Introduction to Electromagnetic Waves with Maxwell's Equations

Dedicated to Oliver Heaviside

Introduction to Electromagnetic Waves with Maxwell's Equations

Özgür Ergül

Middle East Technical University, Ankara, Turkey

This edition first published 2022

Registered Offices
John Wiley & Sons, Inc., 111 River Street, Hoboken, NJ 07030, USA
John Wiley & Sons Ltd, The Atrium, Southern Gate, Chichester, West Sussex, PO19 8SQ, UK

Editorial Office
The Atrium, Southern Gate, Chichester, West Sussex, PO19 8SQ, UK

For details of our global editorial offices, customer services, and more information about Wiley products visit us at www.wiley.com.

Wiley also publishes its books in a variety of electronic formats and by print-on-demand. Some content that appears in standard print versions of this book may not be available in other formats.

Library of Congress Cataloging-in-Publication Data

Names: Ergül, Özgür, author.
Title: Introduction to electromagnetic waves with Maxwell's equations / Özgür Ergül.
Description: Hoboken, NJ : Wiley, 2022. | Includes bibliographical references and index.
Identifiers: LCCN 2021014304 (print) | LCCN 2021014305 (ebook) | ISBN 9781119626725 (cloth) | ISBN 9781119626732 (adobe pdf) | ISBN 9781119626749 (epub)
Subjects: LCSH: Electromagnetic waves. | Maxwell equations.
Classification: LCC QC661 .E694 2021 (print) | LCC QC661 (ebook) | DDC 537/.12–dc23
LC record available at https://lccn.loc.gov/2021014304
LC ebook record available at https://lccn.loc.gov/2021014305

Cover Design: Wiley
Cover Image: © Pobytov/iStock/Getty Images Plus/Getty Images

Set in 11/13pt CMR10 by Straive, Chennai, India

SKYA7042C16-C873-4941-846B-CB84784EEFD9_082621

Contents

Preface

I took my first *serious* course on electromagnetics 23 years ago. My favorite professor introduced us to our main textbook – a very famous primer, as I learned several years later. Even though I am not good at remembering conversations, particularly scientific ones, I still recall the following, with some dramatic visual details probably contributed by my imagination:

– This book is *actually* quite good.

– But?

– Well... It is the starting point... Electrostatics... Magnetostatics... I would be happier to start with Maxwell's...

– Then why don't you write your own book, Professor? (This was a bit provocative, I admit.)

– Maybe someday...

I completed my BSc and PhD at the same university, spending a total of eight years on top of five undergraduate years, and I had a chance to be a teaching assistant for numerous electromagnetics courses. We used the same book in all the undergraduate courses. In fact, I am still addicted to it, without any regret!

Even Feynman Could Not Do It!

After some adventures in mathematics abroad, I returned to my current location in 2013 as a member of the faculty. The Middle East Technical University has a deep-rooted history, and the Department of Electrical and Electronics Engineering is one of the best in the country. In addition, the electromagnetics group has always been strong in this university, with 10+ full professors when I arrived. I, a 33-year-old man, was like a kid in a candy store, where "candy" refers to things I could do. For the undergraduate courses, I was happy to accept materials from my seniors, Prof. Özlem Aydın Çivi and Prof. Nilgün Günalp, as they were well prepared and optimized (some notes were older than me). In fact, they saved my life. But, thinking about teaching electromagnetics, I always shared the feeling with others that something was not right. Certainly it was difficult to describe; it seemed to be felt only by humans who taught electromagnetics. I had a chance to speak to many professors, instructors, and teaching assistants, not only from my university but also from diverse universities all over the world during scientific conferences. Conversations related to teaching materials often reached the same point, already familiar to me:

– We could teach electromagnetics better.

– How?

– Well... What if we could start with Maxwell's...

The *force* that made me write this book appeared in one of these conversations, with remarkable words from a respected professor.

– But even Feynman could not do it!

To be honest, I still do not know if Feynman ever tried to teach electromagnetics in this way, which I now call the *correct* order. Whenever I start to investigate, I find myself studying electromagnetics in the amazing world of this extraordinary scientist. In fact, whether Feynman taught this way is not important. The message I got was that this is a task so important that even Feynman – a man who can teach everything – might not complete it. This is a task that I must do.

All Electromagneticians Are Authors

If an instructor is reading these sentences, I am 100% sure that s/he has an idea for a new book on electromagnetics, if it is not already written. And this is amazing, considering that the material has been more or less the same for decades. Any instructor of an electromagnetics course at some point thinks that the material given to her/him (usually by her/his senior instructors) could be taught in a better way. But any attempt to change the classical order of basic/fundamental electromagnetics surprisingly evolves into a complicated task, discouraging the instructor from embarking on an adventure. We have to accept that writing a book on *basic* electromagnetics is an ambitious task, for several reasons:

- There are many potential readers – i.e. undergraduate students who will become engineers and researchers – so one must consider their variety (age and background) as well as the new habits of the young students of the twenty-first century.
- Many instructors are simply not happy with the available books (not related to their quality), and they have competitive projects, often having the spirit of this book.
- Ironically, books are available, some of which are excellent (discouraging potential authors), but none of which actually present the topic as the instructors wish.

This book is the end product of an attempt to complete this ambitious task. It presents the well-known topics that all instructors would naturally include in their syllabus, while using the correct order (a novel approach). What I found was stranger than the problem itself. This correct order becomes possible only with a necessary pedagogic approach, ironically considering the typical learning behaviors of the students of the new millennium instead of the traditional mechanisms.

What This Book Is

Before explaining the exact structure of this book with the so-called correct order of topics, I should indicate for whom it is written. This book is for undergraduate students who wish to or have to learn electromagnetism for the first time. It is also for senior undergraduate and graduate students who wish to go deeper in primary topics. In addition to Maxwell's equations and their basic applications for canonical problems, the material presented in this book also contains information on the transmission of electromagnetic waves with fundamental details on antennas, waveguides, and transmission lines. These topics must be covered for completeness of the basic knowledge, but readers should understand that each is the topic of individual books that cannot be merged into a single one. Vector calculus and similar fundamental information are also provided and discussed where necessary, but readers are expected have a background in basic calculus.

As already revealed, this book presents basic electromagnetics from the perspective of Maxwell's equations, which is possible via a new pedagogic approach. To further clarify, we may consider a particular example, such as the topic of *gradient*. In this book, *gradient* is introduced in Section 3.6.1 (Chapter 3), after Gauss' law, Ampere's law, and much more material.

In typical textbooks, the concept of *gradient* is explained much earlier – usually at the very beginning – since it is a fundamental mathematical tool. Readers may find very specific books where *gradient* is explained in later chapters, but this is merely an example. In this book, the same strategy is used for all **tools**: e.g. for dot product, cross product, divergence, curl, etc. To describe the approach in one sentence, in this book, a tool is presented *whenever* it is required, not earlier or later. Continuing with the same example, if *gradient* was explained earlier (e.g. in the preliminary (zero) chapter of this book), then we would encounter the two major issues of the traditional approach:

- Related examples to teach *gradient* can be superficial and indirectly related to electromagnetism, since it is actually *not* the time to use *gradient*.
- When it is the right time to use *gradient* in the context of Maxwell's equations, *gradient* becomes an old topic (often not well understood), and revision is needed.

It should be noted that *gradient* is not used in Gauss' law and Ampere's law (at least, not directly), so it can be placed in Faraday's law. This is the idea.

Based on the described approach, the sequence of material in this book is *optimized*. Specifically, readers will frequently find text such as the following (from Chapter 3):

To understand the integral form of Faraday's law, we need the concepts of differential length with direction (Section 2.2.1) and circulation of vector fields (Section 2.1.2). These are in addition to the concepts of dot product (Section 1.1.2), differential surface with direction (Section 1.1.1), and flux of vector fields (Section 1.1.3). Therefore, no new tool is required before we focus on the meaning and application of the integral form of Faraday's law.

All tools and concepts are carefully defined so that readers will know exactly which tools are needed to understand the topic ahead.

As also mentioned above, the structure of this book is suitable for young students of the new millennium. As some instructors might have realized, and I will describe directly, students of the new generation have (good or bad) a pragmatic learning behavior. According to this strategy, learning is a recursive approach of opening boxes inside larger boxes until the required information is reached. Continuing with examples, a student who desires to learn the differential form of Faraday's law needs Stokes' theorem. If s/he does not know Stokes' theorem, s/he must learn it (that is, immediately under the section of the differential form of Faraday's law in this book). But at the second level, Stokes' theorem needs the dot product, curl, circulation, and flux concepts. As all these concepts are also used in Gauss' and Ampere's laws, so the student is forwarded to these tools (in case information is missing). Of course, by adapting this strategy, my aim is not to generate lazy students. The arrangement of topics, connections between them, and the tree-like organization of tools and concepts are designed for effectively learning and teaching electromagnetics. To avoid any misunderstanding, I should emphasize that with the correct order of topics, this book should be studied straightforwardly from the first page to the last page (as the order is already correct). Most instructors will be pleased to learn that, with the approach I have used, electrostatics and magnetostatics become special cases of electromagnetics in this book, as they should always be.

Why Do We Learn Electromagnetics?

I believe I am not the only instructor who is asked, "Why do we learn electromagnetics?" Whether this is a real question or something to test the instructor's patience in the fifth week of the course, I find it an extremely important question that must be handled carefully. This is exactly why I reflect the question as, "Why do you learn electromagnetics?" In my teaching career so far, I have received spectacular answers, including, "To understand how microwave ovens work" when I was a teaching assistant 18 years ago. Today, answers focus more on mobile communication, while some students differentiate themselves by replying, "To understand light." Of course, we can mention a plethora of applications by visualizing the whole electromagnetic spectrum from radio waves to microwaves, THz waves, optical waves, X-rays, and gamma rays. For example, at radio and microwave frequencies, one can immediately list communication (e.g. cellular phones), identification (e.g. radars), tracking (e.g. RFID), and many other applications of modern life. Optical frequencies at which all optical phenomena occur, including human vision, are merely high-frequency samples of electromagnetism.

In general, to understand how electronic devices and nature work, one needs the knowledge of electromagnetism: i.e. how electromagnetic waves emerge, behave, and interact with matter. All technological areas, including aviation, automotive, space technology, energy harvesting, communication, security, electronic circuits, biomedical imaging, and power systems, to name a few, involve electromagnetic waves. On the other hand, at the early stages of learning electromagnetism, students are often overwhelmed by even a brief list of these applications, as electromagnetic waves themselves may not make sense in the first place. Therefore, I find it more useful to provoke students with questions, some of which are as follows:

- How you can stand on the earth if there is a gravitational force to its center?
- How can you touch? What is touching?
- How do you see? Why don't you see your text messages traveling through the air?
- How does the Sun warm us and other objects?

- Why is the sky blue? Why are oceans blue, while the water you drink from a glass is not?
- Why can't you see atoms directly via a microscope with super magnification, e.g. putting all available lenses one after another?
- How does your mobile phone receive signals behind walls? Behind your friend's body?
- How does humanity survive under solar winds?
- Why are X-rays so dangerous, but visible light is not?
- How do astronomers learn information about distant stars?
- Why is a sunset red? Why is an apple red? In fact, what is red?
- How can we see inside a human body without slicing into it?
- How do traditional compasses work? Why does the earth have its own magnetic field in the first place?

This list is instantly and randomly generated, and it is incomplete; all instructors will have many other questions (questioning strangely basic things) based on their students' background. Students reading this book may start with this list, asking the questions of themselves, their friends, and even their pets, without great expectations. I also use similar questions to change the topic of conversations with my wife, who is a molecular biologist. It works. Lucky instructors like me occasionally have a student who questions learning something from 160 years ago. This is the beauty of Maxwell's equations; we are using equations written more than a century ago to describe a plethora of natural and technological phenomena today.

Content, Structure, and Style

Following this preface are mathematical notation, a list of symbols, special functions, frequently used identities, tools to understand Maxwell's equations (a list of tools and concepts), and a preliminary chapter (Chapter 0) that contains some basic information, at which readers may frequently look. The main body of the book contains eight chapters. The first four are devoted to Maxwell's equations, in accordance with the structure of the book. After completing these four chapters, readers are expected to fully understand Maxwell's equations and be ready to use them to understand all electromagnetic phenomena. In addition, all tools and concepts that are necessary to understand Maxwell's equations are completed in the first four chapters. Chapter 5 puts Maxwell's equations into practice to understand electromagnetic waves, as well as their basic behaviors: propagation, reflection, and transmission. Instructors may be shocked to see that conductors are topics of Chapter 6. This is a natural consequence of the correct ordering; Maxwell's equations can be fully understood without metals that behave oddly, particularly *perfect* conductors that often push students into a dark world of confusion. Naturally, capacitance, resistance, and inductance are all covered together in Chapter 6. In Chapter 7, transmission of electromagnetic waves is discussed, considering three fundamental mechanisms: wireless links (using antennas), waveguides, and cables via transmission-line modeling. The concluding chapter provides essential information on the electromagnetic spectrum and history of electromagnetism, in addition to various examples of electromagnetic phenomena in nature and technology. I believe this is when students are able to accept and appreciate all this information.

The book is written in an unusual landscape format, with the main text covering about three-quarters of the page, exactly as it is here. Page sides are reserved for hundreds of figures, as well as hundreds of side notes like this.[1] While reading this book, it is possible to completely ignore the side notes; this is why they are not included in the main text. However, readers may find it interesting that these side notes are based on students' questions; i.e. they contain very critical information that we (instructors) do not necessarily include in our materials, while students are enthusiastically asking these questions to complete their learning process. There are also red side notes like this (without numbering, as shown on the right-hand side), which provide information about important variables and their units; the color makes them easier to find. I believe and hope that the remaining empty spaces will be filled with students' own precious notes.

[1]This is how side notes appear.

This is how red side notes appear.

Maxwell's equations cannot be understood without practice, while I wanted to be careful to avoid making this book a question-solving reference. There are a total of 100 *examples* distributed over the entire book, each with a very detailed solution to help understand the material. In addition, a total of 100 *exercises* are placed at the end of the chapters; detailed solutions to these exercises are also available in a supplementary document on the book's website. Finally, each chapter ends with a set of *questions* – this time, without publicly available solutions. Readers may find it interesting that some of these questions, despite looking trivial, are known to confuse even professors, which may provide a chance for students to test their instructors.

I was thinking about how to complete this preface, when I remembered one question that I received five years ago. My students do not ask this question, as I always ask them first. "Where do we use *this* in real life?" Probably this question is not special to electromagnetism; but when it is directed to us, *this* refers to Maxwell's equations or electromagnetism. There can be alternative answers, but I prefer the short one: "This *is* life!"

Mathematical Notation

In the following, $*$ represents a variable/constant to which the operation is applied.

- $|*|$: Absolute value
- $* \approx *$: Approximates
- $\oint *$: Closed integral
- $[*, *]$: Closed range
- $* \cup *$: Combine
- $*^*$: Complex conjugate
- $*_0$: Constant value
- $* \times *$: Cross product
- $*^\circ$: Degree
- $\bar{\nabla}$: Del operator
- $\frac{d*}{d*}$: Derivative
- d : Differential indicator
  - ✓ df : Differential f
  - ✓ $\overline{df}$: Differential f with direction
  - ✓ $d\bar{f}$: Differential vector $\bar{f}$

- $*/*$ or $\frac{*}{*}$: Division
- $* \cdot *$: Dot product
- $* = *$: Equal to
- $* \stackrel{?}{=} *$: Equality to be checked
- $* \to *$: Goes to
- $* >$: Greater than
- $* \geq$: Greater than or equal to
- $[*, *), (*, *]$: Half open range
- $* \longrightarrow *$: Implies, from/to
- $* \longleftrightarrow *$: Implies both ways
- $* \in *$: In, inside
- ∞ : Infinity
- $[*]_*^*$: Insert values (e.g. after integration)
- $\displaystyle\int *$: Integral
- $*^-$: Left side of number
- $* <$: Less than
- $* \leq$: Less than or equal to
- $|*|$: Magnitude or determinant (if matrix)
- $\begin{bmatrix} * & * & * \\ * & * & * \\ * & * & * \end{bmatrix}$: Matrix (3-by-3 in this example)
- $\mp$: Minus or plus
- $\begin{bmatrix} * & * \\ * & * \end{bmatrix} \begin{bmatrix} * \\ * \end{bmatrix}$: Multiplication of a matrix and a vector (2-by-2 in this example)
- $\tilde{*}$: Normalization
- $* \neq *$: Not equal

- $\not\rightarrow$: Not implies
- $(*, *)$: Open range
- $\frac{\partial *}{\partial *}$: Partial derivative
- $* \perp *$: Perpendicular to
- $\angle *$: Phase
- $\pm$: Plus or minus
- $*^n$: (to the) Power of n
- $* \propto *$: Proportional to
- $*^+$: Right side of number
- $\sqrt{*}$: Square root
- $-$: Subtract, minus
- $+$: Sum, plus
- $\bar{*}$: Vector
- $\begin{bmatrix} * \\ * \\ * \end{bmatrix}$: Vector of elements (3 elements in this example)

List of Symbols

- a : Radius (m), size (m), constant, scalar
- $\hat{a}$: Unit vector
- A : Area (m^2), effective area (m^2), constant
- $\bar{A}$: Magnetic vector potential (Wb/m)
- b : Radius (m), size (m), constant
- B : Constant
- $\bar{B}$: Magnetic flux density (Wb/m^2)
- c : Radius (m), size (m), constant, speed of light (m/s)
- C : Curve, capacitance (F), constant
- d : Distance (m), size (m)
- D : Directive gain, directivity, dimension, constant
- $\bar{D}$: Electric flux density (C/m^2)
- e : Constant, Euler's number, antenna efficiency
- $\bar{e}$: Vector, vector function, vector field
- $\bar{E}$: Electric field intensity (N/C or V/m)
- f : Frequency (Hz), scalar, scalar function, scalar field, scale factor, flux (Wb or C)
- $\bar{f}$: Vector, vector function, vector field

- $\bar{F}$: Force (N)
- g : Green's function, scalar, scalar function, scalar field, scale factor
- $\bar{g}$: Vector, vector function, vector field
- G : Conductance (S), antenna gain
- h : Scalar, scalar function, scalar field, scale factor, height (m), constant, waveguide constant (1/m), Planck's constant (Js)
- $\bar{h}$: Vector, vector function, vector field
- $\bar{H}$: Magnetic field intensity (A/m)
- i : Index
- I : Electric current (A), dipole moment (Cm, Am^2, or Am), integral value
- $\bar{I}$: Dipole moment (Cm, Am^2, or Am), electric current with direction (A)
- j : Imaginary unit ($j = \sqrt{-1}$)
- $\bar{J}$: Electric current density (A/m or A/m^2)
- k : Wavenumber (rad/m), number of items/terms, Coulomb's constant (m/F), magnetic force constant (A/m), constant
- $\bar{k}$: Wave vector (rad/m)
- l : Line (m), length (m), number of items/terms
- L : Inductance (H)
- m : Index
- M : Number of items/terms
- $\bar{M}$: Magnetization (A/m)
- n : Number of items/terms, index
- N : Number of items/terms
- p : Power (W)
- P : Location, point
- $\bar{P}$: Polarization (C/m^2)
- q : Electric charge density (C/m, C/m^2, or C/m^3)

- Q : Electric charge (C)
- $\bar{r}$: Position vector (m)
- R : Spherical variable, distance, resistance (Ω)
- $\bar{R}$: Distance vector (m)
- s : Scalar, surface, skin depth (m)
- S : Surface
- $\bar{S}$: Poynting vector (W/m^2)
- t : Time (s)
- T : Time period (s)
- u : Velocity (m/s), coordinate variable
- $\bar{u}$: Velocity vector (m/s)
- U : Radiation intensity (W/sr), radiation pattern
- v : Volume, coordinate variable, phase (rad)
- V : Volume, voltage (V)
- w : Electric potential energy (J), magnetic potential energy (J)
- x : Cartesian variable (m)
- X : Reactance (Ω)
- y : Cartesian variable (m)
- z : Cartesian/cylindrical variable (m)
- Z : Impedance (Ω)
- α : Attenuation constant (Np/m)
- β : Phase constant (rad/m)
- γ : Propagation constant (1/m)
- Γ : Reflection coefficient
- δ : Dirac delta function, loss-tangent argument

- Δ : Small difference, Laplace operator
- ε : Permittivity (F/m)
- ζ : Excitation coefficient
- η : Wave/intrinsic impedance (Ω)
- θ : Spherical variable (rad), angle (rad)
- λ : Wavelength (m)
- μ : Permeability (H/m)
- π : Pi number, dissipated/generated power flux density (W/m^2)
- ρ : Cylindrical variable (m)
- $\bar{\rho}$: Cylindrical position vector (m)
- σ : Conductivity (S/m)
- τ : Transmission coefficient
- ϕ : Cylindrical/spherical variable, phase (rad)
- Φ : Electromotive force (V), electric scalar potential (V)
- φ : Angle (rad), phase (rad)
- χ : Electric susceptibility, magnetic susceptibility, complex exponential
- ω : Angular frequency (rad/s)

Special Functions

- $\cos^{-1}$: Arccosine
- $\sin^{-1}$: Arcsine
- $\tan^{-1}$: Arctangent
- $\cos$: Cosine
- Ci : Cosine integral
- $\exp$: Exponential
- $\tanh$: Hyperbolic tangent
- Im : Imaginary (part)
- $\lim$: Limit
- $\log$: Logarithm
- $\max$: Maximum
- $\ln$: Natural logarithm
- Re : Real (part)
- $\sin$: Sine
- $\tan$: Tangent

Frequently Used Identities

Cartesian-Cylindrical and Cartesian-Spherical Vector Transformations

Given vector $\bar{f} = \hat{a}_x f_x + \hat{a}_y f_y + \hat{a}_z f_z = \hat{a}_\rho f_\rho + \hat{a}_\phi f_\phi + \hat{a}_z f_z = \hat{a}_R f_R + \hat{a}_\theta f_\theta + \hat{a}_\phi f_\phi$:

$$\begin{bmatrix} f_x \\ f_y \\ f_z \end{bmatrix} = \begin{bmatrix} \cos\phi & -\sin\phi & 0 \\ \sin\phi & \cos\phi & 0 \\ 0 & 0 & 1 \end{bmatrix} \begin{bmatrix} f_\rho \\ f_\phi \\ f_z \end{bmatrix}, \qquad \begin{bmatrix} f_\rho \\ f_\phi \\ f_z \end{bmatrix} = \begin{bmatrix} \cos\phi & \sin\phi & 0 \\ -\sin\phi & \cos\phi & 0 \\ 0 & 0 & 1 \end{bmatrix} \begin{bmatrix} f_x \\ f_y \\ f_z \end{bmatrix}$$

$$\begin{bmatrix} f_x \\ f_y \\ f_z \end{bmatrix} = \begin{bmatrix} \sin\theta\cos\phi & \cos\theta\cos\phi & -\sin\phi \\ \sin\theta\sin\phi & \cos\theta\sin\phi & \cos\phi \\ \cos\theta & -\sin\theta & 0 \end{bmatrix} \begin{bmatrix} f_R \\ f_\theta \\ f_\phi \end{bmatrix}, \qquad \begin{bmatrix} f_R \\ f_\theta \\ f_\phi \end{bmatrix} = \begin{bmatrix} \sin\theta\cos\phi & \sin\theta\sin\phi & \cos\theta \\ \cos\theta\cos\phi & \cos\theta\sin\phi & -\sin\theta \\ -\sin\phi & \cos\phi & 0 \end{bmatrix} \begin{bmatrix} f_x \\ f_y \\ f_z \end{bmatrix}$$

Dot and Cross Products

Given vectors $\bar{f}$, $\bar{g}$, and $\bar{h}$:

$$\bar{f} \cdot (\bar{g} \times \bar{h}) = \bar{g} \cdot (\bar{h} \times \bar{f}) = \bar{h} \cdot (\bar{f} \times \bar{g})$$

$$\bar{f} \times (\bar{g} \times \bar{h}) = \bar{g}(\bar{f} \cdot \bar{h}) - \bar{h}(\bar{f} \cdot \bar{g})$$

Gradient and Laplace

Given scalar field $f(\bar{r},t)$:

$$\begin{aligned}
\bar{\nabla} f(\bar{r},t) &= \hat{a}_x \frac{\partial f(\bar{r},t)}{\partial x} + \hat{a}_y \frac{\partial f(\bar{r},t)}{\partial y} + \hat{a}_z \frac{\partial f(\bar{r},t)}{\partial z} \\
&= \hat{a}_\rho \frac{\partial f(\bar{r},t)}{\partial \rho} + \hat{a}_\phi \frac{1}{\rho} \frac{\partial f(\bar{r},t)}{\partial \phi} + \hat{a}_z \frac{\partial f(\bar{r},t)}{\partial z} \\
&= \hat{a}_R \frac{\partial f(\bar{r},t)}{\partial R} + \hat{a}_\theta \frac{1}{R} \frac{\partial f(\bar{r},t)}{\partial \theta} + \hat{a}_\phi \frac{1}{R\sin\theta} \frac{\partial f(\bar{r},t)}{\partial \phi} \\
\bar{\nabla}^2 f(\bar{r},t) &= \frac{\partial^2 f(\bar{r},t)}{\partial x^2} + \frac{\partial^2 f(\bar{r},t)}{\partial y^2} + \frac{\partial^2 f(\bar{r},t)}{\partial z^2} \\
&= \frac{1}{\rho} \frac{\partial}{\partial \rho} \left(\rho \frac{\partial f(\bar{r},t)}{\partial \rho} \right) + \frac{1}{\rho^2} \frac{\partial^2 f(\bar{r},t)}{\partial \phi^2} + \frac{\partial^2 f(\bar{r},t)}{\partial z^2} \\
&= \frac{1}{R^2} \frac{\partial}{\partial R} \left(R^2 \frac{\partial f(\bar{r},t)}{\partial R} \right) + \frac{1}{R^2 \sin\theta} \frac{\partial}{\partial \theta} \left(\sin\theta \frac{\partial f(\bar{r},t)}{\partial \theta} \right) + \frac{1}{R^2 \sin^2\theta} \frac{\partial^2 f(\bar{r},t)}{\partial \phi^2}
\end{aligned}$$

Divergence

Given vector field $\bar{f}(\bar{r},t) = \hat{a}_x f_x(\bar{r},t) + \hat{a}_y f_y(\bar{r},t) + \hat{a}_z f_z(\bar{r},t) = \hat{a}_\rho f_\rho(\bar{r},t) + \hat{a}_\phi f_\phi(\bar{r},t) + \hat{a}_z f_z(\bar{r},t) = \hat{a}_R f_R(\bar{r},t) + \hat{a}_\theta f_\theta(\bar{r},t) + \hat{a}_\phi f_\phi(\bar{r},t)$:

$$\begin{aligned}
\bar{\nabla} \cdot \bar{f}(\bar{r},t) &= \frac{\partial f_x(\bar{r},t)}{\partial x} + \frac{\partial f_y(\bar{r},t)}{\partial y} + \frac{\partial f_z(\bar{r},t)}{\partial z} \\
&= \frac{1}{\rho} \frac{\partial}{\partial \rho} [\rho f_\rho(\bar{r},t)] + \frac{1}{\rho} \frac{\partial f_\phi(\bar{r},t)}{\partial \phi} + \frac{\partial f_z(\bar{r},t)}{\partial z} \\
&= \frac{1}{R^2} \frac{\partial}{\partial R} [R^2 f_R(\bar{r},t)] + \frac{1}{R\sin\theta} \frac{\partial}{\partial \theta} [\sin\theta f_\theta(\bar{r},t)] + \frac{1}{R\sin\theta} \frac{\partial f_\phi(\bar{r},t)}{\partial \phi}
\end{aligned}$$

Curl

Given vector field $\bar{f}(\bar{r},t) = \hat{a}_x f_x(\bar{r},t) + \hat{a}_y f_y(\bar{r},t) + \hat{a}_z f_z(\bar{r},t) = \hat{a}_\rho f_\rho(\bar{r},t) + \hat{a}_\phi f_\phi(\bar{r},t) + \hat{a}_z f_z(\bar{r},t) = \hat{a}_R f_R(\bar{r},t) + \hat{a}_\theta f_\theta(\bar{r},t) + \hat{a}_\phi f_\phi(\bar{r},t)$:

$$\bar{\nabla} \times \bar{f}(\bar{r},t) = \begin{vmatrix} \hat{a}_x & \hat{a}_y & \hat{a}_z \\ \partial/\partial x & \partial/\partial y & \partial/\partial z \\ f_x(\bar{r},t) & f_y(\bar{r},t) & f_z(\bar{r},t) \end{vmatrix}$$

$$= \hat{a}_x \left(\frac{\partial f_z(\bar{r},t)}{\partial y} - \frac{\partial f_y(\bar{r},t)}{\partial z} \right) + \hat{a}_y \left(\frac{\partial f_x(\bar{r},t)}{\partial z} - \frac{\partial f_z(\bar{r},t)}{\partial x} \right) + \hat{a}_z \left(\frac{\partial f_y(\bar{r},t)}{\partial x} - \frac{\partial f_x(\bar{r},t)}{\partial y} \right)$$

$$\bar{\nabla} \times \bar{f}(\bar{r},t) = \frac{1}{\rho} \begin{vmatrix} \hat{a}_\rho & \hat{a}_\phi \rho & \hat{a}_z \\ \partial/\partial \rho & \partial/\partial \phi & \partial/\partial z \\ f_\rho(\bar{r},t) & \rho f_\phi(\bar{r},t) & f_z(\bar{r},t) \end{vmatrix}$$

$$= \hat{a}_\rho \left(\frac{1}{\rho} \frac{\partial f_z(\bar{r},t)}{\partial \phi} - \frac{\partial f_\phi(\bar{r},t)}{\partial z} \right) + \hat{a}_\phi \left(\frac{\partial f_\rho(\bar{r},t)}{\partial z} - \frac{\partial f_z(\bar{r},t)}{\partial \rho} \right) + \frac{1}{\rho} \hat{a}_z \left(\frac{\partial}{\partial \rho} [\rho f_\phi(\bar{r},t)] - \frac{\partial f_\rho(\bar{r},t)}{\partial \phi} \right)$$

$$\bar{\nabla} \times \bar{f}(\bar{r},t) = \frac{1}{R^2 \sin\theta} \begin{vmatrix} \hat{a}_R & \hat{a}_\theta R & \hat{a}_\phi R \sin\theta \\ \partial/\partial R & \partial/\partial \theta & \partial/\partial \phi \\ f_R(\bar{r},t) & R f_\theta(\bar{r},t) & R \sin\theta f_\phi(\bar{r},t) \end{vmatrix}$$

$$= \frac{1}{R\sin\theta} \hat{a}_R \left(\frac{\partial}{\partial \theta} [\sin\theta f_\phi(\bar{r},t)] - \frac{\partial f_\theta(\bar{r},t)}{\partial \phi} \right) + \frac{1}{R} \hat{a}_\theta \left(\frac{1}{\sin\theta} \frac{\partial f_R(\bar{r},t)}{\partial \phi} - \frac{\partial}{\partial R} [R f_\phi(\bar{r},t)] \right) + \frac{1}{R} \hat{a}_\phi \left(\frac{\partial}{\partial R} [R f_\theta(\bar{r},t)] - \frac{\partial f_R(\bar{r},t)}{\partial \theta} \right)$$

Operations on Multiple Fields

Given scalar fields $f(\bar{r},t)$ and $g(\bar{r},t)$ and vector fields $\bar{h}(\bar{r},t)$ and $\bar{e}(\bar{r},t)$:

$$\bar{\nabla}[f(\bar{r},t) g(\bar{r},t)] = f(\bar{r},t) \bar{\nabla} g(\bar{r},t) + g(\bar{r},t) \bar{\nabla} f(\bar{r},t)$$

$$\bar{\nabla} \cdot [f(\bar{r},t) \bar{h}(\bar{r},t)] = f(\bar{r},t) \bar{\nabla} \cdot \bar{h}(\bar{r},t) + [\bar{\nabla} f(\bar{r},t)] \cdot \bar{h}(\bar{r},t)$$

$$\bar{\nabla} \times [f(\bar{r},t) \bar{h}(\bar{r},t)] = f(\bar{r},t) \bar{\nabla} \times \bar{h}(\bar{r},t) + [\bar{\nabla} f(\bar{r},t)] \times \bar{h}(\bar{r},t)$$

$$\bar{\nabla} \cdot [\bar{h}(\bar{r},t) \times \bar{e}(\bar{r},t)] = [\bar{\nabla} \times \bar{h}(\bar{r},t)] \cdot \bar{e}(\bar{r},t) - \bar{h}(\bar{r},t) \cdot [\bar{\nabla} \times \bar{e}(\bar{r},t)]$$

Multiple Operators

Given scalar field $f(\bar{r},t)$ and vector field $\bar{g}(\bar{r},t) = \hat{a}_x g_x(\bar{r},t) + \hat{a}_y g_y(\bar{r},t) + \hat{a}_z g_z(\bar{r},t)$:

$$\bar{\nabla} \cdot [\bar{\nabla} f(\bar{r},t)] = \bar{\nabla}^2 f(\bar{r},t)$$

$$\bar{\nabla} \times [\bar{\nabla} f(\bar{r},t)] = 0$$

$$\bar{\nabla} \cdot [\bar{\nabla} \times \bar{g}(\bar{r},t)] = 0$$

$$\bar{\nabla} \times \bar{\nabla} \times \bar{g}(\bar{r},t) = \bar{\nabla}[\bar{\nabla} \cdot \bar{g}(\bar{r},t)] - \bar{\nabla}^2 \bar{g}(\bar{r},t)$$

$$\bar{\nabla}^2 \bar{g}(\bar{r},t) = \hat{a}_x \bar{\nabla}^2 g_x(\bar{r},t) + \hat{a}_y \bar{\nabla}^2 g_y(\bar{r},t) + \hat{a}_z \bar{\nabla}^2 g_z(\bar{r},t)$$

Tools to Understand Maxwell's Equations

- Differential surface with direction (Section 1.1.1)
- Dot product (Section 1.1.2)
- Flux of vector fields (Section 1.1.3)
- Electric charge density (Section 1.3.1)
- Divergence of vector fields (Section 1.3.2)
- Divergence theorem (Section 1.3.3)
- Electric dipole (Section 1.7.1)
- Polarization (Section 1.7.2)
- Equivalent polarization charges (Section 1.7.3)
- Differential length with direction (Section 2.1.1)
- Circulation of vector fields (Section 2.1.2)
- Electric current density (Section 2.3.1)
- Cross product (Section 2.3.2)
- Curl of vector fields (Section 2.3.3)
- Stokes' theorem (Section 2.3.4)
- Magnetic dipole (Section 2.8.1)
- Magnetization (Section 2.8.2)

- Equivalent magnetization currents (Section 2.8.3)
- Gradient of scalar fields (Section 3.6.1)
- Gradient theorem (Section 3.6.3)
- Electric potential energy (Section 3.6.5)
- Poisson's equation and Laplace's equation (Section 3.6.7)
- Electrostatic boundary value problems (Section 3.6.11)
- Coulomb's gauge (Section 4.3.1)
- Magnetic potential energy (Section 4.3.3)

About the Companion Website

The companion website for this book is at

www.wiley.com/go/ergulmax 

The website contains:

- PDF & Latex source file of exercise solutions for Students and Instructors

Scan this QR code to visit the companion website.

Preliminary

In this preliminary chapter, we consider some basic materials as the background for the rest of the book. The topic of electromagnetics is based on fields, specifically scalar and vector fields, and their manipulation via basic operations: e.g. multiplication, differentiation, and integration. To describe fields and perform operations on them to explain electromagnetic phenomena, we need to define coordinate systems and their variables. We consider only the most basic systems, namely, Cartesian, cylindrical, and spherical systems, each of which has its own advantages. Using these systems requires knowledge of their variables, differential lengths, surfaces, and volumes. In addition to ordinary vectors defined in these systems, the important concept of a *position vector* is also described in this chapter.

In accordance with this book's format, this chapter is not a summary of vector calculus, which is usually considered the first topic in many elementary texts on electromagnetics. For example, gradient, divergence, and curl operations are considered not in this chapter but later, since they are inseparable parts of Maxwell's equations. Similarly, special integrals called *flux* and *circulation* are defined and explained whenever they are needed. The purpose of this chapter is only to prepare the reader before jumping into Maxwell's equations. At the end of the chapter, we briefly describe electrostatics and magnetostatics as the study of static electric charges and steady electric currents, as well as electromagnetics as the generalized name for both static and dynamic events. Finally, we discuss *time* as a major parameter while considering its absence in static events only a special case in electromagnetics.

Introduction to Electromagnetic Waves with Maxwell's Equations, First Edition. Özgür Ergül.

Companion website: www.wiley.com/go/ergulmax

0.1 Scalar and Vector Fields

It would be impossible to understand electromagnetics without mathematics, and particularly calculus. As a starting point, we need to first consider coordinate systems, as well as scalar and vector fields. Since we are practically living in a three-dimensional space that requires three variables to indicate any arbitrary position, we construct our understanding of real and abstract objects on coordinates. Hence, we need to define a system, namely a coordinate system, to identify the positions of objects with respect to each other. From this perspective, time can be considered the fourth dimension to describe electromagnetic events. Coordinate systems are abstract, and we can define them as we wish, but only a few systems have useful properties that make mathematical operations relatively easy to practice.

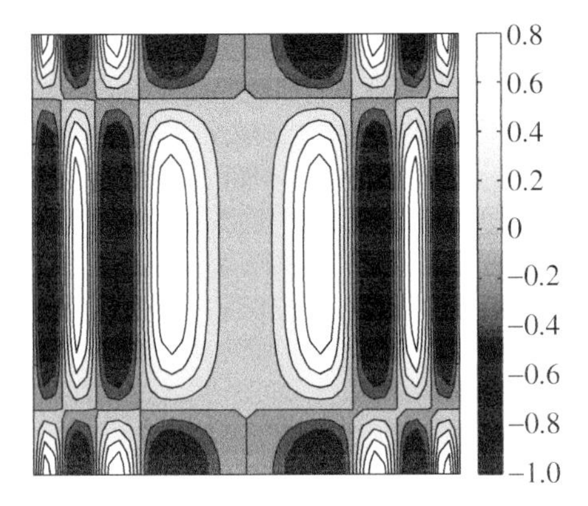

Figure 0.1 A two-dimensional scalar field plotted via contours.

A scalar field is a function defined with respect to position, which can be illustrated in alternative ways (Figure 0.1). Let a three-dimensional space be represented by three variables: i.e. x, y, and z. Then, a scalar field f is naturally a function of these variables, which can be shown as $f(x, y, z)$. For example, the temperature in a room can be seen as a scalar field (Figure 0.2). While it may be constant in the entire domain, it may also depend on the position: e.g. $f(0, 1, 2) = 24$°C and $f(0, 2, 2) = 25$°C for the temperature in the room. If a scalar field $f(x, y, z)$ is changing with position, we may need to know how it changes: i.e. the derivative of $f(x, y, z)$ with respect to position. Since there are multiple variables (x, y, and z in our example), the definition of the derivative may not be as straightforward as a function of a single variable. Later, we will define the *gradient* of a scalar field, taking into account all variables at the same time and adding directional information via vectors. Obviously, a scalar field may also depend on time, making it a function of four variables $f(x, y, z, t)$, while its time derivative may also have a physical meaning.

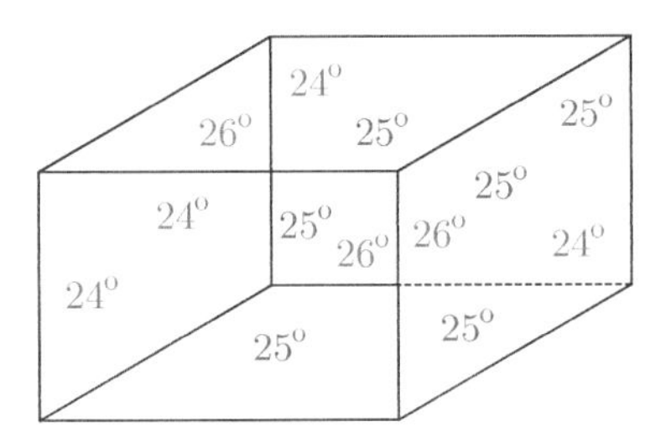

Figure 0.2 Temperature in a room can be considered a scalar field.

The temperature in a room can be represented as a scalar field. But what about air flowing in the room? To make the scenario easier, we can assume that time is frozen. In this case, in addition to the strength of the flow, which can be seen as a scalar field, we need direction information (Figure 0.3). In other words, we need to know in which directions the air is flowing at different positions. In the three-dimensional space with the same variables, we can define $g_x(x, y, z)$, $g_y(x, y, z)$, $g_z(x, y, z)$, each of which is a scalar field representing the strength of the flow in x, y, and z directions, respectively. Instead of considering these components separately, we can use vector calculus to unify them as

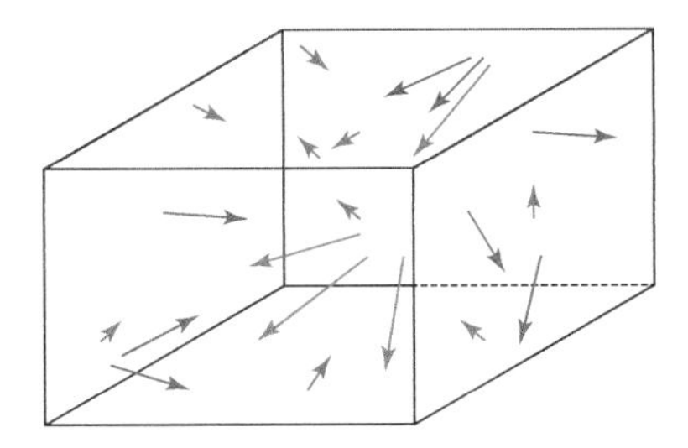

Figure 0.3 Air flow in a room can be considered a vector field.

$$\bar{g}(x, y, z) = \hat{a}_x g_x(x, y, z) + \hat{a}_y g_y(x, y, z) + \hat{a}_z g_z(x, y, z), \tag{1}$$

where $\hat{a}_x$, $\hat{a}_y$, and $\hat{a}_z$ represent the primary directions (unit vectors) in the coordinate system. Therefore, when we define a vector field $\bar{g}(x, y, z)$, we immediately understand that it depends on the position (x, y, z) and has a direction that can be decomposed into $g_x(x, y, z)$, $g_y(x, y, z)$, and

$g_z(x, y, z)$. This compact form has many advantages. For example, imagine that we must describe the following scenario. *We are in a* $4 \times 4 \times 4$ *square room* $(0 \le x \le 4,\ 0 \le y \le 4,\ 0 \le z \le 4)$ *where the air is flowing in a strange manner. At a fixed time* t_0*, it does not flow in the* y *direction, but it flows in the* x *and* z *directions. The flow in the* x *direction does not depend on* x *and* y*, but it depends on* z *quadratically. The flow in the* z *direction does not depend on* z*, but it depends on the multiplication of the values of* x *and* y*. In addition, there is no airflow in one of the corners* $(x = 0,\ y = 0,\ z = 0)$ *of the room.* Instead of all this text, we can simply write

$$\bar{f}(x, y, z, t_0) = \hat{a}_x z^2 + \hat{a}_z xy, \tag{2}$$

and all information is squeezed into a single expression.

Similar to scalar fields, vector fields can be illustrated in alternative ways (Figure 0.4). In addition, we often need to know how a vector field is changing with respect to position. Special differentiation operations, called *divergence* and *curl*, are considered later when they are needed.

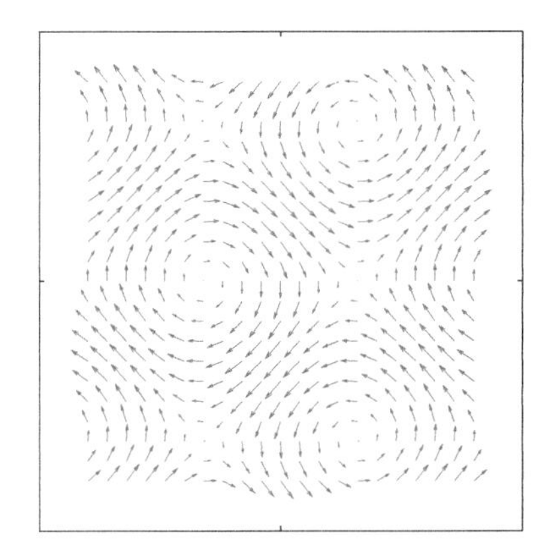

Figure 0.4 A two-dimensional vector field plotted via arrows. Each arrow shows the direction of the field, while its size is the magnitude of the field (exactly at the position of the arrow).

0.2 Cartesian Coordinate Systems

The most natural coordinate systems are Cartesian types (Figure 0.5), defined with perpendicular axes: i.e. x, y, and z, as already used. Then, $\bar{f} = \hat{a}_x f_x + \hat{a}_y f_y + \hat{a}_z f_z$ represents a vector in a Cartesian coordinate system, where f_x, f_y, and f_z are the lengths of the projections of $\bar{f}$ on the x, y, and z axes, respectively. This representation does not change with the location of the vector in a three-dimensional space. Specifically, $\bar{f}$ has the same expression,[1] independent of its location with respect to the origin of the coordinate system. The magnitude of a vector $\bar{f}$ can be found as

$$f = |\bar{f}| = \sqrt{f_x^2 + f_y^2 + f_z^2}. \tag{3}$$

Hence, the unit vector (that has a magnitude of one) in the direction of $\bar{f}$ is (Figure 0.6)

$$\hat{a}_f = \frac{\bar{f}}{f} = \frac{\hat{a}_x f_x + \hat{a}_y f_y + \hat{a}_z f_z}{\sqrt{f_x^2 + f_y^2 + f_z^2}}. \tag{4}$$

The variables of a Cartesian coordinate system (x, y, z) are all metric and related to distances in straight paths.

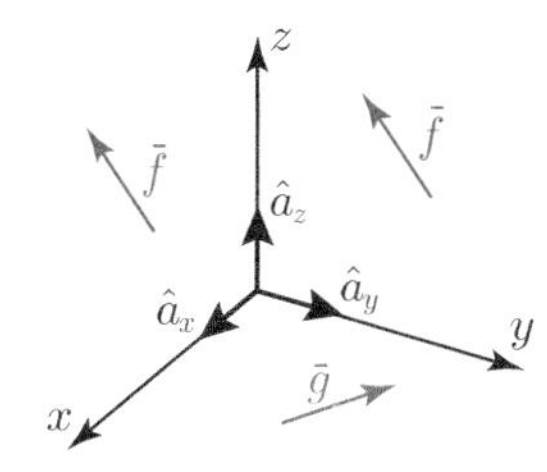

Figure 0.5 A Cartesian coordinate system with unit vectors $\hat{a}_x$, $\hat{a}_y$, and $\hat{a}_z$. Two arbitrary vectors $\bar{f}$ and $\bar{g}$ are defined inside the system. Note that a vector definition is independent of its position: i.e. it is fixed if the magnitude and direction of the vector do not change.

[1] This is not true for all coordinate systems. For example, the representation of a vector in a spherical coordinate system generally depends on the position.

0.3 Basic Vector Operations

In vector calculus, vectors often need to be added, subtracted (Figure 0.7), and multiplied. As shown below, multiplication may occur as a scaling of a vector, while we can also define two types of vector-vector multiplications, namely, dot product ($\cdot$) and cross product ($\times$), to be considered later. Let $\bar{f} = \hat{a}_x f_x + \hat{a}_y f_y + \hat{a}_z f_z$ and $\bar{g} = \hat{a}_x g_x + \hat{a}_y g_y + \hat{a}_z g_z$ be two vectors. Then, we define

$$-\bar{f} = -\hat{a}_x f_x - \hat{a}_y f_y - \hat{a}_z f_z \tag{5}$$

$$\bar{h} = \bar{f} + \bar{g} = \hat{a}_x(f_x + g_x) + \hat{a}_y(f_y + g_y) + \hat{a}_z(f_z + g_z) \tag{6}$$

$$\bar{e} = s\bar{f} = \hat{a}_x(sf_x) + \hat{a}_y(sf_y) + \hat{a}_z(sf_z) \tag{7}$$

as the negative of a vector, addition of two vectors, and multiplication of a vector by a scalar s, respectively.

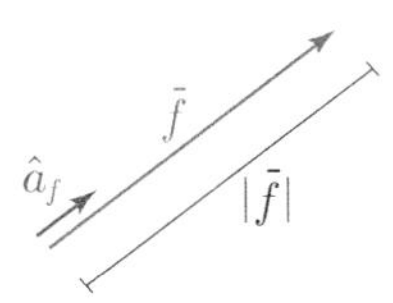

Figure 0.6 A vector $\bar{f}$ can be represented by its direction (unit vector $\hat{a}_f$) and its magnitude $|\bar{f}|$.

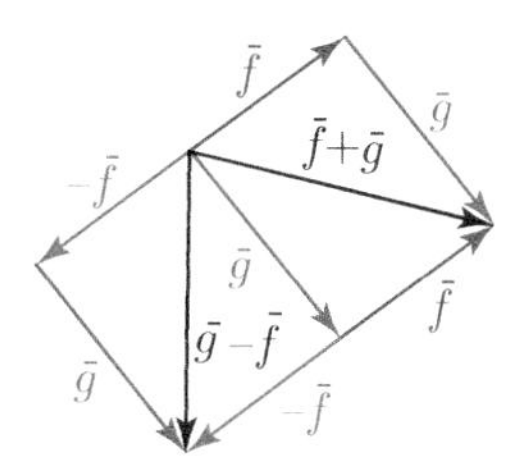

Figure 0.7 A geometric demonstration of addition and subtraction of vectors. It is usually easier to visualize these operations when the starting points of the vectors coincide.

0.4 Orthogonal Coordinate Systems

In a three-dimensional space, an orthogonal coordinate system (Figure 0.8) consists of three mutually orthogonal surfaces defined by variables u_1, u_2, and u_3. In particular, if $\hat{a}_1$, $\hat{a}_2$, and $\hat{a}_3$ are the unit vectors that are normal to these surfaces, they are perpendicular to each other at each location in an orthogonal system. In this book, we consider Cartesian, cylindrical, and spherical coordinate systems of orthogonal type.

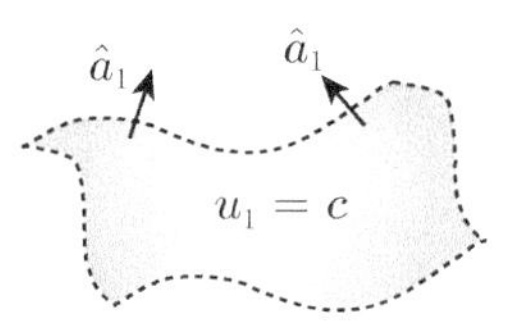

Figure 0.8 For a given variable u_1, equation $u_1 =$ constant corresponds to a surface in a three-dimensional space. Depending on the variable, this surface may extend to infinity (for example, if it is a Cartesian variable) or close upon itself. The unit normal a_1 is always perpendicular to the constant surface; hence, its direction may change, again depending on the properties of the variable.

0.4.1 Properties of a Cartesian Coordinate System

In a Cartesian coordinate system, the variables are defined as $u_1 = x$, $u_2 = y$, and $u_3 = z$, and the corresponding unit vectors are $\hat{a}_x$, $\hat{a}_y$, and $\hat{a}_z$, which do not depend on the position (Figure 0.9). A vector is represented as $\bar{f} = \hat{a}_x f_x + \hat{a}_y f_y + \hat{a}_z f_z$, which also does not depend on the location of the vector. A position vector, which is defined as a vector from the origin to a position (x, y, z), can be shown as[2]

$$\bar{r} = \hat{a}_x x + \hat{a}_y y + \hat{a}_z z, \qquad \bar{r} = |\bar{r}|\hat{a}_r, \qquad |\bar{r}| = \sqrt{x^2 + y^2 + z^2}. \tag{8}$$

When evaluating integrals in a Cartesian system, differential lengths, surfaces, and volumes are needed (Figure 0.9). In particular, for integrals along lines in the x, y, and z directions, we have the differential lengths as $dl = dx$, $dl = dy$, and $dl = dz$, respectively. For a general contour,

[2] A position vector is not a single vector. For a given location (position) in space, the corresponding position vector is always from the *origin* to the location. Therefore, the direction is different for different locations.

these differentials may be combined based on the dependency between the variables. In the most basic cases, differential surfaces are in the form of $ds = dydz$, $ds = dxdz$, or $ds = dxdy$, when x, y, or z is constant, respectively, on the surface. For volume integrals, the basic differential volume is $dv = dxdydz$.

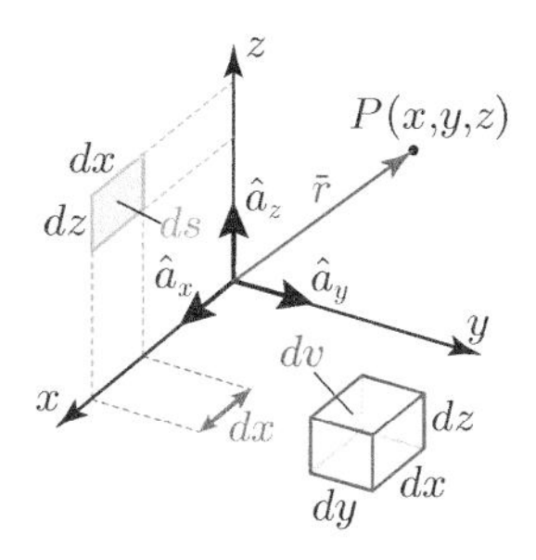

Figure 0.9 Position vector shown in a Cartesian coordinate system, including a geometric representation of differentials.

0.4.2 Cylindrical Coordinate System

In a cylindrical coordinate system, the variables are defined as $u_1 = \rho$, $u_2 = \phi$, and $u_3 = z$, where ρ is the radial distance from the z axis and ϕ is the angle between the projection on the x-y plane and the x axis (Figure 0.10). Note that $\rho \geq 0$ and $\phi \in [0, 2\pi)$, while z may have any value as usual. The unit vectors are $\hat{a}_\rho$, $\hat{a}_\phi$, and $\hat{a}_z$, where $\hat{a}_\rho$ and $\hat{a}_\phi$ depend on the position. This is because the unit vectors in those directions to *increase* the corresponding variables, i.e. ρ and ϕ, depend on the location with respect to the origin. Due to the changing directions of the unit vectors, a representation of a vector in a cylindrical coordinate system, i.e. $\bar{f} = \hat{a}_\rho f_\rho + \hat{a}_\phi f_\phi + \hat{a}_z f_z$, also depends on its location and is not unique (Figure 0.11).

In a cylindrical coordinate system, a position vector, which is defined as a vector from the origin to a position (ρ, ϕ, z), is written[3]

$$\bar{r} = \hat{a}_\rho \rho + \hat{a}_z z, \qquad \bar{r} = \hat{a}_r |\bar{r}|, \qquad |\bar{r}| = \sqrt{\rho^2 + z^2}. \tag{9}$$

[3]Note that a position vector does not contain any ϕ component.

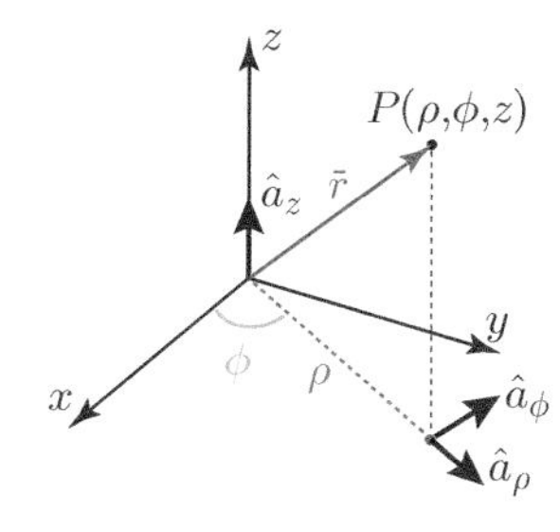

Figure 0.10 A cylindrical coordinate system with cylindrical variables.

Given a point $P = (x, y, z)$ in a Cartesian coordinate system, the same point can be represented in a cylindrical coordinate system as $P = (\rho, \phi, z)$, where

$$\rho = \sqrt{x^2 + y^2} \tag{10}$$

$$\phi = \tan^{-1}(y/x) = \sin^{-1}\left(\frac{y}{\sqrt{x^2 + y^2}}\right) = \cos^{-1}\left(\frac{x}{\sqrt{x^2 + y^2}}\right). \tag{11}$$

Similarly, given a point $P = (\rho, \phi, z)$ in a cylindrical coordinate system, the same point can be represented in a Cartesian coordinate system as $P = (x, y, z)$, where

$$x = \rho \cos\phi \tag{12}$$

$$y = \rho \sin\phi. \tag{13}$$

Based on these, the differential lengths can be written in terms of each other as

$$d\rho = \frac{xdx}{\sqrt{x^2 + y^2}} + \frac{ydy}{\sqrt{x^2 + y^2}} = \cos\phi dx + \sin\phi dy \tag{14}$$

$$\rho d\phi = \frac{-\rho y dx}{x^2 + y^2} + \frac{\rho x dy}{x^2 + y^2} = -\sin\phi dx + \cos\phi dy \tag{15}$$

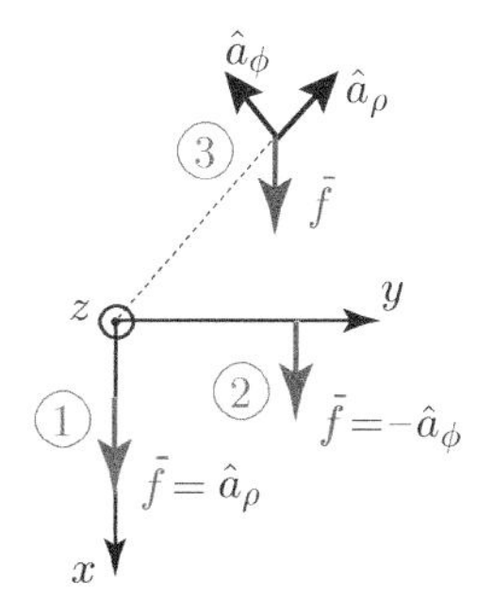

Figure 0.11 A demonstration of how the representation of a vector depends on its location in a cylindrical coordinate system: $\bar{f} = \hat{a}_x$ becomes $\hat{a}_\rho$ and $\hat{a}_\phi$ at positions 1 and 2, respectively, while it needs to be written as a combination of $\hat{a}_\rho$ and $\hat{a}_\phi$ at position 3.

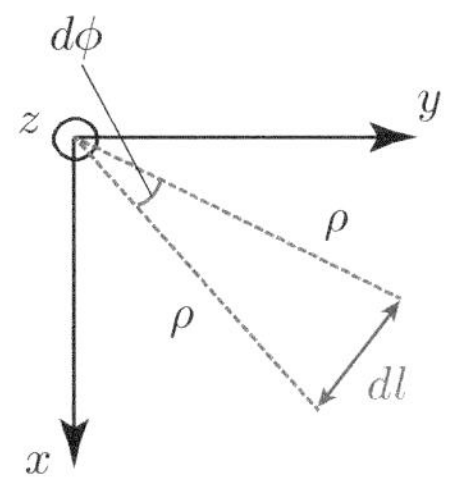

Figure 0.12 Conversion of a differential angle $d\phi$ into a differential length dl in a cylindrical coordinate system.

and

$$dx = \cos\phi d\rho - \sin\phi \rho d\phi \tag{16}$$

$$dy = \sin\phi d\rho + \cos\phi \rho d\phi. \tag{17}$$

It is important that $d\phi$ is (often) used together with ρ (as $\rho d\phi$ above) since ϕ is an angular variable and needs to be converted into a metric variable to be a length (Figure 0.12). Similarly, the basic differential surfaces in a cylindrical coordinate system can be written $ds = \rho d\phi dz$, $ds = d\rho dz$, and $ds = \rho d\rho d\phi$ when ρ, ϕ, or z is constant, respectively, on the surface. Finally, the basic differential volume in a cylindrical coordinate system is simply $dv = \rho d\rho d\phi dz$.

0.4.3 Spherical Coordinate System

In a spherical coordinate system (Figure 0.13), the variables are defined as $u_1 = R$, $u_2 = \theta$, and $u_3 = \phi$, where R is the radial distance from the origin, θ is the angle from the positive z axis, and ϕ is the angle between the projection on the x-y plane and the x axis (same as the one in a cylindrical coordinate system). Note that $R \geq 0$, $\theta \in [0, \pi]$, and $\phi \in [0, 2\pi)$. It is remarkable that θ cannot (and does not have to) be larger than π. Otherwise, a location in a spherical system could not be written uniquely in terms of R, θ, and ϕ. The unit vectors are $\hat{a}_R$, $\hat{a}_\theta$, and $\hat{a}_\phi$, where all these vectors depend on the position. Therefore, a representation of a vector, $\bar{f} = \hat{a}_R f_R + \hat{a}_\theta f_\theta + \hat{a}_\phi f_\phi$, depends on its location.

In a spherical coordinate system, a position vector, which is defined as a vector from the origin to a position (R, θ, ϕ), is written[4]

$$\bar{r} = \hat{a}_R R, \qquad \bar{r} = \hat{a}_r |\bar{r}|, \qquad |\bar{r}| = R. \tag{18}$$

Given a point $P = (x, y, z)$ in a Cartesian coordinate system, the same point can be represented in a spherical coordinate system as $P = (R, \theta, \phi)$, where[5]

$$R = \sqrt{x^2 + y^2 + z^2} \tag{19}$$

$$\theta = \tan^{-1}\left(\frac{\sqrt{x^2 + y^2}}{z}\right) = \cos^{-1}\left(\frac{z}{\sqrt{x^2 + y^2 + z^2}}\right) \tag{20}$$

$$\phi = \tan^{-1}(y/x). \tag{21}$$

Similarly, given a point $P = (R, \theta, \phi)$ in a spherical coordinate system, the same point can be represented in a Cartesian coordinate system as $P = (x, y, z)$, where[6]

$$x = R \sin\theta \cos\phi \tag{22}$$

[4]Note that the expression for a position vector does not contain any θ or ϕ component.

[5]These inverse trigonometric expressions may lead to some numerical problems, but for good reasons. For example, if we try to find ϕ for $x = 0$, we arrive at $\tan^{-1}(\infty)$ (if $y > 0$), which may not be a good idea to calculate on a computer. Indeed, $\tan^{-1}(\infty) = \pi/2$, indicating that the location is on the y-z plane. But what if $y = 0$ too? Then $\tan^{-1}(y/x)$ becomes not well-defined: i.e. we cannot evaluate it without knowing how x and y values go to zero in the limit. In this case, the point is located on the z axis.

[6]The equality $z = R\cos\theta$ is relatively easy to understand, considering that θ is the angle of the location from the positive z axis. While the equality $x = R\sin\theta\cos\phi$ may not be so obvious, one can consider $x = \rho\cos\phi$ in a cylindrical coordinate system. Furthermore, we have $\rho = R\sin\theta$, leading to $x = (R\sin\theta)\cos\phi$. Similarly, we have $y = (R\cos\theta)\cos\phi$.

$$y = R\sin\theta\sin\phi \tag{23}$$

$$z = R\cos\theta. \tag{24}$$

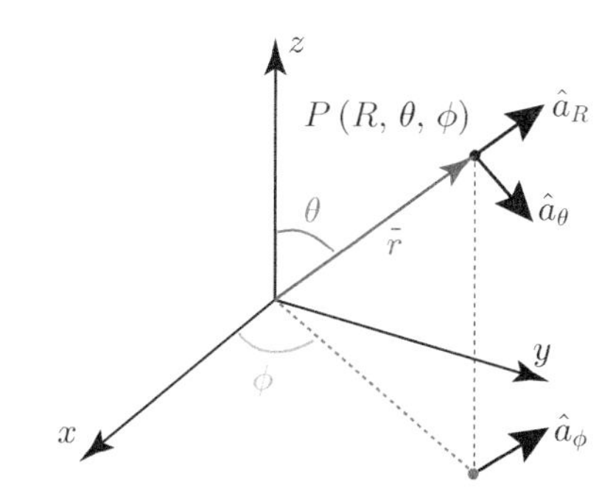

Figure 0.13 A spherical coordinate system with spherical variables.

Using these expressions, differential lengths in Cartesian and spherical coordinate systems can be written in terms of each other as

$$dR = \sin\theta\cos\phi dx + \sin\theta\sin\phi dy + \cos\theta dz \tag{25}$$

$$Rd\theta = \cos\theta\cos\phi dx + \cos\theta\sin\phi dy - \sin\theta dz \tag{26}$$

$$R\sin\theta d\phi = -\sin\phi dx + \cos\phi dy \tag{27}$$

and

$$dx = \sin\theta\cos\phi dR + R\cos\theta\cos\phi d\theta - R\sin\theta\sin\phi d\phi \tag{28}$$

$$dy = \sin\theta\sin\phi dR + R\cos\theta\sin\phi d\theta + R\sin\theta\cos\phi d\phi \tag{29}$$

$$dz = \cos\theta dR - R\sin\theta d\theta. \tag{30}$$

We note how R and $R\sin\theta$ (corresponding to ρ in a cylindrical system) are used with $d\theta$ and $d\phi$ to convert them into differential lengths. Similarly, the basic differential surfaces in a spherical coordinate system can be written $ds = R^2\sin\theta d\theta d\phi$, $ds = R\sin\theta dRd\phi$, and $ds = RdRd\theta$ when R, θ, or ϕ is constant, respectively, on the surface. In a spherical coordinate system, the basic differential volume is $dv = R^2\sin\theta dRd\theta d\phi$.

0.5 Electrostatics, Magnetostatics, and Electromagnetics

Electric charge is a useful concept to describe interactions (specifically *force at distance*) between electric particles. A proton and an electron are assumed to have positive and negative electric charges, respectively, with equal amounts as

Q: electric charge
Unit: coulomb (C)

$$Q_p = -Q_e \approx 1.6022 \times 10^{-19} \quad \text{coulomb (C)}. \tag{31}$$

If these particles are assumed to be static (stationary and time-invariant), Coulomb's law provides the strength of the attractive electric force between them,[7] as considered in Section 1.6. One can claim that this force is the observed quantity, leading to the definition of electric charges. Whichever comes first, the concept of electric charges enables the analyses of complex electrostatic, magnetostatic, and electromagnetic interactions, not only between single particles but also at large scales.

[7]This electric force is very large (more than 10^{36} times) in comparison to the gravity force between a proton and an electron.

I: Electric current
Unit: ampere (A) (coulombs/second)

Electrostatics is a study of static (stationary and time-invariant) electric charges: i.e. understanding interactions between electric charges when they are at rest (not moving) and not changing with respect to time. As formulated via Coulomb's law, electric forces between static electric charges can be interpreted by electric fields created by electric charges affecting each other. When electric charges move or change, however, additional interaction mechanisms must be included.

A continuous movement of large numbers of electric charges is interpreted as electric current (Figure 0.14). In conductors, electric currents are generally created by drifting free electrons. Considering that a derivative with respect to time indicates movement, we have

$$I(t) = -\frac{dQ(t)}{dt}, \tag{32}$$

where the unit is the ampere (A): i.e. coulombs/second (C/s)[8]. Magnetostatics is a study of steady (stationary and time-invariant) electric currents: i.e. when $I(t) = I_0$ itself does not change with respect to time and does not change position. In addition, magnetostatics include time-invariant electric fields; hence, electric charges do not accumulate anywhere, and the electric charge density values (if nonzero) do not change with respect to time.[9] This generally requires time-invariant electric currents that flow in stationary loops. Such loops that carry electric currents apply magnetic force to each other, which can be formulated via Ampere's force law or the Biot-Savart law, similar to Coulomb's law formulating electric force between electric charges. In fact, if current loops are neutral but still carrying electric currents (e.g. drifting free electrons and fixed protons), they should not apply any electric force to each other, and the force between them is only of magnetic type. Similar to electric fields to interpret electric forces, magnetic fields are defined to describe interactions between current-carrying bodies. Specially, an (electric)-current-carrying object generates a magnetic field that affects other current-carrying objects around it.

Electromagnetics is a general name for the study of electric charges and currents. Hence, not only static cases but also dynamic cases are included in electromagnetics.[10] Acceleration of electric charges leads to electromagnetic interactions that cannot be modeled via the laws of electrostatics and magnetostatics.[11] In such a dynamic case, electric and magnetic fields are coupled into electromagnetic waves that propagate in space. Maxwell's equations describe all necessary interactions between electric and magnetic fields and sources (electric charges and currents). These four equations, which are considered basic topics of this book, can be summarized as follows:

- Gauss' law: Describes the generation of electric fields due to electric charges
- Faraday's law: Describes the generation of electric fields due to time-variant magnetic fields

[8] The derivative relationship between the electric charge and electric current can be well understood if one considers a closed volume including an electric charge of $Q(t)$. If the amount of the electric charge is decreasing with respect to time, there should be some net electric current flow outward from the bounding surface of the volume due to the conservation of electric charges. Then, this can be generalized to any arbitrary surface, where we count the total number of electric charges ΔQ passing through the surface per unit time Δt, leading to $\lim_{t\to 0}(-\Delta Q/\Delta t) = -dQ(t)/dt$ for the instantaneous value of the electric current.

[9] Hence, for a given finite volume without any charge variation, the net electric current flowing inward and outward is zero, as also indicated by Kirchhoff's current law.

[10] From this perspective, electrostatics and magnetostatics are special cases of electromagnetics.

[11] In fact, even without acceleration of electric charges, one can see the entanglement of electric and magnetic fields/forces. For example, similar to the freedom in the selection of a coordinate system, we can choose a moving observation point instead of a fixed one. Then, depending on the frame, a given electric charge can be stationary or moving. This means electric and magnetic forces are not completely distinct, and actually, they can be considered different observations or interpretations of physical interactions between electric charges. As a simple case, according to the Lorentz force law, even a moving electric charge exposed to a static magnetic field is affected by a magnetic force. If the observation frame is also moving with the electric charge, making it stationary, there should not be any magnetic force, while the electric charge should still be affected by the same amount of force since the underlying physics has not changed. In fact, according to Lorentz transformations, the force becomes electrical type of the same amount. Based on this interchangeability, all electrical interactions can be named in general as electromagnetic forces, including both static and dynamic cases.

- Gauss' law for magnetism: Describes the nature of magnetic fields and nonexistence of magnetic charges
- Ampere's law: Describes the generation of magnetic fields due to electric currents and time-variant electric fields

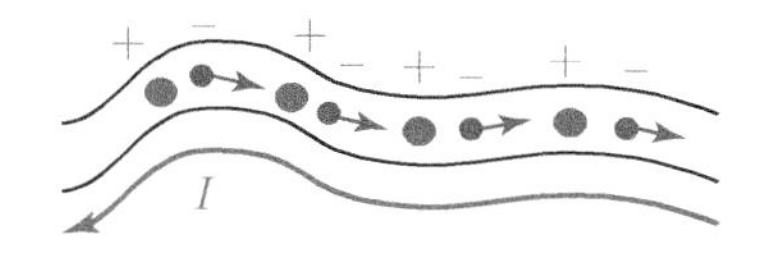

Figure 0.14 In a conductor, drifting electrons due to an applied voltage (electric field) lead to electric current. Since electrons have negative electric charges, by convention, the electric current is in the opposite direction of the electron drift.

The final forms of these laws with necessary extensions were written by Maxwell; hence, they are called Maxwell's equations.

0.6 Time in Electromagnetics

Time is a major parameter in electromagnetics. Electromagnetic propagation is based on synchronized oscillations of electric and magnetic fields due to fluctuating sources with respect to time. As in other areas of engineering, time dependency can be in transient or steady state. In a transient state, which is mainly triggered by an irregular change in sources, the generated fields fluctuate until they settle into regular forms (e.g. zero values in the absence of time-harmonic sources). Specially, in the absence of external effects, all systems enter into steady state after a sufficient time.[12] Frequency is the main parameter in steady-state electromagnetics by describing how sources and fields regularly oscillate with respect to time.[13] Specifically, in many dynamic cases, sources and fields are assumed to have a time-harmonic behavior: i.e. they have a form of

$$f(\bar{r},t) = f_0(\bar{r})\cos(\omega t + \phi_0) \tag{33}$$

for scalar fields and

$$\bar{f}(\bar{r},t) = \bar{f}_0(\bar{r})\cos(\omega t + \phi_0) \tag{34}$$

for vector fields.[14] In these expressions, $\omega = 2\pi f$ is the angular frequency with unit radians/second (rad/s) and f is the frequency with unit hertz (Hz). From this perspective, a static (electrostatic or magnetostatic) problem can be considered a limit case, where $\omega \to 0$.

Time-harmonic problems are best solved in the phasor domain. A given time harmonic vector function $\bar{f}(t) = \bar{f}_0\cos(\omega t + \phi_0)$ can be represented as

$$\bar{f}(t) = \text{Re}\{\bar{f}_0 \exp(j\omega t + j\phi_0)\}, \tag{35}$$

where $j = \sqrt{-1}$ is the imaginary unit, while the real part of the complex exponential function is used to represent the cosine function. Defining a phasor of $\bar{f}(t)$ as

$$\bar{f} = \bar{f}_0 \exp(j\phi_0) = \bar{f}_0\underline{/\phi_0}, \tag{36}$$

[12] In many cases, mathematical models estimate infinite time to reach steady state, while, in practice, systems are assumed be in steady state after a sufficient time for transient fluctuations to become negligible.

[13] Electrostatics and magnetostatics can be seen as steady states with zero frequency.

[14] In these expressions, $f_0(\bar{r})$ and $|\bar{f}_0(\bar{r})|$ are called amplitudes of the fields.

f: Frequency
Unit: hertz (Hz) (1/second)

ω: Angular frequency
Unit: radians/second (rad/s)

we obtain

$$\bar{f}(t) = \text{Re}\{\bar{f} \exp(j\omega t)\}, \tag{37}$$

where all information regarding the function is included in $\bar{f}$. Hence, instead of the function itself, its phasor can be used in mathematical manipulations.

An important advantage of the phasor representation appears when considering time derivatives. Specifically, for a vector function $\bar{f}(t)$ and its phasor $\bar{f}$, we have

$$\frac{\partial \bar{f}(t)}{\partial t} = \frac{\partial}{\partial t}\text{Re}\{\bar{f}_0 \exp(j\omega t + j\phi_0)\} = \text{Re}\{\bar{f}_0 j\omega \exp(j\omega t + j\phi_0)\}, \tag{38}$$

leading to the phasor representation of the derivative itself as[15]

$$\frac{\partial \bar{f}(t)}{\partial t} \longrightarrow j\omega \bar{f}_0 \exp(j\phi_0) = j\omega \bar{f}_0 \underline{/\phi_0} = j\omega \bar{f}. \tag{39}$$

[15]Similarly, we have

$$\frac{\partial^2 \bar{f}(t)}{\partial t^2} \longrightarrow (j\omega)^2 \bar{f} = -\omega^2 \bar{f}.$$

Time integrals and shifts can also be performed easily as

$$\int \bar{f}(t)dt \longrightarrow \frac{1}{j\omega}\bar{f} \tag{40}$$

$$\bar{f}(t - t_0) \longrightarrow \bar{f} \exp(-j\omega t_0). \tag{41}$$

0.7 Final Remarks

In this chapter, we have considered a few important concepts that are useful while discussing Maxwell's equations in the next chapters. Scalar and vector fields, as well as coordinate systems, let us understand and study electromagnetic scenarios that are formulated with Maxwell's equations. In the following (Chapters 1–4), each chapter is devoted to one of Maxwell's equations that are listed in Section 0.5. The mathematical basis and tools to understand each Maxwell equation will also be discussed, but only when they are required. In all discussions, the considered functions (scalar and vector fields) are written in general as $f(\bar{r}, t)$ or $\bar{f}(\bar{r}, t)$ to indicate their dependency on both position $\bar{r}$ and time t. For time-harmonic cases and in steady states, phasor expressions do not contain time dependency, and they are shown as $f(\bar{r})$ or $\bar{f}(\bar{r})$. These expressions are also used in static (zero-frequency) cases. We emphasize once again that electrostatics and magnetostatics are considered special categories of electromagnetics in this book.

1

Gauss' Law

$$\oint_S \bar{D}(\bar{r},t)\cdot\overline{ds} = Q_{\text{enc}}(t)$$

$$\bar{\nabla}\cdot\bar{D}(\bar{r},t) = q_v(\bar{r},t)$$

We start with Gauss' law, which describes how electric fields are created by electric charges. There are two different forms of Gauss' law. First, in the integral form, it states that the flux of the electric field through a surface is equal to the enclosed electric charge. This is a very strong theorem, especially considering that it does not matter how the electric charge is distributed inside the volume. At the same time, it is useful only in symmetric cases to find electric fields from electric charges. Second, in the differential form, Gauss' law provides pointwise information: i.e. the divergence of the electric flux density is the electric charge density. While the divergence operation (as a kind of derivative) is a loss of information, this is also an important tool to relate electric fields to electric charges. Application of Gauss' law at boundaries further leads to important boundary conditions that must be satisfied by electric fields across boundaries. In general, both forms of Gauss' law are valid for all cases, including static and dynamic scenarios. On the other hand, Gauss' law alone is not sufficient to analyze most dynamic cases, and other of Maxwell's equations are required to completely investigate electromagnetic radiation. However, in static cases, Gauss' law is an excellent tool, hand-in-hand with Coulomb's law in a vacuum.

1.1 Integral Form of Gauss' Law

Gauss' law provides the relationship between electric charges and electric fields: i.e. it shows that electric fields are created by electric charges. Consider a surface S bounding an electric

Introduction to Electromagnetic Waves with Maxwell's Equations, First Edition. Özgür Ergül.

Companion website: www.wiley.com/go/ergulmax

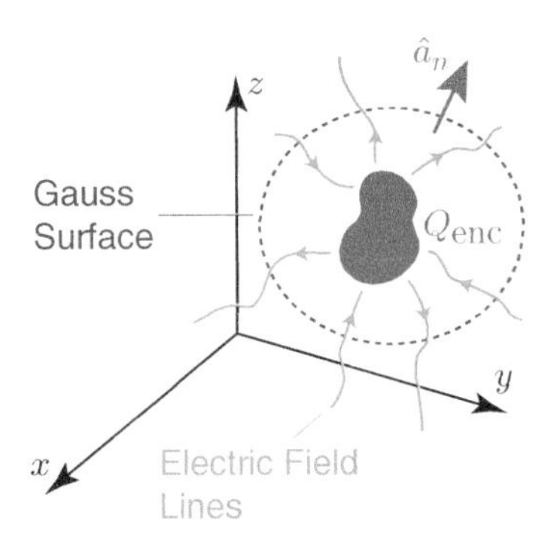

Figure 1.1 A representation of Gauss' law. Depending on the electric charge distribution, electric field lines may enter or leave a mathematical (Gauss) surface at different locations. However, the total net flux is always proportional to the total amount of the net electric charge enclosed. The geometry of the surface does not matter.

$\bar{D}$: electric flux density
Unit: coulombs/meter^2 (C/m^2)

$\bar{E}$: electric field intensity
Unit: newtons/coulomb (N/C)
or volts/meter (V/m)

ε: permittivity
Unit: farads/meter (F/m)

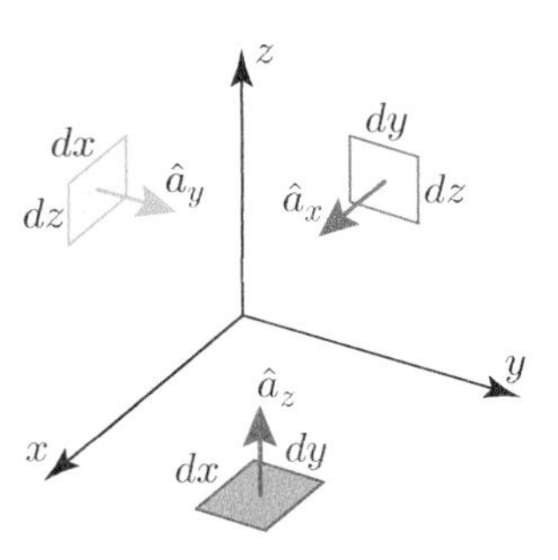

Figure 1.2 Geometric representation of differential surfaces in a Cartesian coordinate system.

charge distribution of arbitrary shape (Figure 1.1). Let the total amount of the electric charge be $Q_{\mathrm{enc}}(t)$. Then Gauss' law states that

$$\oint_S \bar{D}(\bar{r},t)\cdot\overline{ds} = Q_{\mathrm{enc}}(t), \tag{1.1}$$

i.e. the integral of the normal component of the electric flux density $\bar{D}(\bar{r},t)$ over the entire surface (that is, flux) is equal to the total amount of the enclosed electric charge. If the medium is *linear* and *isotropic*, as considered throughout this book, the electric flux density is related to the electric field intensity as

$$\bar{D}(\bar{r},t) = \varepsilon(\bar{r})\bar{E}(\bar{r},t), \tag{1.2}$$

where $\varepsilon(\bar{r})$ is the permittivity at location $\bar{r}$. In a *homogeneous* medium, where ε does not depend on the position, we further have

$$\oint_S \bar{E}(\bar{r},t)\cdot\overline{ds} = \frac{Q_{\mathrm{enc}}(t)}{\varepsilon}, \tag{1.3}$$

where ε is taken out of the integral. In a vacuum (the absence of any material), $\varepsilon = \varepsilon_0 \approx 8.85418782\times10^{-12}$ F/m. In this book, both electric flux density and electric field intensity are referred to as *electric field*. To understand the integral form of Gauss' law, we need the following tools and concepts:

- Differential surface with direction
- Dot product
- Flux of vector fields

1.1.1 Differential Surfaces with Direction

In Gauss' law in Eq. (1.1), we use $\overline{ds}$, which is a differential surface with a direction. To be specific, $\overline{ds}$ involves a differential surface ds and a direction $\hat{a}_n$ that is normal to the surface: i.e. $\overline{ds} = \hat{a}_n ds$. We note that there are two choices for $\hat{a}_n$ for a given surface, while it is selected as the outward direction when Gauss' law is used for a closed surface. In a Cartesian coordinate system, some of the common differential surfaces are as follows (Figure 1.2):

- On a constant-x surface: $\overline{ds} = \hat{a}_n ds = \pm\hat{a}_x dydz$

- On a constant-y surface: $\overline{ds} = \hat{a}_n ds = \pm\hat{a}_y dxdz$
- On a constant-z surface: $\overline{ds} = \hat{a}_n ds = \pm\hat{a}_z dxdy$

When y and z are changing (describing the differential surface $ds = dydz$), the normal is in the $\pm x$ direction ($\hat{a}_n = \pm\hat{a}_x$). Similarly, for $ds = dxdz$ and $ds = dxdy$, we have $\hat{a}_n = \pm\hat{a}_y$ and $\hat{a}_n = \pm\hat{a}_z$, respectively. We also note that constant-x, constant-y, and constant-z surfaces are planes.

In a cylindrical coordinate system, the following differential surfaces are commonly used:

- On a constant-ρ surface: $\overline{ds} = \pm\hat{a}_\rho \rho d\phi dz$
- On a constant-ϕ surface: $\overline{ds} = \pm\hat{a}_\phi d\rho dz$
- On a constant-z surface: $\overline{ds} = \pm\hat{a}_z \rho d\rho d\phi$

We note that $d\phi$ is always multiplied with ρ to change this differential angle into a differential length (Figure 12). As indicated above for a Cartesian coordinate system, a constant-z surface is a plane. In addition, a constant-ϕ surface is also a plane with changing variables ρ and z (such a plane can be spanned completely by changing ρ and z). On the other hand, a constant-ρ surface is a cylinder (Figure 1.3) with a radius of the constant (if the constant is not zero). On such a surface, which closes upon itself, ϕ and z are changing variables, while $\pm\hat{a}_\rho$ is the normal direction.

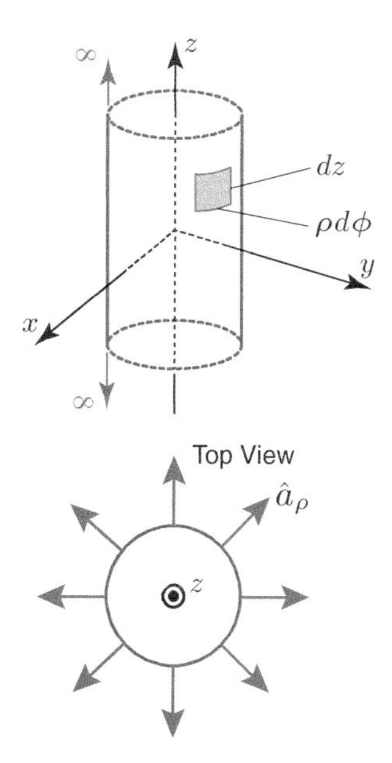

Figure 1.3 A constant-ρ surface in a cylindrical coordinate system and a differential surface on it. The normal direction is selected as outward in this case.

In a spherical coordinate system, commonly used differential surfaces are as follows:

- On a constant-R surface: $\overline{ds} = \pm\hat{a}_R R^2 \sin\theta d\theta d\phi$
- On a constant-θ surface: $\overline{ds} = \pm\hat{a}_\theta R \sin\theta dR d\phi$
- On a constant-ϕ surface: $\overline{ds} = \pm\hat{a}_\phi R dR d\theta$

In the above, $d\theta$ and $d\phi$ are accompanied by R and $R\sin\theta$, respectively, to change these differential angles into differential lengths. A constant-R surface is a sphere of radius R, where θ and ϕ are changing variables. By selecting $\theta \in [0, \pi]$ and $\phi \in [0, 2\pi)$, a full sphere can be spanned. On the other hand, a constant-θ surface is a cone (Figure 1.4), which can be spanned by variables R and ϕ. As also described for a cylindrical coordinate system, a constant-ϕ surface is a plane with changing variables R and θ (instead of ρ and z for a cylindrical system).

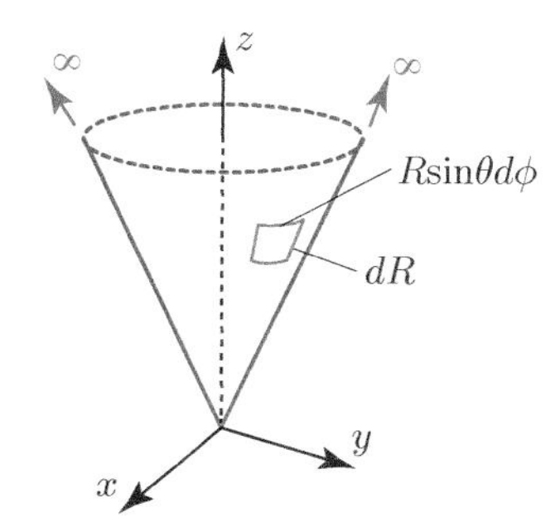

Figure 1.4 A constant-θ surface in a spherical coordinate system and a differential surface on it.

1.1.2 Dot Product

The *dot product* of two vectors $\bar{f}$ and $\bar{g}$ is defined as the length of the projection of the first vector in the direction of the second vector, amplified with the length of the second vector (Figure 1.5). Here, $\bar{f}$ can be projected on $\bar{g}$ or $\bar{g}$ can be projected on $\bar{f}$: i.e. the order does not matter. Using trigonometry, one can derive

$$\bar{f} \cdot \bar{g} = |\bar{f}||\bar{g}| \cos \theta_{fg}, \tag{1.4}$$

where $\theta_{fg} \in [0°, 180°]$ is the angle between $\bar{f}$ and $\bar{g}$. We note some special cases

$$\bar{f} \cdot \bar{g} = |\bar{f}||\bar{g}| \quad \text{if} \quad \theta_{fg} = 0° \tag{1.5}$$

$$\bar{f} \cdot \bar{g} = -|\bar{f}||\bar{g}| \quad \text{if} \quad \theta_{fg} = 180° \tag{1.6}$$

$$\bar{f} \cdot \bar{g} = 0 \quad \text{if} \quad \theta_{fg} = 90° \tag{1.7}$$

and possible ranges of values (Figure 1.6):[1]

$$\bar{f} \cdot \bar{g} > 0 \quad \text{if} \quad 0° \leq \theta_{fg} < 90° \tag{1.8}$$

$$\bar{f} \cdot \bar{g} < 0 \quad \text{if} \quad 90° < \theta_{fg} \leq 180°. \tag{1.9}$$

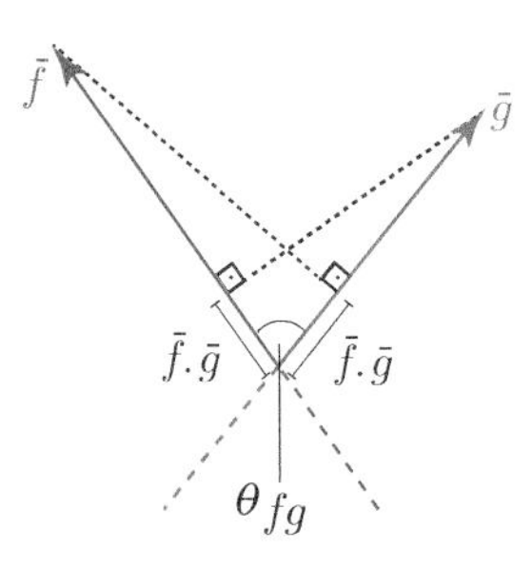

Figure 1.5 A geometric demonstration of the dot product of two vectors $\bar{f}$ and $\bar{g}$. It is usually easier to visualize a dot product when the starting points of the vectors coincide.

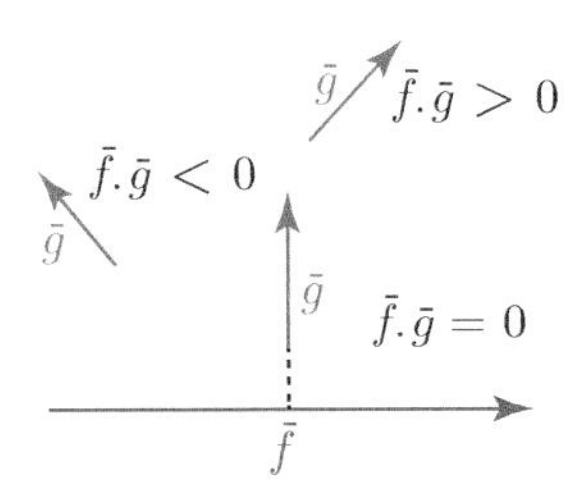

Figure 1.6 Different ranges of values that a dot product can have.

Equations (1.5), (1.6), and (1.7) tell us that the dot product of two vectors is maximized in magnitude when they are aligned, while the projection of a vector on another vector is zero ($\bar{f} \cdot \bar{g} = 0$) if they are perpendicular.[2] We note that the dot product of two vectors is a scalar; therefore, the dot product of two vector fields is a scalar field.[3]

Considering that the dot product gives projection between two vectors, we have

$$f_x = \hat{a}_x \cdot \bar{f}, \quad f_y = \hat{a}_y \cdot \bar{f}, \quad f_z = \hat{a}_z \cdot \bar{f}, \tag{1.10}$$

since the Cartesian components of a vector correspond to its projections on the corresponding Cartesian axes. In addition, we have $\hat{a}_x \cdot \hat{a}_x = 1$, $\hat{a}_y \cdot \hat{a}_y = 1$, $\hat{a}_z \cdot \hat{a}_z = 1$, while $\hat{a}_x \cdot \hat{a}_y = \hat{a}_y \cdot \hat{a}_x = 0$, $\hat{a}_y \cdot \hat{a}_z = \hat{a}_z \cdot \hat{a}_y = 0$, and $\hat{a}_z \cdot \hat{a}_x = \hat{a}_x \cdot \hat{a}_z = 0$. This is the reason for calling a Cartesian system *orthogonal*; its unit vectors $\hat{a}_x$, $\hat{a}_y$, and $\hat{a}_z$ are perpendicular to each other. Using Cartesian representations of vectors, one can derive an expression for the dot product as

$$\bar{f} \cdot \bar{g} = \bar{g} \cdot \bar{f} = (\hat{a}_x f_x + \hat{a}_y f_y + \hat{a}_z f_z) \cdot (\hat{a}_x g_x + \hat{a}_y g_y + \hat{a}_z g_z) \tag{1.11}$$

$$= f_x g_x + f_y g_y + f_z g_z, \tag{1.12}$$

[1] Note that the angle between two vectors is considered to be the smaller angle: i.e. it can be maximum 180°.

[2] Also note that the result of the dot product of two vectors is a scalar, and there is no dot product between a vector and a scalar ($\bar{f} \cdot g =$ undefined). This also means there is no consecutive dot products: i.e. $\bar{f} \cdot \bar{g} \cdot \bar{h} =$ undefined.

[3] Obviously, the dot product reduces the information (some information is lost); therefore, it does not have an inverse function (i.e. a dot division). For example, consider $h = \bar{f} \cdot \bar{g}$. Given h and $\bar{f}$, we cannot find a unique $\bar{g}$, as there are infinitely many vectors satisfying the equality.

considering that the dot product is distributive over summation.[4] Furthermore, using this result for $\bar{g} = \bar{f}$, we have[5]

$$\bar{f} \cdot \bar{f} = f_x^2 + f_y^2 + f_z^2 = |\bar{f}|^2, \tag{1.13}$$

i.e. the dot product of a vector with itself gives information about its length. A related and useful equation can be written

$$(\bar{f} + \bar{g}) \cdot (\bar{f} + \bar{g}) = |\bar{f} + \bar{g}|^2 = |\bar{f}|^2 + |\bar{g}|^2 + 2\bar{f} \cdot \bar{g}. \tag{1.14}$$

Then, using $\bar{f} \cdot \bar{g} \leq |\bar{f}||\bar{g}|$, leading to

$$|\bar{f}|^2 + |\bar{g}|^2 + 2\bar{f} \cdot \bar{g} \leq |\bar{f}|^2 + |\bar{g}|^2 + 2|\bar{f}||\bar{g}| = \left(|\bar{f}| + |\bar{g}|\right)^2, \tag{1.15}$$

we obtain the triangular equality

$$|\bar{f} + \bar{g}| \leq |\bar{f}| + |\bar{g}|, \tag{1.16}$$

which is one of the basic rules in vector calculus (Figure 1.7).

The dot product of two vectors can easily be evaluated in a Cartesian system as in Eq. (1.12). On the other hand, it can be confusing when using other coordinate systems. For example, consider $\hat{a}_\rho \cdot \hat{a}_\phi$, which should be zero since a cylindrical coordinate system is also an orthogonal system. However, we have $\hat{a}_\rho = \hat{a}_y$ on the y axis, while $\hat{a}_\phi = \hat{a}_y$ on the x axis, leading to $\hat{a}_\rho \cdot \hat{a}_\phi = 1$ (by mistake). The problem is that $\hat{a}_\rho$ and $\hat{a}_\phi$ do not represent unique vectors; they depend on where they are defined. For a *given fixed* point, however, we have $\hat{a}_\rho \cdot \hat{a}_\phi = 0$, $\hat{a}_\phi \cdot \hat{a}_z = 0$, and $\hat{a}_z \cdot \hat{a}_\rho = 0$. Similarly, in a spherical coordinate system, $\hat{a}_R \cdot \hat{a}_\theta = 0$, $\hat{a}_\theta \cdot \hat{a}_\phi = 0$, and $\hat{a}_\phi \cdot \hat{a}_R = 0$ at any given point.

The projection property of the dot product provides interesting transformations between vectors. First, we consider the dot products of unit vectors of Cartesian and cylindrical systems. We have (Figure 1.8)

$$\hat{a}_x \cdot \hat{a}_\rho = \cos\phi \tag{1.17}$$

$$\hat{a}_x \cdot \hat{a}_\phi = -\sin\phi \tag{1.18}$$

$$\hat{a}_y \cdot \hat{a}_\rho = \sin\phi \tag{1.19}$$

$$\hat{a}_y \cdot \hat{a}_\phi = \cos\phi, \tag{1.20}$$

where ϕ is the angular position at which the dot products are evaluated. For example, at a given point, the angle between $\hat{a}_x$ and $\hat{a}_\rho$ is ϕ, leading to $\cos\phi$ as their dot product. Similarly, the

[4] $\bar{f} \cdot (\bar{g} + \bar{h}) = \bar{f} \cdot \bar{g} + \bar{f} \cdot \bar{h}$.

[5] Then we also have $\hat{a}_f \cdot \hat{a}_f = |\hat{a}_f|^2 = 1$ for a unit vector $\hat{a}_f$.

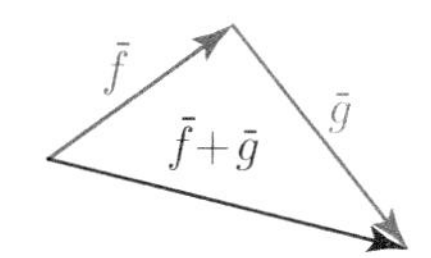

Figure 1.7 According to the triangular inequality, $|\bar{f} + \bar{g}|$ cannot be larger than $|\bar{f}| + |\bar{g}|$, due to the properties of a triangle.

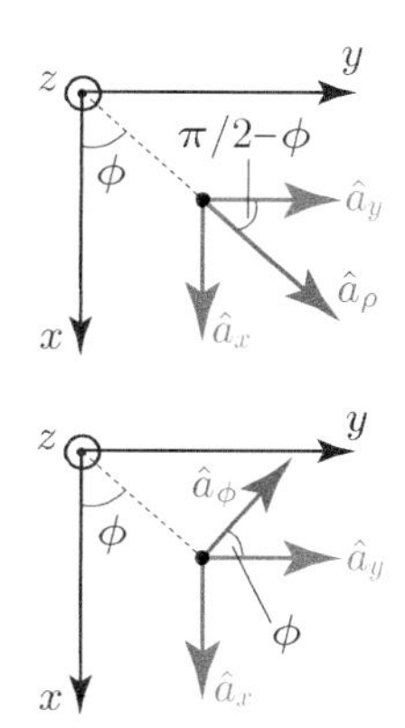

Figure 1.8 Angles between unit vectors, which depend on the position.

angle between $\hat{a}_x$ and $\hat{a}_\phi$ is $90° + \phi$, so that their dot product leads to $\cos(90° + \phi) = -\sin\phi$. These general expressions can be evaluated at any specific point; for example, when $\phi = 90°$ (on the positive y axis), we have $\cos\phi = 0$ and $\sin\phi = 1$, leading to $\hat{a}_x \cdot \hat{a}_\rho = 0$, $\hat{a}_x \cdot \hat{a}_\phi = -1$, $\hat{a}_y \cdot \hat{a}_\rho = 1$, and $\hat{a}_y \cdot \hat{a}_\phi$. Obviously, on a positive y axis, the cylindrical unit vectors become $\hat{a}_\rho = \hat{a}_y$ and $\hat{a}_\phi = -\hat{a}_x$, verifying these values.

Using the results above, we can further write Cartesian unit vectors in terms of cylindrical unit vectors and vice versa. Consider a position represented by ϕ. Since $\hat{a}_z$ is the common unit vector and is perpendicular to all other unit vectors, $\hat{a}_x$ can be written in terms of $\hat{a}_\rho$ and $\hat{a}_\phi$ as[6]

$$\hat{a}_x = (\hat{a}_x \cdot \hat{a}_\rho)\hat{a}_\rho + (\hat{a}_x \cdot \hat{a}_\phi)\hat{a}_\phi = \hat{a}_\rho \cos\phi - \hat{a}_\phi \sin\phi. \tag{1.21}$$

This can be interpreted as $\hat{a}_x$ is decomposed into $\hat{a}_\rho$ and $\hat{a}_\phi$ components, where its projections onto these vectors ($\hat{a}_x \cdot \hat{a}_\rho$ and $\hat{a}_x \cdot \hat{a}_\phi$) provide the values of the components. Similarly, we have[7]

$$\hat{a}_y = \hat{a}_\rho \sin\phi + \hat{a}_\phi \cos\phi \tag{1.22}$$

$$\hat{a}_\rho = \hat{a}_x \cos\phi + \hat{a}_y \sin\phi \tag{1.23}$$

$$\hat{a}_\phi = -\hat{a}_x \sin\phi + \hat{a}_y \cos\phi. \tag{1.24}$$

Dot products between unit vectors further allow us to transform the representation of any vector between different coordinate systems. Considering Cartesian and cylindrical coordinate systems, we have[8]

$$\begin{bmatrix} f_x \\ f_y \\ f_z \end{bmatrix} = \begin{bmatrix} \cos\phi & -\sin\phi & 0 \\ \sin\phi & \cos\phi & 0 \\ 0 & 0 & 1 \end{bmatrix} \begin{bmatrix} f_\rho \\ f_\phi \\ f_z \end{bmatrix} \tag{1.25}$$

and

$$\begin{bmatrix} f_\rho \\ f_\phi \\ f_z \end{bmatrix} = \begin{bmatrix} \cos\phi & \sin\phi & 0 \\ -\sin\phi & \cos\phi & 0 \\ 0 & 0 & 1 \end{bmatrix} \begin{bmatrix} f_x \\ f_y \\ f_z \end{bmatrix} \tag{1.26}$$

for a vector $\bar{f} = \hat{a}_x f_x + \hat{a}_y f_y + \hat{a}_z f_z = \hat{a}_\rho f_\rho + \hat{a}_\phi f_\phi + \hat{a}_z f_z$. In these matrices (namely, transformation matrices), ϕ is again the angle of the position vector showing the *location* of the transformed vector (it is not the angle related to the transformed vector itself). Obviously, we

[6] Hence, depending on the position in terms of ϕ, the unit vector $\hat{a}_x$ may contain both $\hat{a}_\rho$ and $\hat{a}_\phi$, while $\hat{a}_x = \hat{a}_\rho$ when $\phi = 0$ and $\hat{a}_x = -\hat{a}_\phi$ when $\phi = 90°$. But we are in trouble at the origin since the angle ϕ is not defined. In fact, $\hat{a}_x$ is uniquely defined at the origin, while $\hat{a}_\rho$ and $\hat{a}_\phi$ are not unique. This kind of problem arises due to the definition of a coordinate variable. For example, none of the Cartesian coordinate variables has a finite minimum and maximum: i.e. $-\infty < x, y, z < \infty$. However, the cylindrical variable ρ has an end point (minimum) at the origin.

[7] Obviously, using these expressions together should lead to consistent results. For example, we have

$$\begin{aligned} \hat{a}_x &= \hat{a}_\rho \cos\phi - \hat{a}_\phi \sin\phi \\ &= \hat{a}_x \cos^2\phi + \hat{a}_y \cos\phi \sin\phi \\ &\quad + \hat{a}_x \sin^2\phi - \hat{a}_y \sin\phi \cos\phi = \hat{a}_x, \end{aligned}$$

as expected.

[8] Zero terms in these matrices correspond to components with no relationship. Since z is the variable of both Cartesian and cylindrical systems, it is independent of all other variables (x, y, ρ, and ϕ).

should have

$$\begin{bmatrix} \cos\phi & -\sin\phi & 0 \\ \sin\phi & \cos\phi & 0 \\ 0 & 0 & 1 \end{bmatrix} \begin{bmatrix} \cos\phi & \sin\phi & 0 \\ -\sin\phi & \cos\phi & 0 \\ 0 & 0 & 1 \end{bmatrix} = \begin{bmatrix} 1 & 0 & 0 \\ 0 & 1 & 0 \\ 0 & 0 & 1 \end{bmatrix}, \tag{1.27}$$

since transformation from Cartesian to cylindrical and then back to Cartesian should not change the representation of a vector.

It is useful to repeat the discussion above for Cartesian and spherical coordinate systems. First, we note that

$$\hat{a}_x \cdot \hat{a}_R = \sin\theta\cos\phi \tag{1.28}$$
$$\hat{a}_x \cdot \hat{a}_\theta = \cos\theta\cos\phi \tag{1.29}$$
$$\hat{a}_x \cdot \hat{a}_\phi = -\sin\phi \tag{1.30}$$
$$\hat{a}_y \cdot \hat{a}_R = \sin\theta\sin\phi \tag{1.31}$$
$$\hat{a}_y \cdot \hat{a}_\theta = \cos\theta\sin\phi \tag{1.32}$$
$$\hat{a}_y \cdot \hat{a}_\phi = \cos\phi \tag{1.33}$$
$$\hat{a}_z \cdot \hat{a}_R = \cos\theta \tag{1.34}$$
$$\hat{a}_z \cdot \hat{a}_\theta = -\sin\theta \tag{1.35}$$
$$\hat{a}_z \cdot \hat{a}_\phi = 0, \tag{1.36}$$

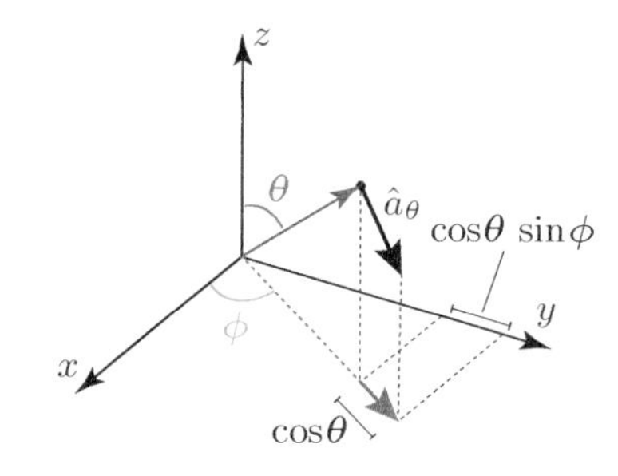

Figure 1.9 Projection of $\hat{a}_\theta$ onto the y axis.

where θ and ϕ represent the position where the dot products are evaluated. These expressions can again be obtained by considering the projection of one unit vector onto the other one.[9] For example, to find the coupling term between $\hat{a}_y$ and $\hat{a}_\theta$, one needs to evaluate $\hat{a}_y \cdot \hat{a}_\theta$ (Figure 1.9). This means it is sufficient to project $\hat{a}_\theta$ onto the y axis. This can be achieved in two steps. First, $\hat{a}_\theta$ can be projected onto the x-y plane. This brings a multiplication with $\cos\theta$. Next, we need to evaluate the projection of the result on y. This brings a second multiplication with $\cos(\pi/2 - \phi) = \sin\phi$. Then we get the overall term as $\cos\theta\sin\phi$.

[9] We are already familiar with the terms $\hat{a}_x \cdot \hat{a}_\phi$, $\hat{a}_y \cdot \hat{a}_\phi$, and $\hat{a}_z \cdot \hat{a}_\phi$ from cylindrical coordinate systems.

We can now write the Cartesian and spherical unit vectors in terms of each other as

$$\hat{a}_x = \hat{a}_R \sin\theta\cos\phi + \hat{a}_\theta \cos\theta\cos\phi - \hat{a}_\phi \sin\phi \tag{1.37}$$
$$\hat{a}_y = \hat{a}_R \sin\theta\sin\phi + \hat{a}_\theta \cos\theta\sin\phi + \hat{a}_\phi \cos\phi \tag{1.38}$$

$$\hat{a}_z = \hat{a}_R \cos\theta - \hat{a}_\theta \sin\theta \tag{1.39}$$

$$\hat{a}_R = \hat{a}_x \sin\theta\cos\phi + \hat{a}_y \sin\theta\sin\phi + \hat{a}_z\cos\theta \tag{1.40}$$

$$\hat{a}_\theta = \hat{a}_x \cos\theta\cos\phi + \hat{a}_y \cos\theta\sin\phi - \hat{a}_z\sin\theta \tag{1.41}$$

$$\hat{a}_\phi = -\hat{a}_x \sin\phi + \hat{a}_y\cos\phi. \tag{1.42}$$

When writing these, we simply consider the dot products (listed above) between the unit vectors. A generalization of these expressions leads to transformation between coordinate systems as[10]

$$\begin{bmatrix} f_x \\ f_y \\ f_z \end{bmatrix} = \begin{bmatrix} \sin\theta\cos\phi & \cos\theta\cos\phi & -\sin\phi \\ \sin\theta\sin\phi & \cos\theta\sin\phi & \cos\phi \\ \cos\theta & -\sin\theta & 0 \end{bmatrix} \begin{bmatrix} f_R \\ f_\theta \\ f_\phi \end{bmatrix} \tag{1.43}$$

and

$$\begin{bmatrix} f_R \\ f_\theta \\ f_\phi \end{bmatrix} = \begin{bmatrix} \sin\theta\cos\phi & \sin\theta\sin\phi & \cos\theta \\ \cos\theta\cos\phi & \cos\theta\sin\phi & -\sin\theta \\ -\sin\phi & \cos\phi & 0 \end{bmatrix} \begin{bmatrix} f_x \\ f_y \\ f_z \end{bmatrix}, \tag{1.44}$$

where $\bar{f} = \hat{a}_x f_x + \hat{a}_y f_y + \hat{a}_z f_z = \hat{a}_R f_R + \hat{a}_\theta f_\theta + \hat{a}_\phi f_\phi$ is any vector.[11]

Different coordinate systems are defined since each of them can be more useful for a given scenario. For example, to represent $\hat{a}_x$ in a spherical coordinate system, all three unit vectors – $\hat{a}_R$, $\hat{a}_\theta$, and $\hat{a}_\phi$ – are needed, as shown in Eq. (1.37). Obviously, a spherical coordinate system is not the most suitable to represent $\hat{a}_x$. On the other hand, a radial direction from the origin is best represented in a spherical system as $\hat{a}_R$. This is particularly useful in such cases with spherical symmetries.[12] Similarly, scenarios with cylindrical symmetries are usually easier to analyze in a cylindrical coordinate system.

1.1.3 Flux of Vector Fields

So far, we have discussed differential surfaces (with directions) and dot product that are necessary tools for Gauss' law in Eq. (1.1). However, to understand Gauss' law, we need the concept of *flux*: i.e. surface integrals in the form of

$$I_f = \int_S \bar{f}(\bar{r}) \cdot \overline{ds} = \int_S \bar{f}(\bar{r}) \cdot \hat{a}_n ds, \tag{1.45}$$

[10]Once again, the multiplication of the transformation matrices is an identity matrix.

[11]Similar to ambiguities in some special positions, defining a vector in a spherical coordinate system may lead to confusions on the z axis. Considering a transformation from Cartesian to spherical representation when $\theta = 0°$, we have

$$f_R = f_z$$
$$f_\theta = \cos\phi f_x + \sin\phi f_y$$
$$f_\phi = -\sin\phi f_x + \cos\phi f_y.$$

But what is ϕ on the z axis? In fact, it can be selected arbitrarily (or depending on the context). For example, if $\phi = 0°$, then $f_\theta = f_x$ while $f_\phi = f_y$. This can be visualized as if we are on the z-x plane ($\phi = 0°$). If $\phi = 90°$, however, $f_\theta = f_y$ and $f_\phi = -f_x$, as we are now assuming that we are on the y-z plane.

[12]We can consider the position vector in different coordinate systems: i.e. $\bar{r} = \hat{a}_x x + \hat{a}_y y + \hat{a}_z z = \hat{a}_\rho \rho + \hat{a}_z z = \hat{a}_R R$. It is best described in a spherical system.

where $\bar{f}(\bar{r})$ is a vector field and S is a surface. This equation can be read as the flux of $\bar{f}(\bar{r})$ through S. The value of the integral (flux) must be independent of the chosen coordinate system; however, given $\bar{f}(\bar{r})$, it may be wiser to choose one coordinate system than another.

In a Cartesian system, if the surface is planar and its normal is in the $\pm x$, $\pm y$, or $\pm z$ directions, we have

$$I_f = \pm \int_S f_x(\bar{r}) dy dz, \quad I_f = \pm \int_S f_y(\bar{r}) dx dz, \quad I_f = \pm \int_S f_z(\bar{r}) dx dy, \tag{1.46}$$

respectively. For a cylindrical surface with radius ρ located around the z axis (normal in the $\pm\rho$ direction), it is best to evaluate the flux as (Figure 1.3)

$$I_f = \pm \int_S f_\rho(\bar{r}) \rho d\phi dz \tag{1.47}$$

using a cylindrical system. Obviously, if not already given, the ρ component of $\bar{f}(\bar{r})$ should be found (using a coordinate transformation) to evaluate this integral. If the normal of the surface is in the $\pm z$ direction, while a cylindrical coordinate system needs to be used, we have (Figure 1.10)

$$I_f = \pm \int_S f_z(\bar{r}) \rho d\rho d\phi. \tag{1.48}$$

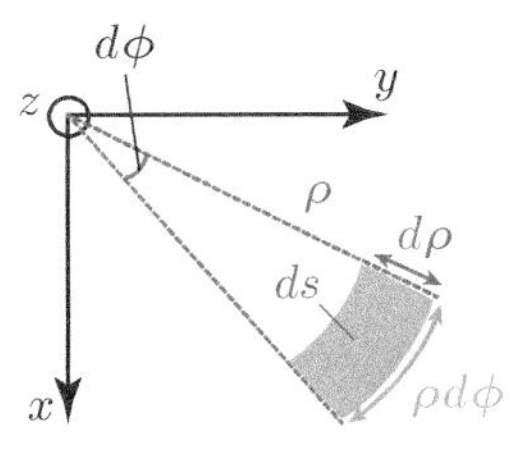

Figure 1.10 A differential surface on a constant-z plane using cylindrical parameters.

In addition, if the surface is planar but its normal is in a $\pm\phi$ direction (not in the x or y direction), one can use

$$I_f = \pm \int_S f_\phi(\bar{r}) dz d\rho. \tag{1.49}$$

On a spherical surface with a normal $\hat{a}_R$, the spherical coordinate system can be employed as (Figure 1.11)

$$I_f = \pm \int_S f_R(\bar{r}) R^2 \sin\theta d\theta d\phi. \tag{1.50}$$

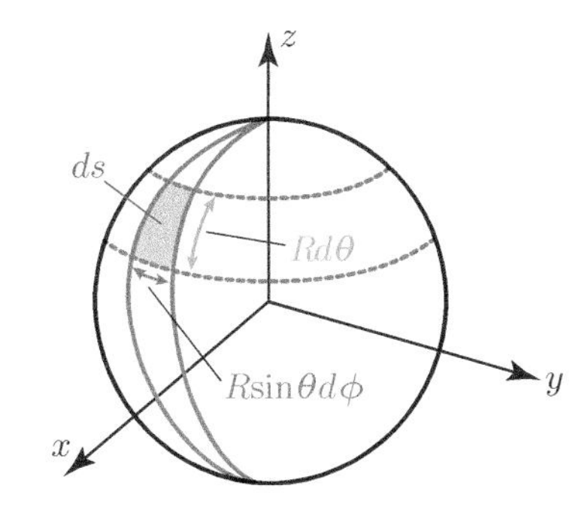

Figure 1.11 A differential surface on a sphere.

If the normal of a planar surface is in a $\pm\phi$ direction, while using a spherical system, an option to evaluate the flux is

$$I_f = \pm \int_S f_\phi(\bar{r}) R dR d\theta. \tag{1.51}$$

Finally, on a conical surface with radial center at the origin (Figure 1.4), it may be an advantage to use a spherical system as

$$I_f = \pm \int_S f_\theta(\bar{r}) R \sin\theta d\phi dR. \tag{1.52}$$

Hence, given $\bar{f}(\bar{r})$ and the surface S, there are many options to evaluate the flux.[13]

As a simple example (Figure 1.12), consider a vector field $\bar{f}(\bar{r}) = \hat{a}_z f_0$, where f_0 is constant, an upper hemisphere is defined as $R \leq a$, $0 \leq \theta \leq \pi/2$, and $0 \leq \phi < 2\pi$. Taking the normal in the R direction, we have

$$I_f = \int_S \bar{f}(\bar{r}) \cdot \overline{ds} = f_0 \int_0^{\pi/2} \int_0^{2\pi} \hat{a}_z \cdot \hat{a}_R R^2 \sin\theta d\theta d\phi. \tag{1.53}$$

Then, using $\hat{a}_z \cdot \hat{a}_R = \cos\theta$ and considering that $R = a$, one can obtain

$$I_f = f_0 a^2 \int_0^{\pi/2} \int_0^{2\pi} \cos\theta \sin\theta d\theta d\phi. \tag{1.54}$$

Evaluating the integral above, we have[14,15]

$$I_f = f_0 2\pi a^2 \int_0^{\pi/2} \cos\theta \sin\theta d\theta = f_0 \pi a^2 \int_0^{\pi/2} \sin(2\theta) d\theta \tag{1.55}$$

$$= f_0 \pi a^2 \left[-\frac{\cos(2\theta)}{2} \right]_0^{\pi/2} = f_0 \pi a^2. \tag{1.56}$$

Note that a larger a (larger surface) leads to more flux. This is similar to collecting more water using a larger cup in the rain.

Consider again $\bar{f}(\bar{r}) = \hat{a}_z f_0$, but a lower hemisphere defined as $R \leq a$, $\pi/2 \leq \theta \leq \pi$, and $0 \leq \phi < 2\pi$. If the normal direction is again selected as $\hat{a}_R$, we have the flux as

$$I_f = f_0 2\pi a^2 \int_{\pi/2}^{\pi} \cos\theta \sin\theta d\theta = f_0 \pi a^2 \left[-\frac{\cos(2\theta)}{2} \right]_{\pi/2}^{\pi} = -f_0 \pi a^2. \tag{1.57}$$

Since $\bar{f}(\bar{r})$ and $\overline{ds}$ are always in the opposite directions for this surface, it is natural to obtain a negative value as the flux. More importantly, if the flux over a full sphere is considered, we obtain

$$I = \oint_S \bar{f}(\bar{r}) \cdot \overline{ds} = f_0 \pi a^2 - f_0 \pi a^2 = 0. \tag{1.58}$$

[13] In all these integrals, the integration is simply shown as $\int_S$, while it is indeed a double integral involving two variables. Integration limits often lead to confusions, especially in cylindrical and spherical coordinate systems. In the general expressions, $\pm$ is used to show the dependency on the normal direction of the surface. If the normal direction is selected correctly, a general rule is to select the limits from smaller to larger values. These selections are better clarified in the examples and exercises.

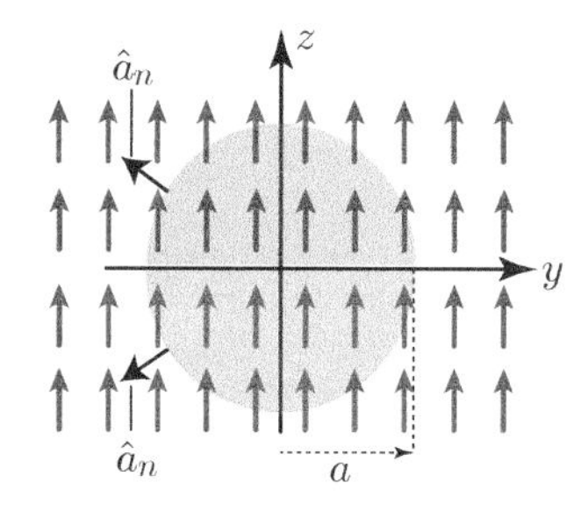

Figure 1.12 A constant vector field through hemispherical surfaces.

[14] We note how the value of $\bar{f}(\bar{r}) \cdot \overline{ds}$ – i.e. the integrand of the flux – changes over the surface. When $\theta = 0°$ and $\cos\theta = 1$, $\bar{f}(\bar{r})$ is aligned with $\overline{ds}$, and the contribution to the flux is large. When $\theta = 90°$ and $\cos\theta = 0$, however, $\bar{f}(\bar{r})$ is tangent to the surface, and no contribution occurs. Hence, depending on the location on the surface, different contributions are added to the overall integral.

[15] $\sin\theta\cos\theta = \sin(2\theta)/2$.

A flux over a closed surface, which is expressed as a closed surface integral, is often called *net flux*. For this example, the net flux is zero, which can be interpreted as all inward and outward $\bar{f}(\bar{r})$ over the entire surface being perfectly balanced, leading to zero net flow of the vector field. In Gauss' law, zero net flux means the algebraic sum of all sources and sinks (negative sources) enclosed in the considered surface is zero.

Next, we investigate the net flux of the constant vector field $\bar{f}(\bar{r}) = \hat{a}_z f_0$ through an upper closed hemisphere (Figure 1.13). Specifically, the upper hemisphere in the previous example is now closed by using a disk at the bottom. Dividing the net flux into two parts, we have

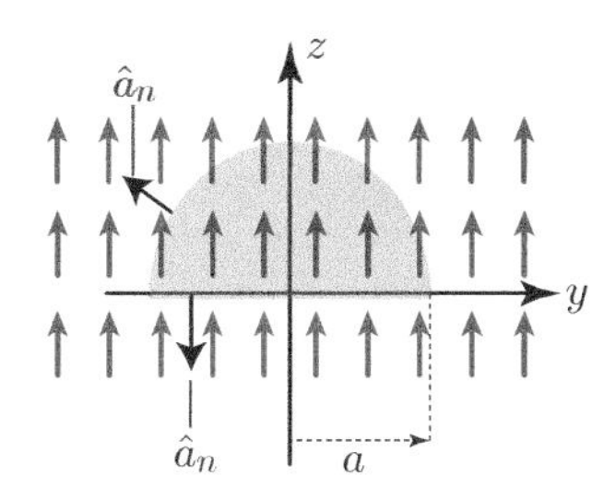

Figure 1.13 A constant vector field through a closed hemisphere.

$$I = \oint_S \bar{f}(\bar{r}) \cdot \overline{ds} = \int_{\text{Hemishere}} \bar{f}(\bar{r}) \cdot \overline{ds} + \int_{\text{Disk}} \bar{f}(\bar{r}) \cdot \overline{ds}, \tag{1.59}$$

where

$$\int_{\text{Hemishere}} \bar{f}(\bar{r}) \cdot \overline{ds} = f_0 \pi a^2 \tag{1.60}$$

as found previously. For the disk surface, we have $\hat{a}_n = -\hat{a}_z$ as the outward direction and $\hat{a}_n \cdot \bar{f}(\bar{r}) = -f_0$. This means the integral on the disk is simply $-f_0$ times the area of the disk:

$$\int_{\text{Disk}} \bar{f}(\bar{r}) \cdot \overline{ds} = -f_0 \pi a^2, \tag{1.61}$$

leading to

$$I = \oint_S \bar{f}(\bar{r}) \cdot \overline{ds} = 0. \tag{1.62}$$

Once again, all entering and leaving $\bar{f}(\bar{r})$ through the surface cancel each other.

In many cases, the net flux through a surface is nonzero. As a simple case (Figure 1.14), we consider $\bar{f}(\bar{r}) = \hat{a}_z z$ through a sphere of radius a. In this case, we can evaluate the net flux as a single surface integral. First, we note that

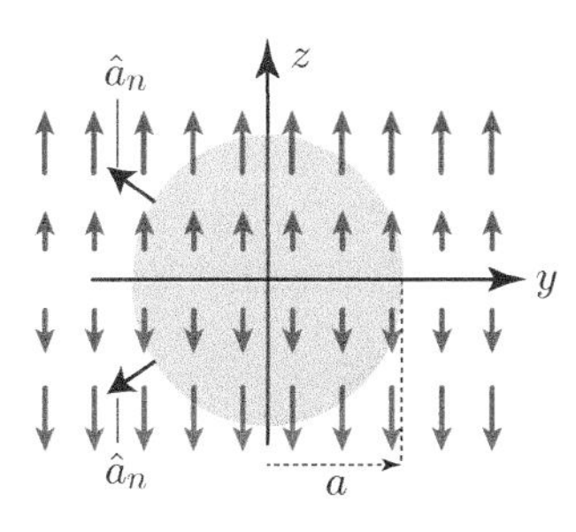

Figure 1.14 A varying vector field $\bar{f}(\bar{r}) = \hat{a}_z z$ through a sphere.

$$\bar{f}(\bar{r}) \cdot \overline{ds} = z\hat{a}_z \cdot \hat{a}_R R^2 \sin\theta d\theta d\phi = R^3 \cos^2\theta \sin\theta d\theta d\phi, \tag{1.63}$$

using $z = R\cos\theta$ and $\hat{a}_z \cdot \hat{a}_R = \cos\theta$. Furthermore, $R = a$ on the sphere, leading to

$$\bar{f}(\bar{r}) \cdot \overline{ds} = a^3 \cos^2\theta \sin\theta d\theta d\phi. \tag{1.64}$$

Then, evaluating the integral, we obtain

$$I_f = \int_S \bar{f}(\bar{r}) \cdot \overline{ds} = \int_0^{\pi} \int_0^{2\pi} a^3 \cos^2\theta \sin\theta d\theta d\phi \tag{1.65}$$

$$= 2\pi a^3 \left[-\frac{\cos^3(\theta)}{3} \right]_0^{\pi} = f_0 \pi a^2 = \frac{4\pi}{3} a^3. \tag{1.66}$$

Hence, in this case, we obtain a positive net flux. In fact, as also shown in Figure 1.14, $\bar{f}(\bar{r})$ is flowing outward on the entire surface of the sphere, and this is numerically verified in Eq. (1.66). To create this net flux, there should be some net source inside the sphere.

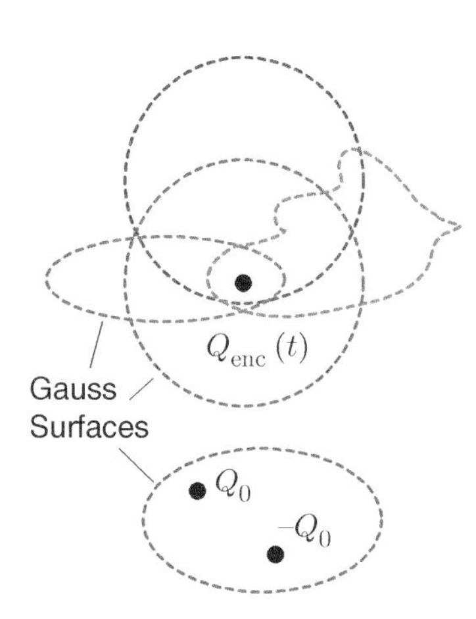

Figure 1.15 According to Gauss' law, the net electric flux through all surfaces enclosing the same electric charge is the same, independent of the shape of the surface. In addition, the flux through a surface enclosing zero net electric charge is zero, even when there are electric charges inside the surface.

1.1.4 Meaning of Gauss' Law and Its Application

According to Gauss' law in Eq. (1.1), the net flux of the electric flux density (this is why $\bar{D}(\bar{r}, t)$ is called flux *density*) through a surface is proportional to the amount of the net electric charge enclosed (at a fixed time t). At first glance, this may seem strange. For example, according to Gauss' law, the net flux through a surface enclosing a static positive electric charge $+Q_0$ and a static negative electric charge $-Q_0$ at two distinct locations (inside the surface) must be zero (Figure 1.15). Indeed, the flux is really zero for this case, but this does not mean $\bar{D}(\bar{r})$ is zero on the surface. It indicates that the vector sum of the electric flux density, which may be outward on some portion of the surface and inward on the other, is algebraically zero.

To understand the physical meaning of Gauss' law, we consider an analogy of a full water tank that is filled and/or emptied by water sources/sinks located inside of it (Figure 1.16). It does not matter where the source and sink are located, provided that they are inside the tank. If the source pumps in more water than is removed by the sink, the amount of water increases, and the tank overflows (positive net flux). On the other hand, if the sink removes more water than the source provides, the amount of water decreases (negative net flux). The volume of water is steady only if the source and sink balance each other, which corresponds to zero net flux. Interestingly, Gauss' law (in this form) does not provide information about where the sources/sinks are located.

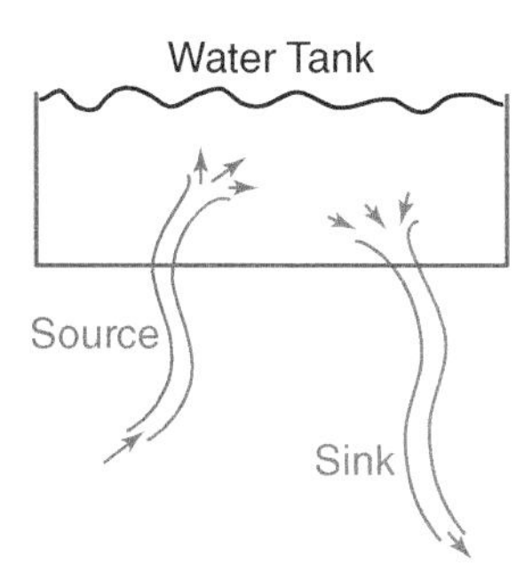

Figure 1.16 An analogy for describing Gauss' law. A water tank is filled by a source and emptied by a sink. The response (increasing/decreasing/steady volume of water) depends on the amounts of water pumped in and out by the source and the sink, respectively.

It is remarkable that, similar to its differential form considered in Section 1.3, the integral form of Gauss' law holds for dynamic electric charges/fields. Indeed, the integral from Eq. (1.1) is written by considering time-variant total enclosed electric charge $Q_{\text{enc}}(t)$ and time-variant net flux:

$$f_D(t) = \oint_S \bar{D}(\bar{r}, t) \cdot \overline{ds}, \tag{1.67}$$

where $f_D(t) = Q_{\text{enc}}(t)$ is indicated by Gauss' law. This means if $Q_{\text{enc}}(t)$ changes due to some electric charges passing through the surface, then the net flux through the surface changes simultaneously.[16]

[16] It should be noted that an electric charge's movement does not necessarily indicate a change in the total enclosed electric charge or the net flux. For example, electric charges inside the surface may move (without leaving the volume enclosed by the surface), leading to a change of the flux density distribution over the surface, while the net flux remains constant.

1.1.5 Examples

Example 1: Let $\bar{f} = \hat{a}_x + \hat{a}_y 4 - \hat{a}_z 5$ and $\bar{g} = -\hat{a}_x 2 - \hat{a}_y 4 + \hat{a}_z 4$ (Figure 1.17). Find the angle between the vectors.

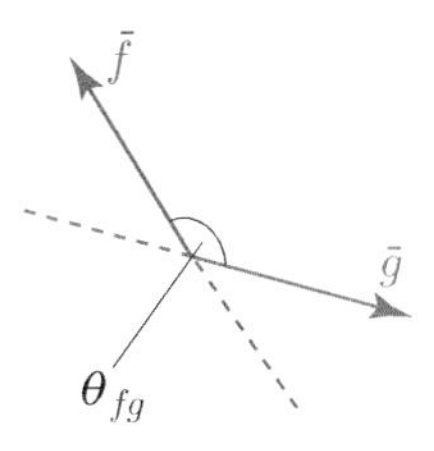

Figure 1.17 Two vectors $\bar{f}$ and $\bar{g}$ with an angle of θ_{fg} between them.

Solution: Using the properties of the dot product, we derive

$$\bar{f} \cdot \bar{g} = -2 - 16 - 20 = -38 \tag{1.68}$$

$$|\bar{f}| = \sqrt{1^2 + 4^2 + (-5)^2} = \sqrt{42} \tag{1.69}$$

$$|\bar{g}| = \sqrt{(-2)^2 + (-4)^2 + 4^2} = \sqrt{36} = 6 \tag{1.70}$$

$$\bar{f} \cdot \bar{g} = |\bar{f}||\bar{g}| \cos\theta_{fg} \longrightarrow -38 = 6\sqrt{42}\cos\theta_{fg} \longrightarrow \cos\theta_{fg} = -\frac{38}{6\sqrt{42}} \tag{1.71}$$

and

$$\theta_{fg} = \cos^{-1}\left(-\frac{38}{6\sqrt{42}}\right) \approx 168°. \tag{1.72}$$

A negative value indicates that $\theta_{fg} > 90°$.

Example 2: Find the net outward flux of $\bar{f}(\bar{r}) = \hat{a}_z z + \hat{a}_y x$ through a cup-shaped closed surface involving a hemisphere of radius b inside another hemisphere of radius a, and a top surface (Figure 1.18).

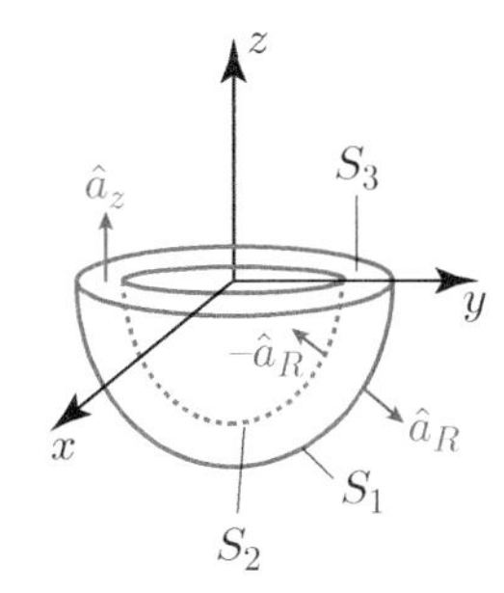

Figure 1.18 A cup-shaped object involving a hemisphere (S_2) of radius b inside another hemisphere (S_1) of radius a, and a top surface (S_3).

Solution: We can find the net outward flux by dividing the surface into three parts: two spherical surfaces and the cylindrical top surface. We have[17,18]

$$I_f = \oint_S \bar{f}(\bar{r}) \cdot \overline{ds} = \int_{S_1} \bar{f}(\bar{r}) \cdot \overline{ds} + \int_{S_2} \bar{f}(\bar{r}) \cdot \overline{ds} + \int_{S_3} \bar{f}(\bar{r}) \cdot \overline{ds} \tag{1.73}$$

$$= \int_0^{2\pi} \int_{\pi/2}^{\pi} (\hat{a}_z z + \hat{a}_y x) \cdot \hat{a}_R R^2 \sin\theta d\theta d\phi$$

$$+ \int_0^{2\pi} \int_{\pi/2}^{\pi} (\hat{a}_z z + \hat{a}_y x) \cdot (-\hat{a}_R) R^2 \sin\theta d\theta d\phi$$

[17] Using the divergence theorem, we simply have $\bar{\nabla} \cdot \bar{f}(\bar{r}) = 1$ and $\int_V 1 dv = \frac{2}{3}\pi a^3 - \frac{2}{3}\pi b^3$ considering the volume for a hemisphere.

[18] Note that for the inner hemisphere, the outward normal is $-\hat{a}_R$. When we are going *outward* from the object on the inner surface, $-\hat{a}_R$ is the direction of movement.

$$+\int_0^{2\pi}\int_b^a (\hat{a}_z z + \hat{a}_y x)\cdot \hat{a}_z \rho d\rho d\phi \tag{1.74}$$

$$= \int_0^{2\pi}\int_{\pi/2}^{\pi} (a\cos^2\theta + a\sin^2\theta\cos\phi\sin\phi)a^2\sin\theta d\theta d\phi$$

$$- \int_0^{2\pi}\int_{\pi/2}^{\pi} (b\cos^2\theta + b\sin^2\theta\cos\phi\sin\phi)b^2\sin\theta d\theta d\phi$$

$$+\int_0^{2\pi}\int_b^a z\rho d\rho d\phi \tag{1.75}$$

$$= (a^3 - b^3)\int_0^{2\pi}\int_{\pi/2}^{\pi} \cos^2\theta\sin\theta d\theta d\phi \tag{1.76}$$

$$= 2\pi(a^3-b^3)\left[-\frac{\cos^3\theta}{3}\right]_{\pi/2}^{\pi} = \frac{2\pi}{3}(a^3-b^3). \tag{1.77}$$

Obviously, when $a = b$, there is a perfect cancellation of the contributions from the spherical surfaces, leading to zero net outward flux.

1.2 Using the Integral Form of Gauss' Law

Gauss' law provides the amount of net flux through a mathematical surface for a given electric charge distribution. However, its usage to find the flux density at desired locations is usually limited to symmetric cases. In addition, Gauss' law usually is not sufficient to find electric fields in dynamic cases. This is not because of the invalidity of Gauss' law (indeed, it is valid for all cases), but because electric fields are not only generated by electric charges when time variance exists. Therefore, we need other of Maxwell's equations for analyses in dynamic cases.

To demonstrate how Gauss' law can be used directly to find the electric flux density, we consider a static point electric charge[19] Q_0 located at the origin in a vacuum (Figure 1.19). Then a symmetrically located spherical surface enclosing the electric charge can be defined, leading to

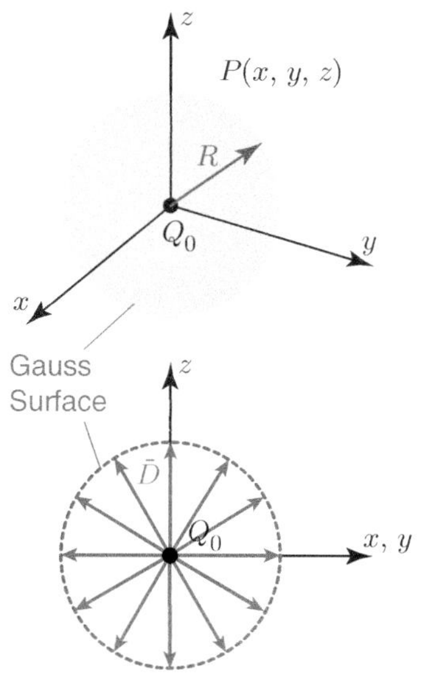

$$\oint_S \bar{D}(\bar{r})\cdot\overline{ds} = \int_0^{2\pi}\int_0^{\pi} \bar{D}(\bar{r})\cdot\hat{a}_R R^2\sin\theta d\theta d\phi = Q_0, \tag{1.78}$$

where R is the radius of the sphere. In the integral form, we are unable to put $\bar{D}(\bar{r})$ directly outside the integral since we do not know how it depends on position $\bar{r}$. On the other hand,

Figure 1.19 Application of Gauss' law for a static point electric charge at the origin. Due to the symmetry, the electric field created by the electric charge is in the radial direction. The arrows show only the directions.

[19]Point electric charge means a finite amount of electric charge is squeezed into a single point. Hence, the electric charge density at that location is actually infinite.

using the symmetry of the Gauss surface, we are able to deduce the required information. First, we can assume that $\bar{D}(\bar{r})$ depends only on R since each observation point on the surface has the same distance from the source.[20] In addition, we further assume that $\bar{D}(\bar{r})$ has only an R component. This is because the point electric charge produces outward (or inward if Q_0 is negative) electric field lines. Then we have

$$\bar{D}(\bar{r}) = \hat{a}_R D_R(\bar{r}) + \hat{a}_\theta D_\theta(\bar{r}) + \hat{a}_\phi D_\phi(\bar{r}) \tag{1.79}$$

$$= \hat{a}_R D_R(R, \theta, \phi) = \hat{a}_R D_R(R). \tag{1.80}$$

[20] If the point electric charge is rotated, it is still a point electric charge.

Inserting the final expression into the integral, we derive

$$\int_0^{2\pi}\int_0^{\pi} \bar{D}(\bar{r}) \cdot \hat{a}_R R^2 \sin\theta d\theta d\phi = \int_0^{2\pi}\int_0^{\pi} D_R(R) R^2 \sin\theta d\theta d\phi \tag{1.81}$$

$$= D_R(R) R^2 \int_0^{2\pi}\int_0^{\pi} \sin\theta d\theta d\phi \tag{1.82}$$

$$= D_R(R) 4\pi R^2 = Q_0, \tag{1.83}$$

leading to $D_R(R) = Q_0/(4\pi R^2)$ or

$$\bar{D}(\bar{r}) = \hat{a}_R \frac{Q_0}{4\pi R^2}. \tag{1.84}$$

While this expression is derived by considering a surface with radius R, we do not have any assumptions about the value of R, and it is valid for any arbitrary location (Figure 1.20). Using $\bar{D}(\bar{r}) = \varepsilon_0 \bar{E}(\bar{r})$, the electric field intensity created by a static point electric charge can be written

$$\bar{E}(\bar{r}) = \hat{a}_R \frac{Q_0}{4\pi\varepsilon_0 R^2}. \tag{1.85}$$

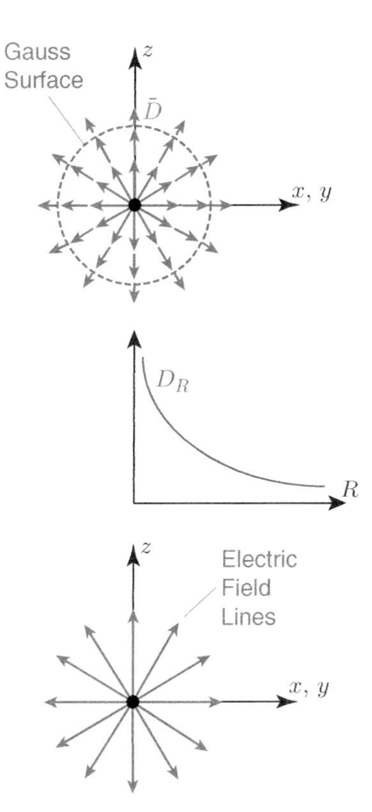

Figure 1.20 The electric field created by a static point electric charge decays quadratically with respect to distance. The arrows show both the direction and the magnitude (size of the arrows). Sometimes it is more illustrative to use electric field lines that show only the direction but not the magnitude.

It is remarkable that the electric field decays quadratically ($1/R^2$) with respect to distance from the point electric charge. This a typical behavior of far-distance (far-zone) electric fields created by most finite-size static electric charge distributions.

As another application of the integral form of Gauss' law, we consider a static line electric charge density (see Section 1.3.1) defined as $q_l(\bar{r}) = q_0$ (C/m) on the z axis from $-\infty$ to ∞ in a vacuum (Figure 1.21). Since the electric charge density is infinitely long, we can assume that $\bar{D}(\bar{r})$ depends only on ρ (cylindrical radial variable) and has only a ρ component: i.e. $\bar{D}(\bar{r}) = \hat{a}_\rho D_\rho(\rho)$. Then, considering a cylindrical Gauss surface, we have

$$\oint_S \bar{D}(\bar{r}) \cdot \overline{ds} = 2\pi\rho D_\rho(\rho) h = Q_{\text{enc}} = q_0 h, \tag{1.86}$$

where h is the height of the surface. In this integral, we simply consider the curved surface while omitting the top and bottom surfaces since $\bar{D}(\bar{r})$ does not have a normal component on these surfaces. Then we obtain

$$\bar{D}(\bar{r}) = \hat{a}_\rho \frac{q_0}{2\pi\rho}. \tag{1.87}$$

In addition, the electric field intensity can be obtained as

$$\bar{E}(\bar{r}) = \frac{\bar{D}(\bar{r})}{\varepsilon_0} = \hat{a}_\rho \frac{q_0}{2\pi\varepsilon_0\rho}. \tag{1.88}$$

It is remarkable that, as opposed to electric fields due to electric charge distributions with finite geometries, the electric field due to an infinitely long electric charge distribution decays with the distance ρ (i.e. not its square).

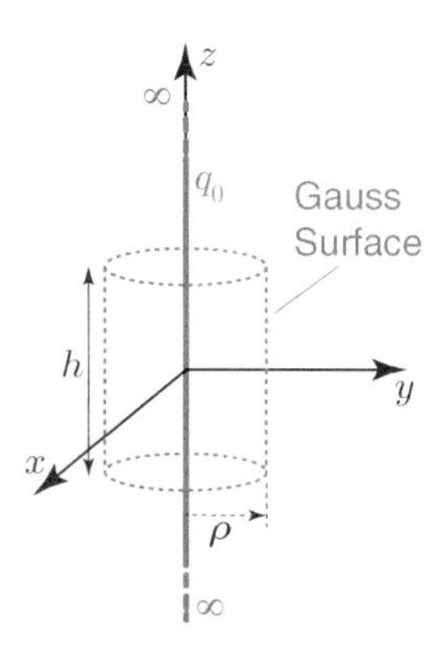

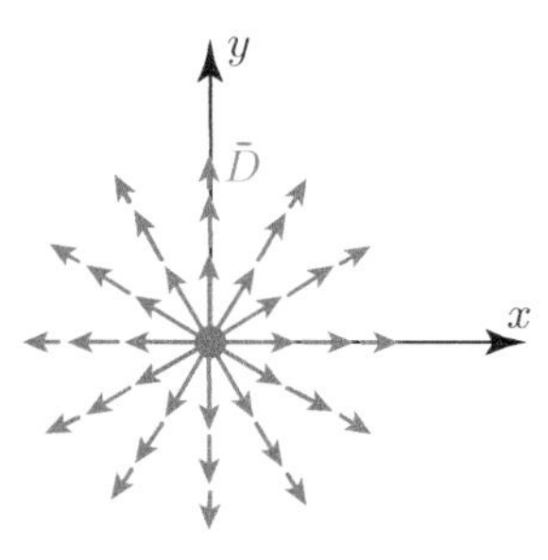

Figure 1.21 A static line electric charge density, which is infinitely long, creates a radial electric field that decays with the distance (proportional to $1/\rho$). For a point electric charge (Figure 1.20), the decay is quadratic (proportional to $1/R^2$).

1.2.1 Examples

Example 3: Consider a static volume electric charge distribution (see Section 1.3.1) $q_v(\bar{r}) = q_0 R$ (C/m^3) defined as a sphere of radius a in a vacuum (Figure 1.22). Find the electric flux density $\bar{D}(\bar{r})$ everywhere, as well as the locations where it becomes maximum.

Solution: Using symmetry, we can assume that $\bar{D}$ depends only on R and has only an R component: i.e. $\bar{D}(\bar{r}) = \hat{a}_R D_R(R)$. Considering a spherical Gauss surface with radius R, we have

$$\oint_S \bar{D}(\bar{r}) \cdot \overline{ds} = 4\pi R^2 D_R = Q_{\text{enc}}. \tag{1.89}$$

If $R < a$, the total amount of the enclosed electric charge can be found as

$$Q_{\text{enc}} = \int_V q_v(\bar{r}')dv' = \int_0^R \int_0^\pi \int_0^{2\pi} q_v(R')^2 \sin\theta' dR' d\theta' d\phi' \tag{1.90}$$

$$= \int_V q_0 (R')^3 \sin\theta' dR' d\theta' d\phi' = 4\pi q_0 \frac{R^4}{4}, \tag{1.91}$$

leading to[21]

$$\bar{D}(\bar{r}) = \hat{a}_R \frac{q_0 R^2}{4} \qquad (R < a). \tag{1.92}$$

[21] One can check the differential form of Gauss' law (see Section 1.3) as

$$\bar{\nabla} \cdot \bar{D}(\bar{r}) = \frac{1}{R^2} \frac{\partial}{\partial R} \left(R^2 \frac{q_0 R^2}{4} \right) = q_0 R.$$

Note that primed coordinates are used as integration variables to distinguish them from observation variables (R in this case). If $R > a$, the total amount of the enclosed electric charge becomes $Q_{\text{enc}} = \pi q_0 a^4$, leading to[22]

$$\bar{D}(\bar{r}) = \hat{a}_R \frac{Q_{\text{enc}}}{4\pi R^2} = \hat{a}_R \frac{q_0 a^4}{4R^2} \qquad (R > a). \tag{1.93}$$

The electric field intensity can be found as

$$\bar{E}(\bar{r}) = \frac{\bar{D}(\bar{r})}{\varepsilon_0} = \hat{a}_R \frac{q_0}{4\varepsilon_0} \begin{cases} R^2 & (R < a) \\ a^4/R^2 & (R > a). \end{cases} \tag{1.94}$$

In addition, the maximum electric flux density and electric field intensity occur at $R = a$, leading to

$$\bar{D}(a, \theta, \phi) = \hat{a}_R \frac{q_0 a^2}{4}. \tag{1.95}$$

We note that the expressions for the electric flux density at $R = a^-$ and $R = a^+$ are consistent: i.e. there is no jump since there is no surface electric charge density (see Section 1.5).

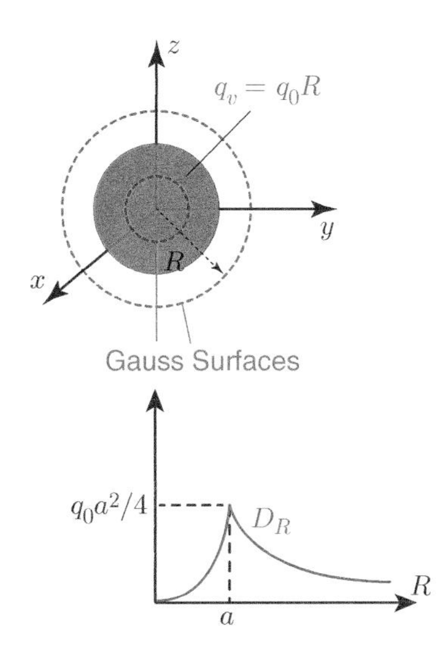

Figure 1.22 A static volume electric charge distribution with a spherical shape and the corresponding electric flux density with respect to R.

[22]We note the similarity of the expression for $R > a$ to the electric field of a point electric charge.

1.3 Differential Form of Gauss' Law

In a differential form, Gauss' law can be written

$$\bar{\nabla} \cdot \bar{D}(\bar{r}, t) = q_v(\bar{r}, t), \tag{1.96}$$

where $\bar{\nabla} \cdot \bar{f}(\bar{r})$ represents the divergence of a vector field $\bar{f}(\bar{r})$ and $q_v(\bar{r}, t)$ is the volume electric charge density (C/m^3). In a homogeneous medium, where ε does not depend on the position, we further have

q_v: volume electric charge density
Unit: coulombs/meter^3 (C/m^3)

$$\bar{\nabla} \cdot \bar{E}(\bar{r}, t) = \frac{q_v(\bar{r}, t)}{\varepsilon}, \tag{1.97}$$

where ε is taken out of the derivative. To understand the differential form of Gauss' law, we need the following tools and concepts:

- Electric charge density
- Divergence of vector fields
- Divergence theorem

These are in addition to previous tools: i.e. differential surface with direction (Section 1.1.1), dot product (Section 1.1.2), and flux of vector fields (Section 1.1.3).

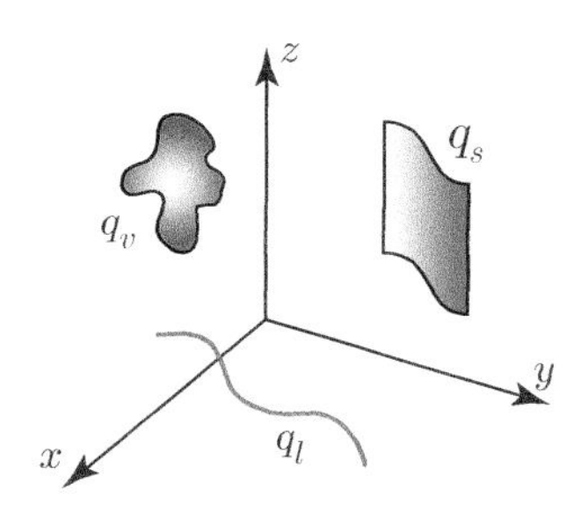

Figure 1.23 Volume, surface, and line electric charge densities.

1.3.1 Electric Charge Density

As briefly discussed in Section 0.5, electric charge is a useful definition to describe electrical interactions. For example, electric fields are created by electric charges, and Gauss' law describes how this occurs. The integral form of Gauss' law in Eq. (1.1) directly uses the electric charge on the right-hand side. In the differential form, however, the volume electric charge density is used. It can be written (Figure 1.23)

$$q_v(\bar{r}, t) = \lim_{\Delta v \to 0} \frac{\Delta Q(t)}{\Delta v}, \tag{1.98}$$

where $\Delta Q(t)$ is the total electric charge enclosed in a volume Δv. Hence, $q_v(\bar{r}, t)$ is a point-wise quantity, providing the amount of the enclosed electric charge as the volume shrinks into a single point. Obviously, a static point electric charge Q_0 indicates an infinite volume electric charge density: i.e. $q_v(\bar{r}) = \lim_{\Delta v \to 0} (Q_0/\Delta v) = \infty$, where $\bar{r}$ is the location of the electric charge.

Similar to the volume electric charge density, surface and line electric charge densities can be defined (Figure 1.23). For example, in some cases, electric charges may be distributed over a surface (zero thickness). Then it is more useful to define a surface electric charge density.[23] Similarly, for electric charge distributions defined as curves, it is useful to define line electric charge density. We have

[23] Surface electric charge densities are particularly important when studying boundaries between two different media.

$$q_s(\bar{r}, t) = \lim_{\Delta s \to 0} \frac{\Delta Q(t)}{\Delta s} \tag{1.99}$$

and

$$q_l(\bar{r}, t) = \lim_{\Delta l \to 0} \frac{\Delta Q(t)}{\Delta l}, \tag{1.100}$$

where Δs and Δl represent a surface and line, respectively, which shrink to zero.

q_s: surface electric charge density
Unit: coulombs/meter^2 (C/m^2)

q_l: Line electric charge density
Unit: coulombs/meter (C/m)

It is remarkable that the right-hand side of the differential form of Gauss' law can only be in the form of a volume electric charge density. If there is a surface electric charge density $q_s(\bar{r}, t)$, it can be seen as a volume density $q_v(\bar{r}, t) = q_s(\bar{r}, t)\delta(u)$, where u is a coordinate variable perpendicular to the associated surface and δ is the Dirac delta (impulse) function. Then enforcing the differential form of Gauss' law by defining a surface charge distribution in terms of a volume charge distribution would require working with the derivative of a discontinuous function (electric field). On the other hand, we can find better expressions for electric fields in the vicinity of surface electric charges: i.e. the boundary conditions discussed in Sections 1.5 and 3.4.

1.3.2 Divergence of Vector Fields

Given a vector field $\bar{f}(\bar{r}, t)$, $g(\bar{r}, t) = \bar{\nabla} \cdot \bar{f}(\bar{r}, t)$ is a scalar field defined as

$$g(\bar{r}, t) = \frac{\partial f_x(\bar{r}, t)}{\partial x} + \frac{\partial f_y(\bar{r}, t)}{\partial y} + \frac{\partial f_z(\bar{r}, t)}{\partial z}. \tag{1.101}$$

Using the definition of the *del* operator

$$\bar{\nabla} = \hat{a}_x \frac{\partial}{\partial x} + \hat{a}_y \frac{\partial}{\partial y} + \hat{a}_z \frac{\partial}{\partial z}, \tag{1.102}$$

the divergence of a vector field may be interpreted as a dot product:[24]

$$\bar{\nabla} \cdot \bar{f}(\bar{r}, t) = \left(\hat{a}_x \frac{\partial}{\partial x} + \hat{a}_y \frac{\partial}{\partial y} + \hat{a}_z \frac{\partial}{\partial z} \right) \cdot \bar{f}(\bar{r}, t). \tag{1.103}$$

Basically, the divergence of a vector field is its rate of change depending on the position (Figure 1.24). Since a vector field has an arbitrary direction at a given point, all possible directions are considered, leading to a dot product with the $\bar{\nabla}$ operator. For example, consider a time-invariant vector field $\bar{f}(\bar{r}) = \hat{a}_x x$ (Figure 1.25). One can find $\bar{\nabla} \cdot \bar{f}(\bar{r}) = 1$ everywhere, indicating that $\bar{f}(\bar{r})$ is changing with a constant rate with respect to position. On the other hand, the divergence result does not give full information about $\bar{f}(\bar{r})$: i.e. we lose the information that the change in $\bar{f}(\bar{r})$ is in fact in the x direction (similar to losing information in a dot product). For example, if $\bar{f}(\bar{r}) = \hat{a}_y y$ or $\bar{f}(\bar{r}) = \hat{a}_z z$, we also have $\bar{\nabla} \cdot \bar{f}(\bar{r}) = 1$.

But what if $\bar{f}(\bar{r}) = \hat{a}_x y$? The value of this vector field changes in the y direction (Figure 1.25). However, the vector field itself is in the x direction. Taking the divergence, we have $\bar{\nabla} \cdot \bar{f}(\bar{r}) = 0$ even though $\bar{f}(\bar{r})$ is in fact changing with respect to position. This is because the divergence operation partially measures the variations in a vector field, while rotational changes are ignored.[25] But why is such a metric important? As electric fields, some vector fields are created by well-defined *direct* sources (e.g. electric charges). Given a vector field, its sources

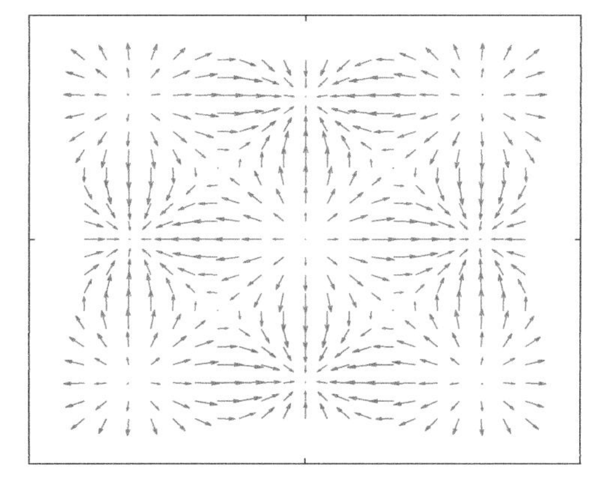

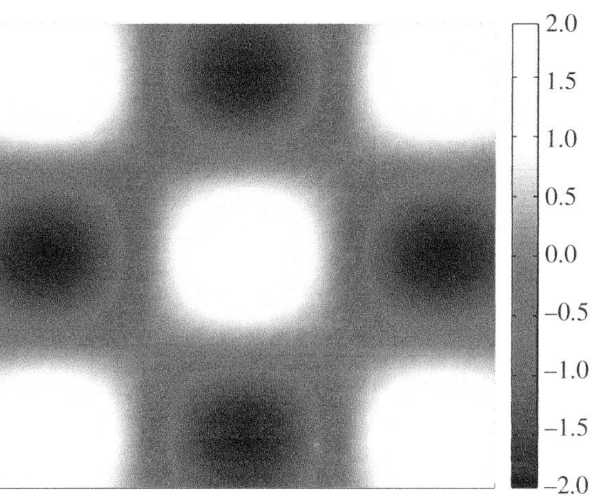

Figure 1.24 A vector field and its divergence (colors). Positive and negative values indicate positive sources and negative sources (sinks), respectively.

[24] The dot product with $\bar{\nabla}$ is an interpretation of divergence, but it should be noted that $\bar{\nabla}$ is an operator. Therefore, not all rules about dot products may be used.

[25] As discussed later, by measuring rotational changes, the curl operation complements the divergence operation.

can be found by evaluating the divergence because if the field is changing in its own direction, then there should be some source to create this change. It would not be wrong to say that the divergence operation is a source detector, and therefore, it is an important kind of differentiation used for vector fields.

The divergence operation can easily be defined in a Cartesian system, as shown in Eq. (1.101). For a cylindrical or spherical system, however, it is not trivial. The problem is that some of the unit vectors change with respect to the position and need to be considered while evaluating derivatives. Specifically, application of the $\bar{\nabla}$ operator to the components of a vector field requires extra care when working with position-dependent unit vectors. First, we consider a cylindrical system as[26]

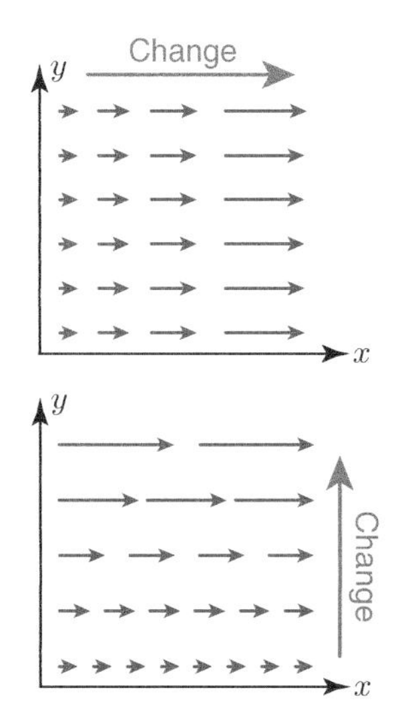

Figure 1.25 A vector field $f(\bar{r}) = \hat{a}_x x$ has a nonzero divergence (top figure), while $f(\bar{r}) = \hat{a}_x y$ has a zero divergence (bottom figure).

$$\bar{\nabla} \cdot \bar{f}(\bar{r}, t) = \left(\hat{a}_\rho \frac{\partial}{\partial \rho} + \hat{a}_\phi \frac{1}{\rho} \frac{\partial}{\partial \phi} + \hat{a}_z \frac{\partial}{\partial z} \right) \cdot \bar{f}(\bar{r}, t) \tag{1.104}$$

$$= \left(\hat{a}_\rho \frac{\partial}{\partial \rho} + \hat{a}_\phi \frac{1}{\rho} \frac{\partial}{\partial \phi} + \hat{a}_z \frac{\partial}{\partial z} \right) \cdot \left(\hat{a}_\rho f_\rho(\bar{r}, t) + \hat{a}_\phi f_\phi(\bar{r}, t) + \hat{a}_z f_z(\bar{r}, t) \right), \tag{1.105}$$

where the divergence operation is written again as a dot product while $\bar{f}(\bar{r})$ is represented by using cylindrical variables. Using the terms one by one, we have[27]

$$\begin{aligned} \bar{\nabla} \cdot \bar{f}(\bar{r}, t) &= \hat{a}_\rho \cdot \frac{\partial}{\partial \rho} \left(\hat{a}_\rho f_\rho(\bar{r}, t) \right) + \hat{a}_\rho \cdot \frac{\partial}{\partial \rho} \left(\hat{a}_\phi f_\phi(\bar{r}, t) \right) + \hat{a}_\rho \cdot \frac{\partial}{\partial \rho} \left(\hat{a}_z f_z(\bar{r}, t) \right) \\ &+ \hat{a}_\phi \cdot \frac{1}{\rho} \frac{\partial}{\partial \phi} \left(\hat{a}_\rho f_\rho(\bar{r}, t) \right) + \hat{a}_\phi \cdot \frac{1}{\rho} \frac{\partial}{\partial \phi} \left(\hat{a}_\phi f_\phi(\bar{r}, t) \right) + \hat{a}_\phi \cdot \frac{\partial}{\rho \partial \phi} \left(\hat{a}_z f_z(\bar{r}, t) \right) \\ &+ \hat{a}_z \cdot \frac{\partial}{\partial z} \left(\hat{a}_\rho f_\rho(\bar{r}, t) \right) + \hat{a}_z \cdot \frac{\partial}{\partial z} \left(\hat{a}_\phi f_\phi(\bar{r}, t) \right) + \hat{a}_z \cdot \frac{\partial}{\partial z} \left(\hat{a}_z f_z(\bar{r}, t) \right). \end{aligned} \tag{1.106}$$

Since $\hat{a}_z$ does not depend on the position, we obtain

$$\hat{a}_\rho \cdot \frac{\partial}{\partial \rho} \left(\hat{a}_z f_z(\bar{r}, t) \right) = \hat{a}_\rho \cdot \hat{a}_z \frac{\partial f_z(\bar{r}, t)}{\partial \rho} = 0 \tag{1.107}$$

$$\hat{a}_\phi \cdot \frac{1}{\rho} \frac{\partial}{\partial \phi} \left(\hat{a}_z f_z(\bar{r}, t) \right) = \hat{a}_\phi \cdot \hat{a}_z \frac{1}{\rho} \frac{\partial f_z(\bar{r}, t)}{\partial \phi} = 0 \tag{1.108}$$

$$\hat{a}_z \cdot \frac{\partial}{\partial z} \left(\hat{a}_z f_z(\bar{r}, t) \right) = \hat{a}_z \cdot \hat{a}_z \frac{\partial f_z(\bar{r}, t)}{\partial z} = \frac{\partial f_z(\bar{r}, t)}{\partial z}. \tag{1.109}$$

[26] Even though the following derivation is pretty long, it is interesting to see how the derivatives of some unit vectors are not zero and how they lead to the final expressions for the divergence in a cylindrical coordinate system. During the derivation, we obtain $\partial \hat{a}_\rho / \partial \phi = \hat{a}_\phi$ and $\partial \hat{a}_\phi / \partial \phi = -\hat{a}_\rho$.

[27] Note again ρ besides $d\phi$.

In addition, we can evaluate the other four terms as

$$\hat{a}_\rho \cdot \frac{\partial}{\partial \rho}\left(\hat{a}_\rho f_\rho(\bar{r},t)\right) = \hat{a}_\rho \cdot \hat{a}_\rho \frac{\partial f_\rho(\bar{r},t)}{\partial \rho} = \frac{\partial f_\rho(\bar{r},t)}{\partial \rho} \tag{1.110}$$

$$\hat{a}_\rho \cdot \frac{\partial}{\partial \rho}\left(\hat{a}_\phi f_\phi(\bar{r},t)\right) = \hat{a}_\rho \cdot \hat{a}_\phi \frac{\partial f_\phi(\bar{r},t)}{\partial \rho} = 0 \tag{1.111}$$

$$\hat{a}_z \cdot \frac{\partial}{\partial z}\left(\hat{a}_\rho f_\rho(\bar{r},t)\right) = \hat{a}_z \cdot \hat{a}_\rho \frac{\partial f_\rho(\bar{r},t)}{\partial z} = 0 \tag{1.112}$$

$$\hat{a}_z \cdot \frac{\partial}{\partial z}\left(\hat{a}_\phi f_\phi(\bar{r},t)\right) = \hat{a}_z \cdot \hat{a}_\phi \frac{\partial f_\phi(\bar{r},t)}{\partial z} = 0, \tag{1.113}$$

leading to

$$\bar{\nabla} \cdot \bar{f}(\bar{r},t) = \frac{\partial f_\rho(\bar{r},t)}{\partial \rho} + \hat{a}_\phi \cdot \frac{1}{\rho}\frac{\partial}{\partial \phi}\left(\hat{a}_\rho f_\rho(\bar{r},t)\right) + \hat{a}_\phi \cdot \frac{1}{\rho}\frac{\partial}{\partial \phi}\left(\hat{a}_\phi f_\phi(\bar{r},t)\right) + \frac{\partial f_z(\bar{r},t)}{\partial z}. \tag{1.114}$$

The second and third terms on the right-hand side of Eq. (1.114) are interesting since the unit vectors $\hat{a}_\rho$ and $\hat{a}_\phi$ depend on ϕ (Figure 1.26). By using the derivative of the multiplication of two functions, we have[28]

$$\hat{a}_\phi \cdot \frac{1}{\rho}\frac{\partial}{\partial \phi}\left(\hat{a}_\rho f_\rho(\bar{r},t)\right) = \hat{a}_\phi \cdot f_\rho(\bar{r},t)\frac{1}{\rho}\frac{\partial \hat{a}_\rho}{\partial \phi} \tag{1.115}$$

and[29]

$$\hat{a}_\phi \cdot \frac{1}{\rho}\frac{\partial}{\partial \phi}\left(\hat{a}_\phi f_\phi(\bar{r},t)\right) = \frac{1}{\rho}\frac{\partial f_\phi(\bar{r},t)}{\partial \phi} + \hat{a}_\phi \cdot f_\rho(\bar{r},t)\frac{1}{\rho}\frac{\partial \hat{a}_\phi}{\partial \phi}. \tag{1.116}$$

To find the derivatives of the unit vectors $\hat{a}_\rho$ and $\hat{a}_\phi$, we can convert them into Cartesian unit vectors (see Eqs. (1.23) and (1.24)) as

$$\frac{\partial \hat{a}_\rho}{\partial \phi} = \frac{\partial}{\partial \phi}\left(\hat{a}_x \cos\phi + \hat{a}_y \sin\phi\right) = -\hat{a}_x \sin\phi + \hat{a}_y \cos\phi = \hat{a}_\phi \tag{1.117}$$

$$\frac{\partial \hat{a}_\phi}{\partial \phi} = \frac{\partial}{\partial \phi}\left(-\hat{a}_x \sin\phi + \hat{a}_y \cos\phi\right) = -\hat{a}_x \cos\phi - \hat{a}_y \sin\phi = -\hat{a}_\rho. \tag{1.118}$$

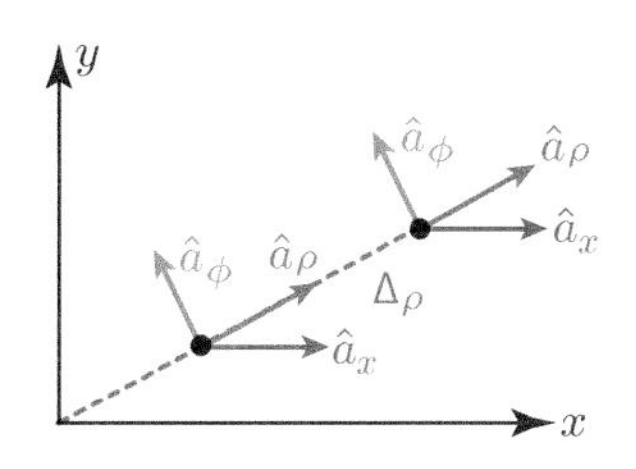

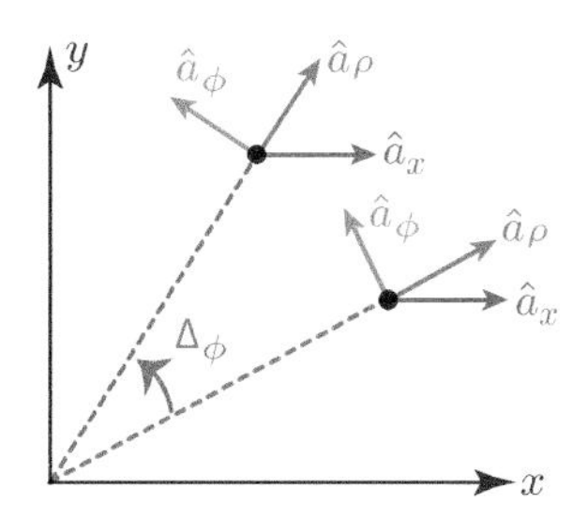

Figure 1.26 Changes in some of the unit vectors with respect to cylindrical variables ρ and ϕ. When ρ is changing, none of the unit vectors changes: i.e. they keep the same directions. When ϕ is changing, however, $\hat{a}_\rho$ and $\hat{a}_\phi$ change rotationally. Therefore, $\partial\hat{a}_\rho/\partial\phi \neq 0$ and $\partial\hat{a}_\phi/\partial\phi \neq 0$. On the other hand, $\hat{a}_x$ does not change also in the second case, showing the position-independent characteristics of this unit vector.

[28] Note that

$$\hat{a}_\phi \cdot \frac{1}{\rho}\frac{\partial}{\partial \phi}\left(\hat{a}_\rho f_\rho(\bar{r},t)\right) = \hat{a}_\phi \cdot \hat{a}_\rho \frac{1}{\rho}\frac{\partial f_\rho(\bar{r},t)}{\partial \phi} + \hat{a}_\phi \cdot f_\rho(\bar{r},t)\frac{1}{\rho}\frac{\partial \hat{a}_\rho}{\partial \phi}.$$

[29] Note that

$$\hat{a}_\phi \cdot \frac{1}{\rho}\frac{\partial}{\partial \phi}\left(\hat{a}_\phi f_\phi(\bar{r},t)\right) = \hat{a}_\phi \cdot \hat{a}_\phi \frac{1}{\rho}\frac{\partial f_\phi(\bar{r},t)}{\partial \phi} + \hat{a}_\phi \cdot f_\rho(\bar{r},t)\frac{1}{\rho}\frac{\partial \hat{a}_\phi}{\partial \phi}.$$

We finally obtain the expression for the divergence in a cylindrical coordinate system as[30]

$$\bar{\nabla} \cdot \bar{f}(\bar{r},t) = \frac{\partial f_\rho(\bar{r},t)}{\partial \rho} + \frac{f_\rho(\bar{r},t)}{\rho} + \frac{1}{\rho}\frac{\partial f_\phi(\bar{r},t)}{\partial \phi} + \frac{\partial f_z(\bar{r},t)}{\partial z}. \tag{1.119}$$

It can also be written[31]

$$\bar{\nabla} \cdot \bar{f}(\bar{r},t) = \frac{1}{\rho}\frac{\partial}{\partial \rho}\left(\rho f_\rho(\bar{r},t)\right) + \frac{1}{\rho}\frac{\partial f_\phi(\bar{r},t)}{\partial \phi} + \frac{\partial f_z(\bar{r},t)}{\partial z}. \tag{1.120}$$

A similar derivation for the divergence operation in a spherical coordinate system leads to

$$\begin{aligned}\bar{\nabla} \cdot \bar{f}(\bar{r},t) &= \frac{1}{R^2}\frac{\partial}{\partial R}\left(R^2 f_R(\bar{r},t)\right) \\ &+ \frac{1}{R\sin\theta}\frac{\partial}{\partial \theta}\left(\sin\theta f_\theta(\bar{r},t)\right) + \frac{1}{R\sin\theta}\frac{\partial f_\phi(\bar{r},t)}{\partial \phi}.\end{aligned} \tag{1.121}$$

We note that the first derivative (with respect to R) is applied to the multiplication $\left(R^2 f_R(\bar{r},t)\right)$, while the second derivative (with respect to θ) is on $(\sin\theta f_\theta(\bar{r},t))$.

Confusion very commonly occurs in a cylindrical or a spherical coordinate system when dealing with vector fields of constant magnitude. For example, consider a time-invariant vector field $\bar{f}(\bar{r}) = \hat{a}_R$ (Figure 1.27). Since $|\bar{f}(\bar{r})| = 1$ everywhere, one may think that $\bar{\nabla} \cdot \bar{f}(\bar{r}) = 0$, while it is not. In fact, using the expression above, we have[32]

$$\bar{\nabla} \cdot \bar{f}(\bar{r}) = \frac{1}{R^2}\frac{\partial}{\partial R}(R^2) = \frac{2R}{R^2} = \frac{2}{R}. \tag{1.122}$$

This result tells us that the divergence of $\bar{f}(\bar{r}) = \hat{a}_R$ is nonzero (at the same time, it is changing with R). But what does this mean? The source of the confusion is usually thinking a vector field as a single vector, while it is actually a collection of infinitely many vectors. When it is difficult to interpret the result of a divergence, one can consider its physical meaning: i.e. the divergence of a vector field indicates the sources that create the field. For example, if we consider the net flux $f(\bar{r}) = \hat{a}_R$ through a spherical surface of radius R, it increases quadratically with R (since the magnitude of $\bar{f}(\bar{r})$ is constant while the area of the surface is $4\pi R^2$). Therefore, there should be sources at each and every point in space to account for the increase in the flux. In other words, such a vector field that has a magnitude of 1 everywhere can only be created by infinitely many sources distributed as $2/R$ in space.

[30] Note that

$$\begin{aligned}\bar{\nabla} \cdot \bar{f}(\bar{r},t) &= \frac{\partial f_\rho(\bar{r},t)}{\partial \rho} + \hat{a}_\phi \cdot f_\rho(\bar{r},t)\frac{1}{\rho}\frac{\partial \hat{a}_\rho}{\partial \phi} \\ &+ \frac{1}{\rho}\frac{\partial f_\phi(\bar{r},t)}{\partial \phi} + \hat{a}_\phi \cdot f_\rho(\bar{r},t)\frac{1}{\rho}\frac{\partial \hat{a}_\phi}{\partial \phi} \\ &+ \frac{\partial f_z(\bar{r},t)}{\partial z} \\ &= \frac{\partial f_\rho(\bar{r},t)}{\partial \rho} + \frac{f_\rho(\bar{r},t)}{\rho} \\ &+ \frac{1}{\rho}\frac{\partial f_\phi(\bar{r},t)}{\partial \phi} + \frac{\partial f_z(\bar{r},t)}{\partial z}.\end{aligned}$$

[31] As commonly shown in formula tables.

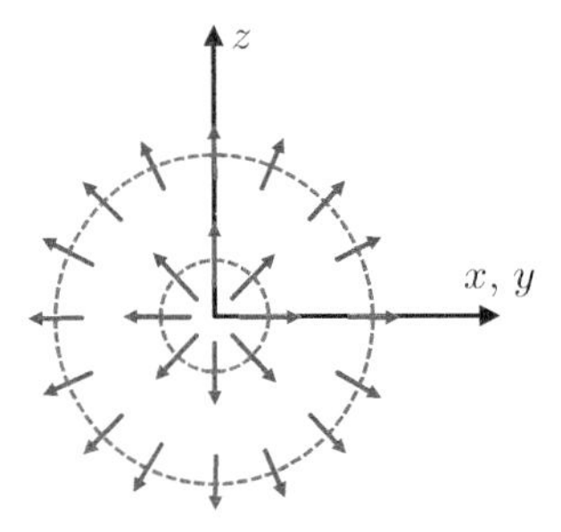

Figure 1.27 The net flux of vector field $\bar{f}(\bar{r}) = \hat{a}_R$ through a spherical surface increases with the size of the sphere. This means more *sources* of this vector field are included as the sphere grows.

[32] This can be verified in a Cartesian system as

$$\bar{f}(\bar{r}) = \hat{a}_R = \hat{a}_r = \frac{\bar{r}}{|\bar{r}|} = \frac{\hat{a}_x x + \hat{a}_y y + \hat{a}_z z}{\sqrt{x^2+y^2+z^2}},$$

and

$$\begin{aligned}\bar{\nabla} \cdot \bar{f}(\bar{r}) &= \frac{\partial}{\partial x}\left(\frac{x}{\sqrt{x^2+y^2+z^2}}\right) \\ &+ \frac{\partial}{\partial y}\left(\frac{y}{\sqrt{x^2+y^2+z^2}}\right) + \frac{\partial}{\partial z}\left(\frac{z}{\sqrt{x^2+y^2+z^2}}\right) \\ &= \frac{3}{\sqrt{x^2+y^2+z^2}} - \frac{x^2+y^2+z^2}{[x^2+y^2+z^2]^{3/2}} \\ &= \frac{2}{\sqrt{x^2+y^2+z^2}}.\end{aligned}$$

Now, we consider some important vector fields and their divergences. First, the divergence of a time-invariant vector field $\bar{f}(\bar{r}) = \bar{r}$ can be found as[33,34]

$$\bar{\nabla} \cdot \bar{f}(\bar{r}) = \bar{\nabla} \cdot (\hat{a}_x x + \hat{a}_y y + \hat{a}_z z) = \frac{\partial x}{\partial x} + \frac{\partial y}{\partial y} + \frac{\partial z}{\partial z} = 1 + 1 + 1 = 3. \tag{1.123}$$

[33]Note that $\bar{f}(\bar{r})$ is not a position vector; it is a vector field whose value and direction at a point is the same as the position vector for that location.

[34]Obviously, $\bar{\nabla} \cdot [\bar{f}(\bar{r}) + \bar{g}(\bar{r})] = \bar{\nabla} \cdot \bar{f}(\bar{r}) + \bar{\nabla} \cdot \bar{g}(\bar{r})$, for any two vector fields $\bar{f}(\bar{r})$ and $\bar{g}(\bar{r})$.

This result should be independent of the coordinate system. Therefore, the same equality can be shown in a cylindrical or a spherical coordinate system. For example, in a spherical coordinate system, we have

$$\bar{f}(\bar{r}) = R\hat{a}_R \longrightarrow \bar{\nabla} \cdot \bar{f}(\bar{r}) = \frac{1}{R^2}\frac{\partial}{\partial R}(R^2 R) = \frac{1}{R^2} 3R^2 = 3, \tag{1.124}$$

as expected. To sum up, a constant source distribution (with a magnitude of 3) is needed in the *whole* space to create a vector field $\bar{f}(\bar{r}) = \bar{r}$.

As another important identity, we derive the divergence of a time-invariant vector field $\bar{f}(\bar{r}) = \bar{r}/|\bar{r}|^3$ as (Figure 1.28)

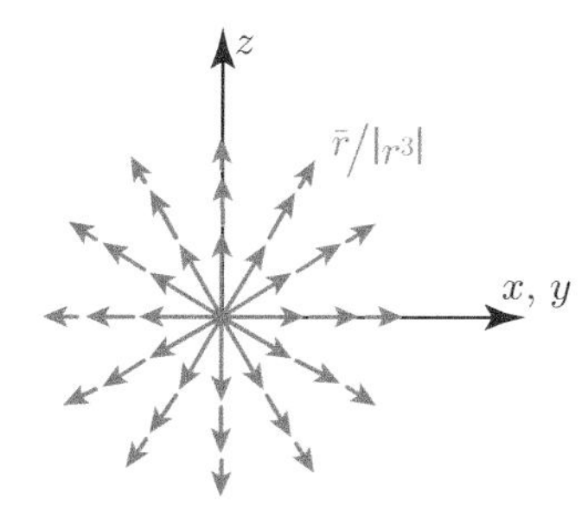

Figure 1.28 A vector field $\bar{f}(\bar{r}) = \bar{r}/|\bar{r}|^3$ seems to change with respect to position. But indeed, its divergence is zero except at the origin. This means if an arbitrary differential volume does not contain the origin, the algebraic sum of all entering/leaving fields is zero. In addition, all spheres centered at the origin contain the same net flux, independent of their radii. When the sphere size grows, the magnitude of the field passing through the surface decreases, but the surface area increases at the same rate (which balances the overall flux).

$$\bar{\nabla} \cdot \left(\frac{\bar{r}}{|\bar{r}|^3}\right) = \bar{\nabla} \cdot \left(\frac{\hat{a}_x x + \hat{a}_y y + \hat{a}_z z}{[x^2 + y^2 + z^2]^{3/2}}\right) \tag{1.125}$$

$$= \frac{\partial}{\partial x}\left(\frac{x}{[x^2 + y^2 + z^2]^{3/2}}\right) + \frac{\partial}{\partial y}\left(\frac{y}{[x^2 + y^2 + z^2]^{3/2}}\right) + \frac{\partial}{\partial z}\left(\frac{z}{[x^2 + y^2 + z^2]^{3/2}}\right) \tag{1.126}$$

$$= \frac{3}{[x^2 + y^2 + z^2]^{3/2}} + \frac{x(-3/2)2x}{[x^2 + y^2 + z^2]^{5/2}} + \frac{y(-3/2)2y}{[x^2 + y^2 + z^2]^{5/2}} + \frac{z(-3/2)2z}{[x^2 + y^2 + z^2]^{5/2}} \tag{1.127}$$

$$= \frac{3(x^2 + y^2 + z^2) - 3x^2 - 3y^2 - 3z^2}{[x^2 + y^2 + z^2]^{5/2}} = 0. \tag{1.128}$$

This result is interesting because, as shown in Eq. (1.84), $\bar{r}/|\bar{r}|^3 = \hat{a}_R/R^2$ represents the electric field created by a static point electric charge. But if its divergence is zero, and keeping in mind that divergence is a *source detector*, where is the source? In fact, the derivation above indicates that there is no source anywhere *except* at the origin. Specifically, when $x = 0$, $y = 0$, and $z = 0$, the derivation above becomes invalid. It can be shown that[35]

$$\lim_{R\to 0} \bar{\nabla} \cdot \left(\frac{\hat{a}_R}{R^2}\right) = 4\pi\delta(R), \tag{1.129}$$

where $\delta(R)$ is the Dirac delta function, indicating a point source located at the origin. Hence, an impulse source at the origin is sufficient (no other electric charges are needed) to create a vector field $\bar{f}(\bar{r}) = \bar{r}/|\bar{r}|^3$.

[35] In the derivation of the differential form of Gauss' law, we assume that the total electric charge can be written as a volume integral of a volume electric charge density. A point electric charge can be considered as infinite electric charge density squeezed into a zero volume. From this perspective, the divergence of the electric flux density due to a point electric charge Q_0 can be said to be infinite at the origin, which can be represented as an impulse function, e.g. $q_v(\bar{r}) = Q_0\delta(R)$.

1.3.3 Divergence Theorem and the Differential Form of Gauss' Law

The divergence theorem is the key to writing the differential form of Gauss' law. As discussed in Section 1.2, the integral form of Gauss' law can be used to derive the electric field due to an electric charge distribution for symmetric cases. However, the differential form provides pointwise information at any point, although this information is partial (indicating only the divergence of the electric field). In a general form, given a surface S enclosing a volume V and a vector field $\bar{f}(\bar{r}, t)$, the divergence theorem states that

$$\int_V \bar{\nabla} \cdot \bar{f}(\bar{r}, t) dv = \oint_S \bar{f}(\bar{r}, t) \cdot \overline{ds}, \tag{1.130}$$

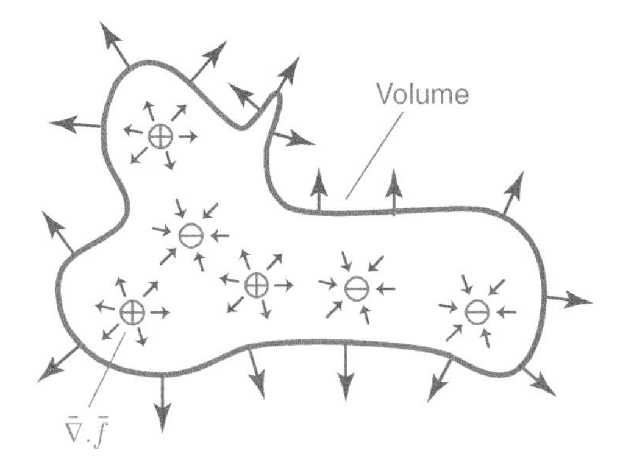

Figure 1.29 Since the divergence $\bar{\nabla} \cdot \bar{f}(\bar{r}, t)$ represents the source/sink of a vector field $\bar{f}(\bar{r}, t)$, where the vector field emerges (positive divergence) or disappears (negative divergence), its volume integral (the sum of all sources/sinks) corresponds to the net flux of the vector field. Some of the fields generated by sources may end at sinks, but unbalanced fields must go outside the volume through its surface. Similarly, if there are more/stronger sinks than sources, there should be net flux inward on the surface.

where $\overline{ds} = \hat{a}_n ds$ and $\hat{a}_n$ is the outward normal. This is a very strong theorem, converting a volume integral into a surface integral and vice versa. To understand how this equality holds, we recall that the divergence is an indicator of a source creating the vector field or a sink where the field diminishes. If we integrate this source/sink distribution in a volume, it gives the net flux of $\bar{f}(\bar{r}, t)$ through the surface enclosing the integration domain. There can be cancellations of positive sources and sinks (negative sources) inside the volume integration, but the net flux is created by the net amount of the sources (Figure 1.29). This is perfectly consistent with the integral form of Gauss' law. Indeed, Gauss' law can be seen as the direct application of the divergence theorem for electric charges and fields.

At this stage, it should be emphasized that the divergence theorem in Eq. (1.130) is still not pointwise: it represents the equality of two integrals. For example, if both integrals are zero, we have

$$\oint_S \bar{f}(\bar{r}, t) \cdot \overline{ds} = 0. \tag{1.131}$$

As discussed in Section 1.1.4, this does not indicate that $\bar{f}(\bar{r}, t)$ is zero on the surface. In addition, it does not indicate that $\bar{\nabla} \cdot \bar{f}(\bar{r}, t)$ is completely zero in the volume enclosed by the surface.[36]

[36] For example, for a static point electric charge at the origin, the net flux through any closed surface that does not include the origin is zero, while it may be difficult to prove this for a surface with an arbitrary shape.

Indeed, there can be nonzero $\bar{f}(\bar{r},t)$, but its integration should give zero flux overall. This also means the sum of all divergences inside the related volume should be zero, whereas the value of $\bar{\nabla}\cdot\bar{f}(\bar{r},t)$ can have positive and negative values that balance each other. This is again consistent with the discussions on the integral form of Gauss' law.[37]

Using the divergence theorem, we are now able to derive the differential form of Gauss' law from its integral form. According to the divergence theorem, we have

$$\oint_S \bar{D}(\bar{r},t)\cdot\overline{ds} = \int_V \bar{\nabla}\cdot\bar{D}(\bar{r},t)dv, \tag{1.132}$$

while the integral form of Gauss' law in Eq. (1.1) can be used to write

$$\int_V \bar{\nabla}\cdot\bar{D}(\bar{r},t)dv = Q_{\text{enc}}(t). \tag{1.133}$$

At the same time, the net enclosed electric charge can be written as the volume integral of the volume electric charge density, leading to

$$\int_V \bar{\nabla}\cdot\bar{D}(\bar{r},t)dv = \int_V q_v(\bar{r},t)dv. \tag{1.134}$$

An important point is that this equation is valid for any arbitrary volume. Specifically, since the volume V above can be anything, one can claim that the integrand should be the same, leading to the differential form of Gauss' law. Alternatively, letting the volume shrink to zero at a selected point $\bar{r}$, we have the differential form of Gauss' law in Eq. (1.96): i.e. $\bar{\nabla}\cdot\bar{D}(\bar{r},t) = q_v(\bar{r},t)$.

[37] It is remarkable that there are many vector fields with zero divergence, and it is often questionable whether such a vector field can represent a physical electric field. For example, consider $\bar{f}(\bar{r},t) = \hat{a}_x\cos(t)$, with a divergence $\bar{\nabla}\cdot\bar{f}(\bar{r},t) = 0$. At a fixed time, this vector field is constant with the same value everywhere. There is probably no physical electric charge distribution, even located at infinity, to create such an electric field. But what about $\bar{f}(\bar{r},t) = \hat{a}_x\cos(t-z)$, where z is the Cartesian variable? In this case, we again have $\bar{\nabla}\cdot\bar{f}(\bar{r},t) = 0$, indicating there is not any detectable electric charge. However, as discussed later, this expression for $\bar{f}(\bar{r},t)$ represents a plane wave that is frequently defined and used in electromagnetic analyses. Plane waves are idealizations of physical waves, and this (idealization) is why we cannot define physical electric charges for them. From another perspective, such electric charges that create plane waves can be considered to have an infinite extent and be located at infinity. It can be shown that an infinitely large planar electric charge density oscillating with time can create symmetric half plane waves.

1.3.4 Examples

Example 4: Consider a vector field $\bar{f}(\bar{r}) = \hat{a}_R R$ and a sphere of radius a (Figure 1.30). Verify the divergence theorem.

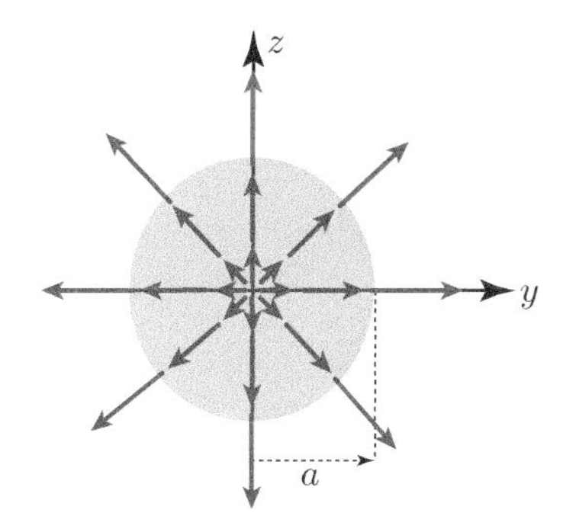

Figure 1.30 A a vector field $\bar{f}(\bar{r}) = \hat{a}_R R$ and a sphere of radius a.

Solution: To verify the divergence theorem, we first evaluate the divergence of $\bar{f}(\bar{r})$ as[38]

$$\bar{\nabla}\cdot\bar{f}(\bar{r}) = \frac{1}{R^2}\frac{\partial}{\partial R}(R^2 R) = \frac{3R^2}{R^2} = 3. \tag{1.135}$$

Then the volume integral can be found as

$$\int_V \bar{\nabla}\cdot\bar{f}(\bar{r})dv = \frac{4}{3}\pi a^3 \times 3 = 4\pi a^3. \tag{1.136}$$

[38] Obviously, this vector field does not correspond to a physical electric field since a static volume electric charge density must fill the whole space to create such a field. Even when such an electric charge distribution is assumed to exist, the electric field created would not be unique.

On the other side, the surface integral can be evaluated as

$$\oint_S \bar{f}(\bar{r}) \cdot \overline{ds} = \int_0^{2\pi} \int_0^{\pi} \hat{a}_R R \cdot \hat{a}_R R^2 \sin\theta d\theta d\phi = 4\pi a^3, \tag{1.137}$$

verifying the divergence theorem.

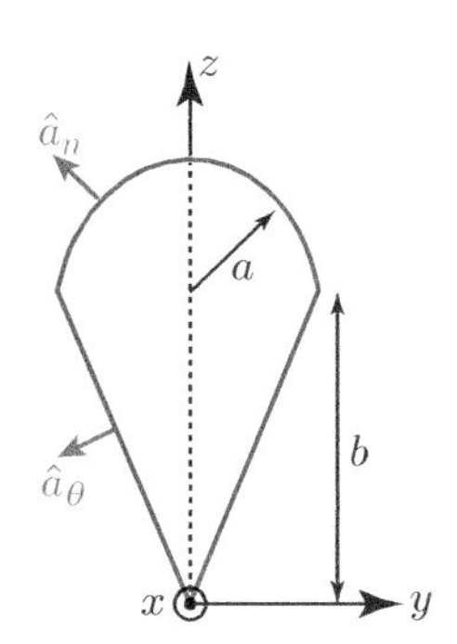

Figure 1.31 A closed surface involving a shifted hemisphere combined with an inverted conic section.

Example 5: Consider a vector field $\bar{f}(\bar{r}) = \hat{a}_z z$ and its net outward flux through a closed surface involving a shifted hemisphere combined with an inverted conic section (Figure 1.31). The height of the cone is b, while the radius of the hemisphere is a. Verify the divergence theorem.

Solution: For the direct calculation of the flux, we have

$$\oint_S \bar{f}(\bar{r}) \cdot \overline{ds} = \int_{\text{hemisphere}} \bar{f}(\bar{r}) \cdot \overline{ds} + \int_{\text{cone}} \bar{f}(\bar{r}) \cdot \overline{ds},$$

where[39]

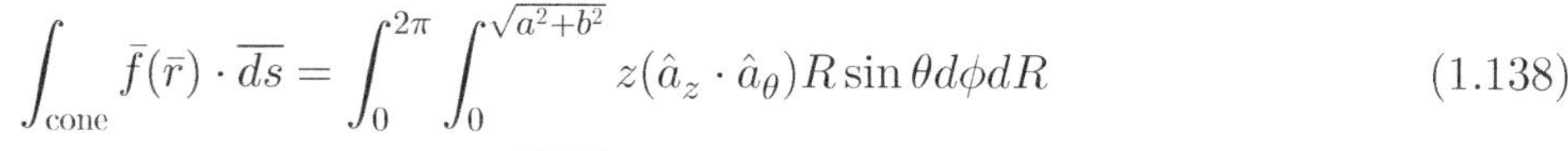

$$\int_{\text{cone}} \bar{f}(\bar{r}) \cdot \overline{ds} = \int_0^{2\pi} \int_0^{\sqrt{a^2+b^2}} z(\hat{a}_z \cdot \hat{a}_\theta) R \sin\theta d\phi dR \tag{1.138}$$

$$= \int_0^{2\pi} \int_0^{\sqrt{a^2+b^2}} R\cos\theta(-\sin\theta) R \sin\theta d\phi dR \tag{1.139}$$

$$= -\int_0^{2\pi} \int_0^{\sqrt{a^2+b^2}} R^2 \cos\theta \sin^2\theta d\phi dR \tag{1.140}$$

$$= -\frac{b}{\sqrt{a^2+b^2}} \frac{a^2}{a^2+b^2} \int_0^{2\pi} \int_0^{\sqrt{a^2+b^2}} R^2 d\phi dR \tag{1.141}$$

$$= -\frac{2\pi a^2 b}{[a^2+b^2]^{3/2}} \left[\frac{R^3}{3}\right]_0^{\sqrt{a^2+b^2}} = -\frac{2\pi a^2 b}{3}.$$

For finding the contribution from the hemispherical surface, one can assume that the hemisphere is located at the origin but $\bar{f} = \hat{a}_z(z+b)$. Then we can evaluate the flux through the surface as[40]

[39] On the cone, the changing spherical variables are R and ϕ. The variable R changes from 0 to $\sqrt{a^2+b^2}$ since the height of the cone is b, while the radius at its top is a. The variable ϕ changes from 0 to 2π since the cone is full around the z axis. Then dS is written $R\sin\theta d\phi dR$, where $R\sin\theta$ is accompanying $d\phi$ to change this differential angle into a differential length. In addition, the outward normal on the cone is $\hat{a}_\theta$. During the evaluations, we further use

$$\hat{a}_z \cdot \hat{a}_\theta = -\sin\theta$$
$$z = R\cos\theta.$$

[40] Without shifting the hemispherical surface, it would be difficult to write $\overline{ds}$. This can be understood by investigating the outward normal in the original case, which is not $\hat{a}_R$ (see Figure 1.32).

$$\int_{\text{hemisphere}} \bar{f}(\bar{r}) \cdot \overline{ds} = \int_0^{2\pi} \int_0^{\pi/2} (z+b)(\hat{a}_z \cdot \hat{a}_R) R^2 \sin\theta d\theta d\phi \tag{1.142}$$

$$= \int_0^{2\pi} \int_0^{\pi/2} (R\cos\theta + b)\cos\theta R^2 \sin\theta d\theta d\phi \tag{1.143}$$

$$= a^3 \int_0^{2\pi} \int_0^{\pi/2} \cos^2\theta \sin\theta d\theta d\phi$$

$$+ ba^2 \int_0^{2\pi} \int_0^{\pi/2} \cos\theta \sin\theta d\theta d\phi \tag{1.144}$$

$$= 2\pi a^3 \left[-\frac{\cos^3\theta}{3}\right]_0^{\pi/2} + 2\pi b a^2 \left[-\frac{\cos(2\theta)}{4}\right]_0^{\pi/2} \tag{1.145}$$

$$= \frac{2\pi a^3}{3} + \pi b a^2. \tag{1.146}$$

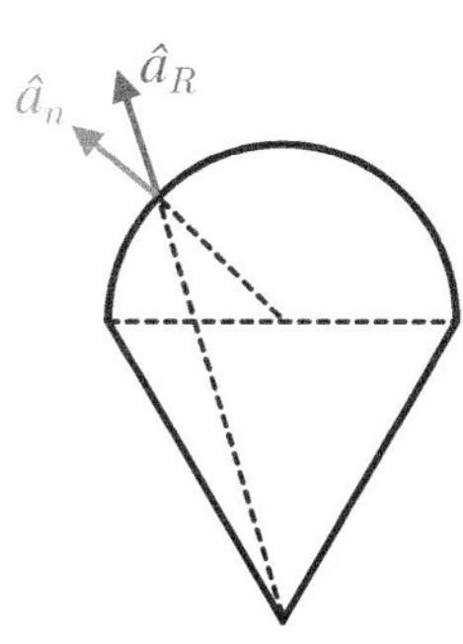

Figure 1.32 An illustration to show that the normal of the hemispherical surface in Figure 1.31 is not $\hat{a}_R$.

Hence, we have

$$\oint_S \bar{f}(\bar{r}) \cdot \overline{ds} = -\frac{2\pi a^2 b}{3} + \frac{2\pi a^3}{3} + \pi b a^2 = \frac{\pi a^2 b}{3} + \frac{2\pi a^3}{3} \tag{1.147}$$

$$= \frac{\pi a^2}{3}(2a+b). \tag{1.148}$$

On the other side, we have $\bar{\nabla} \cdot \bar{f}(\bar{r}) = 1$ and

$$\int_V \bar{\nabla} \cdot \bar{f}(\bar{r}) dv = \frac{4}{6}\pi a^3 + \pi a^2 \frac{b}{3} \tag{1.149}$$

$$= \frac{\pi a^2}{3}(2a+b). \tag{1.150}$$

1.4 Using the Differential Form of Gauss' Law

The differential form of Gauss' law is pointwise: i.e. it provides the equality of the divergence of the electric flux density and the electric charge density at every point. On the other hand, given a distribution of electric charge density, the electric field may not be found uniquely since the information is not complete (divergence of a vector field does not provide full information about the vector field). Further information may be obtained from other properties of the electric field: e.g. its curl and/or boundary conditions. At the same time, given an electric field distribution, the volume electric charge density can be found completely by using the differential form of Gauss' law.

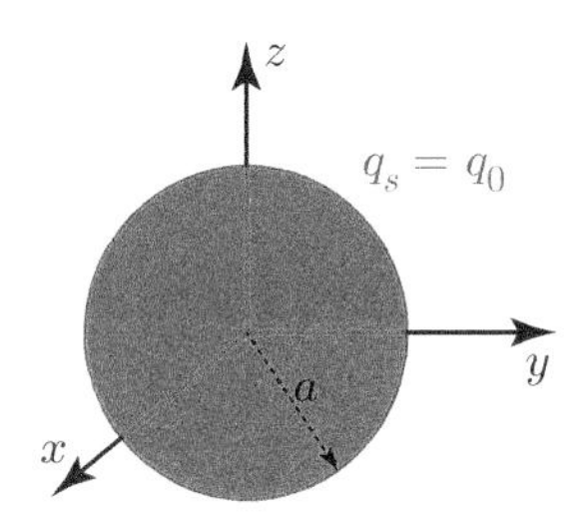

Figure 1.33 A static surface electric charge density $q_s(\bar{r}) = q_0$ (C/m^2) on a sphere of radius a.

An important point, as discussed several times so far, is that the right-hand side of the differential form of Gauss' law in Eq. (1.96) is the *volume* electric charge density. Hence, there are mathematical issues when surface or line electric charge distributions are considered. As an example, we consider a static surface electric charge density $q_s(\bar{r}) = q_0$ (C/m^2) as a sphere of radius a (Figure 1.33) in a vacuum. We assume that there is nothing else in the medium: i.e. $\bar{D}(\bar{r}) = \varepsilon_0 \bar{E}(\bar{r})$, where ε_0 is the permittivity of vacuum. Since the object (spherical charge distribution) and the overall scenario are symmetric,[41] the integral form of Gauss' law can be used to derive the electric flux density everywhere. On the other hand, to use Gauss' law, we need assumptions so that $\bar{D}(\bar{r})$ can be placed outside the integral. For this case, we can assume that $\bar{D}(\bar{r})$ depends only on R and it has only an R component:

[41] It is important that the overall scenario (source and space) is symmetric to use the integral form of Gauss' law.

$$\bar{D}(\bar{r}) = \hat{a}_R D_R(\bar{r}) + \hat{a}_\theta D_\theta(\bar{r}) + \hat{a}_\phi D_\phi(\bar{r}) = \hat{a}_R D_R(R, \theta, \phi) = \hat{a}_R D_R(R). \tag{1.151}$$

Now, considering a spherical Gauss surface with radius R, we have (Figure 1.34)

$$\oint_S \bar{D}(\bar{r}) \cdot \overline{ds} = \int_0^{2\pi} \int_0^{\pi} \hat{a}_R D_R(R) \cdot \hat{a}_R R^2 \sin\theta d\theta d\phi = 4\pi R^2 D_R(R) = Q_{\text{enc}} \tag{1.152}$$

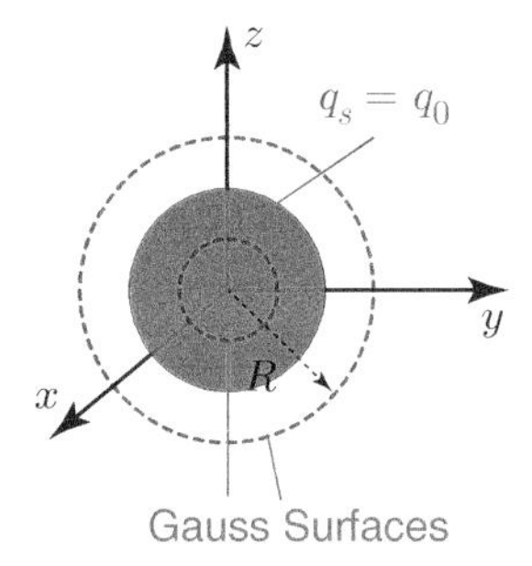

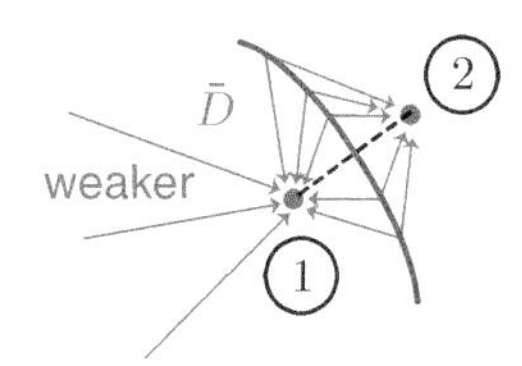

Figure 1.34 Using the integral form of Gauss' law for the surface electric charge density in Figure 1.33. The electric field can change significantly at short distances, which can be understood by examining the position of the observation point with respect to charges.

or $\bar{D}(\bar{r}) = \hat{a}_R Q_{\text{enc}}/(4\pi R^2)$, where Q_{enc} is the amount of the electric charge enclosed by the Gauss surface. If $R < a$, no electric charge is enclosed so that $Q_{\text{enc}} = 0$ and $\bar{D}(\bar{r}) = 0$. Otherwise, if $R > a$, we have $Q_{\text{enc}} = q_0 4\pi a^2$, leading to

$$\bar{D}(\bar{r}) = \hat{a}_R \frac{Q_{\text{enc}}}{4\pi R^2} = \hat{a}_R \frac{q_0 a^2}{R^2} \qquad (R > a). \tag{1.153}$$

Interestingly, for $R > a$, the electric field created by a spherical surface electric charge density is the same as the field created by a point electric charge. Using $\bar{D}(\bar{r}) = \varepsilon_0 \bar{E}(\bar{r})$, the electric field intensity can also be written

$$\bar{E}(\bar{r}) = \hat{a}_R \frac{q_0 a^2}{\varepsilon_0 R^2} \qquad (R > a). \tag{1.154}$$

At this stage, if we consider the differential form of Gauss' law, we have

$$q_v(\bar{r}) = \bar{\nabla} \cdot \bar{D}(\bar{r}) = \frac{1}{R^2}\frac{\partial}{\partial R}[R^2 D_R(R)] = 0 \quad (R \neq a). \tag{1.155}$$

This means there is no *volume* electric charge density in the whole space. While this was expected, we observe that the differential form of Gauss' law does not provide clear information about the electric charge density in this scenario.

It is remarkable that the electric field for the surface electric charge density in Figure 1.33 is discontinuous at $R = a$. Specifically, $\bar{D}(\bar{r})$ in Eq. (1.153) jumps from 0 to $q_0 a^2/a^2 = q_0$ when R changes from a^- to a^+. As shown in the next section, this discontinuity is directly related to the existence of a surface electric charge distribution that becomes clear when considering the boundary conditions. Therefore, even though the differential form of Gauss' law provides clear information about volume electric charge density distributions, application of Gauss' law (specifically, its integral form) at boundaries (leading to the boundary conditions for normal components of electric fields) enables the detection of surface electric charge densities. As also shown in Figure 1.34, such a jump is mainly caused by suddenly changing directions of electric fields created by electric charges that are close to the observation point.[42]

As a proper application of the differential form of Gauss' law, we now consider a static volume electric charge density $q_v(\bar{r}) = q_0$ (C/m^3) as a sphere of radius a (Figure 1.36). Specifically, the electric charge is distributed uniformly in the spherical region $R \leq a$ with a charge density of q_0. Using the integral form of Gauss' law, we again have

$$\bar{D}(\bar{r}) = \hat{a}_R \frac{Q_{\text{enc}}}{4\pi R^2}, \tag{1.156}$$

where Q_{enc} is the amount of the electric charge enclosed by the Gauss surface (with radius R). If $R > a$, we have $Q_{\text{enc}} = q_0 4\pi a^3/3$, leading to

$$\bar{D}(\bar{r}) = \hat{a}_R \frac{q_0 a^3}{3R^2} \qquad (R > a). \tag{1.157}$$

If $R < a$, however, we have $Q_{\text{enc}} = q_0 4\pi R^3/3$ and

$$\bar{D}(\bar{r}) = \hat{a}_R \frac{q_0 R^3}{3R^2} = \hat{a}_R \frac{q_0 R}{3} \qquad (R < a). \tag{1.158}$$

Hence, the electric flux density increases linearly for $R < a$, and then it decays quadratically with the distance for $R > a$. At this stage, knowing the expression for $\bar{D}(\bar{r})$, we can use the

[42]From another perspective, considering the derivative of the step function (jump), one can claim that the value of the divergence $\bar{\nabla} \cdot \bar{D}(\bar{r})$ at $R = a$ is in fact an impulse function (specifically, $q_0\delta(R - a)$) corresponding to a spherical surface electric charge distribution with a radius of a (see Figure 1.35). In this interpretation, a volume electric charge distribution is assumed to be squeezed into a zero volume, becoming a spherical surface with zero thickness in the limit.

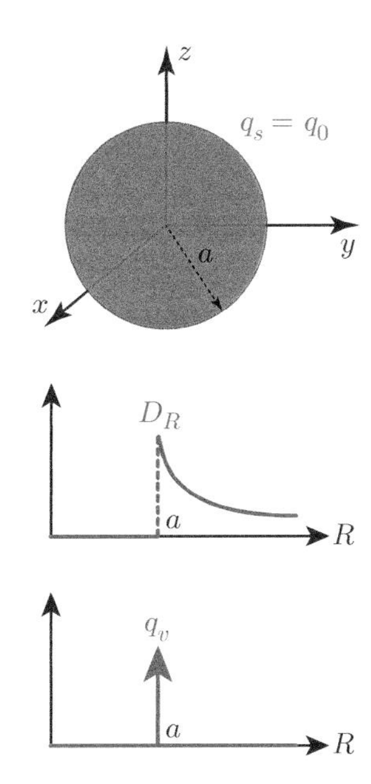

Figure 1.35 The electric flux density due to a static spherical surface electric charge density has a jump at $R = a$, where a is the radius of the electric charge density. From one perspective, the volume electric charge density, which is the divergence of the electric flux density, is an impulse function at $R = a$.

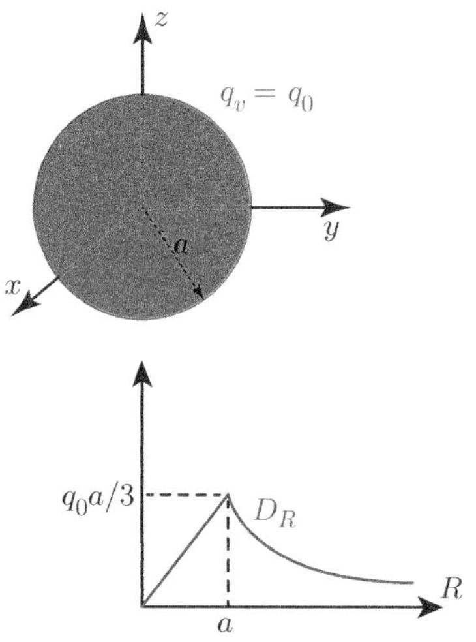

differential form of Gauss' law. For $R > a$, we again have zero divergence and zero volume electric charge density. On the other hand, for $R < a$, we have

$$q_v(\bar{r}) = \bar{\nabla} \cdot \bar{D}(\bar{r}) = \frac{1}{R^2}\frac{\partial}{\partial R}\left(R^2 D_R(\bar{r})\right) = \frac{1}{R^2}\frac{\partial}{\partial R}\left(R^2 \frac{q_0 R}{3}\right) \tag{1.159}$$

$$= \frac{q_0}{3R^2}\frac{\partial}{\partial R}(R^3) = q_0 \qquad (R < a). \tag{1.160}$$

Hence, we correctly find the volume electric charge density for $R < a$ by using the differential form of Gauss' law.

1.4.1 Examples

Example 6: Consider a static spherical volume electric charge density $q_v(\bar{r})$ (C/m^3). The electric charge is limited to $R < a$ (Figure 1.37). It is also given that the electric flux density is

$$\bar{D}(\bar{r}) = \hat{a}_R \frac{q_0 a^2}{R} \qquad (R < a). \tag{1.161}$$

Find the electric flux density for $R > a$.

Solution: First, we need to find the electric charge distribution[43] with respect to R. Using the differential form of Gauss' law, we have

$$q_v(\bar{r}) = \bar{\nabla} \cdot \bar{D}(\bar{r}) = \frac{1}{R^2}\frac{\partial}{\partial R}\left(R^2 D_R(R)\right) = \frac{q_0 a^2}{R^2}. \tag{1.162}$$

Using this, the total electric charge can be found as[44]

$$Q_{\text{total}} = \int_V \frac{q_0 a^2}{R^2} dv = 4\pi q_0 a^2 \int_0^a \frac{1}{R^2} R^2 dR = 4\pi q_0 a^3 \tag{1.163}$$

by considering the whole spherical volume. Then, using the integral form of Gauss' law (with a spherical Gauss surface of radius R), the electric flux density for $R > a$ can be found as

$$4\pi R^2 D_R(R) = 4\pi q_0 a^3, \tag{1.164}$$

leading to

$$\bar{D}(\bar{r}) = \hat{a}_R \frac{q_0 a^3}{R^2} \qquad (R > a). \tag{1.165}$$

Figure 1.36 A static volume electric charge density $q_v(\bar{r}) = q_0$ (C/m^3) in the shape of a sphere of radius a and the electric flux density with respect to R. The electric flux density is continuous at $R = a$: i.e. $\bar{D}(R = a^+) = \bar{D}(R = a^-) = \hat{a}_R q_0 a/3$ since there is no surface electric charge density that could create a discontinuity in $\bar{D}(\bar{r})$.

[43] When considering this question, the region for the charge distribution should be defined clearly as a part of the question. In this case, it is given that the volume charge distribution is limited to the region $R \leq a$. In addition, the volume charge distribution is assumed to be the only charge distribution. For example, if there were surface charge density at $R = a$, then the electric flux density would be discontinuous.

[44] Alternatively, considering a spherical Gauss surface with radius $R < a$, we have

$$\oint_S \bar{D}(\bar{r}) \cdot \overline{ds} = \int_0^{\pi}\int_0^{2\pi} \hat{a}_R \frac{q_0 a^2}{R} \cdot \hat{a}_R R^2 \sin\theta d\theta d\phi$$
$$= 4\pi q_0 a^2 R = Q_{\text{enc}}.$$

Hence, $Q_{\text{enc}} = 4\pi q_0 a^2 R$ for all values of $R < a$, leading to $Q_{\text{total}} = 4\pi q_0 a^3$ in the limit of $R \rightarrow a$.

1.5 Boundary Conditions for Normal Electric Fields

Application of Gauss' law at boundaries between two different media provides some useful conditions that must be satisfied by electric fields. Specifically, Gauss' law provides how the normal component of the electric flux density and the normal component of the electric field intensity should behave at a boundary. To derive the boundary conditions, we consider a boundary S between two media with permittivities $\varepsilon_1(\bar{r})$ and $\varepsilon_2(\bar{r})$, and a normal direction $\hat{a}_n$ pointing into the first medium. We also assume that $\{\bar{E}_1(\bar{r},t), \bar{D}_1(\bar{r},t)\}$ and $\{\bar{E}_2(\bar{r},t), \bar{D}_2(\bar{r},t)\}$ represent electric fields in the first and second media, respectively, where $\bar{D}_1(\bar{r},t) = \varepsilon_1(\bar{r})\bar{E}_1(\bar{r},t)$ and $\bar{D}_2(\bar{r},t) = \varepsilon_2(\bar{r})\bar{E}_2(\bar{r},t)$. For using Gauss' law locally at the boundary, we consider a rectangular prism Gauss surface[45] enclosing a differential part of the boundary around the observation point (Figure 1.38). Using the integral form of Gauss' law, as repeated here, we have

$$\oint_S \bar{D}(\bar{r},t) \cdot \overline{ds} = Q_{\text{enc}}(t), \tag{1.166}$$

where $Q_{\text{enc}}(t)$ is the electric charge density enclosed in the Gauss surface. Now, if we consider a limit case as the height of the Gauss surface goes to zero ($h \to 0$), we derive

$$\oint_S \bar{D}(\bar{r},t) \cdot \overline{ds} \to \Delta A[\hat{a}_n \cdot \bar{D}_1(\bar{r},t) - \hat{a}_n \cdot \bar{D}_2(\bar{r},t)] \tag{1.167}$$

$$Q_{\text{enc}}(t) \to (\Delta A) q_s(\bar{r},t), \tag{1.168}$$

where ΔA is the top/bottom area of the Gauss surface. Hence, the surface integral consists of the contributions from the top and bottom surfaces (with different signs considering the normal of the Gauss surface), while the enclosed electric charge reduces into the area multiplied by the surface electric charge density $q_s(\bar{r},t)$ that possibly exists at the boundary surface. Inserting the limit cases into Gauss' law (Eq. (1.166)), we derive

$$\Delta A[\hat{a}_n \cdot \bar{D}_1(\bar{r},t) - \hat{a}_n \cdot \bar{D}_2(\bar{r},t)] = (\Delta A) q_s(\bar{r},t) \tag{1.169}$$

$$\hat{a}_n \cdot [\bar{D}_1(\bar{r},t) - \bar{D}_2(\bar{r},t)] = q_s(\bar{r},t) \qquad (\bar{r} \in S) \tag{1.170}$$

as the boundary condition for the normal electric flux density. According to this equation, the normal component of the electric flux density may be discontinuous depending on the existence of a surface electric charge density at the boundary (Figure 1.39). If there is no electric charge, we have

$$\hat{a}_n \cdot \bar{D}_1(\bar{r},t) = \hat{a}_n \cdot \bar{D}_2(\bar{r},t) \qquad (\bar{r} \in S), \tag{1.171}$$

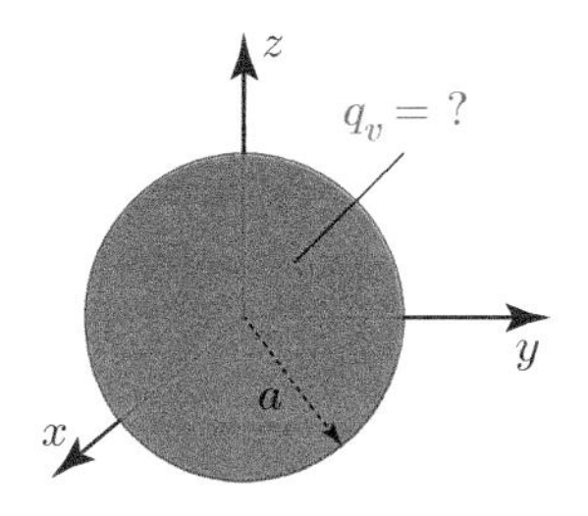

Figure 1.37 A static spherical volume electric charge density with an unknown distribution.

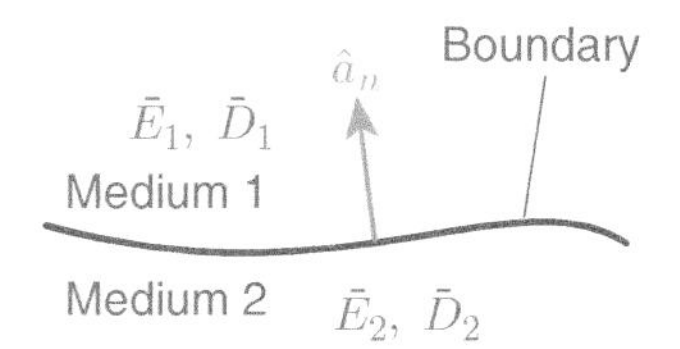

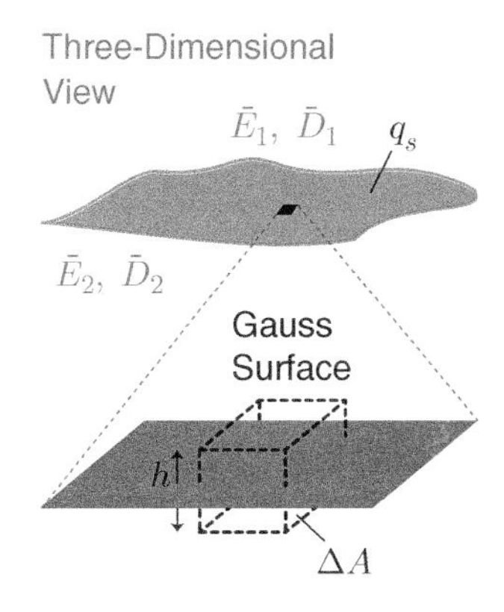

Figure 1.38 An illustration of the derivation of the boundary conditions for normal components of electric fields across an interface between two different media. We note that the derivation holds for observation points on smooth surfaces, and not for observation points located at the edges and corners.

[45]We note that the Gauss surface is assumed to be infinitesimal such that the enclosed boundary is perfectly flat. In addition, it is symmetrically located such that the boundary always remain inside as $h \to 0$.

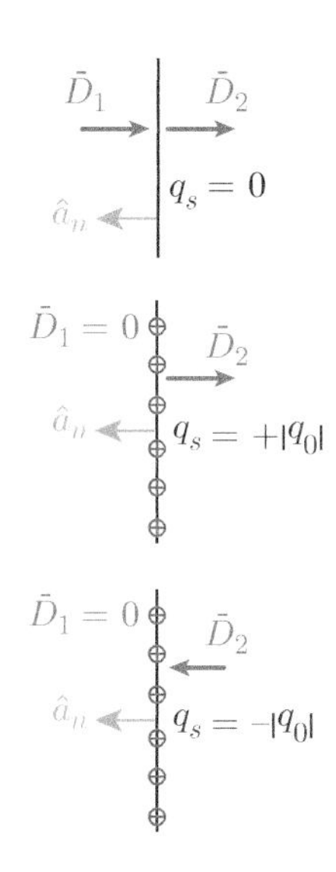

Figure 1.39 Some special cases about the boundary condition for the normal component of the electric flux density. In the first case (top), the surface electric charge density is zero so that the normal component of the electric flux density must be continuous. This can be interpreted to mean that any electric flux entering the boundary (normal component) must leave the boundary. In the second case (middle), $\bar{D}_1(\bar{r},t) = 0$, and there is a static positive electric charge density so that $\hat{a}_n \cdot \bar{D}_2(\bar{r},t) = -|q_0|$: i.e. the flux is in the right direction. In the third case (bottom), $\bar{D}_1(\bar{r},t) = 0$, and there is a negative electric charge density so that $\hat{a}_n \cdot \bar{D}_2(\bar{r},t) = +|q_0|$: i.e. the flux is in the left direction.

i.e. the normal component of the electric flux is continuous. When the dielectric media have zero conductivity, it is common to assume that there is no electric charge at the boundary, if not particularly inserted.

We can also derive the boundary condition for the electric field intensity values. In the general form, we have

$$\hat{a}_n \cdot [\varepsilon_1(\bar{r})\bar{E}_1(\bar{r},t) - \varepsilon_2(\bar{r})\bar{E}_2(\bar{r},t)] = q_s(\bar{r},t) \qquad (\bar{r} \in S), \tag{1.172}$$

which becomes

$$\hat{a}_n \cdot [\varepsilon_1(\bar{r})\bar{E}_1(\bar{r},t) - \varepsilon_2(\bar{r})\bar{E}_2(\bar{r},t)] = 0 \qquad (\bar{r} \in S) \tag{1.173}$$

without electric charge: i.e. when $q_s(\bar{r},t) = 0$. Hence, the normal component of the electric field intensity across a boundary of two different media is discontinuous, even when there is no electric charge. This can be explained by the polarization effects, as discussed in Section 1.7.

The boundary conditions provide important information about electric/magnetic fields by showing their behavior across boundaries. As an interesting example for normal components of electric fields, we consider an infinite sheet of static surface electric charge density defined as $q_s(\bar{r}) = q_0$ (C/m^2) lying on the x-y plane in a vacuum (Figure 1.40). Due to the symmetry and infinity, the electric flux density should not depend on x and y, while it may depend on z. In addition, it should be in opposite directions for[46] $z > 0$ and $z < 0$. Consequently, we can write the electric flux density as

$$\bar{D}(\bar{r}) = \hat{a}_z \begin{cases} f(|z|) & (z > 0) \\ -f(|z|) & (z < 0), \end{cases} \tag{1.174}$$

where $f(|z|)$ represents a possible z dependency. Then, using the boundary condition, we have

$$\hat{a}_z \cdot [\bar{D}_1(\bar{r}) - \bar{D}_2(\bar{r})] = q_0 \qquad (\bar{r} \in S), \tag{1.175}$$

leading to $2f(|z| = 0) = q_0$ or

$$f(|z| = 0) = q_0/2. \tag{1.176}$$

In the above, we have found the value of $f(|z|)$, and hence the electric flux density, only at the boundary ($z = 0$). But for this special case, we can further find the electric flux density everywhere. For this purpose, one can consider a symmetrically located Gauss surface (see Figure 1.40 for the side view). The contributions are only from[47] the top surface (at $z = h$) and the bottom surface (at $z = -h$). But these contributions must be the same considering that $\bar{D}(\bar{r})$ and the

[46] If q_0 is positive, $\bar{D}(\bar{r})$ should be in the $+z$ and $-z$ directions, respectively, for $z > 0$ and $z < 0$.

[47] There is no contribution from the four side surfaces, as $\bar{D}(\bar{r})$ is parallel to them.

normal of these surfaces are aligned, while the surfaces are the same distance from the source. On the other hand, no matter what the value of h is, the same amount of electric charge is enclosed inside the Gauss surface. This indicates that the electric flux density does not depend on z, and we have $f(|z|) = f(|z| = 0)$, leading to

$$\bar{D}(\bar{r}) = \hat{a}_z \frac{q_0}{2} \begin{cases} 1 & (z > 0) \\ -1 & (z < 0). \end{cases} \tag{1.177}$$

To sum up, the electric flux density has a constant magnitude everywhere.[48]

With the latest example, we may list some different behaviors of electric fields created by different configurations:[49]

- A static point electric charge creates an electric field decaying with the square of the distance $(1/R^2)$.
- An infinitely long static line electric charge density creates an electric field decaying with the distance $(1/\rho)$.
- The electric field created by an infinitely large static planar surface electric charge density does not decay.

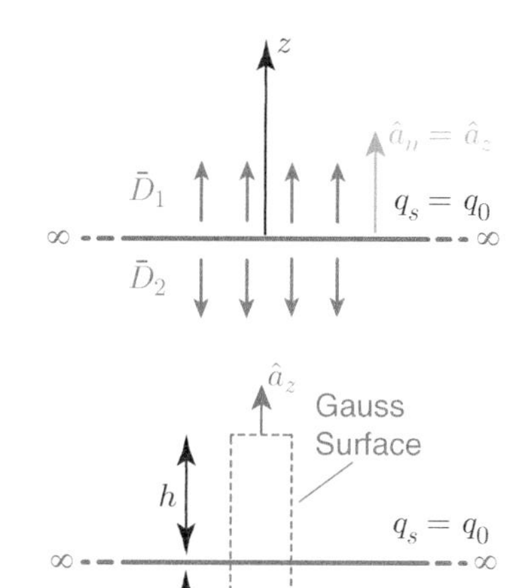

Figure 1.40 The electric field created by an infinitely large static surface electric charge density on the x-y plane. The electric flux density should be in the opposite $(\pm z)$ directions for $z > 0$ and $z < 0$, while it should not depend on x and y. An application of the integral form of Gauss' law further shows that the electric flux density does not depend on z; hence it must be constant everywhere.

[48] At this stage we understand that the magnitude of the electric flux density does not depend on the distance from the plate. Therefore, it was not necessary to place the Gauss surface symmetrically. Nevertheless, we *did not know* this before analyzing the scenario.

[49] As shown in Section 1.7.1, there is also a static electric charge configuration (electric dipole) that creates an electric field decaying with $1/R^3$ in the far zone.

1.6 Static Cases and Coulomb's Law

Like all Maxwell's equations, Gauss' law in Eqs. (1.1) and (1.96) is valid for both static and dynamic cases. When time-harmonic sources are involved, the integral and differential forms can be written in the frequency domain as

$$\oint_S \bar{D}(\bar{r}) \cdot \overline{ds} = Q_{\text{enc}} \tag{1.178}$$

$$\bar{\nabla} \cdot \bar{D}(\bar{r}) = q_v(\bar{r}), \tag{1.179}$$

where $\bar{D}(\bar{r})$, Q_{enc}, and $q_v(\bar{r})$ are phasor-domain representations of the corresponding quantities. On the other hand, as briefly discussed in Section 1.2, Gauss' law is not sufficient alone to analyze a dynamic electromagnetic scenario. This is because it gives partial information about electric fields – i.e. how they are created by electric charges – while electric fields are also produced by time-variant magnetic fields as indicated by Faraday's law. However, in a static case, the electric flux density and the electric field intensity created by an electric charge distribution can be found via Gauss' law. In fact, in a static case, Gauss' law is consistent with Coulomb's law, which is an important tool to analyze static electric charges. In this section, we further investigate Gauss' law for static cases.

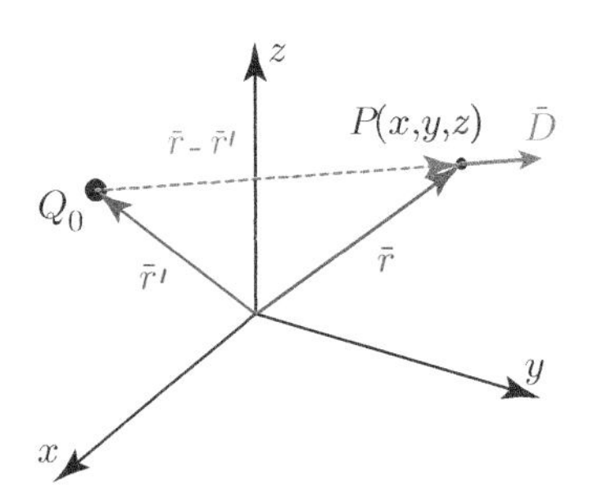

Figure 1.41 The electric field created by a static electric charge Q_0 located at $\bar{r}' = \hat{a}_x x' + \hat{a}_y y' + \hat{a}_z z'$ at another point $\bar{r} = \hat{a}_x x + \hat{a}_y y + \hat{a}_z z$.

1.6.1 Superposition Principle

Considering Eq. (1.84), the electric flux density created by a static point electric charge Q_0 located at the origin in a vacuum can be written

$$\bar{D}(\bar{r}) = \hat{a}_R \frac{Q_0}{4\pi R^2} = \frac{Q_0}{4\pi} \frac{\bar{r}}{|\bar{r}|^3}, \tag{1.180}$$

where $\bar{r}$ is the position vector. In a more general case (Figure 1.41), considering an electric charge Q_0 located at $\bar{r}' = \hat{a}_x x' + \hat{a}_y y' + \hat{a}_z z'$, the electric flux density at an arbitrary position $\bar{r} = \hat{a}_x x + \hat{a}_y y + \hat{a}_z z$ can be written

$$\bar{D}(\bar{r}) = \frac{Q_0}{4\pi} \frac{(\bar{r} - \bar{r}')}{|\bar{r} - \bar{r}'|^3}. \tag{1.181}$$

Several points must be emphasized:

- The electric flux density is proportional to the amount of the source electric charge Q_0.
- The direction of the electric field is from the source to the observation point (if Q_0 is positive) or from the observation point to the source point (if Q_0 is negative). This direction information is included in the term $(\bar{r} - \bar{r}')$ in the numerator of the second fraction.
- The electric flux density decays quadratically with the distance from the source: i.e. it is proportional to $1/|\bar{r} - \bar{r}'|^2$, where $|\bar{r} - \bar{r}'|$ is the distance between the source and observation points. We note that one of $|\bar{r} - \bar{r}'|$ in the denominator cancels with the magnitude of $(\bar{r} - \bar{r}')$ in the numerator of the second fraction.[50]
- It appears that the electric field created by a point electric charge is infinite at the location of the electric charge. We recall that the point electric charge itself is an idealization (a total amount of electric charge Q_0 squeezed into a single point), as discussed in Section 1.3.2. Therefore, having an infinite electric field at the electric charge location is perfectly consistent with the source definition.

[50] We also note that $\bar{r} - \bar{r}' = \hat{a}_x(x - x') + \hat{a}_y(y - y') + \hat{a}_z(z - z')$ in a Cartesian coordinate system.

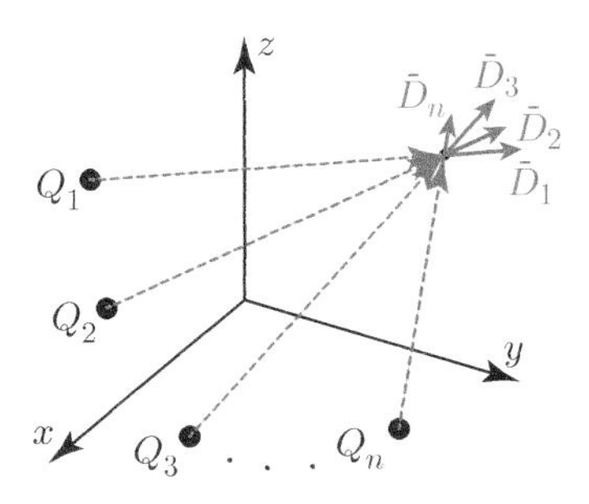

Figure 1.42 Electric fields created by a set of static electric charges. The vector sum of these electric fields gives the total electric field at the observation point.

Electric charges and field values are linearly related to each other. According to the superposition principle, linear effects can be added. For example, if there are n point electric charges

$Q_1, Q_2, \ldots, Q_n$ in a vacuum (Figure 1.42), the total electric field created by them at $\bar{r}$ can be obtained as[51]

$$\bar{D}(\bar{r}) = \sum_{i=1}^{n} \frac{Q_i}{4\pi} \frac{(\bar{r} - \bar{r}_i)}{|\bar{r} - \bar{r}_i|^3}, \tag{1.182}$$

where $\bar{r}_i$ represents the location of the electric charge Q_i. If there is a continuous volumetric distribution of electric charges as defined in Eq. (1.98), then the electric field created by this electric charge distribution (Figure 1.43) can be found via integration as[52]

$$\bar{D}(\bar{r}) = \frac{1}{4\pi} \int_V q_v(\bar{r}') \frac{(\bar{r} - \bar{r}')}{|\bar{r} - \bar{r}'|^3} d\bar{r}'. \tag{1.183}$$

Depending on the definition of the electric charge distribution – i.e. surface electric charges (C/m^2) and line electric charges (C/m), as described in Eqs. (1.99) and (1.100), respectively – the domain of the integration may be changed as

$$\bar{D}(\bar{r}) = \frac{1}{4\pi} \int_S q_s(\bar{r}') \frac{(\bar{r} - \bar{r}')}{|\bar{r} - \bar{r}'|^3} d\bar{r}' \tag{1.184}$$

$$\bar{D}(\bar{r}) = \frac{1}{4\pi} \int_C q_l(\bar{r}') \frac{(\bar{r} - \bar{r}')}{|\bar{r} - \bar{r}'|^3} d\bar{r}'. \tag{1.185}$$

These integrals[53,54,55] are commonly used as tools to find the electric field (electric flux density $\bar{D}(\bar{r})$ and electric field intensity $\bar{E}(\bar{r}) = \bar{D}(\bar{r})/\varepsilon_0$) for given electric charge distributions in a vacuum.

As an example to understand the superposition principle, we consider two static point electric charges Q_1 and Q_2 located at $\bar{r}_1$ and $\bar{r}_2$, respectively, in a vacuum (Figure 1.44). We also consider two spherical Gauss surfaces S_1 and S_2 centered at $\bar{r}_1$ and $\bar{r}_2$. According to Gauss' law, we have

$$\oint_{S_1} \bar{D}_1(\bar{r}) \cdot \overline{ds} = Q_1 \longrightarrow \bar{D}_1(\bar{r}) = \frac{Q_1}{4\pi} \frac{(\bar{r} - \bar{r}_1)}{|\bar{r} - \bar{r}_1|^3} \tag{1.186}$$

and

$$\oint_{S_2} \bar{D}_2(\bar{r}) \cdot \overline{ds} = Q_2 \longrightarrow \bar{D}_2(\bar{r}) = \frac{Q_2}{4\pi} \frac{(\bar{r} - \bar{r}_2)}{|\bar{r} - \bar{r}_2|^3}, \tag{1.187}$$

where $\bar{r}$ represents the observation point. Specifically, using Gauss' law individually, we are able to obtain the electric fields created by Q_1 and Q_2, assuming that Q_1 does not change the point characteristics Q_2 and vice versa. According to the superposition principle, the electric fields

[51] We *assume* that electric charges do not affect each other: i.e. the existence of an electric charge does not change the point properties of other electric charges or the electric fields created by them. This also means a point source creates the same electric field, whether there are other electric charges or not.

[52] One can take the divergence of the electric flux density as $\bar{\nabla} \cdot \bar{D}(\bar{r})$ to derive the differential form of Gauss' law for static cases, keeping in mind that the $\bar{\nabla}$ operator is applied to *unprimed* coordinates and using $\lim_{R\to 0} \bar{\nabla} \cdot \hat{a}_R/R^2 = 4\pi\delta(R)$.

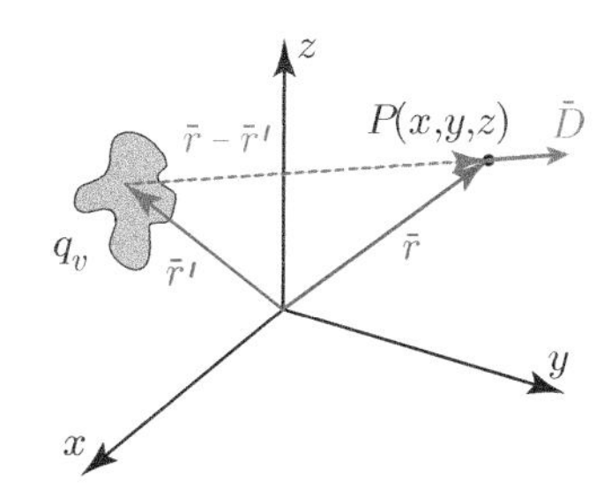

Figure 1.43 The electric field created by a volume electric charge distribution. The total electric field at the observation point can be found by integration that accounts for all possible source locations $\bar{r}'$.

[53] In these integrals, $\bar{r}$ represents the observation point, while $\bar{r}'$ is the source location and the integration variable. In addition, $d\bar{r}'$ indicates that the integral is over the primed coordinates: i.e. (x', y', z'), (ρ', ϕ', z'), or (R', θ', ϕ') depending on the used coordinate system. From this perspective, $d\bar{r}' = dv'$, $d\bar{r}' = ds'$, or $d\bar{r}' = dl'$, depending on the integration domain (source type).

[54] These integral expressions can be verified and formulated more completely via the electric scalar potential.

[55] These integrals can be called Coulomb's law when the electric force is related to the electric field.

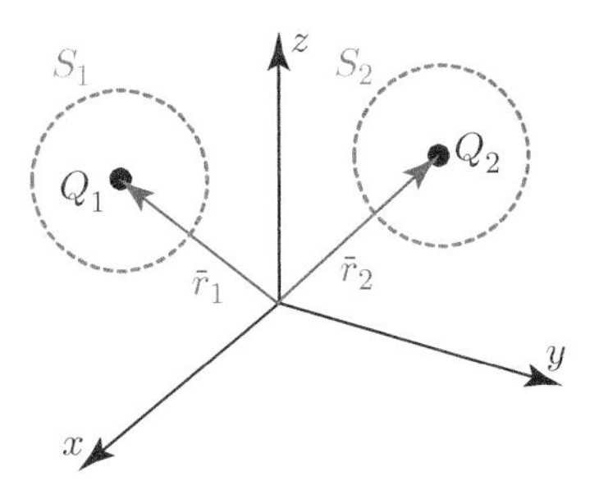

Figure 1.44 Two static point electric charges located at different positions and spherical Gauss surfaces S_1 and S_2 enclosing them.

created by individual electric charges can be superposed: i.e. the total electric flux density can be found anywhere as

$$\bar{D}(\bar{r}) = \bar{D}_1(\bar{r}) + \bar{D}_2(\bar{r}) = \frac{Q_1}{4\pi}\frac{(\bar{r}-\bar{r}_1)}{|\bar{r}-\bar{r}_1|^3} + \frac{Q_2}{4\pi}\frac{(\bar{r}-\bar{r}_2)}{|\bar{r}-\bar{r}_2|^3}. \tag{1.188}$$

Furthermore, since Gauss' law is valid for all electric charge distributions and configurations, we have[56]

$$\oint_{S_1} \bar{D}(\bar{r}) \cdot \overline{ds} = \oint_{S_1} [\bar{D}_1(\bar{r}) + D_2(\bar{r})] \cdot \overline{ds} = Q_1 \tag{1.189}$$

and

$$\oint_{S_2} \bar{D}(\bar{r}) \cdot \overline{ds} = \oint_{S_2} [\bar{D}_1(\bar{r}) + D_2(\bar{r})] \cdot \overline{ds} = Q_2. \tag{1.190}$$

But we emphasize that $\bar{D}(\bar{r})$ cannot be put outside the integrals in Eqs. (1.189) and (1.190) since $\bar{D}(\bar{r})$ (caused by both Q_1 and Q_2) changes with position on S_1 and S_2. Obviously, without the superposition principle, it would be difficult to find the electric flux density in Eq. (1.188) using Gauss' law directly.

The superposition principle is naturally valid and inherently used in all cases,[57] and it can be employed alone (with some previous knowledge) to analyze some interesting cases. As an example, we consider a simplified model of an ideal parallel-plate capacitor (Figure 1.45) that consists of two plates, each with area A_p, separated by a distance of d. We assume that the space between the plates is a vacuum. When a constant voltage (electric potential) is applied between the plates, positive and negative electric charges with equal amounts are accumulated on the plates. For example, if the top plate has a higher voltage, we can assume Q_0 on the top plate (at $z = d$) and $-Q_0$ on the bottom plate (at $z = 0$). In this ideal case, the plates have zero thicknesses, and they are very large such that electric charges are distributed uniformly, leading to surface electric charge densities $q_s(\bar{r}) = Q_0/A_p$ and $q_s(\bar{r}) = -Q_0/A_p$ on the top and bottom plate, respectively. Then, further considering that the plates are very large, we can assume that the electric field is only in the $\pm z$ direction, and we can use Eq. (1.177) to write[58]

$$\bar{D}_{\text{top}}(\bar{r}) = \hat{a}_z \frac{Q_0}{2A} \begin{cases} 1 & (z > d) \\ -1 & (z < d) \end{cases} \tag{1.191}$$

and

$$\bar{D}_{\text{bottom}}(\bar{r}) = -\hat{a}_z \frac{Q_0}{2A} \begin{cases} 1 & (z > 0) \\ -1 & (z < 0) \end{cases} \tag{1.192}$$

[56]Note that we have

$$\oint_{S_1} D_2(\bar{r}) \cdot \overline{ds} = 0 \quad \text{and} \quad \oint_{S_2} D_1(\bar{r}) \cdot \overline{ds} = 0.$$

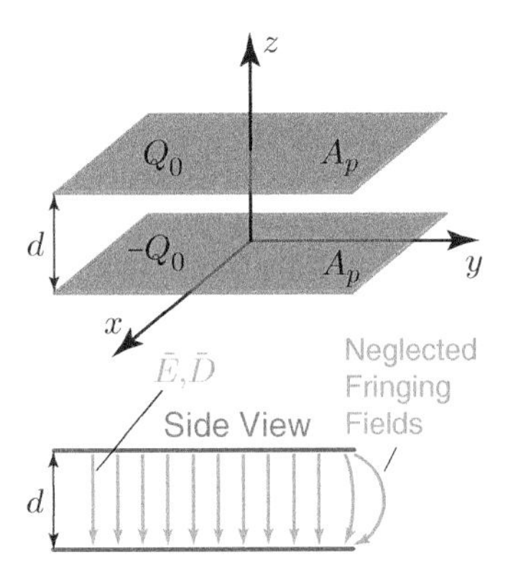

Figure 1.45 A parallel-plate capacitor model.

[57]In time-variant cases, the time (and delay) is involved when superposing the effects of multiple sources.

[58]The assumption of large plates is used twice. In the first, the electric charges are considered to be distributed uniformly: i.e. no electric charges are accumulated at the edges. In the second, we directly use the expression for the electric flux density created by an infinitely large planar electric charge density. These also mean no fringing fields (Figure 1.45), which occur due to the finite sizes of the planar electric charge distributions (edge effects).

for the electric flux density created by the top and bottom plates.[59] Combining these contributions (using the superposition principle), we arrive at the overall expression for the electric flux density as

$$\bar{D}(\bar{r}) = \bar{D}_{\text{top}}(\bar{r}) + \bar{D}_{\text{bottom}}(\bar{r}) = \hat{a}_z \frac{Q_0}{2A} \begin{cases} 0 & (z > d) \\ -2 & (0 < z < d) \\ 0 & (z < 0). \end{cases} \tag{1.193}$$

This can be written simply as

$$\bar{D}(\bar{r}) = -\hat{a}_z \frac{Q_0}{A} \begin{cases} 1 & \text{(between the plates)} \\ 0 & \text{(elsewhere)}. \end{cases} \tag{1.194}$$

Hence, the contributions from the top and bottom plates perfectly cancel outside the capacitor, while they lead to $-z$-directed electric flux density inside the capacitor (Figure 1.46). Finally, the electric field intensity can be written

$$\bar{E}(\bar{r}) = -\hat{a}_z \frac{Q_0}{A\varepsilon_0} \begin{cases} 1 & \text{(between the plates)} \\ 0 & \text{(elsewhere)}. \end{cases} \tag{1.195}$$

The source integrals in Eqs. (1.183), (1.184), and (1.185), which can be called Coulomb's law (see Section 1.6.2), can be used directly to find the electric field due to an arbitrary (but static) electric charge distributions. They are particularly useful when there is no symmetry such that Gauss' law cannot be used. As an example, we consider a static surface electric charge density with a disk shape of radius a located on the x-y plane in a vacuum (Figure 1.47). The electric charge density is constant on the entire disk as $q_s(\bar{r}) = q_0$ (C/m^2). By using Eq. (1.184), we can find the electric flux density at $\bar{r} = \hat{a}_z z$ for $z > 0$ (on the positive z axis). We have

$$\bar{D}(\bar{r}) = \frac{q_0}{4\pi} \int_0^{2\pi} \int_0^a \frac{(\hat{a}_z z - \hat{a}_x x' - \hat{a}_y y')}{[(x')^2 + (y')^2 + z^2]^{3/2}} \rho' d\rho' d\phi' \tag{1.196}$$

since[60] $\bar{r}' = \hat{a}_x x' + \hat{a}_y y'$. Then, inserting $x' = \rho' \cos\phi'$, $y' = \rho' \sin\phi'$, and $(x')^2 + (y')^2 = (\rho')^2$, we obtain[61]

$$\bar{D}(\bar{r}) = \hat{a}_z \frac{q_0 z}{2} \int_0^a \frac{1}{[(\rho')^2 + z^2]^{3/2}} \rho' d\rho'. \tag{1.197}$$

[59] In these expressions, only z is used to define the position, but we recall that the plates are actually finite and the electric field is zero outside the capacitor: i.e. for large values of x and y in Figure 1.45.

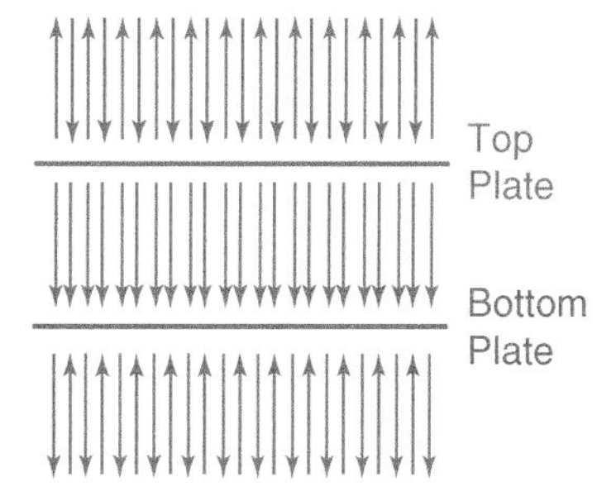

Figure 1.46 Electric fields created by a positively charged (top) plate and negatively charged (bottom) plate.

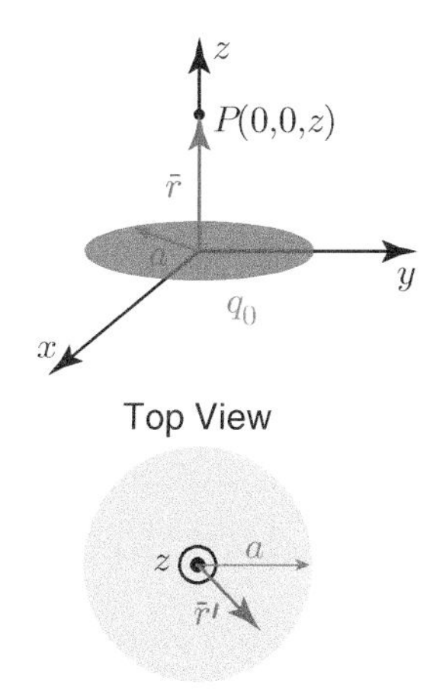

Figure 1.47 A static surface electric charge distribution with a disk shape.

[60] We note that $\bar{r} = \hat{a}_z z$, $\bar{r}' = \hat{a}_x x' + \hat{a}_y y'$, and $d\bar{r}' = \rho' d\rho' d\phi'$, while the integral is over a full disk so that $\rho' : 0 \to a$ and $\phi' : 0 \to 2\pi$ in the integration limits.

[61] Note that

$$\int_0^{2\pi} \cos\phi' d\phi' = 0 \quad \text{and} \quad \int_0^{2\pi} \sin\phi' d\phi' = 0.$$

Evaluating the integral, we have[62,63]

$$\bar{D}(\bar{r}) = \hat{a}_z \frac{q_0 z}{2} \left[-\frac{1}{\sqrt{(\rho')^2 + z^2}} \right]_0^a = \hat{a}_z \frac{q_0 z}{2} \left(\frac{1}{z} - \frac{1}{\sqrt{a^2 + z^2}} \right). \tag{1.198}$$

[62]Note that $z > 0$ and $1/\sqrt{z^2} = 1/z$.

[63]Since the electric charge distribution is in a vacuum, we can also write the electric field intensity as

$$\bar{E}(\bar{r}) = \hat{a}_z \frac{q_0 z}{2\varepsilon_0} \left(\frac{1}{z} - \frac{1}{\sqrt{a^2 + z^2}} \right).$$

Once an electric field due to an electric charge distribution is found in terms of geometric quantities, several different limits can be interesting. In the case of the disk-shaped charge distribution, we rewrite the expression for the electric flux density as

$$\bar{D}(\bar{r}) = \hat{a}_z \frac{q_0}{2} \left(1 - \frac{z}{\sqrt{a^2 + z^2}} \right). \tag{1.199}$$

Then we have the following special cases:

- When $z \to 0$, we have

$$\lim_{z \to 0} \{\bar{D}(\bar{r})\} = \hat{a}_z \frac{q_0}{2}. \tag{1.200}$$

  Interestingly, this is the same as the electric flux density due to an infinitely large planar surface electric charge density.[64]

- When $a \to 0$, we have

$$\lim_{a \to 0} \{\bar{D}(\bar{r})\} = \hat{a}_z \frac{q_0}{2} \left(1 - \frac{z}{\sqrt{z^2}} \right) = 0. \tag{1.201}$$

  Unsurprisingly, the electric field goes to zero since the electric charge distribution disappears as a goes to zero.

- Fixing $q_0 \pi a^2 = Q_0$ constant, and letting $a \to 0$, we have[65]

$$\lim_{a \to 0} \{\bar{D}(\bar{r})\} = \lim_{a \to 0} \left\{ \hat{a}_z \frac{Q_0}{2\pi a^2} \left(1 - \frac{z}{\sqrt{a^2 + z^2}} \right) \right\} \tag{1.202}$$

$$= \hat{a}_z \frac{Q_0}{4\pi z^2}, \tag{1.203}$$

  i.e. the expression becomes the one for a static point electric charge.[66]

[64]The same expression can be obtained by using $z > 0$, while $a \to \infty$: i.e. when the electric charge distribution becomes really infinitely large.

[65]The limit is evaluated by using the L'Hôpital rule, as follows:

$$\begin{aligned}
\lim_{a \to 0} \left\{ \frac{1}{a^2} - \frac{z}{a^2 \sqrt{a^2 + z^2}} \right\} &= \lim_{a \to 0} \left\{ \frac{\sqrt{a^2 + z^2} - z}{a^2 \sqrt{a^2 + z^2}} \right\} \\
&= \lim_{a \to 0} \left\{ \frac{a/\sqrt{a^2 + z^2}}{2a\sqrt{a^2 + z^2} + a^3/\sqrt{a^2 + z^2}} \right\} \\
&= \lim_{a \to 0} \left\{ \frac{1/\sqrt{a^2 + z^2}}{2\sqrt{a^2 + z^2} + a^2/\sqrt{a^2 + z^2}} \right\} \\
&= \lim_{a \to 0} \left\{ \frac{1}{2(a^2 + z^2)} \right\} = \frac{1}{2z}.
\end{aligned}$$

[66]Without fixing the total electric charge as $Q_0 = q_0 \pi a^2$, the electric field intensity would go to zero as $a \to 0$. When fixing $Q_0 = q_0 2\pi a$ as constant, the total electric charge is always Q_0; hence, we arrive at the expression for the point electric charge in the limit $a \to 0$. In this case, the electric charge density q_0 goes to infinity.

1.6.2 Coulomb's Law and Electric Force

Coulomb's law describes the interaction between static (stationary and time-invariant) electric charges in a vacuum. Force by an electric charge Q_1 on an electric charge Q_2 can be written[67]

$$\bar{F}_{e,12} = k_e \frac{Q_1 Q_2}{(R_{12})^2} \hat{a}_{12}, \tag{1.204}$$

where R_{12} is the distance between the electric charges and $\hat{a}_{12}$ is the unit vector directed from Q_1 to Q_2 (Figure 1.48). Values of Q_1 and Q_2 can be positive or negative, making the force attractive (negative) or repulsive (positive). The constant k_e is defined as

$$k_e = \frac{1}{4\pi\varepsilon_0}, \tag{1.205}$$

where ε_0 is permittivity of vacuum.

Force by an electric charge Q_1 on an electric charge Q_2 can be interpreted as the electric field intensity created by Q_1 acting on Q_2 (Figure 1.49). This electric field *is assumed to* exist even when there is no Q_2, and hence

$$\bar{E}_{12} = \lim_{Q_2 \to 0} \frac{\bar{F}_{e,12}}{Q_2} = \frac{1}{4\pi\varepsilon_0} \frac{Q_1}{(R_{12})^2} \hat{a}_{12}. \tag{1.206}$$

The equation above represents the electric field intensity created by Q_1 at a location described by a vector $\bar{R}_{12} = \hat{a}_{12} R_{12}$, which is defined from the source (electric charge location) to the observation point (test charge location). With this definition, the electric field intensity created by a point electric charge or multiple electric charges is a vector field. Obviously, with a well-defined electric field $\bar{E}_0$ created by some sources (electric charges), force applied to any electric charge Q_0 can be found as

$$\bar{F}_{e,0} = Q_0 \bar{E}_0. \tag{1.207}$$

As expected, the electric field intensity of a point electric charge indicated by Coulomb's law is consistent with the expression in Eq. (1.85) found by using Gauss' law.

Since the electric force is a measurable quantity, Coulomb's law provides a physical interpretation for electric fields. A force applied by a point electric charge (source) to another point electric charge (test) can be interpreted as the electric field created by the source electric charge acting on the test electric charge. Even when there is no test electric charge, we *assume* that this electric field exists. On the other hand, Coulomb's law is valid only for static cases. Dynamic cases can be described by Gauss' law, and they can be fully analyzed when Gauss' law is supported by other of Maxwell's equations.

$\bar{F}_e$: electric force
Unit: newton (N)

[67] Obviously, the electric force applied by Q_2 to Q_1 is equal in strength but in the opposite direction: i.e. $\bar{F}_{e,21} = -\bar{F}_{e,12}$.

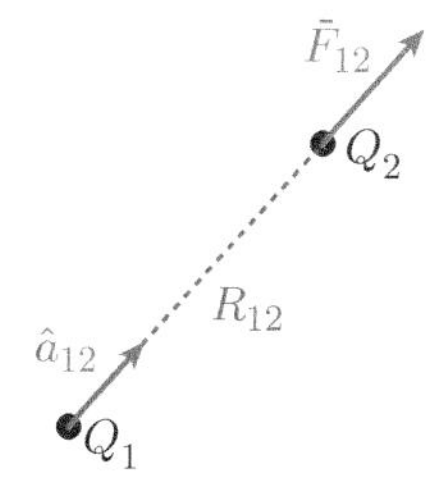

Figure 1.48 Force applied by an electric charge Q_1 to another electric charge Q_2.

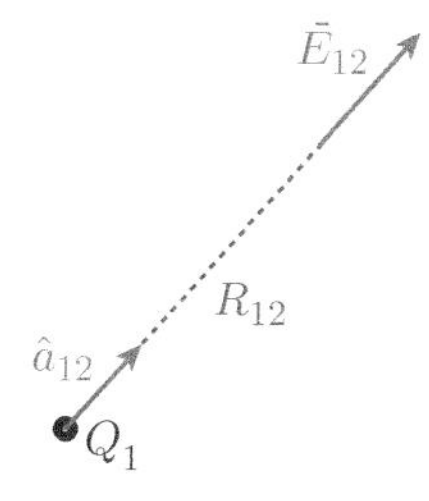

Figure 1.49 The electric field created by an electric charge Q_1 at a location described by a vector $\bar{R}_{12} = \hat{a}_{12} R_{12}$ with respect to the electric charge. The electric field is *assumed* to exist even when there is not any test electric charge: e.g. Q_2.

1.6.3 Examples

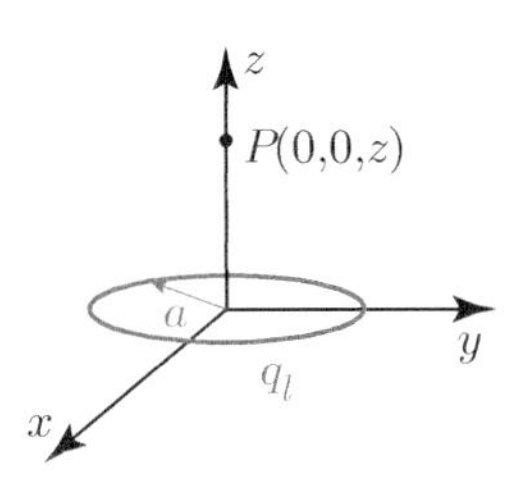

Figure 1.50 A static line electric charge distribution of a circular shape.

Example 7: Consider a static line electric charge distribution $q_l(\bar{r}) = q_0$ (C/m) of a circular shape with radius a located on the x-y plane in a vacuum (Figure 1.50). Find the electric field intensity at $\bar{r} = \hat{a}_z z$ for $z > 0$ (on the positive z axis). Also evaluate the expression when $a \to 0$ while the total electric charge is constant.

Solution: Using source integration (Coulomb's law), we have

$$\bar{E}(\bar{r}) = \frac{1}{4\pi\varepsilon_0}\int_C q_l(\bar{r})\frac{(\bar{r}-\bar{r}')}{|\bar{r}-\bar{r}'|^3}d\bar{r}' = \frac{q_0}{4\pi\varepsilon_0}\int_C \frac{(\bar{r}-\bar{r}')}{|\bar{r}-\bar{r}'|^3}d\bar{r}', \tag{1.208}$$

where $\bar{r} = \hat{a}_z z$, $\bar{r}' = \hat{a}_x x' + \hat{a}_y y'$, $d\bar{r}' = dl' = \rho d\phi' = a d\phi'$, and

$$(\bar{r}-\bar{r}') = \hat{a}_z z - \hat{a}_x x' - \hat{a}_y y' \tag{1.209}$$

$$|\bar{r}-\bar{r}'| = \sqrt{(x')^2 + (y')^2 + z^2} = \sqrt{a^2+z^2}. \tag{1.210}$$

Then we obtain[68]

$$\bar{E}(0,0,z) = \frac{q_0}{4\pi\varepsilon_0}\int_0^{2\pi}\frac{\hat{a}_z z - \hat{a}_x x' - \hat{a}_y y'}{(a^2+z^2)^{3/2}} a d\phi' \tag{1.211}$$

$$= \frac{q_0}{4\pi\varepsilon_0}\int_0^{2\pi}\frac{\hat{a}_z z - \hat{a}_x a\cos\phi' - \hat{a}_y a\sin\phi'}{(a^2+z^2)^{3/2}} a d\phi' \tag{1.212}$$

$$= \frac{q_0}{4\pi\varepsilon_0}\frac{\hat{a}_z z a}{(a^2+z^2)^{3/2}}\int_0^{2\pi} d\phi' - \frac{q_0}{4\pi\varepsilon_0}\frac{\hat{a}_x a^2}{(a^2+z^2)^{3/2}}\int_0^{2\pi}\cos\phi' d\phi' - \frac{q_0}{4\pi\varepsilon_0}\frac{\hat{a}_y a^2}{(a^2+z^2)^{3/2}}\int_0^{2\pi}\sin\phi' d\phi', \tag{1.213}$$

leading to

$$\bar{E}(0,0,z) = \hat{a}_z\frac{q_0 2\pi a}{4\pi\varepsilon_0}\frac{z}{(a^2+z^2)^{3/2}}. \tag{1.214}$$

Considering that $q_0 2\pi a = Q_0$ is the total electric charge, we have

$$\bar{E}(0,0,z) = \hat{a}_z\frac{Q_0}{4\pi\varepsilon_0}\frac{z}{(a^2+z^2)^{3/2}}. \tag{1.215}$$

[68] We note that it is safe to write the observation point $\bar{r}$ and the source point $\bar{r}'$ using the Cartesian unit vectors. This way, when evaluating the integrals, the unit vectors $\hat{a}_x$, $\hat{a}_y$, and $\hat{a}_z$ can be put outside the integrals. For example, in this case, one could also write $\bar{r}' = a\hat{a}_\rho$. But Then one must be careful: i.e. $\int_0^{2\pi}\hat{a}_\rho d\phi' \neq 2\pi\hat{a}_\rho$.

Then, when $a \to 0$, we obtain[69]

$$\bar{E}(0,0,z) \to \hat{a}_z \frac{Q_0}{4\pi\varepsilon_0}\frac{1}{z^2}, \tag{1.216}$$

which is an electric field intensity created by a point electric charge.

[69] We note that if the limit $a \to 0$ is evaluated by assuming constant q_0 before defining $Q_0 = q_0 2\pi a$, we will obtain zero electric field. In such a case, the total electric charge goes to zero, naturally leading to a vanishing field. When fixing $Q_0 = q_0 2\pi a$ as constant, the total electric charge is always Q_0, and hence we arrive at the expression for the point electric charge in the limit $a \to 0$. In this case, the electric charge density q_0 goes to infinity.

Example 8: Consider a static line electric charge distribution $q_l(\bar{r})$ (C/m) of a circular shape with radius a located on the x-y plane in a vacuum (Figure 1.51). Find the electric field intensity at $\bar{r} = \hat{a}_z z$ for $z > 0$ (on the positive z axis) if $q_l(\bar{r}) = q_l(\phi) = q_0 \cos\phi$ (C/m).

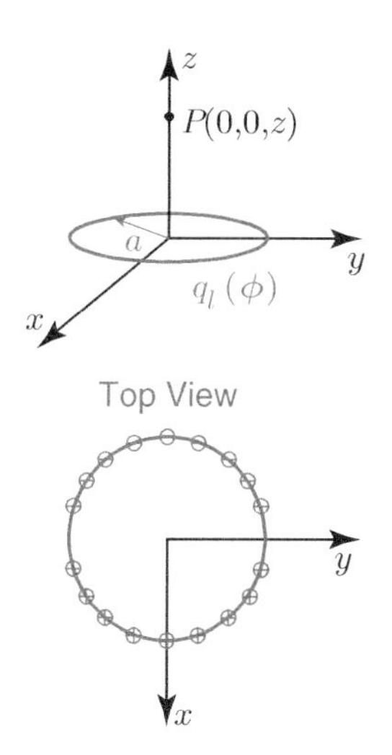

Figure 1.51 A static line electric charge distribution of a circular shape, where the line electric charge density depends on ϕ as $q_l(\bar{r}) = q_0 \cos\phi$. If $q_0 > 0$, then positive and negative electric charges are located at points with $x > 0$ and $x < 0$, respectively.

Solution: Using source integration (Coulomb's law), we have

$$\bar{E}(\bar{r}) = \frac{1}{4\pi\varepsilon_0}\int_C q_l(\bar{r})\frac{(\bar{r}-\bar{r}')}{|\bar{r}-\bar{r}'|^3}d\bar{r}' \tag{1.217}$$

$$= \frac{q_0}{4\pi\varepsilon_0}\int_C \cos\phi' \frac{(\bar{r}-\bar{r}')}{|\bar{r}-\bar{r}'|^3}d\bar{r}', \tag{1.218}$$

where

$$\bar{r} = \hat{a}_z z \tag{1.219}$$

$$\bar{r}' = \hat{a}_x x' + \hat{a}_y y' \tag{1.220}$$

$$d\bar{r}' = dl' = \rho d\phi' = a d\phi' \tag{1.221}$$

$$(\bar{r}-\bar{r}') = \hat{a}_z z - \hat{a}_x x' - \hat{a}_y y' \tag{1.222}$$

$$|\bar{r}-\bar{r}'| = \sqrt{(x')^2 + (y')^2 + z^2} = \sqrt{a^2+z^2}. \tag{1.223}$$

Then we obtain[70]

$$\bar{E}(0,0,z) = \frac{q_0}{4\pi\varepsilon_0}\int_0^{2\pi} \frac{\hat{a}_z z - \hat{a}_x x' - \hat{a}_y y'}{(a^2+z^2)^{3/2}} a\cos\phi' d\phi' \tag{1.224}$$

$$= \frac{q_0}{4\pi\varepsilon_0}\int_0^{2\pi} \frac{\hat{a}_z z - \hat{a}_x a\cos\phi' - \hat{a}_y a\sin\phi'}{(a^2+z^2)^{3/2}} \cos\phi' a d\phi' \tag{1.225}$$

$$= -\hat{a}_x \frac{q_0 a^2}{4\pi\varepsilon_0}\frac{1}{(a^2+z^2)^{3/2}}\int_0^{2\pi}\cos^2\phi' d\phi' \tag{1.226}$$

$$= -\hat{a}_x \frac{q_0 a^2}{4\varepsilon_0}\frac{1}{(a^2+z^2)^{3/2}}. \tag{1.227}$$

[70] We note that (if $q_0 > 0$) the electric charge density is positive for $x > 0$, while it is negative for $x < 0$. Therefore, there is no z component of the electric field intensity due to cancellation effects. In the x direction, however, the contributions from the semicircles ($x > 0$ and $x < 0$) are in the same direction ($-x$.)

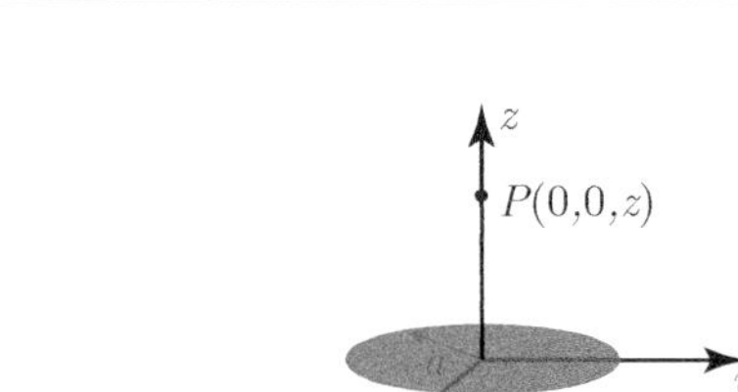

Figure 1.52 A static surface electric charge distribution with a disk shape.

Example 9: Consider a static surface electric charge distribution $q_s(\bar{r}) = q_0$ (C/m^2) of a disk shape with radius a located on the x-y plane in a vacuum (Figure 1.52). Find the electric field intensity at $\bar{r} = \hat{a}_z$ for $z > 0$ (on the positive z axis) using the electric field intensity contributions created by circular loops.

Solution: In Section 1.6.1, the electric field created by this charge distribution was found by using source integration (Coulomb's law). Alternatively, we can consider the electric field intensity as the sum of contributions created by infinitely many circular loops. Specifically, let the disk be formed of infinitely many loops of radius $\rho' \in [0, a]$. A loop with radius ρ' creates an electric field intensity as

$$d\bar{E}(0,0,z) = \hat{a}_z \frac{dQ}{4\pi\varepsilon_0} \frac{z}{[(\rho')^2 + z^2]^{3/2}}, \tag{1.228}$$

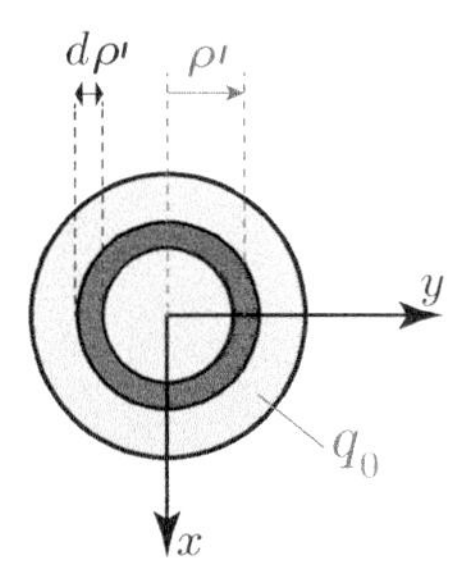

Figure 1.53 Top view of Figure 1.52 to demonstrate the integration strategy.

where dQ is the amount of electric charge represented by the loop (Figure 1.53). If the loop has a differential thickness of $d\rho'$, then we have $dQ = q_0 2\pi\rho' d\rho'$ since q_0 is the surface electric charge density and $2\pi\rho' d\rho'$ is the differential area of the loop. Then the electric field intensity created by the disk can be found as

$$\bar{E}(0,0,z) = \int_0^a \hat{a}_z \frac{q_0 2\pi\rho' d\rho'}{4\pi\varepsilon_0} \frac{z}{[(\rho')^2 + z^2]^{3/2}} = \hat{a}_z \frac{2\pi q_0 z}{4\pi\varepsilon_0} \int_0^a \frac{\rho' d\rho'}{[(\rho')^2 + z^2]^{3/2}} \tag{1.229}$$

$$= \hat{a}_z \frac{2\pi q_0 z}{4\pi\varepsilon_0} \left[\frac{-1}{\sqrt{(\rho')^2 + z^2}} \right]_0^a = \hat{a}_z \frac{2\pi q_0 z}{4\pi\varepsilon_0} \left(\frac{1}{z} - \frac{1}{\sqrt{a^2 + z^2}} \right) \tag{1.230}$$

$$= \hat{a}_z \frac{q_0}{2\varepsilon_0} \left(1 - \frac{z}{\sqrt{a^2 + z^2}} \right). \tag{1.231}$$

Considering the total electric charge $Q_0 = q_0 \pi a^2$, the electric field intensity can also be written[71]

$$\bar{E}(0,0,z) = \hat{a}_z \frac{Q_0}{2\pi\varepsilon_0} \left(\frac{\sqrt{a^2 + z^2} - z}{a^2\sqrt{a^2 + z^2}} \right). \tag{1.232}$$

[71] The total electric charge is the electric charge density multiplied by the area of the disk.

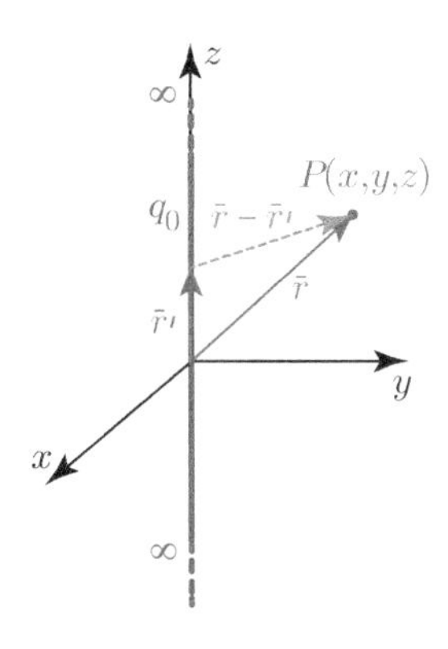

Figure 1.54 An infinitely long static line electric charge with constant density.

Example 10: Consider a static line electric charge density defined as $q_l(\bar{r}) = q_0$ (C/m) on the z axis from $-\infty$ to ∞ in a vacuum (Figure 1.54). Find the electric field intensity $\bar{E}(\bar{r})$ everywhere via source integration (Coulomb's law).

Solution: This problem can easily be solved with the integral form of Gauss' law, as shown in Section 1.2. In this case, we may try source integration (Coulomb's law) to find the electric field intensity. Using

$$\bar{r} = \hat{a}_x x + \hat{a}_y y + \hat{a}_z z \quad \text{and} \quad \bar{r}' = \hat{a}_z z' \tag{1.233}$$

$$d\bar{r}' = dl' = dz' \tag{1.234}$$

$$(\bar{r} - \bar{r}') = \hat{a}_x x + \hat{a}_y y + \hat{a}_z z - \hat{a}_z z' \tag{1.235}$$

$$|\bar{r} - \bar{r}'| = \sqrt{x^2 + y^2 + (z - z')^2} = \sqrt{\rho^2 + (z - z')^2}, \tag{1.236}$$

we have

$$\bar{E}(\bar{r}) = \frac{1}{4\pi\varepsilon_0} \int_C q_l(\bar{r}) \frac{(\bar{r} - \bar{r}')}{|\bar{r} - \bar{r}'|^3} d\bar{r}' = \frac{q_0}{4\pi\varepsilon_0} \int_C \frac{(\bar{r} - \bar{r}')}{|\bar{r} - \bar{r}'|^3} d\bar{r}' \tag{1.237}$$

$$= \frac{q_0}{4\pi\varepsilon_0} \int_{-\infty}^{\infty} \frac{\hat{a}_x x + \hat{a}_y y + \hat{a}_z z - \hat{a}_z z'}{[\rho^2 + (z - z')^2]^{3/2}} dz' \tag{1.238}$$

$$= \frac{q_0}{4\pi\varepsilon_0} (\hat{a}_x x + \hat{a}_y y) \int_{-\infty}^{\infty} \frac{1}{[\rho^2 + (z - z')^2]^{3/2}} dz'$$
$$+ \frac{q_0}{4\pi\varepsilon_0} \hat{a}_z \int_{-\infty}^{\infty} \frac{(z - z')}{[\rho^2 + (z - z')^2]^{3/2}} dz'. \tag{1.239}$$

The first integral can be evaluated as[72]

$$\int_{-\infty}^{\infty} \frac{1}{[\rho^2 + (z - z')^2]^{3/2}} dz' = \frac{2}{\rho^2}. \tag{1.240}$$

For the second integral, we have[73]

$$\int_{-\infty}^{\infty} \frac{(z - z')}{[\rho^2 + (z - z')^2]^{3/2}} dz' = \left[\frac{1}{[\rho^2 + (z - z')^2]^{1/2}} \right]_{-\infty}^{\infty} = 0, \tag{1.241}$$

leading to

$$\bar{E}(\bar{r}) = \frac{q_0}{4\pi\varepsilon_0} (\hat{a}_x x + \hat{a}_y y) \frac{2}{\rho^2} = \hat{a}_\rho \frac{q_0}{2\pi\varepsilon_0 \rho}. \tag{1.242}$$

Hence the electric field created by an infinitely long static line electric charge decays with the distance, as found before.

[72] This type of integral is also common in magnetostatics.

[73] Using $\hat{a}_x x + \hat{a}_y y = \hat{a}_\rho \rho$.

1.7 Gauss' Law and Dielectrics

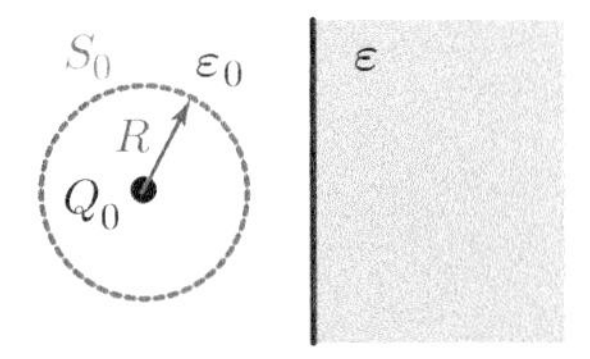

Figure 1.55 A static point electric charge in a vacuum near a dielectric medium.

As shown in Eq. (1.2), the electric flux density $\bar{D}(\bar{r},t)$ and the electric field intensity $\bar{E}(\bar{r},t)$ are related to each other as $\bar{D}(\bar{r},t)=\varepsilon(\bar{r})\bar{E}(\bar{r},t)$. Then, if ε does not change with position $\bar{r}$, we have

$$\oint_S \varepsilon\bar{E}(\bar{r},t)\cdot\overline{ds}=Q_{\text{enc}}(t)\longrightarrow\oint_S \bar{E}(\bar{r},t)\cdot\overline{ds}=\frac{Q_{\text{enc}}(t)}{\varepsilon} \tag{1.243}$$

$$\bar{\nabla}\cdot\left(\varepsilon\bar{E}(\bar{r},t)\right)=q_v(\bar{r},t)\longrightarrow\bar{\nabla}\cdot\bar{E}(\bar{r},t)=\frac{q_v(\bar{r},t)}{\varepsilon}. \tag{1.244}$$

In a vacuum, $\varepsilon=\varepsilon_0\approx 8.85418782\times 10^{-12}$ F/m, while the permittivity has diverse values for different materials.

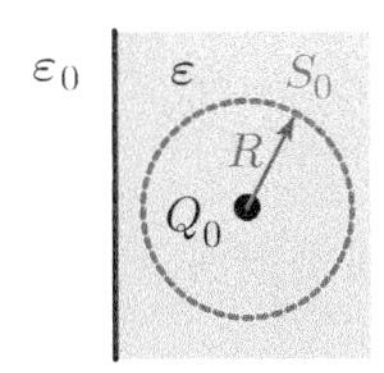

Figure 1.56 A static point electric charge inside a dielectric medium.

Dielectrics are a class of materials that are used as insulators since they do not conduct – or weakly conduct – electricity due to a lack or sparsity of free electrons or ions responsible for electrical conduction. As discussed in Section 6.6.3, being a *good* dielectric (insulator) depends on many conditions, including the frequency. Nevertheless, some well-known dielectrics in practice are different kinds of plastics, paper, wood, glass, and porcelain. Pure water possesses good dielectric properties, while seawater can behave as a good dielectric or a good conductor in practical frequency regimes. Without any charge carrier, dry air is often considered a perfect dielectric (until dielectric breakdown). In general, most dielectrics are imperfect: i.e. they contain charge carriers and conduct small amounts of electric currents. Such materials with non-negligible conductivity are called *lossy* since electrical energy is consumed (*lost* from one perspective) and converted into other forms (e.g. heat) as conduction occurs.[74]

[74] A vacuum is an ultimate insulator with zero conductivity. But without any material properties (existence of atoms and *bound* charges), it should not be considered as a dielectric.

A good dielectric may conduct electricity poorly, but this does not mean it weakly interacts with electric fields. In fact, when exposed to an electric field, a typical dielectric material is polarized. Polarization is due to the response of molecules – or, more specifically, bound charges – to an external electric field. Such a response may involve the separation of positive and negative electric charges (stretching) and/or orientation of polar molecules due to the external electric field. The polarization ability[75] (polarizability) is basically represented by the permittivity ε; larger the value of the permittivity, higher the polarization effect. More specifically, permittivity values close to ε_0 represent poor polarization ability, while $\varepsilon=\varepsilon_0$ corresponds to no polarizability. Some dielectrics, such as dry air, have very low polarization ability ($\varepsilon\approx\varepsilon_0$), while those with polar molecules have very high polarization ability (and hence permittivity).

[75] High polarization ability should not be confused with being a good dielectric (having low conductivity).

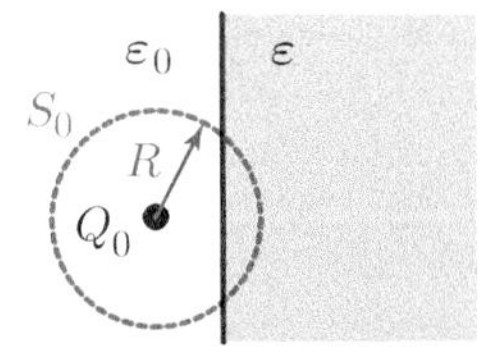

Figure 1.57 A static point electric charge in a vacuum near a dielectric medium. The Gauss surface encloses both the vacuum and part of the dielectric medium.

Considering Eqs. (1.243) and (1.244), dielectric effects are already included in Gauss' law, while the use of the equations becomes more complicated. To understand how Gauss' law should be used, we may investigate a series of different cases depicted in Figure 1.55–1.57. First, we consider the integral form of Gauss' law for a static point electric charge, which is always located at the origin and enclosed in a spherical Gauss surface:

- Case 1: Consider a static point electric charge Q_0 located in a vacuum near a dielectric medium with permittivity $\varepsilon \neq \varepsilon_0$. If the Gauss surface S_0 encloses only vacuum in addition to the electric charge (Figure 1.55), we can write

$$\oint_{S_0} \bar{D}(\bar{r}) \cdot \overline{ds} = Q_0 \longrightarrow \oint_{S_0} \bar{E}(\bar{r}) \cdot \overline{ds} = \frac{Q_0}{\varepsilon_0} \not\rightarrow 4\pi R^2 D_R(R) = Q_0. \tag{1.245}$$

  Specifically, the integral form of Gauss' law can be written for both $\bar{D}(\bar{r})$ and $\bar{E}(\bar{r})$, while they cannot be placed outside the Gauss integral. This is because the electric field is not in the R direction (Figure 1.58) and it does not only depend on R due to the lack of symmetry (the nearby dielectric medium).

- Case 2: Consider a static point electric charge Q_0 located in a dielectric medium with permittivity $\varepsilon \neq \varepsilon_0$. If the Gauss surface S_0 encloses only dielectric material in addition to the electric charge (Figure 1.56), we can write

$$\oint_{S_0} \bar{D}(\bar{r}) \cdot \overline{ds} = Q_0 \longrightarrow \oint_{S_0} \bar{E}(\bar{r}) \cdot \overline{ds} = \frac{Q_0}{\varepsilon} \not\rightarrow 4\pi R^2 D_R(R) = Q_0. \tag{1.246}$$

  This is very similar to Case 1.

- Case 3: Consider again a static point electric charge Q_0 located in a vacuum near a dielectric medium with permittivity $\varepsilon \neq \varepsilon_0$. In this case, the Gauss surface encloses both the vacuum and a part of the dielectric medium (Figure 1.57). Hence, we have

$$\oint_{S_0} \bar{D}(\bar{r}) \cdot \overline{ds} = Q_0 \longrightarrow \varepsilon_0 \oint_{S_v} \bar{E}(\bar{r}) \cdot \overline{ds} + \varepsilon \oint_{S_d} \bar{E}(\bar{r}) \cdot \overline{ds} = Q_0 \tag{1.247}$$

$$\not\rightarrow 4\pi R^2 D_R(R) = Q_0. \tag{1.248}$$

  Once again, we are able to write Gauss' law for the electric flux density, while Gauss' law for the electric field intensity needs to be written in two parts by dividing the integral for the vacuum (S_v) and dielectric (S_d) portions. In addition, the lack of symmetry again inhibits the relocation of the electric field outside the integral.

It is remarkable that when using the integral form of Gauss' law, the locations on the surface (not the volume) determines which permittivity values should be used. For example, if the Gauss

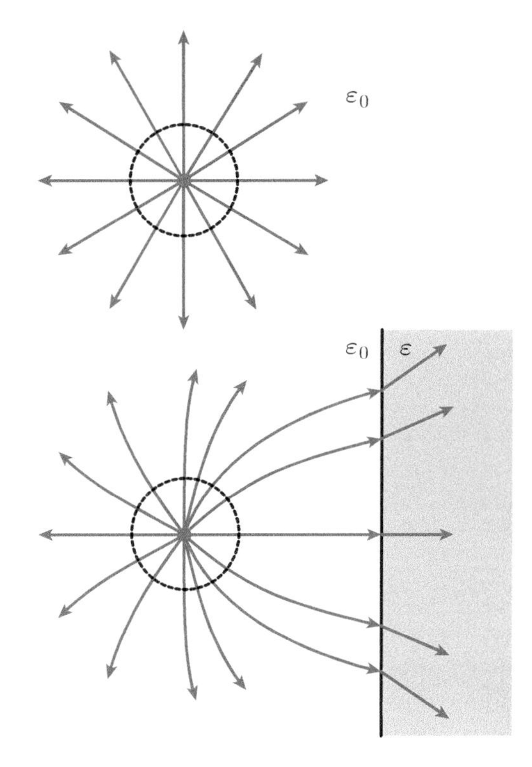

Figure 1.58 Dielectrics interact with external electric fields. For example, electric field lines created by a static point electric charge are deformed near a dielectric medium, while the electric field penetrates the dielectric. We note that considering a point electric charge with or without a nearby dielectric medium, the integral form of Gauss' law does not change (flux is the enclosed electric charge), while the electric flux density cannot be placed outside the integral in the latter due to the lack of symmetry.

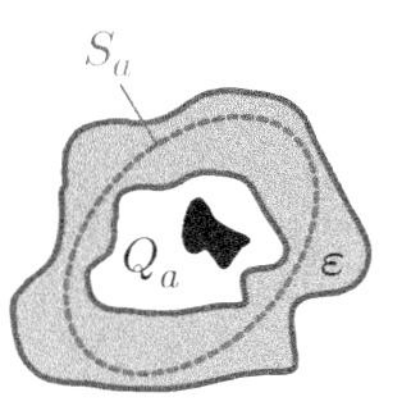

Figure 1.59 An application of the integral form of Gauss' law when all locations on the surface are completely inside a dielectric medium. After writing

$$\oint_{S_a} \bar{D}(\bar{r},t) \cdot \overline{ds} = Q_a(t)$$

and

$$\oint_{S_a} \varepsilon(\bar{r})\bar{E}(\bar{r},t) \cdot \overline{ds} = Q_a(t),$$

we note that $\varepsilon(\bar{r}) = \varepsilon$ can be put outside the integral since it is the same everywhere on the surface. Then the integral form of Gauss' law can be written for the electric field intensity as

$$\oint_{S_a} \bar{E}(\bar{r},t) \cdot \overline{ds} = Q_a(t)/\varepsilon.$$

surface S_a is entirely located in a single medium with constant permittivity ε (no matter what is inside or outside), one can write

$$\oint_{S_a} \bar{D}(\bar{r},t) \cdot \overline{ds} = Q_a(t) \longrightarrow \oint_{S_a} \bar{E}(\bar{r},t) \cdot \overline{ds} = \frac{Q_a(t)}{\varepsilon}, \tag{1.249}$$

where Q_a is the enclosed electric charge (Figure 1.59). In symmetric cases, where there are valid assumptions on the symmetry of the electric flux density and the electric field intensity, this equation can be useful to find electric fields (Figure 1.60).

Next, we consider the differential form of Gauss' law and its application in the presence of dielectrics. For this purpose, we consider different locations (points) around interfaces between a vacuum and dielectric media, as depicted in Figure 1.61. We have the following cases:

- If the point is located in a vacuum (e.g. location 1 in Figure 1.61), whether there is a dielectric in the vicinity or not, we have

$$\bar{\nabla} \cdot \bar{D}(\bar{r},t) = q_v(\bar{r},t) \longrightarrow \bar{\nabla} \cdot \bar{E}(\bar{r},t) = \frac{q_v(\bar{r},t)}{\varepsilon_0}, \tag{1.250}$$

  where $q_v(\bar{r},t)$ is the volume electric charge density at the location.

- If the point is located in a dielectric medium with permittivity $\varepsilon \neq \varepsilon_0$ (e.g. location 3 in Figure 1.61), whether there is another medium (vacuum or another type of dielectric) in the vicinity or not, we have

$$\bar{\nabla} \cdot \bar{D}_d(\bar{r},t) = q_v(\bar{r},t) \longrightarrow \bar{\nabla} \cdot \bar{E}_d(\bar{r},t) = \frac{q_v(\bar{r},t)}{\varepsilon}, \tag{1.251}$$

  where $q_v(\bar{r},t)$ is the volume electric charge density at the location.

- At the boundary between a vacuum and a dielectric medium (e.g. location 2 in Figure 1.61), the boundary conditions for normal electric fields (that is obtained by using Gauss' law at the boundary) can be used: i.e.

$$\hat{a}_n \cdot [\bar{D}_0(\bar{r},t) - \bar{D}_d(\bar{r},t)] = q_s(\bar{r},t) \tag{1.252}$$

$$\longrightarrow \hat{a}_n \cdot [\varepsilon_0\bar{E}_0(\bar{r},t) - \varepsilon\bar{E}_d(\bar{r},t)] = q_s(\bar{r},t) \qquad (\bar{r} \in S), \tag{1.253}$$

  where $q_s(\bar{r},t)$ is the surface electric charge density at the location. If the dielectric medium is lossless, one can select $q_s(\bar{r},t) = 0$: i.e. the electric charge density is zero if no electric charge is particularly inserted.

- If the point is located in a dielectric medium with permittivity $\varepsilon(\bar{r})$ that depends on position (e.g. location 4 in Figure 1.61), we have

$$\bar{\nabla}\cdot\bar{D}_d(\bar{r},t)=q_v(\bar{r},t)\longrightarrow\bar{\nabla}\cdot\big(\varepsilon(\bar{r})\bar{E}_d(\bar{r},t)\big)=q_v(\bar{r},t) \tag{1.254}$$

$$\not\longrightarrow\bar{\nabla}\cdot\bar{E}_d(\bar{r},t)=\frac{q_v(\bar{r},t)}{\varepsilon(\bar{r})}. \tag{1.255}$$

In this case, we are able to use Gauss' law for the electric flux density, while one cannot place the permittivity outside the derivative (divergence operation). In such a case, the derivative of the permittivity (more specifically, its gradient) may be used to write the complete and correct equation for $\bar{\nabla}\cdot\bar{E}_d(\bar{r},t)$.

Before further discussion of electrical responses of dielectrics, we note that a scalar definition of $\varepsilon(\bar{r})$ is possible for a *linear* and *isotropic* medium. Briefly, here is what would happen when these conditions are not satisfied:

- Anisotropy: If the medium is not isotropic, the cross components of $\bar{D}(\bar{r},t)$ and $\bar{E}(\bar{r},t)$ can be related to each other. In this context, using a Cartesian coordinate system, $\bar{D}(\bar{r},t)$ and $\bar{E}(\bar{r},t)$ for a linear and isotropic medium can be written

$$\begin{bmatrix} D_x(\bar{r},t)\\ D_y(\bar{r},t)\\ D_z(\bar{r},t)\end{bmatrix}=\begin{bmatrix}\varepsilon(\bar{r}) & 0 & 0\\ 0 & \varepsilon(\bar{r}) & 0\\ 0 & 0 & \varepsilon(\bar{r})\end{bmatrix}\begin{bmatrix} E_x(\bar{r},t)\\ E_y(\bar{r},t)\\ E_z(\bar{r},t)\end{bmatrix},$$

where $\varepsilon(\bar{r})$ is the permittivity.[76] In an anisotropic (but still linear) medium, however, we generally have

$$\begin{bmatrix} D_x(\bar{r},t)\\ D_y(\bar{r},t)\\ D_z(\bar{r},t)\end{bmatrix}=\begin{bmatrix}\varepsilon_{xx}(\bar{r}) & \varepsilon_{xy}(\bar{r}) & \varepsilon_{xz}(\bar{r})\\ \varepsilon_{yx}(\bar{r}) & \varepsilon_{yy}(\bar{r}) & \varepsilon_{yz}(\bar{r})\\ \varepsilon_{zx}(\bar{r}) & \varepsilon_{zy}(\bar{r}) & \varepsilon_{zz}(\bar{r})\end{bmatrix}\begin{bmatrix} E_x(\bar{r},t)\\ E_y(\bar{r},t)\\ E_z(\bar{r},t)\end{bmatrix}, \tag{1.256}$$

where each component of $\bar{D}(\bar{r},t)$ may be related to each component of $\bar{E}(\bar{r},t)$. Crystals are naturally anisotropic, while anisotropy can also be achieved synthetically for different applications.

- Linearity: If the the medium is not linear, the relationship between $\bar{D}(\bar{r},t)$ and $\bar{E}(\bar{r},t)$ may depend on the value of $\bar{E}(\bar{r},t)$. Specifically, for a nonlinear (but still isotropic) medium, we may have

$$\bar{D}(\bar{r},t)=\varepsilon\big(\bar{r},\bar{E}(\bar{r},t)\big)\bar{E}(\bar{r},t),$$

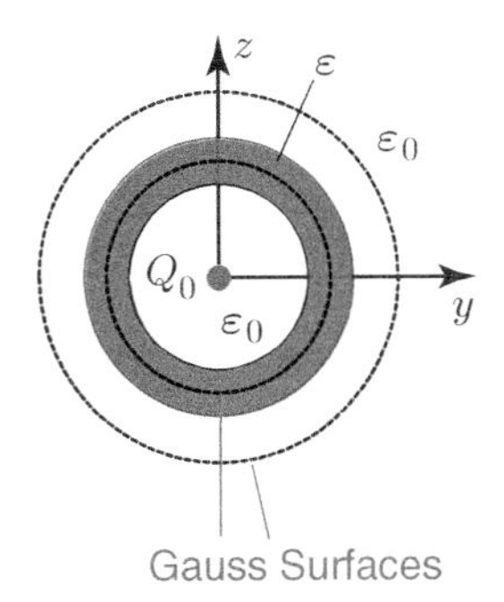

Figure 1.60 The integral form of Gauss' law can be used to find electric fields in static cases involving symmetric dielectric objects. For example, consider a static point electric charge Q_0 located inside a dielectric shell. Gauss' law can be written

$$\oint_S \bar{D}(\bar{r})\cdot\overline{ds}=Q_0,$$

leading to $\bar{D}(\bar{r})=\hat{a}_R Q_0/(4\pi R^2)$ everywhere. This is because, due to symmetry, $\bar{D}(\bar{r})$ has only R component and it only depends on R. This also means $\bar{E}(\bar{r})=\hat{a}_R Q_0/(4\pi\varepsilon_0 R^2)$ when $\bar{r}$ is in a vacuum and $\bar{E}(\bar{r})=\hat{a}_R Q_0/(4\pi\varepsilon R^2)$ when $\bar{r}$ is in the dielectric shell. Considering that $\varepsilon>\varepsilon_0$, the value of the electric field intensity drops inside the dielectric. This can be interpreted as the response of the dielectric material to the electric field created by the point source.

[76]Consequently, we have $D_x(\bar{r},t)=\varepsilon(\bar{r})E_x(\bar{r},t)$, $D_y(\bar{r},t)=\varepsilon(\bar{r})E_y(\bar{r},t)$, and $D_z(\bar{r},t)=\varepsilon(\bar{r})E_z(\bar{r},t)$, leading to $\bar{D}(\bar{r},t)=\varepsilon(\bar{r})\bar{E}(\bar{r},t)$.

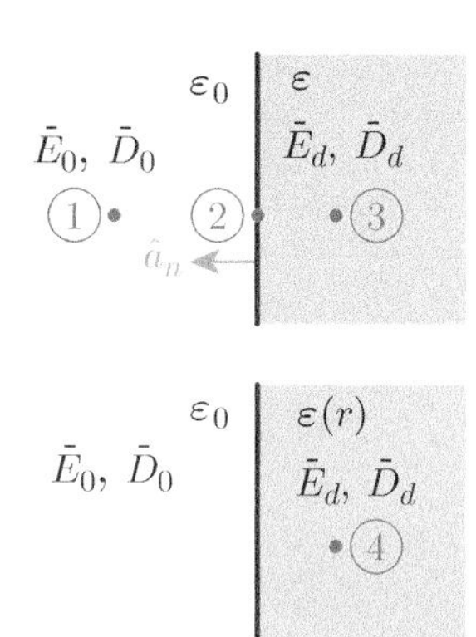

Figure 1.61 Different locations at the interface between the vacuum and a dielectric medium.

where the permittivity is a function of $\bar{E}(\bar{r},t)$. This complex behavior can be simplified into a polynomial form, where $\bar{D}(\bar{r},t)$ depends on not only $\bar{E}(\bar{r},t)$ but also its powers: e.g. $|\bar{E}(\bar{r},t)|\bar{E}(\bar{r},t)$, $|\bar{E}(\bar{r},t)|^2\bar{E}(\bar{r},t)$, etc. Nonlinearity is an important property of ferroelectric materials.

It is remarkable that almost all dielectrics possess weak anisotropy and nonlinearity, while these are negligible for most of them. Nevertheless, even for simple dielectrics, the relationship between $\bar{D}(\bar{r},t)$ and $\bar{E}(\bar{r},t)$ strongly depends on other conditions:[77] e.g. frequency. For example, at room temperature and at low frequencies, the permittivity of pure water is as large as $80\varepsilon_0$. If the frequency is very high[78] – e.g. in the visible light ranges – its permittivity drops to around $1.8\varepsilon_0$.

Understanding the mathematical basis of how dielectrics interact with electric fields (particularly in the context of Gauss' law) requires the following tools and concepts:

- Electric dipole
- Polarization
- Equivalent polarization charges

These are in addition to previous tools: i.e. differential surface with direction (Section 1.1.1), dot product (Section 1.1.2), flux of vector fields (Section 1.1.3), electric charge density (Section 1.3.1), divergence of vector fields (Section 1.3.2), and divergence theorem (Section 1.3.3).

[77] Given an application, these external conditions are often assumed to be fixed to define a unique $\varepsilon(\bar{r})$ for analyses.

[78] In general, the permittivity of all materials depends on the frequency. This is because the response of a dielectric material to a varying external electric field is not instantaneous: i.e. it is delayed. Hence, this delay leads to different responses when the variation rate (frequency) is different.

1.7.1 Electric Dipole

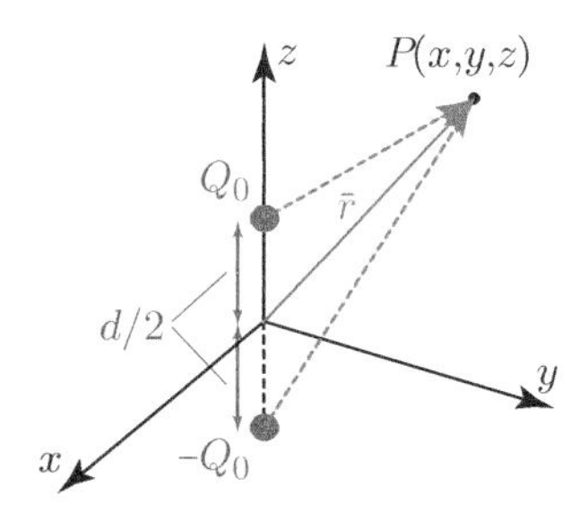

Figure 1.62 Electric dipole involving positive and negative static (i.e. stationary and time-invariant) point electric charges separated by a distance of d.

A special arrangement of a static positive point electric charge and a static negative point electric charge with equal magnitudes is called an *electric dipole* (Figure 1.62). As an example, we consider $+Q_0$ and $-Q_0$ electric charges located at $z = d/2$ and $z = -d/2$, respectively, in a vacuum. The electric flux density at an arbitrary location can be written

$$\bar{D}_d(\bar{r}) = \frac{Q_0}{4\pi}\frac{(\bar{r}-\hat{a}_z d/2)}{|\bar{r}-\hat{a}_z d/2|^3} - \frac{Q_0}{4\pi}\frac{(\bar{r}+\hat{a}_z d/2)}{|\bar{r}+\hat{a}_z d/2|^3}$$
$$= \frac{Q_0}{4\pi}\left(\frac{\bar{r}-\hat{a}_z d/2}{|\bar{r}-\hat{a}_z d/2|^3} - \frac{\bar{r}+\hat{a}_z d/2}{|\bar{r}+\hat{a}_z d/2|^3}\right), \tag{1.257}$$

where $\bar{r} = \hat{a}_x x + \hat{a}_y y + \hat{a}_z z$ is the position vector, as usual. Now, we particularly focus on the case $|\bar{r}| \gg d$: i.e. when the observation point is far away from the dipole (far zone). Using a first-order approximation,[79] we obtain

$$|\bar{r} - \hat{a}_z d/2|^3 = [(\bar{r} - \hat{a}_z d/2) \cdot (\bar{r} - \hat{a}_z d/2)]^{3/2} = \left(|\bar{r}|^2 - \bar{r} \cdot \hat{a}_z d + d^2/4\right)^{3/2} \tag{1.258}$$

$$\approx \left(|\bar{r}|^2 - \bar{r} \cdot \hat{a}_z d\right)^{3/2} \approx |\bar{r}|^3 \left(1 - \hat{a}_R \cdot \hat{a}_z \frac{3d}{2|\bar{r}|}\right). \tag{1.259}$$

[79] If $f = d/|\bar{r}|$, we have $f^2 \approx 0$ and $(1 \pm f)^{3/2} \approx 1 \pm 3f/2$.

Similarly, one can obtain

$$|\bar{r} + \hat{a}_z d/2|^3 \approx |\bar{r}|^3 \left(1 + \hat{a}_R \cdot \hat{a}_z \frac{3d}{2|\bar{r}|}\right). \tag{1.260}$$

Using these approximations and rearranging the terms, we derive[80]

$$\bar{D}_d(\bar{r}) = \frac{Q_0}{4\pi}\bar{r}\left(\frac{1}{|\bar{r} - \hat{a}_z d/2|^3} - \frac{1}{|\bar{r} + \hat{a}_z d/2|^3}\right)$$

$$- \frac{Q_0}{4\pi}\hat{a}_z d/2 \left(\frac{1}{|\bar{r} - \hat{a}_z d/2|^3} + \frac{1}{|\bar{r} + \hat{a}_z d/2|^3}\right) \tag{1.261}$$

$$\approx \frac{Q_0}{4\pi}\frac{\hat{a}_R(\hat{a}_R \cdot \hat{a}_z)3d}{|\bar{r}|^3} - \hat{a}_z \frac{Q_0}{4\pi}\frac{d}{|\bar{r}|^3} = \frac{Q_0 d}{4\pi|\bar{r}|^3}[3\hat{a}_R(\hat{a}_R \cdot \hat{a}_z) - \hat{a}_z]. \tag{1.262}$$

[80] We note some specific cases: (i) When the observation point is located on the x-y plane, $\hat{a}_R$ is a combination of $\hat{a}_x$ and $\hat{a}_y$, leading to $\hat{a}_R \cdot \hat{a}_z = 0$ and an overall electric field is in the $-z$ direction. This is because the electric field in the x and/or y directions created by $+Q_0$ and $-Q_0$ cancel each other. (ii) When the observation point is located on the z axis, the overall electric field is in the z direction.

At this stage, we define an important quantity called *dipole moment* as

$$\bar{I}_{DM,e} = \hat{a}_z Q_0 d, \tag{1.263}$$

$\bar{I}_{DM,e}$: dipole moment
Unit: coulombs × meters (Cm)

which contains all the information regarding the arrangement of a dipole: i.e. magnitude Q_0, the distance between the electric charges d, and the dipole direction (unit vector from the negative electric charge to the positive electric charge). Then we have

$$\bar{D}_d(\bar{r}) \approx \frac{1}{4\pi|\bar{r}|^3}\left(3\hat{a}_R\hat{a}_R \cdot \bar{I}_{DM,e} - \bar{I}_{DM,e}\right), \tag{1.264}$$

which is valid for arbitrary $\bar{I}_{DM,e}$. It is remarkable that an electric dipole's electric field has a cubic decay with the distance $|\bar{r}|$, while it decays quadratically for a single electric charge. Figure 1.63 depicts electric field lines for an electric dipole (not only in the far zone but also in the near zone) when the charges are located as in the original case (on the z axis).[81]

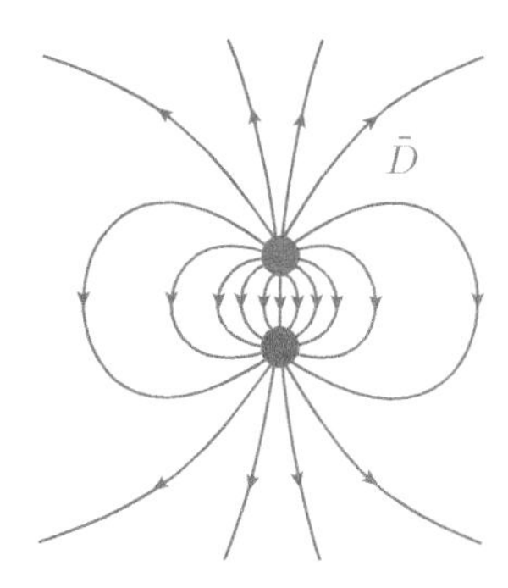

Figure 1.63 Electric field lines for an electric dipole. Electric dipoles are useful to describe various phenomena, such as polarization in dielectrics.

[81] As also used in Figure 1.20, electric field lines are useful to visualize how the electric field created by a source changes its *direction* with respect to position. If one starts from the positive electric charge where the electric field emerges and follow a line, the electric field is always in the direction along the line until it ends at a negative electric charge or infinity.

1.7.2 Polarization

To understand electric fields in a dielectric medium, we need the concept of polarization. In dielectric objects, electrons are strongly bonded to their nuclei.[82] Hence, these molecules can be considered to be electric dipoles. When no electric field is applied, electric dipoles are oriented in random directions (no polarization). However, when an external electric field (e.g. created by some electric charges) is applied, the dipoles are oriented in the electric field direction. This orientation can be due to the separation of positive and negative electric charges (stretching) or the physical orientation of polar molecules. In any case, dipole orientations lead to positional inequalities, which may balance each other at specific points. But due to possible imbalances, the object's overall response affects the electric field inside of it.

[82]When it is a perfect dielectric, all electrons are bound.

As discussed at the beginning of this section, when writing the relationship between the electric flux density and the electric field intensity, we are already using dielectric effects via permittivity. Specifically, we use $\bar{D}(\bar{r},t) = \varepsilon(\bar{r})\bar{E}(\bar{r},t)$, where $\varepsilon(\bar{r})$ is the permittivity that accounts for dielectric properties. In general, permittivity can further be written

$$\varepsilon(\bar{r}) = \varepsilon_0\varepsilon_r(\bar{r}), \tag{1.265}$$

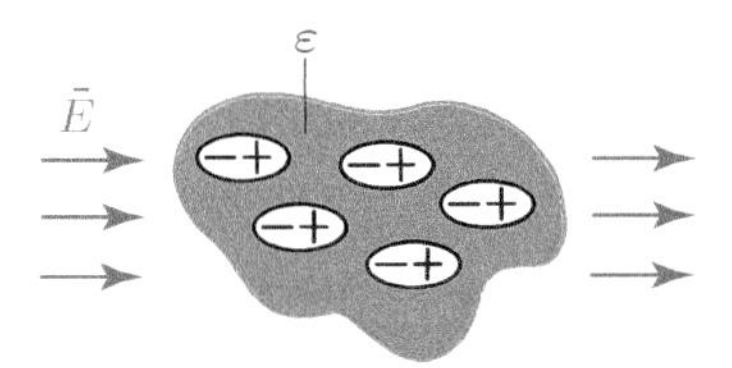

Figure 1.64 A simple illustration of polarization inside a dielectric body in an electric field created by external sources.

where $\varepsilon_r(\bar{r})$ is the relative permittivity, which typically is larger than one. Furthermore, one can define $\chi_e(\bar{r}) = \varepsilon_r(\bar{r}) - 1$ as the electric susceptibility. We note that, in a vacuum, $\chi_e(\bar{r}) = 0$ and $\varepsilon_r(\bar{r}) = 1$ everywhere. Now we consider a dielectric object with permittivity $\varepsilon(\bar{r})$, leading to

$$\bar{E}(\bar{r},t) = \frac{\bar{D}(\bar{r},t)}{\varepsilon_0\varepsilon_r(\bar{r})}. \tag{1.266}$$

χ_e: electric susceptibility
Unitless

Therefore, if $\bar{D}(\bar{r},t)$ is considered to have a constant magnitude, the electric field intensity is reduced by $\varepsilon_r(\bar{r})$ in comparison to a vacuum. As shown in Figure 1.60 for a symmetric case, this reduction is the response of the dielectric (namely, polarization) to the external field. Specifically, when the electric dipoles inside the dielectric are oriented in the electric field direction, they create a *net* secondary electric field intensity in the opposite direction (Figure 1.64).

To formulate the polarization, we define it as a vector field as

$\bar{P}$: polarization
Unit: coulombs/meter^2 (C/m^2)

$$\bar{P}(\bar{r},t) = [\varepsilon(\bar{r}) - \varepsilon_0]\bar{E}(\bar{r},t) = \bar{D}(\bar{r},t) - \varepsilon_0\bar{E}(\bar{r},t), \tag{1.267}$$

which accounts for the difference in $\bar{E}(\bar{r},t)$ when dielectric material exists instead of a vacuum. Since $\chi_e(\bar{r}) = \varepsilon_r(\bar{r}) - 1$, the polarization can also be written

$$\bar{P}(\bar{r},t) = \varepsilon_0[\varepsilon_r(\bar{r}) - 1]\bar{E}(\bar{r},t) = \varepsilon_0\chi_e(\bar{r})\bar{E}(\bar{r},t). \tag{1.268}$$

Consequently, in the absence of a dielectric at a position $\bar{r}$, we have $\chi_e(\bar{r}) = 0$ and $\bar{P}(\bar{r}) = 0$ (no polarization).

At this stage, we may revisit Gauss' law in the presence of dielectrics. Using the differential form and the vector identity[83]

$$\bar{\nabla} \cdot [h(\bar{r},t)\bar{f}(\bar{r},t)] = \bar{\nabla} h(\bar{r},t) \cdot \bar{f}(\bar{r},t) + h(\bar{r},t)\bar{\nabla} \cdot \bar{f}(\bar{r},t) \tag{1.269}$$

for any scalar field $h(\bar{r},t)$ and vector field $\bar{f}(\bar{r},t)$, we have[84]

$$\bar{\nabla} \cdot \bar{D}(\bar{r},t) = \bar{\nabla} \cdot [\varepsilon(\bar{r})\bar{E}(\bar{r},t)] = \bar{\nabla}\varepsilon(\bar{r}) \cdot \bar{E}(\bar{r},t) + \varepsilon(\bar{r})\bar{\nabla} \cdot \bar{E}(\bar{r},t) = q_v(\bar{r},t). \tag{1.270}$$

Rearranging the terms, we further have[85]

$$\bar{\nabla} \cdot \bar{E}(\bar{r},t) = \frac{q_v(\bar{r},t)}{\varepsilon(\bar{r})} - \frac{\bar{\nabla}\varepsilon(\bar{r})}{\varepsilon(\bar{r})} \cdot \bar{E}(\bar{r},t). \tag{1.271}$$

Then, if and only if the permittivity does not change with respect to position at $\bar{r}$, we can write[86]

$$\bar{\nabla} \cdot \bar{E}(\bar{r},t) = \frac{q_v(\bar{r},t)}{\varepsilon}. \tag{1.272}$$

Otherwise (when $\bar{\nabla}\varepsilon(\bar{r}) \neq 0$), the divergence of the electric field intensity contains an extra term. We note that $\bar{\nabla}\varepsilon(\bar{r}) \neq 0$, not only when $\varepsilon(\bar{r})$ changes smoothly but also when there is a jump in $\varepsilon(\bar{r})$: i.e. across a boundary.

[83] We can define $\bar{\nabla}\varepsilon(\bar{r})$ as the change in $\varepsilon(\bar{r})$ with respect to position; this important operator (gradient) is studied later in more detail.

[84] We note that, without using this identity, we cannot find an expression for $\bar{\nabla} \cdot \bar{E}(\bar{r},t)$, as also shown in Eq. (1.255).

[85] The second term explains why the normal component of the electric field intensity can be discontinuous across the boundary between two different media.

[86] In some special cases, the change in permittivity may be perpendicular to the electric field: i.e. $\bar{\nabla}\varepsilon(\bar{r},t) \perp \bar{E}(\bar{r},t)$. For such a case, one can also use Eq. (1.272).

1.7.3 Equivalent Polarization Charges

While dielectric effects can purely be represented by the permittivity, an equivalence theorem can be used to model any polarization in terms of some fictitious electric charges. For this purpose, we again consider the divergence of the electric field intensity using Eq. (1.267) as

$$\bar{\nabla} \cdot \bar{E}(\bar{r},t) = \frac{1}{\varepsilon_0}\bar{\nabla} \cdot \bar{D}(\bar{r},t) - \frac{1}{\varepsilon_0}\bar{\nabla} \cdot \bar{P}(\bar{r},t), \tag{1.273}$$

leading to

$$\bar{\nabla} \cdot \bar{E}(\bar{r},t) = \frac{q_v(\bar{r},t) - \bar{\nabla} \cdot \bar{P}(\bar{r},t)}{\varepsilon_0}. \tag{1.274}$$

Therefore, if the divergence of the polarization is nonzero, it acts like a volume electric charge density that generates $\bar{E}(\bar{r})$ in a vacuum (ε_0 everywhere). This can be written

$$\bar{\nabla} \cdot \bar{E}(\bar{r}, t) = \frac{q_v(\bar{r}, t) + q_{pv}(\bar{r}, t)}{\varepsilon_0}, \tag{1.275}$$

where

$$q_{pv}(\bar{r}, t) = -\bar{\nabla} \cdot \bar{P}(\bar{r}, t) \tag{1.276}$$

is called the *volume polarization charge density* ($\mathrm{C/m^3}$). These *fictitious* electric charges can be used to account for the imbalance of oriented dipoles inside a dielectric object, and they are specifically nonzero due to the inhomogeneity of a material.

Volume polarization charges generally are not sufficient to fully model dielectric effects.[87] Specifically, when considering a dielectric object, the value of the permittivity jumps across its surface S, which is not directly visible in the differential form of Gauss' law. But, similar to the divergence of the electric field intensity, we can now check the boundary conditions for normal electric fields. Using the condition for the electric flux density in Eq. (1.170), we obtain[88]

$$\hat{a}_n \cdot [\varepsilon_0 \bar{E}_0(\bar{r}, t) - \varepsilon_0 \bar{E}_2(\bar{r}, t) - \bar{P}_2(\bar{r}, t)] = q_s(\bar{r}, t) \qquad (\bar{r} \in S), \tag{1.277}$$

where $\hat{a}_n$ is the unit normal vector outward from the dielectric object. This equation can be rearranged as

$$\hat{a}_n \cdot [\varepsilon_0 \bar{E}_0(\bar{r}, t) - \varepsilon_0 \bar{E}_2(\bar{r}, t)] = q_s(\bar{r}, t) + \hat{a}_n \cdot \bar{P}_2(\bar{r}, t) \qquad (\bar{r} \in S). \tag{1.278}$$

Therefore, $q_{ps}(\bar{r}, t) = \hat{a}_n \cdot \bar{P}_2(\bar{r}, t)$ represents a fictitious electric charge ($\mathrm{C/m^2}$) that acts like a surface electric charge density and causes a jump in the electric field intensity in a vacuum (Figure 1.65). We note that, due to these fictitious charges, in the absence of *true* surface electric charges (when $q_s(\bar{r}, t) = 0$), the normal component of the electric field intensity is still discontinuous:

$$\hat{a}_n \cdot [\bar{E}_0(\bar{r}, t) - \bar{E}_2(\bar{r}, t)] = \frac{\hat{a}_n \cdot \bar{P}_2(\bar{r}, t)}{\varepsilon_0} \qquad (\bar{r} \in S). \tag{1.279}$$

This is consistent with the discontinuity expressed in Eq. (1.172).

In summary, if the polarization of a dielectric object is known, the object can be represented by equivalent electric charges that are located entirely in a *vacuum* (Figure 1.66). These are

[87] Indeed, they do not describe the response of a simple homogeneous dielectric object.

[88] If an interface between two dielectric objects is considered and the equivalence theorem is used to model both objects, then two sets of surface polarization charges should be used (one for each object) with the correct direction for $\hat{a}_n$ (always outward from the considered object).

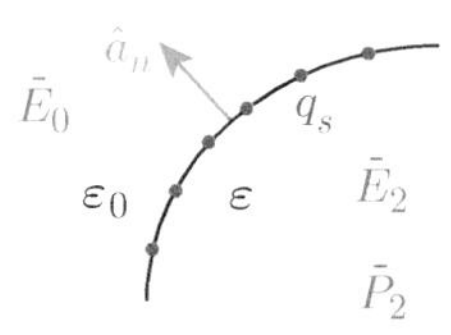

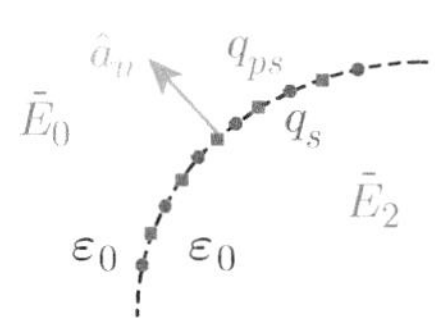

Figure 1.65 Replacing the polarization effect at the interface between a dielectric medium and a vacuum with equivalent surface polarization charges. We note that polarization charges $q_{ps}(\bar{r}, t)$ are added to true electric charges $q_s(\bar{r}, t)$ that may exist.

volume polarization and surface polarization charges, defined as

$$q_{pv}(\bar{r},t) = -\bar{\nabla}\cdot\bar{P}(\bar{r},t) \tag{1.280}$$

$$q_{ps}(\bar{r},t) = \hat{a}_n\cdot\bar{P}(\bar{r},t), \tag{1.281}$$

where $\hat{a}_n$ is the outward unit normal vector. These electric charges can be used to find the electric field intensity $\bar{E}(\bar{r},t)$: e.g. by using

$$\oint_S \bar{E}(\bar{r},t)\cdot\overline{ds} = \frac{Q_{\text{enc}}(t)}{\varepsilon_0} \tag{1.282}$$

or

$$\bar{\nabla}\cdot\bar{E}(\bar{r},t) = \frac{q_v(\bar{r},t)}{\varepsilon_0}. \tag{1.283}$$

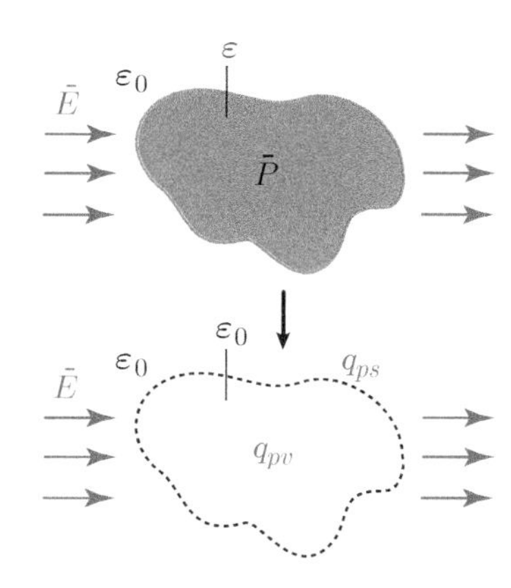

Figure 1.66 Modeling polarization in terms of equivalent electric charges that are located in the host medium (vacuum).

Once the electric field intensity is found, the electric flux density can be found as[89] $\bar{D}(\bar{r},t) = \varepsilon_0\bar{E}(\bar{r},t) + \bar{P}(\bar{r},t)$. Equivalent polarization charges are particularly useful when the polarization is already known (e.g. in electrets).

Since equivalent electric charges are introduced to model a dielectric object, they should not change the total amount of the electric charge in the considered system to obey the conservation of electric charges. This means the sum of polarization charges must be zero:

$$Q_{\text{net}}(t) = \int_V q_{pv}(\bar{r},t)d\bar{r} + \oint_S q_{ps}(\bar{r},t)d\bar{r} = 0. \tag{1.284}$$

This can be shown (for any arbitrary geometry) by using the expressions for the polarization charges as

$$Q_{\text{net}}(t) = \int_V [-\bar{\nabla}\cdot\bar{P}(\bar{r},t)]d\bar{r} + \oint_S \hat{a}_n\cdot\bar{P}(\bar{r},t)d\bar{r} \tag{1.285}$$

and using the divergence theorem, leading to

$$Q_{\text{net}}(t) = -\oint_S \hat{a}_n\cdot\bar{P}(\bar{r},t)d\bar{r} + \oint_S \hat{a}_n\cdot\bar{P}(\bar{r},t)d\bar{r} = 0. \tag{1.286}$$

Therefore, the distribution of these charges (rather than their total amount) effectively models the behavior of a dielectric object.

As an example for equivalent polarization charges, we reconsider an ideal parallel-plate capacitor that consists of two plates, each with area A_p, separated by a distance d (Figure 1.67).

[89] Obviously, one cannot use $\bar{D}(\bar{r},t) = \varepsilon_0\bar{E}(\bar{r},t)$.

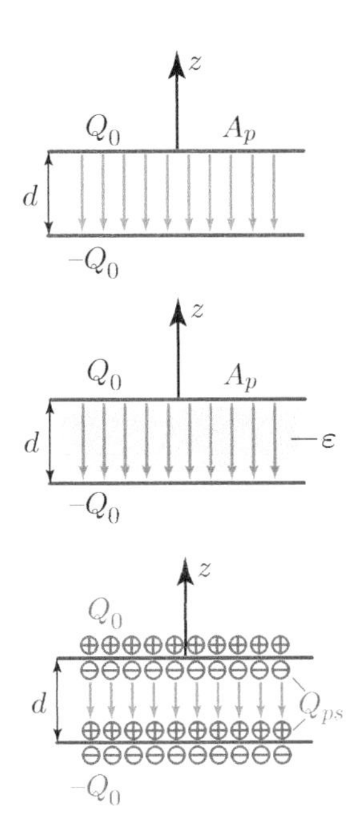

Figure 1.67 A parallel-plate capacitor and a dielectric inserted between the plates. In the equivalent problem, the surface polarization charges reduce the total effective electric charges on the top and bottom plates. This leads to decreasing electric field intensity due to the dielectric. In this scenario, the true electric charges $\pm Q_0$ are assumed to be constant (no connection to a battery).

As discussed in Section 1.6.1, when the space between the plates is a vacuum, the electric flux density and electric field intensity can be written

$$\bar{D}(\bar{r}) = -\hat{a}_z \frac{Q_0}{A} \begin{cases} 1 & \text{(between the plates)} \\ 0 & \text{(elsewhere)} \end{cases} \tag{1.287}$$

and

$$\bar{E}(\bar{r}) = -\hat{a}_z \frac{Q_0}{A\varepsilon_0} \begin{cases} 1 & \text{(between the plates)} \\ 0 & \text{(elsewhere),} \end{cases} \tag{1.288}$$

respectively, assuming that the plates are very large. If the electric charges accumulated on the plates ($\pm Q_0$) are constant (no connection to a battery) and the space between the plates is filled by a dielectric material with permittivity ε, then the electric flux density does not change. This can be understood by considering the boundary condition for the electric flux density (on the top/bottom plates). On the other hand, the electric field intensity changes as

$$\bar{E}(\bar{r}) = -\hat{a}_z \frac{Q_0}{A\varepsilon} \begin{cases} 1 & \text{(between the plates)} \\ 0 & \text{(elsewhere).} \end{cases} \tag{1.289}$$

Considering $\varepsilon > \varepsilon_0$, the electric field intensity decreases when the dielectric material is inserted. This decrease can be explained by the response of the dielectric to the external field (created by the electric charges located on the plates): i.e. the polarization

$$\bar{P}(\bar{r}) = \bar{D}(\bar{r}) - \varepsilon_0 \bar{E}(\bar{r}) = -\hat{a}_z \frac{Q_0}{A}\left(1 - \frac{\varepsilon_0}{\varepsilon}\right) \tag{1.290}$$

between the plates. Using the equivalence theorem, the dielectric can be replaced with surface polarization charges[90]

$$q_{ps}(\bar{r}) = \hat{a}_n \cdot \bar{P}(\bar{r}) = \begin{cases} \hat{a}_z \cdot \bar{P}(\bar{r}) & (\bar{r} \in \text{top plate}) \\ -\hat{a}_z \cdot \bar{P}(\bar{r}) & (\bar{r} \in \text{bottom plate}) \end{cases} \tag{1.291}$$

$$= \frac{Q_0}{A}\left(1 - \frac{\varepsilon_0}{\varepsilon}\right) \begin{cases} -1 & \text{(top plate)} \\ 1 & \text{(bottom plate).} \end{cases} \tag{1.292}$$

Specifically, in the vicinity of the top plate (at $z = d^-$), negative surface polarization charges exist, while positive surface polarization charges exist in the vicinity of the bottom plate

[90] We note how the outward normal direction is taken as $\hat{a}_z$ on the top surface of the dielectric and $-\hat{a}_z$ on the bottom surface of the dielectric.

(at $z = 0^+$) (Figure 1.67). In the equivalent problem, these polarization charges can be combined with the *true* electric charges,[91] leading to

$$q_{s,\text{total}}(\bar{r}) = q_s(\bar{r}) + q_{ps}(\bar{r}) = \frac{Q_0}{A}\frac{\varepsilon_0}{\varepsilon}\begin{cases} 1 & \text{(top plate)} \\ -1 & \text{(bottom plate).} \end{cases} \tag{1.293}$$

[91] *Free* charge is also a common name to distinguish true electric charges from polarization charges (that correspond to bound charges) in equivalent problems.

Therefore, the total electric charges are reduced by $\varepsilon_0/\varepsilon$ when the dielectric material is inserted between the plates. In the equivalent problem, the reduced amount of electric charges in Eq. (1.293) can be assumed to be located in a vacuum and create the electric field intensity in Eq. (1.289).

1.7.4 Examples

Example 11: Consider a static point electric charge Q_0 (C) located at the origin inside a dielectric shell of inner radius a and outer radius b (Figure 1.68). Let the relative permittivity of the shell be ε_r. Find the electric flux density $\bar{D}(\bar{r})$ and the electric field intensity $\bar{E}(\bar{r})$ everywhere. Also find the polarization inside the shell, as well as the equivalent polarization charges to represent the dielectric in a vacuum.

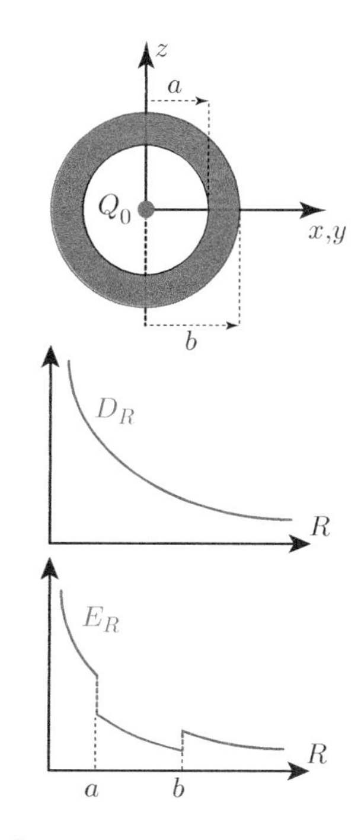

Figure 1.68 A static point electric charge Q_0 located at the origin inside a dielectric shell and the corresponding electric field with respect to R.

Solution: Due to the spherical symmetry, we can assume that the electric flux density $\bar{D}(\bar{r})$ depends only on R and has only an R component: i.e. $\bar{D}(\bar{r}) = \hat{a}_R D_R(R)$. Considering a spherical Gauss surface with radius R, we have

$$\oint_S \bar{D}(\bar{r}) \cdot \overline{ds} = 4\pi R^2 D_R = Q_{\text{enc}} = Q_0. \tag{1.294}$$

Therefore, we obtain

$$\bar{D}(\bar{r}) = \hat{a}_R \frac{Q_0}{4\pi R^2} \tag{1.295}$$

everywhere. Hence, the electric flux density is the same as that due to a point electric charge.[92] After finding $\bar{D}(\bar{r})$, it is straightforward to derive $\bar{E}(\bar{r})$ as

[92] We emphasize that this does not mean $\bar{D}(\bar{r})$ is never affected by dielectrics. In this case, the symmetry leads to an unaffected $\bar{D}(\bar{r})$. Note that, without symmetry, we would not be able to employ Gauss' law (while it is still valid) to find $\bar{D}(\bar{r})$.

$$\bar{E}(\bar{r}) = \hat{a}_R \frac{Q_0}{4\pi\varepsilon_0 R^2} \qquad (R < a \quad \text{and} \quad b < R) \tag{1.296}$$

and

$$\bar{E}(\bar{r}) = \hat{a}_R \frac{Q_0}{4\pi\varepsilon_0\varepsilon_r R^2} \qquad (a < R < b). \tag{1.297}$$

Note that $\bar{E}(\bar{r})$ is discontinuous (has jumps), while $\bar{D}(\bar{r})$ is continuous at the boundaries. Also, considering the expressions above, it is remarkable that the electric field intensity for $R < a$ and $b < R$ is the same as that due to a point electric charge, again as a result of the symmetry. As a verification for $R > b$, if the equivalent problem is considered by replacing the dielectric with equivalent polarization charges, we have

$$\oint_S \bar{E}(\bar{r}) \cdot \overline{ds} = \frac{Q_{\text{enc}}}{\varepsilon_0} = \frac{Q_0 + Q_{pv} + Q_{ps}}{\varepsilon_0}, \quad (1.298)$$

where Q_{pv} and Q_{ps} are the total volume and surface polarization charges enclosed by the Gauss surface. But since $Q_{pv} + Q_{ps} = 0$, we simply have

$$\bar{E}(\bar{r}) = \hat{a}_R \frac{Q_0}{4\pi\varepsilon_0 R^2} \qquad (R > b).$$

Finally, one can find the polarization vector $\bar{P}(\bar{r})$, as well as the polarization charges. We have

$$\bar{P}(\bar{r}) = \bar{D}(\bar{r}) - \varepsilon_0 \bar{E}(\bar{r}) \quad (1.299)$$

$$= \hat{a}_R \frac{Q_0}{4\pi R^2} - \hat{a}_R \frac{Q_0}{4\pi\varepsilon_r R^2} \quad (1.300)$$

$$= \hat{a}_R \frac{Q_0}{4\pi R^2}\left(1 - \frac{1}{\varepsilon_r}\right) \qquad (a < R < b).$$

Then we further obtain[93]

$$q_{pv} = -\bar{\nabla} \cdot \bar{P} = 0 \qquad \text{(everywhere)} \quad (1.301)$$

$$q_{ps} = \hat{a}_n \cdot \bar{P} = \frac{Q_0}{4\pi b^2}\left(1 - \frac{1}{\varepsilon_r}\right) \qquad (R = b) \quad (1.302)$$

$$q_{ps} = \hat{a}_n \cdot \bar{P} = -\frac{Q_0}{4\pi a^2}\left(1 - \frac{1}{\varepsilon_r}\right) \qquad (R = a) \quad (1.303)$$

as the equivalent polarization charges (Figure 1.69). Hence, there is no volume polarization charge (due to homogeneity), while there are surface polarization charge distributions at $R = a$ and $R = b$. One can further show that the total polarization charge (in this particular case, the sum of only surface polarization charges) is zero.

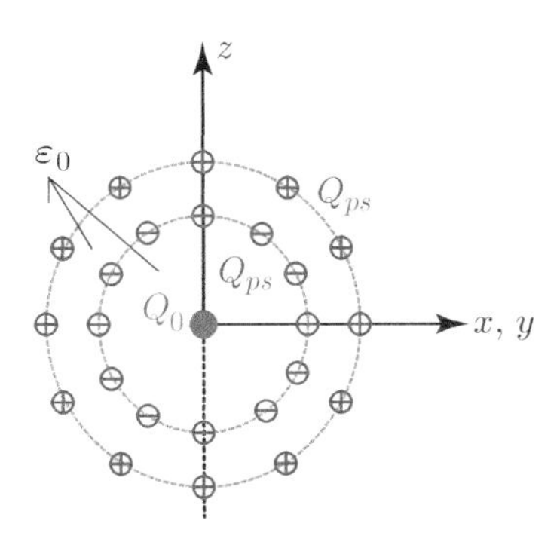

Figure 1.69 In the equivalent problem for the scenario in Figure 1.68, equivalent polarization charges exist on the inner and outer surfaces at $R = a$ and $R = b$ that replace the dielectric layer. If $Q_0 > 0$, then a uniformly distributed negative polarization charge density exists at $R = a$, where the total amount of the polarization charge is $Q_{ps}(R = a) = -Q_0(1 - 1/\varepsilon_r)$. In this case, a positive polarization charge distribution exists with a total amount of $Q_{ps}(R = b) = Q_0(1 - 1/\varepsilon_r)$ at $R = b$. We further note that everywhere is a vacuum in the equivalent problem. Then the jump in the electric field intensity values at $R = a$ and $R = b$ can be explained by the equivalent polarization charges $Q_{ps}(R = a)$ and $Q_{ps}(R = b)$.

[93] When the equivalent problem is used to analyze a dielectric problem, the electric field intensity can be found by assuming all charges (true and equivalent charges) in a vacuum. However, this does not mean $\bar{D}(\bar{r}, t) = \varepsilon_0 \bar{E}(\bar{r}, t)$. Either the original permittivity values should be used or (more safely) we have $\bar{D}(\bar{r}, t) = \varepsilon_0 \bar{E}(\bar{r}, t) + \bar{P}(\bar{r}, t)$.

Example 12: Consider a dielectric sphere of radius a, which is electrically charged with a static volume electric charge density $q_v(\bar{r}) = Q_0 R/a^4$ (C/m^3), located at the origin

(Figure 1.70). Let the relative permittivity of the sphere be ε_r. Find the electric flux density $\bar{D}(\bar{r})$ and the electric field intensity $\bar{E}(\bar{r})$, as well as the polarization and equivalent polarization charges, everywhere.

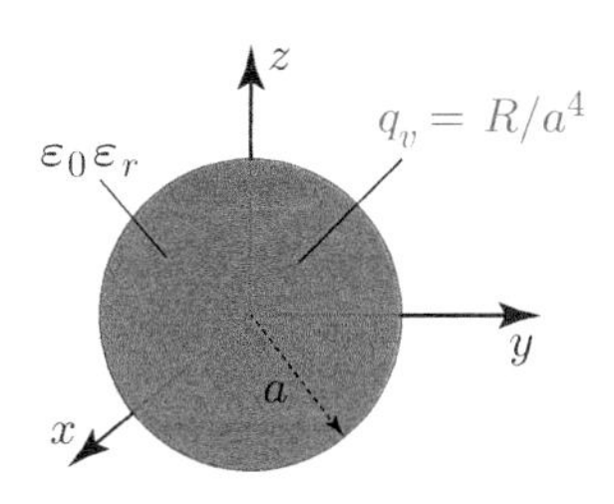

Figure 1.70 An electrically charged dielectric sphere.

Solution: Once again, considering the spherical symmetry, we can use Gauss' law as $\bar{D}(\bar{r}) = \hat{a}_R Q_{\text{enc}}/(4\pi R^2)$, where

$$Q_{\text{enc}} = Q_0 \int_0^R \int_0^\pi \int_0^{2\pi} \frac{R'}{a^4}(R')^2 \sin\theta dR' d\theta' d\phi' \tag{1.304}$$

$$= Q_0 \pi \left[\frac{(R')^4}{a^4}\right]_0^R = Q_0 \frac{\pi R^4}{a^4} \qquad (R < a). \tag{1.305}$$

When $R > a$, we further have $Q_{\text{enc}} = Q_0 \pi$. Therefore, we obtain the electric flux density as

$$\bar{D}(\bar{r}) = \frac{Q_0}{4}\hat{a}_R \begin{cases} R^2/a^4 & (R < a) \\ 1/R^2 & (R > a), \end{cases} \tag{1.306}$$

which is again continuous at $R = a$. The electric field intensity can be found as

$$\bar{E}(\bar{r}) = \frac{Q_0}{4\varepsilon_0}\hat{a}_R \begin{cases} R^2/(\varepsilon_r a^4) & (R < a) \\ 1/R^2 & (R > a), \end{cases} \tag{1.307}$$

which is discontinuous. Obviously, the polarization is nonzero inside the dielectric (assuming $\varepsilon_r \neq 1$):

$$\bar{P}(\bar{r}) = \bar{D}(\bar{r}) - \varepsilon_0 \bar{E}(\bar{r}) = \hat{a}_R \frac{Q_0 R^2}{4a^4} - \hat{a}_R \frac{Q_0 R^2}{4\varepsilon_r a^4} \tag{1.308}$$

$$= \hat{a}_R \frac{Q_0 R^2}{4a^4}\left(1 - \frac{1}{\varepsilon_r}\right) \qquad (R < a). \tag{1.309}$$

The surface polarization charge density at $R = a$ can be found as

$$q_{ps}(\bar{r}) = \hat{a}_n \cdot \bar{P}(\bar{r}) = \frac{Q_0}{4a^2}\left(1 - \frac{1}{\varepsilon_r}\right) \qquad (R = a). \tag{1.310}$$

Similarly, the volume polarization charge density can be found as[94]

$$q_{pv}(\bar{r}) = -\bar{\nabla} \cdot \bar{P}(\bar{r}) = -\frac{Q_0 R}{a^4}\left(1 - \frac{1}{\varepsilon_r}\right) \qquad (R < a). \tag{1.311}$$

[94] If we integrate the polarization charges, we obtain

$$Q_{ps} = Q_0 4\pi a^2 \frac{1}{4a^2}\left(1 - \frac{1}{\varepsilon_r}\right) = Q_0 \pi \left(1 - \frac{1}{\varepsilon_r}\right)$$

$$Q_{pv} = -Q_0 4\pi \int_0^a \frac{R}{a^4}\left(1 - \frac{1}{\varepsilon_r}\right) R^2 dR$$

$$= -Q_0 4\pi \left(1 - \frac{1}{\varepsilon_r}\right)\frac{a^4}{4a^4} = -Q_0 \pi \left(1 - \frac{1}{\varepsilon_r}\right),$$

verifying that $Q_{ps} + Q_{pv} = 0$.

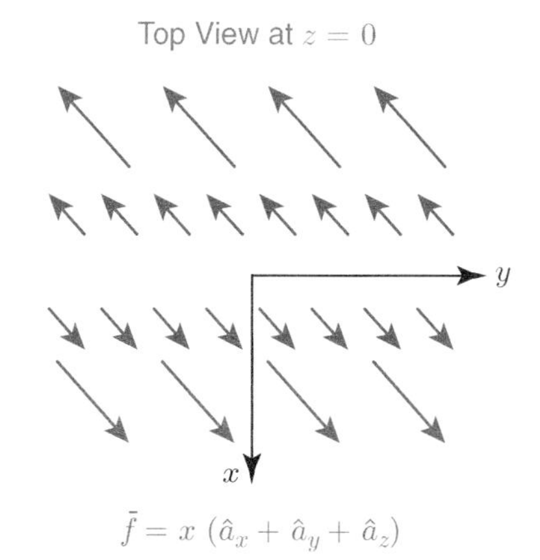

Figure 1.71 A view of a vector field $\bar{f}(\bar{r}) = (\hat{a}_x + \hat{a}_y + \hat{a}_z)x$ on the x-y plane.

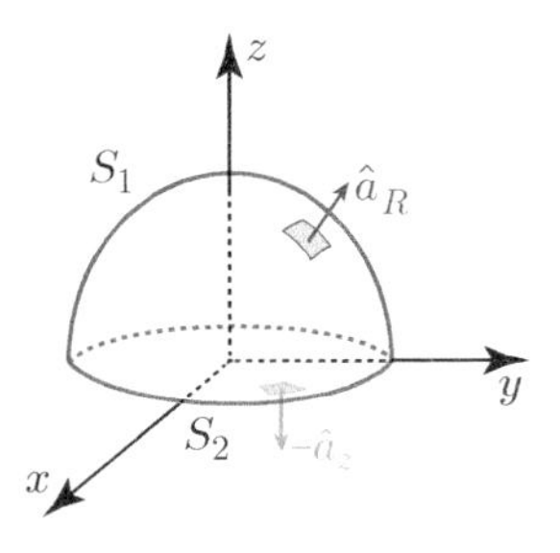

Figure 1.72 A closed hemisphere and outward normal directions.

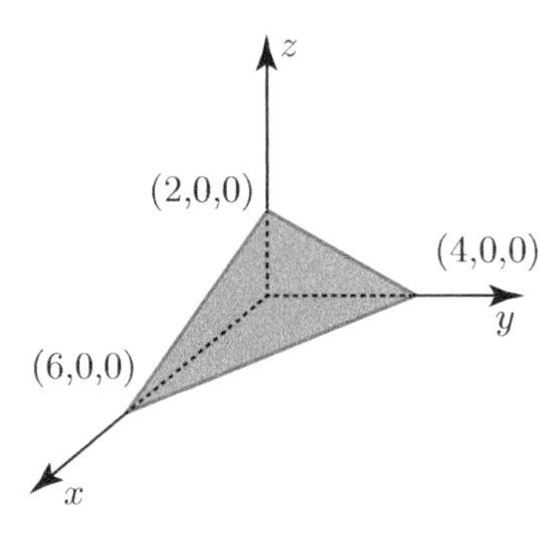

Figure 1.73 A surface defined as $S:\ 2x+3y+6z=12$ in the first octant.

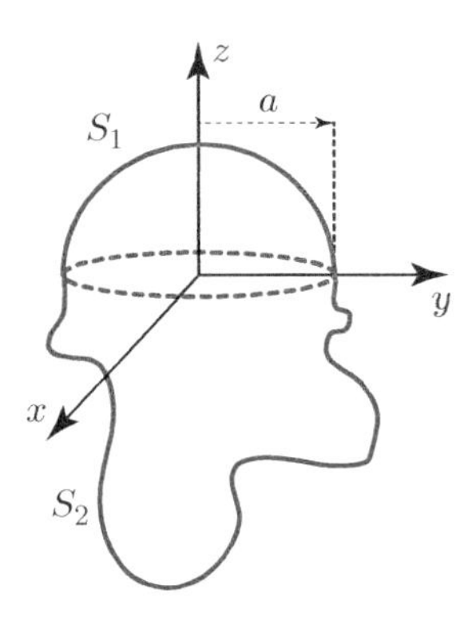

Figure 1.74 A closed surface involving an upper hemisphere and an arbitrary shape.

1.8 Final Remarks

As shown in this chapter, Gauss' law is a fundamental concept that describes the relationship between electric charges and electric fields. To understand Gauss' law for different cases, including those involving dielectrics, we have discussed different tools and concepts, such as a differential surface with direction, dot product, flux of vector fields, electric charge density, divergence of vector fields, divergence theorem, electric dipole, polarization, and equivalent polarization charges. We have seen that Gauss' law alone can be used to find electric fields in static cases. However, in dynamic cases, where sources and fields may depend on time, we need the other of Maxwell's equations discussed in the following chapters.

1.9 Exercises

Exercise 1: Consider a vector $\bar{f}=\hat{a}_x 2+\hat{a}_y-\hat{a}_z 3$. Find an expression for $\bar{f}$ in a cylindrical coordinate system.

Exercise 2: Consider a vector field $\bar{f}(\bar{r})=(\hat{a}_x+\hat{a}_y+\hat{a}_z)x$, which has all three components of the Cartesian system (Figure 1.71). Find the net outward flux of this field through a unit cube limited with surfaces at $x=0$, $x=1$, $y=0$, $y=1$, $z=0$, and $z=1$.

Exercise 3: Find the net outward flux of $\bar{f}(\bar{r})=\hat{a}_x x-\hat{a}_y+\hat{a}_z yz$ through a closed hemisphere of radius 2 with $z>0$ (Figure 1.72).

Exercise 4: Find the flux of $\bar{f}=\hat{a}_x 18z-\hat{a}_y 12+\hat{a}_z 3y$ through a surface defined as $S:\ 2x+3y+6z=12$ in the first octant, where the normal of the surface $\hat{a}_n$ is in the positive direction (Figure 1.73).

Exercise 5: Consider a closed surface $S=S_1\cup S_2$, where S_1 is an upper hemisphere of radius a and S_1 is an arbitrary surface (Figure 1.74). Note that there is no internal surface between the hemisphere and the arbitrary surface. A vector field is given as

$$\bar{f}(\bar{r})=\begin{cases}\hat{a}_z(z+a) & z<0\\ \hat{a}_z(z^2/a+a) & z\geq 0.\end{cases} \tag{1.312}$$

In addition, the net outward flux of the vector field through $S = S_1 \cup S_2$ is

$$\oint_S \bar{f}(\bar{r}) \cdot \overline{ds} = 2\pi a^3. \quad (1.313)$$

- Find the outward flux through S_2.
- Find the total volume enclosed by $S = S_1 \cup S_2$.

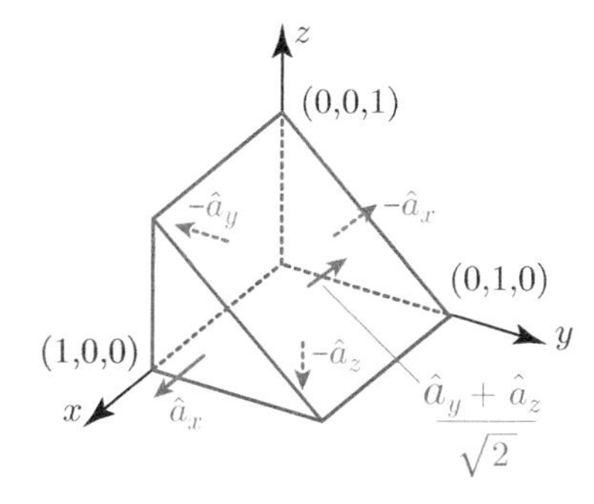

Figure 1.75 A closed surface in the shape of a triangular prism.

Exercise 6: With the help of the divergence theorem, find the net outward flux of $\bar{f}(\bar{r}) = \hat{a}_x x - \hat{a}_y + \hat{a}_z yz$ through a closed hemisphere of radius 2 with $z > 0$ (Figure 1.72).

Exercise 7: Consider a vector field $\bar{f}(\bar{r}) = \bar{r} = \hat{a}_x x + \hat{a}_y y + \hat{a}_z z$ and its net outward flux through a closed surface in the shape of a triangular prism defined with 1×1 squares on the z-x and x-y planes and an oblique surface between them (Figure 1.75). Verify the divergence theorem.

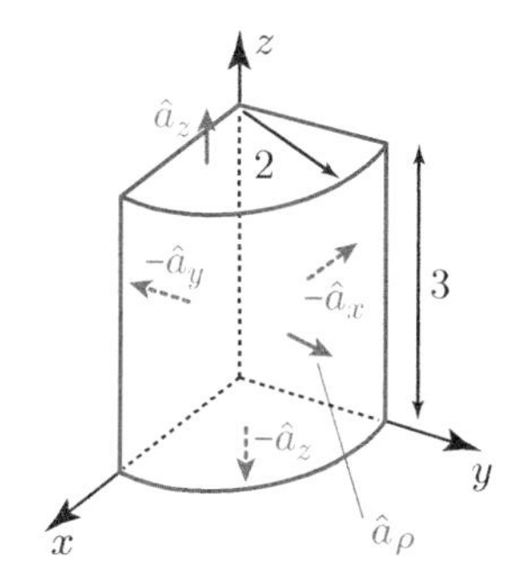

Figure 1.76 A closed surface in the shape of a quarter of a cylinder.

Exercise 8: Consider a vector field $\bar{f}(\bar{r}) = \hat{a}_x y + \hat{a}_y x + \hat{a}_z z$ and its net outward flux through a closed surface in the shape of a quarter of a cylinder (with radius 2 and aligned from $z = 0$ to $z = 3$) in the first octant (Figure 1.76). Verify the divergence theorem.

Exercise 9: A static volume electric charge density (C/m^3) is defined in a cylindrical region $\rho \le a$ and $|z| \le a$ in a vacuum, and there is no electric charge elsewhere (Figure 1.77). The electric flux density in the same region is given as

$$\bar{D}(\bar{r}) = \frac{Q_0}{a^2}\left[\frac{\rho^2}{a^2}\hat{a}_\rho + \sin\left(\frac{\pi z}{2a}\right)\hat{a}_z\right] \qquad (\rho \le a, |z| \le a).$$

Find the total amount of the electric charge.

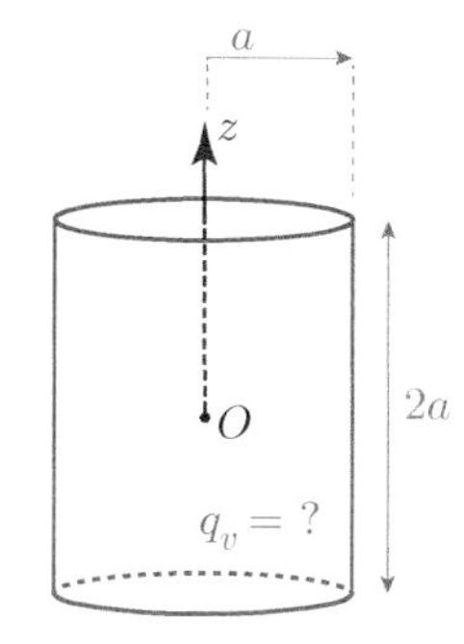

Figure 1.77 A cylindrical region, where the electric charge density is not known.

Exercise 10: The electric flux intensity $\bar{D}(\bar{r})$ in a vacuum is given as

$$\bar{D}(\bar{r}) = \hat{a}_R \frac{q_0}{5a^2R^2}\begin{cases} R^5 & (R < a) \\ a^5 & (R > a), \end{cases} \quad (1.314)$$

where q_0 (C/m^3) and a (m) are constants. Find the static electric charges creating this flux density.

Exercise 11: A static ring-shaped surface electric charge density is defined on the x-y plane in a vacuum (Figure 1.78). The electric charge density is given as

$$q_s(\bar{r}) = \frac{aq_0}{\rho}\cos\phi \tag{1.315}$$

for $b \leq \rho \leq a$. Find the electric field intensity at $\bar{r} = \hat{a}_z z$ for $z > 0$ (on the positive z axis).

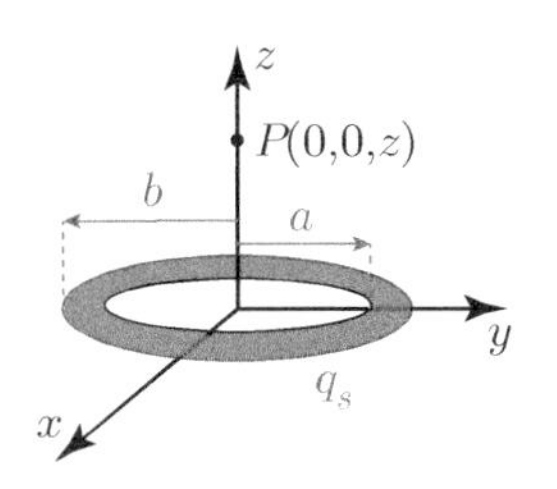

Figure 1.78 A static ring-shaped surface electric charge density.

Exercise 12: Consider positive and negative static electric charge distributions with sector shapes placed on the x-y plane in a vacuum (Figure 1.79): i.e. q_0 (C/m^2) for $-\phi_0 < \phi < \phi_0$ and $-q_0$ (C/m^2) for $\pi - \phi_0 < \phi < \pi + \phi_0$. The radius of the sectors is a. Find the electric field intensity at $z = a$.

Exercise 13: A system of static electric charge distributions consists of a surface electric charge density $q_s(\bar{r}) = q_{s0}\rho/a$ (C/m^2) with a sector shape ($\phi_0 < \phi < 2\pi - \phi_0$) and a static line electric charge density $q_l(\bar{r}) = q_{l0}$ (C/m) with an complementary arch shape (Figure 1.80). Find the constant q_{l0} in terms of the constant q_{s0} and the radius a, if the electric field intensity at the origin is given as zero.

Exercise 14: Consider a static surface electric charge density $q_s(\bar{r}) = q_0 z\cos\phi$ (C/m^2) defined on a half cylinder $\rho = a$, $\pi \leq \phi \leq 2\pi$, and $0 \leq z \leq h$ in a vacuum (Figure 1.81). Find the electric field intensity at the origin.

Exercise 15: Consider a surface involving a combination of a symmetrically placed spherical surface ($R = a$, $\theta < \pi - \varphi$) and a circular surface at the bottom (Figure 1.82). Let a static surface electric charge density q_0 be distributed uniformly on the entire surface in a vacuum. Find the electric field intensity at the origin, and simplify the expression for small φ.

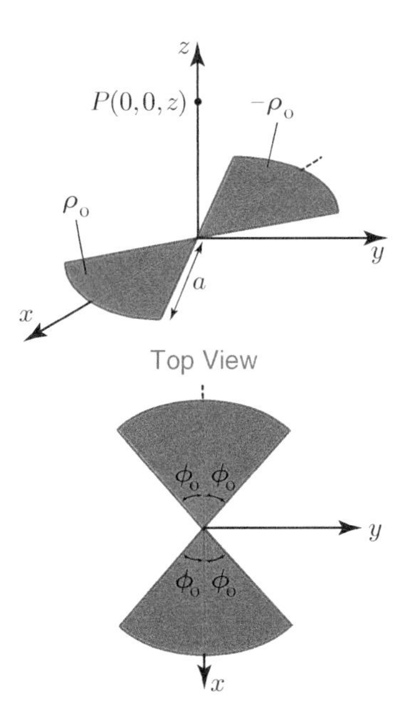

Figure 1.79 Positive and negative static electric charge distributions with sector shapes.

Exercise 16: Consider a static point electric charge Q_0 (C) located at the origin and enclosed in a spherical dielectric region with a constant permittivity of ε_2 and radius a (Figure 1.83). A dielectric shell exists for $a < R < b$ with a varying permittivity defined as

$$\varepsilon_1 = \frac{\varepsilon_2(b - R) + \varepsilon_0(R - a)}{b - a}. \tag{1.316}$$

Find the electric flux density $\bar{D}(\bar{r})$ and the electric field intensity $\bar{E}(\bar{r})$, as well as the polarization and the equivalent polarization charges, everywhere.

1.10 Questions

Question 1: Consider standard Cartesian, cylindrical, and spherical coordinate systems.

- Prove that the position vector $\bar{r} = \hat{a}_x x + \hat{a}_y y + \hat{a}_z z$ can be written $\bar{r} = \rho\hat{a}_\rho + z\hat{a}_z = R\hat{a}_R$ in cylindrical and spherical systems, using the relationships between the coordinate variables and unit vectors given as

$$\hat{a}_x = \hat{a}_\rho \cos\phi - \hat{a}_\phi \sin\phi = \hat{a}_R \sin\theta\cos\phi + \hat{a}_\theta \cos\theta\cos\phi - \hat{a}_\phi \sin\phi \tag{1.317}$$

$$\hat{a}_y = \hat{a}_\rho \sin\phi + \hat{a}_\phi \cos\phi = \hat{a}_R \sin\theta\sin\phi + \hat{a}_\theta \cos\theta\sin\phi + \hat{a}_\phi \cos\phi \tag{1.318}$$

$$\hat{a}_z = \hat{a}_R \cos\theta - \hat{a}_\theta \sin\theta \tag{1.319}$$

$$x = \rho\cos\phi = R\sin\theta\cos\phi \tag{1.320}$$

$$y = \rho\sin\phi = R\sin\theta\sin\phi \tag{1.321}$$

$$z = R\cos\theta. \tag{1.322}$$

  Note that the position vector is always from the origin to a position. Therefore, it naturally fits into $\bar{R} = \hat{a}_R R$ in the spherical coordinate system.

- Prove that[95]

$$\frac{\partial\bar{r}}{\partial\phi} = \hat{a}_\phi\rho = \hat{a}_\phi R\sin\theta \tag{1.323}$$

$$\frac{\partial\bar{r}}{\partial\theta} = \hat{a}_\theta R \tag{1.324}$$

$$\frac{\partial\bar{r}}{\partial R} = \hat{a}_R. \tag{1.325}$$

- Show that

$$\int_S \bar{r}ds = 0 \tag{1.326}$$

  for any spherical surface S centered at the origin.

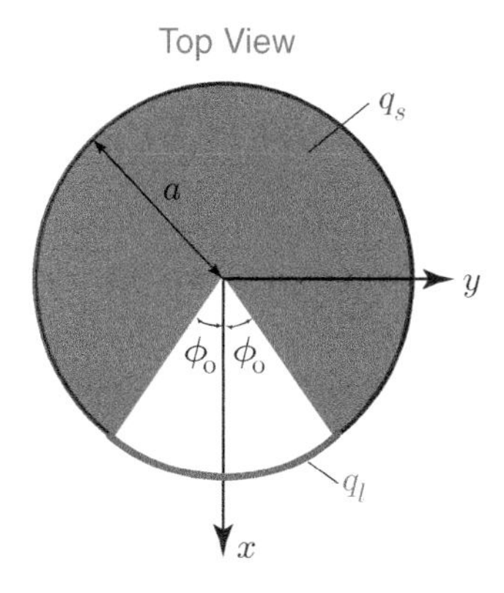

Figure 1.80 A system of static electric charge distributions involving a sector shape and a complementary arch shape.

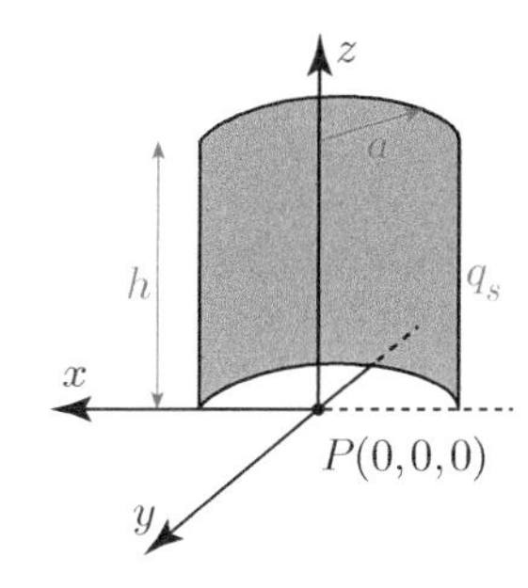

Figure 1.81 A static surface electric charge distribution with a cylindrical shape.

[95]The following are required:

$$\hat{a}_\rho = \hat{a}_x \cos\phi + \hat{a}_y \sin\phi$$

$$\hat{a}_R = \hat{a}_x \sin\theta\cos\phi + \hat{a}_y \sin\theta\sin\phi + \hat{a}_z \cos\theta$$

$$\hat{a}_\theta = \hat{a}_x \cos\theta\cos\phi + \hat{a}_y \cos\theta\sin\phi - \hat{a}_z \sin\theta$$

$$\hat{a}_\phi = -\hat{a}_x \sin\phi + \hat{a}_y \cos\phi.$$

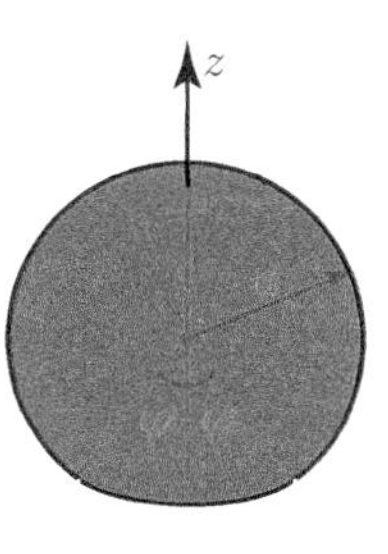

Figure 1.82 A surface involving a combination of a symmetrically placed spherical surface and a circular surface at the bottom.

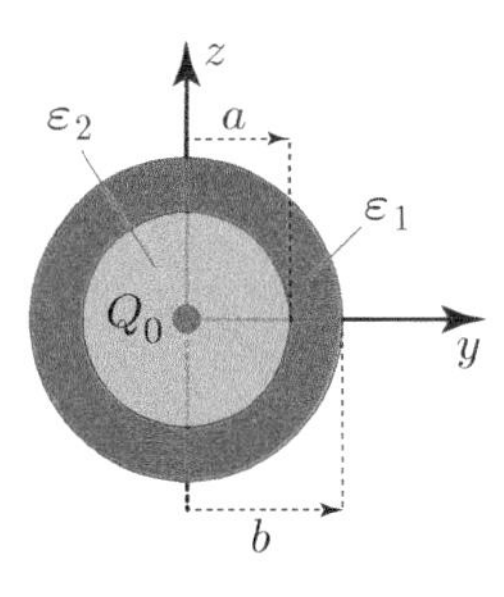

Figure 1.83 A static point electric charge located inside layered spherical dielectrics.

Question 2: There is an unknown static electric charge distribution in a vacuum. The electric flux density is given as

$$\bar{D}(\bar{r}) = \hat{a}_R \begin{cases} aq_0/2 & (R < a) \\ 0 & (R > a), \end{cases} \tag{1.327}$$

where a (m) and q_0 (C/m^3) are constants. Find all electric charges.

Question 3: Consider static surface electric charge distributions with different geometries as described below.

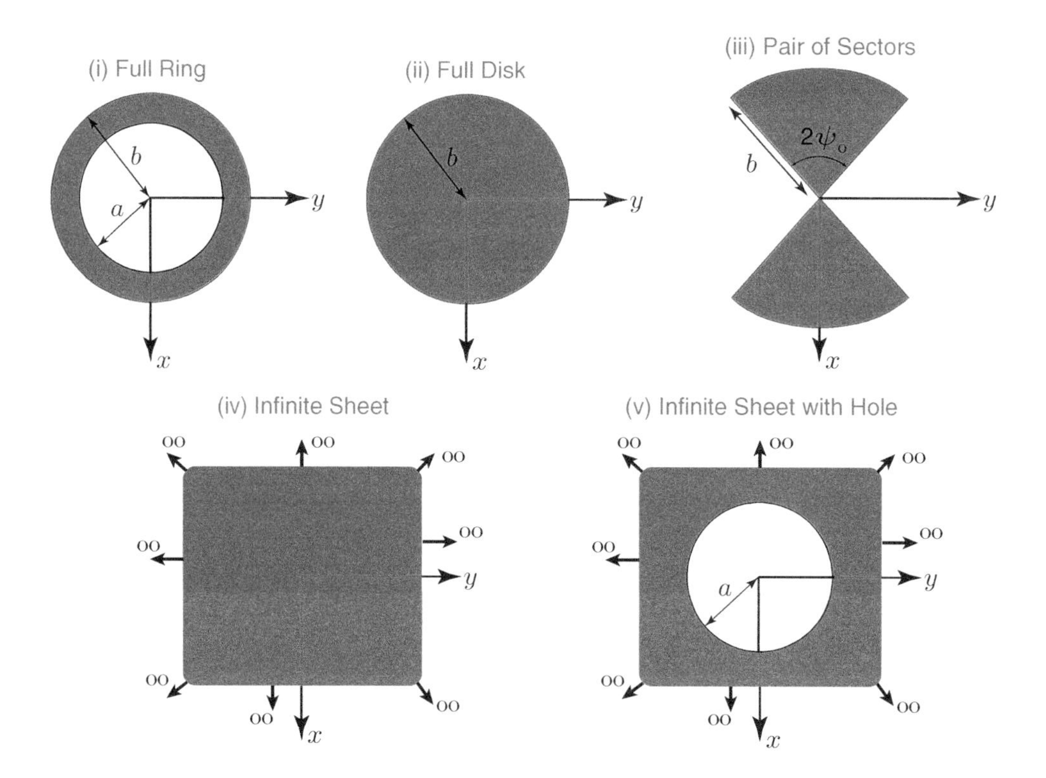

In each case, the surface electric charge density is constant [q_0 (C/m^2)], and it is required to find the electric field intensity at an arbitrary position on the z axis. Instead of dealing with each geometry one by one, find the electric field intensity for the electric charge distribution in Figure 1.84 and then use appropriate limits to reach the expressions for the given geometries.

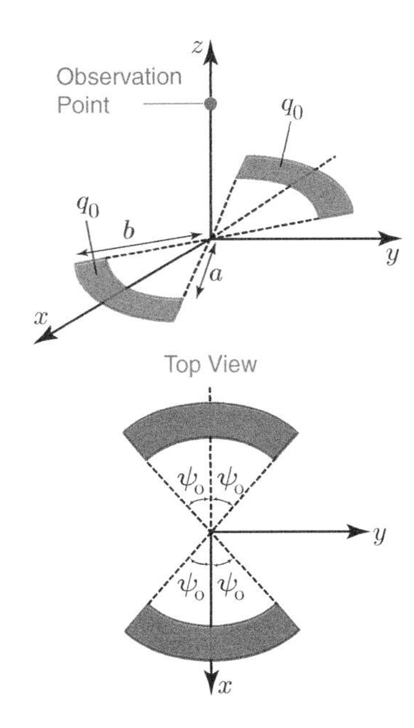

Figure 1.84 A static sectoral surface electric charge distribution.

Question 4: Consider a static line electric charge distribution on the z axis defined as

$$q_l(\bar{r}) = \begin{cases} q_0 & z \in (0, a) \\ -q_0 & z \in (-a, 0), \end{cases} \tag{1.328}$$

where q_0 is a constant, in a vacuum (Figure 1.85).

- Find the electric field intensity at an arbitrary position on the x-y plane for $\rho > 0$.
- Find the electric field intensity when $a \to \infty$.
- Find the electric field intensity when $a \to 0$.
- Find the electric field intensity at the origin for a finite a.
- Find the electric field intensity at the origin when $a \to \infty$.

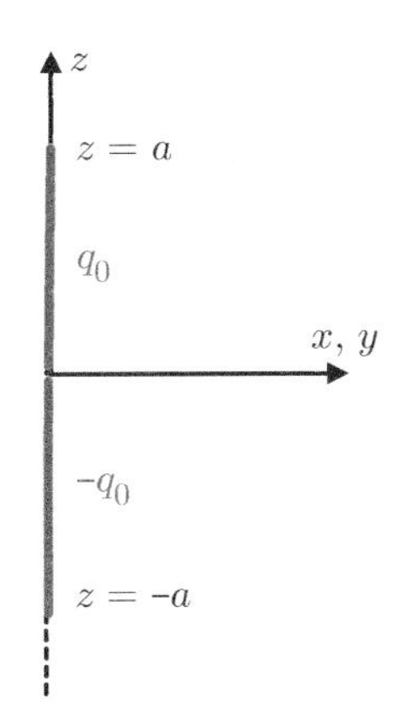

Figure 1.85 A static line electric charge distribution on the z axis.

Question 5: A spherical shell with inner radius a and outer radius $b > a$ is permanently polarized as $\bar{P}(\bar{r}) = \hat{a}_R P_0$, where P_0 is a constant. The shell is located in a vacuum, and there is no (true) electric charge density anywhere (Figure 1.86).

- Find all equivalent polarization charges.
- Show that the total polarization charge is zero.
- Find the electric field intensity everywhere.
- Show that the electric flux density is zero everywhere.

Question 6: An infinitely long static line electric charge density q_0 (C/m) on the z axis is placed inside an infinitely long cylindrical shell with inner radius a and outer radius $b > a$ (Figure 1.87). The permittivity of the shell is given as $\varepsilon_0 a^2/\rho^2$, where ρ is the cylindrical radial variable.

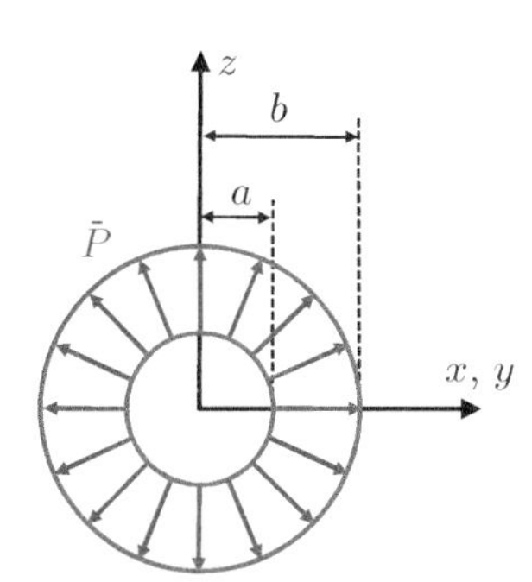

Figure 1.86 A spherical shell with permanent polarization.

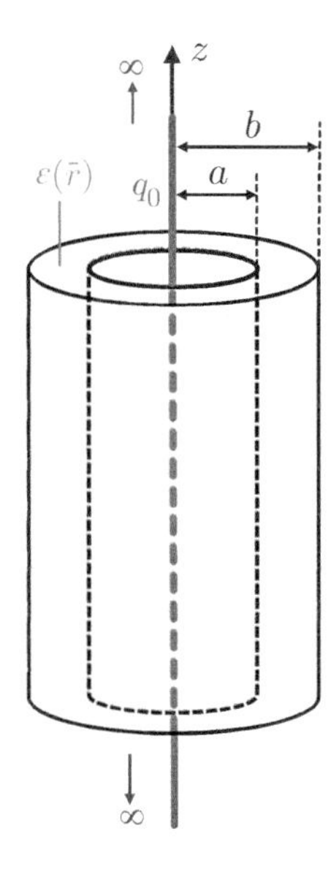

Figure 1.87 An infinitely long static line electric charge density inside an infinitely long cylindrical dielectric shell.

- Find the electric flux density everywhere.
- Find the electric field intensity everywhere.
- Find the polarization inside the dielectric.
- Find all equivalent polarization charges.
- Show that the total polarization charge is zero.
- Using the equivalent problem – i.e. vacuum everywhere – verify the expression for the electric field intensity for $a < \rho < b$.

2

Ampere's Law

$$\oint_C \bar{H}(\bar{r},t) \cdot \overline{dl} = I_{\text{enc}}(t) + \int_S \frac{\partial \bar{D}(\bar{r},t)}{\partial t} \cdot \overline{ds}$$

$$\bar{\nabla} \times \bar{H}(\bar{r},t) = \bar{J}_v(\bar{r},t) + \frac{\partial \bar{D}(\bar{r},t)}{\partial t}$$

In this chapter, we continue with Ampere's law, which describes how magnetic fields are created by electric currents, as well as time-variant electric fields. Similar to Gauss' law, there are two forms of Ampere's law. The integral form relates the magnetic field's circulation around a curve to the enclosed free and displacement currents. Again, this form is useful mostly for symmetric cases, especially for static (stationary and time-invariant) scenarios. On the other hand, in the differential form, we write the curl of the magnetic field intensity as the electric current density and the time-derivative of the electric flux density at every point. Similar to the divergence operation, the curl operation is a kind of derivative, but it measures rotational variance with respect to coordinate variables. This means the generation of magnetic fields from their sources (electric currents and changing electric fields) has a different mechanism than the generation of electric fields from electric charges. Application of Ampere's law at boundaries leads to the boundary conditions that must be satisfied by magnetic fields across boundaries. While Ampere's law is again valid for both static and dynamic cases, it basically needs the other Maxwell's equations for the complete solution of a dynamic problem.

2.1 Integral Form of Ampere's Law

Consider a curve C bounding an open surface[1] S. According to Ampere's law, the magnetic field intensity $\bar{H}(\bar{r},t)$ along C is related to the enclosed electric current $I_{\text{enc}}(t)$ and the integral of the time derivative of the electric flux density over the surface as (Figure 2.1)

$$\oint_C \bar{H}(\bar{r},t) \cdot \overline{dl} = I_{\text{enc}}(t) + \int_S \frac{\partial \bar{D}(\bar{r},t)}{\partial t} \cdot \overline{ds}. \tag{2.1}$$

[1]An open surface can be described by a surface that does not completely enclose a volume.

$\bar{H}$: magnetic field intensity
Unit: amperes/meter (A/m)

$\bar{B}$: magnetic flux density
Unit: webers/meter^2 (Wb/m^2)

μ: permeability
Unit: henrys/meter (H/m)

Introduction to Electromagnetic Waves with Maxwell's Equations, First Edition. Özgür Ergül.

Companion website: www.wiley.com/go/ergulmax

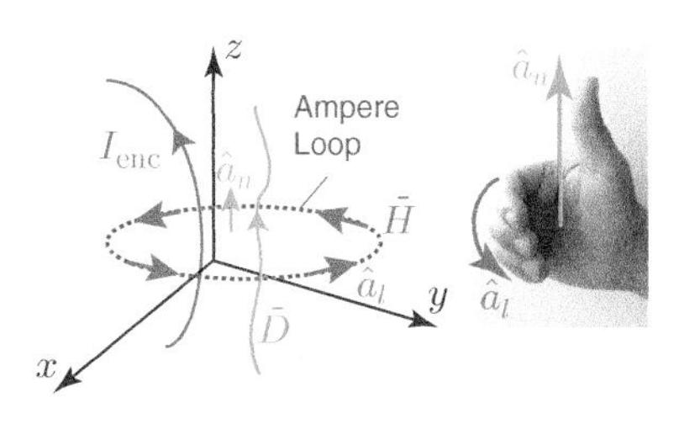

Figure 2.1 An illustration of Ampere's law. The circulation of the magnetic field intensity around a curve is proportional to the net electric current passing through the associated surface in addition to the surface integral of the time derivative of the electric flux density. The geometries of the curve and the bounded surface do not matter. The normal of the surface $\hat{a}_n$ for surface integral of the flux density, which is also the *selected direction* for the electric current, and the circulation direction are related via the right-hand rule.

The direction of $\overline{dl}$ and the electric current/flux directions are related via the right-hand rule. Ampere's law describes how magnetic fields are created, especially considering that magnetic charges do not exist. The first contribution is the electric current $I_{\text{enc}}(t)$, as also described by Biot-Savart law in static cases. The second contribution, which exists only for dynamic cases (when there is a time-variant electric field), is due to the electric flux density. Hence, Ampere's law also provides a relationship between electric and magnetic fields in dynamic cases.

If the medium is (magnetically) linear and isotropic, which is considered throughout this book, the magnetic field intensity and the magnetic flux density are related as

$$\bar{B}(\bar{r},t) = \mu(\bar{r})\bar{H}(\bar{r},t), \tag{2.2}$$

where $\mu(\bar{r})$ is the permeability at location $\bar{r}$. Then in a (magnetically) homogeneous medium, where μ does not depend on the position, we have

$$\oint_C \bar{B}(\bar{r},t) \cdot \overline{dl} = \mu I_{\text{enc}}(t) + \mu \int_S \frac{\partial \bar{D}(\bar{r},t)}{\partial t} \cdot \overline{ds}. \tag{2.3}$$

We note that, if the medium is also electrically homogeneous, ε does not depend on the position and we further have

$$\oint_C \bar{B}(\bar{r},t) \cdot \overline{dl} = \mu I_{\text{enc}}(t) + \mu\varepsilon \int_S \frac{\partial \bar{E}(\bar{r},t)}{\partial t} \cdot \overline{ds}. \tag{2.4}$$

In a vacuum (absence of any material), $\mu = \mu_0 = 4\pi \times 10^{-7}$ H/m. Similar to electric fields, both the magnetic flux density and the magnetic field intensity are called magnetic fields, in general. To understand Ampere's law, we need the following tools and concepts:

- Differential length with direction
- Circulation of vector fields

Circulation of vector fields further needs the concept of dot product in Section 1.1.2. In the addition, on the right-hand side of Ampere's law in Eq. (2.1), two required concepts are differential surface with direction (Section 1.1.1) and flux of vector fields (Section 1.1.3), which were discussed before.

2.1.1 Differential Length With Direction

In Ampere's law in Eq. (2.1), we use $\overline{dl}$, which is a differential length with a direction. Specifically, it contains a differential length dl and a direction $\hat{a}_l$ that is aligned in the direction of the considered line (Figure 2.2). Some of the common differential lengths with direction are

- $\overline{dl} = \pm\hat{a}_x dx$, $\overline{dl} = \pm\hat{a}_y dy$, $\overline{dl} = \pm\hat{a}_z dz$
- $\overline{dl} = \pm\hat{a}_\rho d\rho$, $\overline{dl} = \pm\hat{a}_\phi \rho d\phi$, $\overline{dl} = \pm\hat{a}_z dz$
- $\overline{dl} = \pm\hat{a}_R dR$, $\overline{dl} = \pm\hat{a}_\theta R d\theta$, $\overline{dl} = \hat{a}_\phi R \sin\theta d\phi$

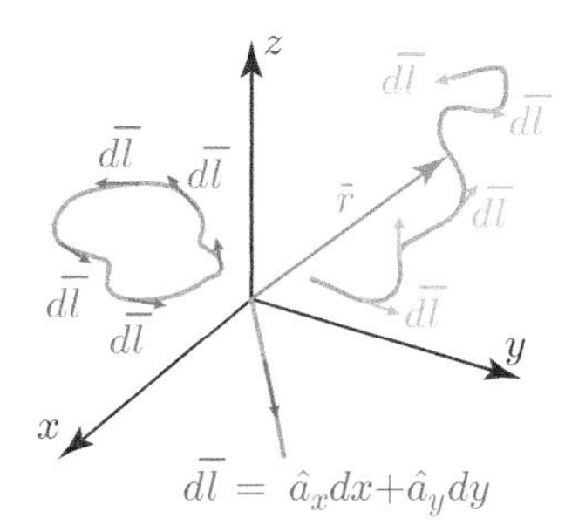

Figure 2.2 Differential length vectors on different curves.

in the Cartesian, cylindrical, and spherical coordinate systems, respectively. Similar to the differential surfaces in Section 1.1.1 (and as discussed in Section 0.4), $d\phi$ and $d\theta$ are accompanied by $\rho = R\sin\theta$ and R, respectively, to convert these variables into lengths. In general, a differential length with a direction can be a combination of the common ones listed above. As an example, we consider $x = y$ line and the direction toward the positive x and y (Figure 2.3). Then we have

$$\overline{dl} = \hat{a}_l dl = \hat{a}_x dx + \hat{a}_y dy, \tag{2.5}$$

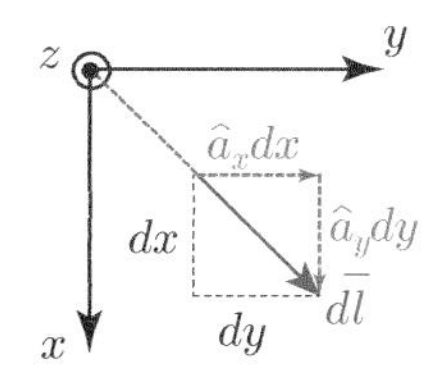

Figure 2.3 A representation of $\overline{dl} = \hat{a}_x x + \hat{a}_y y$.

which can be written by considering that the vector $\overline{dl}$ is a combination of two vectors $\hat{a}_x dx$ and $\hat{a}_y dy$. This also means that

$$\hat{a}_l = \frac{\overline{dl}}{dl} = \frac{\hat{a}_x dx + \hat{a}_y dy}{\sqrt{(dx)^2 + (dy)^2}} = \frac{1}{\sqrt{2}}(\hat{a}_x + \hat{a}_y), \tag{2.6}$$

considering that $dx = dy = dl/\sqrt{2}$.

In the next subsection, we consider a type of line integral: the circulation, involving differential lengths. Similar to surface integrals, line integrals involving cylindrical and spherical variables and unit vectors need to be handled carefully. For example, consider a line integral (Figure 2.4)

$$\bar{f} = \oint \hat{a}_\phi d\phi, \tag{2.7}$$

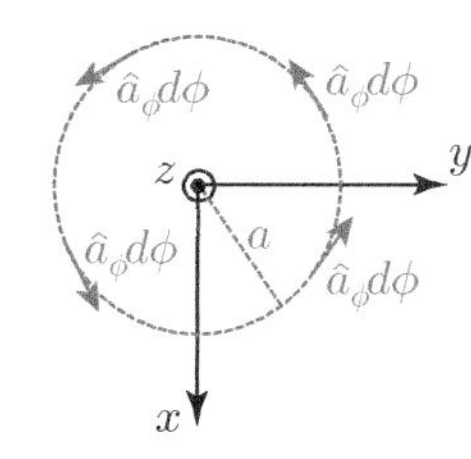

Figure 2.4 A differential $\overline{dl} = \hat{a}_\phi d\phi$ on a circle.

which is evaluated over a circle of radius a. This integral can be interpreted as the sum of all possible vectors $\overline{dl} = \hat{a}_\phi d\phi$. But if $\hat{a}_\phi$ is taken out of the integral, we have

$$\bar{f} \stackrel{?}{=} \hat{a}_\phi \int_0^{2\pi} d\phi = \hat{a}_\phi 2\pi, \tag{2.8}$$

which is obviously incorrect. Since, $\hat{a}_\phi$ depends on the position, the integral should be evaluated as

$$\bar{f} = \int_0^{2\pi} (-\hat{a}_x \sin\phi + \hat{a}_y \cos\phi) d\phi = -\hat{a}_x \int_0^{2\pi} \sin\phi d\phi + \hat{a}_y \int_0^{2\pi} \cos\phi d\phi = 0 \tag{2.9}$$

by passing to the Cartesian representation.

2.1.2 Circulation of Vector Fields

To understand Ampere's law, we need line integrals in the form of

$$I_c = \oint_C \bar{f}(\bar{r}) \cdot \overline{dl} = \oint_C \bar{f}(\bar{r}) \cdot \hat{a}_l dl, \tag{2.10}$$

where $\bar{f}(\bar{r})$ is a vector field and C is a curve. The value of the integral must be independent of the chosen coordinate system; however, given $\bar{f}(\bar{r})$, choosing one coordinate system may be wiser than choosing another. In a Cartesian system, if the curve is linear and in a direction $\pm x$, $\pm y$, or $\pm z$, we have

$$I_c = \pm \int_C f_x(\bar{r})dx, \quad I_c = \pm \int_C f_y(\bar{r})dy, \quad I_c = \pm \int_C f_z(\bar{r})dz, \tag{2.11}$$

respectively, where $f_x(\bar{r})$, $f_y(\bar{r})$, $f_z(\bar{r})$ are the Cartesian components of the vector field. In many cases, it may be easier to evaluate the line integral in cylindrical and spherical coordinate systems and using their specific variables:

$$I_c = \pm \int_C f_\rho(\bar{r})d\rho, \quad I_c = \pm \int_C f_\phi(\bar{r})\rho d\phi \tag{2.12}$$

$$I_c = \pm \int_C f_R(\bar{r})dR, \quad I_c = \pm \int_C f_\theta(\bar{r})Rd\theta, \quad I_c = \pm \int_C f_\phi(\bar{r})R\sin\theta d\phi, \tag{2.13}$$

where $f_\rho(\bar{r})$, $f_\phi(\bar{r})$, $f_R(\bar{r})$, and $f_\theta(\bar{r})$ are the components of the vector field $\bar{f}(\bar{r})$. It can be seen that the line integral in Ampere's law – Eq. (2.1) – is a closed integral:

$$I_c = \oint_C \bar{f}(\bar{r}) \cdot \overline{dl}. \tag{2.14}$$

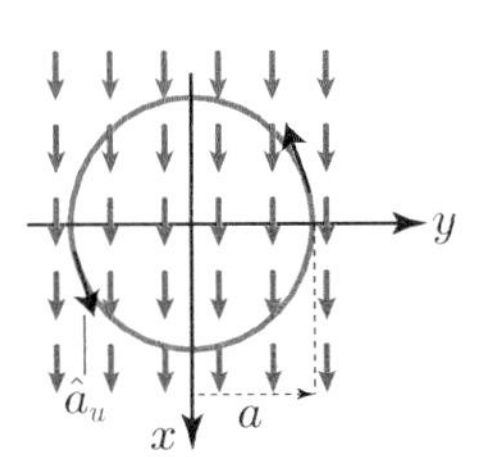

Figure 2.5 Circulation of $\bar{f}(\bar{r}) = \hat{a}_x f_0$ around a circle on the $z = 0$ plane.

This is called *circulation* of the vector field $\bar{f}(\bar{r})$ around the closed curve C.

As an example, consider a vector field $\bar{f}(\bar{r}) = \hat{a}_x f_0$, where f_0 is a constant (Figure 2.5). The circulation of $\bar{f}(\bar{r})$ around a curve in the shape of a circle of radius a on the $z = 0$ plane can be found as[2]

$$I_c = \int_0^{2\pi} f_0 \hat{a}_x \cdot \hat{a}_\phi \rho d\phi = f_0 a \int_0^{2\pi} (-\sin\phi)d\phi = 0. \tag{2.15}$$

Hence, the circulation of $\bar{f}(\bar{r})$ around this curve is zero. This is because the value of $\bar{f}(\bar{r}) \cdot \overline{dl}$ has positive and negative values that perfectly balance each other considering the whole circle.

[2] The direction of $\overline{dl}$ is arbitrarily selected as $\hat{a}_\phi$.

Next, we consider the circulation of the same vector field ($\bar{f}(\bar{r}) = \hat{a}_x f_0$) around a triangular curve, as depicted in Figure 2.6. In this case, we can divide the circulation integral into three parts as

$$I_c = \int_{C_1} f_0 \hat{a}_x \cdot \hat{a}_x dx + \int_{C_2} f_0 \hat{a}_x \cdot (-\hat{a}_x dx + \hat{a}_y dy) + \int_{C_3} f_0 \hat{a}_x \cdot (-\hat{a}_y) dy. \quad (2.16)$$

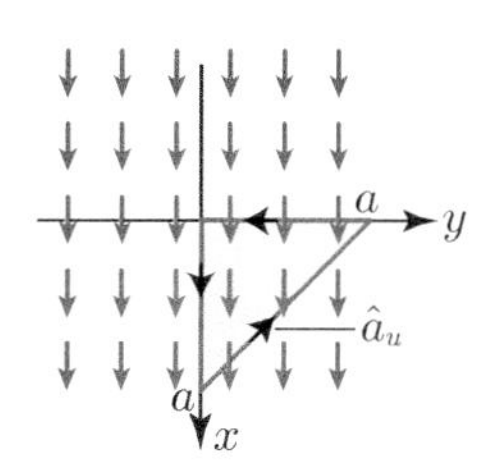

Figure 2.6 Circulation of $\bar{f}(\bar{r}) = \hat{a}_x f_0$ around a triangle on the $z = 0$ plane.

We note the following properties of the curves:

- On C_1: $\overline{dl} = \hat{a}_x dx$, $x = a$, and $y \in [0, a]$
- On C_2: $\overline{dl} = -\hat{a}_x dx + \hat{a}_y dy$, $x + y = a$, and $x, y \in [0, a]$
- On C_3: $\overline{dl} = -\hat{a}_y dy$, $x = 0$, and $y \in [0, a]$

Based on these, the first integral can be evaluated as

$$\int_{C_1} f_0 \hat{a}_x \cdot \hat{a}_x dx = f_0 \int_0^a \hat{a}_x \cdot \hat{a}_x dx = f_0 \int_0^a dx = f_0 a. \quad (2.17)$$

Similarly, the second integral can be evaluated as

$$\int_{C_2} f_0 \hat{a}_x \cdot (-\hat{a}_x dx + \hat{a}_y dy) = f_0 \int_{C_2} \hat{a}_x \cdot (-\hat{a}_x dx + \hat{a}_y dy) \quad (2.18)$$

$$= -f_0 \int_0^a dx = -f_0 a. \quad (2.19)$$

Finally, for the third integral, we have

$$\int_{C_3} f_0 \hat{a}_x \cdot (-\hat{a}_y) dy = f_0 \int_0^a \hat{a}_x \cdot (-\hat{a}_y) dy = f_0 \int_0^a 0 dy = 0. \quad (2.20)$$

Hence, the circulation can be found as

$$I_c = f_0 a - f_0 a + 0 = 0. \quad (2.21)$$

Interestingly, we find the circulation zero again. In fact, using Stokes' theorem, one can show that the circulation of $\bar{f}(\bar{r}) = \hat{a}_x f_0$ should be zero for *any* curve located anywhere.[3]

[3]This is not because $\bar{f}(\bar{r})$ has a constant magnitude. In fact, circulation values can be zero for many vector fields, such as for $\bar{f}(\bar{r}) = \hat{a}_x x$. Stokes' theorem can show these null circulations very clearly.

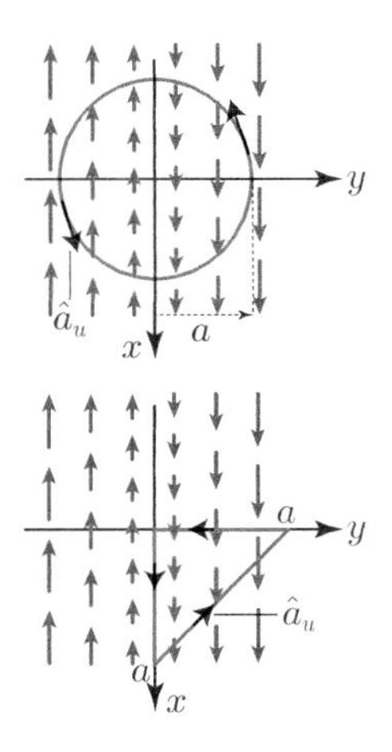

Figure 2.7 Circulation of $\bar{f}(\bar{r}) = \hat{a}_x y$ around a circle and a triangle on the $z = 0$ plane.

As another example (Figure 2.7), we now consider the circulation of a vector field $\bar{f}(\bar{r}) = \hat{a}_x y$ around the circle and the triangle of the previous examples. For the circle, we have

$$I_c = \int_0^{2\pi} y\hat{a}_x \cdot \hat{a}_\phi \rho d\phi = \int_0^{2\pi} \rho \sin\phi(-\sin\phi)\rho d\phi \tag{2.22}$$

$$= -a^2 \int_0^{2\pi} \sin^2\phi d\phi = -\pi a^2. \tag{2.23}$$

For the triangular curve, we can write

$$I_c = \int_{C_1} y\hat{a}_x \cdot \hat{a}_x dx + \int_{C_2} y\hat{a}_x \cdot (-\hat{a}_x dx + \hat{a}_y dy) + \int_{C_3} y\hat{a}_x \cdot (-\hat{a}_y) dy, \tag{2.24}$$

where[4]

$$\int_{C_1} y\hat{a}_x \cdot \hat{a}_x dx = \int_{C_1} y dx = \int_{C_1} 0 dx = 0 \tag{2.25}$$

$$\int_{C_2} y\hat{a}_x \cdot (-\hat{a}_x dx + \hat{a}_y dy) = -\int_{C_2} y dx = -\int_0^a (a - x) dx \tag{2.26}$$

$$= -\left[ax - x^2/2\right]_0^a = -a^2/2 \tag{2.27}$$

$$\int_{C_3} y\hat{a}_x \cdot (-\hat{a}_y) dy = \int_{C_3} 0 dy = 0. \tag{2.28}$$

Hence, we obtain

$$I_c = 0 - a^2/2 + 0 = -a^2/2. \tag{2.29}$$

It is not a coincidence that, due to the properties of $\bar{f}(\bar{r}) = \hat{a}_x y$, we obtain the negative of the area enclosed by the considered curve in both cases.

2.1.3 Meaning of Ampere's Law and Its Application

To understand Ampere's law, first we may consider steady (stationary and time-invariant) electric currents flowing through an imaginary surface formed by an Ampere loop[5,6] (Figure 2.8). If the electric currents are in opposite directions, these magnetic fields tend to cancel each other. Using the right-hand rule and considering the direction of $\overline{dl}$, these electric currents can be called positive and negative. We do not expect that these cancellations lead to zero fields everywhere (unless the electric currents have the same magnitude and they perfectly coincide

[4] Note that the integration limits are always from the smaller value to the larger value. For example, in the second integral, the differential length with direction is used as $-\hat{a}_x dx + \hat{a}_y dy$, by already including the negative sign in the x direction. Then the variable x is changing from 0 to a (not from a to 0).

[5] In most cases, we assume that an electric current is flowing, without directly indicating how it is created. If they are conduction currents, these electric currents are actually carried by a *metal* that is excited by some voltage source.

[6] Existence of an electric current (moving electric charges) does not necessarily mean there is a net electric charge. In most static cases, we assume that moving electrons and stationary protons are perfectly balanced. Then a magnetic field can be created due to the electric current, while there is no electric field in the absence of a net electric charge and changing magnetic field.

in space). For example, at some locations that are close to the positive electric current, the magnetic field intensity tends to be in the counterclockwise direction (looking from the top). However, at other locations, the effect of the negative electric current will dominate, leading to a clockwise magnetic field. In general, there are positive and negative $\bar{H}(\bar{r})$ values (with respect to the position and direction of $\overline{dl}$) along the Ampere loop. Ampere's law enters the picture at this stage. Wherever electric currents are located (provided that they pass through the surface), the circulation of $\bar{H}(\bar{r})$ depends only on their *net* amount. In fact, if the magnitudes of the electric currents are equal, the circulation of $\bar{H}(\bar{r})$ is zero, even though $\bar{H}(\bar{r})$ itself may have different nonzero values on the curve.

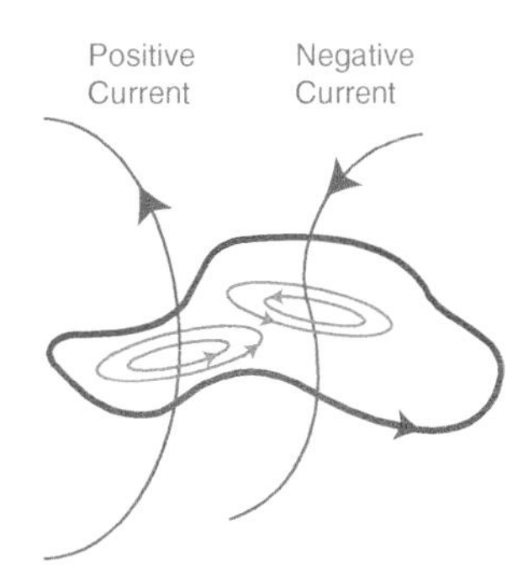

Figure 2.8 An illustration of Ampere's law, where the magnetic fields created by electric currents in opposite directions tend to cancel each other.

Like Gauss' law, Ampere's law is valid for all (static and dynamic) cases. In most dynamic cases, the second term on the right-hand side of Eq. (2.1) – i.e. the surface integral of the time derivative of the electric flux density – cannot be ignored and must be included for a proper analysis. But before including this term, we can consider a simple scenario, where a constant electric current flow I_0 is moving from inside to outside of a given Ampere loop (Figure 2.9). Assuming that there is no changing electric flux density, we have

$$\oint_C \bar{H}(\bar{r},t)\cdot\overline{dl} = I_{\text{enc}}(t), \tag{2.30}$$

where $I_{\text{enc}}(t) = I_0$ before the movement and $I_{\text{enc}}(t) = 0$ after the movement. Therefore, the circulation of the magnetic field intensity changes from I_0 to 0 when the electric current is moved.

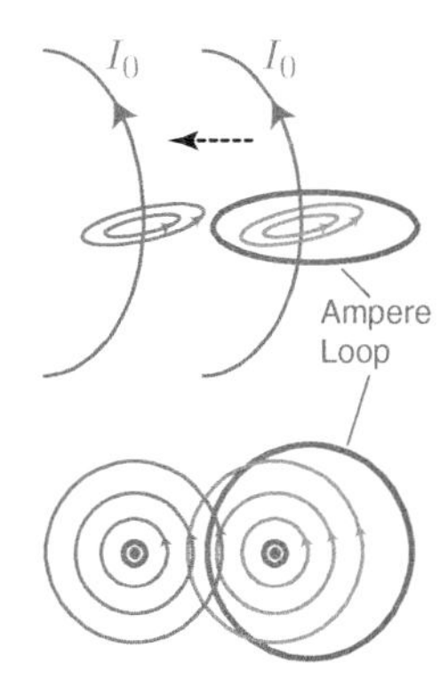

Figure 2.9 A movement of a constant electric current I_0, leading to a changing circulation of the magnetic field intensity around an Ampere loop.

In general dynamic cases, Ampere's law is not complete with only electric currents, and the time-varying electric flux density must be added, as shown in Eq. (2.1). Specifically, if the electric flux passing though a surface changes with respect to time, its effect is the same as an electric current: i.e. it creates a magnetic field and a nonzero circulation around the curve bounding the surface. As an example, consider again an Ampere loop enclosing a time-variant electric current: i.e. $I_{\text{enc}}(t)$ (Figure 2.10). If the time-varying electric flux was not included, we would write

$$\oint_C \bar{H}(\bar{r},t)\cdot\overline{dl} = I_{\text{enc}}(t). \tag{2.31}$$

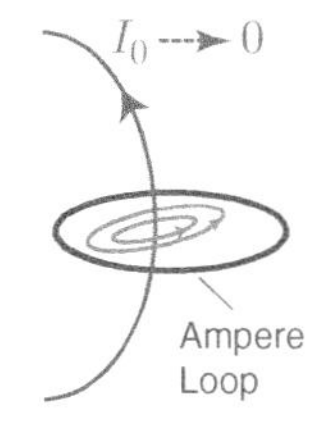

Figure 2.10 An Ampere loop enclosing a time-variant electric current flow that drops from I_0 to zero.

On the other hand, this equation is defective if the electric current changes with respect to time. For example, if the electric current suddenly changes from I_0 to 0, this equation indicates that the circulation of the magnetic field intensity suddenly drops to zero.[7] However, this (infinite speed of *information*) is not feasible and should require some time for the circulation to drop to zero. For a correct formulation, the surface integral of the time derivative of the electric flux density must be added to the right-hand side of Eq. (2.31), which contributes to the time-variant circulation. We further note this time-variant electric flux density is also related to a time-variant magnetic field (as formulated by Faraday's law).

[7] Note that this is different from the movement of a constant electric current I_0 from the inside to the outside of the curve, which needs finite time.

Following the discussion above for dynamic cases, we should write Ampere's law fully as

$$\oint_C \bar{H}(\bar{r},t)\cdot\overline{dl} = I_{\text{enc}}(t) + \int_S \frac{\partial \bar{D}(\bar{r},t)}{\partial t}\cdot\overline{ds}. \tag{2.32}$$

Then we can define the displacement current density as

$$\bar{J}_{vd}(\bar{r},t) = \frac{\partial \bar{D}(\bar{r},t)}{\partial t}, \tag{2.33}$$

which acts like a volume electric current density (see Section 2.3.1), whose surface integral gives the displacement current through the surface. Then we have

$$\oint_C \bar{H}(\bar{r},t)\cdot\overline{dl} = I_{\text{enc}}(t) + I_d(t), \tag{2.34}$$

where[8]

$$I_d(t) = \int_S \bar{J}_{vd}(\bar{r},t)\cdot\overline{ds} \tag{2.35}$$

is the displacement current. Obviously, in a static case, we have $I_d(t) = 0$, and

$$\oint_C \bar{H}(\bar{r})\cdot\overline{dl} = I_{\text{enc}} \tag{2.36}$$

as Ampere's law. Hence, in static scenarios, electric fields do not contribute to the generation of magnetic fields. This can be interpreted as *decoupling* of electric and magnetic fields.[9]

It is remarkable that, if an Ampere loop is stationary,[10] one can write

$$\oint_C \bar{H}(\bar{r},t)\cdot\overline{dl} = I_{\text{enc}}(t) + \frac{d}{dt}f_D(t), \tag{2.37}$$

where

$$f_D(t) = \oint_S \bar{D}(\bar{r},t)\cdot\overline{ds} \tag{2.38}$$

is the electric flux.

The complete form of Ampere's law in Eqs. (2.1) and (2.34) containing both *true* and displacement currents is a crucial component of understanding electromagnetics and, more specifically, how electric and magnetic fields are coupled in dynamic cases. In fact, the concept of displacement current can be used for more complete formulations of basic scenarios.

[8] We note that the unit of the electric flux density $\bar{D}(\bar{r},t)$ is Coulomb/meter^2; hence, its time derivative leads to Coulomb/(second × meter^2) or Ampere/meter^2.

J_v: volume electric current density
Unit: amperes/meter^2 (A/m^2)

[9] In a dynamic case, electric and magnetic fields create each other: i.e. time-variant electric fields create magnetic fields, and time-variant magnetic fields create electric fields. This explains how electromagnetic waves, which contain time-variant electric and magnetic fields, propagate in space. In electrostatics, electric fields are created only by static electric charges. In magnetostatics, magnetic fields are created only by steady electric currents, while there can be time-invariant electric fields without contributing to magnetic fields. Therefore, in both electrostatics and magnetostatics, electric and magnetic fields are decoupled: i.e. they do not interfere with each other.

[10] If an Ampere loop is not stationary (just mathematically or physically in a given application), then the time derivative cannot be placed outside the surface integral. This is because the integration domain itself depends on the time.

A popular example is a charging or discharging capacitor (Figure 2.11). According to Ampere's law, there is a magnetic field intensity $\bar{H}(\bar{r},t)$ around the current-carrying wires:

$$\oint_C \bar{H}(\bar{r},t) \cdot \overline{dl} = I_0, \tag{2.39}$$

where I_0 is constant and the surface bounded by the curve C can be a simple disk. However since the shape of the bounded surface can be arbitrary, it can be deformed such that it passes between the plates without changing C. Without a displacement current, this leads to inconsistencies as no true current passes through the deformed surface and the circulation of $\bar{H}$ should be zero. Using the complete form of Ampere's law, however, we have

$$\oint_C \bar{H}(\bar{r},t) \cdot \overline{dl} = \int_S \frac{\partial \bar{D}(\bar{r},t)}{\partial t} \cdot \overline{ds} = \frac{d}{dt}\int_S \bar{D}(\bar{r},t) \cdot \overline{ds}, \tag{2.40}$$

for the second (deformed surface) case, where $\bar{D}(\bar{r},t)$ is the electric flux density between the plates. If the capacitor is ideal, we further have a uniform $\bar{D}(\bar{r},t)$ so that

$$\frac{d}{dt}\int_S \bar{D}(\bar{r},t) \cdot \overline{ds} = \frac{d}{dt}\left[A_p \frac{Q(t)}{A_p}\right] = \frac{dQ(t)}{dt}, \tag{2.41}$$

where $Q(t)$ is the time-dependent total amount of the electric charge at one of the plates, while A_p represents the surface area of each plate. Obviously, using the two equations, we arrive at

$$\oint_C \bar{H}(\bar{r},t) \cdot \overline{dl} = I_0 = \frac{dQ(t)}{dt} \tag{2.42}$$

as the most basic definition of the electric current in terms of the electric charge. This result also helps us to generalize Kirchhoff's current law for dynamic cases.

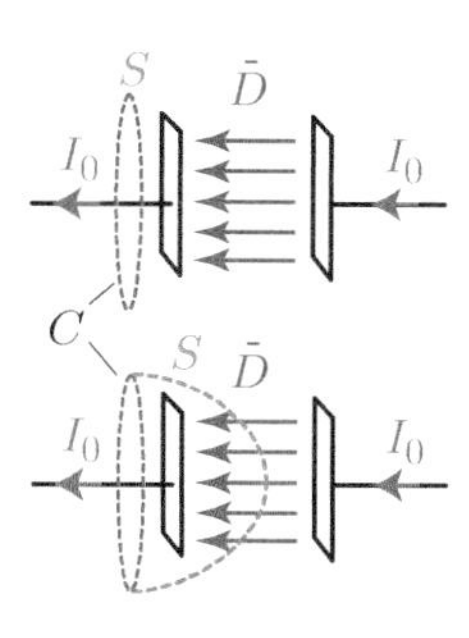

Figure 2.11 Application of Ampere's law for a charging or discharging capacitor. Since the shape of the surface bounded by a given curve C can be arbitrary, displacement currents must be included to avoid inconsistencies.

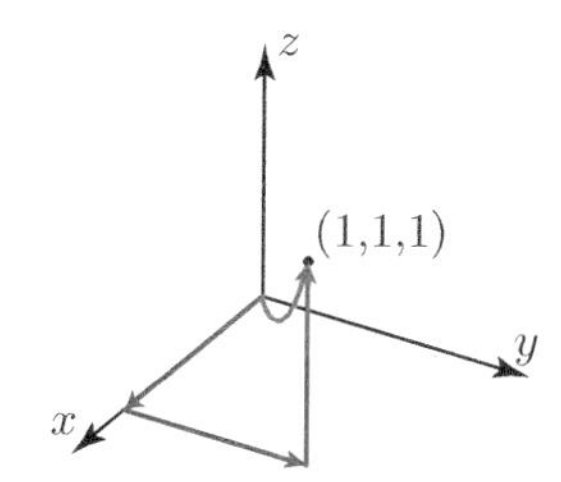

Figure 2.12 Different paths from the origin to $(x,y,z) = (1,1,1)$.

2.1.4 Examples

Example 13: Find the integral of the projection of $\bar{f}(\bar{r}) = \hat{a}_x(3x^2+6y) - \hat{a}_y 14yz + \hat{a}_z 20xz^2$ from $(x,y,z) = (0,0,0)$ to $(x,y,z) = (1,1,1)$ along C defined as $\{y = x^2, z = x^3\}$.

Solution: The integration path is a curved line from the origin to the location $(x,y,z) = (1,1,1)$ (Figure 2.12). We can evaluate the integral in a Cartesian coordinate system by using[11] $dy = d(x^2) = 2xdx$, $dz = d(x^3) = 3x^2dx$, and

$$\overline{dl} = \hat{a}_x dx + \hat{a}_y dy + \hat{a}_z dz = \hat{a}_x dx + \hat{a}_y 2xdx + \hat{a}_z 3x^2 dx. \tag{2.43}$$

[11] We note how these differential identities are found. If $y = x^2$, then $dy/dx = 2x$ via derivative. This means that $dy = 2xdx$. This relationship states that a differential change in x (Δx) creates a differential change in y: $\Delta y = 2x\Delta x$, which depends on the value of x.

Then we have

$$\int_C \bar{f}(\bar{r})\cdot\overline{dl} = \int_{(0,0,0)}^{(1,1,1)} \bar{f}\cdot(\hat{a}_x dx + \hat{a}_y 2x dx + \hat{a}_z 3x^2 dx). \tag{2.44}$$

Using $\bar{f}(\bar{r}) = \hat{a}_x 9x^2 - \hat{a}_y 14x^5 + \hat{a}_z 20x^7$, we obtain

$$\int_C \bar{f}(\bar{r})\cdot\overline{dl} = \int_0^1 \left[\hat{a}_x 9x^2 - \hat{a}_y 14x^5 + \hat{a}_z 20x^7\right]\cdot(\hat{a}_x dx + \hat{a}_y 2x dx + \hat{a}_z 3x^2 dx) \tag{2.45}$$

$$= \int_0^1 (9x^2 - 28x^6 + 60x^9)dx = \left[3x^3 - 4x^7 + 6x^{10}\right]_0^1 = 5. \tag{2.46}$$

At this stage, it is interesting to find the integral $\bar{f}(\bar{r})$ from $(0,0,0)$ to $(1,1,1)$ when it is projected on another path: e.g. along three straight lines (Figure 2.12) as $C_1 : (0,0,0) \longrightarrow (1,0,0)$, $C_2 : (1,0,0) \longrightarrow (1,1,0)$, and $C_3 : (1,1,0) \longrightarrow (1,1,1)$. Then we have $\overline{dl} = \hat{a}_x dx$ on C_1, $\overline{dl} = \hat{a}_y dy$ on C_2, and $\overline{dl} = \hat{a}_z dz$ on C_3, leading to

$$\int_C \bar{f}(\bar{r})\cdot\overline{dl} = \int_{C_1} \bar{f}\cdot\overline{dl} + \int_{C_2} \bar{f}\cdot\overline{dl} + \int_{C_3} \bar{f}\cdot\overline{dl} \tag{2.47}$$

$$= \int_0^1 (3x^2 + 6y)dx + \int_0^1 (-14yz)dy + \int_0^1 (20xz^2)dz \tag{2.48}$$

$$= \int_0^1 3x^2 dx + \int_0^1 0 dy + \int_0^1 20z^2 dz = \left[x^3\right]_0^1 + \left[\frac{20}{3}z^3\right]_0^1 = \frac{23}{3}. \tag{2.49}$$

Comparing the results, it is understood that the integral of the projection of $\bar{f}$ depends on the path.[12]

[12]For conservative fields, the value of a line integral is independent of the path, as discussed in Chapter 3. For example, the electric field intensity in a static case is conservative, while it is not conservative in a dynamic case.

Example 14: Find the circulation of $\bar{f}(\bar{r}) = xy\hat{a}_x$ along a closed loop formed by straight lines from $(x,y,z) = (0,0,0)$ to $(x,y,z) = (1,1,0)$, then from $(x,y,z) = (1,1,0)$ to $(x,y,z) = (1,0,0)$, and finally from $(x,y,z) = (1,0,0)$ to $(x,y,z) = (0,0,0)$ (Figure 2.13).

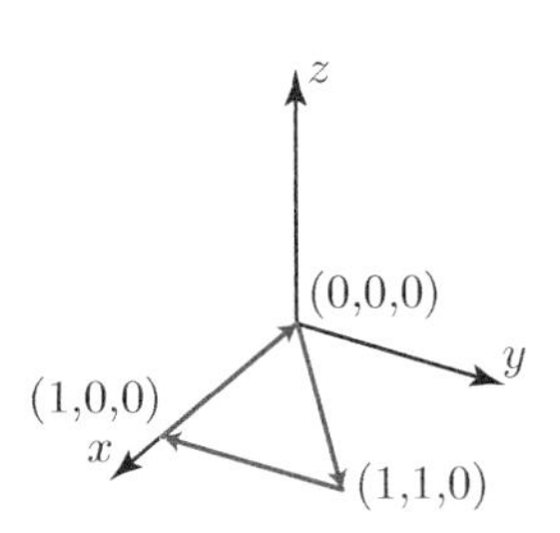

Figure 2.13 A closed path on the x-y plane defined by three straight lines from $(x,y,z) = (0,0,0)$ to $(x,y,z) = (0,0,0)$.

Solution: We have

$$\int_C \bar{f}(\bar{r})\cdot\overline{dl} = \int_{C_1} \bar{f}(\bar{r})\cdot\overline{dl} + \int_{C_2} \bar{f}(\bar{r})\cdot\overline{dl} + \int_{C_3} \bar{f}(\bar{r})\cdot\overline{dl}, \tag{2.50}$$

where $\overline{dl} = \hat{a}_x x + \hat{a}_y y$ on C_1, $\overline{dl} = -\hat{a}_y y$ on C_2, $\overline{dl} = -\hat{a}_x x$ on C_3, and[13]

[13]We note that integral limits are from small values to large values (even though C_2 and C_3 are traced in negative directions) since the direction information is already included in the definition of $\overline{dl}$.

$$\int_{C_1} \bar{f} \cdot \overline{dl} = \int_0^1 xydx = \int_0^1 x^2 dx = \frac{1}{3} \tag{2.51}$$

$$\int_{C_2} \bar{f} \cdot \overline{dl} = \int_0^1 0dy = 0 \tag{2.52}$$

$$\int_{C_3} \bar{f} \cdot \overline{dl} = \int_0^1 (-xy)dx = \int_0^1 0dx = 0. \tag{2.53}$$

Therefore, the circulation can be obtained as

$$\int_C \bar{f}(\bar{r}) \cdot \overline{dl} = \frac{1}{3}. \tag{2.54}$$

2.2 Using the Integral Form of Ampere's Law

Similar to the integral form of Gauss' law, the integral form of Ampere's law is mostly useful in symmetric cases, and particularly for static problems. The symmetry is needed to carry the magnetic field intensity outside the integral such that it can be written in terms of the electric current. Furthermore, for a static case, the displacement current disappears and the magnetic field intensity can be written only in terms of the true electric current. While the existence of the displacement current makes Ampere's law valid for dynamic cases, the electric flux density is often unknown (as opposed to the special cases, such as an ideal capacitor). Therefore, a further equation that relates electric fields to magnetic fields (specifically, Faraday's law) is needed for a complete analysis of a dynamic scenario.

As an example of a case that can be analyzed by using only Ampere's law, we consider an infinitely long line with zero thickness that carries a steady electric current I_0 along the z axis in a vacuum.[14] Using a circular Ampere loop with radius ρ (Figure 2.14), we have

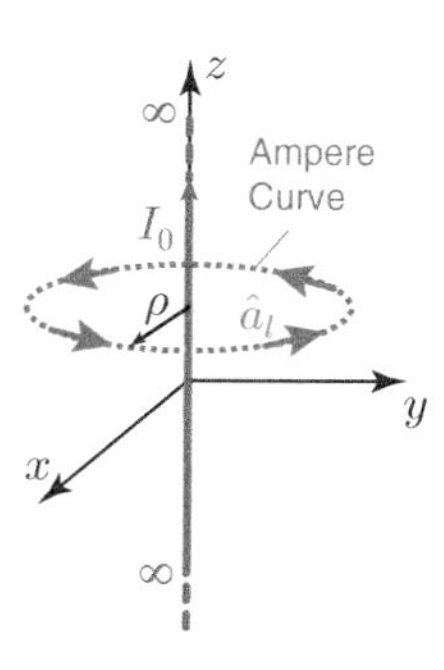

Figure 2.14 An infinitely long wire carrying a steady electric current flowing in the z direction in a vacuum.

[14]Similar to the point electric charge, a line electric current with a zero thickness is an idealization. In this case, a finite amount of electric current I_0 is assumed to be squeezed into a zero volume.

$$\oint_C \bar{H}(\bar{r}) \cdot \overline{dl} = \int_0^{2\pi} \bar{H}(\bar{r}) \cdot \hat{a}_\phi \rho d\phi = I_0. \tag{2.55}$$

Using symmetry, we can assume that $\bar{H}(\bar{r})$ only depends on ρ, and it must be in the ϕ direction (counterclockwise when looking at the loop from the top: i.e. using the right-hand rule shown in Figure 2.1):

$$\bar{H}(\bar{r}) = \hat{a}_\rho H_\rho(\bar{r}) + \hat{a}_\phi H_\phi(\bar{r}) + \hat{a}_z H_z(\bar{r}) = \hat{a}_\phi H_\phi(\rho, \phi, z) = \hat{a}_\phi H_\phi(\rho). \tag{2.56}$$

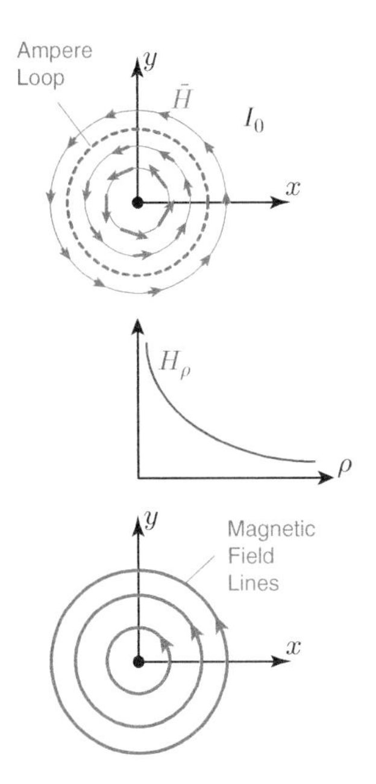

Figure 2.15 The magnetic field created by a steady line electric current decays with the distance from the line. The arrows show the direction and magnitude (size of arrows). Sometimes it is more illustrative to use magnetic field lines that only shows the direction but not magnitude.

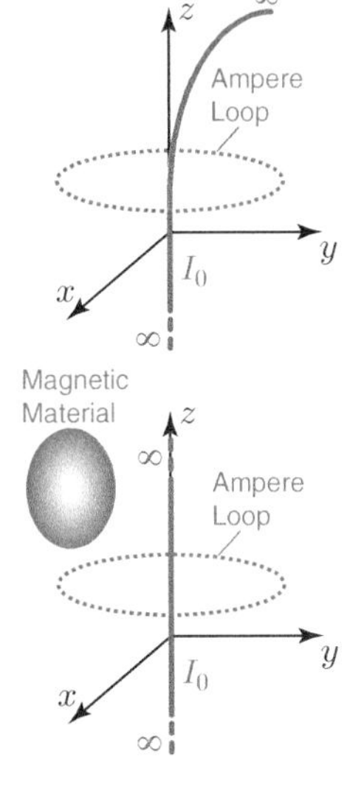

Figure 2.16 Two different examples, where magnetic field intensity cannot be put outside the integral due to the lack of symmetry when using Ampere's law.

Hence, we obtain

$$\int_0^{2\pi} \bar{H}(\bar{r}) \cdot \hat{a}_\phi \rho d\phi = \int_0^{2\pi} H_\phi(\rho)\rho d\phi = 2\pi\rho H_\phi(\rho) = I_0, \tag{2.57}$$

leading to

$$H_\phi(\rho) = \frac{I_0}{2\pi\rho} \tag{2.58}$$

or

$$\bar{H}(\bar{r}) = \hat{a}_\phi \frac{I_0}{2\pi\rho}. \tag{2.59}$$

Therefore, the magnetic field intensity decays with the distance ρ from the wire (Figure 2.15).

Considering a line electric current passing through a circular Ampere loop, several things must be emphasized (Figure 2.16):

- If the electric current is not located symmetrically, we can still write Eq. (2.55), but we are unable to put $\bar{H}(\bar{r})$ outside the integral since it depends on the position on the loop.
- If the wire is curved, the symmetry collapses, and we are again unable to put $\bar{H}(\bar{r})$ outside the integral while Eq. (2.55) is still valid.
- If there is a magnetic material with arbitrary geometry (not symmetric and symmetrically located), the overall symmetry collapses again. Hence, Eq. (2.55) is valid, but $\bar{H}(\bar{r})$ cannot be put outside the integral.

2.2.1 Examples

Example 15: Consider an infinitely long coaxial structure with an inner cylinder of radius a inside a shell defined as $b \leq \rho \leq c$, all located in a vacuum (Figure 2.17). We assume that steady electric currents I_0 (A) and $-I_0$ (A) flow in the cylinder ($\rho < a$) and shell ($b \leq \rho \leq c$), respectively, while they are uniformly distributed. Find the magnetic field intensity $\bar{H}(\bar{r})$ everywhere.

Solution: Considering the symmetry and using Ampere's law on circular loops, we have

$$\oint_C \bar{H}(\bar{r}) \cdot \overline{dl} = 2\pi\rho H_\phi(\rho) = I_{\text{enc}}. \tag{2.60}$$

Now, we need to consider different cases. For $\rho < a$, we have

$$I_{\text{enc}} = \frac{\pi\rho^2}{\pi a^2} I_0 = \frac{\rho^2}{a^2} I_0, \tag{2.61}$$

leading to

$$\bar{H}(\bar{r}) = \hat{a}_\phi \frac{\rho I_0}{2\pi a^2} \qquad (\rho < a). \tag{2.62}$$

In the case of $a < \rho < b$, I_0 flowing in the inner cylinder is completely enclosed by the Ampere loop so that we have

$$\bar{H}(\bar{r}) = \hat{a}_\phi \frac{I_0}{2\pi\rho} \qquad (a < \rho < b). \tag{2.63}$$

We note that this is the same as the magnetic field intensity of a line electric current I_0. Next, we consider the case $b < \rho < c$, where a part of the negative electric current (in the $-z$ direction) in the outer shell is enclosed, in addition to the whole I_0 in the inner cylinder. We have[15]

$$I_{\text{enc}} = I_0 - I_0 \left(\frac{\rho^2 - b^2}{c^2 - b^2}\right) = I_0 \left(\frac{c^2 - \rho^2}{c^2 - b^2}\right), \tag{2.64}$$

leading to

$$\bar{H}(\bar{r}) = \hat{a}_\phi \frac{I_0}{2\pi\rho} \left(\frac{c^2 - \rho^2}{c^2 - b^2}\right) \qquad (b < \rho < c). \tag{2.65}$$

Finally, for $\rho > c$, the effects of the positive and negative electric currents completely cancel each other, and we have

$$\bar{H}(\bar{r}) = 0 \qquad (\rho > c). \tag{2.66}$$

We note that there is no discontinuity in the value of $\bar{H}(\bar{r})$: i.e. it does not jump at $\rho = a$, $\rho = b$, or $\rho = c$ (Figure 2.18). This is because there is no surface electric current density (I_0 and $-I_0$ consist of volume electric current density).

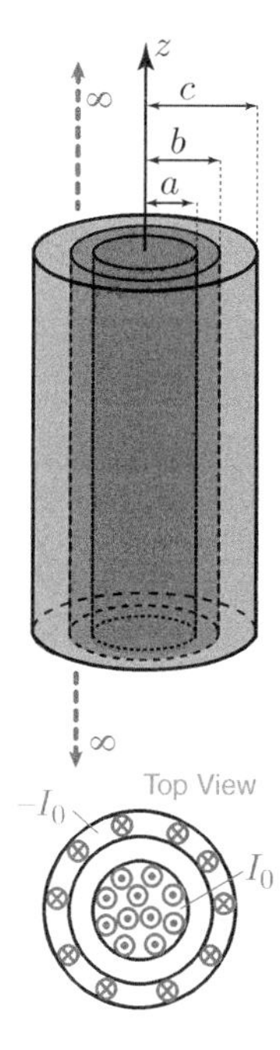

Figure 2.17 A coaxial structure that carries steady electric currents. Electric currents I_0 and $-I_0$ flow in the inner cylinder ($\rho < a$) and in the shell ($b \leq \rho \leq c$), respectively.

[15]Note that the area of a ring with inner/outer radii b/c is $\pi(c^2 - b^2)$.

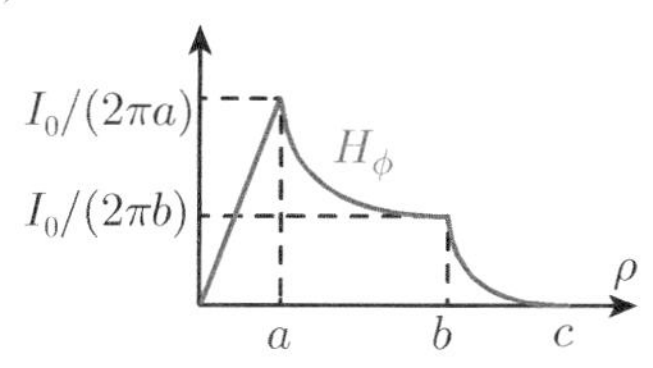

Figure 2.18 The magnitude (ϕ component) of the magnetic field intensity with respect to ρ for the coaxial structure in Figure 2.17.

Example 16: Consider an infinitely long ideal cylindrical inductor (Figure 2.19) involving N_l turns/m and steady electric current I_0 (A). Find the magnetic field intensity $\bar{H}(\bar{r})$ at the center line of the inductor.

Solution: To solve this problem via the integral form of Ampere's law, we consider a frame-like (rectangular) Ampere loop enclosing a part of the electric current (Figure 2.20). The right-hand side of the Ampere loop passes through the center of the inductor, where we look

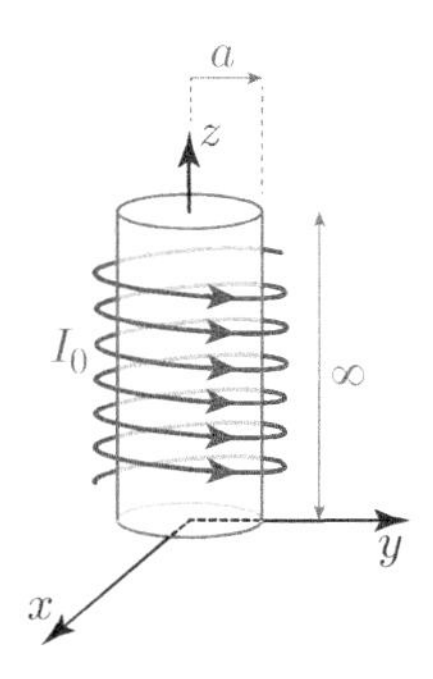

Figure 2.19 An infinitely long ideal cylindrical inductor.

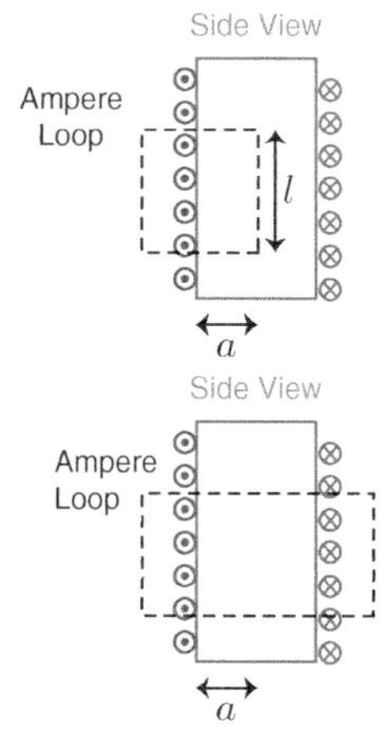

Figure 2.20 Ampere loops to find the magnetic field intensity for an ideal inductor.

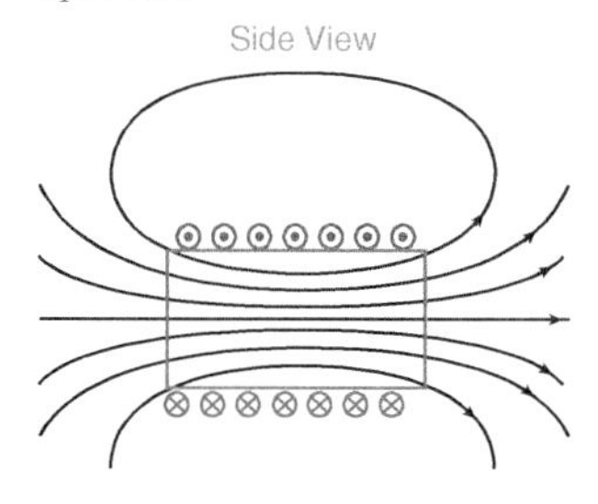

Figure 2.21 The magnetic field created by a cylindrical inductor tends to deviate from the central axis toward the ends of the inductor. In fact, the magnetic field lines close upon themselves, as shown for one of them.

for the value of $\bar{H}(\bar{r})$. Given its length l, we have

$$\oint_C \bar{H}(\bar{r}) \cdot \overline{dl} = I_0 N_l l, \tag{2.67}$$

where $I_0 N_l l$ represents the total enclosed electric current. At this stage, we need assumptions in order to simplify the integral and find an expression for $\bar{H}(\bar{r})$. Since the inductor is ideal and infinitely long, and also considering the direction of the electric current, the magnetic field intensity can be considered to be in the z direction,[16] while it should not depend on z. Therefore, in the circulation of $\bar{H}(\bar{r})$, there is no contribution from the horizontal lines, leading to

$$lH_{\text{center}} - lH_{\text{outside}} = I_0 N_l l \tag{2.68}$$

or

$$H_{\text{center}} - H_{\text{outside}} = I_0 N_l. \tag{2.69}$$

But what about the magnetic field intensity outside? In the case of an ideal capacitor, we know that the electric flux density is zero outside, as discussed in Section 1.6.1. Similarly, for an ideal inductor, the magnetic field intensity is zero outside the inductor (see Figure 2.21 for a non-ideal case). This can be shown by considering a larger and symmetrically located Ampere loop enclosing both outward and inward electric currents, leading to zero net electric current flowing through the bounded surface (Figure 2.20). Consequently, we simply obtain

$$H_{\text{center}} = I_0 N_l \tag{2.70}$$

or more formally

$$\bar{H}(0, 0, z) = \hat{a}_z I_0 N_l. \tag{2.71}$$

If the center of the inductor is a vacuum, we also have

$$\bar{B}(0, 0, z) = \hat{a}_z \mu_0 I_0 N_l. \tag{2.72}$$

If a magnetic material with $\mu > \mu_0$ is inserted inside the inductor while keeping the electric current the same (I_0), we have

$$\bar{B}(0, 0, z) = \hat{a}_z \mu I_0 N_l, \tag{2.73}$$

which means that the magnetic flux density (hence magnetic flux inside the inductor) increases. Alternatively, if the magnetic flux (hence the magnetic flux density $B(\bar{r}) = \hat{a}_z \mu_0 I_0 N_l$) inside the inductor is kept constant, the magnetic field intensity should change as $\bar{H}(\bar{r}) : \hat{a}_z I_0 N_l \rightarrow \hat{a}_z I_0 N_l / \mu$, indicating that the electric current must change (decrease) as $I_0 \rightarrow I_0/\mu$.

[16] This is similar to assuming no fringing electric field for an ideal capacitor.

Example 17: Consider a steady electric current I_0 (A) circulating around a toroid of radius b and cylindrical radius a (Figure 2.22). There are a total of N turns. Find the magnetic field intensity everywhere.

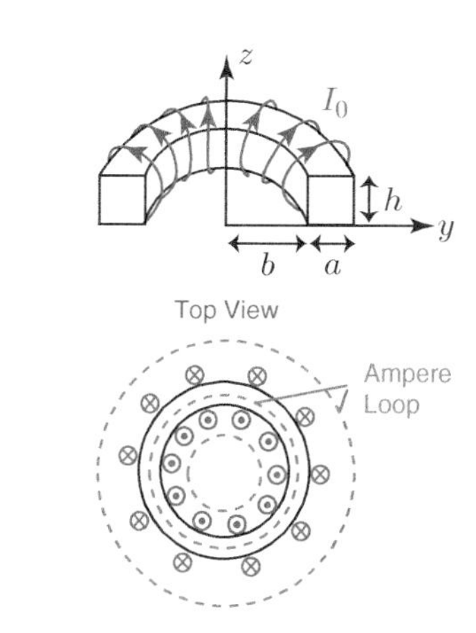

Figure 2.22 A toroid structure and an Ampere loop with different radii to find the magnetic field intensity.

Solution: Using the symmetry, we can use the integral form of Ampere's law to solve this problem. We consider a circular Ampere loop that is oriented similar to the toroid. If the radius of Ampere loop is ρ, we have

$$\oint_C \bar{H}(\bar{r}) \cdot \overline{dl} = I_{\text{enc}}, \tag{2.74}$$

where I_{enc} depends on the size of the curve. At this stage, we assume that $\bar{H}(\bar{r})$ only depends on ρ and is in the ϕ direction. Then for $\rho < b$ or $\rho > b + a$, we have $I_{\text{enc}} = 0$, leading to $\bar{H}(\bar{r}) = 0$. Otherwise – i.e. when $b < \rho < b + a$ – we obtain

$$\oint_C \bar{H}(\bar{r}) \cdot \overline{dl} = 2\pi\rho H_\phi(\rho) = NI_0 \qquad (b < \rho < b + a) \tag{2.75}$$

or

$$\bar{H}(\bar{r}) = \hat{a}_\phi \frac{NI_0}{2\pi\rho} \qquad (b < \rho < b + a). \tag{2.76}$$

We note that the magnetic field intensity is proportional to the number of turns: i.e. given a constant current I_0, it can be increased by using more turns.

2.3 Differential Form of Ampere's Law

Similar to Gauss' law, Ampere's law has a differential form. It can be written[17]

$$\bar{\nabla} \times \bar{H}(\bar{r}, t) = \bar{J}_v(\bar{r}, t) + \frac{\partial \bar{D}(\bar{r}, t)}{\partial t}, \tag{2.77}$$

where $\bar{\nabla} \times \bar{f}(\bar{r})$ represents the curl of a vector field $\bar{f}(\bar{r})$, and $\bar{J}_v(\bar{r}, t)$ is the volume electric current density (A/m^2). Like the differential form of Gauss' law, the differential form of Ampere's law provides a pointwise relationship between the magnetic field intensity, the volume electric current density, and the time derivative of the electric flux density. Specifically, it describes how magnetic fields are created by electric currents and changing electric fields. In a homogeneous medium, where μ does not depend on the position, we further have

$$\bar{\nabla} \times \bar{B}(\bar{r}, t) = \mu \bar{J}_v(\bar{r}, t) + \mu \frac{\partial \bar{D}(\bar{r}, t)}{\partial t}. \tag{2.78}$$

[17] Similar to the differential form of Gauss' law, the differential form of Ampere's law contains the volume density of the electric current (not the surface density or the electric current itself).

Once again, we list the tools and concepts to understand the differential form of Ampere's law:

- Electric current density
- Cross product
- Curl of vector fields
- Stokes' theorem

These are in addition to some of the previous tools: i.e. differential surface with direction (Section 1.1.1), dot product (Section 1.1.2), flux of vector fields (Section 1.1.3), divergence of vector fields (Section 1.3.2), differential length with direction (Section 2.1.1), and circulation of vector fields (Section 2.1.2).

2.3.1 Electric Current Density

As mentioned in Section 0.5, the electric current is important to describe magnetic interactions. As described by Ampere's law, magnetic fields are created by electric currents. Similar to the definition of electric fields for the interpretation of electric forces (measurable quantities) between electrically charged bodies, magnetic fields are defined to interpret magnetic forces (measurable quantities) between electric-current-carrying bodies.

The integral form of Ampere's law in Eq. (2.1) directly uses the electric current, while the differential form in Eq. (2.77) involves the volume electric current density $\bar{J}_v(\bar{r},t)$ as well as the displacement current density:

$$\bar{J}_{vd}(\bar{r},t) = \frac{\partial \bar{D}(\bar{r},t)}{\partial t} \tag{2.79}$$

that is also a kind of volume electric current density. A volume electric current density can be written (Figure 2.23)

$$\bar{J}_v(\bar{r},t) = \hat{a}_J \lim_{\Delta s \to 0} \frac{\Delta I}{\Delta s}, \tag{2.80}$$

where $\hat{a}_J$ is the vector direction of the electric current flow and ΔI is the total electric current flowing through a differential surface Δs as its area goes to zero. We note that, even though

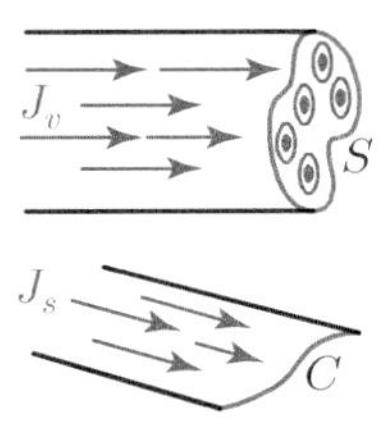

Figure 2.23 Volume electric current density (electric current flowing in nonzero volume) has a unit of A/m^2, and its surface integral (over S) gives the total electric current. Surface electric current density (electric current flowing on surface) has a unit of A/m, and its line integral (along C) corresponds to the electric current.

it is called volume electric current density (since the flow is confined in a volume), the total electric current through a surface S is the flux (surface integral) of the volume electric current density (Figure 2.24):

$$I(t) = \int_S \bar{J}_v(\bar{r}, t) \cdot \overline{ds} = \int_S \bar{J}_v(\bar{r}, t) \cdot \hat{a}_n ds, \tag{2.81}$$

where $\hat{a}_n$ is the unit normal to the surface.

If an electric current is flowing over a surface with zero thickness, a proper definition of the electric current density can be (Figure 2.23)

$$\bar{J}_s(\bar{r}, t) = \hat{a}_J \lim_{\Delta l \to 0} \frac{\Delta I}{\Delta l}, \tag{2.82}$$

where ΔI is the total electric current flowing through a differential length Δl as it goes to zero. To find the total electric current crossing a curve C by using the surface electric current density $\bar{J}_s(\bar{r}, t)$, we need a line integral as

$$I(t) = \int_C \bar{J}_s(\bar{r}, t) \cdot \hat{a}_n dl, \tag{2.83}$$

where $\hat{a}_n$ is the unit normal to the curve (Figure 2.24). If there is a surface electric current density $\bar{J}_s(\bar{r}, t)$, it can be seen as a volume density $\bar{J}_v(\bar{r}, t) = \bar{J}_s(\bar{r}, t)\delta(u)$, where u is a coordinate variable normal to the surface on which the electric current is flowing and δ is the Dirac delta function. While Ampere's law could be used for this kind of expression to evaluate the magnetic field due to a surface electric current density, there is again a better way to deal with surface quantities: i.e. the boundary conditions that are derived by using Ampere's law at boundaries.

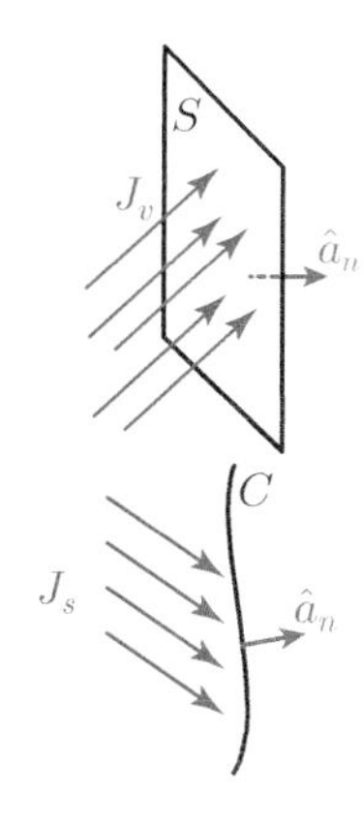

Figure 2.24 To find the total electric current, the surface integral of the volume electric current density or the line integral of the surface electric current density is needed, while considering the direction of the normal to the surface/line.

J_s: surface electric current density
Unit: amperes/meter (A/m)

2.3.2 Cross Product

So far, we have seen the dot product as the only type of multiplication between two vectors. The divergence operation that is mainly used in Gauss' law is also related to the dot product: i.e. it can be interpreted as a dot product between $\bar{\nabla}$ operator and a vector field at all locations. In Ampere's law, we have another kind of operation: the curl operation, which can be interpreted as a cross product between $\bar{\nabla}$ operator and a vector field, again at all locations. In this subsection, we consider the cross product of two vectors in detail before discussing the curl operation.

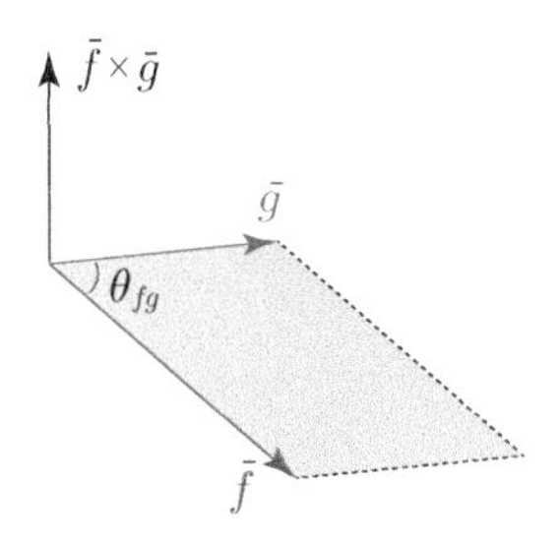

Figure 2.25 A geometric demonstration of the cross product of two vectors $\bar{f}$ and $\bar{g}$.

The cross product of two vectors is another vector that is perpendicular to both vectors and has a magnitude equal to the area of the parallelogram (Figure 2.25) formed by the vectors,[18]

$$\bar{f} \times \bar{g} = \begin{vmatrix} \hat{a}_x & \hat{a}_y & \hat{a}_z \\ f_x & f_y & f_z \\ g_x & g_y & g_z \end{vmatrix} \tag{2.84}$$

$$= \hat{a}_x(f_y g_z - f_z g_y) + \hat{a}_y(f_z g_x - f_x g_z) + \hat{a}_z(f_x g_y - f_y g_x) \tag{2.85}$$

$$= \hat{a}_n |\bar{f}||\bar{g}| \sin\theta_{fg}, \tag{2.86}$$

where $\hat{a}_n \cdot \bar{f} = 0$, $\hat{a}_n \cdot \bar{g} = 0$, and $\theta_{fg} \in [0°, 180°]$ is the angle[19] between $\bar{f}$ and $\bar{g}$. Note that the area $|\bar{f}||\bar{g}|\sin\theta_{AB}$ cannot be negative, while the direction of $\hat{a}_n$ follows the right-hand rule (Figure 2.26). Therefore, we have

$$\bar{f} \times \bar{g} = 0 \quad \text{if } \theta_{fg} = 0° \text{ or } \theta_{fg} = 180° \tag{2.87}$$

$$\bar{f} \times \bar{g} = -\bar{g} \times \bar{f}. \tag{2.88}$$

As indicated in Eq. (2.87), the cross product of two vectors is zero if the vectors are aligned: i.e. when the parallelogram formed by the two vectors vanishes. We also note that

$$\hat{a}_x \times \hat{a}_y = -\hat{a}_y \times \hat{a}_x = \hat{a}_z \tag{2.89}$$

$$\hat{a}_y \times \hat{a}_z = -\hat{a}_z \times \hat{a}_y = \hat{a}_x \tag{2.90}$$

$$\hat{a}_z \times \hat{a}_x = -\hat{a}_x \times \hat{a}_z = \hat{a}_y, \tag{2.91}$$

which can be memorized by following a cycle (Figure 2.27). Similar relations exist for other coordinate systems:

$$\hat{a}_\rho \times \hat{a}_\phi = -\hat{a}_\phi \times \hat{a}_\rho = \hat{a}_z \tag{2.92}$$

$$\hat{a}_\phi \times \hat{a}_z = -\hat{a}_z \times \hat{a}_\phi = \hat{a}_\rho \tag{2.93}$$

$$\hat{a}_z \times \hat{a}_\rho = -\hat{a}_\rho \times \hat{a}_z = \hat{a}_\phi, \tag{2.94}$$

for a cylindrical system and

$$\hat{a}_R \times \hat{a}_\theta = -\hat{a}_\theta \times \hat{a}_R = \hat{a}_\phi \tag{2.95}$$

$$\hat{a}_\theta \times \hat{a}_\phi = -\hat{a}_\phi \times \hat{a}_\theta = \hat{a}_R \tag{2.96}$$

$$\hat{a}_\phi \times \hat{a}_R = -\hat{a}_R \times \hat{a}_\phi = \hat{a}_\theta \tag{2.97}$$

for a spherical system.

[18] As indicated in this formulation, one of the most basic ways to show the cross product is the determinant of a 3×3 matrix, where the first row contains the unit vectors, the second row contains the components of the first vector, and the third row contains the components of the second vector. On the other hand, considering orthogonal coordinate systems, where the cross product of two unit vectors leads to third unit vector, it may be easier to perform the product by considering the components one by one.

[19] Note what these expressions indicate: zero dot products show that $\bar{f} \times \bar{g}$ is perpendicular to both $\bar{f}$ and $\bar{g}$. One can find only two such directions; the correct one is obtained by using the counterclockwise direction (right-hand rule).

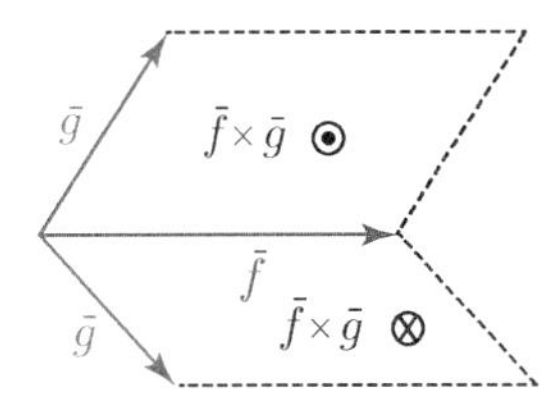

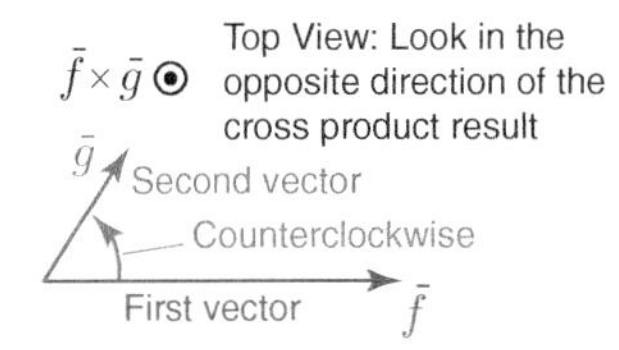

Figure 2.26 Cross products of vectors, where the results are vectors out from the paper or into the paper. The direction is determined by considering the counterclockwise behavior (right-hand rule).

Similar to the dot product, the cross product does not have an inverse operation (division). This is because, given $\bar{h} = \bar{f} \times \bar{g}$ and $\bar{f}$, there are infinitely many vectors that satisfy the cross product result and one cannot find a unique $\bar{g}$. On the other hand, as opposed to the dot product, the result of the cross product is a vector so that multiple cross products can be used consecutively. In fact, the cross products of three vectors have a famous identity, known as *BAC-CAB*:[20]

$$\bar{f} \times (\bar{g} \times \bar{h}) = \bar{g}(\bar{f} \cdot \bar{h}) - \bar{h}(\bar{f} \cdot \bar{g}). \tag{2.98}$$

We note that the parentheses must be used carefully in consecutive cross products. For example,

$$(\bar{f} \times \bar{g}) \times \bar{h} = -\bar{h} \times (\bar{f} \times \bar{g}) = -\bar{f}(\bar{h} \cdot \bar{g}) + \bar{g}(\bar{h} \cdot \bar{f}) \tag{2.99}$$

$$\neq \bar{f} \times (\bar{g} \times \bar{h}). \tag{2.100}$$

When two of the vectors in consecutive cross products are the same, we have

$$\bar{f} \times (\bar{f} \times \bar{h}) = \bar{f}(\bar{f} \cdot \bar{h}) - \bar{h}(\bar{f} \cdot \bar{f}) = \bar{f}(\bar{f} \cdot \bar{h}) - |\bar{f}|^2\bar{h}. \tag{2.101}$$

Furthermore, if the first vector is a unit vector – i.e. $\bar{f} = \hat{a}_f$ – we further have $|\bar{f}| = 1$ and

$$-\hat{a}_f \times (\hat{a}_f \times \bar{h}) = \bar{h} - \hat{a}_f(\hat{a}_f \cdot \bar{h}). \tag{2.102}$$

The final equality can be interpreted as follows:[21] a component of a given vector ($\bar{h}$) in a direction (i.e. the dot product operation $\hat{a}_f \cdot \bar{h}$) can be removed (i.e. the subtraction of the term $-\hat{a}_f(\hat{a}_f \cdot \bar{h})$) by evaluating two cross products with the related unit vector[22] (Figure 2.28).

Since the cross product is related to a surface that is perpendicular to two vectors involved in the product, it provides a new insight on the differential lengths and surfaces with directions. Let $\overline{dl}_1$, $\overline{dl}_2$, and $\overline{dl}_3$ be the basic differential lengths in a coordinate system with variables u_1, u_2, and u_3. Then the differential surfaces can be found:[23]

- On a constant u_1 surface: $\overline{ds}_1 = \overline{dl}_2 \times \overline{dl}_3$
- On a constant u_2 surface: $\overline{ds}_2 = \overline{dl}_3 \times \overline{dl}_1$
- On a constant u_3 surface: $\overline{ds}_3 = \overline{dl}_1 \times \overline{dl}_2$

For example, in a Cartesian system, we have $\overline{dl}_1 = \hat{a}_x dx$, $\overline{dl}_2 = \hat{a}_y dy$, $\overline{dl}_3 = \hat{a}_z dz$, leading to

$$\overline{ds}_1 = \hat{a}_y dy \times \hat{a}_z dz = \hat{a}_x dydz \tag{2.103}$$

$$\overline{ds}_2 = \hat{a}_z dz \times \hat{a}_x dx = \hat{a}_y dxdz \tag{2.104}$$

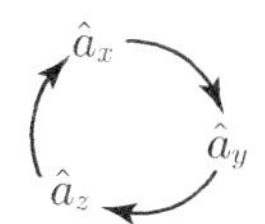

Figure 2.27 A cycle of unit vectors for the correct order in their cross products.

[20]This name is meaningful when $\bar{f} = \bar{A}$, $\bar{g} = \bar{B}$, $\bar{h} = \bar{C}$; but we reserve A and B for representing two important vector fields.

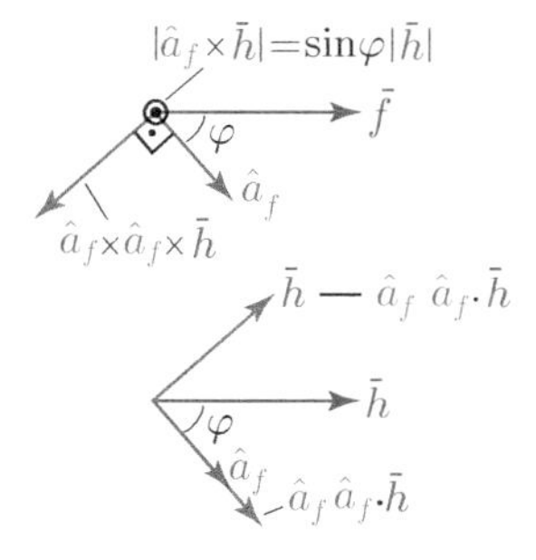

Figure 2.28 Taking a cross product of a vector $\bar{h}$ twice with $\hat{a}_f$ removes a component of $\bar{h}$ in the direction of $\hat{a}_f$. Note that there is also a negative sign.

[21]Note that, $\hat{a}_f(\hat{a}_f \cdot \bar{h})$ is a vector in the direction of $\hat{a}_f$, whose magnitude is $|\hat{a}_f \cdot \bar{h}|$: i.e. $|\bar{h}|\cos\varphi$, where φ is the angle between $\hat{a}_f$ and $\bar{h}$. Obviously, if $\hat{a}_f \perp \bar{h}$, then $\hat{a}_f \cdot \bar{h} = 0$ and $-\hat{a}_f \times (\hat{a}_f \times \bar{h}) = \bar{h}$.

[22]Interestingly, taking a third cross product with a negative sign, we have

$$-\hat{a}_f \times [\hat{a}_f \times (\hat{a}_f \times \bar{h})] = \hat{a}_f \times \bar{h}.$$

[23]Note the cyclic relationship:

$$1 \times 2 \rightarrow 3$$
$$2 \times 3 \rightarrow 1$$
$$3 \times 1 \rightarrow 2.$$

$$\overline{ds}_3 = \hat{a}_x dx \times \hat{a}_y dy = \hat{a}_z dxdy. \tag{2.105}$$

Similarly, for a cylindrical coordinate system, we have $\overline{dl}_1 = \hat{a}_\rho d\rho$, $\overline{dl}_2 = \hat{a}_\phi \rho d\phi$, and $\overline{dl}_3 = \hat{a}_z dz$, leading to

$$\overline{ds}_1 = \hat{a}_\phi \rho d\phi \times \hat{a}_z dz = \hat{a}_\rho \rho d\phi dz \tag{2.106}$$

$$\overline{ds}_2 = \hat{a}_z dz \times \hat{a}_\rho d\rho = \hat{a}_\phi d\rho dz \tag{2.107}$$

$$\overline{ds}_3 = \hat{a}_\rho d\rho \times \hat{a}_\phi \rho d\phi = \hat{a}_z \rho d\rho d\phi. \tag{2.108}$$

Finally, for a spherical coordinate system, using $\overline{dl}_1 = \hat{a}_R dR$, $\overline{dl}_2 = \hat{a}_\theta R d\theta$, and $\overline{dl}_3 = \hat{a}_\phi R \sin\theta d\phi$, we obtain

$$\overline{ds}_1 = \hat{a}_\theta R d\theta \times \hat{a}_\phi R \sin\theta d\phi = \hat{a}_R R^2 \sin\theta d\theta d\phi \tag{2.109}$$

$$\overline{ds}_2 = \hat{a}_\phi R \sin\theta d\phi \times \hat{a}_R dR = \hat{a}_\theta R \sin\theta dR d\phi \tag{2.110}$$

$$\overline{ds}_3 = \hat{a}_R dR \times \hat{a}_\theta R d\theta = \hat{a}_\phi R dR d\theta. \tag{2.111}$$

A differential volume in different coordinate systems can also be written as a product of differential lengths.[24]

[24] A triple product involving a dot product and a cross product can be defined as follows. Let $\bar{f}$, $\bar{g}$, and $\bar{h}$ be three nonzero vectors. Then we have

$$\bar{f} \cdot (\bar{g} \times \bar{h}) = \bar{g} \cdot (\bar{h} \times \bar{f}) = \bar{h} \cdot (\bar{f} \times \bar{g})$$

and $|\bar{f} \cdot (\bar{g} \times \bar{h})|$ is the volume of the solid formed by $\bar{f}$, $\bar{g}$, and $\bar{h}$ (Figure 2.29). If $\bar{f} \cdot \bar{g} \times \bar{h}$ is given without parenthesis, it must be understood that the dot product is evaluated after the cross product since $(\bar{f} \cdot \bar{g}) \times \bar{h}$ is not defined. As expected, $\bar{f} \cdot (\bar{g} \times \bar{f}) = 0$ since there is no solid formed by this combination. Using the geometric definition of the volume above, a differential volume in a coordinate system is defined as

$$dv = |\overline{dl}_1 \cdot \overline{dl}_2 \times \overline{dl}_3|.$$

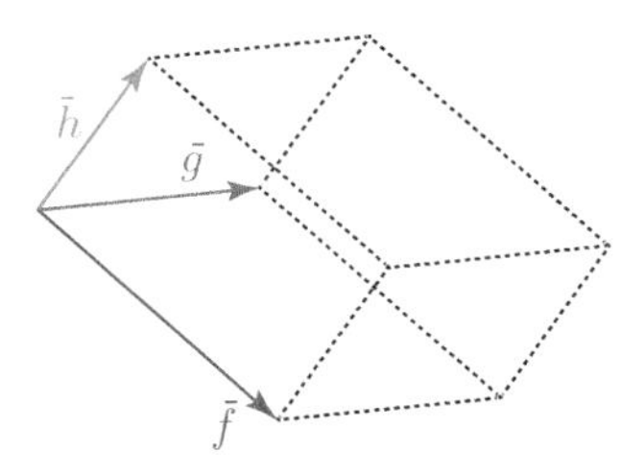

Figure 2.29 A solid formed by three vectors, whose volume equals to the absolute value of $\bar{f} \cdot (\bar{g} \times \bar{h})$.

2.3.3 Curl of Vector Fields

Given a vector field $\bar{f}(\bar{r}, t)$, $\bar{g}(\bar{r}, t) = \bar{\nabla} \times \bar{f}(\bar{r}, t)$ is a vector field defined as

$$\bar{g}(\bar{r}, t) = \begin{vmatrix} \hat{a}_x & \hat{a}_y & \hat{a}_z \\ \partial/\partial x & \partial/\partial y & \partial/\partial z \\ f_x(\bar{r}, t) & f_y(\bar{r}, t) & f_z(\bar{r}, t) \end{vmatrix}, \tag{2.112}$$

which leads to

$$g_x(\bar{r}, t) = \frac{\partial f_z(\bar{r}, t)}{\partial y} - \frac{\partial f_y(\bar{r}, t)}{\partial z} \tag{2.113}$$

$$g_y(\bar{r}, t) = \frac{\partial f_x(\bar{r}, t)}{\partial z} - \frac{\partial f_z(\bar{r}, t)}{\partial x} \tag{2.114}$$

$$g_z(\bar{r}, t) = \frac{\partial f_y(\bar{r}, t)}{\partial x} - \frac{\partial f_x(\bar{r}, t)}{\partial y}. \tag{2.115}$$

Using the definition of the del operator, the curl of a vector field can be interpreted as a cross product:

$$\bar{\nabla} \times \bar{f}(\bar{r}, t) = \left(\hat{a}_x \frac{\partial}{\partial x} + \hat{a}_y \frac{\partial}{\partial y} + \hat{a}_z \frac{\partial}{\partial z} \right) \times \bar{f}(\bar{r}, t). \tag{2.116}$$

The curl of a vector field is a kind of measurement of how much the field is changing. On the other hand, as opposed to the divergence, which measures changes in the field direction, the curl is related to the *rotational* behavior (Figure 2.30). For example, consider $\bar{f}(\bar{r}) = \hat{a}_x x$, whose curl is zero. This vector field changes only in the x direction, while it is constant in the y and z directions. Since the field itself is in the x direction, its curl is zero. On the other hand, consider now $\bar{f}(\bar{r}) = x\hat{a}_y$. We have $\bar{\nabla} \times \bar{f}(\bar{r}) = \hat{a}_z$, indicating a change in $\bar{f}(\bar{r})$ perpendicular to its direction. Note that $\bar{\nabla} \cdot \bar{f}(\bar{r}) = 0$, and in this case, the divergence does not give any information about the variation (and sources) of $\bar{f}(\bar{r})$.

Similar to the divergence, the curl of a vector field has a physical meaning. In general, to change a vector field $\bar{f}(\bar{r})$, we need some kind of a source. This source is not the same as the one that creates nonzero divergence. Instead, it creates a rotational behavior. This relationship describes how magnetic fields are related to electric and displacement currents. Specifically, the relationship between magnetic fields and electric currents is established via the curl operation, indicating the rotational creation of magnetic fields from their sources.

What about the direction of the curl? To understand this, consider a vector field $\bar{f}(\bar{r}) = -\hat{a}_x y + \hat{a}_y x$ whose rotational behavior is more obvious (Figure 2.31). We have

$$\bar{\nabla} \times \bar{f}(\bar{r}) = \hat{a}_z 2, \tag{2.117}$$

i.e. a constant curl in the z direction. The constant value of 2 indicates that the *rate* of the rotation is the same everywhere. On the other hand, the direction z represents the rotation direction via the right-hand rule (Figure 2.1): i.e. using the right hand, if the thumb points at the curl direction, then the fingers show how the field is rotating.

The curl operation is relatively easy to define in a Cartesian coordinate system, as shown in Eqs. (2.112)–(2.115). In a cylindrical or a spherical system, however, one needs to consider the change of unit vectors with respect to the position. Considering all dependencies in a cylindrical system, we have

$$\bar{\nabla} \times \bar{f}(\bar{r}, t) = \frac{1}{\rho} \begin{vmatrix} \hat{a}_\rho & \hat{a}_\phi \rho & \hat{a}_z \\ \partial/\partial\rho & \partial/\partial\phi & \partial/\partial z \\ f_\rho(\bar{r}, t) & \rho f_\phi(\bar{r}, t) & f_z(\bar{r}, t) \end{vmatrix} \tag{2.118}$$

$$= \frac{1}{\rho}\hat{a}_\rho \left(\frac{\partial f_z(\bar{r}, t)}{\partial \phi} - \frac{\partial}{\partial z}[\rho f_\phi(\bar{r}, t)] \right) + \frac{1}{\rho}\hat{a}_\phi \rho \left(\frac{\partial f_\rho(\bar{r}, t)}{\partial z} - \frac{\partial f_z(\bar{r}, t)}{\partial \rho} \right)$$

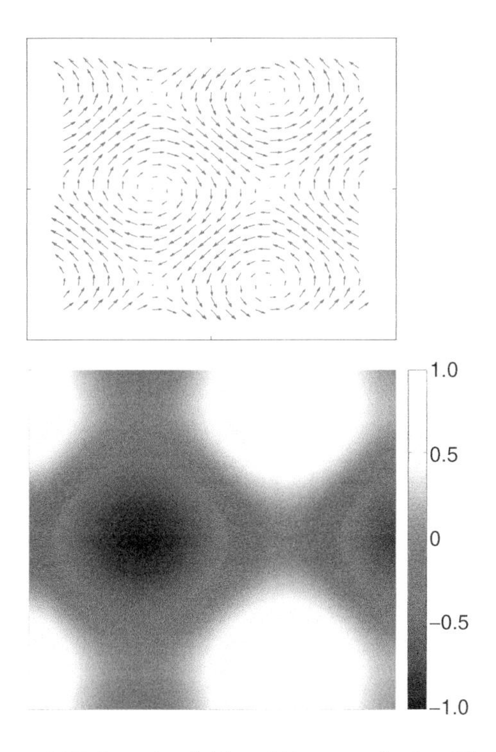

Figure 2.30 A vector field and the magnitude of its curl (colors).

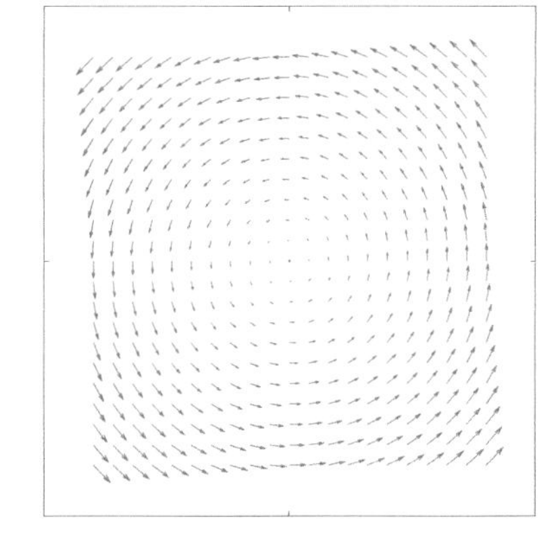

Figure 2.31 A vector field $\bar{f}(\bar{r}) = -\hat{a}_x y + \hat{a}_y x$.

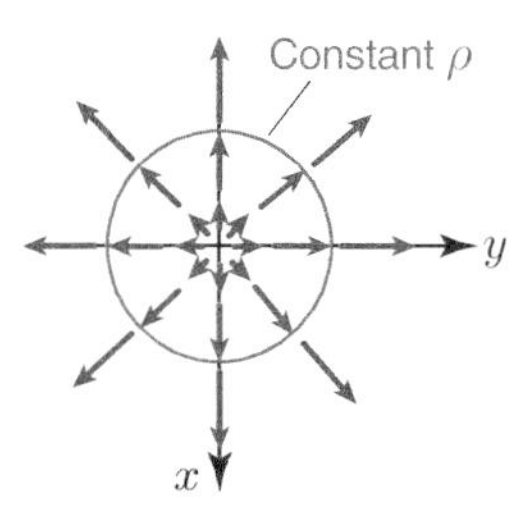

Figure 2.32 A vector field $\bar{f}(\bar{r}) = \hat{a}_\rho \rho$, shown on the x-y plane. This vector field does not possess a rotational behavior.

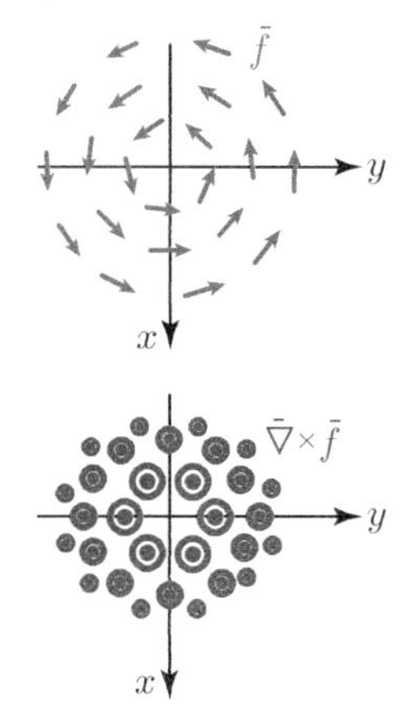

Figure 2.33 A vector field $\bar{f}(\bar{r}) = \hat{a}_\phi$, shown on the x-y plane. As opposed to a vector field $\hat{a}_\phi/\rho$, which is produced by a single source located on the z axis, this vector field can be created by an infinite source distribution represented by $\bar{\nabla} \times \bar{f}(\bar{r}) = \hat{a}_z/\rho$. We note that the magnitude of $\bar{f}(\bar{r})$ is constant, while its curl is nonzero everywhere.

$$+\frac{1}{\rho}\hat{a}_z\left(\frac{\partial}{\partial\rho}[\rho f_\phi(\bar{r},t)] - \frac{\partial f_\rho(\bar{r},t)}{\partial\phi}\right) \tag{2.119}$$

$$= \hat{a}_\rho\left(\frac{1}{\rho}\frac{\partial f_z(\bar{r},t)}{\partial\phi} - \frac{\partial f_\phi(\bar{r},t)}{\partial z}\right) + \hat{a}_\phi\left(\frac{\partial f_\rho(\bar{r},t)}{\partial z} - \frac{\partial f_z(\bar{r},t)}{\partial\rho}\right)$$
$$+\frac{1}{\rho}\hat{a}_z\left(\frac{\partial}{\partial\rho}[\rho f_\phi(\bar{r},t)] - \frac{\partial f_\rho(\bar{r},t)}{\partial\phi}\right). \tag{2.120}$$

The final expression is quite lengthy, but we can consider some special cases to see how it works. First, we consider a vector field $\bar{f}(\bar{r}) = \hat{a}_\rho \rho$ (Figure 2.32). Considering that $f_\rho(\bar{r}) = \rho$, $f_\phi(\bar{r}) = 0$, and $f_z(\bar{r}) = 0$, we obtain

$$\bar{\nabla} \times \bar{f}(\bar{r}) = \hat{a}_\phi\left(\frac{\partial f_\rho(\bar{r})}{\partial z}\right) + \frac{1}{\rho}\hat{a}_z\left(-\frac{\partial f_\rho(\bar{r})}{\partial\phi}\right) = 0. \tag{2.121}$$

Hence, this vector field does not have any rotational behavior. Next, we consider $\bar{f}(\bar{r}) = \hat{a}_\phi$ (Figure 2.33). Considering that $f_\rho(\bar{r}) = 0$, $f_\phi(\bar{r}) = 1$, and $f_z(\bar{r}) = 0$, we have

$$\bar{\nabla} \times \bar{f}(\bar{r}) = \hat{a}_\rho\left(-\frac{\partial f_\phi(\bar{r})}{\partial z}\right) + \frac{1}{\rho}\hat{a}_z\left(\frac{\partial}{\partial\rho}[\rho f_\phi(\bar{r})]\right) = \frac{1}{\rho}\hat{a}_z\frac{\partial\rho}{\partial\rho} = \hat{a}_z\frac{1}{\rho}. \tag{2.122}$$

Even though the magnitude of $\bar{f}(\bar{r})$ is constant 1, its curl is nonzero since this vector field possesses a rotational behavior that is the same everywhere. Specifically, in order to create this rotational vector field, there should be a type of source that can lead a constant rotational behavior everywhere in space. The result of the curl operation indicates that this is possible if the *whole* space is filled with a source distribution of magnitude $1/\rho$. Note that, checking the divergence of the same vector field, we have

$$\bar{\nabla} \cdot \bar{f}(\bar{r}) = \frac{1}{\rho}\frac{\partial}{\partial\rho}\left(\rho f_\rho(\bar{r})\right) + \frac{1}{\rho}\frac{\partial f_\phi(\bar{r})}{\partial\phi} + \frac{\partial f_z(\bar{r})}{\partial z} = \frac{1}{\rho}\frac{\partial f_\phi(\bar{r})}{\partial\phi} = 0, \tag{2.123}$$

indicating that the vector field $\bar{f}(\bar{r}) = \hat{a}_\phi$ is created *only* by rotational sources.[25]

What about a vector field $\bar{f}(\bar{r}) = \hat{a}_\rho \rho + \hat{a}_\phi$, which is a combination of the vector fields in the previous two examples? In this case, we have[26]

$$\bar{\nabla} \times \bar{f}(\bar{r}) = \hat{a}_z\frac{1}{\rho} \tag{2.124}$$

and

$$\bar{\nabla} \cdot \bar{f}(\bar{r}) = \frac{1}{\rho}\frac{\partial}{\partial\rho}\left(\rho^2\right) = 2. \tag{2.125}$$

[25] In this context, rotational sources do not mean the sources themselves have rotational behaviors. They are sources that can create rotational fields, such as a static line electric current flow that creates a rotational magnetic field.

[26] In general, $\bar{\nabla} \times [\bar{f}(\bar{r}) + \bar{g}(\bar{r})] = \bar{\nabla} \times \bar{f}(\bar{r}) + \bar{\nabla} \times \bar{g}(\bar{r})$ for any two vector fields $\bar{f}(\bar{r})$ and $\bar{g}(\bar{r})$.

Hence, this vector field is created by two different kinds of sources. The first source distribution, which can be found by the curl operation $\bar{\nabla} \times \bar{f}(\bar{r})$, creates the rotational behavior in $\bar{f}(\bar{r})$: i.e. $\bar{f}_{df}(\bar{r}) = \hat{a}_{\phi}$. On the other side, the second source distribution, which can be found by the divergence operation $\bar{\nabla} \cdot \bar{f}(\bar{r})$, creates the non-rotational part of $\bar{f}(\bar{r})$: i.e. $\bar{f}_{cf}(\bar{r}) = \hat{a}_{\rho}\rho$. It is remarkable that the two different parts of the vector field $\bar{f}(\bar{r})$ are clearly *decomposed*. Specifically, we have

$$\bar{f}(\bar{r}) = \bar{f}_{df}(\bar{r}) + \bar{f}_{cf}(\bar{r}), \tag{2.126}$$

where $\bar{f}_{df}(\bar{r}) = \hat{a}_{\phi}$ is the divergence-free part:

$$\bar{\nabla} \cdot \bar{f}_{df}(\bar{r}) = 0, \tag{2.127}$$

while $\bar{f}_{cf}(\bar{r}) = \hat{a}_{\rho}\rho$ is the curl-free part:

$$\bar{\nabla} \times \bar{f}_{cf}(\bar{r}) = 0. \tag{2.128}$$

This type of decomposition of a vector field is called *Helmholtz decomposition*.[27,28] In dynamic cases, electric fields have both types of sources: i.e. electric charges that create curl-free parts of electric fields (as formulated by Gauss' law) and time-variant magnetic fields that create divergence-free parts of electric fields (as formulated by Faraday's law).

At this stage, we may further consider a vector field

$$\bar{f}(\bar{r}) = \hat{a}_{\phi}/\rho, \tag{2.129}$$

which represents the magnetic field due to an infinitely long steady electric current on the z axis (line) in a vacuum. Since this vector field has a rotational behavior, one may think that its curl should be nonzero. But we should keep in mind that the curl is a *pointwise* quantity and gives the source at each point where it is evaluated. When there is no source at a particular point, the curl operation should give zero, while the vector field can be created by a source at another location where the curl is not zero. For this vector field, we have

$$\bar{\nabla} \times \bar{f}(\bar{r}) = \frac{1}{\rho}\hat{a}_z \left(\frac{\partial}{\partial \rho}[\rho f_{\phi}(\bar{r}, t)] \right) \tag{2.130}$$

$$= \frac{1}{\rho}\hat{a}_z \left(\frac{\partial}{\partial \rho}\left(\frac{\rho}{\rho}\right) \right) = 0. \tag{2.131}$$

Hence, the magnetic field's curl due to a steady line electric current seems to be zero everywhere, while this evaluation is valid if $\rho \neq 0$. One can show that, when $\rho = 0$ (when the observation point

[27] More formally, Helmholtz decomposition exists for a vector field that is twice continuously differentiable.

[28] We immediately note that the decomposition is not unique for all vector fields. For example, consider a vector field $\bar{f}(\bar{r}) = \hat{a}_x$, which has a constant magnitude. Then we have $\bar{\nabla} \cdot \bar{f}(\bar{r}) = 0$ and $\bar{\nabla} \times \bar{f}(\bar{r}) = 0$. Therefore, this vector field itself is both divergence-free and curl-free. We can write it as $\bar{f}(\bar{r}) = \bar{f}_{df}(\bar{r}) + \bar{f}_{cf}(\bar{r})$, where $\bar{f}_{df}(\bar{r}) = \hat{a}_x$ and $\bar{f}_{cf}(\bar{r}) = 0$ or $\bar{f}_{df}(\bar{r}) = 0$ and $\bar{f}_{cf}(\bar{r}) = \hat{a}_x$. In fact, there are infinitely many choices: e.g. $\bar{f}_{df}(\bar{r}) = \hat{a}_x + \hat{a}_y$ and $\bar{f}_{cf}(\bar{r}) = -\hat{a}_y$, giving $\bar{f}(\bar{r}) = \bar{f}_{df}(\bar{r}) + \bar{f}_{cf}(\bar{r})$. Obviously, adding a constant vector field to any vector field leads to the same non-uniqueness. For example, for a vector field $\bar{f}(\bar{r}) = \hat{a}_{\rho}\rho + \hat{a}_{\phi} + \hat{a}_x$, one can put $\hat{a}_x$ in $\bar{f}_{df}$ or $\bar{f}_{cf}$. There are also vector fields that vary with position while having a non-unique Helmholtz decomposition. A common property of all these vector fields (with non-unique Helmholtz decomposition) is that they do not vanish at infinity. In fact, if a vector field vanishes at infinity, it can be *uniquely* found by knowing its divergence and curl.

is located on the z axis), the curl operation gives an impulse function,[29] correctly indicating the existence of the source on the z axis. Hence, an impulse source located on the z axis[30] (without any other source at other locations) is sufficient to create a vector field $\bar{f}(\bar{r}) = \hat{a}_\phi/\rho$.

For a spherical coordinate system, the curl operation can be performed as

$$\bar{\nabla} \times \bar{f}(\bar{r}, t) = \frac{1}{R^2 \sin\theta} \begin{vmatrix} \hat{a}_R & \hat{a}_\theta R & \hat{a}_\phi R \sin\theta \\ \partial/\partial R & \partial/\partial\theta & \partial/\partial\phi \\ f_R(\bar{r}, t) & R f_\theta(\bar{r}, t) & R\sin\theta f_\phi(\bar{r}, t) \end{vmatrix} \tag{2.132}$$

leading to

$$\begin{aligned} \bar{\nabla} \times \bar{f}(\bar{r}, t) = & \frac{1}{R^2 \sin\theta} \hat{a}_R \left(\frac{\partial}{\partial\theta} [R\sin\theta f_\phi(\bar{r}, t)] - \frac{\partial}{\partial\phi} [R f_\theta(\bar{r}, t)] \right) \\ & + \frac{1}{R\sin\theta} \hat{a}_\theta \left(\frac{\partial f_R(\bar{r}, t)}{\partial\phi} - \frac{\partial}{\partial R} [R\sin\theta f_\phi(\bar{r}, t)] \right) \\ & + \frac{1}{R} \hat{a}_\phi \left(\frac{\partial}{\partial R} [R f_\theta(\bar{r}, t)] - \frac{\partial f_R(\bar{r}, t)}{\partial\theta} \right) \end{aligned} \tag{2.133}$$

$$\begin{aligned} = & \frac{1}{R\sin\theta} \hat{a}_R \left(\frac{\partial}{\partial\theta} [\sin\theta f_\phi(\bar{r}, t)] - \frac{\partial f_\theta(\bar{r}, t)}{\partial\phi} \right) \\ & + \frac{1}{R} \hat{a}_\theta \left(\frac{1}{\sin\theta} \frac{\partial f_R(\bar{r}, t)}{\partial\phi} - \frac{\partial}{\partial R} [R f_\phi(\bar{r}, t)] \right) \\ & + \frac{1}{R} \hat{a}_\phi \left(\frac{\partial}{\partial R} [R f_\theta(\bar{r}, t)] - \frac{\partial f_R(\bar{r}, t)}{\partial\theta} \right). \end{aligned} \tag{2.134}$$

Once again, we can evaluate this lengthy expressions for several basic examples. First, we consider $\bar{f}(\bar{r}) = \hat{a}_R$, whose divergence was found before as[31]

$$\bar{\nabla} \cdot \bar{f}(\bar{r}) = \frac{2}{R}, \tag{2.135}$$

even though the magnitude is constant 1 everywhere. The curl of the same vector field can be found as

$$\bar{\nabla} \times \bar{f}(\bar{r}) = \frac{1}{R} \hat{a}_\theta \left(\frac{1}{\sin\theta} \frac{\partial f_R(\bar{r})}{\partial\phi} \right) + \frac{1}{R} \hat{a}_\phi \left(-\frac{\partial f_R(\bar{r})}{\partial\theta} \right) = 0 \tag{2.136}$$

[29] In the derivation of the differential form of Ampere's law, we assume that the total electric current can be written as a surface integral of a volume electric current density. A line electric current can be considered infinite electric current density squeezed into a zero volume. From this perspective, the curl of the magnetic field intensity due to a line electric current can be said to be infinite on the z axis, which can be represented as an impulse function: e.g. $\bar{J}_v(\bar{r}) = \hat{a}_z I_0 \delta(z)$.

[30] We further note that $\bar{\nabla} \cdot \bar{f}(\bar{r}) = 0$: i.e. the rotational source on the z axis is the only source that creates $\bar{f}(\bar{r})$.

[31] This was found simply as

$$\bar{\nabla} \cdot \bar{f}(\bar{r}) = \frac{1}{R^2} \frac{\partial}{\partial R} (R^2) = \frac{2R}{R^2} = \frac{2}{R}.$$

This means that a vector field $\bar{f}(\bar{r}) = \hat{a}_R$, whose magnitude is constant everywhere, can be created by a distribution of non-rotational sources in the whole space.

everywhere. Obviously, this vector field does not have a rotational source or, hence, any rotational behavior. Next, we consider $\bar{f}(\bar{r}) = \hat{a}_\phi$, whose curl was found in a cylindrical coordinate system as $\hat{a}_z(1/\rho)$. Now, using the expression for a spherical coordinate system, we derive[32]

$$\bar{\nabla} \times \bar{f}(\bar{r}) = \frac{1}{R\sin\theta}\hat{a}_R\left(\frac{\partial}{\partial\theta}[\sin\theta f_\phi(\bar{r})]\right) + \frac{1}{R}\hat{a}_\theta\left(-\frac{\partial}{\partial R}[Rf_\phi(\bar{r})]\right) \tag{2.137}$$

$$= \frac{1}{R\sin\theta}\hat{a}_R\frac{\partial}{\partial\theta}(\sin\theta) + \frac{1}{R}\hat{a}_\theta\left(-\frac{\partial R}{\partial R}\right) \tag{2.138}$$

$$= \hat{a}_R\frac{\cos\theta}{R\sin\theta} - \hat{a}_\theta\frac{1}{R} \tag{2.139}$$

$$= \frac{\hat{a}_R\cos\theta - \hat{a}_\theta\sin\theta}{R\sin\theta} = \frac{\hat{a}_z}{\rho}, \tag{2.140}$$

as expected.

Finally, we may consider a vector field $\bar{f}(\bar{r}) = \bar{r}/|\bar{r}|^3$ (Figure 1.28), which describes the electric field created by a static point electric charge at the origin.[33] Using a spherical coordinate system, the vector field can be rewritten $\bar{f}(\bar{r}) = \hat{a}_R/R^2$ so that $f_R = 1/R^2$ and

$$\bar{\nabla} \times \bar{f}(\bar{r}) = \frac{1}{R}\hat{a}_\theta\left(\frac{1}{\sin\theta}\frac{\partial f_R(\bar{r})}{\partial\phi}\right) + \frac{1}{R}\hat{a}_\phi\left(-\frac{\partial f_R(\bar{r})}{\partial\theta}\right) = 0. \tag{2.141}$$

Hence, the curl of the electric field due to a static point electric charge is zero. While this may be unsurprising, it can be generalized (via *superposition*) to any arbitrary static electric charge distribution (Figure 2.34). Specifically, the curl of *any* static electric field is zero, as it is verified by Faraday's law.

[32] Recognizing that $\hat{a}_z = \hat{a}_R\cos\theta - \hat{a}_\theta\sin\theta$ and $\rho = R\sin\theta$.

[33] We recall that $\bar{\nabla}\cdot\bar{f}(\bar{r}) \neq 0$ at the origin, indicating a non-rotational point source at this location.

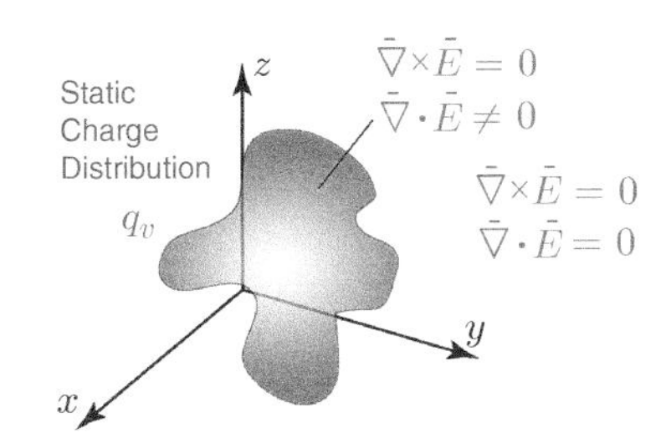

Figure 2.34 Electric field created by any static electric charge distribution has a zero curl everywhere. This can be simply verified by considering the electric field created by a static point electric charge (proportional to $\bar{r}/|\bar{r}|^3$) and using the superposition.

2.3.4 Stokes' Theorem and the Differential Form of Ampere's Law

Stokes' theorem is the main tool to derive the differential form of Ampere's law from its integral form.[34] Given a curve C around a surface S and a vector field $\bar{f}(\bar{r}, t)$, Stokes' theorem states that (Figure 2.35)

$$\int_S \bar{\nabla} \times \bar{f}(\bar{r}, t) \cdot \overline{ds} = \oint_C \bar{f}(\bar{r}, t) \cdot \overline{dl}, \tag{2.142}$$

[34] This is similar to the divergence theorem that is used to derive the differential form of Gauss' law from its integral form. In fact, Stokes' theorem and the divergence theorem are mathematically related to each other.

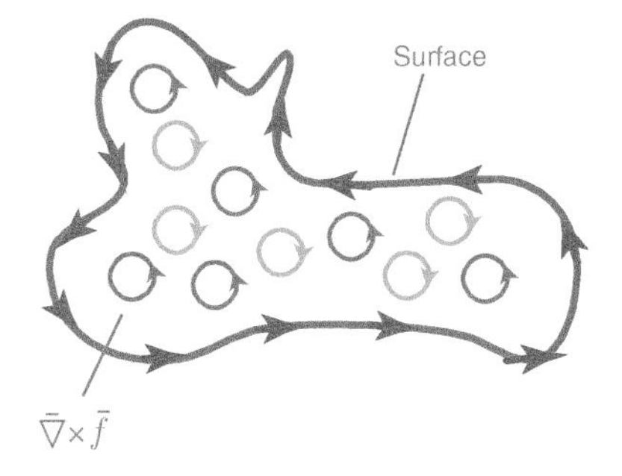

Figure 2.35 Since the curl $\bar{\nabla} \times \bar{f}(\bar{r}, t)$ represents rotational positive/negative (counterclockwise and clockwise) sources of a vector field $\bar{f}(\bar{r}, t)$, its surface integral (the sum of all positive and negative sources) corresponds to the net circulation of the vector field along the curve bounding the open surface. Some of the rotations may cancel each other, but unbalanced rotations appear along the bounding curve. If positive/negative sources dominate, the net circulation becomes positive/negative, while the value of $\bar{f}(\bar{r}, t) \cdot \overline{dl}$ may be positive or negative depending on the position along the line.

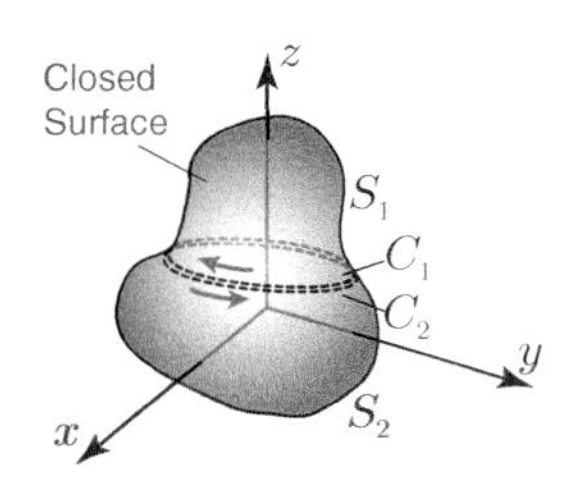

Figure 2.36 According to Stokes' theorem,

$$\oint_S \bar{\nabla} \times \bar{f}(\bar{r}, t) \cdot \overline{ds} = 0$$

for any close surface since there is no any curve bounding the surface. This can be understood by considering the surface S as a combination of two open surfaces S_1 and S_2. Then the line integrals

$$\oint_{C_1} \bar{f}(\bar{r}, t) \cdot \overline{dl} \text{ and } \oint_{C_2} \bar{f}(\bar{r}, t) \cdot \overline{dl}$$

have equal contributions with opposite signs, if the right-hand rule is used consistently (e.g. considering the normal of the surfaces outward).

where $\overline{ds} = \hat{a}_n ds$, $\overline{dl} = \hat{a}_l dl$, while $\hat{a}_n$ and $\hat{a}_l$ are related via the right-hand rule (Figure 2.1). If S is closed, there is not any curve that bounds the surface (Figure 2.36), and

$$\oint_S \bar{\nabla} \times \bar{f}(\bar{r}, t) \cdot \overline{ds} = 0. \tag{2.143}$$

Like the divergence theorem, Stokes' theorem is a very useful tool; it converts a surface integral into a line integral and vice versa. The idea behind the equality above is the fact that the curl is an indicator of rotational sources. When we consider all such sources on a surface via integration, we end up with the net circulation of $\bar{f}(\bar{r}, t)$ along the surface boundary. There can be cancellations of rotational sources with opposite sides, but the net circulation is created by the net amount of rotational sources (Figure 2.35). Ampere's law can be seen as a direct application of Stokes' theorem when we consider magnetic fields as vector fields and currents (both electric and displacement) as rotational sources.

Similar to the divergence theorem, Stokes' theorem in Eq. (2.142) is not pointwise: i.e. it indicates the equality of two integrals. For example, if the line integral is zero:

$$\oint_C \bar{f}(\bar{r}, t) \cdot \overline{dl} = 0, \tag{2.144}$$

this does not indicate that $\bar{f}(\bar{r}, t)$ is zero or $\bar{\nabla} \times \bar{f}(\bar{r}, t)$ is zero. There can be nonzero $\bar{\nabla} \times \bar{f}(\bar{r}, t)$ inside the surface, while the positive and negative values cancel each other, leading to zero overall source and zero circulation. This is perfectly consistent with the discussions of Ampere's law, when the source in Stokes' theorem ($\bar{\nabla} \times \bar{f}(\bar{r}, t)$) is replaced with electric and displacement current densities (sources of magnetic fields).

With the help of Stokes' theorem, we can now convert the integral form of Ampere's law into the differential form. According to Stokes' theorem, which is valid for all vector fields,[35] we have

$$\int_S \bar{\nabla} \times \bar{H}(\bar{r}, t) \cdot \overline{ds} = \oint_C \bar{H}(\bar{r}, t) \cdot \overline{dl}, \tag{2.145}$$

for any S and C. At the same time, using the integral form of Ampere's law in Eq. (2.1), we have

$$\int_S \bar{\nabla} \times \bar{H}(\bar{r}, t) \cdot \overline{ds} = I_{\text{enc}}(t) + \int_S \frac{\partial \bar{D}(\bar{r}, t)}{\partial t} \cdot \overline{ds}, \tag{2.146}$$

Rewriting the right-hand side using the volume electric current density, we have

$$\int_S \bar{\nabla} \times \bar{H}(\bar{r}, t) \cdot \overline{ds} = \int_S \bar{J}_v(\bar{r}, t) \cdot \overline{ds} + \int_S \frac{\partial \bar{D}(\bar{r}, t)}{\partial t} \cdot \overline{ds}. \tag{2.147}$$

[35] Mathematically, all theorems have limitations. For example, vector fields should be differentiable for Stokes' theorem. However, in the context of electromagnetics, we omit such constraints for simplicity.

Once again (as in Gauss' law), the key is that the final equation is valid for any arbitrary surface, including the limit $S \rightarrow 0$ for which the equation becomes pointwise: i.e. $\bar{\nabla} \times \bar{H}(\bar{r},t) = \bar{J}_v(\bar{r},t) + \bar{J}_{vd}(\bar{r},t)$, which is the differential form in Eq. (2.77). We note that $\bar{J}_{vd}(\bar{r},t) = \partial\bar{D}(\bar{r},t)/\partial t$ represents the displacement current, as defined before in Eq. (2.79).

2.3.5 Examples

Example 18: Consider the circulation of $\bar{f}(\bar{r}) = -y^2\hat{a}_x + xy\hat{a}_y + z\hat{a}_z$ through a loop of segments defined as a combination of a straight line C_1 from the origin to $(x,y,z) = (1,0,0)$, a circular line C_2 from $(x,y,z) = (1,0,0)$ to $(x,y,z) = (0,1,0)$, a straight line C_3 from $(x,y,z) = (0,1,0)$ to $(x,y,z) = (0,1,1)$, and a straight line C_4 from $(x,y,z) = (0,1,1)$ to the origin (Figure 2.37). Verify Stokes' theorem.

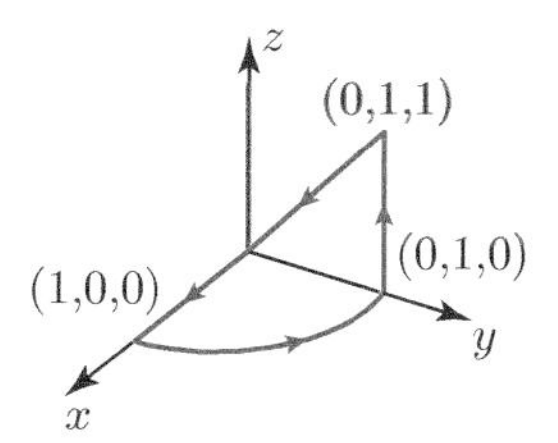

Figure 2.37 A closed path involving a combination of a straight line C_1 from the origin to $(x,y,z) = (1,0,0)$, a circular line C_2 from $(x,y,z) = (1,0,0)$ to $(x,y,z) = (0,1,0)$, a straight line C_3 from $(x,y,z) = (0,1,0)$ to $(x,y,z) = (0,1,1)$, and a straight line C_4 from $(x,y,z) = (0,1,1)$ to the origin.

Solution: Considering the circulation, we have

$$\oint_C \bar{f}(\bar{r}) \cdot \overline{dl} = \int_{C_1} \bar{f}(\bar{r}) \cdot \overline{dl} + \int_{C_2} \bar{f}(\bar{r}) \cdot \overline{dl} + \int_{C_3} \bar{f}(\bar{r}) \cdot \overline{dl} + \int_{C_4} \bar{f}(\bar{r}) \cdot \overline{dl}, \tag{2.148}$$

where

$$\int_{C_1} \bar{f}(\bar{r}) \cdot \overline{dl} = \int_0^1 (-y^2)dx = 0 \tag{2.149}$$

$$\int_{C_2} \bar{f}(\bar{r}) \cdot \overline{dl} = \int_0^{\pi/2} (-y^2\hat{a}_x + xy\hat{a}_y + z\hat{a}_z) \cdot \hat{a}_\phi 1 d\phi \tag{2.150}$$

$$= \int_0^{\pi/2} (\sin^3\phi + \cos^2\phi \sin\phi)d\phi = 1 \tag{2.151}$$

$$\int_{C_3} \bar{f}(\bar{r}) \cdot \overline{dl} = \int_0^1 z dz = \frac{1}{2} \tag{2.152}$$

and

$$\int_{C_4} \bar{f}(\bar{r}) \cdot \overline{dl} = \int_{(0,0,0)}^{(0,1,1)} (-y^2\hat{a}_x + xy\hat{a}_y + z\hat{a}_z) \cdot (-\hat{a}_y dy - \hat{a}_z dz) \tag{2.153}$$

$$= \int_{(0,0,0)}^{(0,1,1)} (-xydy - zdz) = -\int_0^1 zdz = -\frac{1}{2}. \tag{2.154}$$

Therefore, we obtain $\oint_C \bar{f}(\bar{r}) \cdot \overline{dl} = 1$. On the other side, it can be shown that[36] $\bar{\nabla} \times \bar{f}(\bar{r}) = 3y\hat{a}_z$ and

$$\int_S \bar{\nabla} \times \bar{f}(\bar{r}) \cdot \overline{ds} = 1, \tag{2.155}$$

verifying Stokes' theorem.

[36] We have

$$\nabla \times (-y^2\hat{a}_x + xy\hat{a}_y + z\hat{a}_z)$$
$$= \left(\hat{a}_x \frac{\partial}{\partial x} + \hat{a}_y \frac{\partial}{\partial y} + \hat{a}_z \frac{\partial}{\partial z}\right) \times (-y^2\hat{a}_x + xy\hat{a}_y + z\hat{a}_z)$$
$$= \hat{a}_z \frac{\partial}{\partial x}(xy) - \hat{a}_z \frac{\partial}{\partial y}(-y^2)$$
$$= \hat{a}_z y + \hat{a}_z(2y) = \hat{a}_z 3y.$$

Example 19: Consider a mushroom-shaped contour defined on the x-y plane (Figure 2.38). Find the line integral of $\bar{f}(\bar{r}) = \hat{a}_x y$ counterclockwise along the contour.

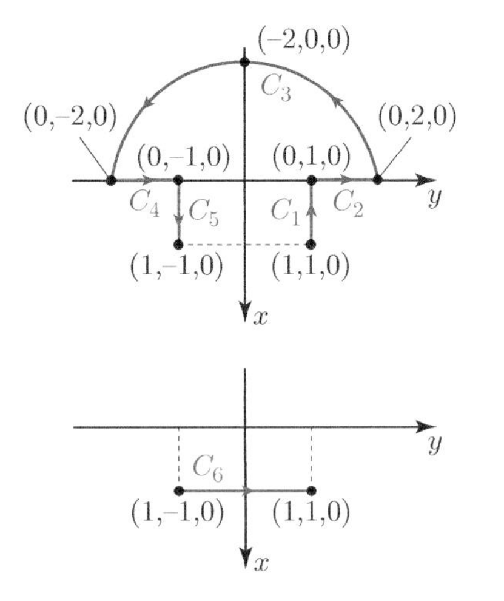

Figure 2.38 A mushroom-shaped contour defined on the x-y plane and a line complementing the contour for a closed one.

Solution: As a brute-force approach, we can evaluate the line integral as

$$\int_C \bar{f}(\bar{r}) \cdot \overline{dl} = \int_{C_1} \bar{f}(\bar{r}) \cdot \overline{dl} + \int_{C_2} \bar{f}(\bar{r}) \cdot \overline{dl} + \int_{C_3} \bar{f}(\bar{r}) \cdot \overline{dl} \tag{2.156}$$

$$+ \int_{C_4} \bar{f}(\bar{r}) \cdot \overline{dl} + \int_{C_5} \bar{f}(\bar{r}) \cdot \overline{dl}, \tag{2.157}$$

where

$$\int_{C_1} \bar{f}(\bar{r}) \cdot \overline{dl} = \int_0^1 \hat{a}_x y \cdot (-\hat{a}_x) dx = -1 \tag{2.158}$$

$$\int_{C_2} \bar{f}(\bar{r}) \cdot \overline{dl} = \int_1^2 \hat{a}_x y \cdot \hat{a}_y dx = 0 \tag{2.159}$$

$$\int_{C_3} \bar{f}(\bar{r}) \cdot \overline{dl} = \int_{\pi/2}^{3\pi/2} \hat{a}_x y \cdot \hat{a}_\phi \rho d\phi = \int_{\pi/2}^{3\pi/2} (-\sin\phi)\rho \sin\phi \rho d\phi \tag{2.160}$$

$$= -\int_{\pi/2}^{3\pi/2} \sin^2\phi \rho^2 d\phi = -2\pi \tag{2.161}$$

$$\int_{C_4} \bar{f}(\bar{r}) \cdot \overline{dl} = \int_{-2}^{-1} \hat{a}_x y \cdot \hat{a}_y dy = 0 \tag{2.162}$$

$$\int_{C_5} \bar{f}(\bar{r}) \cdot \overline{dl} = \int_0^1 \hat{a}_x y \cdot \hat{a}_x dx = -1. \tag{2.163}$$

Then we have

$$\int_C \bar{f}(\bar{r}) \cdot \overline{dl} = -2 - 2\pi. \tag{2.164}$$

On the other hand, the same integral can be found via Stokes' theorem. For this purpose, we can use a line C_6 from $(1,-1,0)$ to $(1,1,0)$ that complements the original contour $C = C_1 \cup C_2 \cup C_3 \cup C_4 \cup C_5$, leading to a closed contour (Figure 2.38). Then we have

$$\int_C \bar{f}(\bar{r}) \cdot \overline{dl} = \oint_{C+C_6} \bar{f}(\bar{r}) \cdot \overline{dl} - \int_{C_6} \bar{f}(\bar{r}) \cdot \overline{dl}, \tag{2.165}$$

where the closed line integral can be found via Stokes' theorem as

$$\oint_{C+C_6} \bar{f}(\bar{r}) \cdot \overline{dl} = \int_S \bar{\nabla} \times \bar{f}(\bar{r}) \cdot \overline{ds} = -A = -\frac{\pi 2^2}{2} - 2 = -2 - 2\pi \tag{2.166}$$

considering that $\bar{\nabla} \times \bar{f}(\bar{r}) = \hat{a}_y \times \hat{a}_x \partial y / \partial y = -\hat{a}_z$. It is relatively straightforward to show that the line integral over C_6 is zero:

$$\int_{C_6} \bar{f}(\bar{r}) \cdot \overline{dl} = \int_{-1}^{1} \hat{a}_x y \cdot \hat{a}_y dy = 0. \tag{2.167}$$

Consequently, we have

$$\int_C \bar{f}(\bar{r}) \cdot \overline{dl} = -2 - 2\pi - 0 = -2 - 2\pi, \tag{2.168}$$

verifying the earlier result.

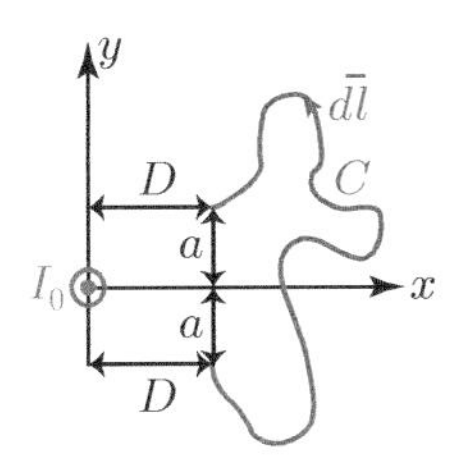

Figure 2.39 An arbitrary curve located close to an infinitely long steady line electric current I_0 in the z direction.

Example 20: An arbitrary curve (C) is located close to an infinitely long steady line electric current I_0 (A) in the z direction (Figure 2.39). The curve starts from $(x, y) = (D, -a)$ and ends at $(x, y) = (D, a)$. Find an expression for the line integral

$$\int_C \bar{H}_0(\bar{r}) \cdot \overline{dl}, \tag{2.169}$$

where $\bar{H}_0(\bar{r})$ is the magnetic field intensity created by I_0.

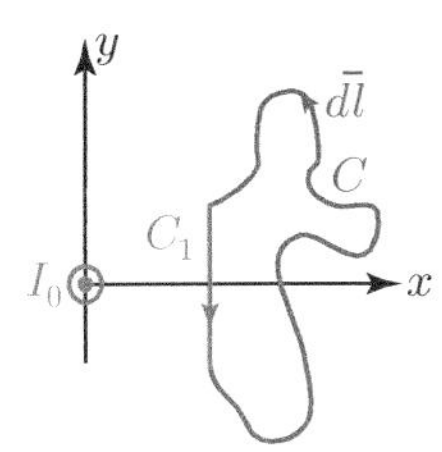

Figure 2.40 Adding a straight line C_1 to the curve C in Figure 2.39 such that a closed contour is obtained.

Solution: Interestingly, even though the shape of the curve is arbitrary and is not given, the line integral has a unique value. To find the value of the integral, we consider a straight line connecting the end points of C to create a closed contour (Figure 2.40). Then we have

$$\oint_{C \cup C_1} \bar{H}_0(\bar{r}) \cdot \overline{dl} = \int_C \bar{H}_0(\bar{r}) \cdot \overline{dl} + \int_{C_1} \bar{H}_0(\bar{r}) \cdot \overline{dl} = 0 \tag{2.170}$$

since there is no any current enclosed by the combined contour $C \cup C_1$. Hence, the required line integral can be found as

$$\int_C \bar{H}_0(\bar{r}) \cdot \overline{dl} = -\int_{C_1} \bar{H}_0(\bar{r}) \cdot \overline{dl} = -\int_{-a}^{a} \bar{H}_0(\bar{r}) \cdot (-\hat{a}_y) dy, \tag{2.171}$$

where $\bar{H}_0(\bar{r})$ is evaluated at observation points $\bar{r}$ on C_1. Using Eq. (2.59), we can obtain the expression for $\bar{H}_0(\bar{r})$ as[37] (Figure 2.41)

$$\bar{H}_0(\bar{r}) = \frac{I_0}{2\pi}\left(-\hat{a}_x \frac{y}{D^2+y^2} + \hat{a}_y \frac{D}{D^2+y^2}\right) \tag{2.172}$$

leading to

$$\int_C \bar{H}_0(\bar{r}) \cdot \overline{dl} = \int_{-a}^{a} \bar{H}_0(\bar{r}) \cdot \hat{a}_y dy = \frac{I_0}{2\pi}\int_{-a}^{a} \frac{D}{D^2+y^2} dy = \frac{I_0}{\pi}\tan^{-1}(a/D). \tag{2.173}$$

[37] Considering Figure 2.41, we have $\tan\alpha = y/D$ for a given observation point (y value) on C_1. The magnetic field intensity $\bar{H}_0$ has x and y components as

$$\begin{aligned} \bar{H}_0 &= -\hat{a}_x H_0 \sin\alpha + \hat{a}_y H_0 \cos\alpha \\ &= -\hat{a}_x H_0 \frac{y}{\sqrt{D^2+y^2}} + \hat{a}_y H_0 \frac{D}{\sqrt{D^2+y^2}}. \end{aligned}$$

Finally, the magnitude of the magnetic field intensity can be written $H_0 = I_0/(2\pi\sqrt{D^2+y^2})$.

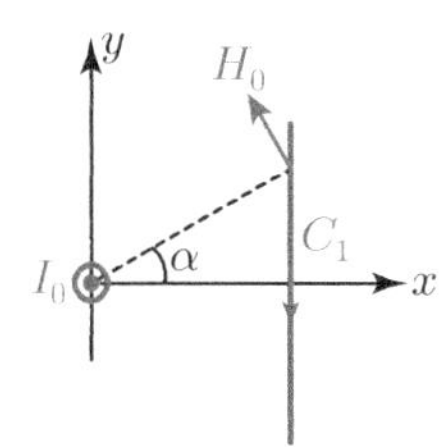

Figure 2.41 Magnetic field intensity created by the infinitely long steady line electric current I_0 on C_1.

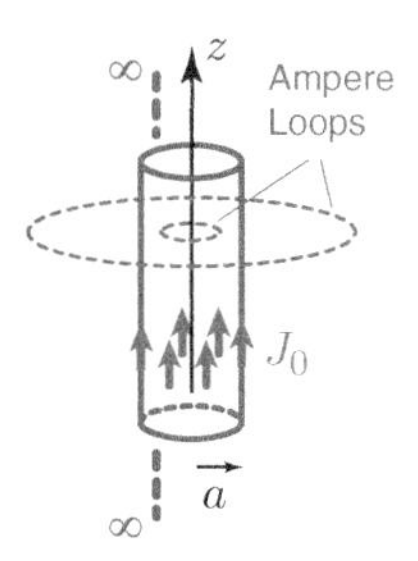

Cross Sectional View

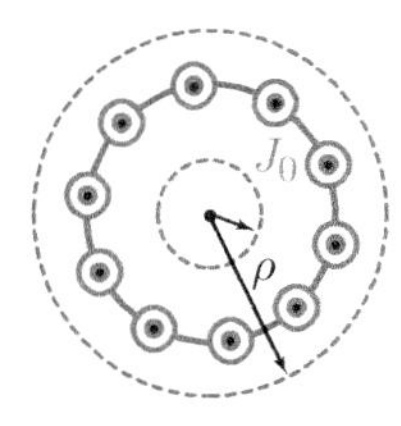

Figure 2.42 A steady surface electric current density $J_s(\bar{r}) = J_0$ (A/m) on an infinitely long cylindrical surface aligned in the z direction. The electric current is assumed to flow only on the surface (surface electric current density).

2.4 Using the Differential Form of Ampere's Law

The differential form of Ampere's law is pointwise and provides information about how magnetic fields and electric currents are related to each other at every point in space. Obviously, knowing the electric current density and/or the time derivative of the electric flux density (displacement current density) does not provide *full* information about the magnetic field intensity. This is because, knowing the curl of a vector field, we cannot find the vector field uniquely, and other properties – e.g. its divergence and/or boundary conditions – are needed.

The differential form of Gauss' law provides the electric charge density for a known electric field intensity. Similarly, if the magnetic field intensity is known, the differential form of Ampere's law can be used to find the right-hand side: i.e. the electric current density plus the time derivative of the electric flux density. On the other hand, if the electric flux density is not known, the electric current density cannot be found by using Ampere's law alone. But in a static case, the electric current density can be found from the magnetic field intensity via Ampere's law since the electric current density is the only type of sources that creates the magnetic field in this regime.

We emphasize that the electric current term on the right-hand side of Ampere's law is a volume density $\bar{J}_v(\bar{r}, t)$, not a surface density $\bar{J}_s(\bar{r}, t)$ or line density (simply electric current $I(t)$). To show how this is critical, we consider a steady surface electric current density J_0 (A/m) in the shape of an infinitely long cylinder of radius a in a vacuum (Figure 2.42). The electric current density is uniformly distributed so that the total amount of electric current flowing along the cylinder is $J_0 2\pi a$ (A) (see Section 2.3.1). Using the integral form of Ampere's law for a circular Ampere loop (considering the symmetry), we have

$$\int_0^{2\pi} \bar{H}(\bar{r}) \cdot \hat{a}_\phi \rho d\phi = \int_0^{2\pi} H_\phi(\rho)\rho d\phi = 2\pi\rho H_\phi(\rho) = I_{\text{enc}}, \tag{2.174}$$

assuming that the magnetic field intensity has only ϕ component and only depends on ρ due to the symmetry. If the radius of the Ampere loop is smaller than a: i.e. for $\rho < a$, we have $I_{\text{enc}} = 0$ and $\bar{H}(\bar{r}) = 0$. Otherwise, for $\rho > a$, we have $I_{\text{enc}} = 2\pi a J_0$, leading to[38]

$$\bar{H}(\bar{r}) = \hat{a}_\phi \frac{I_{\text{enc}}}{2\pi\rho} = \hat{a}_\phi \frac{J_0 2\pi a}{2\pi\rho} = \hat{a}_\phi \frac{J_0 a}{\rho} \qquad (\rho > a). \tag{2.175}$$

Furthermore, using $\bar{B}(\bar{r}) = \mu_0 \bar{H}(\bar{r})$, we obtain

$$\bar{B}(\bar{r}) = \hat{a}_\phi \frac{\mu_0 J_0 a}{\rho} \qquad (\rho > a), \tag{2.176}$$

while it is again zero for $\rho < a$. Now, using the differential form of Ampere's law, we have

$$\bar{J}_v(\bar{r}) = \bar{\nabla} \times \bar{H}(\bar{r}), \tag{2.177}$$

where

$$\bar{\nabla} \times \bar{H}(\bar{r}) = -\hat{a}_\rho \frac{\partial H_\phi(\bar{r})}{\partial z} + \frac{1}{\rho}\hat{a}_z \frac{\partial}{\partial \rho}[\rho H_\phi(\bar{r})] \tag{2.178}$$

$$= \frac{1}{\rho}\hat{a}_z \frac{\partial}{\partial \rho}(J_0 a) = 0. \tag{2.179}$$

Hence, there is no volume electric current density, as expected. This means the differential form of Ampere's law does not provide information about the electric current density in this scenario.[39]

We note that, for the surface electric current density in the above, the magnetic field intensity is discontinuous at $\rho = a$: i.e. its magnitude jumps from 0 to J_0 when ρ changes from a^- to a^+ (Figure 2.43). This jump is due to the surface electric current density at $\rho = a$. Similar to the boundary conditions for normal components of electric fields, which are derived by applying Gauss' law at boundaries, the application of Ampere's law at boundaries leads to the boundary conditions for tangential components of magnetic fields. Instead of the differential form of Ampere's law, these boundary conditions can be used to detect and find surface electric current distributions. The specific cause of the jump across a surface electric current density is the suddenly changing directions of magnetic fields created by the electric currents at the boundary and close to the observation point (Figure 2.44).

[38]It is remarkable that, for $\rho > a$, the magnetic field intensity is the same as that created by an infinitely long steady line electric current $I_0 = 2\pi a J_0$ on the z axis.

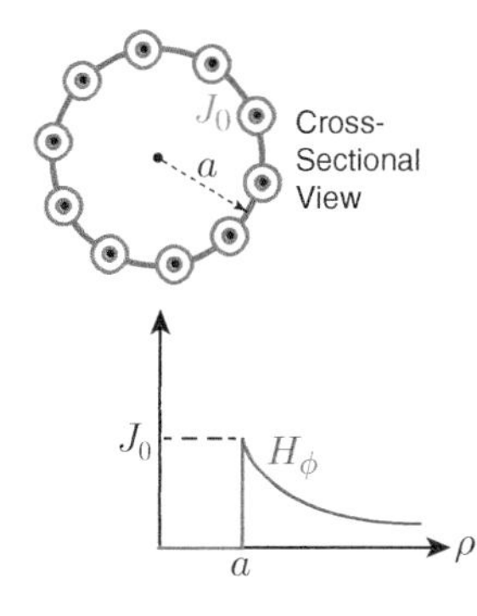

Figure 2.43 The magnetic field intensity created by a steady surface electric current density $J_s(\bar{r}) = J_0$ (A/m) on an infinitely long cylindrical surface (Figure 2.42). The magnetic field intensity is discontinuous at $\rho = a$.

[39]From a mathematical point of view, the curl operation should not be used at $\rho = a$ since the magnetic field intensity is discontinuous.

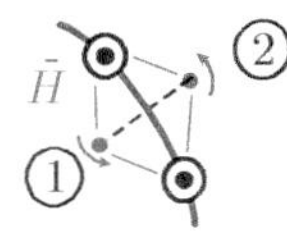

Figure 2.44 The magnetic field may jump at short distances when an observation point moves across a surface electric current density. This is due to the suddenly changing direction of the magnetic field created by nearby sources.

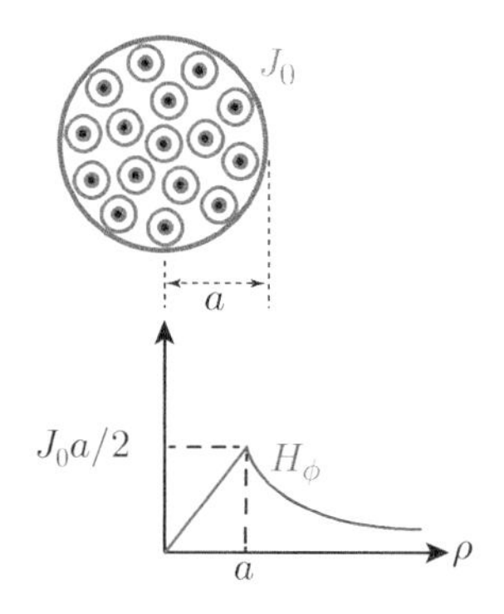

Figure 2.45 A steady volume electric current density $J_v = J_0$ (A/m^2) flowing in a cylindrical region of radius a in the z direction. The electric current is assumed to be uniformly distributed over the cross-sectional surface (disk). It is remarkable that the magnetic field intensity is continuous at $\rho = a$: i.e. $\bar{H}(\rho = a^+) = \bar{H}(\rho = a^-) = \hat{a}_\phi J_0 a/2$. This is because there is no any surface electric current density to create a discontinuity in $\bar{H}(\bar{r})$.

Next, we focus on another scenario, which involves a volume electric current density (Figure 2.45). We consider a steady volume electric current density J_0 (A/m^2), again in the shape of an infinitely long cylinder of radius a. In this case, the electric current is uniformly distributed over the cross-section of the cylinder, leading to a total of $I_0 = J_0\pi a^2$ (A) electric current flowing along the cylinder. The integral form of Ampere's law leads to

$$\int_0^{2\pi} \bar{H}(\bar{r}) \cdot \hat{a}_\phi \rho d\phi = \int_0^{2\pi} H_\phi(\rho)\rho d\phi = 2\pi\rho H_\phi(\rho) = I_{\text{enc}}, \tag{2.180}$$

again considering the symmetry. For $\rho > a$, we have

$$\bar{H}(\bar{r}) = \hat{a}_\phi \frac{I_{\text{enc}}}{2\pi\rho} = \hat{a}_\phi \frac{J_0\pi a^2}{2\pi\rho} = \hat{a}_\phi \frac{J_0 a^2}{2\rho} \qquad (\rho > a), \tag{2.181}$$

which is the same as the magnetic field intensity due to an infinitely long steady line electric current $I_0 = \pi a^2 J_0$ on the z axis. For $\rho < a$, we need to use the total enclosed electric current as $I_{\text{enc}} = J_0\pi\rho^2$, leading to

$$\bar{H}(\bar{r}) = \hat{a}_\phi \frac{I_{\text{enc}}}{2\pi\rho} = \hat{a}_\phi \frac{J_0\pi\rho^2}{2\pi\rho} = \hat{a}_\phi \frac{J_0\rho}{2} \qquad (\rho < a). \tag{2.182}$$

Hence, inside the cylinder, the magnetic field intensity increases linearly with ρ. Using the differential form of Ampere's law, we still have $\bar{J}_v(\bar{r}) = 0$ for $\rho > a$, while it is nonzero for $\rho < a$ as

$$\bar{\nabla} \times \bar{H}(\bar{r}) = -\hat{a}_\rho \frac{\partial H_\phi(\bar{r})}{\partial z} + \frac{1}{\rho}\hat{a}_z \frac{\partial}{\partial\rho}[\rho H_\phi(\bar{r})] = \frac{1}{\rho}\hat{a}_z \frac{\partial}{\partial\rho}\left(\frac{J_0\rho^2}{2}\right) = \hat{a}_z J_0. \tag{2.183}$$

To sum up, the differential form of Ampere's law correctly finds the volume electric current density. It is remarkable that the magnetic field intensity is continuous with respect to ρ and does not have any jump since there is no any surface electric current density in this case.

2.4.1 Examples

Example 21: An infinitely long cylindrical object of radius a is located in a vacuum (Figure 2.46). A steady volume electric current density with an unknown distribution flows in the z direction inside the cylinder, and there is no electric current elsewhere. The magnetic field intensity inside the cylinder is given as

$$\bar{H}(\bar{r}) = \hat{a}_\phi H_0 \rho^3 \qquad (\rho < a). \tag{2.184}$$

Find the magnetic field intensity for $\rho > a$.

Solution: First, we can use the differential form of Ampere's law to find the distribution of the electric current density as

$$\bar{J}_v(\bar{r}) = \bar{\nabla} \times \bar{H}(\bar{r}) = \hat{a}_z \frac{H_0}{\rho} \frac{\partial}{\partial \rho}(\rho^4) = \hat{a}_z 4H_0\rho^2. \tag{2.185}$$

Hence the total electric current can be found via integration as

$$I_0 = H_0 \int_0^{2\pi} \int_0^a 4\rho^2 \rho d\rho d\phi = 2\pi H_0 a^4. \tag{2.186}$$

Using the integral form of Ampere's law, we can find the magnetic field intensity outside the cylinder as

$$2\pi\rho H_\phi(\rho) = I_0 = 2\pi H_0 a^4 \longrightarrow \bar{H}(\bar{r}) = \hat{a}_\phi \frac{H_0 a^4}{\rho} \qquad (\rho > a). \tag{2.187}$$

We also note that the magnetic field intensity is continuous (due to the lack of surface electric current density) at $\rho = a$.

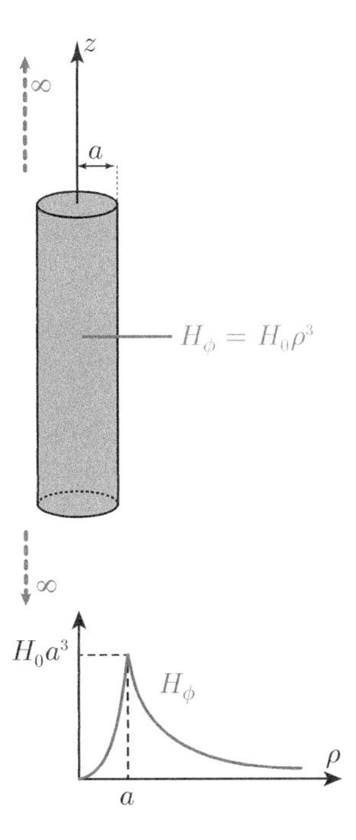

Figure 2.46 A cylindrical structure with unknown electric current distribution but known magnetic field intensity for $\rho < a$. The overall magnetic field intensity is plotted with respect to ρ.

2.5 Boundary Conditions for Tangential Magnetic Fields

Application of Ampere's law at boundaries between two different media with different magnetic properties provides the boundary conditions that must be satisfied by magnetic fields. While Gauss' law leads to the boundary conditions for normal components of electric fields, Ampere's law provides information about tangential components of magnetic fields. Now, we consider a boundary S between two media with permeability values μ_1 and μ_2, and a normal direction $\hat{a}_n$ pointing into the first medium. We further write $\{\bar{H}_1(\bar{r},t), \bar{B}_1(\bar{r},t)\}$ and $\{\bar{H}_2(\bar{r},t), \bar{B}_2(\bar{r},t)\}$ to represent magnetic fields in the first medium and the second medium, respectively, where $\bar{B}_1(\bar{r},t) = \mu_1(\bar{r})\bar{H}_1(\bar{r},t)$ and $\bar{B}_2(\bar{r},t) = \mu_2(\bar{r})\bar{H}_2(\bar{r},t)$. To use Ampere's law locally at the boundary, we consider a frame (Ampere loop) enclosing a differential part of the boundary around the observation point (Figure 2.47). We repeat the integral form of Ampere's law as

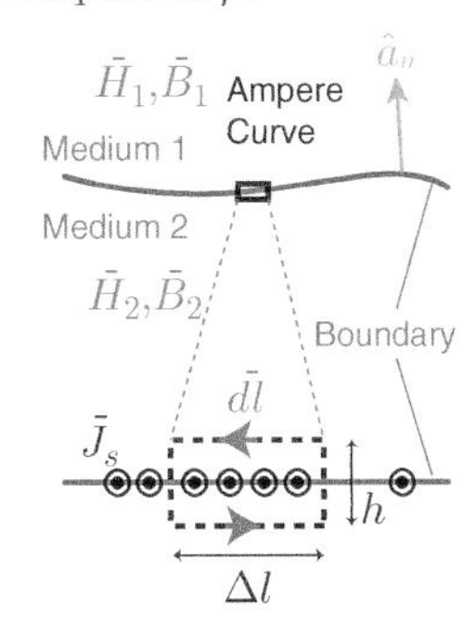

Figure 2.47 An illustration of the derivation of the boundary conditions for tangential magnetic fields across an interface between two different media. We note that the derivation holds for observation points on smooth surfaces, and not for observation points located at the edges and corners.

$$\oint_C \bar{H}(\bar{r},t) \cdot \overline{dl} = I_{\text{enc}}(t) + \int_S \frac{\partial \bar{D}(\bar{r},t)}{\partial t} \cdot \overline{ds}, \tag{2.188}$$

where $I_{\text{enc}}(t)$ is the electric current flowing through the surface bounded by the frame. We assume that the frame has a differential width Δl, while its height h is shrinking to zero. Hence, we derive the limit case as[40]

$$\Delta l[\hat{a}_n \times \bar{H}_1(\bar{r},t) - \hat{a}_n \times \bar{H}_2(\bar{r},t)] = (\Delta l)\bar{J}_s(\bar{r},t) + (\Delta l)\lim_{h\to 0}\left[h\frac{\partial \bar{D}(\bar{r},t)}{\partial t}\right], \tag{2.189}$$

or

$$\hat{a}_n \times [\bar{H}_1(\bar{r},t) - \bar{H}_2(\bar{r},t)] = \bar{J}_s(\bar{r},t) + \lim_{h\to 0}\left[h\frac{\partial \bar{D}(\bar{r},t)}{\partial t}\right], \tag{2.190}$$

where $\bar{J}_s(\bar{r},t)$ is the surface electric current density. The electric flux density term on the right-hand side goes to zero if the time derivative of the electric flux density is not infinite. Hence, we arrive at

$$\hat{a}_n \times [\bar{H}_1(\bar{r},t) - \bar{H}_2(\bar{r},t)] = \bar{J}_s(\bar{r},t) \qquad (\bar{r} \in S) \tag{2.191}$$

as the boundary condition for the magnetic field intensity at the boundary.

Obviously, the magnetic field intensity can be discontinuous if there are surface electric currents flowing on the surface, which create magnetic fields and lead to imbalance across the surface (Figure 2.44). If there is no surface electric current density, however, we have

$$\hat{a}_n \times \bar{H}_1(\bar{r},t) = \hat{a}_n \times \bar{H}_2(\bar{r},t) \qquad (\bar{r} \in S), \tag{2.192}$$

i.e. the tangential component of the magnetic field intensity is continuous across the boundary.

But what about the magnetic flux density? Using $\bar{B}_1(\bar{r},t) = \mu_1(\bar{r})\bar{H}_1(\bar{r},t)$ and $\bar{B}_2(\bar{r},t) = \mu_2(\bar{r})\bar{H}_2(\bar{r},t)$, we have

$$\hat{a}_n \times [\bar{B}_1(\bar{r},t)/\mu_1(\bar{r}) - \bar{B}_2(\bar{r},t)/\mu_2(\bar{r})] = \bar{J}_s(\bar{r}) \qquad (\bar{r} \in S) \tag{2.193}$$

and

$$\hat{a}_n \times \bar{B}_1(\bar{r},t)/\mu_1(\bar{r}) = \hat{a}_n \times \bar{B}_2(\bar{r},t)/\mu_2(\bar{r}) \qquad (\bar{r} \in S) \tag{2.194}$$

if there is no surface electric current density. Therefore, at the boundary between two different magnetic media, the tangential magnetic flux density is discontinuous *even* when there is no electric current density. This can be explained by the magnetization effects that are discussed in Section 2.8.

In addition to providing important information about how magnetic fields should behave across boundaries, the boundary conditions for magnetic fields can be used directly to analyze some magnetic scenarios. As an example, we consider a infinite sheet of steady surface electric current density $\bar{J}_s(\bar{r}) = \hat{a}_z J_0$ (A/m) on the z-x plane in a vacuum (Figure 2.48). The electric current flows in the z direction, and we would like to find the magnetic field intensity everywhere.

[40]We note that $\bar{H}(\bar{r},t)$ and $\bar{J}_s(\bar{r},t)$ are related via the right-hand rule, as represented by the cross products with $\hat{a}_n$.

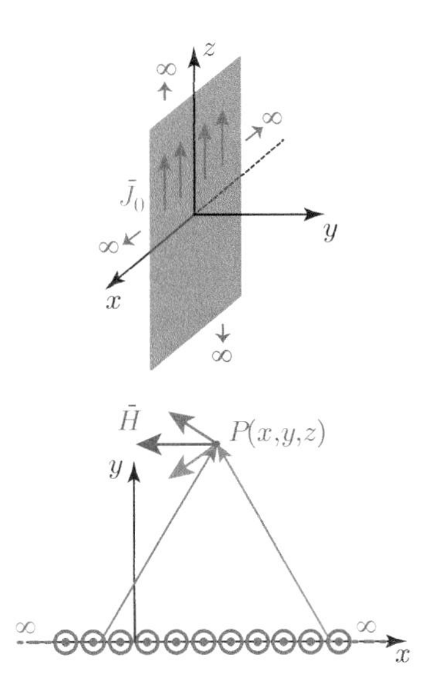

Figure 2.48 An infinitely large steady surface electric current density on the z-x plane, flowing in the z direction. The magnetic field should not depend on x and z, and it should have only an x component. The direction of the magnetic field can be understood by considering that, for any point in space, one can find two line electric currents (infinitesimal parts of the infinite sheet) creating opposite but equal magnetic fields in the $\pm y$ direction. Note that we are also using the superposition principle: i.e. magnetic fields created by different electric currents can be added.

First, by using infinity and symmetry, we claim that the magnetic field does not depend on x and z, while it should be in the $-x$ and $+x$ directions for $y > 0$ and $y < 0$, respectively. Then the magnetic field intensity can be written

$$\bar{H}(\bar{r}) = -\hat{a}_x \begin{cases} f(|y|) & (y > 0) \\ -f(|y|) & (y < 0), \end{cases} \tag{2.195}$$

where $f(|y|)$ is a function of $|y|$: i.e. the distance from the surface electric current density. If one uses the boundary condition for the tangential magnetic field intensity, we have

$$\hat{a}_y \times [\bar{H}_1(\bar{r}) - \bar{H}_2(\bar{r})] = \bar{J}_s(\bar{r}) \qquad (\bar{r} \in S), \tag{2.196}$$

leading to $\hat{a}_z 2f(|y| = 0) = \hat{a}_z J_0$ or

$$f(|y| = 0) = J_0/2. \tag{2.197}$$

Interestingly, similar to the electric field due to an infinitely large surface electric charge density, the function $f(|y|)$ in the above and the magnetic field intensity do not depend on y. To see this, one can consider an Ampere loop in the shape of a rectangle located symmetrically (Figure 2.49). The rectangle's height is $2h$: i.e. the distance between the top/bottom edge and the electric current density is h. The contributions from the side edges are zero (since the magnetic field intensity is in the $\pm x$ direction), while the top and bottom edges' contributions are the same.[41] At the same time, no matter what the value of h is, the same amount of electric current is enclosed inside the Ampere surface. This indicates that the magnetic field intensity does not depend on y, and we have $f(|y|) = f(|y| = 0)$, leading to

$$\bar{H}(\bar{r}) = -\hat{a}_x \frac{J_0}{2} \begin{cases} 1 & (y > 0) \\ -1 & (y < 0). \end{cases} \tag{2.198}$$

We note that an infinitely long steady line electric current creates a magnetic field decaying with the distance $(1/\rho)$, while there is no decay for a surface electric current density of infinite size.

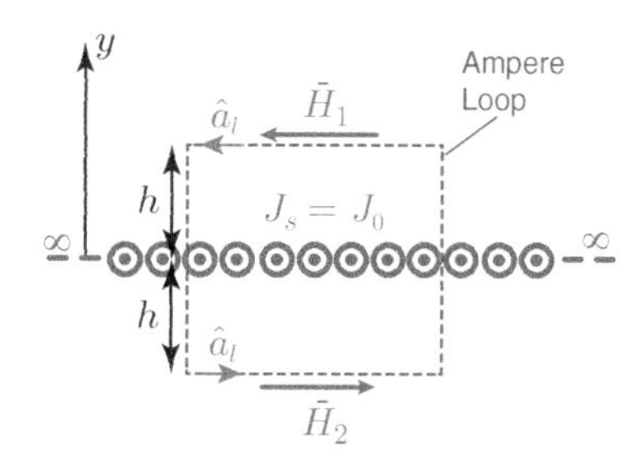

Figure 2.49 An application of the integral form of Ampere's law shows that the magnetic field intensity due to an infinitely large surface electric current density does not depend on the distance from the current density.

[41] We note that the direction of the line integration $\hat{a}_l$ and the direction of the magnetic field intensity are aligned at both $y = -h$ and $y = +h$ in Figure 2.49.

2.6 Gauss' Law and Ampere's Law

Gauss' law and Ampere's law provide information about how electric and magnetic fields are created. According to Gauss' law, electric charge distributions create electric fields. According to Ampere's law, electric current distributions create magnetic fields. Ampere's law further states that electric fields are also sources of magnetic fields in *dynamic* cases. On the other hand, it does not provide a complete relationship between electric and magnetic fields. For a complete picture, we need Faraday's law and the information that magnetic fields are solenoidal (Gauss' law for magnetic fields). These are discussed in the following chapters.

While Gauss' law and Ampere's law are valid for all cases – i.e. they describe both static and dynamic scenarios – they are particularly useful in electrostatics and magnetostatics, respectively. This is because when they are used as isolated tools (without the link provided by Faraday's law), they do not fully describe a dynamic event: e.g. electromagnetic propagation. On the other hand, combining Gauss' law, Ampere's law, Faraday's law, and Gauss' law for magnetic fields, we can fully formulate both static and dynamic electromagnetic scenarios.

At this stage, an interesting practice is to combine Gauss' law and Ampere's law, as much as we can. In electromagnetics, different differential equations are combined by using derivatives such that similar elements arise. For example, considering the divergence of the differential form of Ampere's law, we have

$$\bar{\nabla} \cdot \bar{\nabla} \times \bar{H}(\bar{r},t) = \bar{\nabla} \cdot \bar{J}_v(\bar{r},t) + \bar{\nabla} \cdot \frac{\partial \bar{D}(\bar{r},t)}{\partial t} = \bar{\nabla} \cdot \bar{J}_v(\bar{r},t) + \frac{\partial}{\partial t}\bar{\nabla} \cdot \bar{D}(\bar{r},t). \tag{2.199}$$

Further using Gauss' law, we obtain

$$\bar{\nabla} \cdot \bar{\nabla} \times \bar{H}(\bar{r},t) = \bar{\nabla} \cdot \bar{J}_v(\bar{r},t) + \frac{\partial q_v(\bar{r},t)}{\partial t}. \tag{2.200}$$

At this point, we can introduce the null identity[42]

$$\bar{\nabla} \cdot \bar{\nabla} \times \bar{f}(\bar{r},t) = 0 \tag{2.201}$$

for any vector field $\bar{f}(\bar{r},t)$. Therefore, we reach a relationship between the volume electric current density and the volume electric charge density:[43]

$$\bar{\nabla} \cdot \bar{J}_v(\bar{r},t) = -\frac{\partial q_v(\bar{r},t)}{\partial t}. \tag{2.202}$$

This equality, which is called *continuity* equation, is one of most basic equations in electromagnetics and related areas.[44] It states that the electric current density is related to the negative time derivative of the electric charge density. Specifically, if the electric charge density is changing with respect to time (decreasing/increasing) at a position $\bar{r}$, then there should be a *change* in the electric current density with respect to position (due to a source/sink) exactly at the

[42] This can be shown easily in a Cartesian coordinate system as

$$\begin{aligned}
&\bar{\nabla} \cdot \bar{\nabla} \times \bar{f}(\bar{r},t)\\
&= \left(\hat{a}_x\frac{\partial}{\partial x} + \hat{a}_y\frac{\partial}{\partial y} + \hat{a}_z\frac{\partial}{\partial z}\right) \cdot \bar{\nabla} \times \bar{f}(\bar{r},t)\\
&= \frac{\partial}{\partial x}\left(\frac{\partial f_z(\bar{r},t)}{\partial y} - \frac{\partial f_y(\bar{r},t)}{\partial z}\right)\\
&+ \frac{\partial}{\partial y}\left(\frac{\partial f_x(\bar{r},t)}{\partial z} - \frac{\partial f_z(\bar{r},t)}{\partial x}\right)\\
&+ \frac{\partial}{\partial z}\left(\frac{\partial f_y(\bar{r},t)}{\partial x} - \frac{\partial f_x(\bar{r},t)}{\partial y}\right)\\
&= \frac{\partial^2 f_z(\bar{r},t)}{\partial x \partial y} - \frac{\partial^2 f_y(\bar{r},t)}{\partial x \partial z}\\
&+ \frac{\partial^2 f_x(\bar{r},t)}{\partial y \partial z} - \frac{\partial^2 f_z(\bar{r},t)}{\partial y \partial x}\\
&+ \frac{\partial^2 f_y(\bar{r},t)}{\partial z \partial x} - \frac{\partial^2 f_x(\bar{r},t)}{\partial z \partial y}\\
&= 0.
\end{aligned}$$

[43] In the phasor domain, the continuity equation can be written

$$\bar{\nabla} \cdot \bar{J}(\bar{r}) = -j\omega q_v(\bar{r}).$$

[44] So, from one perspective, the continuity equation is a basic identity so that Gauss' law and Ampere's law are in fact related to each other via the continuity equation.

same point. For a further physical interpretation (Figure 2.50), we may integrate the expression in an arbitrary stationary volume and use the divergence theorem as

$$\int_V \bar{\nabla} \cdot \bar{J}_v(\bar{r}, t) dv = -\int_V \frac{\partial q_v(\bar{r}, t)}{\partial t} dv = \frac{d}{dt} \int_V q_v(\bar{r}, t) dv \tag{2.203}$$

$$\oint_S \bar{J}_v(\bar{r}, t) \cdot \overline{ds} = -\frac{dQ(t)}{dt} \longrightarrow I_{\text{net}}(t) = -\frac{dQ(t)}{dt}, \tag{2.204}$$

leading to the basic definition of the electric current in terms of the electric charge, as also given in Eq. (32).

The expression in Eq. (2.204) can be seen as a generalized Kirchhoff's current law (Figure 2.50). It states that, given a volume V, there is a net electric current $I_{\text{net}}(t)$ through its surface if the electric charge enclosed in the volume changes. Obviously, if $dQ(t)/dt = 0$ (no time dependency in the electric charge), then

$$I_{\text{net}}(t) = \oint_S \bar{J}_v(\bar{r}, t) \cdot \overline{ds} = 0. \tag{2.205}$$

Specifically, electric currents entering and leaving a volume (which can be a node in a circuit) balance each other. In magnetostatics, we have steady electric currents and no electric charge accumulation (no time-variant electric fields) so that

$$\oint_S \bar{J}_v(\bar{r}) \cdot \overline{ds} = 0. \tag{2.206}$$

As a pointwise equality, this can be written

$$\bar{\nabla} \cdot \bar{J}_v(\bar{r}) = 0 \tag{2.207}$$

when considering steady electric currents.

As a common practice to be discussed later, one can take the curl of Ampere's law as

$$\bar{\nabla} \times \bar{\nabla} \times \bar{H}(\bar{r}, t) = \bar{\nabla} \times \bar{J}_v(\bar{r}, t) + \bar{\nabla} \times \frac{\partial \bar{D}(\bar{r}, t)}{\partial t}. \tag{2.208}$$

This equation is not informative yet since we do not know the curl of the electric flux density. But we will see how this equation is extremely useful to describe electromagnetic events.

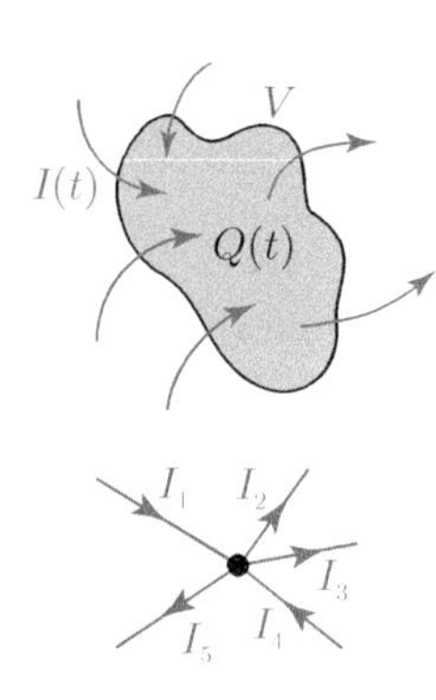

Figure 2.50 When the continuity equation is used for an arbitrary volume, we obtain the relationship between electric currents and electric charges. Specifically, if the net electric current leaving the volume – i.e. $\oint_S \bar{J}_v(\bar{r}, t) \cdot \overline{ds}$ – is positive/negative, then the total electric charge inside volume should decrease/increase (conservation of electric charges). This can be seen as the generalized version of Kirchhoff's current law. In circuit analyses, Kirchhoff's current law is used at nodes, where multiple wires are connected. Since there cannot be electric charge accumulation at nodes, the algebraic sum of electric currents entering/leaving a node should be zero. In the above, we have $I_1(t) - I_2(t) - I_3(t) + I_4(t) - I_5(t) = 0$.

2.7 Static Cases, Biot-Savart Law, and Ampere's Force Law

Ampere's law is valid for both static and dynamic cases. If time-harmonic sources are involved, Ampere's law can be written[45]

$$\oint_C \bar{H}(\bar{r}) \cdot \overline{dl} = I_{\text{enc}} + \int_S j\omega \bar{D}(\bar{r}) \cdot \overline{ds} \tag{2.209}$$

$$\bar{\nabla} \times \bar{H}(\bar{r}) = \bar{J}_v(\bar{r}) + j\omega \bar{D}(\bar{r}), \tag{2.210}$$

where $\bar{H}(\bar{r})$, I_{enc}, $\bar{J}_v(\bar{r})$, and $\bar{D}(\bar{r})$ are phasor-domain representations, while ω is the angular frequency. Nevertheless, as discussed in Section 2.2, Ampere's law is not sufficient alone to analyze a dynamic scenario. This is because it provides information about how magnetic fields are created from electric currents and time-variant electric fields, while such electric fields may be created by changing magnetic fields as formulated by Faraday's law. In a static case, where steady electric currents are involved and nothing depends on time, Ampere's law can be used directly to find magnetic fields from electric currents. This was shown before for symmetric electric current distributions (using the integral form of Ampere's law or its application at boundaries). In this section, we further focus on static cases.[46] and show the relationship between Ampere's law, Biot-Savart law, and Ampere's force law.[47]

Before proceeding, we note that $\omega = 0$ represents static cases. Obviously, when ω is set to zero, the electric flux term in the equations above drops such that we reach the equations for magnetostatics. A very interesting (but difficult to interpret) limit occurs when $\omega \to \infty$. In this case, to have a finite magnetic field intensity, the electric flux density should go to zero. In the case of a capacitor discussed in Section 2.1.3, this means the charges accumulated on the capacitor terminals should vanish. This is consistent with the fact that capacitors in time-harmonic circuits behave like short circuits as the frequency goes to infinity. When the frequency is too high, charges *do not have time* to accumulate.

[45] We recall that, using the $\exp(j\omega t)$ time convention, a time derivative is replaced with $j\omega$, as formulated in Section 0.6. The exponential $\exp(j\omega t)$ means all sources and fields in the considered electromagnetic problem has the form of $f(\bar{r}, t) = f_0(\bar{r}) \cos(\omega t + \phi_0)$.

[46] In electrostatics, we consider static electric charges: i.e. stationary and time-invariant electric charges and electric charge distributions. In magnetostatics, we consider steady electric currents: i.e. stationary and time-invariant electric currents and electric current distributions. The term *static electric current* is not common since an electric current means a movement of electric charges and there is actually something moving.

[47] Biot-Savart law describes how magnetic fields are created by steady electric currents, while Ampere's force law describes forces between steady electric currents. Hence, they correspond to Coulomb's law in electrostatics.

2.7.1 Superposition Principle

As described in Eq. (2.59), the magnetic field intensity created by an infinitely long steady electric current I_0 along the z axis in a vacuum can be written (Figure 2.51)

$$\bar{H}(\bar{r}) = \hat{a}_\phi \frac{I_0}{2\pi\rho}. \tag{2.211}$$

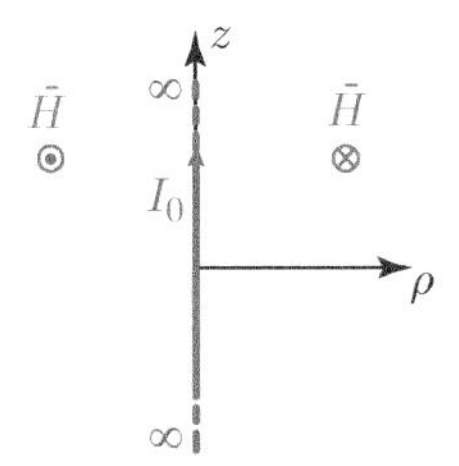

Figure 2.51 An infinitely long steady electric current flowing in the z direction in a vacuum. The magnetic field intensity decays with the distance ($1/\rho$), while its direction can be found via the right-hand rule.

Similar to the application of the superposition principle for static point electric charges, we can add the magnetic field intensity created by multiple electric current distributions. Hence, we *assume* that the existence of an electric current distribution does not change the magnetic field created by another electric current distribution.[48] First, we can consider an infinitely long steady electric current oriented in the z direction, but at an arbitrary location (x and y). Since the right-hand rule does not change, we have[49]

$$\bar{H}(\rho) = \left(\hat{a}_z \times \frac{(\bar{\rho} - \bar{\rho}')}{|\bar{\rho} - \bar{\rho}'|}\right) \frac{I_0}{2\pi|\bar{\rho} - \bar{\rho}'|} = \frac{I_0}{2\pi}\hat{a}_z \times \frac{(\bar{\rho} - \bar{\rho}')}{|\bar{\rho} - \bar{\rho}'|^2}, \tag{2.212}$$

where $\bar{\rho} = \hat{a}_x x + \hat{a}_y y$ is a cylindrical position vector (Figure 2.52). We note that primed and unprimed coordinates are used to represent the source and observation locations, respectively. Also note that $\hat{a}_z$ is the electric current direction and the cross product

$$\hat{a}_H = \hat{a}_z \times \frac{(\bar{\rho} - \bar{\rho}')}{|\bar{\rho} - \bar{\rho}'|} \tag{2.213}$$

is an application of the right-hand rule to find the direction of the magnetic field intensity. When Eq. (2.212) is compared with Eq. (1.181), it can be observed that the cube of the distance ($|\bar{r} - \bar{r}'|^3$) is replaced with the square of the distance ($|\bar{\rho} - \bar{\rho}'|^2$) in the denominator (other than the directional difference). This is because the line electric current is infinitely long and the decay is one-order less (inversely proportional to the distance rather than its square).

Next, we can consider multiple infinitely long steady electric currents, all oriented in the z direction (Figure 2.53). Given an observation point $\bar{\rho}$, we have

$$\bar{H}(\bar{\rho}) = \sum_{i=1}^{n} \frac{I_i}{2\pi}\hat{a}_z \times \frac{(\bar{\rho} - \bar{\rho}_i)}{|\bar{\rho} - \bar{\rho}_i|^2}, \tag{2.214}$$

where $\bar{\rho}_i$ represents the location of the electric current I_i. Equation (2.214) can be compared with the electric flux density due to multiple static point electric charges in Eq. (1.182).

Like the electric field calculations, Eq. (2.214) can be generalized to surface and volume electric current densities. However, these electric current distributions are restricted to infinitely long geometries with fixed cross-sections. This is because, in the above, the superposition principle is used for an infinitely long electric current as the most basic element. For an arbitrary electric current distribution, we need to know the magnetic field created by an infinitesimal electric current filament $I_0\overline{dl'}$ (Figure 2.54). According to Biot-Savart law, we have

$$\bar{H}(\bar{r}) = \frac{I_0}{4\pi} \frac{\overline{dl'} \times (\bar{r} - \bar{r}')}{|\bar{r} - \bar{r}'|^3}. \tag{2.215}$$

This equation is very similar to the expression for the electric field created by a static point electric charge in Eq. (1.181), which can be repeated here as

$$\bar{D}(\bar{r}) = \frac{Q_0}{4\pi} \frac{(\bar{r} - \bar{r}')}{|\bar{r} - \bar{r}'|^3}. \tag{2.216}$$

[48] Of course, we also assume that an electric current distribution does not change the shape of the other electric current distribution. But the superposition principle involves a *further* assumption that the magnetic field intensity created by an electric current distribution remains the same when it is not located alone: i.e. when there are other electric current distributions located in the same space.

[49] Since there is no z variance, we can write the magnetic field intensity as $\bar{H}(\rho)$, where $\hat{a}_\rho \rho = \hat{a}_x x + \hat{a}_y y = \bar{r} - \hat{a}_z z$.

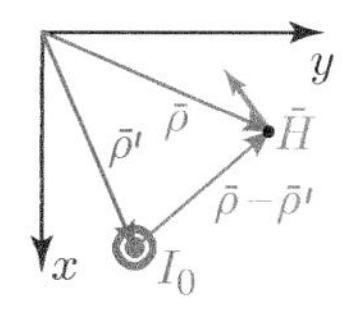

Figure 2.52 An infinitely long steady electric current aligned in the z direction at an arbitrary location. The right-hand rule between the electric current direction and the magnetic field intensity direction should not change, while the magnitude of the magnetic field intensity should be again inversely proportional to the distance.

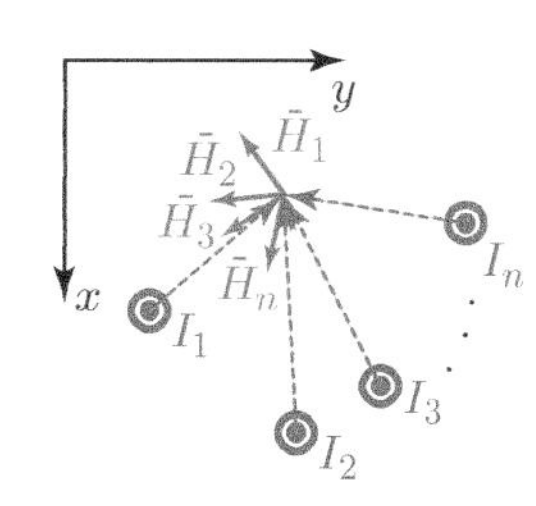

Figure 2.53 Magnetic fields created by a set of infinitely long steady electric currents oriented in the z direction. The vector sum of these magnetic fields gives the total magnetic field at the observation point.

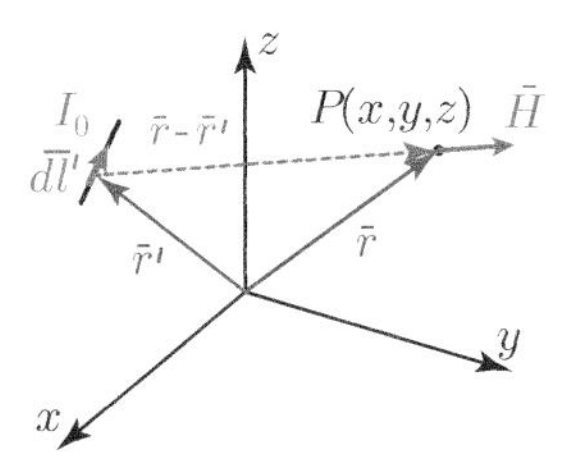

Figure 2.54 The magnetic field created by a steady electric current filament $I_0\overline{dl'}$ located at $\bar{r}' = \hat{a}_x x' + \hat{a}_y y' + \hat{a}_z z'$ at another point $\bar{r} = \hat{a}_x x + \hat{a}_y y + \hat{a}_z z$.

However, as opposed to Gauss' law that leads to Eq. (2.216) for a static point electric charge, it cannot be directly derived from Ampere's law since an electric current filament (even when it is aligned in the z direction) does not possess a symmetry that is useful to employ Ampere's law.[50] On the other hand, we can use it to derive the expressions for the magnetic field intensity due to arbitrary steady electric current distributions. Specifically, for volume (A/m^2), surface (A/m), and line (A) electric current distributions, we have[51]

$$\bar{H}(\bar{r}) = \frac{1}{4\pi}\int_V \frac{\bar{J}_v(\bar{r}') \times (\bar{r}-\bar{r}')}{|\bar{r}-\bar{r}'|^3} d\bar{r}' \tag{2.217}$$

$$\bar{H}(\bar{r}) = \frac{1}{4\pi}\int_S \frac{\bar{J}_s(\bar{r}') \times (\bar{r}-\bar{r}')}{|\bar{r}-\bar{r}'|^3} d\bar{r}' \tag{2.218}$$

$$\bar{H}(\bar{r}) = \frac{1}{4\pi}\int_C \frac{I_0\overline{dl'} \times (\bar{r}-\bar{r}')}{|\bar{r}-\bar{r}'|^3} \tag{2.219}$$

where the integration region depends on the type of the distribution.[52]

At this stage, it may be useful to check these expressions (Biot-Savart law) with Ampere's law for static cases. For this purpose, we consider the curl of the magnetic field intensity due to a volume electric current distribution:

$$\bar{\nabla} \times \bar{H}(\bar{r}) = \frac{1}{4\pi}\bar{\nabla} \times \int_V \frac{\bar{J}_v(\bar{r}') \times (\bar{r}-\bar{r}')}{|\bar{r}-\bar{r}'|^3} d\bar{r}' \tag{2.220}$$

$$= \frac{1}{4\pi}\int_V \bar{\nabla} \times \left[\bar{J}_v(\bar{r}') \times \left(\frac{(\bar{r}-\bar{r}')}{|\bar{r}-\bar{r}'|^3}\right)\right] d\bar{r}'. \tag{2.221}$$

Using the vector identity

$$\bar{\nabla} \times [\bar{f}(\bar{r}) \times \bar{g}(\bar{r})] = \bar{f}(\bar{r})[\bar{\nabla} \cdot \bar{g}(\bar{r})] - \bar{g}(\bar{r})[\bar{\nabla} \cdot \bar{f}(\bar{r})] + (\bar{g}(\bar{r}) \cdot \bar{\nabla})\bar{f}(\bar{r}) - (\bar{f}(\bar{r}) \cdot \bar{\nabla})\bar{g}(\bar{r}), \tag{2.222}$$

and considering that the del operator in Eq. (2.221) is on *primed* coordinates,[53] we have

$$\bar{\nabla} \times \bar{H}(\bar{r}) = \frac{1}{4\pi}\int_V \bar{J}_v(\bar{r}')\bar{\nabla} \cdot \left(\frac{(\bar{r}-\bar{r}')}{|\bar{r}-\bar{r}'|^3}\right) d\bar{r}' \tag{2.223}$$

$$= \frac{1}{4\pi}\int_V \bar{J}_v(\bar{r}')4\pi\delta(|\bar{r}-\bar{r}'|)d\bar{r}' = \bar{J}_v(\bar{r}) \tag{2.224}$$

[50] In fact, the concept of an electric current filament can cause other theoretical problems: e.g. in magnetic force (see below). However, by defining such a current filament, we can construct an analogy between electric and magnetic fields due to point sources.

[51] All these expressions can be called Biot-Savart law, in general.

[52] Similar to the integral forms in Eqs. (1.183), (1.184), and (1.185) for the electric flux density due to static electric charge distributions, these expressions can be derived more completely via the magnetic vector potential.

[53] Writing in open form, we have

$$\bar{\nabla}' = \hat{a}_x\frac{\partial}{\partial x'} + \hat{a}_y\frac{\partial}{\partial y'} + \hat{a}_z\frac{\partial}{\partial z'}$$

so that $\bar{\nabla}' \cdot \bar{f}(\bar{r}) = 0$ and $\bar{\nabla}' \times \bar{f}(\bar{r}) = 0$ since $\bar{f}(\bar{r})$ uses unprimed coordinates: i.e. $\bar{r} = \hat{a}_x x + \hat{a}_y y + \hat{a}_z z$. Specifically, the curl operator in Eq. (2.220) is used on the magnetic field intensity that is a function of the observation point (unprimed coordinates). When we place it inside the integral in Eq. (2.221), it should not be applied to the electric current density.

using the limit in Eq. (1.129). The final equality is simply the differential form of Ampere's law for static cases.

To verify Biot-Savart law and the integral form of Ampere's law with each other, we may reconsider an infinitely long steady line electric current in the z direction (Figure 2.55). The overall electric current can be assumed to be formed of infinitely many electric current filaments defined as

$$\overline{dl'} = \hat{a}_z dz'. \tag{2.225}$$

In addition, the source and observation points can be shown with position vectors as

$$\bar{r} = \hat{a}_x x + \hat{a}_y y + \hat{a}_z z \quad \text{and} \quad \bar{r}' = \hat{a}_z z'. \tag{2.226}$$

We note that only the z component exists for the primed position vector since the source is located on the z axis. However, for the unprimed position vector, we keep all components to find an expression for a general observation point. Then using Eq. (2.219), we have[54]

$$\bar{H}(\bar{r}) = \frac{I_0}{4\pi} \int_C \frac{\hat{a}_z \times (\hat{a}_x x + \hat{a}_y y + \hat{a}_z z - \hat{a}_z z')}{[x^2 + y^2 + (z - z')^2]^{3/2}} dz' \tag{2.227}$$

$$= \frac{I_0}{4\pi} \int_C \frac{(\hat{a}_y x - \hat{a}_x y)}{[x^2 + y^2 + (z - z')^2]^{3/2}} dz' \tag{2.228}$$

$$= \hat{a}_\phi \frac{I_0}{4\pi} \rho \int_{-\infty}^{\infty} \frac{1}{[\rho^2 + (z - z')^2]^{3/2}} dz' \tag{2.229}$$

using $\hat{a}_\phi \rho = \hat{a}_y x - \hat{a}_x y$. For the evaluation of the integral, we may first consider it from $-l$ to l as[55]

$$\int_{-l}^{l} \frac{1}{[\rho^2 + (z - z')^2]^{3/2}} dz' = \left[\frac{z' - z}{\rho^2 \sqrt{\rho^2 + (z - z')^2}} \right]_{-l}^{l} \tag{2.230}$$

$$= \frac{l - z}{\rho^2 \sqrt{\rho^2 + (z - l)^2}} - \frac{-l + z}{\rho^2 \sqrt{\rho^2 + (z + l)^2}}. \tag{2.231}$$

Then evaluating the limit for $l \to \infty$, we arrive at[56]

$$\lim_{l \to \infty} \int_{-l}^{l} \frac{1}{[\rho^2 + (z - z')^2]^{3/2}} dz' = \frac{l - z}{\rho^2 \sqrt{\rho^2 + (z - l)^2}} - \frac{-l + z}{\rho^2 \sqrt{\rho^2 + (z + l)^2}} \tag{2.232}$$

$$= \frac{2}{\rho^2}, \tag{2.233}$$

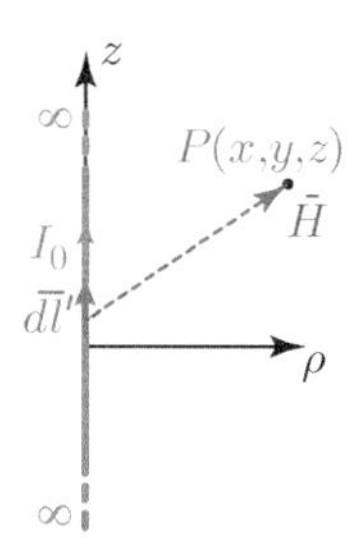

Figure 2.55 The magnetic field intensity created by an infinitely long steady line electric current can be found via Biot-Savart law.

[54]In many texts, this line integral is written as a closed loop ($\oint_C$) since steady electric currents must flow along closed loops. For this case (infinitely long line electric current), one can still claim the existence of a closed loop, where the two ends of the line are connected at infinity.

[55]It is remarkable that, for $z = 0$, this integral result reduces into

$$\frac{l}{\rho^2\sqrt{\rho^2 + l^2}} - \frac{-l}{\rho^2\sqrt{\rho^2 + l^2}} = \frac{2l}{\rho^2\sqrt{\rho^2 + l^2}}$$

leading to an expression for the magnetic field intensity as

$$\bar{H}(\bar{r}) = \hat{a}_\phi \frac{I_0}{4\pi} \rho \frac{2l}{\rho^2\sqrt{\rho^2 + l^2}} = \hat{a}_\phi \frac{I_0 l}{2\pi\rho\sqrt{\rho^2 + l^2}}.$$

This expression is the magnetic field intensity due to a symmetrically located segment of steady line electric current with $2l$ length.

[56]As expected, the magnetic field intensity does not change in the z direction since the source is infinitely long and symmetric such that a movement of the observer in the z direction does not make any difference.

leading to the magnetic field intensity as

$$\bar{H}(\bar{r}) = \bar{H}(\bar{\rho}) = \hat{a}_\phi \frac{I_0}{4\pi}\rho\frac{2}{\rho^2} = \hat{a}_\phi \frac{I_0}{2\pi\rho}. \tag{2.234}$$

This expression is the same as the one found by using the integral form of Ampere's law in Eq. (2.59).

To further practice the superposition principle in the context of Ampere's law, we consider two infinitely long lines, both aligned in the z direction, carrying steady electric currents I_1 and I_2 (Figure 2.56). Using Ampere's law and employing Ampere loops centered at $\bar{\rho}_1$ and $\bar{\rho}_2$ (positions of the line electric currents), we have[57]

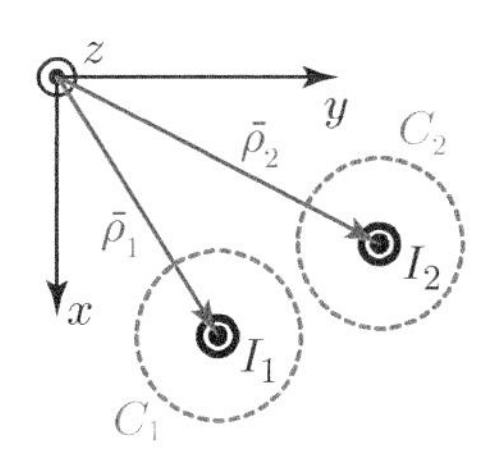

Figure 2.56 Two steady line electric currents (both oriented in the z direction) located at different positions and circular Ampere loops C_1 and C_2.

[57] We simply neglect z since the magnetic field intensity does not change with respect to z. Hence the locations are represented by a position vector $\bar{\rho} = \hat{a}_x x + \hat{a}_y y$.

$$\oint_{C_1} \bar{H}_1(\bar{\rho}) \cdot \overline{dl} = I_1 \longrightarrow \bar{H}_1(\bar{\rho}) = \frac{I_1}{2\pi}\hat{a}_z \times \frac{(\bar{\rho}-\bar{\rho}_1)}{|\bar{\rho}-\bar{\rho}_1|^2} \tag{2.235}$$

and

$$\oint_{C_2} \bar{H}_2(\bar{\rho}) \cdot \overline{dl} = I_2 \longrightarrow \bar{H}_2(\bar{\rho}) = \frac{I_2}{2\pi}\hat{a}_z \times \frac{(\bar{\rho}-\bar{\rho}_2)}{|\bar{\rho}-\bar{\rho}_2|^2}, \tag{2.236}$$

where $\bar{\rho}$ represents the observation point. Hence, $\bar{H}_1(\bar{\rho})$ and $\bar{H}_2(\bar{\rho})$ that are due to the first and the second lines, respectively, can be found via Ampere's law. Then the total magnetic field intensity at any arbitrary position $\bar{\rho}$ can be written[58]

$$\bar{H}(\bar{\rho}) = \bar{H}_1(\bar{\rho}) + \bar{H}_2(\bar{\rho}) = \frac{I_1}{2\pi}\hat{a}_z \times \frac{(\bar{\rho}-\bar{\rho}_1)}{|\bar{\rho}-\bar{\rho}_1|^2} + \frac{I_2}{2\pi}\hat{a}_z \times \frac{(\bar{\rho}-\bar{\rho}_2)}{|\bar{\rho}-\bar{\rho}_2|^2}. \tag{2.237}$$

[58] An interesting case is when the observation point is between the electric current lines: i.e. $\bar{\rho} = (\bar{\rho}_1 + \bar{\rho}_2)/2$, while $I_1 = I_2 = I_0$. We have

$$\begin{aligned}\bar{H}(\bar{\rho}) &= \frac{I_0}{2\pi}\hat{a}_z \\ &\times \left(\frac{(\bar{\rho}_2/2-\bar{\rho}_1/2)}{|\bar{\rho}_2/2-\bar{\rho}_1/2|^2} + \frac{(\bar{\rho}_1/2-\bar{\rho}_2/2)}{|\bar{\rho}_1/2-\bar{\rho}_2/2|^2}\right) \\ &= 0,\end{aligned}$$

i.e. the magnetic fields created by the lines perfectly cancel each other.

Since Ampere's law is valid for all cases, we can also write

$$\oint_{C_1} \bar{H}(\bar{\rho}) \cdot \overline{dl} = \oint_{C_1} [\bar{H}_1(\bar{\rho}) + \bar{H}_2(\bar{\rho})] \cdot \overline{dl} = I_1 \tag{2.238}$$

$$\oint_{C_2} \bar{H}(\bar{\rho}) \cdot \overline{dl} = \oint_{C_2} [\bar{H}_1(\bar{\rho}) + \bar{H}_2(\bar{\rho})] \cdot \overline{dl} = I_2, \tag{2.239}$$

while $\bar{H}(\bar{\rho})$ cannot be placed outside the integrals due to the lack of symmetry of the overall scenario (two electric current lines). Obviously, we further have

$$\oint_{C_1} \bar{H}_2(\bar{\rho}) \cdot \overline{dl} = 0 \quad \text{and} \quad \oint_{C_2} \bar{H}_1(\bar{\rho}) \cdot \overline{dl} = 0. \tag{2.240}$$

The superposition principle is naturally valid and inherently used for all cases.[59] As a direct illustration, we consider a coaxial structure involving two cylindrical surfaces with radii a and b, where $b > a$ (Figure 2.57). There are surface electric currents that are uniformly distributed over the cylinders in opposite directions. Using Eq. (2.175), we have

$$\bar{H}_{\text{inner}}(\bar{\rho}) = \hat{a}_\phi \frac{I_0}{2\pi\rho} \begin{cases} 0 & (\rho < a) \\ 1 & (\rho > a) \end{cases} \tag{2.241}$$

and

$$\bar{H}_{\text{outer}}(\bar{\rho}) = -\hat{a}_\phi \frac{I_0}{2\pi\rho} \begin{cases} 0 & (\rho < b) \\ 1 & (\rho > b) \end{cases} \tag{2.242}$$

as the magnetic field intensity distributions created by the electric currents on the inner and outer cylinders. Combining these fields (via superposition principle), we have

$$\bar{H}(\bar{\rho}) = \bar{H}_{\text{inner}}(\bar{\rho}) + \bar{H}_{\text{outer}}(\bar{\rho}) = \hat{a}_\phi \frac{I_0}{2\pi\rho} \begin{cases} 0 & (\rho < a) \\ 1 & (a < \rho < b) \\ 0 & (\rho > b). \end{cases} \tag{2.243}$$

Specifically, nonzero magnetic field exists only between the cylinders, while it is zero both inside the inner cylinder and outside the overall structure.

As another illustration of the superposition principle, we now consider two infinitely long electric current lines (cylinders), each of radius a (Figure 2.58). The lines are aligned in the z direction. One of them is located at the origin, while the other is at $x = d$, where $a \ll d$. We also assume that the electric currents are $\pm I_0$ – i.e. in opposite directions – and they are flowing only through surfaces of the cylinders. If the magnetic field intensity at any location on the $y = 0$ plane (and for $x > 0$ for simplicity) needs to be found, we may consider the cylinders one by one and add their effects via superposition. The magnetic field intensity created by the first cylinder can be found as

$$\bar{H}_1(x, y = 0, z) = \hat{a}_y \frac{I_0}{2\pi x} \begin{cases} 1 & (x > a) \\ 0 & (0 < x < a) \end{cases}. \tag{2.244}$$

Similarly, the magnetic field intensity created by the second cylinder can be written

$$\bar{H}_2(x, y = 0, z) = \hat{a}_y \frac{I_0}{2\pi|d - x|} \begin{cases} 1 & (0 < x < d - a) \\ 0 & (d - a < x < d + a) \\ -1 & (x > d + a) \end{cases}. \tag{2.245}$$

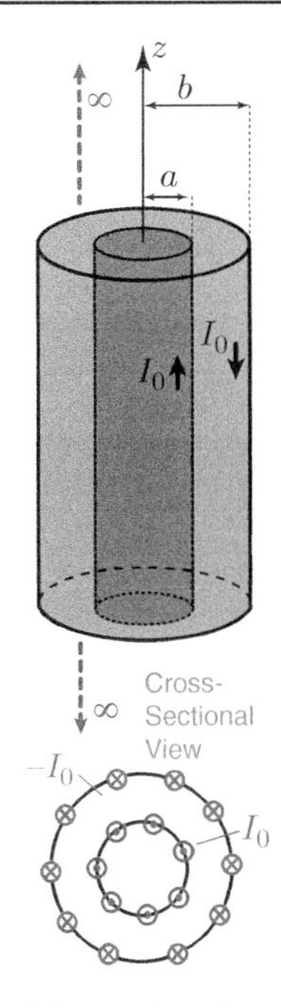

Figure 2.57 A coaxial structure involving two cylindrical surfaces with radii a and $b > a$ aligned in the z direction. Surface electric currents that are uniformly distributed over the cylindrical surfaces flow in opposite directions.

[59]In time-variant cases, the time (and delay) is involved when superposing effects of sources.

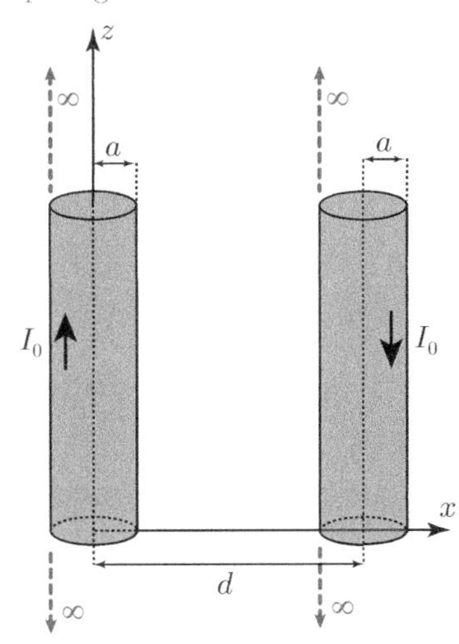

Figure 2.58 Two infinitely long cylinders carrying steady electric currents in opposite directions.

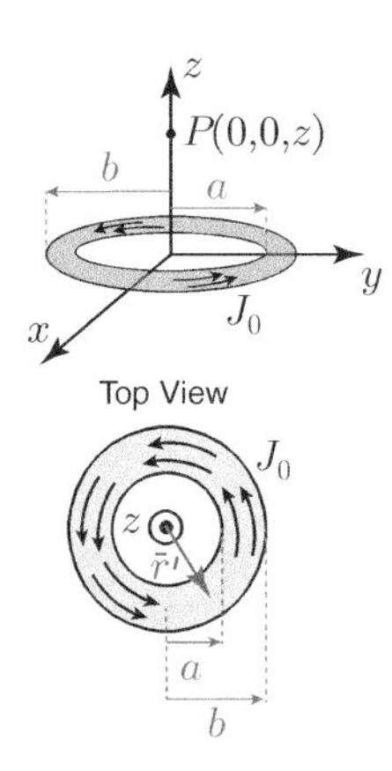

Figure 2.59 A steady surface electric current distribution with a ring shape.

Then the total magnetic field intensity can be obtained as

$$\bar{H}(x, y=0, z) = \bar{H}_1(x, y=0, z) + \bar{H}_2(x, y=0, z). \tag{2.246}$$

We note that, if $x = d/2$, the expressions become

$$\bar{H}_1(d/2, y=0, z) = \bar{H}_2(d/2, y=0, z) = \hat{a}_y \frac{I_0}{\pi d} \tag{2.247}$$

and

$$\bar{H}(d/2, y=0, z) = \hat{a}_y \frac{2I_0}{\pi d}. \tag{2.248}$$

Biot-Savart law – i.e. the source integrals in Eqs. (2.217), (2.218), and (2.219) – can be used directly to find the magnetic field created by an arbitrary (but steady) electric current distribution, especially when Ampere's law cannot be used due to a lack of symmetry. As an example, we consider ring-shaped surface electric current density in a vacuum and the magnetic field intensity created on its axis (at an arbitrary point on the z axis). A constant steady electric current with density J_0 (A/m) flows in the ϕ direction and can be written $\bar{J}_s(\bar{r}) = \hat{a}_\phi J_0$ for $a < \rho < b$. By using Eq. (2.218), we have

$$\bar{H}(\bar{r}) = \frac{J_0}{4\pi} \int_0^{2\pi} \int_a^b \frac{\hat{a}_\phi \times (\hat{a}_z z - \hat{a}_x x' - \hat{a}_y y')}{[(x')^2 + (y')^2 + z^2]^{3/2}} \rho' d\rho' d\phi' \tag{2.249}$$

using $\bar{r} = \hat{a}_z z$ and $\bar{r}' = \hat{a}_x x' + \hat{a}_y y'$. Furthermore, using $x' = \rho' \cos\phi'$, $y' = \rho' \sin\phi'$, and

$$\hat{a}_\phi \times \hat{a}_z = \hat{a}_\rho = \hat{a}_x \cos\phi' + \hat{a}_x \sin\phi' \tag{2.250}$$

$$\hat{a}_\phi \times \hat{a}_x = (-\hat{a}_x \sin\phi' + \hat{a}_y \cos\phi') \times \hat{a}_x = -\hat{a}_z \cos\phi' \tag{2.251}$$

$$\hat{a}_\phi \times \hat{a}_y = (-\hat{a}_x \sin\phi' + \hat{a}_y \cos\phi') \times \hat{a}_y = -\hat{a}_z \sin\phi', \tag{2.252}$$

we arrive at[60,61]

$$\bar{H}(\bar{r}) = \frac{J_0}{4\pi} \int_0^{2\pi} \int_a^b \frac{\hat{a}_z \rho' \cos^2\phi' + \hat{a}_z \rho' \sin^2\phi'}{[(\rho')^2 + z^2]^{3/2}} \rho' d\rho' d\phi' \tag{2.253}$$

$$= \hat{a}_z \frac{J_0}{4\pi} \int_0^{2\pi} \int_a^b \frac{(\rho')^2}{[(\rho')^2 + z^2]^{3/2}} d\rho' d\phi' \tag{2.254}$$

$$= \hat{a}_z \frac{J_0}{2} \left[\ln\left(\sqrt{(\rho')^2 + z^2} + \rho'\right) - \frac{\rho'}{\sqrt{(\rho')^2 + z^2}} \right]_a^b \tag{2.255}$$

[60]Using

$$\int \frac{(\rho')^2}{[(\rho')^2 + z^2]^{3/2}} d\rho' = \ln\left(\sqrt{(\rho')^2 + z^2} + \rho'\right) - \frac{\rho'}{\sqrt{(\rho')^2 + z^2}}.$$

[61]We note that $\bar{H}(z = -z_0) = \bar{H}(z = z_0)$. Hence, if $J_0 > 0$, the magnetic field is in the z direction for all values of $z \in (-\infty, \infty)$.

$$= \hat{a}_z \frac{J_0}{2} \left[\ln \left(\frac{\sqrt{b^2+z^2}+b}{\sqrt{a^2+z^2}+a} \right) + \frac{a}{\sqrt{a^2+z^2}} - \frac{b}{\sqrt{b^2+z^2}} \right]. \quad (2.256)$$

While the final expression is lengthy, we can check several interesting limits:

- When b becomes the same as a, we have

$$\lim_{b \to a} \{\bar{H}(\bar{r})\} = 0 \quad (2.257)$$

since the source disappears.

- When the observation point goes to infinity, we have

$$\lim_{z \to \infty} \{\bar{H}(\bar{r})\} = 0 \quad (2.258)$$

since the magnetic field intensity of a finite electric current is measured at an infinitely long distance.

- When the observation point is at the origin, we have

$$\lim_{z \to 0} \{\bar{H}(\bar{r})\} = \hat{a}_z \frac{J_0}{2} \ln \left(\frac{b}{a} \right). \quad (2.259)$$

- Fixing the total electric current constant – $(b-a)J_0 = I_0$ (Figure 2.60)– and considering the limit $b \to a$, we have[62]

$$\lim_{b \to a} \left\{ \hat{a}_z \frac{I_0}{2(b-a)} \left[\ln \left(\frac{\sqrt{b^2+z^2}+b}{\sqrt{a^2+z^2}+a} \right) + \frac{a}{\sqrt{a^2+z^2}} - \frac{b}{\sqrt{b^2+z^2}} \right] \right\}$$
$$= \hat{a}_z \frac{I_0}{2} \frac{a^2}{[a^2+z^2]^{3/2}}. \quad (2.260)$$

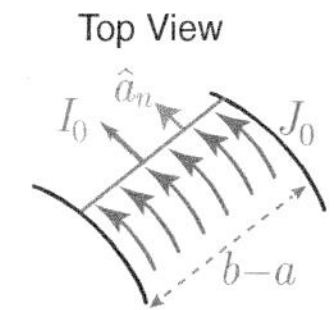

Figure 2.60 Considering the ring-shaped surface electric current density in Figure 2.59, the total electric current flow I_0 can be found by integrating the electric current density as

$$I_0 = \int_C \bar{J}_s(\bar{r}) \cdot \hat{a}_n dl,$$

where C is the line crossed by the electric current and $\hat{a}_n$ is the normal direction. Since the magnitude of $\bar{J}_s$ is constant J_0 (independent of position) and the electric current is in the ϕ direction, we further have $I_0 = J_0(b-a)$.

[62]This limit can be evaluated by using the L'Hôpital's rule (it may take several pages).

The final expression above, which corresponds to the magnetic field intensity of a circular line electric current, is particularly important. As discussed in Section 2.8.1, a circular line electric current (particularly with small a) is called a magnetic dipole.[63] Defining a dipole moment $\bar{I}_{DM,m} = \hat{a}_z I_0 \pi a^2$, the magnetic field intensity on the axis of a magnetic dipole (centered at the origin) can be written

$$\bar{H}_d(\bar{r}) = \bar{I}_{DM,m} \frac{1}{2\pi(a^2+z^2)^{3/2}}. \quad (2.261)$$

Magnetic dipoles are useful to describe magnetization (similar to electric dipoles to describe polarization) as considered in Section 2.8.

[63]We note that, as opposed to an electric dipole, a magnetic dipole does not consist of two magnetic charges as there is no magnetic charge at all. But since the magnetic field created by a circular line electric current is *similar* to the electric field created by an electric dipole (when the observation point is far away from the sources), it is generally considered to be a magnetic dipole.

2.7.2 Ampere's Force Law and Magnetic Force

Similar to Coulomb's law, which describes interactions between static electric charges, Ampere's force law describes interactions between steady electric currents in a vacuum. Consider two loops C_1 and C_2 carrying steady electric currents I_1 and I_2, respectively (Figure 2.61), in a vacuum. The magnetic force applied by C_1 to C_2 can be written

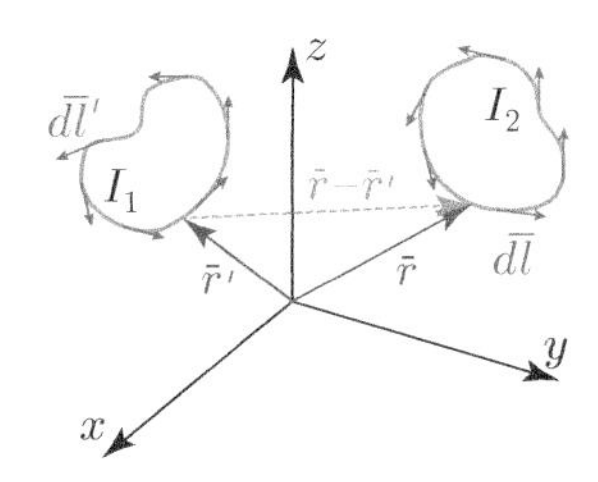

Figure 2.61 Two loops carrying electric currents apply equal amounts of force to each other.

$$\bar{F}_{m,12} = \frac{\mu_0}{4\pi} \oint_{C_2} I_2 \overline{dl} \times \oint_{C_1} \frac{I_1 \overline{dl'} \times (\bar{r} - \bar{r}')}{|\bar{r} - \bar{r}'|^3}, \tag{2.262}$$

where $\overline{dl}$ and $\overline{dl'}$ represent the differential lengths with directions on the test loop C_2 and on the source loop C_1, respectively. In addition, in the line integrals, $\bar{r}$ and $\bar{r}'$ are observation (on C_1) and source (on C_2) points, as usual. Considering that I_2 and I_1 are consistent with $\overline{dl}$ and $\overline{dl'}$ (Figure 2.61), the cross products in Eq. (2.262) determine the direction of the force.[64] As defined before, $\mu_0 = 4\pi \times 10^{-7}$ H/m.

[64]For example, if both loops are located on the x-y plane, $(\bar{r} - \bar{r}')$ is on the x-y plane, while $\overline{dl'} \times (\bar{r} - \bar{r}')$ is in the z direction. Hence, $\overline{dl} \times \overline{dl'} \times (\bar{r} - \bar{r}')$ is nonzero and contains x and y components.

The force applied by an electric current loop C_1 to another electric current loop C_2 can be interpreted as the magnetic field created by C_1 acting on C_2 (Figure 2.62). Specifically, we have[65]

$$\bar{F}_{m,12} = \oint_{C_2} I_2 \overline{dl} \times \bar{B}_1(\bar{r}), \tag{2.263}$$

where

$$\bar{B}_1(\bar{r}) = \frac{\mu_0}{4\pi} \oint_{C_1} \frac{I_1 \overline{dl'} \times (\bar{r} - \bar{r}')}{|\bar{r} - \bar{r}'|^3} \tag{2.264}$$

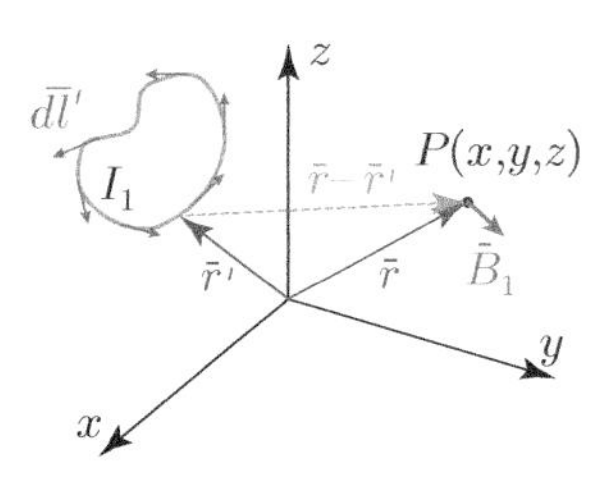

Figure 2.62 Magnetic field created by an electric current loop. The total magnetic field at the observation point can be found by integration that accounts for all possible source locations. The magnetic field is *assumed* to exist even when there is not any test electric current.

[65]Hence, we assume that, whether I_2 exists or not, a magnetic flux density is created by I_1.

is the magnetic flux density created by C_1 at an observation point $\bar{r}$. This expression is consistent with Eq. (2.219) using $\bar{B}(\bar{r}) = \mu_0 \bar{H}(\bar{r})$.

Investigating Coulomb's law in Eq. (1.204), it is tempting to write Ampere's force law between two differential electric currents (instead of loops). Consider two differential steady electric currents $I_1 \overline{dl}_1$ and $I_2 \overline{dl}_2$. The force applied by $I_1 \overline{dl}_1$ to $I_2 \overline{dl}_2$ may be written

$$\bar{F}_{m,12} = k_m I_2 \overline{dl}_2 \times \left[I_1 \overline{dl}_1 \times \frac{\hat{a}_{12}}{(R_{12})^2} \right], \tag{2.265}$$

where R_{12} is the distance between the differential electric currents and $\hat{a}_{12}$ is the unit vector directed from the location of I_1 to the location of I_2. In addition, we have the magnetic force constant as

$$k_m = \frac{\mu_0}{4\pi}, \tag{2.266}$$

where μ_0 is the permeability of a vacuum.

While the expression in Eq. (2.265) is similar to Eq. (1.204) and may describe how the integrals in Eq. (2.262) are constructed, it is in fact *defective* (see below). But first, we may consider an unproblematic case: i.e. when the electric currents are aligned in opposite directions such that $\overline{dl}_1 \cdot \overline{dl}_2 = -dl_1 dl_2$ (Figure 2.63). Evaluating the cross products in Eq. (2.265) as[66]

$$I_2\overline{dl}_2 \times \left[I_1\overline{dl}_1 \times \frac{\hat{a}_{12}}{(R_{12})^2}\right] = I_1\overline{dl}_1\left[I_2\overline{dl}_2 \cdot \frac{\hat{a}_{12}}{(R_{12})^2}\right] - \frac{\hat{a}_{12}}{(R_{12})^2}\left[I_2\overline{dl}_2 \cdot I_1\overline{dl}_1\right] \tag{2.267}$$

$$= \frac{\hat{a}_{12}}{(R_{12})^2}(I_1 dl_1)(I_2 dl_2), \tag{2.268}$$

we have[67]

$$\bar{F}_{m,12} = k_m \frac{(I_1 dl_1)(I_2 dl_2)}{(R_{12})^2}\hat{a}_{12}, \tag{2.269}$$

where the values of I_1 and I_2 can be positive or negative. Specifically, if I_1 and I_2 have the same sign, the force is positive, indicating a repulsion.[68] If I_1 and I_2 have different signs, leading to electric currents in the same direction, there is an attractive force between the differential electric currents. Obviously, the magnetic force applied by the second differential electric current to the first differential electric current can be written

$$\bar{F}_{m,21} = k_m \frac{(I_2 dl_2)(I_1 dl_1)}{(R_{21})^2}\hat{a}_{21} = -\bar{F}_{m,12}, \tag{2.270}$$

considering that $R_{21} = R_{12}$ and $\hat{a}_{21} = -\hat{a}_{12}$. Hence, as a natural force, the magnetic force seems to obey the *antisymmetry* in this case.

Now, we focus on Eq. (2.265) for a general case, where the differential electric currents are oriented arbitrarily (Figure 2.65). Using Eq. (2.265), we have

$$\bar{F}_{m,12} = k_m\left\{I_1\overline{dl}_1\left[I_2\overline{dl}_2 \cdot \frac{\hat{a}_{12}}{(R_{12})^2}\right] - \frac{\hat{a}_{12}}{(R_{12})^2}\left[I_2\overline{dl}_2 \cdot I_1\overline{dl}_1\right]\right\} \tag{2.271}$$

$$\bar{F}_{m,21} = k_m\left\{I_2\overline{dl}_2\left[I_1\overline{dl}_1 \cdot \frac{\hat{a}_{21}}{(R_{21})^2}\right] - \frac{\hat{a}_{21}}{(R_{21})^2}\left[I_1\overline{dl}_1 \cdot I_2\overline{dl}_2\right]\right\} \tag{2.272}$$

$$= k_m\left\{-I_2\overline{dl}_2\left[I_1\overline{dl}_1 \cdot \frac{\hat{a}_{12}}{(R_{12})^2}\right] + \frac{\hat{a}_{12}}{(R_{12})^2}\left[I_1\overline{dl}_1 \cdot I_2\overline{dl}_2\right]\right\}. \tag{2.273}$$

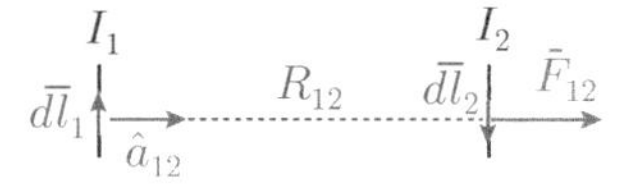

Figure 2.63 Force applied by a differential electric current $I_1\overline{dl}_1$ to another differential electric current $I_2\overline{dl}_2$ aligned in opposite direction.

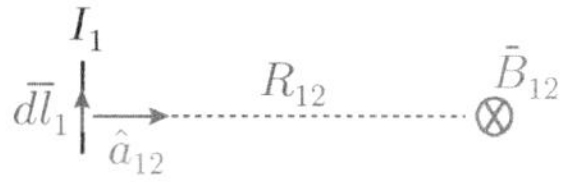

Figure 2.64 Magnetic field created by a differential electric current $I_1\overline{dl}_1$ at a location described by a vector $\bar{R}_{12} = \hat{a}_{12}R_{12}$ with respect to the electric current. In this arrangement, the direction of the magnetic field is into the paper.

[66]See Eq. (2.98) for the BAC-CAB rule.

[67]Hence, in this configuration, we have (Figure 2.64)

$$\bar{B}_{12} = \frac{\mu_0}{4\pi}\frac{I_1\overline{dl}_1 \times \hat{a}_{12}}{(R_{12})^2}.$$

[68]We note that, in this case, due to the different directions of dl_1 and dl_2, the electric currents are indeed flowing in opposite directions.

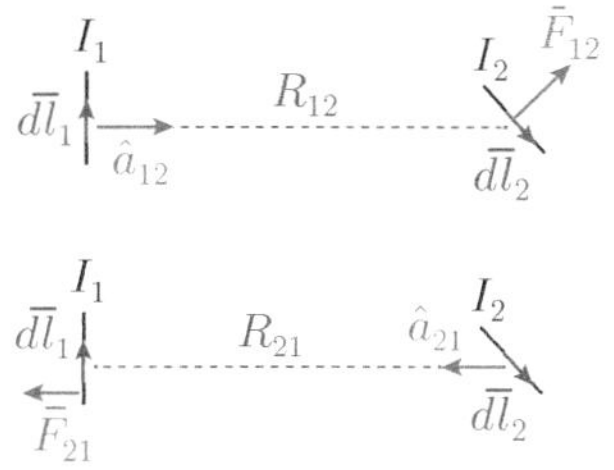

Figure 2.65 Forces between two arbitrarily oriented differential electric currents may not be antisymmetric due to the unphysical nature of the system. If there is a steady electric current on a different length, it must *flow* to somewhere. Hence, the differential electric currents do not represent a complete system.

Therefore, since $I_1\overline{dl}_1\left[I_2\overline{dl}_2\cdot\hat{a}_{12}/(R_{12})^2\right]\neq I_2\overline{dl}_2\left[I_1\overline{dl}_1\cdot\hat{a}_{12}/(R_{12})^2\right]$, there is no antisymmetry:[69]

$$\bar{F}_{m,21}\neq-\bar{F}_{m,12}. \tag{2.274}$$

The source of this problem is the step from Eq. (2.262) to Eq. (2.265). When we are considering the magnetic force, we actually assume that a differential current is a part of a full electric current circulating around a loop or along a wire from infinity to infinity (as in an infinitely long straight wire).[70] Hence, dealing with individual electric current pieces and considering the forces between them may not correspond to a valid interaction, as opposed to interacting static point electric charges.

At this stage, we may ask if the complete expression in Eq. (2.262) possesses antisymmetry or not for two general loops. Using the BAC-CAB rule, it can be rewritten

$$\bar{F}_{m,12}=\frac{\mu_0}{4\pi}\oint_{C_2}\oint_{C_1}I_2\overline{dl}\times\left[I_1\overline{dl'}\times\frac{(\bar{r}-\bar{r}')}{|\bar{r}-\bar{r}'|^3}\right] \tag{2.275}$$

$$=\frac{\mu_0}{4\pi}\oint_{C_2}\oint_{C_1}I_1\overline{dl'}\left[I_2\overline{dl}\cdot\frac{(\bar{r}-\bar{r}')}{|\bar{r}-\bar{r}'|^3}\right]-\frac{\mu_0}{4\pi}\oint_{C_2}\oint_{C_1}\left[I_1\overline{dl'}\cdot I_2\overline{dl}\right]\frac{(\bar{r}-\bar{r}')}{|\bar{r}-\bar{r}'|^3}. \tag{2.276}$$

The first term can be manipulated as

$$\frac{\mu_0}{4\pi}\oint_{C_2}\oint_{C_1}I_1\overline{dl'}\left[I_2\overline{dl}\cdot\frac{(\bar{r}-\bar{r}')}{|\bar{r}-\bar{r}'|^3}\right]=\frac{\mu_0}{4\pi}\oint_{C_1}I_1\overline{dl'}\oint_{C_2}I_2\overline{dl}\cdot\left(\frac{(\bar{r}-\bar{r}')}{|\bar{r}-\bar{r}'|^3}\right) \tag{2.277}$$

$$=\frac{\mu_0}{4\pi}\oint_{C_1}I_1\overline{dl'}\oint_{S_2}I_2\bar{\nabla}\times\left(\frac{(\bar{r}-\bar{r}')}{|\bar{r}-\bar{r}'|^3}\right)\cdot\overline{ds}, \tag{2.278}$$

where Stokes' theorem is used for the second integral. But since the curl of $(\bar{r}-\bar{r}')/|\bar{r}-\bar{r}'|^3$ is zero as shown in Eq. (2.141), we have

$$\frac{\mu_0}{4\pi}\oint_{C_2}\oint_{C_1}I_1\overline{dl'}\left[I_2\overline{dl}\cdot\frac{(\bar{r}-\bar{r}')}{|\bar{r}-\bar{r}'|^3}\right]=0 \tag{2.279}$$

and

$$\bar{F}_{m,12}=-\frac{\mu_0}{4\pi}\oint_{C_2}\oint_{C_1}\left[I_1\overline{dl'}\cdot I_2\overline{dl}\right]\frac{(\bar{r}-\bar{r}')}{|\bar{r}-\bar{r}'|^3} \tag{2.280}$$

as the magnetic force between two arbitrary loops.[71] Obviously, the force is antisymmetric:

$$\bar{F}_{m,21}=-\frac{\mu_0}{4\pi}\oint_{C_1}\oint_{C_2}\left[I_2\overline{dl}\cdot I_1\overline{dl'}\right]\frac{(\bar{r}'-\bar{r})}{|\bar{r}'-\bar{r}|^3}=-\bar{F}_{m,12}. \tag{2.281}$$

[69]In general, natural forces are expected to be antisymmetric. Specifically, consider two objects 1 and 2 and the forces $\bar{F}_{12}=\hat{a}_{F,12}|\bar{F}_{12}|$ (applied by 1 to 2) and $\bar{F}_{21}=\hat{a}_{F,21}|\bar{F}_{21}|$ (applied by 2 to 1) between them. Then we have $\hat{a}_{F,21}=-\hat{a}_{F,12}$ (directions are opposite) and $|\bar{F}_{21}|=|\bar{F}_{12}|$ (magnitudes are the same). For example, Coulomb's law in Eq. (1.204) shows that two static point electric charges apply equal amounts of forces to each other with opposite (*anti*) directions.

[70]Therefore, if there is a steady differential electric current, it must flow to somewhere.

[71]When discussing electrostatics and magnetostatics, the following is a common confusion: a steady electric current creates a magnetic field. But if an electric current is a movement of electric charges, then why don't we consider the electric field created by these charges? For example, when considering the force between two electric-current-carrying wires, we only use Ampere's force law, not Coulomb's law, for the electric charges flowing through the wires. The answer is hidden in the fact that we simply consider *steady* electric currents. When the electric current is steady, electrons move (more specifically drift due to an applied voltage/electric field) through wires with constant rates such that there is no electric charge acceleration or accumulation. Considering that there are also positive electric charges (protons) in the wires, constant movement of electrons does not remove the electric charge neutrality macroscopically anywhere. Hence, being electrically charge-neutral, wires do apply electric force to each other, and the only type of force between them is magnetic.

As one of the most common scenario for Ampere's force law, we now consider a steady line electric current I_2 of unit length[72] (1 m) placed near an infinitely long line electric current I_1, both aligned in the z direction in a vacuum (Figure 2.66). The electric current I_2 is located at $y = d$, hence the distance between the lines is d. To find the applied magnetic force to I_2, we can start by writing the magnetic flux density created by the infinitely long wire as

[72] The force between two infinitely long steady electric currents is infinite. Therefore, we consider a unit length for one of the electric currents.

$$\bar{B}_1(y = d) = -\hat{a}_x \frac{\mu_0 I_1}{2\pi d} \tag{2.282}$$

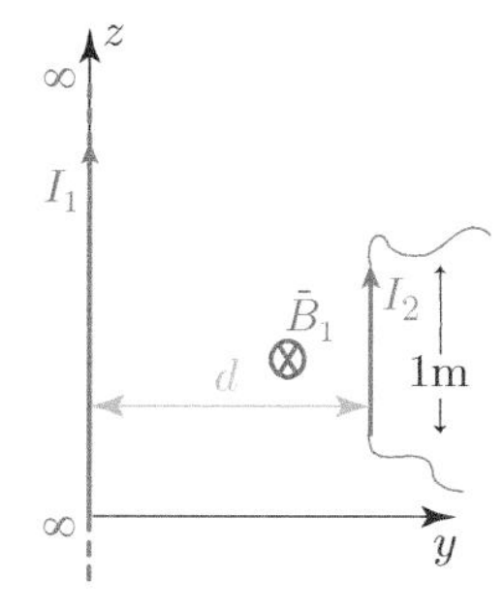

Figure 2.66 A line electric current of unit length placed near an infinitely long line electric current.

at all observation points on I_2. Then the force on line I_2 can be found as[73]

$$\bar{F}_{12} = \int_{C_2} I_2 \overline{dl}_2 \times \bar{B}_1 = \int_0^1 I_2 dz \hat{a}_z \times \left(-\hat{a}_x \frac{\mu_0 I_1}{2\pi d}\right) = -\hat{a}_y \frac{\mu_0 I_1 I_2}{2\pi d}, \tag{2.283}$$

which indicates that there is an attraction if I_1 and I_2 have the same sign. The final expression for a special case of $I_1 = 1\text{A}$, $I_2 = 1\text{A}$, and $d = 1\text{m}$ leads to[74]

$$|\bar{F}_{12}| = \frac{\mu_0}{2\pi} = 2 \times 10^{-7} \text{ N}. \tag{2.284}$$

[73] We note that $\bar{F}_{21} \neq -\bar{F}_{12}$ since I_2 is a part of a electric current loop, and in this case, we are not considering the rest of the loop.

It is remarkable that, even though the distance of 1 m and the current magnitude of 1 A are physically reasonable values, the amount of the force is quite small. In fact, the infinitely long current of 1A creates a magnetic flux density of 2×10^{-7} Wb/m^2, which is at least 100 times smaller than the Earth's magnetic field. To compare, the magnetic flux density is more than 5 Wb/m^2 in the state-of-the-art magnetic resonance imaging (MRI) machines, while it may reach billions on the surface of a neutron star.

[74] We note that the same amount of force exists between two static point electric charges (again separated by a distance of 1 m) with only $Q_1 = Q_2 \approx 4.72 \times 10^{-9}$ C.

2.7.3 Examples

Example 22: Consider a steady electric current sheet (strip) located at $y = 0$ from $x = -d/2$ to $x = d/2$ (Figure 2.67). The surface electric current density $\bar{J}_s(\bar{r})$ is constant and in the z direction: i.e. $\bar{J}_s(\bar{r}) = \hat{a}_z J_0$ (A/m). Find the magnetic field intensity at an arbitrary position $\bar{r} = \hat{a}_x x + \hat{a}_y y + \hat{a}_z z$, where $y > 0$.

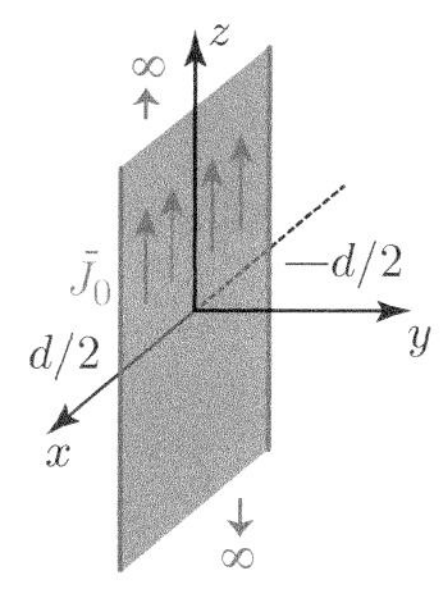

Figure 2.67 An electric current sheet that is infinitely long in the $-z$ and z directions.

Solution: This problem can be solved by directly using Biot-Savart law. Alternatively, we can use the magnetic field intensity created by an infinitely long line electric current to reach the magnetic field due to the electric current sheet as follows (Figure 2.68). First, we can start by considering a differential part of the sheet described by dx'. The total amount of electric

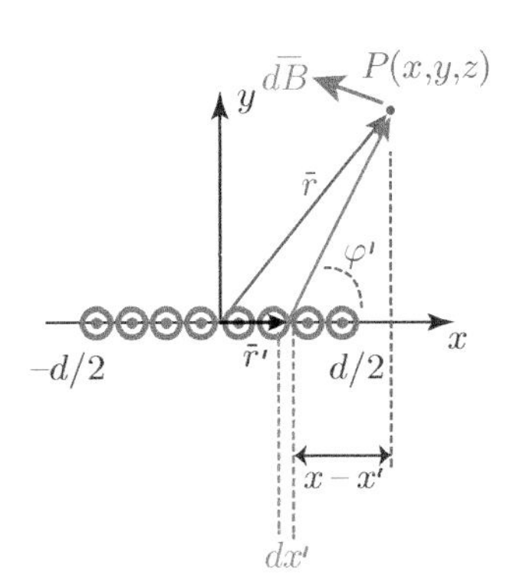

Figure 2.68 Another view of the infinite electric current sheet in Figure 2.67.

current in the z direction passing through this differential length is $dI = J_0 dx'$. Therefore, its contribution at $\bar{r} = \hat{a}_x x + \hat{a}_y y + \hat{a}_z z$ can be found as[75]

$$d\bar{H}(\bar{r}) = \frac{J_0 dx'}{2\pi[(x-x')^2+y^2]^{1/2}}(-\hat{a}_x \sin\varphi' + \hat{a}_y \cos\varphi'), \tag{2.285}$$

where

$$\sin\varphi' = \frac{y}{[(x-x')^2+y^2]^{1/2}} \tag{2.286}$$

$$\cos\varphi' = \frac{x-x'}{[(x-x')^2+y^2]^{1/2}}. \tag{2.287}$$

Inserting the expressions for the trigonometric functions, we derive

$$d\bar{H}(\bar{r}) = \frac{J_0 dx'}{2\pi[(x-x')^2+y^2]}[-\hat{a}_x y + \hat{a}_y(x-x')]. \tag{2.288}$$

Then the magnetic field intensity of the overall sheet can be found by integration as

$$\bar{H}(\bar{r}) = \frac{J_0}{2\pi}\int_{-d/2}^{d/2} \frac{-\hat{a}_x y + \hat{a}_y(x-x')}{[(x-x')^2+y^2]} dx'. \tag{2.289}$$

Obviously, $\bar{B}$ has only x and y directions. Evaluating the integral, one can find that[76]

$$H_x(\bar{r}) = \frac{J_0}{2\pi}\left\{\tan^{-1}\left(\frac{x-d/2}{y}\right) - \tan^{-1}\left(\frac{x+d/2}{y}\right)\right\} \tag{2.290}$$

$$H_y(\bar{r}) = \frac{J_0}{2\pi}\ln\left[\frac{(x-d/2)^2+y^2}{(x+d/2)^2+y^2}\right]. \tag{2.291}$$

[75]We use the expression for the magnetic field due to an infinitely long line electric current: e.g. see Eq. (2.59).

[76]An interesting practice is to check these expressions in the limit $d \to \infty$. We have

$$\lim_{d\to\infty}\left\{\tan^{-1}\left(\frac{x-d/2}{y}\right)\right\} = -\pi/2$$

$$\lim_{d\to\infty}\left\{\tan^{-1}\left(\frac{x+d/2}{y}\right)\right\} = \pi/2$$

$$\lim_{d\to\infty}\left\{\ln\left[\frac{(x-d/2)^2+y^2}{(x+d/2)^2+y^2}\right]\right\} = 0$$

and

$$\lim_{d\to\infty}\{H_x(\bar{r})\} = -J_0/2$$

$$\lim_{d\to\infty}\{H_y(\bar{r})\} = 0,$$

which are consistent with Eq. (2.198).

Example 23: Consider a loop of radius a carrying a steady electric current I_0 (A) located on the x-y plane symmetrically at the origin (Figure 2.69). Find the magnetic field intensity at $\bar{r} = \hat{a}_z z$ for $z > 0$ (on the positive z axis).

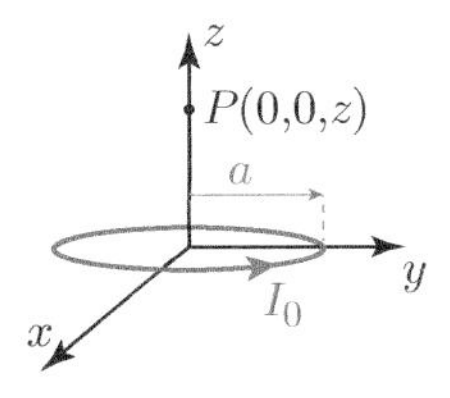

Figure 2.69 A steady electric current flowing through a circular loop.

Solution: For this structure – i.e. a magnetic dipole – the expression for the magnetic field intensity is found via limit in Section 2.7.1. Now, using Biot-Savart law directly, we have

$$\bar{H}(\bar{r}) = \frac{1}{4\pi}\oint_C \frac{I_0 \overline{dl'} \times (\bar{r}-\bar{r}')}{|\bar{r}-\bar{r}'|^3} = \frac{\mu_0 I_0}{4\pi}\oint_C \frac{\overline{dl'}\times(\bar{r}-\bar{r}')}{|\bar{r}-\bar{r}'|^3}, \tag{2.292}$$

where

$$\bar{r} = z\hat{a}_z \qquad \text{(observation point)} \tag{2.293}$$

$$\bar{r}' = a\hat{a}_\rho = a\cos\phi'\hat{a}_x + a\sin\phi'\hat{a}_y \qquad \text{(source point)} \tag{2.294}$$

$$\overline{dl'} = ad\phi'\hat{a}_\phi = ad\phi'(-\hat{a}_x\sin\phi' + \hat{a}_y\cos\phi') \tag{2.295}$$

$$(\bar{r} - \bar{r}') = -\hat{a}_x x' - \hat{a}_y y' + \hat{a}_z z = -\hat{a}_x a\cos\phi' - \hat{a}_y a\sin\phi' + \hat{a}_z z \tag{2.296}$$

$$|\bar{r} - \bar{r}'| = \sqrt{(x')^2 + (y')^2 + z^2} = \sqrt{a^2 + z^2} \tag{2.297}$$

$$\overline{dl'} \times (\bar{r} - \bar{r}') = (a^2\sin^2\phi'\hat{a}_z + az\sin\phi'\hat{a}_y + a^2\cos^2\phi'\hat{a}_z + az\cos\phi'\hat{a}_x)d\phi' \tag{2.298}$$

$$= (a^2\hat{a}_z + az\sin\phi'\hat{a}_y + az\cos\phi'\hat{a}_x)d\phi'. \tag{2.299}$$

Then we derive[77]

$$\bar{H}(0,0,z) = \frac{I_0}{4\pi}\int_0^{2\pi} \frac{az\cos\phi'\hat{a}_x + az\sin\phi'\hat{a}_y + a^2\hat{a}_z}{(a^2+z^2)^{3/2}}d\phi' \tag{2.300}$$

$$= \hat{a}_z\frac{I_0}{2\pi}\frac{\pi a^2}{(a^2+z^2)^{3/2}}, \tag{2.301}$$

[77]We note that $\bar{B}(0,0,z) \to 0$ when $a \to 0$, provided that $z > 0$. If $z = 0$, then shrinking loop (as $a \to 0$) means the observation point is located on a vanishingly small source. Then we have $\bar{B}(0,0,0) \to \infty$ when $a \to 0$.

considering the zero integrals of sin and cos from 0 to 2π. The final expression is consistent with Eq. (2.261). In addition, when $z \gg a$, we have

$$\bar{H}(0,0,z) = \hat{a}_z\frac{I_0}{2\pi}\frac{\pi a^2}{(a^2+z^2)^{3/2}} \approx \hat{a}_z\frac{I_0}{2\pi}\frac{\pi a^2}{z^3}, \tag{2.302}$$

which is consistent with Eq. (2.351).

Example 24: Consider a cylindrical inductor involving N turns aligned in the z direction from $z = 0$ to $z = l$ (Figure 2.70). The inductor is symmetrically located with radius a. Assuming that N is very large, find the magnetic field intensity on the z axis inside the inductor.

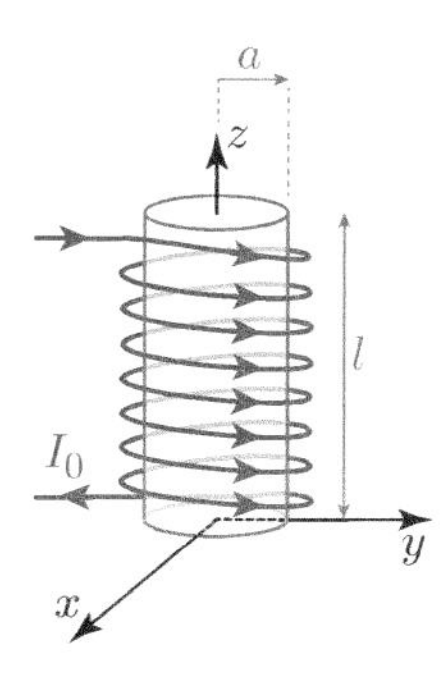

Figure 2.70 A cylindrical inductor of finite size aligned in the z direction.

Solution: Considering that N is very large, we can assume that the inductor consists of infinitely many perfect loops. Hence, the expression for the magnetic field intensity created by a loop can be used. Let dz' be a differential length along the z direction. The number of loops within dz' can be found as $dN = (N/l)dz'$. Then considering a differential amount of these loops located at z' (source location), the magnetic field intensity created at an observation point z can be written[78]

[78]We are using Eq. (2.301).

$$d\bar{H}(0,0,z) = \hat{a}_z \frac{I_0}{2\pi} \frac{\pi a^2 (N/l) dz'}{[a^2 + (z - z')^2]^{3/2}}. \tag{2.303}$$

We note that $z - z'$ is the distance between the source loop and the observation point, and the expression is valid for both $z - z' \geq 0$ and $z - z' < 0$. To find the overall magnetic field intensity, we write the integral as

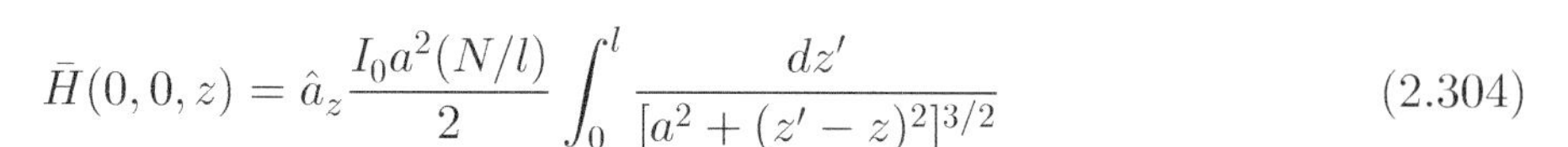

$$\bar{H}(0,0,z) = \hat{a}_z \frac{I_0 a^2 (N/l)}{2} \int_0^l \frac{dz'}{[a^2 + (z' - z)^2]^{3/2}} \tag{2.304}$$

$$= \hat{a}_z \frac{I_0 a^2 (N/l)}{2} \frac{1}{a^2} \left[\frac{z' - z}{\sqrt{a^2 + (z' - z)^2}} \right]_0^l \tag{2.305}$$

leading to

$$\bar{H}(0,0,z) = \hat{a}_z \frac{I_0 (N/l)}{2} \left[\frac{l - z}{\sqrt{a^2 + (l - z)^2}} - \frac{-z}{\sqrt{a^2 + z^2}} \right] \tag{2.306}$$

$$= \frac{I_0 N}{2l} \left(\frac{l - z}{\sqrt{a^2 + (l - z)^2}} + \frac{z}{\sqrt{a^2 + z^2}} \right). \tag{2.307}$$

At the center – i.e. at $z = l/2$ – we have

$$H_z(0,0,l/2) = \frac{I_0 N}{2} \frac{1}{\sqrt{a^2 + (l/2)^2}}. \tag{2.308}$$

At the ends of the inductor, we further have

$$H_z(0,0,0) = H_z(0,0,l) = \frac{I_0 N}{2} \frac{1}{\sqrt{a^2 + l^2}}. \tag{2.309}$$

If $l \gg a$, these expressions can be simplified as

$$H_z(0,0,L/2) \approx I_0 N_l \tag{2.310}$$

$$H_z(0,0,0) = H_z(0,0,L) \approx \frac{I_0 N_l}{2}, \tag{2.311}$$

where N_l is the number of turns per length. We note that Eq. (2.310) is perfectly consistent with the expression in Eq. (2.71). Obviously, the magnetic field intensity decreases toward the ends of the inductor (Figure 2.71), which cannot be predicted via an ideal (infinite) model.

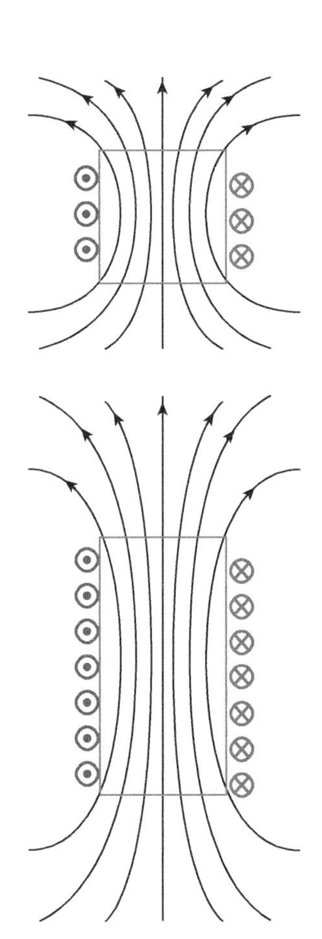

Figure 2.71 Depending on the size of the inductor, magnetic field lines tend to deviate from the center, unless the inductor is not infinite. This deviation is consistent with the fact that the magnetic field intensity decreases toward the ends.

Example 25: Consider an infinitely long steady line electric current on the z axis I_0 (A). A triangular loop, which also carries steady electric current I_0, is placed on the z-x plane at a distance of d (Figure 2.72). The structures are located in a vacuum. Find the magnetic force acting on the loop.

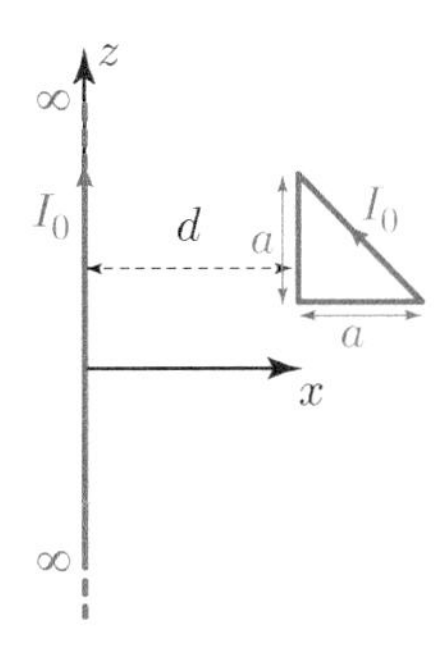

Figure 2.72 An infinitely long wire carrying a steady electric current flowing in the z direction and a triangular loop that also carries steady electric current I_0 located on the z-x plane in a vacuum.

Solution: To find the total magnetic force, we can consider the three segments of the triangular loop and the magnetic force acting on them separately. First, the magnetic flux density created by the infinitely long line can be written $\bar{B}(\bar{r}) = \hat{a}_y \mu_0 I_0/(2\pi x)$ on the z-x plane. For the vertical line, we have

$$\bar{F}_{m1} = \int_{C_1} I_0 \overline{dl} \times \bar{B}(\bar{r}) = \int_0^a I_0 \hat{a}_z \times \bar{B}(\bar{r}) dz = -\hat{a}_x \frac{\mu_0 I_0^2}{2\pi d} a = -\hat{a}_x \frac{\mu_0 I_0^2 a}{2\pi d}. \tag{2.312}$$

For the horizontal line, the magnetic force can be found as

$$\bar{F}_{m3} = \int_{C_3} I_0 \overline{dl} \times \bar{B}(\bar{r}) = \int_d^{d+a} I_0(-\hat{a}_x) \times \bar{B}(\bar{r}) dx \tag{2.313}$$

$$= -\hat{a}_z \frac{\mu_0 I_0^2}{2\pi} [\ln(x)]_d^{d+a} = -\hat{a}_z \frac{\mu_0 I_0^2}{2\pi} \ln\left(\frac{d+a}{a}\right). \tag{2.314}$$

Finally, for the diagonal line, $\overline{dl} = \hat{a}_x dx - \hat{a}_z dz$ and

$$\overline{dl} \times \bar{B}(\bar{r}) = [\hat{a}_x dx - \hat{a}_z dz] \times \hat{a}_y \frac{\mu_0 I_0}{2\pi x} = \hat{a}_z \frac{\mu_0 I_0}{2\pi x} dx + \hat{a}_x \frac{\mu_0 I_0}{2\pi x} dz, \tag{2.315}$$

leading to

$$\bar{F}_{m2} = \int_{C_3} I_0 \overline{dl} \times \bar{B}(\bar{r}) = \int \left[\hat{a}_z \frac{\mu_0 I_0^2}{2\pi x} dx + \hat{a}_x \frac{\mu_0 I_0^2}{2\pi x} dz\right] \tag{2.316}$$

$$= \int_d^{d+a} \hat{a}_z \frac{\mu_0 I_0^2}{2\pi x} dx + \int_0^a \hat{a}_x \frac{\mu_0 I_0^2}{2\pi(a+d-z)} dz \tag{2.317}$$

$$= \hat{a}_z \frac{\mu_0 I_0^2}{2\pi} \ln\left(\frac{d+a}{a}\right) + \hat{a}_x \frac{\mu_0 I_0^2}{2\pi} \ln\left(\frac{d+a}{a}\right). \tag{2.318}$$

Hence, the total magnetic force can be found as[79]

$$\bar{F}_m = \bar{F}_{m1} + \bar{F}_{m2} + \bar{F}_{m3} = -\hat{a}_x \frac{\mu_0 I_0^2 a}{2\pi d} + \hat{a}_x \frac{\mu_0 I_0^2}{2\pi} \ln\left(\frac{d+a}{a}\right) \tag{2.319}$$

[79] As shown in Chapter 6, this problem can also be solved via the magnetic potential energy and virtual displacement.

$$= \hat{a}_x \frac{\mu_0 I_0^2}{2\pi} \left[\ln \left(\frac{d+a}{a} \right) - \frac{a}{d} \right]. \tag{2.320}$$

2.8 Ampere's Law and Magnetic Materials

The magnetic flux density $\bar{B}(\bar{r}, t)$ and the magnetic field intensity $\bar{H}(\bar{r}, t)$ are related to each other, as also given in Eq. (2.2), as $\bar{B}(\bar{r}, t) = \mu(\bar{r})\bar{H}(\bar{r}, t)$, where $\mu(\bar{r}) \neq \mu_0 = 4\pi \times 10^{-7}$ H/m for a magnetic material. If $\mu(\bar{r})$ does not depend on position $\bar{r}$, we have

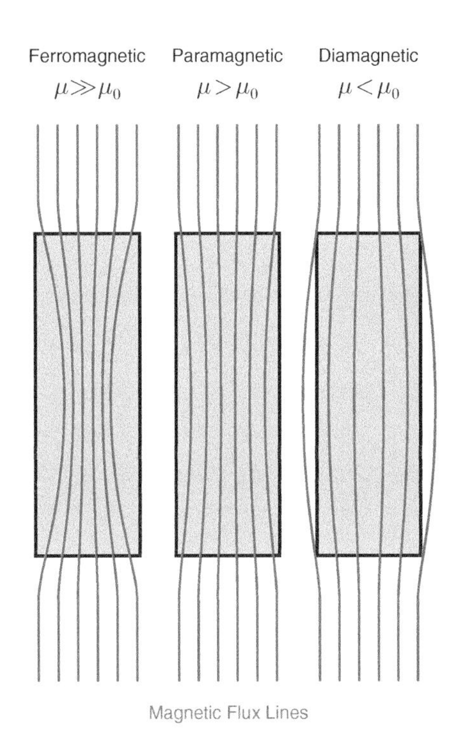

Figure 2.73 Various responses of magnetic materials to external magnetic fields.

$$\oint_C \frac{1}{\mu} \bar{B}(\bar{r}, t) \cdot \overline{dl} = I_{\text{enc}}(t) + \int_S \frac{\partial \bar{D}(\bar{r}, t)}{\partial t} \cdot \overline{ds}$$

$$\longrightarrow \oint_C \bar{B}(\bar{r}, t) \cdot \overline{dl} = \mu I_{\text{enc}}(t) + \mu \int_S \frac{\partial \bar{D}(\bar{r}, t)}{\partial t} \cdot \overline{ds} \tag{2.321}$$

$$\bar{\nabla} \times \left(\frac{1}{\mu} \bar{B}(\bar{r}, t) \right) = \bar{J}_v(\bar{r}, t) + \frac{\partial \bar{D}(\bar{r}, t)}{\partial t}$$

$$\longrightarrow \bar{\nabla} \times \bar{B}(\bar{r}, t) = \mu \bar{J}_v(\bar{r}, t) + \mu \frac{\partial \bar{D}(\bar{r}, t)}{\partial t} \tag{2.322}$$

as the integral and differential forms of Ampere's law for the magnetic flux density. We note that the permeability can be placed outside the integral and curl operations if and only if it does not depend on position (at related observation points).

Magnetic materials are those having various responses to magnetic fields (Figure 2.73), similar to the response of dielectrics to electric fields. This response can be modeled by the permeability, which can be inhomogeneous (depends on the position), nonlinear (depends on the magnetic field itself), and/or anisotropic (depends on direction). In this book, we naturally consider inhomogeneity: i.e. by writing the permeability as $\mu(\bar{r})$, since the existence of a magnetic material (even when it is itself homogeneous) located in a vacuum can be interpreted as an inhomogeneity. On the other hand, similar to dielectrics, we omit anisotropy: i.e. we assume

$$\begin{bmatrix} B_x(\bar{r}, t) \\ B_y(\bar{r}, t) \\ B_z(\bar{r}, t) \end{bmatrix} = \begin{bmatrix} \mu(\bar{r}) & 0 & 0 \\ 0 & \mu(\bar{r}) & 0 \\ 0 & 0 & \mu(\bar{r}) \end{bmatrix} \cdot \begin{bmatrix} H_x(\bar{r}, t) \\ H_y(\bar{r}, t) \\ H_z(\bar{r}, t) \end{bmatrix} \tag{2.323}$$

and there is no directional dependency or cross coupling between the components of the magnetic flux density and the magnetic field intensity. Besides, in most cases (e.g. free-space propagation, waveguides, etc.), nonlinearity is omitted. On the other hand, nonlinearity is needed to explain magnets that possess permanent magnetization and that are more common than electrets (with permanent polarization) in practice.

Before proceeding with the application of Ampere's law in the presence of magnetic materials, we will list various kinds of magnetic materials that are well categorized. As an important class, ferromagnetic materials have strong responses to magnetic fields ($\mu(\bar{r}) \gg \mu_0$). The response of a ferromagnetic object can be explained by its magnetization: i.e. alignment of magnetic dipoles.[80] Ferromagnetic materials are typically anisotropic and nonlinear (with strong dependency on the temperature), which explains how they can become permanently magnetized.[81] Well-known ferromagnetic materials are cobalt, iron, and nickel, as well as their alloys. Paramagnetic materials have weaker (but still nonzero) responses to magnetic fields, and they generally cannot be permanently magnetized (as opposed to ferromagnetic materials). Macroscopically, they can be represented by a permeability $\mu(\bar{r}) > \mu_0$ that does not depend on the applied magnetic field (linear). Some examples of paramagnetic materials are aluminum, oxygen, and titanium. Since a paramagnetic material's response to an external magnetic field is based on the alignment of magnetic dipoles in the same direction, it is typically attracted to the applied magnetic field. Finally, diamagnetism is the opposite of paramagnetism and refers to the negative response of a material to an external magnetic field. Hence, a purely diamagnetic material is repelled by the applied magnetic field since magnetic dipoles are aligned in the opposite direction. This can be modeled as $\mu(\bar{r}) < \mu_0$, while diamagnetism is usually very weak and $\mu \approx \mu_0$. In fact, all materials possess different levels of diamagnetism (as a natural property), while paramagnetism and ferromagnetism (if they exist) are dominant behaviors.

At this stage, we may now focus on applying Ampere's law in different cases when magnetic materials exist. For this purpose, we consider two-dimensional scenarios involving a steady line electric current I_0 aligned in the z direction and always located on the z axis being enclosed by a circular Ampere loop (Figure 2.74–2.76). We have the following different cases for the integral form:

- Case 1: If a line electric current is located in a vacuum near a magnetic medium $\mu \neq \mu_0$ and the Ampere loop C encloses only a vacuum in addition to the electric current (Figure 2.74), we have

$$\oint_C \bar{H}(\bar{r}) \cdot \overline{dl} = I_0 \longrightarrow \oint_C \bar{B}(\bar{r}) \cdot \overline{dl} = \mu_0 I_0 \not\rightarrow 2\pi\rho H_\phi(\rho) = I_0. \tag{2.324}$$

  Hence, we are able to write and use the integral form of Ampere's law for both $\bar{H}(\bar{r})$ and $\bar{B}(\bar{r})$. However, the existence of the magnetic material destroys the symmetry such that the magnetic field term cannot be placed outside the integral even though the loop is symmetric around the electric current. Specifically, the magnetic field does not have only ϕ component and does not depend on only ρ (Figure 2.77).

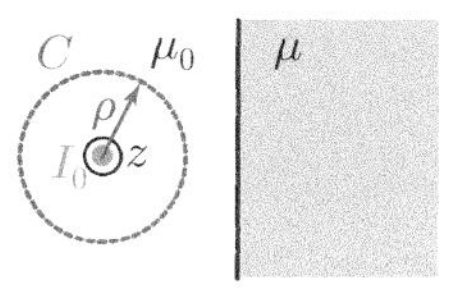

Figure 2.74 A steady line electric current in a vacuum near a magnetic medium.

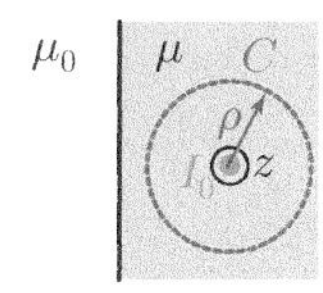

Figure 2.75 A steady line electric current inside a magnetic medium.

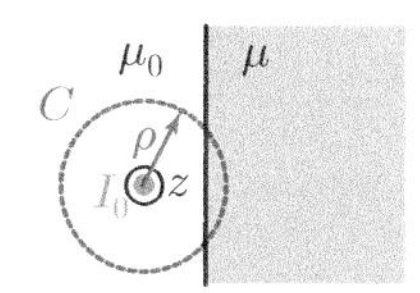

Figure 2.76 A steady line electric current in a vacuum near a magnetic medium. The Ampere loop encloses both a vacuum and a part of the magnetic medium.

[80] This is similar to the polarization: i.e. alignment of electric dipoles in the presence of an external electric field.

[81] In such a case of permanent magnetization, magnetic fields are created (that can be modeled as fields created by magnetization currents), even though there is no true electric current.

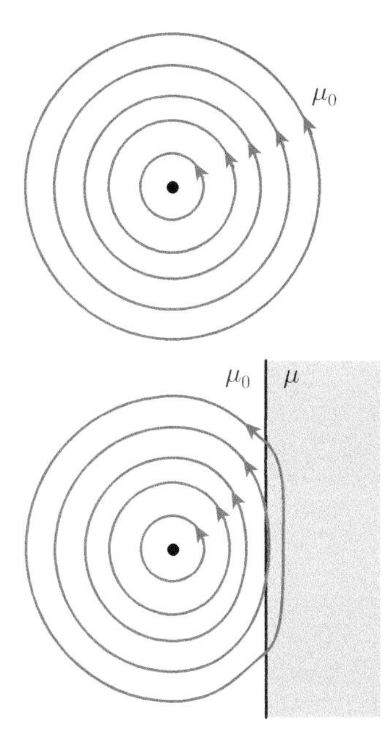

Figure 2.77 Magnetic materials interact with external magnetic fields. For example, magnetic field lines created by a steady line electric current are deformed near a magnetic medium, while a magnetic field penetrates the medium. We note that, considering a steady electric current in a vacuum with or without a nearby magnetic medium, the integral form of Ampere's law does not change (circulation is the enclosed electric current), while the magnetic field intensity cannot be placed outside the integral in the latter due to the lack of symmetry.

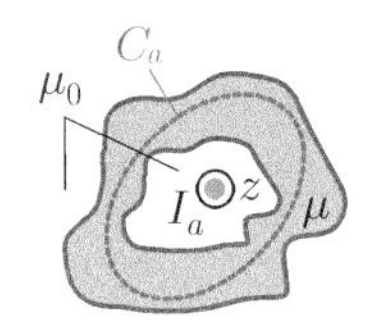

Figure 2.78 An application of the integral form of Ampere's law when all locations on the Ampere loop are completely inside a magnetic medium. As shown in Eq. (2.330), the permeability $\mu(\bar{r}) = \mu$ can be placed outside the integral. If there is an overall symmetry, the magnetic flux density can also be put outside the integral, which can be very useful in a static case (without the time derivative of the electric flux density).

- Case 2: If a line electric current is located inside a magnetic medium and the Ampere loop C encloses only magnetic material in addition to the electric current (Figure 2.75), we have

$$\oint_C \bar{H}(\bar{r}) \cdot \overline{dl} = I_0 \longrightarrow \oint_C \bar{B}(\bar{r}) \cdot \overline{dl} = \mu I_0 \not\rightarrow 2\pi\rho H_\phi(\rho) = I_0. \tag{2.325}$$

Hence, similar to Case 1, Ampere's integral is directly applicable to $\bar{H}(\bar{r})$ and $\bar{B}(\bar{r})$, while the lack of symmetry (the vacuum part of the space) inhibits the relocation of the magnetic field outside the integral.

- Case 3: If a line electric current is located in the vicinity of a boundary between a vacuum and magnetic medium and the Ampere loop C encloses both the vacuum and a part of the magnetic medium (Figure 2.76), we have

$$\oint_C \bar{H}(\bar{r}) \cdot \overline{dl} = I_0 \longrightarrow \frac{1}{\mu_0}\oint_{C_v} \bar{B}(\bar{r}) \cdot \overline{dl} + \frac{1}{\mu}\oint_{C_m} \bar{B}(\bar{r}) \cdot \overline{dl} = I_0 \tag{2.326}$$

$$\not\rightarrow 2\pi\rho H_\phi(\rho) = I_0. \tag{2.327}$$

In this case, Ampere's law is directly written for $\bar{H}(\bar{r})$, while two separate terms are needed for $\bar{B}(\bar{r})$ in order to account for the contributions from the part of the loop in the vacuum (C_v) and the rest of the loop in magnetic medium (C_m). Obviously, the magnetic field cannot be placed outside the integral.

Interestingly, when Ampere's law is written for the magnetic flux density, the permeability values on the Ampere loop should be used. For example, we consider a two-dimensional case (infinitely long in the z direction), where a line electric current $I_0(t)$ is located inside a cavity enclosed by a magnetic shell with constant permeability μ (Figure 2.78). If the Ampere loop is completely located inside the magnetic shell, we have

$$\oint_{C_a} \bar{H}(\bar{r},t) \cdot \overline{dl} = I_a(t) + \int_{S_a} \frac{\partial \bar{D}(\bar{r},t)}{\partial t} \cdot \overline{ds} \tag{2.328}$$

$$\longrightarrow \oint_{C_a} \bar{B}(\bar{r},t) \cdot \overline{dl} = \mu I_a(t) + \mu \int_{S_a} \frac{\partial \bar{D}(\bar{r},t)}{\partial t} \cdot \overline{ds}, \tag{2.329}$$

where μ is safely put outside the integral. If the permittivity on the *surface* (disk) bounded by the Ampere loop is fixed (ε), we further have

$$\oint_{C_a} \bar{B}(\bar{r},t) \cdot \overline{dl} = \mu I_a(t) + \mu\varepsilon \int_{S_a} \frac{\partial \bar{E}(\bar{r},t)}{\partial t} \cdot \overline{ds}, \tag{2.330}$$

where ε is placed outside the flux integral. In the case of perfect symmetry (when everything including the magnetic material is symmetric), the magnetic field (and electric field) can further be moved outside the integrals. Then in a static case – i.e. $I_0(t) = I_0$ – Ampere's law can be used directly to find magnetic fields.

Next, we consider the differential form of Ampere's law in the presence of magnetic materials. For this purpose, we consider different observation points in the vicinity of a boundary between a vacuum and a magnetic material (Figure 2.79).

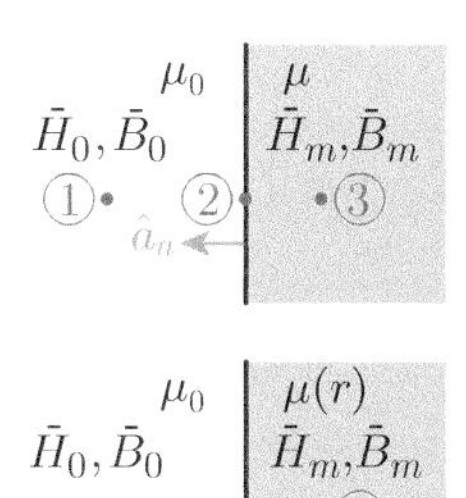

Figure 2.79 Different locations around the interface between a vacuum and a magnetic medium.

- If the point is located in a vacuum (e.g. location 1 in Figure 2.79), whether there is a magnetic material in the vicinity or not, we have

$$\bar{\nabla} \times \bar{H}(\bar{r}, t) = \bar{J}_v(\bar{r}, t) + \frac{\partial \bar{D}(\bar{r}, t)}{\partial t} \tag{2.331}$$

$$\longrightarrow \bar{\nabla} \times \bar{B}(\bar{r}, t) = \mu_0 \bar{J}_v(\bar{r}, t) + \mu_0 \varepsilon_0 \frac{\partial \bar{E}(\bar{r}, t)}{\partial t}, \tag{2.332}$$

  where $J_v(\bar{r}, t)$ is the volume electric current density and $\bar{E}(\bar{r}, t)$ is the electric field intensity at the location.

- If the point is located in a magnetic medium with permeability $\mu \neq \mu_0$ (e.g. location 3 in Figure 2.79), whether there is another medium (a vacuum or another type of magnetic material) in the vicinity or not, we have

$$\bar{\nabla} \times \bar{H}_m(\bar{r}, t) = \bar{J}_v(\bar{r}, t) + \frac{\partial \bar{D}_m(\bar{r}, t)}{\partial t} \tag{2.333}$$

$$\longrightarrow \bar{\nabla} \times \bar{B}_m(\bar{r}, t) = \mu \bar{J}_v(\bar{r}, t) + \mu \frac{\partial \bar{D}_m(\bar{r}, t)}{\partial t}, \tag{2.334}$$

  where $J_v(\bar{r}, t)$ is the volume electric current density and $\bar{D}_m(\bar{r}, t)$ is the electric flux density at the location.

- At the boundary between a vacuum and a magnetic medium (e.g. location 2 in Figure 2.79), the boundary conditions for tangential magnetic fields (that are obtained by using Ampere's law at the boundary) can be used:

$$\hat{a}_n \times [\bar{H}_0(\bar{r}, t) - \bar{H}_m(\bar{r}, t)] = \bar{J}_s(\bar{r}, t) \qquad (\bar{r} \in S) \tag{2.335}$$

$$\longrightarrow \hat{a}_n \times [\bar{B}_0(\bar{r}, t)/\mu_0 - \bar{B}_m(\bar{r}, t)/\mu] = \bar{J}_s(\bar{r}, t) \qquad (\bar{r} \in S), \tag{2.336}$$

  where $\bar{J}_s(\bar{r}, t)$ is the surface electric current density at the location. If the magnetic medium is lossless, one can select $J_s(\bar{r}, t) = 0$: i.e. the electric current density is zero if no electric current is particularly inserted.

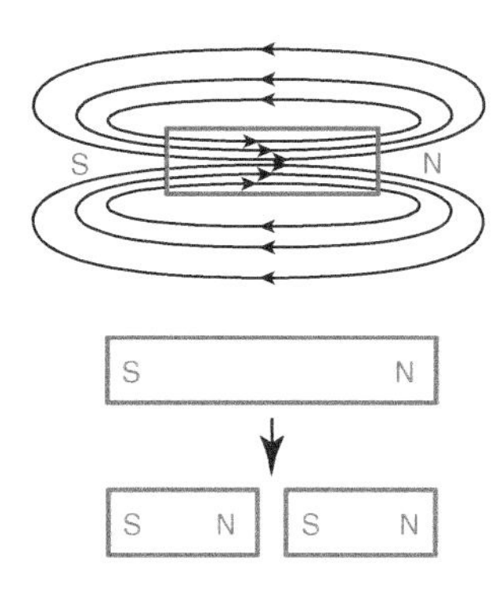

Figure 2.80 There is no magnetic charge (magnetic *monopole*) in nature. This means magnetic field lines do not start from a location or end at a location; they close upon themselves. Considering a bar magnet (a permanently magnetized bar), magnetic field lines emerge from north side (N) and enter into south side (S). On the other hand, magnetic fields continue inside the bar, closing the loops. One cannot obtain a magnetic monopole (isolated N or S) by dividing the bar magnet into two.

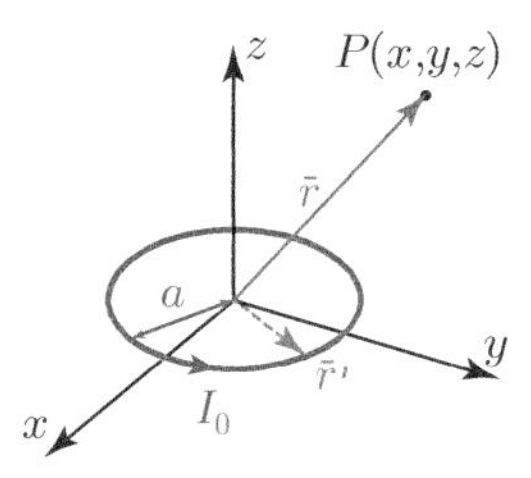

Figure 2.81 Magnetic dipole: i.e. a steady electric current flowing through a loop of radius a.

- If the point is located in a magnetic medium with permeability $\mu(\bar{r})$ that depends on position (e.g. location 4 in Figure 2.79), we have

$$\bar{\nabla} \times \bar{H}_m(\bar{r},t) = \bar{J}_v(\bar{r},t) + \frac{\partial \bar{D}_m(\bar{r},t)}{\partial t} \tag{2.337}$$

$$\longrightarrow \bar{\nabla} \times \left(\bar{B}_m(\bar{r},t)/\mu(\bar{r})\right) = \bar{J}_v(\bar{r},t) + \frac{\partial \bar{D}_m(\bar{r},t)}{\partial t} \tag{2.338}$$

$$\not\longrightarrow \bar{\nabla} \times \bar{B}_m(\bar{r},t) = \mu(\bar{r})\bar{J}_v(\bar{r},t) + \mu(\bar{r})\frac{\partial \bar{D}_m(\bar{r},t)}{\partial t}. \tag{2.339}$$

In other words, we are able to use Ampere's law for the magnetic flux density, while we cannot place the permeability outside the derivative (curl operation). In such a case, the derivative of the permeability (its gradient) may be used to write the complete and correct equation for $\bar{\nabla} \times \bar{B}_m(\bar{r},t)$.

We may now construct the mathematical basis for the interaction of magnetic fields with magnetic materials, particularly in the context of Ampere's law. As usual, we list the tools and concepts for this purpose:

- Magnetic dipole
- Magnetization
- Equivalent magnetization currents

These are in addition to previous tools: differential length with direction (Section 2.1.1), cross product (Section 2.3.2), circulation of vector fields (Section 2.1.2), electric current density (Section 2.3.1), curl of vector fields (Section 2.3.3), and Stokes' theorem (Section 2.3.4). Also, we need to know differential surface with direction (Section 1.1.1), dot product (Section 1.1.2), and flux of vector fields (Section 1.1.3).

2.8.1 Magnetic Dipole

As opposed to the existence of electric charges, there is no magnetic charge in nature (Figure 2.80). On the other hand, a magnetic dipole, which creates a magnetic field that is similar[82] to the electric field created by an electric dipole, can be defined as a circular loop carrying a constant electric current (Figure 2.81). Consider a steady I_0 flowing

[82]This similarity is particularly for the far zone: i.e. away from the dipole.

along a loop of radius a located at the origin in a vacuum. The magnetic field intensity at an arbitrary location $\bar{r}$ can be found as

$$\bar{H}_d(\bar{r}) = \frac{I_0}{4\pi}\oint_C \frac{\overline{dl'} \times (\bar{r}-\bar{r}')}{|\bar{r}-\bar{r}'|^3} = \frac{I_0}{4\pi}\int_0^{2\pi} \frac{a\hat{a}_\phi \times (\bar{r}-\bar{r}')}{|\bar{r}-\bar{r}'|^3} d\phi'. \tag{2.340}$$

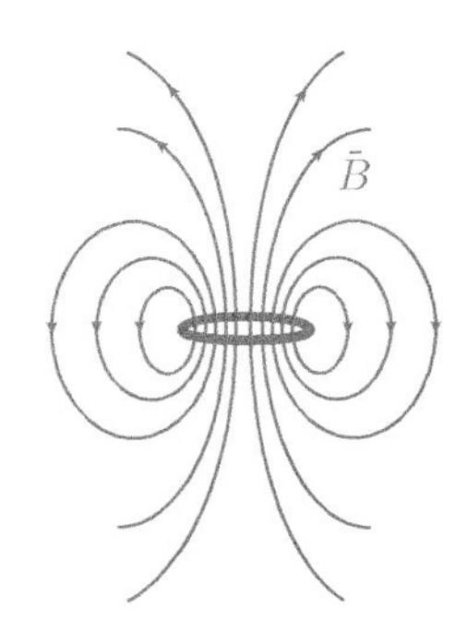

Figure 2.82 Magnetic field lines for a magnetic dipole. Magnetic dipoles are very useful to describe various phenomena, such as magnetization. We note that magnetic field lines close upon themselves (since there is no magnetic monopole/charge), while electric field lines in Figure 1.63 emerge from a positive electric charge and end at a negative electric charge.

Similar to the electric dipole, we are particularly interested in the far-zone expression: i.e. $\bar{H}_d(\bar{r})$ for $|\bar{r}| \ll a$. Using a first-order approximation, we have

$$|\bar{r}-\bar{r}'|^{-3} = \left[(x - a\cos\phi')^2 + (y - a\sin\phi')^2 + z^2\right]^{-3/2} \tag{2.341}$$

$$\approx [|\bar{r}|^2 - 2a(x\cos\phi' + y\sin\phi')]^{-3/2} \tag{2.342}$$

$$= |\bar{r}|^{-3}\left[1 - \frac{2a(x\cos\phi' + y\sin\phi')}{|\bar{r}|^2}\right]^{-3/2} \tag{2.343}$$

$$\approx |\bar{r}|^{-3}\left[1 + \frac{3a(x\cos\phi' + y\sin\phi')}{|\bar{r}|^2}\right]. \tag{2.344}$$

In addition, in the numerator, we have

$$a\hat{a}_\phi \times (\bar{r}-\bar{r}') = az(\hat{a}_x\cos\phi' + \hat{a}_y\sin\phi') - \hat{a}_z\left[a\sin\phi'(y - a\sin\phi') + a\cos\phi'(x - a\cos\phi')\right] \tag{2.345}$$

$$= az(\hat{a}_x\cos\phi' + \hat{a}_y\sin\phi') - \hat{a}_z\left[ay\sin\phi' + ax\cos\phi' - a^2\right]. \tag{2.346}$$

Considering the integrals,[83] we derive

$$\bar{H}_d(\bar{r}) \approx \frac{I_0}{2|\bar{r}|^3}\left\{\hat{a}_z a^2\left(1 - \frac{3x^2}{2|\bar{r}|^2} - \frac{3y^2}{2|\bar{r}|^2}\right) + \hat{a}_x a^2\frac{3xz}{2|\bar{r}|^2} + \hat{a}_y a^2\frac{3yz}{2|\bar{r}|^2}\right\} \tag{2.347}$$

$$= \frac{I_0a^2}{2|\bar{r}|^3}\hat{a}_z\left(1 - \frac{3x^2}{2|\bar{r}|^2} - \frac{3y^2}{2|\bar{r}|^2} - \frac{3z^2}{2|\bar{r}|^2}\right) + \frac{I_0a^2}{2|\bar{r}|^3}\left(\hat{a}_x\frac{3xz}{2|\bar{r}|^2} + \hat{a}_y\frac{3yz}{2|\bar{r}|^2} + \hat{a}_z\frac{3z^2}{2|\bar{r}|^2}\right) \tag{2.348}$$

$$= -\hat{a}_z\frac{I_0a^2}{4|\bar{r}|^3} + \frac{\mu_0 I_0 a^2}{4|\bar{r}|^5}3z\bar{r} = \frac{I_0a^2}{4|\bar{r}|^3}\left(-\hat{a}_z + \frac{3z}{|\bar{r}|}\hat{a}_R\right). \tag{2.349}$$

[83] We repeat these important symmetry integrals as

$$\int_0^{2\pi}\cos\phi' d\phi' = \int_0^{2\pi}\sin\phi' d\phi' = 0$$

$$\int_0^{2\pi}\sin\phi'\cos\phi' d\phi' = 0$$

$$\int_0^{2\pi}\sin^2\phi' d\phi' = \int_0^{2\pi}\cos^2\phi' d\phi' = \pi.$$

$\bar{I}_{DM,m}$: magnetic dipole moment
Unit: amperes × meter² (Am²)

Defining the magnetic dipole moment as $\bar{I}_{DM,m} = \hat{a}_z I_0 \pi a^2$, which contains the normal direction ($\hat{a}_z$), the magnitude of the electric current, and the surface area of the loop, we obtain[84]

$$\bar{H}_d(\bar{r}) \approx -\frac{\bar{I}_{DM,m}}{4\pi|\bar{r}|^3} + \frac{\bar{I}_{DM,m}}{4\pi|\bar{r}|^3} 3\hat{a}_R \cdot \bar{I}_{DM,m} \tag{2.350}$$

$$= \frac{1}{4\pi|\bar{r}|^3}\left(3\hat{a}_R\hat{a}_R \cdot \bar{I}_{DM,m} - \bar{I}_{DM,m}\right). \tag{2.351}$$

[84] Note the similarity of this expression to the electric field intensity created by an electric dipole in Eq. (1.264).

The final expression, which contains the dipole moment $\bar{I}_{DM,m}$, is independent of the orientation of the dipole and is valid for all cases. As the distance $\bar{r}$ becomes large, the magnetic field intensity decays with $1/|\bar{r}|^3$, similar to the behavior of the electric field of an electric dipole. At the same time, the near-zone behavior of the magnetic dipole is very different from that of the electric dipole (Figure 2.82 in comparison to Figure 1.63).

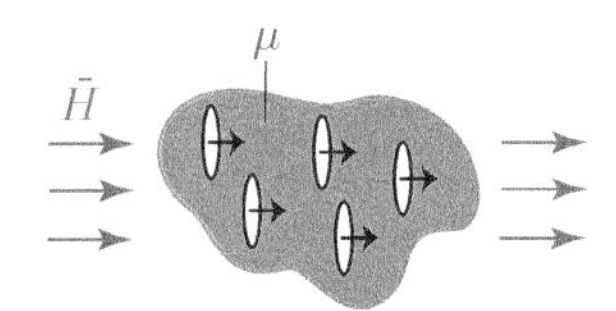

Figure 2.83 A simple illustration of a magnetization inside a magnetic body in a magnetic field created by external sources.

2.8.2 Magnetization

To understand how magnetic fields and magnetic materials interact with each other, we need the concept of magnetization. Since there are different kinds of magnetism (basically, ferromagnetism, paramagnetism, and diamagnetism[85]), it is useful to briefly describe different dynamics in magnetization. First, all materials possess diamagnetism: i.e. negative responses of materials to external magnetic fields. Considering the atoms in a material, paired electrons spinning around nuclei can be considered circulating electric currents and hence magnetic dipoles oriented randomly without any external field. When a magnetic field is applied, these dipoles tend to orient in the negative direction (opposing the external field), as predicted by Lenz's law.[86] On the other hand, this response is very small and usually undetectable in practice. Paramagnetism occurs when the atoms of a material have free electrons that can be modeled as relatively large magnetic dipoles. These dipoles are again randomly oriented, while they become aligned when an external magnetic field is applied (Figure 2.83). The net alignment is in the direction of the field, as opposed to diamagnetism.[87] Ferromagnets behave similar to paramagnets with much larger effects, but interestingly, they can also possess permanent magnetization due to the alignment of magnetic dipoles even when there is no external field (Figure 2.84). This *self-alignment* is obviously due to the interaction of magnetic dipoles with each other, while the physics behind the ferromagnetism is explained thoroughly via quantum mechanics.

[85] There are also other types, such as antiferromagnetism and ferrimagnetism.

[86] This is the reason for $\mu < \mu_0$ macroscopically.

[87] Hence, we have $\mu > \mu_0$.

In any case, the magnetic property of a material can be represented by a permeability $\mu(\bar{r})$ such that we have $\bar{B}(\bar{r},t) = \mu(\bar{r})\bar{H}(\bar{r},t)$ or

$$\bar{B}(\bar{r},t) = \mu_0\mu_r(\bar{r})\bar{H}(\bar{r},t), \tag{2.352}$$

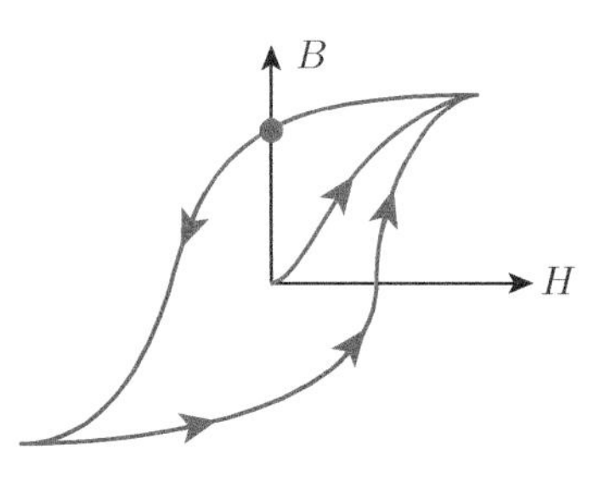

Figure 2.84 When a ferromagnetic material is magnetized, a well-known B-H (hysteresis) curve is formed, which can be seen as a nonlinear permeability. Then for some cases, the value of $\mu(\bar{r})$ can be zero or infinity. Hence, in some cases, it may be better to use the magnetization vector field $\bar{M}(\bar{r},t)$ for the relationship between $\bar{H}(\bar{r},t)$ and $B(\bar{r},t)$.

where $\mu_r(\bar{r})$ is the relative permeability. In addition, we define $\chi_m(\bar{r}) = \mu_r(\bar{r}) - 1$ as the magnetic susceptibility. We note that, for paramagnetic materials, we have $\chi_m(\bar{r}) > 0$, while diamagnetism can be represented with $\chi_m(\bar{r}) < 0$. At the same time, for most materials, $\chi_m(\bar{r}) \approx 0$, and they can simply be considered nonmagnetic. Considering a magnetic material with $\chi_m(\bar{r}) > 0$ and $\mu_r(\bar{r}) > 1$, Eq. (2.352) indicates that the magnetic flux density is increased for a fixed magnetic field intensity. This increase can be considered the response of the material to the magnetic field (magnetization).

χ_m: magnetic susceptibility
Unitless

To formulate magnetization, we define it as a vector field as[88]

$$\bar{M}(\bar{r},t) = (\mu_r(\bar{r}) - 1)\bar{H}(\bar{r},t) = \frac{1}{\mu_0}\bar{B}(\bar{r},t) - \bar{H}(\bar{r},t), \tag{2.353}$$

which accounts for the difference in $\bar{B}(\bar{r},t)$ when a magnetic material exists instead of a vacuum. Further using $\chi_m(\bar{r}) = \mu_r(\bar{r}) - 1$, we have

$$\bar{M}(\bar{r},t) = \chi_m(\bar{r})\bar{H}(\bar{r},t) \tag{2.354}$$

under the assumption that the magnetic material is linear.

[88] When considering dielectrics, we write $\bar{P}(\bar{r},t) = \bar{D}(\bar{r},t) - \varepsilon_0\bar{E}(\bar{r},t)$ as the general expression, while $\bar{D}(\bar{r},t) = \varepsilon(\bar{r})\bar{E}(\bar{r},t)$ can be used for linear (and isotropic) cases. For magnetic materials, the general expression is $\bar{M}(\bar{r},t) = \bar{B}(\bar{r},t)/\mu_0 - \bar{H}(\bar{r},t)$, while $\bar{B}(\bar{r},t) = \mu(\bar{r})\bar{H}(\bar{r},t)$ assumes linear (and isotropic) materials. For example, inside a magnet, there is a nonzero magnetization, but $\mu(\bar{r})$ is not well defined.

$\bar{M}$: magnetization
Unit: amperes/meter (A/m)

Now, we can revisit Ampere's law, particularly its differential form, while considering magnetic material. Using the vector identity[89]

$$\bar{\nabla} \times [h(\bar{r},t)\bar{f}(\bar{r},t)] = \bar{\nabla}h(\bar{r},t) \times \bar{f}(\bar{r},t) + h(\bar{r},t)\bar{\nabla} \times \bar{f}(\bar{r},t) \tag{2.355}$$

for any scalar field $h(\bar{r},t)$ and vector field $\bar{f}(\bar{r},t)$, we have

[89] We note that $\bar{\nabla}h(\bar{r},t)$ represents the gradient of a scalar field $h(\bar{r},t)$, discussed in Section 3.6.1.

$$\bar{\nabla} \times \bar{H}(\bar{r},t) = \bar{\nabla} \times \left[\bar{B}(\bar{r},t)/\mu(\bar{r})\right] \tag{2.356}$$

$$= \bar{\nabla}\left(\frac{1}{\mu(\bar{r})}\right) \times \bar{B}(\bar{r},t) + \frac{1}{\mu(\bar{r})}\bar{\nabla} \times \bar{B}(\bar{r},t) \tag{2.357}$$

$$= \bar{J}_v(\bar{r},t) + \frac{\partial \bar{D}(\bar{r},t)}{\partial t}. \tag{2.358}$$

Hence, we obtain a general expression for the curl of the magnetic flux density as

$$\bar{\nabla} \times \bar{B}(\bar{r},t) = \mu(\bar{r})\bar{J}_v(\bar{r},t) + \mu(\bar{r})\frac{\partial \bar{D}(\bar{r},t)}{\partial t} - \mu(\bar{r})\bar{\nabla}\left(\frac{1}{\mu(\bar{r})}\right) \times \bar{B}(\bar{r},t) \tag{2.359}$$

or[90]

$$\bar{\nabla} \times \bar{B}(\bar{r},t) = \mu(\bar{r})\bar{J}_v(\bar{r},t) + \mu(\bar{r})\frac{\partial \bar{D}(\bar{r},t)}{\partial t} + \frac{\bar{\nabla}\mu(\bar{r})}{\mu(\bar{r})} \times \bar{B}(\bar{r},t). \tag{2.360}$$

[90] We use the identity

$$\bar{\nabla}\left(\frac{1}{h(\bar{r},t)}\right) = -\frac{\bar{\nabla}h(\bar{r},t)}{[h(\bar{r},t)]^2}$$

for any scalar field $h(\bar{r},t)$.

In the above, $\bar{\nabla}\mu(\bar{r})$ represent the change in $\mu(\bar{r})$ with respect to position, including a jump across a boundary. Obviously, if the permeability does not change with position at $\bar{r}$, we have $\mu(\bar{r}) = \mu$, $\bar{\nabla}\mu(\bar{r}) = 0$, and

$$\bar{\nabla} \times \bar{B}(\bar{r},t) = \mu \bar{J}_v(\bar{r},t) + \mu \frac{\partial \bar{D}(\bar{r},t)}{\partial t}. \tag{2.361}$$

2.8.3 Equivalent Magnetization Currents

Similar to the equivalent polarization charges to represent polarization, we can define equivalent magnetization currents to represent magnetization. This way, for a given magnetic material, an equivalent problem can be found, where magnetization currents exist in a vacuum. To formulate the equivalence, we reconsider the curl of the magnetic flux density as

$$\bar{\nabla} \times \bar{B}(\bar{r},t) = \mu_0 \bar{\nabla} \times \bar{H}(\bar{r},t) + \mu_0 \bar{\nabla} \times \bar{M}(\bar{r},t), \tag{2.362}$$

leading to

$$\bar{\nabla} \times \bar{B}(\bar{r},t) = \mu_0 \bar{J}_v(\bar{r},t) + \mu_0 \frac{\partial \bar{D}(\bar{r},t)}{\partial t} + \mu_0 \bar{\nabla} \times \bar{M}(\bar{r},t) \tag{2.363}$$

$$= \mu_0 \left[\bar{J}_v(\bar{r},t) + \bar{\nabla} \times \bar{M}(\bar{r},t)\right] + \mu_0 \frac{\partial \bar{D}(\bar{r},t)}{\partial t}. \tag{2.364}$$

The final equality can be written

$$\bar{\nabla} \times \bar{B}(\bar{r},t) = \mu_0 \left[\bar{J}_v(\bar{r},t) + \bar{J}_{mv}(\bar{r},t)\right] + \mu_0 \frac{\partial \bar{D}(\bar{r},t)}{\partial t}, \tag{2.365}$$

where $\bar{J}_{mv}(\bar{r},t) = \bar{\nabla} \times \bar{M}(\bar{r},t)$ is the volume magnetization current density (A/m^2). Hence, $\bar{\nabla} \times \bar{M}(\bar{r},t)$ behaves like a *fictitious* electric[91] current density to represent magnetization effects.

[91] We note that equivalent magnetization currents are *electric*-type currents: i.e. they are not magnetic currents.

Similar to the volume polarization charge density $q_{pv}(\bar{r},t) = -\bar{\nabla} \cdot \bar{P}(\bar{r},t)$, the volume magnetization current density is nonzero when the magnetization changes inside a magnetic object. This particularly occurs when the permeability changes with respect to position such that there is a variance in the magnitudes of the aligned magnetic dipoles and a lack of perfect cancellation. At the same time, magnetic dipoles at the boundaries of a magnetic object generally are not canceled, and they also need to be represented by equivalent magnetization currents (specifically by surface electric currents). To derive the expression for the surface magnetization current density, we consider the boundary conditions for tangential magnetic fields in the presence of magnetization (Figure 2.85):

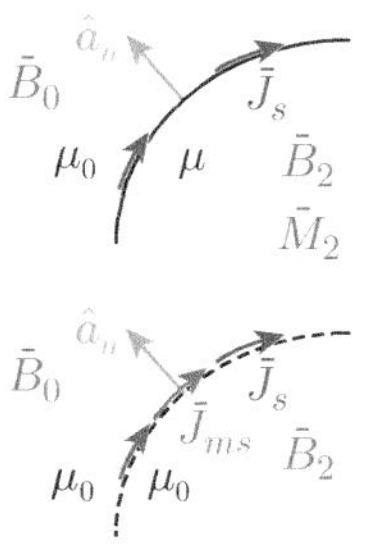

Figure 2.85 Replacing the magnetization effect at the interface between a magnetic medium and a vacuum with equivalent surface magnetization currents. Note that magnetization currents $\bar{J}_{ms}(\bar{r},t)$ are added to true electric currents $\bar{J}_s(\bar{r},t)$ that may possibly exist.

$$\hat{a}_n \times [\bar{B}_0(\bar{r},t)/\mu_0 - \bar{B}_2(\bar{r},t)/\mu_0 + \bar{M}_2(\bar{r},t)] = \bar{J}_s(\bar{r},t) \qquad (\bar{r} \in S), \tag{2.366}$$

where $\hat{a}_n$ is the unit normal vector out from the magnetic object. This equation can be rearranged as

$$\hat{a}_n \times [\bar{B}_0(\bar{r},t)/\mu_0 - \bar{B}_2(\bar{r},t)/\mu_0] = \bar{J}_s(\bar{r},t) - \hat{a}_n \times \bar{M}_2(\bar{r},t) \qquad (\bar{r} \in S) \tag{2.367}$$

or[92]

$$\hat{a}_n \times [\bar{B}_0(\bar{r},t) - \bar{B}_2(\bar{r},t)] = \mu_0\bar{J}_s(\bar{r},t) - \mu_0\hat{a}_n \times \bar{M}_2(\bar{r},t) \qquad (\bar{r} \in S). \tag{2.368}$$

In the above, the term $-\hat{a}_n \times \bar{M}_2(\bar{r},t)$ acts like a surface electric current density that creates a jump in the magnetic flux density, even when there is no true electric current density $\bar{J}_s(\bar{r},t)$.

To sum up, a magnetized object can be represented by the surface magnetization current density and the volume magnetization current density defined as (Figure 2.86)

$$\bar{J}_{ms}(\bar{r},t) = -\hat{a}_n \times \bar{M}(\bar{r},t) \tag{2.369}$$

$$\bar{J}_{mv}(\bar{r},t) = \bar{\nabla} \times \bar{M}(\bar{r},t), \tag{2.370}$$

where $\bar{M}(\bar{r},t)$ is the magnetization. If these equivalent currents are used, we assume that they generate magnetic flux density in *a vacuum*. For example, differential and integral forms of Ampere's law can be written

$$\bar{\nabla} \times \bar{B}(\bar{r},t) = \mu_0\bar{J}_v(\bar{r},t) + \mu_0\frac{\partial\bar{D}(\bar{r},t)}{\partial t} \tag{2.371}$$

$$\oint_C \bar{B}(\bar{r},t) \cdot \overline{dl} = \mu_0 I_{\text{enc}}(t) + \mu_0 \int_S \frac{\partial\bar{D}(\bar{r},t)}{\partial t} \cdot \overline{ds}, \tag{2.372}$$

where $\bar{J}_v(\bar{r},t)$ is the total volume electric current density (including true volume electric current density and volume magnetization current density), while $I_{\text{enc}}(t)$ is the total enclosed electric current (including true electric current and magnetization current). Once the magnetic flux density is found, the magnetic field intensity can be found as $\bar{H}(\bar{r},t) = \bar{B}(\bar{r},t)/\mu_0 - \bar{M}(\bar{r},t)$.

Since equivalent currents are not true currents, their introduction in a problem should not change the total electric current (conservation of electric charges). In fact, we have

$$\bar{J}_{\text{net}}(t) = \int_V \bar{J}_{mv}(\bar{r})dv + \oint_S \bar{J}_{ms}(\bar{r})ds = 0, \tag{2.373}$$

where V and S are the volume and surface where magnetization is defined. This can be shown clearly by inserting the expressions for the magnetization currents as

$$\bar{J}_{\text{net}}(t) = \int_V \bar{\nabla} \times \bar{M}(\bar{r},t)dv - \oint_S \hat{a}_n \times \bar{M}(\bar{r},t)ds = 0, \tag{2.374}$$

[92] This is consistent with the discontinuity in the magnetic flux density shown in Eq. (2.194). Specifically, the magnetic flux density is discontinuous at the boundary between two (magnetically) different media, even when there is no surface electric current density $\bar{J}_s(\bar{r},t)$. This can be explained by the unbalanced magnetization at the boundary, which can be represented by a fictitious surface electric current density in the equivalent problem.

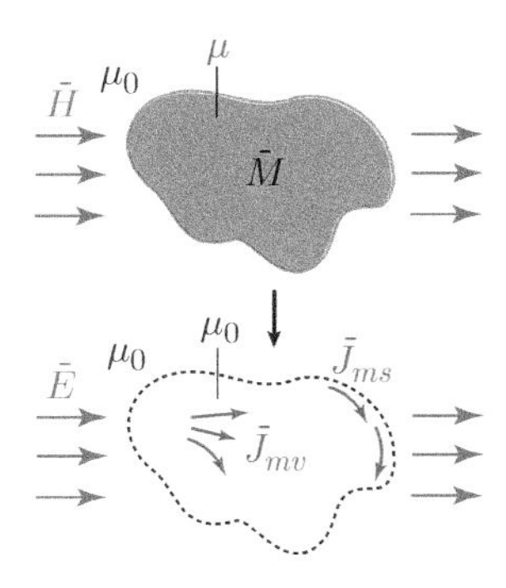

Figure 2.86 Modeling of magnetization in terms of equivalent electric currents that are located in the host medium (a vacuum). Surface magnetization currents account for the discontinuity of the permeability between the object and the vacuum. On the other hand, volume magnetization currents are generally needed to model permeability variations inside the object.

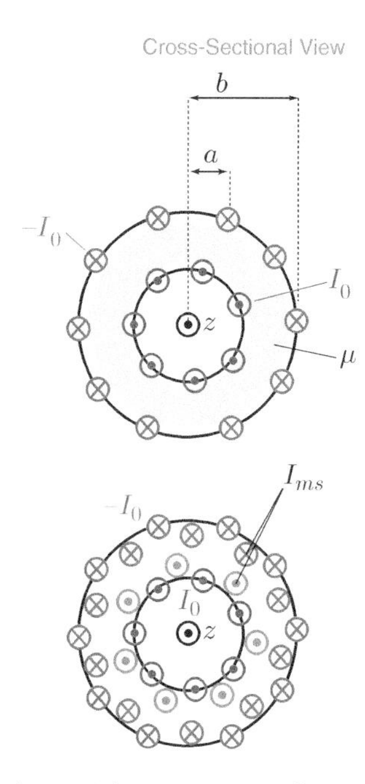

Figure 2.87 A coaxial structure involving two cylindrical surfaces with radii a and $b > a$ aligned in the z direction. Surface electric currents that are uniformly distributed over the cylindrical surfaces flow in opposite directions. If a magnetic material is inserted between the cylinders, surface magnetization currents exist in the equivalent problem. These electric currents reduce the magnetic flux density between the cylinders. In this scenario, the true electric currents $\pm I_0$ are assumed to be constant.

where $\hat{a}_n$ is the unit normal vector out from the magnetic object. Using the vector identity

$$\bar{\nabla} \cdot [\bar{f}(\bar{r},t) \times \bar{g}(\bar{r},t)] = \bar{g}(\bar{r},t) \cdot \bar{\nabla} \times \bar{f}(\bar{r},t) - \bar{f}(\bar{r},t) \cdot \bar{\nabla} \times \bar{g}(\bar{r},t) \tag{2.375}$$

for any vector fields $\bar{f}(\bar{r},t)$ and $\bar{g}(\bar{r},t)$, we have

$$\bar{\nabla} \cdot [\bar{f} \times \bar{M}(\bar{r},t)] = \bar{M}(\bar{r},t) \cdot \bar{\nabla} \times \bar{f} - \bar{f} \cdot \bar{\nabla} \times \bar{M}(\bar{r},t) = \bar{f} \cdot \bar{\nabla} \times \bar{M}(\bar{r},t), \tag{2.376}$$

where $\bar{f}$ is *any constant* vector field. Taking a volume integral of both sides, we arrive at

$$\int_V \bar{\nabla} \cdot [\bar{f} \times \bar{M}(\bar{r},t)] dv = \int_V \bar{f} \cdot \bar{\nabla} \times \bar{M}(\bar{r},t) dv = \bar{f} \cdot \int_V \bar{\nabla} \times \bar{M}(\bar{r},t) dv \tag{2.377}$$

$$\oint_S \bar{f} \times \bar{M}(\bar{r},t) \cdot \overline{ds} = \bar{f} \cdot \int_V \bar{\nabla} \times \bar{M}(\bar{r},t) dv, \tag{2.378}$$

where the divergence theorem is used, while $\bar{f}$ is taken out of the integral. Further using $\bar{f} \times \bar{M}(\bar{r},t) \cdot \hat{a}_n = -\bar{f} \cdot \hat{a}_n \times \bar{M}(\bar{r},t)$, we have

$$\bar{f} \cdot \oint_S \hat{a}_n \times \bar{M}(\bar{r},t) \cdot \overline{ds} = \bar{f} \cdot \int_V \bar{\nabla} \times \bar{M}(\bar{r},t) dv. \tag{2.379}$$

Finally, since $\bar{f}$ is arbitrary, the integrals should be equal to each other, verifying the equality in Eq. (2.374).

As an example of the equivalent magnetization currents, we reconsider an infinitely long coaxial structure involving two cylindrical surfaces with radii a and $b > a$ aligned in the z direction (Figure 2.57). There are steady surface electric currents that are uniformly distributed over cylinders in opposite directions. As shown in Section 2.7.1, the magnetic field intensity can be written

$$\bar{H}(\bar{\rho}) = \hat{a}_\phi \frac{I_0}{2\pi\rho} \begin{cases} 0 & (\rho < a) \quad \text{or} \quad (\rho > b) \\ 1 & (a < \rho < b) \end{cases} \tag{2.380}$$

In addition, if everywhere is a vacuum, we have the magnetic flux density as

$$\bar{B}(\bar{\rho}) = \hat{a}_\phi \frac{\mu_0 I_0}{2\pi\rho} \begin{cases} 0 & (\rho < a) \quad \text{or} \quad (\rho > b) \\ 1 & (a < \rho < b). \end{cases} \tag{2.381}$$

If a magnetic material is inserted between the cylinders (Figure 2.87) while keeping the electric currents $\pm I_0$ constant (e.g. by using an electric current source),[93] the magnetic field intensity

[93] When discussing dielectric effects in a parallel-plate capacitor in Section 1.7.3 (e.g. see Figure 1.67) we assumed that true electric charges are constant so that insertion of a dielectric reduces the electric field intensity. Alternatively, one can assume constant voltage (electric field intensity), which can be possible if the capacitor is connected to a voltage source. In such a case, the electric field intensity remains constant while the electric flux density increases as a dielectric is inserted. In the case of a coaxial structure (or an inductor), there are again two options. Keeping the electric current fixed leads to a constant magnetic field intensity but changing magnetic flux density since a magnetic material is inserted. Alternative, the magnetic flux (hence the magnetic flux density) can be fixed, while the electric current and the magnetic field intensity change.

does not change (considering Ampere's law). On the other hand, the magnetic flux density is changed as

$$\bar{B}(\bar{\rho}) = \hat{a}_\phi \frac{\mu I_0}{2\pi\rho} \begin{cases} 0 & (\rho < a) \quad \text{or} \quad (\rho > b) \\ 1 & (a < \rho < b), \end{cases} \tag{2.382}$$

i.e. it increases if $\mu > \mu_0$. This increase can be explained as the response of the magnetic material to the external field (created by the true electric currents on the cylinders). One can find the magnetization for $a < \rho < b$ as

$$\bar{M}(\bar{\rho}) = \bar{B}(\bar{\rho})/\mu_0 - \bar{H}(\bar{\rho}) = \hat{a}_\phi \frac{I_0}{2\pi\rho}\left(\frac{\mu}{\mu_0} - 1\right). \tag{2.383}$$

Hence, the equivalent magnetization currents can be written

$$\bar{J}_{mv}(\bar{r}) = \bar{\nabla} \times \bar{M}(\bar{r}) = 0 \tag{2.384}$$

and[94]

$$\bar{J}_{ms}(\bar{r}) = -\hat{a}_n \times \bar{M}(\bar{r}) = -\begin{cases} -\hat{a}_\rho \times \bar{M}(\bar{r}) & (\bar{r} \in \text{inner cylinder}) \\ \hat{a}_\rho \times \bar{M}(\bar{r}) & (\bar{r} \in \text{outer cylinder}) \end{cases} \tag{2.385}$$

$$= \hat{a}_z \frac{I_0}{2\pi\rho}\left(\frac{\mu}{\mu_0} - 1\right) \begin{cases} 1 & (\bar{r} \in \text{inner cylinder}) \\ -1 & (\bar{r} \in \text{outer cylinder}). \end{cases} \tag{2.386}$$

It is remarkable that the surface magnetization current density is added positively to the true[95] electric current density on both inner and outer cylinders. Therefore, the increasing value of the magnetic flux density is verified in the equivalent problem.

As mentioned in Section 2.8.2, ferromagnetic materials can possess permanent magnetization (Figure 2.88). Since permanently magnetized objects (magnets) are commonly observed in real life (as opposed to electrets with permanent polarization), we consider an example of their electromagnetic analyses. We assume that a cylindrical object is permanently magnetized uniformly as $\bar{M}(\bar{r}) = \hat{a}_z M_0$ and located in a vacuum (Figure 2.89). The object has a radius of a and is oriented from $z = 0$ to $z = l$. Even though there is no any external magnetic field, this magnet itself creates a magnetic field, and our purpose is to find it on the z axis for $z > l$. Using the definitions of the magnetization currents, we have

$$\bar{J}_{mv}(\bar{r}) = \bar{\nabla} \times \bar{M}(\bar{r}) = 0 \tag{2.387}$$

$$\bar{J}_{ms}(z = 0) = \bar{J}_{ms}(z = L) = 0 \tag{2.388}$$

$$\bar{J}_{ms}(\rho = a) = -\hat{a}_n \times \bar{M}(\rho = a) = -\hat{a}_\rho \times \hat{a}_z M_0 = \hat{a}_\phi M_0. \tag{2.389}$$

[94] We note that $\hat{a}_n = -\hat{a}_\rho$ on the surface of the inner cylinder, while $\hat{a}_n = \hat{a}_\rho$ on the surface of the outer cylinder. This is because we always consider the *outward* normal with respect to the magnetic material.

[95] *Free electric current* is also a common name to distinguish true electric currents from magnetization (bound) currents.

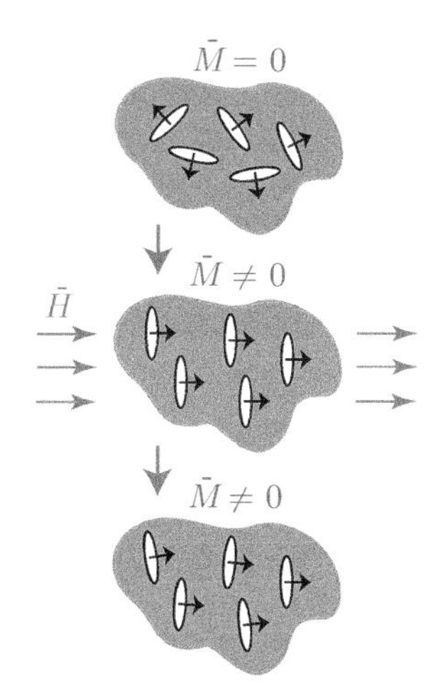

Figure 2.88 A simple illustration of permanent magnetization.

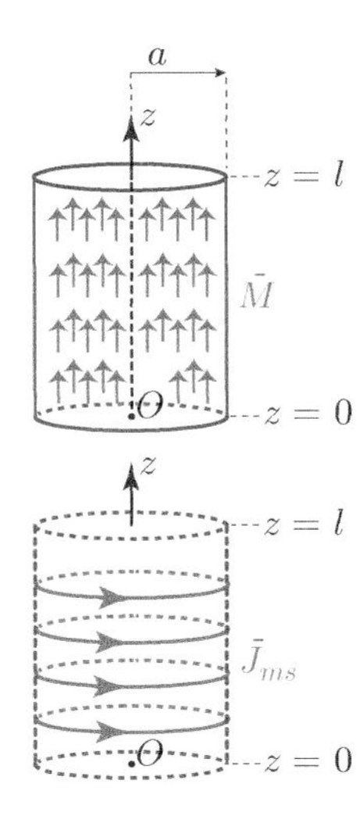

Figure 2.89 A permanently magnetized cylindrical object and its model in terms of equivalent currents.

Therefore, nonzero surface magnetization currents exist only on the cylindrical surface. At this stage, the object can be replaced by circular electric current loops lying from $z = 0$ to $z = l$ (Figure 2.89). Then using the expression for the magnetic flux density due to a loop electric current in Eq. (2.260), we have

$$dB_z(\bar{r}) = \frac{\mu_0 M_0 dz'}{2} \frac{a^2}{[a^2 + (z - z')^2]^{3/2}} \tag{2.390}$$

as the magnetic field created by loops inside a differential length dz'. We note that inside such a differential length, the total amount of electric current is $I_0 = M_0 dz'$. Then an integration (superposing the effects of all loops from $z = 0$ to $z = l$) leads to

$$B_z = \frac{\mu_0 M_0 a^2}{2} \int_0^l \frac{dz'}{[a^2 + (z - z')^2]^{3/2}} \tag{2.391}$$

$$= \frac{\mu_0 M_0}{2} \left[\frac{z}{\sqrt{z^2 + a^2}} + \frac{l - z}{\sqrt{(z - l)^2 + a^2}} \right]. \tag{2.392}$$

Hence, we obtain

$$\bar{B}(0, 0, z) = \hat{a}_z \frac{\mu_0 M_0}{2} \left[\frac{z}{\sqrt{z^2 + a^2}} + \frac{l - z}{\sqrt{(z - l)^2 + a^2}} \right] \qquad (z > l). \tag{2.393}$$

Considering *a vacuum*, we also have[96]

$$\bar{H}(0, 0, z) = \hat{a}_z \frac{M_0}{2} \left[\frac{z}{\sqrt{z^2 + a^2}} + \frac{l - z}{\sqrt{(z - l)^2 + a^2}} \right] \qquad (z > l). \tag{2.394}$$

Interestingly, Eq. (2.393) is also valid for $0 < z < l$, while for this region, the magnetic field intensity can be found as[97]

$$\bar{H}(0, 0, z) = \bar{B}(0, 0, z)/\mu_0 - \bar{M}(0, 0, z) \tag{2.395}$$

$$= \hat{a}_z \frac{M_0}{2} \left[\frac{z}{\sqrt{z^2 + a^2}} + \frac{l - z}{\sqrt{(z - l)^2 + a^2}} \right] - \hat{a}_z M_0. \tag{2.396}$$

Hence, at the center of the magnet, we have

$$\bar{H}(0, 0, l/2) = \hat{a}_z \frac{M_0}{2} \left[\frac{l/2}{\sqrt{l^2/4 + a^2}} + \frac{l/2}{\sqrt{l^2/4 + a^2}} \right] - \hat{a}_z M_0 \tag{2.397}$$

[96] This result may be confusing. Since there is no true electric current, we may ask why there is nonzero magnetic field intensity $\bar{H}(\bar{r})$. In fact, since there is no true electric current, $\bar{\nabla} \times \bar{H}(\bar{r}) = 0$ everywhere; however, this does *not* indicate that $\bar{H}(\bar{r})$ itself is zero.

[97] It is remarkable that the magnetic field intensity has a jump at $z = l$. As discussed in Chapter 4, the boundary conditions for normal magnetic fields enforce continuity in the normal component of the magnetic flux density, while the normal component of the magnetic field intensity jumps when the permeability changes across a boundary. In an equivalent problem, this can be explained by magnetization charges.

$$= \hat{a}_z \frac{M_0}{2} \frac{l}{\sqrt{l^2/4 + a^2}} - \hat{a}_z M_0. \qquad (2.398)$$

It is remarkable that, for $l \gg a$, we have $\bar{H}(0, 0, l/2) = 0$. Finally, Eq. (2.394) can also be used for a magnet with small height: i.e. when $l \ll a$ (Figure 2.90). In this case, we have

$$\bar{H}(0, 0, z) = \hat{a}_z \frac{M_0 l a^2}{2[z^2 + a^2]^{3/2}} \qquad (2.399)$$

using a first-order approximation.

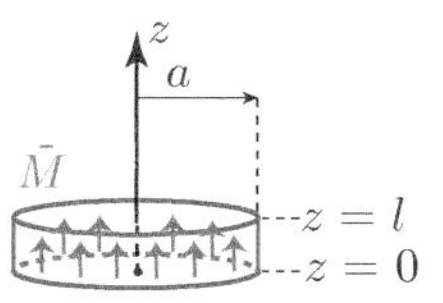

Figure 2.90 A cylindrical magnet with small height.

2.8.4 Examples

Example 26: Consider an infinitely long cylindrical magnet with given magnetization

$$\bar{M}(\bar{r}) = \hat{a}_\phi M_0 \frac{(\rho - a)}{a} \qquad (2.400)$$

for $\rho < a$ (Figure 2.91). The magnet is located in a vacuum. Find the magnetic field intensity $\bar{H}(\bar{r})$ and the magnetic flux density $\bar{B}(\bar{r})$ everywhere.

Solution: First, we can find the magnetization currents as (Figure 2.91)

$$\bar{J}_{mv}(\bar{r}) = \bar{\nabla} \times \bar{M}(\bar{r}) = \frac{M_0}{\rho} \hat{a}_z \frac{\partial}{\partial \rho} \left(\frac{\rho(\rho - a)}{a} \right) = \hat{a}_z M_0 \frac{(2\rho - a)}{\rho a}. \qquad (2.401)$$

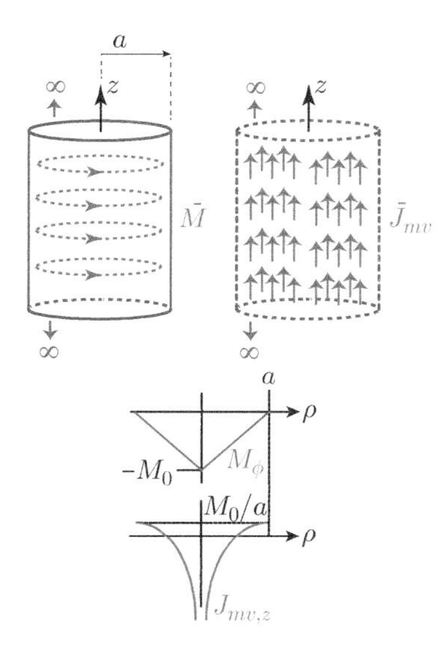

Figure 2.91 A magnetized cylindrical object and the volume magnetization current density to represent it.

We note that there is a volume magnetization current distribution to represent the magnet, while there is no need for any surface magnetization current (i.e. $\bar{M}(\bar{r}) \times \hat{a}_n = 0$) in this case. The magnetic flux density can be found via Ampere's law (in *a vacuum*) using a circular Ampere loop as

$$\bar{B}(\bar{r}) = \hat{a}_\phi \frac{\mu_0 I_{\text{enc}}}{2\pi\rho}, \qquad (2.402)$$

where ρ is the radius of the loop. For $\rho < a$, we have

$$I_{\text{enc}} = M_0 \int_0^\rho \int_0^{2\pi} \frac{(2\rho' - a)}{\rho' a} \rho' d\phi' d\rho' \qquad (2.403)$$

$$= \frac{2\pi M_0}{a} \left[(\rho')^2 - a\rho' \right]_0^\rho = \frac{2\pi M_0 \rho(\rho - a)}{a}, \qquad (2.404)$$

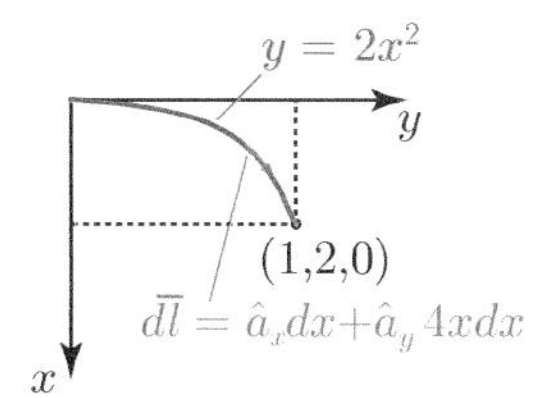

Figure 2.92 A path from the origin to $(x, y, z) = (1, 2, 0)$, where $y = 2x^2$.

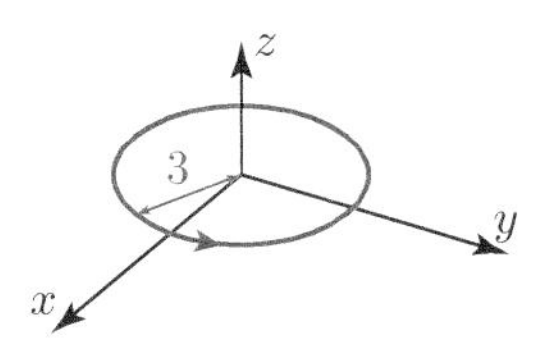

Figure 2.93 A closed circular path on the x-y plane.

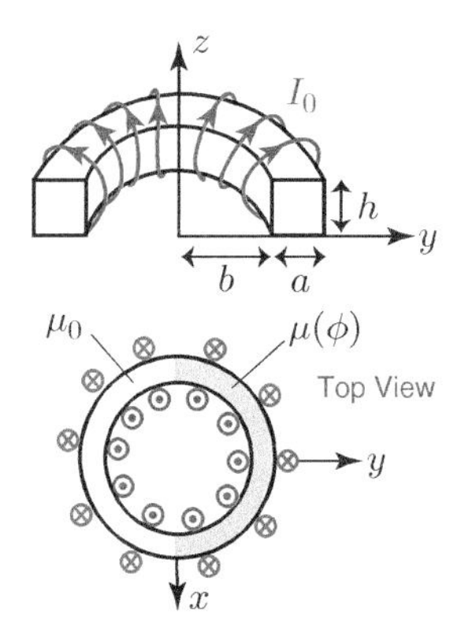

Figure 2.94 A toroid structure, half of which is filled by an inhomogeneous magnetic material.

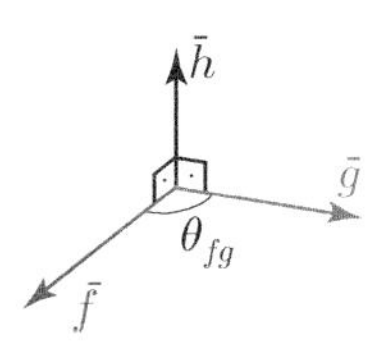

Figure 2.95 Cross product of two vectors: i.e. $\bar{f} \times \bar{g} = \bar{h}$.

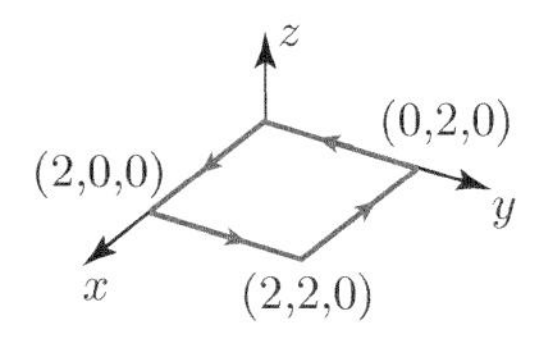

Figure 2.96 A square closed path on the x-y plane.

leading to

$$\bar{B}(\bar{r}) = \hat{a}_\phi \frac{\mu_0}{2\pi\rho} \frac{2\pi M_0 \rho(\rho - a)}{a} = \hat{a}_\phi \mu_0 M_0 \frac{(\rho - a)}{a} \qquad (\rho < a). \tag{2.405}$$

On the other hand, for $\rho > a$, we have $I_{\text{enc}} = 0$; hence,

$$\bar{B}(\bar{r}) = 0 \qquad (\rho > a). \tag{2.406}$$

At this stage, we can find $\bar{H}(\bar{r})$ outside the magnet by simply using $\bar{H}(\bar{r}) = \bar{B}(\bar{r})/\mu_0$, leading to

$$\bar{H}(\bar{r}) = 0 \qquad (\rho > a). \tag{2.407}$$

On the other hand, inside the magnet, we may use $\bar{H}(\bar{r}) = \bar{B}(\bar{r})/\mu_0 - \bar{M}(\bar{r})$, leading to

$$\bar{H}(\bar{r}) = \hat{a}_\phi M_0 \frac{(\rho - a)}{a} - \hat{a}_\phi M_0 \frac{(\rho - a)}{a} = 0 \qquad (\rho < a). \tag{2.408}$$

Hence, the magnetic field intensity is zero everywhere. In fact, we are expecting this result since the symmetry allows us to solve the original problem as

$$\oint_C \bar{H}(\bar{r}) \cdot \overline{dl} = 2\pi\rho H_\phi = I_{\text{enc}}, \tag{2.409}$$

while there is no true electric current ($I_{\text{enc}} = 0$) and $H_\phi = 0$ for any ρ. We emphasize that we are unable to use $\bar{B}(\bar{r}) = \mu(\bar{r})\bar{H}(\bar{r})$ inside the magnet, not only because $\mu(\bar{r})$ is not clearly defined; from the results, we can deduce that $\chi_m(\bar{r}) = \bar{H}/\bar{M} \to \infty$ when $\rho < a$.

2.9 Final Remarks

In this chapter, we have focused on Ampere's law, which describes the relationship between magnetic fields and electric currents and changing electric fields. In addition to the tools introduced in Chapter 1, we understand Ampere's law by considering new tools and concepts, such as differential length with direction, circulation of vector fields, electric current density, cross product, curl of vector fields, Stokes' theorem, magnetic dipole, magnetization, and equivalent magnetization currents. All these tools will also be useful in the next chapters when discussing Faraday's law and Gauss' law for magnetic fields. Ampere's law is valid for all cases, but, similar to Gauss' law, it is particularly useful when finding magnetic fields in static cases. For general dynamic cases, we still need Faraday's law and Gauss' law for magnetic fields that are considered in later chapters.

2.10 Exercises

Exercise 17: Find the integral of the projection of $\bar{f}(\bar{r}) = \hat{a}_x 3xy - \hat{a}_y y^2$ from $(x, y, z) = (0, 0, 0)$ to $(x, y, z) = (1, 2, 0)$ along C defined as $y = 2x^2$ (Figure 2.92).

Exercise 18: Find the circulation of $\bar{f}(\bar{r}) = \hat{a}_x(2x - y + z) + \hat{a}_y(x + y - z^2) + \hat{a}_z(3x - 2y + 4z)$ along a circle on the x-y plane with radius $a = 3$ (Figure 2.93).

Exercise 19: Consider a steady electric current I_0 (A) circulating around a toroid of radius b and cylindrical radius a (Figure 2.94). There are a total of N turns. Half of the toroid ($\pi < \phi < 2\pi$) is filled with a vacuum, while the other half is filled with an inhomogeneous material with $\mu(\bar{r}) = \mu_0 + \mu_0 \sin\phi$ for $\phi \in [0, \pi]$. Find the magnetic flux density everywhere.

Exercise 20: Consider two vectors $\bar{f} = \hat{a}_x + \hat{a}_y 4 - \hat{a}_z 5$ and $\bar{g} = -\hat{a}_x 2 - \hat{a}_y 4 + \hat{a}_z 4$. Find $\bar{h} = \bar{f} \times \bar{g}$ and show that $\bar{h}$ is perpendicular to both $\bar{f}$ and $\bar{g}$ (Figure 2.95).

Exercise 21: Consider three vectors $\bar{f} = \hat{a}_x + \hat{a}_y 4 - \hat{a}_z 5$, $\bar{g} = -\hat{a}_x 2 - \hat{a}_y 4 + \hat{a}_z 4$, and $\bar{h} = -\hat{a}_x + \hat{a}_y - \hat{a}_z$. Find $\bar{f} \cdot \bar{g} \times \bar{h}$.

Exercise 22: Consider three vectors $\bar{f} = \hat{a}_x + \hat{a}_y 4 - \hat{a}_z 5$, $\bar{g} = -\hat{a}_x 2 - \hat{a}_y 4 + \hat{a}_z 4$, and $\bar{h} = -\hat{a}_x + \hat{a}_y - \hat{a}_z$. Evaluate and show that $\bar{f} \times (\bar{g} \times \bar{h}) \neq (\bar{f} \times \bar{g}) \times \bar{h}$.

Exercise 23: Consider the circulation of $\bar{f}(\bar{r}) = x\hat{a}_x + x^2 y\hat{a}_y + y^2 z\hat{a}_z$ around a square formed of straight lines as $(x, y) = (0, 0) \longrightarrow (2, 0) \longrightarrow (2, 2) \longrightarrow (0, 2) \longrightarrow (0, 0)$ on the x-y plane (Figure 2.96). Verify Stokes' theorem.

Exercise 24: Consider the circulation of $\bar{f}(\bar{r}) = \hat{a}_z \cos\phi/\rho$ through a cylindrical surface defined as $\rho = 2$, $\pi/3 \leq \phi \leq \pi/2$, and $0 \leq z \leq 3$ (Figure 2.97). Verify Stokes' theorem.

Exercise 25: Consider the coaxial structure in Figure 2.17. An inner cylinder of radius a is located inside a shell defined as $b \leq \rho \leq c$, while we assume that uniformly distributed steady electric currents I_0 (A) and $-I_0$ (A) flow in the cylinder ($\rho < a$) and in the shell ($b \leq \rho \leq c$), respectively. Show that the differential form of Ampere's law holds everywhere.

Exercise 26: A distribution of the magnetic field intensity with respect to position is given as (Figure 2.98)

$$\bar{H}(\bar{r}) = \hat{a}_\phi H_0 \begin{cases} -\rho^2/a^2 & (\rho < a) \\ a/\rho & (\rho > a). \end{cases} \tag{2.410}$$

Find the electric current everywhere.

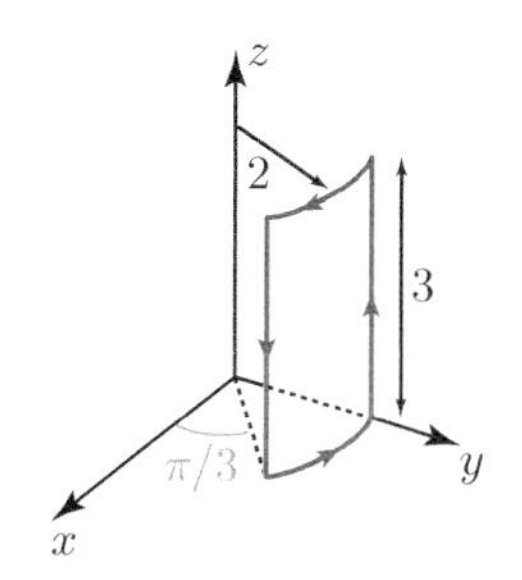

Figure 2.97 A cylindrical surface defined as $\rho = 2$, $\pi/3 \leq \phi \leq \pi/2$, and $0 \leq z \leq 3$, and a contour around it.

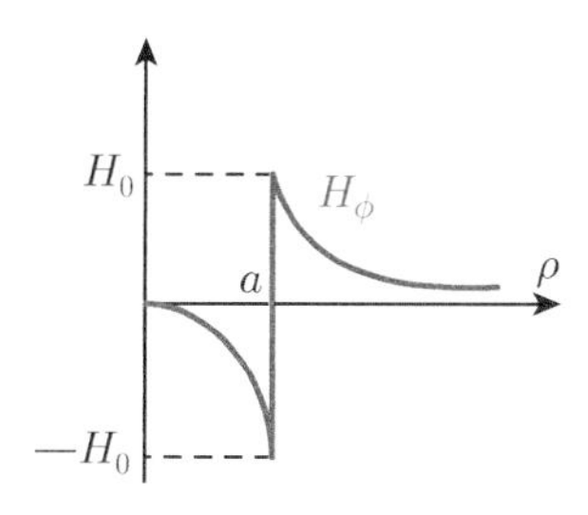

Figure 2.98 A static magnetic field intensity with respect to ρ. Nonzero surface electric current density is responsible for the jump at $\rho = a$.

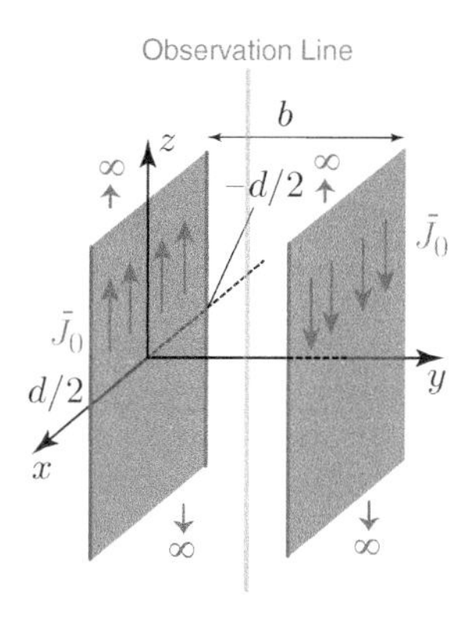

Figure 2.99 Two infinitely long strips carrying steady electric currents.

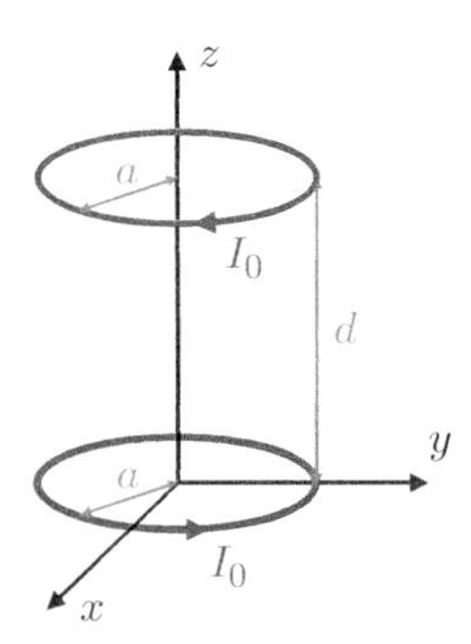

Figure 2.100 A loop over another loop, carrying currents in opposite directions. This structure is commonly known as Helmholtz coil.

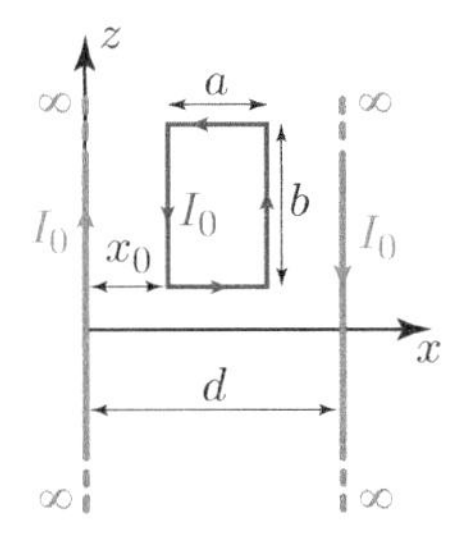

Figure 2.101 A rectangular current-carrying loop between two infinitely long wires.

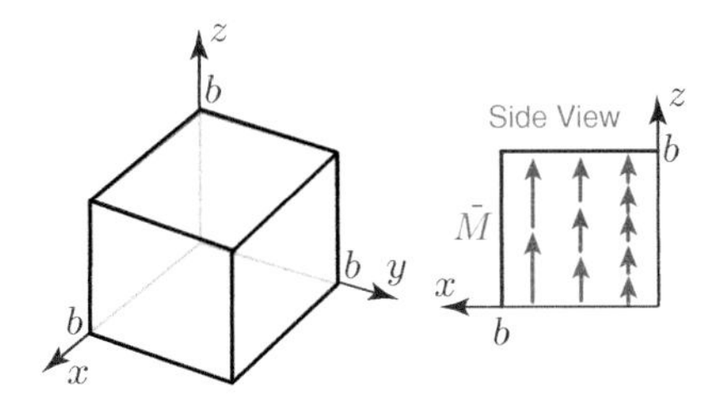

Figure 2.102 A magnetized cube (magnet) with a given magnetization $\bar{M}(\bar{r}) = \hat{a}_z M_0 x/b$ (A/m).

Exercise 27: Consider two infinitely long strips carrying steady electric currents $\bar{J}_s(\bar{r}) = \pm\hat{a}_z J_0$ (A/m) in reverse directions (Figure 2.99). One of the strips is at $y = 0$ from $x = -d/2$ to $x = d/2$, and the other is located at $y = b$ from $x = -d/2$ to $x = d/2$. Find the magnetic field intensity on a line at $x = 0$ and $y = b/2$.

Exercise 28: Consider a loop of radius a carrying a steady electric current I_0 (A) located at the origin. Another loop of radius a is located at $z = d$, and it also carries a steady electric current I_0 (A) in the opposite direction (Figure 2.100). The loops are located in a vacuum. Assuming that $d \gg a$, find the magnetic force between the loops.

Exercise 29: Consider an $a \times b$ rectangular loop carrying a steady electric current I_0 (A) between two infinitely long lines that also carry I_0 (A) in opposite directions in a vacuum (Figure 2.101). The distance between the infinitely long lines is d, while the distance between the left vertical side of the rectangular loop to one of the lines is $x_0 < (d - a)$. Find the magnetic force acting on the loop.

Exercise 30: Consider a permanently magnetized cube defined as $0 \leq x \leq b$, $0 \leq y \leq b$, and $0 \leq z \leq b$ (Figure 2.102). Magnetization inside the cube is given as $\bar{M}(\bar{r}) = \hat{a}_z M_0 x/b$ (A/m). Find all magnetization currents.

Exercise 31: Consider a conical magnet defined as $\theta \in [0, \theta_0]$ and $z \in [a, b]$, which has a permanent magnetization $\bar{M}(\bar{r}) = \hat{a}_z M_0$ (Figure 2.103). Find the magnetic flux density and the magnetic field intensity at the origin.

2.11 Questions

Question 7: Prove the Jacobi identity

$$\bar{f} \times (\bar{g} \times \bar{h}) + \bar{g} \times (\bar{h} \times \bar{f}) + \bar{h} \times (\bar{f} \times \bar{g}) = 0 \tag{2.411}$$

for any vectors $\bar{f}$, $\bar{g}$, and $\bar{h}$.

Question 8: Magnetic field intensity with respect to position is given as (Figure 2.104)

$$\bar{H}(\bar{r}) = \hat{a}_\phi H_0 \begin{cases} 1 & (\rho < a) \\ a/\rho & (a < \rho < b) \\ -a/\rho & (\rho > b). \end{cases} \tag{2.412}$$

Find the electric current (expressions and types) everywhere.

Question 9: Consider a current-carrying wire involving two infinitely long lines connected via a semicircle (Figure 2.105). Find the magnetic field intensity at the center point of the semicircle, if steady electric current I_0 (A) flows along the wire.

Question 10: A cylindrical inductor of length l and radius a has nonuniform wiring (Figure 2.106), and the number of wires per unit length is given as

$$N_l = N_0 \frac{|l/2 - z|}{l^2}. \tag{2.413}$$

Note that the wiring is dense at the ends of the inductor, while it becomes sparse at the middle.

- Find the magnetic flux density at the middle of the inductor: i.e. at $z = l/2$.
- Compare your result with an ideally wired inductor (still with length l and radius a, as shown in Figure 2.70) using the *same amount* of wire.

Question 11: Consider an infinitely long cylindrical magnetic layer with a constant permeability μ, which is located symmetrically around an infinitely long wire carrying a steady electric current I_0 (A) (Figure 2.107).

- Find the magnetic field intensity everywhere.
- Find the magnetic flux density everywhere.
- Find the equivalent magnetization currents if the magnetic material is to be replaced with equivalent currents.
- Verify the expression for the magnetic flux density considering the equivalent problem (equivalent magnetization currents).

Question 12: Consider a permanently magnetized cylindrical layer with inner radius a and outer radius b located around the z axis from $z = -l/2$ to $z = l/2$ in a vacuum (Figure 2.108). The magnetization is given as

$$\bar{M}(\bar{r}) = \hat{a}_\rho \frac{M_0 a}{\rho}, \tag{2.414}$$

where ρ is the cylindrical variable. Show that the magnetic flux density on the z axis for $z > l$ can be written

$$\bar{B}(\bar{r}) = \hat{a}_z \frac{\mu_0 M_0 a}{2} \left\{ \frac{1}{\sqrt{z^2 + a^2}} - \frac{1}{\sqrt{z^2 + b^2}} \right\}$$

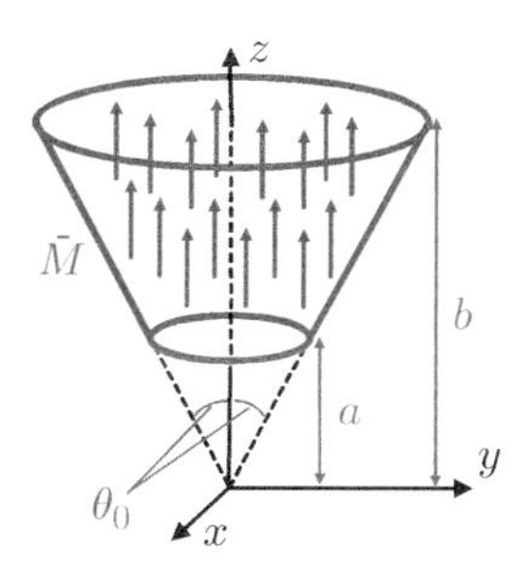

Figure 2.103 A conical magnet.

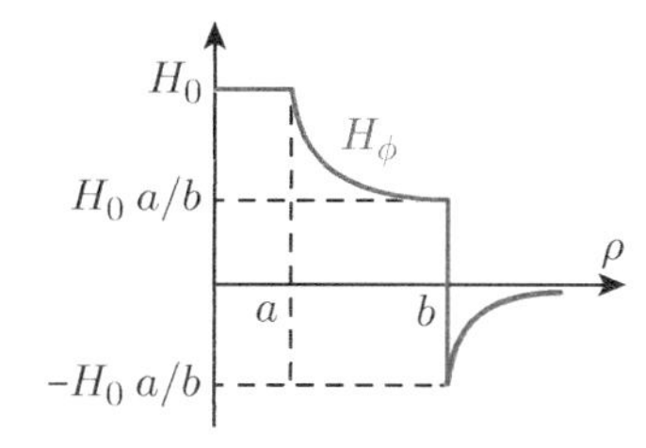

Figure 2.104 A magnetic field intensity with respect to ρ generated by an unknown electric current distribution.

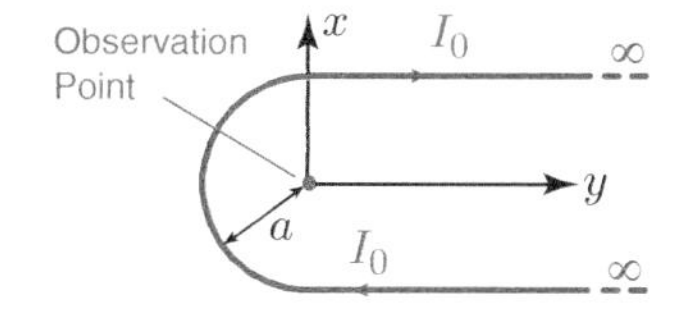

Figure 2.105 A current-carrying wire involving two infinitely long lines and a semicircle.

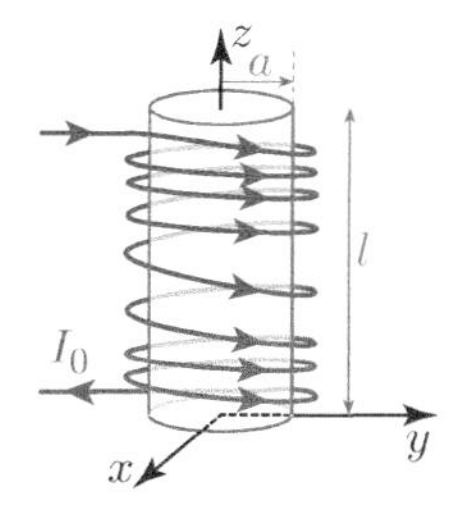

Figure 2.106 A cylindrical inductor with nonuniform wiring.

$$- \hat{a}_z \frac{\mu_0 M_0 a}{2} \left\{ \frac{1}{\sqrt{(z-l)^2 + a^2}} - \frac{1}{\sqrt{(z-l)^2 + b^2}} \right\}. \tag{2.415}$$

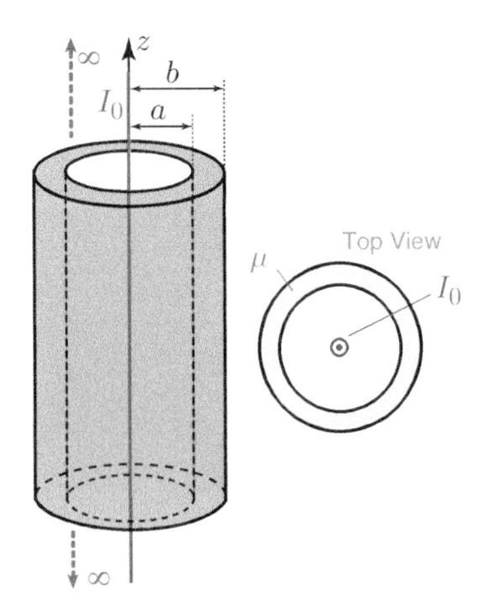

Figure 2.107 An infinitely long cylindrical magnetic layer around an infinitely long wire carrying a steady electric current I_0 (A).

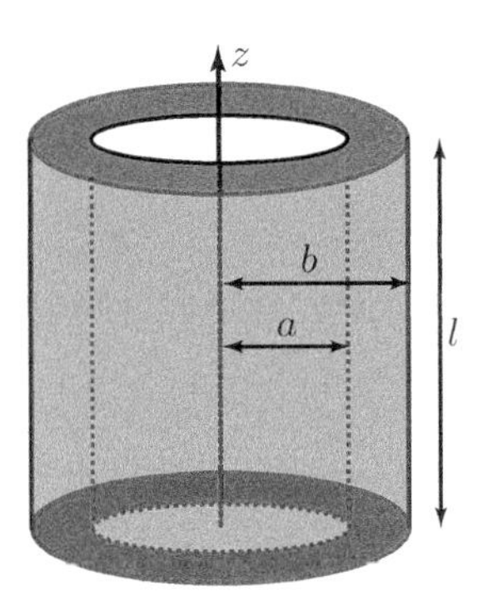

Figure 2.108 A magnetized cylindrical layer.

Question 13: An infinitely long cylindrical structure is defined as follows:

- Steady volume electric current density $\bar{J}_v(\bar{r}) = \hat{a}_z J_0$ (A/m^2) for $0 < \rho < a$ flowing in a vacuum
- A vacuum for $a < \rho < b$ and $\rho > c$
- A magnetic material with $\mu(\bar{r}) = \mu_0 \rho / a$ for $b < \rho < c$

Do the following:

- Find the magnetic field intensity everywhere.
- Find the magnetic flux density everywhere.
- Find the magnetization inside the magnetic material.
- Find all equivalent magnetization currents and show that the net magnetization current is zero.
- How would the magnetic field intensity and magnetic flux density change if the magnetic material was changed to a permanent magnet with magnetization $\bar{M}(\bar{r}) = -J_0 a^2/(2\rho)$ (A/m)?

3

Faraday's Law

So far, we have discussed two of the four Maxwell's equations: i.e. Gauss' law and Ampere's law. These fundamental laws describe how electric and magnetic fields are created by electric charges and currents, respectively. In addition, Ampere's law has an extra term, showing the relationship between electric and magnetic fields in dynamic cases. Specifically, if there is a time-variant electric flux, it creates magnetic fields, even in the absence of electric currents. In this sense, Faraday's law, which is considered in this chapter, complements Ampere's law by showing that changing magnetic flux creates electric fields, even when there is no electric charge. This mutual relationship between electric and magnetic fields – generation of electric/magnetic fields due to changing magnetic/electric fields – is the major mechanism to describe how electromagnetic waves and electromagnetic energy propagate in space, including a vacuum in the absence of a material.

Like the previous two chapters, we first focus on the integral form of Faraday's law and then discuss its differential form. Faraday's law, especially when used together with Gauss' and Ampere's laws, is particularly useful in dynamic cases, while it also provides important information on static electric fields: i.e. their conservative nature.

$$\oint_C \bar{E}(\bar{r},t) \cdot \overline{dl} = -\int_S \frac{\partial \bar{B}(\bar{r},t)}{\partial t} \cdot \overline{ds}$$

$$\bar{\nabla} \times \bar{E}(\bar{r},t) = -\frac{\partial \bar{B}(\bar{r},t)}{\partial t}$$

Introduction to Electromagnetic Waves with Maxwell's Equations, First Edition. Özgür Ergül.

Companion website: www.wiley.com/go/ergulmax

3.1 Integral Form of Faraday's Law

Electric charges are direct sources of electric fields as given by Gauss' law as $\bar{\nabla} \cdot \bar{D}(\bar{r}, t) = q_v(\bar{r}, t)$ or

$$\oint_S \bar{D}(\bar{r}, t) \cdot \overline{ds} = Q_{\text{enc}}(t). \tag{3.1}$$

In fact, in static cases, electric charges form the only type of sources to generate electric fields. On the hand, in dynamic scenarios, we have another type of sources: changing magnetic fields. Consider a curve C bounding an open surface S (Figure 3.1). According to Faraday's law (of *induction*), we have

$$\oint_C \bar{E}(\bar{r}, t) \cdot \overline{dl} = -\int_S \frac{\partial \bar{B}(\bar{r}, t)}{\partial t} \cdot \overline{ds}, \tag{3.2}$$

where the circulation of $\bar{E}(\bar{r}, t)$ along the closed loop is the negative of the total magnetic flux change through the bounded surface. Similar to Ampere's law in Eq. (2.1), the direction of $\overline{dl}$ and the flux direction are related via the right-hand rule. If the medium is magnetically linear and isotropic, Faraday's law can be written as

$$\oint_C \bar{E}(\bar{r}, t) \cdot \overline{dl} = -\int_S \mu(\bar{r}) \frac{\partial \bar{H}(\bar{r}, t)}{\partial t} \cdot \overline{ds}. \tag{3.3}$$

Furthermore, if we consider a homogeneous medium, we have $\mu(\bar{r}) = \mu$ and

$$\oint_C \bar{E}(\bar{r}, t) \cdot \overline{dl} = -\mu \int_S \frac{\partial \bar{H}(\bar{r}, t)}{\partial t} \cdot \overline{ds}, \tag{3.4}$$

where $\mu = \mu_0$ for a vacuum or any magnetically inactive (nonmagnetic) medium. In addition, if the medium is also electrically linear, isotropic, and homogeneous, we have

$$\oint_C \bar{D}(\bar{r}, t) \cdot \overline{dl} = -\mu\varepsilon \int_S \frac{\partial \bar{H}(\bar{r}, t)}{\partial t} \cdot \overline{ds}, \tag{3.5}$$

where $\varepsilon = \varepsilon_0$ for a vacuum or for any electrically inactive medium. Note that Eq. (3.2) using *the electric field intensity* $\bar{E}(\bar{r}, t)$ is the main form of Faraday's law that is valid for inhomogeneous cases, in comparison to Gauss' law that is generally written for *the electric flux density* $\bar{D}(\bar{r}, t)$.

To understand the integral form of Faraday's law, we need the concepts of differential length with direction (Section 2.2.1) and circulation of vector fields (Section 2.1.2). These are

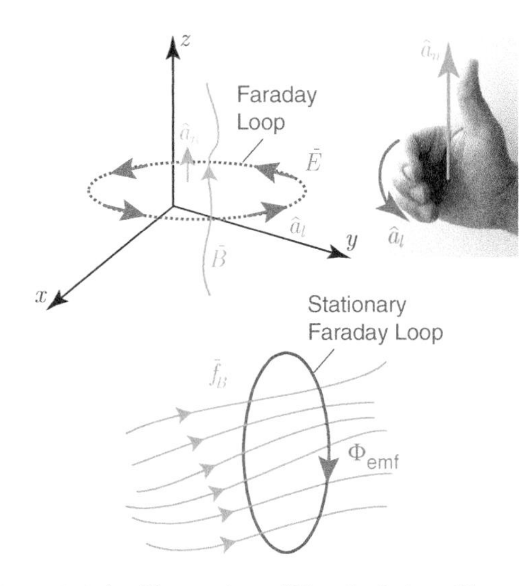

Figure 3.1 An illustration of Faraday's law. For general cases, Eq. (3.2) is valid: i.e. the change in the magnetic flux density with respect to time should be integrated, and the result corresponds to the circulation of the electric field intensity. On the other hand, if the loop is stationary – i.e. if the integration domains S and C are fixed – then the time derivative operation can be evaluated after the integration. In this case, the circulation of the electric field intensity is the time derivative of the magnetic flux (integral of the magnetic flux density). Furthermore, in this stationary (does not mean static) case, since the time derivative of the magnetic flux is *always* the electromotive force induced along the loop, the circulation of the electric field intensity becomes the only source of the electromotive force.

in addition to the concepts of dot product (Section 1.1.2), differential surface with direction (Section 1.1.1), and flux of vector fields (Section 1.1.3). Therefore, no new tool is required before we focus on the meaning and application of the integral form of Faraday's law.

3.1.1 Meaning and Application of Faraday's Law

We already know the meaning of the circulation and flux to interpret Faraday's law in Eq. (3.2) from a mathematical point of view. Integrating the time derivative of the magnetic flux density means all contributions crossing the bounded surface should be combined. The effects of changing magnetic flux density in opposite directions may cancel each other. But the circulation of the electric field intensity depends on the *net* value, while it is not directly related to the shape of the loop and the distribution of the magnetic flux density. Similar to the integral forms of Gauss' and Ampere's laws, the integral form of Faraday's law may be useful to find the electric field intensity $\bar{E}(\bar{r}, t)$ for symmetric and special cases: i.e. when $\bar{E}(\bar{r}, t)$ can be put outside the integral. On the other hand, the integral form of Faraday's law is often written in such a way that an important physical quantity, the electromotive force, is used and evaluated without dealing with the circulation integral. According to this form, we have[1]

$$\Phi_{\text{emf}}(t) = -\frac{df_B(t)}{dt}, \tag{3.6}$$

where $\Phi_{\text{emf}}(t)$ is the electromotive force (voltage) induced in the loop (Figure 3.2) and

$$f_B(t) = \int_S \bar{B}(\bar{r}, t) \cdot \overline{ds} \tag{3.7}$$

is the magnetic flux. Hence, we can write

$$\Phi_{\text{emf}}(t) = -\frac{d}{dt}\int_S \bar{B}(\bar{r}, t) \cdot \overline{ds}. \tag{3.8}$$

According to this equation, changing magnetic flux (with respect to time) through the surface of a loop creates an electromotive[2] force along the loop.

The difference between the right-hand sides of Eqs. (3.2) and (3.8) is important. In Eq. (3.2), the time derivative is inside the integral and directly applied to the magnetic flux density. In Eq. (3.8), however, the time derivative is applied to the magnetic flux: i.e. after the integral is evaluated. These two equations are valid for all cases; but the location of the time derivative indicates the difference between the circulation of the electric field intensity and the electromotive force. When the loop under investigation is not stationary, the time derivative in Eq. (3.8)

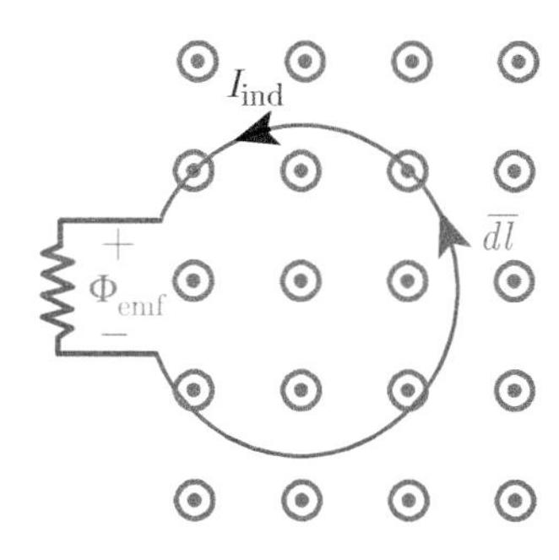

Figure 3.2 A representation of the electromotive force due to a changing magnetic flux. The normal of the surface (to be used to evaluate the flux integral) and $\overline{dl}$ are defined by using the right-hand rule. The induced electric current $I_{\text{ind}}(t)$ along the loop is selected in the same direction as $\overline{dl}$. Then the polarization of the electromotive force is chosen accordingly as if a resistor is connected to measure the voltage. With these directions, if $df_B(t)/dt > 0$ (due to a time variance of the magnetic flux density and/or the movement of the loop under a position-dependent magnetic flux density), then $\Phi_{\text{emf}}(t)$ is negative. Hence, the actual direction of $I_{ind}(t)$ is clockwise, creating a magnetic field $\bar{B}_{\text{ind}}(\bar{r}, t)$ inward that opposes the positive change in $\bar{B}(\bar{r}, t)$.

[1] The negative sign on the right-hand side of Faraday's law has a special interpretation. Due to this sign, a change in the net magnetic flux through a surface creates a negative electromotive force (voltage). Specifically, the created electromotive force *opposes* the change in the magnetic flux, which is called Lenz's law.

f_B: magnetic flux
Unit: weber (Wb) (volts × seconds)

[2] This naming becomes more meaningful when the loop is defined not only mathematically but also physically as a conducting wire.

should be evaluated carefully by considering the time-dependent integration domain S. For example, considering a loop moving with a velocity of $\bar{u}(t)$, it can be shown that[3]

$$\frac{d}{dt}\int_S \bar{B}(\bar{r},t)\cdot\overline{ds} = \int_S \frac{\partial\bar{B}(\bar{r},t)}{\partial t}\cdot\overline{ds} - \oint_C \bar{u}(t)\times\bar{B}(\bar{r},t)\cdot\overline{dl}. \tag{3.9}$$

[3]While a moving loop in an electric field is less considered in the context of Ampere's law, a moving loop in a magnetic field is of particular interest (in the context of Faraday's law) since it corresponds to generating electric voltage (electromotive force) from kinetic energy.

Therefore, in general, an electromotive force along a loop is caused by a time-varying magnetic flux density (first component) and a movement of the loop under a magnetic field (second component), both changing the *magnetic flux* through the loop with respect to time.[4] Then using Eq. (3.2), we have

[4]A movement under a magnetic flux density does not necessarily mean electromotive force is induced. In fact, if $\bar{B}(\bar{r},t)$ does not change with respect to position, we have

$$\oint_C \bar{u}(t)\times\bar{B}(\bar{r},t)\cdot\overline{dl} = \int_S \bar{\nabla}\times[\bar{u}(t)\times\bar{B}(\bar{r},t)]\cdot\overline{ds} = 0.$$

In general, Eq. (3.8) may be easier to use for all cases instead of resorting to Eq. (3.9).

$$\Phi_{\text{emf}}(t) = \oint_C \bar{E}(\bar{r},t)\cdot\overline{dl} + \oint_C \bar{u}(t)\times\bar{B}(\bar{r},t)\cdot\overline{dl} \tag{3.10}$$

or

$$\Phi_{\text{emf}}(t) = \oint_C [\bar{E}(\bar{r},t) + \bar{u}(t)\times\bar{B}(\bar{r},t)]\cdot\overline{dl}. \tag{3.11}$$

Obviously, if the loop is stationary, the electromotive force is the same as the circulation of the electric field intensity. Furthermore, if the magnetic flux does not change with respect to time, we simply have

$$\Phi_{\text{emf}}(t) = \oint_C \bar{E}(\bar{r},t)\cdot\overline{dl} = -\frac{df_B(t)}{dt} = 0. \tag{3.12}$$

Vector fields with zero circulation everywhere are called *conservative*. Hence, when a distribution of the electric field intensity is only due to electric charges (electrostatics), it is a conservative vector field.

3.1.2 Lorentz Force Law

The expression for the electromotive force in Eq. (3.11) is directly related to Lorentz force law that describes the forces acting on a moving electric charge under electric and magnetic fields.[5] First, we recall the types of forces we are already familiar with. Coulomb's law states that there is an electric force between two static electric charges. This is formulated as an electric field created by one of the electric charges to affect the other electric charge. Therefore, as a generalization, we assume that an electric field (created somehow) applies force to electric charges. Ampere's force law states that there is a counterpart magnetic force between two steady electric-current-carrying wires. Again, as a generalization, we assume that an electric-current-carrying wire creates a magnetic field (Biot-Savart law), and there is a magnetic force on an electric-current-carrying

[5]Specifically, the electromotive force induced along a contour is formed of two different components. One of them is the electric field intensity, whose circulation can be nonzero only if there is a changing magnetic flux density. The second is a possible movement of the loop under magnetic flux density, and it is nonzero even when the magnetic flux density itself does not change with respect to time. If the integrand of Eq. (3.11) is multiplied by a test charge Q_0 (which may be an electron moving along the path C), we simply have $Q_0\bar{E}(\bar{r},t) + Q_0\bar{u}(t)\times\bar{B}(\bar{r},t)$ as the total force acting on Q_0.

wire if it is exposed to a magnetic field. Since an electric current is a movement of electric charges, this indicates a magnetic force acting on a moving electric charge in a magnetic field (Figure 3.3). Overall, electric and magnetic forces applied to a moving electric charge Q_0 can be written as

$$\bar{F}_0(t) = Q_0\bar{E}_0(t) + Q_0\bar{u}_0(t) \times \bar{B}_0(t), \tag{3.13}$$

where $\bar{E}_0(t)$ and $\bar{B}_0(t)$ are the electric field intensity and the magnetic flux density, respectively, at the position of the electric charge, whereas $\bar{u}_0(t)$ represents the velocity vector. This equality is called the Lorentz force law.

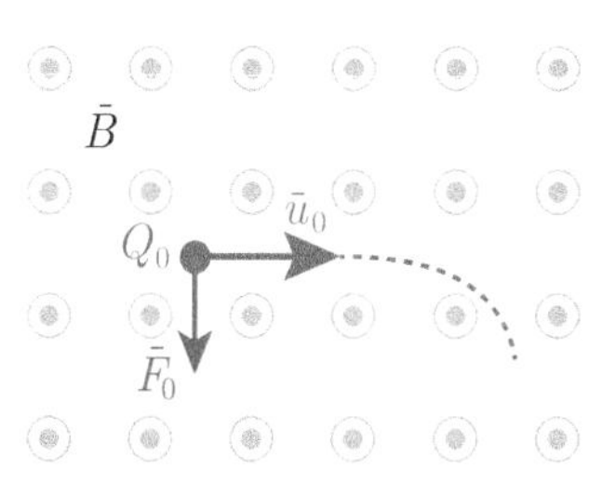

Figure 3.3 According to Lorentz force law, there is a force on a moving electric charge in a magnetic field. In this figure, the electric charge is assumed to be positive so that its trajectory is downward when the magnetic field is out of page.

According to Eq. (3.13), if the electric charge is not moving, there is only electric force, and there is no magnetic force despite the existence of a magnetic field. On the other hand, if the electric charge is moving under a magnetic field, there is a magnetic force described as

$$\bar{F}_{m,0}(t) = Q_0\bar{u}_0(t) \times \bar{B}_0(t). \tag{3.14}$$

Similar to Ampere's force law, the direction of the force is perpendicular to both the velocity vector (electric current direction) and the magnetic field. These directions follow the right-hand rule as follows. If the velocity of a positive electric charge is in the x direction and the magnetic field is in the y direction, then the force should be in the z direction. Obviously, fixing the velocity in the x direction, the magnetic field does not have to be in the y direction; but its y component is responsible for creating the force in the z direction. Under the same scenario, any component of the magnetic field in the z direction creates a force in the $-y$ direction.

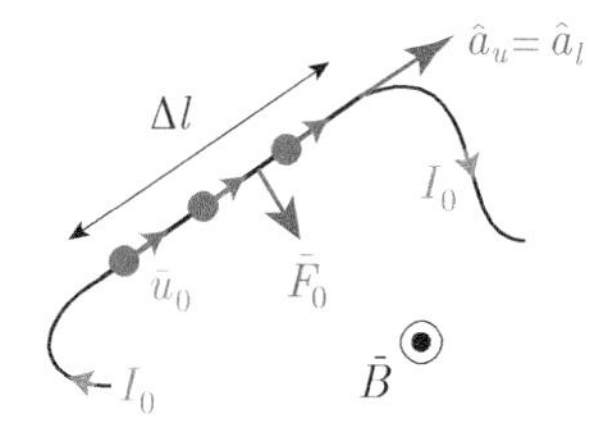

Figure 3.4 A segment of wire under a magnetic field.

Coulomb's law inside Lorentz force law is clearly visible. To complete the picture, we now consider an electric current flowing through a segment of wire[6] (Figure 3.4):

$$I_0(t) = \frac{\Delta Q_0(t)}{\Delta t}. \tag{3.15}$$

Multiplying both sides with the length of the segment, we arrive at

$$I_0(t)\Delta l = \Delta Q_0(t)\frac{\Delta l}{\Delta t} = \Delta Q_0(t)u_0(t), \tag{3.16}$$

where $u_0(t)$ represent the speed of the electric charges. Then, if the electric current segment is exposed to a magnetic field, there is a force on the moving electric charges as predicted by Lorentz force law as[7]

$$F_{m,0}(t) = \Delta Q_0(t)u_0(t)B_0(t). \tag{3.17}$$

[6] In this equation, a positive sign is used (as opposed to Eq. (32)) since we focus on the amount of electric current flowing along a line in terms of the amount of moving electric charges, rather than defining electric current as the movement of electric charges out of a volume.

[7] We assume that the electric current segment is small enough such that magnetic flux density is constant over it.

If the direction of the movement is selected as $\hat{a}_u = \hat{a}_l$, which is the same as the alignment of the electric current segment,[8] we have

[8] We assume that the electric current segment is small enough that its direction is fixed.

$$\bar{F}_{m,0}(t) = \Delta Q_0(t)\bar{u}_0(t) \times \bar{B}_0(t). \tag{3.18}$$

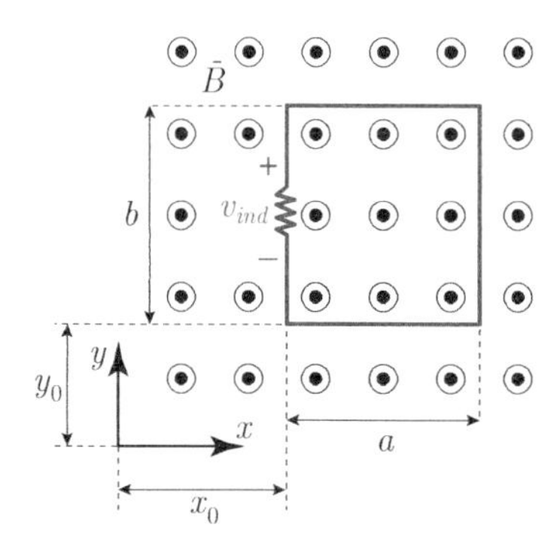

Figure 3.5 A rectangular loop on the x-y plane exposed to a magnetic flux density in the z direction.

Using electric current instead of electric charge, the same expression can be written as

$$\bar{F}_{m,0}(t) = I_0(t)\Delta l\hat{a}_l \times \bar{B}_0(t) \tag{3.19}$$

or

$$\bar{F}_{m,0}(t) = I_0(t)\overline{\Delta l} \times \bar{B}_0(t) \tag{3.20}$$

as the force applied by the magnetic field to the electric current segment. This is perfectly in agreement with Ampere's force law: e.g. see the integrand of Eq. (2.263).

3.2 Using the Integral Form of Faraday's Law

The integral form of Faraday's law is rarely useful to directly find electric fields due to changing magnetic fields.[9] On the other hand, it is commonly used to find the electromotive force (which sometimes corresponds to the circulation of the electric field intensity) along a loop. As examples, we consider $a \times b$ rectangular loops located on the x-y plane and exposed to a z-directed magnetic flux density (Figures 3.5 and 3.6). We assume that the corners of the loop are at (x_0, y_0), $(x_0 + a, y_0)$, $(x_0, y_0 + b)$, and $(x_0 + a, y_0 + b)$. We consider both stationary (Figure 3.5) and moving loops (Figure 3.6) under time-invariant and time-variant magnetic flux densities (one by one) as

[9] In fact, Ampere's law can be more useful for this purpose.

$$\bar{B}_1(\bar{r}, t) = \hat{a}_z B_0 \tag{3.21}$$

$$\bar{B}_2(\bar{r}, t) = \hat{a}_z B_0 \sin(\omega t - kx), \tag{3.22}$$

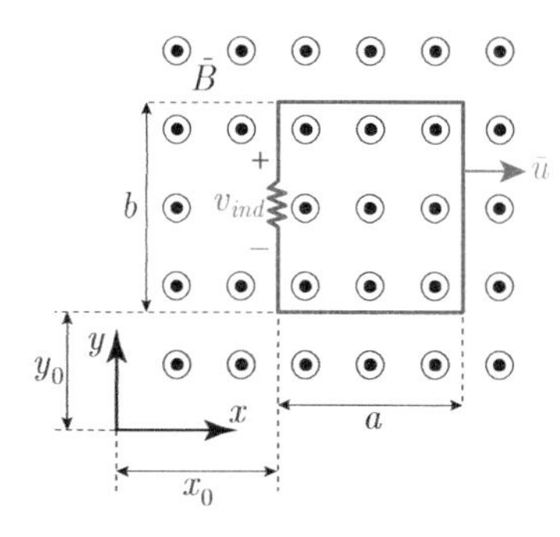

Figure 3.6 A rectangular loop on the x-y plane exposed to a magnetic flux density in the z direction. The loop is moving in the x direction,

where ω and k are constants with units rad/s and rad/m. In all cases, the loops are located in a vacuum. Hence the given magnetic fields are created by some sources at infinity, which are simply omitted in the analysis.

Before using Faraday's law, we can employ Ampere's law to find the corresponding electric field intensity expressions. We have[10]

$$\frac{\partial \bar{E}_1(\bar{r},t)}{\partial t} = \frac{1}{\mu_0\varepsilon_0}\bar{\nabla}\times\bar{B}_1(\bar{r},t) = 0 \tag{3.23}$$

$$\frac{\partial \bar{E}_2(\bar{r},t)}{\partial t} = \frac{1}{\mu_0\varepsilon_0}\bar{\nabla}\times\bar{B}_2(\bar{r},t) = \hat{a}_y\frac{kB_0}{\mu_0\varepsilon_0}\cos(\omega t - kx) \tag{3.24}$$

leading to[11] $\bar{E}_1(\bar{r},t) = 0$ and

$$\bar{E}_2(\bar{r},t) = \hat{a}_y\frac{kB_0}{\omega\mu_0\varepsilon_0}\sin(\omega t - kx). \tag{3.25}$$

Now we consider several cases:

- Case 1: First, we focus on loops exposed to the static magnetic field $\bar{B}_1(\bar{r},t) = \hat{a}_z B_0$. The magnetic flux can be found as

$$f_B(t) = \int_S \bar{B}_1(\bar{r},t)\cdot\overline{ds} = abB_0 \longrightarrow \Phi_{\text{emf}}(t) = -\frac{df_B(t)}{dt} = 0. \tag{3.26}$$

Hence, there is no electromotive force, whether the loop is moving or not. Considering that $\bar{E}_1(\bar{r},t) = 0$ and the magnetic flux density is time-invariant, we further have the circulation of the electric field intensity as

$$\oint_C \bar{E}_1(\bar{r},t)\cdot\overline{dl} = -\int_S \frac{\partial \bar{B}_1(\bar{r},t)}{\partial t}\cdot\overline{ds} = 0 \tag{3.27}$$

when the loop is moving or not. On the other hand, in the case of the moving loop, the electromotive force is zero because

$$\oint_C \bar{u}(t)\times\bar{B}_1(\bar{r},t)\cdot\overline{dl} = 0 \tag{3.28}$$

is satisfied[12] *in addition* to the zero circulation of the electric field intensity.

- Case 2: Now, we consider a stationary loop (Figure 3.5) exposed to $\bar{B}_2(\bar{r},t) = \hat{a}_z B_0\sin(\omega t - kx)$. The magnetic flux can be found as

$$f_B(t) = \int_S \bar{B}_2(\bar{r},t)\cdot\overline{ds} = \frac{bB_0}{k}\Big[\cos(\omega t - kx)\Big]_{x_0}^{x_0+a} \tag{3.29}$$

[10] We are using

$$\frac{\partial \bar{E}(\bar{r},t)}{\partial t} = \frac{1}{\varepsilon_0}\bar{\nabla}\times\bar{H}(\bar{r},t) = \frac{1}{\mu_0\varepsilon_0}\bar{\nabla}\times\bar{B}(\bar{r},t)$$

[11] We omit possible time-invariant components of the electric field intensity.

[12] According to Eqs. (3.26) and (3.27), this (zero integral in Eq. (3.28)) must hold for any type of movement since

$$\Phi_{\text{emf}}(t) = \oint_C \bar{E}_1(\bar{r},t)\cdot\overline{dl} + \oint_C \bar{u}(t)\times\bar{B}_1(\bar{r},t)\cdot\overline{dl}$$

$$0 = 0 + \oint_C \bar{u}(t)\times\bar{B}_1(\bar{r},t)\cdot\overline{dl}.$$

For example, if $\bar{u} = \hat{a}_x u_0$ (movement in the x direction with a constant velocity), we have

$$\bar{u}(t)\times\bar{B}_1(\bar{r},t) = -\hat{a}_y u_0 B_0.$$

Then the horizontal edges give no contribution, since $\overline{dl} = \pm\hat{a}_x$ and $\bar{u}(t)\times\bar{B}_1(\bar{r},t)\cdot\overline{dl} = 0$ for them. The vertical edges give contributions, but they cancel each other, leading to the zero integral in Eq. (3.28).

or[13]

$$f_B(t) = \frac{2bB_0}{k}\sin(\omega t - kx_0 - ka/2)\sin(ka/2) \tag{3.30}$$

leading to the expression for the electromotive force[14] as

$$\Phi_{\text{emf}}(t) = -\frac{df_B(t)}{dt} = -\frac{2b\omega B_0}{k}\cos(\omega t - kx_0 - ka/2)\sin(ka/2). \tag{3.31}$$

Note that, when $t = kx_0/\omega + ka/(2\omega)$, the magnetic flux through the loop is zero, while the electromotive force has its maximum value (with a negative sign) as

$$\Phi_{\text{emf}}(t = x_0/\omega) = -\frac{2b\omega B_0}{k}\sin(ka/2). \tag{3.32}$$

In fact, there is a typical 90° phase difference between the electromotive force and the magnetic flux (Figure 3.7). Since the loop in this case is stationary, the electromotive force above corresponds to the circulation of the electric field intensity. Using Eq. (3.25), we have[15]

$$\oint_C \bar{E}_2(\bar{r},t)\cdot\overline{dl} = \left[\hat{a}_y \frac{kB_0}{\omega\mu_0\varepsilon_0}\sin(\omega t - kx)\cdot\hat{a}_y\right]_{y_0}^{y_0+b}\Bigg|_{x=x_0+a} + \left[\hat{a}_y \frac{kB_0}{\omega\mu_0\varepsilon_0}\sin(\omega t - kx)\cdot(-\hat{a}_y)\right]_{y_0}^{y_0+b}\Bigg|_{x=x_0} \tag{3.33}$$

$$= -\frac{2bkB_0}{\omega\mu_0\varepsilon_0}\cos(wt - kx_0 - ka/2)\sin(ka/2). \tag{3.34}$$

Note that only the vertical edges of the rectangular loop contribute to the circulation since the electric field intensity is in the y direction.[16]

- Case 3: Finally, we consider a moving loop with velocity $\bar{u} = \hat{a}_x u_0$ exposed to $\bar{B}_2(\bar{r},t) = \hat{a}_z B_0 \sin(\omega t - kx)$. Even though the loop is moving, we would like to investigate the case at the instant when one of the corners is at (x_0, y_0) (Figure 3.6). Then the magnetic flux is the same as for the stationary loop (Case 2):

$$f_B(t) = \frac{2bB_0}{k}\sin(\omega t - kx_0 - ka/2)\sin(ka/2). \tag{3.35}$$

On the other hand, we must be careful with the time derivative since $\partial x_0/\partial t = u_0 \neq 0$. We obtain

$$\Phi_{\text{emf}}(t) = -\frac{2bB_0}{k}(\omega - ku_0)\cos(\omega t - kx_0 - ka/2)\sin(ka/2). \tag{3.36}$$

[13] Using

$$\cos(a) - \cos(b) = -2\sin\left(\frac{a+b}{2}\right)\sin\left(\frac{a-b}{2}\right).$$

[14] This electromotive force can be interpreted as the voltage across a resistor (with the direction shown in Figure 3.5).

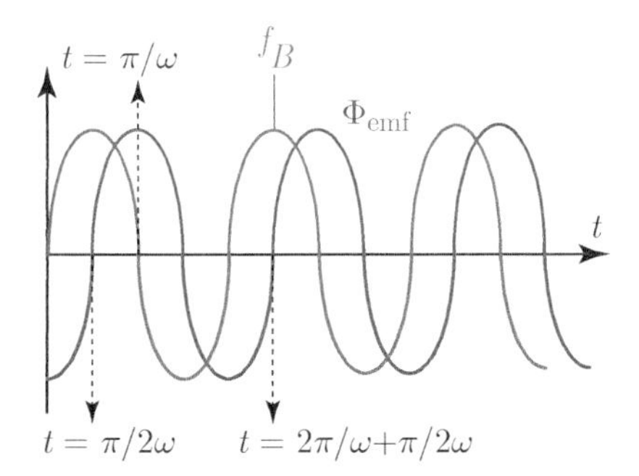

Figure 3.7 Time variation of the magnetic flux $f_B(t)$ and the corresponding electromotive force for a stationary rectangular loop (see Figure 3.5) when $\bar{B}_2(\bar{r},t) = \hat{a}_z B_0 \sin(\omega t - kx)$. In this particular plot, we assume $x_0 = -a/2$ (symmetrically located loop) such that $f_B(t)$ is zero when $t = 0$.

[15] Using

$$\sin(a) - \sin(b) = 2\cos\left(\frac{a+b}{2}\right)\sin\left(\frac{a-b}{2}\right).$$

[16] Interestingly, to make Eqs. (3.31) and (3.34) equal, we must have

$$\frac{2b\omega B_0}{k} = \frac{2bkB_0}{\omega\mu_0\varepsilon_0} \rightarrow k^2 = \omega^2\mu_0\varepsilon_0$$

Hence, even though we start with arbitrary constants k and ω, they *must* be related to each other if Ampere's law and Faraday's law are to be satisfied. This is because magnetic and electric fields in this example represent an electromagnetic wave: i.e. coupled electric and magnetic fields propagating in space. This wave must satisfy certain conditions, such as a fixed velocity, as we discuss later in more detail.

Obviously, this electromotive force has two components. One of them is due to the time variance of the magnetic flux density (not magnetic flux):

$$\Phi_{\mathrm{emf},B}(t) = -\frac{2b\omega B_0}{k}\cos(\omega t - kx_0 - ka/2)\sin(ka/2), \tag{3.37}$$

which is the same as the one in Eq. (3.31). We also note that

$$\Phi_{\mathrm{emf},B}(t) = \oint_C \bar{E}_2(\bar{r},t)\cdot\overline{dl}, \tag{3.38}$$

i.e. the electromotive force due to the time variance of the magnetic flux density corresponds to the circulation of the electric field intensity. The second component of the electromotive force is due to the movement of the loop under a *position-dependent* magnetic flux density:[17,18]

$$\Phi_{\mathrm{emf},M}(t) = -2bB_0u_0\cos(\omega t - kx_0 - ka/2)\sin(ka/2). \tag{3.39}$$

[17]As expected, we have

$$\begin{aligned}\Phi_{\mathrm{emf},M}(t) &= \oint_C \bar{u}(t)\times\bar{B}_2(\bar{r},t)\cdot\overline{dl}\\ &= \oint_C -\hat{a}_yB_0u_0\sin(\omega t - kx)\cdot\overline{dl}\\ &= \left[-\hat{a}_yB_0u_0\sin(\omega t-kx)\cdot\hat{a}_y\right]_{y_0}^{y_0+b}\Big|_{x=x_0+a}\\ &\quad+\left[\hat{a}_yB_0u_0\sin(\omega t-kx)\cdot\hat{a}_y\right]_{y_0}^{y_0+b}\Big|_{x=x_0}\\ &= -2bB_0u_0\cos(\omega t - kx_0 - ka/2)\sin(ka/2).\end{aligned}$$

[18]Interestingly, it is possible to make the total electromotive force zero. Considering Eq. (3.36), we simply need $u_0 = \omega/k$. This is actually the case when the loop is moving with the speed of the electromagnetic wave (*surfing* on the wave) such that the magnetic flux becomes fixed.

Before more examples of the integral form of Faraday's law, we now consider a structure that is quite confusing: Faraday's disk generator.[19] A metallic disk of radius a is rotating under a static magnetic flux density $\hat{a}_zB_0$ that is normal to the surface of the disk (Figure 3.8). The angular frequency of the rotation is ω, while the induced electromotive force (voltage) is measured from the center line to a *fixed* point at the edge using a stationary circuit. Considering the rotation of free charges on the metal disk, it is expected that positive electric charges are attracted to the edge while the negative electric charges tend to accumulate at the center. Therefore, there must be an electromotive force. However, as the source of the confusion, integration of the magnetic flux density through the disk is constant: i.e. $df_B(t)/dt = 0$. But if the magnetic flux does not change, how is an electromotive force generated? The confusion is due to the incorrect interpretation of the disk, as if it is a circular loop along which the electromotive force is to be measured. In fact, in such a case (a simple rotating circular loop), the electromotive force must really be zero. But if we look carefully at the original scenario (the electromotive force is measured by using a fixed circuit), we can conclude that the *loop* to be considered is, in fact, *not* a circular loop.

[19]That can be considered a steady (direct) current (DC) generator.

To explain the analysis of Faraday's disk generator, we can consider a differential part of the disk: i.e. a rod of length a that is rotating with an angular velocity of ω. This rod is part of the closed *Faraday loop*, along which the electromotive force is calculated (Figure 3.9). As the rod rotates, the loop expands and includes increasingly more magnetic flux.[20] More specifically, the magnetic flux is increasing as

[20]Note that if the circuit was not stationary and it was moving together with the disk, the Faraday loop would not be expanding.

$$\Delta f_B = B_0A_s \tag{3.40}$$

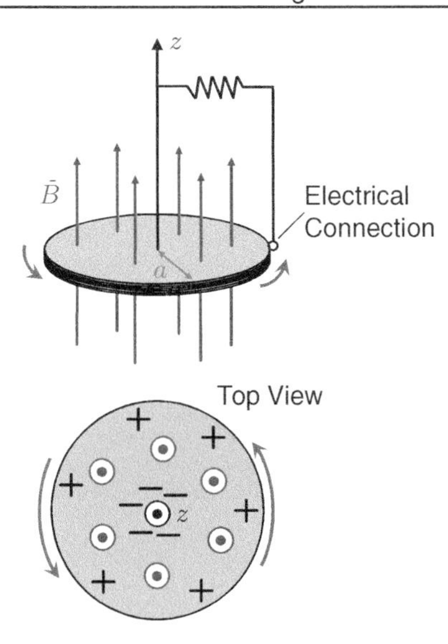

Figure 3.8 Faraday's disk generator that involves a rotating metallic disk under a static magnetic field to generate DC current. The electromotive force is measured from the center line to the edge of the disk by using a stationary circuit: i.e. the wire of the circuit touches the side of the disk with only an electrical connection.

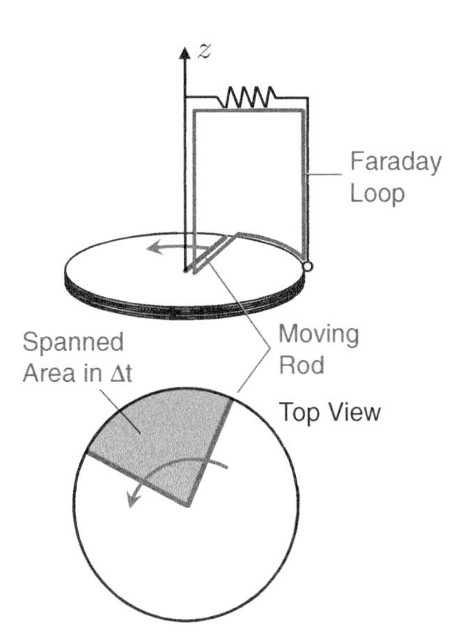

Figure 3.9 Illustration of the loop to find the electromotive force measured by the stationary circuit in the Faraday's disk generator. Rotation of the rod leads to the expansion of the loop.

where A_s is the sectoral area spanned in time Δt. It can be found that $A_s = \omega \Delta t a^2/2$, leading to the electromotive force as

$$\Phi_{\text{emf}}(t) = \frac{\Delta f_B}{\Delta t} = \omega B_0 a^2/2. \tag{3.41}$$

Alternatively, if the magnetic flux density does not change with time, we can use

$$\Phi_{\text{emf}}(t) = \oint_C \bar{u}(t) \times \bar{B}(\bar{r}, t) \cdot \overline{dl}, \tag{3.42}$$

leading to

$$\Phi_{\text{emf}}(t) = \int_0^a \omega \rho \hat{a}_\phi \times \hat{a}_z B_0 \cdot \hat{a}_\rho d\rho \tag{3.43}$$

since the only contribution is from the rotating rod. Note that the velocity vector is $u_t = \hat{a}_\phi \omega \rho$ on the rod. Evaluating the integral, we reach again

$$\Phi_{\text{emf}}(t) = \omega B_0 [\rho^2/2]_0^a = \omega B_0 a^2/2 \tag{3.44}$$

as the electromotive force.

3.2.1 Examples

Example 27: Consider an alternating current (AC) generator involving a loop rotating with an angular frequency of ω in a static magnetic field $\bar{B}(\bar{r}) = \hat{a}_z B_0$ (Figure 3.10). Find the induced voltage between the terminals.

Solution: First, we can write the magnetic flux as

$$f_B(t) = abB_0 \cos(\omega t), \tag{3.45}$$

assuming that the normal of the loop is in the z direction at $t = 0$. Then the electromotive force (the induced voltage between the terminals) can be found simply as

$$\Phi_{\text{emf}}(t) = -\frac{df_B(t)}{dt} = abB_0 \omega \sin(\omega t), \tag{3.46}$$

considering the given polarity (Figure 3.10).

Example 28: Consider an $a \times b$ rectangular loop on the z-x plane moving away from an infinitely long wire carrying a steady electric current I_0 (A) in the z direction (Figure 3.11). The velocity of the loop is u_0. Find the electromotive force induced in the loop (when it is at position x_0).

Solution: With the given coordinate system, the magnetic flux density created by the infinitely long wire can be found as

$$\bar{B}(\bar{r}) = \hat{a}_y \frac{\mu_0 I_0}{2\pi x}. \tag{3.47}$$

If the position of the loop (distance between the left vertical wire and the infinitely long wire) is assumed to be x_0 at time t, the total magnetic flux can be found as

$$f_B(t) = \int_S \bar{B}(\bar{r}) \cdot \overline{ds} = \int_S \bar{B}(\bar{r}) \cdot \hat{a}_y ds = \int_{x_0}^{x_0+a} \int_0^b \frac{\mu_0 I_0}{2\pi x} dz dx \tag{3.48}$$

$$= \frac{\mu_0 I_0}{2\pi} b \ln\left(\frac{x_0 + a}{x_0}\right), \tag{3.49}$$

where the direction of $\overline{ds}$ is selected in the y direction, leading to $\overline{dl}$ and I_{ind} in the clockwise direction. Then the polarity of the electromotive force $\Phi_{\text{emf}}(t)$ is selected accordingly (Figure 3.11). To find the value of $\Phi_{\text{emf}}(t)$, we use Faraday's law as

$$\Phi_{\text{emf}}(t) = -\frac{df_B(t)}{dt} = -\frac{b\mu_0 I_0}{2\pi}\frac{d}{dt}\left[\ln\left(1 + \frac{a}{x_0}\right)\right]$$

$$= -\frac{b\mu_0 I_0}{2\pi}\left(\frac{x_0}{x_0 + a}\right)\left(-\frac{a}{(x_0)^2}\right)\frac{dx_0}{dt} = \frac{ab\mu_0 I_0 u_0}{2\pi x_0(x_0 + a)}. \tag{3.50}$$

Note that, despite the magnetic flux density $\bar{B}(\bar{r})$ does not change with respect to time, the movement of the loop creates an electromotive force that acts against the decreasing magnetic flux.

Alternatively, this problem can be solved by using

$$\Phi_{\text{emf}}(t) = \oint_C \bar{u}(t) \times \bar{B}(\bar{r}) \cdot \overline{dl}, \tag{3.51}$$

where

$$\bar{u}(t) \times \bar{B}(\bar{r}) = \hat{a}_x u_0 \times \hat{a}_y B(\bar{r}) = \hat{a}_z u_0 B(\bar{r}) \tag{3.52}$$

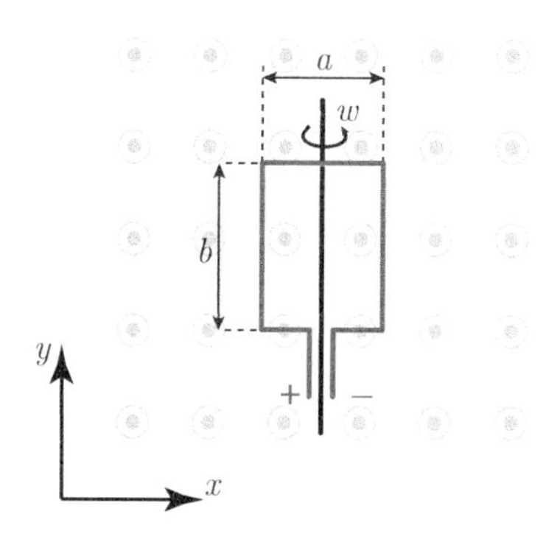

Figure 3.10 An AC generator: A loop rotating in a static magnetic field.

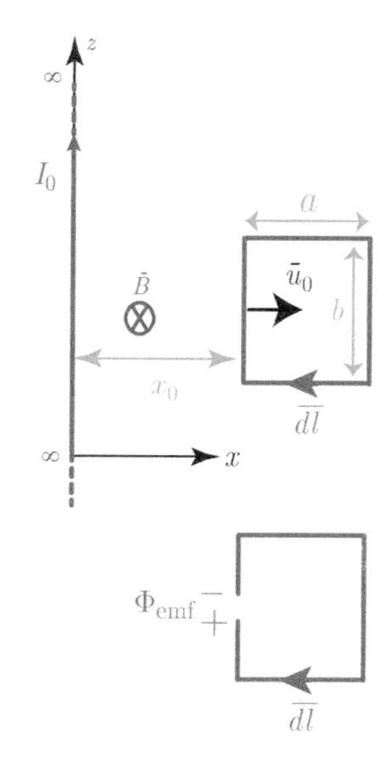

Figure 3.11 A rectangular loop moving away from an infinitely long wire carrying a steady electric current. The polarity of the electromotive force for the selected direction of $\overline{dl}$ is also shown.

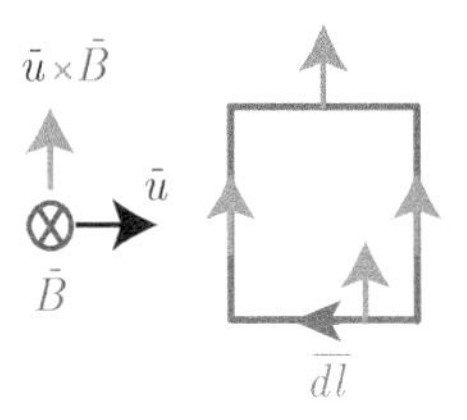

Figure 3.12 For the scenario in Figure 3.11, $\bar{u}(t) \times \bar{B}(\bar{r})$ is parallel to the vertical (side) lines, leading to contributions to the circulation.

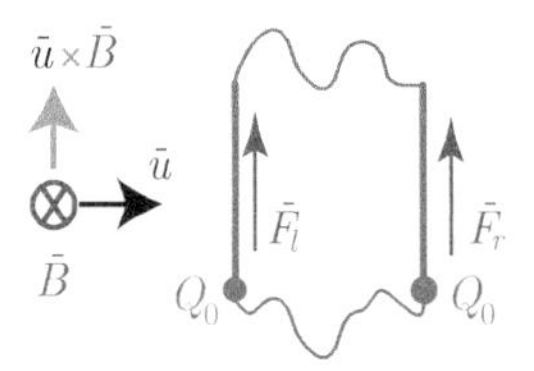

Figure 3.13 Considering the scenario in Figure 3.11, charges located on the side lines of the loop are exposed to force (see the Lorentz force law in Eq. (3.13)). Specifically, a positive charge Q_0 is pushed in the z direction. On the other hand, the force on the left line ($\bar{F}_l$) is larger than the force on the right line ($\bar{F}_r$) since it is closer to the infinitely long line current. This leads to an overall clockwise electromotive force, which is consistent with the demonstration in Figure 3.11 with $\Phi_{\text{emf}} > 0$ according to Eq. (3.50).

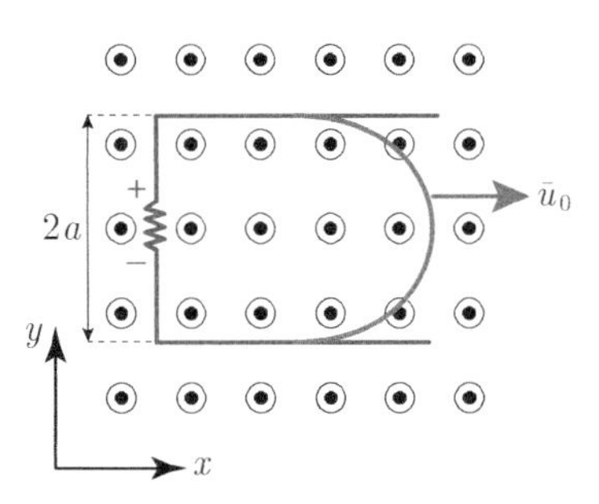

Figure 3.14 A closed loop involving a stationary part and a moving semicircular contour under a static magnetic flux density.

so that $\bar{u}(t) \times \bar{B}(\bar{r}) \cdot \overline{dl}$ is nonzero only on the vertical sides of the loop (Figure 3.12). Therefore, we obtain

$$\Phi_{\text{emf}}(t) = \int_0^b u_0 B(x_0) dz - \int_0^b u_0 B(x_0 + a) dz = \frac{b\mu_0 I_0 u_0}{2\pi x_0} - \frac{b\mu_0 I_0 u_0}{2\pi(x_0 + a)} \tag{3.53}$$

$$= \frac{b\mu_0 I_0 u_0}{2\pi}\left(\frac{1}{x_0} - \frac{1}{x_0 + a}\right) = \frac{ab\mu_0 I_0 u_0}{2\pi x_0(x_0 + a)}, \tag{3.54}$$

verifying the result (Figure 3.13).

Example 29: Consider a closed loop exposed to a static magnetic flux density $\bar{B}(\bar{r}) = \hat{a}_z B_0$. A semicircular contour that completes the closed loop slides with a velocity $\bar{u}_0 = \hat{a}_x u_0$ in the x direction (Figure 3.14). Find an expression for the induced electromotive force.

Solution: To find the electromotive force, we can consider the instant when the moving semicircle is at $x = x_0$ while the left vertical side of the loop is at $x = 0$ (Figure 3.15). Then the magnetic flux can be found as

$$f_B(t) = \int_S \bar{B}_0(\bar{r}, t) \cdot \overline{ds} = 2ax_0 B_0 + \frac{\pi a^2}{2} B_0, \tag{3.55}$$

where the first term comes from the rectangular part while the second term is the contribution of the semicircular region. The electromotive force can be obtained as

$$\Phi_{\text{emf}}(t) = -\frac{df_B(t)}{dt} = 2aB_0 \frac{\partial x_0}{\partial t} = 2aB_0 u_0, \tag{3.56}$$

considering the given polarity (Figure 3.14). Note that the shape of the moving contour (semicircle) has no effect on the electromotive force, which is also time-invariant.

Alternatively, since $\bar{B}(\bar{r}, t) = \hat{a}_z B_0$ is static (hence there is no electric field intensity), we can find the electromotive force as

$$\Phi_{\text{emf}}(t) = -\int_C \bar{u}(t) \times \bar{B}(\bar{r}) \cdot \overline{dl}, \tag{3.57}$$

where the line integral is reduced to the semicircular contour C. Then we have

$$\Phi_{\text{emf}}(t) = -\int_{-\pi/2}^{\pi/2} (-\hat{a}_y) u_0 B_0 \cdot \hat{a}_\phi a d\phi' \tag{3.58}$$

assuming that the center of the semicircle is located at the origin of a primed coordinate system (ρ', ϕ', z'). Then we have

$$\Phi_{\text{emf}}(t) = aB_0u_0 \int_{-\pi/2}^{\pi/2} \cos\phi' d\phi' = aB_0u_0 [\sin\phi']_{-\pi/2}^{\pi/2} = 2aB_0u_0, \tag{3.59}$$

verifying the earlier solution.

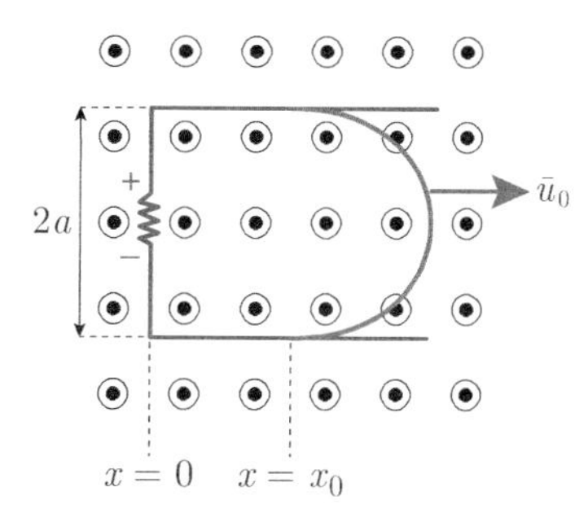

Figure 3.15 An instant when the moving contour in Figure 3.14 is at $x = x_0$.

Example 30: Consider a loop located at the origin expanding radially with velocity u_0 (Figure 3.16). The loop is exposed to a time-varying magnetic flux density in the z direction defined as $\bar{B}(\bar{r}, t) = \hat{a}_z b_0 t$, for a constant b_0 [Wb/(m²s)]. Assuming that the radius of the loop is a at $t = 0$, find the electromotive force induced in the loop.

Solution: First, we can evaluate the total flux with respect to time as

$$f_B(t) = \int_S \bar{B}(\bar{r}, t) \cdot \overline{ds} = \int_0^{2\pi} \int_0^{a+u_0t} b_0 t \rho d\rho d\phi = \pi b_0 t (a + u_0 t)^2. \tag{3.60}$$

Then the electromotive force can be found as usual via time derivative as

$$\Phi_{\text{emf}}(t) = -\frac{df_B(t)}{dt} = -\pi b_0 (a + u_0 t)^2 - 2\pi b_0 t (a + u_0 t) u_0 \tag{3.61}$$

$$= -\pi b_0 (a + u_0 t)(a + 3u_0 t). \tag{3.62}$$

Hence, if $u_0 = 0$, we have $\Phi_{\text{emf}}(t) = -\pi b_0 a^2$, which is only caused by the time variance of the magnetic flux density.[21]

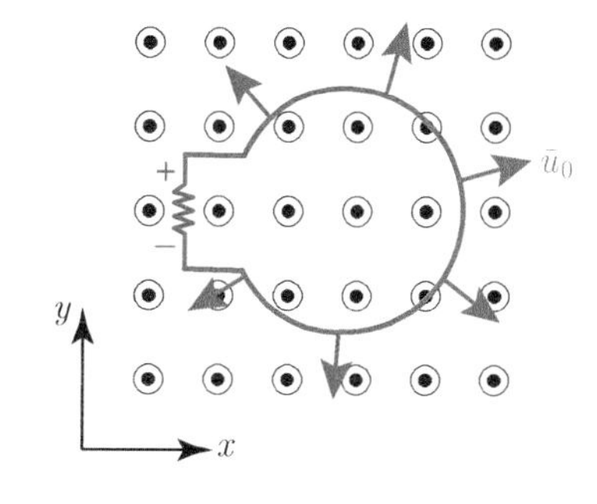

Figure 3.16 An expanding loop under a time-variant magnetic flux density.

[21] We can also find the electromotive force by using

$$\Phi_{\text{emf}}(t) = -\int_S \frac{\partial \bar{B}(\bar{r}, t)}{\partial t} \cdot \overline{ds} + \oint_C \bar{u}(t) \times \bar{B}(\bar{r}, t) \cdot \overline{dl}.$$

For the first term, we have $\partial B(\bar{r}, t)/\partial t = \hat{a}_z b_0$, leading to

$$-\int_S \frac{\partial \bar{B}(\bar{r}, t)}{\partial t} \cdot \overline{ds} = -\pi b_0 (a + u_0 t)^2.$$

For the second term, the cross product can be evaluated as $\bar{u}(t) \times \bar{B}(\bar{r}, t) = \hat{a}_\rho u_0 \times \hat{a}_z b_0 t = -\hat{a}_\phi b_0 u_0 t$. Then integration on the circle (using $\overline{dl} = \hat{a}_\phi \rho d\phi$) leads to

$$\oint_C \bar{u}(t) \times \bar{B}(\bar{r}, t) \cdot \overline{dl} = -2\pi (a + u_0 t) b_0 u_0 t.$$

Finally, we can add the two contributions to confirm the earlier result.

3.3 Differential Form of Faraday's Law

As we did for Gauss' and Ampere's laws, we now focus on the differential form of Faraday's law. For this purpose, we need the concepts of the cross product (Section 2.3.2), curl of vector fields (Section 2.3.3), and Stokes' theorem (Section 2.3.4). These are in addition to those required to understand the integral form of Faraday's law: i.e. differential surface with direction (Section 1.1.1), dot product (Section 1.1.2), flux of vector fields (Section 1.1.3), differential length with direction (Section 2.1.1), and circulation of vector fields (Section 2.1.2). Therefore, no new tools are needed to understand the differential form of Faraday's law.

Basically, using Stokes' theorem:

$$\int_S \bar{\nabla} \times \bar{f}(\bar{r},t) \cdot \overline{ds} = \oint_C \bar{f}(\bar{r},t) \cdot \overline{dl}, \tag{3.63}$$

the integral form of Faraday's law can be written as[22]

$$\int_S \bar{\nabla} \times \bar{E}(\bar{r},t) \cdot \overline{ds} = -\int_S \frac{\partial \bar{B}(\bar{r},t)}{\partial t} \cdot \overline{ds}, \tag{3.64}$$

where both sides of the equation involves surface integrals. Since this equation can be written for any surface, the integrands should be equal everywhere:

$$\bar{\nabla} \times \bar{E}(\bar{r},t) = -\frac{\partial \bar{B}(\bar{r},t)}{\partial t}, \tag{3.65}$$

which is the differential form of Faraday's law. Using $\bar{B}(\bar{r},t) = \mu(\bar{r})\bar{H}(\bar{r})$, we also have

$$\bar{\nabla} \times \bar{E}(\bar{r},t) = -\mu(\bar{r})\frac{\partial \bar{H}(\bar{r},t)}{\partial t}. \tag{3.66}$$

According to Eq. (3.65), changing magnetic fields are sources for electric fields. This complements direct sources for electric fields – i.e. electric charges – according to Gauss' law $\bar{\nabla} \cdot \bar{D}(\bar{r},t) = \rho_v(\bar{r},t)$. Note that electric charges create electric fields in both static and dynamic cases. On the other hand, magnetic fields become sources of electric fields only in dynamic cases: i.e. when the time derivative in Eq. (3.65) is nonzero.

Using the differential form of Faraday's law, we can also write the curl of $\bar{D}(\bar{r},t)$ in a dielectric medium with permittivity $\varepsilon(\bar{r})$ as[23]

$$\bar{\nabla} \times \bar{D}(\bar{r},t) = \bar{\nabla} \times [\varepsilon(\bar{r})\bar{E}(\bar{r},t)] = \varepsilon(\bar{r})\bar{\nabla} \times \bar{E}(\bar{r},t) + \bar{\nabla}\varepsilon(\bar{r}) \times \bar{E}(\bar{r},t) \tag{3.67}$$

$$= -\mu(\bar{r})\varepsilon(\bar{r})\frac{\partial \bar{H}(\bar{r},t)}{\partial t} + \bar{\nabla}\varepsilon(\bar{r}) \times \bar{E}(\bar{r},t), \tag{3.68}$$

where a position derivative of the permittivity (more specifically its gradient) is used. Using polarization, we also have

$$\bar{\nabla} \times \bar{D}(\bar{r},t) = \bar{\nabla} \times [\varepsilon_0\bar{E}(\bar{r},t) + \bar{P}(\bar{r},t)] = \varepsilon_0\bar{\nabla} \times \bar{E}(\bar{r},t) + \bar{\nabla} \times \bar{P}(\bar{r},t) \tag{3.69}$$

$$= -\varepsilon_0\frac{\partial \bar{B}(\bar{r},t)}{\partial t} + \bar{\nabla} \times \bar{P}(\bar{r},t) = -\varepsilon_0\frac{\partial \bar{B}(\bar{r},t)}{\partial t} + \varepsilon_0\bar{J}_{pv}(\bar{r},t), \tag{3.70}$$

[22] Note that if there is a closed surface S, we have

$$\oint_S \bar{\nabla} \times \bar{E}(\bar{r},t) \cdot \overline{ds} = 0,$$

which is valid not only for static electric fields but also for dynamic cases. This means

$$\oint_S \frac{\partial \bar{B}(\bar{r},t)}{\partial t} \cdot \overline{ds} = 0.$$

For a stationary closed surface S, we further have

$$\frac{d}{dt}\oint_S \bar{B}(\bar{r},t) \cdot \overline{ds} = \frac{df_B(t)}{dt} = 0.$$

Hence the net magnetic flux through a stationary closed surface does not change with respect to time.

[23] This equation indicates that, even in a static case, the curl of the electric flux density can be nonzero: i.e. $\bar{\nabla} \times \bar{D}(\bar{r}) = \bar{\nabla}\varepsilon(\bar{r}) \times \bar{E}(\bar{r})$.

which can be nonzero even when there is no changing magnetic flux density. In the above, $\bar{J}_{pv}(\bar{r},t) = \bar{\nabla} \times \bar{P}(\bar{r},t)/\varepsilon_0$ acts like a volume magnetic current density[24] that contributes to the nonzero curl of the electric flux density.

[24]Not to be confused with the volume magnetization current density, which is a kind of electric current density.

As discussed in Chapters 1 and 2, Gauss' law and Ampere's law are useful alone for static problems. For general solutions to dynamic problems, we further need Faraday's law and a property of magnetic fields (specifically, the absence of magnetic charges). These solutions are discussed thoroughly in Chapter 4, where Maxwell's equations are completed. At this stage, we further use Faraday's law to solve electrostatic problems. For known electric charge distributions, Gauss' law is the primary tool to find static electric fields. On the other hand, Faraday's law brings a new insight into the characteristics of static electric fields: i.e. their irrotational behavior.

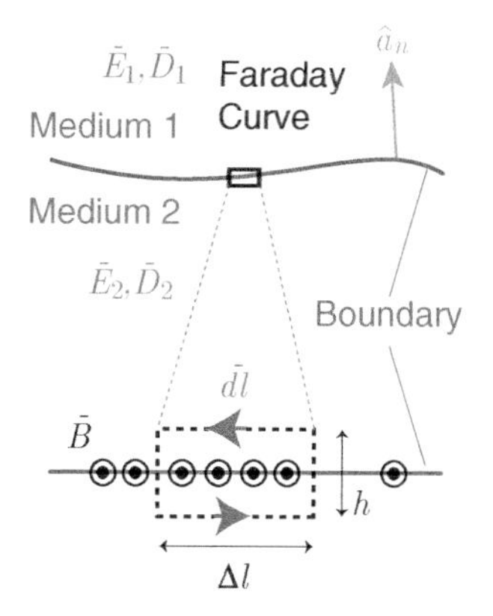

Figure 3.17 An illustration of the derivation of the boundary conditions for tangential magnetic fields across an interface between two different media. There can be nonzero tangential magnetic flux density (which usually corresponds to the magnetic field intensity, hence surface electric current density) flowing through the frame, while the integral of the time derivative of the magnetic flux density goes to zero as the frame shrinks to a line. Note that the derivation holds for observation points on smooth surfaces (not at the edges and corners).

3.4 Boundary Conditions for Tangential Electric Fields

We now derive the boundary conditions for tangential electric fields. The procedure is very similar to the boundary conditions for tangential magnetic fields using Ampere's law. In this case, we use Faraday's law on a rectangular frame enclosing part of the boundary between two dielectric media (Figure 3.17). Specifically, we consider a boundary S between two media with permittivity values $\varepsilon_1(\bar{r})$ and $\varepsilon_2(\bar{r})$, and a normal direction $\hat{a}_n$ pointing into the first medium. Using the integral form of Faraday's law, we have

$$\oint_C \bar{E}(\bar{r},t) \cdot \overline{dl} = -\int_S \frac{\partial \bar{B}(\bar{r},t)}{\partial t} \cdot \overline{ds}, \tag{3.71}$$

where the circulation of the electric field intensity along the frame is equal to the integral of the time derivative of the magnetic flux density through its surface. Letting the height of the frame h go to zero, we have a limit case[25]

$$\Delta l[\hat{a}_n \times \bar{E}_1(\bar{r},t) - \hat{a}_n \times \bar{E}_2(\bar{r},t)] = (\Delta l) \lim_{h\to 0} \left[h \frac{\partial \bar{B}(\bar{r},t)}{\partial t} \right], \tag{3.72}$$

where Δl is the width of the frame. Further canceling the differential width of the frame, we have

$$\hat{a}_n \times [\bar{E}_1(\bar{r},t) - \bar{E}_2(\bar{r},t)] = \lim_{h\to 0} \left[h \frac{\partial \bar{B}(\bar{r},t)}{\partial t} \right]. \tag{3.73}$$

[25]Note that performing a cross product with $\hat{a}_n$ means projecting the vector onto the boundary such that we are considering its tangential component.

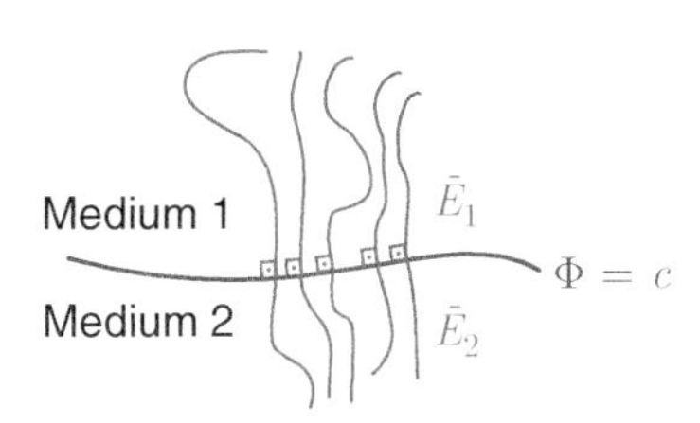

Figure 3.18 Considering the boundary condition for the tangential component of the electric field intensity immediately leads to interesting results that must be satisfied by electric fields. For example, if the electric field intensity is perpendicular to a surface (zero tangential components) on one side, then it must also be perpendicular on the other side. Hence, that portion of the surface must be an *equipotential* (constant electric scalar potential) surface. Note that this boundary condition does not provide any information on normal components.

The right-hand side of this equation is zero, if we assume that the time derivative of the magnetic flux density is not infinite. Hence, we obtain the boundary condition as

$$\hat{a}_n \times [\bar{E}_1(\bar{r},t) - \bar{E}_2(\bar{r},t)] = 0 \qquad (\bar{r} \in S) \tag{3.74}$$

$$\hat{a}_n \times \bar{E}_1(\bar{r},t) = \hat{a}_n \times \bar{E}_2(\bar{r},t) \qquad (\bar{r} \in S) \tag{3.75}$$

Therefore, the tangential component of the electric field intensity is *always* continuous across a boundary separating two media (Figure 3.18).

At this stage, we can also derive the boundary condition for the electric flux density as[26]

$$\hat{a}_n \times \bar{D}_1(\bar{r},t)/\varepsilon_1(\bar{r}) = \hat{a}_n \times \bar{D}_2(\bar{r},t)/\varepsilon_2(\bar{r}) \qquad (\bar{r} \in S). \tag{3.76}$$

Obviously, the tangential electric flux density is discontinuous if $\varepsilon_1(\bar{r}) \neq \varepsilon_2(\bar{r})$. Note that this discontinuity is not directly related to the discontinuity of the normal component of $\bar{D}(\bar{r},t)$ that is caused by electric charges at the boundary. In fact, even in the absence of electric charges, the tangential component of $\bar{D}(\bar{r},t)$ is discontinuous across a boundary between two different dielectrics. We can explain this by considering the relationship between $\bar{E}(\bar{r},t)$ and $\bar{D}(\bar{r},t)$ in terms of polarization and rewriting the boundary condition for $\bar{E}(\bar{r},t)$ as[27]

$$\hat{a}_n \times \bar{D}_1(\bar{r},t)/\varepsilon_0 - \hat{a}_n \times \bar{P}_1(\bar{r},t)/\varepsilon_0 = \hat{a}_n \times \bar{D}_2(\bar{r},t)/\varepsilon_0 - \hat{a}_n \times \bar{P}_2(\bar{r},t)/\varepsilon_0 \tag{3.77}$$

for $\bar{r} \in S$, leading to

$$\hat{a}_n \times [\bar{D}_1(\bar{r},t) - \bar{D}_2(\bar{r},t)] = \hat{a}_n \times [\bar{P}_1(\bar{r},t) - \bar{P}_2(\bar{r},t)] \qquad (\bar{r} \in S) \tag{3.78}$$

$$= -\hat{a}_n \times \bar{P}(\bar{r},t) = \varepsilon_0 \bar{J}_{ps}(\bar{r},t), \qquad (\bar{r} \in S) \tag{3.79}$$

where $\bar{P}(\bar{r},t) = \bar{P}_2(\bar{r},t) - \bar{P}_1(\bar{r},t)$ is the net polarization. Hence, the quantity $\bar{J}_{ps}(\bar{r},t) = -\hat{a}_n \times \bar{P}(\bar{r},t)/\varepsilon_0$ acts like a surface magnetic current density[28] that is responsible for the discontinuity of the electric flux density. This equation is consistent with Eq. (3.70), where the curl of the electric flux density can be nonzero as $\bar{\nabla} \times \bar{P}(\bar{r},t)$ even in the absence of a time-variant magnetic flux density.

[26] Considering this boundary condition – i.e. $D_{1t}(\bar{r},t) = \varepsilon_1(\bar{r})D_{2t}(\bar{r},t)/\varepsilon_2(\bar{r},t)$, together with $D_{1n}(\bar{r},t) = D_{2n}(\bar{r},t)$ in the absence of electric charges, where subscripts t and n indicate tangential and normal, respectively, we have

$$\tan\varphi_1 = [\varepsilon_1(\bar{r})/\varepsilon_2(\bar{r})]\tan\varphi_2,$$

where φ_1 and φ_2 are the angles describing the ratios of the components (Figure 3.19).

[27] Using $\bar{E}(\bar{r},t) = \bar{D}(r,t)/\varepsilon_0 - \bar{P}(r,t)/\varepsilon_0$.

[28] Not to be confused with the surface magnetization current density, which is a kind of electric current density.

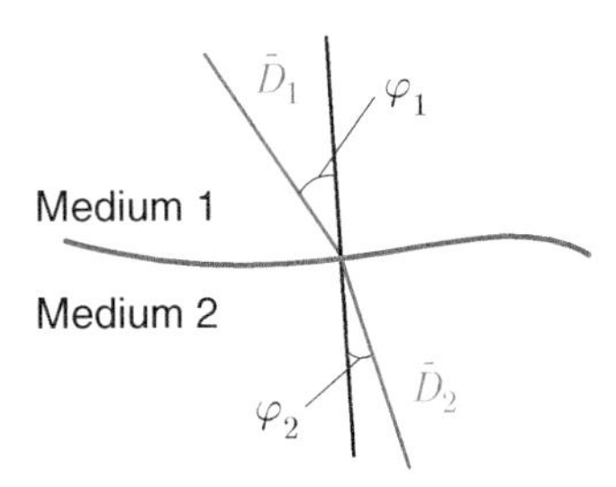

Figure 3.19 Demonstration of the angular relationship that must be satisfied by the electric flux density at a boundary.

3.5 Combining Faraday's Law with Gauss' and Ampere's Laws

At this stage, we have three of four Maxwell's equations: i.e. Gauss' law, Ampere's law, and Faraday's law. As shown in Section 2.6, using Gauss' law and Ampere's law together leads to the continuity Eq. (2.202) and another equation:

$$\bar{\nabla} \times \bar{\nabla} \times \bar{H}(\bar{r},t) = \bar{\nabla} \times \bar{J}_v(\bar{r},t) + \bar{\nabla} \times \frac{\partial \bar{D}(\bar{r},t)}{\partial t}. \tag{3.80}$$

At that point, we stopped with this expression since we did not know the curl of the electric field. But now, with the help of Faraday's law, we can proceed a little further. First, assuming an electrically homogeneous medium,[29] we have $\bar{D}(\bar{r},t) = \varepsilon\bar{E}(\bar{r},t)$, leading to[30]

$$\bar{\nabla} \times \bar{\nabla} \times \bar{H}(\bar{r},t) = \bar{\nabla} \times \bar{J}_v(\bar{r},t) + \varepsilon\frac{\partial}{\partial t}\bar{\nabla} \times \bar{E}(\bar{r},t). \tag{3.81}$$

Then, using Faraday's law, we derive[31]

$$\bar{\nabla} \times \bar{\nabla} \times \bar{H}(\bar{r},t) = \bar{\nabla} \times \bar{J}_v(\bar{r},t) - \varepsilon\frac{\partial^2}{\partial t^2}\bar{B}(\bar{r},t). \tag{3.82}$$

If the medium is magnetically homogeneous (for simplicity), we further have $\bar{B}(\bar{r},t) = \mu\bar{H}(\bar{r},t)$. Then, rearranging the terms, we obtain[32]

$$\bar{\nabla} \times \bar{\nabla} \times \bar{H}(\bar{r},t) + \mu\varepsilon\frac{\partial^2}{\partial t^2}\bar{H}(\bar{r},t) = \bar{\nabla} \times \bar{J}_v(\bar{r},t). \tag{3.83}$$

This equation is noteworthy since it provides an expression for the magnetic field intensity only in terms of a source $\bar{J}_v(\bar{r},t)$, while it is written for both static and dynamic cases.[33]

Until now, we have discussed what happens when the del operator ($\bar{\nabla}$) is applied to Ampere's law, either as the divergence ($\bar{\nabla}\cdot$) or the curl ($\bar{\nabla}\times$) operation.[34] Similar procedures can be applied to Faraday's law to derive useful formulations towards solutions of Maxwell's equations. First, we start with the divergence operation as

$$\bar{\nabla} \cdot \bar{\nabla} \times \bar{E}(\bar{r},t) = -\bar{\nabla} \cdot \frac{\partial \bar{B}(\bar{r},t)}{\partial t} = -\frac{\partial}{\partial t}\bar{\nabla} \cdot \bar{B}(\bar{r},t), \tag{3.84}$$

where the left-hand side is zero due to the null identity $\bar{\nabla} \cdot \bar{\nabla} \times \bar{f}(\bar{r},t) = 0$ for any vector field $\bar{f}(\bar{r},t)$. Therefore, we obtain an interesting equality as

$$\frac{\partial}{\partial t}\bar{\nabla} \cdot \bar{B}(\bar{r},t) = 0. \tag{3.85}$$

Even though it is short, this is a very strong statement and related to an important property of magnetic fields. Specifically, the divergence of the magnetic flux density is zero, and it is a solenoidal (divergence-free) vector field (in both static and dynamic cases), while this physically means[35] there is no magnetic charge. Gauss' law for magnetic fields (divergence of the magnetic flux density) is extensively discussed in Chapter 4.

[29] If the medium is electrically inhomogeneous with a position-dependent permittivity $\varepsilon(\bar{r})$, then an extra term involving the gradient of the permittivity appears.

[30] Also by changing the order of the time derivative and the curl operation.

[31] Using

$$\bar{\nabla} \times \bar{\nabla} \times \bar{H}(\bar{r},t) = \bar{\nabla} \times \bar{J}_v(\bar{r},t) - \varepsilon\frac{\partial}{\partial t}\left(\frac{\partial \bar{B}(\bar{r},t)}{\partial t}\right).$$

[32] Using

$$\bar{\nabla} \times \bar{\nabla} \times \bar{H}(\bar{r},t) = \bar{\nabla} \times \bar{J}_v(\bar{r},t) - \mu\varepsilon\frac{\partial^2}{\partial t^2}\bar{H}(\bar{r},t).$$

[33] Obviously, for a static case, this equation reduces to $\bar{\nabla} \times \bar{\nabla} \times \bar{H}(\bar{r}) = \bar{\nabla} \times \bar{J}_v(\bar{r})$, which is simply the curl of Ampere's law.

[34] For the first case, we arrive at the continuity equation as Eq. (2.202). In the second case, we derive an equation for the magnetic field as Eq. (3.83).

[35] Be careful: this is not a general proof, since a zero time derivative does not imply a zero function.

Next, we consider using the curl operation on Faraday's law. We have

$$\bar{\nabla} \times \bar{\nabla} \times \bar{E}(\bar{r},t) = -\bar{\nabla} \times \frac{\partial \bar{B}(\bar{r},t)}{\partial t} = -\frac{\partial}{\partial t}\bar{\nabla} \times \bar{B}(\bar{r},t). \tag{3.86}$$

For a magnetically homogeneous medium[36] – i.e. for $\bar{B}(\bar{r},t) = \mu\bar{H}(\bar{r},t)$ – we further derive

$$\bar{\nabla} \times \bar{\nabla} \times \bar{E}(\bar{r},t) = -\mu\frac{\partial}{\partial t}\bar{\nabla} \times \bar{H}(\bar{r},t). \tag{3.87}$$

For the right-hand side, we can insert Ampere's law, leading to

$$\bar{\nabla} \times \bar{\nabla} \times \bar{E}(\bar{r},t) = -\mu\frac{\partial \bar{J}_v(\bar{r},t)}{\partial t} - \mu\frac{\partial^2}{\partial t^2}\bar{D}(\bar{r},t). \tag{3.88}$$

If the medium is electrically homogeneous (for simplicity), we have $\bar{D}(\bar{r},t) = \varepsilon\bar{E}(\bar{r},t)$, further leading to[37]

$$\bar{\nabla} \times \bar{\nabla} \times \bar{E}(\bar{r},t) + \mu\varepsilon\frac{\partial^2}{\partial t^2}\bar{E}(\bar{r},t) = -\mu\frac{\partial \bar{J}_v(\bar{r},t)}{\partial t}. \tag{3.89}$$

This equation,[38] which is very similar to Eq. (3.83), has an isolated electric field intensity (there is no magnetic field intensity or magnetic flux density) that depends on a source: i.e. $\bar{J}_v(\bar{r},t)$.

To sum up, by using three of four Maxwell's equations, we can derive equations Eqs. (3.83) and (3.89), which provide information on electric and magnetic fields in terms of the volume electric current density. As discussed in Chapter 4, these equations can further be modified to derive wave equations that are fundamentals of electromagnetics.

[36] If the medium is magnetically inhomogeneous with a position-dependent permeability $\mu(\bar{r})$, then an extra term involving the gradient of the permeability appears.

[37] Using

$$\bar{\nabla} \times \bar{\nabla} \times \bar{E}(\bar{r},t) = -\mu\frac{\partial \bar{J}_v(\bar{r},t)}{\partial t} - \mu\varepsilon\frac{\partial^2}{\partial t^2}\bar{E}(\bar{r},t).$$

[38] Note that, for a static case, we have $\bar{\nabla} \times \bar{\nabla} \times \bar{E}(\bar{r}) = 0$.

3.6 Static Cases and Electric Scalar Potential

Like all Maxwell's equations, Faraday's law is valid for both static and dynamic cases. When time-harmonic sources (with an angular frequency of ω) are involved in a dynamic case, we have[39]

$$\oint_C \bar{E}(\bar{r}) \cdot \overline{dl} = -\int_S j\omega\bar{B}(\bar{r}) \cdot \overline{ds} \tag{3.90}$$

$$\bar{\nabla} \times \bar{E}(\bar{r}) = -j\omega\bar{B}(\bar{r}), \tag{3.91}$$

[39] Then, as we will practice later, we have

$$\begin{aligned}\bar{\nabla} \times \bar{\nabla} \times \bar{E}(\bar{r}) &= -j\omega\mu\bar{\nabla} \times \bar{H}(\bar{r}) \\ &= -j\omega\mu\bar{J}_v(\bar{r}) + (-j\omega\mu)j\omega\bar{D}(\bar{r}) \\ &= -j\omega\mu\bar{J}_v(\bar{r}) + \omega^2\mu\bar{D}(\bar{r})\end{aligned}$$

for a magnetically homogeneous medium. In addition, if the medium is electrically homogeneous, we arrive at a very fundamental equation:

$$\bar{\nabla} \times \bar{\nabla} \times \bar{E}(\bar{r}) - \omega^2\mu\varepsilon\bar{E}(\bar{r}) = -j\omega\mu\bar{J}_v(\bar{r}).$$

where $\bar{E}(\bar{r})$ and $\bar{B}(\bar{r})$ are phasor domain representations. On the other hand, in a static case ($\omega = 0$), we have

$$\oint_C \bar{E}(\bar{r}) \cdot \overline{dl} = 0 \tag{3.92}$$

$$\bar{\nabla} \times \bar{E}(\bar{r}) = 0. \tag{3.93}$$

Based on these equations, we can form two important conclusions about the behavior of static electric fields:

- The circulation of the static electric field intensity is always zero, independent of the closed path C. This means a line integral of the static electric field intensity is path-independent:

$$\int_C \bar{E}(\bar{r}) \cdot \overline{dl} = \text{constant}, \tag{3.94}$$

  where the constant depends on the limits of the integral (not the path between them). Vector fields with zero circulations are called *conservative*.[40,41]

- The curl of the static electric field intensity is zero. Hence, such an electric field is created solely by electric charges. A vector field with zero curl everywhere is called *irrotational* (curl-free).

Hence, the static electric field intensity is *both* conservative and irrotational, as indicated by Faraday's law. In fact, from a mathematical point of view, being conservative guarantees that the vector field is irrotational. This relationship comes from the null identity[42]

$$\bar{\nabla} \times \bar{\nabla} g(\bar{r}, t) = 0 \tag{3.95}$$

for any scalar field $g(\bar{r}, t)$. As discussed in Section 3.6.3, being conservative means the vector field can be written as the gradient of a scalar field. Then this makes the vector field automatically irrotational. On the other hand, the converse is not generally true: i.e. irrotational means conservative only for simply connected spaces.

Being conservative, the electric field intensity is the gradient of a scalar field, the electric scalar potential[43]:

$$\bar{E}(\bar{r}) = -\bar{\nabla}\Phi(\bar{r}). \tag{3.96}$$

In the following, we mainly focus on the electric scalar potential and its usage for analyzing electrostatic cases, as well as energy due to electric interactions. For this purpose, we need the following new tools and concepts:

[40] Formally, a vector field is conservative if it can be written as the gradient of a scalar field. But as discussed in Section 3.6.3, this is equivalent to path independence.

[41] The flexibility on the choice of the path is useful when calculating the electric scalar potential from known electric field intensity. Integrals related to some paths may be easier to evaluate than others.

[42] This can be shown in a Cartesian coordinate system as

$$\begin{aligned}
&\bar{\nabla} \times \bar{\nabla} f(\bar{r}, t) \\
&= \bar{\nabla} \times \left(\hat{a}_x \frac{\partial f(\bar{r}, t)}{\partial x} + \hat{a}_y \frac{\partial f(\bar{r}, t)}{\partial y} + \hat{a}_z \frac{\partial f(\bar{r}, t)}{\partial z} \right) \\
&= \hat{a}_x \left(\frac{\partial^2 f(\bar{r}, t)}{\partial y \partial z} - \frac{\partial^2 f(\bar{r}, t)}{\partial z \partial y} \right) \\
&+ \hat{a}_y \left(\frac{\partial^2 f(\bar{r}, t)}{\partial z \partial x} - \frac{\partial^2 f(\bar{r}, t)}{\partial x \partial z} \right) \\
&+ \hat{a}_z \left(\frac{\partial^2 f(\bar{r}, t)}{\partial x \partial y} - \frac{\partial^2 f(\bar{r}, t)}{\partial y \partial x} \right) = 0.
\end{aligned}$$

[43] It is also called electrostatic potential since it is the only contribution to the electric field intensity in a static case. The negative sign is merely a selected convention, which is useful to define the electric scalar potential with respect to infinity.

Φ: electric scalar potential
Unit: volt (V)

- Gradient of scalar fields
- Gradient theorem
- Electric potential energy
- Poisson's equation and Laplace's equation
- Electrostatic boundary value problems

These are in addition to the previous tools and concepts: i.e. differential surface with direction (Section 1.1.1), dot product (Section 1.1.2), flux of vector fields (Section 1.1.3), electric charge density (Section 1.3.1), divergence of vector fields (Section 1.3.2), divergence theorem (Section 1.3.3), electric dipoles (Section 1.7.1), differential length with direction (Section 2.1.1), circulation of vector fields (Section 2.1.2), cross product (Section 2.3.2), and curl of vector fields (Section 2.3.3).

3.6.1 Gradient of Scalar Fields

Given a scalar field $f(\bar{r},t)$, $\bar{g}(\bar{r},t) = \bar{\nabla}f(\bar{r},t)$ is a vector field defined as

$$\bar{g}(\bar{r},t) = \hat{a}_x \frac{\partial f(\bar{r},t)}{\partial x} + \hat{a}_y \frac{\partial f(\bar{r},t)}{\partial y} + \hat{a}_z \frac{\partial f(\bar{r},t)}{\partial z} \tag{3.97}$$

in a Cartesian coordinate system. Obviously, this is simply an application of the del operator[44]: i.e. $\bar{\nabla} = \hat{a}_x\partial/\partial x + \hat{a}_y\partial/\partial y + \hat{a}_z\partial/\partial z$ on a scalar field $f(\bar{r},t)$. As a major property, $\bar{\nabla}f(\bar{r},t)$ is always perpendicular to the equipotential surfaces of $f(\bar{r},t)$: i.e. surfaces S where $f(\bar{r},t)$ is constant. In addition, we have[45]

$$\frac{\partial f(\bar{r},t)}{\partial x} = \hat{a}_x \cdot \bar{\nabla}f(\bar{r},t), \quad \frac{\partial f(\bar{r},t)}{\partial y} = \hat{a}_y \cdot \bar{\nabla}f(\bar{r},t), \quad \frac{\partial f(\bar{r},t)}{\partial z} = \hat{a}_z \cdot \bar{\nabla}f(\bar{r},t) \tag{3.98}$$

and

$$\frac{\partial f(\bar{r},t)}{\partial n} = \hat{a}_n \cdot \bar{\nabla}f(\bar{r},t) \tag{3.99}$$

for any direction $\hat{a}_n$.

In general, the gradient of a scalar field represents its derivative: i.e. its rate of change with respect to position (Figure 3.20). Since we are considering scalar fields in three-dimensional spaces, the derivatives need to be known in all possible directions. For example, in a

[44] Note that this basic operation has not been required until now, other than being mentioned in a few places.

[45] Also note that

$$\bar{\nabla}[f(\bar{r},t) + g(\bar{r},t)] = \bar{\nabla}f(\bar{r},t) + \bar{\nabla}g(\bar{r},t)$$
$$\bar{\nabla}[f(\bar{r},t)g(\bar{r},t)] = g(\bar{r},t)\bar{\nabla}f(\bar{r},t) + f(\bar{r},t)\bar{\nabla}g(\bar{r},t)$$

for any scalar fields $f(\bar{r},t)$ and $g(\bar{r},t)$.

Cartesian coordinate system, finding derivatives in the x, y, and z directions gives all the necessary information about how the field is changing. Combinations of the derivatives with respect to x, y, and z can be represented as a vector field: the gradient. Alternatively, in cylindrical and spherical coordinate systems, it is required to find the rate of change in (ρ, ϕ, z) and (R, θ, ϕ) directions, respectively. Of course, the gradient of a scalar field does not change (only its representation changes) with the coordinate system in which it is evaluated.

Consider a scalar field $f(\bar{r}) = x$ whose gradient is simply $\bar{g}(\bar{r}) = \bar{\nabla} f(\bar{r}) = \hat{a}_x$. This expression indicates that the field changes in the x direction (because of $\hat{a}_x$), it does not change in the y and z directions (because of the absence of $\hat{a}_y$ and $\hat{a}_z$), and the rate of change in the x direction is unity (since $|\hat{a}_x| = 1$). Now, consider another field $f(\bar{r}) = x + \sqrt{3}y$ (Figure 3.21). In this case, $\bar{g}(\bar{r}) = \bar{\nabla} f(\bar{r}) = \hat{a}_x + \sqrt{3}\hat{a}_y$, indicating that the value of the field changes in both x and y directions but still with a constant rate of $\sqrt{1+3} = 2$. The direction of $\bar{g}$ is 60° from the x axis; this is the direction perpendicular to the equipotential lines ($x + \sqrt{3}y = c$): i.e. lines with constant function values. In fact, the gradient of a scalar field points into the direction with maximum rate of change, which may of course depend on the position.

Consider another example with a scalar field $f(\bar{r}) = xy$ (Figure 3.22). In this case, we have $\bar{g}(\bar{r}) = \hat{a}_x y + \hat{a}_y x$. For large values of x, $g_y > g_x$: i.e. the y component of the gradient is dominant and the scalar field changes *more* in the y direction. For large values of y, however, the x component of the gradient becomes dominant. When $x = y = c$ (a location on the x-y plane), we have $\bar{g}(\bar{r}) = (\hat{a}_x + \hat{a}_y)c$, and the rate of change is the same in the x and y directions. But what about the z direction? The scalar field $f(\bar{r}) = xy$ does not change with respect to z: i.e. it is constant in this direction. Therefore, its gradient does not have a z component.

As practiced for the divergence and curl operations, we can derive the gradient operation in cylindrical and spherical coordinate systems. Using the definition of the del operator in a cylindrical system, we have

$$\bar{\nabla} f(\bar{r}, t) = \hat{a}_\rho \frac{\partial f(\bar{r}, t)}{\partial \rho} + \hat{a}_\phi \frac{1}{\rho} \frac{\partial f(\bar{r}, t)}{\partial \phi} + \hat{a}_z \frac{\partial f(\bar{r}, t)}{\partial z}. \tag{3.100}$$

In the above, $\partial\phi$ is accompanied by ρ to change the angular derivative into a metric derivative. The expression for the gradient in a spherical coordinate system can be written as

$$\bar{\nabla} f(\bar{r}, t) = \hat{a}_R \frac{\partial f(\bar{r}, t)}{\partial R} + \hat{a}_\theta \frac{1}{R} \frac{\partial f(\bar{r}, t)}{\partial \theta} + \hat{a}_\phi \frac{1}{R\sin\theta} \frac{\partial f(\bar{r}, t)}{\partial \phi}, \tag{3.101}$$

where $\partial\theta$ and $\partial\phi$ are used with R and $R\sin\theta$, respectively.

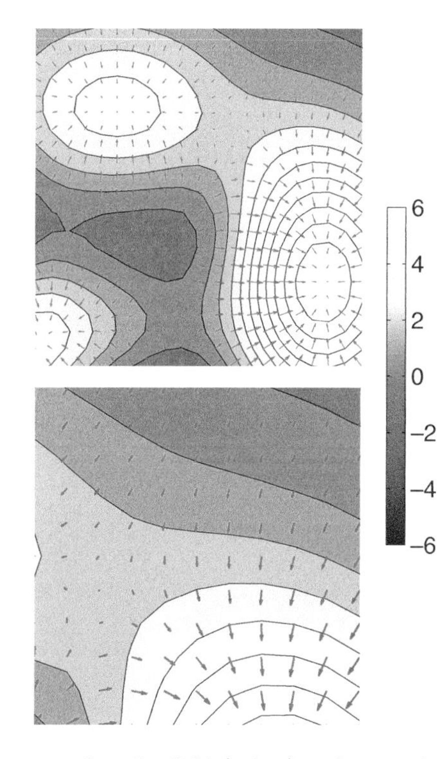

Figure 3.20 A scalar field (colors) and its gradient (arrows).

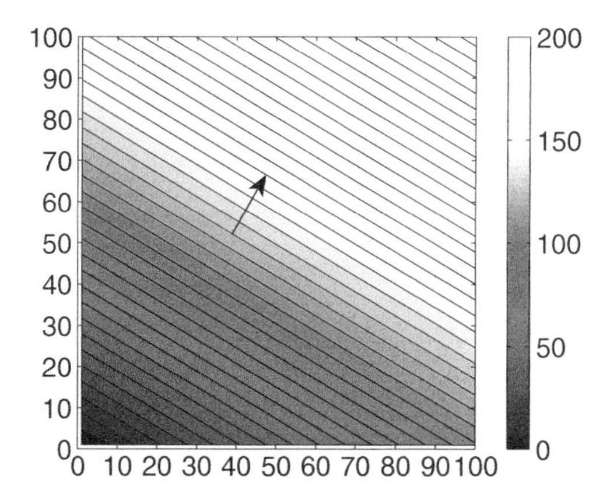

Figure 3.21 A scalar field $f(\bar{r}) = x + \sqrt{3}y$ (colors) and its gradient (vector) at a single point.

As an important case, the gradient of the magnitude of the position vector field $f(\bar{r}) = |\bar{r}|$ can be found as[46]

$$\bar{\nabla}(|\bar{r}|) = \bar{\nabla}\left(\sqrt{x^2+y^2+z^2}\right) \tag{3.102}$$

$$= \left(\hat{a}_x\frac{\partial}{\partial x} + \hat{a}_y\frac{\partial}{\partial y} + \hat{a}_z\frac{\partial}{\partial z}\right)\left(\sqrt{x^2+y^2+z^2}\right) \tag{3.103}$$

$$= \frac{\hat{a}_x 2x + \hat{a}_y 2y + \hat{a}_z 2z}{2\sqrt{x^2+y^2+z^2}} = \frac{\bar{r}}{|\bar{r}|} = \hat{a}_R. \tag{3.104}$$

We emphasize that $f(\bar{r}) = |\bar{r}|$ is a scalar field, which is defined everywhere (not simply a magnitude of a vector). The gradient result as $\bar{\nabla}\bar{f}(\bar{r}) = \hat{a}_R$ indicates that the equipotential surfaces of $f = |\bar{r}|$ are spheres[47] (Figure 3.23). As another important case, we have

$$\bar{\nabla}\left(\frac{1}{|\bar{r}|}\right) = -\frac{\bar{r}}{|\bar{r}|^3} = -\frac{\hat{a}_R}{|\bar{r}|^2} = -\frac{\hat{a}_R}{R^2}. \tag{3.105}$$

Note that the result of this gradient operation is similar to the electric field created by a static point electric charge. Hence, the scalar field $f(\bar{r}) = 1/|\bar{r}|$ should represent the electric scalar potential due to a point source.

Sometimes, the gradient operation provides useful information on a scalar field that is not obvious at the first glance. For example, we consider a scalar field $\bar{f}(\bar{r}) = \rho\sin\phi$ for $0 < \phi < \pi$. The gradient of the field can be found as

$$\bar{\nabla}f(\bar{r},t) = \hat{a}_\rho\frac{\partial}{\partial\rho}(\rho\sin\phi) + \hat{a}_\phi\frac{1}{\rho}\frac{\partial}{\partial\phi}(\rho\sin\phi) + \hat{a}_z\frac{\partial}{\partial z}(\rho\sin\phi) \tag{3.106}$$

using a cylindrical coordinate system. Then we have

$$\bar{\nabla}f(\bar{r},t) = \hat{a}_\rho\sin\phi + \hat{a}_\phi\frac{1}{\rho}\rho\cos\phi = \hat{a}_\rho\sin\phi + \hat{a}_\phi\cos\phi = \hat{a}_y. \tag{3.107}$$

This result indicates that the scalar field changes only in the y direction, the rate of change is unity at all locations, and the equipotential lines[48] on the x-y plane are perpendicular to $\hat{a}_y$. This simply means $f(\bar{r},t) = y + c$, where c is a constant.[49] In fact, one can recognize that

$$f(\bar{r}) = \rho\sin\phi = \sqrt{(x^2+y^2)}\sin\left(\tan^{-1}(y/x)\right) \tag{3.108}$$

$$= \sqrt{(x^2+y^2)}\frac{y}{\sqrt{(x^2+y^2)}} = y, \tag{3.109}$$

which fits into our finding with $c = 0$.

[46]As expected, the final equality is independent of the coordinate system. Therefore, the same equality can be shown in a cylindrical or a spherical coordinate system. In fact, it is relatively easy in a spherical system:

$$\bar{\nabla}|f(\bar{r})| = \bar{\nabla}(R) = \hat{a}_R.$$

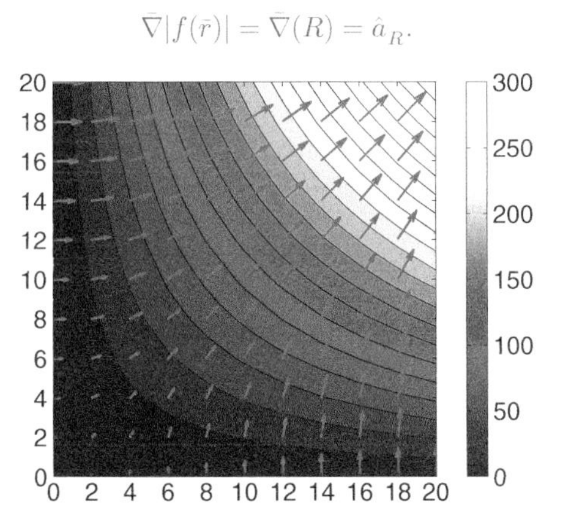

Figure 3.22 A scalar field $f(\bar{r}) = xy$ (colors) and its gradient (arrows).

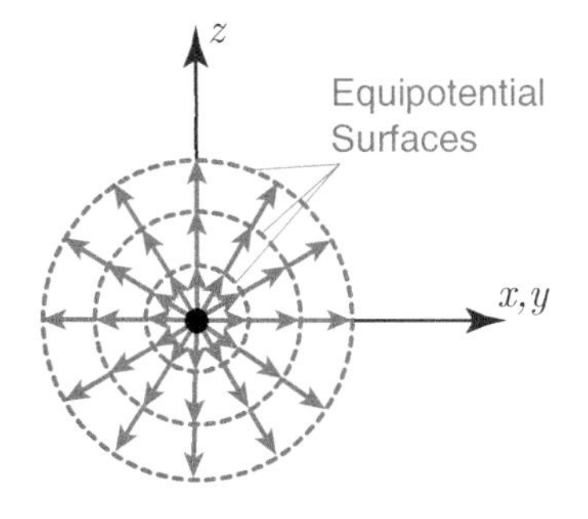

Figure 3.23 Vector field $\bar{g}(\bar{r}) = \hat{a}_R$, which is the gradient of the scalar field $f(\bar{r}) = |\bar{r}|$. Note that $f(\bar{r}) = |\bar{r}| = c$ (constant) surfaces (spheres) are equipotential surfaces of $\bar{g}(\bar{r})$.

[47]This is because $\hat{a}_R$ is perpendicular to spherical surfaces. The value $|\bar{\nabla}\bar{f}(\bar{r})| = |\hat{a}_R| = 1$ indicates that the rate of change is constant in the entire space.

[48]The term *equipotential surfaces* is better, if the z direction is also included.

[49]Note that, similar to the divergence and curl operations, the gradient operation leads to a loss of information. In this case, this is c, which may be zero or nonzero to satisfy the same gradient value.

3.6.2 Examples

Example 31: Consider a distribution of the static electric field intensity $\bar{E}(\bar{r}) = e_0(\hat{a}_x 2y + \hat{a}_y ax - \hat{a}_z 3z)$, where e_0 is a constant with unit V/m^2, while a is a unitless constant. Find the value of a.

Solution: The static electric field intensity is irrotational[50] and must satisfy $\bar{\nabla} \times \bar{E}(\bar{r}) = 0$:

$$\begin{vmatrix} \hat{a}_x & \hat{a}_y & \hat{a}_z \\ \partial/\partial x & \partial/\partial y & \partial/\partial z \\ 2y & ax & -3z \end{vmatrix} = \hat{a}_x \left[\frac{\partial(-3z)}{\partial y} - \frac{\partial(ax)}{\partial z}\right] - \hat{a}_y \left[\frac{\partial(-3z)}{\partial x} - \frac{\partial(2y)}{\partial z}\right]$$
$$+ \hat{a}_z \left[\frac{\partial(ax)}{\partial x} - \frac{\partial(2y)}{\partial y}\right] = \hat{a}_z(a-2) = 0. \tag{3.110}$$

[50] Note that if this field is in a vacuum, the associated volume electric current density can be found as $q_v(\bar{r}) = \bar{\nabla} \cdot \bar{E}(\bar{r})/\varepsilon_0 = -3/\varepsilon_0$, defined everywhere.

Therefore,[51] $a = 2$ and $\bar{E}(\bar{r}) = \hat{a}_x 2y + \hat{a}_y 2x - \hat{a}_z 3z$.

[51] Note that $a = 2$ is the only option to make $\bar{E}(\bar{r})$ a valid expression to represent a distribution of the static electric field intensity.

Example 32: Let a surface be defined as $2xz^2 - 3xy - 4x = 7$. Find an equation for the tangent plane (Figure 3.24) at $(x, y, z) = (1, -1, 2)$.

Solution: Defining a scalar field[52] $f(\bar{r}) = 2xz^2 - 3xy - 4x$, we have

[52] Hence, $f(\bar{r}) = 7$ is an equipotential surface of $f(\bar{r})$.

$$\bar{\nabla} f(\bar{r}) = \hat{a}_x(2z^2 - 3y - 4) + \hat{a}_y(-3x) + \hat{a}_z(4xz) \tag{3.111}$$

and

$$[\bar{\nabla} f]_{(1,-1,2)} = 7\hat{a}_x - 3\hat{a}_y + 8\hat{a}_z \tag{3.112}$$

as the normal direction at the given location. To find an equation for the tangent plane at $(1, -1, 2)$, we need to solve $\overline{\Delta l} \cdot \hat{a}_n = 0$, where

$$\overline{\Delta l} = \hat{a}_x(x-1) + \hat{a}_y(y+1) + \hat{a}_z(z-2) \tag{3.113}$$

$$\hat{a}_n = \frac{\hat{a}_x 7 - \hat{a}_y 3 + \hat{a}_z 8}{\sqrt{7^2 + 3^2 + 8^2}}. \tag{3.114}$$

Then we obtain $7(x-1) - 3(y+1) + 8(z-2) = 0$ or

$$7x - 3y + 8z = 26 \tag{3.115}$$

as a formula for the tangent plane.

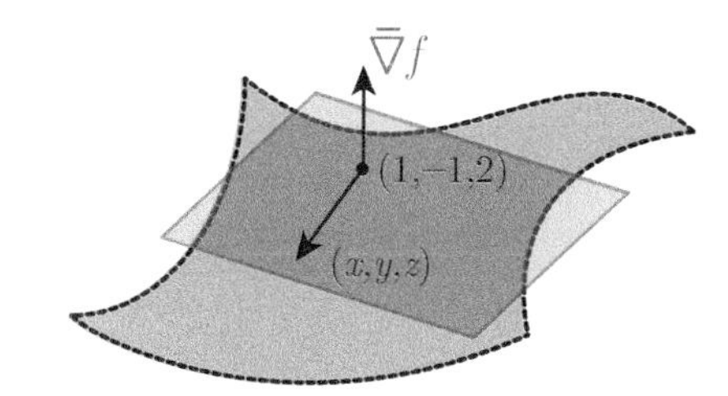

Figure 3.24 A surface defined as $2xz^2 - 3xy - 4x = 7$ and the tangent plane at $(x, y, z) = (1, -1, 2)$. To find an equation for the tangent plane, one can consider an arbitrary location (x, y, z) on the plane. Then a vector from $(1, -1, 2)$ (which is known to be located on the plane) to (x, y, z) must be perpendicular to the normal direction of the surface $2xz^2 - 3xy - 4x = 7$ at $(1, -1, 2)$. The required expression can be found by setting the dot product to zero.

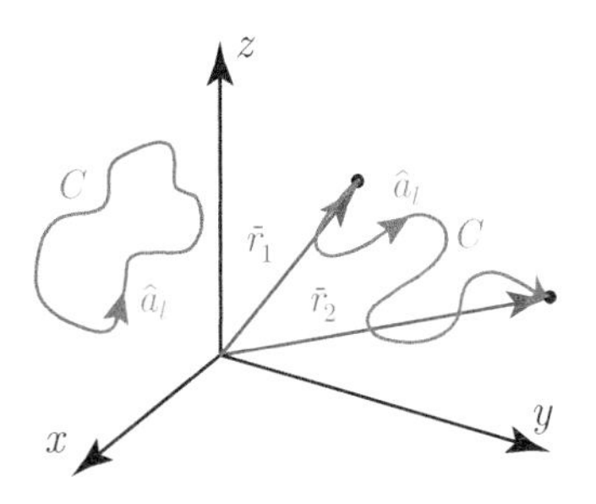

Figure 3.25 A demonstration of the gradient theorem. If a vector field is conservative – i.e. if it can be written as $\bar{g}(\bar{r},t)=\bar{\nabla}f(\bar{r},t)$ – then its line integral along a curve C only depends on the end points as $f(\bar{r}_2,t)-f(\bar{r}_1,t)$. This means if the curve is closed, the line integral of $\bar{g}(\bar{r},t)$ is zero.

3.6.3 Gradient Theorem

Consider a scalar field $f(\bar{r},t)$, whose gradient is $\bar{g}(\bar{r},t)=\bar{\nabla}f(\bar{r},t)$, and its line integral from a position $\bar{r}_1$ to $\bar{r}_2$. Then the gradient theorem states that (Figure 3.25)

$$\int_{\bar{r}_1}^{\bar{r}_2}\bar{g}(\bar{r},t)\cdot\overline{dl}=\int_{\bar{r}_1}^{\bar{r}_2}\bar{\nabla}f(\bar{r},t)\cdot\overline{dl}=f(\bar{r}_2,t)-f(\bar{r}_1,t), \tag{3.116}$$

i.e. the line integral of the gradient of a scalar field is simply the difference between the value of the field at the end points ($\bar{r}_1$ and $\bar{r}_2$), while it does not depend on the path of the integration. The path independence can easily be shown as

$$\int_{\bar{r}_1}^{\bar{r}_2}\bar{\nabla}f(\bar{r},t)\cdot\overline{dl}=\int_{\bar{r}_1}^{\bar{r}_2}\bar{\nabla}f(\bar{r},t)\cdot\hat{a}_l dl=\int_{\bar{r}_1}^{\bar{r}_2}\frac{\partial f(\bar{r},t)}{\partial l}dl=f(\bar{r}_2,t)-f(\bar{r}_1,t). \tag{3.117}$$

Then a vector field $\bar{g}(\bar{r},t)$ that can be written as the gradient of a scalar field $[\bar{g}(\bar{r},t)=\bar{\nabla}f(\bar{r},t)]$ has

$$\int_{\bar{r}_1}^{\bar{r}_1}\bar{g}(\bar{r},t)\cdot\overline{dl}=\int_{\bar{r}_1}^{\bar{r}_1}\bar{\nabla}f(\bar{r},t)\cdot\overline{dl}=f(\bar{r}_1,t)-f(\bar{r}_1,t)=0, \tag{3.118}$$

i.e. its closed line integral (circulation) for any path is zero. Formally, a *conservative* vector field is defined as one that can be written as the gradient of a scalar field. The gradient theorem states that such a vector field has a zero circulation for all closed paths. Although not shown here, the gradient theorem also has the converse: i.e. a vector field that demonstrates path independency (zero circulation) can be written as the gradient of a scalar field. Consequently, with a zero-circulation property (based on the integral form of Faraday's law), the static electric field intensity is conservative, and it can be written as[53] $\bar{E}(\bar{r})=-\bar{\nabla}\Phi(\bar{r})$.

It is interesting that a combination of the gradient theorem and Stokes' theorem verifies the null identity $\bar{\nabla}\times\bar{\nabla}f(\bar{r},t)=0$. Specifically, considering $\bar{g}(\bar{r},t)=\bar{\nabla}f(\bar{r},t)$, we have

$$\int_S\bar{\nabla}\times\bar{g}(\bar{r},t)\cdot\overline{ds}=\int_S\bar{\nabla}\times\bar{\nabla}f(\bar{r},t)\cdot\overline{ds}, \tag{3.119}$$

where S is an arbitrary open surface. Then, using Stokes' theorem and the gradient theorem subsequently, we derive

$$\int_S\bar{\nabla}\times\bar{\nabla}f(\bar{r},t)\cdot\overline{ds}=\oint_C\bar{\nabla}f(\bar{r},t)\cdot\overline{dl}=0 \tag{3.120}$$

as the line integral is along a closed curve. Since the selected surface S and the bounding curve C can be arbitrary, the integrand $\bar{\nabla}\times\bar{\nabla}f(\bar{r},t)$ should be zero.

[53]This is consistent with the fact that a static electric field is irrotational. Specifically, we have the following:

- The integral form of Faraday's law states that the circulation of the static electric field intensity is zero. Hence, it can be written as $\bar{E}(\bar{r})=-\bar{\nabla}\Phi(r)$, and it is a conservative field.
- All conservative fields are irrotational. Then $\bar{E}(\bar{r})$ must be irrotational, and it should satisfy $\bar{\nabla}\times\bar{E}(\bar{r})=0$. This is stated by the differential form of Faraday's law.

3.6.4 Gradient in Gauss' Law, Ampere's Law, and Faraday's Law

The gradient operation is not directly involved in Maxwell's equations.[54] We have already mentioned it several times when discussing dielectric and magnetic materials. With knowledge of the gradient, we can now revisit the differential forms of the laws and the related boundary conditions. We consider the differential laws at locations where permittivity $\varepsilon(\bar{r})$ or permeability $\mu(\bar{r})$ may depend on the position. We consider locations at a boundary S, which may separate two dielectric or magnetic media labeled 1 and 2.

[54]This subsection requires previous knowledge of Maxwell's equations and some related content. However, it is not directly related to understanding electrostatic cases and the electric scalar potential.

- According to Gauss' law, we have

$$\bar{\nabla} \cdot \bar{D}(\bar{r},t) = q_v(\bar{r},t) \tag{3.121}$$

$$\bar{\nabla} \cdot \bar{E}(\bar{r},t) = \frac{q_v(\bar{r},t)}{\varepsilon(\bar{r})} - \frac{\bar{\nabla}\varepsilon(\bar{r}) \cdot \bar{E}(\bar{r},t)}{\varepsilon(\bar{r})} = \frac{q_v(\bar{r},t)}{\varepsilon_0} - \frac{\bar{\nabla} \cdot \bar{P}(\bar{r},t)}{\varepsilon_0}, \tag{3.122}$$

where $\bar{P}(\bar{r},t)$ is the polarization. In the above, $\bar{\nabla}\varepsilon(\bar{r})$ represents the variation of the permittivity with respect to position. The dot product with $\bar{E}(\bar{r},t)$ indicates that we are particularly interested in the variation in the direction of the electric field intensity. Hence, we observe that $\bar{\nabla}\varepsilon(\bar{r}) \cdot \bar{E}(\bar{r},t)$ or $\bar{\nabla} \cdot \bar{P}(\bar{r},t)$ acts like a volume electric charge density,[55] depending on the form used. In fact, in some cases (e.g. electrets), $\bar{\nabla} \cdot \bar{P}(\bar{r},t)$ (or $\bar{\nabla}\varepsilon(\bar{r}) \cdot \bar{E}(\bar{r},t)$) is the only type of source that directly creates electric fields:

[55]For these quantities, we note the units as $(1/\text{m}) \times (\text{F/m}) \times (\text{V/m}) = \text{C/m}^3$ and $(1/\text{m}) \times (\text{C/m}^2) = \text{C/m}^3$, respectively, where the del operator brings $1/\text{m}$.

$$q_v(\bar{r},t) = 0 \longrightarrow \bar{\nabla} \cdot \bar{E}(\bar{r},t) = -\frac{\bar{\nabla}\varepsilon(\bar{r}) \cdot \bar{E}(\bar{r},t)}{\varepsilon(\bar{r})} = -\frac{\bar{\nabla} \cdot \bar{P}(\bar{r},t)}{\varepsilon_0}. \tag{3.123}$$

- According to the boundary conditions for normal electric fields, we have

$$\hat{a}_n \cdot [\bar{D}_1(\bar{r},t) - \bar{D}_2(\bar{r},t)] = q_s(\bar{r},t) \qquad (\bar{r} \in S) \tag{3.124}$$

$$\hat{a}_n \cdot [\bar{E}_1(\bar{r},t) - \bar{E}_2(\bar{r},t)] = \frac{q_s(\bar{r},t)}{\varepsilon_0} - \frac{\hat{a}_n \cdot [\bar{P}_1(\bar{r},t) - \bar{P}_2(\bar{r},t)]}{\varepsilon_0} \qquad (\bar{r} \in S). \tag{3.125}$$

Therefore, $\hat{a}_n \cdot [\bar{P}_1(\bar{r},t) - \bar{P}_2(\bar{r},t)]$ acts like a surface electric charge density that creates a discontinuity in the normal electric field intensity even in the absence of a true surface electric charge density. It accounts for the jump in the permittivity across the boundary.[56]

[56]When there is a jump in the permittivity, the function $\varepsilon(\bar{r})$ is discontinuous across the boundary. Hence, $\bar{\nabla}\varepsilon(\bar{r})$ is an impulse function with a direction normal to the boundary. This leads to *infinitely large* volume electric charge density $\bar{\nabla}\varepsilon(\bar{r}) \cdot \bar{E}(\bar{r},t)$ located on the surface that can be interpreted as a finite surface electric charge density.

- According to Ampere's law, we have

$$\bar{\nabla} \times \bar{H}(\bar{r},t) = \bar{J}_v(\bar{r},t) + \frac{\partial \bar{D}(\bar{r},t)}{\partial t} \tag{3.126}$$

$$\bar{\nabla} \times \bar{B}(\bar{r},t) = \mu(\bar{r})\bar{J}_v(\bar{r},t) + \mu(\bar{r})\frac{\partial \bar{D}(\bar{r},t)}{\partial t} + \mu(\bar{r})\frac{\bar{\nabla}\mu(\bar{r}) \times \bar{B}(\bar{r},t)}{[\mu(\bar{r})]^2} \tag{3.127}$$

$$= \mu_0 \bar{J}_v(\bar{r},t) + \mu_0 \frac{\partial \bar{D}(\bar{r},t)}{\partial t} + \mu_0 \bar{\nabla} \times \bar{M}(\bar{r},t), \tag{3.128}$$

where $\bar{M}(\bar{r},t)$ is the magnetization. In this case, $\bar{\nabla}\mu(\bar{r})$ represents the variation of the permeability with respect to position. Specifically, a variation perpendicular to the magnetic flux density (due to the cross product) may lead to an additional nonzero term on the right-hand side. This means $\bar{\nabla}\mu(\bar{r}) \times \bar{B}(\bar{r},t)/[\mu(\bar{r})]^2$ acts like a volume electric current density.[57] Similarly, in the above, $\bar{\nabla} \times \bar{M}(\bar{r},t)$ acts like a volume electric current density[58] if that form is preferred.

- According to the boundary conditions for tangential magnetic fields, we have

$$\hat{a}_n \times [\bar{H}_1(\bar{r},t) - \bar{H}_2(\bar{r},t)] = \bar{J}_s(\bar{r},t) \qquad (\bar{r} \in S) \tag{3.129}$$

$$\hat{a}_n \times [\bar{B}_1(\bar{r},t) - \bar{B}_2(\bar{r},t)] = \mu_0 \bar{J}_s(\bar{r},t) + \mu_0 \hat{a}_n \times [\bar{M}_1(\bar{r},t) - \bar{M}_2(\bar{r},t)] \qquad (\bar{r} \in S). \tag{3.130}$$

Hence, $\hat{a}_n \times [\bar{M}_1(\bar{r},t) - \bar{M}_2(\bar{r},t)]$ acts like a surface electric current density that creates a discontinuity in the tangential magnetic flux density even in the absence of a true surface electric current density.

- According to Faraday's law, we have[59]

$$\bar{\nabla} \times \bar{E}(\bar{r},t) = -\frac{\partial \bar{B}(\bar{r},t)}{\partial t} \tag{3.131}$$

$$\bar{\nabla} \times \bar{D}(\bar{r},t) = -\varepsilon(\bar{r})\frac{\partial \bar{B}(\bar{r},t)}{\partial t} + \varepsilon(\bar{r})\frac{\bar{\nabla}\varepsilon(\bar{r}) \times \bar{D}(\bar{r},t)}{[\varepsilon(\bar{r})]^2} \tag{3.132}$$

$$= -\varepsilon_0 \frac{\partial \bar{B}(\bar{r},t)}{\partial t} + \varepsilon_0 \frac{\bar{\nabla} \times \bar{P}(\bar{r},t)}{\varepsilon_0}, \tag{3.133}$$

where the change in the permittivity perpendicular to the electric flux density appears on the right-hand side of $\bar{\nabla} \times \bar{D}(\bar{r},t)$.[60] In this case, $\bar{\nabla}\varepsilon(\bar{r}) \times \bar{D}(\bar{r},t)/[\varepsilon(\bar{r})]^2$ or $\bar{\nabla} \times \bar{P}(\bar{r},t)/\varepsilon_0$ acts like a volume *magnetic* current density.[61]

[57] Once again, we can check the units as $(1/\text{m}) \times (\text{H/m}) \times (\text{Wb/m}^2)/(\text{H/m})^2 = \text{A/m}^2$.

[58] We can check the units as $(1/\text{m}) \times \text{A/m} = \text{A/m}^2$.

[59] Note that

$$\bar{\nabla} \times [h(\bar{r},t)\bar{f}(\bar{r},t)] = \bar{\nabla}h(\bar{r},t) \times \bar{f}(\bar{r},t) + h(\bar{r},t)\bar{\nabla} \times \bar{f}(\bar{r},t)$$

and

$$\bar{\nabla}\left(\frac{1}{h(\bar{r},t)}\right) = -\frac{1}{[h(r,t)]^2}\bar{\nabla}h(\bar{r},t).$$

[60] Hence, the static electric flux density is not necessarily irrotational in dielectric media. For example, in electrets, where no free charge but polarization exists, nonzero electric fields may be created via this mechanism.

[61] The units for these magnetic sources can be checked as $(1/\text{m}) \times (\text{F/m}) \times (\text{C/m}^2)/(\text{F/m})^2 = \text{V/m}^2$ for $\bar{\nabla}\varepsilon(\bar{r}) \times \bar{D}(\bar{r},t)/[\varepsilon(\bar{r})]^2$ and as $(1/\text{m}) \times (\text{C/m}^2)/(\text{F/m}) = \text{V/m}^2$ for $\bar{\nabla} \times \bar{P}(\bar{r},t)/\varepsilon_0$.

- According to the boundary conditions for tangential electric fields, we have

$$\hat{a}_n \times [\bar{E}_1(\bar{r},t) - \bar{E}_2(\bar{r},t)] = 0 \qquad (\bar{r} \in S) \tag{3.134}$$

$$\hat{a}_n \times [\bar{D}_1(\bar{r},t) - \bar{D}_2(\bar{r},t)] = \varepsilon_0 \frac{\hat{a}_n \times [\bar{P}_1(\bar{r},t) - \bar{P}_2(\bar{r},t)]}{\varepsilon_0} \qquad (\bar{r} \in S). \tag{3.135}$$

  Hence, $\hat{a}_n \times [\bar{P}_1(\bar{r},t) - \bar{P}_2(\bar{r},t)]/\varepsilon_0$ acts like a surface magnetic current density that creates a discontinuity in the tangential electric flux density.[62]

[62] The units can be checked as $(\mathrm{C/m^2})/(\mathrm{F/m}) = \mathrm{V/m}$.

In the next chapter, this discussion is completed when considering the divergence of the magnetic flux density and the boundary conditions for normal magnetic fields.

3.6.5 Electric Potential Energy

As indicated in Eq. (3.96), the static electric field intensity can be written as the negative gradient of the electric scalar potential as $\bar{E}(\bar{r}) = -\bar{\nabla}\Phi(\bar{r})$. This also means

$$-\int_{\bar{r}_1}^{\bar{r}_2} \bar{E}(\bar{r}) \cdot \overline{dl} = \int_{\bar{r}_1}^{\bar{r}_2} \bar{\nabla}\Phi(\bar{r}) \cdot \overline{dl} = \Phi(\bar{r}_2) - \Phi(\bar{r}_1), \tag{3.136}$$

i.e. the line integral of the static electric field intensity depends only on the values of the electric scalar potential at the end points $\bar{r}_1$ and $\bar{r}_2$ and not the integration path. Since the electric field intensity is related to the electric force (as indicated in Eq. 1.207) as $\bar{F}_{e,0}(\bar{r}) = Q_0\bar{E}(\bar{r})$, where Q_0 is a test charge that is exposed to the electric field, the line integral of the electric field intensity is related to the electric potential energy. Specifically, we have

w_e: electric potential energy
Unit: joule (J) (coulombs × volts or newtons × meters)

$$w_{e,0}(\bar{r}_1,\bar{r}_2) = w_{e,0}(\bar{r}_2) - w_{e,0}(\bar{r}_1) = -Q_0\int_{\bar{r}_1}^{\bar{r}_2} \bar{E}(\bar{r}) \cdot \overline{dl}, \tag{3.137}$$

which represents the amount of the electric potential energy gained by the test charge Q_0 when it is theoretically moved from $\bar{r}_1$ to $\bar{r}_2$. Hence, it also corresponds to the work done (by external forces) to theoretically move the charge from $\bar{r}_1$ to $\bar{r}_2$. Obviously, using the electric scalar potential, we have

$$w_{e,0}(\bar{r}_1,\bar{r}_2) = Q_0\Phi(\bar{r}_1,\bar{r}_2) = Q_0[\Phi(\bar{r}_2) - \Phi(\bar{r}_1)], \tag{3.138}$$

where $\Phi(\bar{r}_1,\bar{r}_2)$ is the electric scalar potential at $\bar{r}_2$ *with respect to* $\bar{r}_1$.

It is often useful to define a reference point and the corresponding electric scalar potential.[63]

[63] This is similar to defining a ground with zero potential in circuit analysis.

A common approach is assigning zero electric scalar potential at infinity: i.e. $\Phi(\infty) = 0$. Then the electric scalar potential at a position $\bar{r}$ can be defined uniquely (Figure 3.26) as[64]

$$\Phi(\bar{r}) = -\int_{\infty}^{\bar{r}} \bar{E}(\bar{r}') \cdot \overline{dl'}. \tag{3.139}$$

[64]Note that primed coordinates are used inside the integral to distinguish the integration variables from the observation point $\bar{r}$, where the electric scalar potential is evaluated.

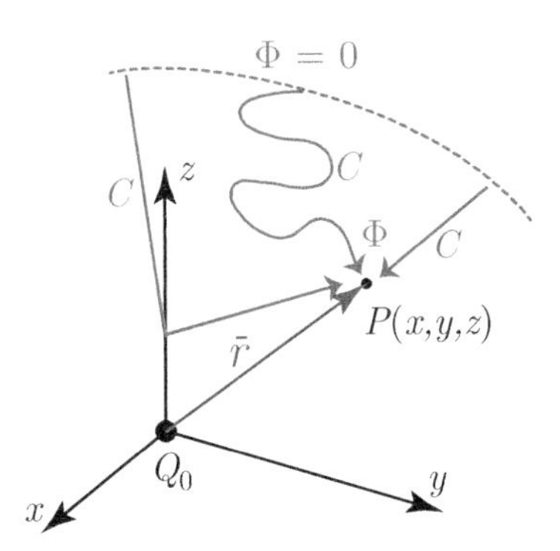

Figure 3.26 When finding the electric scalar potential from the static electric field intensity, any arbitrary path can be chosen from the reference point (e.g. infinity) to the observation point. For a static point electric charge, the best path may be a radial one, for which the line integral can easily be evaluated.

Depending on the case, the static electric field intensity may be known partially or entirely. If it is possible to evaluate the line integral between two points (if the electric field is known on *any* path between the points), one can find the electric scalar potential at these points relative to each other. As an example, we consider the electric field generated by a static point electric charge at the origin:

$$\bar{E}(\bar{r}) = \hat{a}_R \frac{Q_0}{4\pi\varepsilon_0 R^2}. \tag{3.140}$$

First, we can check the curl of this electric field intensity as

$$\bar{\nabla} \times \bar{E}(\bar{r}) = \frac{1}{R}\hat{a}_\theta \left(\frac{1}{\sin\theta}\frac{\partial E_R(\bar{r})}{\partial \phi}\right) - \frac{1}{R}\hat{a}_\phi \left(\frac{\partial E_R(\bar{r})}{\partial \theta}\right) = 0. \tag{3.141}$$

Hence, the electric field intensity due to a static point electric charge is irrotational, as expected.[65] Next, by using the expression for the electric field intensity, we can find the electric scalar potential as a function of position. Assigning $\Phi(\infty) = 0$, all we need to do is approaching an arbitrary (observation) point $\bar{r}$ from infinity.[66] We have[67]

$$\Phi(\bar{r}) = -\int_{\infty}^{\bar{r}} \hat{a}_R \frac{Q_0}{4\pi\varepsilon_0 (R')^2} \cdot \overline{dl'}, \tag{3.142}$$

where we can choose the path ($\overline{dl}$). The best option seems to be moving radially from infinity to the position: i.e. $\overline{dl'} = \hat{a}_R dR'$. Then we have[68]

$$\Phi(\bar{r}) = -\int_{\infty}^{R} \hat{a}_R \frac{Q_0}{4\pi\varepsilon_0 (R')^2} \cdot \hat{a}_R dR' = -\frac{Q_0}{4\pi\varepsilon_0}\int_{\infty}^{R} \frac{1}{(R')^2} dR' \tag{3.143}$$

$$= \frac{Q_0}{4\pi\varepsilon_0}\left[\frac{1}{R'}\right]_{\infty}^{R} = \frac{Q_0}{4\pi\varepsilon_0 R}. \tag{3.144}$$

This can be rewritten as

$$\Phi(\bar{r}) = \frac{Q_0}{4\pi\varepsilon_0 |\bar{r}|}, \tag{3.145}$$

[65]This means, by superposition, the electric field intensity due to any static charge distribution is irrotational. In addition, a vector field $\bar{f}(\bar{r})$ with $\bar{\nabla} \times \bar{f}(\bar{r}) \neq 0$ cannot be an expression for the static electric field intensity.

[66]Once again, we emphasize that, since the static electric field intensity is conservative, the integral should not depend on the path. Therefore, we can select any path from infinity to the location $\bar{r}$. This selection can be made by considering the electric field direction to simplify the integral (Figure 3.26).

[67]Primed coordinates are used for the integrand to avoid any confusion when considering spherical components of the observation point $\bar{r}$.

[68]While the movement is from infinity to the location $\bar{r}$, we select $\overline{dl'}$ in the outward direction since the integral limits already contain the direction information.

where the magnitude of the position vector is used for R. It should be emphasized that the electric scalar potential of a static point electric charge is inversely proportional to R, while its electric field intensity behaves as $1/R^2$. Specifically, the electric scalar potential decays slower than the electric field intensity, which is more or less expected since the potential is the derivative of the intensity. In addition, the equipotential surfaces ($\Phi(\bar{r})$ = constant surfaces) for a static point electric charge are spherical such that the electric field intensity (the gradient of the electric scalar potential) is perpendicular to them at all points (Figure 3.27).

The superposition principle is also applicable to the electric scalar potential. For example, we can reconsider the electric field created by an electric dipole in a vacuum (Figure 3.28). Let a positive charge Q_0 and a negative charge $-Q_0$ be located at $z = d/2$ and $z = -d/2$, respectively. Using Eq. (3.145), the electric scalar potential at an arbitrary point $\bar{r} = (R, \theta, \phi)$ can be found as

$$\Phi_d(\bar{r}) = \frac{Q_0}{4\pi\varepsilon_0}\left\{\frac{1}{|\bar{r} - \hat{a}_z d/2|} - \frac{1}{|\bar{r} + \hat{a}_z d/2|}\right\}. \tag{3.146}$$

When the observation point is far away from the dipole – i.e. when $|\bar{r}| \gg d$ – we have

$$|\bar{r} - \hat{a}_z d/2| = \sqrt{(\bar{r} - \hat{a}_z d/2) \cdot (\bar{r} - \hat{a}_z d/2)} \tag{3.147}$$

$$= \sqrt{R^2 + d^2/4 - 2\bar{r} \cdot \hat{a}_z d/2} \approx \sqrt{R^2 - \bar{r} \cdot \hat{a}_z d} \tag{3.148}$$

$$= \sqrt{R^2 - Rd\cos\theta} = R\sqrt{1 - \frac{d\cos\theta}{R}}$$

$$\approx R - \frac{d\cos\theta}{2} = |\bar{r}| - \frac{d\cos\theta}{2} \tag{3.149}$$

using a first-order approximation. Similarly, one can derive

$$|\bar{r} + \hat{a}_z d/2| \approx |\bar{r}| + \frac{d\cos\theta}{2}, \tag{3.150}$$

leading to

$$\Phi_d(\bar{r}) \approx \frac{Q_0 d\cos\theta}{4\pi\varepsilon_0 |\bar{r}|^2} \tag{3.151}$$

or

$$\Phi_d(\bar{r}) \approx \frac{Q_0 d\cos\theta}{4\pi\varepsilon_0 R^2}. \tag{3.152}$$

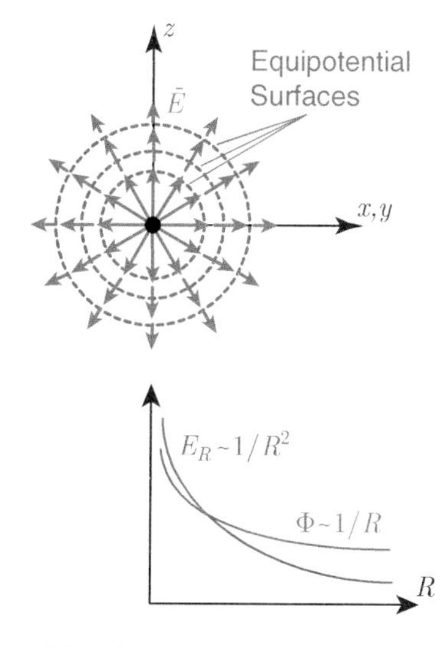

Figure 3.27 The electric scalar potential created by a static point electric charge has spherical equipotential surfaces and decays with the distance as $1/R$.

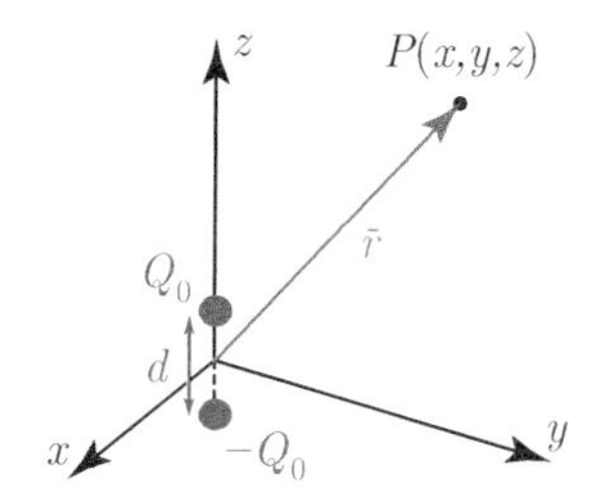

Figure 3.28 Electric dipole located in a vacuum.

Using this expression for the electric scalar potential, the corresponding electric field intensity can be found as[69]

$$\bar{E}_d(\bar{r}) = -\bar{\nabla}\Phi(\bar{r}) \approx \frac{Q_0 d}{4\pi\varepsilon_0|\bar{r}|^3}(\hat{a}_R 2\cos\theta + \hat{a}_\theta \sin\theta), \tag{3.153}$$

where the approximation is due to the assumption $|\bar{r}| \gg d$. The expression in Eq. (3.153) is consistent with the one in Eq. (1.262) where

$$\hat{a}_R 2\cos\theta + \hat{a}_\theta \sin\theta = \hat{a}_R 3\cos\theta + \hat{a}_\theta \sin\theta - \hat{a}_R \cos\theta = \hat{a}_R 3(\hat{a}_R \cdot \hat{a}_z) - \hat{a}_z. \tag{3.154}$$

[69]When evaluating derivatives, particularly in the gradient operation, we should be careful about the generality of the expression for the electric scalar potential. Specifically, if the expression for the potential is found for special cases, evaluating the derivative may lead to incorrect results due to loss of information. To see this problem, consider the electric field intensity at a given observation point $\bar{r}_0$. We have

$$\bar{E}(\bar{r}_0) = -[\bar{\nabla}\Phi(\bar{r})]_{\bar{r}=\bar{r}_0},$$

i.e. $\bar{r} = \bar{r}_0$ should be inserted *after* the gradient is evaluated and the expression is found in terms of the variable $\bar{r}$. From this perspective, an expression

$$\bar{E}(\bar{r}_0) = -\bar{\nabla}[\Phi(\bar{r}_0)],$$

is generally incorrect since inserting $\bar{r} = \bar{r}_0$ *before* the gradient operation can lead to a loss of information. In the case of the electric dipole, the electric scalar potential is found for large $|\bar{r}|$ values, but it is still general – i.e. valid for all $\bar{r}$ values satisfying the criteria $|\bar{r}| \gg d$ – so that the evaluation of the gradient to find the electric field intensity is correct.

3.6.5.1 Electric Potential Energy of Discrete Charge Distributions

When multiple charges are arranged in space, electric potential energy is involved in the system since each charge is exposed to the electric field intensity created by other charges. Considering two static point electric charges Q_1 and Q_2 in a vacuum, one can assume that Q_2 is brought from infinity to a location r_2 such that the energy required for this theoretical movement can be written as

$$w_e = Q_2\Phi_2, \tag{3.155}$$

where Φ_2 is the electric scalar potential created by Q_1 at the location of Q_2. Based on the expression for the electric scalar potential due to a static point electric charge, we also have

$$\Phi_2 = \frac{Q_1}{4\pi\varepsilon_0 R_{12}}, \tag{3.156}$$

where $R_{12} = |\bar{r}_1 - \bar{r}_2|$ is the distance between the charges, leading to

$$w_e = \frac{Q_1 Q_2}{4\pi\varepsilon_0 R_{12}}. \tag{3.157}$$

This energy is spent to bring the charge Q_2 from infinity to its final position.

If there is a positive electric potential energy in a system of discrete (point) charges, this means there is an external force to keep them at their positions. Recall from Section 1.6.2 that, for a pair of charges Q_1 and Q_2, the force between them is

$$F_e = \frac{Q_1 Q_2}{4\pi\varepsilon_0 (R_{12})^2} \tag{3.158}$$

if we omit the direction information. The external force to keep the system stationary is the negative of this force. When Q_1 and Q_2 have the same sign (Figure 3.29), the force F_e is positive,

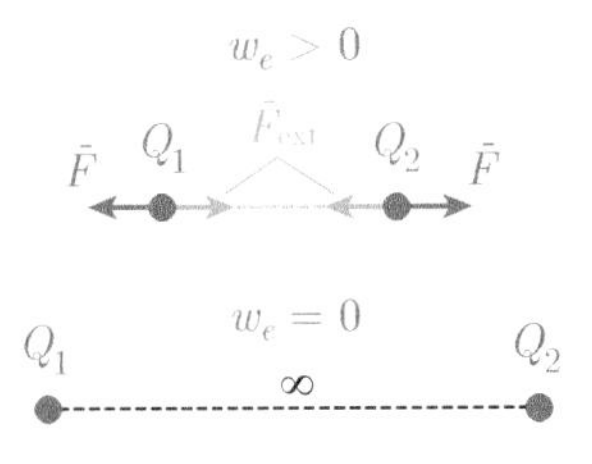

Figure 3.29 If there are two point electric charges with the same sign, there is a repulsive force between them. Therefore, to keep them stationary, an external force is needed. The energy stored by the system is positive, which can be $w_{e,\text{stored}} = \infty$ if the charges are brought together with zero distance between them. If the external force is removed, the system will tend to converge to an equilibrium state (zero energy), which corresponds to charges with an infinite distance between them.

indicating that the charges repel each other. Hence, an external force should be applied such that the charges do not move away from each other. Then the expression in Eq. (3.157), which is the energy that is provided externally to the system while bringing Q_2 from infinity, is the stored energy due to the interaction of charges. This energy can be released if the external force is removed and the charges are allowed to move away from each other. The energy becomes zero when the distance between them becomes infinite such that there is no force between the charges.

Following the discussion above, it is interesting to consider the case when Q_1 and Q_2 have opposite signs (Figure 3.30). The force between them is negative (attractive), the external force is needed to keep them separated, and the energy to bring Q_2 from infinity to its final position is negative. This negative sign indicates that bringing Q_2 from infinity does not consume external energy. In fact, energy is released due to such movement. In the final position, if the external force is removed, the charges will attract each other until they collapse. This is the case when the electric potential energy due to the interaction of the charges becomes minus infinite.[70,71]

In the expression given in Eq. (3.155), we assume that Q_2 is the charge that is brought from infinity while Q_1 was already in its position. Alternatively, if Q_1 is brought from infinity while Q_2 is fixed, we obtain

$$w_e = Q_1 \Phi_1, \tag{3.159}$$

where

$$\Phi_1 = \frac{Q_2}{4\pi\varepsilon_0 R_{12}} \tag{3.160}$$

is the electric scalar potential created by Q_2 at the position of Q_1. Obviously, we have

$$w_e = \frac{Q_1 Q_2}{4\pi\varepsilon_0 R_{12}}, \tag{3.161}$$

which is the same as Eq. (3.157). Hence, as long as the final status is the same, it does not matter how the charges are moved to reach it (Figure 3.31). For systems with more charges, however, it may be difficult to construct the final scenario by moving the charges one by one. Therefore, as a more generalized expression, the electric potential energy to construct a system of pair of charges (again by bringing them from infinity) can be written as

$$w_e = \frac{1}{2} Q_1 \Phi_1 + \frac{1}{2} Q_2 \Phi_2 \tag{3.162}$$

without thinking about how the charges are actually brought together.

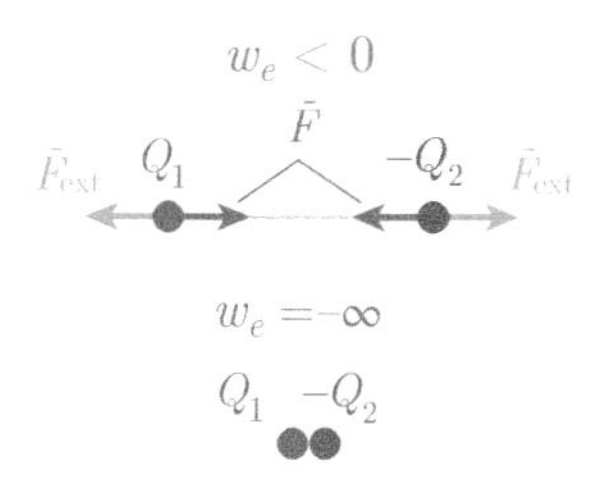

Figure 3.30 If there are two point electric charges with opposite signs, there is an attractive force between them. Therefore, to keep them stationary, an external force is needed. The energy stored by the system is negative, which can be $w_{e,\text{stored}} = 0$ if the charges are separated by an infinite distance. If the external force is removed, the system convergences into an equilibrium state (negative infinite energy), which corresponds to charges with zero distance between them.

[70] Hence for both cases (Q_1 and Q_2 with the same or opposite signs), the force between them tries to minimize the electric potential energy due to their interaction.

[71] Following this discussion, one may think that a static point electric charge should store infinite energy by itself, since it is a collection of charge particles that are brought together at a single point. This is correct if we consider the total stored energy in the system. On the other hand, here, we consider only the electric potential energy due to the interaction of point charges, not the energy to create each of them. For example, in Eq. (3.155), the electric scalar potential created by Q_2 at its position (which is infinite) is not considered.

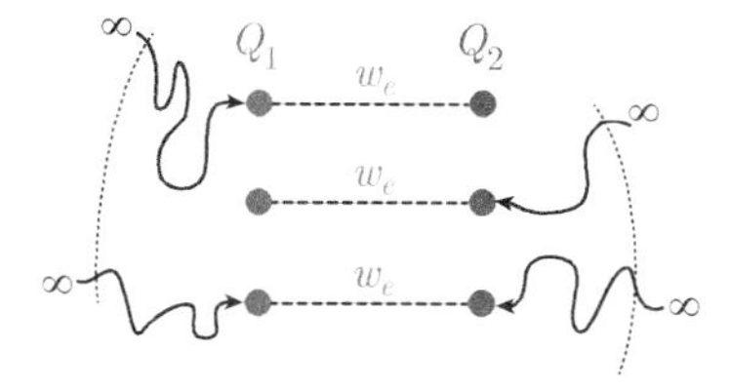

Figure 3.31 As long as the final status is the same, the stored energy does not depend on how two charges are brought together. If one of the charges is stationary, the energy corresponds to the line integral of the force applied by the stationary charge to the moved charge. When both charges are moving, one needs to consider the force between them dynamically as they move along their paths.

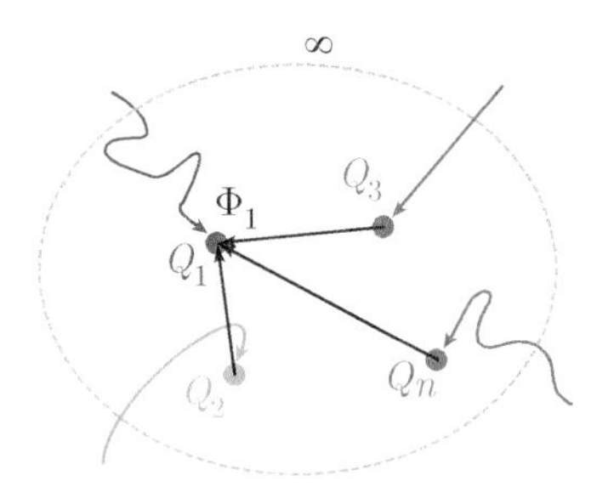

Figure 3.32 When considering the energy stored by a system multiple charges, it may be difficult to consider how charges are brought from infinity while electric forces between them change depending on their locations. Instead, Eq. (3.163) can be used by simply considering the electric scalar potential created at the final position of each charge by other charges.

Next, we consider the case when multiple static point electric charges $Q_1, Q_2, \ldots, Q_n$ are located at positions $\bar{r}_1, \bar{r}_2, \ldots, \bar{r}_n$. By extending Eq. (3.162), we have

$$w_e = \frac{1}{2}\sum_{k=1}^{n} Q_k \Phi_k, \tag{3.163}$$

where Φ_k is the total potential created by other $n-1$ charges at the position of the kth charge (Figure 3.32). Once again, this is the energy due to the interaction of point charges, without taking into account how these point charges are formed at the beginning.

3.6.5.2 Electric Potential Energy Stored by an Electric Dipole

When an electric dipole is exposed to an external electric field, it stores electric potential energy. This kind of scenario (electric dipole in an external electric field) is important since electric dipoles are useful to describe dielectrics and their responses to external electric fields, as discussed in Section 1.7.2. We now consider an electric dipole involving static point electric charges $-Q_0$ and $+Q_0$. If the distance between the charges is d, this dipole can be represented by a dipole moment as (Figure 3.33)

$$\bar{I}_{DM,e} = \hat{a}_d Q_0 d, \tag{3.164}$$

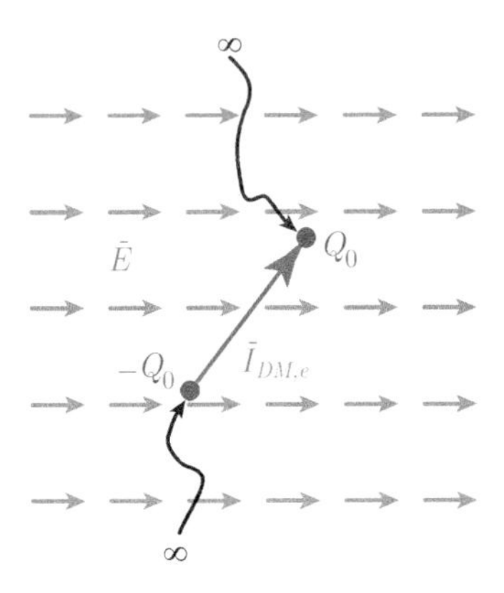

Figure 3.33 The energy stored by an electric dipole refers to the electric potential energy for bringing positive and negative charges together against an external electric field. The force between the electric charges is not included.

where $\hat{a}_d$ is the unit vector from the negative charge $(-Q_0)$ to the positive charge $(+Q_0)$. The stored electric potential energy for an electric dipole is generally defined as due to the interaction of the dipole and the external electric field, *excluding* the interaction between the charges of the dipole. Let us consider a static electric field intensity $\bar{E}(\bar{r})$. For bringing the positive charge $+Q_0$ from infinity to its position $\bar{r}_+$, the required (hence stored) energy can be found as

$$w^+_{ed,\text{stored}} = -Q_0 \int_{\infty}^{\bar{r}_+} \bar{E}(\bar{r}') \cdot \overline{dl'}, \tag{3.165}$$

where the path is not important. Similarly, for bringing the negative charge $-Q_0$ from infinity to its position $\bar{r}_-$, we have

$$w^-_{ed,\text{stored}} = Q_0 \int_{\infty}^{\bar{r}_-} \bar{E}(\bar{r}') \cdot \overline{dl'}. \tag{3.166}$$

Combining these energies, we have the total stored energy as

$$w_{ed,\text{stored}} = -Q_0 \int_{\infty}^{\bar{r}_+} \bar{E}(\bar{r}') \cdot \overline{dl'} + Q_0 \int_{\infty}^{\bar{r}_-} \bar{E}(\bar{r}') \cdot \overline{dl'} \tag{3.167}$$

$$= Q_0 \int_{\bar{r}_+}^{\infty} \bar{E}(\bar{r}') \cdot \overline{dl'} + Q_0 \int_{\infty}^{\bar{r}_-} \bar{E}(\bar{r}') \cdot \overline{dl'} \tag{3.168}$$

$$= Q_0 \int_{\bar{r}_+}^{\bar{r}_-} \bar{E}(\bar{r}') \cdot \overline{dl'}. \tag{3.169}$$

For the line integral from $\bar{r}_+$ to $\bar{r}_-$, we can choose a straight path. In addition, assuming that d is small such that the electric field intensity is constant on this path ($\bar{E}(\bar{r}') = \bar{E}_0$), we have

$$w_{ed,\text{stored}} = -Q_0 d\hat{a}_d \cdot \bar{E}(\bar{r}) = -\bar{I}_{DM,e} \cdot \bar{E}_0, \tag{3.170}$$

as the stored energy by the electric dipole.[72]

As discussed in Section 1.7.2, electric dipoles tend to orient in the direction of an external electric field. From this perspective, an electric dipole is a compact (rigid) structure that can rotate while the charges are stationary with respect to each other. Then the expression for the stored electric potential energy in Eq. (3.170) represents a correct interpretation of this physical phenomenon.[73] Using the expression in Eq. (3.170), zero electric potential energy occurs then $\bar{I}_{DM,e} \cdot \bar{E}_0 = 0$: i.e. when the dipole moment is perpendicular to the external electric field intensity. This is the case when the torque (force to rotate) applied by the external electric field to the electric dipole is maximum. On the other hand, the stored electric potential energy becomes maximum as $w_{ed,\text{stored}} = |\bar{I}_{DM,e}||\bar{E}_0|$ when there is 180° between $\bar{I}_{DM,e}$ and $\bar{E}_0$: i.e. when the electric dipole is oriented in the opposite direction of the electric field. This is the case when the torque applied by the external electric field to the electric dipole is zero, while the dipole orientation is *unstable*. Finally, if there is 0° between $\bar{I}_{DM,e}$ and $\bar{E}_0$ – i.e. when the electric dipole is perfectly aligned with the electric field – the minimum energy occurs as $w_{ed,\text{stored}} = -|\bar{I}_{DM,e}||\bar{E}_0|$. This is an equilibrium state: i.e. the electric dipole tends to maintain its orientation whenever there is no external force (other than the applied electric field).

3.6.5.3 Stored Electric Potential Energy in Charge Distributions

The stored electric potential energy in a distribution of charges can be defined as the amount of the electric potential energy to form the overall system. This energy is measured with respect the equilibrium case: i.e. the status when there is no external force to keep the system together and when the energy is assumed to be zero. For a volumetric (and static) electric charge distribution,

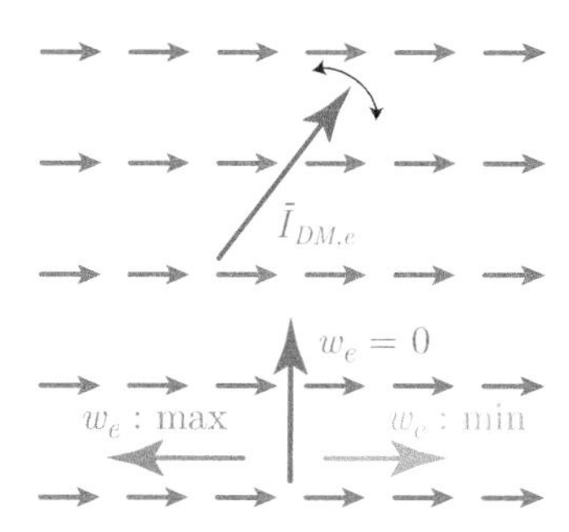

Figure 3.34 The energy stored by an electric dipole is the electric potential energy due to the orientation of the dipole in an external electric field. Depending on the orientation that can be represented by the dipole moment, the energy can be zero, positive, or negative.

[72]Note the other energies in the system, considering that this expression only represents the energy due to constructing a dipole *against* the external electric field intensity. Specifically, due to the interaction between the point electric charges, we further have

$$w_{eq,\text{stored}} = -\frac{Q_0^2}{4\pi\varepsilon_0 d}.$$

In addition, we recall that forming a point electric charge requires infinite energy: i.e. $w_{e+} = w_{e-} = \infty$. Therefore, when considering an energy in a system, it should be stated clearly which part of the system is considered.

[73]Note that Eq. (3.170) is obtained by imaginary movements of charges from infinity to their positions, while it does not matter how to interpret the corresponding stored energy. The same energy is stored when there is an electric dipole with rigidly connected charges (with constant dipole-moment magnitude) while it is free to rotate with changing dipole-moment direction (Figure 3.34).

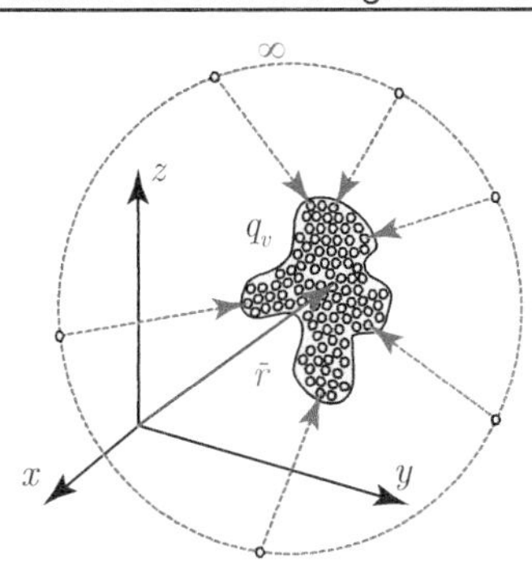

Figure 3.35 For a given charge distribution, infinitesimal particles can be assumed to be brought from infinity to find the stored electric potential energy.

the stored electric potential energy can be found as (Figure 3.35)

$$w_{e,\text{stored}} = \frac{1}{2}\int_V q_v(\bar{r})\Phi(\bar{r})dv, \tag{3.171}$$

where the integration is used to represent the summation of infinitesimal charges (represented by the volume electric charge density $q_v(\bar{r})$) multiplied by the corresponding electric scalar potential values. We emphasize that, being the total energy to form the overall system, the expression in Eq. (3.171) includes not only the electric potential energy due to interactions of charges but also self-interactions of infinitesimal charges.[74] As an example, considering a single point electric charge located at the origin, we have $q_v(\bar{r}) = Q_0\delta(R)$ and

$$w_{e,\text{stored}} = \frac{1}{2}\int_V Q_0\delta(R)\Phi(\bar{r})dv = Q_0\Phi(R=0) = \infty. \tag{3.172}$$

Naturally, squeezing electric charges at a single point (point electric charge) requires an infinite amount of work, leading to the storage of an infinite amount of energy in the system.

The expression in Eq. (3.171) can be used for arbitrary electric charge distributions, including surface and line distributions where the integral should be evaluated on surfaces and curves, respectively.[75] In some cases, however, integration on or inside the source region may be difficult either due to the expression or because of the definition of charges. For this purpose, there is an alternative way to obtain the stored electric potential energy when the electric field created by the charge distribution is known. Assuming a volumetric (and static) electric charge distribution and using Eq. (3.171), we first obtain

$$w_{e,\text{stored}} = \frac{1}{2}\int_V [\bar{\nabla}\cdot\bar{D}(\bar{r})]\Phi(\bar{r})dv \tag{3.173}$$

using Gauss' law. Furthermore, we have[76]

$$\bar{\nabla}\cdot[\Phi(\bar{r})\bar{D}(\bar{r})] = \bar{\nabla}\Phi(\bar{r})\cdot\bar{D}(\bar{r}) + \Phi(\bar{r})\bar{\nabla}\cdot\bar{D}(\bar{r}), \tag{3.174}$$

leading to

$$w_{e,\text{stored}} = \frac{1}{2}\int_V \bar{\nabla}\cdot[\Phi(\bar{r})\bar{D}(\bar{r})]dv - \frac{1}{2}\int_V \bar{\nabla}\Phi(\bar{r})\cdot\bar{D}(\bar{r})dv. \tag{3.175}$$

[74] Therefore, it naturally gives a positive value, as opposed to Eqs. (3.162), (3.163), and (3.170).

[75] We have

$$w_{e,\text{stored}} = \frac{1}{2}\int_S q_s(\bar{r})\Phi(\bar{r})ds$$

or

$$w_{e,\text{stored}} = \frac{1}{2}\int_C q_l(\bar{r})\Phi(\bar{r})dl.$$

[76] Using the identity

$$\bar{\nabla}\cdot[h(\bar{r},t)\bar{f}(\bar{r},t)] = \bar{\nabla}h(\bar{r},t)\cdot\bar{f}(\bar{r},t) + h(\bar{r},t)\bar{\nabla}\cdot\bar{f}(\bar{r},t)$$

for any scalar field $h(\bar{r},t)$ and vector field $\bar{f}(\bar{r},t)$.

Using the divergence theorem and $\bar{E}(\bar{r}) = -\bar{\nabla}\Phi(\bar{r})$, we have

$$w_{e,\text{stored}} = \frac{1}{2}\oint_S \Phi(\bar{r})\bar{D}(\bar{r}) \cdot \overline{ds} - \frac{1}{2}\int_V \bar{\nabla}\Phi(\bar{r}) \cdot \bar{D}(\bar{r})dv \tag{3.176}$$

$$= \frac{1}{2}\oint_S \Phi(\bar{r})\bar{D}(\bar{r}) \cdot \overline{ds} + \frac{1}{2}\int_V \bar{E}(\bar{r}) \cdot \bar{D}(\bar{r})dv, \tag{3.177}$$

where the first integral is converted into a surface integral enclosing the electric charge density.[77] The final equality becomes useful when assuming the integration domains are made infinite (Figure 3.36): i.e. when

$$\int_\infty \Phi(\bar{r})\bar{D}(\bar{r}) \cdot \overline{ds} \propto \oint (1/R)(1/R^2)ds \propto (1/R), \tag{3.178}$$

leading to

$$w_{e,\text{stored}} = \frac{1}{2}\int_\infty \bar{E}(\bar{r}) \cdot \bar{D}(\bar{r})dv. \tag{3.179}$$

Although its derivation is performed under the assumption of electrostatics, this final expression for the electric potential energy in terms of electric fields is valid for *both* static and dynamic cases. Note that, in Eq. (3.178), the surface integral brings a multiplication with R^2, while the integrand is overall proportional to $1/R^3$. Therefore, as the surface grows, the integral goes to zero with order $1/R$. The final expression in Eq. (3.179) indicates that the integral should be evaluated at *everywhere* (specifically where the electric field is nonzero). Using $\bar{D}(\bar{r}) = \varepsilon(\bar{r})\bar{E}(\bar{r})$, we further have

$$w_{e,\text{stored}} = \frac{1}{2}\int_\infty \varepsilon(\bar{r})\bar{E}(\bar{r}) \cdot \bar{E}(\bar{r})dv, \tag{3.180}$$

which becomes

$$w_{e,\text{stored}} = \frac{\varepsilon}{2}\int_\infty |\bar{E}(\bar{r})|^2 dv \tag{3.181}$$

for a homogeneous medium and

$$w_{e,\text{stored}} = \frac{\varepsilon_0}{2}\int_\infty |\bar{E}(\bar{r})|^2 dv \tag{3.182}$$

for a vacuum.

In general, capacitors are devices that can store electric potential energy (commonly stated as electrical energy). An interesting example is a parallel-plate capacitor (Figure 3.37) involving

[77] This conversion is valid for all cases except a point electric charge that has zero volume and surface. Specifically, the final expression in Eq. (3.179) does not give the infinite internal energy stored by a point electric charge; it simply gives zero, assuming that the charge is *already* formed but it does not interact with any other charge.

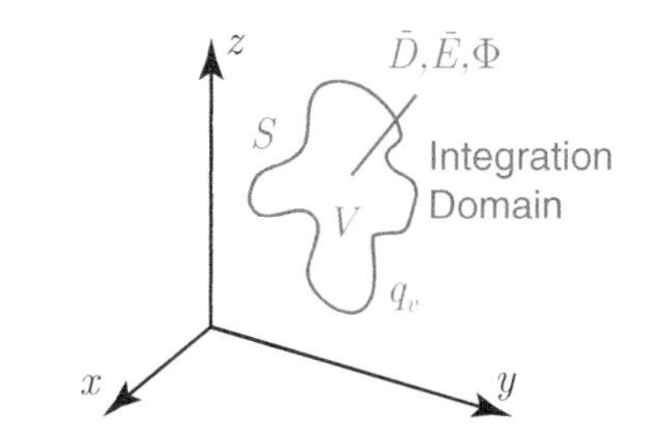

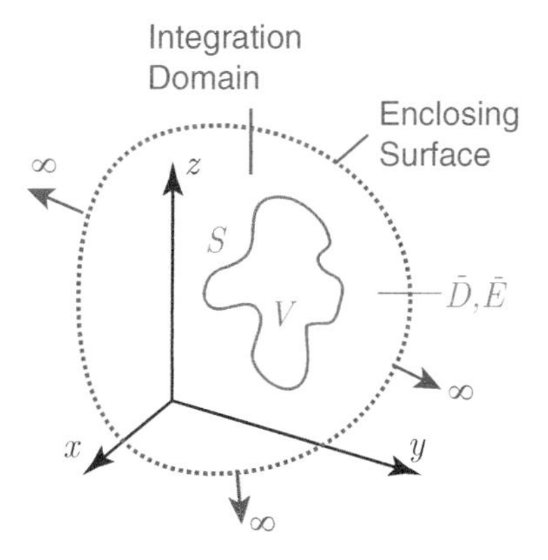

Figure 3.36 The electric potential energy stored by a volume electric charge distribution can be found by integration on the source. However, since the energy stored in an enclosing surface is the same and can be selected as infinity, the expression in Eq. (3.179) can be used, while the first term in Eq. (3.177) decays to zero in the limit. In the new form that only uses electric fields, the integral should be evaluated everywhere in space.

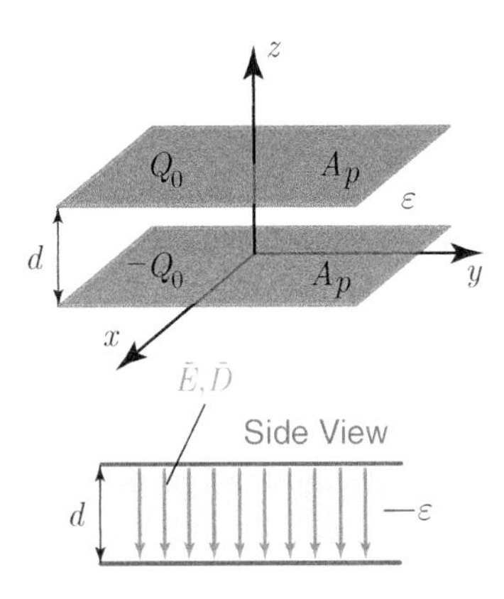

Figure 3.37 An ideal parallel-plate capacitor with a dielectric material between the plates.

two perfectly conducting plates separated by a distance d. We assume that each plate's area is A_p and a dielectric material with permittivity ε exists between the plates. In addition, the electric voltage (potential difference) between the plates is V_0 with the higher potential at the top plate, leading to Q_0 electric charge on the top plate (at $z = d$) and $-Q_0$ electric charge on the bottom plate. Using Eq. (3.179), the stored electric potential energy can be found as

$$w_{e,\text{stored}} = \frac{1}{2}\int_{\infty} \bar{E}(\bar{r}) \cdot \bar{D}(\bar{r})dv = \frac{1}{2}\int_{\text{between plates}} \bar{E}(\bar{r}) \cdot \bar{D}(\bar{r})dv, \tag{3.183}$$

where the integration domain is restricted to the volume between the plates, assuming that the capacitor is ideal and there is no electric field outside. Then, using Eq. (1.194), we have

$$w_{e,\text{stored}} = \frac{1}{2}\int_{\text{between plates}} \frac{Q_0}{\varepsilon A}\frac{Q_0}{A}dv = \frac{1}{2}\frac{Q_0^2}{\varepsilon A^2}Ad = \frac{Q_0^2 d}{2\varepsilon A}. \tag{3.184}$$

Defining the capacitance of the capacitor[78] as

$$C_0 = \frac{Q_0}{V_0} = \frac{A\varepsilon}{d}, \tag{3.185}$$

the stored electric potential energy can be written as[79,80]

$$w_{e,\text{stored}} = \frac{Q_0^2}{2C_0} = \frac{1}{2}C_0 V_0^2. \tag{3.186}$$

Note that this expression is always positive, and it corresponds to the energy stored by the capacitor.

Obviously, the expression in Eq. (3.186) should also be obtained by using the electric scalar potential. In fact, the applied voltage of V_0 means the electric scalar potentials at the bottom and top plates are c and $c + V_0$, respectively, where c is an arbitrary constant (depending on the selected reference point, at which the electric scalar potential is zero). Using the surface version of Eq. (3.171), we have

$$w_{e,\text{stored}} = \frac{1}{2}\int_S q_s(\bar{r})\Phi(\bar{r})ds \tag{3.187}$$

$$= \frac{1}{2}\int_{\text{top plate}} q_s(\bar{r})\Phi(\bar{r})ds + \frac{1}{2}\int_{\text{bottom plate}} q_s(\bar{r})\Phi(\bar{r})ds, \tag{3.188}$$

[78] Capacitance calculations are detailed in Section 6.7.

[79] In fact, these expressions are true for all capacitors.

[80] An interestingly confusing question: does the energy stored in a capacitor increase or decrease with the capacitance? The answer is both. If the charges on the plates do not change, increasing the capacitance (by increasing the area A_p, decreasing the distance d, and/or increasing the permittivity ε) reduces energy. If the voltage is constant (e.g. by keeping the capacitance connected to a battery), however, increasing the capacitance increases the stored energy.

leading to

$$w_{e,\text{stored}} = \frac{1}{2}Q_0(a + V_0) + \frac{1}{2}(-Q_0)a = \frac{1}{2}Q_0V_0, \tag{3.189}$$

which is consistent with the expressions in Eq. (3.186).

3.6.5.4 Electric Potential Energy and Electric Force

One may easily write the relationship between the electric force and the electric potential energy for static point electric charges. Considering two point electric charges Q_1 and Q_2 located at $\bar{r}_1$ and $\bar{r}_2$, respectively, we have

$$w_{e,\text{stored}} = \frac{Q_1Q_2}{4\pi\varepsilon_0 R_{12}} \tag{3.190}$$

as the electric potential energy due to the interaction of Q_1 and Q_2, while

$$\bar{F}_{e,12} = \frac{Q_1Q_2}{4\pi\varepsilon_0(R_{12})^2}\hat{a}_{12} \tag{3.191}$$

$$\bar{F}_{e,21} = \frac{Q_2Q_1}{4\pi\varepsilon_0(R_{21})^2}\hat{a}_{21} = -\bar{F}_{e,12} \tag{3.192}$$

represent the electric force applied by the electric charges to each other. In these equations, we have $\bar{r}_1 - \bar{r}_2 = \hat{a}_{12}R_{12} = -\hat{a}_{21}R_{21}$ and $R_{12} = R_{21}$. Note the consistency that (i) the electric scalar potential is the line integral of the electric field intensity, (ii) the electric force is the electric field intensity multiplied by the electric charge, (iii) the electric potential energy is the electric scalar potential multiplied by the electric charge, and (iv) energy is the line integral of the corresponding force. In the above, Eq. (3.190) can be seen as the line integral of Eq. (3.191) or Eq. (3.192), while Eq. (3.191) or Eq. (3.192) can be seen as the derivative of Eq. (3.190), if the distance R_{12} is the variable.

For a general system of electric charges/fields, however, the variable and the acting force due to the stored electric potential energy may not be obvious as in the case of two point electric charges. In such a case, we can rely on the derivative of scalar fields – i.e. gradient – as[81]

$$\bar{F}_e = \pm\bar{\nabla}w_{e,\text{stored}}, \tag{3.193}$$

[81] Recall that the stored electric potential energy $w_{e,\text{stored}}$ is positive, by definition.

where the sign depends on what is constant – i.e. the electric charge or the electric scalar potential – when the variable of the system is changing.[82] Specifically, we have[83]

$$\bar{F}_e = -\bar{\nabla} w_{e,\text{stored}} \qquad \text{(constant charge)} \tag{3.194}$$

and

$$\bar{F}_e = \bar{\nabla} w_{e,\text{stored}} \qquad \text{(constant potential).} \tag{3.195}$$

[82] This equation is general and also valid for static cases. Specifically, when we consider electrostatics, there is no movement with respect to time. But to interpret the force, we consider an imaginary action as if there is a movement leading to a change in the stored electric potential energy. This approach is often called *virtual displacement*.

[83] There is a formal derivation of the sign of the gradient expression relating the stored electric potential energy and the electric force. Basically, the case of constant charge occurs when the system is isolated from external sources. On the other hand, keeping constant voltage sources corresponds to an external voltage source that provides energy. When this energy is taken into account, the sign becomes a plus.

As an example, we consider again a parallel-plate capacitor (Figure 3.38) involving perfectly conducting plates separated by a distance d. As found above, the stored electric potential energy can be written as

$$w_{e,\text{stored}} = \frac{Q_0^2}{2C_0^2} = \frac{Q_0^2 d}{2\varepsilon A} \tag{3.196}$$

or

$$w_{e,\text{stored}} = \frac{V_0^2 C_0}{2} = \frac{V_0^2 \varepsilon A}{2d}. \tag{3.197}$$

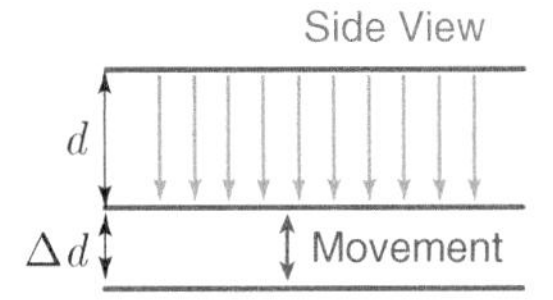

Figure 3.38 Virtual displacement (imaginary movement) to find the electric force between the plates of a capacitor.

To find the force between the plates, we can consider d as the variable. If the *charge* is assumed to be constant, we have

$$\bar{F}_e = -\bar{\nabla} w_{e,\text{stored}} = -\hat{a}_d \frac{\partial}{\partial d}\left(\frac{Q_0^2 d}{2\varepsilon A}\right) = -\hat{a}_d \frac{Q_0^2}{2\varepsilon A}. \tag{3.198}$$

Alternatively, if the *voltage* is assumed to be constant, we have[84]

$$\bar{F}_e = \bar{\nabla} w_{e,\text{stored}} = \hat{a}_d \frac{\partial}{\partial d}\left(\frac{V_0^2 \varepsilon A}{2d}\right) = -\hat{a}_d \frac{V_0^2 \varepsilon A}{2d^2}. \tag{3.199}$$

The equality of these force expressions can be shown by using $Q_0 = C_0 V_0$ as

$$\bar{F}_e = -\hat{a}_d \frac{Q_0^2}{2\varepsilon A} = -\hat{a}_d \frac{V_0^2 C_0^2}{2\varepsilon A} = -\hat{a}_d \frac{V_0^2 \varepsilon A}{2d^2}. \tag{3.200}$$

The negative value for $V_0 > 0$ and $Q_0 > 0$ indicates that the force *tries to reduce* the value of d: i.e. the plates attract each other.

[84] Considering these expressions, a confusing question is about how the force is related to the geometric properties. According to one expression, the electric force should increase if the area of each plate (A) increases, while it should decrease using the other expression. In fact, both of them are correct. If the charges on the plates are constant ($\pm Q_0$), enlarging the plates leads to smaller electric charge density (as the total charges will be distributed on larger surfaces), leading to a reduced electric force between the plates. If the voltage between the plates is fixed (V_0), however, enlarging the plates means collecting more charges on the plates. This also means there should be an external source to keep the voltage the same while the capacitor is enlarged.

Another interesting example involves a dielectric slab inserted partially into a parallel-plate capacitor (Figure 3.39). We again consider the distance between the plates as d, while the area of each plate is $A = a \times b$. The dielectric slab with a permittivity of ε is inserted by an amount of x, while the rest of the capacitor is a vacuum. The electric potential energy stored in the system can be found as[85]

$$w_{e,\text{stored}} = \frac{1}{2}\left(\frac{\varepsilon x b}{d}\right)V_0^2 + \frac{1}{2}\left(\frac{\varepsilon_0(a-x)b}{d}\right)V_0^2. \tag{3.201}$$

Then, assuming constant V_0, we have[86]

$$\bar{F}_e = \bar{\nabla} w_{e,\text{stored}} = \hat{a}_x\frac{1}{2}\left(\frac{\varepsilon b}{d}\right)V_0^2 - \hat{a}_x\frac{1}{2}\left(\frac{\varepsilon_0 b}{d}\right)V_0^2 \tag{3.202}$$

$$= \hat{a}_x\frac{b}{2d}V_0^2(\varepsilon - \varepsilon_0). \tag{3.203}$$

The positive value (in the direction of $\hat{a}_x$, assuming that $\varepsilon > \varepsilon_0$) indicates that the force tries to increase x, hence the dielectric slab is attracted into the capacitor.

[85] This can be found by considering the dielectric and vacuum parts separately as different capacitors and adding the energy stored in them.

[86] The alternative solution (using electric charges) is more difficult since the charges are not distributed uniformly on the plates.

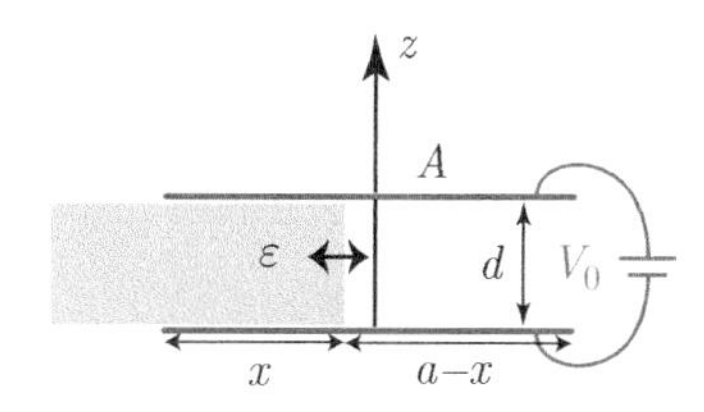

Figure 3.39 A dielectric slab that is partially inserted into a parallel-plate capacitor. The slab is allowed to move horizontally.

3.6.6 Examples

Example 33: Consider a static spherical electric charge distribution with a constant volume electric charge density $q_v(\bar{r}) = q_0$ (C/m^3) and radius a in a vacuum (Figure 3.40). Find the stored electric potential energy in the system.

Solution: Given the charge density, the electric field intensity can be found via Gauss' law:

$$\bar{E}(\bar{r}) = \hat{a}_R\frac{q_0}{3\varepsilon_0 R^2}\begin{cases} a^3 & (R > a) \\ R^3 & (R < a). \end{cases} \tag{3.204}$$

Then the electric scalar potential can be found via integration as

$$\Phi(\bar{r}) = -\int_\infty^R \frac{q_0 a^3}{3\varepsilon_0(R')^2}dR' = \frac{q_0 a^3}{3\varepsilon_0 R} \qquad (R > a) \tag{3.205}$$

and[87]

$$\Phi(\bar{r}) = \frac{q_0 a^3}{3\varepsilon_0 a} - \int_a^R \frac{q_0 R'}{3\varepsilon_0}dR' = \frac{q_0 a^3}{3\varepsilon_0 a} - \left[\frac{q_0(R')^2}{6\varepsilon_0}\right]_a^R \tag{3.206}$$

$$= \frac{q_0 a^2}{2\varepsilon_0} - \frac{q_0 R^2}{6\varepsilon_0} = \frac{q_0}{2\varepsilon_0}\left(a^2 - \frac{R^2}{3}\right) \qquad (R < a). \tag{3.207}$$

[87] Note that, to find the electric scalar potential for $R < a$, we first consider an integral from infinity to the surface of the charge distribution using the electric field intensity defined for $R > a$. This is the electric scalar potential at the surface of the sphere: i.e. the amount of energy to bring a test charge from infinity to the surface at $R = a$. Then a further penetration inside the charge distribution is a second integration from the surface to the position with $R < a$.

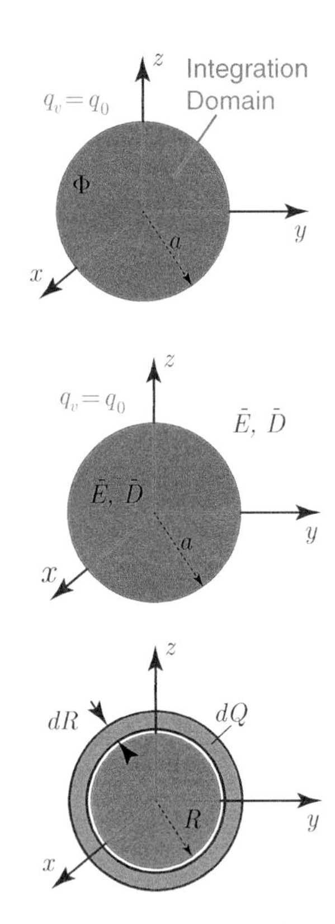

Figure 3.40 Three different ways to find the electric potential energy stored by a static spherical electric charge distribution: (i) the electric scalar potential inside the distribution is multiplied by the density and integrated inside the sphere; (ii) the multiplication of the electric field intensity and the electric flux density is integrated everywhere; (iii) the charge distribution is assumed to be formed by the accumulation of charges brought from infinity.

Then the stored electric potential energy can be found via integration (Figure 3.40) as

$$w_{e,\text{stored}} = \frac{1}{2}\int_V q_v(\bar{r})\Phi(\bar{r})dv = \frac{1}{2}\int_{R<a} q_0 \left[\frac{q_0 a^2}{2\varepsilon_0} - \frac{q_0 R^2}{6\varepsilon_0}\right] dv \tag{3.208}$$

$$= \frac{1}{2}\int_0^a \int_0^{2\pi} \int_0^{\pi} q_0 \frac{q_0}{2\varepsilon_0}\left(a^2 - \frac{R^2}{3}\right) R^2 \sin\theta d\theta d\phi dR \tag{3.209}$$

$$= \frac{q_0^2}{4\varepsilon_0} 4\pi \left[\frac{a^2 R^3}{3} - \frac{R^5}{15}\right]_0^a = \frac{4\pi q_0^2 a^5}{15\varepsilon_0}. \tag{3.210}$$

Alternatively, using the expression for the electric field intensity (Figure 3.40), one can find

$$w_{e,\text{stored}} = \frac{1}{2}\int_\infty \bar{E}(\bar{r}) \cdot \bar{D}(\bar{r})dv = \frac{\varepsilon_0}{2}\int_\infty |\bar{E}(\bar{r})|^2 dv \tag{3.211}$$

$$= \frac{\varepsilon_0}{2}\int_0^a \int_0^{2\pi} \int_0^{\pi} \frac{q_0^2 R^2}{9\varepsilon_0^2} R^2 \sin\theta d\theta d\phi dR \tag{3.212}$$

$$+ \frac{\varepsilon_0}{2}\int_a^\infty \int_0^{2\pi} \int_0^{\pi} \frac{q_0^2 a^6}{9\varepsilon_0^2 R^4} R^2 \sin\theta d\theta d\phi dR \tag{3.213}$$

$$= \frac{\varepsilon_0}{2} 4\pi \frac{q_0^2}{9\varepsilon_0^2}\left[\frac{R^5}{5}\right]_0^a + \frac{\varepsilon_0}{2} 4\pi \frac{q_0^2 a^6}{9\varepsilon_0^2}\left[-\frac{1}{R}\right]_a^\infty \tag{3.214}$$

$$= \frac{2\pi q_0^2 a^5}{45\varepsilon_0} + \frac{2\pi q_0^2 a^5}{9\varepsilon_0} = \frac{4\pi q_0^2 a^5}{15\varepsilon_0}, \tag{3.215}$$

which is the same as the expression above.

There is even another way to find the stored electric potential energy in this system (Figure 3.40). In this case, the spherical charge distribution can be assumed to be built by adding layers of differential thickness dR. When the radius is R, the potential created by the charge distribution on its surface can be found as

$$\Phi(R) = \frac{Q(R)}{4\pi\varepsilon_0 R} = \frac{4}{3}\pi R^3 q_0 \frac{1}{4\pi\varepsilon_0 R} = \frac{q_0 R^2}{3\varepsilon_0}. \tag{3.216}$$

Adding a new layer with a total electric charge of $dQ = 4\pi R^2 q_0 dR$, we have

$$dw_{e,\text{stored}}(R) = \frac{q_0 R^2}{3\varepsilon_0} 4\pi R^2 q_0 dR = \frac{4\pi q_0^2}{3\varepsilon_0} R^4 dR \tag{3.217}$$

as the electric potential energy added into the system. Then, integrating this expression from 0 to a (considering the whole process of building the final distribution), we arrive at

$$w_{e,\text{stored}} = \frac{4\pi q_0^2}{3\varepsilon_0}\int_0^a R^4 dR = \frac{4\pi q_0^2 a^5}{15\varepsilon_0}, \tag{3.218}$$

which is consistent with the earlier results.

Example 34: The electric scalar potential inside an $a \times a \times a$ parallel-plate structure filled with a dielectric material (constant ε permittivity) is given as $\Phi(\bar{r}) = V_0 z^2/a^2$ for $z \in [0, a]$ (Figure 3.41). The electric field intensity outside the structure (a vacuum) is zero. Find the force on the top plate.

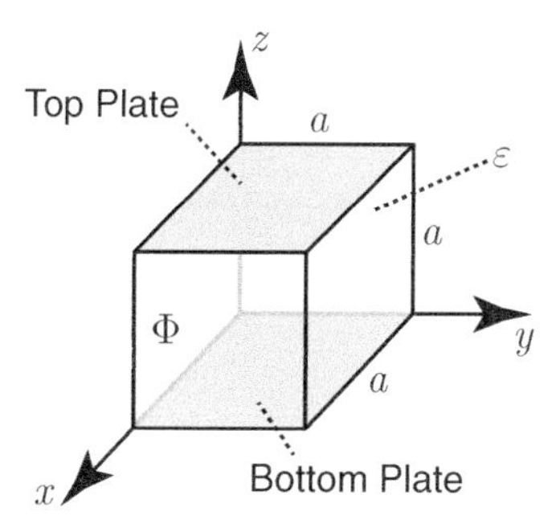

Figure 3.41 An $a \times a \times a$ parallel-plate structure with a given distribution of the electric scalar potential between the top and bottom plates.

Solution: First, we can find the electric field intensity between the plates as

$$\bar{E}(\bar{r}) = -\bar{\nabla}\Phi(\bar{r}) = -\hat{a}_z V_0 \frac{2z}{a^2}. \tag{3.219}$$

Then, considering $\bar{D}(\bar{r}) = \varepsilon\bar{E}(\bar{r})$, the volume electric charge density can be found as[88]

$$q_v(\bar{r}) = \bar{\nabla}\cdot\bar{D}(\bar{r}) = -\frac{2\varepsilon V_0}{a^2}. \tag{3.220}$$

The electric potential energy between the plates can be found as

$$w_{e,\text{stored}}^{\text{between}} = \frac{1}{2}\int_V q_v(\bar{r})\Phi(\bar{r})dv = \frac{a^2}{2}\int_0^a\left(-\frac{2\varepsilon V_0}{a^2}\right)\frac{V_0 z^2}{a^2}dz \tag{3.221}$$

$$= -\frac{\varepsilon V_0^2}{a^2}\frac{a^3}{3} = -\frac{\varepsilon V_0^2 a}{3}. \tag{3.222}$$

But this is not the total energy stored in the system since the electric charges on the plates have not been considered. Specifically, the electric scalar potential on the top plate is V_0, while the total accumulated charge on the plate can be found as[89,90]

$$Q_{\text{top}} = -a^2\hat{a}_z\cdot\bar{D}(\bar{r}) = 2\varepsilon V_0 a. \tag{3.223}$$

Then the contribution by the top plate is

$$w_{e,\text{stored}}^{\text{top}} = \frac{1}{2}V_0 Q_{\text{top}} = \varepsilon V_0^2 a, \tag{3.224}$$

leading to[91]

[88] The nonzero value, despite the constant permittivity, indicates that the dielectric is charged a priori.

[89] We use the boundary condition for the normal electric flux density while the electric flux density is zero outside the structure.

[90] We do not need to consider the bottom plate since the corresponding electric scalar potential is zero. In addition, there is no charge accumulation on the bottom plate so that, even when the electric scalar potential was not zero on it, its contribution to the electric potential energy would be zero.

[91] Note that the stored energy must be positive.

$$w_{e,\text{stored}} = w_{e,\text{stored}}^{\text{between}} + w_{e,\text{stored}}^{\text{top}} = 2\varepsilon V_0^2 a/3. \tag{3.225}$$

Alternatively, following the field approach, we simply have

$$w_{e,\text{stored}} = \frac{\varepsilon}{2}\int_{\infty} |\bar{E}(\bar{r})|^2 dv = \frac{\varepsilon}{2}\int_{\text{between}} |\bar{E}(\bar{r})|^2 dv \tag{3.226}$$

$$= \frac{\varepsilon}{2} a^2 \frac{V_0^2}{a^4}\int_0^a 4z^2 dz = 2\varepsilon V_0^2 a/3. \tag{3.227}$$

Finally, considering a as the virtual displacement variable, the force acting on the top plate can be found as

$$\bar{F}_e = \bar{\nabla} w_{e,\text{stored}} = \hat{a}_a 2\varepsilon_0 V_0^2/3, \tag{3.228}$$

indicating a repulsive force (a force that tries to increase a).

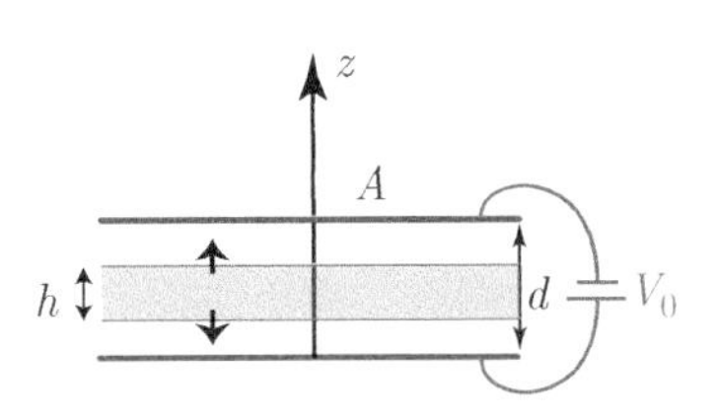

Figure 3.42 A dielectric slab that is completely inserted into a parallel-plate capacitor. A force is applied on the slab to stretch it.

Example 35: Consider a dielectric slab that is completely inserted into a parallel-plate capacitor (Figure 3.42). The distance between the plates is d, while the area of each plate is A. The slab has a thickness of h, and the rest of the capacitor is filled with a vacuum. Find the force that tries to stretch the slab.

Solution: First, we can find the electric potential energy stored in the capacitor. We can write the electric flux density as

$$\bar{D}(\bar{r}) = \hat{a}_z Q_0/A, \tag{3.229}$$

assuming that $\pm Q_0$ charges are accumulated on the plates, while the bottom plate has a higher electric scalar potential. Then the electric field intensity can be written as

$$\bar{E}(\bar{r}) = \bar{D}(\bar{r})/\varepsilon(\bar{r}) = \hat{a}_z \frac{Q_0}{A}\begin{cases} 1/\varepsilon & \text{(in slab)} \\ 1/\varepsilon_0 & \text{(otherwise).}\end{cases} \tag{3.230}$$

The stored electric potential energy can be found as[92]

$$w_{e,\text{stored}} = \frac{1}{2}\int_{\text{between plates}} \bar{E}(\bar{r})\cdot\bar{D}(\bar{r})dv = \frac{Q_0^2}{2A^2\varepsilon}hA + \frac{Q_0^2}{2A^2\varepsilon_0}(d-h)A. \tag{3.231}$$

Then the electric force acting on the slab can be found by considering h as the changing variable:

$$\bar{F}_e = -\bar{\nabla} w_{e,\text{stored}} = -\hat{a}_h \frac{Q_0^2}{2A\varepsilon} + \hat{a}_h \frac{Q_0^2}{2A\varepsilon_0} = -\hat{a}_h \frac{Q_0^2}{2A}(1/\varepsilon - 1/\varepsilon_0). \tag{3.232}$$

[92] Using $C_0 = Q_0/V_0$, this expression can also be written in terms of the potential difference between the plates, while the capacitance value in terms of geometric parameters needs to be found.

The positive value (for $\varepsilon > \varepsilon_0$) indicates that the force tries to stretch the dielectric (tries to increase h).

3.6.7 Poisson's Equation and Laplace's Equation

Electric scalar potential is useful since it provides alternative and sometimes easier solutions of static and dynamic problems. In a static case, if the electric scalar potential is found (as shown below), then the electric field intensity can be found without directly employing Gauss' law. To derive an equation for the electric scalar potential in a static case, we repeat that the electric field intensity can be written as $\bar{E}(\bar{r}) = -\bar{\nabla}\Phi(\bar{r})$, where $\Phi(\bar{r})$ is the electric scalar potential. Then, evaluating the divergence of both sides, we have

$$\bar{\nabla} \cdot \bar{E}(\bar{r}) = -\bar{\nabla} \cdot \bar{\nabla}\Phi(\bar{r}) = -\bar{\nabla}^2\Phi(\bar{r}) = -\Delta\Phi(\bar{r}), \tag{3.233}$$

where $\Delta = \bar{\nabla}^2 = \bar{\nabla} \cdot \bar{\nabla}$ is called the Laplace operator.[93] Considering Gauss' law, we also have

$$\bar{\nabla} \cdot \bar{D}(\bar{r}) = \bar{\nabla}\varepsilon(\bar{r}) \cdot \bar{E}(\bar{r}) + \varepsilon(\bar{r})\bar{\nabla} \cdot \bar{E}(\bar{r}) = q_v(\bar{r}), \tag{3.234}$$

where $\bar{\nabla}\varepsilon(\bar{r})$ accounts for inhomogeneity, as also discussed in Section 3.6.4. Then we have

$$\bar{\nabla}\varepsilon(\bar{r}) \cdot \bar{E}(\bar{r}) - \varepsilon(\bar{r})\bar{\nabla}^2\Phi(\bar{r}) = q_v(\bar{r}) \tag{3.235}$$

or

$$\bar{\nabla}^2\Phi(\bar{r}) = -\frac{q_v(\bar{r})}{\varepsilon(\bar{r})} + \frac{\bar{\nabla}\varepsilon(\bar{r}) \cdot \bar{E}(\bar{r})}{\varepsilon(\bar{r})}, \tag{3.236}$$

where $\bar{\nabla}\varepsilon(\bar{r}) \cdot \bar{E}(\bar{r})$ acts like a volume electric charge density. Equation (3.236) can be categorized as a Poisson's equation: i.e. an equation involving Laplace of a scalar field with a nonzero right-hand side.

In general, solutions of Poisson's equation are well studied in mathematics. In the context of electrostatics, we also consider different cases, where the electric scalar potential needs to be found for given conditions. First, in a homogeneous medium,[94] we have

$$\bar{\nabla}^2\Phi(\bar{r}) = -\frac{q_v(\bar{r})}{\varepsilon}. \tag{3.237}$$

Therefore, if we know the electric charge density, we can find the electric scalar potential.[95] In

[93] The Laplace operator is very important in electromagnetics as it appears in numerous equations. It is basically the divergence of the gradient of a scalar field. At the same time, it can be defined as a standalone operator. For example, in a Cartesian coordinate system, we have

$$\bar{\nabla}^2 f(\bar{r},t) = \frac{\partial^2 f(\bar{r},t)}{\partial x^2} + \frac{\partial^2 f(\bar{r},t)}{\partial y^2} + \frac{\partial^2 f(\bar{r},t)}{\partial z^2}$$

In a cylindrical coordinate system, the expression becomes

$$\bar{\nabla}^2 f(\bar{r},t) = \frac{1}{\rho}\frac{\partial}{\partial \rho}\left(\rho\frac{\partial f(\bar{r},t)}{\partial \rho}\right) + \frac{1}{\rho^2}\frac{\partial^2 f(\bar{r},t)}{\partial \phi^2} + \frac{\partial^2 f(\bar{r},t)}{\partial z^2}$$

Finally, if a Laplace operation needs to be evaluated in a spherical coordinate system, we have

$$\bar{\nabla}^2 f(\bar{r},t) = \frac{1}{R^2}\frac{\partial}{\partial R}\left(R^2\frac{\partial f(\bar{r},t)}{\partial R}\right) + \frac{1}{R^2\sin\theta}\frac{\partial}{\partial \theta}\left(\sin\theta\frac{\partial f(\bar{r},t)}{\partial \theta}\right) + \frac{1}{R^2\sin^2\theta}\frac{\partial^2 f(\bar{r},t)}{\partial \phi^2}.$$

[94] Having $\bar{\nabla}\varepsilon(\bar{r},t) \perp \bar{E}(\bar{r},t)$ also leads to the same equation. Specifically, if the change in the permittivity is perpendicular to the electric field intensity at the observation point $\bar{r}$, then the extra term $\bar{\nabla}\varepsilon(\bar{r}) \cdot \bar{E}(\bar{r})/\varepsilon(\bar{r})$ again vanishes.

[95] For some cases, this may be easier than using Gauss' law directly.

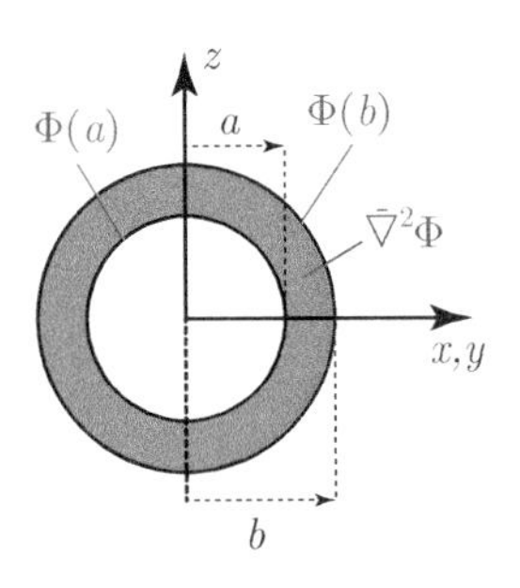

Figure 3.43 If there is a spherical symmetry, Laplace's equation is best solved in a spherical coordinate system, leading to $\Phi(R) = -c_1/R + c_2$, where c_1 and c_2 needs to be found via the boundary conditions $\Phi(a)$ and $\Phi(b)$. It is possible that $a \to 0$ and/or $b \to \infty$, while Φ is then required at the origin and/or at infinity.

a vacuum, we further have

$$\bar{\nabla}^2\Phi(\bar{r}) = -\frac{q_v(\bar{r})}{\varepsilon_0}. \tag{3.238}$$

If there is no electric charge density at $\bar{r}$, we arrive at Laplace's equation (Poisson's equation with a zero right-hand side)

$$\bar{\nabla}^2\Phi(\bar{r}) = 0, \tag{3.239}$$

which is another fundamental equation.

It can be observed that, with or without the volume electric charge density $q_v(\bar{r})$, we obtain equations to find the electric scalar potential. The usage may depend on the context, but we can also study some general solutions. First, we consider Laplace's equation in a spherical coordinate system (Figure 3.43). Assuming that the electric scalar potential only depends on R, we have

$$\bar{\nabla}^2\Phi(\bar{r}) = 0 \longrightarrow \frac{1}{R^2}\frac{\partial}{\partial R}\left(R^2\frac{\partial\Phi(R)}{\partial R}\right) = 0, \tag{3.240}$$

using the definition of Laplace operator. Then we derive[96]

$$\frac{1}{R^2}\frac{\partial}{\partial R}\left(R^2\frac{\partial\Phi(R)}{\partial R}\right) = 0 \longrightarrow R^2\frac{\partial\Phi(R)}{\partial R} = c_1 \longrightarrow \frac{\partial\Phi(R)}{\partial R} = \frac{c_1}{R^2} \tag{3.241}$$

$$\longrightarrow \Phi(R) = -\frac{c_1}{R} + c_2, \tag{3.242}$$

where c_1 and c_2 are constants (with appropriate units). If $\Phi(R)$ decays at infinity (often selected as the reference for the electric scalar potential), we further have $c_2 = 0$ and

$$\Phi(\bar{r}) = \Phi(R) = -\frac{c_1}{R}, \tag{3.243}$$

which satisfy Laplace's equation for any c_1.

Next, we can consider Poisson's equation for an impulse function: i.e. $q_v(\bar{r}) = Q_0\delta(R)$, representing a point source (a static point electric charge) at the origin.[97] Considering spherical symmetry and a vacuum, we need to solve

$$\bar{\nabla}^2\Phi(R) = -Q_0\delta(R)/\varepsilon_0, \tag{3.244}$$

where the solution should be in the form of[98] $\Phi(R) = -c/R$, where c is a constant (with unit V×m). To find constant c, letting $R \to 0$ directly is not useful due to the singularity. A proper

[96] If the derivative of a function is zero, it must be constant for the variable of the derivative.

[97] The unit of Q_0 is C, while $Q_0\delta(R)$ has units of C/m³. More formally, to represent the impulse function at the origin, the Dirac delta function should be used as $\delta(\bar{r})$, while we simply write it as $\delta(R)$ considering the spherical symmetry.

[98] This comes from the solution of Laplace's equation for an R-dependent scalar field in a spherical coordinate system. Note that the homogeneous equation is still *valid* if the observation point is not exactly on the source.

way is to write Laplace operator as $\bar{\nabla}^2 = \bar{\nabla} \cdot \bar{\nabla}$ and study a spherical volume centered at the origin (Figure 3.44). Then we have

$$\int_V \bar{\nabla}^2 \Phi(R) dv = \int_V \bar{\nabla} \cdot \bar{\nabla}\Phi(R) dv = \int_V [-Q_0 \delta(R)/\varepsilon_0] dv = -Q_0/\varepsilon_0, \tag{3.245}$$

where the integral of the impulse function is unity. The divergence theorem lets us rewrite the equality as

$$\oint_S \bar{\nabla}\Phi(R) \cdot \overline{ds} = -Q_0/\varepsilon_0. \tag{3.246}$$

In addition, inserting $\Phi(R) = -c/R$ or $\bar{\nabla}\Phi(R) = c/R^2$ we derive

$$\oint_S \bar{\nabla}\Phi(R) \cdot \overline{ds} = 4\pi c = -Q_0/\varepsilon_0, \tag{3.247}$$

leading to $c = -Q_0/(4\pi\varepsilon_0)$. Therefore, the solution of Poisson's equation for a point source can be written as

$$\Phi(R) = \frac{Q_0}{4\pi\varepsilon_0 R}. \tag{3.248}$$

The general solution of Poisson's equation for an impulse function provides insight about the electric scalar potential for static electric charge distributions. First, considering a static point electric charge Q_0 at the origin, the electric scalar potential can be rewritten as

$$\Phi(\bar{r}) = \frac{Q_0}{4\pi\varepsilon_0|\bar{r}|}. \tag{3.249}$$

This equation is the same as the one in Eq. (3.145), where the electric scalar potential is found from the electric field intensity of a static point electric charge. If the charge is located at $\bar{r}'$, the electric scalar potential can be written as (Figure 3.45)

$$\Phi(\bar{r}) = \frac{Q_0}{4\pi\varepsilon_0|\bar{r} - \bar{r}'|}. \tag{3.250}$$

In addition, for a given charge distribution $q_v(\bar{r})$, the superposition principle can be used to derive the electric scalar potential[99] as (Figure 3.45)

$$\Phi(\bar{r}) = \frac{1}{4\pi\varepsilon_0} \int_V \frac{q_v(\bar{r}')}{|\bar{r} - \bar{r}'|} d\bar{r}'. \tag{3.251}$$

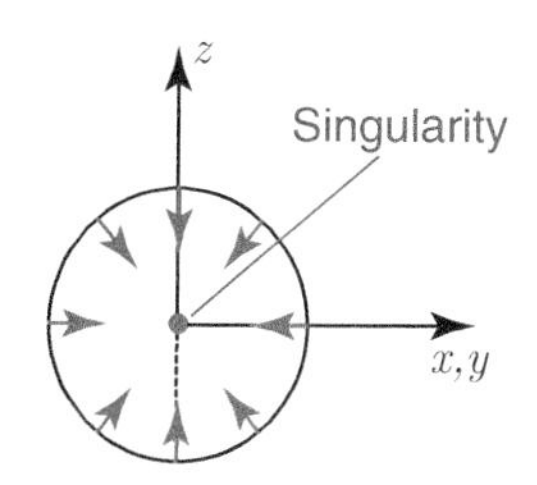

Figure 3.44 It can be difficult to find the constants of the electric scalar potential, if there is a singularity at the origin when solving Laplace's equation. Then a volume integral can be considered and converted into a surface integral that can be evaluated without problems.

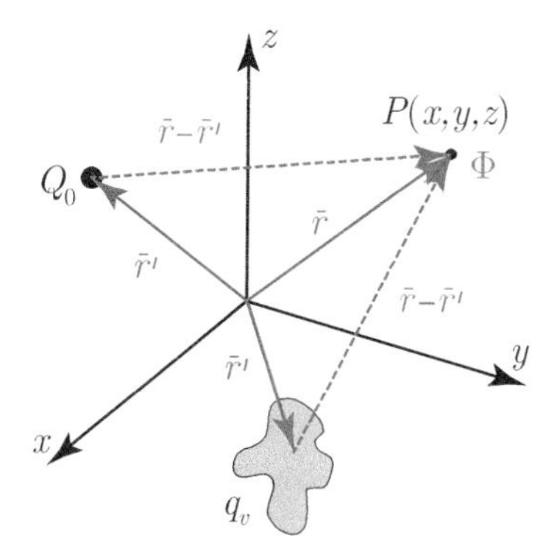

Figure 3.45 Electric scalar potential created by a point charge or a distribution of electric charges. For the charge distribution, the total electric scalar potential at the observation point can be found by integration that accounts for all possible source locations.

[99]Recall that the effects of different electric charges can be added (superposed) directly since we consider *static* cases.

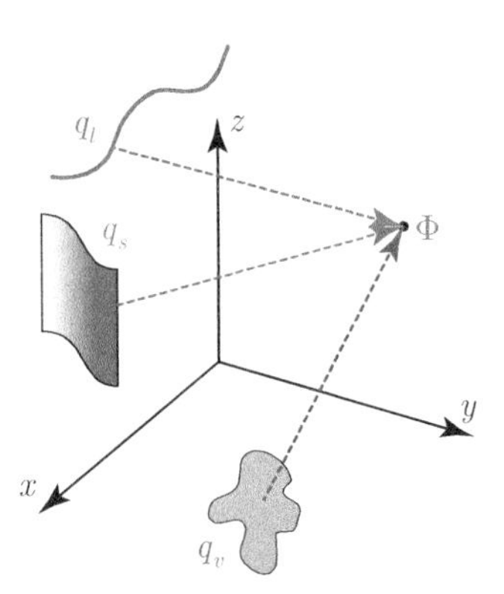

Figure 3.46 Electric scalar potential created by volume, surface, and line charge distributions can be found via superposition integrals. Note that, as opposed to the electric field intensity, the electric scalar potential values are directly added without any vector operation.

Obviously, as in the case of the electric field intensity, the superposition principle works for any charge distribution (Figure 3.46):

$$\Phi(\bar{r}) = \frac{1}{4\pi\varepsilon_0}\int_S \frac{q_s(\bar{r}')}{|\bar{r}-\bar{r}'|}d\bar{r}' \tag{3.252}$$

for a surface electric charge density and

$$\Phi(\bar{r}) = \frac{1}{4\pi\varepsilon_0}\int_C \frac{q_l(\bar{r}')}{|\bar{r}-\bar{r}'|}d\bar{r}' \tag{3.253}$$

for a line electric charge density, respectively.

At this stage, we can perform a consistency check by finding the electric field intensity from a general expression for the electric scalar potential. For a volume electric charge distribution, we have[100,101]

$$\bar{E}(\bar{r}) = -\bar{\nabla}\Phi(\bar{r}) = -\frac{1}{4\pi\varepsilon_0}\bar{\nabla}\int_V \frac{q_v(\bar{r}')}{|\bar{r}-\bar{r}'|}d\bar{r}' \tag{3.254}$$

$$= -\frac{1}{4\pi\varepsilon_0}\int_V \bar{\nabla}\left(\frac{q_v(\bar{r}')}{|\bar{r}-\bar{r}'|}\right)d\bar{r}' \tag{3.255}$$

$$= -\frac{1}{4\pi\varepsilon_0}\int_V [\bar{\nabla}q_v(\bar{r}')]\frac{1}{|\bar{r}-\bar{r}'|}d\bar{r}' - \frac{1}{4\pi\varepsilon_0}\int_V q_v(\bar{r}')\bar{\nabla}\left(\frac{1}{|\bar{r}-\bar{r}'|}\right)d\bar{r}' \tag{3.256}$$

$$= \frac{1}{4\pi\varepsilon_0}\int_V q_v(\bar{r}')\frac{(\bar{r}-\bar{r}')}{|\bar{r}-\bar{r}'|^3}d\bar{r}'. \tag{3.257}$$

[100] The del operator is put directly inside the integral since the integral is using primed coordinates ($\bar{r}'$) while the derivative (del operator) is in unprimed coordinates ($\bar{r}$).

[101] We have $\bar{\nabla}q_v(\bar{r}') = 0$, since the operator is in unprimed coordinates while $q_v(\bar{r}')$ is in primed coordinates.

The final expression is perfectly consistent with the one in Eq. (1.183). Note that, in a static case, electric fields are created only by electric charges.

The source integrals in Eqs. (3.251), (3.252), and (3.253) can be used directly to find the electric scalar potential due to given electric charge distributions. As an example, we consider a finite line electric charge density $q_l(\bar{r}) = q_0$ (C/m) of length $2a$ located at $x = a$ and $z = 0$, from $y = -a$ to $y = a$ (Figure 3.47). To find the electric scalar potential at $\bar{r} = \hat{a}_z a$, we have

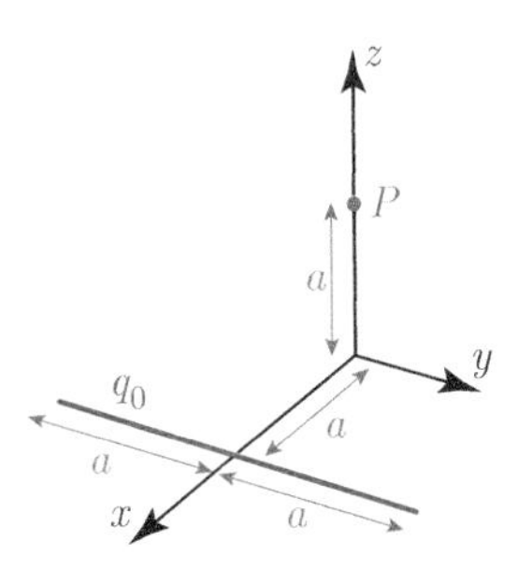

Figure 3.47 A finite static line electric charge density.

$$\Phi(0,0,a) = \frac{q_0}{4\pi\varepsilon_0}\int_C \frac{1}{|\bar{r}-\bar{r}'|}d\bar{r}' = \frac{q_0}{4\pi\varepsilon_0}\int_{-a}^{a}\frac{1}{\sqrt{2a^2+(y')^2}}dy' \tag{3.258}$$

$$= \frac{1}{4\pi\varepsilon_0}\left[\ln\left(y'+\sqrt{2a^2+(y')^2}\right)\right]_{-a}^{a} \tag{3.259}$$

leading to

$$\Phi(0,0,a) = \frac{1}{4\pi\varepsilon_0}\left[\ln\left(a+\sqrt{3a^2}\right) - \ln\left(-a+\sqrt{3a^2}\right)\right] \tag{3.260}$$

$$= \frac{1}{4\pi\varepsilon_0}\ln\left(\frac{a+\sqrt{3a^2}}{-a+\sqrt{3a^2}}\right) = \frac{1}{4\pi\varepsilon_0}\ln\left(\frac{1+\sqrt{3}}{-1+\sqrt{3}}\right). \tag{3.261}$$

The final expression is free of a since both the size of the electric charge distribution and the distance of the observation point from the distribution are proportional to a.

Although there are some ambiguities due to the actual behaviors of electrical particles (electrons and protons), the electric scalar potential can be related to the electric voltage that is a measurable quantity in electrical circuits. For example, we consider a simple capacitor involving parallel plates (Figure 3.48), as considered several times so far (e.g. in Section 1.6.1). A voltage source with a value of V_0 is connected to its terminals. Assuming electric scalar potential values $\Phi(z=0)=0$ and $\Phi(z=d)=V_0$ at the terminals, the values between the plates must obey

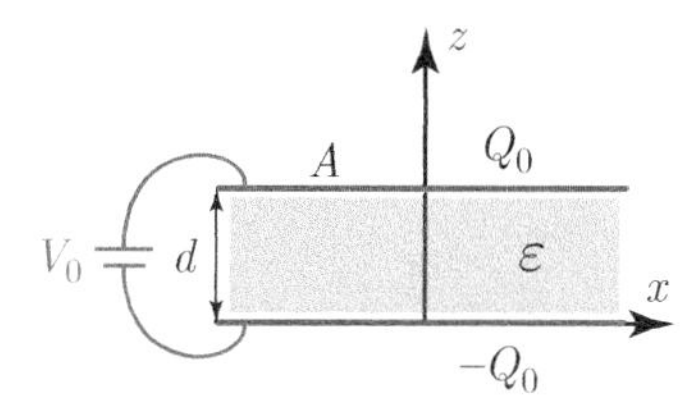

Figure 3.48 A voltage source connected to a parallel-plate capacitor.

$$\bar{\nabla}^2\Phi(\bar{r}) = 0 \tag{3.262}$$

if there is no volume electric charge density for $0 < z < d$. Then we have

$$\bar{\nabla}^2\Phi(\bar{r}) = \frac{\partial^2\Phi(\bar{r})}{\partial x^2} + \frac{\partial^2\Phi(\bar{r})}{\partial y^2} + \frac{\partial^2\Phi(\bar{r})}{\partial z^2} = 0. \tag{3.263}$$

If the electric scalar potential only depends on z, we further have[102]

$$\frac{\partial^2\Phi(\bar{r})}{\partial z^2} = 0. \tag{3.264}$$

Then the electric scalar potential can be written as

$$\frac{\partial\Phi(\bar{r})}{\partial z} = c_1 \longrightarrow \Phi(\bar{r}) = c_1 z + c_2, \tag{3.265}$$

where c_1 and c_2 are constant (with appropriate units). If we assign $\Phi(z=0)=0$, we must have $c_2 = 0$ (Figure 3.49). Further using $\Phi(z=d)=V_0$ leads to $c_1 = V_0/d$ and

$$\Phi(\bar{r}) = \frac{V_0 z}{d}. \tag{3.266}$$

[102] This is equivalent to assuming that there is no electric field (not even a fringing field) outside the capacitor.

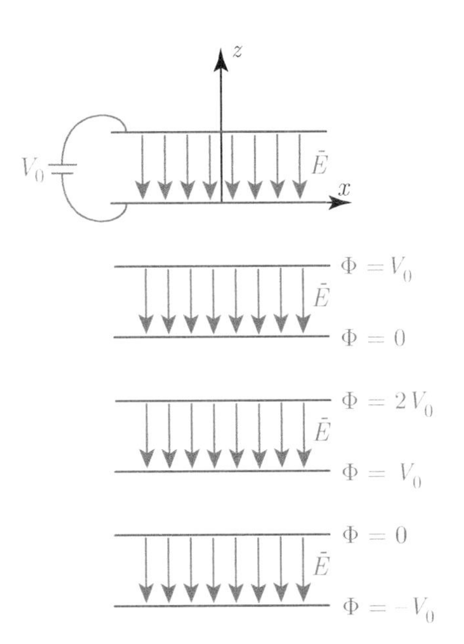

Figure 3.49 If the electric scalar potential changes only in one direction (z) in a Cartesian coordinate system, then it must change linearly according to Laplace's equation. Since the electric potential is a relative quantity that involves the difference of two values, one can assign any electric scalar potential values to the boundaries in accordance with the given voltage. The electric field intensity is not affected by the selection.

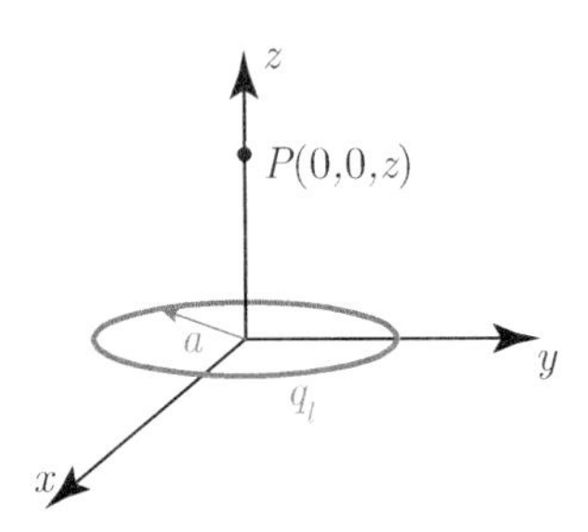

Figure 3.50 A static line electric charge distribution with a circular shape.

The corresponding electric field intensity can be found as

$$\bar{E}(\bar{r}) = -\bar{\nabla}\Phi(\bar{r}) = -\hat{a}_z\frac{V_0}{d} \qquad (0 \leq z \leq d). \tag{3.267}$$

Note that if this expression is the same as the one in Eq. (1.194), we must have

$$-\hat{a}_z\frac{V_0}{d} = -\hat{a}_z\frac{Q_0}{A\varepsilon}, \tag{3.268}$$

where Q_0 is the total charge on the top plate (while the bottom plate has $-Q_0$), A is the area of each plate, and ε is the permittivity of the dielectric material between the plates. This leads to

$$C_0 = \frac{Q_0}{V_0} = \frac{A\varepsilon}{d}, \tag{3.269}$$

which is simply the capacitance of the capacitor. Refer to Section 6.7 for detailed capacitance calculations.

3.6.8 Examples

Example 36: Consider a static line electric charge distribution $q_l(\bar{r}) = q_0$ (C/m) with a circular shape of radius a located on the x-y plane in a vacuum (Figure 3.50). Find the electric scalar potential at $\bar{r} = \hat{a}_z z$ for $z > 0$ (on the positive z axis).

Solution: Using the definition of the electric scalar potential in terms of electric charge density, we have

$$\Phi(0,0,z) = \frac{1}{4\pi\varepsilon_0}\int_C \frac{q_l(\bar{r})}{|\bar{r}-\bar{r}'|}d\bar{r}' = \frac{q_0}{4\pi\varepsilon_0}\int_0^{2\pi}\frac{1}{\sqrt{a^2+z^2}}ad\phi' \tag{3.270}$$

$$= \frac{q_0 2\pi a}{4\pi\varepsilon_0}\frac{1}{\sqrt{a^2+z^2}}. \tag{3.271}$$

Considering that $q_0 2\pi a = Q_0$ is the total charge, we can write the potential also as

$$\Phi(0,0,z) = \frac{Q_0}{4\pi\varepsilon_0}\frac{1}{\sqrt{a^2+z^2}}. \tag{3.272}$$

At this stage, one may ask if we can find the electric field intensity $\bar{E}(\bar{r})$ from $\Phi(\bar{r})$ using $\bar{E}(\bar{r}) = -\bar{\nabla}\Phi(\bar{r})$. In general, the gradient should be used only when $\Phi(\bar{r})$ is known as a full expression in terms of the coordinate variables. Therefore, when it is known for a specific case – e.g. just on the z axis – evaluating the gradient can lead to incorrect expressions.[103]

[103]While this is not suggested, $\bar{E}(\bar{r}) = -\bar{\nabla}\Phi(\bar{r})$ may still hold for partially known $\Phi(\bar{r})$ in some symmetric cases, including this example, for which $\bar{E}(0,0,z) = \bar{\nabla}\Phi(0,0,z)$ gives the correct expression.

3.6.9 Finding Electric Scalar Potential from Electric Field Intensity

In general, the electric scalar potential is defined to facilitate solutions to Maxwell's equations. Specifically, if potentials are known, electric and magnetic fields can be derived from them. In an electrostatic case, the electric scalar potential can be used to find the electric field intensity via $\bar{E}(\bar{r}) = -\bar{\nabla}\Phi(\bar{r})$ since it is the only type of potential contributing to $\bar{E}(\bar{r})$. To find the electric scalar potential, one may use source integration: e.g. Eq. (3.251) for a volume electric charge density or solve the associated boundary-value problem via Poisson's equation or Laplace's equation (see Section 3.6.11). However, in some cases, the electric scalar potential may be needed while the electric field intensity is already known or easy to find. In such a case, we can use the line integral in Eq. (3.139). Since the value of the integral is independent of the path, it is sufficient to know the electric field intensity on *any* path from the reference point to the observation location where the electric scalar potential needs to be found.[104]

[104]Remember that this is valid only for static cases.

As an example, we can revisit the static line electric charge distribution $q_l(\bar{r}) = q_0$ (C/m) with a circular shape (Figure 3.50). Using Coulomb's law, the electric field intensity on the z axis can be found as Eq. (1.215):

$$\bar{E}(0,0,z) = \hat{a}_z \frac{Q_0}{4\pi\varepsilon_0} \frac{z}{(a^2+z^2)^{3/2}}. \tag{3.273}$$

Then the electric scalar potential at any point on the z axis can be found as[105] (Figure 3.51)

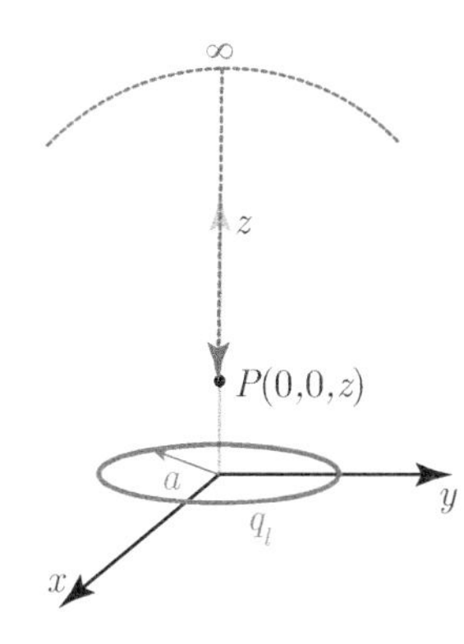

Figure 3.51 To find the electric scalar potential at an arbitrary point on the z axis, it is sufficient to know the electric field intensity only on the z axis for the integration.

$$\Phi(0,0,z) = -\int_{\infty}^{z} \bar{E}(0,0,z') \cdot \hat{a}_z dz' = -\frac{Q_0}{4\pi\varepsilon_0} \int_{\infty}^{z} \frac{z'}{[a^2+(z')^2]^{3/2}} dz' \tag{3.274}$$

$$= \frac{Q_0}{4\pi\varepsilon_0} \left[\frac{1}{\sqrt{a^2+(z')^2}}\right]_{\infty}^{z} = \frac{Q_0}{4\pi\varepsilon_0} \frac{1}{\sqrt{a^2+z^2}}, \tag{3.275}$$

[105]Note that the primed coordinate variable (i.e. z') is used to distinguish the integration variable from the arbitrary observation point z.

which is consistent with the earlier result in Eq. (3.272). Note that, at the origin, we have $\bar{E}(0,0,0) = 0$ and[106]

$$\Phi(0,0,0) = \frac{2\pi a q_0}{4\pi\varepsilon_0\sqrt{a^2}} = \frac{q_0}{2\varepsilon_0}. \tag{3.276}$$

[106]Note that the electric scalar potential at the origin does not depend on the radius a.

Hence, a zero electric field intensity does not indicate a zero electric scalar potential, and vice versa.

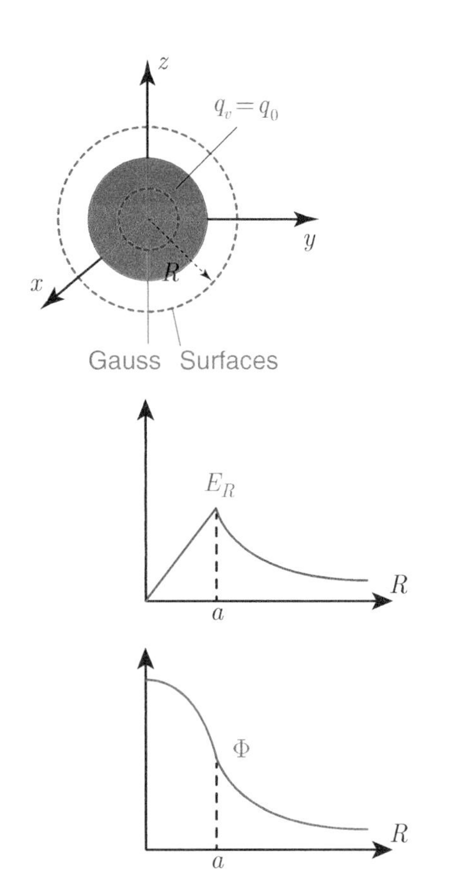

Figure 3.52 A static volume electric charge density $q_v(\bar{r}) = q_0$ (C/m^3) in the shape of a sphere of radius a. The electric field intensity and the electric scalar potential are plotted with respect to R.

As another example, we consider a static volume electric charge distribution $q_v(\bar{r}) = q_0$ (C/m^3) defined as a sphere of radius a in a vacuum (Figure 3.52). To find the electric scalar potential, we can first find the electric field intensity since it is easy to obtain via Gauss' law. Assuming that $\bar{E}(\bar{r})$ depends only on R and has only R component – i.e. $\bar{E}(\bar{r}) = \hat{a}_R E_R(R)$ – we have

$$\oint_S \bar{E}(\bar{r}) \cdot \overline{ds} = 4\pi R^2 E_R(R) = \frac{Q_{\text{enc}}}{\varepsilon_0}. \tag{3.277}$$

If $R < a$, the total amount of the enclosed electric charge can be found as

$$Q_{\text{enc}} = q_0 \frac{4}{3}\pi R^3, \tag{3.278}$$

leading to

$$\bar{E}(\bar{r}) = \hat{a}_R \frac{q_0 R}{3\varepsilon_0} \qquad (R < a). \tag{3.279}$$

On the other hand, if $R > a$, the total amount of the enclosed electric charge becomes

$$Q_{\text{enc}} = q_0 \frac{4}{3}\pi a^3, \tag{3.280}$$

leading to

$$\bar{E}(\bar{r}) = \hat{a}_R \frac{q_0 a^3}{3\varepsilon_0 R^2} \qquad (R > a). \tag{3.281}$$

The electric scalar potential can be found by using the electric field intensity expressions. For $|\bar{r}| > a$, we have

$$\Phi(\bar{r}) = -\int_\infty^{\bar{r}} \bar{E}(\bar{r}') \cdot \overline{dl'} = -\frac{q_0 a^3}{3\varepsilon_0} \int_\infty^{\bar{r}} \frac{\hat{a}_R}{(R')^2} \cdot \hat{a}_R dR' \tag{3.282}$$

$$= \frac{q_0 a^3}{3\varepsilon_0 |\bar{r}|} \qquad (|\bar{r}| > a). \tag{3.283}$$

For $|\bar{r}| < a$, we need to account for both expressions of the electric field intensity to derive[107]

$$\Phi(\bar{r}) = -\int_{\infty}^{a} \bar{E}(\bar{r}') \cdot \overline{dl'} - \int_{a}^{\bar{r}} \bar{E}(\bar{r}') \cdot \overline{dl'} \tag{3.284}$$

$$= \frac{q_0 a^3}{3\varepsilon_0 a} - \frac{q_0}{3\varepsilon_0}\int_{a}^{\bar{r}} R'\hat{a}_R \cdot \hat{a}_R dR' = \frac{q_0 a^3}{3\varepsilon_0 a} - \frac{q_0}{3\varepsilon_0}\left[\frac{(R')^2}{2}\right]_a^{\bar{r}} \tag{3.285}$$

$$= \frac{q_0 a^2}{2\varepsilon_0} - \frac{q_0|\bar{r}|^2}{6\varepsilon_0} \qquad (|\bar{r}| < a). \tag{3.286}$$

Hence, the value of the electric scalar potential at the origin is $\Phi(0,0,0) = q_0 a^2/2\varepsilon_0$. The positive value of $q_0 > 0$ indicates that it requires energy to bring a positive test charge from infinity to the origin.

[107]The two terms are as follows: first, we bring a test charge from infinity to the surface of the sphere. Then we further move it to the center (Figure 3.53).

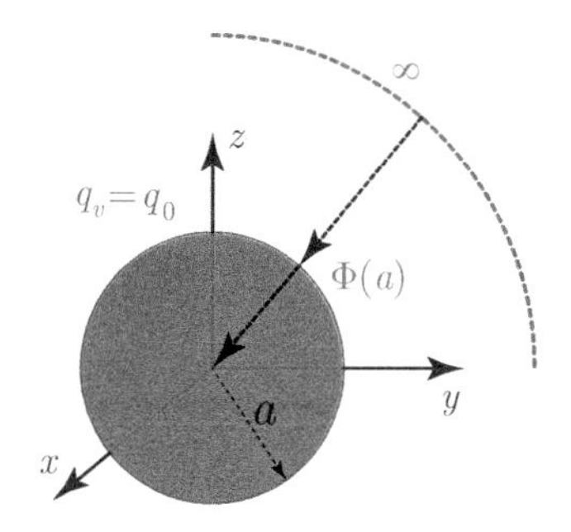

Figure 3.53 When finding the electric scalar potential at the center of a spherical volume electric charge distribution, the test charge can be brought from infinity in two steps since the expression for the electric field intensity is different inside and outside the charge distribution. First, it can be brought to the surface of the sphere to find the electric scalar potential for $R = a$. Then it can be further moved to the origin.

3.6.10 Examples

Example 37: Consider a static surface electric charge distribution $q_s(\bar{r}) = q_0$ (C/m^2) as a sphere with radius a in a vacuum (Figure 3.54). Find the electric scalar potential for $|\bar{r}| < a$.

Solution: Since the charge distribution is symmetric, it is relatively easy to find the electric field intensity. Using the integral form of Gauss' law, one can derive

$$\bar{E}(\bar{r}) = \hat{a}_R \frac{q_0 a^2}{\varepsilon_0 |\bar{r}|^2} \qquad (|\bar{r}| > a), \tag{3.287}$$

while $\bar{E}(\bar{r}) = 0$ for $|\bar{r}| < a$. At this stage, since the electric field intensity is known, we can evaluate the electric scalar potential via line integral. For $|\bar{r}| > a$, we have

$$\Phi(\bar{r}) = -\int_{\infty}^{\bar{r}} \bar{E}(\bar{r}') \cdot \overline{dl'} = -\frac{q_0 a^2}{\varepsilon_0}\int_{\infty}^{\bar{r}} \frac{\hat{a}_R}{(R')^2} \cdot \hat{a}_R dR' = \frac{q_0 a^2}{\varepsilon_0 |\bar{r}|} \qquad (|\bar{r}| > a) \tag{3.288}$$

by approaching radially from infinity. Then, for $|\bar{r}| < a$, we further have

$$\Phi(\bar{r}) = -\int_{\infty}^{a} \bar{E}(\bar{r}') \cdot \overline{dl'} - \int_{a}^{\bar{r}} \bar{E}(\bar{r}') \cdot \overline{dl'} \tag{3.289}$$

$$= \frac{q_0 a^2}{\varepsilon_0 a} - \int_{a}^{\bar{r}} 0 \cdot \overline{dl'} = \frac{q_0 a}{\varepsilon_0} \qquad (|\bar{r}| < a), \tag{3.290}$$

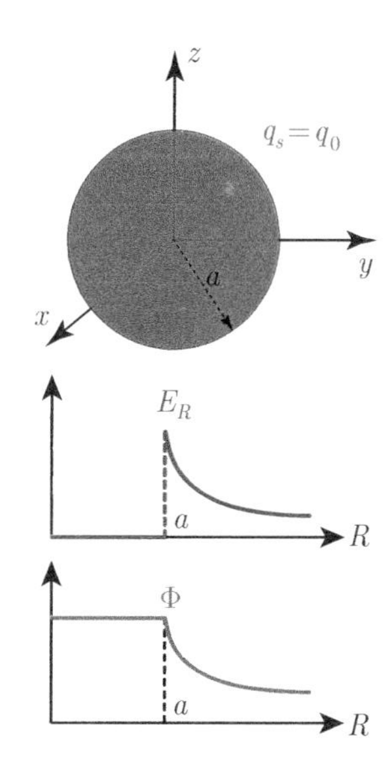

Figure 3.54 The electric scalar potential due to a static spherical surface electric charge density is constant inside the sphere. Note that the electric scalar potential should always be continuous if there is no unphysical jump in the electric potential energy.

which is constant. This means a test charge can be moved inside the spherical region without using/releasing electric potential energy.[108]

[108]This is true assuming that the test charge does change the uniform distribution of the original electric charges on the sphere.

Example 38: Consider a dielectric sphere of radius a, which is charged with a static volume electric charge density $q_v(\bar{r}) = R/a^4$ (C/m^3), located at the origin in a vacuum (Figure 3.55). Let the relative permittivity of the sphere be ε_r. Find the electric scalar potential at the origin.

Solution: Using the symmetry, the electric field intensity can be found as

$$\bar{E}(\bar{r}) = \frac{1}{4\varepsilon_0}\hat{a}_R \begin{cases} R^2/(\varepsilon_r a^4) & (R < a) \\ 1/R^2 & (R > a). \end{cases}$$

Hence, the electric scalar potential at the origin can be found via line integral involving two consecutive paths[109] as

[109]This is done in two parts since the expression for the electric field intensity is different for $R < a$ and $R > a$.

$$\Phi(0,0,0) = -\int_{\infty}^{a} \bar{E}(\bar{r}') \cdot \overline{dl'} - \int_{a}^{0} \bar{E}(\bar{r}') \cdot \overline{dl'} \tag{3.291}$$

$$= -\frac{1}{4\varepsilon_0}\int_{\infty}^{a} \frac{1}{(R')^2} dR' - \frac{1}{4a^4\varepsilon_0\varepsilon_r}\int_{a}^{0} (R')^2 dR' \tag{3.292}$$

$$= \frac{1}{4a\varepsilon_0} + \frac{1}{12a\varepsilon_0\varepsilon_r}. \tag{3.293}$$

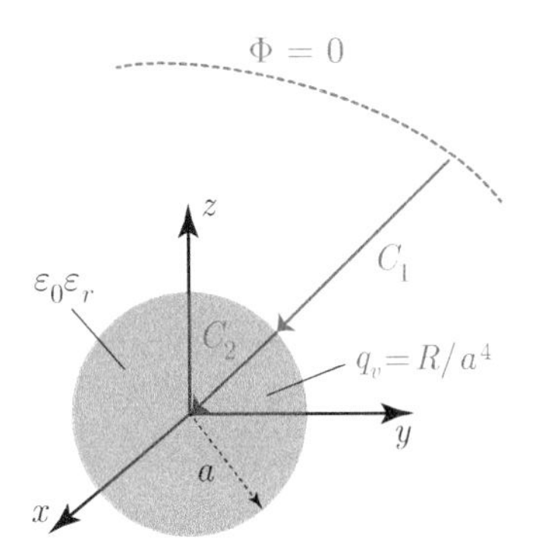

Figure 3.55 A charged dielectric sphere of radius a. When finding the electric scalar potential at the origin, one can first move from infinity to the surface of the sphere (C_1), and then from the surface of the sphere to the origin (C_2). For simple integration, these paths can be selected as radial.

Note that this electric scalar potential is caused by both electric charges $q_v(\bar{r})$ and polarization effects. For example, the pure contribution of the electric charges can easily be obtained by setting $\varepsilon_r = 1$ as

$$\Phi^{\text{charge}}(0,0,0) = \frac{1}{4a\varepsilon_0} + \frac{1}{12a\varepsilon_0} = \frac{1}{3a\varepsilon_0}. \tag{3.294}$$

Then the polarization (dielectric) effect can be found as[110]

[110]Note that the polarization effect is negative: i.e. the dielectric decreases the electric scalar potential at the origin.

$$\Phi^{\text{polarization}}(0,0,0) = \frac{1}{4a\varepsilon_0} + \frac{1}{12a\varepsilon_0\varepsilon_r} - \frac{1}{3a\varepsilon_0} = \frac{1}{12a\varepsilon_0\varepsilon_r} - \frac{1}{12a\varepsilon_0}. \tag{3.295}$$

3.6.11 Electrostatic Boundary Value Problems

In electrostatics and magnetostatics, electric fields and magnetic fields are decoupled such that they can be solved independently for given static electric charges and steady electric currents. For symmetric problems, Gauss' law and Ampere's law can also be used directly to obtain electric and magnetic fields. In some cases, however, resorting first to potentials may be useful. Especially in bounded regions, using potentials may lead to Poisson's equation or Laplace's equation that can be relatively easy to solve. As an example, we consider an infinitely long cylindrical region (Figure 3.56) aligned in the z direction and bounded by $\rho = a$ and $\rho = b$ surfaces with $b > a$. The value of the electric scalar potential at the boundaries are given as $\Phi(\rho = a) = V_0$ and $\Phi(\rho = b) = 0$. These values may be created by a voltage source connected to cylindrical perfect electric conductors. Assuming that the region is a vacuum, we have

$$\bar{\nabla}^2\Phi(\bar{r}) = \frac{1}{\rho}\frac{\partial}{\partial\rho}\left(\rho\frac{\partial\Phi(\bar{r})}{\partial\rho}\right) + \frac{1}{\rho^2}\frac{\partial^2\Phi(\bar{r})}{\partial\phi^2} + \frac{\partial^2\Phi(\bar{r})}{\partial z^2} \tag{3.296}$$

$$= \frac{1}{\rho}\frac{\partial}{\partial\rho}\left(\rho\frac{\partial\Phi(\bar{r})}{\partial\rho}\right) = 0 \qquad (a < \rho < b), \tag{3.297}$$

where Laplace's equation is written in a cylindrical coordinate system. Hence, we derive

$$\rho\frac{\partial\Phi(\bar{r})}{\partial\rho} = c_1 \longrightarrow \frac{\partial\Phi(\bar{r})}{\partial\rho} = \frac{c_1}{\rho} \longrightarrow \Phi(\bar{r}) = c_1\ln(\rho) + c_2, \tag{3.298}$$

where c_1 and c_2 are constants. To find these constants, we need to check and satisfy the boundary conditions. Specifically, we have

$$\Phi(\rho = b) = 0 = c_1\ln(b) + c_2 \longrightarrow c_2 = -c_1\ln(b) \tag{3.299}$$

$$\Phi(\rho = a) = V_0 = c_1\ln(a) + c_2 = c_1\ln(a) - c_1\ln(b) = c_1\ln(a/b) \tag{3.300}$$

and we obtain $c_1 = V_0/\ln(a/b)$, leading to

$$\Phi(\bar{r}) = V_0\ln(\rho)/\ln(a/b) - V_0\ln(b)/\ln(a/b) = V_0\frac{\ln(\rho/b)}{\ln(a/b)}. \tag{3.301}$$

The corresponding electric field intensity can be found as

$$\bar{E}(\bar{r}) = -\bar{\nabla}\Phi(\bar{r}) = -\frac{\partial\Phi(\bar{r})}{\partial\rho} = \hat{a}_\rho\frac{V_0}{\rho\ln(b/a)} \qquad (a < \rho < b). \tag{3.302}$$

Now, for the same cylindrical structure (Figure 3.56), we consider the region $\rho < a$. Equation $\Phi(\bar{r}) = c_1\ln(\rho) + c_2$ still holds for the electric scalar potential. Using the boundary condition at $\rho = a$, we have

$$\Phi(\rho = a) = c_1\ln(a) + c_2 = V_0 \longrightarrow c_2 = V_0 - c_1\ln(a) \tag{3.303}$$

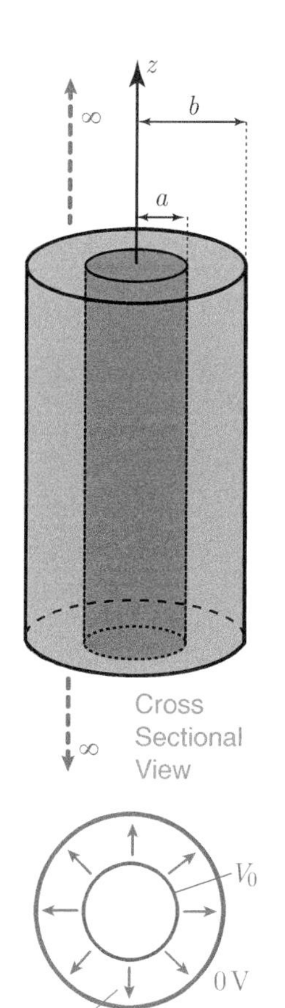

Figure 3.56 An infinitely long cylindrical region with 0 and V_0 electric scalar potential values at the boundaries.

leading to

$$\Phi(\bar{r}) = c_1 \ln(\rho) + c_2 = c_1 \ln(\rho) + V_0 - c_1 \ln(a)$$
$$= V_0 + c_1 \ln(\rho/a). \tag{3.304}$$

But there is no any other boundary to find the constant c_1. On the other hand, if the electric scalar potential at $\rho = 0$ is finite,[111] we must have $c_1 = 0$, leading to (Figure 3.57)

[111] We assume that there is no source at $\rho = 0$ that might create an infinite electric scalar potential.

$$\Phi(\bar{r}) = V_0 \qquad (\rho < a). \tag{3.305}$$

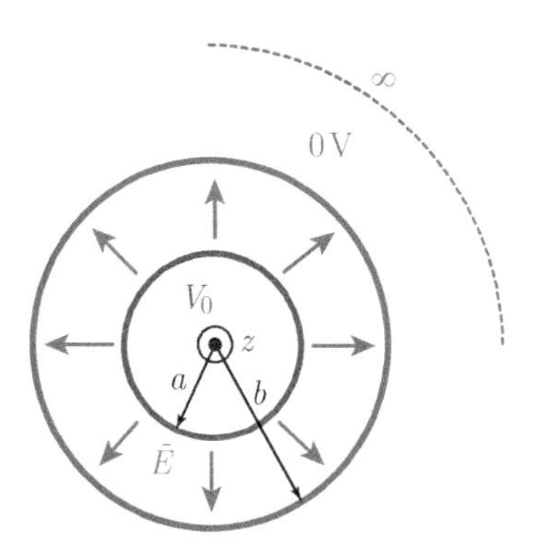

Figure 3.57 If the electric scalar potential changes only radially in a cylindrical coordinate system, it must change with a logarithmic decay between two boundaries with different potential values. Interestingly, if the electric scalar potential is finite on the z axis ($\rho = 0$), then its values for $\rho < a$ must be the same as the electric scalar potential at the given boundary $\rho = a$. In addition, if the electric scalar potential is finite (e.g. zero) at infinity, then its values for $\rho > b$ must be the same as the electric scalar potential at the given boundary $\rho = b$.

Similarly, for $\rho > b$, we again have $\Phi = c_1 \ln(\rho) + c_2$, and

$$c_1 \ln(b) + c_2 = 0 \longrightarrow c_2 = -c_1 \ln(b) \tag{3.306}$$
$$\Phi(\bar{r}) = c_1 \ln(\rho) + c_2 = c_1 \ln(\rho/b). \tag{3.307}$$

But if the electric scalar potential at infinity is finite (zero), we must have $c_1 = 0$, and

$$\Phi(\bar{r}) = 0 \qquad (\rho > b). \tag{3.308}$$

Therefore, the electric scalar potential is zero for $\rho > b$. This is consistent with the fact that there is no electric field intensity for $\rho > b$, which can be verified by the integral form of Gauss' law.

As another example, we now consider a spherical region (Figure 3.58) bounded by $R = a$ and $R = b$ surfaces with $b > a$. The electric scalar potential is given at the boundaries[112] as $\Phi(R = a) = V_0$ and $\Phi(R = b) = 0$. Assuming that the region is a vacuum, we have

[112] Similar to the cylindrical case, this automatically means $\Phi(R < a) = V_0$ and $\Phi(R > b) = 0$.

$$\bar{\nabla}^2 \Phi(\bar{r}) = \frac{1}{R^2} \frac{\partial}{\partial R} \left(R^2 \frac{\partial \Phi(\bar{r})}{\partial R} \right) = 0 \tag{3.309}$$
$$\longrightarrow \frac{\partial}{\partial R} \left(R^2 \frac{\partial \Phi(\bar{r})}{\partial R} \right) = 0 \longrightarrow R^2 \frac{\partial \Phi(\bar{r})}{\partial R} = c_1 \tag{3.310}$$
$$\longrightarrow \frac{\partial \Phi(\bar{r})}{\partial R} = \frac{c_1}{R^2} \longrightarrow \Phi(\bar{r}) = -\frac{c_1}{R} + c_2 \tag{3.311}$$

in a spherical coordinate system. Then, to find c_1 and c_2, we use the boundary conditions as

$$\Phi(R = b) = -\frac{c_1}{b} + c_2 = 0 \longrightarrow c_2 = \frac{c_1}{b} \tag{3.312}$$

and

$$\Phi(R=a) = -\frac{c_1}{a} + c_2 = -\frac{c_1}{a} + \frac{c_1}{b} = V_0 \tag{3.313}$$

$$\longrightarrow c_1 = V_0 \frac{ab}{(a-b)}. \tag{3.314}$$

Then we derive

$$\Phi(\bar{r}) = -V_0 \frac{ab}{(a-b)} \frac{1}{R} + V_0 \frac{ab}{(a-b)} \frac{1}{b} = V_0 \frac{ab}{(a-b)} \left(\frac{1}{b} - \frac{1}{R} \right) \tag{3.315}$$

$$= V_0 \frac{a}{(a-b)} \left(1 - \frac{b}{R} \right). \tag{3.316}$$

Finally, the corresponding electric field intensity can be found as

$$\bar{E}(\bar{r}) = -\bar{\nabla}\Phi(\bar{r}) = -\hat{a}_R \frac{\partial \Phi(\bar{r})}{\partial R} = -\hat{a}_R V_0 \frac{a}{(a-b)} \frac{b}{R^2} \tag{3.317}$$

$$= \hat{a}_R V_0 \frac{ab}{(b-a)} \frac{1}{R^2}. \tag{3.318}$$

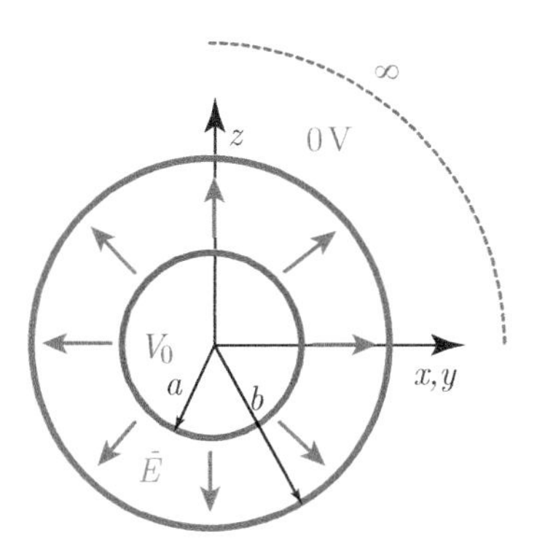

Figure 3.58 If the electric scalar potential changes only radially in a spherical coordinate system, it must decay with the radial distance R between two boundaries with different potential values. Interestingly, if the electric scalar potential is finite at the origin ($R = 0$), then its values for $R < a$ must be the same as the electric scalar potential at the given boundary $R = a$. In addition, if the electric scalar potential is finite (e.g. zero) at infinity, then its values for $R > b$ must be the same as the electric scalar potential at the given boundary $R = b$.

As an interesting practice, by using the related boundary condition, we can further find the electric charge density on the inner sphere.[113] Assuming that the region is a vacuum, we have

$$q_s(\bar{r}) = \hat{a}_R \cdot \bar{D}(R=a) = \varepsilon_0 \hat{a}_R \cdot \bar{E}(R=a) = \varepsilon_0 V_0 \frac{b}{a(b-a)}. \tag{3.319}$$

[113]For example, it may be a spherical perfect electric conductor.

3.6.12 Examples

Example 39: Consider an electrostatic case involving a spherical region (Figure 3.59) bounded by $R = a$ and $R = b$ surfaces with $b > a$. Let the electric scalar potential values at the boundaries be $\Phi(R = a) = V_0$ (V/m) and $\Phi(R = b) = 0$, respectively. Also there is an electric charge distribution with density $q_v(\bar{r}) = q_v(R) = -q_0/R^2$ for $a < R < b$. Find the electric scalar potential everywhere.

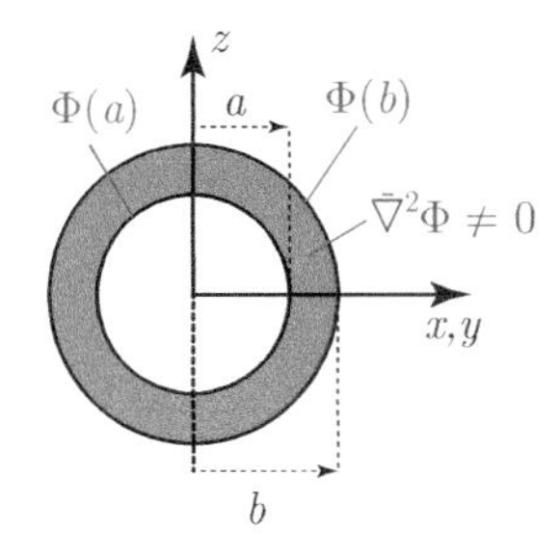

Figure 3.59 A spherical region bounded by $R = a$ and $R = b$ surfaces (with $b > a$) and filled with an electric charge distribution so that Laplace's equation (Poisson's equation with zero right-hand side) is not valid.

Solution: Writing Poisson's equation, we have[114]

$$\bar{\nabla}^2\Phi(\bar{r}) = \frac{1}{R^2}\frac{\partial}{\partial R}\left(R^2\frac{\partial\Phi(\bar{r})}{\partial R}\right) = \frac{q_0}{\varepsilon_0 R^2} \tag{3.320}$$

$$\longrightarrow \frac{\partial}{\partial R}\left(R^2\frac{\partial\Phi(\bar{r})}{\partial R}\right) = \frac{q_0}{\varepsilon_0} \longrightarrow R^2\frac{\partial\Phi(\bar{r})}{\partial R} = \frac{q_0}{\varepsilon_0}R + c_1 \tag{3.321}$$

$$\longrightarrow \frac{\partial\Phi(\bar{r})}{\partial R} = \frac{q_0}{\varepsilon_0}\frac{1}{R} + \frac{c_1}{R^2} \tag{3.322}$$

$$\longrightarrow \Phi(\bar{r}) = \frac{q_0}{\varepsilon_0}\ln(R) - \frac{c_1}{R} + c_2 \qquad (a < R < b). \tag{3.323}$$

[114]Recall that, if there were no volume electric charge distribution, the solution of Laplace's equation would lead to

$$\Phi(\bar{r}) = V_0\frac{ab}{(b-a)}\left(\frac{1}{R} - \frac{1}{b}\right) \qquad (a < R < b).$$

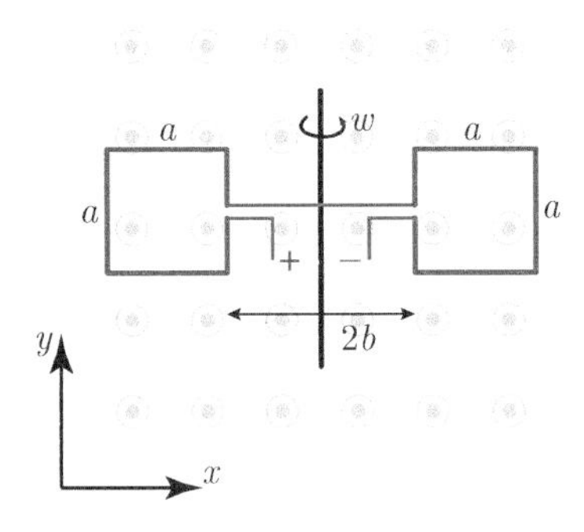

Figure 3.60 A system of two loops rotating in a static magnetic field.

Since $\Phi(R = a) = V_0$ and $\Phi(R = b) = 0$, the constants c_1 and c_2 can be derived as

$$\frac{q_0}{\varepsilon_0}\ln(b) - \frac{c_1}{b} + c_2 = 0 \longrightarrow c_2 = \frac{c_1}{b} - \frac{q_0}{\varepsilon_0}\ln(b) \tag{3.324}$$

$$\frac{q_0}{\varepsilon_0}\ln(a) - \frac{c_1}{a} + c_2 = \frac{q_0}{\varepsilon_0}\ln(a) - \frac{c_1}{a} + \frac{c_1}{b} - \frac{q_0}{\varepsilon_0}\ln(b) = V_0 \tag{3.325}$$

$$\longrightarrow c_1 = -\frac{ab}{b-a}\left[V_0 + \frac{q_0}{\varepsilon_0}\ln(b/a)\right]. \tag{3.326}$$

Then we obtain

$$\Phi(\bar{r}) = \frac{q_0}{\varepsilon_0}\ln(R) + \frac{ab}{b-a}\left[V_0 + \frac{q_0}{\varepsilon_0}\ln(b/a)\right]\frac{1}{R} - \frac{ab}{b-a}\left[V_0 + \frac{q_0}{\varepsilon_0}\ln(b/a)\right]\frac{1}{b} - \frac{q_0}{\varepsilon_0}\ln b, \tag{3.327}$$

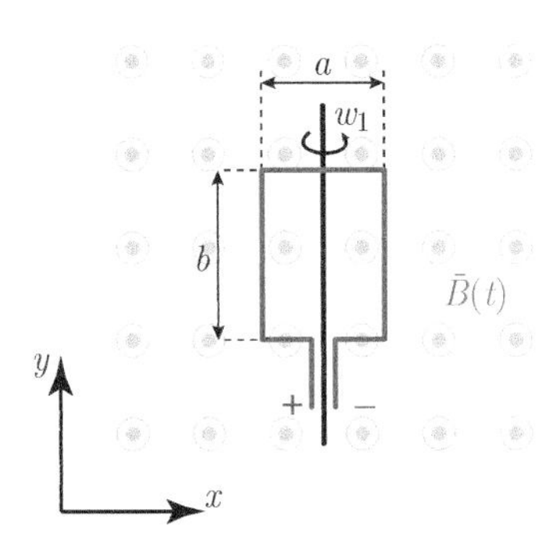

Figure 3.61 An AC generator: a loop rotating in a time-variant magnetic flux density.

leading to

$$\Phi(\bar{r}) = \frac{q_0}{\varepsilon_0}\ln(R/b) + \frac{ab}{b-a}\left[V_0 + \frac{q_0}{\varepsilon_0}\ln(b/a)\right]\left(\frac{1}{R} - \frac{1}{b}\right) \qquad (a < R < b). \tag{3.328}$$

One can check that the final expression satisfies the boundary values as

$$\Phi(R=a) = \frac{q_0}{\varepsilon_0}\ln(a/b) + \frac{ab}{b-a}\left[V_0 + \frac{q_0}{\varepsilon_0}\ln(b/a)\right]\left(\frac{1}{a}-\frac{1}{b}\right) \tag{3.329}$$

$$= \frac{q_0}{\varepsilon_0}\ln(a/b) + V_0 + \frac{q_0}{\varepsilon_0}\ln(b/a) = V_0 \tag{3.330}$$

$$\Phi(R=b) = \frac{q_0}{\varepsilon_0}\ln(b/b) + \frac{ab}{b-a}[V_0 + \frac{q_0}{\varepsilon_0}\ln(b/a)]\left(\frac{1}{b}-\frac{1}{b}\right) \tag{3.331}$$

$$= \frac{q_0}{\varepsilon_0}\ln(b/b) = 0. \tag{3.332}$$

Considering the final expression, we have

$$\bar{E}(\bar{r}) = -\bar{\nabla}\Phi(\bar{r}) = -\hat{a}_R\frac{\partial\Phi(\bar{r})}{\partial R} \tag{3.333}$$

$$= -\hat{a}_R\frac{q_0}{\varepsilon_0}\frac{1}{R} + \hat{a}_R\frac{ab}{b-a}\left[V_0 + \frac{q_0}{\varepsilon_0}\ln(b/a)\right]\frac{1}{R^2} \tag{3.334}$$

$$q_v(\bar{r}) = \frac{1}{\varepsilon_0}\bar{\nabla}\cdot\bar{E}(\bar{r}) = \frac{1}{\varepsilon_0}\frac{1}{R^2}\frac{\partial}{\partial R}\left(R^2E_R(\bar{r})\right) \tag{3.335}$$

$$= -\frac{1}{\varepsilon_0}\frac{1}{R^2}\frac{\partial}{\partial R}\left(\frac{R^2}{R}\right) \tag{3.336}$$

$$= -\frac{1}{\varepsilon_0}\frac{1}{R^2} \tag{3.337}$$

as expected.

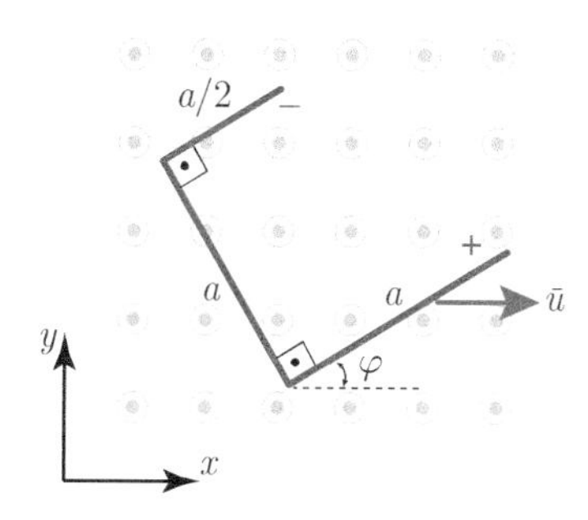

Figure 3.62 A three-segment wire moving in a static magnetic field.

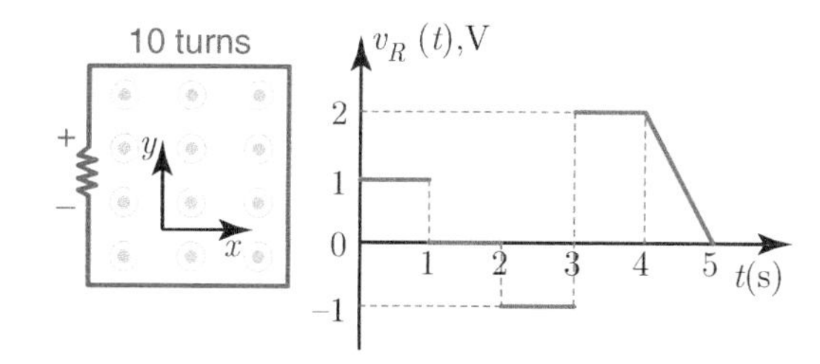

Figure 3.63 A loop of 10 turns with area 0.1 m^2 under a time-variant magnetic field.

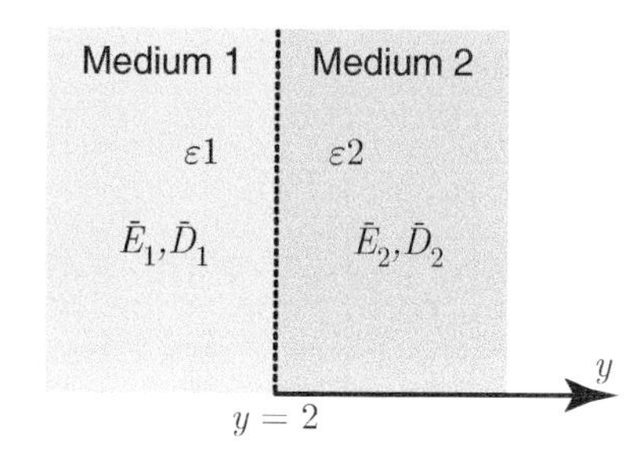

Figure 3.64 A infinitely large boundary at $y = 2$, separating two dielectric media.

3.7 Final Remarks

As the third Maxwell's equations considered in this book, Faraday's law shows how electric fields are created by changing (time-variant) magnetic fields. Apart from being a fundamental key to understand the propagation of electromagnetic ways, this law also describes how kinetic and/or magnetic potential energy (movements of loops under static or time-variant magnetic flux density) can be converted into the electromotive force (voltage). However, to fully understand electromagnetic phenomena, we need the final Maxwell's equation: Gauss' law for magnetic fields, discussed in the next chapter. We will then see how Maxwell's equations can be used to solve and analyze dynamic electromagnetic problems.

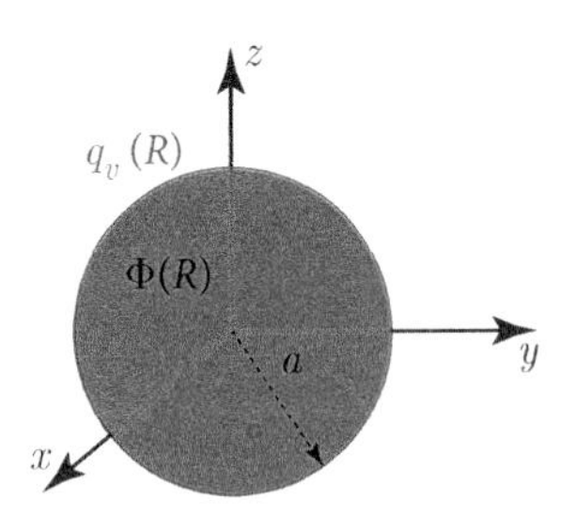

Figure 3.65 A static spherical electric charge distribution of radius a represented by a radially changing volume electric charge density $q_v(R)$.

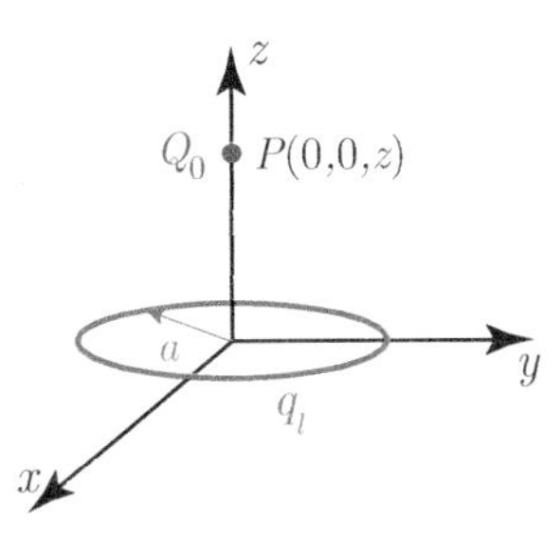

Figure 3.66 A static point electric charge located over a static line electric charge distribution of a circular shape.

[115]Hence, we assume that this magnetic flux density and the related electric field intensity are created by some sources located at infinity.

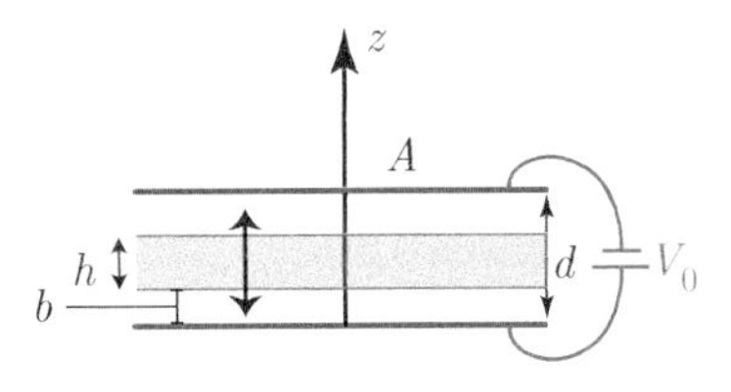

Figure 3.67 A dielectric slab (which can vertically move) inserted into a parallel-plate capacitor.

3.8 Exercises

Exercise 32: A system of two loops is rotating with an angular frequency of ω in a static magnetic field $\bar{B}(\bar{r}) = \hat{a}_z B_0$ (Figure 3.60). Ignoring the wires between the loops, find the induced voltage between the terminals.

Exercise 33: Consider an AC generator that involves a rotating loop with an angular frequency of ω_1 in a time-variant magnetic flux density $\bar{B}(t) = B_0 \sin(\omega_2 t)$ (Figure 3.61). Find the induced voltage between the terminals.

Exercise 34: A three-segment wire is moving with a velocity of $\bar{u} = \hat{a}_x u_0$ in a static magnetic field $\bar{B}(\bar{r}) = \hat{a}_z B_0$ (Figure 3.62). Find an expression for the angle φ such that the induced voltage between the ends of the wire is zero.

Exercise 35: A loop of 10 turns with area 0.1 m^2 is connected to a resistor to measure a time-variant but position-independent magnetic field in the z direction. Keeping the loop stationary, the measured voltage across the resistor is shown in Figure 3.63, while it is zero for $t > 5$ s. Find and plot the magnetic flux density $B_z(t)$ if $B_z(t) = 0$ for $t < 0$.

Exercise 36: The magnetic flux density in a vacuum[115] with permittivity ε_0 and permeability μ_0 is given as $\bar{B}(\bar{r}, t) = \hat{a}_x B_0 \cos(\omega t - kz)$. Find the relationship between ω and k.

Exercise 37: Consider two dielectric media separated by a boundary surface defined as $y = 2$ (Figure 3.64). Let the relative permittivity of the first ($y < 2$) and second ($y > 2$) media be $\varepsilon_{r1} = 2$ and $\varepsilon_{r2} = 3$, respectively. If the electric field intensity for the second medium is given as $\bar{E}_2(\bar{r}) = e_0(\hat{a}_x x^2 + \hat{a}_y y^2 + \hat{a}_z z^2)$, where e_0 is a constant with unit V/m^3, find $\bar{E}_1(\bar{r})$ and $\bar{D}_1(\bar{r})$ at the boundary ($y = 2^-$).

Exercise 38: Let a surface be defined as $x^2 + 3yz + z^3 = 5$. Find the unit normal vector that is perpendicular to the surface at $(x, y, z) = (1, 1, 1)$.

Exercise 39: Consider a scalar field $f(\bar{r}) = x^2yz + 4xz^2$. Find the directional derivative of this field in the direction of $\bar{g} = \hat{a}_x 2 - \hat{a}_y - \hat{a}_z 2$ at $(x, y, z) = (1, -2, -1)$.

Exercise 40: Consider a static spherical electric charge distribution of radius a represented by a radially changing volume electric charge density $q_v(R)$ (Figure 3.65) in a vacuum. The electric scalar potential for $R < a$ is given as

$$\Phi(R) = V_0[1 - R/(2a)] \qquad (R < a), \tag{3.338}$$

where V_0 (V) is a constant. Find the electric potential energy stored in the system.

Exercise 41: Consider a static point electric charge Q_0 (C) located on the z axis and a static line electric charge distribution $q_l(\bar{r}) = q_0$ (C/m) of a circular shape with radius a on the x-y plane (Figure 3.66). Find the position at which the electric force applied on the point charge by the circular charge distribution is maximum.

Exercise 42: Consider a dielectric slab inserted into a parallel-plate capacitor (Figure 3.67). The distance between the plates is d, while the area of each plate is A. The slab has a thickness of h, and the rest of the capacitor is filled with a vacuum. Find the force that tries to move the slab *vertically*.

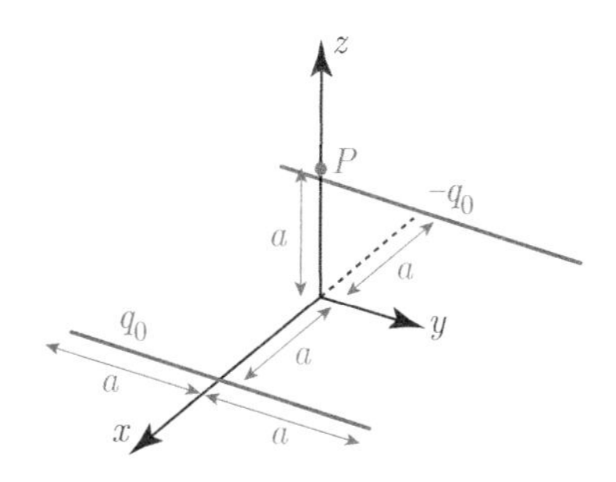

Figure 3.68 Finite static line electric charge densities located at $(x = a, z = 0)$ and $(x = -a, z = 0)$.

Exercise 43: Consider finite static line electric charge densities $q_l(\bar{r}) = q_0$ (C/m^3) and $q_l(\bar{r}) = -q_0$ located at $(x = a, z = 0)$ and $(x = -a, z = 0)$, respectively (Figure 3.68). The length of both lines is $2a$: i.e. they are aligned from $y = -a$ to $y = a$. Find the electric scalar potential and the electric field intensity at $\bar{r} = \hat{a}_z a$.

Exercise 44: Consider a cylindrical dielectric (Figure 3.69) of radius a aligned from $z = 0$ to $z = d$. The dielectric is permanently polarized with $\bar{P} = \hat{a}_z P_0$ (C/m^2). Find the electric scalar potential at $\bar{r} = \hat{a}_z z$ for $z > d$ with the help of the equivalent problem.

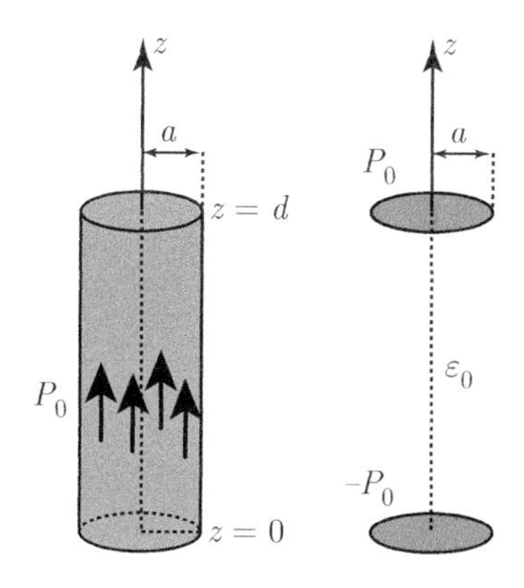

Figure 3.69 A permanently polarized dielectric (electret) with a constant polarization and the equivalent problem.

Exercise 45: Consider a static electric field intensity $\bar{E}(\bar{r}) = e_0[\hat{a}_x 2y + \hat{a}_y 2x - \hat{a}_z 2a]$, where e_0 (V/m^2) and a (m) are constants. Find the electric scalar potential at $\bar{r} = (x, y, z) = (5, 3, 20)$ m, if the electric scalar potential at a reference point $\bar{r}_{\text{ref}} = (x, y, z) = (0, -2, 8)$ m is defined as zero.

Exercise 46: Consider again a static spherical electric charge distribution of radius a represented by a radially changing volume electric charge density $q_v(R)$ (Figure 3.65) in a vacuum. The electric scalar potential for $R < a$ is given as

$$\Phi(R) = V_0[1 - R/(2a)] \qquad (R < a), \tag{3.339}$$

where V_0 (V) is a constant. Find the electric scalar potential for $R > a$.

3.9 Questions

Question 14: A rectangular loop is moving with a velocity $\bar{u} = \hat{a}_x u_0$ in a dynamic magnetic flux density $\bar{B}(\bar{r}, t) = \hat{a}_z B_0 \exp(kx - \omega t)$, where k and ω are constants and t represents the time (Figure 3.70). Given the dimensions of the loop in the figure, find an expression for u_0 such that no electromotive force is induced in the loop.

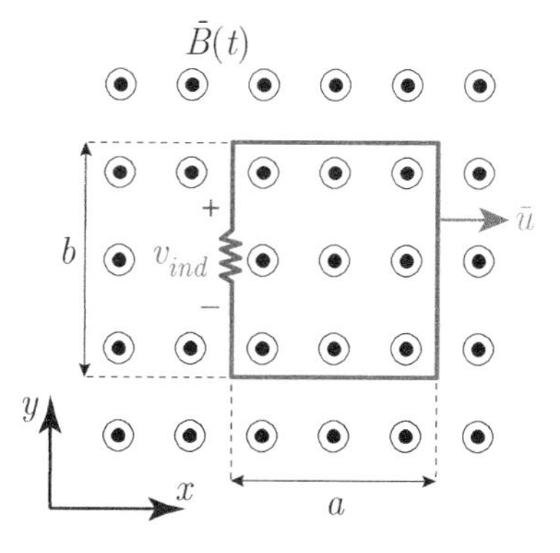

Figure 3.70 A rectangular loop on the x-y plane exposed to a magnetic flux density in the z direction. The loop is moving in the x direction.

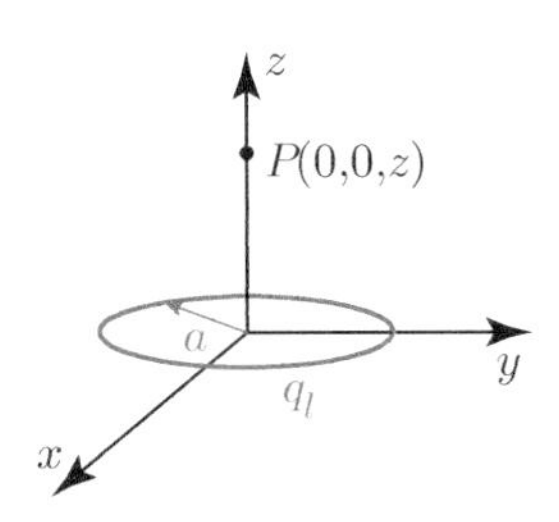

Figure 3.71 A static line electric charge distribution with a circular shape.

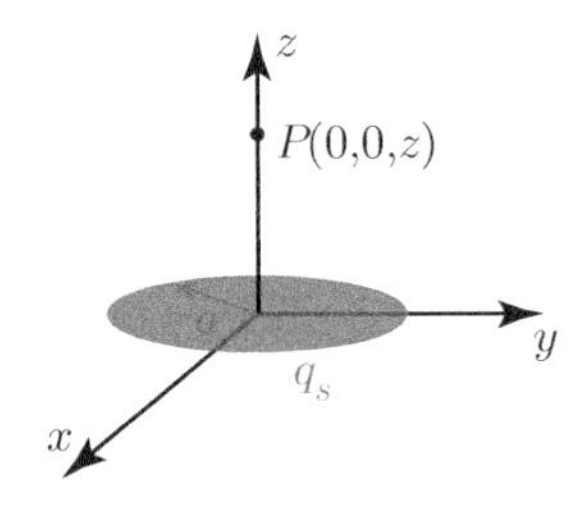

Figure 3.72 A static surface electric charge distribution with a disk shape.

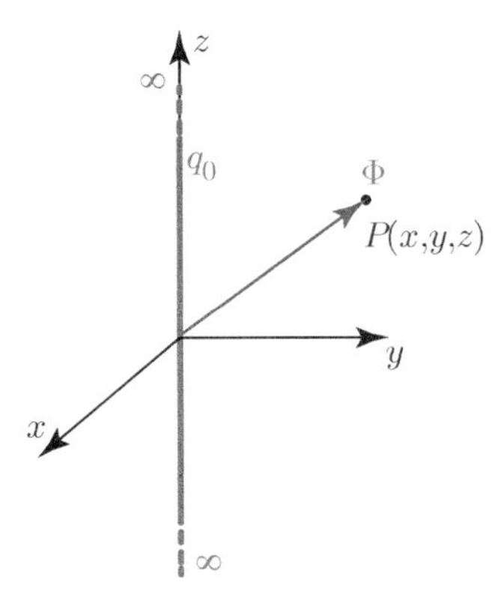

Figure 3.73 An infinitely long static line electric charge density.

Question 15: Prove the following identities:

$$\bar{\nabla}(\bar{f}\cdot\bar{g}) = (\bar{f}\cdot\bar{\nabla})\bar{g} + (\bar{g}\cdot\bar{\nabla})\bar{f} + \bar{f}\times(\bar{\nabla}\times\bar{g}) + \bar{g}\times(\bar{\nabla}\times\bar{f}) \tag{3.340}$$

$$\frac{1}{2}\bar{\nabla}(\bar{f}\cdot\bar{f}) = (\bar{f}\cdot\bar{\nabla})\bar{f} + \bar{f}\times(\bar{\nabla}\times\bar{f}). \tag{3.341}$$

Question 16: Consider a static line electric charge distribution $q_l(\bar{r}) = q_0$ (C/m) of a circle shape with radius a located on the x-y plane in a vacuum (Figure 3.71). Show that the electric scalar potential at $\bar{r} = \hat{a}_z z$ for $z > 0$ (on the positive z axis) is zero if $q_l(\bar{r}) = q_0\cos\phi$. Also show that this value is consistent with the electric field intensity, which is zero on the z axis.

Question 17: Consider a static surface electric charge density $q_s(\bar{r}) = q_0$ (C/m^2) having a disk shape of radius a located at the origin in a vacuum (Figure 3.72). Show that the electric scalar potential at $\bar{r} = \hat{a}_z z$ for $z > 0$ (on the positive z axis) can be written as

$$\Phi(0,0,z) = \frac{q_0}{2\varepsilon_0}\left(\sqrt{z^2+a^2} - z\right). \tag{3.342}$$

Question 18: Consider an infinitely long static line electric charge density defined as $q_l(\bar{r}) = q_0$ (C/m^3) on the z axis from $-\infty$ to ∞ (Figure 3.73). Show that the electric scalar potential difference between two points $\bar{r}_a = (\rho_a, \phi_a, z_a)$ and $\bar{r}_b = (\rho_b, \phi_b, z_b)$ is

$$\Phi(\bar{r}_b) - \Phi(\bar{r}_a) = \frac{q_0}{2\pi\varepsilon_0}\ln\left(\frac{\rho_a}{\rho_b}\right). \tag{3.343}$$

Is it possible to define a unique potential at an arbitrary point $\bar{r}$?

Question 19: Consider a static electric field intensity defined as

$$\bar{E}(\bar{r}) = e_0(\hat{a}_x axyz + \hat{a}_y x^2 z + \hat{a}_z x^2 y), \tag{3.344}$$

where e_0 is a constant with unit V/m^4. Show that $a = 2$ and the electric scalar potential at $(x,y,z) = (2,2,2)$ m is $-16e_0$ (V) if the potential at the origin is zero.

Question 20: In a three-dimensional space, the electric scalar potential is given as $\Phi(R) = V_0\exp(-R/a)$ for $R > a$, while it is constant $\Phi(R) = V_0\exp(-1)$ for $R < a$.

- Is there a volume electric charge density? If yes, find its distribution.
- Is there a surface electric charge density? If yes, find its distribution.
- Is there a line electric charge density? If yes, find its distribution.

4

Gauss' Law for Magnetic Fields

The final Maxwell's equation that we consider in this chapter is Gauss' law for magnetic fields. Compared to the other three equations – i.e. Gauss' law (for electric fields), Ampere's law, and Faraday's law – this law is usually considered in less detail. In fact, it only indicates that there is no magnetic charge; hence, magnetic fields must be solenoidal. However, this information is crucial. Without Gauss' law for magnetic fields, electromagnetic problems (particularly dynamic cases) cannot be solved systematically. This chapter starts with the law and proceeds with the boundary conditions for normal magnetic fields. A counterpart of the electric scalar potential – the magnetic vector potential – is also discussed, leading to the concept of magnetic potential energy. Finally, we practice combining all of Maxwell's equations before considering the solutions to Maxwell's equations in detail in the following chapters.

$$\oint_S \bar{B}(\bar{r},t) \cdot \overline{ds} = 0$$

$$\bar{\nabla} \cdot \bar{B}(\bar{r},t) = 0$$

4.1 Integral and Differential Forms of Gauss' law for Magnetic Fields

Gauss' law for electric fields shows that electric charges create electric fields, e.g. $\bar{\nabla} \cdot \bar{D}(\bar{r},t) = q_v(\bar{r},t)$. On the other hand, as discussed in the previous chapters, the generation of magnetic fields is different; electric currents and changing electric flux density values create magnetic fields, but there is a rotational dependency that is obviously different than the relationship between electric charges and electric fields. We do not consider a direct generation of magnetic fields simply because there is no magnetic charge in nature.

Magnetic charges are only used conceptually for numerical purposes, e.g. in equivalence problems, and they do not exist in nature.[1] This means magnetic fields do not start from a point (source) or end at a point (sink or negative source). Then if there is a magnetic field in a medium, the associated field lines must close upon themselves. They are still generated

[1] As discussed in Section 2.8, a hypothetical experiment of dividing a magnet into two leads to two new magnets such that a pole of a magnet cannot be isolated. We also note that, given a magnet, the generated magnetic field lines do not emerge from its north pole (or end at its south pole); they are closed loops passing through the inside of the magnet.

Introduction to Electromagnetic Waves with Maxwell's Equations, First Edition. Özgür Ergül.

Companion website: www.wiley.com/go/ergulmax

via electric currents and/or changing electric flux density values, but the relationship is always rotational, leading to this *solenoidal* property.

To understand the integral and differential forms of Gauss' law for magnetic fields, we need the tools and concepts (that are considered in previous chapters), such as differential surface with direction (Section 1.1.1), dot product (Section 1.1.2), flux of vector field (Section 1.1.3), divergence of vector field (Section 1.3.2), electric current density (Section 2.3.1), cross product (Section 2.3.2), curl of vector field (Section 2.3.3), magnetization (Section 2.8.2), and gradient of scalar fields (Section 3.6.1). First, from a mathematical point of view, if we write Gauss' law for magnetic fields, we arrive at[2]

[2]Note that Gauss' law for magnetic fields is written in terms of the magnetic flux density. In comparison to this, Gauss' law for electric fields is written for the electric flux density, Ampere's law is written for the magnetic field intensity, and Faraday's law is written for the electric field intensity. Then the constitutive relationships $\bar{D}(\bar{r},t) = \varepsilon(\bar{r})\bar{E}(\bar{r},t)$ and $\bar{B}(\bar{r},t) = \mu(\bar{r})\bar{H}(\bar{r},t)$ allow us to use them for the other quantities.

$$\oint_S \bar{B}(\bar{r},t) \cdot \overline{ds} = 0 \tag{4.1}$$

and

$$\bar{\nabla} \cdot \bar{B}(\bar{r},t) = 0 \tag{4.2}$$

with zero right-hand sides. Even though these equations seem trivial, one needs to be careful when writing them for the magnetic field intensity. For the integral form, we generally have

$$\oint_S \mu(\bar{r})\bar{H}(\bar{r},t) \cdot \overline{ds} = 0, \tag{4.3}$$

assuming isotropic and linear medium, as usual. If the integration is carried out though a surface, on which the permeability is fixed as $\mu(\bar{r}) = \mu$, then we have (Figure 4.1)

$$\oint_S \bar{H}(\bar{r},t) \cdot \overline{ds} = 0. \tag{4.4}$$

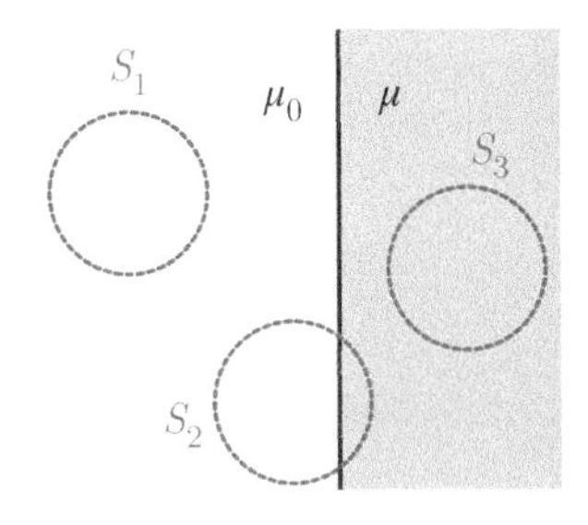

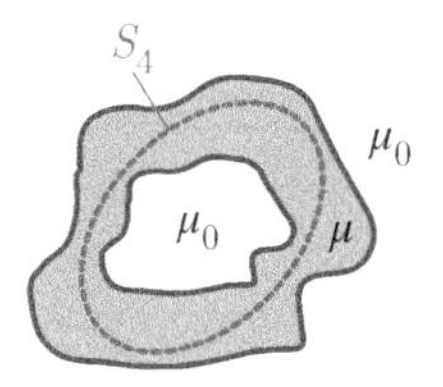

Figure 4.1 Given the integral form of Gauss' law for magnetic fields, the closed surface integral (net flux) of the magnetic field intensity is zero if all points on the surface are in the same magnetic medium (S_1, S_3, and S_4). Otherwise – i.e. if the surface encloses multiple types of volumes with different magnetic properties (S_2) – the position-dependent $\mu(\bar{r})$ cannot be placed outside the integral such that net flux of $\bar{H}(\bar{r},t)$ (alone) may not be zero.

On the other side, the differential form can be written

$$\bar{\nabla} \cdot [\mu(\bar{r})\bar{H}(\bar{r},t)] = \bar{\nabla}\mu(\bar{r}) \cdot \bar{H}(\bar{r},t) + \mu(\bar{r})\bar{\nabla} \cdot \bar{H}(\bar{r}) = 0 \tag{4.5}$$

or

$$\bar{\nabla} \cdot \bar{H}(\bar{r},t) = -\frac{\bar{\nabla}\mu(\bar{r}) \cdot \bar{H}(\bar{r},t)}{\mu(\bar{r})}, \tag{4.6}$$

where $\bar{\nabla}\mu(\bar{r})$ is the gradient of the permeability that accounts for possible variations in magnetic properties. If the variation is zero at $\bar{r}$ – that is, if $\bar{\nabla}\mu(\bar{r}) = 0$ – then we have

$$\bar{\nabla} \cdot \bar{H}(\bar{r},t) = 0, \tag{4.7}$$

i.e., there is no direct source for the magnetic field intensity.

The expression in Eq. (4.6) indicates that, when the permeability varies locally, $\bar{\nabla}\mu(\bar{r}) \cdot \bar{H}(\bar{r},t)$ acts like a volume magnetic charge density.[3] Note that the existence of this type of a *fictitious* magnetic charge density depends on the existence of $\bar{H}(\bar{r},t)$ itself: i.e. it does not represent a true magnetic charge distribution. Still, $\bar{\nabla}\mu(\bar{r}) \cdot \bar{H}(\bar{r},t)$ can create a discontinuity (step up or down) in the magnetic field intensity, as if there is a magnetic charge. Naturally, this is directly related to the magnetization. To see this, we rewrite the differential form as

[3]Note the units: $(1/\text{m}) \times (\text{H/m}) \times (\text{A/m}) = \text{Wb/m}^3$.

$$\bar{\nabla} \cdot \bar{B}(\bar{r},t) = 0 \longrightarrow \bar{\nabla} \cdot [\bar{H}(\bar{r},t)/\mu_0 + \bar{M}(\bar{r},t)/\mu_0] = 0, \tag{4.8}$$

leading to

$$\bar{\nabla} \cdot \bar{H}(\bar{r},t) = -\bar{\nabla} \cdot \bar{M}(\bar{r},t) = \frac{1}{\mu_0} q_{mv}(\bar{r},t). \tag{4.9}$$

Hence, from this perspective, $q_{mv}(\bar{r},t) = -\mu_0 \bar{\nabla} \cdot \bar{M}(\bar{r},t)$ behaves like a volume magnetic charge density that creates the nonzero divergence of the magnetic field intensity.[4]

[4]Recall the corresponding equation $q_{pv}(\bar{r},t) = -\bar{\nabla} \cdot \bar{P}(\bar{r},t)$ that relates the volume polarization charge density to the polarization.

4.1.1 Meaning of Gauss' Law for Magnetic Fields

According to Gauss' law for magnetic fields, the magnetic flux density is divergence-free: i.e. $\bar{\nabla} \cdot \bar{B}(\bar{r},t) = 0$. This was partially predicted[5] in Section 3.5 when the divergence of Faraday's law is evaluated and the null identity $\bar{\nabla} \cdot \bar{\nabla} \times \bar{f}(\bar{r},t) = 0$ is used. Zero divergence means there is no magnetic charge: i.e. magnetic field lines do not vanish/emerge anywhere, and they close upon themselves. We already know that these types of vector fields are called *solenoidal*. But according to the null identity[6] $\bar{\nabla} \cdot \bar{\nabla} \times \bar{f}(\bar{r},t) = 0$, a solenoidal vector field $\bar{g}(\bar{r},t)$ with $\bar{\nabla} \cdot \bar{g}(\bar{r},t) = 0$ can be written $\bar{g}(\bar{r},t) = \bar{\nabla} \times \bar{f}(\bar{r},t)$: i.e. the curl of another vector field. This leads to the definition of the magnetic vector potential, as discussed in Section 4.3. This is similar to the electric field intensity in a *static* case: i.e. it is an irrotational field with zero curl $\bar{\nabla} \times \bar{E}(\bar{r}) = 0$, and it can be written as a gradient of a scalar field $[\bar{E}(\bar{r}) = -\bar{\nabla}\Phi(\bar{r})]$. On the other hand, the magnetic flux density is solenoidal in *both* static and dynamic cases. Physically, the integral form of Gauss' law for magnetic fields indicates that magnetic field lines entering an arbitrary surface S must leave, leading to an overall zero net flux. This is a natural consequence of the absence of magnetic charges: i.e. magnetic fields cannot vanish/emerge at a location to unbalance the total flux through a surface.

[5]It is a partial prediction due to the time derivative in Eq. (3.85). Only according to Gauss' law for magnetic fields, the divergence of the magnetic flux density is zero in all cases (static or dynamic).

[6]Interestingly, this null identity can be verified by combining the divergence theorem and Stokes' theorem. According to the divergence theorem for a vector field $\bar{g}(\bar{r},t) = \bar{\nabla} \times \bar{f}(\bar{r},t)$, we have

$$\begin{aligned}\int_V \bar{\nabla} \cdot \bar{g}(\bar{r},t) dv &= \int_V \bar{\nabla} \cdot \bar{\nabla} \times \bar{f}(\bar{r},t) dv \\ &= \oint_S \bar{g}(\bar{r},t) \cdot \overline{ds} = \oint_S \bar{\nabla} \times \bar{f}(\bar{r},t) \cdot \overline{ds}\end{aligned}$$

for any arbitrary volume V enclosed by a closed surface S. Further using Stokes' theorem, we have

$$\oint_S \bar{\nabla} \times \bar{f}(\bar{r},t) \cdot \overline{ds} = 0$$

since the surface is closed and there is no line to evaluate the circulation of $\bar{f}(\bar{r},t)$. Therefore, we can deduce that

$$\int_V \bar{\nabla} \cdot \bar{\nabla} \times \bar{f}(\bar{r},t) dv = 0.$$

For satisfying this equality for any arbitrary volume V, the integrand must be zero, leading to the null identity.

Even though Gauss' law for magnetic fields is valid for both static and dynamic cases, we can easily check it for static cases involving steady electric currents. As an example, we consider an infinitely long line with zero thickness that carries a steady electric current I_0 along

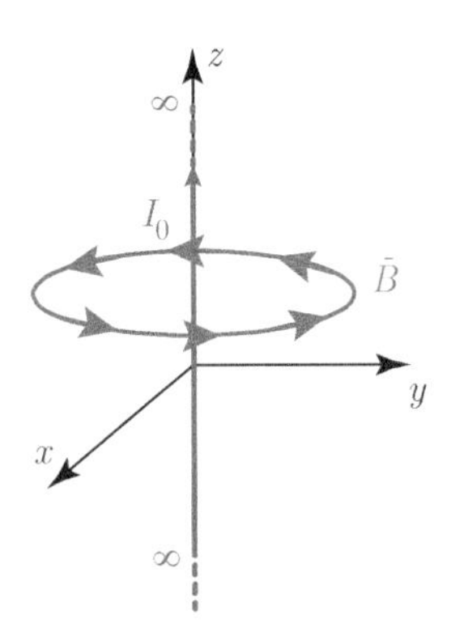

Figure 4.2 An infinitely long wire carrying a steady electric current flowing in the z direction in a vacuum.

the z axis in a vacuum (Figure 4.2). The magnetic flux density at an arbitrary point can be written

$$\bar{B}(\bar{r}) = \hat{a}_\phi \frac{\mu_0 I_0}{2\pi\rho} \tag{4.10}$$

in a cylindrical coordinate system. Then we have

$$\bar{\nabla} \cdot \bar{B}(\bar{r}) = \frac{1}{\rho}\frac{\partial}{\partial \rho}\left(\rho B_\rho(\bar{r})\right) + \frac{1}{\rho}\frac{\partial B_\phi(\bar{r})}{\partial \phi} + \frac{\partial B_z(\bar{r})}{\partial z} = \frac{1}{\rho}\frac{\partial B_\phi(\bar{r})}{\partial \phi} = 0 \tag{4.11}$$

everywhere, as expected.[7] But what about an arbitrary steady current distribution? According to the superposition principle, we have

$$\bar{B}(\bar{r}) = \frac{\mu_0}{4\pi}\int_V \frac{\bar{J}_v(\bar{r}') \times (\bar{r} - \bar{r}')}{|\bar{r} - \bar{r}'|^3} d\bar{r}' \tag{4.12}$$

for a steady volume electric current density $\bar{J}_v(\bar{r})$ in a vacuum. Evaluating the divergence, we have

$$\bar{\nabla} \cdot \bar{B}(\bar{r}) = \frac{\mu_0}{4\pi}\int_V \bar{\nabla} \cdot \left(\bar{J}_v(\bar{r}') \times \frac{(\bar{r} - \bar{r}')}{|\bar{r} - \bar{r}'|^3}\right) d\bar{r}'. \tag{4.13}$$

Then using the vector identity

$$\bar{\nabla} \cdot [\bar{f}(\bar{r},t) \times \bar{g}(\bar{r},t)] = \bar{g}(\bar{r},t) \cdot \bar{\nabla} \times \bar{f}(\bar{r},t) - \bar{f}(\bar{r},t) \cdot \bar{\nabla} \times \bar{g}(\bar{r},t) \tag{4.14}$$

and given that the divergence operation is evaluated in unprimed coordinates, we obtain[8]

$$\bar{\nabla} \cdot \bar{B}(\bar{r}) = \frac{\mu_0}{4\pi}\int_V \bar{J}_v(\bar{r}') \cdot \bar{\nabla} \times \left(\frac{(\bar{r} - \bar{r}')}{|\bar{r} - \bar{r}'|^3}\right) d\bar{r}' = 0. \tag{4.15}$$

4.1.2 Examples

Example 40: In a vacuum region defined as $0 \le x \le a$, $0 \le y \le b$, $-\infty < z < \infty$ (an infinitely long waveguide), the magnetic field intensity is given in the phasor domain as[9]

$$\bar{H}(\bar{r}) = \hat{a}_z c_1 \cos\left(\frac{\pi y}{b}\right)\exp(-\gamma z) + \hat{a}_y c_2 \sin\left(\frac{\pi y}{b}\right)\exp(-\gamma z), \tag{4.16}$$

where c_1, c_2, γ are constant with appropriate units. Find the relationship between c_1 and c_2.

[7] Note that the magnetic flux density is singular on the z axis: i.e. it becomes infinite at $\rho = 0$. On the other hand, $\bar{\nabla} \cdot \bar{B}(\bar{r}) = 0$ everywhere, including the positions on the z axis.

[8] Note that

$$\bar{\nabla} \times \left(\frac{(\bar{r} - \bar{r}')}{|\bar{r} - \bar{r}'|^3}\right) = 0.$$

This is consistent with the fact that the static electric field intensity due to a static point electric charge is irrotational (curl-free).

[9] As discussed in Chapter 7, this expression represents a TE_{10} mode.

Solution: According to Gauss' law for magnetic fields, we have

$$\bar{\nabla} \cdot \bar{B}(\bar{r}) = \mu_0 \bar{\nabla} \cdot \bar{H}(\bar{r}) = 0, \tag{4.17}$$

leading to

$$c_1 \cos\left(\frac{\pi y}{b}\right)(-\gamma)\exp(-\gamma z) + c_2\left(\frac{\pi}{b}\right)\cos\left(\frac{\pi y}{b}\right)\exp(-\gamma z) = 0. \tag{4.18}$$

Then we obtain[10]

$$c_1 = c_2\left(\frac{\pi}{\gamma b}\right) \longleftrightarrow c_2 = c_1\left(\frac{\gamma b}{\pi}\right). \tag{4.19}$$

[10] As a further practice, one can find the corresponding electric field intensity (using Ampere's law) as

$$\begin{aligned}
\bar{E}(\bar{r}) &= \frac{1}{j\omega\varepsilon_0}\bar{\nabla} \times \bar{H}(\bar{r}) \\
&= -\hat{a}_x c_1\left(\frac{\pi}{b}\right)\sin\left(\frac{\pi y}{b}\right)\exp(-\gamma z) \\
&\quad - \hat{a}_x c_2 \sin\left(\frac{\pi y}{b}\right)(-\gamma)\exp(-\gamma z) \\
&= \hat{a}_x \frac{1}{j\omega\varepsilon_0}\left[c_2\gamma - c_1\left(\frac{\pi}{b}\right)\right]\sin\left(\frac{\pi y}{b}\right)\exp(-\gamma z) \\
&= \hat{a}_x \frac{c_1}{j\omega\varepsilon_0}\left[\gamma^2\frac{b}{\pi} - \frac{\pi}{b}\right]\sin\left(\frac{\pi y}{b}\right)\exp(-\gamma z).
\end{aligned}$$

Example 41: Consider a static magnetic field defined as

$$\bar{B}(\bar{r}) = -\hat{a}_x B_0\left(\frac{x}{b}\right) + \hat{a}_z B_0\left(\frac{z+a}{a}\right), \tag{4.20}$$

where B_0 (Wb/m^2), a (m), and b (m) are constants. Find the magnetic flux through a hemispherical surface of radius a with normal direction $\hat{a}_R$ (Figure 4.3).

Solution: First, we can find apply Gauss' law to find the relationship between a and b as

$$\bar{\nabla} \cdot \bar{B}(\bar{r}) = -\frac{B_0}{b} + \frac{B_0}{a} = 0 \longrightarrow b = a. \tag{4.21}$$

Then the magnetic flux can be obtained:[11]

$$f_B = \int_S \bar{B}(\bar{r}) \cdot \overline{ds} = \int_S \bar{B}(\bar{r}) \cdot \hat{a}_R ds = \int_0^{2\pi}\int_0^{\pi/2} \bar{B}(\bar{r}) \cdot \hat{a}_R a^2 \sin\theta d\theta d\phi \tag{4.22}$$

$$\begin{aligned}
&= -\int_0^{2\pi}\int_0^{\pi/2} B_0\left(\frac{x}{a}\right)\hat{a}_x \cdot \hat{a}_R a^2 \sin\theta d\theta d\phi \\
&+ \int_0^{2\pi}\int_0^{\pi/2} B_0\left(\frac{z+a}{a}\right)\hat{a}_z \cdot \hat{a}_R a^2 \sin\theta d\theta d\phi
\end{aligned} \tag{4.23}$$

[11] We use

$$\begin{aligned}
x &= a\sin\theta\cos\phi \\
\hat{a}_x \cdot \hat{a}_R &= \sin\theta\cos\phi \\
z &= a\cos\theta \\
\hat{a}_z \cdot \hat{a}_R &= \cos\theta \\
\int_0^{2\pi}\cos^2\phi d\phi &= \frac{1}{2}\int_0^{2\pi}[\cos(2\phi)+1]d\phi = \pi \\
\sin\theta\cos\theta &= \frac{1}{2}\sin(2\theta) \\
\int_0^{\pi/2}\sin^3\theta d\theta &= \int_0^{\pi/2}\sin\theta(1-\cos^2\theta)d\theta \\
&= \left[-\cos\theta\right]_0^{\pi/2} + \frac{1}{3}\left[\cos^3\theta\right]_0^{\pi/2} \\
&= \frac{2}{3} \\
\int_0^{\pi/2}\cos^2\theta\sin\theta d\theta &= -\frac{1}{3}\left[\cos^3\theta\right]_0^{\pi/2} = \frac{1}{3}.
\end{aligned}$$

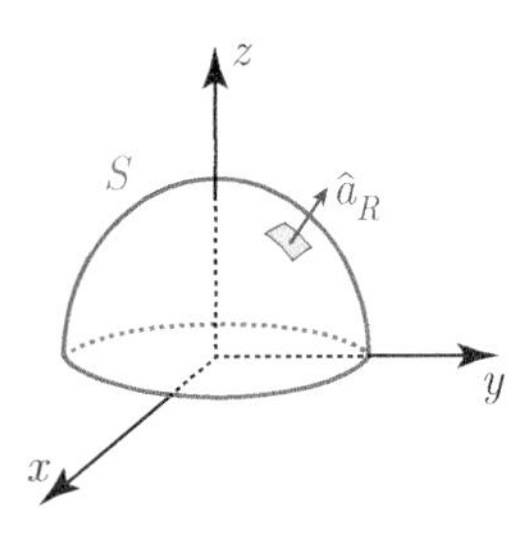

Figure 4.3 A hemispherical surface with normal direction $\hat{a}_R$. The surface is open: i.e. it is not closed from the bottom with a disk.

$$= -B_0a^2 \int_0^{2\pi}\int_0^{\pi/2} \sin^3\theta\cos^2\phi d\theta d\phi$$

$$+ B_0a^2 \int_0^{2\pi}\int_0^{\pi/2} \cos^2\theta\sin\theta d\theta d\phi + B_0a^2 \int_0^{2\pi}\int_0^{\pi/2} \cos\theta\sin\theta d\theta d\phi \tag{4.24}$$

$$= -B_0\pi a^2 \int_0^{\pi/2} \sin^3\theta d\theta$$

$$+ B_0 2\pi a^2 \int_0^{\pi/2} \cos^2\theta\sin\theta d\theta + B_0\pi a^2 \int_0^{\pi/2} \sin(2\theta)d\theta \tag{4.25}$$

$$= -\frac{2B_0\pi a^2}{3} + \frac{2B_0\pi a^2}{3} + B_0\pi a^2 = B_0\pi a^2. \tag{4.26}$$

On the other hand, instead of these lengthy integrals, one can use the solenoidal property of the magnetic flux density and evaluate the magnetic flux through a disk of radius a that complements the hemispherical surface.[12] Specifically, we have

$$f_B = \int_0^{2\pi}\int_0^{a} \bar{B}(z=0)\cdot\hat{a}_z\rho d\rho d\phi = \int_0^{2\pi}\int_0^{a} B_0\rho d\rho d\phi = B_0\pi a^2, \tag{4.27}$$

which is the same as the expression above.

[12]The disk of radius a on the x-y plane complements the hemispherical surface to a closed one. Since the net flux through the closed surface is zero, we should have

$$\int_S \bar{B}(\bar{r})\cdot\overline{ds} + \int_{\text{disk}} \bar{B}(\bar{r})\cdot\overline{ds} = 0,$$

when the normal direction is outward ($-z$ on the disk). Alternatively, we can claim that, since the hemispherical surface and the disk share the same circular edge, any magnetic flux passing through the disk must also pass through the hemispherical surface, given that magnetic field lines do not vanish/emerge at any location.

4.2 Boundary Conditions for Normal Magnetic Fields

Since there is no magnetic charge, the boundary condition for the normal component of the magnetic flux density can easily be obtained as (Figure 4.4)

$$\hat{a}_n\cdot[\bar{B}_1(\bar{r},t) - \bar{B}_2(\bar{r},t)] = 0 \qquad (\bar{r}\in S) \tag{4.28}$$

or

$$\hat{a}_n\cdot\bar{B}_1(\bar{r},t) = \hat{a}_n\cdot\bar{B}_2(\bar{r},t) \qquad (\bar{r}\in S), \tag{4.29}$$

where S is the boundary separating medium 1 and medium 2, and $\hat{a}_n$ is the normal direction pointing into the first medium (as in all boundary conditions in this book). Hence, the normal component of the magnetic flux density is continuous across boundaries. Interestingly, this is not directly true for the magnetic field intensity. If the permeability values of the first and second media are $\mu_1(\bar{r})$ and $\mu_2(\bar{r})$, respectively, we have

$$\hat{a}_n \cdot [\mu_1(\bar{r})\bar{H}_1(\bar{r},t) - \mu_2(\bar{r})\bar{H}_2(\bar{r},t)] = 0 \qquad (\bar{r} \in S) \tag{4.30}$$

or

$$\mu_1(\bar{r})\hat{a}_n \cdot \bar{H}_1(\bar{r},t) = \mu_2(\bar{r})\hat{a}_n \cdot \bar{H}_2(\bar{r},t) \qquad (\bar{r} \in S). \tag{4.31}$$

Hence, if $\mu_1(\bar{r}) \neq \mu_2(\bar{r})$, the normal component of the magnetic field intensity is discontinuous across the boundary. This is even in the absence of any electric current density at the boundary: i.e. even when the tangential magnetic field intensity is continuous according to the related boundary condition. One can explain the discontinuity by rewriting the boundary condition for the magnetic flux density (using $\bar{B}(\bar{r},t) = \mu_0\bar{H}(\bar{r},t) + \mu_0\bar{M}(\bar{r},t)$) as

$$\begin{aligned}&\hat{a}_n \cdot \mu_0\bar{H}_1(\bar{r},t) + \hat{a}_n \cdot \mu_0\bar{M}_1(\bar{r},t)\\ &\quad = \hat{a}_n \cdot \mu_0\bar{H}_2(\bar{r},t) + \hat{a}_n \cdot \mu_0\bar{M}_2(\bar{r},t) \qquad (\bar{r} \in S)\end{aligned} \tag{4.32}$$

or

$$\hat{a}_n \cdot [\bar{H}_1(\bar{r},t) - \bar{H}_2(\bar{r},t)] = \hat{a}_n \cdot [\bar{M}_2(\bar{r},t) - \bar{M}_1(\bar{r},t)] \qquad (\bar{r} \in S) \tag{4.33}$$

$$= \hat{a}_n \cdot \bar{M}(\bar{r},t) = \frac{q_{ms}(\bar{r},t)}{\mu_0} \qquad (\bar{r} \in S), \tag{4.34}$$

where $\bar{M}(\bar{r},t) = \bar{M}_2(\bar{r},t) - \bar{M}_1(\bar{r},t)$ is the net magnetization. In the above,

$$q_{ms}(\bar{r},t) = \mu_0\hat{a}_n \cdot \bar{M}(\bar{r},t) \tag{4.35}$$

with units Wb/m^2 acts like a surface magnetic charge density. This fictitious surface charge density is consistent with the fictitious volume charge density $q_{mv}(\bar{r},t)$ in Eq. (4.9).

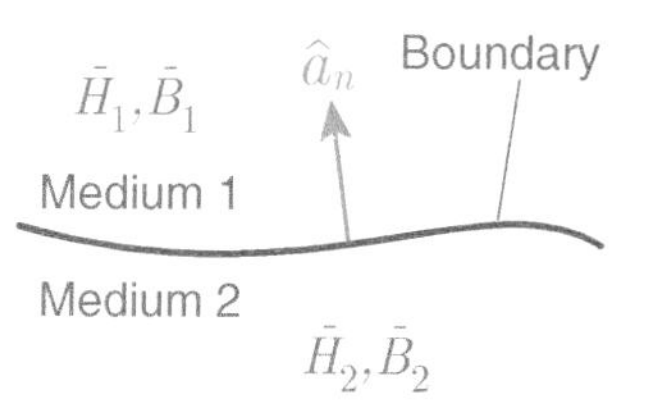

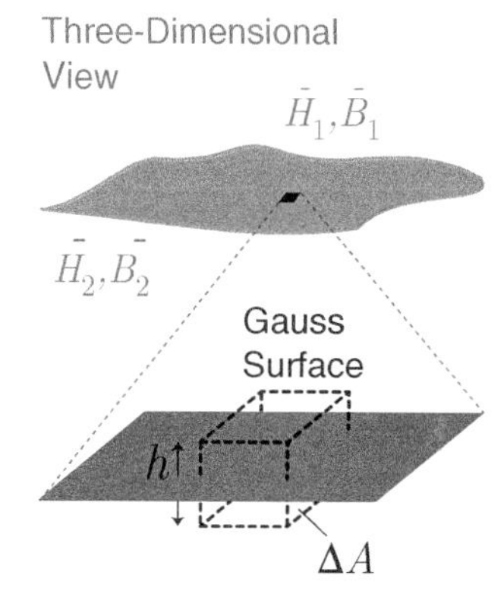

Figure 4.4 An illustration of the derivation of the boundary conditions for normal components of magnetic fields across an interface between two different media. Similar to the practice in the boundary conditions for normal electric fields (Section 1.5), one can use a rectangular prism around the boundary while its height shrinks to zero. Since there is no magnetic charge, the total magnetic flux through the bottom and top surfaces must balance each other, leading to the continuity of the normal magnetic flux density.

4.2.1 Examples

Example 42: Consider an interface at $x = 0$ with a steady surface electric current density $\bar{J}_s(\bar{r}) = \hat{a}_y 200$ (A/m). The permeability of the first ($x < 0$) and second ($x > 0$) media is $2\mu_0$ and $5\mu_0$, respectively (Figure 4.5). In addition, the magnetic field intensity in the first medium is given as $\bar{H}_1(\bar{r}) = \hat{a}_x 150 - \hat{a}_y 400 + \hat{a}_z 250$. Find the magnetic field intensity and the magnetic flux density at the interface on the side of the second medium.

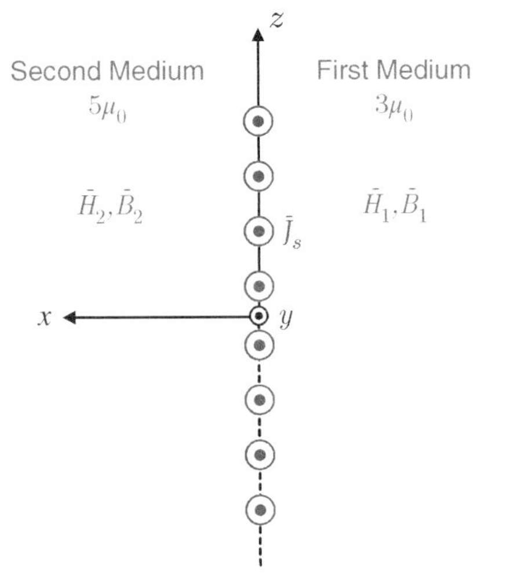

z

Second Medium $5\mu_0$ | First Medium $3\mu_0$

$\bar{H}_2, \bar{B}_2$ | $\bar{H}_1, \bar{B}_1$ | $\bar{J}_s$

x | y

Figure 4.5 An interface at $x = 0$ that separate two different media. A steady surface electric current density is defined in the y direction.

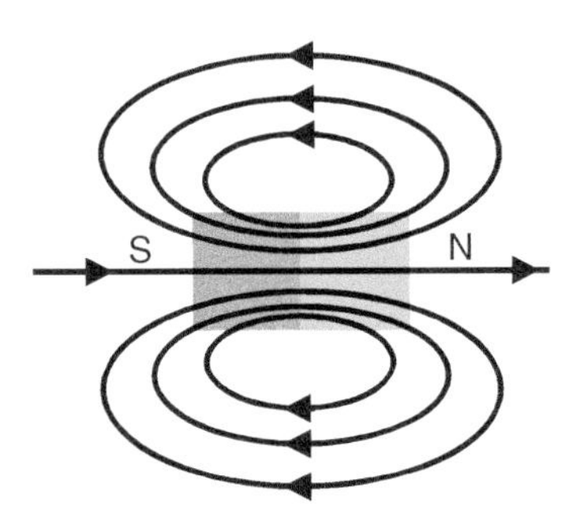

Figure 4.6 Due to the solenoidal nature of the magnetic flux density, the magnetic field lines should always close upon themselves. This is also true for a magnet, where the field lines continue inside the material.

Solution: First, the magnetic flux density in the first medium can be written

$$\bar{B}_1(\bar{r}) = \hat{a}_x 300\mu_0 - \hat{a}_y 800\mu_0 + \hat{a}_z 500\mu_0 \tag{4.36}$$

everywhere, including $x = 0^-$. Using the boundary conditions, we have

$$\bar{B}_{2n}(x=0) = \bar{B}_{1n}(x=0) \longrightarrow \bar{B}_{2n}(x=0) = \hat{a}_x 300\mu_0 \tag{4.37}$$

and

$$-\hat{a}_x \times [\bar{H}_1(\bar{r}) - \bar{H}_2(\bar{r})] = \bar{J}_s(\bar{r}) = \hat{a}_y 200 \tag{4.38}$$

$$\longrightarrow H_{2y}(x=0) = H_{1y}(x=0) = -400 \tag{4.39}$$

$$\longrightarrow H_{1z}(x=0) - H_{2z}(x=0) = 200 \tag{4.40}$$

$$\longrightarrow H_{2z}(x=0) = H_{1z}(x=0) - 200 = 50. \tag{4.41}$$

Then we derive

$$\bar{H}_2(x=0) = \hat{a}_x 60 - \hat{a}_y 400 + \hat{a}_z 50 \tag{4.42}$$

$$\bar{B}_2(x=0) = 5\mu_0 \bar{H}_2(x=0) = \hat{a}_x 300\mu_0 - \hat{a}_y 2000\mu_0 + \hat{a}_z 250\mu_0. \tag{4.43}$$

Note that $\bar{H}_2(\bar{r})$ and $\bar{B}_2(\bar{r})$ are known only at the boundary: i.e. at $x = 0$.

4.3 Static Cases and Magnetic Vector Potential

For time-harmonic dynamic cases (with an angular frequency of ω), Gauss' law for magnetic fields can be written in the phasor domain as

$$\bar{\nabla} \cdot \bar{B}(\bar{r}) = 0, \tag{4.44}$$

where $\bar{B}(\bar{r})$ is the phasor representation of $\bar{B}(\bar{r}, t)$. In a static case, we also have

$$\bar{\nabla} \cdot \bar{B}(\bar{r}) = 0. \tag{4.45}$$

The two equations above seem exactly the same; they both show the solenoidal nature of the magnetic flux density (Figures 4.6 and 4.7) that is valid for both dynamic and static cases. In this section, we particularly consider the solution of magnetostatic problems using the magnetic vector potential. This is similar to solving electrostatic problems via the electric scalar potential, as discussed in Section 3.6. To understand the magnetic vector potential and its use in magnetostatic problems, as well as energy due to magnetic interactions, we need the following tools and concepts.

- Magnetic potential energy
- Coulomb gauge

These are in addition to earlier tools and concepts:[13] differential surfaces with direction (Section 1.1.1), dot products (Section 1.1.2), flux of vector fields (Section 1.1.3), divergence of vector fields (Section 1.3.2), divergence theorem (Section 1.3.3), differential length with direction (Section 2.1.1), circulation of vector fields (Section 2.1.2), electric current density (Section 2.3.1), cross products (Section 2.3.2), curl of vector fields (Section 2.3.3), magnetic dipoles (Section 2.8.1), magnetization (Section 2.8.2), gradient of scalar fields (Section 3.6.1), gradient theorem (Section 3.6.3), and Poisson's and Laplace's equations (Section 3.6.7).

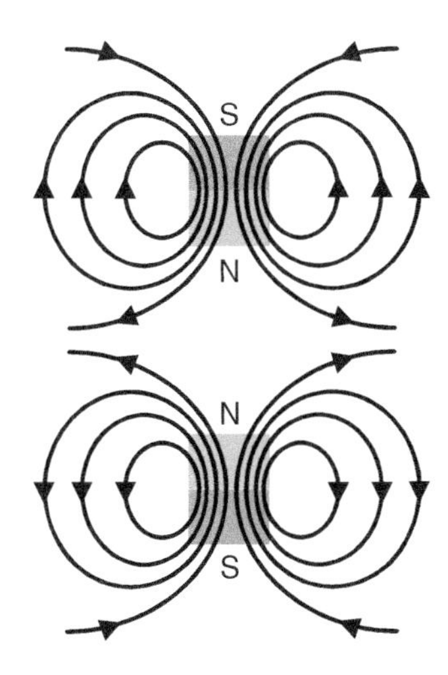

Figure 4.7 When two magnets are brought together, the magnetic field lines depend on their interaction, while the solenoidal nature of the magnetic flux density does not change.

4.3.1 Magnetic Vector Potential and Coulomb Gauge

Since $\bar{\nabla} \cdot \bar{B}(\bar{r},t) = 0$, the null identity states that[14]

$$\bar{B}(\bar{r},t) = \bar{\nabla} \times \bar{A}(\bar{r},t) \tag{4.46}$$

for a vector field $\bar{A}(\bar{r},t)$, which is defined as the magnetic vector potential. Taking the curl of the magnetic flux density, we have

$$\bar{\nabla} \times \bar{B}(\bar{r},t) = \bar{\nabla} \times \bar{\nabla} \times \bar{A}(\bar{r},t). \tag{4.47}$$

Using the curl operation twice on a vector field can be written[15]

$$\bar{\nabla} \times \bar{\nabla} \bar{f}(\bar{r},t) = \bar{\nabla}[\bar{\nabla} \cdot \bar{f}(\bar{r},t)] - \bar{\nabla} \cdot \bar{\nabla} \bar{f}(\bar{r},t) = \bar{\nabla}[\bar{\nabla} \cdot \bar{f}(\bar{r},t)] - \bar{\nabla}^2 \bar{f}(\bar{r},t), \tag{4.48}$$

where $\Delta = \bar{\nabla}^2 = \bar{\nabla} \cdot \bar{\nabla}$ is called the Laplace operator, as also used in Section 3.6.7. Hence, we obtain

$$\bar{\nabla} \times \bar{B}(\bar{r},t) = \bar{\nabla}[\bar{\nabla} \cdot \bar{A}(\bar{r},t)] - \bar{\nabla}^2 \bar{A}(\bar{r},t). \tag{4.49}$$

At this stage, we can use Ampere's law (the general form in Eq. (2.360)):

$$\bar{\nabla} \times \bar{B}(\bar{r},t) = \mu(\bar{r}) \bar{J}_v(\bar{r},t) + \mu(\bar{r}) \frac{\partial \bar{D}(\bar{r},t)}{\partial t} + \frac{\bar{\nabla}\mu(\bar{r})}{\mu(\bar{r})} \times \bar{B}(\bar{r},t) \tag{4.50}$$

[13] Note that, at this stage, we need almost all of the tools and concepts.

[14] Note that this is valid in general (for both dynamic and static cases).

$\bar{A}$: magnetic vector potential
Unit: webers/meter (Wb/m)

[15] This is similar to the BAC-CAB rule that is used for the cross product between three vectors.

to arrive at

$$\bar{\nabla}[\bar{\nabla} \cdot \bar{A}(\bar{r}, t)] - \bar{\nabla}^2 \bar{A}(\bar{r}, t) = \mu(\bar{r}) \bar{J}_v(\bar{r}, t) + \mu(\bar{r}) \frac{\partial \bar{D}(\bar{r}, t)}{\partial t} + \frac{\bar{\nabla}\mu(\bar{r})}{\mu(\bar{r})} \times \bar{B}(\bar{r}, t). \tag{4.51}$$

For a magnetically homogeneous medium, we further have

$$\bar{\nabla}^2 \bar{A}(\bar{r}, t) - \bar{\nabla}[\bar{\nabla} \cdot \bar{A}(\bar{r}, t)] = -\mu \bar{J}_v(\bar{r}, t) - \mu \frac{\partial \bar{D}(\bar{r}, t)}{\partial t}, \tag{4.52}$$

which turns into

$$\bar{\nabla}^2 \bar{A}(\bar{r}, t) - \bar{\nabla}[\bar{\nabla} \cdot \bar{A}(\bar{r}, t)] = -\mu_0 \bar{J}_v(\bar{r}, t) - \mu_0 \varepsilon_0 \frac{\partial \bar{E}(\bar{r}, t)}{\partial t} \tag{4.53}$$

in a vacuum.[16]

[16] Note that this equation is valid for all (static and dynamic) cases.

At this stage, we focus on the solution of the magnetic vector potential for static cases, similar to the solution of the electric scalar potential in Section 3.6. Dropping the time derivative of the electric field intensity, we have

$$\bar{\nabla}^2 \bar{A}(\bar{r}) - \bar{\nabla}[\bar{\nabla} \cdot \bar{A}(\bar{r})] = -\mu_0 \bar{J}_v(\bar{r}). \tag{4.54}$$

We already know how to solve Poisson's equation, while in the above, we have an extra term $\bar{\nabla}[\bar{\nabla} \cdot \bar{A}(\bar{r})]$ that prevents us from proceeding. But what is the gradient of the divergence of the magnetic vector potential? The *trick* to simplify the equation above is to *select* $\bar{\nabla} \cdot \bar{A}(\bar{r})$. According to Helmholtz decomposition, a vector field $\bar{f}(\bar{r}, t)$ is uniquely defined by its divergence $\bar{\nabla} \cdot \bar{f}(\bar{r}, t)$ and its curl $\bar{\nabla} \times \bar{f}(\bar{r}, t)$. Therefore, when we write $\bar{B}(\bar{r}, t) = \bar{\nabla} \times \bar{A}(\bar{r}, t)$ (including dynamic cases), there are infinitely many vector fields $\bar{A}(\bar{r}, t)$ that satisfy this equality. Consequently, we are free to choose $\bar{\nabla} \cdot \bar{A}(\bar{r}, t)$. This selection, which can be considered a process of eliminating redundant degrees of freedom, is called *gauge fixing*. To facilitate the solution of the magnetic vector potential for a static case, one of the best selection seems to be $\bar{\nabla} \cdot \bar{A}(\bar{r}) = 0$, which is called Coulomb gauge.[17] Then we obtain

[17] For dynamic cases, there are better selections for $\bar{\nabla} \cdot \bar{A}(\bar{r}, t)$.

$$\bar{\nabla}^2 \bar{A}(\bar{r}) = -\mu_0 \bar{J}_v(\bar{r}), \tag{4.55}$$

where the magnetic vector potential is written only in terms of the volume electric current density.

When the Laplace operator is applied to a vector field, it is applied to its components individually. In a Cartesian coordinate system,[18] we have

$$\bar{\nabla}^2 A_x(\bar{r}) = -\mu_0 J_{vx}(\bar{r}), \quad \bar{\nabla}^2 A_y(\bar{r}) = -\mu_0 J_{vy}(\bar{r}), \quad \bar{\nabla}^2 A_z(\bar{r}) = -\mu_0 J_{vz}(\bar{r}) \tag{4.56}$$

with separate equations for the x, y, and z components of $\bar{A}(\bar{r})$ and $\bar{J}_v(\bar{r})$. Then the solutions are obtained (as discussed in Section 3.6) as

$$A_x(\bar{r}) = \frac{\mu_0}{4\pi}\int_V \frac{J_{vx}(\bar{r}')}{|\bar{r}-\bar{r}'|}d\bar{r}' \tag{4.57}$$

$$A_y(\bar{r}) = \frac{\mu_0}{4\pi}\int_V \frac{J_{vy}(\bar{r}')}{|\bar{r}-\bar{r}'|}d\bar{r}' \tag{4.58}$$

$$A_z(\bar{r}) = \frac{\mu_0}{4\pi}\int_V \frac{J_{vz}(\bar{r}')}{|\bar{r}-\bar{r}'|}d\bar{r}', \tag{4.59}$$

leading to (Figure 4.8)

$$\bar{A}(\bar{r}) = \frac{\mu_0}{4\pi}\int_V \frac{\bar{J}_v(\bar{r}')}{|\bar{r}-\bar{r}'|}d\bar{r}'. \tag{4.60}$$

Similarly, for the surface electric current density and a line electric current (Figure 4.9), we have

$$\bar{A}(\bar{r}) = \frac{\mu_0}{4\pi}\int_S \frac{\bar{J}_s(\bar{r}')}{|\bar{r}-\bar{r}'|}d\bar{r}' \tag{4.61}$$

and[19]

$$\bar{A}(\bar{r}) = \frac{\mu_0}{4\pi}\oint_C \frac{I_0\overline{dl'}}{|\bar{r}-\bar{r}'|}, \tag{4.62}$$

respectively, where $\overline{dl'}$ contains the direction information.

Whenever a magnetic vector potential is obtained from an electric current distribution, one can find the magnetic flux density as[20] $\bar{B}(\bar{r}) = \bar{\nabla}\times\bar{A}(\bar{r})$. The relationship between $\bar{A}(\bar{r})$ and $\bar{J}_v(\bar{r})$ may be easier to use than the Biot-Savart law, which contains cross products in the numerator of the integrand. Therefore, in some cases, finding the magnetic vector

[18]This may be tricky when using other coordinate systems.

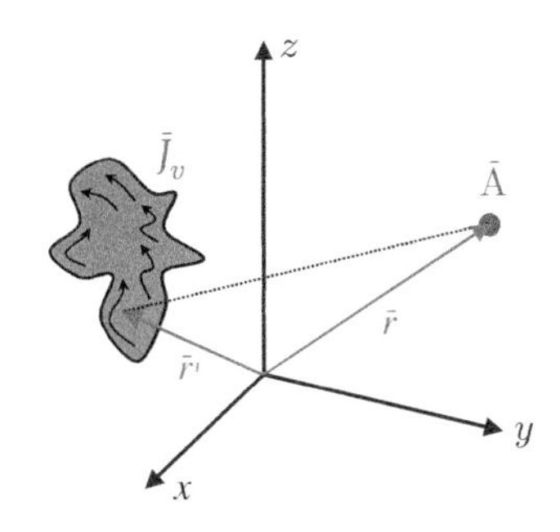

Figure 4.8 The magnetic vector potential due to a steady volume electric current density can be calculated via superposition. There is a direct relationship between the vector directions of $\bar{J}_v(\bar{r})$ and $\bar{A}(\bar{r})$.

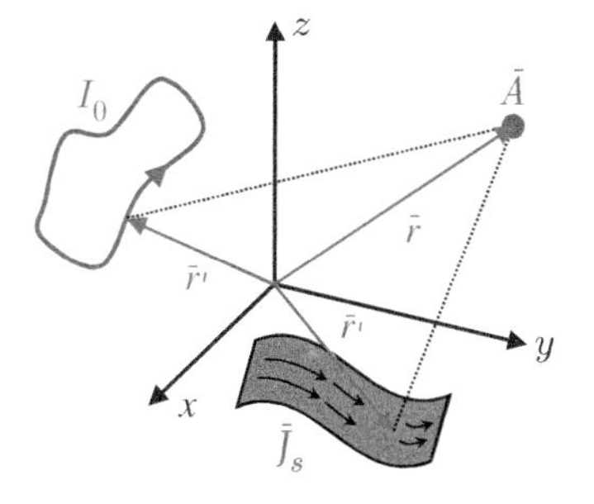

Figure 4.9 The magnetic vector potential due to surface and line current distributions can also be found via appropriate integration.

[19]We generally prefer closed line integrals in the case of a line electric current in static cases.

[20]Note that this is generally valid, including dynamic cases, as $\bar{B}(\bar{r},t) = \bar{\nabla}\times\bar{A}(\bar{r},t)$.

potential as the first step can be helpful to find the magnetic flux density caused by a current source.

Since the derivation above is done for a static case, the result should be consistent with the Biot-Savart law. Indeed, we can verify this as[21]

[21] Recall that

$$\bar{\nabla}\left(\frac{1}{|\bar{r}-\bar{r}'|}\right) = \frac{(\bar{r}-\bar{r}')}{|\bar{r}-\bar{r}'|^3}.$$

$$\bar{B}(\bar{r}) = \bar{\nabla}\times\bar{A}(\bar{r}) = \bar{\nabla}\times\frac{\mu_0}{4\pi}\int_V \frac{\bar{J}_v(\bar{r}')}{|\bar{r}-\bar{r}'|}d\bar{r}' \tag{4.63}$$

$$= \frac{\mu_0}{4\pi}\int_V \bar{\nabla}\times\left(\frac{\bar{J}_v(\bar{r}')}{|\bar{r}-\bar{r}'|}\right)d\bar{r}' \tag{4.64}$$

$$= \frac{\mu_0}{4\pi}\int_V [\bar{\nabla}\times\bar{J}_v(\bar{r}')]\frac{1}{|\bar{r}-\bar{r}'|}d\bar{r}' - \frac{\mu_0}{4\pi}\int_V \bar{J}_v(\bar{r}')\times\bar{\nabla}\left(\frac{1}{|\bar{r}-\bar{r}'|}\right)d\bar{r}' \tag{4.65}$$

$$= \frac{\mu_0}{4\pi}\int_V \bar{J}_v(\bar{r}')\times\frac{(\bar{r}-\bar{r}')}{|\bar{r}-\bar{r}'|^3}d\bar{r}'. \tag{4.66}$$

In the above, $\bar{\nabla}\times\bar{J}_v(\bar{r}')=0$ because $\bar{J}_v(\bar{r}')$ is in primed coordinates while the curl operation is in unprimed coordinates.

At this stage, we can revisit some static cases involving steady electric currents. First, we consider the magnetic field due to a magnetic dipole with dipole moment $\bar{I}_{DM,m}$. For this purpose, we consider a loop of radius a carrying a steady electric current I_0 located on the x-y plane (Figure 4.10). We have

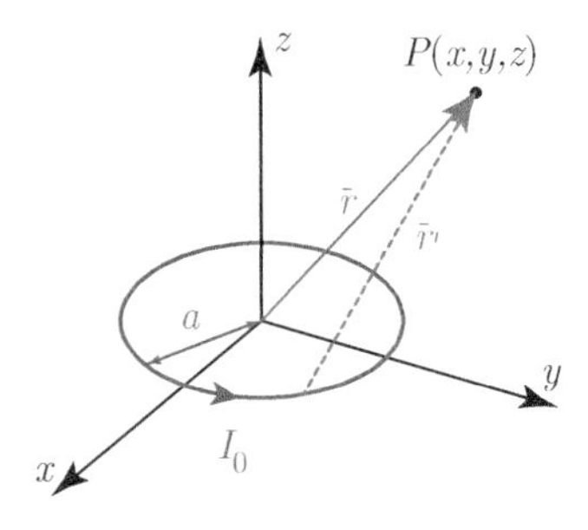

Figure 4.10 Magnetic dipole: i.e. a steady electric current flowing through a loop of radius a.

$$\bar{r} = \hat{a}_x x + \hat{a}_y y + \hat{a}_z z, \quad \bar{r}' = \hat{a}_x x' + \hat{a}_y y' \tag{4.67}$$

$$|\bar{r}-\bar{r}'| = \sqrt{(x-x')^2+(y-y')^2+z^2}, \quad \overline{dl'} = \hat{a}_\phi a d\phi', \tag{4.68}$$

leading to[22]

[22] We use

$$\hat{a}_\phi = -\hat{a}_x\sin\phi' + \hat{a}_y\cos\phi'$$
$$x' = a\cos\phi'$$
$$y' = a\sin\phi'.$$

$$\bar{A}_d(\bar{r}) = \frac{\mu_0 I_0 a}{4\pi}\int_0^{2\pi}\frac{\hat{a}_\phi d\phi'}{\sqrt{(x-x')^2+(y-y')^2+z^2}} \tag{4.69}$$

$$= \frac{\mu_0 I_0 a}{4\pi}\int_0^{2\pi}\frac{(-\hat{a}_x\sin\phi'+\hat{a}_y\cos\phi')d\phi'}{\sqrt{x^2-2xx'+(x')^2+y^2-2yy'+(y')^2+z^2}} \tag{4.70}$$

$$= \frac{\mu_0 I_0 a}{4\pi}\int_0^{2\pi}\frac{(-\hat{a}_x\sin\phi'+\hat{a}_y\cos\phi')d\phi'}{\sqrt{R^2+a^2-2xa\cos\phi'-2ya\sin\phi'}}, \tag{4.71}$$

where R is the spherical variable. If $R = |\bar{r}| \gg a$ – i.e. if the observation point is far from the dipole – one can approximate the denominator as

$$\sqrt{R^2 + a^2 - 2xa\cos\phi' - 2ya\sin\phi'} \approx \frac{1}{R}\left[1 - \frac{2xa\cos\phi'}{R^2} - \frac{2ya\sin\phi'}{R^2}\right]^{-1/2} \tag{4.72}$$

$$\approx \frac{1}{R}\left[1 + \frac{xa\cos\phi'}{R^2} + \frac{ya\sin\phi'}{R^2}\right]. \tag{4.73}$$

Then we obtain[23] (Figure 4.11)

$$\bar{A}_d(\bar{r}) \approx -\hat{a}_x\frac{\mu_0 I_0 a}{4\pi R}\int_0^{2\pi}\sin\phi'\left[1 + \frac{xa\cos\phi'}{R^2} + \frac{ya\sin\phi'}{R^2}\right]d\phi' \tag{4.74}$$

$$+\hat{a}_y\frac{\mu_0 I_0 a}{4\pi R}\int_0^{2\pi}\cos\phi'\left[1 + \frac{xa\cos\phi'}{R^2} + \frac{ya\sin\phi'}{R^2}\right]d\phi' \tag{4.75}$$

$$= -\hat{a}_x\frac{\mu_0 I_0 a^2}{4\pi R}\int_0^{2\pi}\frac{y\sin^2\phi'}{R^2}d\phi' + \hat{a}_y\frac{\mu_0 I_0 a^2}{4\pi R}\int_0^{2\pi}\frac{x\cos^2\phi'}{R^2}d\phi' \tag{4.76}$$

$$= -\hat{a}_x\frac{\mu_0 I_0 a^2 y}{4R^3} + \hat{a}_y\frac{\mu_0 I a^2 x}{4R^3} \tag{4.77}$$

$$= -\hat{a}_x\frac{\mu_0 I_0 a^2\sin\theta\sin\phi}{4R^2} + \hat{a}_y\frac{\mu_0 I_0 a^2\sin\theta\cos\phi}{4R^2} \tag{4.78}$$

$$= \hat{a}_\phi\frac{\mu_0 I_0 a^2}{4R^2}\sin\theta. \tag{4.79}$$

Using the expression for $\bar{A}_d(\bar{r})$, the magnetic flux density in the far zone ($|\bar{r}| \gg a$) can be found as[24]

$$\bar{B}_d(\bar{r}) = \bar{\nabla}\times\bar{A}_d(\bar{r}) = \frac{\mu_0 I_0 a^2}{4R^3}\left(\hat{a}_R 2\cos\theta + \hat{a}_\theta\sin\theta\right). \tag{4.80}$$

Furthermore, using $\hat{a}_z = (\hat{a}_R\cos\theta - \hat{a}_\theta\sin\theta)$, we have

$$\bar{B}_d(\bar{r}) = \frac{\mu_0 I_0 a^2}{4R^3}\left(\hat{a}_R 3\cos\theta - \hat{a}_z\right). \tag{4.81}$$

Finally, using the dipole moment $\bar{I}_{DM,m} = \hat{a}_z I_0\pi a^2$ and the identities $\cos\theta = \hat{a}_z\cdot\hat{a}_R$ and $R = |\bar{r}|$, we derive a generalized form as

$$\bar{B}_d(\bar{r}) = \frac{\mu_0}{4\pi|\bar{r}|^3}(3\hat{a}_R\hat{a}_R\cdot\bar{I}_{DM,m} - \bar{I}_{DM,m}). \tag{4.82}$$

This expression is consistent with the one in Eq. (2.351) that is found via the Biot-Savart law.

[23] We repeat these important symmetric integrals:

$$\int_0^{2\pi}\cos\phi' d\phi' = 0, \quad \int_0^{2\pi}\sin\phi' d\phi' = 0$$

$$\int_0^{2\pi}\sin\phi'\cos\phi' d\phi' = 0$$

$$\int_0^{2\pi}\sin^2\phi' d\phi' = \pi, \quad \int_0^{2\pi}\cos^2\phi' d\phi' = \pi.$$

[24] Note that to use $\bar{B}(\bar{r},t) = \bar{\nabla}\times\bar{A}(\bar{r},t)$, the expression for $\bar{A}(\bar{r},t)$ should be known fully with respect to position. Knowing $\bar{A}(\bar{r},t)$ for special cases may lead to incorrect expressions for the magnetic flux density. For example, if $\bar{A}(\bar{r})$ is found only on the z axis (again in the far zone), we have

$$\bar{A}(z) \approx 0$$

for $z \gg a$. Obviously $\bar{B}(z) = \bar{\nabla}\times\bar{A}(z) = 0$ is incorrect. Formally, if the magnetic flux density needs to be found at $\bar{r} = \bar{r}_0$, we should have (inserting $\bar{r} = \bar{r}_0$ after evaluating the curl)

$$\bar{B}(\bar{r} = \bar{r}_0) = [\bar{\nabla}\times\bar{A}(\bar{r})]_{\bar{r}=\bar{r}_0}$$

while

$$\bar{B}(\bar{r} = \bar{r}_0) \neq \bar{\nabla}\times\bar{A}(\bar{r} = \bar{r}_0)$$

in general.

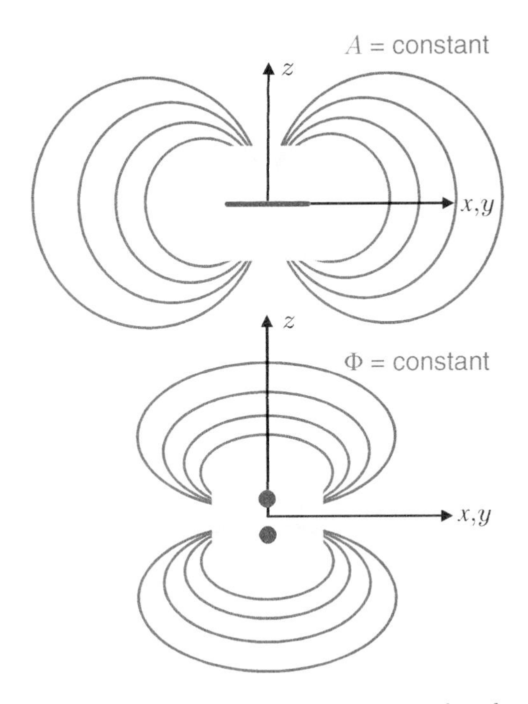

Figure 4.11 In the far zone, the equipotential surfaces for the magnetic vector potential due to a magnetic dipole satisfy $R^2 = \sin\theta$. As a comparison, the equipotential surfaces for the electric scalar potential due to an electric dipole satisfy $R^2 = \cos\theta$.

Using the magnetic vector potential is not always trivial, even for some simple cases. For example, we consider an infinitely long steady line electric current I_0 along the z axis located in a vacuum (Figure 4.2). Using Ampere's law,[25] the magnetic flux density can be obtained everywhere as

$$\bar{B}(\bar{r}) = \hat{a}_\phi \frac{\mu_0 I_0}{2\pi\rho}. \tag{4.83}$$

On the other side, the magnetic vector potential can be found as

$$\bar{A}(\bar{r}) = \frac{\mu_0}{4\pi}\int_C \frac{\hat{a}_z I_0}{|\bar{r}-\bar{r}'|}dl' = \hat{a}_z I_0 \frac{\mu_0}{4\pi}\int_{-\infty}^{\infty} \frac{1}{\sqrt{\rho^2+(z-z')^2}}dz'. \tag{4.84}$$

If the current line is infinitely long, the magnetic vector potential (and the magnetic flux density) should not depend on z, so we can just set $z = 0$. On the other hand, even when using this simplification, as well as the even symmetry of the integrand, we have

$$\bar{A}(\bar{r}) = \hat{a}_z I_0 \frac{\mu_0}{2\pi}\int_0^{\infty} \frac{1}{\sqrt{\rho^2+(z')^2}}dz' = \hat{a}_z \frac{\mu_0 I_0}{2\pi}\left[\ln\left(\sqrt{\rho^2+(z')^2}+z'\right)\right]_0^{\infty} \tag{4.85}$$

$$= \hat{a}_z \frac{\mu_0 I_0}{2\pi}\lim_{z_0\to\infty}\left[\ln\left(\sqrt{\rho^2+(z')^2}+z'\right)\right]_0^{z_0} \tag{4.86}$$

$$= \hat{a}_z \frac{\mu_0 I_0}{2\pi}\lim_{z_0\to\infty}\left[\ln\left(\frac{\sqrt{\rho^2+z_0^2}+z_0}{\rho}\right)\right]. \tag{4.87}$$

The final expression is infinite if $z_0 \to \infty$ – i.e. when the current line becomes infinitely long.[26] Specifically, the magnetic vector potential due to an infinitely long steady line electric current seems to be infinite. On the other hand, as shown above, the magnetic flux density is finite. To construct the relationship between the expressions, we can still evaluate the magnetic vector potential for $z_0 \gg z$ as

$$\bar{A}(\bar{r}) \approx \hat{a}_z \frac{\mu_0 I_0}{2\pi}\ln\left(\frac{2z_0}{\rho}\right). \tag{4.88}$$

Then the curl operation can be used as

$$\bar{B}(\bar{r}) = \bar{\nabla}\times\bar{A}(\bar{r}) \approx -\hat{a}_\phi \frac{\mu_0 I_0}{2\pi}\frac{\partial}{\partial\rho}\left[\ln\left(\frac{2z_0}{\rho}\right)\right] \tag{4.89}$$

[25] Note that this is a special case, where Ampere's law is applicable without resorting to the Biot-Savart law.

[26] Note that this is similar to the infinite electric scalar potential due to an infinitely long static line electric charge density.

leading to

$$\bar{B}(\bar{r}) \approx -\hat{a}_\phi \frac{\mu_0 I_0}{2\pi} \frac{\rho}{2z_0} \left(\frac{-2z_0}{\rho^2} \right) = \hat{a}_\phi \frac{\mu_0 I_0}{2\pi\rho}. \tag{4.90}$$

The final expression does not contain z_0. But formally, we still need to evaluate the limit $z_0 \to \infty$, leading to

$$\bar{B}(\bar{r}) = \lim_{z_0 \to 0} \left[\hat{a}_\phi \frac{\mu_0 I_0}{2\pi\rho} \right] = \hat{a}_\phi \frac{\mu_0 I_0}{2\pi\rho}. \tag{4.91}$$

As another example, we consider an infinitely long cylinder with a steady volume electric current density $\bar{J}_v(\bar{r}) = \hat{a}_z J_0$ (A/m^2) in a vacuum (Figure 4.12). Using the differential form $\bar{\nabla}^2 \bar{A}(\bar{r}) = -\mu_0 \bar{J}_v(\bar{r})$ (while keeping in mind the integral form[27]), we must have

$$\bar{\nabla}^2 A_z(\bar{r}) = -\mu_0 J_0 \qquad (\rho < a) \tag{4.92}$$

while Laplace's equation holds for $\rho > a$:

$$\bar{\nabla}^2 \bar{A}_z(\bar{r}) = 0 \qquad (\rho > a). \tag{4.93}$$

Assuming that $\bar{A}(\bar{r})$ only depends on ρ, we have[28]

$$\frac{1}{\rho} \frac{\partial}{\partial \rho} \left(\rho \frac{\partial A_z(\rho)}{\partial \rho} \right) = -\mu_0 J_0, \tag{4.94}$$

leading to

$$\frac{\partial}{\partial \rho} \left(\rho \frac{\partial A_z(\rho)}{\partial \rho} \right) = -\mu_0 J_0 \rho \longrightarrow \frac{\partial A_z(\rho)}{\partial \rho} = -\mu_0 J_0 \frac{\rho}{2} + \frac{c_1}{\rho} \tag{4.95}$$

$$\longrightarrow A_z(\rho) = -\mu_0 J_0 \frac{\rho^2}{4} + c_1 \ln(\rho) + c_2, \tag{4.96}$$

where c_1 and c_2 are constants. As in the case of the electric scalar potential, constant c_1 and c_2 depend on the selection of boundary values. In this case, if $A_z(\rho)$ is *selected* finite for $\rho = 0$, we must have $c_1 = 0$. In addition, a selection $A_z(\rho = 0) = 0$ leads to

$$A_z(\rho) = -\mu_0 J_0 \frac{\rho^2}{4} \qquad (\rho < a). \tag{4.97}$$

Cross-Sectional View

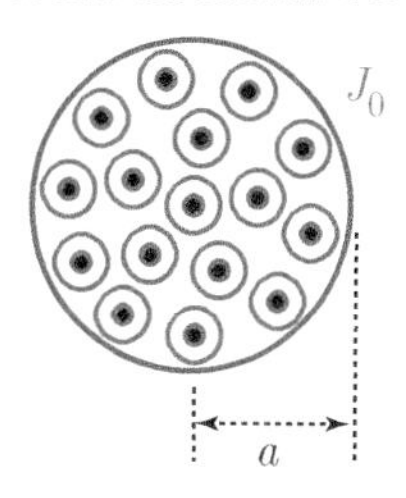

Figure 4.12 A steady volume electric current density $\bar{J}_v(\bar{r}) = \hat{a}_z J_0$ (A/m^2) flowing in a cylindrical region of radius a in the z direction. The electric current is assumed to be uniformly distributed over the cross-sectional surface (disk).

[27] Specifically, based on the fact that the current is in the z direction, we claim that the magnetic vector potential has only z component.

[28] Based on the symmetry, $\bar{A}(\bar{r})$ should not vary in the ϕ and z directions.

This also means $A_z(\rho = a) = -\mu_0 J_0 a^2/4$. Now, for $\rho > a$, we have

$$\frac{1}{\rho}\frac{\partial}{\partial\rho}\left(\rho\frac{\partial A_z(\rho)}{\partial\rho}\right) = 0 \longrightarrow A_z(\rho) = c_3\ln(\rho) + c_4, \tag{4.98}$$

where c_3 and c_4 are constants to be determined. Similar to the electric scalar potential, the magnetic vector potential must be continuous.[29] This means

[29] A discontinuity in the magnetic vector potential would create infinite magnetic flux density.

$$A_z(a^+) = A_z(a^-) \longrightarrow c_3\ln(a) + c_4 = -\mu_0 J_0 a^2/4, \tag{4.99}$$

while we are still unable to find c_3 and c_4 uniquely. In fact, these constants depend on other boundary conditions that we must reconsider. Since there is no surface electric current density at $\rho = a$, we must further have a continuity in the tangential component of the magnetic field intensity: i.e. $\hat{a}_\rho \times \bar{H}(\bar{r})$ must be continuous at $\rho = a$. Furthermore, since there is no magnetic material, the tangential component of the magnetic flux density: i.e. $\hat{a}_\rho \times \bar{B}(\bar{r})$, is also continuous.[30] This brings a new condition on $\bar{A}(\bar{r})$ as

[30] If the cylinder had a magnetic material, the tangential component of the magnetic flux density would be discontinuous, and this would simply change the expressions for the constants.

$$\bar{J}_s(\bar{r} = a) = 0 \quad \text{and} \quad \mu(\rho = a^+) = \mu(\rho = a^-) \tag{4.100}$$

$$\longrightarrow [\hat{a}_\rho \times \bar{\nabla} \times \bar{A}(\bar{r})]_{\rho=a^+} = [\hat{a}_\rho \times \bar{\nabla} \times \bar{A}(\bar{r})]_{\rho=a^-}. \tag{4.101}$$

Evaluating the curl operations, we have

$$[\hat{a}_\rho \times \bar{\nabla} \times \bar{A}(\bar{r})]_{\rho=a^-} = \left[\hat{a}_\rho \times \hat{a}_\phi \frac{\mu_0 J_0 \rho}{2}\right]_{\rho=a} = \hat{a}_z \frac{\mu_0 J_0 a}{2} \tag{4.102}$$

$$[\hat{a}_\rho \times \bar{\nabla} \times \bar{A}(\bar{r})]_{\rho=a^+} = \left[\hat{a}_\rho \times (-\hat{a}_\phi)\frac{c_3}{\rho}\right]_{\rho=a} = -\hat{a}_z \frac{c_3}{a}, \tag{4.103}$$

leading to $c_3 = -\mu_0 J_0 a^2/2$ and

$$c_3\ln(a) + c_4 = -\mu_0 J_0 a^2/4 \longrightarrow c_4 = -\mu_0 J_0 a^2/4 + \mu_0 J_0 \ln(a) a^2/2. \tag{4.104}$$

This leads to the magnetic vector potential expression[31]

[31] Note that inserting $\rho = a$ in the final expression leads to

$$A_z(\rho = a) = \frac{\mu_0 J_0 a^2}{2}[\ln(a/a) - 1/2] = -\frac{\mu_0 J_0 a^2}{4},$$

as desired due to the continuity of the magnetic vector potential. On the other side, note that

$$\lim_{\rho\to\infty} A_z(\rho) = -\infty.$$

$$A_z(\rho) = -\frac{\mu_0 J_0 a^2}{2}\ln(\rho) - \frac{\mu_0 J_0 a^2}{4} + \frac{\mu_0 J_0 \ln(a) a^2}{2} \tag{4.105}$$

$$= \frac{\mu_0 J_0 a^2}{2}[\ln(a/\rho) - 1/2] \qquad (\rho > a). \tag{4.106}$$

Finally, we can check these expressions by finding $\bar{B}(\bar{r}) = \bar{\nabla} \times \bar{A}(\bar{r})$. For $\rho < a$, we have[32]

$$\bar{B}(\bar{r}) = \bar{\nabla} \times \bar{A}(\bar{r}) = -\hat{a}_\phi \frac{\partial}{\partial \rho} A_z(\rho) = \hat{a}_\phi \frac{\mu_0 J_0 \rho}{2} \qquad (\rho < a). \tag{4.107}$$

[32] Note that the expressions for $\bar{A}(\bar{r})$ depend on a specific selection: i.e. its value when $\rho = 0$. On the other hand, independent of the selection of $\bar{A}(\rho = 0)$, the expressions for $\bar{B}(\bar{r})$ are unique.

This is consistent with the integral form of Ampere's law as

$$2\pi\rho H_\phi(\rho) = I_{\text{enc}} = J_0 \pi \rho^2 \longrightarrow \bar{H}(\bar{r}) = \hat{a}_\phi \frac{J_0 \pi \rho^2}{2\pi\rho} = \hat{a}_\phi \frac{J_0 \rho}{2} \qquad (\rho < a). \tag{4.108}$$

Similarly, for $\rho > a$, we have

$$\bar{B}(\bar{r}) = \bar{\nabla} \times \bar{A}(\bar{r}) = -\hat{a}_\phi \frac{\partial}{\partial \rho} A_z(\rho) = \hat{a}_\phi \frac{\mu_0 J_0 a^2}{2} \frac{\rho}{a} \frac{a}{\rho^2} = \hat{a}_\phi \frac{\mu_0 J_0 a^2}{2\rho}. \tag{4.109}$$

Obviously, using the integral form of Ampere's law, we obtain a consistent result as

$$2\pi\rho H_\phi(\rho) = I_{\text{enc}} = J_0 \pi a^2 \longrightarrow \bar{H}(\bar{r}) = \hat{a}_\phi \frac{J_0 \pi a^2}{2\pi\rho} = \hat{a}_\phi \frac{J_0 a^2}{2\rho} \qquad (\rho > a). \tag{4.110}$$

4.3.2 Examples

Example 43: Consider a ring-shaped (with inner radius a and outer radius b) surface current distribution in a vacuum (Figure 4.13). The surface electric current density is given as

$$\bar{J}_s(\bar{r}) = \hat{a}_\phi J_0 b / \rho, \tag{4.111}$$

where J_0 (A/m) is a constant. Find the magnetic vector potential and the magnetic flux density at an arbitrary point in the z axis.

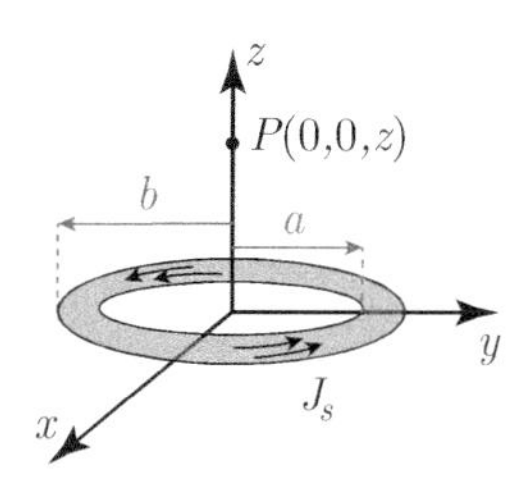

Figure 4.13 A steady surface electric current distribution with a ring shape.

Solution: Using the source integral Eq. (4.61), we have

$$\bar{A}(\bar{r}) = \frac{\mu_0}{4\pi} \int_S \frac{\bar{J}_s(\bar{r})}{|\bar{r} - \bar{r}'|} d\bar{r}' = \frac{\mu_0 J_0 b}{4\pi} \int_S \frac{\hat{a}_\phi}{\rho' |\bar{r} - \bar{r}'|} d\bar{r}'. \tag{4.112}$$

Further inserting $\bar{r} = \hat{a}_z z$, $\bar{r}' = \hat{a}_\rho \rho'$, and $d\bar{r}' = \rho' d\rho' d\phi'$, we obtain

$$\bar{A}(\bar{r}) = \frac{\mu_0 J_0 b}{4\pi} \int_0^{2\pi} \int_a^b \frac{\hat{a}_\phi}{\rho' [(\rho')^2 + z^2]^{1/2}} \rho' d\rho' d\phi'. \tag{4.113}$$

But the integral gives zero[33] – i.e. $\bar{A}(\bar{r}) = 0$, since $\hat{a}_\phi = -\hat{a}_x \sin\phi' + \hat{a}_y \cos\phi'$, while the integral is from 0 to 2π. On the other hand, a zero magnetic vector potential does not indicate a zero

[33] This can be interpreted as the perfect cancellation of contributions from electric currents in different directions.

magnetic flux density. Mathematically, this can be written

$$\bar{B}(\bar{r}_0) = [\bar{\nabla} \times \bar{A}(\bar{r})]_{\bar{r}=\bar{r}_0}, \tag{4.114}$$

i.e., the specific evaluation of the magnetic flux density at a position $\bar{r}_0$ can be done by inserting $\bar{r} = \bar{r}_0$ *after* the curl operation is evaluated.[34] Given that the electric current distribution is steady, the magnetic flux density can be found via the Biot-Savart law as

$$\bar{B}(\bar{r}) = \frac{\mu_0 J_0 b}{4\pi} \int_0^{2\pi} \int_a^b \frac{\hat{a}_\phi \times (\hat{a}_z z - \hat{a}_\rho \rho')}{\rho'[(\rho')^2 + z^2]^{3/2}} \rho' d\rho' d\phi', \tag{4.115}$$

leading to[35]

$$\bar{B}(\bar{r}) = \frac{\mu_0 J_0 b}{4\pi} \int_0^{2\pi} \int_a^b \frac{\hat{a}_\rho z + \hat{a}_z \rho'}{[(\rho')^2 + z^2]^{3/2}} d\rho' d\phi' \tag{4.116}$$

$$= \hat{a}_z \frac{\mu_0 J_0 b}{4\pi} \int_0^{2\pi} \int_a^b \frac{\rho'}{[(\rho')^2 + z^2]^{3/2}} d\rho' d\phi' \tag{4.117}$$

$$= \hat{a}_z \frac{\mu_0 J_0 b}{4\pi} 2\pi \left[-\frac{1}{\sqrt{(\rho')^2 + z^2}} \right]_a^b = \hat{a}_z \frac{\mu_0 J_0 b}{2} \left[\frac{1}{\sqrt{a^2 + z^2}} - \frac{1}{\sqrt{b^2 + z^2}} \right]. \tag{4.118}$$

Furthermore, if the total current flowing through the ring is defined as $I_0 = J_0(b - a)$, we have[36]

$$\bar{B}(\bar{r}) = \hat{a}_z \frac{\mu_0 I_0}{2} \frac{b}{b - a} \left[\frac{1}{\sqrt{a^2 + z^2}} - \frac{1}{\sqrt{b^2 + z^2}} \right] \tag{4.119}$$

$$= \hat{a}_z \frac{\mu_0 I_0}{2} \frac{b}{b - a} \left[\frac{\sqrt{b^2 + z^2} - \sqrt{a^2 + z^2}}{\sqrt{(a^2 + z^2)(b^2 + z^2)}} \right]. \tag{4.120}$$

Example 44: Consider an infinitely long cylinder with radius a carrying a steady electric current in the form of a volume electric current distribution in a vacuum (Figure 4.14). The magnetic vector potential for $\rho < a$ is given as

$$\bar{A}(\bar{r}) = -\hat{a}_z \frac{\mu J_0}{a} \rho^3, \tag{4.121}$$

where μ is the permeability (constant) of the cylinder and J_0 (A/m²) is a constant. Find the magnetic field intensity outside the cylinder.

[34] This is similar to $\bar{E}(\bar{r}_0) = -[\bar{\nabla}\Phi(\bar{r})]_{\bar{r}=\bar{r}_0}$. When the magnetic vector potential is found for a particular case (a specific position), the variance information (curl) may be lost.

[35] Using $\hat{a}_\rho = \hat{a}_x \cos\phi' + \hat{a}_y \sin\phi'$ so that its integral leads to zero.

[36] In this final form, it is possible to take a limit $a \to b$ as

$$\lim_{b\to a}\{\bar{B}(\bar{r})\}$$
$$= \hat{a}_z \frac{\mu_0 I_0}{2} \frac{a}{a^2 + z^2} \lim_{b\to a} \left\{ \frac{\sqrt{b^2 + z^2} - \sqrt{a^2 + z^2}}{b - a} \right\}$$
$$= \hat{a}_z \frac{\mu_0 I_0}{2} \frac{a}{a^2 + z^2} \lim_{b\to a} \left\{ \frac{b/\sqrt{b^2 + z^2}}{1} \right\}$$
$$= \hat{a}_z \frac{\mu_0 I_0}{2} \frac{a}{a^2 + z^2} \frac{a}{\sqrt{a^2 + z^2}}$$
$$= \hat{a}_z \frac{\mu_0 I_0}{2} \frac{a^2}{[a^2 + z^2]^{3/2}},$$

which is consistent with Eq. (2.260).

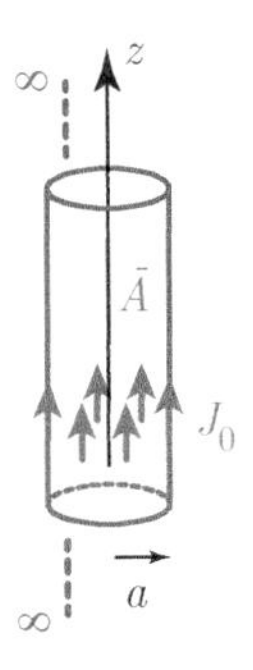

Figure 4.14 A steady volume electric current density flowing in a cylindrical region of radius a in the z direction. The magnetic vector potential is known inside the cylinder.

Solution: Given the magnetic vector potential, we can obtain the magnetic flux density inside the cylinder as

$$\bar{B}(\bar{r}) = \bar{\nabla} \times \bar{A}(\bar{r}) = -\hat{a}_\phi \frac{\partial A_z(\bar{r})}{\partial \rho} = \hat{a}_\phi \frac{3\mu J_0}{a}\rho^2 \qquad (\rho < a). \tag{4.122}$$

Then the volume electric current density can be obtained:

$$\bar{J}_v(\bar{r}) = \frac{1}{\mu}\bar{\nabla} \times \bar{B}(\bar{r}) = \hat{a}_z \frac{1}{\rho}\frac{\partial}{\partial \rho}[\rho B_\phi(\bar{r})] = \hat{a}_z \frac{1}{\rho}\frac{3J_0}{a}3\rho^2 = \hat{a}_z \frac{9J_0\rho}{a}. \tag{4.123}$$

The magnetic field intensity outside the cylinder can be found via the integral form of Ampere's law. First, we can find the total current flowing through the cylinder as

$$I_0 = \int_0^{2\pi}\int_0^a \frac{9J_0\rho}{a}\rho d\rho d\phi = \frac{9J_0}{a}2\pi\frac{a^3}{3} = 6\pi J_0 a^2. \tag{4.124}$$

Hence, we have $2\pi\rho H_\phi(\rho) = I_{\text{enc}} = 6\pi J_0 a^2$, leading to[37]

$$\bar{H}(\bar{r}) = \hat{a}_\phi \frac{3J_0a^2}{\rho}. \tag{4.125}$$

[37]Note that the magnetic field intensity is continuous across the surface of the cylinder since it is tangential and there is no surface electric current density.

4.3.3 Magnetic Potential Energy

Similar to the electric potential energy, which is related to the electric force, electric scalar potential, and electric fields, one can define the magnetic potential energy: i.e. a form of energy stored by means of magnetic fields. While there are alternative ways to approach the physical interpretation of the magnetic potential energy, we construct analogies with the electric potential energy and related concepts. First, we may consider two infinitesimal electric currents: $I_1\overline{dl}_1$ and $I_2\overline{dl}_2$ located in a vacuum, and the magnetic force between them, (Figure 4.15) as

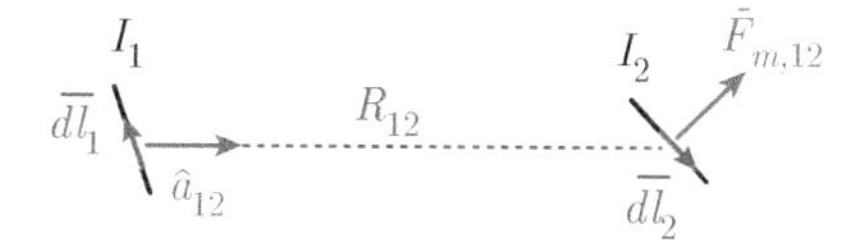

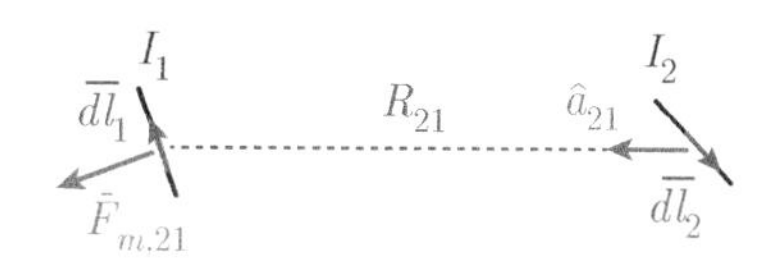

Figure 4.15 Magnetic force between two arbitrarily oriented differential electric currents. Recall that there is no antisymmetry in general.

$$\bar{F}_{m,12} = \frac{\mu_0}{4\pi}I_2\overline{dl}_2 \times \left[I_1\overline{dl}_1 \times \frac{\hat{a}_{12}}{(R_{12})^2}\right] \tag{4.126}$$

$$\bar{F}_{m,21} = \frac{\mu_0}{4\pi}I_1\overline{dl}_1 \times \left[I_2\overline{dl}_2 \times \frac{\hat{a}_{21}}{(R_{21})^2}\right], \tag{4.127}$$

where $R_{12} = R_{21} = |\bar{r}_1 - \bar{r}_2|$ is the distance between currents, $\hat{a}_{12}$ is the unit vector directed from the location of I_1 to the location of I_2, and $\hat{a}_{21} = -\hat{a}_{12}$. As discussed in Section 2.7.2, $|\bar{F}_{m,12}|$ may not be the same as $|\bar{F}_{m,21}|$: i.e. the magnetic force between infinitesimal currents is not necessarily antisymmetric.[38] In addition, it is possible to have $\bar{F}_{m,12} = 0$ or $\bar{F}_{m,21} = 0$ by

[38]Also recall that this is due to the unphysical nature of an isolated infinitesimal current.

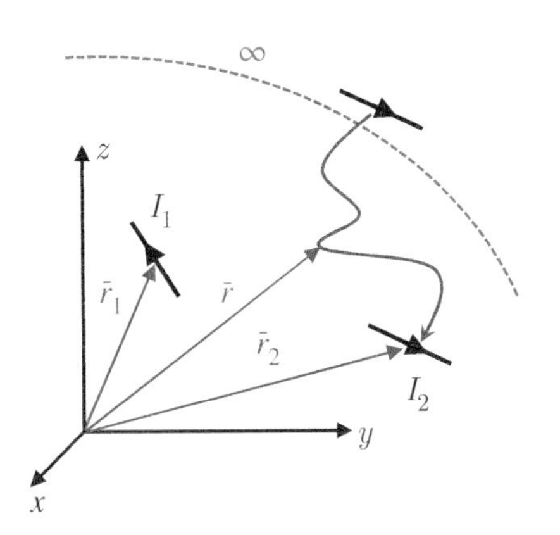

Figure 4.16 The magnetic potential energy stored due to the interaction of two infinitesimal currents can be found by integrating the magnetic force applied by the first current to the second one when it is brought from infinity to its final position.

[39]Here, the critical point is that there is a dot product between $\overline{dl}_1$ and $\overline{dl}$, which can be enforced to zero by selecting the path (integration variable $\overline{dl}$) accordingly. This is not generally possible for the second term, where the dot product is between $\overline{dl}_1$ and $\overline{dl}_2$: the directions of the infinitesimal currents.

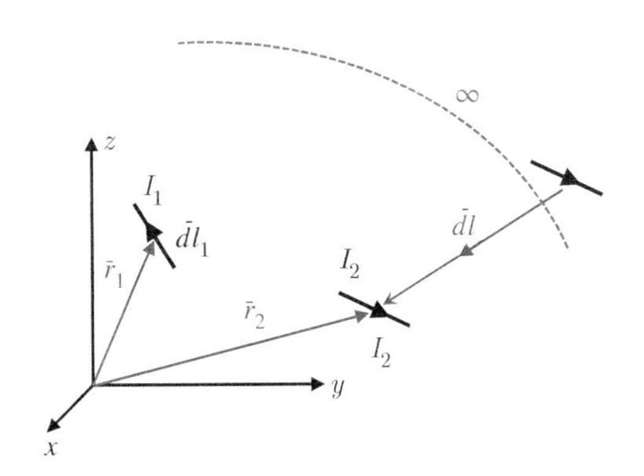

Figure 4.17 The movement path in Figure 4.16 can be selected as a straight line that is perpendicular to the direction of the first infinitesimal current ($\overline{dl}_1$), leading to $\overline{dl} \cdot \overline{dl}_1 = 0$.

aligning the differential currents without separating them by an infinite distance. Obviously, the magnetic force is quite different from the electric force, for which the distance is the major factor without any orientation concerns.

But how do we find the magnetic potential energy in current distributions? Similar to the electric potential energy, we can consider the amount of work done *against* magnetic forces. We should also have an equilibrium state to define zero energy such that the energy stored by a current distribution can be uniquely defined. As it is done for the electric potential energy for a system of electric charges, we can consider imaginary movements to bring electric currents together. In this case, however, we must be careful about the directions (of the electric currents).

For example, we consider again infinitesimal electric currents – i.e. $I_1\overline{dl}_1$ and $I_2\overline{dl}_2$ – and assume that $I_2\overline{dl}_2$ is brought from infinity (the case of zero energy) to its position $\bar{r}_2$ (Figure 4.16). The required energy should be the line integral of the magnetic force applied by $I_1\overline{dl}_1$ to $I_2\overline{dl}_2$ along the path of movement. In addition, it should not depend on the path itself. Specifically, we must integrate

$$\bar{F}_{m,12}(\bar{r}) = \frac{\mu_0}{4\pi} I_2\overline{dl}_2 \times \left[I_1\overline{dl}_1 \times \frac{(\bar{r}_1 - \bar{r})}{|\bar{r}_1 - \bar{r}|^3} \right] \tag{4.128}$$

along a suitable path C, where the location of the second infinitesimal current ($\bar{r}$) changes from infinity to $\bar{r}_2$. Further using the BAC-CAB rule, we have

$$\bar{F}_{m,12}(\bar{r}) = \frac{\mu_0}{4\pi} I_1\overline{dl}_1 \left[I_2\overline{dl}_2 \cdot \frac{(\bar{r}_1 - \bar{r})}{|\bar{r}_1 - \bar{r}|^3} \right] - \left[I_2\overline{dl}_2 \cdot I_1\overline{dl}_1 \right] \frac{(\bar{r}_1 - \bar{r})}{|\bar{r}_1 - \bar{r}|^3}, \tag{4.129}$$

which leads to

$$w_m = \frac{\mu_0}{4\pi} I_1\overline{dl}_1 \cdot \int_{\infty}^{\bar{r}_2} \left[I_2\overline{dl}_2 \cdot \frac{(\bar{r}_1 - \bar{r})}{|\bar{r}_1 - \bar{r}|^3} \right] \overline{dl} \tag{4.130}$$

$$- \frac{\mu_0}{4\pi} \left[I_2\overline{dl}_2 \cdot I_1\overline{dl}_1 \right] \int_{\infty}^{\bar{r}_2} \frac{(\bar{r}_1 - \bar{r})}{|\bar{r}_1 - \bar{r}|^3} \cdot \overline{dl} \tag{4.131}$$

as the work done (hence the stored magnetic potential energy) to bring $I_2\overline{dl}_2$ to the location $\bar{r}_2$ under the effect of $I_1\overline{dl}_1$.

The first integral above is not trivial to evaluate for an arbitrary path; but we can *select* a particular path with $\overline{dl} \perp \overline{dl}_1$ to see that this integral is in fact zero[39] (Figure 4.17). Therefore, the magnetic potential energy can be updated as

$$w_m = -\frac{\mu_0}{4\pi}\left[I_2\overline{dl}_2 \cdot I_1\overline{dl}_1\right]\int_{\infty}^{\bar{r}_2}\frac{(\bar{r}_1 - \bar{r})}{|\bar{r}_1 - \bar{r}|^3}\cdot\overline{dl} \tag{4.132}$$

$$= \frac{\mu_0}{4\pi}\left[I_2\overline{dl}_2 \cdot I_1\overline{dl}_1\right]\int_{\infty}^{\bar{r}_2}\bar{\nabla}\left(\frac{1}{|\bar{r}_1 - \bar{r}|}\right)\cdot\overline{dl} \tag{4.133}$$

$$= \frac{\mu_0}{4\pi}\left[I_2\overline{dl}_2 \cdot I_1\overline{dl}_1\right]\frac{1}{|\bar{r}_1 - \bar{r}_2|}, \tag{4.134}$$

where the gradient theorem is used[40]. Recognizing

$$\bar{A}_2 = \frac{\mu_0}{4\pi}\frac{I_1\overline{dl}_1}{|\bar{r}_1 - \bar{r}_2|} = \frac{\mu_0}{4\pi}\frac{I_1\overline{dl}_1}{R_{12}} \tag{4.135}$$

as the magnetic vector potential created by $I_1\overline{dl}_1$ at $\bar{r}_2$, we obtain the magnetic potential energy as

$$w_m = I_2\overline{dl}_2 \cdot \bar{A}_2. \tag{4.136}$$

This final expression is remarkably similar to the one in Eq. (3.155) for the electric potential energy involved in a pair of static point electric charges.[41]

[40]In addition, we use

$$\bar{\nabla}\left(\frac{1}{|\bar{r}_1 - \bar{r}|}\right) = \frac{(\bar{r}_1 - \bar{r})}{|\bar{r}_1 - \bar{r}|^3}.$$

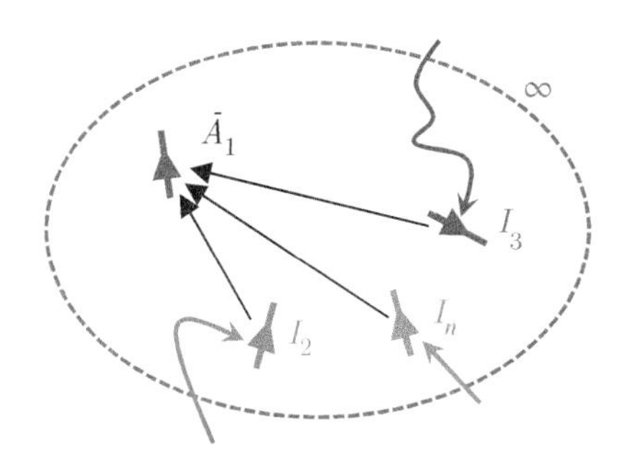

Figure 4.18 Finding the magnetic potential energy stored by a system of multiple infinitesimal currents.

[41]Specifically, we have $w_e = Q_2\Phi_2$, where Φ_2 is the electric scalar potential created by Q_1 at the location of Q_2.

4.3.3.1 Magnetic Potential Energy of Discrete Current Distributions

When there are more than two infinitesimal electric currents distributed in a vacuum, the magnetic potential energy in the system can be found by considering the total magnetic vector potential created at the location of each current (Figure 4.18). For this purpose, we generalize Eq. (4.136) as[42]

$$w_m = \frac{1}{2}I_1\overline{dl}_1 \cdot \bar{A}_1 + \frac{1}{2}I_2\overline{dl}_2 \cdot \bar{A}_2 \tag{4.137}$$

as the magnetic potential energy due to the interaction of two infinitesimal electric currents.[43] In the above, A_1 is the magnetic vector potential created by the second current at the location of the first current ($\bar{r}_1$) and A_2 is the magnetic vector potential created by the first current at the location of the second current ($\bar{r}_2$). Then for a system of n infinitesimal electric currents $I_1\overline{dl}_1, I_2\overline{dl}_2, \ldots, I_n\overline{dl}_n$, we have

$$w_m = \frac{1}{2}\sum_{k=1}^{n} I_k\overline{dl}_k \cdot \bar{A}_k, \tag{4.138}$$

[42]This is similar to the generalization in Section 3.6.5.1 for the electric scalar potential.

[43]Note that it does not matter how the currents are brought together. The magnetic potential energy depends only on the final positions and orientations.

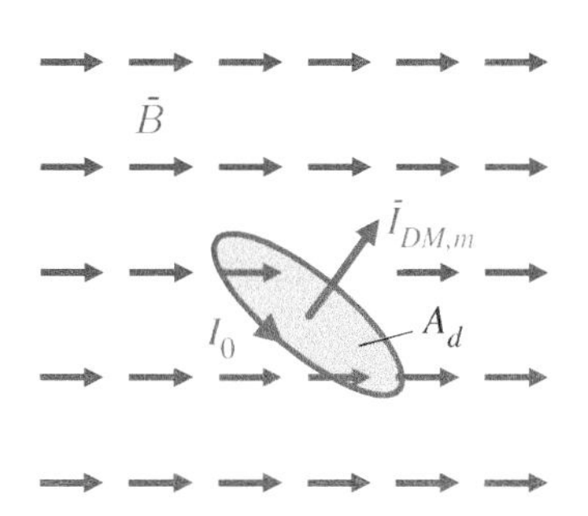

Figure 4.19 Magnetic potential energy is stored by a magnetic dipole when it is exposed to an external magnetic field.

where $\bar{A}_k$ is the total magnetic vector potential created by other $n-1$ currents at the position of the kth current. Similar to the expression in Eq. (3.163) for the electric potential energy due to a system of static point electric charges, the expression in Eq. (4.138) is the magnetic potential energy due to the interaction of n currents: i.e. it does not include the self energies of the currents (that is in fact infinity). Hence, we assume that these infinitesimal steady line electric currents are somehow formed initially before they are brought to their positions to interact.

4.3.3.2 Magnetic Potential Energy Stored by a Magnetic Dipole

We now consider the magnetic potential energy stored by a magnetic dipole. When a magnetic dipole (a circular line electric current) is exposed to an external magnetic field (Figure 4.19), it stores magnetic potential energy.[44] This is similar to the electric dipole storing electric potential energy under an external electric field. We further note that this phenomenon is important to explain the magnetization (Section 2.8): i.e. responses of magnetic materials to external magnetic fields.

[44]We emphasize that this is not the stored energy due to the self-interaction of the dipole (interactions between current segments). We are considering the stored energy due to the positioning and orientation of the dipole in an external magnetic field.

We start by considering a magnetic dipole with dipole moment (Figure 4.19)

$$\bar{I}_{DM,m} = \hat{a}_d I_0 A_d, \tag{4.139}$$

where I_0 is the steady line electric current flowing through the associated loop enclosing an area A_d, while $\hat{a}_d$ is the normal of the enclosed surface in accordance with the right-hand rule. To find the stored magnetic potential energy, we can consider each piece of the electric current brought from infinity (zero-energy location) to its position. To make this easier, we may assume that the loop has a square shape and it is formed of very short edges of length Δd numbered 1, 2, 3, and 4 (Figure 4.20). To bring edge 1 from infinity to its position $\bar{r}_1$, work must be done as

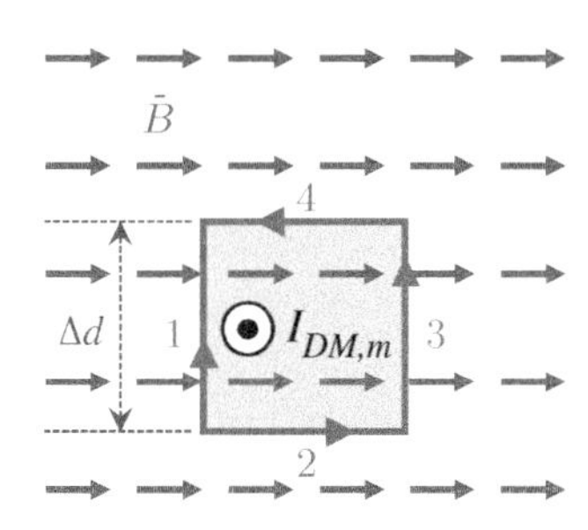

Figure 4.20 A magnetic dipole with a square shape can be considered to derive the stored magnetic potential energy.

$$w^{(1)}_{md,\text{stored}} = -I_0 \Delta d \int_{\infty}^{\bar{r}_1} \hat{a}_1 \times \bar{B}(\bar{r}') \cdot \overline{dl'}, \tag{4.140}$$

which is simply added to the stored magnetic potential energy. In this expression, a_1 is the unit vector in the direction of edge 1, whereas $\bar{r}_1$ represents its final position (e.g. the final position of its center). Similarly, we have

$$w^{(2)}_{md,\text{stored}} = -I_0 \Delta d \int_{\infty}^{\bar{r}_2} \hat{a}_2 \times \bar{B}(\bar{r}') \cdot \overline{dl'} \tag{4.141}$$

$$w^{(3)}_{md,\text{stored}} = -I_0 \Delta d \int_{\infty}^{\bar{r}_3} \hat{a}_3 \times \bar{B}(\bar{r}') \cdot \overline{dl'} \tag{4.142}$$

$$w^{(4)}_{md,\text{stored}} = -I_0 \Delta d \int_{\infty}^{\bar{r}_4} \hat{a}_4 \times \bar{B}(\bar{r}') \cdot \overline{dl'}. \tag{4.143}$$

for edges 2, 3, and 4. Furthermore, the square shape allows us to insert $\hat{a}_3 = -\hat{a}_1$, leading to[45]

$$w_{md,\text{stored}}^{(1)} + w_{md,\text{stored}}^{(3)}$$

$$= -I_0 \Delta d \int_{\infty}^{\bar{r}_1} \hat{a}_1 \times \bar{B}(\bar{r}') \cdot \overline{dl'} - I_0 \Delta d \int_{\bar{r}_3}^{\infty} \hat{a}_1 \times \bar{B}(\bar{r}') \cdot \overline{dl'} \tag{4.144}$$

$$= -I_0 \Delta d \int_{\bar{r}_3}^{\bar{r}_1} \hat{a}_1 \times \bar{B}(\bar{r}') \cdot \overline{dl'} \tag{4.145}$$

$$= I_0 \Delta d \int_{\bar{r}_3}^{\bar{r}_1} \bar{B}(\bar{r}') \cdot \hat{a}_1 \times \overline{dl'}. \tag{4.146}$$

[45]We also use

$$\hat{a}_1 \times \bar{B}(\bar{r}') \cdot \overline{dl'} = \overline{dl'} \cdot [\hat{a}_1 \times \bar{B}(\bar{r}')]$$
$$= \hat{a}_1 \cdot [\bar{B}(\bar{r}') \times \overline{dl'}] = \bar{B}(\bar{r}') \cdot [\overline{dl'} \times \hat{a}_1]$$
$$= -\bar{B}(\bar{r}') \cdot [\hat{a}_1 \times \overline{dl'}].$$

Because that the dipole is very small, the magnetic flux density can be assumed to be constant ($\bar{B}(\bar{r}') = \bar{B}_0$) over it so that

$$w_{md,\text{stored}}^{(1)} + w_{md,\text{stored}}^{(3)} = -I_0 \Delta d \bar{B}_0 \cdot \hat{a}_d (\Delta d/2) = -I_0 (\Delta d)^2 \bar{B}_0 \cdot \hat{a}_d/2. \tag{4.147}$$

Since we have

$$w_{md,\text{stored}}^{(2)} + w_{md,\text{stored}}^{(4)} = w_{md,\text{stored}}^{(1)} + w_{md,\text{stored}}^{(3)}, \tag{4.148}$$

we arrive at the total work done to form a magnetic dipole at its position, and hence the stored magnetic potential energy, as

$$w_{md,\text{stored}} = -I_0 (\Delta d)^2 \bar{B}_0 \cdot \hat{a}_d = -\bar{I}_{DM,m} \cdot \bar{B}_0. \tag{4.149}$$

This is very similar to the electric potential energy stored by an electric dipole[46] in Eq. (3.170).

According to Eq. (4.149), the magnetic potential energy stored by a magnetic dipole is zero when the dipole moment is perpendicular to the magnetic field direction[47] (Figure 4.21). In addition, the maximum energy is $|\bar{I}_{DM,m}||\bar{B}_0|$ when the dipole moment is the opposite of the magnetic field. If the dipole is in this position, it will *tend* to rotate (minimize its energy) if no external force is applied to keep it as it is.[48,49] Finally, the minimum energy occurs when the dipole moment is in the direction of the magnetic field. This is the *equilibrium* state, which is the preferred orientation if no external force (other than the applied magnetic field) exists. When a magnetic material is placed under an external magnetic field, magnetization occurs due to such magnetic dipoles oriented in the direction of the magnetic field.

[46]Specifically, we have $w_{ed,\text{stored}} = -\bar{I}_{DM,e} \cdot \bar{E}_0$ as the electric potential energy stored by an electric dipole with dipole moment $\bar{I}_{DM,e}$ in an external electric field intensity $\bar{E}_0$.

[47]This is also the case when the torque (force to rotate) is maximum.

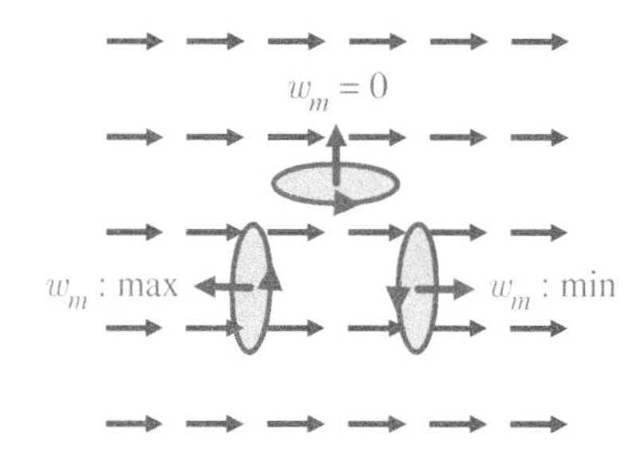

Figure 4.21 The magnetic potential energy stored by a magnetic dipole depends on its orientation.

[48]We assume that the dipole is a rigid body.

[49]If fact, the torque is zero when the dipole moment is perfectly aligned in the opposite direction, while it is unstable: i.e. any small perturbation makes it rotate to minimize its energy.

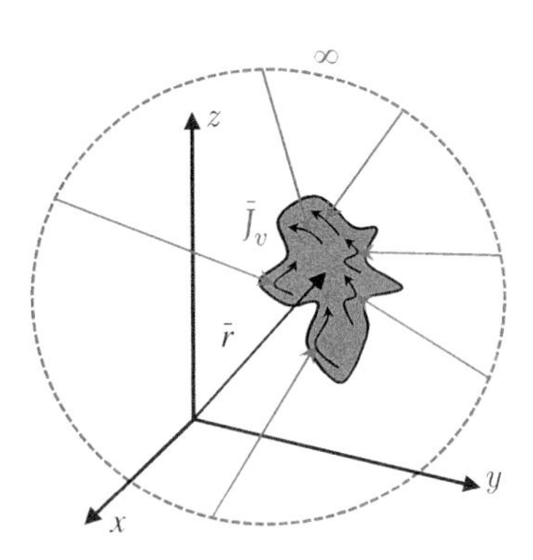

Figure 4.22 For a given current distribution, infinitesimal portions can be assumed to be brought from infinity to find the stored magnetic potential energy.

4.3.3.3 Stored Magnetic Potential Energy in Current Distributions

At the stage, we can find the stored magnetic potential energy in arbitrary electric current distributions, similar to the electric potential energy stored in electric charge distributions (discussed in Section 3.6.5.3). The stored magnetic potential energy can be defined as the total amount of work to bring electric currents together with respect to an equilibrium state[50] (with zero energy) (Figure 4.22). For a volume electric current density, the summation in Eq. (4.138) can be changed to an integral over the source as[51]

$$w_{m,\text{stored}} = \frac{1}{2}\int_V \bar{J}_v(\bar{r}) \cdot \bar{A}(\bar{r})dv. \tag{4.150}$$

This expression can be modified such that the stored magnetic potential energy can be found via field quantities. Given steady volume electric current density, we have

$$w_{m,\text{stored}} = \frac{1}{2}\int_V [\bar{\nabla} \times \bar{H}(\bar{r})] \cdot \bar{A}(\bar{r})dv \tag{4.151}$$

using Ampere's law. Furthermore, using the identity

$$\bar{\nabla} \cdot [\bar{f}(\bar{r},t) \times \bar{g}(\bar{r},t)] = \bar{g}(\bar{r},t) \cdot [\bar{\nabla} \times \bar{f}(\bar{r},t)] - \bar{f}(\bar{r},t) \cdot [\bar{\nabla} \times \bar{g}(\bar{r},t)] \tag{4.152}$$

for any vector fields $\bar{f}(\bar{r},t)$ and $\bar{g}(\bar{r},t)$, we obtain

$$\bar{A}(\bar{r}) \cdot [\bar{\nabla} \times \bar{H}(\bar{r})] = \bar{H}(\bar{r}) \cdot [\bar{\nabla} \times \bar{A}(\bar{r})] - \bar{\nabla} \cdot [\bar{A}(\bar{r}) \times \bar{H}(\bar{r})], \tag{4.153}$$

leading to

$$w_{m,\text{stored}} = \frac{1}{2}\int_V \bar{H}(\bar{r}) \cdot [\bar{\nabla} \times \bar{A}(\bar{r})]dv - \frac{1}{2}\int_V \bar{\nabla} \cdot [\bar{A}(\bar{r}) \times \bar{H}(\bar{r})]dv. \tag{4.154}$$

Then using the divergence theorem and $\bar{B}(\bar{r}) = \bar{\nabla} \times \bar{A}(\bar{r})$, we have

$$w_{m,\text{stored}} = \frac{1}{2}\int_V \bar{H}(\bar{r}) \cdot \bar{B}(\bar{r})dv - \frac{1}{2}\oint_S \bar{A}(\bar{r}) \times \bar{H}(\bar{r}) \cdot \overline{ds}, \tag{4.155}$$

where the second integral is converted into a surface integral enclosing the current density.

[50] Hence, it includes all work done to bring electric currents together, including their interactions with each other. Therefore, this energy is *always* positive.

[51] Similar integrals can be written for surface and line current distributions:

$$w_{m,\text{stored}} = \frac{1}{2}\int_S \bar{J}_s(\bar{r}) \cdot \bar{A}(\bar{r})ds$$

and

$$w_{m,\text{stored}} = \frac{1}{2}\oint_C I\overline{dl} \cdot \bar{A}(\bar{r}),$$

assuming closed contours for line currents.

The final equality above becomes useful when assuming that the integration domains are made infinity (Figure 4.23). We have[52]

$$\oint_{\infty} \bar{A}(\bar{r}) \times \bar{H}(\bar{r}) \cdot \overline{ds} \propto \oint (1/R)(1/R^2)ds \propto (1/R), \tag{4.156}$$

leading to[53]

$$w_{m,\text{stored}} = \frac{1}{2}\int_{\infty} \bar{H}(\bar{r}) \cdot \bar{B}(\bar{r})dv. \tag{4.157}$$

This integral should be evaluated *everywhere* (where the magnetic field is nonzero). Using $\bar{B}(\bar{r}) = \mu(\bar{r})\bar{H}(\bar{r})$, we further have

$$w_{m,\text{stored}} = \frac{1}{2}\int_{\infty} \mu(\bar{r})\bar{H}(\bar{r}) \cdot \bar{H}(\bar{r})dv, \tag{4.158}$$

which becomes

$$w_{m,\text{stored}} = \frac{\mu}{2}\int_{\infty} |\bar{H}(\bar{r})|^2 dv \tag{4.159}$$

for a homogeneous medium and

$$w_{m,\text{stored}} = \frac{\mu_0}{2}\int_{\infty} |\bar{H}(\bar{r})|^2 dv \tag{4.160}$$

for a vacuum.[54]

Like capacitors that can store electric potential energy, inductors are devices that can store magnetic potential energy. As an example, we can consider an ideal inductor that has a cylindrical shape of radius a (Figure 4.24). The inductor's length is l, and the material inside the cylinder (the core of the inductor) has a permeability of μ. Assuming a steady electric current I_0 flowing through a total of N turns, the magnetic field intensity inside the inductor can be approximated as

$$\bar{H}(\bar{r}) = \hat{a}_z I_0 N_l, \tag{4.161}$$

where $N_l = N/l$ is the number of turns per length. Note that there is no magnetic field outside the inductor. Therefore, the stored magnetic potential energy can be found as

$$w_{m,\text{stored}} = \frac{\mu}{2}\int_{\infty} |\bar{H}(\bar{r})|^2 dv = \frac{\mu}{2}\int_{\text{inside cylinder}} I_0^2 N_l^2 dv = \frac{\mu I_0^2 N_l^2}{2}(\pi a^2)l \tag{4.162}$$

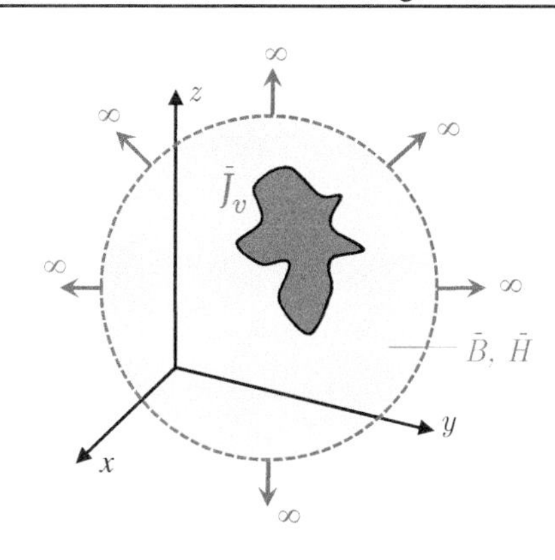

Figure 4.23 When finding the magnetic potential energy using only magnetic fields, the integral should be evaluated everywhere in space.

[52]Note that the surface integral brings a multiplication with R^2, while the integrand is overall proportional to $1/R^3$. Therefore, as the surface grows, the integral goes to zero with order $1/R$.

[53]Although the derivation here is based on magnetostatics, this final form in terms of magnetic fields is valid for *both* static and dynamic cases. This is similar to the derivation in Section 3.6.5.3, where the electric potential energy is written in terms of electric fields (that is again valid for both static and dynamic cases, although the derivation is based on electrostatics).

[54]Note the similarity of these equations with the corresponding ones in Section 3.6.5.3. Specifically, we have

$$w_{e,\text{stored}} = \frac{1}{2}\int_{\infty} \varepsilon(\bar{r})\bar{E}(\bar{r}) \cdot \bar{E}(\bar{r})dv,$$

which becomes

$$w_{e,\text{stored}} = \frac{\varepsilon}{2}\int_{\infty} |\bar{E}(\bar{r})|^2 dv$$

for a homogeneous medium and

$$w_{e,\text{stored}} = \frac{\varepsilon_0}{2}\int_{\infty} |\bar{E}(\bar{r})|^2 dv$$

for a vacuum.

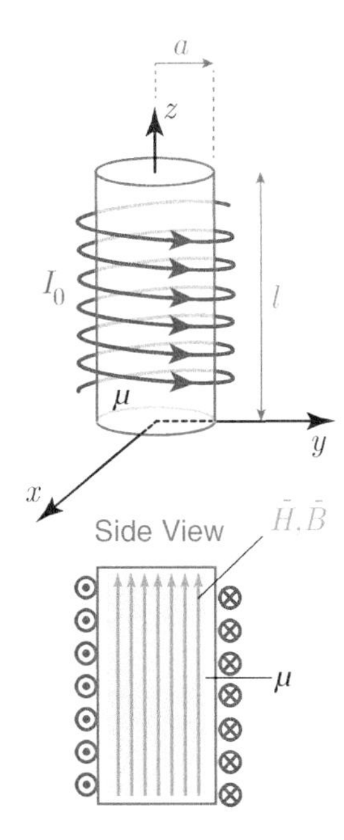

Figure 4.24 An ideal cylindrical inductor with a magnetic material inside.

or

$$w_{m,\text{stored}} = \frac{\mu I_0^2 N_l^2}{2}(\pi a^2) l = \frac{\mu I_0^2 N^2}{2l}(\pi a^2). \tag{4.163}$$

Defining the inductance of the inductor[55] as

$$L_0 = \frac{N f_{B0}}{I_0} = \frac{\mu N^2 \pi a^2}{l}, \tag{4.164}$$

where f_{B0} is the magnetic flux, we have

$$w_{m,\text{stored}} = \frac{1}{2} L_0 I_0^2 = \frac{1}{2}\frac{N^2 f_{B0}^2}{L_0}. \tag{4.165}$$

Obviously, this expression is always positive and corresponds to the energy stored by the inductor.

[55] Inductance calculations are detailed in Section 6.9.

4.3.3.4 Magnetic Potential Energy and Magnetic Force

When the stored magnetic potential energy in a system is known in terms of geometric and other physical parameters, the corresponding magnetic forces can be found based on the parameter to be considered the variable. Specifically, as discussed in Section 3.6.5.3 for the electric potential energy and the electric force, we visualize an imaginary movement (virtual displacement) for the *parameter* that is used to define the applied force. The force is then the gradient of the energy:

$$\bar{F}_m = \pm \bar{\nabla} w_{m,\text{stored}}, \tag{4.166}$$

where the sign depends on which quantity (magnetic flux or electric current) is taken as the constant of the system. If the magnetic flux is constant, we have[56]

$$\bar{F}_m = -\bar{\nabla} w_{m,\text{stored}} \qquad \text{(constant magnetic flux)}. \tag{4.167}$$

On the other hand, if the electric current is constant, the sign becomes plus:

$$\bar{F}_m = \bar{\nabla} w_{m,\text{stored}} \qquad \text{(constant electric current)}. \tag{4.168}$$

This is similar to taking electric charge or electric potential as fixed when finding the electric force from the electric potential energy.

[56] There is a formal derivation of the sign of the gradient expression relating the stored magnetic potential energy and the magnetic force. Basically, the case of constant flux occurs when the system is isolated from external sources. On the other hand, keeping constant current corresponds to an external current source that provides energy. When this energy is taken into account, the sign becomes plus.

As an example, we consider again an ideal inductor with a cylindrical shape of radius a and length l (Figure 4.24). The stored magnetic potential energy can be written

$$w_{m,\text{stored}} = \frac{1}{2}\frac{N^2 f_{B0}^2}{L_0} = \frac{N^2 f_{B0}^2}{2\mu N^2 \pi a^2/l} = \frac{f_{B0}^2 l}{2\mu\pi a^2} \tag{4.169}$$

or

$$w_{m,\text{stored}} = \frac{1}{2}L_0 I_0^2 = \frac{(\mu N^2 \pi a^2/l)I_0^2}{2} = \frac{\mu N^2 \pi a^2 I_0^2}{2l}, \tag{4.170}$$

where N is the number of turns. Given a as the virtual displacement variable (Figure 4.25), we have

$$\bar{F}_m = -\bar{\nabla} w_{m,\text{stored}} = -\hat{a}_a \frac{\partial}{\partial a}\left(\frac{f_{B0}^2 l}{2\mu\pi a^2}\right) = \hat{a}_a \frac{f_{B0}^2 l}{\mu\pi a^3}, \tag{4.171}$$

when the magnetic flux is assumed to be constant. Alternatively, assuming that the electric current is constant, we have

$$\bar{F}_m = \bar{\nabla} w_{m,\text{stored}} = \hat{a}_a \frac{\partial}{\partial a}\left(\frac{\mu N^2 \pi a^2 I_0^2}{2l}\right) = \hat{a}_a \frac{\mu N^2 \pi a I_0^2}{l}. \tag{4.172}$$

Obviously, these force expressions must be equal, which can be shown by using $L_0 = N f_{B0}/I_0 = \mu N^2 \pi a^2/l$ as

$$\bar{F}_m = \hat{a}_a \frac{f_{B0}^2 l}{\mu\pi a^3} = \hat{a}_a \frac{L_0^2 I_0^2 l/(N^2)}{\mu\pi a^3} = \hat{a}_a \frac{\mu^2 N^4 \pi^2 a^4 I_0^2 l}{l^2 N^2 \mu\pi a^3} = \hat{a}_a \frac{\mu N^2 \pi a I_0^2}{l}. \tag{4.173}$$

The positive value of the force indicates that it tries to increase a: i.e. to *expand* the inductor.[57]

An interesting example involves a magnetic core (with permeability μ) inserted partially into an ideal inductor (with radius a, length l, and number of turns N). The core is inserted by an amount of b, while the rest of the inductor is a vacuum (Figure 4.26). The energy stored in the system can be written

$$w_{m,\text{stored}} = \frac{1}{2}\int_{\infty} \bar{H}(\bar{r}) \cdot \bar{B}(\bar{r}) dv \tag{4.174}$$

$$= \frac{1}{2}\pi a^2 \int_0^b H_c B_c dz + \frac{1}{2}\pi a^2 \int_b^l H_0 B_0 dz, \tag{4.175}$$

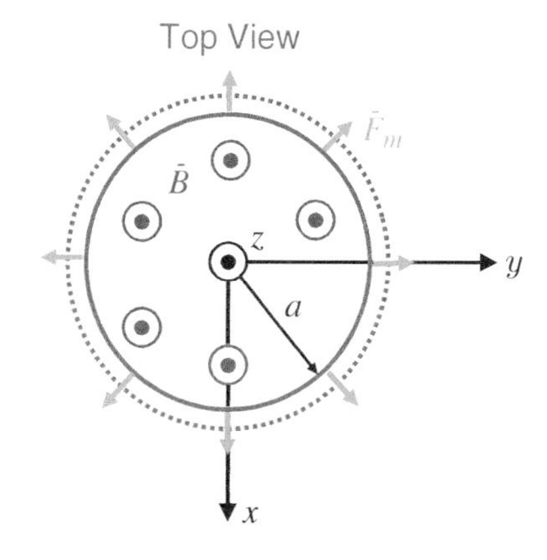

Figure 4.25 By considering the radius as the virtual displacement variable, we are able to find the magnetic force that tries to expand an ideal inductor.

[57] Hence, when being connected to a source, an inductor tries to break down by expanding (against physical forces that keep it intact). This is similar to a capacitor that tries to break down by collapsing.

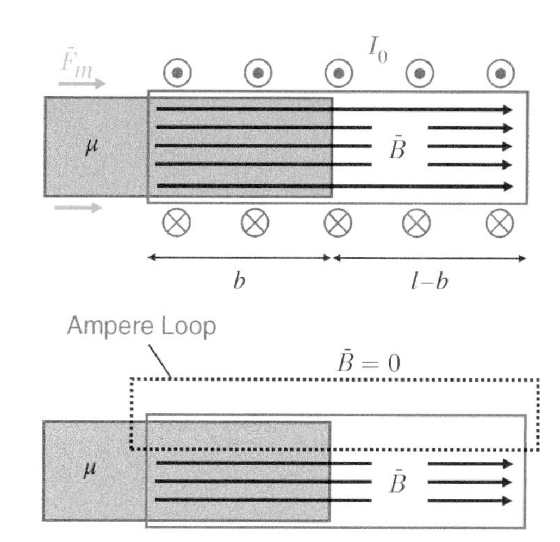

Figure 4.26 When it is partially inserted, a magnetic core is attracted into an inductor. The magnetic flux density can be found by using an Ampere loop, assuming that the inductor is ideal (there is zero magnetic flux density outside, while the magnetic field lines are parallel inside the inductor).

where

$$B_c = B_0 = \frac{NI_0}{b/\mu + (l-b)/\mu_0} \tag{4.176}$$

and

$$H_c = \frac{NI_0/\mu}{b/\mu + (l-b)/\mu_0} \tag{4.177}$$

$$H_0 = \frac{NI_0/\mu_0}{b/\mu + (l-b)/\mu_0} \tag{4.178}$$

are the magnitudes of the magnetic flux density and the magnetic field intensity,[58] in a vacuum (B_0 and H_0) and inside the core (B_c and H_c). For this example, using constant I_0 leads to lengthy expressions. Instead, one can choose the magnetic flux and the magnitude of the magnetic flux density constant (independent of the variable b), leading to[59]

$$w_{m,\text{stored}} = \frac{1}{2}\pi a^2 \int_0^b \frac{B_0^2}{\mu} dz + \frac{1}{2}\pi a^2 \int_b^l \frac{B_0^2}{\mu_0} dz \tag{4.179}$$

$$= \frac{1}{2}\pi a^2 B_0^2 \left(\frac{b}{\mu} + \frac{l-b}{\mu_0}\right). \tag{4.180}$$

Then given b as the variable, we have

$$\bar{F}_m = -\bar{\nabla} w_{m,\text{stored}} = -\hat{a}_b \frac{\partial w_{m,\text{stored}}}{\partial b} = \hat{a}_b \frac{1}{2}\pi a^2 B_0^2 \left(\frac{\mu-\mu_0}{\mu_0\mu}\right) \tag{4.181}$$

$$= \hat{a}_b \frac{1}{2}\pi a^2 \frac{N^2 I^2}{[b/\mu + (l-b)/\mu_0]^2} \left(\frac{\mu-\mu_0}{\mu_0\mu}\right), \tag{4.182}$$

which indicates that the force tries to increase[60] b (assuming $\mu > \mu_0$).

4.3.4 Examples

Example 45: Consider an infinitely long cylinder of radius a carrying a steady electric current with a constant volume electric current density $\bar{J}_v(\bar{r}) = \hat{a}_z J_0$ (A/m^2) in a vacuum (Figure 2.45). The permeability is μ_0 everywhere. Find the magnetic potential energy stored *inside* the cylinder per length.[61].

[58] These expressions can be found via Ampere's law. Specifically, using the Ampere loop in Figure 4.26, we have

$$\int_C \bar{H}(\bar{r}) \cdot \overline{dl} = bH_c + (l-b)H_0 = NI_0.$$

Then, inserting $H_c = B_c/\mu = B_0/\mu$ and $H_0 = B_0/\mu$, we have

$$bB_0/\mu + (l-b)B_0/\mu_0 = NI_0.$$

[59] Considering the expression in Eq. (4.176), it may be confusing to assume that the magnetic flux density is independent of b, as its expression is given particularly in terms of b. In fact, this is possible by assuming that I_0 is a function of b. Specifically, I_0 depends on b such that any change in b also changes I_0 accordingly, keeping B_0 constant.

[60] Hence, the magnetic core is attracted into the inductor.

[61] Note that the magnetic potential energy per length stored outside the cylinder would be

$$w_{ml,\text{external}} = \frac{1}{2}\int_{\rho>a} \bar{H}(\bar{r}) \cdot \bar{B}(\bar{r}) ds = \infty.$$

Solution: The magnetic field intensity can be found via the integral form of Ampere's law (as already practiced in Section 2.4) as

$$\bar{H}(\bar{r}) = \hat{a}_\phi \frac{I_{\text{enc}}}{2\pi\rho} = \hat{a}_\phi \frac{J_0 \pi \rho^2}{2\pi\rho} = \hat{a}_\phi \frac{J_0 \rho}{2} \qquad (\rho < a). \tag{4.183}$$

Then the magnetic potential energy stored inside the cylinder per length can be found as

$$w_{ml,\text{internal}} = \frac{1}{2}\int_{\rho<a} \bar{H}(\bar{r}) \cdot \bar{B}(\bar{r}) ds = \frac{1}{2}\int_0^a \int_0^{2\pi} \frac{J_0\rho}{2} \frac{\mu_0 J_0 \rho}{2} \rho d\phi d\rho \tag{4.184}$$

$$= \frac{1}{2} \frac{2\pi\mu_0 J_0^2}{4} \frac{a^4}{4} = \frac{\pi\mu_0 J_0^2 a^4}{16}. \tag{4.185}$$

If the total current $I_0 = J_0 \pi a^2$ is inserted, this expression becomes

$$w_{ml,\text{internal}} = \frac{\mu_0 I_0^2}{16\pi}. \tag{4.186}$$

This means the internal inductance per length of a nonmagnetic cylindrical cable (with uniform volume current distribution) is $\mu_0/(8\pi)$.

Example 46: Consider a coaxial structure involving two infinitely long cylindrical surfaces with radii a and b, where $b > a$, and a homogeneous magnetic material with permeability μ between them (Figure 4.27). Assuming that steady electric currents flow in opposite directions on the cylindrical surfaces with uniform distributions, find the magnetic potential energy stored per length.

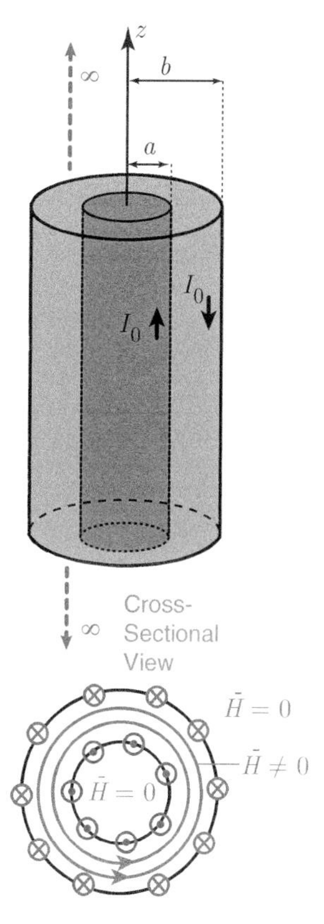

Figure 4.27 A coaxial structure involving two cylindrical surfaces with radii a and $b > a$ aligned in the z direction. Steady surface electric currents that are uniformly distributed over the cylindrical surfaces flow in opposite directions.

Solution: Using the integral form of Ampere's law, we can find the magnetic field intensity and the magnetic flux density as

$$\bar{H}(\bar{r}) = \hat{a}_\phi \frac{I_0}{2\pi\rho}, \quad \bar{B}(\bar{r}) = \hat{a}_\phi \frac{\mu I_0}{2\pi\rho} \qquad (a < \rho < b), \tag{4.187}$$

while $\bar{H}(\bar{r}) = 0$ and $\bar{B}(\bar{r}) = 0$ for $\rho < a$ and $\rho > b$. Then the energy per length can be found as

$$w_{ml,\text{stored}} = \frac{1}{2}\int_{\infty} \bar{H}(\bar{r}) \cdot \bar{B}(\bar{r}) ds = \frac{1}{2}\int_a^b \int_0^{2\pi} \frac{I_0}{2\pi\rho} \frac{\mu I_0}{2\pi\rho} \rho d\phi d\rho \tag{4.188}$$

$$= \frac{1}{2} \frac{\mu I_0^2}{4\pi^2} 2\pi \int_a^b \frac{1}{\rho} d\rho = \frac{\mu I_0^2}{4\pi} \ln(b/a). \tag{4.189}$$

Note that this is nothing but $w_{ml,\text{stored}} = L_{0l}I_0^2/2$, where $L_{0l} = \mu \ln(b/a)/(2\pi)$ is the inductance per length for a coaxial structure with uniform surface current distributions.[62]

[62]Inductance calculations are detailed in Section 6.9.

4.4 Combining All of Maxwell's Equations

At this stage,[63] by including Gauss' law for magnetic fields, we have all four Maxwell's equations to analyze all static and dynamic problems. This section focuses on the derivation of secondary equations to solve for electric and magnetic fields from sources. Before this, we revise Maxwell's equations as follows:

[63]In the rest of the book, including this section, all the tools and concepts introduced so far are extensively used.

- Electric fields are directly generated via electric charges (Gauss' law), as opposed to magnetic fields in the absence of magnetic charges (Gauss' law for magnetic fields). Faraday's law describes another kind of source for electric fields: changing magnetic fields that create rotational electric fields.
- Magnetic fields are directly generated via electric currents (Ampere's law). In addition, changing electric fields also create magnetic fields (Ampere's law). As a symmetric relationship, changing magnetic fields create electric fields (Faraday's law). This explains how electric and magnetic fields are coupled (generate each other) so that electromagnetic waves occur.
- There is no magnetic charge (Gauss' law for magnetic fields) so that the magnetic flux density is solenoidal (divergence-free).
- There is no magnetic current[64,65] that would create rotational electric fields. Therefore, in a static case, the electric field intensity is irrotational (curl-free).

[64]Magnetic currents should not be confused with magnetization currents, which are *electric* currents defined to represent magnetization effects.

[65]Magnetic currents can be defined *mathematically* for symmetric applications of Maxwell's equations.

With the knowledge of all of Maxwell's equations, we can now proceed to find secondary equations for electric and magnetic fields in terms of electric currents and charges (main sources[66]). In Section 3.5, Faraday's Law, Gauss' Law, and Ampere's Law are combined to arrive at

[66]By main sources, we mean the origins of electric and magnetic fields. Once these main sources create fields, electric and magnetic fields continue to create each other in dynamic cases.

$$\bar{\nabla} \times \bar{\nabla} \times \bar{E}(\bar{r}, t) + \mu\varepsilon \frac{\partial^2}{\partial t^2} \bar{E}(\bar{r}, t) = -\mu \frac{\partial \bar{J}_v(\bar{r}, t)}{\partial t} \tag{4.190}$$

$$\bar{\nabla} \times \bar{\nabla} \times \bar{H}(\bar{r}, t) + \mu\varepsilon \frac{\partial^2}{\partial t^2} \bar{H}(\bar{r}, t) = \bar{\nabla} \times \bar{J}_v(\bar{r}, t) \tag{4.191}$$

if the considered medium is homogeneous both electrically and magnetically.[67,68] Now, using the vector identity $\bar{\nabla} \times \bar{\nabla} \times \bar{f}(\bar{r}, t) = \bar{\nabla}[\bar{\nabla} \cdot \bar{f}(\bar{r}, t)] - \bar{\nabla}^2 \bar{f}(\bar{r}, t)$, we obtain

$$\bar{\nabla}^2 \bar{E}(\bar{r}, t) - \mu\varepsilon \frac{\partial^2}{\partial t^2} \bar{E}(\bar{r}, t) - \bar{\nabla}\bar{\nabla} \cdot \bar{E}(\bar{r}, t) = \mu \frac{\partial \bar{J}_v(\bar{r}, t)}{\partial t}. \quad (4.192)$$

The third term on the left-hand side can be further written in terms of the volume electric charge density using Gauss' law. Specifically, we have

$$\bar{\nabla}^2 \bar{E}(\bar{r}, t) - \mu\varepsilon \frac{\partial^2}{\partial t^2} \bar{E}(\bar{r}, t) = \mu \frac{\partial \bar{J}_v(\bar{r}, t)}{\partial t} + \frac{1}{\varepsilon} \bar{\nabla} q_v(\bar{r}, t). \quad (4.193)$$

It can be observed that Gauss' law for magnetic fields is not used to derive this equation for the electric field intensity. Using a similar procedure, we have

$$\bar{\nabla}^2 \bar{H}(\bar{r}, t) - \mu\varepsilon \frac{\partial^2}{\partial t^2} \bar{H}(\bar{r}, t) - \bar{\nabla}\bar{\nabla} \cdot \bar{H}(\bar{r}, t) = -\bar{\nabla} \times \bar{J}_v(\bar{r}, t). \quad (4.194)$$

Without knowing the divergence of the magnetic field intensity – i.e. $\bar{\nabla} \cdot \bar{H}(\bar{r}, t)$ – we cannot proceed on this equation. However, according to Gauss' law for magnetic fields, $\bar{\nabla} \cdot \bar{H}(\bar{r}, t) = 0$, and we obtain

$$\bar{\nabla}^2 \bar{H}(\bar{r}, t) - \mu\varepsilon \frac{\partial^2}{\partial t^2} \bar{H}(\bar{r}, t) = -\bar{\nabla} \times \bar{J}_v(\bar{r}, t). \quad (4.195)$$

It is interesting that Eqs. (4.193) and (4.195) are very similar to each other; each contains Laplace ($\bar{\nabla}^2$) of the related field intensity, a double time derivative term with a multiplier $\mu\varepsilon$, and a right-hand side involving sources $\bar{J}_v(\bar{r}, t)$ and $q_v(\bar{r}, t)$.

4.4.1 Wave Equations

Equations (4.193) and (4.195) are called non-homogeneous[69] wave equations since they describe how electric and magnetic fields behave like waves. From this perspective, these equations describe how electromagnetic waves are generated by sources: i.e. electric charges and currents. Specifically, circulation of the volume electric current density creates the magnetic field intensity that behaves like a wave. The gradient of the volume electric charge density and/or the time derivative of the volume electric current density creates the electric field intensity that behaves like a wave. These equations are important since they involve isolated electric and magnetic field intensities, while Maxwell's equations state that they are not completely independent from each other in dynamic cases.

[67] If the medium is not homogeneous, extra terms arise involving the gradient of the permittivity and/or the gradient of the permeability.

[68] Recall the derivation and sequence of assumptions as follows:

- Curl of Ampere's and Faraday's laws:

$$\bar{\nabla} \times \bar{\nabla} \times \bar{H}(\bar{r}, t) = \bar{\nabla} \times \bar{J}_v(\bar{r}, t) + \frac{\partial}{\partial t} \bar{\nabla} \times \bar{D}(\bar{r}, t)$$

$$\bar{\nabla} \times \bar{\nabla} \times \bar{E}(\bar{r}, t) = -\frac{\partial}{\partial t} \bar{\nabla} \times \bar{B}(\bar{r}, t)$$

- Constant permittivity/permeability:

$$\bar{\nabla} \times \bar{\nabla} \times \bar{H}(\bar{r}, t) = \bar{\nabla} \times \bar{J}_v(\bar{r}, t) + \varepsilon \frac{\partial}{\partial t} \bar{\nabla} \times \bar{E}(\bar{r}, t)$$

$$\bar{\nabla} \times \bar{\nabla} \times \bar{E}(\bar{r}, t) = -\mu \frac{\partial}{\partial t} \bar{\nabla} \times \bar{H}(\bar{r}, t)$$

- Inserting Faraday's law and Ampere's law:

$$\bar{\nabla} \times \bar{\nabla} \times \bar{H}(\bar{r}, t) = \bar{\nabla} \times \bar{J}_v(\bar{r}, t) - \varepsilon \frac{\partial^2}{\partial t^2} \bar{B}(\bar{r}, t)$$

$$\bar{\nabla} \times \bar{\nabla} \times \bar{E}(\bar{r}, t) = -\mu \frac{\partial \bar{J}_v(\bar{r}, t)}{\partial t} - \mu \frac{\partial^2}{\partial t^2} \bar{D}(\bar{r}, t)$$

- Constant permeability/permittivity:

$$\bar{\nabla} \times \bar{\nabla} \times \bar{H}(\bar{r}, t) + \mu\varepsilon \frac{\partial^2}{\partial t^2} \bar{H}(\bar{r}, t) = \bar{\nabla} \times \bar{J}_v(\bar{r}, t)$$

$$\bar{\nabla} \times \bar{\nabla} \times \bar{E}(\bar{r}, t) + \mu\varepsilon \frac{\partial^2}{\partial t^2} \bar{E}(\bar{r}, t) = -\mu \frac{\partial \bar{J}_v(\bar{r}, t)}{\partial t}$$

[69] In this context, the term *non-homogeneous* is used to indicate a nonzero right-hand side so that the corresponding *homogeneous* equation means the one with a zero right-hand side.

General solutions of non-homogeneous wave Eqs. (4.193) and (4.195) in terms of sources are slightly complicated due to the right-hand sides that contain derivative operations. In fact, as shown in Section 4.4.2, there are relatively easier equations involving potentials. On the other hand, we can still investigate wave equations for the fields and understand how they represent waves propagating in space. For this purpose, we consider homogeneous versions (with zero right-hand sides) of the equations as[70]

$$\bar{\nabla}^2\bar{E}(\bar{r},t) - \mu\varepsilon\frac{\partial^2}{\partial t^2}\bar{E}(\bar{r},t) = 0 \tag{4.196}$$

$$\bar{\nabla}^2\bar{H}(\bar{r},t) - \mu\varepsilon\frac{\partial^2}{\partial t^2}\bar{H}(\bar{r},t) = 0. \tag{4.197}$$

Note that wave equations are *pointwise*. Thus the homogeneous equations are valid whenever a source does not exist at the particular position $\bar{r}$, even when electric and magnetic fields are nonzero at that position, as they can be created by sources at other locations. The solutions of the homogeneous equations further need boundary conditions; hence, the resulting electric and magnetic fields naturally depend on the problem. But as an example, we consider an electric field intensity that has only an x component. The corresponding equation that must be solved becomes

$$\bar{\nabla}^2 E_x(\bar{r},t) - \mu\varepsilon\frac{\partial^2}{\partial t^2}E_x(\bar{r},t) = 0 \tag{4.198}$$

$$\frac{\partial^2}{\partial x^2}E_x(\bar{r},t) + \frac{\partial^2}{\partial y^2}E_x(\bar{r},t) + \frac{\partial^2}{\partial z^2}E_x(\bar{r},t) - \mu\varepsilon\frac{\partial^2}{\partial t^2}E_x(\bar{r},t) = 0, \tag{4.199}$$

while[71] $E_y(\bar{r},t) = 0$ and $E_z(\bar{r},t) = 0$. Note that having only an x component does not indicate that the electric field intensity depends only on x; it may also depend on y and z. Therefore, to further simplify the analysis, we assume that $E_x(\bar{r},t)$ depends only on z, leading to

$$\frac{\partial^2}{\partial z^2}E_x(z,t) - \mu\varepsilon\frac{\partial^2}{\partial t^2}E_x(z,t) = 0. \tag{4.200}$$

Hence, we arrive at an equation with only $E_x(z,t)$ that depends on two variables: position z and time t.

[70] In a Cartesian system, these can be written as six different equations:

$$\bar{\nabla}^2 E_x(\bar{r},t) - \mu\varepsilon\frac{\partial^2}{\partial t^2}E_x(\bar{r},t) = 0$$
$$\bar{\nabla}^2 E_y(\bar{r},t) - \mu\varepsilon\frac{\partial^2}{\partial t^2}E_y(\bar{r},t) = 0$$
$$\bar{\nabla}^2 E_z(\bar{r},t) - \mu\varepsilon\frac{\partial^2}{\partial t^2}E_z(\bar{r},t) = 0$$
$$\bar{\nabla}^2 H_x(\bar{r},t) - \mu\varepsilon\frac{\partial^2}{\partial t^2}H_x(\bar{r},t) = 0$$
$$\bar{\nabla}^2 H_y(\bar{r},t) - \mu\varepsilon\frac{\partial^2}{\partial t^2}H_y(\bar{r},t) = 0$$
$$\bar{\nabla}^2 H_z(\bar{r},t) - \mu\varepsilon\frac{\partial^2}{\partial t^2}H_z(\bar{r},t) = 0,$$

where

$$\bar{E}(\bar{r},t) = \hat{a}_x E_x(\bar{r},t) + \hat{a}_y E_y(\bar{r},t) + \hat{a}_z E_z(\bar{r},t)$$
$$\bar{H}(\bar{r},t) = \hat{a}_x H_x(\bar{r},t) + \hat{a}_y H_y(\bar{r},t) + \hat{a}_z H_z(\bar{r},t).$$

[71] Depending on how $E_x(\bar{r},t)$ behaves, other coordinate systems can be used. For example, in a spherical coordinate system, we have

$$\frac{1}{R^2}\frac{\partial}{\partial R}\left[R^2\frac{\partial E_x(\bar{r},t)}{\partial R}\right] + \frac{1}{R^2\sin\theta}\frac{\partial}{\partial\theta}\left[\sin\theta\frac{\partial E_x(\bar{r},t)}{\partial\theta}\right] + \frac{1}{R^2\sin^2\theta}\frac{\partial^2 E_x(\bar{r},t)}{\partial\phi^2} = 0.$$

The final form of the equation can be solved easily for given boundary conditions.[72] But in all cases, $E_x(z,t)$ behaves as a wave. To see this, note the trivial solution in the form of[73]

$$\bar{E}_x(z,t) = \hat{a}_x E_x(z,t) = \hat{a}_x E_x(t - z/u_p) \tag{4.201}$$

where u_p is a constant. This can be verified by inserting $E_x(z,t)$ into the equation as

$$\frac{\partial^2}{\partial z^2} E_x(t - z/u_p) - \mu\varepsilon \frac{\partial^2}{\partial t^2} E_x(t - z/u_p) = 0, \tag{4.202}$$

where

$$\frac{\partial^2}{\partial z^2} E_x(t - z/u_p) = \frac{1}{u_p^2} \frac{\partial^2}{\partial v^2} E_x(v) \tag{4.203}$$

and[74]

$$\frac{\partial^2}{\partial t^2} E_x(t - z/u_p) = \frac{\partial^2}{\partial v^2} E_x(v) \tag{4.204}$$

using a substitution[75] $v = t - z/u_p$. Then we obtain

$$\frac{1}{u_p^2} \frac{\partial^2}{\partial v^2} E_x(v) - \mu\varepsilon \frac{\partial^2}{\partial v^2} E_x(v) = 0, \tag{4.205}$$

which can be satisfied if

$$u_p = \frac{1}{\sqrt{\mu\varepsilon}}. \tag{4.206}$$

Hence, $E_x(z,t) = E_x(t - z/u_p)$ is really a solution,[76] but if this solution is selected, then u_p must be fixed based on the medium parameters μ and ε.

At first glance, the solution of the electric field intensity $\bar{E}(z,t) = \hat{a}_x E_x(t - z/u_p)$ seems to be independent of the magnetic field intensity. On the other hand, even though the electric and magnetic fields appear separately in non-homogeneous wave equations (4.193) and (4.195), they are related to each other via the root (Maxwell's) equations for dynamic cases. Specifically,

[72] To this point, we have made many assumptions that one may question the generality of the final equation. In fact, there are conceptually well-known waves: plane waves, with electric fields in fixed directions (e.g. x) while varying with respect to perpendicular directions (e.g. z). Plane waves do not exist in real life, but they are good interpretations of the wave propagation. In realistic cases, more complex waves can be visualized as combinations of many plane waves or similar fundamental waves with complex field directions and dependencies on positions.

[73] This does not prove that this is the only form of solution. For example, one can easily claim that a combination of such functions is also a solution, and that combination itself may not be simply a function of $t - z/u_p$.

[74] The derivation of the first equation can be shown in detail as

$$\begin{aligned}
\frac{\partial^2}{\partial z^2} E_x(t - z/u_p) &= \frac{\partial}{\partial z}\frac{\partial}{\partial z} E_x(t - z/u_p) \\
&= \frac{\partial}{\partial z}\left[\frac{\partial E_x(v)}{\partial v}\frac{\partial}{\partial z}(t - z/u_p)\right] \\
&= -\frac{1}{u_p}\frac{\partial}{\partial z}\left[\frac{\partial E_x(v)}{\partial v}\right] \\
&= -\frac{1}{u_p}\frac{\partial}{\partial z}(t - z/u_p)\frac{\partial}{\partial v}\left[\frac{\partial E_x(v)}{\partial v}\right] \\
&= \frac{1}{u_p^2}\frac{\partial^2}{\partial v^2} E_x(v).
\end{aligned}$$

[75] One can also show that $E_x(t + z/u_p)$ is another form of solution.

[76] We can also see that Gauss' law is satisfied as $\bar{\nabla} \cdot \bar{E}(\bar{r}, t) = 0$.

using Faraday's law for $\bar{E}(z,t) = \hat{a}_x E_x(t - z/u_p)$, we have[77]

$$\frac{\partial \bar{H}(\bar{r},t)}{\partial t} = -\frac{1}{\mu}\bar{\nabla} \times \bar{E}(\bar{r},t). \tag{4.207}$$

[77] We already assume that there is no source at $\bar{r}$.

Evaluating the curl as

$$\bar{\nabla} \times \bar{E}(\bar{r},t) = \hat{a}_z \times \hat{a}_x \frac{\partial}{\partial z} E_x(t - z/u_p) = \hat{a}_y \left(-\frac{1}{u_p}\right) \frac{\partial E_x(v)}{\partial v}, \tag{4.208}$$

we obtain[78]

$$\frac{\partial \bar{H}(\bar{r},t)}{\partial t} = \hat{a}_y \frac{1}{\mu u_p} \frac{\partial E_x(v)}{\partial v}. \tag{4.209}$$

[78] Hence, by selecting the electric field intensity in the x direction *and* assuming that it depends only on z, we automatically assume that the magnetic field intensity is in the y direction.

This means, given the electric field intensity $\bar{E}(z,t) = \hat{a}_x E_x(t - z/u_p)$, the magnetic field intensity must be in the form[79]

$$\bar{H}(z,t) = \hat{a}_y H_y(t - z/u_p), \tag{4.210}$$

[79] Note that this form also satisfies Gauss' law for magnetic fields: $\bar{\nabla} \cdot \bar{H}(\bar{r},t) = 0$. Recall that all derivations until this point are done for a homogeneous medium and a source-free location $\bar{r}$.

where the function $H_y(v)$ should be related to the function $E_x(v)$. To find this relationship, we use the time derivative of the magnetic field intensity as

$$\frac{\partial H_y(z,t)}{\partial t} = \frac{\partial H_y(v)}{\partial v} \tag{4.211}$$

with $v = t - z/u_p$, leading to

$$\frac{\partial H_y(v)}{\partial v} = \frac{1}{\mu u_p} \frac{\partial E_x(v)}{\partial v}. \tag{4.212}$$

Assuming that there is no constant (that is, independent of z and t) term in the magnetic field intensity, we have

$$H_y(t - z/u_p) = \frac{1}{\mu u_p} \frac{\partial}{\partial v} E_x(t - z/u_p) = \frac{1}{\eta} E_x(t - z/u_p), \tag{4.213}$$

and

$$\bar{H}(t - z/u_p) = \hat{a}_y \frac{1}{\eta} E_x(t - z/u_p), \tag{4.214}$$

where $\eta = \sqrt{\mu/\varepsilon}$ is another constant.[80]

[80] Note the units of η: Given the magnetic field intensity and the electric field intensity, we have

$$(\mathrm{V/m})/(\mathrm{A/m}) = \mathrm{V/A} = \Omega.$$

Similarly, given the permittivity and permeability, we have

$$\begin{aligned}\sqrt{(\mathrm{H/m})/(\mathrm{F/m})} &= \sqrt{\mathrm{H/F}} = \sqrt{(\mathrm{Wb/A})/(\mathrm{C/V})} \\ &= \sqrt{(\mathrm{Vs/A})/(\mathrm{C/V})} \\ &= \sqrt{\mathrm{V}^2/(\mathrm{AC/s})} \\ &= \sqrt{\mathrm{V}^2/\mathrm{A}^2} = \Omega.\end{aligned}$$

In fact, η is a kind of impedance.

Obviously, the expression for the magnetic field intensity in Eq. (4.214) satisfies the corresponding wave equation in Eq. (4.197). In addition, during its derivation, we have used Faraday's law, while Gauss' law holds for both electric and magnetic fields. At this stage, we can now check Ampere's law as

$$\bar{\nabla} \times \bar{H}(\bar{r}, t) = \hat{a}_z \times \hat{a}_y \frac{\partial}{\partial z}\left[\frac{1}{\eta} E_x(t - z/u_p)\right] = -\hat{a}_x \frac{1}{\eta}\left(-\frac{1}{u_p}\right)\frac{\partial E_x(v)}{\partial v} \tag{4.215}$$

that must be

$$\bar{\nabla} \times \bar{H}(\bar{r}, t) = \varepsilon \frac{\partial \bar{E}(\bar{r}, t)}{\partial t} = \hat{a}_x \varepsilon \frac{\partial E_x(v)}{\partial v}, \tag{4.216}$$

leading to $1/(\eta u_p) = \varepsilon$. This is correct since $\eta = \sqrt{\mu/\varepsilon}$ and $u_p = 1/\sqrt{\mu\varepsilon}$.

To sum up, it is obvious that a pair

$$\bar{E}(\bar{r}, t) = \hat{a}_x E_x(t - z/u_p) \tag{4.217}$$

$$\bar{H}(\bar{r}, t) = \hat{a}_y \frac{E_x}{\eta}(t - z/u_p) \tag{4.218}$$

satisfies all of Maxwell's equations and wave equations (that are in fact derived from Maxwell's equations) in a homogeneous medium and source-free region. These expressions represent an electromagnetic wave traveling along the z direction (Figure 4.28). To see this, we can consider the value of $E_x(z, t) = E_x(v) = E_x(t - z/u_p)$ at specific times, e.g. at $t = 0$, $t = 1/u_p$, and $t = 2/u_p$. We have

$$t = 0 \longrightarrow E_x(z, t) = E_x(-z/u_p) \tag{4.219}$$

$$t = 1/u_p \longrightarrow E_x(z, t) = E_x(1/u_p - z/u_p) \tag{4.220}$$

$$t = 2/u_p \longrightarrow E_x(z, t) = E_x(2/u_p - z/u_p). \tag{4.221}$$

We observe that, as the time progress, the function shifts (travels) in the z direction. For example, a function

$$g(z) = E_x(1/u_p - z/u_p) \tag{4.222}$$

is a right-shifted version of a function

$$f(z) = E_x(-z/u_p) \tag{4.223}$$

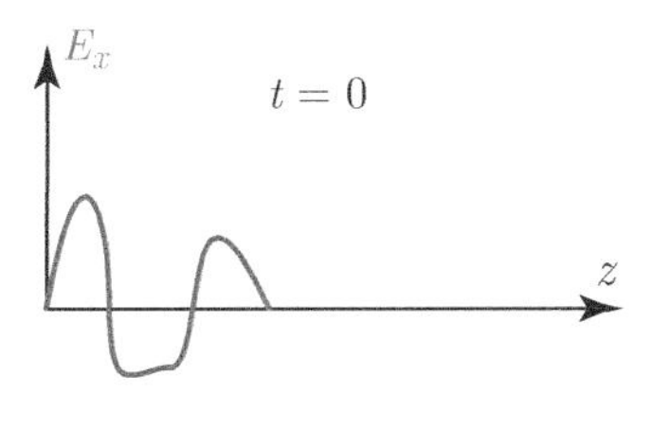

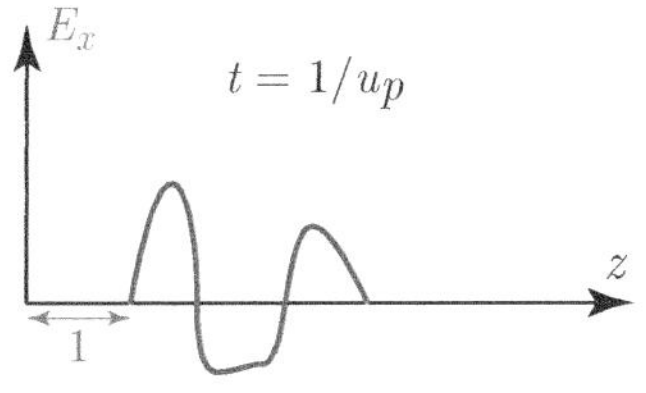

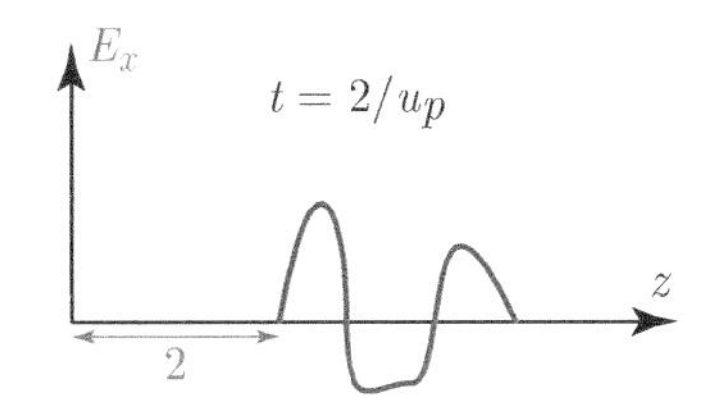

Figure 4.28 An electric field function in the form $E_x(t - z/u_p)$ represents a wave traveling in the z direction. If we draw the function with respect to z, the same shape moves to the right as time progresses. The speed of this movement is u_p, which corresponds to the speed of light.

u_p: phase velocity
Unit: meters/second (m/s)

as $g(z) = f(z-1)$. Obviously, the velocity of the wave is $\Delta z/\Delta t = u_p$, given that $\Delta z = 1$ for $\Delta t = 1/u_p$. In general, u_p is called the *phase* velocity to distinguish it from the *group* velocity of combined waves involving multiple frequencies. In a vacuum, u_p turns into the well-known speed of light:

$$c_0 = \frac{1}{\sqrt{\mu_0\varepsilon_0}} \approx \frac{1}{\sqrt{8.85418782 \times 10^{-12} \times 4\pi \times 10^{-7}}} \text{ m/s} \tag{4.224}$$

or[81]

$$c_0 = 2.99792458 \times 10^8 \text{ m/s}. \tag{4.225}$$

[81]The speed of light in a vacuum is exactly $299,792,458$ m/s – this is not an approximation. In fact, one meter can be defined as 1/299,792,458 of the distance that light travels in one second in a vacuum. The approximate equality in Eq. (4.224) is due to the approximation of the vacuum permittivity as $8.85418782 \times 10^{-12}$ F/m.

Hence, in a vacuum, electromagnetic waves should propagate with a constant speed of c_0 if they must satisfy Maxwell's equations.

Even though the natural solutions of wave equations (4.196) and (4.197) are traveling waves with a single argument in the form $t - z/u_p$, there are also solutions that do not possess a positional movement. Given the electric field intensity, any function in the form

$$E_x(z,t) = E_x(\omega t \pm kz) \tag{4.226}$$

with a constant k (1/m) satisfies the reduced equation (hence the general form of the wave equation)

$$\frac{\partial^2}{\partial z^2}E_x(z,t) - \mu\varepsilon\frac{\partial^2}{\partial t^2}E_x(z,t) = 0 \tag{4.227}$$

provided that $1/u_p = k/\omega$ or $u_p = \omega/k$. Each of these functions corresponds to a *traveling* wave with a velocity $u_p = \omega/k$ in either the $+z$ direction or $-z$ direction. Note that this set also includes time-harmonic functions (Figure 4.29). Specifically, an electric field intensity in the form

$$E_x(z,t) = E_0 \cos(\omega t \pm kz + \phi_0), \tag{4.228}$$

where $\omega = 2\pi f$ is the angular frequency in radian per second and ϕ_0 is an arbitrary phase in radian, is a traveling-wave solution. But in general, a solution may also be a combination of individual waves that satisfy wave equations. To see that such a solution may not represent a traveling wave by itself, we consider two individual solutions with the same frequency and

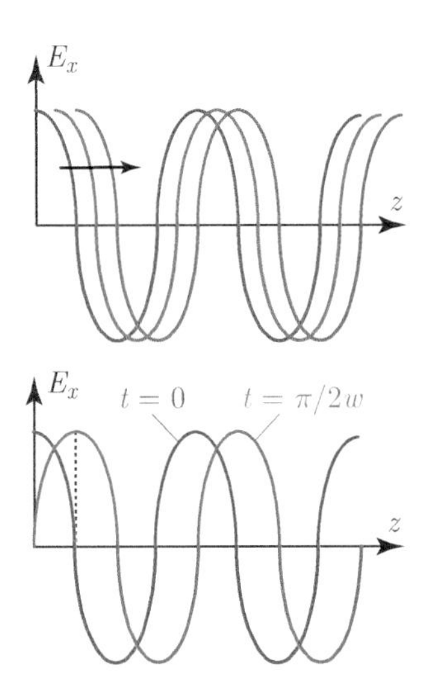

Figure 4.29 A time-harmonic electric field intensity $E_x(z,t) = E_0\cos(\omega t - kz)$ represents a sinusoidal wave traveling in the $+z$ direction. For example, at $t = 0$, it is $E_x(z,0) = E_0\cos(-kz) = E_0\cos(kz)$ with respect to z. When $t = \pi/(2\omega)$, it becomes $E_x(z,\pi/2\omega) = E_0\cos(\pi/2 - kz) = E_0\sin(kz)$, which is a right-shifted version of $E_x(z,t=0)$. Note that time samples should be selected carefully (with small intervals) for a correct interpretation of the traveling direction.

amplitude and zero phases as (Figure 4.30)

$$E_{x1}(z,t) = E_0\cos(\omega t - kz) \tag{4.229}$$

$$E_{x2}(z,t) = E_0\cos(\omega t + kz) \tag{4.230}$$

and their combination $E_x(z,t) = E_{x1}(z,t) + E_{x2}(z,t)$. In the above, $u_p = 1/\sqrt{\mu\varepsilon} = \omega/k$ so that $k = \omega\sqrt{\mu\varepsilon}$ to satisfy the wave equation (4.196). These functions represent right-traveling (traveling in the $+z$ direction) and left-traveling (traveling in the $-z$ direction) waves, respectively, while their combination can be written

$$E_x(z,t) = E_0\left[\cos(\omega t - kz) + \cos(\omega t + kz)\right] \tag{4.231}$$

$$= E_0\cos\left(\frac{\omega t + kz + \omega t - kz}{2}\right)\cos\left(\frac{\omega t + kz - \omega t + kz}{2}\right) \tag{4.232}$$

$$= 2E_0\cos(\omega t)\cos(kz). \tag{4.233}$$

This final expression (which is also a time-harmonic function) is not a traveling wave since its argument is not in the form $t - z/u_p$. But we can check that it satisfies the wave equation as

$$\frac{\partial^2}{\partial z^2}E_x(z,t) - \mu\varepsilon\frac{\partial^2}{\partial t^2}E_x(z,t) \tag{4.234}$$

$$= -2E_0k^2\cos(\omega t)\cos(kz) + \mu\varepsilon 2w^2E_0k^2\cos(\omega t)\cos(kz) = 0, \tag{4.235}$$

given that $k = \omega\sqrt{\mu\varepsilon}$. Waves in the form of Eq. (4.233) are called *standing* waves as they oscillate with time (ωt) and change with position (kz), while they do not represent a wave that moves. Standing waves are important concepts particularly when electromagnetic waves encounter reflections (from reflective bodies) so that waves traveling in opposite directions naturally exist together (superposed).

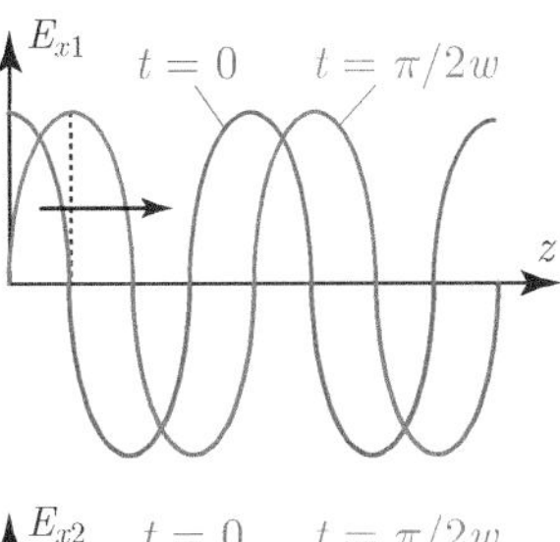

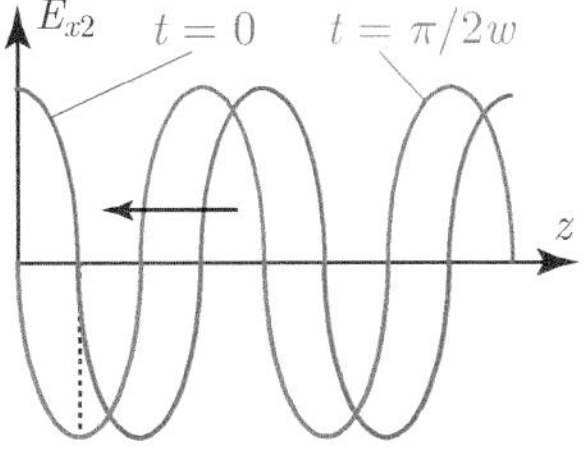

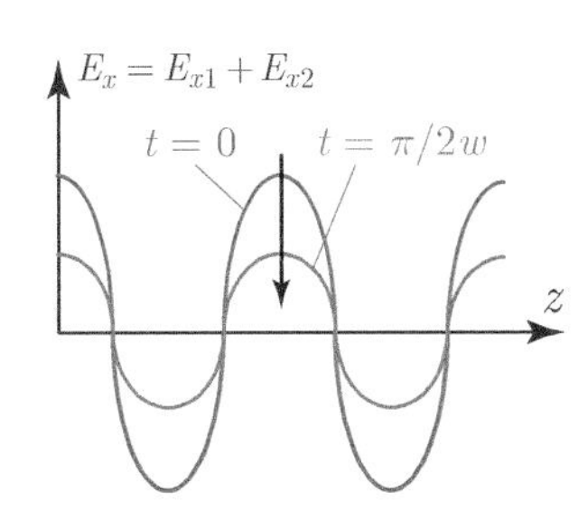

Figure 4.30 Time-harmonic electric field intensities $E_{x1}(z,t) = E_0\cos(\omega t - kz)$ and $E_{x2}(z,t) = E_0\cos(\omega t + kz)$ represent right-traveling and left-traveling waves, respectively, with the same frequency and amplitude. But their sum $E_x(z,t) = E_{x1}(z,t) + E_{x2}(z,t)$ is a time-harmonic electric field intensity that does not travel (standing wave), while simply oscillating with respect to time.

4.4.2 Wave Equations for Potentials

While homogeneous wave equations (equations with zero right-hand sides) in Eqs. (4.196) and (4.197) provide important information on solutions of Maxwell's equations and they can be used in various cases for deriving electric and magnetic fields, non-homogeneous versions in Eqs. (4.193) and (4.195) are usually not employed directly due to their relatively complicated right-hand sides. Instead, one can resort to potentials (the electric scalar potential and the magnetic vector potential) whose solutions are easier. In fact, once potentials are found, it is generally straightforward to find fields.[82] First, recall that the magnetic flux density can be

[82] This can be seen as follows: instead of solving equations involving derivatives of sources, we can solve the versions involving sources and then take derivatives.

written as the curl of the magnetic vector potential as

$$\bar{B}(\bar{r},t) = \bar{\nabla} \times \bar{A}(\bar{r},t) \tag{4.236}$$

for both static and dynamic cases. On the other hand, in a general dynamic case, the electric field intensity is not irrotational $[\bar{\nabla} \times \bar{E}(\bar{r},t) \neq 0]$ and it cannot be written as a gradient of a scalar function[83]: the electric scalar potential $[\bar{E}(\bar{r},t) \neq -\bar{\nabla}\Phi(\bar{r},t)]$. Considering Faraday's law, we have

[83] This also means the closed line integral of the electric field intensity is not identically zero:

$$\oint_C \bar{E}(\bar{r},t) \cdot \overline{dl} \neq 0.$$

$$\bar{\nabla} \times \bar{E}(\bar{r},t) = -\frac{\partial \bar{B}(\bar{r},t)}{\partial t} = -\frac{\partial}{\partial t}\bar{\nabla} \times \bar{A}(\bar{r},t), \tag{4.237}$$

leading to

$$\bar{\nabla} \times \left[\bar{E}(\bar{r},t) + \frac{\partial \bar{A}(\bar{r},t)}{\partial t}\right] = 0. \tag{4.238}$$

This means a vector field $\bar{E}(\bar{r},t) + \partial\bar{A}(\bar{r},t)/\partial t$ is irrotational rather than $\bar{E}(\bar{r},t)$ itself. Therefore, we have

$$\bar{E}(\bar{r},t) + \frac{\partial \bar{A}(\bar{r},t)}{\partial t} = -\bar{\nabla}\Phi(\bar{r},t), \tag{4.239}$$

where $\Phi(\bar{r},t)$ is the electric scalar potential.[84] Hence, we obtain the electric field intensity in terms of potentials as

[84] The irrotational property of the combined vector field $\bar{E}(\bar{r},t) + \partial\bar{A}(\bar{r},t)/\partial t$ means it can be written as the gradient of a scalar function. Selecting this function as the electric scalar potential, which satisfies $\bar{E}(\bar{r}) = -\bar{\nabla}\Phi(\bar{r})$ in static cases, is a kind of choice among many.

$$\bar{E}(\bar{r},t) = -\bar{\nabla}\Phi(\bar{r},t) - \frac{\partial \bar{A}(\bar{r},t)}{\partial t}. \tag{4.240}$$

Note that, in a static case, $\partial\bar{A}(\bar{r},t)/\partial t = 0$ and $\bar{E}(\bar{r}) = -\bar{\nabla}\Phi(\bar{r})$, as required.

In the following, our aim is to use Maxwell's equations to derive useful equations for the potentials. First, using Ampere's law, we have[85]

[85] Here we also include inhomogeneous cases.

$$\bar{\nabla} \times \bar{H}(\bar{r},t) = \bar{\nabla} \times [\bar{B}(\bar{r},t)/\mu(\bar{r})] = \frac{1}{\mu(\bar{r})}\bar{\nabla} \times \bar{B}(\bar{r},t) + \bar{\nabla}\left(\frac{1}{\mu(\bar{r})}\right) \times \bar{B}(\bar{r},t) \tag{4.241}$$

$$= \bar{J}_v(\bar{r},t) + \frac{\partial \bar{D}(\bar{r},t)}{\partial t} \tag{4.242}$$

or[86]

[86] In a magnetically homogeneous case, this can be written

$$\bar{\nabla} \times \bar{B}(\bar{r},t) = \mu\bar{J}_v(\bar{r},t) + \mu\varepsilon(\bar{r})\frac{\partial \bar{E}(\bar{r},t)}{\partial t}.$$

$$\bar{\nabla} \times \bar{B}(\bar{r},t) - \frac{\bar{\nabla}\mu(\bar{r})}{\mu(\bar{r})} \times \bar{B}(\bar{r},t) = \mu(\bar{r})\bar{J}_v(\bar{r},t) + \mu(\bar{r})\varepsilon(\bar{r})\frac{\partial \bar{E}(\bar{r},t)}{\partial t}. \tag{4.243}$$

Inserting the potentials as $\bar{B}(\bar{r},t) = \bar{\nabla} \times \bar{A}(\bar{r},t)$ and $\bar{E}(\bar{r},t) = -\bar{\nabla}\Phi(\bar{r},t) - \partial\bar{A}(\bar{r},t)/\partial t$, we derive

$$\bar{\nabla} \times \bar{\nabla} \times \bar{A}(\bar{r},t) - \frac{\bar{\nabla}\mu(\bar{r})}{\mu(\bar{r})} \times \bar{\nabla} \times \bar{A}(\bar{r},t)$$
$$= \mu(\bar{r})\bar{J}_v(\bar{r},t) - \mu(\bar{r})\varepsilon(\bar{r})\frac{\partial}{\partial t}\bar{\nabla}\Phi(\bar{r},t) - \mu(\bar{r})\varepsilon(\bar{r})\frac{\partial^2}{\partial t^2}\bar{A}(\bar{r},t). \tag{4.244}$$

Rearranging the terms,[87] we obtain

$$\bar{\nabla}^2\bar{A}(\bar{r},t) - \frac{1}{u_p^2(\bar{r})}\frac{\partial^2}{\partial t^2}\bar{A}(\bar{r},t) - \bar{\nabla}\left[\bar{\nabla}\cdot\bar{A}(\bar{r},t) + \frac{1}{u_p^2(\bar{r})}\frac{\partial\Phi(\bar{r},t)}{\partial t}\right]$$
$$+ \frac{\bar{\nabla}\mu(\bar{r})}{\mu(\bar{r})} \times \bar{\nabla} \times \bar{A}(\bar{r},t) = -\mu(\bar{r})\bar{J}_v(\bar{r},t), \tag{4.245}$$

where $u_p^2(\bar{r}) = 1/[\mu(\bar{r})\varepsilon(\bar{r})]$. In a homogeneous medium, this equation becomes

$$\bar{\nabla}^2\bar{A}(\bar{r},t) - \frac{1}{u_p^2}\frac{\partial^2}{\partial t^2}\bar{A}(\bar{r},t) - \bar{\nabla}\left[\bar{\nabla}\cdot\bar{A}(\bar{r},t) + \frac{1}{u_p^2}\frac{\partial\Phi(\bar{r},t)}{\partial t}\right] = -\mu\bar{J}_v(\bar{r},t), \tag{4.246}$$

where $u_p = 1/\sqrt{\mu\varepsilon}$ is the phase velocity inside the medium.

In static cases (see Section 4.3.1), we use the Coulomb gauge to simplify the equation for $\bar{A}(\bar{r})$. In this case, however, a good selection to eliminate $\Phi(\bar{r},t)$ and write the equation only in terms of $\bar{A}(\bar{r},t)$ can be written

$$\bar{\nabla}\cdot\bar{A}(\bar{r},t) = -\frac{1}{u_p^2}\frac{\partial\Phi(\bar{r},t)}{\partial t}, \tag{4.247}$$

which is called the *Lorenz gauge*.[88] Using this gauge condition, a non-homogeneous wave equation is obtained:

$$\bar{\nabla}^2\bar{A}(\bar{r},t) - \frac{1}{u_p^2}\frac{\partial^2}{\partial t^2}\bar{A}(\bar{r},t) = -\mu\bar{J}_v(\bar{r},t), \tag{4.248}$$

which can be used to find $\bar{A}(\bar{r},t)$.

To derive an equation for the electric scalar potential $\Phi(\bar{r},t)$, we first consider Gauss' law as

$$\bar{\nabla}\varepsilon(\bar{r})\cdot\bar{E}(\bar{r},t) + \varepsilon(\bar{r})\bar{\nabla}\cdot\bar{E}(\bar{r},t) = q_v(\bar{r},t). \tag{4.249}$$

[87] We insert

$$\bar{\nabla}\times\bar{\nabla}\times\bar{A}(\bar{r},t) = \bar{\nabla}[\bar{\nabla}\cdot\bar{A}(\bar{r},t)] - \bar{\nabla}^2\bar{A}(\bar{r},t),$$

leading to

$$\bar{\nabla}[\bar{\nabla}\cdot\bar{A}(\bar{r},t)] - \bar{\nabla}^2\bar{A}(\bar{r},t) - \frac{\bar{\nabla}\mu(\bar{r})}{\mu(\bar{r})}\times\bar{\nabla}\times\bar{A}(r,t)$$
$$= \mu(\bar{r})\bar{J}_v(\bar{r},t) - \mu(\bar{r})\varepsilon(\bar{r})\frac{\partial}{\partial t}\bar{\nabla}\Phi(\bar{r},t)$$
$$- \mu(\bar{r})\varepsilon(\bar{r})\frac{\partial^2}{\partial t^2}\bar{A}(\bar{r},t).$$

Then the terms are rearranged.

[88] Obviously, this does not mean we cannot use other gauges. In fact, we are free to define the divergence of $\bar{A}(\bar{r},t)$. But the Lorenz gauge may be the best selection if one needs a simple formula for the magnetic vector potential. If we use the Coulomb gauge, the equation for the magnetic vector potential becomes

$$\bar{\nabla}^2\bar{A}(\bar{r},t) - \frac{1}{u_p^2}\frac{\partial^2}{\partial t^2}\bar{A}(\bar{r},t) - \bar{\nabla}\left[\frac{1}{u_p^2}\frac{\partial\Phi(\bar{r},t)}{\partial t}\right]$$
$$= -\mu\bar{J}_v(\bar{r},t),$$

where the electric scalar potential is not eliminated.

Inserting the potentials, we obtain[89]

$$\bar{\nabla}^2\Phi(\bar{r},t) + \frac{\partial}{\partial t}\bar{\nabla}\cdot\bar{A}(\bar{r},t) = -\frac{q_v(\bar{r},t)}{\varepsilon(\bar{r})} + \frac{\bar{\nabla}\varepsilon(\bar{r})\cdot\bar{E}(\bar{r},t)}{\varepsilon(\bar{r})}. \tag{4.250}$$

[89] As an intermediate step, we have

$$\bar{\nabla}\varepsilon(\bar{r})\cdot\bar{E}(\bar{r},t) + \varepsilon(\bar{r})\bar{\nabla}\cdot\left[-\bar{\nabla}\Phi(\bar{r},t) - \frac{\partial\bar{A}(\bar{r},t)}{\partial t}\right] = q_v(\bar{r},t).$$

In an electrically homogeneous medium, we further derive

$$\bar{\nabla}^2\Phi(\bar{r},t) + \frac{\partial}{\partial t}\bar{\nabla}\cdot\bar{A}(\bar{r},t) = -\frac{q_v(\bar{r},t)}{\varepsilon}. \tag{4.251}$$

At this stage, using the Lorenz gauge, we arrive at a non-homogeneous wave equation as[90]

$$\bar{\nabla}^2\Phi(\bar{r},t) - \frac{1}{u_p^2}\frac{\partial^2}{\partial t^2}\Phi(\bar{r},t) = -\frac{q_v(\bar{r},t)}{\varepsilon}. \tag{4.252}$$

[90] Note that using the Coulomb gauge leads to

$$\bar{\nabla}^2\Phi(\bar{r},t) = -\frac{q_v(\bar{r},t)}{\varepsilon},$$

while the corresponding equation for $\bar{A}(\bar{r},t)$ is complicated, as shown above.

Note the similarity of this equation and the one for the magnetic vector potential in Eq. (4.248).

Wave equations (4.248) and (4.252) can be solved for arbitrary electric current and charge distributions. Note that, in static cases, these equations reduce into the corresponding Poisson's equations as

$$\bar{\nabla}^2\Phi(\bar{r}) = -q_v(\bar{r})/\varepsilon \tag{4.253}$$

$$\bar{\nabla}^2\bar{A}(\bar{r}) = -\mu\bar{J}_v(\bar{r}) \tag{4.254}$$

whose solutions are shown in Section 3.6.7 and Section 4.3.1 as[91]

$$\Phi(\bar{r}) = \frac{1}{4\pi\varepsilon}\int_V \frac{q_v(\bar{r}')}{|\bar{r}-\bar{r}'|}d\bar{r}' \tag{4.255}$$

$$\bar{A}(\bar{r}) = \frac{\mu}{4\pi}\int_V \frac{\bar{J}_v(\bar{r}')}{|\bar{r}-\bar{r}'|}d\bar{r}'. \tag{4.256}$$

[91] Recall that these integral forms can be written for surface and line sources as

$$\Phi(\bar{r}) = \frac{1}{4\pi\varepsilon}\int_S \frac{q_s(\bar{r}')}{|\bar{r}-\bar{r}'|}d\bar{r}'$$

$$\bar{A}(\bar{r}) = \frac{\mu}{4\pi}\int_S \frac{\bar{J}_s(\bar{r}')}{|\bar{r}-\bar{r}'|}d\bar{r}'$$

for $d\bar{r}' = ds'$ and

$$\Phi(\bar{r}) = \frac{1}{4\pi\varepsilon}\int_C \frac{q_l(\bar{r}')}{|\bar{r}-\bar{r}'|}d\bar{r}'$$

$$\bar{A}(\bar{r}) = \frac{\mu}{4\pi}\oint_C \frac{I_0\overline{dl'}}{|\bar{r}-\bar{r}'|}$$

for $d\bar{r}' = dl'$, respectively.

Similar to the derivations in these sections, we may consider the general solution of Eq. (4.252) for the electric scalar potential and then extend it for the magnetic scalar potential. Without considering the details, the general steps are as follows:

- First, as shown Section 3.6.7 for the electric scalar potential in electrostatics, we can consider the homogeneous version of Eq. (4.252):

$$\bar{\nabla}^2\Phi(\bar{r},t) - \frac{1}{u_p^2}\frac{\partial^2}{\partial t^2}\Phi(\bar{r},t) = 0. \tag{4.257}$$

This equation can be solved in a spherical coordinate system with the assumption that the electric scalar potential only depends on R. Specifically, we have

$$\bar{\nabla}^2\Phi(R,t) - \frac{1}{u_p^2}\frac{\partial^2}{\partial t^2}\Phi(R,t) = 0$$
$$\longrightarrow \frac{1}{R^2}\frac{\partial}{\partial R}\left[R^2\frac{\partial\Phi(R,t)}{\partial R}\right] - \frac{1}{u_p^2}\frac{\partial^2}{\partial t^2}\Phi(R,t) = 0. \tag{4.258}$$

If there is no time dependency – i.e. when $\Phi(R,t) = \Phi(R)$ and the second term disappears – we already know the general solution (see Section 3.6.7) as $\Phi(R) = -a/R$, where a is a constant. Then, taking a hint from Section 4.4.1, the solution of Eq. (4.258) should be in the form[92,93]

$$\Phi(R,t) = \frac{f(t - R/u_p)}{R}, \tag{4.259}$$

where $f(v) = f(t - R/u_p)$ is a function to be found depending on the scenario, e.g. the boundary conditions.

- Second, the solution found in a spherical coordinate system can be used directly for a point source where a spherical symmetry exists. Note that, if the observation point is not exactly on the point source, the homogeneous equation and its solution are still valid. Specifically, we may consider a point source (charge) $q_v(\bar{r},t) = Q_0\delta(R)h(t)$, where $h(t)$ is a function of time that is nonzero for $t > 0$. Here, $h(t)$ represents the time dependency of the point source,[94] and we have $q_v(\bar{r},t) = 0$ if $R \neq 0$ (due to the impulse function $\delta(R)$) or $t < 0$ (due to $h(t)$). Then considering the non-homogeneous equation

$$\bar{\nabla}^2\Phi(R,t) - \frac{1}{u_p^2}\frac{\partial^2}{\partial t^2}\Phi(R,t) = -Q_0\delta(R)h(t)/\varepsilon, \tag{4.260}$$

the solution can be found as

$$\Phi(R,t) = \frac{Q_0 h(t - R/u_p)}{4\pi\varepsilon R} \tag{4.261}$$

or[95]

$$\Phi(R,t) = \frac{Q_0 h(t')}{4\pi\varepsilon R}, \tag{4.262}$$

where $t' = t - R/u_p$.

[92] Note that $\Phi(R,t)$ is assumed to be zero at infinity.

[93] This can be verified as

$$\frac{1}{R^2}\frac{\partial}{\partial R}\left[R^2\frac{\partial\Phi(R,t)}{\partial R}\right] - \frac{1}{u_p^2}\frac{\partial^2}{\partial t^2}\Phi(R,t)$$
$$= \frac{1}{R^2}\frac{\partial}{\partial R}\left\{R^2\left[-\frac{1}{R^2}f(v) - \frac{1}{Ru_p}\frac{\partial f(v)}{\partial v}\right]\right\}$$
$$- \frac{1}{Ru_p^2}\frac{\partial^2 f(v)}{\partial v^2}$$
$$= \frac{1}{R^2}\frac{\partial}{\partial R}\left[-f(v) - \frac{R}{u_p}\frac{\partial f(v)}{\partial v}\right] - \frac{1}{Ru_p^2}\frac{\partial^2 f(v)}{\partial v^2}$$
$$= \frac{1}{R^2 u_p}\frac{\partial f(v)}{\partial v} - \frac{1}{R^2 u_p}\frac{\partial f(v)}{\partial v}$$
$$+ \frac{1}{Ru_p^2}\frac{\partial^2 f(v)}{\partial v^2} - \frac{1}{Ru_p^2}\frac{\partial^2 f(v)}{\partial v^2} = 0,$$

where $v = t + R/u_p$. Here, note that

$$\frac{\partial f(v)}{\partial R} = \frac{\partial v}{\partial R}\frac{\partial f(v)}{\partial v} = -\frac{1}{u_p}\frac{\partial f(v)}{\partial v}$$
$$\frac{\partial f(v)}{\partial t} = \frac{\partial v}{\partial t}\frac{\partial f(v)}{\partial v} = \frac{\partial f(v)}{\partial v}$$

via the chain rule.

[94] Time variance of a point charge may be questioned; but note that this point charge can be a part of a time-variant charge distribution so that its value may change depending on time.

[95] This solution can be found by considering a spherical surface and evaluating the limit as the sphere shrinks to zero.

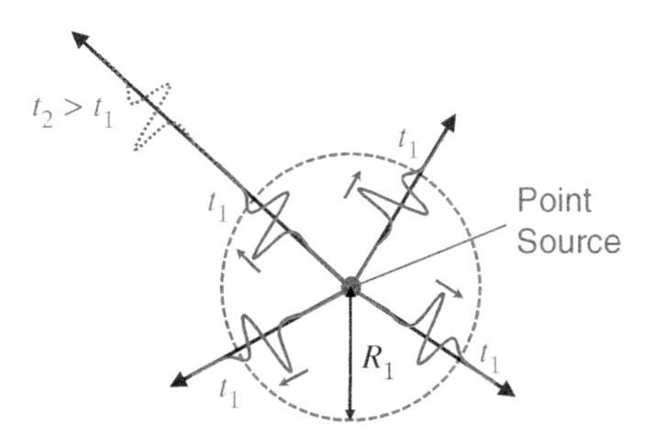

Figure 4.31 A variation in a point source creates a spherical wave propagating in all directions. For observation points located at R_1, the time delay (the time period between a source variation and the arrival of the wave) is $t_1 = R_1/u_p$, where u_p is the phase velocity. Observation points that are farther away from the source receive the same wave (with further decay) at a later time.

- Third, as the final stage, the superposition principle can be used to write a general expression for arbitrary source distributions. If the point charge with $h(t)$ time dependency is located at $\bar{r}'$, the electric scalar potential at an arbitrary position can be written

$$\Phi(\bar{r}, t) = \frac{Q_0 h(t - |\bar{r} - \bar{r}'|/u_p)}{4\pi|\bar{r} - \bar{r}'|}. \tag{4.263}$$

This expression indicates that $\Phi(\bar{r}, t)$ is nonzero if $t > |\bar{r} - \bar{r}'|/u_p$. The critical time $t = |\bar{r} - \bar{r}'|/u_p$ is exactly when the wave (in this case, the electric scalar potential) that emerges from the point source arrives at the observation point $\bar{r}$. As one may expect, the time required for the wave's travel is the distance $|\bar{r} - \bar{r}'|$ divided by the phase velocity u_p. Hence, a variation in the point charge *creates* a *spherical* wave propagating in all directions (Figure 4.31). For an arbitrary volume charge distribution with volume electric charge density $q_v(\bar{r}, t)$, the superposition principle[96] lets us generalize the expression as (Figure 4.32)

$$\Phi(\bar{r}, t) = \frac{1}{4\pi\varepsilon}\int_V \frac{q_v(\bar{r}, t')}{|\bar{r} - \bar{r}'|}d\bar{r}', \tag{4.264}$$

where $t' = t - |\bar{r} - \bar{r}'|/u_p$ is called *retarded* time.[97] As practiced many times, we can further extend the expression as

$$\Phi(\bar{r}, t) = \frac{1}{4\pi\varepsilon}\int_S \frac{q_s(\bar{r}, t')}{|\bar{r} - \bar{r}'|}d\bar{r}' \tag{4.265}$$

and

$$\Phi(\bar{r}, t) = \frac{1}{4\pi\varepsilon}\int_C \frac{q_l(\bar{r}, t')}{|\bar{r} - \bar{r}'|}d\bar{r}' \tag{4.266}$$

for a surface electric charge density and a line electric charge density, respectively.

Based on the results for the electric scalar potential, we can write the expressions for the magnetic vector potential as

$$\bar{A}(\bar{r}, t) = \frac{\mu}{4\pi}\int_V \frac{\bar{J}_v(\bar{r}, t')}{|\bar{r} - \bar{r}'|}d\bar{r}' \tag{4.267}$$

$$\bar{A}(\bar{r}, t) = \frac{\mu}{4\pi}\int_S \frac{\bar{J}_s(\bar{r}, t')}{|\bar{r} - \bar{r}'|}d\bar{r}' \tag{4.268}$$

$$\bar{A}(\bar{r}, t) = \frac{\mu}{4\pi}\int_C \frac{I(t')\overline{dl'}}{|\bar{r} - \bar{r}'|}, \tag{4.269}$$

[96] A source produces the same field whether there is another source or not.

[97] Note that $t' = t - |\bar{r} - \bar{r}'|/u_p$ is different for different locations of the source. Specifically, given an observation point $\bar{r}$, waves (in this case, electric scalar potential values) produced by different parts of the source arrive at different times (Figure 4.32).

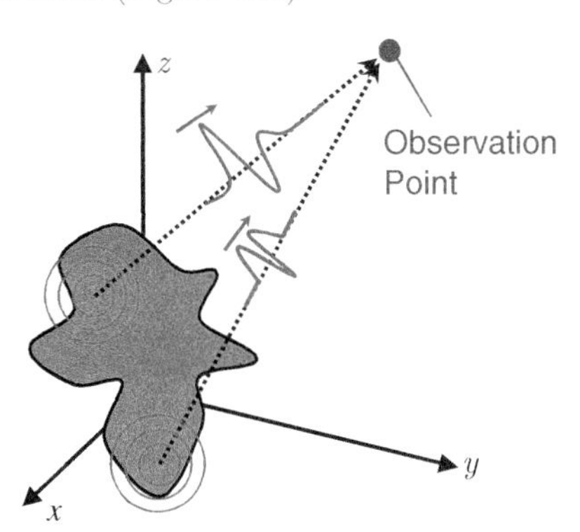

Figure 4.32 Each location in a source distribution can be considered a point source that creates spherical waves. At a given observation point, combinations of these waves (which travel for different time periods, depending on the distances) are measured.

where $t' = t - |\bar{r} - \bar{r}'|/u_p$, for different forms of the electric current distribution. Finally, once the time-dependent potentials are found from sources, electric and magnetic fields can be found:[98,99]

$$\bar{E}(\bar{r}, t) = -\bar{\nabla}\Phi(\bar{r}, t) - \frac{\partial \bar{A}(\bar{r}, t)}{\partial t} \tag{4.270}$$

$$\bar{H}(\bar{r}, t) = \frac{1}{\mu}\bar{\nabla} \times \bar{A}(\bar{r}, t). \tag{4.271}$$

[98]Using these expressions, it is also possible to write the general expressions for electric and magnetic fields in terms of sources.

[99]Note that potentials depend on the selected gauge, while electric and magnetic fields do not.

4.4.3 Time-Harmonic Sources and Helmholtz Equations

When time-harmonic sources are involved in a dynamic case, Maxwell's equations can be written in the phasor domain. Then they can be combined to derive higher-level equations (similar to wave equations) to analyze the given dynamic problem. In a homogeneous medium, the result is Helmholtz equations,[100] which are among the most fundamental tools of electromagnetics. Instead of starting from Maxwell's equations, one can also convert wave equations into Helmholtz equations via proper transformations of time derivatives. Now, we assume that the volume electric current density and the volume electric charge density can be written

[100]In general, Helmholtz equations are in the form $(\bar{\nabla}^2 + k^2)\bar{f} = \bar{g}$ or $(\nabla^2 + k^2)f = g$.

$$\bar{J}_v(\bar{r}, t) = \bar{J}_0(\bar{r})\cos(\omega t + \phi_{j0}) \tag{4.272}$$

$$q_v(\bar{r}, t) = q_0(\bar{r})\cos(\omega t + \phi_{q0}), \tag{4.273}$$

where $\omega = 2\pi f$ is the angular frequency and f is the frequency. In these expressions, if ω is known, the amplitudes ($\bar{J}_0$ and q_0) and phases (ϕ_{j0} and ϕ_{q0}) provide all the information regarding the sources. When the sources are time-harmonic, the resulting electric and magnetic fields are also time-harmonic with the same frequency (Figure 4.33). Using phasor representations for all quantities and given that a single time derivative corresponds to a multiplication with $j\omega$, we obtain Helmholtz equations in homogeneous media (using Eqs. (4.193), (4.195), (4.248), and (4.252):[101]

$$\bar{\nabla}^2\bar{E}(\bar{r}) + \omega^2\mu\varepsilon\bar{E}(\bar{r}) = j\omega\mu\bar{J}_v(\bar{r}) + \frac{1}{\varepsilon}\bar{\nabla}q_v(\bar{r}) \tag{4.274}$$

$$\bar{\nabla}^2\bar{H}(\bar{r}) + \omega^2\mu\varepsilon\bar{H}(\bar{r}) = -\bar{\nabla} \times \bar{J}_v(\bar{r}) \tag{4.275}$$

for fields and

$$\bar{\nabla}^2\bar{A}(\bar{r}) + \omega^2\mu\varepsilon\bar{A}(\bar{r}) = -\mu\bar{J}_v(\bar{r}) \tag{4.276}$$

$$\bar{\nabla}^2\Phi(\bar{r}) + \omega^2\mu\varepsilon\Phi(\bar{r}) = -\frac{q_v(\bar{r})}{\varepsilon} \tag{4.277}$$

[101]Hence, we have

$$\bar{E}(\bar{r}, t) = \bar{E}_0(\bar{r})\cos(\omega t + \phi_{e0})$$
$$\bar{H}(\bar{r}, t) = \bar{H}_0(\bar{r})\cos(\omega t + \phi_{h0})$$
$$\bar{A}(\bar{r}, t) = \bar{A}_0(\bar{r})\cos(\omega t + \phi_{a0})$$
$$\Phi(\bar{r}, t) = \Phi_0(\bar{r})\cos(\omega t + \phi_{\phi0})$$

and

$$\bar{E}(\bar{r}) = \bar{E}_0(\bar{r})\underline{/\phi_{e0}}$$
$$\bar{H}(\bar{r}) = \bar{H}_0(\bar{r})\underline{/\phi_{h0}}$$
$$\bar{A}(\bar{r}) = \bar{A}_0(\bar{r})\underline{/\phi_{a0}}$$
$$\Phi(\bar{r}) = \Phi_0(\bar{r})\underline{/\phi_{\phi0}},$$

as well as

$$\bar{J}_v(\bar{r}) = \bar{J}_0(\bar{r})\underline{/\phi_{j0}}$$
$$q_v(\bar{r}) = q_0(\bar{r})\underline{/\phi_{q0}}.$$

k: wavenumber
Unit: radians/meter (rad/m)

for potentials. An important quantity is the *wavenumber*, which is defined as

$$k = \omega\sqrt{\mu\varepsilon} = \frac{\omega}{u_p} = \frac{2\pi f}{u_p} = \frac{2\pi}{\lambda}, \tag{4.278}$$

λ: wavelength
Unit: meter (m)

where $\lambda = f/u_p$ is the wavelength. Hence, we arrive at the well-known forms as

$$\bar{\nabla}^2\bar{E}(\bar{r}) + k^2\bar{E}(\bar{r}) = j\omega\mu\bar{J}_v(\bar{r}) + \frac{1}{\varepsilon}\bar{\nabla}q_v(\bar{r}) \tag{4.279}$$

$$\bar{\nabla}^2\bar{H}(\bar{r}) + k^2\bar{H}(\bar{r}) = -\bar{\nabla}\times\bar{J}_v(\bar{r}) \tag{4.280}$$

$$\bar{\nabla}^2\bar{A}(\bar{r}) + k^2\bar{A}(\bar{r}) = -\mu\bar{J}_v(\bar{r}) \tag{4.281}$$

$$\bar{\nabla}^2\Phi(\bar{r}) + k^2\Phi(\bar{r}) = -\frac{q_v(\bar{r})}{\varepsilon}. \tag{4.282}$$

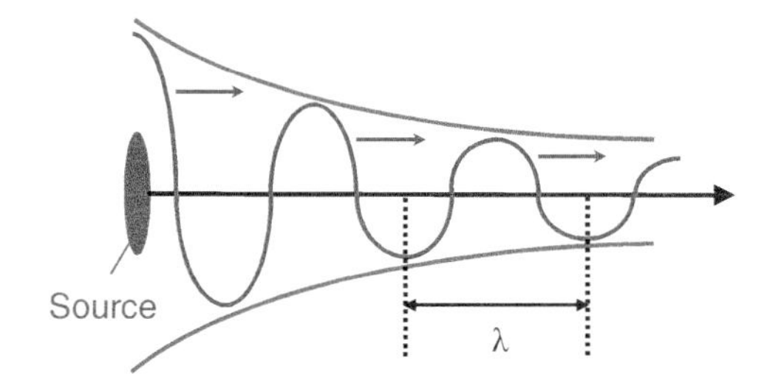

Figure 4.33 A time-harmonic source generates a time-harmonic (oscillatory) wave that travels with the phase velocity. If the source is finite, the wave decays with the distance. At two observation points that are separated by the wavelength (λ), the wave becomes maximum/minimum simultaneously.

We emphasize that Helmholtz equations for potentials are written based on the assumption of the Lorenz gauge

$$\bar{\nabla}\cdot\bar{A}(\bar{r}) = -j\omega\frac{1}{u_p^2}\Phi(\bar{r}) = -j\omega\mu\varepsilon\Phi(\bar{r}) \tag{4.283}$$

in phasor notation.

As shown in the previous section, solutions of wave equations for potentials lead to integrals containing sources with retarded times. Now, we can also convert the solutions into their phasor-domain representations by noting that

$$q_v(\bar{r}, t - |\bar{r}-\bar{r}'|/u_p) \longrightarrow q_v(\bar{r})\exp(-j\omega|\bar{r}-\bar{r}'|/u_p) \tag{4.284}$$

$$\bar{J}_v(\bar{r}, t - |\bar{r}-\bar{r}'|/u_p) \longrightarrow \bar{J}_v(\bar{r})\exp(-j\omega|\bar{r}-\bar{r}'|/u_p). \tag{4.285}$$

Specifically, without directly solving Helmholtz equations, we can obtain

$$\bar{A}(\bar{r}) = \frac{\mu}{4\pi}\int_V \bar{J}_v(\bar{r}')\frac{\exp(-jk|\bar{r}-\bar{r}'|)}{|\bar{r}-\bar{r}'|}d\bar{r}' \tag{4.286}$$

$$\Phi(\bar{r}) = \frac{1}{4\pi\varepsilon}\int_V q_v(\bar{r}')\frac{\exp(-jk|\bar{r}-\bar{r}'|)}{|\bar{r}-\bar{r}'|}d\bar{r}', \tag{4.287}$$

where $k = \omega/u_p$ is used. Once again, these integrals can also be written for surface and line sources. In these expressions, note the important term

$$g(\bar{r},\bar{r}') = \frac{\exp(-jk|\bar{r}-\bar{r}'|)}{4\pi|\bar{r}-\bar{r}'|}, \tag{4.288}$$

which is called Green's function.[102] Then, inserting $g(\bar{r},\bar{r}')$, we derive

$$\bar{A}(\bar{r}) = \mu \int_V \bar{J}_v(\bar{r}')g(\bar{r},\bar{r}')d\bar{r}' \tag{4.289}$$

$$\Phi(\bar{r}) = \frac{1}{\varepsilon} \int_V q_v(\bar{r}')g(\bar{r},\bar{r}')d\bar{r}' \tag{4.290}$$

as the expressions for potentials in time-harmonic cases.

Note that the retardation of time in general time-varying expressions is converted into a simple phase term $\exp(-jk|\bar{r}-\bar{r}'|)$ in phasor-domain expressions. This may be confusing, especially considering different sources at different distances from an observation point. As an example, let us consider two point sources located at $\bar{r}_1$ and $\bar{r}_2$ with[103]

$$|\bar{r}-\bar{r}_1| = |\bar{r}-\bar{r}_2| + a\lambda, \tag{4.291}$$

where λ is the wavelength and $a \neq 0$ is an integer (Figure 4.34). Then the corresponding Green's function values can be written[104]

$$\frac{\exp(-jk|\bar{r}-\bar{r}_1|)}{4\pi|\bar{r}-\bar{r}_1|} = \frac{\exp(-jk|\bar{r}-\bar{r}_2| - jka\lambda)}{4\pi|\bar{r}-\bar{r}_2| + 4\pi a\lambda} \tag{4.292}$$

$$= \frac{\exp(-jk|\bar{r}-\bar{r}_2|)\exp(-j2\pi a)}{4\pi|\bar{r}-\bar{r}_2| + 4\pi a\lambda} = \frac{\exp(-jk|\bar{r}-\bar{r}_2|)}{4\pi|\bar{r}-\bar{r}_2| + 4\pi a\lambda}. \tag{4.293}$$

Hence, the waves emerging from the two sources differ only in terms of magnitude. This means these waves are *in phase*:[105] i.e. they become maximum and minimum simultaneously at the observation point, although their sources are at different distances. Since we are considering a steady state (and time-harmonic sources), the waves emerging from the sources have different traveling times,[106] but their periodicity makes it possible to have synchronization at various locations (including our observation point).

Using Eqs. (4.270) and (4.271), electric and magnetic fields can be obtained in the phasor domain as

$$\bar{E}(\bar{r}) = -\bar{\nabla}\Phi(\bar{r}) - j\omega\bar{A}(\bar{r}) \tag{4.294}$$

$$\bar{H}(\bar{r}) = \frac{1}{\mu}\bar{\nabla}\times\bar{A}(\bar{r}). \tag{4.295}$$

[102] It can be shown that Green's function is the solution of a Helmholtz equation with an impulse right-hand side (point source).

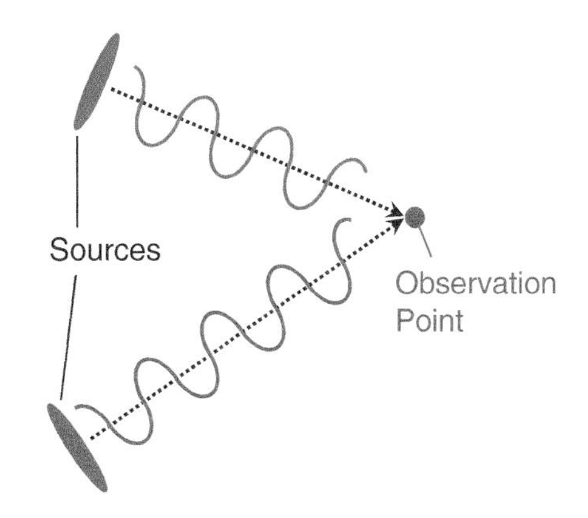

Figure 4.34 Waves created by two time-harmonic sources (at the same frequency) can be added in phase at an observation point depending on the distances between the sources and the observation point.

[103] Using phasor notation, we automatically assume that these two sources have the same time dependency: i.e. they oscillate and create waves at the same frequency.

[104] Using

$$\exp(-j2\pi a) = \cos(2\pi a) - j\sin(2\pi a) = 1.$$

[105] We assume that the sources are already in phase so that it is sufficient to compare the corresponding Green's function values. Otherwise, if the sources are not in phase, extra phase terms are added via $\bar{J}_v(\bar{r})$ and $q_v(\bar{r})$ in Eqs. (4.289) and (4.290). In such a case, one may still find observation points where the waves from the two sources arrive in phase.

[106] Specifically, we have

$$t_2 = |r - \bar{r}_2|/u_p$$

and

$$t_1 = |r - \bar{r}_1|/u_p = t_2 + \frac{a\lambda}{u_p} = t_2 + \frac{a}{f}.$$

in terms of potentials. Given the general expression in Eq. (4.289) for the magnetic vector potential, we have[107]

$$\bar{H}(\bar{r}) = \bar{\nabla} \times \int_V \bar{J}_v(\bar{r}')g(\bar{r},\bar{r}')d\bar{r}' = \int_V \bar{\nabla} \times [\bar{J}_v(\bar{r}')g(\bar{r},\bar{r}')]d\bar{r}' \tag{4.296}$$

$$= \int_V \bar{\nabla} g(\bar{r},\bar{r}') \times \bar{J}_v(\bar{r}')d\bar{r}' + \int_V g(\bar{r},\bar{r}')\bar{\nabla} \times \bar{J}_v(\bar{r}')d\bar{r}' \tag{4.297}$$

$$= \int_V \bar{\nabla} g(\bar{r},\bar{r}') \times \bar{J}_v(\bar{r}')d\bar{r}' \tag{4.298}$$

or[108]

$$\bar{H}(\bar{r}) = \int_V \bar{J}_v(\bar{r}') \times \bar{\nabla}' g(\bar{r},\bar{r}')d\bar{r}'. \tag{4.299}$$

Also using Eq. (4.290), we have the expression for the electric field intensity as

$$\bar{E}(\bar{r}) = -\frac{1}{\varepsilon}\bar{\nabla}\int_V q_v(\bar{r}')g(\bar{r},\bar{r}')d\bar{r}' - j\omega\mu\int_V \bar{J}_v(\bar{r}')g(\bar{r},\bar{r}')d\bar{r}' \tag{4.300}$$

$$= -\frac{1}{\varepsilon}\int_V q_v(\bar{r}')\bar{\nabla} g(\bar{r},\bar{r}')d\bar{r}' - j\omega\mu\int_V \bar{J}_v(\bar{r}')g(\bar{r},\bar{r}')d\bar{r}' \tag{4.301}$$

or[109]

$$\bar{E}(\bar{r}) = \frac{1}{\varepsilon}\int_V q_v(\bar{r}')\bar{\nabla}' g(\bar{r},\bar{r}')d\bar{r}' - j\omega\mu\int_V \bar{J}_v(\bar{r}')g(\bar{r},\bar{r}')d\bar{r}'. \tag{4.302}$$

Finally, we check the expressions for time-harmonic sources in the limit static cases: i.e. when $\omega \to 0$ (hence $k \to 0$) in steady state. We have[110]

$$\bar{\nabla}^2\bar{E}(\bar{r}) = \frac{1}{\varepsilon}\bar{\nabla} q_v(\bar{r}), \qquad \bar{\nabla}^2\bar{H}(\bar{r}) = -\bar{\nabla} \times \bar{J}_v(\bar{r}), \tag{4.303}$$

$$\bar{\nabla}^2\bar{A}(\bar{r}) = -\mu\bar{J}_v(\bar{r}), \qquad \bar{\nabla}^2\Phi(\bar{r}) = -\frac{q_v(\bar{r})}{\varepsilon} \tag{4.304}$$

[107] Note that the differential operator is in unprimed coordinates so that it can be directly carried inside the integral. In addition, we use $\bar{\nabla} \times [h(\bar{r})\bar{f}(\bar{r})] = \bar{\nabla} h(\bar{r}) \times \bar{f}(\bar{r}) + h(\bar{r})\bar{\nabla} \times \bar{f}(\bar{r})$, while $\bar{\nabla} \times \bar{J}(\bar{r}') = 0$, again due to the different coordinate system variables for $\bar{\nabla}$ and $\bar{J}(\bar{r}')$.

[108] Note that $\bar{\nabla} g(\bar{r},\bar{r}') = -\bar{\nabla}' g(\bar{r},\bar{r}')$. For example, we have

$$\frac{\partial g(\bar{r},\bar{r}')}{\partial x} = \frac{\partial}{\partial x}\frac{\exp(-jk|\bar{r}-\bar{r}'|)}{4\pi|\bar{r}-\bar{r}'|}$$
$$= \frac{\partial}{\partial x}(|\bar{r}-\bar{r}'|)\frac{\partial}{\partial R}\left[\frac{\exp(-jkR)}{4\pi R}\right]$$
$$= \frac{(x-x')}{|\bar{r}-\bar{r}'|}\frac{\partial}{\partial R}\left[\frac{\exp(-jkR)}{4\pi R}\right]$$
$$= -\frac{\partial g(\bar{r},\bar{r}')}{\partial x'},$$

where $R = |\bar{r}-\bar{r}'|$.

[109] At this stage, it is a nice practice to consider these expressions in Maxwell's equations. For example, using Faraday's law, we have

$$\bar{\nabla} \times \bar{E}(\bar{r}) = \frac{1}{\varepsilon}\int_V q_v(\bar{r}')\bar{\nabla} \times \bar{\nabla}' g(\bar{r},\bar{r}')d\bar{r}'$$
$$- j\omega\mu\int_V \bar{\nabla} g(\bar{r},\bar{r}') \times \bar{J}_v(\bar{r}')d\bar{r}'$$
$$= -j\omega\mu\int_V \bar{J}_v(\bar{r}') \times \bar{\nabla}' g(\bar{r},\bar{r}')d\bar{r}'$$
$$= -j\omega\mu\bar{H}(\bar{r})$$

correctly. Here, we use $\bar{\nabla} \times \bar{\nabla}' g(\bar{r},\bar{r}') = -\bar{\nabla} \times \bar{\nabla} g(\bar{r},\bar{r}') = 0$ based on the null identity.

[110] Note that Green's function becomes

$$g(\bar{r},\bar{r}') = \frac{\exp(-jk|\bar{r}-\bar{r}'|)}{4\pi|\bar{r}-\bar{r}'|} \to \frac{1}{4\pi|\bar{r}-\bar{r}'|},$$

leading to

$$\bar{\nabla}' g(\bar{r},\bar{r}') \to \frac{(\bar{r}-\bar{r}')}{4\pi|\bar{r}-\bar{r}'|^3}.$$

and[111]

$$\bar{A}(\bar{r}) = \frac{\mu}{4\pi}\int_V \bar{J}_v(\bar{r}')\frac{1}{|\bar{r}-\bar{r}'|}d\bar{r}' \tag{4.305}$$

$$\Phi(\bar{r}) = \frac{1}{4\pi\varepsilon}\int_V q_v(\bar{r}')\frac{1}{|\bar{r}-\bar{r}'|}d\bar{r}' \tag{4.306}$$

$$\bar{E}(\bar{r}) = \frac{1}{4\pi\varepsilon}\int_V q_v(\bar{r}')\frac{(\bar{r}-\bar{r}')}{|\bar{r}-\bar{r}'|^3}d\bar{r}' = -\bar{\nabla}\Phi(\bar{r}) \tag{4.307}$$

$$\bar{H}(\bar{r}) = \frac{1}{4\pi}\int_V \bar{J}_v(\bar{r}') \times \frac{(\bar{r}-\bar{r}')}{|\bar{r}-\bar{r}'|^3}d\bar{r}' = \frac{1}{\mu}\bar{\nabla}\times\bar{A}(\bar{r}). \tag{4.308}$$

All these equations are perfectly consistent with the earlier expressions used in static cases.

[111]Note that the Lorenz gauge

$$\bar{\nabla}\cdot\bar{A}(\bar{r},t) = -j\omega\mu\varepsilon\Phi(\bar{r})$$

becomes the Coulomb gauge

$$\bar{\nabla}\cdot\bar{A}(\bar{r}) = 0$$

in the static limit ($\omega \to 0$).

4.4.4 Examples

Example 47: In a three-dimensional half space $z > 0$ with vacuum properties, the electric field intensity is defined as (Figure 4.35)

$$\bar{E}(\bar{r},t) = \hat{a}_x E_0 \exp(-az)\cos(bt), \tag{4.309}$$

where $a \neq 0$ (1/m) and $b \neq 0$ (1/s) are constants. Find the magnetic field intensity and all sources (electric charges and currents) for $z > 0$. Also check the equality

$$\bar{\nabla}^2\bar{E}(\bar{r},t) - \mu_0\varepsilon_0\frac{\partial^2}{\partial t^2}\bar{E}(\bar{r},t) \stackrel{?}{=} 0. \tag{4.310}$$

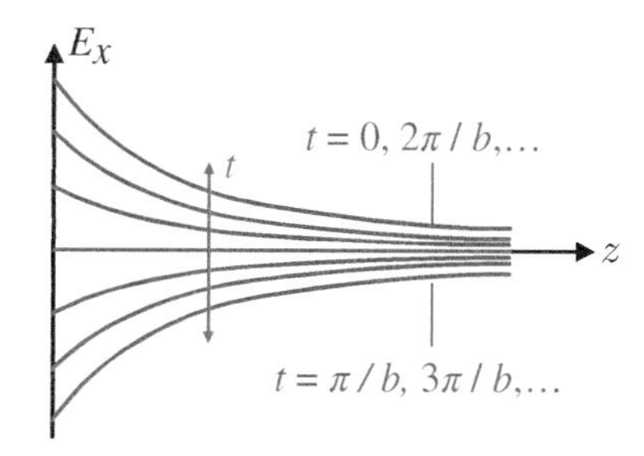

Figure 4.35 The amplitude (corresponding to the x component) of the electric field intensity $\bar{E}(\bar{r},t) = \hat{a}_x E_0 \exp(-az)\cos(bt)$ decays exponentially in the z direction, while it oscillates with time. At specific instants ($t = \pi/(2b), 3\pi/(2b), \ldots$), the electric field intensity is completely zero.

Solution: First, given the electric field intensity, we can find the volume electric charge density as

$$q_v(\bar{r},t) = \varepsilon_0\bar{\nabla}\cdot\bar{E}(\bar{r},t) = \varepsilon_0\frac{\partial}{\partial x}\big[E_0\exp(-az)\cos(bt)\big] = 0. \tag{4.311}$$

For the magnetic field intensity, we can employ Faraday's law, which needs the curl of the electric field intensity:

$$\bar{\nabla}\times\bar{E}(\bar{r},t) = \hat{a}_z\times\hat{a}_x\frac{\partial}{\partial z}\big[E_0\exp(-az)\cos(bt)\big] = -\hat{a}_y a E_0\exp(-az)\cos(bt). \tag{4.312}$$

In the absence of a static magnetic field, we obtain[112,113]

$$-\mu_0\frac{\partial \bar{H}(\bar{r},t)}{\partial t} = \bar{\nabla}\times\bar{E}(\bar{r},t) = -\hat{a}_y aE_0\exp(-az)\cos(bt) \tag{4.313}$$

$$\longrightarrow \bar{H}(\bar{r},t) = \hat{a}_y\frac{aE_0}{b\mu_0}\exp(-az)\sin(bt). \tag{4.314}$$

To find the volume electric current density, we use Ampere's law as

$$\bar{\nabla}\times\bar{H}(\bar{r},t) = \bar{J}_v(\bar{r},t) + \varepsilon_0\frac{\partial \bar{E}(\bar{r},t)}{\partial t} \tag{4.315}$$

$$\hat{a}_z\times\hat{a}_y\frac{aE_0}{b\mu_0}\frac{\partial}{\partial z}\big[\exp(-az)\sin(bt)\big] = \bar{J}_v(\bar{r},t) + \hat{a}_x E_0\varepsilon_0\frac{\partial}{\partial t}\big[\exp(-az)\cos(bt)\big], \tag{4.316}$$

leading to[114]

$$\bar{J}_v(\bar{r},t) = \hat{a}_x\frac{a^2E_0}{b\mu_0}\exp(-az)\sin(bt) + \hat{a}_x bE_0\varepsilon_0\exp(-az)\sin(bt) \tag{4.317}$$

$$= \hat{a}_x\left(\frac{a^2}{b\mu_0} + b\varepsilon_0\right)E_0\exp(-az)\sin(bt). \tag{4.318}$$

Finally, given Eq. (4.193), we have[115]

$$\bar{\nabla}^2\bar{E}(\bar{r},t) - \mu_0\varepsilon_0\frac{\partial^2}{\partial t^2}\bar{E}(\bar{r},t) = \mu_0\frac{\partial \bar{J}_v(\bar{r},t)}{\partial t} + \frac{1}{\varepsilon_0}\bar{\nabla}q_v(\bar{r},t) \tag{4.319}$$

$$= \hat{a}_x\left(a^2 + b^2\mu_0\varepsilon_0\right)E_0\exp(-az)\cos(bt), \tag{4.320}$$

which is definitely not zero.

Example 48: In a three-dimensional space with vacuum properties, the magnetic vector potential is defined in the phasor domain as[116]

$$\bar{A}(\bar{r}) = \hat{a}_x A_0\sin(kx)\exp(-jkz), \tag{4.321}$$

where $k=\omega\sqrt{\mu_0\varepsilon_0}$, assuming the Lorenz gauge. Find the electric field intensity and the magnetic field intensity, as well as all sources (electric charges and currents).

[112]Note that $\bar{\nabla}\cdot\bar{H}(\bar{r},t)=0$ is satisfied.

[113]Note that the magnetic field intensity has the same exponential decay with respect to position ($\exp(-az)$), while its time dependency ($\sin(bt)$) is different from the time dependency of the electric field intensity ($\cos(bt)$). It is particularly noteworthy that, when the electric field intensity is zero (at instants $t=\pi/(2b), 3\pi/(2b),\ldots$), the magnitude of the magnetic field intensity is at a maximum, e.g.

$$H_y\left(z, t=\frac{\pi}{2b}\right) = \frac{aE_0}{b\mu_0}\exp(-az)$$

$$H_y\left(z, t=\frac{3\pi}{2b}\right) = -\frac{aE_0}{b\mu_0}\exp(-az).$$

Similarly, when the magnetic field intensity is zero (at instants $t=0,\pi/b,2\pi/b,\ldots$), the magnitude of the electric field intensity becomes maximum.

[114]Note that $\bar{J}_v(\bar{r},t)$ exists in the absence of $q_v(\bar{r},t)$: i.e. zero electric charge density does not mean zero electric current density. For this case, we can check the continuity equation as

$$\bar{\nabla}\cdot\bar{J}(\bar{r},t) = 0 = -\frac{\partial q_v(\bar{r},t)}{\partial t},$$

which is perfectly satisfied.

[115]Similarly, given Eq. (4.195), we have

$$\bar{\nabla}^2\bar{H}(\bar{r},t) - \mu_0\varepsilon_0\frac{\partial^2}{\partial t^2}\bar{H}(\bar{r},t) = -\bar{\nabla}\times\bar{J}_v(\bar{r},t)$$

$$= \hat{a}_z\times\hat{a}_x\left(\frac{a^2}{b\mu_0} + b\varepsilon_0\right)E_0\sin(bt)\frac{\partial}{\partial z}[\exp(-az)]$$

$$= -\hat{a}_y\left(\frac{a^2}{b\mu_0} + b\varepsilon_0\right)aE_0\sin(bt)\exp(-az).$$

[116]In the time domain, the magnetic vector potential can be written

$$\bar{A}(\bar{r},t) = \hat{a}_x A_0\sin(kx)\cos(\omega t - kz).$$

Solution: First, using the Lorenz gauge, we can find the electric scalar potential:[117]

$$\Phi(\bar{r}) = \frac{j}{\omega\mu_0\varepsilon_0}\bar{\nabla}\cdot\bar{A}(\bar{r}) = \frac{j}{\omega\mu_0\varepsilon_0}A_0 k\cos(kx)\exp(-jkz) \tag{4.322}$$

$$= \frac{jA_0}{\sqrt{\mu_0\varepsilon_0}}\cos(kx)\exp(-jkz). \tag{4.323}$$

Then the electric field intensity can be found via potentials:

$$\bar{E}(\bar{r}) = -\bar{\nabla}\Phi(\bar{r}) - j\omega\bar{A}(\bar{r}) \tag{4.324}$$

$$= \hat{a}_x\frac{jA_0}{\sqrt{\mu_0\varepsilon_0}}k\sin(kx)\exp(-jkz) + \hat{a}_z\frac{jA_0}{\sqrt{\mu_0\varepsilon_0}}\cos(kx)(jk)\exp(-jkz)$$

$$- \hat{a}_x jA_0\omega\sin(kx)\exp(-jkz) \tag{4.325}$$

$$= -\hat{a}_z A_0\omega\cos(kx)\exp(-jkz). \tag{4.326}$$

Similarly, the magnetic field intensity can be obtained from the magnetic vector potential as[118,119]

$$\bar{H}(\bar{r}) = \frac{1}{\mu_0}\bar{\nabla}\times\bar{A}(\bar{r}) = \frac{1}{\mu_0}\hat{a}_z\times\hat{a}_x A_0\sin(kx)(-jk)\exp(-jkz) \tag{4.327}$$

$$= -\hat{a}_y\frac{jA_0\omega}{\eta_0}\sin(kx)\exp(-jkz), \tag{4.328}$$

where $\eta_0 = \sqrt{\mu_0/\varepsilon_0}$. The volume electric charge density can be found from the electric field intensity via Gauss' law:

$$q_v(\bar{r}) = \varepsilon_0\bar{\nabla}\cdot\bar{E}(\bar{r}) = jkA_0\omega\varepsilon_0\cos(kx)\exp(-jkz). \tag{4.329}$$

For the volume electric current density, we can directly use the given magnetic vector potential in the related Helmholtz equation as[120]

$$-\mu_0\bar{J}_v(\bar{r}) = \bar{\nabla}^2\bar{A}(\bar{r}) + k^2\bar{A}(\bar{r}) \tag{4.330}$$

$$= \hat{a}_x\left\{\frac{\partial^2}{\partial x^2} + \frac{\partial^2}{\partial z^2} + k^2\right\}[A_0\sin(kx)\exp(-jkz)] \tag{4.331}$$

$$= \hat{a}_x A_0(-k^2)\sin(kx)\exp(-jkz) + \hat{a}_x A_0\sin(kx)(-jk)^2\exp(-jkz)$$

$$+ \hat{a}_x k^2 A_0\sin(kx)\exp(-jkz) \tag{4.332}$$

$$= -\hat{a}_x A_0 k^2\sin(kx)\exp(-jkz) \tag{4.333}$$

[117] Hence, in the time domain, we have

$$\Phi(\bar{r},t) = \mathrm{Re}\{\Phi(\bar{r})\exp(j\omega t)\}$$
$$= \frac{A_0}{\sqrt{\mu_0\varepsilon_0}}\cos(kx)\mathrm{Re}\{\exp(-jkz)\exp(j\omega t)\}$$
$$= \frac{A_0}{\sqrt{\mu_0\varepsilon_0}}\cos(kx)\cos(\omega t - kz + \pi/2)$$
$$= -\frac{A_0}{\sqrt{\mu_0\varepsilon_0}}\cos(kx)\sin(\omega t - kz),$$

given that $j = \exp(j\pi/2)$.

[118] Note that $\bar{\nabla}\cdot\bar{H}(\bar{r}) = 0$, as expected.

[119] Time-domain expressions can be written

$$\bar{E}(\bar{r},t) = -\hat{a}_z A_0\omega\cos(kx)\cos(\omega t - kz)$$
$$\bar{H}(\bar{r},t) = \hat{a}_y\frac{A_0\omega}{\eta_0}\sin(kx)\sin(\omega t - kz).$$

[120] Alternatively, we can employ Ampere's law to derive the volume electric current density. For this purpose, the curl of the magnetic field intensity can be found as

$$\bar{\nabla}\times\bar{H}(\bar{r}) = -\hat{a}_z\times\hat{a}_y\frac{jA_0\omega}{\eta_0}\sin(kx)(-jk)\exp(-jkz)$$
$$- \hat{a}_x\times\hat{a}_y\frac{jA_0\omega}{\eta_0}k\cos(kx)\exp(-jkz)$$
$$= \hat{a}_x A_0\omega^2\varepsilon_0\sin(kx)\exp(-jkz)$$
$$- \hat{a}_z jA_0\omega^2\varepsilon_0\cos(kx)\exp(-jkz).$$

Then we obtain

$$\bar{J}_v(\bar{r}) = \bar{\nabla}\times\bar{H}(\bar{r}) - j\omega\varepsilon_0\bar{E}(\bar{r})$$
$$= \hat{a}_x A_0\omega^2\varepsilon_0\sin(kx)\exp(-jkz)$$
$$- \hat{a}_z jA_0\omega^2\varepsilon_0\cos(kx)\exp(-jkz)$$
$$+ \hat{a}_z jA_0\omega^2\varepsilon_0\cos(kx)\exp(-jkz)$$
$$= \hat{a}_x A_0\omega^2\varepsilon_0\sin(kx)\exp(-jkz)$$
$$= \hat{a}_x\frac{A_0k^2}{\mu_0}\sin(kx)\exp(-jkz).$$

leading to[121]

$$\bar{J}_v(r) = \hat{a}_x \frac{A_0 k^2}{\mu_0} \sin(kx) \exp(-jkz). \tag{4.334}$$

Example 49: In a three-dimensional space with vacuum properties, the electric field intensity is defined as[122] (Figure 4.36)

$$\bar{E}(\bar{r}, t) = \hat{a}_y E_0 \sin\left(\frac{\pi x}{a}\right) \cos(\omega t - bz). \tag{4.335}$$

Find the magnetic field intensity, as well as the relationship between a (m) and b (1/m), assuming that the volume electric current density is zero.[123]

Solution: In the phasor domain, the electric field intensity can be written

$$\bar{E}(\bar{r}) = \hat{a}_y E_0 \sin\left(\frac{\pi x}{a}\right) \exp(-jbz). \tag{4.336}$$

Then using Faraday's law, we can obtain the magnetic field intensity as

$$\bar{H}(\bar{r}) = -\frac{1}{j\omega\mu_0} \bar{\nabla} \times \bar{E}(\bar{r}) \tag{4.337}$$

$$= -\frac{1}{j\omega\mu_0} \hat{a}_x \times \hat{a}_y E_0 \exp(-jbz) \frac{\partial}{\partial x}\left[\sin\left(\frac{\pi x}{a}\right)\right] - \frac{1}{j\omega\mu_0} \hat{a}_z \times \hat{a}_y E_0 \sin\left(\frac{\pi x}{a}\right) \frac{\partial}{\partial z}[\exp(-jbz)] \tag{4.338}$$

$$= \frac{E_0}{\omega\mu_0} \exp(-jbz)\left[\hat{a}_z j\left(\frac{\pi}{a}\right) \cos\left(\frac{\pi x}{a}\right) - \hat{a}_x b \sin\left(\frac{\pi x}{a}\right)\right]. \tag{4.339}$$

The corresponding expression in the time domain can be written[124]

$$\bar{H}(\bar{r}, t) = -\hat{a}_z \frac{E_0}{\omega\mu_0}\left(\frac{\pi}{a}\right) \cos\left(\frac{\pi x}{a}\right) \sin(\omega t - bz) - \hat{a}_x \frac{E_0}{\omega\mu_0} b \sin\left(\frac{\pi x}{a}\right) \cos(\omega t - bz). \tag{4.340}$$

Continuing in the phasor domain, application of Gauss' law leads to

$$q_v(\bar{r}) = \varepsilon_0 \bar{\nabla} \cdot \bar{E}(\bar{r}) = \varepsilon_0 \frac{\partial}{\partial y}\left[E_0 \sin\left(\frac{\pi x}{a}\right) \exp(-jbz)\right] = 0, \tag{4.341}$$

i.e., there is no volume electric charge density, as expected.[125]

[121] We can check the continuity equation as

$$\nabla \cdot \bar{J}_v(\bar{r}) = A_0 \omega^2 \varepsilon_0 k \cos(kx) \exp(-jkz)$$
$$= -j\omega\left[jkA_0\omega\varepsilon_0 \cos(kx) \exp(-jkz)\right]$$
$$= -j\omega q_v(\bar{r}).$$

[122] As discussed in Chapter 7, this kind of waves can be generated inside waveguides.

[123] Note that, considering time-harmonic sources and fields, zero electric current density means zero volume electric charge density, based on the continuity equation $\bar{\nabla} \cdot \bar{J}(\bar{r}) = -j\omega q(\bar{r})$.

[124] We have

$$\bar{H}(\bar{r}, t) = \text{Re}\{\bar{H}(\bar{r}) \exp(j\omega t)\}$$
$$= \hat{a}_z \frac{E_0}{\omega\mu_0}\left(\frac{\pi}{a}\right) \cos\left(\frac{\pi x}{a}\right) \text{Re}\{\exp(j\omega t - jbz)j\}$$
$$- \hat{a}_x \frac{E_0}{\omega\mu_0} b \sin\left(\frac{\pi x}{a}\right) \text{Re}\{\exp(j\omega t - jbz)\}$$

[125] On the other side, using Gauss' law for magnetic fields, we have

$$\bar{\nabla} \cdot \bar{H}(\bar{r}, t) =$$
$$-\frac{E_0}{\omega\mu_0}\left(\frac{\pi}{a}\right) \cos\left(\frac{\pi x}{a}\right) \frac{\partial}{\partial z}[\sin(\omega t - bz)]$$
$$-\frac{E_0}{\omega\mu_0} b \cos(\omega t - bz) \frac{\partial}{\partial x}\left[\sin\left(\frac{\pi x}{a}\right)\right]$$
$$= \frac{E_0}{\omega\mu_0} b \left(\frac{\pi}{a}\right) \cos\left(\frac{\pi x}{a}\right) \cos(\omega t - bz)$$
$$- \frac{E_0}{\omega\mu_0} b \cos(\omega t - bz) \left(\frac{\pi}{a}\right) \cos\left(\frac{\pi x}{a}\right)$$
$$= 0$$

correctly.

To find the relationship between a and b, we can use either Ampere's law or the Helmholtz equation for the electric field intensity. Considering the latter, we have

$$\bar{\nabla}^2\bar{E}(\bar{r}) + k^2\bar{E}(\bar{r}) = j\omega\mu\bar{J}_v(\bar{r}) + \frac{1}{\varepsilon_0}\bar{\nabla}q_v(\bar{r}) = 0 \tag{4.342}$$

$$\longrightarrow E_0\exp(-jbz)\frac{\partial^2}{\partial x^2}\left[\sin\left(\frac{\pi x}{a}\right)\right] + E_0\sin\left(\frac{\pi x}{a}\right)\frac{\partial^2}{\partial z^2}[\exp(-jbz)]$$
$$+ k^2E_0\sin\left(\frac{\pi x}{a}\right)\exp(-jbz) = 0, \tag{4.343}$$

leading to

$$-\left(\frac{\pi}{a}\right)^2 + (-jb)^2 + k^2 = 0 \tag{4.344}$$

or

$$\left(\frac{\pi}{a}\right)^2 + b^2 = k^2 = \omega^2\mu_0\varepsilon_0 \tag{4.345}$$

as the condition between a and b.

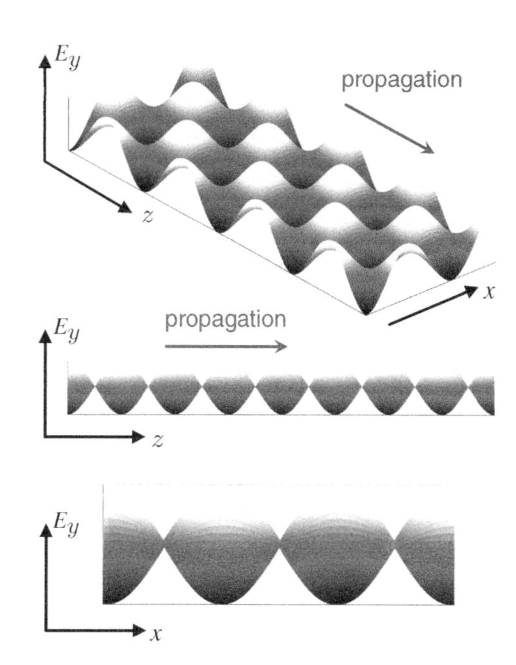

Figure 4.36 A snapshot (plots at a fixed time) of $\bar{E}_y(\bar{r},t) = E_0\sin(\pi x/a)\cos(\omega t - bz)$. The propagation is in the z direction. The plots show the wave in a limited region. It should be noted that the plot is not a three-dimensional view of the wave: i.e. there is no variation in the y direction.

4.5 Final Remarks

With this chapter, we have covered all of Maxwell's equations. While being simple, Gauss' law for magnetic fields states the important property of the magnetic flux density: i.e. its solenoidal nature, which is essential to fully use Maxwell's equations to solve electromagnetic problems. Because it is solenoidal, the magnetic flux density can be written as the curl of a vector field: the magnetic vector potential, which can be seen as a counterpart of the electric scalar potential. We have also considered the magnetic potential energy (counterpart of the electric potential energy) and its relationship with the magnetic force (counterpart of the electric force). In this chapter, we have also practiced the combination of Maxwell's equations and their general solutions. Combinations for both fields and potentials lead to wave equations, which are fundamental tools and the key to explaining why electromagnetic fields and potentials must behave as waves. In the the phasor domain involving time-harmonic sources, wave equations turn into Helmholtz equations that are also fundamentals of electromagnetics. All equations converge into suitable static forms when there is no time dependency.

This chapter also has completed all necessary tools and concepts to understand how Maxwell's equations can be used to understand electromagnetic phenomena. In the next

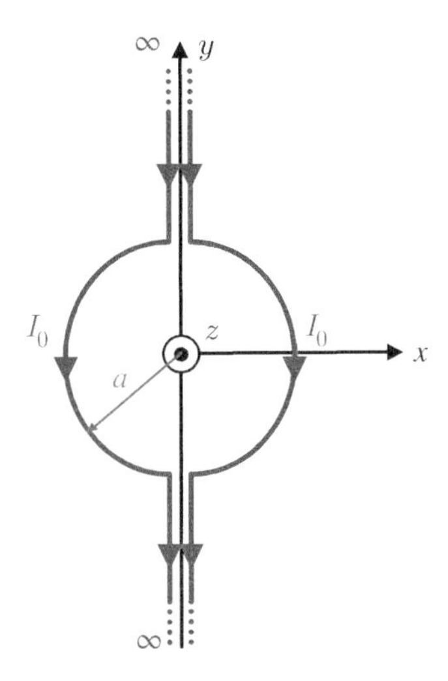

Figure 4.37 Steady electric currents I_0 flowing along two infinitely long wires that make a circular shape. The distance between the straight segments can be assumed to be zero.

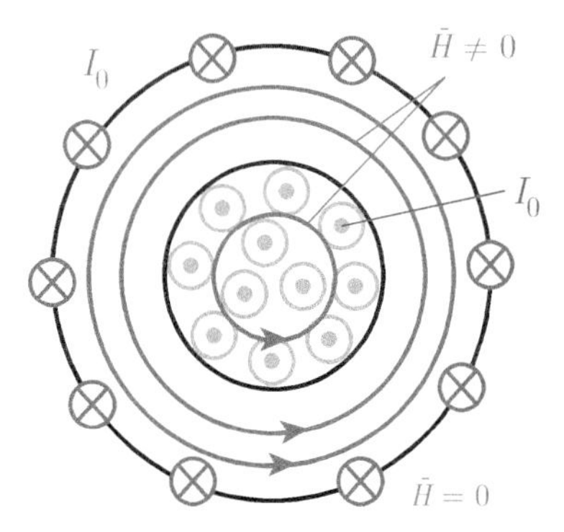

Figure 4.38 A coaxial structure involving two cylindrical surfaces with radii a and $b > a$ aligned in the z direction. A steady surface electric current flows uniformly on the outer cylinder, while the electric current (in the opposite direction) is distributed over the cross section of the inner cylinder (volume electric current density).

chapters, which can be considered to constitute the second part of the book, we use Maxwell's equations to analyze basic electromagnetic scenarios in more detail.

4.6 Exercises

Exercise 47: Show that the magnetic flux density due to a z-directed magnetic dipole,

$$\bar{B}_d(\bar{r}) = \frac{\mu_0 I_{DM}}{4\pi R^3}\left(\hat{a}_R 2\cos\theta + \hat{a}_\theta \sin\theta\right), \tag{4.346}$$

is solenoidal.

Exercise 48: In a three-dimensional space with vacuum properties, the magnetic vector potential is defined as

$$\bar{A}(\bar{r}) = \hat{a}_z A_0 \begin{cases} -\rho/a & (\rho < a) \\ \ln(\rho/a) & (\rho > a), \end{cases} \tag{4.347}$$

where A_0 (Wb/m) is a constant. Find the volume electric current density and the surface electric current density.

Exercise 49: Consider steady electric currents I_0 (A) flowing along two infinitely long wires that make a circular shape in a vacuum (Figure 4.37). Find the magnetic vector potential and the magnetic flux density at the origin, considering (i) only the circular part and (ii) the whole structure.

Exercise 50: Consider a coaxial structure involving infinitely long cylindrical surfaces with radii a and b, where $b > a$, and a homogeneous magnetic material with permeability μ between them (Figure 4.38). Steady electric currents (I_0 (A)) flow in opposite directions. The current is uniformly distributed on the surface of the outer cylinder (surface current density), while it is uniformly distributed over the cross section of the inner (nonmagnetic) cylinder (volume current density). Find the magnetic potential energy stored per length.

Exercise 51: Consider again a coaxial structure involving infinitely long cylindrical surfaces with radii a and b, where $b > a$, and a homogeneous magnetic material with permeability μ between them (Figure 4.38). In this case, the volume electric current density inside the inner cylinder is given as

$$\bar{J}_v(\bar{r}) = \hat{a}_z \frac{2I_0\rho^2}{\pi a^4}. \tag{4.348}$$

The same amount of steady electric current I_0 (A) in the form of surface electric current density flows in the opposite direction on the outer cylinder. Find the magnetic field intensity everywhere, as well as the magnetic potential energy stored per length.

Exercise 52: Consider a wave in free space (source-free vacuum) with an expression for the electric field intensity (Figure 4.39)

$$\bar{E}(\bar{r},t) = \hat{a}_y E_0 \cos(az)\cos(\omega t - bx), \tag{4.349}$$

where E_0 (V/m), a (1/m), and b (1/m) are constants. Find an expression for a and b in terms of ω, ε_0, and μ_0.

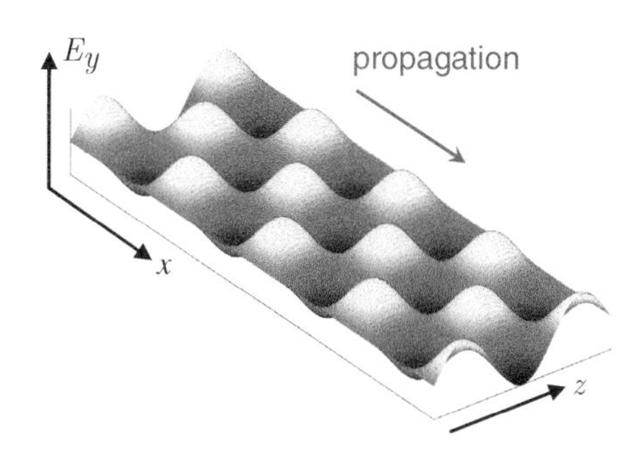

Figure 4.39 A snapshot (plot at a fixed time) of $\bar{E}_y(\bar{r},t) = E_0 \cos(az)\cos(\omega t - bx)$. The propagation is in the x direction. The plot shows the wave in a limited region.

Exercise 53: In a three-dimensional half space $z > 0$ with vacuum properties, the electric field intensity is defined as (Figure 4.40)

$$\bar{E}(\bar{r},t) = \hat{a}_z E_0 \exp(-kz)\cos(\omega t - kx), \tag{4.350}$$

where E_0 (V/m) is a constant and $k = \omega\sqrt{\mu_0\varepsilon_0}$. Assuming no static magnetic field, show that the wave equation is satisfied for $\bar{E}(\bar{r},t)$.

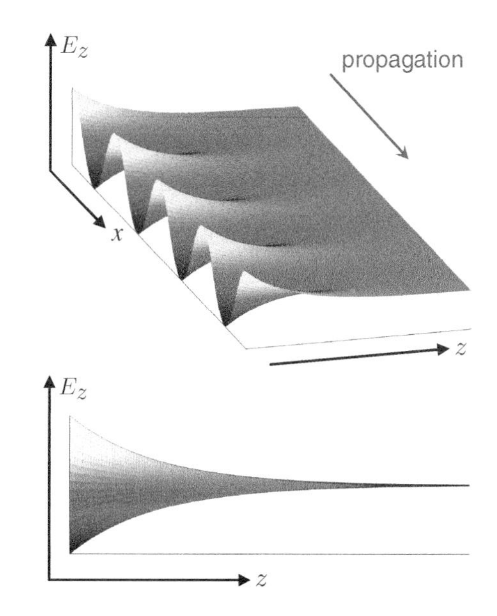

Figure 4.40 A snapshot (plots at a fixed time) of $\bar{E}_z(\bar{r},t) = E_0 \exp(-kz)\cos(\omega t - kx)$. The propagation is in the x direction. The plots show the wave in a limited region. It should be noted that the plot is not a three-dimensional view of the wave: i.e. there is no variation in the y direction.

Exercise 54: In a three-dimensional space with vacuum properties, the magnetic vector potential is defined (assuming the Lorenz gauge) as

$$\bar{A}(\bar{r},t) = \hat{a}_z A_0 \rho \exp(-at), \tag{4.351}$$

where A_0 (Wb/m^2) and a (1/s) are constants. Find the volume electric current density, as well as the electric field intensity and the magnetic field intensity, assuming that there is no static electric field.

Exercise 55: In a three-dimensional space with vacuum properties, the magnetic vector potential is defined (assuming the Lorenz gauge) as

$$\bar{A}(\bar{r}) = \hat{a}_z A_0 \rho \exp(-jkz) \tag{4.352}$$

in the phasor domain, where A_0 (Wb/m^2) is a constant and $k = \omega\sqrt{\mu_0\varepsilon_0}$. Find the volume electric current density, as well as the electric field intensity and the magnetic field intensity.

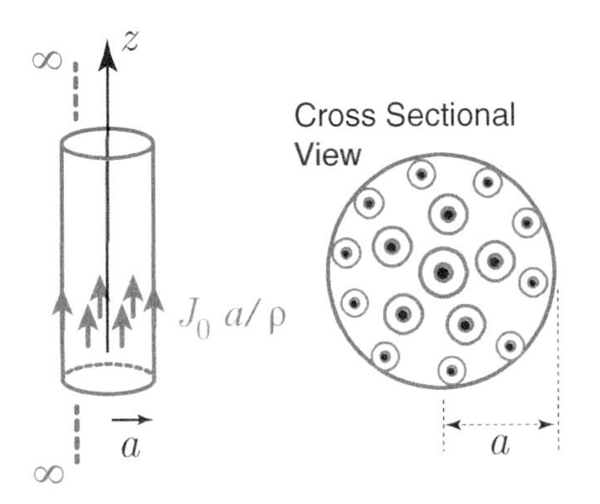

Figure 4.41 A steady volume electric current density $\bar{J}_v(\bar{r}) = \hat{a}_z J_0 a/\rho$ (A/m^2) flowing in a cylindrical region of radius a in the z direction.

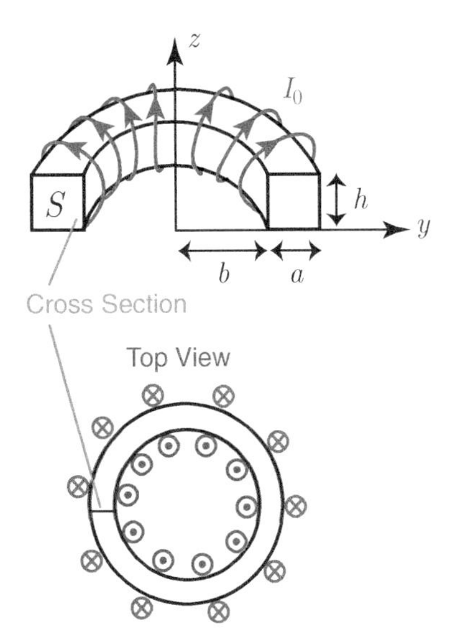

Figure 4.42 A steady electric current I_0 (A) circulating around a toroid of radius b and cylindrical radius a.

Exercise 56: Show that the general phasor-domain expressions for the electric field intensity and the magnetic field intensity for an arbitrary (time-harmonic) source,

$$\bar{E}(\bar{r}) = -\frac{1}{\varepsilon}\int_V q_v(\bar{r}')\bar{\nabla}g(\bar{r},\bar{r}')d\bar{r}' - j\omega\mu\int_V \bar{J}_v(\bar{r}')g(\bar{r},\bar{r}')d\bar{r}' \tag{4.353}$$

$$\bar{H}(\bar{r}) = \int_V \bar{J}_v(\bar{r}') \times \bar{\nabla}'g(\bar{r},\bar{r}')d\bar{r}', \tag{4.354}$$

satisfy Gauss' law.

4.7 Questions

Question 21: Consider a steady volume electric current density $\bar{J}_v(\bar{r}) = \hat{a}_z J_0 a/\rho$ (A/m^2) flowing in a cylindrical region of radius a in the z direction located in a vacuum (Figure 4.41). Find the magnetic vector potential and the magnetic flux density everywhere.

Question 22: Consider a steady electric current I_0 (A) circulating around a toroid of radius b and cylindrical radius a (Figure 4.42). The core is a magnetic material with a constant permeability μ, while there are a total of N turns. Find the stored magnetic potential energy. Describe the magnetic force in the system.

Question 23: The electric field intensity in a three-dimensional space with vacuum properties is defined as

$$\bar{E}(\bar{r},t) = \hat{a}_x E_0\cos(az)\exp(-bx)\cos(\omega t), \tag{4.355}$$

where E_0 (V/m), a (1/m), and b (1/m) are constants.

- Find the magnetic field intensity $\bar{H}(\bar{r},t)$.
- Find the volume electric charge density $q_v(\bar{r},t)$ and the volume electric current density $\bar{J}_v(\bar{r},t)$. Also find a in terms of the given parameters.
- Show that the continuity equation is satisfied.
- Define $\bar{A}(\bar{r},t)$ and $\Phi(\bar{r},t)$ that describe this wave.

Question 24: The magnetic field intensity in a three-dimensional space with vacuum properties is defined in the phasor domain as

$$\bar{H}(\bar{r},t) = \hat{a}_x H_0 \exp(-az)\exp(-jbz), \tag{4.356}$$

where H_0 (A/m), a (1/m), and b (1/m) are constants. Find the electric field intensity and all sources (electric charges and currents).

5

Basic Solutions to Maxwell's Equations

Until this point, we have dedicated each chapter to one of four Maxwell's equations. In addition, we analyzed electromagnetic problems (specifically, static cases) via suitable selections of Maxwell's equations. A general dynamic scenario, however, requires all of Maxwell's equations for a complete analysis. As discussed in Section 4.4, combinations of Maxwell's equations lead to higher-level tools, such as wave equations, to grasp underlying physical mechanisms and solve dynamic problems. This chapter focuses on solutions to Maxwell's equations (or those formulations already derived from them) for various basic electromagnetic scenarios. After understanding electromagnetic propagation and radiation, we will study plane waves (Figure 5.1) – i.e. basic solutions to Maxwell's equations – to understand fundamental wave phenomena: propagation, refraction, and reflection, as well as essential characteristics of waves (e.g. polarization). To proceed further in this chapter, all the tools and concepts introduced and discussed in the previous chapters need to be known.

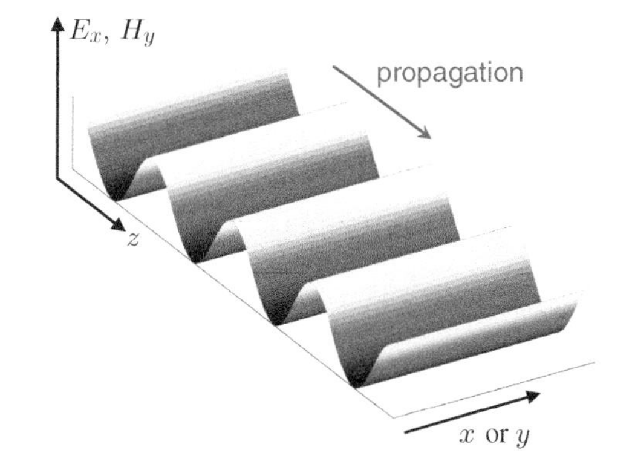

Figure 5.1 A snapshot (plot at a fixed time) of a uniform plane wave propagating in the z direction. Electric and magnetic fields are in the x and y directions, respectively, and they are perpendicular to the propagation direction (z) at each and every point in space. The amplitudes of the electric and magnetic fields are sinusoidal functions with respect to z, while there is no variation in the x and y directions (uniformity). Note that this plot is not a three-dimensional view of the wave: i.e. it shows the dependency of the electric field intensity (E_x) and the magnetic field intensity (H_y) with respect to z (propagation direction) and x/y.

5.1 Summary of Maxwell's Equations, Wave Equations, and Helmholtz Equations

First, in this section, we summarize Maxwell's equations and related concepts, including higher-level equations: i.e. wave equations and Helmholtz equations. Gauss' law, Ampere's law, Faraday's law, and Gauss' law for magnetic fields can be written

$$\bar{\nabla} \cdot \bar{D}(\bar{r}, t) = q_v(\bar{r}, t) \tag{5.1}$$

$$\bar{\nabla} \times \bar{H}(\bar{r}, t) = \bar{J}_v(\bar{r}, t) + \frac{\partial \bar{D}(\bar{r}, t)}{\partial t} \tag{5.2}$$

Introduction to Electromagnetic Waves with Maxwell's Equations, First Edition. Özgür Ergül.

Companion website: www.wiley.com/go/ergulmax

$$\bar{\nabla} \times \bar{E}(\bar{r}, t) = -\frac{\partial \bar{B}(\bar{r}, t)}{\partial t} \tag{5.3}$$

$$\bar{\nabla} \cdot \bar{B}(\bar{r}, t) = 0 \tag{5.4}$$

in differential forms[1,2,3] and

$$\oint_S \bar{D}(\bar{r}, t) \cdot \overline{ds} = Q_{\text{enc}}(t) \tag{5.5}$$

$$\oint_C \bar{H}(\bar{r}, t) \cdot \overline{dl} = I_{\text{enc}}(t) + \int_S \frac{\partial \bar{D}(\bar{r}, t)}{\partial t} \cdot \overline{ds} \tag{5.6}$$

$$\oint_C \bar{E}(\bar{r}, t) \cdot \overline{dl} = -\int_S \frac{\partial \bar{B}(\bar{r}, t)}{\partial t} \cdot \overline{ds} \tag{5.7}$$

$$\oint_S \bar{B}(\bar{r}, t) \cdot \overline{ds} = 0 \tag{5.8}$$

in integral forms.[4,5] In this style, Maxwell's equations are often called *macroscopic*, since material properties (dielectric and magnetic effects) are hidden. These effects can be revealed by using

$$\bar{D}(\bar{r}, t) = \varepsilon_0 \bar{E}(\bar{r}, t) + \bar{P}(\bar{r}, t) \tag{5.9}$$

$$\bar{B}(\bar{r}, t) = \mu_0 \bar{H}(\bar{r}, t) + \mu_0 \bar{M}(\bar{r}, t), \tag{5.10}$$

where $\bar{P}(\bar{r}, t)$ and $\bar{M}(\bar{r}, t)$ represent polarization and magnetization, respectively. The constitutive relations are simplified versions of these expressions as

$$\bar{D}(\bar{r}, t) = \varepsilon(\bar{r}) \bar{E}(\bar{r}, t) \tag{5.11}$$

$$\bar{B}(\bar{r}, t) = \mu(\bar{r}) \bar{H}(\bar{r}, t), \tag{5.12}$$

where $\varepsilon(\bar{r})$ and $\mu(\bar{r})$ are permittivity and permeability at $\bar{r}$. When discussing medium effects, we often resort to representations of polarization/magnetization as equivalent sources. These can be listed as follows:[6]

- Volume and surface polarization charge density (C/m^3 and C/m^2):

$$q_{pv}(\bar{r}, t) = -\bar{\nabla} \cdot \bar{P}(\bar{r}, t) \quad \& \quad q_{ps}(\bar{r}, t) = \hat{a}_n \cdot \bar{P}(\bar{r}, t) \tag{5.13}$$

- Volume and surface magnetization current density (A/m^2 and A/m):

$$\bar{J}_{mv}(\bar{r}, t) = \bar{\nabla} \times \bar{M}(\bar{r}, t) \quad \& \quad \bar{J}_{ms}(\bar{r}, t) = -\hat{a}_n \times \bar{M}(\bar{r}, t) \tag{5.14}$$

[1] Recall that the differential forms represent pointwise identities: i.e. they are satisfied independently at each and every location space. For example, $\bar{\nabla} \cdot \bar{D}(\bar{r}, t) = 0$ if there is no source exactly at $\bar{r}$ (when the sources that create $\bar{D}(\bar{r}, t)$ are located elsewhere).

[2] Recall that, in the differential forms, only *volume* densities appear as sources.

[3] As a complementary one, recall the continuity equation

$$\bar{\nabla} \cdot \bar{J}_v(\bar{r}, t) = -\frac{\partial}{\partial t} q_v(\bar{r}, t).$$

[4] Recall that, in the integral forms, all types of sources that are represented by volume, surface, or line densities are included.

[5] Recall that, if the integration domain does not depend on time (does not move), we have

$$\oint_C \bar{H}(\bar{r}, t) \cdot \overline{dl} = I_{\text{enc}}(t) + \frac{\partial f_D(t)}{\partial t}$$

$$\oint_C \bar{E}(\bar{r}, t) \cdot \overline{dl} = -\frac{\partial f_B(t)}{\partial t},$$

where

$$f_D(\bar{r}, t) = \int_S \bar{D}(\bar{r}, t) \cdot \overline{ds}$$

$$f_B(\bar{r}, t) = \int_S \bar{B}(\bar{r}, t) \cdot \overline{ds}$$

are the electric and magnetic fluxes.

[6] We emphasize that a polarized object should be modeled with *either* polarization (electric) charge densities [$q_{pv}(\bar{r}, t)$ and $q_{ps}(\bar{r}, t)$] *or* magnetic current densities [$\bar{J}_{pv}(\bar{r}, t)$ and $\bar{J}_{ps}(\bar{r}, t)$], not with both sets at the same time (Figure 5.2). Similarly, a magnetized object should be modeled with either magnetization (electric) current densities [$\bar{J}_{mv}(\bar{r}, t)$ and $\bar{J}_{ms}(\bar{r}, t)$] or magnetic charge densities [$q_{mv}(\bar{r}, t)$ and $q_{ms}(\bar{r}, t)$].

- Volume and surface magnetic current density due to polarization (V/m^2 and V/m):

$$\bar{J}_{pv}(\bar{r},t) = \bar{\nabla} \times \bar{P}(\bar{r},t)/\varepsilon_0 \quad \& \quad \bar{J}_{ps}(\bar{r},t) = -\hat{a}_n \times \bar{P}(\bar{r},t)/\varepsilon_0 \tag{5.15}$$

- Volume and surface magnetic charge density due to magnetization (Wb/m^3 and Wb/m^2):

$$q_{mv}(\bar{r},t) = -\mu_0 \bar{\nabla} \cdot \bar{M}(\bar{r},t) \quad \& \quad q_{ms}(\bar{r},t) = \mu_0 \hat{a}_n \cdot \bar{M}(\bar{r},t) \tag{5.16}$$

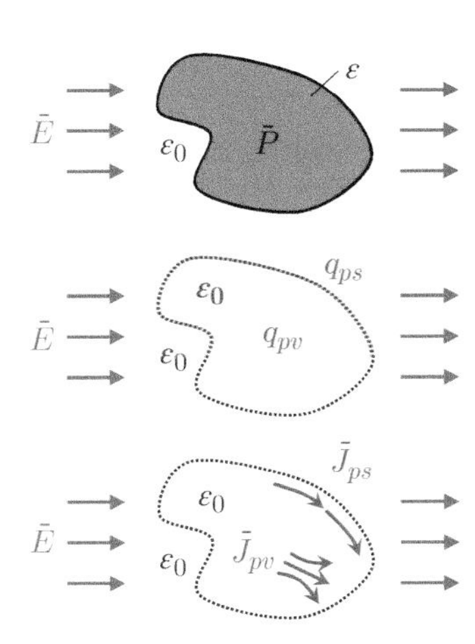

Figure 5.2 Modeling dielectric polarization by either equivalent polarization (electric) charges or equivalent magnetic currents.

Volume equivalent sources describe relationships between flux densities and field intensities, while surface equivalent sources explain discontinuities at material boundaries.

As discussed in Sections 3.6, 4.3, and 4.4.2, potentials are commonly used to facilitate solutions to Maxwell's equations. In dynamic cases, we have

$$\bar{E}(\bar{r},t) = -\bar{\nabla}\Phi(\bar{r},t) - \frac{\partial \bar{A}(\bar{r},t)}{\partial t} \tag{5.17}$$

$$\bar{B}(\bar{r},t) = \bar{\nabla} \times \bar{A}(\bar{r},t), \tag{5.18}$$

where $\Phi(\bar{r},t)$ and $\bar{A}(\bar{r},t)$ are the electric scalar potential and the magnetic vector potential, respectively. The equality for the magnetic flux density is consistent with Gauss' law for magnetic fields: i.e. $\bar{\nabla} \cdot \bar{B}(\bar{r}) = 0$. Specifically, the magnetic flux density is solenoidal (absence of magnetic sources). On the other hand, the magnetic field intensity is not necessarily solenoidal, and we have

$$\bar{\nabla} \cdot \bar{H}(\bar{r},t) = -\bar{\nabla} \cdot \bar{M}(\bar{r},t). \tag{5.19}$$

Together with the corresponding boundary condition, this may explain how magnetic fields are created by *magnets* (permanently magnetized materials) in the absence of (true) electric currents.

In static cases – i.e. when there is no time dependency – we consider stationary electric charges and steady electric currents (Figure 5.3). Since we have zero time derivatives, Maxwell's equations can be written

$$\bar{\nabla} \cdot \bar{D}(\bar{r}) = q_v(\bar{r}), \qquad \oint_S \bar{D}(\bar{r}) \cdot \overline{ds} = Q_{\text{enc}} \tag{5.20}$$

$$\bar{\nabla} \times \bar{H}(\bar{r}) = \bar{J}_v(\bar{r}), \qquad \oint_C \bar{H}(\bar{r}) \cdot \overline{dl} = I_{\text{enc}} \tag{5.21}$$

$$\bar{\nabla} \times \bar{E}(\bar{r}) = 0, \qquad \oint_C \bar{E}(\bar{r}) \cdot \overline{dl} = 0 \tag{5.22}$$

$$\bar{\nabla} \cdot \bar{B}(\bar{r}) = 0, \qquad \oint_S \bar{B}(\bar{r}) \cdot \overline{ds} = 0. \tag{5.23}$$

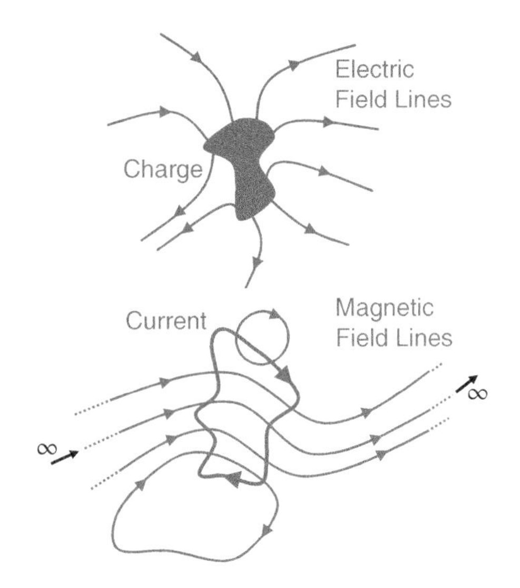

Figure 5.3 In electrostatics and magnetostatics, electric and magnetic fields (hence, their analyses) are completely decoupled from each other. In electrostatics, stationary electric charges create electric fields that can be formulated via Gauss' law, while Faraday's law states that the generated electric field intensity is irrotational. In magnetostatics, steady electric currents are considered without charge accumulation, and they create magnetic fields that can be formulated via Ampere's law. Gauss' law for magnetic fields indicates that the magnetic flux density is solenoidal: i.e. magnetic field lines must close upon themselves (or end at infinity). In any case, electric and magnetic fields do not change with respect to time.

In electrostatics, Gauss' law is consistent with Coulomb's law (Section 1.6). Similarly, in magnetostatics, Ampere's law is consistent with Biot-Savart law and Ampere's force law (Section 2.7). All these laws can be used alone to find static electric and magnetic fields from their sources (electric charges and currents). In electrostatics, the curl of the electric field intensity is zero and is called *irrotational*. This is also evident in the static form of the corresponding potential relationship:

$$\bar{E}(\bar{r}) = -\bar{\nabla}\Phi(\bar{r}) \longrightarrow \bar{\nabla} \times \bar{E}(\bar{r}) = -\bar{\nabla} \times \bar{\nabla}\Phi(\bar{r}) = 0. \tag{5.24}$$

But note that

$$\bar{\nabla} \times \bar{D}(\bar{r}) = -\bar{\nabla} \times \bar{P}(\bar{r}), \tag{5.25}$$

and the electric flux density is not necessarily irrotational. Together with the corresponding boundary condition, this may explain how electric fields can be generated by *electrets* (permanently polarized materials) in the absence of (true) electric charges.[7,8]

Note that for a static case, the magnetic field intensity is irrotational in the absence of the electric current density: i.e. $\bar{\nabla} \times \bar{H}(\bar{r}) = 0$. On the other hand, we have the curl of the magnetic flux density as

$$\bar{\nabla} \times \bar{B}(\bar{r}) = \mu_0\bar{\nabla} \times \bar{H}(\bar{r}) + \mu_0\bar{\nabla} \times \bar{M}(\bar{r}), \tag{5.26}$$

leading to

$$\bar{\nabla} \times \bar{B}(\bar{r}) = \mu_0\bar{\nabla} \times \bar{M}(\bar{r}). \tag{5.27}$$

Hence, inside magnetic materials without any electric currents, the magnetic flux density is not necessarily irrotational, depending on the magnetization. This is also how one may model the generation of magnetic fields by magnets.

The electric flux density can be solenoidal in the absence of electric charge density: i.e. $\bar{\nabla} \cdot \bar{D}(\bar{r}, t) = 0$. On the other hand, we have the divergence of the electric field intensity as

$$\bar{\nabla} \cdot \bar{E}(\bar{r}, t) = \bar{\nabla} \cdot \bar{D}(\bar{r}, t)/\varepsilon_0 - \bar{\nabla} \cdot \bar{P}(\bar{r}, t)/\varepsilon_0, \tag{5.28}$$

leading to

$$\bar{\nabla} \cdot \bar{E}(\bar{r}, t) = -\bar{\nabla} \cdot \bar{P}(\bar{r}, t)/\varepsilon_0. \tag{5.29}$$

Hence, inside dielectrics without any electric charges, the electric field intensity is not necessarily solenoidal, depending on the polarization. Once again, this may explain electric fields created by electrets.

[7] In this context, electrets refer to permanently polarized objects, usually via chemical/electrical processes. Therefore, they involve microscopically oriented electric dipoles, while charge neutrality is generally maintained. There are also so-called real-charge electrets, which basically contain net electric charges to produce electric fields without external sources.

[8] Similar to ferromagnetic materials that can be permanently magnetized by systematically applying magnetic fields, there are also ferroelectric materials that can be permanently polarized by systematically applying electric fields. These materials possess hysteresis behaviors (in terms of the relationship between the electric field intensity and electric flux density), similar to the B-H curves exhibited by ferromagnetic materials (Figure 2.84).

When time-harmonic sources are involved in dynamic cases (Figure 5.4), steady-state analyses become possible in the phasor domain. Maxwell's equations can be written

$$\bar{\nabla}\cdot\bar{D}(\bar{r}) = q_v(\bar{r}), \qquad \oint_S \bar{D}(\bar{r})\cdot\overline{ds} = Q_{\mathrm{enc}} \tag{5.30}$$

$$\bar{\nabla}\times\bar{H}(\bar{r}) = \bar{J}_v(\bar{r}) + j\omega\bar{D}(\bar{r}), \qquad \oint_C \bar{H}(\bar{r})\cdot\overline{dl} = I_{\mathrm{enc}} + \int_S j\omega\bar{D}(\bar{r})\cdot\overline{ds} \tag{5.31}$$

$$\bar{\nabla}\times\bar{E}(\bar{r}) = -j\omega\bar{B}(\bar{r}), \qquad \oint_C \bar{E}(\bar{r})\cdot\overline{dl} = -\int_S j\omega\bar{B}(\bar{r})\cdot\overline{ds} \tag{5.32}$$

$$\bar{\nabla}\cdot\bar{B}(\bar{r}) = 0, \qquad \oint_S \bar{B}(\bar{r})\cdot\overline{ds} = 0 \tag{5.33}$$

in differential and integral forms. In these equations, $\omega = 2\pi f$ is the angular frequency, while all quantities are phasor representations of the corresponding fields and sources. In terms of potentials, we further have

$$\bar{E}(\bar{r}) = -\bar{\nabla}\Phi(\bar{r}) - j\omega\bar{A}(\bar{r}), \qquad \bar{B}(\bar{r}) = \bar{\nabla}\times\bar{A}(\bar{r}). \tag{5.34}$$

Obviously, $\omega = 0$ leads to $\bar{E}(\bar{r}) = -\bar{\nabla}\Phi(\bar{r})$, which mathematically corresponds to electrostatics.

Since electric and magnetic fields directly depend on each other in dynamic cases, Maxwell's equations should be combined for their analyses. Strategies to combine them mainly depend on the electromagnetic problem itself. In general, when a homogeneous medium (with permittivity ε and permeability μ) is considered, combinations of Gauss' law, Ampere's law, and Faraday's law lead to

$$\bar{\nabla}\times\bar{\nabla}\times\bar{E}(\bar{r},t) + \mu\varepsilon\frac{\partial^2}{\partial t^2}\bar{E}(\bar{r},t) = -\mu\frac{\partial\bar{J}_v(\bar{r},t)}{\partial t} \tag{5.35}$$

$$\bar{\nabla}\times\bar{\nabla}\times\bar{H}(\bar{r},t) + \mu\varepsilon\frac{\partial^2}{\partial t^2}\bar{H}(\bar{r},t) = \bar{\nabla}\times\bar{J}_v(\bar{r},t) \tag{5.36}$$

as discussed in Section 3.5. In addition, as shown in Section 4.4, the first equation can be modified to a wave equation as

$$\bar{\nabla}^2\bar{E}(\bar{r},t) - \mu\varepsilon\frac{\partial^2}{\partial t^2}\bar{E}(\bar{r},t) = \mu\frac{\partial\bar{J}_v(\bar{r},t)}{\partial t} + \frac{1}{\varepsilon}\bar{\nabla}q_v(\bar{r},t). \tag{5.37}$$

Further using Gauss' law for magnetic fields, a similar wave equation for the magnetic field intensity can be written:

$$\bar{\nabla}^2\bar{H}(\bar{r},t) - \mu\varepsilon\frac{\partial^2}{\partial t^2}\bar{H}(\bar{r},t) = -\bar{\nabla}\times\bar{J}_v(\bar{r},t). \tag{5.38}$$

These equations describe how electric and magnetic fields behave like *waves*.

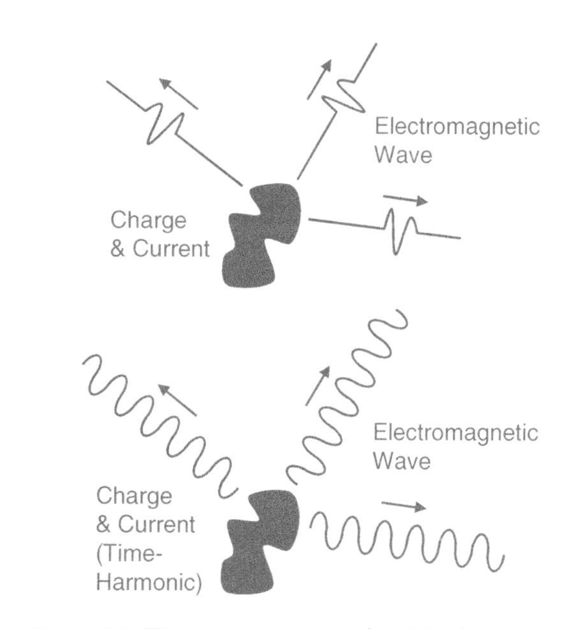

Figure 5.4 Time-variant sources (electric charges and currents, which depend on each other via the continuity equation) create electromagnetic waves (time-variant electric and magnetic fields) that propagate in space. For a time-harmonic source, the generated wave is also time-harmonic: i.e. we observe a continuous movement of peaks and dips with the phase velocity determined by the medium parameters. Note that electric and magnetic fields are inseparable parts of an electromagnetic wave. An electromagnetic wave propagation can be seen as the generation of magnetic fields from electric fields and vice versa. We emphasize that an electromagnetic wave does not need a material to propagate, as opposed to some other types of (e.g. acoustic) waves.

As discussed in Eq. (4.4.2), wave equations can also be derived for potentials. Using a proper gauge, it is possible to derive relatively compact equations to solve for potentials. For example, using the Lorenz gauge[9,10]

$$\bar{\nabla} \cdot \bar{A}(\bar{r}, t) = -\mu\varepsilon \frac{\partial \Phi(\bar{r}, t)}{\partial t}, \tag{5.39}$$

in a homogeneous medium, we have

$$\bar{\nabla}^2 \bar{A}(\bar{r}, t) - \mu\varepsilon \frac{\partial^2}{\partial t^2} \bar{A}(\bar{r}, t) = -\mu \bar{J}_v(\bar{r}, t) \tag{5.40}$$

$$\bar{\nabla}^2 \Phi(\bar{r}, t) - \mu\varepsilon \frac{\partial^2}{\partial t^2} \Phi(\bar{r}, t) = -\frac{q_v(\bar{r}, t)}{\varepsilon}. \tag{5.41}$$

The importance of these equations is that potentials can be solved for arbitrary charge and current distributions, and they can be used to find fields via Eqs. (5.17) and (5.18). Obviously, for static cases, the time derivatives vanish, and we arrive at the corresponding Poisson's equations:

$$\bar{\nabla}^2 \bar{A}(\bar{r}) = -\mu \bar{J}_v(\bar{r}), \qquad \bar{\nabla}^2 \Phi(\bar{r}) = -\frac{q_v(\bar{r})}{\varepsilon}. \tag{5.42}$$

Similarly, for static cases, the Lorenz gauge reduces to the Coulomb gauge as[11] $\bar{\nabla} \cdot \bar{A}(\bar{r}) = 0$. As also discussed in Section 4.4.2, solutions of wave equations for potentials lead to[12] (Figure 5.5)

$$\Phi(\bar{r}, t) = \frac{1}{4\pi\varepsilon} \int_V \frac{q_v(\bar{r}, t')}{|\bar{r} - \bar{r}'|} d\bar{r}', \qquad \bar{A}(\bar{r}, t) = \frac{\mu}{4\pi} \int_V \frac{\bar{J}_v(\bar{r}, t')}{|\bar{r} - \bar{r}'|} d\bar{r}' \tag{5.43}$$

where $t' = t - |\bar{r} - \bar{r}'|/u_p$ (for $u_p = 1/\sqrt{\mu\varepsilon}$) is the *retarded time* that accounts for the time required for the wave to travel from the source point $\bar{r}'$ to the observation point $\bar{r}$.

When time-harmonic sources (steady state) are considered, phasor-domain representations lead to Helmholtz equations:

$$\bar{\nabla}^2 \bar{E}(\bar{r}) + \omega^2 \mu\varepsilon \bar{E}(\bar{r}) = j\omega\mu \bar{J}_v(\bar{r}) + \frac{1}{\varepsilon} \bar{\nabla} q_v(\bar{r}) \tag{5.44}$$

$$\bar{\nabla}^2 \bar{H}(\bar{r}) + \omega^2 \mu\varepsilon \bar{H}(\bar{r}) = -\bar{\nabla} \times \bar{J}_v(\bar{r}) \tag{5.45}$$

$$\bar{\nabla}^2 \bar{A}(\bar{r}) + \omega^2 \mu\varepsilon \bar{A}(\bar{r}) = -\mu \bar{J}_v(\bar{r}) \tag{5.46}$$

$$\bar{\nabla}^2 \Phi(\bar{r}) + \omega^2 \mu\varepsilon \Phi(\bar{r}) = -\frac{q_v(\bar{r})}{\varepsilon}, \tag{5.47}$$

[9] Recall that, without performing a *gauge fixing*, there are infinitely many choices for the pair of the electric scalar potential and magnetic vector potential, all of which lead to the same electric/magnetic fields. Hence, by a gauge fixing, we are removing redundant degrees of freedom. In general, some choices can be better than others in terms of simplification and physical interpretation. For example, the Lorenz gauge leads to symmetric wave equations for the potentials, while it has an advantage of being *Lorentz* invariant: i.e. suitable for relativistic computations (the similarity of the names is a coincidence).

[10] As an alternative to the Lorenz gauge, one could use the Coulomb gauge, $\bar{\nabla} \cdot \bar{A}(\bar{r}, t) = 0$, which would lead to a different set of expressions for potentials but the same set of expressions for fields. The general solution with the Coulomb gauge is out of the scope of this book; but note that the Coulomb gauge is considered a *complete* gauge (unlike the Lorenz gauge that may need extra conditions for a fully unique solution), while the resulting potentials violate causality.

[11] This does not mean that the Coulomb gauge cannot be used for dynamic cases. For a dynamic solution, one can still select $\bar{\nabla} \cdot \bar{A}(\bar{r}, t) = 0$, but this choice leads to more complicated expressions for the magnetic vector potential in terms of sources.

[12] Note that these expressions can be written using surface and line integrals, depending on the source type (for surface or line distributions).

whose solutions can be obtained as[13] (Figure 5.5)

$$\bar{A}(\bar{r}) = \mu \int_V \bar{J}_v(\bar{r}')g(\bar{r},\bar{r}')d\bar{r}', \qquad \Phi(\bar{r}) = \frac{1}{\varepsilon}\int_V q_v(\bar{r}')g(\bar{r},\bar{r}')d\bar{r}' \tag{5.48}$$

$$\bar{H}(\bar{r}) = \int_V \bar{J}_v(\bar{r}') \times \bar{\nabla}'g(\bar{r},\bar{r}')d\bar{r}' \tag{5.49}$$

$$\bar{E}(\bar{r}) = \frac{1}{\varepsilon}\int_V q_v(\bar{r}')\bar{\nabla}'g(\bar{r},\bar{r}')d\bar{r}' - j\omega\mu\int_V \bar{J}_v(\bar{r}')g(\bar{r},\bar{r}')d\bar{r}', \tag{5.50}$$

where $k = \omega\sqrt{\mu\varepsilon} = \omega/u_p$ is the wavenumber and

$$g(\bar{r},\bar{r}') = \exp(-jk|\bar{r}-\bar{r}'|)/(4\pi|\bar{r}-\bar{r}'|) \tag{5.51}$$

is Green's function. Note that the magnetic vector potential has a simple form thanks to the Lorenz gauge,

$$\bar{\nabla}\cdot\bar{A}(\bar{r}) = -j\omega\frac{1}{u_p^2}\Phi(\bar{r}) = -j\omega\mu\varepsilon\Phi(\bar{r}), \tag{5.52}$$

while the field expressions do not depend on the selected gauge. The expression for the electric field intensity can be rewritten purely in terms of the electric current density as

$$\bar{E}(\bar{r}) = \frac{j}{\omega\varepsilon}\int_V \bar{\nabla}'\cdot\bar{J}_v(\bar{r}')\bar{\nabla}'g(\bar{r},\bar{r}')d\bar{r}' - j\omega\mu\int_V \bar{J}_v(\bar{r}')g(\bar{r},\bar{r}')d\bar{r}', \tag{5.53}$$

using the phasor form of the continuity equation:

$$\bar{\nabla}\cdot\bar{J}_v(\bar{r}) = -j\omega q_v(\bar{r}). \tag{5.54}$$

Static cases can be considered time-harmonic sources with zero frequency such that all expressions in the phasor domain reduce to static forms (such as Coulomb's and Biot-Savart laws), as discussed in Section 4.4.3.

When Maxwell's equations are applied at boundaries, the boundary conditions are derived (Figure 5.6). For electric and magnetic fields, we have

$$\hat{a}_n \times [\bar{E}_1(\bar{r},t) - \bar{E}_2(\bar{r},t)] = 0 \qquad (\bar{r}\in S) \tag{5.55}$$

$$\hat{a}_n \cdot [\bar{D}_1(\bar{r},t) - \bar{D}_2(\bar{r},t)] = q_s(\bar{r},t) \qquad (\bar{r}\in S) \tag{5.56}$$

$$\hat{a}_n \times [\bar{H}_1(\bar{r},t) - \bar{H}_2(\bar{r},t)] = \bar{J}_s(\bar{r},t) \qquad (\bar{r}\in S) \tag{5.57}$$

$$\hat{a}_n \cdot [\bar{B}_1(\bar{r},t) - \bar{B}_2(\bar{r},t)] = 0 \qquad (\bar{r}\in S). \tag{5.58}$$

[13] Note that these expressions can be written using surface and line integrals, depending on the source type (for surface or line distributions).

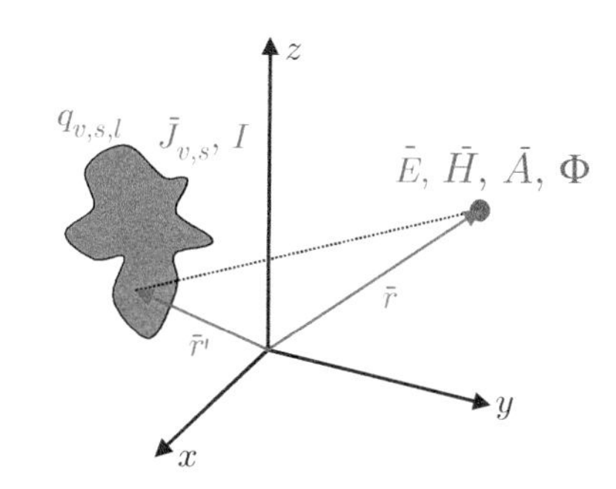

Figure 5.5 Electric and magnetic fields, as well as potentials, can be found via integration on sources (electric charge distributions $[q_v(\bar{r},t)$, $q_s(\bar{r},t)$, $q_l(\bar{r},t)]$ and electric current distributions $[\bar{J}_v(\bar{r},t)$, $\bar{J}_s(\bar{r},t)$, $I(t)]$).

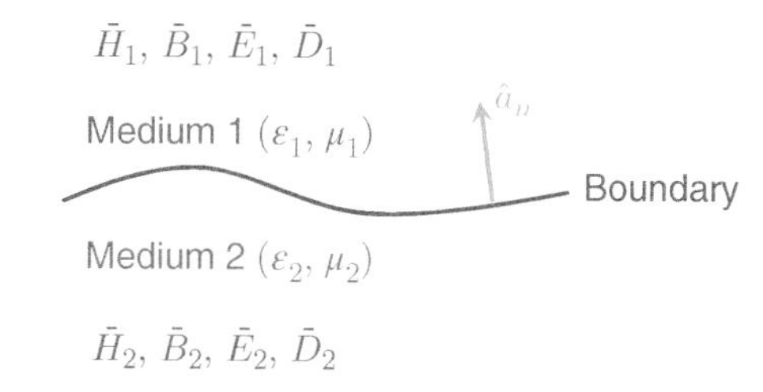

Figure 5.6 Application of Maxwell's equations at a boundary between two different media leads to four different boundary conditions that must be satisfied by field intensities ($\bar{E}(\bar{r},t)$ and $\bar{H}(\bar{r},t)$) and flux densities ($D(\bar{r},t)$ and $\bar{B}(\bar{r},t)$).

Specifically, the tangential electric field intensity and normal magnetic flux density are continuous across boundaries between different media, while the normal electric flux density and tangential magnetic field intensity can be discontinuous. The discontinuity in the normal electric flux density corresponds to the surface electric charge density, while the discontinuity in the tangential magnetic field intensity corresponds to the surface electric current density in the perpendicular direction. Obviously, these boundary conditions can alternatively be written

$$\hat{a}_n \times [\bar{D}_1(\bar{r},t)/\varepsilon_1(\bar{r}) - \bar{D}_2(\bar{r},t)/\varepsilon_2(\bar{r})] = 0 \qquad (\bar{r} \in S) \tag{5.59}$$

$$\hat{a}_n \cdot [\varepsilon_1(\bar{r})\bar{E}_1(\bar{r},t) - \varepsilon_2(\bar{r})\bar{E}_2(\bar{r},t)] = q_s(\bar{r},t) \qquad (\bar{r} \in S) \tag{5.60}$$

$$\hat{a}_n \times [\bar{B}_1(\bar{r},t)/\mu_1(\bar{r}) - \bar{B}_2(\bar{r},t)/\mu_2(\bar{r})] = \bar{J}_s(\bar{r}) \qquad (\bar{r} \in S) \tag{5.61}$$

$$\hat{a}_n \cdot [\mu_1(\bar{r})\bar{H}_1(\bar{r},t) - \mu_2(\bar{r})\bar{H}_2(\bar{r},t)] = 0 \qquad (\bar{r} \in S) \tag{5.62}$$

by interchanging intensity and density quantities. More informative equations can be obtained if polarization and magnetization are used,

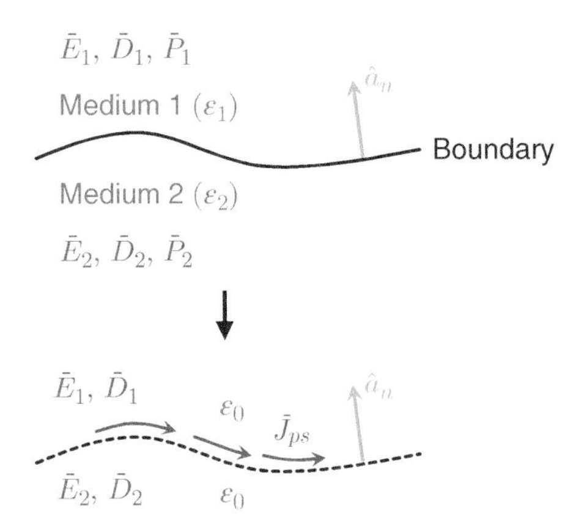

Figure 5.7 A discontinuity in the tangential component of the polarization at a boundary between two different media (e.g. between two different dielectrics or between a dielectric and a vacuum) can be represented by the equivalent magnetic current density $\bar{J}_{ps}(\bar{r},t)$. This surface current density can be used to explain the discontinuity in the tangential component of the electric flux density. Note that the discontinuity in the normal component of the polarization can be represented by the surface polarization charge density $q_{ps}(\bar{r},t)$ in the alternative equivalent problem, which may explain the discontinuity in the normal component of the electric field intensity (even in the absence of a true charge density).

$$\hat{a}_n \times [\bar{D}_1(\bar{r},t) - \bar{D}_2(\bar{r},t)] = -\hat{a}_n \times \bar{P}(\bar{r},t) = \varepsilon_0 \bar{J}_{ps}(\bar{r},t) \qquad (\bar{r} \in S) \tag{5.63}$$

$$\hat{a}_n \cdot [\bar{H}_1(\bar{r},t) - \bar{H}_2(\bar{r},t)] = \hat{a}_n \cdot \bar{M}(\bar{r},t) = q_{ms}(\bar{r},t)/\mu_0 \qquad (\bar{r} \in S), \tag{5.64}$$

where $\bar{P}(\bar{r},t)$ is the *net* polarization $(\bar{P}_2(\bar{r},t) - \bar{P}_1(\bar{r},t))$ and $\bar{M}(\bar{r},t)$ is the *net* magnetization $(\bar{M}_2(\bar{r},t) - \bar{M}_1(\bar{r},t))$ at the boundary. The equivalent sources $\bar{J}_{ps}(\bar{r},t)$ and $q_{ms}(\bar{r},t)$ explain how the tangential electric flux density and normal magnetic field intensity can be discontinuous (Figure 5.7). Similarly, we have

$$\hat{a}_n \cdot [\bar{E}_1(\bar{r},t) - \bar{E}_2(\bar{r},t)] = q_s(\bar{r},t)/\varepsilon_0 + \hat{a}_n \cdot \bar{P}(\bar{r},t)/\varepsilon_0 \tag{5.65}$$

$$= q_s(\bar{r},t)/\varepsilon_0 + q_{ps}(\bar{r},t)/\varepsilon_0 \qquad (\bar{r} \in S) \tag{5.66}$$

$$\hat{a}_n \times [\bar{B}_1(\bar{r},t) - \bar{B}_2(\bar{r},t)] = \mu_0 \bar{J}_s(\bar{r},t) - \mu_0 \hat{a}_n \times \bar{M}(\bar{r},t) \qquad (\bar{r} \in S) \tag{5.67}$$

$$= \mu_0 \bar{J}_s(\bar{r},t) + \mu_0 \bar{J}_{ms}(\bar{r},t) \qquad (\bar{r} \in S). \tag{5.68}$$

In these equations, the equivalent sources $q_{ps}(\bar{r},t)$ and $\bar{J}_{ms}(\bar{r},t)$ explain how the normal electric field intensity and normal magnetic flux density can be discontinuous (Figure 5.7) even when there are no true sources $q_s(\bar{r},t)$ and $\bar{J}_s(\bar{r},t)$. We further note that all boundary conditions are also valid for static cases.

5.1.1 Examples

Example 50: Consider an electrostatic case with a spherical vacuum region bounded by $R = a$ and $R = b$ surfaces with $b > a$. Let the potentials at the boundaries be V_0 and 0, respectively (Figure 5.8). Also, a charge distribution is defined for $a < R < b$ with a volume electric charge density of $q_v(R) = -q_0 a^2/R^2$ (C/m^3). Find the electric scalar potential everywhere.

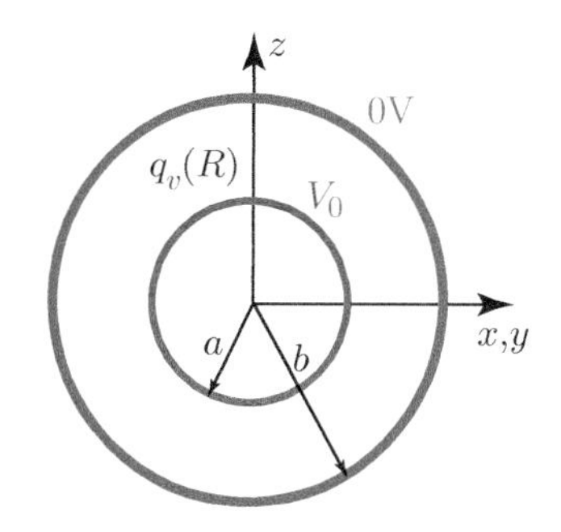

Figure 5.8 A spherical vacuum region bounded by $R = a$ and $R = a$ surfaces with V_0 and 0 electric scalar potential values, respectively. There is a volumetric electric charge distribution between the surfaces.

Solution: Using Poisson's equation, we have[14]

$$\bar{\nabla}^2\Phi(R) = \frac{1}{R^2}\frac{\partial}{\partial R}\left(R^2\frac{\partial\Phi(R)}{\partial R}\right) = \frac{q_0 a^2}{\varepsilon_0 R^2} \tag{5.69}$$

$$\longrightarrow \frac{\partial}{\partial R}\left(R^2\frac{\partial\Phi(R)}{\partial R}\right) = \frac{q_0 a^2}{\varepsilon_0} \longrightarrow R^2\frac{\partial\Phi(R)}{\partial R} = R\frac{q_0 a^2}{\varepsilon_0} + c_1 \tag{5.70}$$

$$\longrightarrow \frac{\partial\Phi(R)}{\partial R} = \frac{q_0 a^2}{\varepsilon_0 R} + \frac{c_1}{R^2} \tag{5.71}$$

$$\longrightarrow \Phi(R) = \frac{q_0 a^2}{\varepsilon_0}\ln(R) - \frac{c_1}{R} + c_2. \tag{5.72}$$

Since $\Phi(R = a) = V_0$ and $\Phi(R = b) = 0$, the constants c_1 and c_2 can be derived as

$$\frac{q_0 a^2}{\varepsilon_0}\ln(b) - \frac{c_1}{b} + c_2 = 0 \longrightarrow c_2 = \frac{c_1}{b} - \frac{q_0 a^2}{\varepsilon_0}\ln(b) \tag{5.73}$$

$$\frac{q_0 a^2}{\varepsilon_0}\ln(a) - \frac{c_1}{a} + c_2 = \frac{q_0 a^2}{\varepsilon_0}\ln(a) - \frac{c_1}{a} + \frac{c_1}{b} - \frac{q_0 a^2}{\varepsilon_0}\ln b = V_0, \tag{5.74}$$

leading to

$$c_1 = -\frac{ab}{b-a}\left[V_0 + \frac{q_0 a^2}{\varepsilon_0}\ln(b/a)\right] \tag{5.75}$$

$$c_2 = -\frac{a}{b-a}\left[V_0 + \frac{q_0 a^2}{\varepsilon_0}\ln(b/a)\right] - \frac{q_0 a^2}{\varepsilon_0}\ln(b). \tag{5.76}$$

[14] The corresponding electric field intensity can be found as

$$\bar{E}(\bar{r}) = -\bar{\nabla}\Phi(\bar{r}) = -\hat{a}_R\frac{q_0 a^2}{\varepsilon_0}\frac{1}{R} - \hat{a}_R\frac{c_1}{R^2}.$$

Then we can verify the volume electric charge density via Gauss' law as

$$q_v(\bar{r}) = \varepsilon_0\bar{\nabla}\cdot\bar{E}(\bar{r}) = \varepsilon_0\frac{1}{R^2}\frac{\partial}{\partial R}\left[-R^2\frac{q_0 a^2}{\varepsilon_0}\frac{1}{R}\right]$$
$$= -\varepsilon_0\frac{1}{R^2}\frac{q_0 a^2}{\varepsilon_0} = -\frac{q_0 a^2}{R^2}.$$

We further note that

$$\bar{\nabla}\times\bar{E}(\bar{r}) = 0,$$

i.e. the electric field intensity is irrotational, as expected.

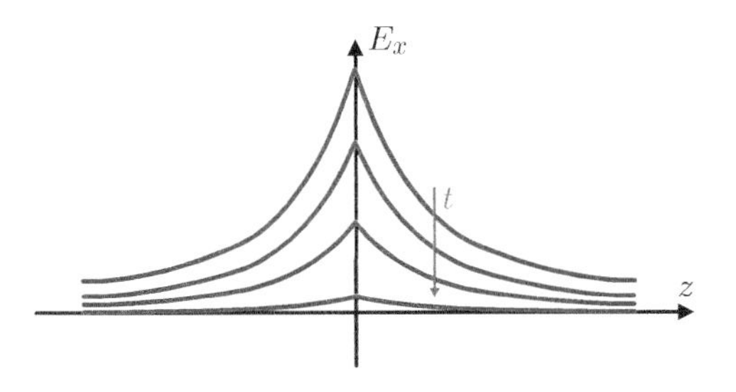

Figure 5.9 The electric field intensity given in Eq. (5.77) with respect to z and t. As time progresses, the electric field intensity decays to zero. This can be interpreted as propagation in the z direction (for $z > 0$) and in the $-z$ direction (for $z < 0$). Note that there is no variation in the x and y directions (for both electric and magnetic fields).

Example 51: In free space (source-free vacuum), the electric field intensity is defined as (Figure 5.9)

$$\bar{E}(\bar{r},t) = \hat{a}_x E_0 \begin{cases} \exp(-t/t_0 + z/z_0) & (z < 0) \\ \exp(-t/t_0 - z/z_0) & (z > 0), \end{cases} \tag{5.77}$$

where E_0 (V/m), t_0 (s), and z_0 (m) are constants.

- Show that there is no volume electric charge density.
- Find the magnetic field intensity.
- Find the relationship between t_0, z_0, and the medium parameters (ε_0, μ_0).

Solution: First, given the expression for the electric field intensity, we have

$$\bar{\nabla} \cdot \bar{D}(\bar{r},t) = \varepsilon_0 \bar{\nabla} \cdot \bar{E}(\bar{r},t) = \varepsilon_0 \frac{\partial E_x(\bar{r},t)}{\partial x} = 0 \longrightarrow q_v(\bar{r},t) = 0, \tag{5.78}$$

i.e. there is no volume electric current density.[15] The magnetic field intensity can be found by considering Faraday's law as

$$-\mu_0 \frac{\partial \bar{H}(\bar{r},t)}{\partial t} = \bar{\nabla} \times \bar{E}(\bar{r},t) = \hat{a}_z \times \hat{a}_x \frac{\partial}{\partial z} [E_0 \exp(-t/t_0 \pm z/z_0)] \tag{5.79}$$

$$= \pm \hat{a}_y \frac{E_0}{z_0} \exp(-t/t_0 \pm z/z_0), \tag{5.80}$$

leading to[16]

$$\frac{\partial \bar{H}(\bar{r},t)}{\partial t} = \mp \hat{a}_y \frac{E_0}{z_0 \mu_0} \exp(-t/t_0 \pm z/z_0) \tag{5.81}$$

$$\bar{H}(\bar{r},t) = \pm \hat{a}_y \frac{E_0 t_0}{z_0 \mu_0} \exp(-t/t_0 \pm z/z_0) \tag{5.82}$$

or

$$\bar{H}(\bar{r},t) = \hat{a}_y \frac{E_0 t_0}{z_0 \mu_0} \begin{cases} \exp(-t/t_0 + z/z_0) & (z < 0) \\ \exp(-t/t_0 - z/z_0) & (z > 0), \end{cases} \tag{5.83}$$

[15] Since there is no discontinuity and infinity in the values of $\bar{D}(\bar{r},t)$, we can state that the surface electric charge density and line electric charge density are also zero everywhere. Hence, this electromagnetic field may be created by some sources that vanish before $t = 0$.

[16] Note that the magnetic field intensity has the same exponential form, while its amplitude and direction (y) are different.

assuming that there is no static magnetic field intensity. At this stage, we can check Ampere's law to find the relationship between the constants. Considering that

$$\bar{\nabla} \times \bar{H}(\bar{r}, t) = \pm \hat{a}_z \times \hat{a}_y \frac{E_0 t_0}{z_0 \mu_0} \frac{\partial}{\partial z} [\exp(-t/t_0 \pm z/z_0)] \tag{5.84}$$

$$= -\hat{a}_x \frac{E_0 t_0}{z_0^2 \mu_0} \exp(-t/t_0 \pm z/z_0) \tag{5.85}$$

and

$$\frac{\partial \bar{D}(\bar{r}, t)}{\partial t} = \varepsilon_0 \frac{\partial \bar{E}(\bar{r}, t)}{\partial t} = -\hat{a}_x \frac{E_0 \varepsilon_0}{t_0} \exp(-t/t_0 \pm z/z_0), \tag{5.86}$$

the equality

$$\bar{\nabla} \times \bar{H}(\bar{r}, t) = \frac{\partial \bar{D}(\bar{r}, t)}{\partial t} \tag{5.87}$$

leads to $z_0/t_0 = \pm 1/\sqrt{\mu_0 \varepsilon_0}$.

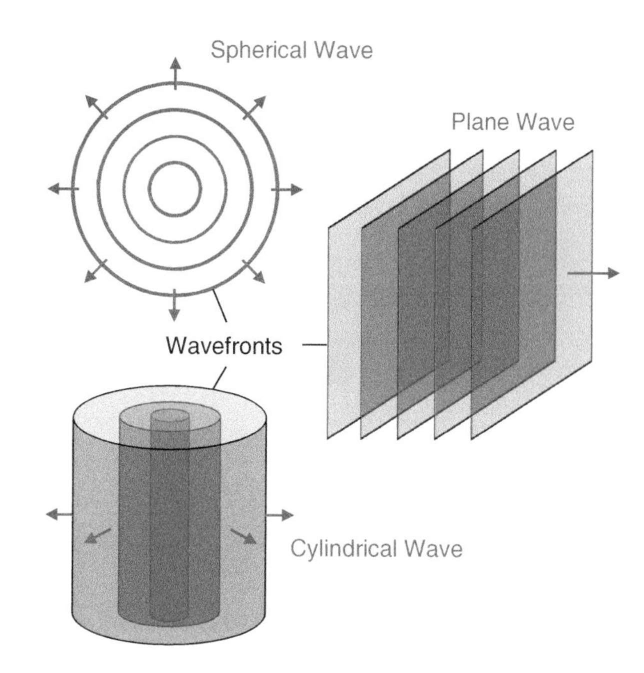

Figure 5.10 Propagation of an electromagnetic wave is mainly characterized, formulated, and visualized by the movement of its wavefronts (constant-phase surfaces) with respect to time. Depending on the source and the resulting wave, the wavefronts can be spherical, planar, cylindrical, or more complex shapes. For some basic types of waves, peaks (maxima) and dips (minima) may directly correspond to wavefronts. Note that wavefronts only describe propagation; the vector directions of electric and magnetic fields, their variations on wavefronts, and the relationship (e.g. phase difference) between them depend on the wave.

5.2 Electromagnetic Propagation and Radiation

As discussed in Sections 4.4.1 and 4.4.2, wave equations are satisfied by functions that represent waves: i.e. dynamic distributions of electric and magnetic fields. Depending on the electromagnetic scenario, there can be traveling and standing waves, while a complex dynamic problem usually involves diverse behaviors of electric and magnetic fields. In general, when time-variant sources create electric and magnetic fields, they tend to propagate: i.e. their functions depend on position and time such that intensity/density distributions with respect to time look like traveling with a phase velocity. More formally, propagation is characterized by movements of *wavefronts*. A wavefront is defined as the surface (plane, sphere, cylinder, etc., depending on the type of the wave) on which the *phase* of the wave is constant (Figures 5.10 and 5.11). If the source and the generated wave are time-harmonic (Figure 5.11), infinitely many wavefronts travel consecutively, and one can define a wavelength – i.e. the period with respect to position:

$$\lambda = \frac{u_p}{f} = \frac{2\pi u_p}{\omega}, \tag{5.88}$$

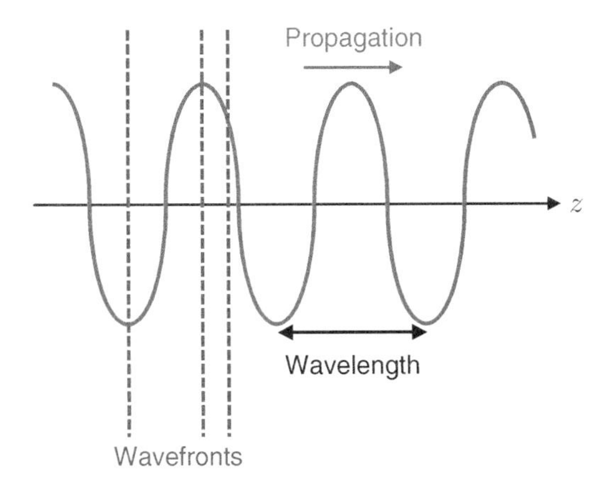

Figure 5.11 For a time-harmonic wave defined as $\bar{f}(\bar{r},t) = \bar{g}(x,y)\cos(\omega t - kz)$, the propagation direction is z, while the wavefronts – i.e. those surfaces on which the phase term $v = \omega t - kz$ is constant – are $z =$ constant planes. Note that the value (amplitude and direction) of the wave may depend on the position (dependency of $\bar{g}(x,y)$ on x and y) on a wavefront. Hence, a wavefront does not necessarily mean a constant value (amplitude and direction) of the wave, while there are waves (e.g. uniform plane waves) that satisfy this property (uniformity).

where f is the frequency and u_p is the phase velocity. The phase velocity generally depends on the medium (not sources), which can be written $u_p = 1/\sqrt{\mu\varepsilon}$ for a homogeneous space. Hence, if the frequency of the source is high, the wavelength of the corresponding wave must be short.

Electromagnetic waves do not need a material to propagate. In fact, the speed of an electromagnetic wave is fixed to $u_p = c_0 = 2.99792458 \times 10^8$ m/s in a vacuum. On the other hand, electromagnetic waves interact with matter. There are basic types of interactions, as follows[17] (Figure 5.12):

- Reflection: When a wave arrives at a boundary, it is reflected back. In many cases, part of the wave is reflected, while the remainder passes through the boundary (transmission).
- Refraction: When a wave or a part of it passes through a boundary, it changes direction.
- Diffraction: When a wave arrives at an edge, tip, or similar geometric discontinuity, it is bent, redirected, and/or reshaped.

[17] In these definitions, *boundary* means a surface separating two different media, one of which may be a vacuum.

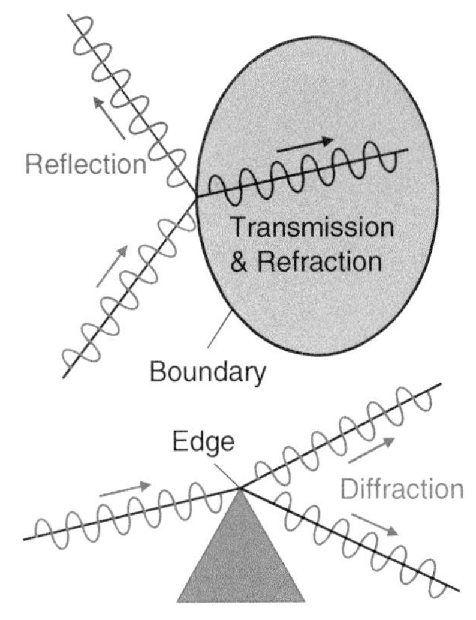

Figure 5.12 Basic interactions of electromagnetic waves and objects.

There are also higher-level definitions that can be considered to define an overall interaction. For example, when a wave hits an object, many reflections, refractions, and/or diffractions may occur such that different secondary waves can be generated (Figure 5.13). These waves can be considered scattered from the object (*scattering*). Note that all these interactions are naturally expected by Maxwell's equations that must be satisfied by electromagnetic waves.

Electromagnetic waves carry electromagnetic energy, which is called electromagnetic *radiation*. This energy/power flow can be seen as the movement of the electric potential energy and magnetic potential energy. A commonly used parameter to describe propagation of energy/power is the Poynting vector (power flux density), which is defined as

$$\bar{S}(\bar{r},t) = \bar{E}(\bar{r},t) \times \bar{H}(\bar{r},t). \tag{5.89}$$

To understand the Poynting vector, we may consider a volume V bounded by a surface S (Figure 5.14). The total power flux – i.e. the electromagnetic power flowing outward the surface – can be found as

$$p_{em}(t) = \oint_S \bar{S}(\bar{r},t) \cdot \overline{ds} = \oint_S \bar{E}(\bar{r},t) \times \bar{H}(\bar{r},t) \cdot \overline{ds}. \tag{5.90}$$

Using the divergence theorem, we can write

$$p_{em}(t) = \int_V \bar{\nabla} \cdot [\bar{E}(\bar{r},t) \times \bar{H}(\bar{r},t)]dv, \tag{5.91}$$

where the integrand can be manipulated as

$$p_{em}(t) = \int_V \bar{H}(\bar{r},t) \cdot \bar{\nabla} \times \bar{E}(\bar{r},t) dv - \int_V \bar{E}(\bar{r},t) \cdot \bar{\nabla} \times \bar{H}(\bar{r},t) dv. \tag{5.92}$$

To derive an expression for the power flux in terms of the electric potential energy and magnetic potential energy, we may first use Faraday's law and Ampere's law:

$$p_{em}(t) = -\int_V \bar{H}(\bar{r},t) \cdot \frac{\partial \bar{B}(\bar{r},t)}{\partial t} dv - \int_V \bar{E}(\bar{r},t) \cdot \bar{J}_v(\bar{r},t) dv - \int_V \bar{E}(\bar{r},t) \cdot \frac{\partial \bar{D}(\bar{r},t)}{\partial t} dv. \tag{5.93}$$

To simplify analysis, we assume that the permittivity and permeability do not depend on the position, leading to

$$p_{em}(t) = -\mu \int_V \bar{H}(\bar{r},t) \cdot \frac{\partial \bar{H}(\bar{r},t)}{\partial t} dv - \int_V \bar{E}(\bar{r},t) \cdot \bar{J}_v(\bar{r},t) dv - \varepsilon \int_V \bar{E}(\bar{r},t) \cdot \frac{\partial \bar{E}(\bar{r},t)}{\partial t} dv \tag{5.94}$$

$$= -\frac{\partial}{\partial t}\left(\frac{\mu}{2}\int_V |\bar{H}(\bar{r},t)|^2 dv\right) - \frac{\partial}{\partial t}\left(\frac{\varepsilon}{2}\int_V |\bar{E}(\bar{r},t)|^2 dv\right) - \int_V \bar{E}(\bar{r},t) \cdot \bar{J}_v(\bar{r},t) dv \tag{5.95}$$

$$= -\frac{\partial w_{m,\text{stored}}(t)}{\partial t} - \frac{\partial w_{e,\text{stored}}(t)}{\partial t} - p_J(t), \tag{5.96}$$

where[18]

$$w_{m,\text{stored}}(t) = \frac{\mu}{2}\int_V |\bar{H}(\bar{r},t)|^2 dv \tag{5.97}$$

$$w_{e,\text{stored}}(t) = \frac{\varepsilon}{2}\int_V |\bar{E}(\bar{r},t)|^2 dv \tag{5.98}$$

$$p_J(t) = \int_V \bar{E}(\bar{r},t) \cdot \bar{J}_v(\bar{r},t) dv. \tag{5.99}$$

In the above, depending on the nature of the volume electric current density, the power term $p_J(t)$ represents dissipated power due to conduction losses and/or generated power by external

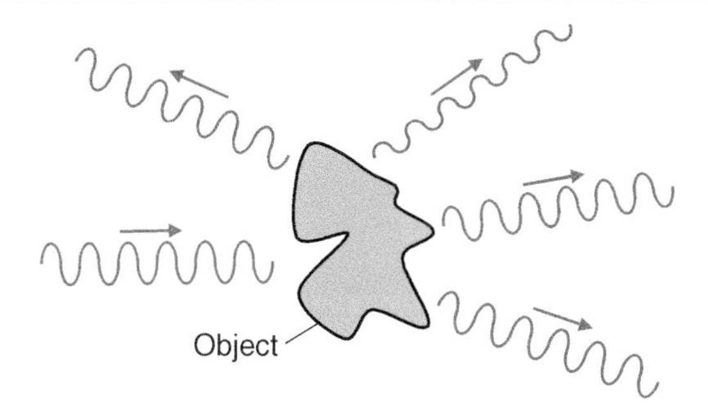

Figure 5.13 When an electromagnetic wave hits an object, secondary waves are generated (more or less in all directions) as a result of many reflections, refractions, and diffractions.

$\bar{S}$: poynting vector
Unit: watts/meter^2 (W/m^2)

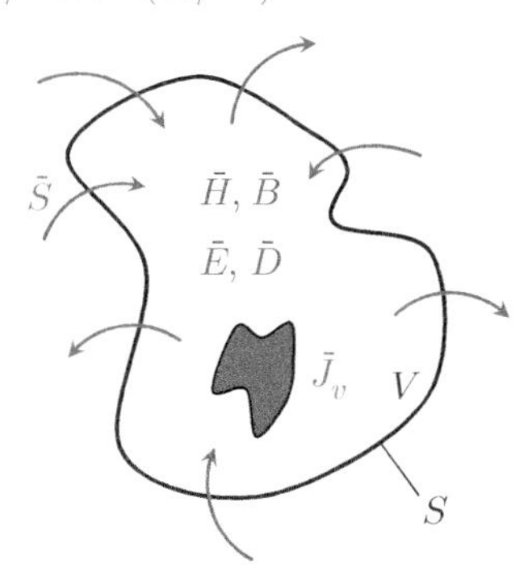

Figure 5.14 Given a volume V bounded by a surface S, the electromagnetic power that flows outward through the surface is related to the time variation of the electric potential energy and magnetic potential energy stored in the volume, as well as to the generated/dissipated power associated with the electric current density. This balance (formulated in Eq. (5.96)) can be seen as a conservation of energy.

[18] As usual, depending on the type of the electric current density, the related integral can be changed to

$$p_J(t) = \int_S \bar{E}(\bar{r},t) \cdot \bar{J}_s(\bar{r},t) ds$$

for a surface current distribution and

$$p_J(t) = I \int_C \bar{E}(\bar{r},t) \cdot \overline{dl}.$$

for a line current.

sources. The final expression can be read as follows: the electromagnetic power propagating through a closed surface in the *outward* direction ($p_{em}(t)$) is equal to (i) decreasing magnetic potential energy ($-\partial w_{m,\text{stored}}(t)/\partial t$), (ii) decreasing electric potential energy ($-\partial w_{e,\text{stored}}(t)/\partial t$), and (iii) negative dissipated power due to conduction losses and/or power supplied by external sources located inside the enclosed volume ($p_J(t)$). This can be seen as a conservation of energy.

In the time domain, the Poynting vector represents the direction of energy flow at a particular time and is often called the *instantaneous* power flux density. The phasor-domain version[19] of the Poynting vector – the complex Poynting vector – is an important quantity when time-harmonic sources are considered. It is defined as

$$\bar{S}_c(\bar{r}) = \frac{1}{2}\bar{E}(\bar{r}) \times [\bar{H}(\bar{r})]^*, \tag{5.100}$$

where $\bar{E}(\bar{r})$ and $\bar{H}(\bar{r})$ are phasor-domain representations of the electric field intensity and magnetic field intensity, respectively. The real part of $\bar{S}_c(\bar{r})$ is the same as the time-average power flux density

$$\bar{S}_{\text{avg}}(\bar{r}) = \frac{1}{T}\int_0^T \bar{S}(\bar{r},t)dt, \tag{5.101}$$

where $T = 1/f = 2\pi/\omega$ is the time period. To see this, we consider $\bar{S}(\bar{r},t) = \bar{E}(\bar{r},t) \times \bar{H}(\bar{r},t)$ as[20]

$$\bar{S}(\bar{r},t) = \text{Re}\{\bar{E}(\bar{r})\exp(j\omega t)\} \times \text{Re}\{\bar{H}(\bar{r})\exp(j\omega t)\} \tag{5.102}$$

$$= \frac{1}{2}\text{Re}\{\bar{E}(\bar{r}) \times \bar{H}(\bar{r})\exp(j2\omega t)\} + \frac{1}{2}\text{Re}\{\bar{E}(\bar{r}) \times [\bar{H}(\bar{r})]^*\} \tag{5.103}$$

or

$$\bar{S}(\bar{r},t) = \frac{1}{2}\text{Re}\{\bar{E}(\bar{r}) \times \bar{H}(\bar{r})\exp(j2\omega t)\} + \text{Re}\{\bar{S}_c(\bar{r})\}. \tag{5.104}$$

Given that $\bar{E}(\bar{r}) \times \bar{H}(\bar{r})$ can be a complex number, we further have

$$\bar{S}(\bar{r},t) = \frac{1}{2}\text{Re}\{\bar{E}(\bar{r}) \times \bar{H}(\bar{r})\}\cos(2\omega t) - \frac{1}{2}\text{Im}\{\bar{E}(\bar{r}) \times \bar{H}(\bar{r})\}\sin(2\omega t) + \text{Re}\{\bar{S}_c(\bar{r})\}. \tag{5.105}$$

Then considering trigonometric functions,[21] the time-average power flux density can be found:

$$\bar{S}_{\text{avg}}(\bar{r}) = \frac{1}{T}\int_0^T \bar{S}(\bar{r},t)dt = \text{Re}\{\bar{S}_c(\bar{r})\}. \tag{5.106}$$

[19] The complex Poynting vector is not directly the phasor representation of the Poynting vector.

[20] Here, we are using

$$\text{Re}\{f\} = \frac{f+f^*}{2}$$
$$\text{Im}\{f\} = \frac{f-f^*}{j2},$$

where $\text{Re}\{f\}$ and $\text{Im}\{f\}$ are real and imaginary parts of a complex number f. Then we have

$$\begin{aligned}\bar{S}(\bar{r},t) &= \text{Re}\{\bar{E}(\bar{r})\exp(j\omega t)\} \times \text{Re}\{\bar{H}(\bar{r})\exp(j\omega t)\}\\ &= \frac{1}{2}\{\bar{E}(\bar{r})\exp(j\omega t) + [\bar{E}(\bar{r})]^*\exp(-j\omega t)\}\\ &\times \frac{1}{2}\{\bar{H}(\bar{r})\exp(j\omega t) + [\bar{H}(\bar{r})]^*\exp(-j\omega t)\}\\ &= \frac{1}{4}\bar{E}(\bar{r}) \times \bar{H}(\bar{r})\exp(j2\omega t)\\ &+ \frac{1}{4}\bar{E}(\bar{r}) \times [\bar{H}(\bar{r})]^* + \frac{1}{4}[\bar{E}(\bar{r})]^* \times \bar{H}(\bar{r})\\ &+ \frac{1}{4}[\bar{E}(\bar{r})]^* \times [\bar{H}(\bar{r})]^*\exp(-j2\omega t)\\ &= \frac{1}{2}\text{Re}\{\bar{E}(\bar{r}) \times \bar{H}(\bar{r})\exp(j2\omega t)\}\\ &+ \frac{1}{2}\text{Re}\{\bar{E}(\bar{r}) \times [\bar{H}(\bar{r})]^*\}.\end{aligned}$$

[21] We have

$$\int_0^T \cos(2\omega t)dt = \left[\frac{\sin(2\omega t)}{2\omega}\right]_0^{2\pi/\omega} = 0$$
$$\int_0^T \sin(2\omega t)dt = \left[-\frac{\cos(2\omega t)}{2\omega}\right]_0^{2\pi/\omega} = 0.$$

In general, the complex Poynting vector provides an overview of the electromagnetic power associated with a time-harmonic field: i.e. its real part corresponds to *net* propagation of the electromagnetic power, while its imaginary part represents oscillatory power (changing direction with respect to time).

5.2.1 Hertzian Dipole

Like electric dipoles in electrostatics and magnetic dipoles (current loops) in magnetostatics, Hertzian dipoles are basic and ideal sources that generate time-harmonic electromagnetic fields. Basically, a Hertzian dipole is an infinitesimal line current with a time-harmonic magnitude. In the time domain, a z-directed Hertzian dipole of length Δz located at the origin can be defined as (Figure 5.15)

$$\bar{J}_{v,d}(\bar{r},t) = \hat{a}_z\delta(x)\delta(y)I_0\cos(\omega t) \qquad (-\Delta z/2 \leq z \leq \Delta z/2), \tag{5.107}$$

which can be written in the phasor domain as

$$\bar{J}_{v,d}(\bar{r}) = \hat{a}_z\delta(x)\delta(y)I_0 \qquad (-\Delta z/2 \leq z \leq \Delta z/2). \tag{5.108}$$

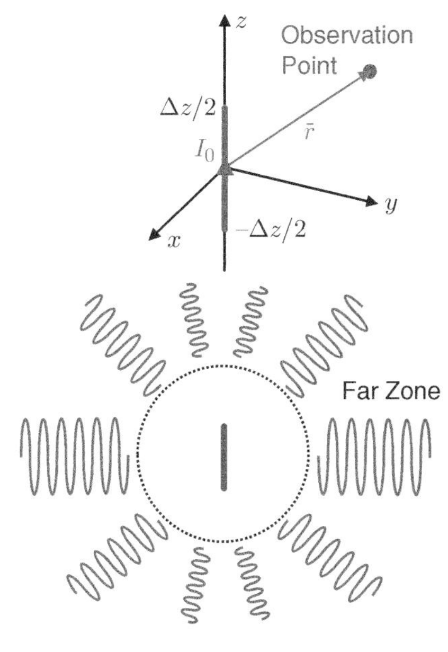

Figure 5.15 A Hertzian dipole is a basic source that creates time-harmonic electromagnetic fields. The volume electric current density for a z-directed Hertzian dipole can be written in the phasor domain as $\bar{J}_{v,d}(\bar{r}) = \hat{a}_z\delta(x)\delta(y)I_0$ from $z = -\Delta z/2$ to $z = \Delta z/2$, where impulse functions $\delta(x)$ and $\delta(y)$ are used to indicate that the thickness of the dipole is zero. In the far zone – i.e. for observation points with $kR \gg 1$ – all electromagnetic waves propagate away (in the $\hat{a}_R$ direction), and the distribution is not uniform. Specifically, electromagnetic radiation is maximum toward observation points at $\theta = \pi/2$ (on the x-y plane), while it is zero at $\theta = 0$ (north pole) and $\theta = \pi$ (south pole).

Given a homogeneous medium, the magnetic vector potential can be evaluated as

$$\bar{A}_d(\bar{r}) = \frac{\mu}{4\pi}\int_V \bar{J}_{v,d}(\bar{r}')\frac{\exp(-jk|\bar{r}-\bar{r}'|)}{|\bar{r}-\bar{r}'|}d\bar{r}' \tag{5.109}$$

$$= \frac{\mu}{4\pi}\int_{-\Delta z/2}^{\Delta z/2}\iint \hat{a}_z\delta(x')\delta(y')I_0\frac{\exp(-jk|\bar{r}-\bar{r}'|)}{|\bar{r}-\bar{r}'|}dx'dy'dz' \tag{5.110}$$

$$= \hat{a}_z\frac{\mu I_0}{4\pi}\int_{-\Delta z/2}^{\Delta z/2}\frac{\exp\left(-jk\sqrt{x^2+y^2+(z-z')^2}\right)}{\sqrt{x^2+y^2+(z-z')^2}}dz'. \tag{5.111}$$

Assuming that Δz is vanishingly small, the integral over z' can be approximated,[22] leading to

$$\bar{A}_d(\bar{r}) = \hat{a}_z\frac{\mu I_0}{4\pi}\Delta z\frac{\exp(-jkR)}{R}, \tag{5.112}$$

where $R = \sqrt{x^2+y^2+z^2}$, as usual. Defining a dipole moment $\bar{I}_{DM,h} = \hat{a}_z I_{DM,h} = \hat{a}_z I_0\Delta z$, we obtain

$$\bar{A}_d(\bar{r}) = \mu\bar{I}_{DM,h}\frac{\exp(-jkR)}{4\pi R} = \mu\bar{I}_{DM,h}g(R), \tag{5.113}$$

[22] Specifically, given that z' is very small, we assume that the integrand is constant as

$$\frac{\exp\left(-jk\sqrt{x^2+y^2+(z-z')^2}\right)}{\sqrt{x^2+y^2+(z-z')^2}} \approx \frac{\exp\left(-jk\sqrt{x^2+y^2+z^2}\right)}{\sqrt{x^2+y^2+z^2}} = \frac{\exp(-jkR)}{R}.$$

where

$$g(R) = \frac{\exp(-jkR)}{4\pi R}. \tag{5.114}$$

In a spherical coordinate system, the magnetic vector potential can also be written[23]

$$\bar{A}_d(\bar{r}) = (\hat{a}_R \cos\theta - \hat{a}_\theta \sin\theta)\mu I_{DM,h} g(R). \tag{5.115}$$

Using $\bar{A}_d(\bar{r})$, we can find the magnetic flux density as[24]

$$\bar{B}_d(\bar{r}) = \bar{\nabla} \times \bar{A}_d(\bar{r}) = \hat{a}_\phi \mu I_{DM,h} \sin\theta \left(jk + \frac{1}{R}\right) g(R). \tag{5.116}$$

The electric field intensity can be found via Ampere's law as[25]

$$\bar{E}_d(\bar{r}) = -\hat{a}_R j\omega\mu I_{DM,h} g(R) \left[2\cos\theta \left(\frac{j}{kR} + \frac{1}{k^2R^2}\right)\right] + \hat{a}_\theta j\omega\mu I_{DM,h} g(R) \left[\sin\theta \left(1 - \frac{j}{kR} - \frac{1}{k^2R^2}\right)\right]. \tag{5.118}$$

At first glance, the magnetic and electric fields, even for an infinitesimal source, have relatively complicated expressions. On the other hand, we can gain more insight by investigating the terms in some specific cases. For example, in the near zone, where $R \ll \lambda$ and $kR \ll 1$, we have

$$\bar{E}_d(\bar{r}) \approx \frac{I_{DM,h}}{j\omega} \frac{1}{4\pi\varepsilon R^3} (\hat{a}_R 2\cos\theta + \hat{a}_\theta \sin\theta) \tag{5.119}$$

$$\bar{B}_d(\bar{r}) \approx \hat{a}_\phi I_{DM,h} \frac{\mu}{4\pi R^2} \sin\theta. \tag{5.120}$$

Hence, in the near zone, the electric field intensity (in both the R and θ directions) decays with $1/R^3$, while the magnetic field intensity (only in the ϕ direction) is proportional to $1/R^2$. On the other hand, in the far zone, where $R \gg \lambda$ and $kR \gg 1$, we have

$$\bar{E}_d(\bar{r}) \approx \hat{a}_\theta j\omega\mu I_{DM,h} \frac{\exp(-jkR)}{4\pi R} \sin\theta \tag{5.121}$$

$$\bar{B}_d(\bar{r}) \approx \hat{a}_\phi jk\mu I_{DM,h} \frac{\exp(-jkR)}{4\pi R} \sin\theta. \tag{5.122}$$

[23]We use $\hat{a}_z = \hat{a}_R \cos\theta - \hat{a}_\theta \sin\theta$.

[24]We have

$$\begin{aligned}\bar{B}_d(\bar{r}) &= \bar{\nabla} \times \bar{A}_d(\bar{r}) \\ &= -\hat{a}_\phi \frac{1}{R}\frac{\partial}{\partial R}\left[R\sin\theta\mu I_{DM,h}\frac{\exp(-jkR)}{4\pi R}\right] \\ &\quad - \hat{a}_\phi \frac{1}{R}\frac{\partial}{\partial \theta}\left[\cos\theta\mu I_{DM,h}\frac{\exp(-jkR)}{4\pi R}\right].\end{aligned}$$

[25]We have (for $R \neq 0$)

$$\begin{aligned}\bar{E}_d(\bar{r}) &= \frac{1}{j\omega\varepsilon\mu}\bar{\nabla} \times \bar{B}_d(\bar{r}) \\ &= \hat{a}_R \frac{\mu I_{DM,h}}{j\omega\varepsilon\mu}\left(jk + \frac{1}{R}\right) g(R) \frac{1}{R\sin\theta}\frac{\partial}{\partial\theta}(\sin^2\theta) \\ &\quad - \hat{a}_\theta \frac{\mu I_{DM,h}}{j\omega\varepsilon\mu}\sin\theta\frac{1}{R}\frac{\partial}{\partial R}\left[R\left(jk + \frac{1}{R}\right) g(R)\right] \\ &= \hat{a}_R \frac{\mu I_{DM,h}}{j\omega\varepsilon\mu}\left(\frac{jk}{R} + \frac{1}{R^2}\right) g(R) 2\cos\theta \\ &\quad - \hat{a}_\theta \frac{\mu I_{DM,h}}{j\omega\varepsilon\mu}\sin\theta\left(-\frac{1}{R^2} - \frac{jk}{R} + k^2\right) g(R),\end{aligned} \tag{5.117}$$

which can be rearranged to get the final expression.

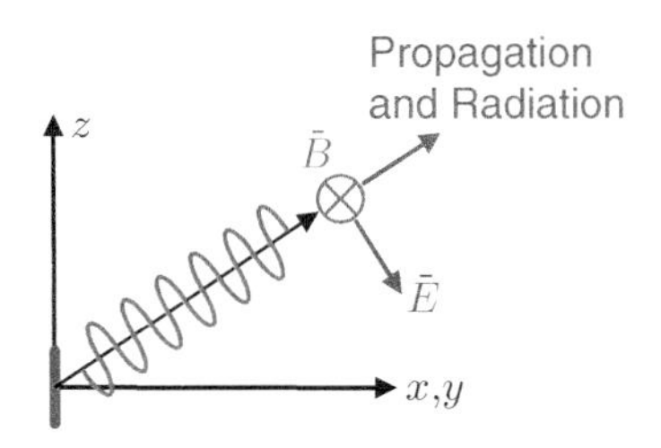

Figure 5.16 A Hertzian dipole creates electromagnetic waves that propagate radially in the far zone: i.e. when $kR \gg 1$. For a z-directed dipole, the electric field is in the θ direction, while the magnetic field is in the ϕ direction. Therefore, the electric field, magnetic field, and propagation (radiation) directions are perpendicular to each other, which is a common property observed in the far-zone electromagnetic waves generated by all finite sources in a homogeneous space.

In the time domain, these expressions lead to[26]

$$\bar{E}_d(\bar{r}) \approx \hat{a}_\theta \omega\mu I_{DM,h} \frac{1}{4\pi R} \sin\theta \cos(\omega t - kR + \pi/2) \quad (5.123)$$

$$\approx -\hat{a}_\theta \omega\mu I_{DM,h} \frac{1}{4\pi R} \sin\theta \sin[\omega(t - R/u_p)] \quad (5.124)$$

$$\bar{B}_d(\bar{r}) \approx \hat{a}_\phi k\mu I_{DM,h} \frac{1}{4\pi R} \sin\theta \cos(\omega t - kR + \pi/2) \quad (5.125)$$

$$\approx -\hat{a}_\phi k\mu I_{DM,h} \frac{1}{4\pi R} \sin\theta \sin[\omega(t - R/u_p)]. \quad (5.126)$$

[26]It is clear in these time-domain expressions (containing a $(\omega t - kR)$ term) that the generated wave is propagating in the radial (R) direction: i.e. away from the dipole, in the far zone.

An important relationship between the electric and magnetic fields in the far zone can be written[27] (Figure 5.16)

$$\bar{H}_d(\bar{r}) \approx \frac{k}{\omega\mu} \hat{a}_R \times \bar{E}_d(\bar{r}) = \frac{\omega\sqrt{\mu\varepsilon}}{\omega\mu} \hat{a}_R \times \bar{E}_d(\bar{r}) = \frac{1}{\sqrt{\mu/\varepsilon}} \hat{a}_R \times \bar{E}_d(\bar{r}) \quad (5.127)$$

or

$$\bar{H}_d(\bar{r}) \approx \frac{1}{\eta} \hat{a}_R \times \bar{E}_d(\bar{r}), \quad (5.128)$$

where η with units volt/ampere (Ω) is called *intrinsic impedance* or *wave impedance*. Note that, in a vacuum, $\eta_0 = \sqrt{\mu_0/\varepsilon_0} \approx 120\pi\ \Omega$. In the above, $\hat{a}_R$ corresponds to the direction of the wave that is generated by the dipole and propagating away from it.

[27]Although we show it for a Hertzian dipole, these expressions are valid for any radiated electromagnetic field from an arbitrary finite source distribution. Specifically, in the far zone, (i) electric and magnetic fields are perpendicular to each other and to the propagation direction (which is always $\hat{a}_R$), and (ii) their amplitudes are related to each other as $|\bar{E}(\bar{r})|/|\bar{H}(\bar{r})| = \eta$.

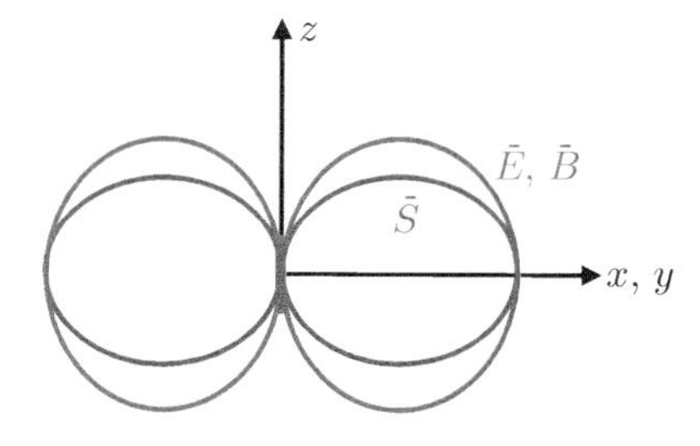

Figure 5.17 Electric and magnetic fields created by a z-directed Hertzian dipole have $\sin\theta$ dependencies in the far zone. Consequently, the Poynting vector, which is purely real in the far zone, behaves as $\sin^2\theta$. These plots that demonstrate relative amplitudes of electric/magnetic fields and the Poynting vector with respect to observation directions are known as radiation patterns.

At this stage, we can consider the electromagnetic power associated with the z-directed Hertzian dipole. In the far zone,[28] we have

$$\bar{S}^{\text{far}}_{c,d}(\bar{r}) = \frac{1}{2}\bar{E}_d(\bar{r}) \times [\bar{H}_d(\bar{r})]^* = \frac{1}{2}\bar{E}_d(\bar{r}) \times [\bar{B}_d(\bar{r})/\mu]^* \quad (5.129)$$

$$= \frac{1}{2}\hat{a}_\theta j\omega\mu I_{DM,h} \frac{\exp(-jkR)}{4\pi R} \sin\theta \times \left[\hat{a}_\phi jkI_{DM,h} \frac{\exp(-jkR)}{4\pi R} \sin\theta\right]^* \quad (5.130)$$

$$= \hat{a}_R |I_{DM,h}|^2 \frac{\omega k\mu}{32\pi^2 R^2} \sin^2\theta. \quad (5.131)$$

Hence, the complex Poynting vector is purely real, indicating that all power has a *propagating* nature in the far zone. The direction of the power flow is $\hat{a}_R$ (away from the Hertzian dipole), which is the same as the direction of wave propagation. At the same time, the power does not radiate uniformly (Figures 5.15 and 5.17): i.e. it is maximum for observation points with $\theta = \pi/2$, while it becomes zero when $\theta = 0$ (north pole) or $\theta = \pi$ (south pole).

η: intrinsic impedance
Unit: volts/ampere (V/A) or ohms (Ω)

[28]The expression for the near zone, which is left as an exercise, contains both oscillatory and propagating power contributions.

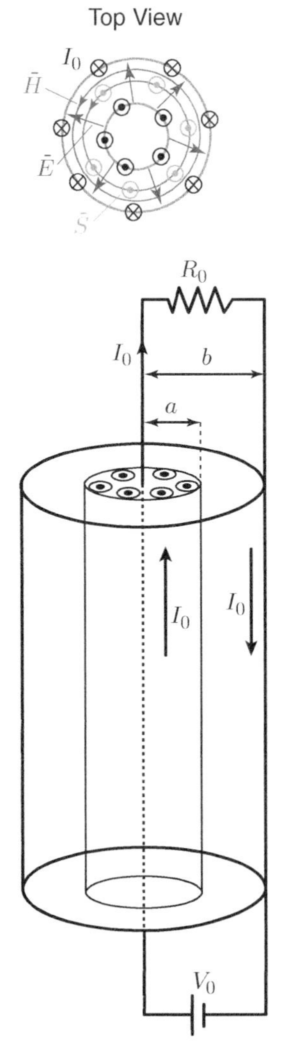

Figure 5.18 A coaxial structure with $\pm I_0$ currents on the inner/outer cylinders.

5.2.2 Examples

Example 52: Consider a very long coaxial cable with a cylindrical surface of radius a inside another cylindrical surface of radius b and dielectric material between them (Figure 5.18). The voltage between the inner and outer cylinders is V_0. The coaxial cable is connected to a resistor R such that a steady electric current $I_0 = V_0/R$ flows uniformly on the inner cylinder, while the same amount of current flows in the opposite direction uniformly on the the outer conductor. Find the Poynting vector between the cylinders.

Solution: First, using Ampere's law (and given that the structure is very long), we can obtain the magnetic field intensity:

$$\bar{H}(\bar{r}) = \hat{a}_\phi \frac{I_0}{2\pi\rho} \qquad (a \leq \rho \leq b). \tag{5.132}$$

In addition, assuming q_l charge density per length on the inner conductor, Gauss' law leads to

$$\bar{E}(\bar{r}) = \hat{a}_\rho \frac{q_l}{2\pi\varepsilon\rho} \qquad (a \leq \rho \leq b).$$

Then the Poynting vector can be obtained

$$\bar{S}(\bar{r}) = \bar{E}(\bar{r}) \times \bar{H}(\bar{r}) = \hat{a}_\rho \frac{q_l}{2\pi\varepsilon\rho} \times \hat{a}_\phi \frac{I_0}{2\pi\rho} = \hat{a}_z \frac{q_l I_0}{4\pi^2\varepsilon\rho^2}, \tag{5.133}$$

which is in the z direction. At this stage, we can use the expression for the per-length capacitance of a coaxial line (see Section 6.7 for capacitance calculations),

$$C_l = \frac{q_l}{V_0} = \frac{2\pi\varepsilon}{\ln(b/a)}, \tag{5.134}$$

leading to

$$\bar{S}(\bar{r}) = \hat{a}_z \frac{2\pi\varepsilon V_0}{\ln(b/a)} \frac{I_0}{4\pi^2\varepsilon\rho^2} = \hat{a}_z \frac{V_0 I_0}{2\pi\rho^2 \ln(b/a)}.$$

Integrating the Poynting vector in the region bounded by the cylinders (e.g. at the end of the coaxial), we have

$$\begin{aligned} p_{em} &= \int_S \bar{S}(\bar{r}) \cdot \overline{ds} = \int_0^{2\pi} \int_a^b \hat{a}_z \frac{V_0 I_0}{2\pi\rho^2 \ln(b/a)} \cdot \hat{a}_z \rho d\rho d\phi \\ &= 2\pi \frac{V_0 I_0}{2\pi \ln(b/a)} \ln(b/a) = V_0 I_0, \end{aligned} \tag{5.135}$$

which is the power carried by the coaxial line and dissipated by the resistor.

Example 53: In free space (source-free vacuum), the electric field intensity is defined as

$$\bar{E}(\bar{r},t) = \hat{a}_z E_0 \cos(ay)\cos(\omega t), \tag{5.136}$$

where E_0 (V/m) and a (1/m) are constants. Consider a unit cube (Figure 5.19) defined as $x \in [0,1]$ (m), $y \in [0,1]$ (m), and $z \in [0,1]$ (m).

- Find the electric potential energy contained inside the cube with respect to time.
- Find the magnetic potential energy contained inside the cube with respect to time.
- Find the Poynting vector everywhere.
- Integrate the Poynting vector on the surface of the unit cube, and show the conservation of energy.

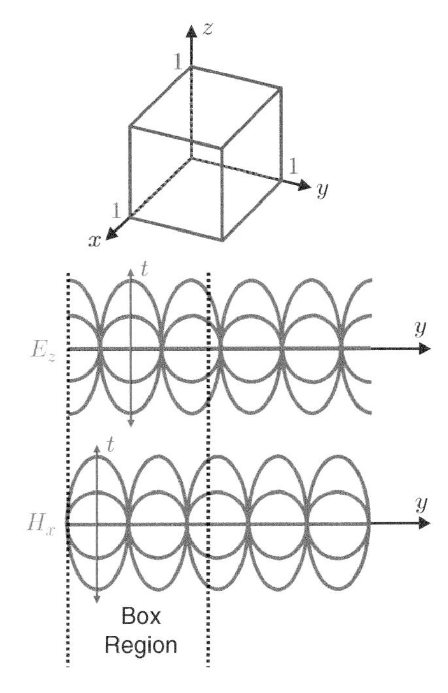

Figure 5.19 A unit box that includes a portion of a standing wave. Since the electric field intensity and magnetic field intensity change with respect to time (with a 90° phase difference), the electric potential energy and magnetic potential energy contained inside the cube depend on time, and there is a time-harmonic power flow through the $y = 1$ surface.

Solution: First, using Faraday's law, we have

$$\bar{\nabla} \times \bar{E}(\bar{r},t) = \hat{a}_x E_0(-a)\sin(ay)\cos(\omega t) = -\mu_0 \frac{\partial \bar{H}(\bar{r},t)}{\partial t}, \tag{5.137}$$

leading to[29]

$$\bar{H}(\bar{r},t) = \hat{a}_x \frac{E_0 a}{\omega\mu_0}\sin(ay)\sin(\omega t). \tag{5.138}$$

To find a, we can use Ampere's law:

$$\bar{\nabla} \times \bar{H}(\bar{r},t) = -\hat{a}_z \frac{E_0 a^2}{\omega\mu_0}\cos(ay)\sin(\omega t) \tag{5.139}$$

$$= \varepsilon_0 \frac{\partial \bar{E}(\bar{r},t)}{\partial t} = -\hat{a}_z E_0 \omega\varepsilon_0 \cos(ay)\sin(\omega t), \tag{5.140}$$

leading to[30] $a^2/(\omega\mu_0) = \omega\varepsilon_0$ or $a = \omega\sqrt{\mu_0\varepsilon_0}$. Then the intensity expressions can be updated as (Figure 5.19)

$$\bar{E}(\bar{r},t) = \hat{a}_z E_0 \cos(k_0 y)\cos(\omega t) \tag{5.141}$$

$$\bar{H}(\bar{r},t) = \hat{a}_x \frac{E_0}{\eta_0}\sin(k_0 y)\sin(\omega t), \tag{5.142}$$

[29] We assume that the magnetic field intensity does not have a time-invariant (static) part.

[30] Hence, a must be nothing but the wavenumber.

where $k_0 = \omega\sqrt{\mu_0\varepsilon_0}$ and $\eta_0 = \sqrt{\mu_0/\varepsilon_0}$, as usual. For the given cubic volume, the contained electric potential energy can be found as[31]

$$w_{e,\text{stored}}(t) = \frac{\varepsilon_0}{2}\int_V |\bar{E}(\bar{r},t)|^2 dv = \frac{\varepsilon_0}{4}E_0^2\cos^2(\omega t)\left[1 + \frac{1}{2k_0}\sin(2k_0)\right]. \quad (5.143)$$

Similarly, the contained magnetic potential energy can be found as[32]

$$w_{m,\text{stored}}(t) = \frac{\mu_0}{2}\int_V |\bar{H}(\bar{r},t)|^2 dv = \frac{\varepsilon_0}{4}E_0^2\sin^2(\omega t)\left[1 - \frac{1}{2k_0}\sin(2k_0)\right]. \quad (5.144)$$

On the other side, we find the Poynting vector as[33]

$$\bar{S}(\bar{r},t) = \bar{E}(\bar{r},t)\times\bar{H}(\bar{r},t) = \hat{a}_y\frac{E_0^2}{\eta_0}\cos(k_0y)\cos(\omega t)\sin(k_0y)\sin(\omega t) \quad (5.145)$$

$$= \hat{a}_y\frac{E_0^2}{4\eta_0}\sin(2k_0y)\sin(2\omega t). \quad (5.146)$$

Integrating the Poynting vector on the surface of the cube gives[34,35]

$$p_{em}(t) = \oint_S \bar{S}(\bar{r},t)\cdot\overline{ds} = \int_{y=1}\frac{E_0^2}{4\eta_0}\sin(2k_0y)\sin(2\omega t)dxdz \quad (5.147)$$

$$= \frac{E_0^2}{4\eta_0}\sin(2k_0)\sin(2\omega t). \quad (5.148)$$

According to the conservation of energy, we must have

$$p_{em}(t) = -\frac{\partial w_{m,\text{stored}}(t)}{\partial t} - \frac{\partial w_{e,\text{stored}}(t)}{\partial t}, \quad (5.149)$$

which can be shown as

$$-\frac{\partial w_{e,\text{stored}}(t)}{\partial t} = \frac{\varepsilon_0}{4}E_0^2\omega\sin(2\omega t)\left[1 + \frac{1}{2k_0}\sin(2k_0)\right] \quad (5.150)$$

$$-\frac{\partial w_{m,\text{stored}}(t)}{\partial t} = -\frac{\varepsilon_0}{4}E_0^2\omega\sin(2\omega t)\left[1 - \frac{1}{2k_0}\sin(2k_0)\right], \quad (5.151)$$

and

$$-\frac{\partial w_{e,\text{stored}}(t)}{\partial t} - \frac{\partial w_{m,\text{stored}}(t)}{\partial t} = \frac{E_0^2}{4\eta_0}\sin(2\omega t)\sin(2k_0). \quad (5.152)$$

[31] We have

$$w_{e,\text{stored}}(t) = \frac{\varepsilon_0}{2}E_0^2\cos^2(\omega t)\int_0^1\cos^2(k_0y)dy$$
$$= \frac{\varepsilon_0}{2}E_0^2\cos^2(\omega t)\frac{1}{2}\left[y + \frac{1}{2k_0}\sin(2k_0y)\right]_0^1.$$

[32] We have

$$w_{m,\text{stored}}(t) = \frac{\mu_0}{2}\frac{E_0^2}{\eta_0^2}\sin^2(\omega t)\int_0^1\sin^2(k_0y)dy$$
$$= \frac{\mu_0}{2}\frac{E_0^2}{\eta_0^2}\sin^2(\omega t)\frac{1}{2}\left[y - \frac{1}{2k_0}\sin(2k_0y)\right]_0^1.$$

[33] In the phasor domain, we have

$$\bar{E}(\bar{r}) = \hat{a}_zE_0\cos(k_0y)$$
$$\bar{H}(\bar{r}) = -\hat{a}_xj\frac{E_0}{\eta_0}\sin(k_0y)$$

and

$$\bar{S}_c(\bar{r}) = \frac{1}{2}\bar{E}(\bar{r})\times[\bar{H}(\bar{r})]^*$$
$$= \frac{1}{2}\hat{a}_zE_0\cos(k_0y)\times\hat{a}_xj\frac{E_0}{\eta_0}\sin(k_0y)$$
$$= \hat{a}_yj\frac{E_0^2}{2\eta_0}\cos(k_0y)\sin(k_0y)$$
$$= \hat{a}_yj\frac{E_0^2}{4\eta_0}\sin(2k_0y),$$

assuming real E_0. Hence, the complex Poynting vector is purely imaginary.

[34] Note that the time-average power flux density can be found as

$$\bar{S}_{\text{avg}}(\bar{r}) = \frac{1}{T}\int_0^T\bar{S}(\bar{r},t)dt = 0.$$

Specifically, at any given location, the electromagnetic power oscillates back and forth in the y direction, while there is no net power flow on average. This is in agreement with the fact that the complex Poynting vector is purely imaginary.

[35] Note that five of the six surfaces (other than the $y = 1$ surface) give zero contribution.

5.3 Plane Waves

A particular set of solutions to wave equations, hence Maxwell equations, are plane waves. These waves are assumed to be produced by infinitely large sources located at infinity. Therefore, at first glance, it may not be obvious why these waves are considered and studied extensively in electromagnetics. In fact, despite their ideal behavior, plane waves can be used to represent almost all waves at large distances from sources (Figure 5.20). In addition, some waves in bounded regions, e.g. in waveguides, may behave like plane waves. Finally, since they are natural solutions to wave equations, more complex waves can be expanded in terms of plane waves, making them important concepts in electromagnetics.

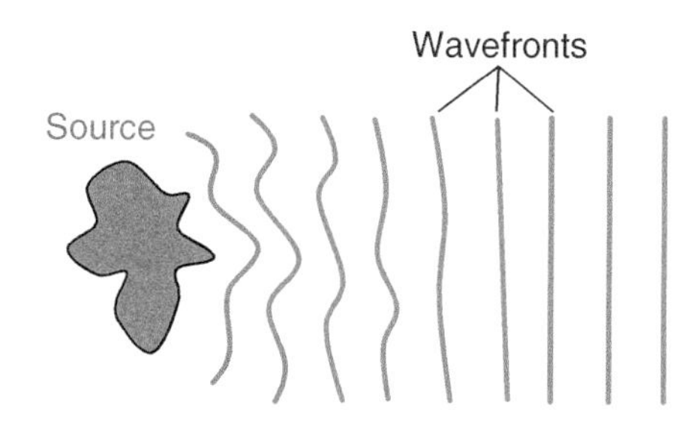

Figure 5.20 Waves generated by an arbitrary source tend to behave like plane waves far away from the source.

Starting with a general expression, a *linearly* polarized *uniform* plane wave[36,37] in a source-free and homogeneous medium with permittivity ε and permeability μ can be expressed via its electric field intensity as

$$\bar{E}(\bar{r}) = \bar{E}_0 \cos(\omega t - \bar{k} \cdot \bar{r}), \tag{5.153}$$

where $\omega = 2\pi f$ represents the angular frequency and $\bar{E}_0$ represents the constant amplitude. In addition, we define the wave vector as

$$\bar{k} = \hat{a}_k k = \hat{a}_k \omega \sqrt{\mu\varepsilon}, \tag{5.154}$$

where $\hat{a}_k$ is the propagation direction and k is the wavenumber defined in Section 4.4.3. Note that Eq. (5.153) is a time-harmonic field and fits into the definition of a wave discussed in Section 4.4.1. Also recall that $k = \omega/u_p = 2\pi/\lambda$, where u_p is the phase velocity (propagation velocity of the plane wave) and λ is the wavelength.

[36]More specifically, plane waves are waves with infinitely large planar wavefronts. For a uniform plane wave (to be considered in more detail), the electric field intensity and magnetic field intensity do not vary with position on the wavefronts: i.e. they only change in the propagation direction. There are also *nonuniform* plane waves, which still have infinitely large planar wavefronts but also have electric/magnetic field intensity distributions that vary with position on the wavefronts.

[37]As discussed in Section 5.3.2, plane waves can have different polarizations: i.e. linear, circular, and elliptical.

A plane wave satisfies the wave equations[38]

$$\bar{\nabla}^2 \bar{E}(\bar{r}, t) - \mu\varepsilon \frac{\partial^2}{\partial t^2} \bar{E}(\bar{r}, t) = 0, \qquad \bar{\nabla}^2 \bar{H}(\bar{r}, t) - \mu\varepsilon \frac{\partial^2}{\partial t^2} \bar{H}(\bar{r}, t) = 0 \tag{5.155}$$

[38]Hence, there is no source at any finite point in the three-dimensional space.

everywhere. Considering the general expressions in Eqs. (5.153) and (5.154), $\hat{a}_k$ is the direction of travel. Specifically, as time progresses, the value of the electric field intensity (and the magnetic field intensity) changes in the direction of $\hat{a}_k$. We emphasize that the phase of a plane wave is constant on a plane (wavefront). This can be seen by considering $\bar{k} \cdot \bar{r} =$ constant (Figure 5.21), which defines a set of positions $\bar{r}$ that are located on a plane perpendicular to $\bar{k}$. In a Cartesian system, these locations satisfy

$$k_x x + k_y y + k_z z = \text{constant}, \tag{5.156}$$

given the components of the wave vector as $\bar{k} = \hat{a}_x k_x + \hat{a}_y k_y + \hat{a}_z k_z$.

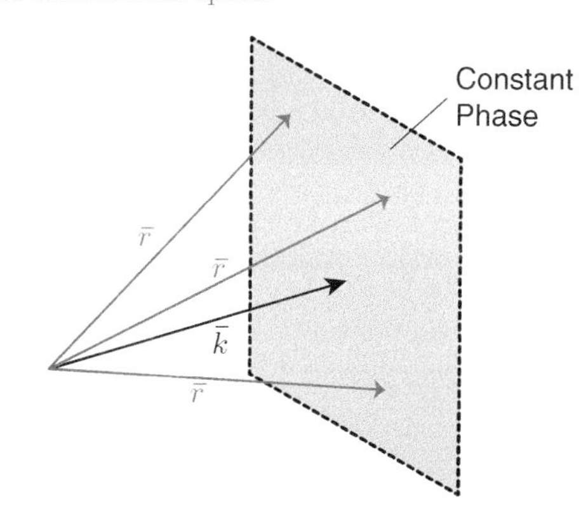

Figure 5.21 Locations that satisfy $\bar{k} \cdot \bar{r} =$ constant define a plane.

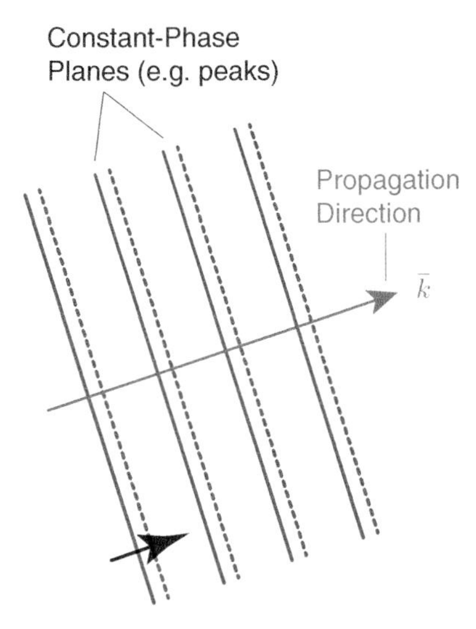

Figure 5.22 A view of a plane wave, where constant-phase planes (wavefronts) that are perpendicular to the propagation direction are shown. As time progresses, the wavefronts move.

Obviously, the overall argument in Eq. (5.153) – i.e. $v = \omega t - \bar{k} \cdot \bar{r}$ – contains both time and position, while the relationship between them defines how the wave propagates (Figure 5.22). For example, at a fixed time $t = t_0$, the electric field intensity is maximum[39] with value E_0 at positions $\bar{r}$ that satisfy $\bar{k} \cdot \bar{r} = \omega t_0 \pm 2\pi$, minimum with value $-E_0$ at positions $\bar{r}$ that satisfy $\bar{k} \cdot \bar{r} = \omega t_0 + \pi \pm 2\pi$, and zero at positions $\bar{r}$ that satisfy $\bar{k} \cdot \bar{r} = \omega t_0 + \pi/2 \pm 2\pi$. The maximum-to-maximum distance d (or the distance between any subsequent planes with identical phases) can be found as

$$\bar{k} \cdot \bar{r}_1 - \bar{k} \cdot \bar{r}_2 = \hat{a}_k \cdot (\bar{r}_1 - \bar{r}_2)k = 2\pi \longrightarrow d = \hat{a}_k \cdot (\bar{r}_1 - \bar{r}_2) = \frac{k}{2\pi} = \lambda, \tag{5.157}$$

which corresponds to the wavelength. As time progresses, these peak-to-peak distances do not change, but the constant-phase planes (wavefronts) move. For example, at $t = t_0 + \Delta t$, the electric field intensity is maximum at positions $\bar{r}$ that satisfy $\bar{k} \cdot \bar{r} = \omega t_0 + \omega \Delta t + \pi \pm 2\pi$: i.e. it is shifted by a distance of $\omega \Delta t/(\omega/u_p) = (\Delta t)u_p$ in Δt time.

At this stage, we can now use Maxwell's equations to further understand plane waves. Given the expression for the electric field intensity in Eq. (5.153), the phasor-domain representation can be found as[40]

$$\bar{E}(\bar{r}) = \bar{E}_0 \exp(-j\bar{k} \cdot \bar{r}) = \bar{E}_0 \exp(-jk_x x - jk_y y - jk_z z). \tag{5.158}$$

Note that this expression satisfies homogeneous Helmholtz equation: i.e. $\bar{\nabla}^2 \bar{E}(\bar{r}) + k^2 \bar{E}(\bar{r}) = 0$. Then Gauss' law states that[41]

$$\bar{\nabla} \cdot \bar{E}(\bar{r}) = \left[\hat{a}_x \frac{\partial}{\partial x} + \hat{a}_y \frac{\partial}{\partial y} + \hat{a}_z \frac{\partial}{\partial z}\right] \cdot \bar{E}_0 \exp(-jk_x x - jk_y y - jk_z z) \tag{5.159}$$

$$= -j\left[\hat{a}_x k_x + \hat{a}_y k_y + \hat{a}_z k_z\right] \cdot \bar{E}_0 \exp(-jk_x x - jk_y y - jk_z z) \tag{5.160}$$

$$= -j\bar{k} \cdot \bar{E}_0 \exp(-j\bar{k} \cdot \bar{r}) = q_v(\bar{r})/\varepsilon = 0. \tag{5.161}$$

But this is possible only when $\bar{k} \cdot \bar{E}_0 = 0$ or $\bar{E}_0 \perp \bar{k}$: i.e. the electric field intensity must be *perpendicular* to the propagation direction.

To find the magnetic field intensity of a plane wave, we can use Faraday's law (again in a source-free region)

$$\bar{\nabla} \times \bar{E}(\bar{r}) = \left[\hat{a}_x \frac{\partial}{\partial x} + \hat{a}_y \frac{\partial}{\partial y} + \hat{a}_z \frac{\partial}{\partial z}\right] \times \bar{E}_0 \exp(-jk_x x - jk_y y - jk_z z) \tag{5.162}$$

$$= -j\left[\hat{a}_x k_x + \hat{a}_y k_y + \hat{a}_z k_z\right] \times \bar{E}_0 \exp(-jk_x x - jk_y y - jk_z z) \tag{5.163}$$

$$= -j\bar{k} \times \bar{E}_0 \exp(-j\bar{k} \cdot \bar{r}) = -j\omega\mu \bar{H}(\bar{r}), \tag{5.164}$$

[39] Assuming that E_0 is positive.

[40] Here $\bar{E}_0$ is considered a real-valued vector since it is coming from the time-domain expression in Eq. (5.153). But in general, the expression in Eq. (5.158) may have a complex constant. This corresponds to an extra phase in the cosine of Eq. (5.153).

[41] Given that a plane wave is defined in a source-free region: i.e. it is generated by sources at infinity.

leading to[42,43,44]

$$\bar{H}(\bar{r}) = \frac{k}{\omega\mu}\hat{a}_k \times \bar{E}_0 \exp(-j\bar{k}\cdot\bar{r}) = \hat{a}_k \times \frac{\bar{E}_0}{\eta}\exp(-j\bar{k}\cdot\bar{r}). \tag{5.165}$$

In the above, $\eta = \sqrt{\mu/\varepsilon}$ is the intrinsic impedance, as defined before.[45,46] Considering the cross product in Eq. (5.165), we find that the magnetic field intensity must be perpendicular to *both* the electric field intensity and the propagation direction: i.e. $\bar{H}(\bar{r}) \perp \bar{E}(\bar{r})$ and $\bar{H}(\bar{r}) \perp \bar{k}$.

On the other side, Ampere's law should verify the relationship between the magnetic field intensity and electric field intensity. We have

$$\bar{\nabla}\times\bar{H}(\bar{r}) = \left[\hat{a}_x\frac{\partial}{\partial x} + \hat{a}_y\frac{\partial}{\partial y} + \hat{a}_z\frac{\partial}{\partial z}\right]\times\bar{H}_0\exp(-jk_x x - jk_y y - jk_z z), \tag{5.166}$$

where $\bar{H}_0 = \hat{a}_k \times \bar{E}_0/\eta$. This leads to[47]

$$\bar{\nabla}\times\bar{H}(\bar{r}) = -j\bar{k}\times\bar{H}_0\exp(-j\bar{k}\cdot\bar{r}) = -j\frac{k}{\eta}\hat{a}_k\times[\hat{a}_k\times\bar{E}_0]\exp(-j\bar{k}\cdot\bar{r}) \tag{5.167}$$

$$= j\frac{k}{\eta}\bar{E}_0\exp(-j\bar{k}\cdot\bar{r}) \tag{5.168}$$

using $\bar{E}(\bar{r}) \perp \bar{k}$. According to Ampere's law, the final equation must be $j\omega\varepsilon\bar{E}(\bar{r}) = j\omega\varepsilon\bar{E}_0\exp(-j\bar{k}\cdot\bar{r})$, which is indeed the case since

$$\frac{k}{\eta} = \frac{\omega\sqrt{\mu\varepsilon}}{\sqrt{\mu/\varepsilon}} = \omega\varepsilon. \tag{5.169}$$

One can further verify that Gauss' law is satisfied for the magnetic field intensity: i.e. $\bar{\nabla}\cdot\bar{H}(\bar{r}) = 0$.

Plane waves are actually natural solutions to wave and Helmholtz equations in a homogeneous and source-free region, if separation of variables is assumed in a Cartesian coordinate system. Starting with Helmholtz equation for the electric field intensity as

$$\bar{\nabla}^2\bar{E}(\bar{r}) + k^2\bar{E}(\bar{r}) = 0, \tag{5.170}$$

where $k = \omega\sqrt{\mu\varepsilon}$, a general solution can be written

$$\bar{E}(\bar{r}) = \hat{a}_x E_x(\bar{r}) + \hat{a}_y E_y(\bar{r}) + \hat{a}_z E_z(\bar{r}). \tag{5.171}$$

[42] Hence, in the time domain, we have

$$\bar{H}(\bar{r},t) = \hat{a}_k \times \frac{\bar{E}_0}{\eta}\cos(\omega t - \bar{k}\cdot\bar{r}).$$

[43] Note that the electric field intensity and magnetic field intensity are *in phase* at all points in space.

[44] This important relationship,

$$\bar{H}(\bar{r},t) = \frac{1}{\eta}\hat{a}_k \times \bar{E}(\bar{r},t)$$

$$\bar{H}(\bar{r}) = \frac{1}{\eta}\hat{a}_k \times \bar{E}(\bar{r}),$$

is valid for *uniform* plane waves.

[45] For waves, η is also called wave impedance.

[46] Hence, the amplitude of the magnetic field intensity is the amplitude of the electric field intensity scaled by η.

[47] We have

$$\hat{a}_k \times [\hat{a}_k \times \bar{E}_0] = \hat{a}_k\hat{a}_k\cdot\bar{E}_0 - \bar{E}_0\hat{a}_k\cdot\hat{a}_k$$

using the BAC-CAB rule. Then given that $\hat{a}_k\cdot\hat{a}_k = 1$ and $\hat{a}_k\cdot\bar{E}_0 = 0$, we obtain

$$\hat{a}_k \times [\hat{a}_k \times \bar{E}_0] = -\bar{E}_0.$$

Now, focusing on the x component, keeping in mind that the same procedure can be applied to the y and z components, we have

$$\frac{\partial^2}{\partial x^2}E_x(\bar{r}) + \frac{\partial^2}{\partial y^2}E_x(\bar{r}) + \frac{\partial^2}{\partial z^2}E_x(\bar{r}) + k^2 E_x(\bar{r}) = 0. \tag{5.172}$$

Assuming that the dependency of the electric field intensity can be separated for x, y, z as $E_x(x, y, z) = f(x)g(y)h(z)$, we obtain[48]

$$\frac{1}{f(x)}\frac{\partial^2}{\partial x^2}f(x) + \frac{1}{g(y)}\frac{\partial^2}{\partial y^2}g(y) + \frac{1}{h(z)}\frac{\partial^2}{\partial z^2}h(z) + k^2 = 0. \tag{5.173}$$

[48] We have

$$g(y)h(z)\frac{\partial^2}{\partial x^2}f(x) + f(x)h(z)\frac{\partial^2}{\partial y^2}g(y) + f(x)g(y)\frac{\partial^2}{\partial z^2}h(z) + k^2 f(x)g(y)h(z) = 0.$$

Then inserting

$$k_x^2 = \frac{1}{f(x)}\frac{\partial^2}{\partial x^2}f(x) \quad k_y^2 = \frac{1}{g(y)}\frac{\partial^2}{\partial y^2}g(y) \quad k_z^2 = \frac{1}{h(z)}\frac{\partial^2}{\partial z^2}h(z), \tag{5.174}$$

where

$$k_x^2 + k_y^2 + k_z^2 = k^2 = \omega^2 \mu\varepsilon, \tag{5.175}$$

we obtain solutions as follows:

$$f(x) = f_0 \exp(-jk_x x), \quad g(y) = g_0 \exp(-jk_y y), \quad h(z) = h_0 \exp(-jk_z z). \tag{5.176}$$

This leads to

$$E_x(\bar{r}) = f_0 g_0 h_0 \exp(-jk_x x)\exp(-jk_y y)\exp(-jk_z z) \tag{5.177}$$

$$= E_{x0}\exp(-jk_x x)\exp(-jk_y y)\exp(-jk_z z) = E_{x0}\exp(-j\bar{k}\cdot\bar{r}). \tag{5.178}$$

Exactly the same procedure can be applied to the y and z components of the electric field intensity, leading to

$$E_y(\bar{r}) = E_{y0}\exp(-j\bar{k}\cdot\bar{r}) \tag{5.179}$$

$$E_z(\bar{r}) = E_{z0}\exp(-j\bar{k}\cdot\bar{r}), \tag{5.180}$$

and[49]

$$\bar{E}(\bar{r}) = (\hat{a}_x E_{x0} + \hat{a}_y E_{y0} + \hat{a}_z E_{z0})\exp(-j\bar{k}\cdot\bar{r}) = \bar{E}_0\exp(-j\bar{k}\cdot\bar{r}). \tag{5.181}$$

[49] In this derivation in the phasor domain, $\bar{E}_0$ can be a real or complex vector. If $\bar{E}_0$ is real, the time-domain expression corresponding to Eq. (5.181) is simply Eq. (5.153):

$$\bar{E}(\bar{r}) = \bar{E}_0\cos(\omega t - \bar{k}\cdot\bar{r}).$$

For complex values of $\bar{E}_0$, the amplitude of the time-domain electric field intensity becomes $|\bar{E}_0|$, and a phase is added into the cosine function.

At this stage, we may consider a special case: i.e. a plane wave propagating in the z direction with the electric field intensity in the x direction. The electric field intensity can be written in the time domain as

$$\bar{E}(\bar{r},t) = \hat{a}_x E_0 \cos(\omega t - kz), \tag{5.182}$$

given that $\bar{k} = \hat{a}_z k_z = \hat{a}_z k = \hat{a}_z \omega\sqrt{\mu\varepsilon}$. Then the magnetic field intensity can be written

$$\bar{H}(\bar{r},t) = \hat{a}_z \times \hat{a}_x \frac{E_0}{\eta} \cos(\omega t - kz) = \hat{a}_y \frac{E_0}{\eta} \cos(\omega t - kz). \tag{5.183}$$

The corresponding representations in the phasor domain are

$$\bar{E}(\bar{r}) = \hat{a}_x E_0 \exp(-jkz) \tag{5.184}$$

$$\bar{H}(\bar{r}) = \hat{a}_y \frac{E_0}{\eta} \exp(-jkz). \tag{5.185}$$

Obviously, we have

$$\bar{\nabla} \cdot \bar{E}(\bar{r}) = \frac{\partial}{\partial x}[E_0 \exp(-jkz)] = 0 \tag{5.186}$$

$$\bar{\nabla} \cdot \bar{H}(\bar{r}) = \frac{\partial}{\partial y}\left[\frac{E_0}{\eta} \exp(-jkz)\right] = 0 \tag{5.187}$$

$$\bar{\nabla} \times \bar{E}(\bar{r}) = \hat{a}_z \times \hat{a}_x \frac{\partial}{\partial z}[E_0 \exp(-jkz)] = \hat{a}_y(-jk)E_0 \exp(-jkz) \tag{5.188}$$

$$= \hat{a}_y(-j)(\omega\sqrt{\mu\varepsilon})\eta\frac{E_0}{\eta} \exp(-jkz) = -j\omega\mu\bar{H}(\bar{r}) \tag{5.189}$$

$$\bar{\nabla} \times \bar{H}(\bar{r}) = \hat{a}_z \times \hat{a}_y \frac{\partial}{\partial z}\left[\frac{E_0}{\eta} \exp(-jkz)\right] = -\hat{a}_x \frac{E_0}{\eta}(-jk)\exp(-jkz) \tag{5.190}$$

$$= \hat{a}_x j(\omega\sqrt{\mu\varepsilon})\frac{\varepsilon}{\mu} E_0 \exp(-jkz) = j\omega\varepsilon\bar{E}(\bar{r}) \tag{5.191}$$

and

$$\bar{\nabla}^2\bar{E}(\bar{r}) + k^2\bar{E}(\bar{r}) = \hat{a}_x(-jk)^2 E_0 \exp(-jkz) + \hat{a}_x k^2 E_0 \exp(-jkz) = 0 \tag{5.192}$$

$$\bar{\nabla}^2\bar{H}(\bar{r}) + k^2\bar{H}(\bar{r}) = \hat{a}_y(-jk)^2 \frac{E_0}{\eta} \exp(-jkz) + \hat{a}_y k^2 \frac{E_0}{\eta} \exp(-jkz) = 0. \tag{5.193}$$

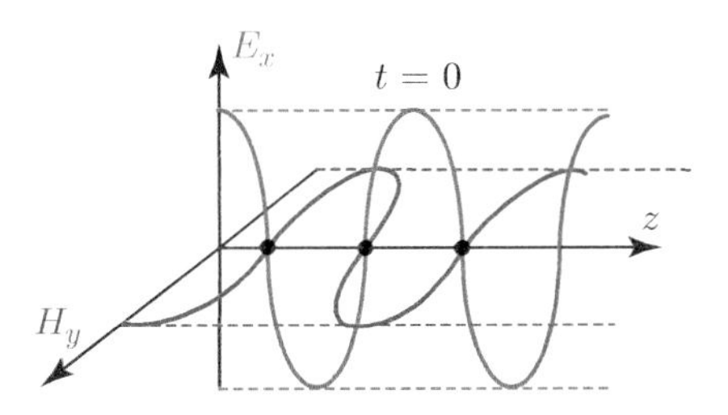

Figure 5.23 The electric field intensity and magnetic field intensity of a plane wave with respect to position at a fixed time.

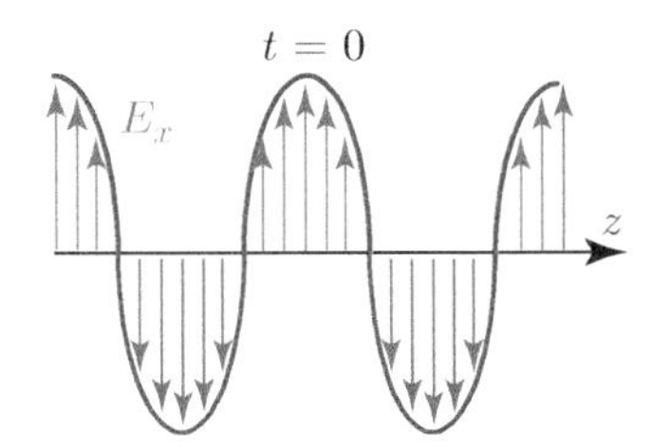

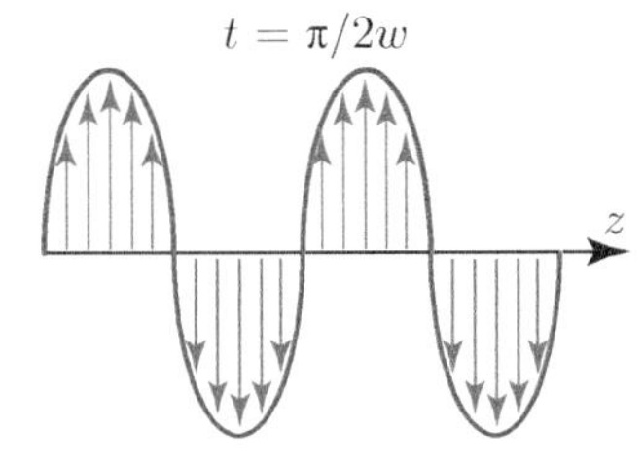

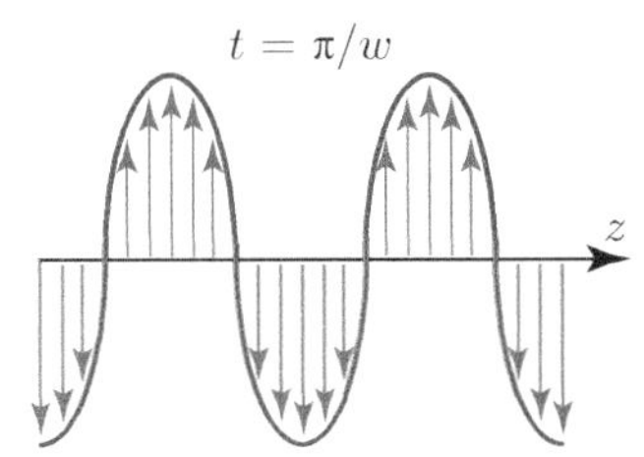

Figure 5.24 The electric field intensity of a plane wave propagating in the z direction at different fixed times. Note that the electric field intensity does not depend on x and y.

One can also show that the time-domain expressions satisfy all Maxwell's equations (in the time domain) and homogeneous wave equations.

Continuing with the special case of a plane wave propagating in the z direction, we may now visualize how the wave looks like. First, we consider the expressions at fixed times: i.e. when $t = t_0$. If $t = 0$, we have (Figure 5.23)

$$\bar{E}(\bar{r}, t = 0) = \hat{a}_x E_0 \cos(kz) \tag{5.194}$$

$$\bar{H}(\bar{r}, t = 0) = \hat{a}_y \frac{E_0}{\eta} \cos(kz), \tag{5.195}$$

i.e. cosine functions with respect to position z. Also note that, being a uniform plane wave, the electric field intensity and magnetic field intensity do not depend on x and y. As time progresses, the cosine functions shift in the z direction. For example, at $t = \pi/(2\omega)$, we have (Figure 5.24)

$$\bar{E}(\bar{r}, t = \pi/(2\omega)) = \hat{a}_x E_0 \sin(kz) \tag{5.196}$$

$$\bar{H}(\bar{r}, t = \pi/(2\omega)) = \hat{a}_y \frac{E_0}{\eta} \sin(kz). \tag{5.197}$$

Similarly, at time π/ω, the expressions become

$$\bar{E}(\bar{r}, t = \pi/\omega) = -\hat{a}_x E_0 \cos(kz) \tag{5.198}$$

$$\bar{H}(\bar{r}, t = \pi/\omega) = -\hat{a}_y \frac{E_0}{\eta} \cos(kz). \tag{5.199}$$

On the other side, we may consider the case when the position is fixed (Figure 5.25), e.g. as $z = 0$. We have

$$\bar{E}(z = 0, t) = \hat{a}_x E_0 \cos(\omega t) \tag{5.200}$$

$$\bar{H}(z = 0, t) = \hat{a}_y \frac{E_0}{\eta} \cos(\omega t), \tag{5.201}$$

i.e. cosine functions with respect to time. For another value of z, e.g. when $z = \pi/(2k)$, we have

$$\bar{E}(z = \pi/(2k), t) = \hat{a}_x E_0 \sin(\omega t) \tag{5.202}$$

$$\bar{H}(z = \pi/(2k), t) = \hat{a}_y \frac{E_0}{\eta} \sin(\omega t). \tag{5.203}$$

Hence, when the field values are maximum at $z = 0$, they are zero at $z = \pi/(2k) = \lambda/4$, and vice versa. Note that these expressions are given for all x and y: i.e. there is no dependency in these directions. In addition, the ratio between the electric field intensity and magnetic field intensity is always η, independent of the time.[50]

Fixing the overall argument of the cosine functions provides another view of plane waves. Specifically, if $v = \omega t - kz$ is fixed, then the electric field intensity and magnetic field intensity simply become $\bar{E}(\bar{r}, t) = \hat{a}_x E_0 \cos(v)$ and $\bar{H}(\bar{r}, t) = \hat{a}_y (E_0/\eta) \cos(v)$, like static fields. This corresponds to *surfing on* the wave: i.e. moving with the speed of the wave such that the electric field intensity and magnetic field intensity do not change with respect to time.

When the electric field is expressed as $\bar{E}(\bar{r}, t) = \hat{a}_x E_0 \cos(\omega t - kz)$, we assume that $\bar{E}(z = 0, t = 0) = \hat{a}_x E_0$ as a kind of initial condition. Obviously, there can be a nonzero phase, leading to more general expressions (for a plane wave propagating in the z direction):[51]

$$\bar{E}(\bar{r}, t) = \hat{a}_x E_0 \cos(\omega t - kz + \varphi) \tag{5.204}$$

$$\bar{H}(\bar{r}, t) = \hat{a}_y \frac{E_0}{\eta} \cos(\omega t - kz + \varphi). \tag{5.205}$$

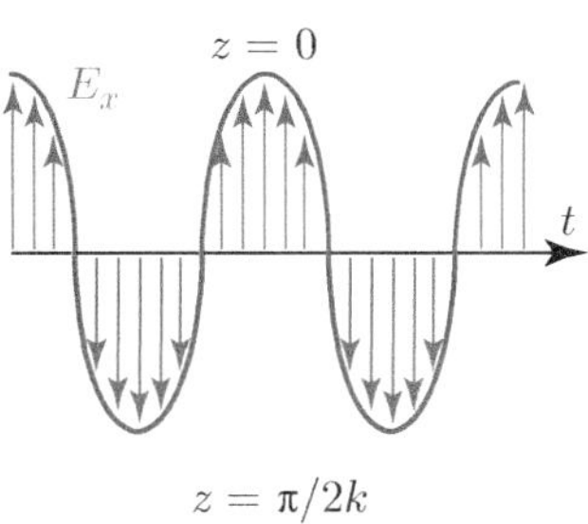

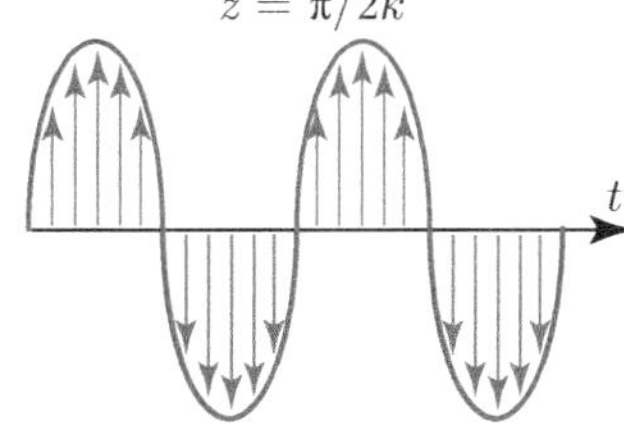

Figure 5.25 The electric field intensity of a plane wave propagating in the z direction at two different sets of positions (z is fixed, while x and y can be anything) with respect to time.

As natural basic solutions to wave equations in a Cartesian coordinate system, plane waves can be considered building blocks of more complex waves that generally exist. As already discussed in Section 4.4.1, a superposition of identical waves in opposite directions leads to standing waves that oscillate with respect to time without traveling. We now consider two plane waves with equal amplitudes and the same electric field direction ($\hat{a}_x$) propagating in the $\pm z$ directions:

$$\bar{E}_1(\bar{r}, t) = \hat{a}_x E_0 \cos(\omega t - kz) \tag{5.206}$$

$$\bar{E}_2(\bar{r}, t) = \hat{a}_x E_0 \cos(\omega t + kz). \tag{5.207}$$

Then the total electric field intensity can be written[52]

$$\bar{E}(\bar{r}, t) = \hat{a}_x E_0 [\cos(\omega t - kz) + \cos(\omega t + kz)] \tag{5.208}$$

$$= \hat{a}_x 2E_0 \cos(\omega t) \cos(kz). \tag{5.209}$$

With completely separated position and time, this is simply a standing wave (Figure 4.30) oscillating with an angular frequency of ω. In the phasor domain, these correspond to

$$\bar{E}_1(\bar{r}) = \hat{a}_x E_0 \exp(-jkz) \tag{5.210}$$

$$\bar{E}_2(\bar{r}) = \hat{a}_x E_0 \exp(jkz) \tag{5.211}$$

$$\bar{E}(\bar{r}) = \hat{a}_x E_0 [\exp(-jkz) + \exp(jkz)] = \hat{a}_x 2E_0 \cos(kz). \tag{5.212}$$

[50] Hence, at a fixed position $\bar{r}$, $\bar{E}(\bar{r}, t)$ and $\bar{H}(\bar{r}, t)$ increase/decrease simultaneously with respect to time.

[51] In the phasor domain, these correspond to

$$\bar{E}(\bar{r}) = \hat{a}_x E_0 \exp(-jkz + j\varphi)$$
$$= \hat{a}_x E_0 \exp(j\varphi) \exp(-jkz)$$
$$\bar{H}(\bar{r}) = \hat{a}_y \frac{E_0}{\eta} \exp(-jkz + j\varphi)$$
$$= \hat{a}_y \frac{E_0}{\eta} \exp(j\varphi) \exp(-jkz),$$

i.e. a phase is added in the exponentials.

[52] Using

$$\cos x + \cos y = 2 \cos\left(\frac{x + y}{2}\right) \cos\left(\frac{x - y}{2}\right).$$

We further note that

$$\bar{H}_1(\bar{r}) = \hat{a}_y(E_0/\eta)\exp(-jkz) \tag{5.213}$$

$$\bar{H}_2(\bar{r}) = -\hat{a}_y(E_0/\eta)\exp(jkz) \tag{5.214}$$

$$\bar{H}(\bar{r}) = \bar{H}_1(\bar{r}) + \bar{H}_2(\bar{r}) = \hat{a}_y\frac{E_0}{\eta}[\exp(-jkz) - \exp(jkz)] \tag{5.215}$$

$$= -\hat{a}_y\frac{j2E_0}{\eta}\sin(kz), \tag{5.216}$$

and[53]

$$\bar{H}(\bar{r},t) = \mathrm{Re}\{\bar{H}(\bar{r})\exp(j\omega t)\} = -\hat{a}_y\frac{2E_0}{\eta}\sin(kz)\mathrm{Re}\{j\exp(j\omega t)\} \tag{5.217}$$

$$= -\hat{a}_y\frac{2E_0}{\eta}\cos(\omega t + \pi/2)\sin(kz) = \hat{a}_y\frac{2E_0}{\eta}\sin(\omega t)\sin(kz). \tag{5.218}$$

Therefore, we have $|\bar{H}(\bar{r},t)|/|\bar{E}(\bar{r},t)| \neq \eta$ even though $|\bar{H}_1(\bar{r},t)|/|\bar{E}_1(\bar{r},t)| = \eta$ and $|\bar{H}_2(\bar{r},t)|/|\bar{E}_2(\bar{r},t)| = \eta$.

But what happens if two plane waves traveling in the same direction and with the same electric field intensity direction are superposed? If the plane waves have the same amplitude E_0 and are in phase, we will simply obtain a plane wave with larger amplitude ($2E_0$). But if they are not in phase, we have

$$\bar{E}_1(\bar{r},t) = \hat{a}_x E_0\cos(\omega t - kz + \varphi_1) \tag{5.219}$$

$$\bar{E}_2(\bar{r},t) = \hat{a}_x E_0\cos(\omega t - kz + \varphi_2) \tag{5.220}$$

and

$$\bar{E}(\bar{r},t) = \bar{E}_1(\bar{r},t) + \bar{E}_2(\bar{r},t) \tag{5.221}$$

$$= \hat{a}_x 2E_0\cos(\omega t - kz + \varphi_1/2 + \varphi_2/2)\cos(\varphi_1/2 - \varphi_2/2). \tag{5.222}$$

In the phasor domain, we have

$$\bar{E}_1(\bar{r}) = \hat{a}_x E_0\exp(-jkz + j\varphi_1) \tag{5.223}$$

$$\bar{E}_2(\bar{r}) = \hat{a}_x E_0\exp(-jkz + j\varphi_2) \tag{5.224}$$

$$\bar{E}(\bar{r}) = \hat{a}_x E_0\exp(-jkz)[\exp(j\varphi_1) + \exp(j\varphi_2)]. \tag{5.225}$$

[53] Using only time-domain expressions, this can be verified as

$$\bar{H}_1(\bar{r},t) = \hat{a}_z \times \bar{E}_1(\bar{r},t)/\eta = \hat{a}_y\frac{E_0}{\eta}\cos(\omega t - kz)$$

$$\bar{H}_2(\bar{r},t) = -\hat{a}_z \times \bar{E}_2(\bar{r},t)/\eta = -\hat{a}_y\frac{E_0}{\eta}\cos(\omega t + kz)$$

and

$$\begin{aligned}\bar{H}(r,t) &= \bar{H}_1(\bar{r},t) + \bar{H}_2(\bar{r},t)\\ &= \hat{a}_y\frac{E_0}{\eta}[\cos(\omega t - kz) - \cos(\omega t + kz)]\\ &= -\hat{a}_y\frac{2E_0}{\eta}\sin(\omega t)\sin(-kz)\\ &= \hat{a}_y\frac{2E_0}{\eta}\sin(\omega t)\sin(kz)\end{aligned}$$

using

$$\cos x - \cos y = -2\sin\left(\frac{x+y}{2}\right)\sin\left(\frac{x-y}{2}\right).$$

This is a plane wave propagating in the z direction, while its amplitude is

$$|\bar{E}(\bar{r})| = 2E_0 \cos(\varphi_1/2 - \varphi_2/2). \tag{5.226}$$

Hence, there is a chance that the amplitude becomes zero, e.g. when $\varphi_1 = \varphi_2 + \pi$, which corresponds to the perfect cancellation of the superposed plane waves.

As another example, we now consider two plane waves propagating in opposite directions with unequal amplitudes. Still assuming that the electric field intensity is in the x direction for both plane waves, we have

$$\bar{E}_1(\bar{r},t) = \hat{a}_x E_1 \cos(\omega t - kz) \tag{5.227}$$

$$\bar{E}_2(\bar{r},t) = \hat{a}_x E_2 \cos(\omega t + kz) \tag{5.228}$$

and[54]

$$\bar{E}(\bar{r},t) = \bar{E}_1(\bar{r},t) + \bar{E}_2(\bar{r},t) \tag{5.229}$$

$$= \hat{a}_x(E_1 - E_2)\cos(\omega t - kz) + \hat{a}_x E_2[\cos(\omega t - kz) + \cos(\omega t + kz)] \tag{5.230}$$

$$= \hat{a}_x(E_1 - E_2)\cos(\omega t - kz) + \hat{a}_x 2E_2 \cos(\omega t)\cos(kz). \tag{5.231}$$

[54]Alternatively, we have

$$\begin{aligned}\bar{E}(\bar{r},t) &= \bar{E}_1(\bar{r},t) + \bar{E}_2(\bar{r},t)\\ &= \hat{a}_x(E_2 - E_1)\cos(\omega t + kz)\\ &\quad + \hat{a}_x E_1[\cos(\omega t - kz) + \cos(\omega t + kz)]\\ &= \hat{a}_x(E_2 - E_1)\cos(\omega t - kz)\\ &\quad + \hat{a}_x 2E_1 \cos(\omega t)\cos(kz).\end{aligned}$$

This sum is interesting since it contains an effective traveling wave

$$\bar{E}_t(\bar{r},t) = \hat{a}_x(E_1 - E_2)\cos(\omega t - kz) \tag{5.232}$$

and an effective standing wave

$$\bar{E}_s(\bar{r},t) = \hat{a}_x 2E_2 \cos(\omega t)\cos(kz). \tag{5.233}$$

5.3.1 Examples

Example 54: Consider a plane wave defined in a medium with permittivity $\varepsilon = 4\varepsilon_0$ and permeability $\mu = \mu_0$

$$\bar{E}(\bar{r},t) = \left(\hat{a}_x \frac{1}{2} - \hat{a}_z \frac{\sqrt{3}}{2}\right) E_0 \cos\left[\omega t - 9\pi\left(\sqrt{3}x + z\right)\right], \tag{5.234}$$

where E_0 (V/m) is a constant. Find the phasor-domain representations of the electric field intensity and magnetic field intensity, as well as the velocity and frequency of the wave (approximately).

Solution: Given that the electric field intensity must be in the form of $\bar{E}(\bar{r},t) = \bar{E}_0 \cos(\omega t - \bar{k}\cdot\bar{r})$, we have

$$\bar{E}_0 = \left(\hat{a}_x \frac{1}{2} - \hat{a}_z \frac{\sqrt{3}}{2}\right) E_0 \tag{5.235}$$

and

$$\bar{k}\cdot\bar{r} = k_x x + k_y y + k_z z = 9\pi\left(\sqrt{3}x + z\right), \tag{5.236}$$

leading to $k_x = 9\sqrt{3}\pi$, $k_y = 0$, $k_z = 9\pi$, and[55,56]

$$\bar{k} = \hat{a}_x 9\sqrt{3}\pi + \hat{a}_z 9\pi, \qquad k = |\bar{k}| = 18\pi \text{ rad/m}. \tag{5.237}$$

Then the phasor-domain representation of the electric field intensity can be written

$$\bar{E}(\bar{r}) = \left(\hat{a}_x \frac{1}{2} - \hat{a}_z \frac{\sqrt{3}}{2}\right) E_0 \exp\left[-j9\pi\left(\sqrt{3}x + z\right)\right]. \tag{5.238}$$

Given that this is a uniform plane wave, we can obtain the phasor-domain representation of the magnetic field intensity as[57]

$$\bar{H}(\bar{r}) = \frac{1}{\eta}\hat{a}_k \times \bar{E}(\bar{r}) \tag{5.239}$$

$$= \frac{2}{\eta_0}\left(\hat{a}_x \frac{\sqrt{3}}{2} + \hat{a}_z \frac{1}{2}\right) \times \left(\hat{a}_x \frac{1}{2} - \hat{a}_z \frac{\sqrt{3}}{2}\right) E_0 \exp\left[-j9\pi\left(\sqrt{3}x + z\right)\right] \tag{5.240}$$

$$= \hat{a}_y \frac{2E_0}{\eta_0} \exp\left[-j9\pi\left(\sqrt{3}x + z\right)\right]. \tag{5.241}$$

Finally, given that $k = 18\pi$ rad/m, we have[58]

$$18\pi = \omega\sqrt{\mu\varepsilon} = \omega\sqrt{\mu_0\varepsilon_0}\sqrt{4} = 2\omega\sqrt{\mu_0\varepsilon_0} \approx \frac{4\pi f}{3\times 10^8}, \tag{5.242}$$

leading to

$$f \approx 13.5 \times 10^8 \text{ Hz} = 1.35 \text{ GHz}. \tag{5.243}$$

Obviously, the wave velocity can be obtained as $u_p = 1/\sqrt{\mu\varepsilon} \approx 1.5 \times 10^8$ m/s.

[55] Hence, we have

$$\bar{E}_0 \cdot \bar{k} = \frac{1}{2}9\sqrt{3}\pi - \frac{\sqrt{3}}{2}9\pi = 0,$$

as required.

[56] The propagation direction can be written

$$\hat{a}_k = \frac{\bar{k}}{k} = \hat{a}_x \frac{\sqrt{3}}{2} + \hat{a}_z \frac{1}{2}.$$

[57] Note that

$$\eta = \sqrt{\frac{\mu}{\varepsilon}} = \frac{1}{\sqrt{4}}\sqrt{\frac{\mu_0}{\varepsilon_0}} = \frac{\eta_0}{2}.$$

[58] We use an approximation:

$$c_0 = \frac{1}{\sqrt{\mu_0\varepsilon_0}} \approx 3 \times 10^8 \text{ m/s}.$$

5.3.2 Polarization of Plane Waves

Polarization[59] is an important property of plane waves, as well as all electromagnetic waves, which describes how fields change as a wave propagates. In general, polarization of a wave affects how it interacts with matter. To understand polarization, we may consider combinations of basic plane waves, as examined above. First, we consider two in-phase plane waves, both propagating in the z direction but with different electric field directions:

[59]Not to be confused with polarization in dielectric materials.

$$\bar{E}_1(\bar{r},t) = \hat{a}_x E_1 \cos(\omega t - kz) \tag{5.244}$$

$$\bar{E}_2(\bar{r},t) = \hat{a}_y E_2 \cos(\omega t - kz). \tag{5.245}$$

Their sum can be written

$$\bar{E}(\bar{r},t) = \bar{E}_1(\bar{r},t) + \bar{E}_2(\bar{r},t) \tag{5.246}$$

$$= (\hat{a}_x E_1 + \hat{a}_y E_2)\cos(\omega t - kz), \tag{5.247}$$

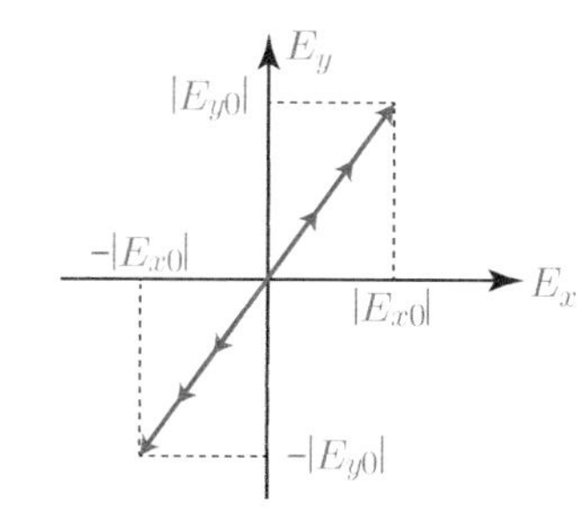

Figure 5.26 A linearly polarized plane wave propagating in the z direction.

which is also a plane wave (with planar wavefronts) propagating in the z direction. The polarization is best viewed in the transverse (phase) plane (on a wavefront), e.g. on the x-y plane if the wave propagates in the $\pm z$ direction. Setting $z = 0$, we have

$$\bar{E}(z=0,t) = (\hat{a}_x E_1 + \hat{a}_y E_2)\cos(\omega t). \tag{5.248}$$

Then as time progresses, we have a sequence as

$$t = 0 \longrightarrow \bar{E}(z=0,t) = (\hat{a}_x E_1 + \hat{a}_y E_2) \tag{5.249}$$

$$t = \pi/(2\omega) \longrightarrow \bar{E}(z=0,t) = 0 \tag{5.250}$$

$$t = \pi/\omega \longrightarrow \bar{E}(z=0,t) = -(\hat{a}_x E_1 + \hat{a}_y E_2) \tag{5.251}$$

$$t = 3\pi/(2\omega) \longrightarrow \bar{E}(z=0,t) = 0 \tag{5.252}$$

$$t = 2\pi/\omega \longrightarrow \bar{E} = (\hat{a}_x E_1 + \hat{a}_y E_2). \tag{5.253}$$

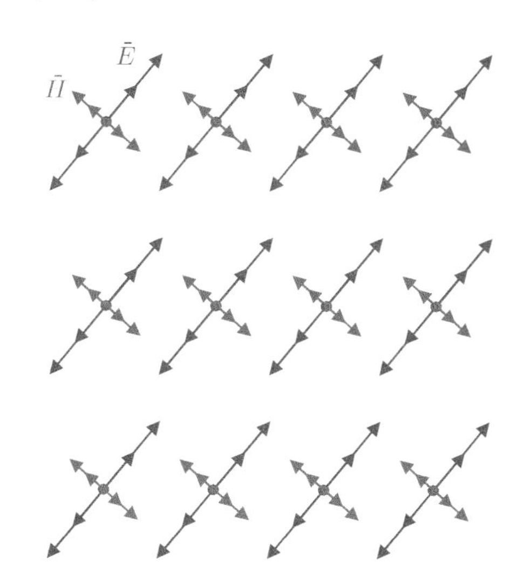

Figure 5.27 A view of a linearly polarized (and uniform) plane wave on a wavefront. At a given location, the tip of the electric field intensity oscillates back and forth on a linear line. Similarly, the tip of the magnetic field intensity oscillates back and forth (in phase) on a perpendicular line. Note that directions/amplitudes are identical on the entire wavefront.

One can further select other instants, e.g. between $t = 0$ and $t = \pi/(2\omega)$, to see that the *tip* of the electric field vector oscillates back and forth on a linear line. This means the plane wave is linearly polarized. The special cases – i.e. an x-polarized plane wave ($E_2 = 0$) and a y-polarized plane wave ($E_1 = 0$) – are also linearly polarized plane waves (Figures 5.26 and 5.27).

Next, we consider the sum of two out-of-phase plane waves with identical amplitudes:

$$\bar{E}_1(\bar{r},t) = \hat{a}_x E_0 \cos(\omega t - kz) \tag{5.254}$$

$$\bar{E}_2(\bar{r},t) = \hat{a}_y E_0 \cos(\omega t - kz - \pi/2), \tag{5.255}$$

as

$$\bar{E}(\bar{r},t) = \bar{E}_1(\bar{r},t) + \bar{E}_2(\bar{r},t) \tag{5.256}$$

$$= \hat{a}_x E_0 \cos(\omega t - kz) + \hat{a}_y E_0 \sin(\omega t - kz). \tag{5.257}$$

This is also a plane wave (with planar wavefronts), but the electric field intensity behaves quite differently in comparison to linearly polarized plane waves. As time progresses on the $z = 0$ plane, we have

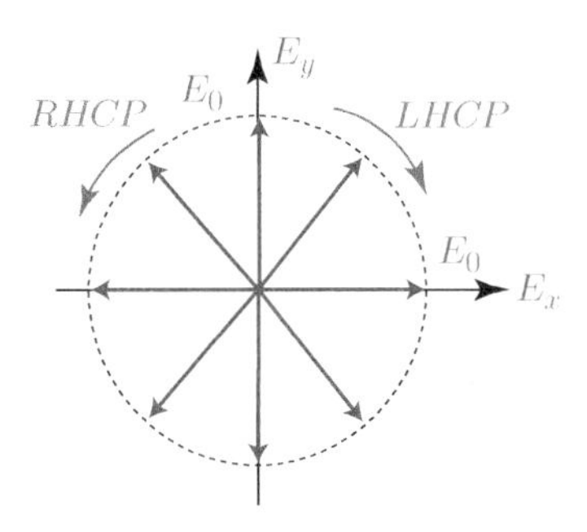

Figure 5.28 Circularly polarized plane waves propagating in the z direction.

$$t = 0 \longrightarrow \bar{E}(\bar{r},t) = \hat{a}_x E_0 \tag{5.258}$$

$$t = \pi/(2\omega) \longrightarrow \bar{E}(\bar{r},t) = \hat{a}_y E_0 \tag{5.259}$$

$$t = \pi/\omega \longrightarrow \bar{E}(\bar{r},t) = -\hat{a}_x E_0 \tag{5.260}$$

$$t = 3\pi/(2\omega) \longrightarrow \bar{E}(\bar{r},t) = -\hat{a}_y E_0 \tag{5.261}$$

$$t = 2\pi/\omega \longrightarrow \bar{E}(\bar{r},t) = \hat{a}_x E_0, \tag{5.262}$$

indicating a *circular* motion. Specifically, the tip of the electric field intensity rotates counterclockwise; this type of plane wave is called a *right-handed circularly polarized* (RHCP) wave (Figure 5.28 and 5.29). Alternatively, if the superposed plane waves are given as

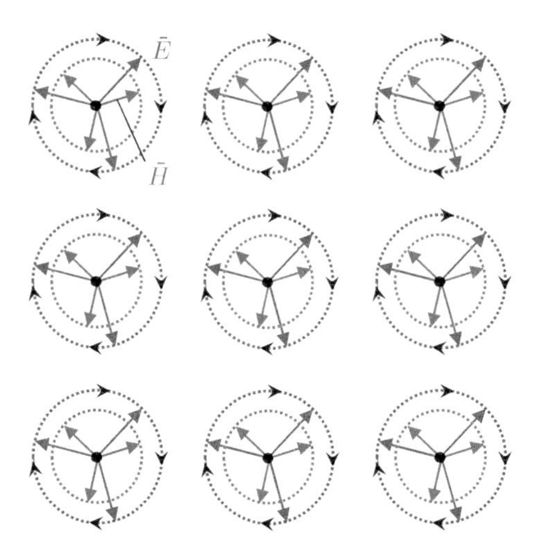

Figure 5.29 A view of an LHCP (and uniform) plane wave on a wavefront. At a given location, the tip of the electric field intensity makes a circular motion with respect to time. Similarly, the tip of the magnetic field intensity makes a circular motion, and there is always 90° angle between the directions of the electric and magnetic fields. Note that directions/amplitudes are identical on the entire wavefront.

$$\bar{E}_1(\bar{r},t) = \hat{a}_x E_0 \cos(\omega t - kz) \tag{5.263}$$

$$\bar{E}_2(\bar{r},t) = \hat{a}_y E_0 \cos(\omega t - kz + \pi/2), \tag{5.264}$$

their sum

$$\bar{E}(\bar{r},t) = \bar{E}_1(\bar{r},t) + \bar{E}_2(\bar{r},t) \tag{5.265}$$

$$= \hat{a}_x E_0 \cos(\omega t - kz) - \hat{a}_y E_0 \sin(\omega t - kz) \tag{5.266}$$

is a *left-handed circularly polarized* (LHCP) plane wave (Figure 5.28). This can be understood by considering the time variance of the electric field intensity, again on the $z = 0$ plane, as

$$t = 0 \longrightarrow \bar{E} = \hat{a}_x E_0 \tag{5.267}$$

$$t = \pi/(2\omega) \longrightarrow \bar{E} = -\hat{a}_y E_0 \tag{5.268}$$

$$t = \pi/\omega \longrightarrow \bar{E} = -\hat{a}_x E_0 \tag{5.269}$$

$$t = 3\pi/(2\omega) \longrightarrow \bar{E} = \hat{a}_y E_0 \tag{5.270}$$

$$t = 2\pi/\omega \longrightarrow \bar{E} = \hat{a}_x E_0, \tag{5.271}$$

indicating a clockwise direction.[60,61,62]

Now we turn our attention to the case when there is a $\pi/2$ phase difference between the x and y components of the electric field intensity (again, considering plane waves propagating in the z direction), while the amplitudes are different. Using

$$\bar{E}_1(\bar{r}, t) = \hat{a}_x E_1 \cos(\omega t - kz) \tag{5.272}$$

$$\bar{E}_2(\bar{r}, t) = \hat{a}_y E_2 \cos(\omega t - kz - \pi/2), \tag{5.273}$$

we have

$$\bar{E}(\bar{r}, t) = \bar{E}_1(\bar{r}, t) + \bar{E}_2(\bar{r}, t) \tag{5.274}$$

$$= \hat{a}_x E_1 \cos(\omega t - kz) + \hat{a}_y E_2 \sin(\omega t - kz). \tag{5.275}$$

In this case, the tip of the electric field intensity rotates on an ellipse, while the direction of the rotation is counterclockwise. Hence, this is a right-handed elliptically polarized (RHEP) plane wave (Figure 5.30). Alternatively, if we have

$$\bar{E}_1(\bar{r}, t) = \hat{a}_x E_1 \cos(\omega t - kz) \tag{5.276}$$

$$\bar{E}_2(\bar{r}, t) = \hat{a}_y E_2 \cos(\omega t - kz + \pi/2), \tag{5.277}$$

then their sum,

$$\bar{E}(\bar{r}, t) = \bar{E}_1(\bar{r}, t) + \bar{E}_2(\bar{r}, t) \tag{5.278}$$

$$= \hat{a}_x E_1 \cos(\omega t - kz) - \hat{a}_y E_2 \sin(\omega t - kz), \tag{5.279}$$

is a left-handed elliptically polarized (LHEP) plane wave (Figure 5.30).

In general, if the phase difference between the components is not $\pi/2$, the result is also an elliptically polarized wave. Specifically, if we have

$$\bar{E}_1(\bar{r}, t) = \hat{a}_x E_1 \cos(\omega t - kz) \tag{5.280}$$

$$\bar{E}_2(\bar{r}, t) = \hat{a}_y E_2 \cos(\omega t - kz + \varphi), \tag{5.281}$$

[60] Note that the magnetic field intensity corresponding to the RHCP and LHCP plane waves described in Eqs. (5.257) and (5.266) can be written

$$\bar{H}(\bar{r}, t) = \frac{1}{\eta}\hat{a}_z \times \hat{a}_x E_0 \cos(\omega t - kz) + \frac{1}{\eta}\hat{a}_z \times \hat{a}_y E_0 \sin(\omega t - kz) = \hat{a}_y \frac{E_0}{\eta}\cos(\omega t - kz) - \hat{a}_x \frac{E_0}{\eta}\sin(\omega t - kz)$$

and

$$\bar{H}(\bar{r}, t) = \hat{a}_y \frac{E_0}{\eta}\cos(\omega t - kz) + \hat{a}_x \frac{E_0}{\eta}\sin(\omega t - kz),$$

respectively.

[61] In many cases (especially for elliptically polarized waves), it may not be trivial to decide whether the wave is right-handed or left-handed. A formal approach can consider the electric field when $t_0 = 0$ (or any value) and $t_1 = \pi/(2\omega)$ (time progress). Then a positive value of the product $\bar{k} \cdot \bar{E}(t_0) \times \bar{E}(t_1)$ indicates a right-handed wave, while a negative value of the same expression indicates a left-handed wave.

[62] A circularly polarized plane wave is a superposition of two linearly polarized plane waves, and a linearly polarized plane wave is a superposition of two circularly polarized plane waves.

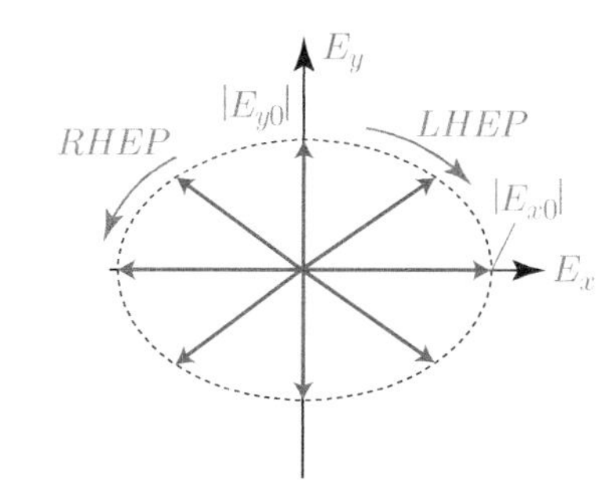

Figure 5.30 Elliptically polarized plane waves propagating in the z direction.

their sum is

$$\bar{E}(\bar{r},t) = \bar{E}_1(\bar{r},t) + \bar{E}_2(\bar{r},t) \tag{5.282}$$

$$= \hat{a}_x E_1 \cos(\omega t - kz) + \hat{a}_y E_2 \cos(\omega t + kz + \varphi). \tag{5.283}$$

Then on the $z = 0$ plane, we obtain

$$\bar{E}(z=0,t) = \hat{a}_x E_1 \cos(\omega t) + \hat{a}_y E_2 \cos(\omega t + \varphi). \tag{5.284}$$

For $\varphi \neq 0$ and $\varphi \neq \pm\pi/2$, the tip of the electric field intensity rotates on a tilted ellipse as time progresses.

At this stage, we now consider a general phasor-domain expression for a plane wave propagating in the z direction

$$\bar{E}(\bar{r}) = (\hat{a}_x E_{x0} + \hat{a}_y E_{y0}) \exp(-jkz), \tag{5.285}$$

where E_{x0} and E_{y0} are complex numbers. At $z = 0$, we further have

$$\bar{E}(z=0) = \hat{a}_x E_{x0} + \hat{a}_y E_{y0}, \tag{5.286}$$

where

$$E_{x0} = |E_{x0}| \exp(j\varphi_{x0}), \qquad E_{y0} = |E_{y0}| \exp(j\varphi_{y0}), \tag{5.287}$$

leading to

$$\bar{E}(z=0) = \hat{a}_x |E_{x0}| \exp(j\varphi_{x0}) + \hat{a}_y |E_{y0}| \exp(j\varphi_{y0}). \tag{5.288}$$

Then in the time domain, we obtain

$$\bar{E}(z=0,t) = \hat{a}_x |E_{x0}| \cos(\omega t + \varphi_{x0}) + \hat{a}_y |E_{y0}| \cos(\omega t + \varphi_{y0}) \tag{5.289}$$

or

$$E_x(z=0,t) = |E_{x0}| \cos(\omega t + \varphi_{x0}) \tag{5.290}$$

$$E_y(z=0,t) = |E_{y0}| \cos(\omega t + \varphi_{y0}) \tag{5.291}$$

$$\bar{E}(z=0,t) = \hat{a}_x E_x(z=0,t) + \hat{a}_y E_y(z=0,t). \tag{5.292}$$

The polarization of the plane wave can be considered the vector $\bar{E}(z=0,t)$ with respect to time, which can be categorized as follows:

- If $\varphi_{x0} = \varphi_0$ and $\varphi_{y0} = \varphi_0$ or $\varphi_{y0} = \varphi_0 \pm \pi$, then we have

$$\bar{E}(z=0,t) = \hat{a}_x |E_{x0}| \cos(\omega t + \varphi_0) \pm \hat{a}_y |E_{y0}| \cos(\omega t + \varphi_0) \tag{5.293}$$

$$= \left[\hat{a}_x |E_{x0}| \pm \hat{a}_y |E_{y0}|\right] \cos(\omega t + \varphi_0). \tag{5.294}$$

As time progresses, $\bar{E}(z=0,t)$ moves in a line with a slope of $\pm|E_{y0}|/|E_{x0}|$, indicating linear polarization for any values of $|E_{x0}|$ and $|E_{y0}|$ (Figure 5.26).

- If $\varphi_{x0} = \varphi_0$, $\varphi_{y0} = \varphi_0 \pm \pi/2$, and $|E_{x0}| = |E_{y0}| = |E_0|$, then we have[63]

$$E_x(z=0,t) = |E_0| \cos(\omega t + \varphi_0) \tag{5.295}$$

$$E_y(z=0,t) = |E_0| \cos(\omega t + \varphi_0 \pm \pi/2) = \mp |E_0| \sin(\omega t + \varphi_0) \tag{5.296}$$

$$\bar{E}(z=0,t) = \hat{a}_x E_x(z=0,t) + \hat{a}_y E_y(z=0,t). \tag{5.297}$$

This means

$$[E_x(z=0,t)]^2 + [E_y(z=0,t)]^2 = |E_0|^2, \tag{5.298}$$

indicating a circle of radius $|E_0|$. Hence, if the components (x and y) have the same amplitude, and there is a $\pi/2$ phase difference between the components, then $\bar{E}(z=0,t)$ moves in a circle as time progresses, representing circular polarization (Figure 5.28).

- If the phase difference is $\pi/2$, while the components do not have the same amplitude, we obtain

$$E_x(z=0,t) = |E_{x0}| \cos(\omega t + \varphi_0) \tag{5.299}$$

$$E_y(z=0,t) = |E_{y0}| \cos(\omega t + \varphi_0 \pm \pi/2) = \mp |E_{y0}| \sin(\omega t + \varphi_0) \tag{5.300}$$

$$\bar{E}(z=0,t) = \hat{a}_x E_x(z=0,t) + \hat{a}_y E_y(z=0,t). \tag{5.301}$$

We have

$$\left(\frac{E_x(z=0,t)}{|E_{x0}|}\right)^2 + \left(\frac{E_y(z=0,t)}{|E_{y0}|}\right)^2 = 1, \tag{5.302}$$

which represents an ellipse. Hence, as time progresses, $\bar{E}(z=0,t)$ moves in an ellipse, passing through $(x,y) = (\pm|E_{x0}|, 0)$ and $(x,y) = (0, \pm|E_{y0}|)$ (Figure 5.30).

- If the components have the same amplitude, but they have a phase difference other than $\pi/2$, we have

$$E_x(z=0,t) = |E_0| \cos(\omega t + \varphi_{x0}) \tag{5.303}$$

$$E_y(z=0,t) = |E_0| \cos(\omega t + \varphi_{y0}) \tag{5.304}$$

$$\bar{E}(z=0,t) = \hat{a}_x E_x(z=0,t) + \hat{a}_y E_y(z=0,t). \tag{5.305}$$

[63]Obviously, the sense of the polarization depends on the sign of $\pi/2$. We have

$$\bar{E}(z,t) = \hat{a}_x |E_0| \cos(\omega t - kz + \varphi_0) + \hat{a}_y |E_0| \sin(\omega t - kz + \varphi_0)$$

for an RHCP wave and

$$\bar{E}(z,t) = \hat{a}_x |E_0| \cos(\omega t - kz + \varphi_0) - \hat{a}_y |E_0| \sin(\omega t - kz + \varphi_0)$$

for an LHCP wave. Note that the sum of two circularly polarized plane waves can lead to a linearly polarized wave. With equal amplitudes/phases and opposite polarization senses, we have

$$\begin{aligned} \bar{E}(z,t) &= \bar{E}_1(z,t) + \bar{E}_2(z,t) \\ &= \hat{a}_x |E_0| \cos(\omega t - kz + \varphi_0) \\ &\quad + \hat{a}_y |E_0| \sin(\omega t - kz + \varphi_0) \\ &\quad + \hat{a}_x |E_0| \cos(\omega t - kz + \varphi_0) \\ &\quad - \hat{a}_y |E_0| \sin(\omega t - kz + \varphi_0) \\ &= \hat{a}_x 2|E_0| \cos(\omega t - kz + \varphi_0), \end{aligned}$$

which represents a linearly polarized plane wave.

While these are not as trivial as the expressions above, the electric field intensity $\bar{E}(z=0,t)$ with these components also follows an elliptical path, while the ellipse is tilted.[64]

[64]The amplitude of the electric field intensity can be written

$$|\bar{E}(z=0,t)|^2 = |E_0|^2\cos^2(\omega t+\varphi_{x0}) + |E_0|^2\cos^2(\omega t+\varphi_{y0}),$$

which becomes maximum/minimum when

$$\cos(\omega t+\varphi_{x0})\sin(\omega t+\varphi_{x0}) + \cos(\omega t+\varphi_{y0})\sin(\omega t+\varphi_{y0}) = 0 \longrightarrow \sin(2\omega t+2\varphi_{x0}) = \sin(-2\omega t-2\varphi_{y0})$$

or

$$2\omega t+2\varphi_{x0} = -2\omega t-2\varphi_{y0}\pm n2\pi \longrightarrow \omega t = \pm n\frac{\pi}{2} - \frac{\varphi_{x0}+\varphi_{y0}}{2},$$

where n is an integer.

5.3.3 Examples

Example 55: Consider the following expressions for the electric field intensity related to three different (uniform) plane waves, where E_0 is a real constant:

- $\bar{E}_1(\bar{r},t) = \hat{a}_x E_0\cos(\omega t-kz+\pi/3) - \hat{a}_y E_0\sin(\omega t-kz-\pi/6)$
- $\bar{E}_2(\bar{r}) = (\hat{a}_y+\hat{a}_z j2)E_0\exp(-jkx)$
- $\bar{E}_3(\bar{r}) = \left[\hat{a}_z\exp(-j\pi/8)+\hat{a}_y\exp(j3\pi/8)\right]E_0\exp(-jkx)$

Find the type and sense of the polarization for each case.

Solution: For the first plane wave, we have

$$\bar{E}_1(\bar{r},t) = \hat{a}_x E_0\cos(\omega t-kz+\pi/3) - \hat{a}_y E_0\cos(\omega t-kz-\pi/6-\pi/2) \tag{5.306}$$

$$= \hat{a}_x E_0\cos(\omega t-kz+\pi/3) + \hat{a}_y E_0\cos(\omega t-kz-2\pi/3) \tag{5.307}$$

$$= (\hat{a}_x-\hat{a}_y)E_0\cos(\omega t-kz+\pi/3). \tag{5.308}$$

Hence, this is a linearly polarized plane wave. For the second plane wave, we can first find the time-domain representation as[65]

$$\bar{E}_2(\bar{r},t) = \hat{a}_y E_0\cos(\omega t-kx) + \hat{a}_z 2E_0\cos(\omega t-kx+\pi/2). \tag{5.309}$$

[65]We have

$$\begin{aligned}\bar{E}_2(\bar{r},t) &= \mathrm{Re}\{\bar{E}_2(\bar{r})\exp(j\omega t)\}\\ &= \mathrm{Re}\{(\hat{a}_y+\hat{a}_z j2)E_0\exp(-jkx)\exp(j\omega t)\}\\ &= \mathrm{Re}\{\hat{a}_y E_0\exp(j\omega t-jkx)\} + \mathrm{Re}\{\hat{a}_z j2E_0\exp(j\omega t-jkx)\}\\ &= \hat{a}_y E_0\cos(\omega t-kx) - \hat{a}_z 2E_0\sin(\omega t-kx)\\ &= \hat{a}_y E_0\cos(\omega t-kx) + \hat{a}_z 2E_0\cos(\omega t-kx+\pi/2).\end{aligned}$$

Therefore, for this plane wave that propagates in the x direction, the components (y and z) have a $\pi/2$ phase difference, while their amplitudes are not equal. Consequently, it is an elliptically polarized plane wave. Investigating the time dependence,[66] one can further find that the wave is left-handed (overall, it is a LHEP plane wave). Finally, for the third plane wave, we have the time-domain representation of the electric field intensity as

[66]When $x=0$ (on a wavefront), we have

$$\bar{E}_2(x=0,t) = \hat{a}_y E_0\cos(\omega t) + \hat{a}_z 2E_0\cos(\omega t+\pi/2).$$

Then at $t_0=0$ and $t_1=\pi/(2\omega)$, we obtain

$$\bar{E}_2(x=0,t_0) = \hat{a}_y E_0$$
$$\bar{E}_2(x=0,t_1) = -\hat{a}_z 2E_0,$$

leading to

$$\bar{k}\cdot\bar{E}(x=0,t_0)\times\bar{E}(x=0,t_1) = -\hat{a}_x\cdot\hat{a}_x 2kE_0^2 < 0,$$

indicating that the wave is left-handed.

$$\bar{E}_3(\bar{r},t) = \hat{a}_y E_0\cos(\omega t-kx+3\pi/8) + \hat{a}_z E_0\cos(\omega t-kx-\pi/8). \tag{5.310}$$

For this plane wave, which also travels in the x direction, the components (y and z) have equal amplitudes and $\pi/2$ phase difference, indicating circular polarization. Given time instants $t_0=0$

and $t_1 = \pi/(2\omega)$ at $x_0 = -\pi/(8k)$, we have

$$\bar{E}_3(x_0, t_0) \times \bar{E}_3(x_0, t_1) = \hat{a}_z E_0 \times (-\hat{a}_y E_0) = \hat{a}_x E_0^2, \tag{5.311}$$

which is in the propagation direction. Consequently, this is an RHCP plane wave.

5.3.4 Power of Plane Waves

Now, we focus on the energy carried by plane waves. As an example, we consider again a plane wave propagating in the z direction with the electric field polarized in the x direction (linear polarization). We have[67]

[67] For simplicity, we assume that there is no additional phase term.

$$\bar{E}(\bar{r}, t) = \hat{a}_x E_0 \cos(\omega t - kz), \qquad \bar{H}(\bar{r}, t) = \hat{a}_y \frac{E_0}{\eta} \cos(\omega t - kz) \tag{5.312}$$

and

$$\bar{S}(\bar{r}, t) = \hat{a}_z \frac{E_0^2}{\eta} \cos^2(\omega t - kz) = \hat{a}_z \frac{E_0^2}{2\eta} \left[\cos(2\omega t - 2kz) + 1\right]. \tag{5.313}$$

Hence, the Poynting vector $\bar{S}(\bar{r}, t)$ represents a time-harmonic *wave* with an angular frequency of 2ω, while its direction indicates that the power flow is in the z direction (Figure 5.31), as expected. The time-average power flux density can be found as

$$\bar{S}_{\text{avg}}(\bar{r}) = \hat{a}_z \frac{E_0^2}{2\eta} \frac{\omega}{\pi} \int_0^{\pi/\omega} \left[\cos(2\omega t - 2\beta z) + 1\right] dt \tag{5.314}$$

$$= \hat{a}_z \frac{E_0^2}{2\eta} \frac{\omega}{\pi} \left[\frac{\sin(2\omega t - 2\beta z)}{2\omega} + t\right]_0^{\pi/\omega} \tag{5.315}$$

$$= \hat{a}_z \frac{E_0^2}{2\eta} \frac{\omega}{\pi} \left[\frac{\sin(2\pi - 2\beta z)}{2\omega} + \frac{\pi}{\omega} - \frac{\sin(-2\beta z)}{2\omega}\right] = \hat{a}_z \frac{E_0^2}{2\eta}. \tag{5.316}$$

Therefore, although it oscillates with respect to time, becoming E_0^2/η (maximum) and zero (minimum) periodically, there is a net power flow in the z direction at *any* position in space.

Investigating the power in the phasor domain, we can find the complex Poynting vector as

$$\bar{S}_c(\bar{r}) = \frac{1}{2}\bar{E}(\bar{r}) \times [\bar{H}(\bar{r})]^* = \frac{1}{2}\hat{a}_x E_0 \exp(-jkz) \times \hat{a}_y \frac{E_0}{\eta} \exp(jkz) = \hat{a}_z \frac{E_0^2}{2\eta}. \tag{5.317}$$

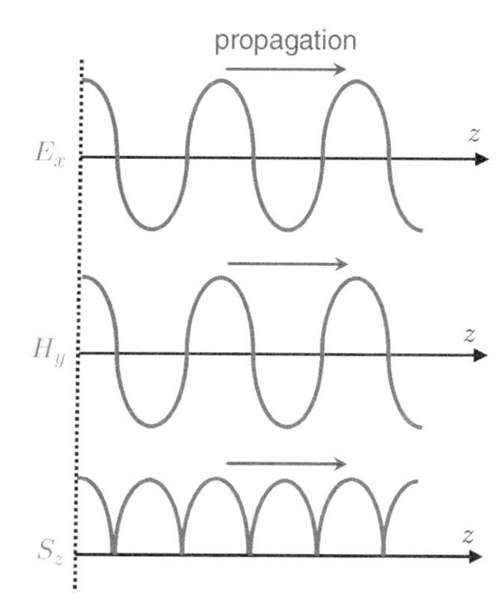

Figure 5.31 The Poynting vector (instantaneous power flux density) for a linearly polarized plane wave represents a time-harmonic wave propagating in the direction of the plane wave (z in this example). Note that the Poynting vector is always in the z direction (positive value): i.e. it does not change to the $-z$ direction at any time/position. At a fixed location, the instantaneous power flux density increases (up to E_0^2/η) and decreases (down to zero) with respect to time with an angular frequency of 2ω.

As we expect, we have

$$\text{Re}\{\bar{S}_c(\bar{r})\} = \hat{a}_z \frac{E_0^2}{2\eta} = S_{\text{avg}}(\bar{r}) \tag{5.318}$$

$$\text{Im}\{\bar{S}_c(\bar{r})\} = 0, \tag{5.319}$$

i.e. the complex Poynting vector is purely real.[68]

[68] This means there is no oscillatory power.

As a further illustration, we know consider an $a \times a \times a$ cubic box centered around the origin. The net outward power flux can be found as[69]

[69] We have

$$p_{em}(t) = \int_{\text{top}} \hat{a}_z \frac{E_0^2}{2\eta} [\cos(2\omega t - 2kz) + 1] \cdot \hat{a}_z dxdy + \int_{\text{bottom}} \hat{a}_z \frac{E_0^2}{2\eta} [\cos(2\omega t - 2kz) + 1] \cdot (-\hat{a}_z) dxdy.$$

$$p_{em}(t) = \oint_S \bar{S}(\bar{r}, t) \cdot \overline{ds} = \oint_S \hat{a}_z \frac{E_0^2}{2\eta} [\cos(2\omega t - 2kz) + 1] \cdot \overline{ds} \tag{5.320}$$

$$= \frac{E_0^2 a^2}{2\eta} [\cos(2\omega t - ka) + 1] - \frac{E_0^2 a^2}{2\eta} [\cos(2\omega t + ka) + 1] \tag{5.321}$$

$$= \frac{E_0^2 a^2}{2\eta} [\cos(2\omega t - ka) - \cos(2\omega t + ka)] = \frac{E_0^2 a^2}{\eta} \sin(2\omega t) \sin(ka). \tag{5.322}$$

Hence, if the dimensions are not selected particularly as $a = n\pi/k = n\lambda/2$ for $n = 1, 2, \ldots$, then there is an imbalance in inward and outward power flows with respect to time.[70] The electric potential energy contained inside the cube at a particular time can be found as[71]

[70] Indicating that energy (in the form of the electric/magnetic potential energy) flows into or out of the cubic region.

[71] We have

$$w_{e,\text{stored}}(t) = \frac{\varepsilon E_0^2}{2} \int_V [\cos(2\omega t - 2kz) + 1] dv = \frac{\varepsilon E_0^2 a^2}{2} \int_{-a/2}^{a/2} [\cos(2\omega t - 2kz) + 1] dz = \frac{\varepsilon E_0^2 a^2}{2} \left[z - \frac{\sin(2\omega t - 2kz)}{2k} \right]_{-a/2}^{a/2}.$$

$$w_{e,\text{stored}}(t) = \varepsilon \int_V |\bar{E}(\bar{r}, t)|^2 dv = \frac{\varepsilon}{2} E_0^2 \int_V \cos^2(\omega t - kz) dv \tag{5.323}$$

$$= \frac{\varepsilon E_0^2 a^2}{2} \left[a - \frac{\sin(2\omega t - ka)}{2k} + \frac{\sin(2\omega t + ka)}{2k} \right]. \tag{5.324}$$

Similarly, the contained magnetic potential energy can be obtained as[72]

[72] Hence, if the dimensions are not selected particularly as $a = n\pi/k = n\lambda/2$ for $n = 1, 2, \ldots$, the electric potential energy and magnetic potential energy contained by the cube oscillate with respect to time.

$$w_{m,\text{stored}}(t) = \mu \int_V |\bar{H}(\bar{r}, t)|^2 dv = \frac{\mu E_0^2}{2\eta^2} \int_V \cos^2(\omega t - kz) dv = w_{e,\text{stored}}(t). \tag{5.325}$$

Consequently, the corresponding power expressions can be written

$$p_{e,\text{stored}}(t) = \frac{\partial w_{e,\text{stored}}(t)}{\partial t} = \frac{\varepsilon E_0^2 a^2 \omega}{4k} [-\cos(2\omega t - ka) + \cos(2\omega t + ka)] \tag{5.326}$$

$$= -\frac{\varepsilon E_0^2 a^2 \omega}{2k} \sin(2\omega t) \sin(ka) = -\frac{E_0^2 a^2}{2\eta} \sin(2\omega t) \sin(ka) \tag{5.327}$$

and

$$p_{m,\text{stored}}(t) = \frac{\partial w_{m,\text{stored}}(t)}{\partial t} = -\frac{E_0^2 a^2}{2\eta}\sin(2\omega t)\sin(ka). \tag{5.328}$$

Obviously, the conservation of energy perfectly holds as

$$p_{em}(t) = -\frac{\partial w_{m,\text{stored}}(t)}{\partial t} - \frac{\partial w_{e,\text{stored}}(t)}{\partial t} = \frac{E_0^2 a^2}{\eta}\sin(2\omega t)\sin(ka). \tag{5.329}$$

When a plane wave is linearly polarized, the value of the instantaneous power flux density oscillates with respect to time (for a fixed position) and with respect to position (for a fixed time). This is not true for all plane waves. As an example, we consider a circularly polarized plane wave propagating in the z direction. When the wave is RHCP, we have

$$\bar{E}(\bar{r},t) = \hat{a}_x E_0\cos(\omega t - kz) + \hat{a}_y E_0 \sin(\omega t - kz) \tag{5.330}$$

$$\bar{H}(\bar{r},t) = \hat{a}_y \frac{E_0}{\eta}\cos(\omega t - kz) - \hat{a}_x \frac{E_0}{\eta}\sin(\omega t - kz). \tag{5.331}$$

Then the Poynting vector can be found as[73]

$$\bar{S}(\bar{r},t) = \bar{E}(\bar{r},t) \times \bar{H}(\bar{r},t) \tag{5.332}$$

$$= \hat{a}_z \frac{E_0^2}{\eta}\cos^2(\omega t - kz) + \hat{a}_z \frac{E_0^2}{\eta}\sin^2(\omega t - kz) = \hat{a}_z \frac{E_0^2}{\eta}. \tag{5.333}$$

Interestingly, the instantaneous power flux density depends on neither position nor time.[74] In the phasor domain, we have

$$\bar{E}(\bar{r}) = (\hat{a}_x - j\hat{a}_y)E_0\exp(-jkz) \tag{5.334}$$

$$\bar{H}(\bar{r}) = (\hat{a}_y + j\hat{a}_x)\frac{E_0}{\eta}\exp(-jkz). \tag{5.335}$$

Hence, the complex Poynting vector can be found as

$$\bar{S}_c(\bar{r}) = \frac{1}{2}\bar{E}(\bar{r}) \times [\bar{H}(\bar{r})]^* = [(\hat{a}_x - j\hat{a}_y) \times (\hat{a}_y - j\hat{a}_x)]\frac{E_0^2}{\eta} = \hat{a}_z \frac{E_0^2}{\eta}, \tag{5.336}$$

which is purely real and the same as the time-average Poynting vector (simply the instantaneous power flux density itself, as it does not change with respect to time).

[73]We gain more insight if we consider that the circularly polarized plane wave is a combination of two linearly polarized plane waves. For the first linearly polarized plane wave, we have

$$\bar{E}_1(\bar{r},t) = \hat{a}_x E_0 \cos(\omega t - kz)$$
$$\bar{H}_1(\bar{r},t) = \hat{a}_y (E_0/\eta)\cos(\omega t - kz)$$
$$\bar{S}_1(\bar{r},t) = \hat{a}_z (E_0^2/\eta)\cos^2(\omega t - kz).$$

Similarly, for the second linearly polarized plane wave, we can obtain

$$\bar{E}_2(\bar{r},t) = \hat{a}_y E_0 \sin(\omega t - kz)$$
$$\bar{H}_2(\bar{r},t) = -\hat{a}_x (E_0/\eta)\sin(\omega t - kz)$$
$$\bar{S}_2(\bar{r},t) = \hat{a}_z (E_0^2/\eta)\sin^2(\omega t - kz).$$

Hence, these two linearly polarized plane waves simultaneously carry energy with oscillatory instantaneous power flux density values (with respect to both position and time). When they are superposed, however, we obtain the total instantaneous power flux density as

$$\bar{S}(\bar{r},t) = \bar{S}_1(\bar{r},t) + \bar{S}_2(\bar{r},t)$$
$$= \hat{a}_z \frac{E_0^2}{\eta}\cos^2(\omega t - kz) + \hat{a}_z \frac{E_0^2}{\eta}\sin^2(\omega t - kz)$$
$$= \hat{a}_z \frac{E_0^2}{\eta},$$

which is independent of position and time.

[74]Hence, at a fixed position, the instantaneous power flux density is continuously $\bar{S}(\bar{r},t) = \hat{a}_z E_0^2/\eta$.

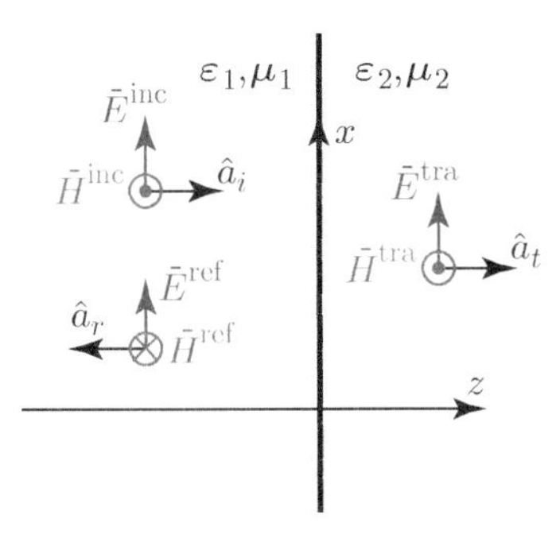

Figure 5.32 Normal incidence of a plane wave on a boundary between two media.

5.3.5 Reflection and Refraction of Plane Waves

Like all electromagnetic waves, plane waves interact with matter. For example, at a boundary between two media, reflection and refraction occur depending on electrical and magnetic properties. Investigating plane waves in ideal cases can be very illustrative to understand these phenomena in general cases. In the following, we consider plane waves and infinitely large planar boundaries to see how reflection and refraction can be formulated.

First, we consider normal incidence: i.e. when the incident plane wave hits the infinitely large boundary normally (Figure 5.32). We assume that the boundary is located at $z = 0$, and it separates the first medium (ε_1, μ_1) defined as $z < 0$ and the second medium (ε_2, μ_2) defined as $z > 0$. In addition to the incident plane wave, reflected and transmitted plane waves are generated, depending on the medium parameters. In general, we can write the incident, reflected, and transmitted waves in the phasor domain as[75]

$$\bar{E}^{\text{inc}}(\bar{r}) = \hat{a}_x E_0^{\text{inc}} \exp(-jk_1 z), \qquad \bar{H}^{\text{inc}}(\bar{r}) = \hat{a}_y \frac{E_0^{\text{inc}}}{\eta_1} \exp(-jk_1 z), \tag{5.337}$$

$$\bar{E}^{\text{ref}}(\bar{r}) = \hat{a}_x E_0^{\text{ref}} \exp(jk_1 z), \qquad \bar{H}^{\text{ref}}(\bar{r}) = -\hat{a}_y \frac{E_0^{\text{ref}}}{\eta_1} \exp(jk_1 z), \tag{5.338}$$

$$\bar{E}^{\text{tra}}(\bar{r}) = \hat{a}_x E_0^{\text{tra}} \exp(-jk_2 z), \qquad \bar{H}^{\text{tra}}(\bar{r}) = \hat{a}_y \frac{E_0^{\text{tra}}}{\eta_2} \exp(-jk_2 z), \tag{5.339}$$

respectively. To relate the electric and magnetic fields to each other, we need the boundary conditions. Assuming no surface electric current density at the boundary,[76] the tangential electric field intensity and tangential magnetic field intensity must be continuous.[77] Then we must have

$$\bar{E}^{\text{inc}}(z = 0) + \bar{E}^{\text{ref}}(z = 0) = \bar{E}^{\text{tra}}(z = 0) \tag{5.340}$$

$$\bar{H}^{\text{inc}}(z = 0) + \bar{H}^{\text{ref}}(z = 0) = \bar{H}^{\text{tra}}(z = 0), \tag{5.341}$$

given that both electric and magnetic fields are tangential to the boundary.[78] This leads to

$$E_0^{\text{inc}} + E_0^{\text{ref}} = E_0^{\text{tra}} \tag{5.342}$$

$$E_0^{\text{inc}}/\eta_1 - E_0^{\text{ref}}/\eta_1 = E_0^{\text{tra}}/\eta_2. \tag{5.343}$$

Setting $E_0^{\text{inc}} = E_0$, we can find others in terms of E_0:

$$E_0^{\text{ref}} = \frac{\eta_2 - \eta_1}{\eta_1 + \eta_2} E_0 \tag{5.344}$$

$$E_0^{\text{tra}} = \frac{2\eta_2}{\eta_1 + \eta_2} E_0. \tag{5.345}$$

[75] At this stage, directions of electric and magnetic fields are selected arbitrarily as the boundary conditions reveal the exact directions. It is only required to be consistent in the initial definition of directions and the corresponding formulation.

[76] If not particularly defined as *impressed* sources, the surface electric current density between two penetrable media is zero.

[77] The boundary conditions also explain why the reflected and transmitted waves are also plane waves when the incident wave is a plane wave.

[78] The electric and magnetic fields are in the x and y directions, respectively, while the boundary normal is in the $\pm z$ direction.

At this stage, we define reflection and transmission coefficients as[79]

$$\Gamma = \frac{\eta_2 - \eta_1}{\eta_2 + \eta_1} \tag{5.346}$$

$$\tau = 1 + \Gamma = \frac{2\eta_2}{\eta_2 + \eta_1} \tag{5.347}$$

so that the reflected and transmitted waves can be written

$$\bar{E}^{\text{ref}}(\bar{r}) = \hat{a}_x \Gamma E_0 \exp(jk_1 z) \tag{5.348}$$

$$\bar{H}^{\text{ref}}(\bar{r}) = -\hat{a}_y \Gamma (E_0/\eta_1) \exp(jk_1 z) \tag{5.349}$$

$$\bar{E}^{\text{tra}}(\bar{r}) = \hat{a}_x \tau E_0 \exp(-jk_2 z) \tag{5.350}$$

$$\bar{H}^{\text{tra}}(\bar{r}) = \hat{a}_y \tau (E_0/\eta_2) \exp(-jk_2 z). \tag{5.351}$$

In addition, expressions for the total fields in the first medium can be obtained as[80]

$$\bar{E}_1(\bar{r}) = \hat{a}_x E_0 \exp(-jk_1 z) + \hat{a}_x \Gamma E_0 \exp(jk_1 z) \tag{5.352}$$

$$\bar{H}_1(\bar{r}) = \hat{a}_y (E_0/\eta_1) \exp(-jk_1 z) - \hat{a}_y \Gamma (E_0/\eta_1) \exp(jk_1 z). \tag{5.353}$$

Assuming real E_0, we have time-domain expressions

$$\bar{E}_1(\bar{r}, t) = \hat{a}_x E_0 \cos(\omega t - k_1 z) + \hat{a}_x \Gamma E_0 \cos(\omega t + k_1 z) \tag{5.354}$$

$$\bar{H}_1(\bar{r}, t) = \hat{a}_y (E_0/\eta_1) \cos(\omega t - jk_1 z) - \hat{a}_y \Gamma (E_0/\eta_1) \cos(\omega t + k_1 z). \tag{5.355}$$

In the transmission region, the total fields are simply the transmitted fields

$$\bar{E}_2(\bar{r}) = \hat{a}_x \tau E_0 \exp(-jk_2 z) \tag{5.356}$$

$$\bar{H}_2(\bar{r}) = \hat{a}_y \tau (E_0/\eta_2) \exp(-jk_2 z) \tag{5.357}$$

in the phasor domain and

$$\bar{E}_2(\bar{r}, t) = \hat{a}_x \tau E_0 \cos(\omega t - k_2 z) \tag{5.358}$$

$$\bar{H}_2(\bar{r}, t) = \hat{a}_y \tau (E_0/\eta_2) \cos(\omega t - k_2 z) \tag{5.359}$$

in the time domain.

[79] Note the special cases $\Gamma = 0$ and $\tau = 1$ if $\eta_2 = \eta_1$ (impedance matching) and $\Gamma = -1$ and $\tau = 0$ if $\eta_2 = 0$ (second medium is perfectly conducting).

[80] Note that the total fields can be decomposed into propagating and standing waves. As an example, for the electric field intensity, we have

$$\begin{aligned}\bar{E}_1(\bar{r}) &= \hat{a}_x E_0 \exp(-jk_1 z) \\ &\quad - \hat{a}_x E_0 \exp(jk_1 z) + \hat{a}_x (1+\Gamma) E_0 \exp(jk_1 z) \\ &= -\hat{a}_x j2E_0 \sin(k_1 z) + \hat{a}_x (1+\Gamma) E_0 \exp(jk_1 z),\end{aligned}$$

leading to

$$\begin{aligned}\bar{E}_1(\bar{r}, t) &= \hat{a}_x 2E_0 \sin(k_1 z) \sin(\omega t) \\ &\quad + \hat{a}_x (1+\Gamma) E_0 \cos(\omega t + k_1 z),\end{aligned}$$

in the time domain. On the right-hand side above, the first term represents a standing wave, while the second term represents a propagating wave in the $-z$ direction. Alternatively, we have

$$\begin{aligned}\bar{E}_1(\bar{r}) &= \hat{a}_x (1+\Gamma) E_0 \exp(-jk_1 z) \\ &\quad - \hat{a}_x \Gamma E_0 \exp(-jk_1 z) + \hat{a}_x \Gamma E_0 \exp(jk_1 z) \\ &= \hat{a}_x (1+\Gamma) E_0 \exp(-jk_1 z) + \hat{a}_x \Gamma j2E_0 \sin(k_1 z)\end{aligned}$$

and

$$\begin{aligned}\bar{E}_1(\bar{r}, t) &= \hat{a}_x (1+\Gamma) E_0 \cos(\omega t - k_1 z) \\ &\quad - \hat{a}_x \Gamma 2E_0 \sin(\omega t) \sin(k_1 z).\end{aligned}$$

Now, on the right-hand side, we have a propagating wave in the z direction and a standing wave.

Next, we can consider electromagnetic energy and power. The complex Poynting vector in the first medium can be written[81]

$$\bar{S}_{c,1}(\bar{r}) = \frac{1}{2}\bar{E}_1(\bar{r}) \times [\bar{H}_1(\bar{r})]^* \tag{5.360}$$

$$= \hat{a}_z\frac{E_0^2}{2\eta_1} - \hat{a}_z\Gamma\frac{E_0^2}{2\eta_1}\exp(-j2k_1z) + \hat{a}_z\Gamma\frac{E_0^2}{2\eta_1}\exp(j2k_1z) - \hat{a}_z\Gamma^2\frac{E_0^2}{2\eta_1} \tag{5.361}$$

$$= \hat{a}_z\frac{E_0^2}{2\eta_1}(1-\Gamma^2) + \hat{a}_z\Gamma\frac{E_0^2}{2\eta_1}j2\sin(2k_1z). \tag{5.362}$$

[81] We have

$$\bar{S}_{c,1}(\bar{r}) = \frac{1}{2}\bar{E}_1(\bar{r}) \times [\bar{H}_1(\bar{r})]^*$$
$$= \frac{1}{2}[\hat{a}_xE_0\exp(-jk_1z) + \hat{a}_x\Gamma E_0\exp(jk_1z)]$$
$$\times\left[\hat{a}_y\frac{E_0}{\eta_1}\exp(-jk_1z) - \hat{a}_y\Gamma\frac{E_0}{\eta_1}\exp(jk_1z)\right]^*$$
$$= \frac{1}{2}[\hat{a}_xE_0\exp(-jk_1z) + \hat{a}_x\Gamma E_0\exp(jk_1z)]$$
$$\times\left[\hat{a}_y\frac{E_0}{\eta_1}\exp(jk_1z) - \hat{a}_y\Gamma\frac{E_0}{\eta_1}\exp(-jk_1z)\right].$$

Obviously, the complex power flux density (Poynting vector) has both real and imaginary (reactive) parts. The real part corresponds to the propagating power, since

$$\bar{S}_{\text{avg},1}(\bar{r}) = \text{Re}\{\bar{S}_{c,1}(\bar{r})\} = \hat{a}_z\frac{E_0^2}{2\eta_1}(1-\Gamma^2), \tag{5.363}$$

which is nonzero[82] if $\Gamma \neq \pm 1$ (if there is no total reflection). The reactive power corresponds to the oscillatory power as a result of the reflection: i.e. when $\Gamma \neq 0$. In the second medium, we have

[82] Specifically, if some nonzero power is transmitted into the second medium, then there should be a net power flow toward the boundary in the first medium.

$$\bar{S}_{c,2}(\bar{r}) = \frac{1}{2}\bar{E}_2(\bar{r}) \times [\bar{H}_2(\bar{r})]^* = \hat{a}_z\tau^2\frac{E_0^2}{2\eta_2}. \tag{5.364}$$

The conservation of energy is satisfied since $\bar{S}_{c,2}(z=0) = \text{Re}\{\bar{S}_{c,1}(z=0)\}$, given that

$$\frac{1}{\eta_1}(1-\Gamma^2) = \frac{1}{\eta_1}\left[1 - \frac{(\eta_2-\eta_1)^2}{(\eta_1+\eta_2)^2}\right] = \frac{1}{\eta_1}\frac{4\eta_1\eta_2}{(\eta_1+\eta_2)^2} \tag{5.365}$$

$$= \frac{1}{\eta_2}\frac{4\eta_2^2}{(\eta_1+\eta_2)^2} = \frac{1}{\eta_2}\tau^2. \tag{5.366}$$

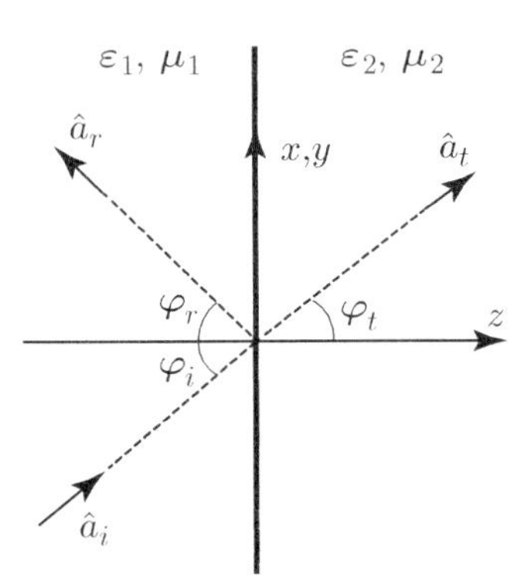

Figure 5.33 Oblique incidence of a plane wave on a boundary between two media. The wave directions and the boundary are shown at the *plane of incidence*: i.e. the plane containing the wave direction vectors and the boundary.

5.3.6 General Case for Reflection and Refraction

At this stage, we now consider a general case of *oblique* incidence, with a boundary located at $z = 0$ separating two media. First, we can omit the polarization and derive the angles for the reflected and refracted (transmitted to the second medium) waves (Figure 5.33). In the phasor domain, the incident plane wave can be written

$$\bar{E}^{\text{inc}}(\bar{r}) = \bar{E}_0^{\text{inc}}\exp(-j\bar{k}_i\cdot\bar{r}), \qquad \bar{H}^{\text{inc}}(\bar{r}) = \frac{1}{\eta_1}\hat{a}_i \times \bar{E}_0^{\text{inc}}\exp(-j\bar{k}_i\cdot\bar{r}), \tag{5.367}$$

where $\hat{a}_i$ is the direction of the wave: i.e. $\bar{k}_i = \hat{a}_i k_1 = \hat{a}_i \omega\sqrt{\mu_1\varepsilon_1}$. Similarly, the reflected and transmitted waves can be written

$$\bar{E}^{\text{ref}}(\bar{r}) = \bar{E}_0^{\text{ref}} \exp(-j\bar{k}_r \cdot \bar{r}), \qquad \bar{H}^{\text{ref}}(\bar{r}) = \frac{1}{\eta_1}\hat{a}_r \times \bar{E}_0^{\text{ref}} \exp(-j\bar{k}_r \cdot \bar{r}) \tag{5.368}$$

and

$$\bar{E}^{\text{tra}}(\bar{r}) = \bar{E}_0^{\text{tra}} \exp(-j\bar{k}_t \cdot \bar{r}), \qquad \bar{H}^{\text{tra}}(\bar{r}) = \frac{1}{\eta_2}\hat{a}_t \times \bar{E}_0^{\text{tra}} \exp(-j\bar{k}_t \cdot \bar{r}), \tag{5.369}$$

respectively, where $\bar{k}_r = \hat{a}_r k_1 = \hat{a}_r \omega\sqrt{\mu_1\varepsilon_1}$ and $\bar{k}_t = \hat{a}_t k_2 = \hat{a}_t \omega\sqrt{\mu_2\varepsilon_2}$.

According to the boundary conditions, the tangential electric field intensity and tangential magnetic field intensity must be continuous across the boundary. Hence, we can write

$$[\bar{E}^{\text{inc}}(z=0) + \bar{E}^{\text{ref}}(z=0)]_{\tan} = [\bar{E}^{\text{tra}}(z=0)]_{\tan} \tag{5.370}$$

$$[\bar{H}^{\text{inc}}(z=0) + \bar{H}^{\text{ref}}(z=0)]_{\tan} = [\bar{H}^{\text{tra}}(z=0)]_{\tan}, \tag{5.371}$$

where subscript "tan" represents the tangential components. The relationships between the amplitudes of the electric field intensity and magnetic field intensity depend on polarization and medium parameters. At this stage, however, we can focus on the exponential parts. If the boundary conditions need to be satisfied at all points at the boundary: i.e. for all x and y when $z = 0$, the exponential term must be the same for all waves. According to this condition, which is called *phase matching*, we must have

$$[\exp(-j\bar{k}_i \cdot \bar{r})]_{z=0} = [\exp(-j\bar{k}_r \cdot \bar{r})]_{z=0} = [\exp(-j\bar{k}_t \cdot \bar{r})]_{z=0} \tag{5.372}$$

or simply

$$[\bar{k}_i \cdot \bar{r}]_{z=0} = [\bar{k}_r \cdot \bar{r}]_{z=0} = [\bar{k}_t \cdot \bar{r}]_{z=0}. \tag{5.373}$$

Given that

$$\bar{k}_i = \hat{a}_x k_{ix} + \hat{a}_y k_{iy} + \hat{a}_z k_{iz} \tag{5.374}$$

$$\bar{k}_r = \hat{a}_x k_{rx} + \hat{a}_y k_{ry} + \hat{a}_z k_{rz} \tag{5.375}$$

$$\bar{k}_t = \hat{a}_x k_{tx} + \hat{a}_y k_{ty} + \hat{a}_z k_{tz}, \tag{5.376}$$

we obtain the necessary conditions:

$$k_{ix} = k_{rx} = k_{tx}, \qquad k_{iy} = k_{ry} = k_{ty}. \tag{5.377}$$

Hence, the *tangential* components of the wave vectors must *match* at the boundary.[83]

[83] This is indeed how the boundary in the plane of incidence in Figure 5.33 must look like a line.

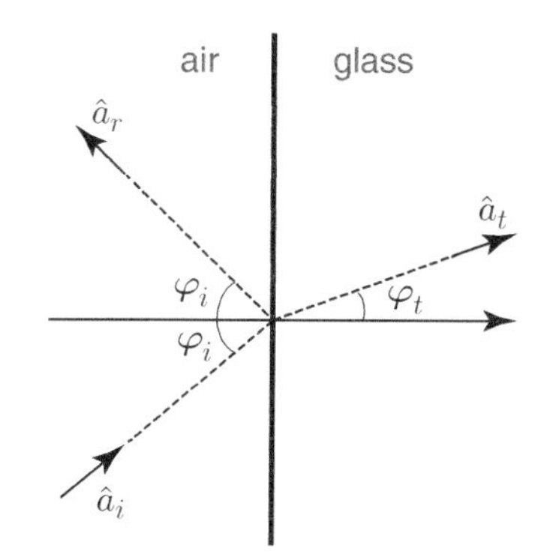

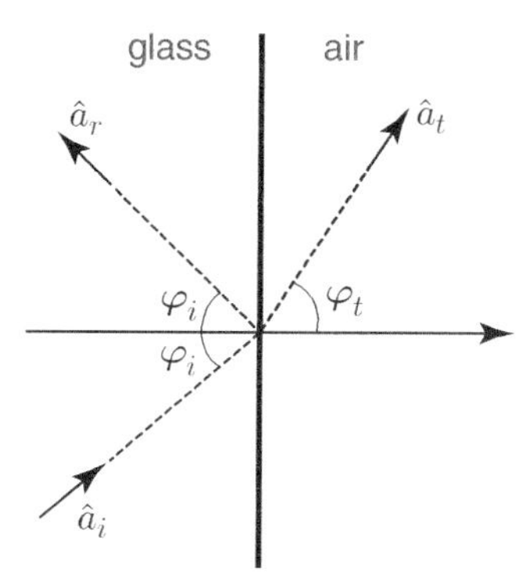

Figure 5.34 Reflection and refraction of waves at boundaries between air and glass.

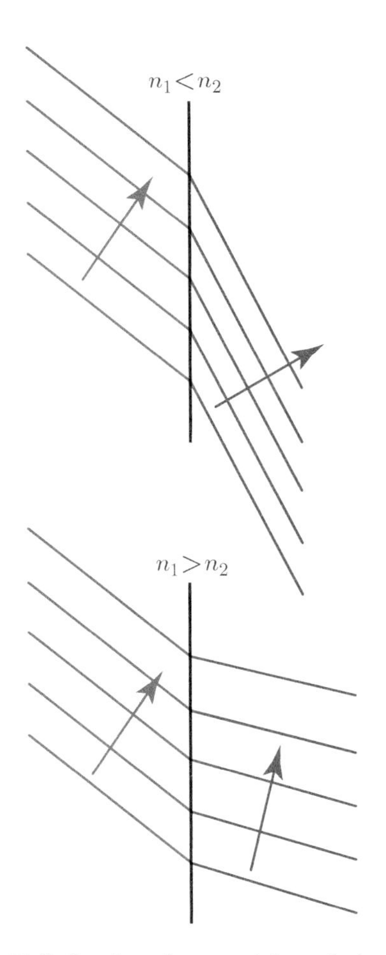

Figure 5.35 Refraction of waves at boundaries between different media. According to phase matching, wavefronts (lines) must match at the boundary. However, the phase velocity and wavelength (the distance between the lines) must change depending on the medium.

n: refractive index
Unitless

The directions of the incident, reflected, and transmitted waves depend on the medium parameters. Specifically, the wave vector $\bar{k}_t$ has a z-component magnitude different than those of $\bar{k}_i$ and $\bar{k}_r$. This is because $|\bar{k}_i| = |\bar{k}_r| \neq |\bar{k}_t|$. Specifically, we have

$$\sin\varphi_i = \frac{[k_{ix}^2 + k_{iy}^2]^{1/2}}{[k_{ix}^2 + k_{iy}^2 + k_{iz}^2]^{1/2}} = \frac{[k_{ix}^2 + k_{iy}^2]^{1/2}}{\omega\sqrt{\mu_1\varepsilon_1}} \tag{5.378}$$

$$\sin\varphi_r = \frac{[k_{rx}^2 + k_{ry}^2]^{1/2}}{[k_{rx}^2 + k_{ry}^2 + k_{rz}^2]^{1/2}} = \frac{[k_{rx}^2 + k_{ry}^2]^{1/2}}{\omega\sqrt{\mu_1\varepsilon_1}} \tag{5.379}$$

$$\sin\varphi_t = \frac{[k_{tx}^2 + k_{ty}^2]^{1/2}}{[k_{tx}^2 + k_{ty}^2 + k_{tz}^2]^{1/2}} = \frac{[k_{tx}^2 + k_{ty}^2]^{1/2}}{\omega\sqrt{\mu_2\varepsilon_2}}, \tag{5.380}$$

so that

$$k_1 \sin\varphi_i = k_1 \sin\varphi_r = k_2 \sin\varphi_t \tag{5.381}$$

or

$$\sqrt{\mu_1\varepsilon_1}\sin\varphi_i = \sqrt{\mu_1\varepsilon_1}\sin\varphi_r = \sqrt{\mu_2\varepsilon_2}\sin\varphi_t. \tag{5.382}$$

Hence, we derive Snell's law as $\varphi_i = \varphi_r$ and

$$\frac{\sin\varphi_t}{\sin\varphi_i} = \frac{\sqrt{\mu_1\varepsilon_1}}{\sqrt{\mu_2\varepsilon_2}} = \frac{n_1}{n_2}, \tag{5.383}$$

where $n = \sqrt{\mu\varepsilon}/\sqrt{\mu_0\varepsilon_0}$ is the refractive index. Note that the wavelength and phase velocity can be written

$$u_p = \frac{\omega}{k} = f\lambda = \frac{c}{n}, \tag{5.384}$$

indicating *reduced* velocity (keeping f and ω constant) of the wave inside a dielectric/magnetic material (in comparison to a vacuum). According to Snell's law, the wave is inclined toward the normal of the boundary plane if the second medium has a higher refractive index than the first medium (Figure 5.34). On the other hand, the wave inclines away from the normal if the second medium has a lower index. This changing direction can be interpreted as the enforcement of phase matching, while the phase velocity must change depending on the medium parameters (Figure 5.35).

5.3.6.1 Perpendicular Polarization

In an oblique incidence, the relationship between the reflected, refracted, and incident waves in terms of field intensities depends on the polarization. We now consider the case when the electric field intensity is perpendicular to the plane of incidence z-x (Figure 5.36). We derive

$$\bar{E}^{\mathrm{inc}}(\bar{r}) = \hat{a}_y E_0^{\mathrm{inc}} \exp[-jk_1(x\sin\varphi_i + z\cos\varphi_i)] \tag{5.385}$$

$$\bar{H}^{\mathrm{inc}}(\bar{r}) = (-\hat{a}_x\cos\varphi_i + \hat{a}_z\sin\varphi_i)\frac{E_0^{\mathrm{inc}}}{\eta_1}\exp[-jk_1(x\sin\varphi_i + z\cos\varphi_i)] \tag{5.386}$$

for the incident wave

$$\bar{E}^{\mathrm{ref}}(\bar{r}) = \hat{a}_y E_0^{\mathrm{ref}} \exp[-jk_1(x\sin\varphi_i - z\cos\varphi_i)] \tag{5.387}$$

$$\bar{H}^{\mathrm{ref}}(\bar{r}) = (\hat{a}_x\cos\varphi_i + \hat{a}_z\sin\varphi_i)\frac{E_0^{\mathrm{ref}}}{\eta_1}\exp[-jk_1(x\sin\varphi_i - z\cos\varphi_i)] \tag{5.388}$$

for the reflected wave, and

$$\bar{E}^{\mathrm{tra}}(\bar{r}) = \hat{a}_y E_0^{\mathrm{tra}} \exp[-jk_2(x\sin\varphi_t + z\cos\varphi_t)] \tag{5.389}$$

$$\bar{H}^{\mathrm{tra}}(\bar{r}) = (-\hat{a}_x\cos\varphi_t + \hat{a}_z\sin\varphi_t)\frac{E_0^{\mathrm{tra}}}{\eta_2}\exp[-jk_2(x\sin\varphi_t + z\cos\varphi_t)] \tag{5.390}$$

for the transmitted wave.[84] According to the boundary condition for the electric field intensity at $z = 0$, we have

$$E_0^{\mathrm{inc}}\exp(-jk_1x\sin\varphi_i) + E_0^{\mathrm{ref}}\exp(-jk_1x\sin\varphi_i) = E_0^{\mathrm{tra}}\exp(-jk_2x\sin\varphi_t), \tag{5.391}$$

while $k_1\sin\varphi_i = k_2\sin\varphi_t$ according to Snell's law, so that

$$E_0^{\mathrm{inc}} + E_0^{\mathrm{ref}} = E_0^{\mathrm{tra}}. \tag{5.392}$$

Further information is needed to express E_0^{ref} and E_0^{tra} in terms of E_0^{inc}. This information is derived from the boundary condition for the magnetic field intensity. Given that there is no surface electric current density at the boundary,[85] the tangential magnetic field intensity must

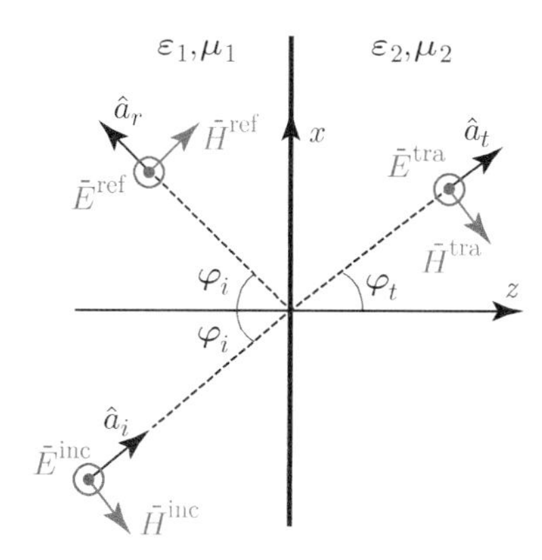

Figure 5.36 Reflection and refraction for perpendicular polarization in oblique incidence.

[84] Note that

$$\bar{k}_i = \hat{a}_i k_1 = (\hat{a}_x\sin\varphi_i + \hat{a}_z\cos\varphi_i)k_1$$
$$\bar{k}_r = \hat{a}_r k_1 = (\hat{a}_x\sin\varphi_i - \hat{a}_z\cos\varphi_i)k_1$$
$$\bar{k}_t = \hat{a}_t k_2 = (\hat{a}_x\sin\varphi_t + \hat{a}_z\cos\varphi_t)k_2$$

in the given setup (Figure 5.36). In addition, we have

$$\bar{H}^{\mathrm{inc}}(\bar{r}) = \hat{a}_i \times \bar{E}^{\mathrm{inc}}(\bar{r})/\eta_1$$
$$\bar{H}^{\mathrm{ref}}(\bar{r}) = \hat{a}_r \times \bar{E}^{\mathrm{ref}}(\bar{r})/\eta_1$$
$$\bar{H}^{\mathrm{tra}}(\bar{r}) = \hat{a}_t \times \bar{E}^{\mathrm{tra}}(\bar{r})/\eta_2.$$

[85] Recall that, if not particularly defined as an impressed source, the surface electric current density between two penetrable media is zero.

be continuous. Hence, we have

$$-\cos\varphi_i \frac{E_0^{\text{inc}}}{\eta_1}\exp(-jk_1x\sin\varphi_i) + \cos\varphi_i \frac{E_0^{\text{ref}}}{\eta_1}\exp(-jk_1x\sin\varphi_i)$$
$$= -\cos\varphi_t \frac{E_0^{\text{tra}}}{\eta_2}\exp(-jk_2x\sin\varphi_t). \tag{5.393}$$

Again, with Snell's law, exponentials are the same, and we obtain

$$-\cos\varphi_i \frac{E_0^{\text{inc}}}{\eta_1} + \cos\varphi_i \frac{E_0^{\text{ref}}}{\eta_1} = -\cos\varphi_t \frac{E_0^{\text{tra}}}{\eta_2}. \tag{5.394}$$

Using Eqs. (5.392) and (5.394) derived above,[86] we arrive at a ratio between the amplitudes of the reflected and incident waves

$$\Gamma_\perp = \frac{E_0^{\text{ref}}}{E_0^{\text{inc}}} = \frac{\eta_2\cos\varphi_i - \eta_1\cos\varphi_t}{\eta_2\cos\varphi_i + \eta_1\cos\varphi_t}, \tag{5.395}$$

which is called the *reflection coefficient* for perpendicular polarization. Then, similarly, we can obtain a ratio between the amplitudes of the transmitted and incident waves as

$$\tau_\perp = \frac{E_0^{\text{tra}}}{E_0^{\text{inc}}} = 1 + \Gamma_\perp = \frac{2\eta_2\cos\varphi_i}{\eta_2\cos\varphi_i + \eta_1\cos\varphi_t}, \tag{5.396}$$

which is the *transmission coefficient* for perpendicular polarization.

We may also consider two interesting special cases, as follows:

- If $\varphi_i = 0$, then $\varphi_r = 0$ (according to Snell's law), and we get

$$\Gamma_\perp = \frac{\eta_2 - \eta_1}{\eta_2 + \eta_1} \tag{5.397}$$

$$\tau_\perp = \frac{2\eta_2}{\eta_2 + \eta_1}, \tag{5.398}$$

which are the reflection and transmission coefficients for a normal incidence case, as defined in Eqs. (5.346) and (5.347).

[86] We insert

$$E_0^{\text{tra}} = E_0^{\text{inc}} + E_0^{\text{ref}}$$

into Eq. (5.394) to derive

$$\cos\varphi_i\left(-\frac{E_0^{\text{inc}}}{\eta_1} + \frac{E_0^{\text{ref}}}{\eta_1}\right) = \cos\varphi_t\left(-\frac{E_0^{\text{inc}}}{\eta_2} - \frac{E_0^{\text{ref}}}{\eta_2}\right)$$
$$E_0^{\text{inc}}\left(\frac{\cos\varphi_i}{\eta_1} - \frac{\cos\varphi_t}{\eta_2}\right) = E_0^{\text{ref}}\left(\frac{\cos\varphi_i}{\eta_1} + \frac{\cos\varphi_t}{\eta_2}\right)$$

or

$$\frac{E_0^{\text{ref}}}{E_0^{\text{inc}}} = \frac{\left(\frac{\cos\varphi_i}{\eta_1} - \frac{\cos\varphi_t}{\eta_2}\right)}{\left(\frac{\cos\varphi_i}{\eta_1} + \frac{\cos\varphi_t}{\eta_2}\right)} = \frac{\eta_1\eta_2\left(\frac{\cos\varphi_i}{\eta_1} - \frac{\cos\varphi_t}{\eta_2}\right)}{\eta_1\eta_2\left(\frac{\cos\varphi_i}{\eta_1} + \frac{\cos\varphi_t}{\eta_2}\right)} = \frac{\eta_2\cos\varphi_i - \eta_1\cos\varphi_t}{\eta_2\cos\varphi_i + \eta_1\cos\varphi_t}.$$

- If $\varepsilon_2 = \varepsilon_1$ and $\mu_2 = \mu_1$, leading to $\eta_2 = \eta_1$ and $\varphi_t = \varphi_i$, we have

$$\Gamma_\perp = \frac{\eta_1 \cos\varphi_i - \eta_1 \cos\varphi_i}{\eta_1 \cos\varphi_i + \eta_1 \cos\varphi_i} = 0 \tag{5.399}$$

$$\tau_\perp = \frac{2\eta_1 \cos\varphi_i}{\eta_1 \cos\varphi_i + \eta_1 \cos\varphi_t} = 1, \tag{5.400}$$

corresponding to full transmission.[87]

[87]Note that $\eta_1 = \eta_2$ (impedance matching) can be satisfied even when $\varepsilon_2 \neq \varepsilon_1$ and $\mu_2 \neq \mu_1$. On the other hand, the conditions $\varepsilon_2 = \varepsilon_1$ and $\mu_2 = \mu_1$ further lead to $\varphi_t = \varphi_i$ and full transmission.

5.3.6.2 Parallel Polarization

We now consider parallel polarization, for which the electric field intensity is *in* the plane of incidence (Figure 5.37). Hence, the magnetic field is in the y or $-y$ direction (normal to the plane of incidence). For the incident wave, we have

$$\bar{E}^{\rm inc}(\bar{r}) = (\hat{a}_x \cos\varphi_i - \hat{a}_z \sin\varphi_i) E_0^{\rm inc} \exp[-jk_1(x\sin\varphi_i + z\cos\varphi_i)] \tag{5.401}$$

$$\bar{H}^{\rm inc}(\bar{r}) = \hat{a}_y \frac{E_0^{\rm inc}}{\eta_1} \exp[-jk_1(x\sin\varphi_i + z\cos\varphi_i)]. \tag{5.402}$$

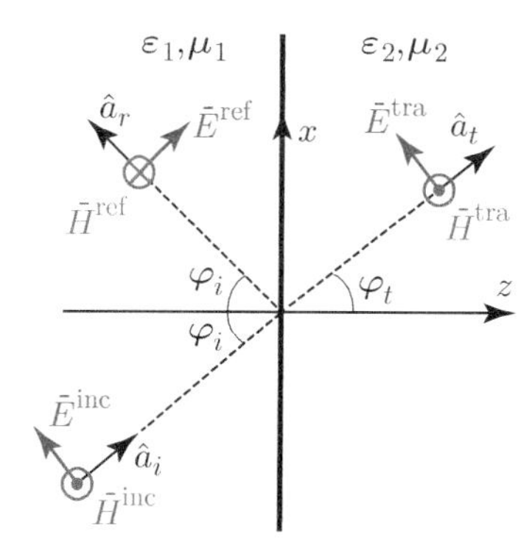

Figure 5.37 Reflection and refraction for parallel polarization in oblique incidence.

Similarly, field intensities for the reflected and transmitted waves can be written[88]

[88]Similar to perpendicular polarization, we have

$$\bar{k}_i = \hat{a}_i k_1 = (\hat{a}_x \sin\varphi_i + \hat{a}_z \cos\varphi_i) k_1$$
$$\bar{k}_r = \hat{a}_r k_1 = (\hat{a}_x \sin\varphi_i - \hat{a}_z \cos\varphi_i) k_1$$
$$\bar{k}_t = \hat{a}_t k_2 = (\hat{a}_x \sin\varphi_t + \hat{a}_z \cos\varphi_t) k_2$$
$$\bar{H}^{\rm inc}(\bar{r}) = \hat{a}_i \times \bar{E}^{\rm inc}(\bar{r})/\eta_1$$
$$\bar{H}^{\rm ref}(\bar{r}) = \hat{a}_r \times \bar{E}^{\rm ref}(\bar{r})/\eta_1$$
$$\bar{H}^{\rm tra}(\bar{r}) = \hat{a}_t \times \bar{E}^{\rm tra}(\bar{r})/\eta_2.$$

$$\bar{E}^{\rm ref}(\bar{r}) = (\hat{a}_x \cos\varphi_i + \hat{a}_z \sin\varphi_i) E_0^{\rm ref} \exp[-jk_1(x\sin\varphi_i - z\cos\varphi_i)] \tag{5.403}$$

$$\bar{H}^{\rm ref}(\bar{r}) = -\hat{a}_y \frac{E_0^{\rm ref}}{\eta_1} \exp[-jk_1(x\sin\varphi_i - z\cos\varphi_i)] \tag{5.404}$$

and

$$\bar{E}^{\rm tra}(\bar{r}) = (\hat{a}_x \cos\varphi_t - \hat{a}_z \sin\varphi_t) E_0^{\rm tra} \exp[-jk_2(x\sin\varphi_t + z\cos\varphi_t)] \tag{5.405}$$

$$\bar{H}^{\rm tra}(\bar{r}) = \hat{a}_y \frac{E_0^{\rm tra}}{\eta_2} \exp[-jk_2(x\sin\varphi_t + z\cos\varphi_t)], \tag{5.406}$$

respectively. Then the boundary conditions lead to

$$\cos\varphi_i E_0^{\rm inc} + \cos\varphi_i E_0^{\rm ref} = \cos\varphi_t E_0^{\rm tra} \tag{5.407}$$

$$\frac{E_0^{\rm inc}}{\eta_1} - \frac{E_0^{\rm ref}}{\eta_1} = \frac{E_0^{\rm tra}}{\eta_2}, \tag{5.408}$$

given that the tangential electric and magnetic field intensities must be continuous at $z = 0$. Using these equations, the reflection and transmission coefficients can be found as[89]

$$\Gamma_{\parallel} = \frac{E_0^{\mathrm{ref}}}{E_0^{\mathrm{inc}}} = \frac{\eta_2 \cos\varphi_t - \eta_1 \cos\varphi_i}{\eta_2 \cos\varphi_t + \eta_1 \cos\varphi_i} \tag{5.409}$$

$$\tau_{\parallel} = \frac{E_0^{\mathrm{tra}}}{E_0^{\mathrm{inc}}} = \frac{\eta_2}{\eta_1}\left(1 - \Gamma_{\parallel}\right) = \frac{2\eta_2 \cos\varphi_i}{\eta_2 \cos\varphi_t + \eta_1 \cos\varphi_i}. \tag{5.410}$$

[89]We insert

$$E_0^{\mathrm{tra}} = \frac{\eta_2}{\eta_1} E_0^{\mathrm{inc}} - \frac{\eta_2}{\eta_1} E_0^{\mathrm{ref}}$$

into Eq. (5.407) to obtain

$$\cos\varphi_i \left(E_0^{\mathrm{inc}} + E_0^{\mathrm{ref}}\right) = \cos\varphi_t \frac{\eta_2}{\eta_1}\left(E_0^{\mathrm{inc}} - E_0^{\mathrm{ref}}\right)$$

$$E_0^{\mathrm{ref}}\left(\cos\varphi_i + \frac{\eta_2}{\eta_1}\cos\varphi_t\right) = E_0^{\mathrm{inc}}\left(\frac{\eta_2}{\eta_1}\cos\varphi_t - \cos\varphi_i\right)$$

or

$$\frac{E_0^{\mathrm{ref}}}{E_0^{\mathrm{inc}}} = \frac{\left(\frac{\eta_2}{\eta_1}\cos\varphi_t - \cos\varphi_i\right)}{\left(\cos\varphi_i + \frac{\eta_2}{\eta_1}\cos\varphi_t\right)} = \frac{(\eta_2 \cos\varphi_t - \eta_1 \cos\varphi_i)}{(\eta_1 \cos\varphi_i + \eta_2 \cos\varphi_t)}.$$

Note that, for $\varphi_i = \varphi_t = 0$, these expressions reduce to

$$\Gamma_{\parallel} = \frac{E_0^{\mathrm{ref}}}{E_0^{\mathrm{inc}}} = \frac{\eta_2 - \eta_1}{\eta_2 + \eta_1} \tag{5.411}$$

$$\tau_{\parallel} = \frac{E_0^{\mathrm{tra}}}{E_0^{\mathrm{inc}}} = \frac{2\eta_2}{\eta_2 + \eta_1}, \tag{5.412}$$

i.e. exactly the same as $\Gamma_{\perp}$ and $\tau_{\perp}$, since polarization does not matter for a normal incidence. Similarly, we obtain

$$\Gamma_{\parallel} = 0 \tag{5.413}$$

$$\tau_{\parallel} = 1, \tag{5.414}$$

independent of the angles, for $\varepsilon_2 = \varepsilon_1$ and $\mu_2 = \mu_1$.

5.3.7 Examples

Example 56: Consider a normal incidence of a circularly polarized plane wave

$$\bar{E}^{\mathrm{inc}}(\bar{r}) = (\hat{a}_x + \hat{a}_y j) E_0 \exp(-jk_0 z) \tag{5.415}$$

onto an interface at $z = 0$ between a vacuum ($z < 0$) and a dielectric medium ($z > 0$) with a relative permittivity of $\varepsilon_r = 4$ (Figure 5.38). Find expressions for the electric field intensities of the reflected and transmitted waves.

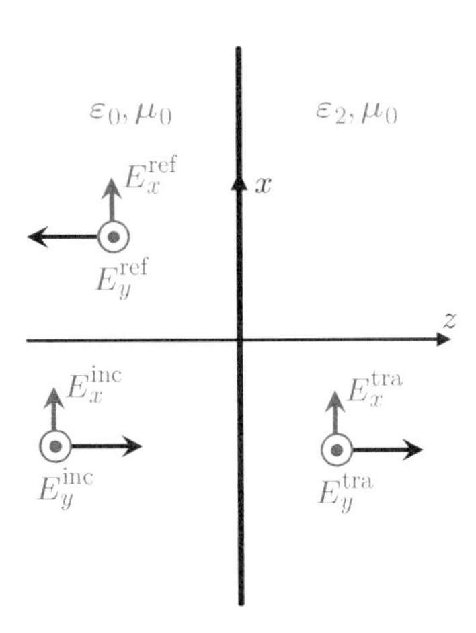

Figure 5.38 Normal incidence of a circularly polarized plane wave onto an interface between a vacuum and a dielectric medium.

Solution: First, note that the reflection and transmission coefficients in Eqs. (5.346) and (5.347) are derived for linearly polarized waves assuming the *setup* in Figure 5.32: i.e. when the electric field intensities of the reflected and transmitted waves are assumed to be in the same direction as the electric field intensity of the incident wave. Therefore, given that a circularly polarized wave is a superposition of two linearly polarized waves, we have[90]

$$\bar{E}^{\mathrm{ref}}(\bar{r}) = (\hat{a}_x + \hat{a}_y j)\Gamma E_0 \exp(jk_0 z) \tag{5.416}$$

$$\bar{E}^{\mathrm{tra}}(\bar{r}) = (\hat{a}_x + \hat{a}_y j)\tau E_0 \exp(-jk_2 z). \tag{5.417}$$

[90]Specifically, multiplying with Γ and τ, the $+x$ component of the incident electric field intensity leads to the $+x$ components of the reflected and transmitted electric field intensities. Similarly, the $+y$ component of the incident electric field intensity leads to the $+y$ components of the reflected and transmitted electric field intensities.

Since $\eta_2 = \sqrt{\mu_2/\varepsilon_2} = \sqrt{\mu_0/\varepsilon_0}/\sqrt{\varepsilon_r} = \eta_0/2$, we obtain the coefficients as

$$\Gamma = \frac{\eta_2 - \eta_1}{\eta_2 + \eta_1} = \frac{\eta_0/2 - \eta_0}{\eta_0/2 + \eta_0} = -1/3, \qquad \tau = 1 + \Gamma = 2/3. \tag{5.418}$$

Hence, we simply obtain[91]

$$\bar{E}^{\text{ref}}(\bar{r}) = -(\hat{a}_x + \hat{a}_y j)\frac{E_0}{3}\exp(jk_0 z) \tag{5.419}$$

$$\bar{E}^{\text{tra}}(\bar{r}) = (\hat{a}_x + \hat{a}_y j)\frac{2E_0}{3}\exp(-j2k_0 z). \tag{5.420}$$

[91]We have

$$k_2 = \omega\sqrt{\mu_2\varepsilon_2} = \omega\sqrt{\mu_0\varepsilon_0}\sqrt{\varepsilon_r} = 2k_0.$$

Note that the incident and transmitted waves are LHCP waves, while the reflected one is an RHCP wave.

Example 57: Consider an oblique incidence of a circularly polarized plane wave

$$\bar{E}^{\text{inc}}(\bar{r}) = (\hat{a}_x/2 - \hat{a}_z\sqrt{3}/2 + \hat{a}_y j)E_0\exp(-jk_0 x\sqrt{3}/2 - jk_0 z/2) \tag{5.421}$$

onto an interface at $z = 0$ between a vacuum ($z < 0$) and a material medium ($z > 0$) with $\varepsilon_r = 3$ relative permittivity and $\mu_r = 4$ relative permeability. Find an expression for the electric field intensity of the reflected wave.

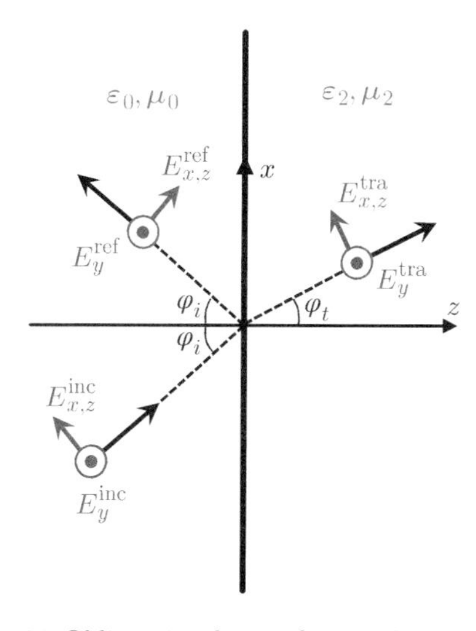

Figure 5.39 Oblique incidence of a circularly polarized plane wave onto an interface between a vacuum and a material medium.

Solution: For the scenarios in the derivations of the reflection/transmission coefficients, we construct our setup (directions) as shown in Figure 5.39. The incident electric field intensity can be decomposed into parallel (x and z) and perpendicular (y) components:

$$\bar{E}^{\text{inc}}_{\perp}(\bar{r}) = \hat{a}_y jE_0\exp(-jk_0 x\sqrt{3}/2 - jk_0 z/2) \tag{5.422}$$

$$\bar{E}^{\text{inc}}_{\parallel}(\bar{r}) = (\hat{a}_x/2 - \hat{a}_z\sqrt{3}/2)E_0\exp(-jk_0 x\sqrt{3}/2 - jk_0 z/2). \tag{5.423}$$

Then the components of the reflected electric field intensity can be written[92,93]

$$\bar{E}^{\text{ref}}_{\perp}(\bar{r}) = \hat{a}_y\Gamma_{\perp} jE_0\exp(-jk_0 x\sqrt{3}/2 + jk_0 z/2) \tag{5.424}$$

$$\bar{E}^{\text{ref}}_{\parallel}(\bar{r}) = (\hat{a}_x/2 + \hat{a}_z\sqrt{3}/2)\Gamma_{\parallel}E_0\exp(-jk_0 x\sqrt{3}/2 + jk_0 z/2). \tag{5.425}$$

To find the reflection coefficients, we can first find the transmission angle via Snell's law as

$$\sqrt{\mu_0\varepsilon_0}\sin\varphi_i = \sqrt{\mu_2\varepsilon_2}\sin\varphi_t = \sqrt{\mu_0\varepsilon_0}2\sqrt{3}\sin\varphi_t \tag{5.426}$$

$$\longrightarrow \sin\varphi_t = \frac{1}{2\sqrt{3}}\sin\varphi_i = \frac{1}{2\sqrt{3}}\frac{\sqrt{3}}{2} = \frac{1}{4} \longrightarrow \varphi_t = \sin^{-1}(1/4), \tag{5.427}$$

[92]We emphasize that $\hat{a}_y$ in the incident electric field intensity *generates* $\hat{a}_y$ in the reflected electric field intensity. On the other hand, $\hat{a}_x/2 - \hat{a}_z\sqrt{3}/2$ in the incident electric field intensity *generates* $\hat{a}_x/2 + \hat{a}_z\sqrt{3}/2$ in the reflected electric field intensity, according to the setup in Figure 5.37 (hence, in Figure 5.39). Only under these conditions can the reflection coefficients be used directly (as they are derived specifically for these setups).

[93]Note how the exponential terms are changed to $\exp(-jk_0 x\sqrt{3}/2 + jk_0 z/2)$, considering the direction of the reflected wave.

given that $\hat{a}_i = \hat{a}_x \sin\varphi_i + \hat{a}_z \cos\varphi_i = \hat{a}_x(\sqrt{3}/2) + \hat{a}_z(1/2)$. Then we have $\cos\varphi_t = [1 - \sin^2\varphi_t]^{1/2} = \sqrt{15}/4$. We also have

$$\eta_2 = \sqrt{\frac{\mu_2}{\varepsilon_2}} = \sqrt{\frac{\mu_r\mu_0}{\varepsilon_r\varepsilon_0}} = \sqrt{\frac{4}{3}}\sqrt{\mu_0\varepsilon_0} = \frac{2}{\sqrt{3}}\eta_0. \tag{5.428}$$

Then we obtain

$$\Gamma_\perp = \frac{\eta_2\cos\varphi_i - \eta_1\cos\varphi_t}{\eta_2\cos\varphi_i + \eta_1\cos\varphi_t} = \frac{\frac{2}{\sqrt{3}}\eta_0\frac{1}{2} - \eta_0\frac{\sqrt{15}}{4}}{\frac{2}{\sqrt{3}}\eta_0\frac{1}{2} + \eta_0\frac{\sqrt{15}}{4}} = \frac{\frac{1}{\sqrt{3}} - \frac{\sqrt{15}}{4}}{\frac{1}{\sqrt{3}} + \frac{\sqrt{15}}{4}} \approx -0.25 \tag{5.429}$$

$$\Gamma_\parallel = \frac{\eta_2\cos\varphi_t - \eta_1\cos\varphi_i}{\eta_2\cos\varphi_t + \eta_1\cos\varphi_i} = \frac{\frac{2}{\sqrt{3}}\eta_0\frac{\sqrt{15}}{4} - \eta_0\frac{1}{2}}{\frac{2}{\sqrt{3}}\eta_0\frac{\sqrt{15}}{4} + \eta_0\frac{1}{2}} = \frac{\frac{\sqrt{15}}{2\sqrt{3}} - \frac{1}{2}}{\frac{\sqrt{15}}{2\sqrt{3}} + \frac{1}{2}} \approx 0.38. \tag{5.430}$$

Consequently, the electric field intensity of the reflected wave can be written approximately as[94]

$$\bar{E}^{\text{ref}}(\bar{r}) \approx \left[\left(\frac{\hat{a}_x}{2} + \frac{\hat{a}_z\sqrt{3}}{2}\right)0.38 - \hat{a}_y j0.25\right] E_0 \exp(-jk_0x\sqrt{3}/2 + jk_0z/2). \tag{5.431}$$

[94]Note that the reflected wave is not circularly polarized: i.e. it is elliptically polarized.

Example 58: Consider an interface at $z = 0$ between two different media, defined as follows:

- Medium 1 (Dielectric): Defined for $z < 0$ with $\varepsilon_1 = 2\varepsilon_0$ and $\mu_1 = \mu_0$
- Medium 2 (Magnetic): Defined for $z > 0$ with $\varepsilon_2 = \varepsilon_0$ and $\mu_2 = 6\mu_0$

An incident wave is defined in medium 1 with the electric field intensity expression

$$\bar{E}^{\text{inc}}(\bar{r}) = \hat{a}_y 2E_0 \exp\left(-j\frac{1}{2}k_1 z\right)\cos\left(\frac{\sqrt{3}}{2}k_1 x\right). \tag{5.432}$$

Find an expression for the electric field intensity of the transmitted wave (in medium 2).

Solution: First, note that the incident wave is not a plane wave, but it can be decomposed into two plane waves as (Figure 5.40)

$$\bar{E}^{\text{inc}}(\bar{r}) = \hat{a}_y E_0 \exp\left(-j\frac{1}{2}k_1 z\right)\left[\exp\left(j\frac{\sqrt{3}}{2}k_1 x\right) + \exp\left(-j\frac{\sqrt{3}}{2}k_1 x\right)\right] \tag{5.433}$$

$$= E_1^{\text{inc}}(\bar{r}) + E_2^{\text{inc}}(\bar{r}), \tag{5.434}$$

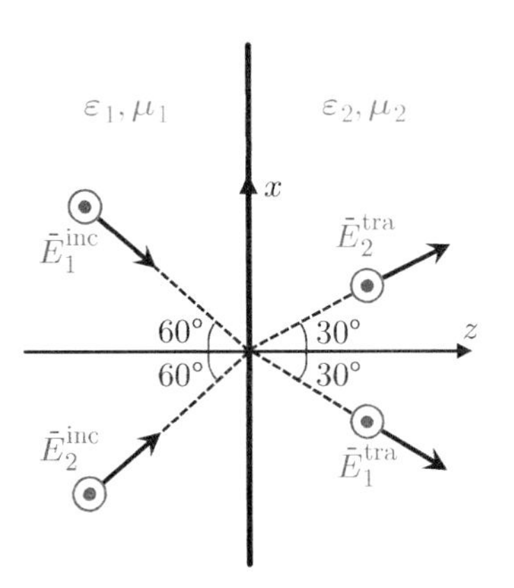

Figure 5.40 An incident wave, which can be decomposed into two plane waves with perpendicular polarizations onto an interface between two material media.

where[95]

$$E_1^{\text{inc}}(\bar{r}) = \hat{a}_y E_0 \exp\left(-jk_1 z/2 + jk_1 x\sqrt{3}/2\right) \tag{5.435}$$

$$E_2^{\text{inc}}(\bar{r}) = \hat{a}_y E_0 \exp\left(-jk_1 z/2 - jk_1 x\sqrt{3}/2\right). \tag{5.436}$$

These linearly (and perpendicularly) polarized waves are incident onto the boundary with 60° angles (Figure 5.40). For the corresponding transmitted waves, the angles can be obtained via Snell's law, e.g. by considering the second plane wave:

$$\sqrt{\mu_1\varepsilon_1}\sin 60^\circ = \sqrt{\mu_2\varepsilon_2}\sin\varphi_t \longrightarrow \sqrt{\mu_0\varepsilon_0}\sqrt{2}\frac{\sqrt{3}}{2} = \sqrt{\mu_0\varepsilon_0}\sqrt{6}\sin\varphi_t \tag{5.437}$$

$$\longrightarrow \sin\varphi_t = \frac{1}{2} \longrightarrow \varphi_t = 30^\circ. \tag{5.438}$$

In addition, the transmission coefficient can be found as[96,97]

$$\tau_\perp = \frac{2\eta_2\cos\varphi_i}{\eta_2\cos\varphi_i + \eta_1\cos\varphi_t} = \frac{2\sqrt{6}\eta_0\frac{1}{2}}{\sqrt{6}\eta_0\frac{1}{2} + \frac{1}{\sqrt{2}}\eta_0\frac{\sqrt{3}}{2}} = \frac{\sqrt{6}}{\sqrt{6}/2 + \sqrt{6}/4} = \frac{4}{3}. \tag{5.439}$$

Hence, the transmitted plane waves can be written

$$E_1^{\text{tra}}(\bar{r}) = \hat{a}_y(4/3)E_0 \exp\left(-jk_2 z\sqrt{3}/2 + jk_2 x/2\right) \tag{5.440}$$

$$E_2^{\text{tra}}(\bar{r}) = \hat{a}_y(4/3)E_0 \exp\left(-jk_2 z\sqrt{3}/2 - jk_2 x/2\right). \tag{5.441}$$

Finally, the total transmitted wave can be obtained:

$$E^{\text{tra}}(\bar{r}) = E_1^{\text{tra}}(\bar{r}) + E_2^{\text{tra}}(\bar{r}) = \hat{a}_y\frac{8E_0}{3}\exp\left(-j\frac{\sqrt{3}}{2}k_2 z\right)\cos\left(\frac{1}{2}k_2 x\right). \tag{5.442}$$

5.3.8 Total Internal Reflection

When the first medium has a higher refractive index than the second medium in the reflection/refraction setups described above, it is possible that all incident power is reflected back to the first medium and there is no *time-average* power crossing the boundary (Figure 5.41).

[95] We use

$$\cos(a) = \frac{\exp(ja) + \exp(-ja)}{2}.$$

[96] Note that the reflection coefficient is the same for the two plane waves since they are incident onto the boundary with the same angle.

[97] We have

$$\eta_1 = \sqrt{\frac{\mu_1}{\varepsilon_1}} = \frac{1}{\sqrt{2}}\eta_0$$

$$\eta_2 = \sqrt{\frac{\mu_2}{\varepsilon_2}} = \sqrt{6}\eta_0.$$

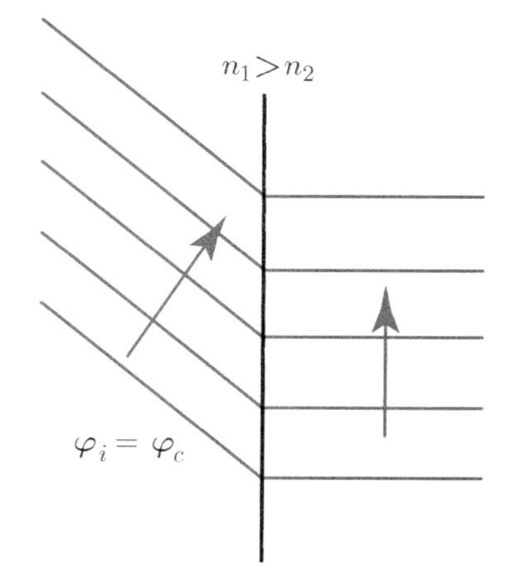

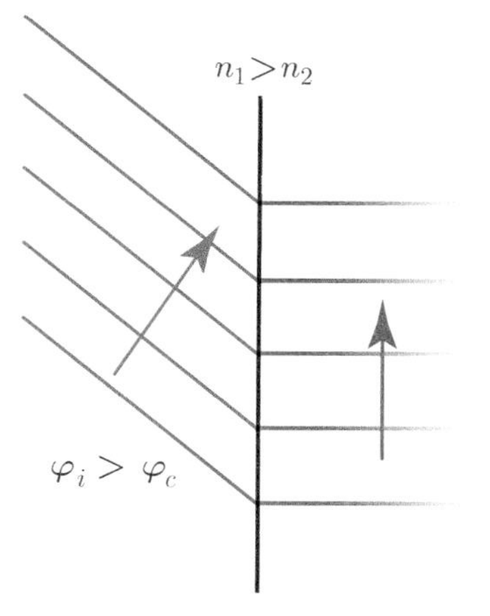

Figure 5.41 Total internal reflection scenarios.

The angle – called the *critical angle* – that leads this scenario can be found simply by considering Snell's law as

$$\sin\varphi_c = \frac{n_2}{n_1}\sin(\pi/2) = \frac{n_2}{n_1} = \frac{\sqrt{\mu_2\varepsilon_2}}{\sqrt{\mu_1\varepsilon_1}} \tag{5.443}$$

or

$$\varphi_c = \sin^{-1}\left(\frac{n_2}{n_1}\right) = \sin^{-1}\left(\frac{\sqrt{\mu_2\varepsilon_2}}{\sqrt{\mu_1\varepsilon_1}}\right). \tag{5.444}$$

When the incident wave hits the boundary exactly at a critical angle ($\varphi_i = \varphi_c$, so that $\varphi_t = \pi/2$), we have[98]

$$\Gamma_\perp = \frac{\eta_2\cos\varphi_c}{\eta_2\cos\varphi_c} = 1, \qquad \tau_\perp = \frac{2\eta_2\cos\varphi_c}{\eta_2\cos\varphi_c} = 2 \tag{5.445}$$

$$\Gamma_\parallel = \frac{-\eta_1\cos\varphi_c}{+\eta_1\cos\varphi_c} = -1, \qquad \tau_\parallel = \frac{2\eta_2\cos\varphi_c}{\eta_1\cos\varphi_c} = \frac{2\eta_2}{\eta_1}. \tag{5.446}$$

Based on these, the electric field intensity in the second medium (transmission region) can be obtained as[99]

$$\bar{E}_\perp^{\text{tra}}(\bar{r}) = \hat{a}_y\tau_\perp E_0^{\text{inc}}\exp(-jk_2x) = \hat{a}_y 2E_0\exp(-jk_2x) \tag{5.447}$$

and

$$\bar{E}_\parallel^{\text{tra}}(\bar{r}) = -\hat{a}_z\tau_\parallel E_0^{\text{inc}}\exp(-jk_2x) = -\hat{a}_z\frac{2\eta_2 E_0}{\eta_1}\exp(-jk_2x) \tag{5.448}$$

for perpendicular and parallel polarizations, respectively. Obviously, these expressions represent plane waves propagating in the x direction: i.e. parallel to the surface.[100] Given the corresponding expressions for the magnetic field intensity as

$$\bar{H}_\perp^{\text{tra}}(\bar{r}) = \hat{a}_z\frac{2E_0}{\eta_2}\exp(-jk_2x), \qquad \bar{H}_\parallel^{\text{tra}}(\bar{r}) = \hat{a}_y\frac{2E_0}{\eta_1}\exp(-jk_2x), \tag{5.449}$$

the complex Poynting vector expressions in the second medium can be found as

$$\bar{S}_{c,\perp}^{\text{tra}}(\bar{r}) = \hat{a}_x\frac{2E_0^2}{\eta_2}, \qquad \bar{S}_{c,\parallel}^{\text{tra}}(\bar{r}) = \hat{a}_x\frac{2\eta_2 E_0^2}{\eta_1^2} \tag{5.450}$$

for perpendicular and parallel polarizations. Hence, the power flux density is purely real and in the x direction: i.e. parallel to the boundary.

[98] *Total reflection* indicates that there is no time-average power flowing into the second medium (transmission region). Transmission coefficients, as well as fields in the transmission region, are generally nonzero.

[99] Setting $E_0^{\text{inc}} = E_0$.

[100] If there is no time-average power transmission, then the origin of the plane wave (which carries energy) may be questioned in terms of the conservation of energy. Specifically, if there is no net power that flows into the transmission region, how are these plane waves generated in the first place? In fact, in these scenarios, we are investigating the *steady state*: i.e. after the transient state that actually involves a power flow into the second medium until the plane wave in this medium is set up.

It becomes interesting when the incidence angle is larger than the critical angle. In this case, according to Snell's law, we have

$$\sin\varphi_t = \frac{n_1}{n_2}\sin\varphi_i = \frac{\sin\varphi_i}{\sin\varphi_c} > 1, \tag{5.451}$$

since $\varphi_i > \varphi_c$. Hence, the angle φ_t is not well-defined; but mathematically, one can still find $\cos\varphi_t$ as

$$\cos\varphi_t = \sqrt{1-\sin^2\varphi_t} = \sqrt{1-\frac{\sin^2\varphi_i}{\sin^2\varphi_c}} \tag{5.452}$$

$$= \sqrt{\frac{\sin^2\varphi_c - \sin^2\varphi_i}{\sin^2\varphi_c}} = -j\frac{1}{\sin\varphi_c}\sqrt{\sin^2\varphi_i - \sin^2\varphi_c}, \tag{5.453}$$

which is purely imaginary:[101]

$$\cos\varphi_t = -j|\cos\varphi_t|. \tag{5.454}$$

Then the reflection and transmission coefficients can be found

$$\Gamma_\perp = \frac{\eta_2\cos\varphi_i + j\eta_1|\cos\varphi_t|}{\eta_2\cos\varphi_i - j\eta_1|\cos\varphi_t|} = \exp(j\Phi_\perp), \qquad \tau_\perp = 1 + \exp(j\Phi_\perp) \tag{5.455}$$

$$\Gamma_\parallel = \frac{-j\eta_2|\cos\varphi_t| - \eta_1\cos\varphi_i}{-j\eta_2|\cos\varphi_t| + \eta_1\cos\varphi_i} = \exp(j\Phi_\parallel), \qquad \tau_\parallel = \frac{\eta_2}{\eta_1}\left[1-\exp(j\Phi_\parallel)\right], \tag{5.456}$$

where $\exp(j\Phi_\perp)$ and $\exp(j\Phi_\parallel)$ represent phase shifts (between incident and reflected waves) introduced by the boundary.[102] The transmitted electric field intensity can be derived as

$$\bar{E}_\perp^{\text{tra}}(\bar{r}) = \hat{a}_y[1+\exp(j\Phi_\perp)]E_0\exp(-jk_2x\sin\varphi_t)\exp(-k_2z|\cos\varphi_t|) \tag{5.457}$$

$$\bar{E}_\parallel^{\text{tra}}(\bar{r}) = (-\hat{a}_x j|\cos\varphi_t| - \hat{a}_z\sin\varphi_t)\frac{\eta_2}{\eta_1}[1-\exp(j\Phi_\parallel)]$$
$$E_0\exp(-jk_2x\sin\varphi_t)\exp(-k_2z|\cos\varphi_t|) \tag{5.458}$$

for perpendicular and parallel polarizations, respectively. The corresponding expressions for the magnetic field intensity are[103]

[101] The choice of the minus sign – i.e. using $-j$ instead of $+j$ – is due to the choice of the correct branch (decaying waves instead of unphysical growing waves) in the second medium.

[102] This phase shift – Goos-Hänchen shift – can be interpreted as the increased path length when a wave is transmitted inside the second medium and then *pushed* back into the first medium due to the total reflection.

[103] We use

$$\bar{H}_{\perp,\parallel}^{\text{tra}}(\bar{r}) = \frac{1}{\eta_2}\hat{a}_t \times \bar{E}_{\perp,\parallel}^{\text{tra}}(\bar{r})$$
$$= \frac{1}{\eta_2}(\hat{a}_x\sin\varphi_t + \hat{a}_z\cos\varphi_t)\times\bar{E}_{\perp,\parallel}^{\text{tra}}(\bar{r})$$
$$= \frac{1}{\eta_2}(\hat{a}_x\sin\varphi_t - \hat{a}_z j|\cos\varphi_t|)\times\bar{E}_{\perp,\parallel}^{\text{tra}}(\bar{r})$$

and

$$\sin^2\varphi_t - |\cos\varphi_t|^2 = \sin^2\varphi_t - (j\cos\varphi_t)^2$$
$$= \sin^2\varphi_t + \cos^2\varphi_t = 1.$$

$$\bar{H}_{\perp}^{\text{tra}}(\bar{r}) = (\hat{a}_x j|\cos\varphi_t| + \hat{a}_z \sin\varphi_t)[1+\exp(j\Phi_{\perp})]$$
$$\frac{E_0}{\eta_2}\exp(-jk_2 x\sin\varphi_t)\exp(-k_2 z|\cos\varphi_t|) \tag{5.459}$$
$$\bar{H}_{\parallel}^{\text{tra}}(\bar{r}) = \hat{a}_y[1-\exp(j\Phi_{\parallel})]\frac{E_0}{\eta_1}\exp(-jk_2 x\sin\varphi_t)\exp(-k_2 z|\cos\varphi_t|). \tag{5.460}$$

Finally, expressions for the complex Poynting vector can be derived as

$$\bar{S}_{c,\perp}^{\text{tra}}(\bar{r}) = (\hat{a}_x \sin\varphi_t + \hat{a}_z j|\cos\varphi_t|)\,[1+\cos\Phi_{\perp}]\,\frac{E_0^2}{\eta_2}\exp(-2k_2 z|\cos\varphi_t|) \tag{5.461}$$
$$\bar{S}_{c,\parallel}^{\text{tra}}(\bar{r}) = (\hat{a}_x \sin\varphi_t - \hat{a}_z j|\cos\varphi_t|)\,\left[1-\cos\Phi_{\parallel}\right]\,\frac{\eta_2 E_0^2}{\eta_1^2}\exp(-2k_2 z|\cos\varphi_t|), \tag{5.462}$$

indicating real power flux density in the x direction and reactive power flux density in the z direction. All these expressions represent propagating waves in the x direction (along the surface) with *decaying* characteristics in the z direction (Figure 5.41). These types of waves are called *surface waves*. The purely imaginary Poynting vector in the z direction means all instantaneous power crossing the boundary is eventually reflected back into the first medium.[104] Therefore, there is no time-average power flow across the boundary. On the other hand, the surface wave propagates in the x direction with a nonzero time-average power flux density.

[104] The transmitted power flux density (which is in the z direction) exactly at the boundary can be written

$$S_{c,z,\perp}^{\text{tra}}(z=0) = \hat{a}_z j[1+\cos\Phi_{\perp}]|\cos\varphi_t|\frac{E_0^2}{\eta_2}$$
$$S_{c,z,\parallel}^{\text{tra}}(z=0) = -\hat{a}_z j[1-\cos\Phi_{\parallel}]|\cos\varphi_t|\frac{\eta_2 E_0^2}{\eta_1^2},$$

indicating zero time-average power flowing across the boundary.

5.3.9 Total Transmission

A conceptual counterpart of total reflection is the total transmission that may occur in special cases: when the incident plane wave hits the boundary at Brewster's angle. Note that a total transmission naturally occurs when two media are already matched: i.e. when $\varepsilon_1 = \varepsilon_2$ and $\mu_1 = \mu_2$. However, our focus here is the total transmission that can occur even when the two media are different.

As opposed to total reflection, total transmission depends on the polarization. First, for perpendicular polarization, we set[105]

$$\Gamma_{\perp} = \frac{\eta_2\cos\varphi_i - \eta_1\cos\varphi_t}{\eta_2\cos\varphi_i + \eta_1\cos\varphi_t} = 0, \tag{5.463}$$

which may be satisfied when $\eta_2\cos\varphi_i - \eta_1\cos\varphi_t = 0$ or

$$\cos\varphi_t = \frac{\eta_2}{\eta_1}\cos\varphi_i = \frac{\sqrt{\mu_2\varepsilon_1}}{\sqrt{\mu_1\varepsilon_2}}\cos\varphi_i. \tag{5.464}$$

[105] By setting the reflection coefficient to zero, we make the reflected wave zero. The corresponding transmission coefficient for perpendicular polarization can be written $\tau_{\perp} = 1 + \Gamma_{\perp} = 1$.

Using Snell's law, we further have

$$\sin\varphi_t = \frac{\sqrt{\mu_1\varepsilon_1}}{\sqrt{\mu_2\varepsilon_2}}\sin\varphi_i. \tag{5.465}$$

Combining the two equations above, we arrive at

$$\cos^2\varphi_t + \sin^2\varphi_t = 1 = \frac{\mu_2\varepsilon_1}{\mu_1\varepsilon_2}\cos^2\varphi_i + \frac{\mu_1\varepsilon_1}{\mu_2\varepsilon_2}\sin^2\varphi_i. \tag{5.466}$$

Further using $\cos^2\varphi_i = 1 - \sin^2\varphi_i$, we have

$$\frac{\mu_2\varepsilon_1}{\mu_1\varepsilon_2} - \frac{\mu_2\varepsilon_1}{\mu_1\varepsilon_2}\sin^2\varphi_i + \frac{\mu_1\varepsilon_1}{\mu_2\varepsilon_2}\sin^2\varphi_i = 1, \tag{5.467}$$

leading to[106]

$$\sin^2\varphi_i = \left[\frac{\mu_1\varepsilon_2}{\mu_2\varepsilon_1} - 1\right] \Big/ \left[\left(\frac{\mu_1}{\mu_2}\right)^2 - 1\right]. \tag{5.468}$$

Hence, Brewster's angle for perpendicular polarization can be written[107]

$$\sin\varphi_{B\perp} = \sqrt{\frac{1 - \dfrac{\mu_1\varepsilon_2}{\mu_2\varepsilon_1}}{1 - \left(\dfrac{\mu_1}{\mu_2}\right)^2}}. \tag{5.469}$$

Obviously, for a nonmagnetic case – i.e. when $\mu_1 = \mu_2 = \mu_0$ – there is no Brewster's angle since $\sin\varphi_{B\perp} = \infty$ cannot be satisfied.

The derivation of Brewster's angle leads to a different result for parallel polarization. In this case, we need

$$\Gamma_\parallel = \frac{\eta_2\cos\varphi_t - \eta_1\cos\varphi_i}{\eta_2\cos\varphi_t + \eta_1\cos\varphi_i} = 0 \tag{5.470}$$

or

$$\cos\varphi_t = \frac{\eta_1}{\eta_2}\cos\varphi_i = \frac{\sqrt{\mu_1\varepsilon_2}}{\sqrt{\mu_2\varepsilon_1}}\cos\varphi_i. \tag{5.471}$$

Combining this equality with Snell's law, we arrive at

$$\cos^2\varphi_t + \sin^2\varphi_t = 1 = \frac{\mu_1\varepsilon_2}{\mu_2\varepsilon_1}\cos^2\varphi_i + \frac{\mu_1\varepsilon_1}{\mu_2\varepsilon_2}\sin^2\varphi_i. \tag{5.472}$$

[106] Grouping the terms, we first obtain

$$\sin^2\varphi_i\left(\frac{\mu_1\varepsilon_1}{\mu_2\varepsilon_2} - \frac{\mu_2\varepsilon_1}{\mu_1\varepsilon_2}\right) = 1 - \frac{\mu_2\varepsilon_1}{\mu_1\varepsilon_2}.$$

[107] Note that, if $\varepsilon_1 = \varepsilon_2$, then we have

$$\sin\varphi_{B\perp} = \sqrt{\frac{1 - \dfrac{\mu_1}{\mu_2}}{1 - \left(\dfrac{\mu_1}{\mu_2}\right)^2}} = \sqrt{\frac{1}{1 + \left(\dfrac{\mu_1}{\mu_2}\right)}} = \sqrt{\frac{\mu_2}{\mu_1 + \mu_2}}.$$

Once again, using $\cos^2\varphi = 1 - \sin^2\varphi$ and rearranging the terms, we have

$$\sin^2\varphi_i\left(\frac{\mu_1\varepsilon_1}{\mu_2\varepsilon_2} - \frac{\mu_1\varepsilon_2}{\mu_2\varepsilon_1}\right) = 1 - \frac{\mu_1\varepsilon_2}{\mu_2\varepsilon_1} \tag{5.473}$$

or

$$\sin^2\varphi_i = \left[\frac{\mu_2\varepsilon_1}{\mu_1\varepsilon_2} - 1\right] \Big/ \left[\left(\frac{\varepsilon_1}{\varepsilon_2}\right)^2 - 1\right]. \tag{5.474}$$

Then Brewster's angle for parallel polarization can be written

$$\sin\varphi_{B\|} = \sqrt{\frac{1 - \dfrac{\mu_2\varepsilon_1}{\mu_1\varepsilon_2}}{1 - \left(\dfrac{\varepsilon_1}{\varepsilon_2}\right)^2}}. \tag{5.475}$$

For a nonmagnetic case, this expression reduces into a neat form (if $\varepsilon_1 \neq \varepsilon_2$) as

$$\sin\varphi_{B\|} = \sqrt{\frac{1 - \dfrac{\varepsilon_1}{\varepsilon_2}}{1 - \left(\dfrac{\varepsilon_1}{\varepsilon_2}\right)^2}} = \sqrt{\frac{1}{1 + \dfrac{\varepsilon_1}{\varepsilon_2}}} = \sqrt{\frac{\varepsilon_2}{\varepsilon_1 + \varepsilon_2}}. \tag{5.476}$$

Hence, for a nonmagnetic case, there exists a Brewster's angle[108] (Figure 5.42).

[108] But in this case (for parallel polarization), there is no Brewster's angle if the two media have the same permittivity.

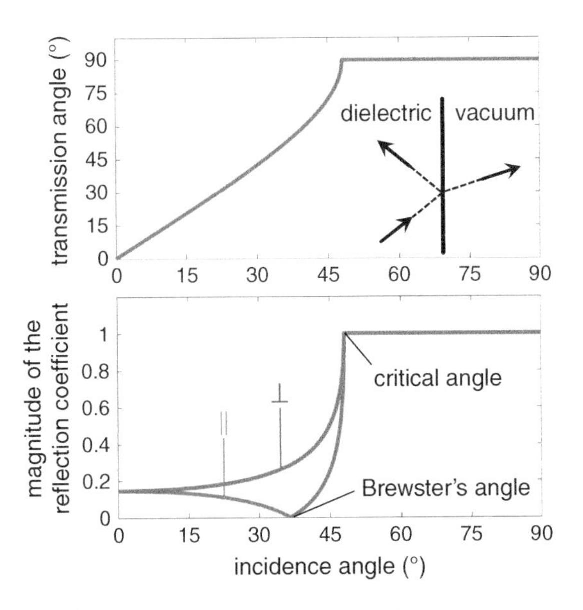

Figure 5.42 Reflection/transmission of plane waves that are incident onto a boundary between a dielectric medium (first medium) and a vacuum (second medium). The relative permittivity of the dielectric is $\varepsilon_r = 1.8$. Different incidence angles are considered from 0° (normal incidence) to 90°. The transmission angle (top plot) and the magnitude of the reflection coefficient (bottom plot, for both perpendicular and parallel polarization) are plotted with respect to the incidence angle. There is Brewster's angle for the parallel polarization, for which the reflection coefficient becomes zero. After the critical angle (which occurs *later* than Brewster's angle), the transmission angle is fixed at 90°, while the reflection coefficient becomes unity for both polarizations.

In the context of total transmission, the different behaviors for different polarizations can be used in practice to separate components of a mixed wave. Assume that a plane wave with both perpendicular and parallel polarization is incident onto a boundary between two dielectrics. If the incidence angle coincides with Brewster's angle (for parallel polarization), then the reflected wave contains only perpendicular polarization due to the total transmission of parallel polarization (Figure 5.43).

There is another equality, which may be called *Brewster's condition*, satisfied when the incident wave with parallel polarization has Brewster's angle in a nonmagnetic case.[109] First, in this case, note that Brewster's angle satisfies

$$\sin^2\varphi_{B\|} = \frac{1}{1 + \varepsilon_1/\varepsilon_2} \tag{5.477}$$

[109] Alternatively, Brewster's condition can be satisfied for perpendicular polarization in a non-dielectric case.

or

$$\sin^2\varphi_{B\|} + \frac{\varepsilon_1}{\varepsilon_2}\sin^2\varphi_{B\|} = 1. \tag{5.478}$$

But according to Snell's law (for a nonmagnetic case), we also have

$$\sin^2\varphi_t = \frac{\varepsilon_1}{\varepsilon_2}\sin^2\varphi_{B\|} \tag{5.479}$$

or $\varepsilon_1/\varepsilon_2 = \sin^2\varphi_t/\sin^2\varphi_{B\|}$. Combining the equations, we obtain

$$\sin^2\varphi_{B\|} + \sin^2\varphi_t = 1. \tag{5.480}$$

For $\varepsilon_1 \neq \varepsilon_2$, this is only possible when

$$\varphi_{B\|} + \varphi_t = \frac{\pi}{2}, \tag{5.481}$$

which is the condition that must be satisfied by incidence and transmission angles. Specifically, the sum of incidence and transmission angles must be 90° in a nonmagnetic case of total transmission.

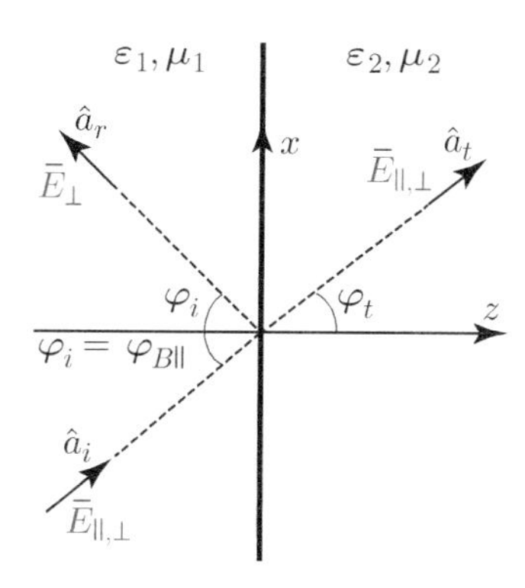

Figure 5.43 Using total transmission (Brewster's angle) to filter out parallel polarization component from an incident wave.

5.3.10 Examples

Example 59: Consider an interface at $z = 0$ that separates a dielectric medium ($z < 0$) with ε_r relative permittivity and a vacuum ($z > 0$). Assuming incidence of plane waves with parallel polarization from the dielectric side onto the boundary (Figure 5.44), compare the critical angle and Brewster's angle for different values of ε_r.

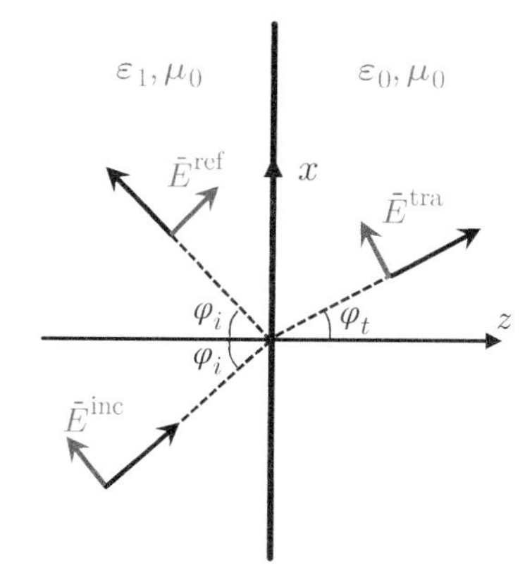

Figure 5.44 Reflection/transmission of plane waves with parallel polarization that are incident onto an interface between a dielectric medium (first medium) and a vacuum (second medium).

Solution: The critical angle with respect to ε_r can be written

$$\sin\varphi_c = \sqrt{\frac{\varepsilon_2}{\varepsilon_1}} = \sqrt{\frac{\varepsilon_0}{\varepsilon_r\varepsilon_0}} = \frac{1}{\sqrt{\varepsilon_r}}, \tag{5.482}$$

while Brewster's angle (for parallel polarization) can be written

$$\sin\varphi_{B\|} = \sqrt{\frac{\varepsilon_2}{\varepsilon_1+\varepsilon_2}} = \sqrt{\frac{\varepsilon_0}{\varepsilon_r\varepsilon_0+\varepsilon_0}} = \frac{1}{\sqrt{1+\varepsilon_r}}. \tag{5.483}$$

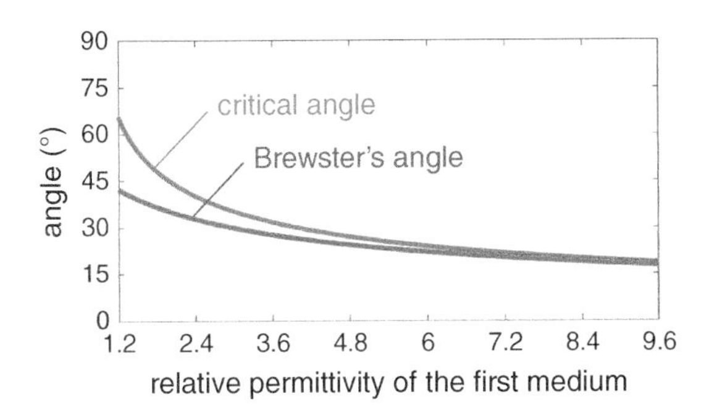

Figure 5.45 Comparison of the critical and Brewster's angles for the setup described in Figure 5.44. The critical angle is always larger than Brewster's angle, while they become similar to each other as the permittivity of the first medium increases.

Therefore, we have

$$\sin\varphi_{B\|} < \sin\varphi_c \longrightarrow \varphi_{B\|} < \varphi_c, \tag{5.484}$$

i.e. the critical angle is always larger than Brewster's angle for any value of ε_r (Figure 5.45). Also note that, for large values of ε_r, we have

$$\varphi_{B\|} \approx \varphi_c \qquad (\varepsilon_r \gg 1). \tag{5.485}$$

Example 60: A plane wave is obliquely incident from air to a dielectric medium with a relative permittivity of ε_r. The electric field intensities for the incident and reflected waves are given as

$$\bar{E}^{\text{inc}}(\bar{r}) = (\hat{a}_x 2 - \hat{a}_y j2 - \hat{a}_z 2\sqrt{3})E_0 \exp[-j(2\sqrt{3}\pi x + 2\pi z)] \tag{5.486}$$

$$\bar{E}^{\text{ref}}(\bar{r}) = \hat{a}_y a E_0 \exp[-j(2\sqrt{3}\pi x - 2\pi z)], \tag{5.487}$$

where a (unitless) and E_0 (V/m) are constants. Find a, ε_r, and the electric field intensity of the transmitted wave.

Solution: First, since there is no parallel component of the reflected wave, the incidence angle should be Brewster's angle for parallel polarization. Then we have[110]

$$\sin\varphi_i = \sin(60°) = \frac{\sqrt{3}}{2} = \sin\varphi_{B\|} = \sqrt{\frac{\varepsilon_2}{\varepsilon_1+\varepsilon_2}} = \sqrt{\frac{\varepsilon_r}{1+\varepsilon_r}}, \tag{5.488}$$

which leads to $\varepsilon_r = 3$. In addition, using Brewster's condition, the transmission angle can be found as[111] $\varphi_t = 90° - \varphi_i = 30°$. The reflection coefficient for the perpendicular polarization can be found as

$$\Gamma_\perp = \frac{\eta_2\cos\varphi_i - \eta_1\cos\varphi_t}{\eta_2\cos\varphi_i + \eta_1\cos\varphi_t} = \frac{\dfrac{\eta_0}{\sqrt{3}}\dfrac{1}{2} - \eta_0\dfrac{\sqrt{3}}{2}}{\dfrac{\eta_0}{\sqrt{3}}\dfrac{1}{2} + \eta_0\dfrac{\sqrt{3}}{2}} = \frac{\sqrt{3}/6 - \sqrt{3}/2}{\sqrt{3}/6 + \sqrt{3}/2} = -\frac{1}{2}, \tag{5.489}$$

leading to $a = j$ and

$$\bar{E}^{\text{ref}}(\bar{r}) = \hat{a}_y j E_0 \exp[-j(2\sqrt{3}\pi x - 2\pi z)]. \tag{5.490}$$

[110] Note that

$$\bar{k}_i \cdot \bar{r} = 2\sqrt{3}\pi x + 2\pi z,$$

leading to

$$\bar{k}_i = \hat{a}_x 2\sqrt{3}\pi + \hat{a}_z 2\pi$$
$$k_{ix} = 2\sqrt{3}\pi$$
$$k_{iy} = 2\pi$$
$$k_1 = |\bar{k}_i| = \sqrt{12\pi^2 + 4\pi^2} = 4\pi$$
$$\hat{a}_i = \frac{\bar{k}_i}{k_1} = \hat{a}_x\frac{\sqrt{3}}{2} + \hat{a}_z\frac{1}{2}$$
$$= \hat{a}_x \sin\varphi_i + \hat{a}_z\cos\varphi_i$$
$$\varphi_i = 60°.$$

[111] Then we have

$$k_2 = \omega\sqrt{\mu_2\varepsilon_2} = \omega\sqrt{\mu_0\varepsilon_0}\sqrt{\varepsilon_r}$$
$$= k_1\sqrt{\varepsilon_r} = 4\sqrt{3}\pi$$
$$\bar{k}_t = \hat{a}_t k_2 = (\hat{a}_x\sin\varphi_t + \hat{a}_z\cos\varphi_t)k_2$$
$$= (\hat{a}_x/2 + \hat{a}_z\sqrt{3}/2)4\sqrt{3}\pi.$$

Finally, the transmission coefficients can be found as[112]

$$\tau_{\|} = \frac{\cos\varphi_i}{\cos\varphi_t}(1+\Gamma_{\|}) = \frac{\cos\varphi_i}{\cos\varphi_t} = \frac{1}{\sqrt{3}}, \qquad \tau_{\perp} = 1 + \Gamma_{\perp} = \frac{1}{2}. \tag{5.491}$$

[112]We emphasize that zero reflection ($\Gamma_{\|} = 0$) does not mean unity transmission: i.e. $\tau_{\|} \neq 1$.

This leads to the electric field intensity of the transmitted wave as[113]

$$\bar{E}^{\text{tra}}(\bar{r}) = \left(\hat{a}_x\tau_{\|}2\sqrt{3} - \hat{a}_y\tau_{\perp}j2 - \hat{a}_z\tau_{\|}2\right)\exp[-j(2\sqrt{3}\pi x + 6\pi z)] \tag{5.492}$$

$$= \left(\hat{a}_x 2 - \hat{a}_y j - \hat{a}_z\frac{2}{\sqrt{3}}\right)\exp[-j(2\sqrt{3}\pi x + 6\pi z)]. \tag{5.493}$$

[113]We use

$$\bar{k}_t \cdot \bar{r} = (x/2 + z\sqrt{3}/2)4\sqrt{3}\pi$$
$$= 2\sqrt{3}\pi x + 6\pi z.$$

Also, note that the parallel component of the electric field intensity of the transmitted wave is in the direction of $\hat{a}_x\sqrt{3}/2 - \hat{a}_z/2$.

Note that the incident and transmitted waves are RHEP, while the reflected wave is linearly polarized.

Example 61: Consider a normal incidence of a plane wave on a triangular right prism (Figure 5.46) with permittivity ε_p and permeability μ_p located in a vacuum (ε_0, μ_0). Assume that all prism dimensions are very large compared to the wavelength, and there is no edge effect. In addition, there is no reflection when the plane wave is passing from the prism side to a vacuum. The angle of the prism is $\varphi_p = \sin^{-1}(1/8)$. For the directions defined in Figure 5.46 (i.e. both incident and reflected electric field intensities are defined in the x direction), it is given that $E^{\text{ref}} = \Gamma E^{\text{inc}} = E^{\text{inc}}/5$. Find the relative permittivity and relative permeability of the prism and the angle φ_t: i.e. the angle between the transmitted wave and the normal of the prism.

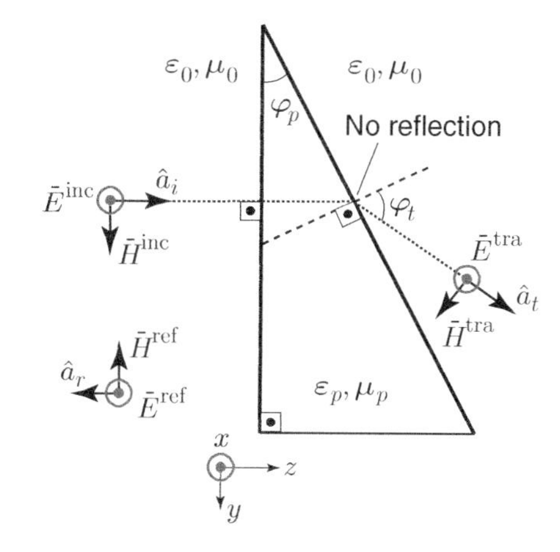

Figure 5.46 Normal incidence of a plane wave on a large triangular right prism.

Solution: Obviously, without a reflection, the plane wave is incident with Brewster's angle onto the boundary from the prism to a vacuum. In addition, this angle is the same as the prism angle: φ_p. Given that the polarization is perpendicular, we have[114]

$$\sin\varphi_p = \sin\varphi_{B\perp} = \sqrt{\frac{1 - \dfrac{\mu_1\varepsilon_2}{\mu_2\varepsilon_1}}{1 - \left(\dfrac{\mu_1}{\mu_2}\right)^2}} = \sqrt{\frac{1 - \dfrac{\mu_p\varepsilon_0}{\mu_0\varepsilon_p}}{1 - \left(\dfrac{\mu_p}{\mu_0}\right)^2}} = \sqrt{\frac{1 - \dfrac{\mu_{pr}}{\varepsilon_{pr}}}{1 - (\mu_{pr})^2}}, \tag{5.494}$$

[114]Note that, on the boundary from the prism to the vacuum, the first and second media are defined as the prism and the vacuum, respectively.

where $\varepsilon_{pr} = \varepsilon_p/\varepsilon_0$ and $\mu_{pr} = \mu_p/\mu_0$. Since $\varphi_p = \sin^{-1}(1/8)$, we obtain

$$\frac{1}{8} = \sqrt{\frac{1 - \dfrac{\mu_{pr}}{\varepsilon_{pr}}}{1 - (\mu_{pr})^2}} \longrightarrow 1 - (\mu_{pr})^2 = 64 - 64\frac{\mu_{pr}}{\varepsilon_{pr}}. \tag{5.495}$$

To derive another relationship between ε_{pr} and μ_{pr}, we can consider the boundary from the vacuum to the prism. Given that $\Gamma = 1/5$, we have[115]

$$\Gamma = \frac{1}{5} = \frac{\eta_2 - \eta_1}{\eta_2 + \eta_1} = \frac{\eta_p - \eta_0}{\eta_p + \eta_0} \longrightarrow 4\eta_p = 6\eta_0 \longrightarrow 4\eta_p^2 = 9\eta_0^2, \tag{5.496}$$

leading to $\mu_{pr}/\varepsilon_{pr} = 9/4$. Inserting this relationship into Eq. (5.495), we obtain[116]

$$\varepsilon_{pr} = 4, \qquad \mu_{pr} = 9. \tag{5.497}$$

The transmission angle can be found via Snell's law:

$$\sqrt{\mu_p \varepsilon_p} \sin\varphi_p = \sqrt{\mu_0 \varepsilon_0} \sin\varphi_t \longrightarrow \sin\varphi_t = \sqrt{\mu_{pr}\varepsilon_{pr}} \sin\varphi_p = 6 \times \frac{1}{8} = \frac{3}{4} \tag{5.498}$$

$$\longrightarrow \varphi_t = \sin^{-1}(3/4). \tag{5.499}$$

[115] Note that, on the boundary from the vacuum to the prism, the first and second media are defined as the vacuum and the prism, respectively.

[116] We have

$$1 - (\mu_{pr})^2 = 64 - 64\frac{\mu_{pr}}{\varepsilon_{pr}} = 64 - 64\frac{9}{4}$$
$$= 64 - 144 = -80$$
$$(\mu_{pr})^2 = 81 \longrightarrow \mu_{pr} = 9.$$

5.3.11 Reflection and Transmission for Two Parallel Interfaces

When there is more than one interface, we need to consider left-traveling and right-traveling waves that satisfy the boundary conditions at each interface. This section presents the idea for a relatively simple case with two interfaces parallel to each other (Figure 5.47). We consider normal incidence and three media, with boundaries at $z = 0$ and $z = d$. The electromagnetic parameters are defined as (ε_1, μ_1), (ε_2, μ_2), and (ε_3, μ_3), for the first (left), second (medium), and third (right) media, respectively. In the first medium, the electric field intensity and magnetic field intensity can be written

$$\bar{E}_1(\bar{r}) = \hat{a}_x E_0^{\text{inc}} \exp(-jk_1 z) + \hat{a}_x E_0^{\text{ref}} \exp(jk_1 z) \tag{5.500}$$

$$\bar{H}_1(\bar{r}) = \hat{a}_y \frac{E_0^{\text{inc}}}{\eta_1} \exp(-jk_1 z) - \hat{a}_y \frac{E_0^{\text{ref}}}{\eta_1} \exp(jk_1 z). \tag{5.501}$$

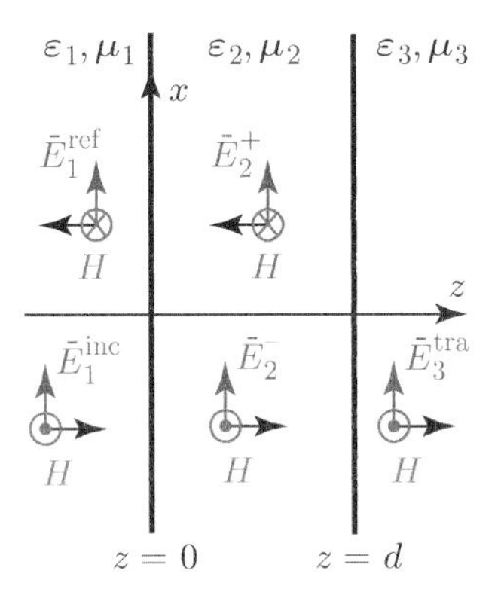

Figure 5.47 Two parallel interfaces separating three media with different electromagnetic parameters.

In these expressions, the incident and reflected waves are shown separately. However, as opposed to single-interface cases considered so far, the reflected wave contains *all* reflections: i.e. the wave reflected from the interface at $z = 0$, as well as the wave reflected from the interface at $z = d$ and then transmitted from the second medium into the first medium, etc. In general, there are infinitely many reflections (inside the second medium) back and forth at the interfaces (Figure 5.48), and the reflected wave in the first medium is a combination of the resulting infinitely many waves that are somehow reflected back into the first medium.

In the second medium, again as a result of infinitely many reflections, there are right-traveling and left-traveling waves, leading to electric and magnetic field expressions:

$$\bar{E}_2(\bar{r}) = \hat{a}_x E_{20}^+ \exp(-jk_2 z) + \hat{a}_x E_{20}^- \exp(jk_2 z) \tag{5.502}$$

$$\bar{H}_2(\bar{r}) = \hat{a}_y \frac{E_{20}^+}{\eta_2} \exp(-jk_2 z) - \hat{a}_y \frac{E_{20}^-}{\eta_2} \exp(jk_2 z). \tag{5.503}$$

Finally, in the third medium, we only have a right-traveling wave

$$\bar{E}_3(\bar{r}) = \hat{a}_x E_0^{\text{tra}} \exp(-jk_3 z), \qquad \bar{H}_3(\bar{r}) = \hat{a}_y \frac{E_0^{\text{tra}}}{\eta_3} \exp(-jk_3 z), \tag{5.504}$$

which is also a combination of infinitely many waves as a result of infinitely many reflections inside the second medium. Basically, the aim is to find constants E_0^{ref}, E_{20}^+, E_{20}^-, and E_0^{tra} in terms of the amplitude of the incident wave: i.e. $E_0^{\text{inc}} = E_0$. Therefore, we need four equations, which can be derived by using the boundary conditions. First, at $z = 0$, the tangential electric and magnetic field intensities must be continuous:

$$\bar{E}_1(z=0) = \bar{E}_2(z=0), \qquad \bar{H}_1(z=0) = \bar{H}_2(z=0). \tag{5.505}$$

Hence, we obtain

$$E_0 + E_0^{\text{ref}} = E_{20}^+ + E_{20}^-, \qquad \frac{E_0}{\eta_1} - \frac{E_0^{\text{ref}}}{\eta_1} = \frac{E_{20}^+}{\eta_2} - \frac{E_{20}^-}{\eta_2}. \tag{5.506}$$

The second set of equations can be obtained by similar continuities at $z = d$ (between the second and third media):

$$E_{20}^+ \exp(-jk_2 d) + E_{20}^- \exp(jk_2 d) = E_0^{\text{tra}} \exp(-jk_3 d) \tag{5.507}$$

$$\frac{E_{20}^+}{\eta_2} \exp(-jk_2 d) - \frac{E_{20}^-}{\eta_2} \exp(jk_2 d) = \frac{E_0^{\text{tra}}}{\eta_3} \exp(-jk_3 d). \tag{5.508}$$

By solving these equations, all amplitudes can be written in terms of E_0.

A particularly important ratio is the one between the reflected and incident waves in the first medium. One can obtain

$$\Gamma = \frac{E_0^{\text{ref}}}{E_0^{\text{inc}}} = \frac{\eta_{eq} - \eta_1}{\eta_{eq} + \eta_1}, \tag{5.509}$$

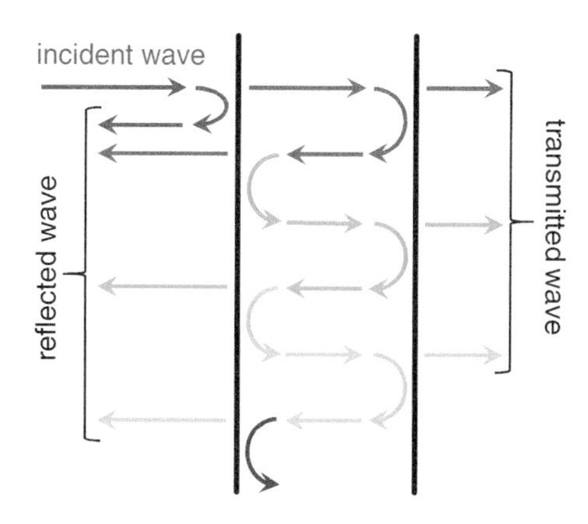

Figure 5.48 An illustration of multiple reflections in a double-interface case. To reach a steady state, infinitely many reflections occur at the interfaces inside the second (middle) medium. Consequently, the reflected wave in the first (left) medium is the superposition of infinitely many waves (the directly reflected wave, as well as those waves that are finally transmitted from the second medium to the first medium after various numbers of internal reflections). Similarly, the transmitted wave in the third (right) medium consists of infinitely many waves (the directly transmitted wave, as well as those waves that are finally transmitted from the second medium to the third medium after various numbers of internal reflections).

where the equivalent impedance can be written[117]

$$\eta_{eq} = \eta_2 \frac{\eta_3 \cos(k_2 d) + j\eta_2 \sin(k_2 d)}{\eta_2 \cos(k_2 d) + j\eta_3 \sin(k_2 d)}. \tag{5.510}$$

[117]Note that for $d = 0$ (i.e. in the absence of the second medium), $\eta_{eq} = \eta_2\eta_3/\eta_2 = \eta_3$. Similarly, if $\eta_2 = \eta_3$ (i.e. if the second medium and the third medium are electromagnetically the same and there is not an actual interface at d), we obtain $\eta_{eq} = \eta_3$.

Interestingly, it is possible to make $\Gamma = 0$: i.e. to obtain zero reflection, which is not possible in the case of a single interface (under normal incidence). For example, if the thickness of the second medium is selected as $d = \lambda_2/4$, where λ_2 is the wavelength (inside the second medium), we have

$$\eta_{eq} = \frac{\eta_2^2}{\eta_3}. \tag{5.511}$$

Then selecting $\eta_{eq} = \eta_1$, we arrive at a condition

$$\eta_2 = \sqrt{\eta_1 \eta_3} \tag{5.512}$$

for zero reflection (without having the same electromagnetic parameters for the first and third media).

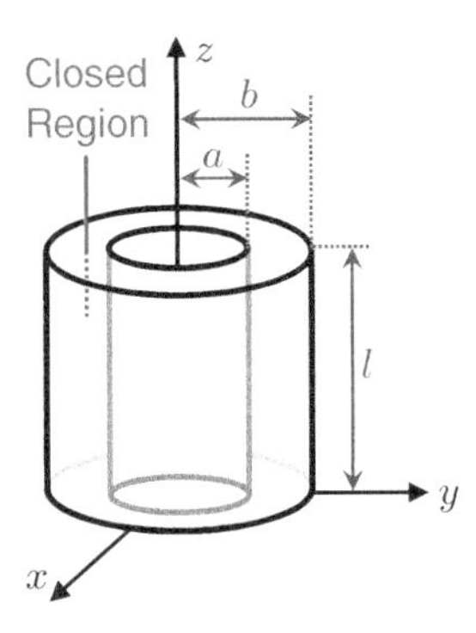

Figure 5.49 A cylindrical closed region defined as $a < \rho < b$ and $0 < z < l$.

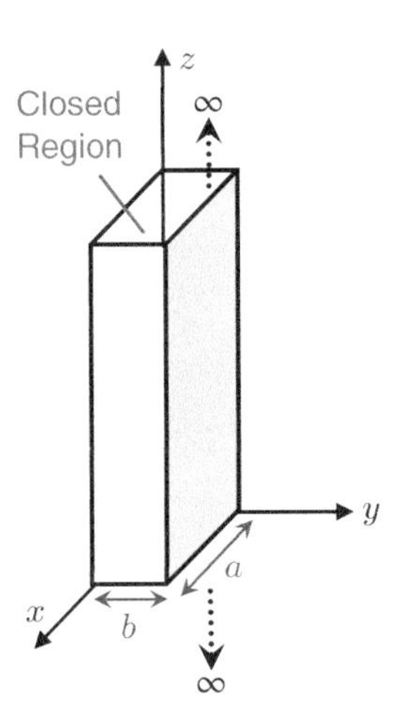

Figure 5.50 An infinitely long rectangular (closed) region defined as $0 \le x \le a$ and $0 \le y \le b$.

5.4 Final Remarks

Each of the first four chapters was devoted to one of four Maxwell's equations, and this chapter has presented basic solutions to Maxwell's equations. Specifically, we have considered plane waves, which are elementary solutions to wave/Helmholtz equations (hence Maxwell's equations) in unbounded regions. Plane waves represent a theoretical set of solutions and provide insight into fundamental characteristics and behaviors of waves, such as propagation, radiation, polarization, reflection, and refraction. Again, it should be emphasized that, away from sources, all waves tend to behave locally like plane waves, which make analyses of plane waves useful and essential to understand electromagnetics in general cases. Finally, more complex waves can usually be decomposed into plane waves (elementary solutions) so that the physics behind complicated scenarios can be explained in terms of the behaviors of their components.

Electromagnetic waves interact with matter, as investigated so far by considering various cases involving dielectric and/or magnetic structures and media. In the next chapter, we focus on the interactions of electromagnetic waves with a very important class of materials: conductors.

5.5 Exercises

Exercise 57: In a cylindrical closed region defined as $a \leq \rho \leq b$ and $0 \leq z \leq l$ (Figure 5.49), the magnetic field intensity is given as

$$\bar{H}(\bar{r},t) = \hat{a}_\phi I_0 \frac{1}{\rho} \cos\left(\frac{\pi z}{l}\right) \cos(\omega t), \tag{5.513}$$

where I_0 (A) is a constant. There is no material or source (electric current/charge) in the region. Find the electric field intensity and the Poynting vector inside the region, as well as the surface electric current density on the inner ($\rho = a$) and outer ($\rho = b$) surfaces, if $\bar{H}(\bar{r},t) = 0$ outside the region.

Exercise 58: In an infinitely long rectangular region defined as $0 \leq x \leq a$ and $0 \leq y \leq b$ (Figure 5.50), the magnetic field intensity is given as

$$\bar{H}(\bar{r}) = H_0 \left[\hat{a}_y jh \left(\frac{b}{\pi}\right) \sin\left(\frac{\pi}{b}y\right) + \hat{a}_z \cos\left(\frac{\pi}{b}y\right)\right] \exp(-jhz), \tag{5.514}$$

where H_0 (A/m) and h (1/m) are constants. There is no material or source (electric current/charge) in the region. Find h and the electric field intensity inside the region.

Exercise 59: Find the complex power flux density (complex Poynting vector) and the time-average power flux density created by a Hertzian dipole without a far-zone approximation.

Exercise 60: Consider the filling stage of a parallel-plate capacitor with circular plates of radii a located in a vacuum (Figure 5.51). The distance between the plates is d. There is a constant current I_0 flowing through the infinitely thin wires connected to the capacitor. Find the Poynting vector between the plates (Figure 5.52), assuming that there are no fringing fields, and show the conservation of energy.

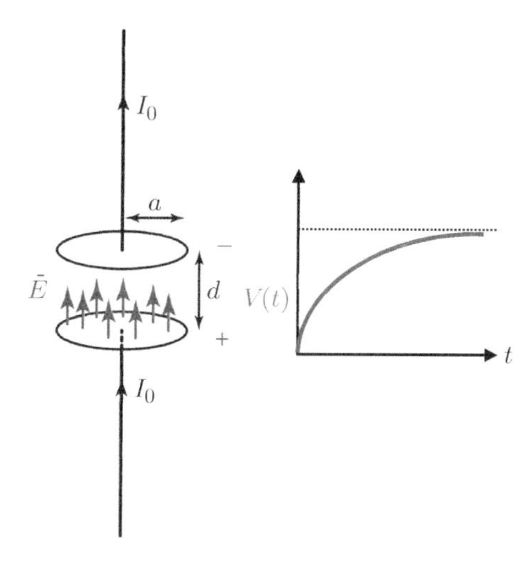

Figure 5.51 Filling stage of a parallel-plate capacitor with circular plates. There is a constant current flow of I_0, while the voltage between the plates increases with respect to time.

Exercise 61: Consider a plane wave with electric field intensity in the phasor domain as

$$\bar{E}(\bar{r}) = [\hat{a}_x 3j + \hat{a}_y 3j + \hat{a}_z 3\sqrt{2}]E_0 \exp[-j60\pi(x-y)], \tag{5.515}$$

where E_0 (V/m) is a constant. Find the polarization of the wave.

Exercise 62: Consider a plane wave with the electric field intensity in the time domain as

$$\bar{E}(\bar{r},t) = \hat{a}_x 2\cos\left(10^8 t - z/\sqrt{3}\right) - \hat{a}_y \sin\left(10^8 t - z/\sqrt{3}\right) \text{ (V/m)}. \tag{5.516}$$

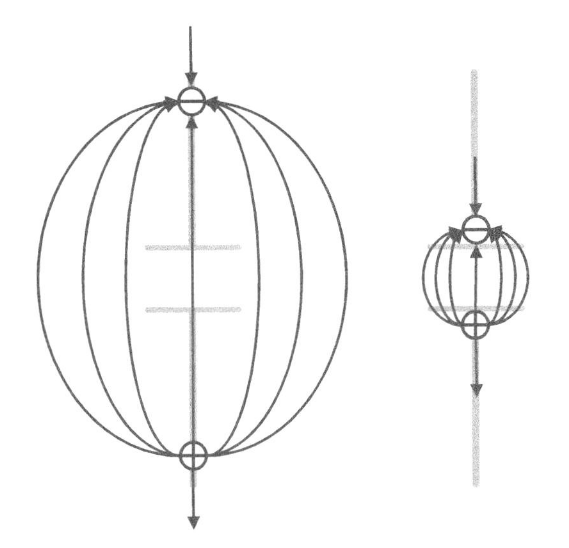

Figure 5.52 Surprisingly, in the case of the filling stage of a parallel-plate capacitor, the Poynting vector appears to be in the negative *radial* direction: i.e. the electromagnetic energy flows into the capacitor from outside. This kind of counter-intuitive result is because an electromagnetic analysis considers all given components (including the host medium in this case) as a whole system. In this particular case, one can consider a pair of positive and negative charges collected on the capacitor plates. When the charges are far away from each other, most of the electric field they create exists outside the capacitor. Then, as they are brought onto the capacitor plates, more and more electric field is confined between the plates. Hence, the electric potential energy indeed flows from outside into the space between the plates.

The medium is dielectric and nonmagnetic ($\mu = \mu_0$). Find the relative permittivity of the medium (approximately), the polarization of the wave, and the magnetic field intensity.

Exercise 63: Consider two (uniform) plane waves described as

$$\bar{E}_1(\bar{r}) = (-\hat{a}_x 6j + \hat{a}_y 4 + \hat{a}_z 3)E_0 \exp(-j6y + j8z) \quad (5.517)$$

$$\bar{E}_2(\bar{r}) = (\hat{a}_x a + \hat{a}_y b + \hat{a}_z 6)E_0 \exp(-j6y + j8z), \quad (5.518)$$

where E_0 (V/m), a, and b are constants. Find the polarization of the first plane wave. In addition, find a and b if the sum of these two plane waves represented by $\bar{E}_3(\bar{r}) = \bar{E}_1(\bar{r}) + \bar{E}_2(\bar{r})$ is an LHCP plane wave.

Exercise 64: Consider an oblique incidence of a (uniform) plane wave

$$\bar{E}^{\text{inc}}(\bar{r}) = (\hat{a}_x\sqrt{3} - \hat{a}_z)E_0 \exp[-j4\pi(\sqrt{3}x + az)], \quad (5.519)$$

where E_0 (V/m) and a are constants, onto an interface at $z = 0$ between a dielectric medium ($z < 0$) with $\varepsilon_1 = 4\varepsilon_0$ permittivity and another dielectric medium ($z > 0$) with $\varepsilon_1 = 2\varepsilon_0$ permittivity (Figure 5.53). Find the phasor-domain expressions for the electric field intensities of the reflected and transmitted waves.

Exercise 65: Consider an oblique incidence of a circularly polarized plane wave

$$\bar{E}^{\text{inc}}(\bar{r}) = (\hat{a}_x\sqrt{3}/2 - \hat{a}_z/2 + \hat{a}_y j)E_0 \exp(-jk_1x/2 - jk_1z\sqrt{3}/2), \quad (5.520)$$

where E_0 (V/m) is a constant, onto an interface at $z = 0$ between a magnetic medium ($z < 0$) with $\mu_1 = 9\mu_0$ permeability and another magnetic medium ($z > 0$) with $\mu_2 = 3\mu_0$ permeability (Figure 5.54). Both media have vacuum permittivity (ε_0). Find an expression for the electric field intensity of the transmitted wave.

Exercise 66: Consider an oblique incidence of a linearly polarized plane wave

$$\bar{E}^{\text{inc}}(\bar{r}) = \hat{a}_y E_0 \exp(-jk_1x/2 - jk_1z\sqrt{3}/2), \quad (5.521)$$

where E_0 (V/m) is a constant, onto an interface at $z = 0$ between a dielectric medium ($z < 0$) with $\varepsilon_1 = 9\varepsilon_0$ permittivity and a vacuum ($z > 0$). Find an expression for the electric field intensity of the transmitted wave.

Exercise 67: Consider an interface at $z = 0$ between a material medium ($z < 0$) with (ε_1,μ_1) permittivity/permeability and a vacuum ($z > 0$). Figure 5.55 depicts the magnitude of the reflection coefficient with respect to the incidence angle from 0° to 30° for both perpendicular and parallel polarizations of (uniform) plane waves. Find ε_1 and μ_1.

Exercise 68: Consider an oblique incidence of a circularly polarized plane wave

$$\bar{E}^{\text{inc}}(\bar{r}) = (\hat{a}_x/2 - \hat{a}_z\sqrt{3}/2 + \hat{a}_y j)E_0 \exp(-jk_1 x\sqrt{3}/2 - jk_1 z/2), \tag{5.522}$$

where E_0 (V/m) is a constant, onto an interface at $z = 0$ between a dielectric medium ($z < 0$) with $\varepsilon_1 = 2\varepsilon_0/\sqrt{3}$ permittivity ($\mu_1 = \mu_0$) and a magnetic medium ($z > 0$) with $\mu_2 = \sqrt{3}\mu_0$ permeability ($\varepsilon_2 = \varepsilon_0$) (Figure 5.56). Find an expression for the electric field intensity of the reflected wave.

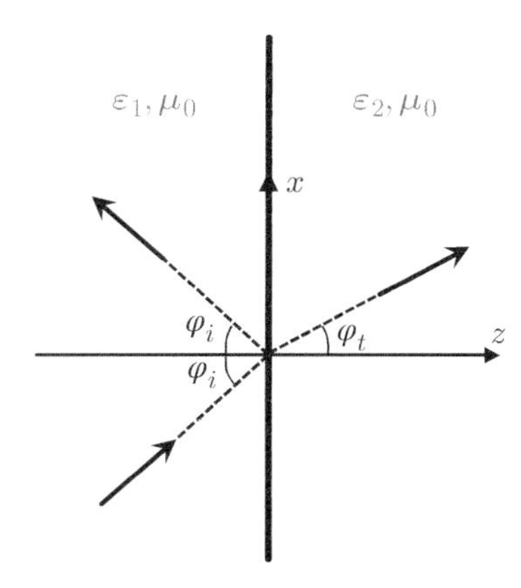

Figure 5.53 Oblique incidence of a plane wave onto an interface between two dielectric media.

5.6 Questions

Question 25: Consider a plane wave propagating in a vacuum with an expression for the magnetic field intensity

$$\bar{H}(\bar{r}) = [\hat{a}_x\sqrt{2}\exp(j\pi/4) - \hat{a}_z(1+j)]H_0 \exp(-jk_0 y), \tag{5.523}$$

where H_0 (A/m) is a constant. Find an expression for the corresponding electric field intensity, as well as the polarization of the wave.

Question 26: Consider a (uniform) plane wave propagating in a dielectric medium with $16\varepsilon_0$ permittivity. The angular frequency of the wave is given as 15×10^8 rad/s, while the electric field intensity is written

$$\bar{E}(\bar{r}) = (\hat{a}_y a + \hat{a}_z j)E_0 \exp[-j(k_x x + 16y)], \tag{5.524}$$

where a, E_0 (V/m), and k_x (rad/m) are constants.

- Find the values of a and k_x, approximating the speed of light in a *vacuum* as $c_0 \approx 3 \times 10^8$ m/s.
- Find an expression for the phasor-domain representation of the magnetic field intensity.
- Find an expression for the time-domain representation of the electric field intensity.

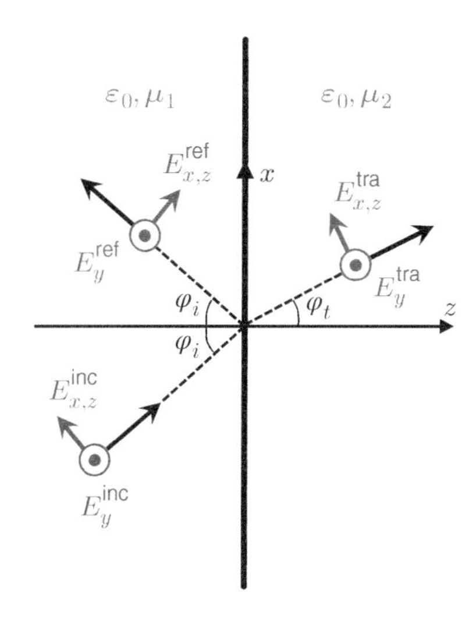

Figure 5.54 Oblique incidence of a circularly polarized plane wave onto an interface between two magnetic media.

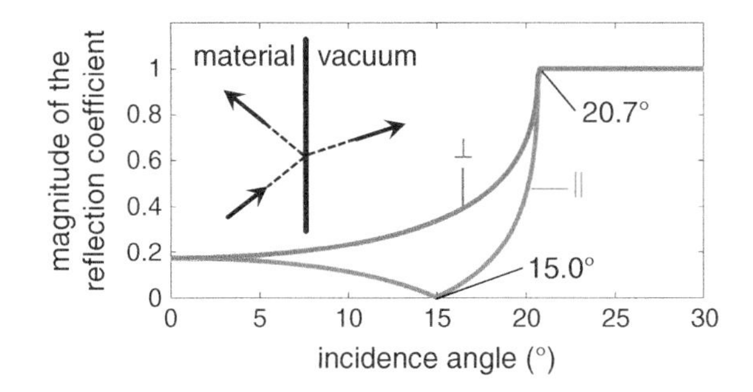

Figure 5.55 Reflection/transmission of plane waves that are incident onto a boundary between a material medium (first medium) and a vacuum (second medium).

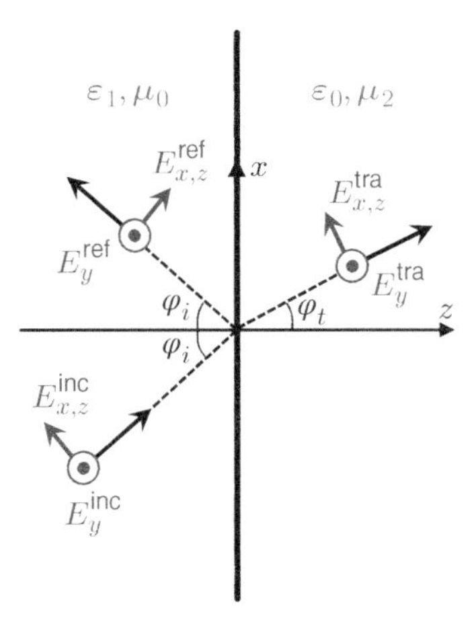

Figure 5.56 Oblique incidence of a circularly polarized plane wave onto an interface between a dielectric medium and a magnetic medium.

Question 27: Consider a plane wave propagating in a dielectric medium with $4\varepsilon_0$ permittivity. The electric field intensity is defined as

$$\bar{E}(\bar{r}) = [\hat{a}_x 4 - \hat{a}_y(1+j) + \hat{a}_z 3]E_0 \exp[-j20\pi(3x - 4z)], \tag{5.525}$$

where E_0 (V/m) is a constant.

- Find the frequency and the propagation direction.
- Find the polarization of the wave.
- Find the time-average power flux density (time-average Poynting vector).

Question 28: A scientist performs an experiment to find the properties (ε_2 and μ_2) of a material. An oblique incidence of a perpendicularly polarized plane wave is considered (Figure 5.57), where the left half-space is a vacuum and the right half-space is filled with the material. When the angle of incidence is 60°, it is measured that the angle of refraction is 30°, while the amplitude of the electric field is reduced by 66.7%: i.e. $|\bar{E}_t| = |\bar{E}_i|/3$. Show that the relative permittivity and relative permeability of the material must be 3/5 and 5, respectively.

Question 29: Consider an oblique incidence of a plane wave

$$\bar{E}^{\text{inc}}(\bar{r}) = [\hat{a}_x a + \hat{a}_y j4 - \hat{a}_z b]E_0 \exp(-j2x - j2\sqrt{3}z), \tag{5.526}$$

where a, b, and E_0 (V/m) are constants, onto an interface at $z = 0$ between a dielectric medium ($z < 0$) with ε_1 permittivity and a vacuum ($z > 0$) (Figure 5.58).

- Find a and b such that the incident wave is circularly polarized. Also find the sense of the polarization.
- Find the permittivity of the first medium (ε_1) such that the reflected wave is linearly polarized.

Question 30: An RHCP plane wave of amplitude $\sqrt{2}$ V/m is sent to an interface between a vacuum and a dielectric medium (Figure 5.59). The frequency is given as 300 MHz.

- Find the angular frequency (ω), wavelengths (λ_0 and λ_2), wave impedances (η_0 and η_2), and wavenumbers (k_0 and k_2) for both the vacuum and the dielectric medium, while approximating the speed of light in a vacuum as $c_0 \approx 3 \times 10^8$ m/s.

- The incident electric field intensity in the phasor domain can be written

$$\bar{E}^{\text{inc}}(\bar{r}) = (\hat{a}_x a_1 + \hat{a}_y a_2 + \hat{a}_z a_3)\exp[j(a_4 x + a_5 z)]. \tag{5.527}$$

  Find all constants a_i, which are possibly complex numbers, for $i = 1, 2, 3, 4, 5$.
- Find the transmission coefficient for perpendicular polarization ($\tau_\perp$).
- Find the transmission coefficient for parallel polarization ($\tau_\|$).
- The transmitted electric field in the phasor domain can be written

$$\bar{E}^{\text{tra}}(\bar{r}) = (\hat{a}_x b_1 + \hat{a}_y b_2 + \hat{a}_z b_3)\exp[j(b_4 x + b_5 z)]. \tag{5.528}$$

  Find all constants b_i, which are possibly complex numbers, for $i = 1, 2, 3, 4, 5$.

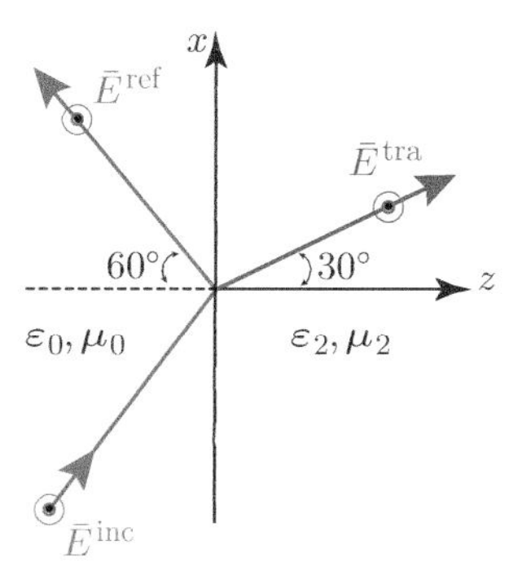

Figure 5.57 Reflection and refraction of a perpendicularly polarized plane wave with oblique incidence onto an interface between a vacuum and a medium with unknown material properties.

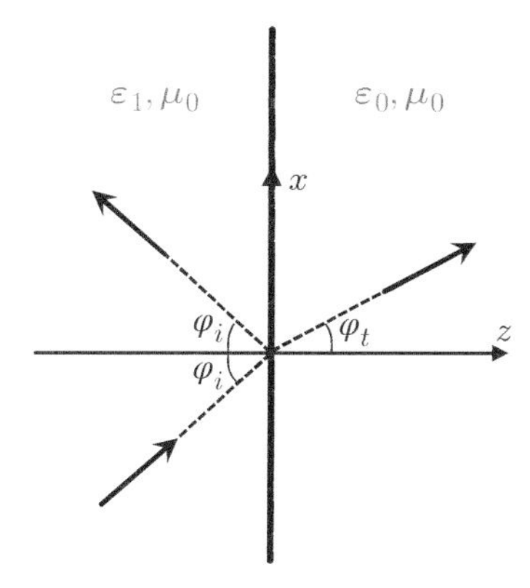

Figure 5.58 Oblique incidence of a plane wave onto an interface between a dielectric medium and a vacuum.

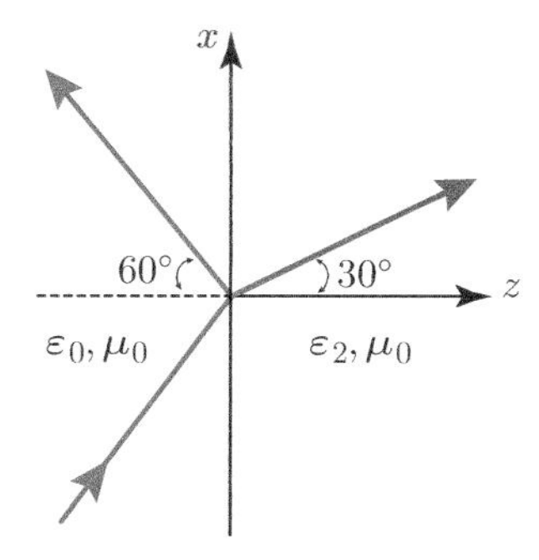

Figure 5.59 Oblique incidence of a plane wave onto an interface between a vacuum and a dielectric medium.

6

Analyses of Conducting Objects

As briefly discussed in Section 0.5, electric current can be considered a collective movement of free electric charges, particularly electrons. This movement can be a specific (but useful) way of injecting electrons to create an electron beam (which can be generated in a vacuum), while the more common form is the conduction current caused by drifting electrons in a conducting object or medium. In Section 1.7, which examines dielectrics, a good dielectric is defined as a material with low conductivity, which is due to a lack of free electrons. In the extreme case of a perfect dielectric (insulator), such as a vacuum,[1] there are no free electrons, so the electrical conduction becomes zero. All of Chapter 2 is devoted to magnetic fields created by electric currents, and we simply omit how these currents are generated when discussing magnetostatics.

Conduction is an important property of all materials, and it is naturally involved in electromagnetics. However, we have not discussed analyses of conducting objects until this stage since conduction is not directly needed to understand Maxwell's equations, while all Maxwell's equations must be understood before thorough analyses of conducting objects. This chapter discusses electromagnetic analyses of conducting objects that are defined macroscopically by conductivity, similar to permittivity and permeability, representing electrical and magnetic properties of materials. Before proceeding, we need to emphasize the following fundamental facts of the macroscopic view of conduction to avoid some common sources of confusion:

- Conduction is caused by drifting (moving) free electrons, but one must be careful about the actual movements of electrons (Figures 6.1 and 6.2). When a voltage is applied across

[1] Air is also considered an excellent insulator. However, high electric field intensity values may cause a dielectric break down that makes the insulator (e.g. air) conductive.

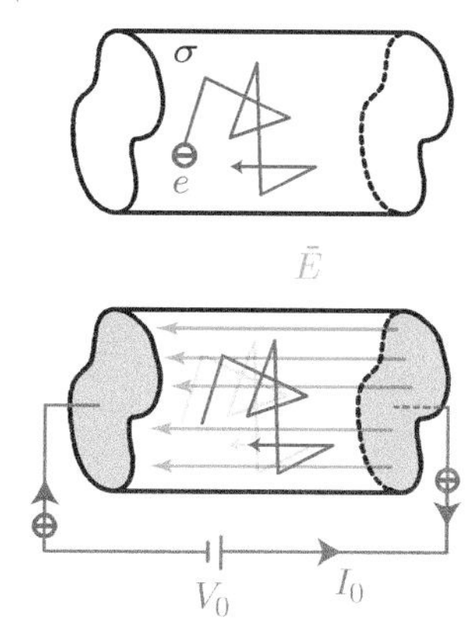

Figure 6.1 In a conducting object, free electrons move at high speeds (thermal movements) – e.g. on the order of 10^6 m/s for some metals – and these movements are random so that there is no net electric current flowing along the object. When an external electric field is applied, average movement of electrons becomes nonzero (electrons drift). The drift is much slower – e.g. on the order of millimeters per second – than thermal movement.

Introduction to Electromagnetic Waves with Maxwell's Equations, First Edition. Özgür Ergül.

Companion website: www.wiley.com/go/ergulmax

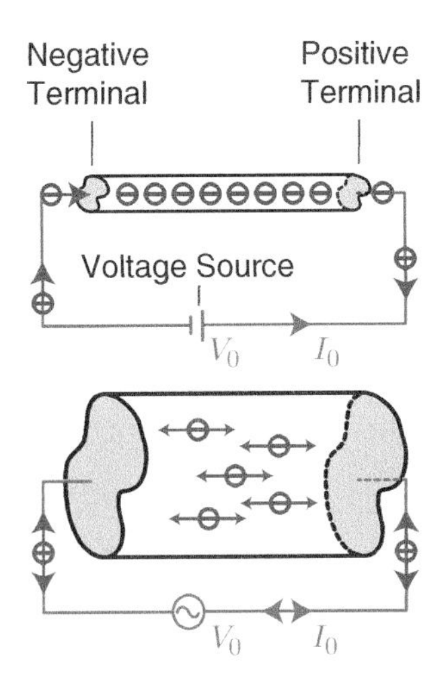

Figure 6.2 For steady currents, the speed of electricity can be considered the electron drift speed, while each electron entering into the conducting object leads to an electron leaving the object on the other side (like a Newton cradle). In dynamic cases involving alternating electric currents, electrons drift back and forth (oscillate), while the speed of electricity corresponds to the *wave* movement of oscillations that can be comparable to the speed of light.

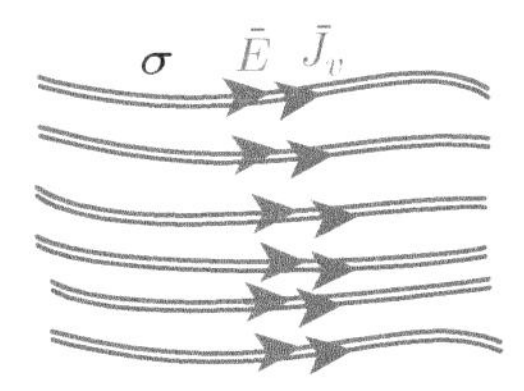

Figure 6.3 According to Ohm's law, electric fields create electric currents in a conducting medium due to free electric charges that drift as a response to electric fields. The electric field intensity and volume electric current density are in the same direction at each and every point, while their ratio is determined by the conductivity.

a conducting object, electric field intensity is formed across it such that free electrons drift in the opposite direction. The drift speed is relatively low in comparison to the much faster thermal (random) movements of electrons. Therefore, drift movements only shift the actual positional variance of electrons. For steady (direct) current (DC), electrons have a constant drift speed, making them effectively migrate from the negative terminal to the positive terminal. In dynamic cases, such as for alternating current (AC), drifting occurs in alternating directions, leading to oscillating positive/negative electric current. The speed of the oscillation *wave* – i.e. the speed of *electricity* but not the drift speed of electrons – can be comparable to the speed of light.

- Conduction, if not perfect, involves a conversion of energy from one form into another form. When free electrons drift in an external electric field, they move from a location with lower potential to another location with higher potential (the opposite of the electric field direction). The *reduced* electric potential energy is converted into thermal energy (like kinetic energy), which is generally released as heat. Obviously, the main source of the energy is that creating the electric field: e.g. a voltage source. This conversion can be purposeful and useful (as in light bulbs), or it may be a side effect to avoid (when transferring energy). In any case, it is considered *loss* (of energy), and an object that demonstrates this kind of loss is called *lossy*. From this perspective, a perfect dielectric with zero conductivity is not lossy (lossless) since an electric field can be formed inside it while there is no free electron and conduction. Ironically, a perfect conductor with infinite conductivity is also lossless since no electric field can be formed inside it, so no voltage can be established across its terminals despite the existence of many drifting electrons.[2,3]

In the following sections, we first consider two fundamental laws (Ohm's law and Joule's law) before using the concept of conductivity in Maxwell's equations to analyze conducting objects.

6.1 Ohm's Law

To integrate the conductivity of conductors into Maxwell's equations, we start with Ohm's law, stating that the volume electric current density due to conduction is proportional to the electric field intensity as (Figure 6.3)

$$\bar{J}_v(\bar{r},t) = \sigma(\bar{r})\bar{E}(\bar{r},t) \tag{6.1}$$

[2]But we will use the common convention that lossless means zero conductivity.

[3]This is perfectly consistent with Joule's law, which indicates that the converted energy is proportional to the voltage times the current. Hence, both voltage (electric field) and current (conduction) must be nonzero for energy conversion.

in a conducting medium. In this simple equation, $\sigma(\bar{r})$ is the conductivity at the observation $\bar{r}$, where the volume electric current density and electric field intensity are evaluated.[4] We also recall that the volume electric current density with unit A/m^2 is related to the electric current through a given surface S as

$$I(t) = \int_S \bar{J}_v(\bar{r}, t) \cdot \overline{ds}. \tag{6.2}$$

In general, conductivity is a property of a material, similar to permittivity and permeability. While it generally provides information about the ohmic losses in an object, two extreme cases should be noted: $\sigma(\bar{r}) = 0$ (perfect insulation without loss) and $\sigma(\bar{r}) = \infty$ (perfect conduction without loss). For $\sigma(\bar{r}) = 0$, we have $\bar{J}(\bar{r}) = 0$: i.e. there is no conduction current despite the existence of electric fields. For $\sigma(\bar{r}) = \infty$, there is a perfect conduction such that any nonzero electric field intensity can create infinite volume electric current density. A physical interpretation is that the electric field intensity must vanish at a location with $\sigma(\bar{r}) = \infty$ so that volume electric current density can be finite.

As a general and illustrative example (Figure 6.4), we consider an object having uniform conductivity $\sigma(\bar{r}) = \sigma$ oriented in the z direction with a cross-section area A and length l. If a constant voltage V_0 is applied across the object, the electric scalar potential can be assumed to be linearly changing[5] (assuming no fringing fields) as[6]

$$\Phi(\bar{r}) = V_0 z/l, \tag{6.3}$$

leading to the electric field intensity

$$\bar{E}(\bar{r}) = -\bar{\nabla}\Phi(\bar{r}) = \hat{a}_z V_0/l. \tag{6.4}$$

Now, using Ohm's law, the volume electric current density is related to the applied voltage as

$$\bar{J}_v(\bar{r}) = \sigma(\bar{r})\bar{E}(\bar{r}) = \hat{a}_z \sigma V_0/l. \tag{6.5}$$

Then the total steady current flowing through the conductor can be found as

$$I_0 = \int_S \bar{J}_v(\bar{r}) \cdot \overline{ds} = \int_S \hat{a}_z \sigma(V_0/l) \cdot \hat{a}_z ds = V_0 \sigma A/l, \tag{6.6}$$

leading to

$$R_0 = \frac{V_0}{I_0} = \frac{l}{\sigma A}, \tag{6.7}$$

σ: conductivity
Unit: siemens/meter (S/m)

[4]Hence, Ohm's law is a pointwise equation.

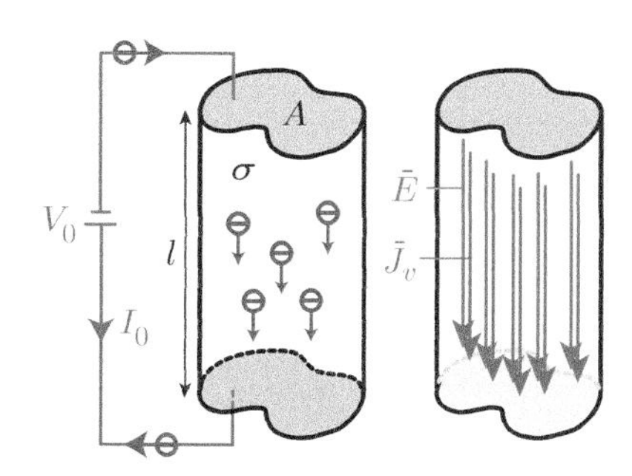

Figure 6.4 An object with uniform conductivity (homogeneous object) with a voltage source connected across it. For finite and nonzero values of the conductivity σ, both electric field intensity and volume electric current density exist inside the object.

[5]Note that, generally, this linear change is a result of the solution of Laplace's equation in a Cartesian coordinate system.

[6]While this derivation provides the general idea, it should be emphasized that we consider steady currents with zero frequency: i.e. static electric and magnetic fields. In dynamic cases, the electric field intensity and volume electric current density may not be approximated as uniform as they tend to be squeezed in the vicinity of surfaces.

R_0: resistance
Unit: ohm (Ω) (volts/ampere)

which is defined as the resistance of the object. This equation indicates that, in a very general and static scenario, the resistance of a conductor is proportional to its length and inversely proportional to the area of its cross section as well as the conductivity. In most cases, however, a more detailed analysis is required to find the resistance of a given structure[7] (even in static cases).

[7] Resistance calculations are detailed in Section 6.8

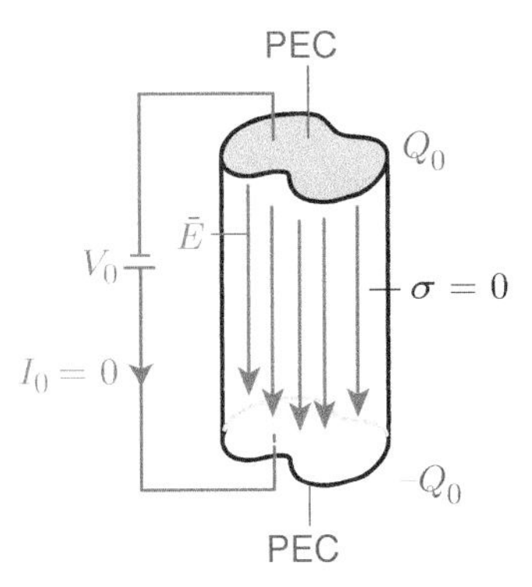

Figure 6.5 When a lossless object is connected to a (voltage) source, a nonzero electric field intensity is created inside, and there is no current flow.

We can now revisit the extreme cases for the resistance expression found above. First, note that $R_0 \to \infty$ as $\sigma \to 0$: i.e. in the case of zero conductivity (perfect insulation without loss) (Figure 6.5). As mentioned several times, an object in this case is described as lossless: e.g. a lossless dielectric. The electric field intensity is created inside the object, but it does not cause a current flow. An ideal capacitor is based on this property: i.e. electric charges accumulated on the plates (which are considered perfectly conducting) create electric fields inside (filled with a vacuum or lossless dielectric), but no current flows through the structure. In the other extreme, we have $R_0 \to 0$ as $\sigma \to \infty$: i.e. in the case of perfect conductivity. Such an object is called a *perfect electric conductor* (PEC). While we can directly claim that the electric field intensity cannot exist inside a PEC (otherwise, there would be infinite volume electric current density), a slightly more detailed view is presented in the following, since the derivation of the resistance above is based on the assumption that there is a voltage across the object (and this is not the case for a PEC).

When considering a PEC connected to a voltage source, a source of confusion is the initial assumption that there should be a voltage across the object. For a PEC, this is not possible since the electric field intensity cannot be established inside the object. Otherwise, an infinite amount of current would pass through the object, leading to an infinite amount of electric power (voltage times current) provided by the voltage source. This conflicting status is well known in circuit theory as a short circuit. However, in the absence of the electric field intensity, the voltage (the electric scalar potential value) everywhere on and inside the PEC is the same. Therefore, visualizing a voltage source across a PEC simply becomes an *invalid* scenario. As an alternative view, one can consider a current source that provides a constant electric current no matter what the voltage is (Figure 6.6). In this case, assuming that the total current flowing through the object is I_0, the volume electric current density can be written[8] $\bar{J}_v(\bar{r}) = \hat{a}_z I_0/A$, while $\bar{E}(\bar{r}) = \bar{J}_v(\bar{r})/\sigma = 0$ is perfectly satisfied.

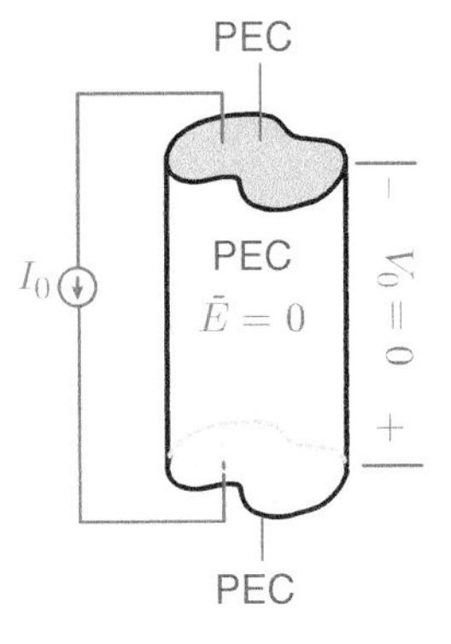

Figure 6.6 When a PEC is connected to a (current) source, no electric field intensity is created inside, and the voltage between its terminals is simply zero.

[8] We still assume that the current is uniformly distributed with steady electric current.

6.2 Joule's Law

As discussed above, energy is converted into heat (it is dissipated) when current flows in a lossy media. Joule's law provides the expression for the amount of converted energy per time.

According to this law, the dissipated power flux density can be written[9,10]

$$\pi_J(\bar{r},t) = \bar{J}_v(\bar{r},t) \cdot \bar{E}(\bar{r},t). \tag{6.8}$$

According to this expression, the dissipated power is proportional to both the volume electric current density and electric field intensity. Therefore, no power is dissipated if one of these quantities is zero.[11] Given that $\bar{J}_v(\bar{r},t) = \sigma(\bar{r})\bar{E}(\bar{r},t)$, we further have

$$\pi_J(\bar{r},t) = \sigma(\bar{r})|\bar{E}(\bar{r},t)|^2 dv. \tag{6.9}$$

Given an object with volume V, the total dissipated power can be obtained as

$$p_J(t) = \int_V \bar{J}_v(\bar{r},t) \cdot \bar{E}(\bar{r},t) dv = \int_V \sigma(\bar{r})|\bar{E}(\bar{r},t)|^2 dv, \tag{6.10}$$

where the density is simply integrated.

To gain more insight into Joule's law, we again consider a static case involving a steady electric current flowing through an object (Figure 6.4) having uniform conductivity $\sigma(\bar{r}) = \sigma$ oriented in the z direction with cross-section area A and length l. Given that $\bar{E}(\bar{r}) = \hat{a}_z V_0/l$ and $\bar{J}(\bar{r}) = \hat{a}_z \sigma V_0/l = \hat{a}_z I_0/A$, the dissipated power can be found:

$$p_J = \int_V \bar{J}_v(\bar{r}) \cdot \bar{E}(\bar{r}) dv = \frac{1}{Al} \int_V V_0 I_0 dv = V_0 I_0. \tag{6.11}$$

This is exactly the application of Joule's law for a resistor

$$p_J = V_0 I_0 = \frac{V_0^2}{R_0} = I_0^2 R_0, \tag{6.12}$$

and it is the same as the power provided (generated) by the voltage source. In dynamic cases, the multiplication

$$p_J(t) = V_0(t) I_0(t) \tag{6.13}$$

gives the instantaneous power, while its average is more frequently used in AC circuits.

[9]Note that instead of dissipated energy, we are more interested in dissipated power. For example, in the case of a steady electric current, the dissipated energy grows monotonously; and especially without reference time information (when the energy is set to zero), the value of the energy does not provide much information. The power provides better information by giving the value of the dissipated energy per second, which can be compared with electric potential energy and magnetic potential energy, or electromagnetic energy in general.

[10]Note that if the electric current density is *impressed* – i.e. is provided by an external source outside the system – the same expression can be considered the *generated* power flux density from the perspective of the system.

[11]For example, $\bar{J}_v(\bar{r},t) = 0$ in a lossless dielectric and $\bar{E}(\bar{r},t) = 0$ in a PEC so that there is no dissipated power in these cases.

p_J: dissipated/generated power
Unit: watt (W) (volts × amperes)

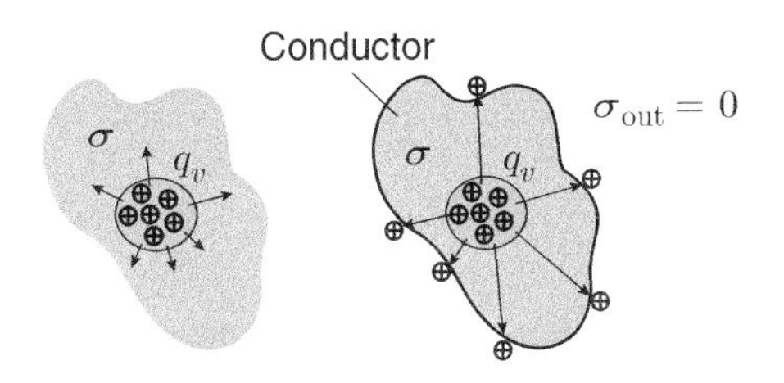

Figure 6.7 If there are electric charges at a location $\bar{r}$ inside a conducting medium ($q_v(\bar{r}) \neq 0$), they must be conducted and moved away ($q_v(t) \to 0$). If we consider a charged conductor located inside a non-conducting (e.g. a vacuum) background, these movements will end when all net electric charge is collected at the boundary: i.e. when they cannot move any further (if they cannot jump out of the object). Theoretically, it takes infinite time to collect all net electric charges on the surface, if the conductivity itself is not infinite.

6.3 Relaxation Time

Relaxation time is an important property of conductors that describes their ability to *remove* electric charge. Although it looks like a variable related to the transient response of conductors, it also provides information about how electric charge is distributed in a steady state (static or dynamic), especially for PECs. To understand this concept in the following, we consider a hypothetical experiment involving a conducting medium with conductivity $\sigma(\bar{r})$ and permittivity $\varepsilon(\bar{r})$. We assume that, at $t = 0$, some electric charge distribution exists with volume electric charge density $q_v(\bar{r})$ (Figure 6.7). First, in a conducting medium, Gauss' law can be manipulated as

$$\bar{\nabla} \cdot \bar{D}(\bar{r}, t) = q_v(\bar{r}, t) \longrightarrow \bar{\nabla} \cdot \left[\frac{\varepsilon(\bar{r})}{\sigma(\bar{r})}\bar{J}_v(\bar{r}, t)\right] = q_v(\bar{r}, t), \tag{6.14}$$

where the current is assumed to be only due to the movement of free charge (conduction current). Then we obtain[12]

$$\bar{\nabla}\left[\varepsilon(\bar{r})/\sigma(\bar{r})\right] \cdot \bar{J}_v(\bar{r}, t) + [\varepsilon(\bar{r})/\sigma(\bar{r})]\bar{\nabla} \cdot \bar{J}_v(\bar{r}, t) = q_v(\bar{r}, t). \tag{6.15}$$

If the change in $\varepsilon(\bar{r})/\sigma(\bar{r})$ is perpendicular[13] to $\bar{J}_v(\bar{r}, t)$ or if it is constant,[14] we derive

$$\bar{\nabla} \cdot \bar{J}_v(\bar{r}, t) = [\sigma(\bar{r})/\varepsilon(\bar{r})]q_v(\bar{r}, t). \tag{6.16}$$

This equation indicates that if there are electric charges at position $\bar{r}$ in a conducting medium, then they *must* move to *generate* electric currents.[15]

At this stage, we can use the relationship between the divergence of the volume electric current density and volume electric charge density: the continuity equation in Eq. (2.202), which can be repeated here as

$$\bar{\nabla} \cdot \bar{J}_v(\bar{r}, t) = -\partial q_v(\bar{r}, t)/\partial t. \tag{6.17}$$

This equation indicates that the generation of electric current requires changing electric charge with respect to time. This basic definition is valid for all cases, and it is not directly related to conduction. On the other hand, combining Eqs. (6.16) and (6.17) leads to an interesting equation:

$$-\frac{\partial q_v(\bar{r}, t)}{\partial t} = [\sigma(\bar{r})/\varepsilon(\bar{r})]q_v(\bar{r}, t). \tag{6.18}$$

[12] Recall the similar derivation for the electric field intensity as

$$\nabla \cdot \bar{D}(\bar{r}, t) = q_v(\bar{r}) \longrightarrow \bar{\nabla} \cdot [\varepsilon(\bar{r})\bar{E}(\bar{r}, t)] = q_v(\bar{r}, t),$$
$$\longrightarrow \bar{\nabla}\varepsilon(\bar{r}) \cdot \bar{E}(\bar{r}, t) + \varepsilon(\bar{r})\bar{\nabla} \cdot \bar{E}(\bar{r}, t) = q_v(\bar{r}, t).$$

If $\bar{J}_v(\bar{r}, t) = \sigma(\bar{r})\bar{E}(\bar{r}, t)$ is used at this stage, we have

$$\frac{1}{\sigma(\bar{r})}\bar{\nabla}\varepsilon(\bar{r}) \cdot \bar{J}_v(\bar{r}, t) + \varepsilon(\bar{r})\bar{\nabla} \cdot \left[\frac{\bar{J}_v(\bar{r}, t)}{\sigma(\bar{r})}\right] = q_v(\bar{r}, t)$$

leading to

$$\frac{1}{\sigma(\bar{r})}\bar{\nabla}\varepsilon(\bar{r}) \cdot \bar{J}(\bar{r}, t) + \varepsilon(\bar{r})\bar{\nabla}\left[\frac{1}{\sigma(\bar{r})}\right] \cdot \bar{J}(\bar{r}, t)$$
$$+ \frac{\varepsilon(\bar{r})}{\sigma(\bar{r})}\bar{\nabla} \cdot \bar{J}(\bar{r}, t) = q_v(\bar{r}, t).$$

This is exactly the same as Eq. (6.15).

[13] This is easily satisfied in a simple medium with constant ε and σ, but it can also be satisfied in some inhomogeneous cases.

[14] It is interesting that $\varepsilon(\bar{r})$ and $\sigma(\bar{r})$ may change with respect to position, while it is sufficient to have a position-independent $\varepsilon(\bar{r})/\sigma(\bar{r})$ to continue this derivation.

[15] Once again, we can check the special case as $\sigma(\bar{r}) = 0 \longrightarrow \bar{\nabla} \cdot \bar{J}_v(\bar{r}, t) = 0$ (no generation), while the $\sigma(\bar{r}) = \infty$ case must be considered in detail.

The general solution of this equation can be written

$$q_v(\bar{r},t) = q_v(\bar{r},t=0)\exp[-t/\tau(\bar{r})], \tag{6.19}$$

where $\tau(\bar{r}) = \varepsilon(\bar{r})/\sigma(\bar{r})$ (with units in seconds) is a time constant called the *relaxation time*.

The expression in Eq. (6.19) indicates that if there is a nonzero conductivity at location $\bar{r}$, the volume electric charge density at that location must decay exponentially as time progresses.[16] In fact, it *must* vanish completely[17] – i.e. $q_v(\bar{r}) \to 0$ – when $t \to \infty$. The case $t \to \infty$ corresponds to the steady state (static or dynamic) where all charge is removed and an equilibrium is reached without any further time dependency. Obviously, after a sufficient time $t \gg \tau(\bar{r})$, we can safely assume that a steady state (equilibrium) is reached; the required time for this assumption is directly related to the time constant.

[16] We assume that no new charges are added externally.

[17] Note the existence of very special cases: i.e. when $\bar{\nabla}[\varepsilon(\bar{r})/\sigma(\bar{r})] \cdot \bar{J}_v(\bar{r},t) \neq 0$, in which nonzero electric charge distributions may remain even in a steady state.

The expression in Eq. (6.19) is pointwise: i.e. it describes how electric charges located at $\bar{r}$ must decay, but it does not indicate where they go. To further exploit the expression, we now consider a conductor located in a non-conducting host medium: e.g. a vacuum (Figure 6.7). If there is a volume electric charge density inside the conductor, Eq. (6.19) indicates that it must vanish. Given that all locations inside the conductor must satisfy the same equality, this tells us that movements should continue until all unbalanced charges (if any) are located conveniently such that they do not move any further. This means they must be located on the conductor's surface in a steady state, if they cannot jump out of the conductor.

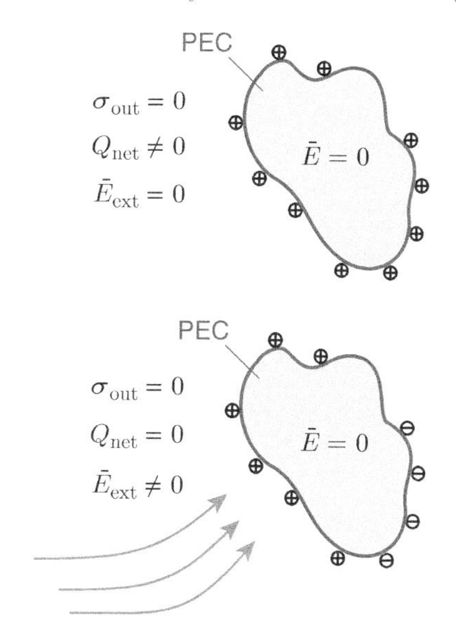

Figure 6.8 For a PEC, (i) all net electric charges must be located at its surface, and (ii) charges should be distributed optimally (whether there is a net amount of electric charge or not) such that the total electric field intensity inside the object is zero (whether there is an external field or not).

Now, we can check what happens when the conductivity of the object is infinite: i.e. we consider a PEC located in a non-conducting medium (Figure 6.8). When $\sigma(\bar{r}) = \infty$, we have $\tau(\bar{r}) = 0$ and $q_v(\bar{r},t) = 0$, independent of $q_v(\bar{r},t=0)$. This strange null equation indicates that any electric charge density that is somehow created inside a PEC must decay *immediately*. In both static and dynamic cases, this means it is impossible to keep an electric charge inside a PEC. Depending on the existence of net electric charge on the surface of a PEC, we may have the following (Figure 6.8):

- If the PEC is charged with some net amount of electric charge, this excessive charge must be located on the surface. Note that this is true for *all* cases: i.e. in transient or steady states (including static steady states and dynamic steady states). This is because the transient time for a PEC is zero.
- If the PEC is not charged, there may not be an accumulation of electric charge on the surface. But if the PEC is exposed to an electromagnetic field, free charges arrange themselves such that the electric field intensity inside the PEC remains zero (which must be satisfied as discussed before[18]). But this arrangement must happen on the surface. This is also true for both transient and steady states.

[18] Two properties of PECs – having zero internal electric field intensity and zero electric charge inside – are often explained together in the literature, but this often causes confusion in dynamic cases. In this book, these two properties are considered separate items to be satisfied, while they complement each other perfectly.

In Section 6.5, electromagnetic analyses of PECs are further detailed, particularly considering two aspects of these objects: i.e. vanishing internal electric field intensity and collection of electric charges on surfaces. These two aspects perfectly complement each other, and they lead to other properties of electromagnetic quantities that must be satisfied in the presence of PECs.

6.4 Boundary Conditions for Conducting Media

Introducing conductivity as a property of materials, we need to revisit boundary conditions. We consider a boundary between two lossy media defined as $\{\varepsilon_1, \sigma_1\}$ and $\{\varepsilon_2, \sigma_2\}$, respectively[19] (Figure 6.9). The boundary conditions for electric fields require

$$\hat{a}_n \cdot [\bar{D}_1(\bar{r}, t) - \bar{D}_2(\bar{r}, t)] = q_s(\bar{r}, t) \quad (\bar{r} \in S), \tag{6.20}$$

where $\hat{a}_n$ is the normal direction towards the first medium, as usual, and $q_s(\bar{r}, t)$ is the surface electric charge density at the observation point $\bar{r}$ on the interface S. Using $\bar{J}_{v,1}(\bar{r}, t) = [\sigma_1(\bar{r})/\varepsilon_1(\bar{r})]\bar{D}_1(\bar{r}, t)$ and $\bar{J}_{v,2}(\bar{r}, t) = [\sigma_2(\bar{r})/\varepsilon_2(\bar{r})]\bar{D}_2(\bar{r}, t)$, we further have[20]

$$\hat{a}_n \cdot \left[\frac{\varepsilon_1(\bar{r})}{\sigma_1(\bar{r})}\bar{J}_{v,1}(\bar{r}, t) - \frac{\varepsilon_2(\bar{r})}{\sigma_2(\bar{r})}\bar{J}_{v,2}(\bar{r}, t)\right] = q_s(\bar{r}, t) \quad (\bar{r} \in S) \tag{6.21}$$

as a boundary condition for the normal component of the electric current density. Inserting relaxation times as $\tau_1(\bar{r}) = \varepsilon_1(\bar{r})/\sigma_1(\bar{r})$ and $\tau_1(\bar{r}) = \varepsilon_1(\bar{r})/\sigma_1(\bar{r})$, this boundary condition can be rewritten

$$\hat{a}_n \cdot \left[\tau_1(\bar{r})\bar{J}_{v,1}(\bar{r}, t) - \tau_2(\bar{r})\bar{J}_{v,2}(\bar{r}, t)\right] = q_s(\bar{r}, t). \qquad (\bar{r} \in S) \tag{6.22}$$

For steady electric currents, there are even more useful boundary conditions for the normal component of the electric current density. Since $\bar{\nabla} \cdot \bar{J}(\bar{r}) = 0$ should be satisfied everywhere, including the boundary, we must have

$$\hat{a}_n \cdot \left[\bar{J}_{v,1}(\bar{r}) - \bar{J}_{v,2}(\bar{r})\right] = 0 \qquad \text{or} \qquad \hat{a}_n \cdot \bar{J}_{v,1}(\bar{r}) = \hat{a}_n \cdot \bar{J}_{v,2}(\bar{r}) \qquad (\bar{r} \in S), \tag{6.23}$$

i.e. the volume electric current density must be continuous.[21] Hence, for steady electric currents, the normal component of the electric current density must be continuous. This can be interpreted as another version of Kirchhoff's current law; currents passing through a boundary must be continuous at each and every point at the boundary. Using the relationship between the electric current density and electric field intensity, we have

$$\hat{a}_n \cdot [\sigma_1(\bar{r})\bar{E}_1(\bar{r}) - \sigma_2(\bar{r})\bar{E}_2(\bar{r})] = 0, \tag{6.24}$$

[19] These two media can also be magnetic, while the following derivation does not involve permeability.

[20] This equation can be found using Eq. (6.16) and the pillbox method used for deriving the boundary conditions for normal electric and magnetic fields (Sections 1.5 and 4.2).

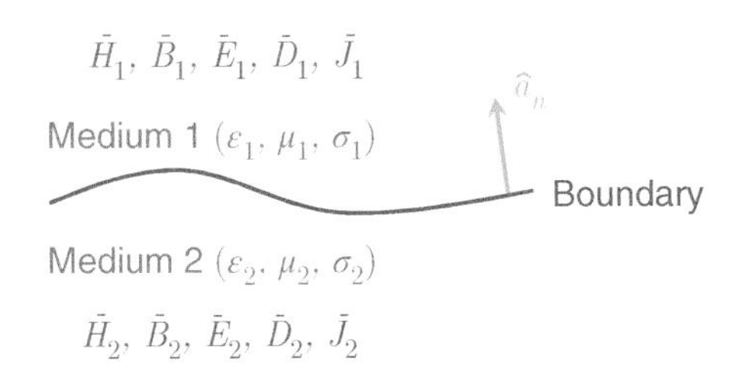

Figure 6.9 A boundary between two general media with different permittivity, permeability, and/or conductivity values.

[21] This is similar to $\bar{\nabla} \cdot \bar{D}(\bar{r}) = 0 \longrightarrow \hat{a}_n \cdot \bar{D}_1(\bar{r}, t) = \hat{a}_n \cdot \bar{D}_2(\bar{r}, t) \qquad (\bar{r} \in S)$.

which can be much more useful than

$$\hat{a}_n \cdot [\varepsilon_1(\bar{r})\bar{E}_1(\bar{r}) - \varepsilon_2(\bar{r})\bar{E}_2(\bar{r})] = q_s(\bar{r}) \quad (\bar{r} \in S), \tag{6.25}$$

if the surface electric charge density is unknown.

Another interesting piece of information can be obtained by inserting Eq. (6.23), which is valid only for steady electric currents, into Eq. (6.22), which is general and valid for all cases. Using $\hat{a}_n \cdot \bar{J}_{v,1}(\bar{r}) = \hat{a}_n \cdot \bar{J}_{v,2}(\bar{r}) = \hat{a}_n \cdot \bar{J}_v(\bar{r})$, we obtain

$$\hat{a}_n \cdot \bar{J}_v(\bar{r}) [\tau_1(\bar{r}) - \tau_2(\bar{r})] = q_s(\bar{r}) \quad (\bar{r} \in S). \tag{6.26}$$

This equation indicates that a difference between the relaxation times of two conducting media causes an accumulation of electric charges at the boundary between them. This is clearly a surface version of Eq. (6.15) with $\bar{\nabla} \cdot \bar{J}_v(\bar{r}) = 0$ as

$$q_v(\bar{r}) = \bar{\nabla}\tau(\bar{r}) \cdot \bar{J}_v(\bar{r}), \tag{6.27}$$

where the gradient term becomes an impulse function when there is a discontinuity in $\tau(\bar{r})$ across the boundary. We finally note that, for steady electric currents, $\hat{a}_n \cdot \bar{J}_{v,1}(\bar{r}) = 0$ if the second medium does not have any conductivity ($\sigma_2(\bar{r}) = 0$). Hence, steady currents must flow tangential to the surface of a conducting object located in a vacuum or in a lossless medium.

6.5 Analyses of Perfectly Conducting Objects

Even though it may sound unphysical, PEC modeling is commonly used as an excellent representation of good conductors in electromagnetics, including both static and dynamic cases. As discussed above, two main properties must be satisfied by a PEC:

- The electric field intensity inside the object must be zero.
- Unbalanced electric charges, if any, must be distributed on the surface of the object. These unbalanced charges may be due to a net amount of electric charges if the PEC is charged or due to the arrangement and localization of electric charges as a response to external effects (e.g. fields).

Surprisingly, these two simple and fundamental properties lead us to proper analysis strategies for PEC structures using Maxwell's equations.

6.5.1 Electric Scalar Potential for PECs

Since the electric field intensity is zero inside a PEC, an electric charge located at a location (on the surface) can be moved to another location (also on the surface) without any decrease or increase in the potential. Therefore, a whole PEC has the same electric scalar potential value, and its surface is equipotential. This is valid in both static and dynamic cases, including transient and steady states, and whether the PEC has a net charge or not. In an electrostatic case, however, the electric field intensity must be perpendicular to the surface of the PEC since it is an equipotential surface[22] and we have $\bar{E}(\bar{r}) = -\bar{\nabla}\Phi(\bar{r})$.

[22]For a dynamic case, this does not directly indicate that the electric field intensity is perpendicular to the surface. However, as discussed below, it is perpendicular *also* in dynamic cases given the boundary conditions.

6.5.2 Boundary Conditions for PECs

For general conducting media, the boundary conditions are discussed in Section 6.4. On the other hand, for PECs, the boundary conditions have very specific and interesting forms since the electric field intensity is zero inside them. Recalling the boundary condition for the tangential electric field intensity, we have[23]

$$\hat{a}_n \times [\bar{E}_1(\bar{r},t) - \bar{E}_2(\bar{r},t)] = 0 \quad (\bar{r} \in S), \tag{6.28}$$

leading to

$$\hat{a}_n \times \bar{E}_1(\bar{r},t) = 0 \quad (\bar{r} \in S) \tag{6.29}$$

if $\bar{E}_2(\bar{r},t) = 0$. Hence, the tangential electric field intensity must be zero on the surface of a PEC. This means the electric field intensity must be normal to the surface of a PEC (Figure 6.10). For a static case, this is already obvious since a PEC is an equipotential object.[24]

[23]For a dynamic case, zero tangential electric field intensity indicates that the tangential component of the magnetic vector potential should also be zero on a PEC.

[24]And, the electric field intensity is simply the gradient of the potential.

Next, we can consider the boundary condition for the normal electric flux density as

$$\hat{a}_n \cdot [\bar{D}_1(\bar{r},t) - \bar{D}_2(\bar{r},t)] = q_s(\bar{r},t) \quad (\bar{r} \in S), \tag{6.30}$$

which becomes

$$\hat{a}_n \cdot \bar{D}_1(\bar{r},t) = q_s(\bar{r},t) \quad (\bar{r} \in S) \tag{6.31}$$

given that[25] $\bar{D}_2(\bar{r},t) = 0$. This means the normal component of the electric flux density must be the surface electric charge density on a PEC.[26] If there is no surface electric charge density at a point $\bar{r} \in S$, we simply have

$$\hat{a}_n \cdot \bar{D}_1(\bar{r},t) = 0. \tag{6.32}$$

[25]Vanishing electric flux density can be discussed in more detail, since it is not straightforward to explain using $\bar{D}(\bar{r},t) = \varepsilon(\bar{r})\bar{E}(\bar{r},t)$ as the permittivity of a PEC is not well defined.

[26]In many cases, $q_s(\bar{r},t)$ is interpreted as the *induced* electric charge density, which is a response by a PEC exposed to an external field.

For a simple host medium, such as a vacuum, using $\bar{D}_1(\bar{r},t) = \varepsilon_1 \bar{E}_1(\bar{r},t)$ leads to $\bar{D}_1(\bar{r},t) = 0$ and $\bar{E}_1(\bar{r},t) = 0$ at $\bar{r} \in S$ when there is no electric charge.

The boundary conditions for magnetic fields can be considered similar to the boundary conditions for electric fields. Since electric and magnetic fields are coupled in dynamic cases, zero electric field intensity means zero magnetic field intensity and magnetic flux density inside a PEC. Consequently, the two main boundary conditions

$$\hat{a}_n \times [\bar{H}_1(\bar{r},t) - \bar{H}_2(\bar{r},t)] = \bar{J}_s(\bar{r},t) \quad (\bar{r} \in S) \tag{6.33}$$

$$\hat{a}_n \cdot [\bar{B}_1(\bar{r},t) - \bar{B}_2(\bar{r},t)] = 0 \quad (\bar{r} \in S) \tag{6.34}$$

turn into

$$\hat{a}_n \times \bar{H}_1(\bar{r},t) = \bar{J}_s(\bar{r},t) \quad (\bar{r} \in S) \tag{6.35}$$

$$\hat{a}_n \cdot \bar{B}_1(\bar{r},t) = 0 \quad (\bar{r} \in S). \tag{6.36}$$

Hence, the normal component of the magnetic flux density must be zero on a PEC surface, while the tangential component of the magnetic field intensity corresponds to the surface electric charge density[27] (Figure 6.10). These boundary conditions are debatable for static cases since vanishing electric field intensity does not necessarily need vanishing magnetic fields. In fact, there is no direct indication that magnetic fields must vanish inside a PEC.[28]

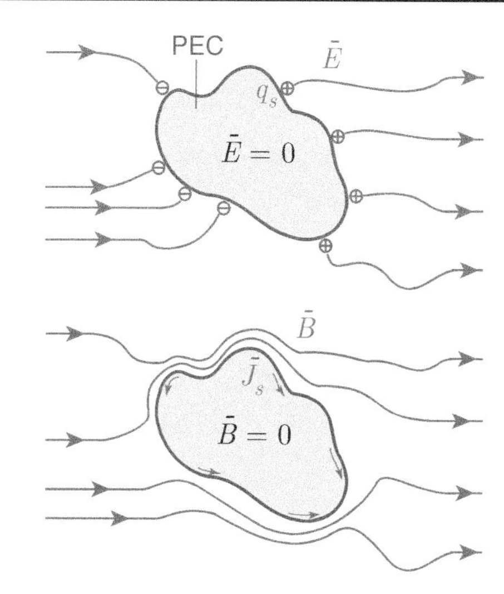

Figure 6.10 The response of a PEC to external fields. Electric field lines must be perpendicular to the surface of the PEC, while magnetic field lines must be parallel according to the boundary conditions. In addition, the discontinuity in the electric flux density and magnetic field intensity must correspond to the induced surface electric charge density and induced surface electric current density, respectively.

[27]In many cases, $J_s(\bar{r},t)$ is interpreted as the induced electric current density, which is a response by a PEC exposed to an external field.

[28]Note that a PEC is not a superconductor, for which magnetic fields are ejected from the interior region due to the Meissner effect.

6.5.3 Basic Responses of PECs

At this stage, with the help of the associated boundary conditions, we have a more complete picture of PECs. We now consider three examples to provide a better understanding of PEC responses:

- Charged PEC: First, we consider a static case involving a PEC with a nonzero net charge located in a vacuum (Figure 6.11). We know that the excess electric charges must accumulate on the surface.[29] Since the electric charge is at rest by definition (electrostatic), it should be distributed on the surface such that it does not move any further. This is consistent with the fact that the tangential electric field intensity is zero on the surface according to the boundary condition.[30] Electric fields are generated outside the PEC due to the electric charges, while the normal electric flux density must be equal to the surface electric charge density on the PEC. Electric field lines can be considered to emerge from the electric charges residing on the surface. Considering the whole scenario, it is obvious that the electric charges must be distributed *optimally* such that the sum of the electric field intensity values created by them is

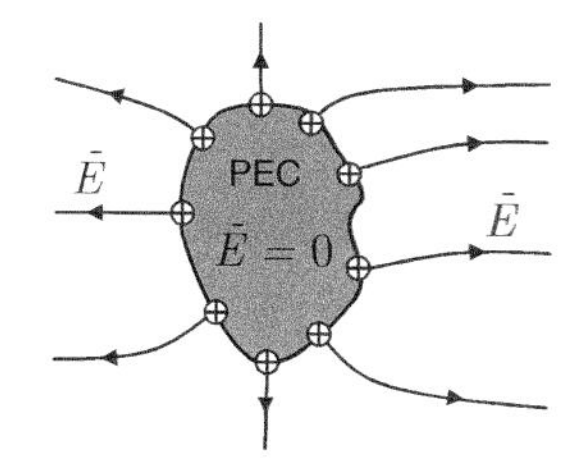

Figure 6.11 A PEC with a nonzero net charge (positive in this case) and the created electric field.

[29]If some electric charge escaped to the inside, it would take zero time for it to accumulate on the surface of the PEC again.

[30]Otherwise, the tangential electric field intensity would apply force to the electric charges and make them move.

zero (perfect cancellation) inside the PEC.[31] For a spherical PEC, which has perfect symmetry, electric charges must be distributed uniformly. For an arbitrary geometry, however, the shape of the surface determines how electric charges should arrange themselves so that the electric field intensity vanishes inside and is normal to the surface just outside the PEC.

[31]This is commonly interpreted as optimal distribution of electric charge for perfect equilibrium while minimizing the energy.

It is commonly known that electric charges tend to accumulate at sharp edges and corners of metallic objects, and our knowledge so far should explain why this occurs. For this purpose, we can consider a PEC with a point (Figure 6.12) where the radius of curvature is small. Recalling that the surface of the PEC is equipotential, the gradient relationship $\bar{E}(\bar{r}) = -\bar{\nabla}\Phi(\bar{r})$ gives large electric field intensity values in the vicinity of the point since the electric scalar potential $\Phi(\bar{r})$ changes quickly with position. The limit of a perfectly sharp tip (zero radius of curvature) leads to a *singularity* in the electric field intensity values.[32] But if the electric field intensity is large, the surface electric charge density must also be large due to the boundary conditions. Therefore, electric charges must accumulate at the tip to satisfy the basic properties of a PEC.

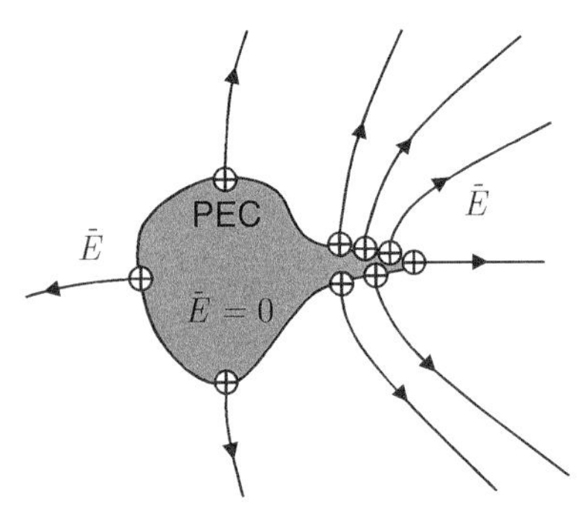

Figure 6.12 A PEC with a nonzero net charge (positive in this case) and the created electric field. The object has a sharp tip where most of the electric charge accumulates.

[32]While a perfectly sharp tip or corner is not physically possible in real life, this is also a common modeling technique (approximation).

- Neutral PEC under an external static electric field: When a PEC is exposed to an external electric field (created by outside sources) (Figure 6.13), free charges of the PEC should arrange themselves such that the basic properties are satisfied. Specifically, although their sum is zero, electric charges appear on the surface to create a secondary electric field that perfectly cancels the external electric field. The *total* (external plus secondary) electric field lines end at negative charges, while they emerge at positive charges. In addition, they must be perpendicular to the surface of the PEC.

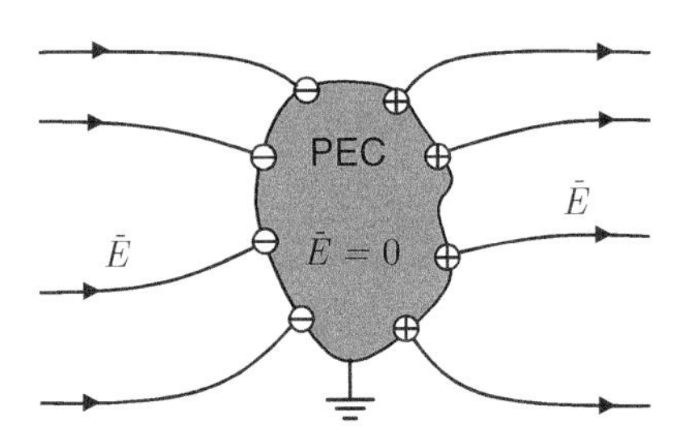

Figure 6.13 A neutral PEC that is exposed to an external electric field. The field lines shown in the figure are for the total (external plus secondary due to the distribution of charges) electric field.

- Neutral PEC under external electromagnetic fields: In dynamic cases, PEC properties do not change, but fields and other quantities change with respect to time. For example, given a PEC under an electromagnetic wave in a vacuum, PEC properties must be satisfied at each time instant. This means when the external electric and magnetic fields change (e.g. due to wave propagation) on the surface, electric charges should arrange themselves accordingly such that electric and magnetic fields are zero inside the PEC, while the boundary conditions are satisfied dynamically as

$$\hat{a}_n \times \bar{E}_1(\bar{r},t) = 0 \quad (\bar{r} \in S) \tag{6.37}$$

$$\hat{a}_n \cdot \bar{D}_1(\bar{r},t) = q_s(\bar{r},t) \quad (\bar{r} \in S) \tag{6.38}$$

$$\hat{a}_n \times \bar{H}_1(\bar{r},t) = \bar{J}_s(\bar{r},t) \quad (\bar{r} \in S) \tag{6.39}$$

$$\hat{a}_n \cdot \bar{B}_1(\bar{r},t) = 0 \quad (\bar{r} \in S) \tag{6.40}$$

for all values of t. We emphasize that the response of a PEC to an external field is *instant* and there is no way to violate the basic properties[33] (null internal fields and charges).

[33]Instant response should not be confused with the time-variant properties of electromagnetic waves that continuously interact with the PEC in accordance with the frequency or external time constants.

6.5.4 Concerns in Geometric Representations of PECs

While the electromagnetic properties of PECs can be well understood, electromagnetic scenarios involving PECs may still be confusing just because it may not be obvious where the PECs are. To illustrate this, we consider cross sections of various perfectly conducting structures (Figure 6.14). When solving electromagnetic problems, the *region of interest* is critical to determine how to approach the problem. When a solid PEC is located in a host medium (I in Figure 6.14), it is typical to omit the interior of the object while the region of interest is outside. In some problems, however, the PEC is simply represented as a surface with *zero* thickness (III in Figure 6.14), while the region of interest is still outside. This problem (III) is theoretically the same as the problem of a solid PEC (I) since the region of interest is outside, and it is not important whether there is an actual material inside the structure.

A theoretical PEC with zero thickness sometimes corresponds to a realistic case with a very thin PEC shell. Such shells (II in Figure 6.14), thin or thick, often represent shield structures that are frequently used in electromagnetic applications. Specifically, if electromagnetic sources are located outside, electric and magnetic fields vanish inside a PEC shell (not just inside the PEC layer). Similarly, if sources are located inside such a shell, electric and magnetic fields disappear outside of it[34] (not just inside the PEC layer). This is a well-known phenomenon of electromagnetic shielding. Obviously, in such a case, both inside and outside of a PEC shell are regions of interest.

In some cases, the inside a PEC is the region of interest, while the outside is out of scope (IV in Figure 6.14). Similar to the opposite case (I in Figure 6.14), the PEC can be represented by an ideal surface with zero thickness. For example, in waveguides and resonators, the inner fields are important, while the outer fields can be vanishingly small or completely zero. For some structures, such as an ideal parallel-plate capacitor (Figure 6.15), the PECs may not even be closed, while the fields are assumed to be confined in certain regions (of interest; e.g. between plates). Specifically, for an ideal parallel-plate capacitor, the plates (PECs) are assumed to be large enough that fringing fields (due to the sharp edges) can be neglected compared to the general behavior of the electric field intensity from the top plate to the bottom plate. This corresponds to assuming that electric charges are uniformly distributed in most parts of the plates. The electric field intensity can then be considered zero outside the region of interest: i.e. the region between the plates. When carefully done, this type of assumption can be extremely useful to facilitate the analysis of the corresponding problem without deforming the true nature of the phenomena.

In another extreme but useful assumption, a current-carrying PEC may be approximated as a wire with zero thickness (Figure 6.16). For example, in an ideal cylindrical inductor, a wire is assumed to be wound on a core material, while it is assumed to carry an electric current squeezed in a zero volume. Obviously, this corresponds to assuming that the wire is perfectly conducting and does not consume energy.

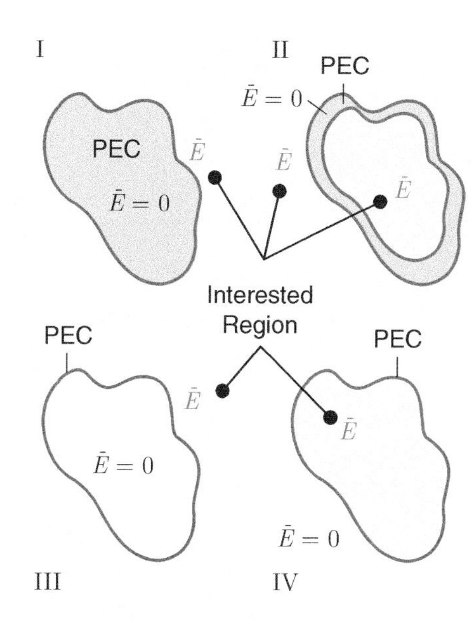

Figure 6.14 Different scenarios involving PECs and various regions of interest.

[34] This is true if the shell is properly grounded such that opposite charges can be collected automatically from the ground to eliminate the effect of sources. For a perfectly neutral shell, inside-to-outside shielding may not work. The most trivial example can be the electric field generated by a static point electric charge enclosed by a spherical PEC (Figure 6.17), where fields are generated outside the shell. In all cases, Maxwell's equations explain what happens.

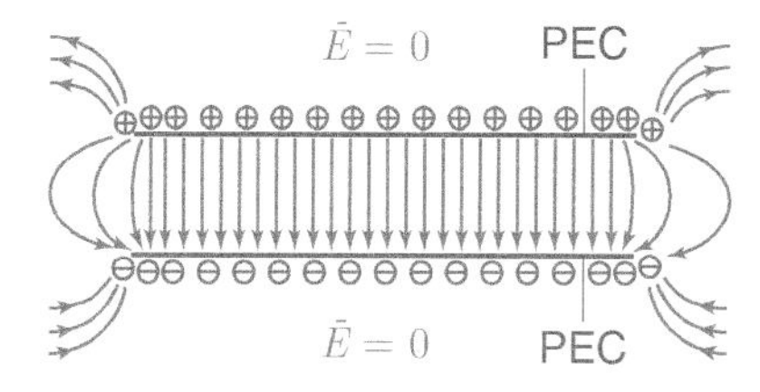

Figure 6.15 A parallel-plate capacitor with large perfectly conducting plates.

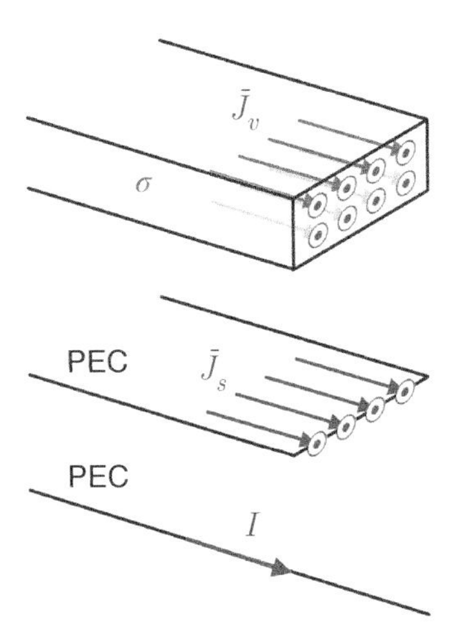

Figure 6.16 In real life, conduction currents have volumetric distributions due to finite conductivity. On the other hand, in the limit of a perfect conductivity, the current flow can be assumed to be squeezed into zero volume, either in the form of a surface current distribution or as a line current, depending on the geometry.

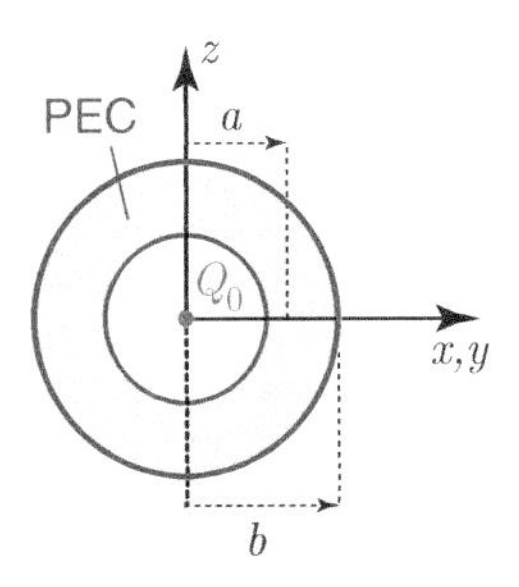

Figure 6.17 A static point electric charge Q_0 located inside a perfectly conducting shell of inner radius a and outer radius b.

6.5.5 Electrostatics for PECs

As discussed in this section, properties of PECs and their responses to external effects can be generalized, including both static and dynamic cases. On the other hand, Maxwell's equations together with our knowledge of PECs let us analyze electrostatic cases in interesting ways. As an example, we consider a static point electric charge Q_0 located at the origin inside a perfectly conducting shell (without any net charge) of inner radius a and outer radius b (Figure 6.17). The entire system is in a vacuum. Due to spherical symmetry, we can assume that $\bar{E}(\bar{r})$ depends only on R and has only an R component: i.e. $\bar{E}(\bar{r}) = \hat{a}_R E_R(R)$. Given a spherical Gauss surface with radius R, we have

$$\oint_S \bar{E}(\bar{r}) \cdot \overline{ds} = 4\pi R^2 E_R(R) = \frac{Q_{\text{enc}}}{\varepsilon_0} \longrightarrow \bar{E}(\bar{r}) = \hat{a}_R \frac{Q_{\text{enc}}}{4\pi\varepsilon_0 R^2}. \tag{6.41}$$

We can now investigate different cases, as follows:

- If $R < a$, $Q_{\text{enc}} = Q_0$ and

$$\bar{E}(\bar{r}) = \hat{a}_R \frac{Q_0}{4\pi\varepsilon_0 R^2} \quad (R < a). \tag{6.42}$$

  Note that the perfect symmetry allows us to employ Gauss' law to find the electric field intensity. Symmetry also indicates that the free charges of the shell should be distributed symmetrically on its inner surface such that their effects at the observation points for $R < a$ should cancel each other.

- If $R > b$, $Q_{\text{enc}} = Q_0$ since the shell is neutral[35] and

$$\bar{E}(\bar{r}) = \hat{a}_R \frac{Q_0}{4\pi\varepsilon_0 R^2} \quad (R > b). \tag{6.43}$$

- If $a < R < b$, we must have $\bar{E}(\bar{r}) = 0$ since the shell is a PEC. But then, according to Gauss' law, the total enclosed charge must be zero: i.e. $Q_{\text{enc}} = 0$. Given the point charge Q_0 at the origin, this is only possible if some negative charge is accumulated on the shell's inner surface. The total amount of this charge must be $-Q_0$ so that the total enclosed charge can be zero when $a < R < b$. Given the spherical symmetry, this charge must be distributed uniformly, leading to the surface charge density:

$$q_{s,\text{inner}}(\bar{r}) = -Q_0/(4\pi a^2). \tag{6.44}$$

[35] Hence, the point electric charge is the only net charge that is enclosed, without any net contribution from the shell.

Indeed, this is consistent with the boundary condition as

$$\bar{D}(R = a^-) = \hat{a}_R \frac{Q_0}{4\pi a^2} \rightarrow \hat{a}_n \cdot \bar{D}(R = a^-) = -\frac{Q_0}{4\pi a^2}, \tag{6.45}$$

given that $\bar{D}(\bar{r}) = 0$ inside the PEC. To complete the picture, one may note that positive charges must accumulate on the shell's outer surface. The total charge should be Q_0 so that the electric field intensity outside the shell becomes only due to the point electric charge. In addition, this satisfies the given neutrality of the conducting shell (Figure 6.18). Obviously, the charge density on the outer surface is different from that on the inner surface:

$$q_{s,\text{outer}}(\bar{r}) = Q_0/(4\pi b^2). \tag{6.46}$$

This is again perfectly consistent with the boundary condition for $\bar{D}(\bar{r})$. Using the expressions for the electric field intensity, the electric scalar potential can be derived via integration, as usual. Given $|\bar{r}| > b$, we have

$$\Phi(\bar{r}) = -\int_{\infty}^{\bar{r}} \bar{E}(\bar{r}) \cdot \overline{dl} = -\frac{Q_0}{4\pi\varepsilon_0} \int_{\infty}^{\bar{r}} \frac{\hat{a}_R}{R^2} \cdot \hat{a}_R dR \tag{6.47}$$

$$= \frac{Q_0}{4\pi\varepsilon_0 |\bar{r}|} \quad (|\bar{r}| > a). \tag{6.48}$$

For $a < |\bar{r}| < b$, there is no electric field intensity, and the potential remains the same inside the shell:

$$\Phi(\bar{r}) = -\int_{\infty}^{b} \bar{E}(\bar{r}) \cdot \overline{dl} = \frac{Q_0}{4\pi\varepsilon_0 b} \quad (a < |\bar{r}| < b). \tag{6.49}$$

This means once a test charge is brought from infinity to the *outer* surface of the shell, it can be moved freely inside the shell without using or releasing energy. Finally, for $|\bar{r}| < a$, we have

$$\Phi(\bar{r}) = \frac{Q_0}{4\pi\varepsilon_0 b} - \int_{a}^{\bar{r}} \bar{E}(\bar{r}) \cdot \overline{dl} = \frac{Q_0}{4\pi\varepsilon_0 b} - \frac{Q_0}{4\pi\varepsilon_0} \int_{a}^{\bar{r}} \frac{\hat{a}_R}{R^2} \cdot \hat{a}_R dR \tag{6.50}$$

$$= \frac{Q_0}{4\pi\varepsilon_0 b} + \frac{Q_0}{4\pi\varepsilon_0 |\bar{r}|} - \frac{Q_0}{4\pi\varepsilon_0 a} \tag{6.51}$$

$$= \frac{Q_0}{4\pi\varepsilon_0} \left(\frac{1}{|\bar{r}|} + \frac{1}{b} - \frac{1}{a} \right) \quad (|\bar{r}| < a). \tag{6.52}$$

Obviously, the electric scalar potential does not have a discontinuity since such a jump would correspond to a jump in the energy and an infinite amount of force (Figure 6.19).

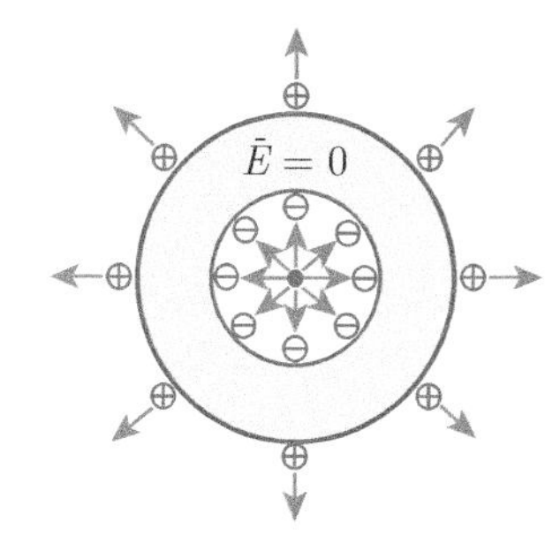

Figure 6.18 Free electric charges of the PEC in Figure 6.17 arrange themselves such that the electric field intensity is zero inside the shell. The charge distributions and electric field lines are shown for a positive Q_0.

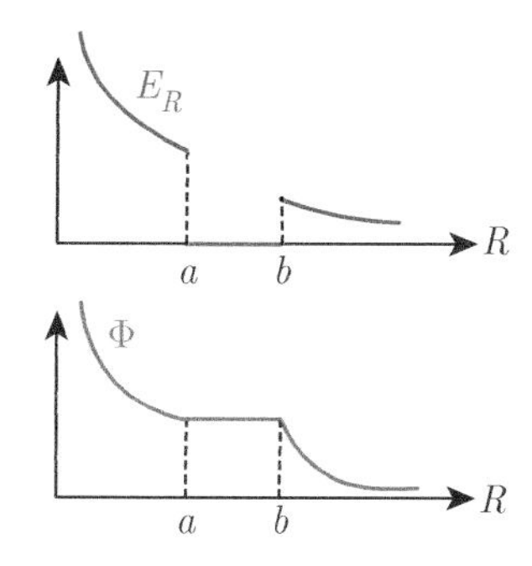

Figure 6.19 For the structure in Figure 6.17, the electric field intensity is zero inside the PEC, while the electric scalar potential possesses a constant value.

6.5.6 Method of Images

When analyzing scenarios involving PECs, it is often useful to replace them via equivalent current and charge sources. This is similar to representing dielectric materials in electric fields via equivalent polarization charges and magnetic materials in magnetic fields via equivalent magnetization charges. In practice, this type of equivalence can be very useful in numerical and computational simulations, where equivalent sources are found approximately at a numerical level.[36] In the context of PECs, the application of the equivalence theorem is often called the *method of images* since the effect of a PEC can be interpreted as creating an image of true sources.

[36]In some simple cases, we may still find equivalent sources analytically, leading to equivalent problems that can be analyzed via Maxwell's equations.

To demonstrate the method of images in electrostatics, we can start with a simple example involving a static point electric charge Q_0 located in a vacuum above an infinitely large planar PEC or simply a perfectly conducting plate (Figure 6.20). We assume that the charge is located at $z = d$, while the planar surface (plate) is on the x-y plane. Heuristically, there should be an image charge below the plate to represent the effect of the PEC. Specifically, there should be a point electric charge Q_i located at (x_i, y_i, z_i) that can replace the plate. But where should we put this charge so that the effect of the plate is represented correctly? To determine this, we can write the electric scalar potential due to the actual charge and its image as[37]

[37]Note that, after the plate is removed, there is a vacuum everywhere.

$$\Phi(\bar{r}) = \frac{Q_0}{4\pi\varepsilon_0}\left[\frac{1}{x^2 + y^2 + (z-d)^2}\right]^{1/2} \tag{6.53}$$

$$+ \frac{Q_i}{4\pi\varepsilon_0}\left[\frac{1}{(x-x_i)^2 + (y-y_i)^2 + (z-z_i)^2}\right]^{1/2}, \tag{6.54}$$

where $\bar{r} = \hat{a}_x x + \hat{a}_y y + \hat{a}_z z$ is the observation point. Now the condition that must be satisfied is $\Phi(z = 0) = \Phi_0$: i.e. the electric scalar potential on the x-y plane must be constant, given the PEC in the original problem.[38] For simplicity, we can assume that the plate is grounded so that $\Phi(z = 0) = 0$. Then we have

[38]Since the surface of a PEC must be an equipotential surface.

$$\frac{Q_i}{4\pi\varepsilon_0}\left[\frac{1}{(x-x_i)^2 + (y-y_i)^2 + z_i^2}\right]^{1/2} = -\frac{Q_0}{4\pi\varepsilon_0}\left[\frac{1}{x^2 + y^2 + d^2}\right]^{1/2} \tag{6.55}$$

to be satisfied for all x and y. This equality can be satisfied if one selects[39] $Q_i = -Q_0$, $x_i = 0$, $y_i = 0$, and $z_i = -d$. This means the image charge should be $-Q_0$ (negative of the actual charge), while it must be placed just opposite of the actual charge at distance d from the plate location.

[39]But this does not prove that this is a unique solution.

If the electric scalar potential is updated accordingly, we have

$$\Phi(\bar{r}) = \frac{Q_0}{4\pi\varepsilon_0}\left\{\left[\frac{1}{x^2+y^2+(z-d)^2}\right]^{1/2} - \left[\frac{1}{x^2+y^2+(z+d)^2}\right]^{1/2}\right\} \tag{6.56}$$

or

$$\Phi(\bar{r}) = \frac{Q_0}{4\pi\varepsilon_0}\left(\frac{1}{R^-} - \frac{1}{R^+}\right), \tag{6.57}$$

where

$$R^- = \sqrt{x^2+y^2+(z-d)^2} \quad \text{and} \quad R^+ = \sqrt{x^2+y^2+(z+d)^2}. \tag{6.58}$$

Using this expression, the electric field intensity can be found:

$$\bar{E}(\bar{r}) = -\bar{\nabla}\Phi(\bar{r}) = -\hat{a}_x\frac{\partial}{\partial x}\Phi(\bar{r}) - \hat{a}_y\frac{\partial}{\partial y}\Phi(\bar{r}) - \hat{a}_z\frac{\partial}{\partial z}\Phi(\bar{r}) \tag{6.59}$$

$$= \hat{a}_x\frac{Q_0}{4\pi\varepsilon_0}\left[\frac{x}{(R^-)^3} - \frac{x}{(R^+)^3}\right] + \hat{a}_y\frac{Q_0}{4\pi\varepsilon_0}\left[\frac{y}{(R^-)^3} - \frac{y}{(R^+)^3}\right]$$
$$+ \hat{a}_z\frac{Q_0}{4\pi\varepsilon_0}\left[\frac{z-d}{(R^-)^3} - \frac{z+d}{(R^+)^3}\right]. \tag{6.60}$$

This is similar the electric field intensity of an electric dipole.

Once a problem is solved via the method of images, it can be interesting to go back to the original problem for a physical interpretation. In the *original* problem, the electric scalar potential and electric field intensity given above are valid only for $z > 0$, while $\Phi(\bar{r}) = 0$ and $\bar{E}(\bar{r}) = 0$ for $z < 0$. Then given the normal component of the electric field,[40] there should be some charge accumulation on the surface of the PEC. This can be found by using the expression for the electric flux density just above the PEC as

$$\bar{D}(z=0^+) = \varepsilon_0\bar{E}(z=0^+) \tag{6.61}$$

$$= \hat{a}_z\varepsilon_0\frac{Q_0}{4\pi\varepsilon_0}\left(\frac{-d}{[x^2+y^2+d^2]^{3/2}} - \frac{d}{[x^2+y^2+d^2]^{3/2}}\right) \tag{6.62}$$

or

$$\bar{D}(z=0^+) = -\hat{a}_z\varepsilon_0\frac{Q_0}{2\pi\varepsilon_0}\frac{d}{[\rho^2+d^2]^{3/2}}. \tag{6.63}$$

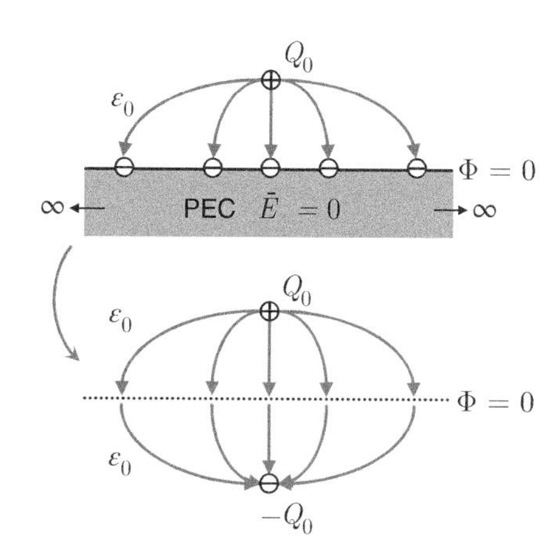

Figure 6.20 A point electric charge located above an infinitely large planar PEC (or simply a perfectly conducting plate) can be modeled via the method of images. In these figures, we assume that the PEC is kept at 0 potential. In the actual problem, the point electric charge attracts negative charges that are accumulated non-uniformly on the surface of the PEC. The electric field intensity is zero for $z < 0$. The total amount of the accumulated charge can be found to be $-Q_0$. In the equivalent problem, the PEC is represented as a single point charge $-Q_0$ (image) located symmetrically at $z = -d$. Both the actual charge and its image are assumed to be located in a vacuum.

[40] Indeed, there is no tangential component due to the PEC.

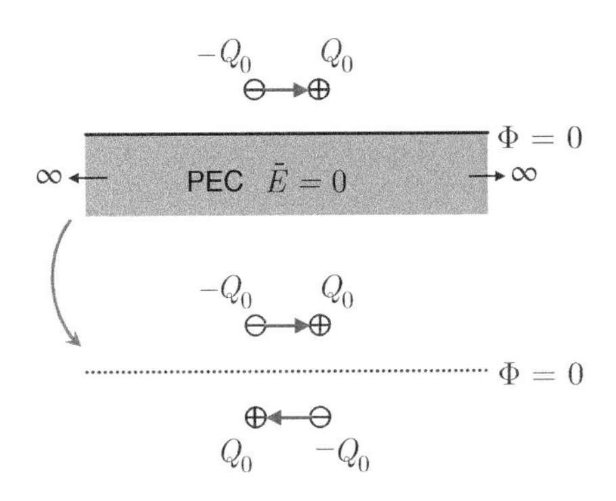

Figure 6.21 Application of the method of images to an electric dipole above an infinitely large planar PEC.

Hence, given the related boundary condition, the surface electric charge density induced on the PEC due to the static point electric charge can be found:

$$q_s(z=0) = \hat{a}_z \cdot \bar{D}(z=0^+) = -\varepsilon_0 \frac{Q_0}{2\pi\varepsilon_0} \frac{d}{[\rho^2 + d^2]^{3/2}}. \tag{6.64}$$

These charges can be assumed to come from the ground connection (or from infinity, since the PEC *touches* infinity) when the point electric charge is brought from infinity. Note that the total induced charge is

$$Q_s(z=0) = -\varepsilon_0 \frac{Q_0}{2\pi\varepsilon_0} \int_0^{2\pi} \int_0^{\infty} \frac{d}{[\rho^2 + d^2]^{3/2}} \rho d\rho d\phi \tag{6.65}$$

$$= -\varepsilon_0 \frac{Q_0}{2\pi\varepsilon_0} 2\pi d \left[-\frac{1}{\sqrt{\rho^2 + d^2}} \right]_0^{\infty} = -\varepsilon_0 \frac{Q_0}{2\pi\varepsilon_0} 2\pi d \frac{1}{d} = -Q_0. \tag{6.66}$$

Hence, the point electric charge causes the accumulation of the same amount of charge with an opposite sign.[41]

[41] We further note that if the point electric charge is brought to the surface of the PEC, we have $\bar{E}(\bar{r}) = 0$ everywhere due to the perfect cancellation of the charge and its image.

The method of images can be used (but may not always be useful) for arbitrary scenarios, including static and dynamic cases, as well as for arbitrary charge and current distributions. For example, an electric dipole located above a planar PEC creates an image (which can replace the effect of the PEC) as an electric dipole in the reverse direction (Figure 6.21). This can be understood by considering the image of the two actual charges separately. Similarly, an electric current flowing in a certain direction creates an image electric current in the reverse direction (Figure 6.22).

An interesting confusion arises when considering electric charges/currents on PECs. For example, according to the method of images, a time-harmonic electric current and its image should cancel each other (considering the equivalent problem: e.g. see Figure 6.22), leading to zero electric and magnetic fields everywhere. This means an externally *impressed* electric current distribution on a PEC does *not* radiate. On the other hand, in scattering problems (particularly in radar scenarios), incident fields create *induced* electric currents on perfectly conducting bodies, which are often employed (allowed to radiate) to find scattered fields. This is because induced electric currents are *responses* of PECs: i.e. they are not impressed currents whose interactions with PECs need to be represented by images.

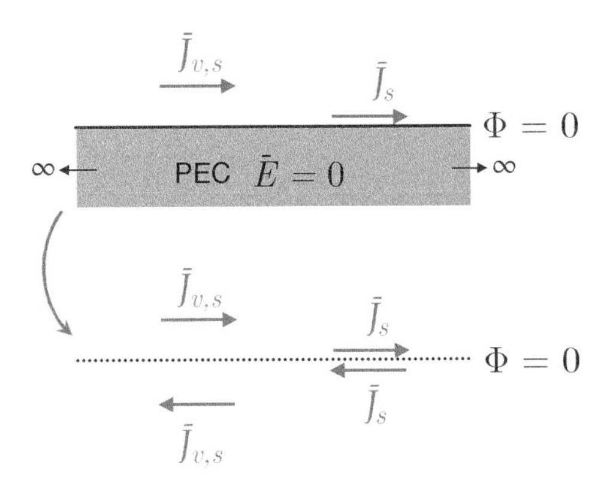

Figure 6.22 Application of the method of images to electric current distributions above an infinitely large planar PEC.

As another example, we can consider a grounded spherical PEC of radius a located at the origin (Figure 6.23). A point charge Q_0 is located at $(b, 0, 0)$, where $b > a$. If the problem is analyzed via the method of images, there should be an image Q_i at $(d, 0, 0)$ with $d < a$ to represent the conductor. Since the electric scalar potential on the surface of the sphere must be zero, we have

$$\frac{Q_0}{4\pi\varepsilon R^+} + \frac{Q_i}{4\pi\varepsilon R^-} = 0 \tag{6.67}$$

$$\frac{Q_0}{\sqrt{(b-x)^2+y^2+z^2}} + \frac{Q_i}{\sqrt{(d-x)^2+y^2+z^2}} = 0, \tag{6.68}$$

where $x^2+y^2+z^2=a^2$. Then we obtain

$$\frac{Q_0}{\sqrt{b^2-2bx+a^2}} + \frac{Q_i}{\sqrt{d^2-2dx+a^2}} = 0 \qquad |x| \leq a. \tag{6.69}$$

A set of selections to satisfy this equation is

$$Q_i = -Q_0 a/b, \qquad d = a^2/b, \tag{6.70}$$

indicating a negative point charge (assuming positive Q_0) located asymmetrically inside the sphere as the image.

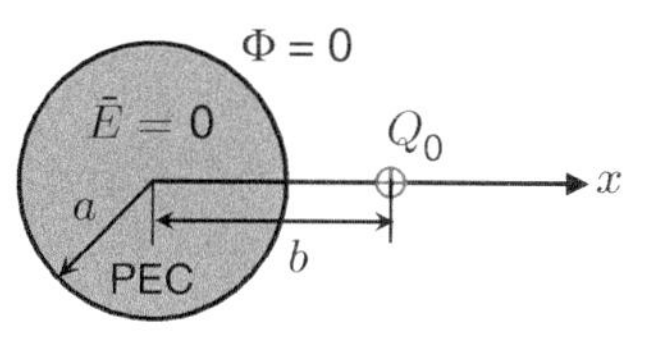

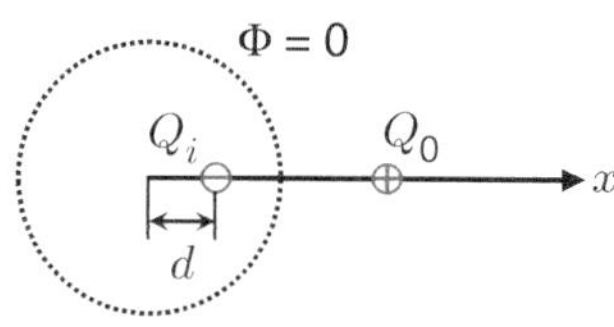

Figure 6.23 Application of the method of images to a point electric charge located near a spherical PEC.

6.5.7 Examples

Example 62: Consider a perfectly conducting cylinder of radius a inside a perfectly conducting cylindrical surface of radius $b > a$, and a vacuum between them (Figure 6.24). The overall structure is infinitely long in the z direction and located in a vacuum, and it has a zero net electric charge. Let the voltage between the inner and outer PECs be constant V_0. Find the electric field intensity everywhere.

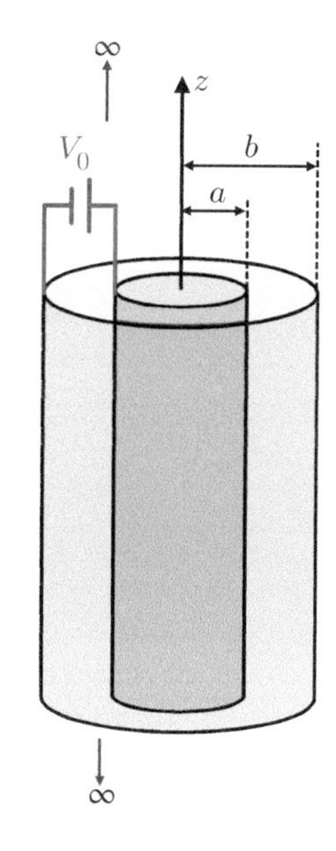

Figure 6.24 An infinitely long coaxial structure consisting of a cylindrical PEC inside a perfectly conducting cylindrical surface.

Solution: We can start by assuming $Q_{l,\text{inner}}$ and $Q_{l,\text{outer}}$ (C/m) charge densities (per length) at $\rho = a$ and $\rho = b$, respectively, given the voltage between the PECs. Since the entire structure is charge-neutral, we must have

$$2\pi a Q_{l,\text{inner}} = 2\pi b Q_{l,\text{outer}}, \tag{6.71}$$

given uniform distributions of charges. Due to the cylindrical symmetry, the electric field intensity depends only on ρ, and it has only ρ component (Figure 6.25): i.e. $\bar{E} = \hat{a}_\rho E_\rho(\rho)$. Then using Gauss' law, we arrive at

$$\oint_S \bar{E}(\bar{r}) \cdot \overline{ds} = 2\pi\rho E_\rho = \frac{Q_{l,\text{inner}}}{\varepsilon_0} \quad (a < \rho < b), \tag{6.72}$$

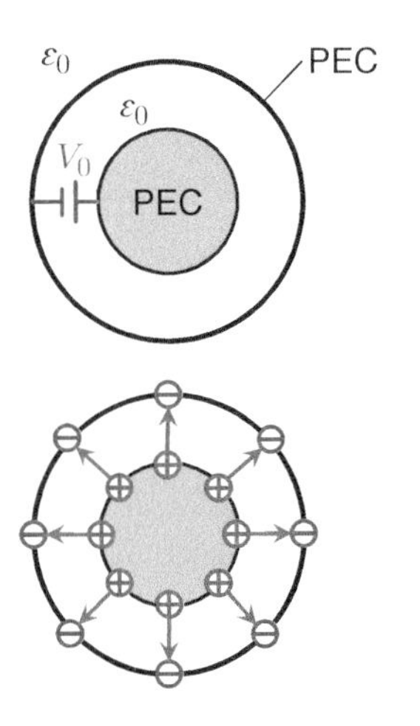

Figure 6.25 Cross-sectional view of the coaxial structure in Figure 6.24. Considering voltage V_0 between the PECs, there should be uniform electric charge accumulation, leading to the electric field intensity in the radial direction.

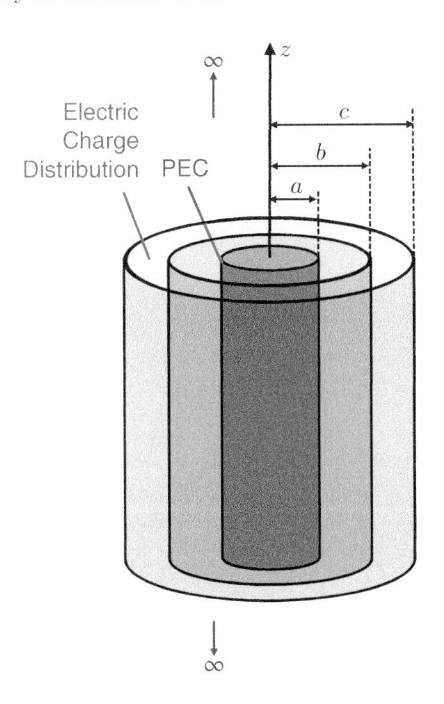

Figure 6.26 A coaxial structure consisting of a perfectly conducting cylinder inside a volume electric charge distribution as a cylindrical layer.

leading to

$$\bar{E}(\bar{r}) = \hat{a}_\rho \frac{Q_{l,\text{inner}}}{2\pi\varepsilon_0\rho} \quad (a < \rho < b). \tag{6.73}$$

In addition, we have $\bar{E}(\bar{r}) = 0$ for $\rho < a$ and $\rho > b$. To write nonzero $\bar{E}(\bar{r})$ for $a < \rho < b$ in terms of voltage V_0, note that the electric scalar potential is the line integral of the electric field intensity. Hence, we derive

$$V_0 = -\int_b^a \bar{E}(\bar{r}) \cdot \overline{dl} = \frac{Q_{l,\text{inner}}}{2\pi\varepsilon_0} \ln\left(\frac{b}{a}\right). \tag{6.74}$$

Therefore, the charge density is related to the voltage as

$$Q_{l,\text{inner}} = \frac{V_0 2\pi\varepsilon_0}{\ln(b/a)}, \tag{6.75}$$

which can be used to obtain

$$\bar{E}(\bar{r}) = \hat{a}_\rho \frac{V_0}{\rho \ln(b/a)} \quad (a < \rho < b). \tag{6.76}$$

Example 63: Consider an infinitely long coaxial structure consisting of a perfectly conducting cylinder of radius a (Figures 6.26 and 6.27). The conductor is charged with Q_{l0} (C/m) static electric charge per length, and it is enclosed in a uniformly distributed volume electric charge density for $b < \rho < c$ (where $b > a$) with $-Q_{l0}$ charge per length. Other than the PEC, there is a vacuum everywhere. Find the electric field intensity everywhere. Also find the electric scalar potential at $\rho = a$, $\rho = b$, and $\rho = c$, if $\Phi(\rho = 0) = 0$.

Solution: Obviously, $\bar{E}(\bar{r}) = 0$ for $\rho < a$: i.e. inside the PEC. For $\rho > a$, using the integral form of Gauss' law, we have

$$2\pi\rho E_\rho(\rho) = \frac{Q_{l,\text{enc}}(\rho)}{\varepsilon_0} \longrightarrow \bar{E}(\bar{r}) = \hat{a}_\rho \frac{Q_{l,\text{enc}}(\rho)}{2\pi\varepsilon_0\rho}, \tag{6.77}$$

where $Q_{l,\text{enc}}(\rho)$ is the enclosed electric charge per length. For $a < \rho < b$, we have $Q_{l,\text{enc}}(\rho) = Q_{l0}$, leading to

$$\bar{E}(\bar{r}) = \hat{a}_\rho \frac{Q_{l0}}{2\pi\varepsilon_0\rho} \quad (a < \rho < b). \tag{6.78}$$

For $b < \rho < c$, the enclosed electric charge needs to be found by considering the volume electric charge density: i.e. $q_v(\bar{r}) = -Q_{l0}/[\pi(c^2 - b^2)]$, leading to

$$Q_{l,\text{enc}}(\rho) = Q_{l0} - Q_{l0}\frac{\pi(\rho^2 - b^2)}{\pi(c^2 - b^2)} = Q_{l0}\left(\frac{c^2 - \rho^2}{c^2 - b^2}\right) \quad (b < \rho < c). \tag{6.79}$$

Hence, we obtain

$$\bar{E}(\bar{r}) = \hat{a}_\rho \frac{Q_{l0}}{2\pi\varepsilon_0 \rho}\left(\frac{c^2 - \rho^2}{c^2 - b^2}\right) \quad (b < \rho < c). \tag{6.80}$$

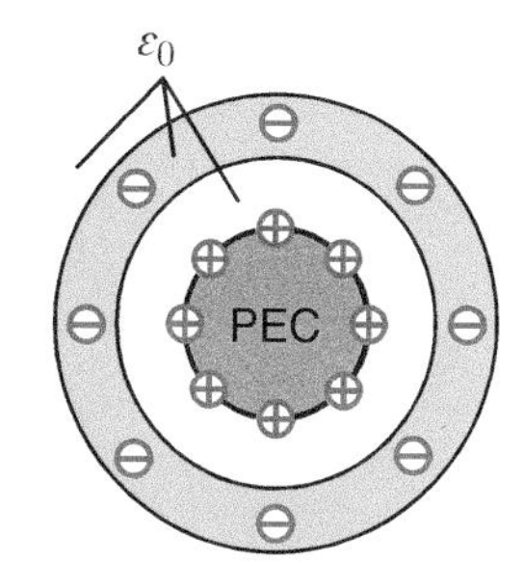

Figure 6.27 Cross-sectional view of the structure in Figure 6.26, assuming a positive value for Q_{l0}.

Finally, for $\rho > c$, we have $Q_{l,\text{enc}}(\rho) = 0$, and we again have $\bar{E}(\rho) = 0$.

Considering the electric scalar potential, we first have $\Phi(\rho = a) = 0$ since $\Phi(\rho = 0) = 0$ and the potential must be the same for the entire PEC. To find the electric scalar potential for $a < \rho < b$, we can consider the line integral of the electric field intensity as

$$\Phi(\rho) = -\int_a^\rho \hat{a}_\rho \frac{Q_{l0}}{2\pi\varepsilon_0 \rho'} \cdot \hat{a}_\rho d\rho' = -\frac{Q_{l0}}{2\pi\varepsilon_0}\ln(\rho/a) \quad (a < \rho < b). \tag{6.81}$$

Similarly, for $b < \rho < c$, we have[42]

$$\Phi(\rho) = \Phi(\rho = b) - \int_b^\rho \hat{a}_\rho \frac{Q_{l0}}{2\pi\varepsilon_0 \rho'}\left[\frac{c^2 - (\rho')^2}{c^2 - b^2}\right] \cdot \hat{a}_\rho d\rho' \tag{6.82}$$

$$= -\frac{Q_{l0}}{2\pi\varepsilon_0}\left[\ln(b/a) + \frac{c^2}{c^2 - b^2}\ln(\rho/b) - \frac{\rho^2 - b^2}{2(c^2 - b^2)}\right] \quad (b < \rho < c). \tag{6.83}$$

Therefore, the electric scalar potential at $\rho = c$ can be found as[43]

$$\Phi(\rho = c) = -\frac{Q_{l0}}{2\pi\varepsilon_0}\left[\ln(b/a) + \frac{c^2}{c^2 - b^2}\ln(c/b) - \frac{1}{2}\right]. \tag{6.84}$$

[42] As intermediate steps, we can write

$$\begin{aligned}\Phi(\rho) &= -\frac{Q_{l0}}{2\pi\varepsilon_0}\ln(b/a) \\ &\quad - \frac{Q_{l0}}{2\pi\varepsilon_0}\int_b^\rho \frac{1}{\rho'}\left[\frac{c^2 - (\rho')^2}{c^2 - b^2}\right] d\rho' \\ &= -\frac{Q_{l0}}{2\pi\varepsilon_0}\ln(b/a) \\ &\quad - \frac{Q_{l0}}{2\pi\varepsilon_0}\left[\frac{c^2}{c^2 - b^2}\ln(\rho') - \frac{(\rho')^2}{2(c^2 - b^2)}\right]_b^\rho.\end{aligned}$$

[43] Also note that $\Phi(\rho) = \Phi(\rho = c)$ for $\rho > c$.

Example 64: A perfectly conducting sphere of radius a is coated with a dielectric of thickness $(b - a)$ having a permittivity of ε_d (Figure 6.28). The conducting sphere is charged: i.e. there is a constant surface electric charge density q_0 (C/m^2) at $R = a$. The entire structure is located in a vacuum. Find the electric field intensity, the electric flux density, the electric potential energy stored by the system, as well as the electric force that tries to change the outer radius b.

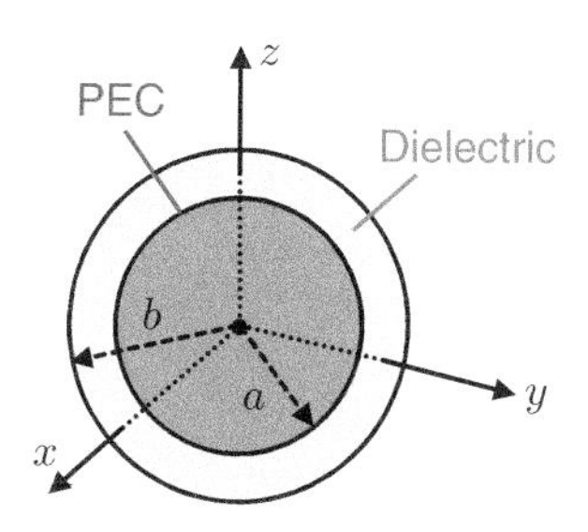

Figure 6.28 A perfectly conducting sphere coated with a dielectric layer.

Solution: First, using Gauss' law, we can find the electric flux density as

$$\bar{D}(\bar{r}) = \hat{a}_R \frac{4\pi a^2 q_0}{4\pi R^2} = \hat{a}_R \frac{a^2 q_0}{R^2} \quad (R > a), \tag{6.85}$$

while $\bar{D}(\bar{r}) = 0$ for $R < a$. Then the electric field intensity can be found as

$$\bar{E}(\bar{r}) = \hat{a}_R \frac{a^2 q_0}{R^2} \begin{cases} 0 & (R < a) \\ 1/\varepsilon_d & (a < R < b) \\ 1/\varepsilon_0 & (R > b). \end{cases} \tag{6.86}$$

The electric potential energy stored by the system can be obtained as[44]

$$w_{e,\text{stored}} = \frac{4\pi}{2} \int_a^b \hat{a}_R \frac{a^2 q_0}{\varepsilon_d R^2} \cdot \hat{a}_R \frac{a^2 q_0}{R^2} R^2 dR + \frac{4\pi}{2} \int_b^\infty \hat{a}_R \frac{a^2 q_0}{\varepsilon_0 R^2} \cdot \hat{a}_R \frac{a^2 q_0}{R^2} R^2 dR \tag{6.87}$$

$$= 2\pi \frac{a^4 q_0^2}{\varepsilon_d} \left[-\frac{1}{R} \right]_a^b + 2\pi \frac{a^4 q_0^2}{\varepsilon_0} \left[-\frac{1}{R} \right]_b^\infty = 2\pi a^4 q_0^2 \left(\frac{1}{a\varepsilon_d} - \frac{1}{b\varepsilon_d} + \frac{1}{b\varepsilon_0} \right). \tag{6.88}$$

We can finally get the expression for the force that tries to change b:

$$\bar{F}_e = -\bar{\nabla} w_e = -\hat{a}_b 2\pi a^4 q_0^2 \frac{\partial}{\partial b} \left(\frac{1}{a\varepsilon_d} - \frac{1}{b\varepsilon_d} + \frac{1}{b\varepsilon_0} \right) = \hat{a}_b \frac{2\pi a^4 q_0^2}{b^2} \left(\frac{1}{\varepsilon_0} - \frac{1}{\varepsilon_d} \right). \tag{6.89}$$

[44] We use

$$w_{e,\text{stored}} = \frac{1}{2} \int_\infty \bar{E}(\bar{r}) \cdot \bar{D}(\bar{r}) dv,$$

leading to

$$w_{e,\text{stored}} = \frac{4\pi}{2} \int_a^b \bar{E}(\bar{r}) \cdot \bar{D}(\bar{r}) R^2 dR + \frac{4\pi}{2} \int_b^\infty \bar{E}(\bar{r}) \cdot \bar{D}(\bar{r}) R^2 dR,$$

considering spherical symmetry.

6.6 Maxwell's Equations in Conducting Media

In Section 4.4, Maxwell's equations, wave equations, and Helmholtz equations are considered by keeping the electric current density $\bar{J}_v(\bar{r}, t)$ (or $\bar{J}_v(\bar{r})$ in the phasor domain) as an impressed source.[45] In a conducting medium, however, electric currents and electric fields are related to each other via Ohm's law. Specifically, Ampere's law can be written

$$\bar{\nabla} \times \bar{H}(\bar{r}, t) = \sigma(\bar{r}) \bar{E}(\bar{r}, t) + \varepsilon(\bar{r}) \frac{\partial \bar{E}(\bar{r}, t)}{\partial t}, \tag{6.90}$$

where the magnetic field intensity is written purely in terms of the electric field intensity, assuming that there is no impressed current.[46] Evaluating the curl of both sides and employing Faraday's law,[47] we derive

$$\bar{\nabla} \times \bar{\nabla} \times \bar{H}(\bar{r}, t) = \bar{\nabla} \times [\sigma(\bar{r}) \bar{E}(\bar{r}, t)] + \varepsilon \frac{\partial}{\partial t} \bar{\nabla} \times \bar{E}(\bar{r}, t) \tag{6.91}$$

$$-\bar{\nabla}^2 \bar{H}(\bar{r}, t) = \bar{\nabla}\sigma(\bar{r}) \times \bar{E}(\bar{r}, t) - \mu\sigma(\bar{r}) \frac{\partial \bar{H}(\bar{r}, t)}{\partial t} - \mu\varepsilon \frac{\partial^2}{\partial t^2} \bar{H}(\bar{r}, t) \tag{6.92}$$

[45] An impressed source can be thought as an externally inserted excitation in an electromagnetic scenario. Hence, this source is considered an unalterable component of the scenario. In reality, there is no impressed source, as all sources must be created and are part of the overall system, where Maxwell's equations must hold.

[46] Otherwise, the general expression should be written

$$\bar{\nabla} \times \bar{H}(\bar{r}, t) = \bar{J}_v^{\text{imp}}(\bar{r}, t) + \sigma(\bar{r}) \bar{E}(\bar{r}, t) + \varepsilon(\bar{r}) \frac{\partial \bar{E}(\bar{r}, t)}{\partial t}.$$

[47] And also using $\bar{\nabla} \cdot \bar{H}(\bar{r}, t) = 0$.

or

$$\bar{\nabla}^2\bar{H}(\bar{r},t) - \mu\varepsilon\frac{\partial^2}{\partial t^2}\bar{H}(\bar{r},t) - \mu\sigma(\bar{r})\frac{\partial\bar{H}(\bar{r},t)}{\partial t} = -\bar{\nabla}\sigma(\bar{r}) \times \bar{E}(\bar{r},t), \tag{6.93}$$

assuming that permittivity and permeability do not depend on position at $\bar{r}$. If the conductivity also does not depend on position at $\bar{r}$, we derive

$$\bar{\nabla}^2\bar{H}(\bar{r},t) - \mu\varepsilon\frac{\partial^2}{\partial t^2}\bar{H}(\bar{r},t) - \mu\sigma\frac{\partial\bar{H}(\bar{r},t)}{\partial t} = 0. \tag{6.94}$$

The third term in this expression, which does not exist in non-conducting media, leads to a *damping* in the wave. This decay is associated with ohmic losses in the conducting medium.

To derive an equation for the electric field intensity, we consider the curl of Faraday's law and use Ampere's law as

$$\bar{\nabla} \times \bar{\nabla} \times \bar{E}(\bar{r},t) = -\mu\frac{\partial}{\partial t}\bar{\nabla} \times \bar{H}(\bar{r},t) = -\mu\varepsilon\frac{\partial^2}{\partial t^2}\bar{E}(\bar{r},t) - \mu\sigma(\bar{r})\frac{\partial\bar{E}(\bar{r},t)}{\partial t}, \tag{6.95}$$

assuming again position-independent ε and μ. Then we further have[48]

$$\frac{1}{\varepsilon}\bar{\nabla}q_v(\bar{r},t) - \bar{\nabla}^2\bar{E}(\bar{r},t) = -\mu\varepsilon\frac{\partial^2}{\partial t^2}\bar{E}(\bar{r},t) - \mu\sigma(\bar{r})\frac{\partial\bar{E}(\bar{r},t)}{\partial t} \tag{6.96}$$

or

$$\bar{\nabla}^2\bar{E}(\bar{r},t) - \mu\varepsilon\frac{\partial^2}{\partial t^2}\bar{E}(\bar{r},t) - \mu\sigma(\bar{r})\frac{\partial\bar{E}(\bar{r},t)}{\partial t} = \frac{1}{\varepsilon}\bar{\nabla}q_v(\bar{r},t). \tag{6.97}$$

If the conductivity does not depend on position, the expression becomes

$$\bar{\nabla}^2\bar{E}(\bar{r},t) - \mu\varepsilon\frac{\partial^2}{\partial t^2}\bar{E}(\bar{r},t) - \mu\sigma\frac{\partial\bar{E}(\bar{r},t)}{\partial t} = \frac{1}{\varepsilon}\bar{\nabla}q_v(\bar{r},t), \tag{6.98}$$

which is very similar to Eq. (6.94).

For time-harmonic sources, and given the phasor domain, similar derivations can be done starting from the phasor forms of Maxwell's equations. Alternatively, wave equations (6.94) and (6.98) can be converted directly to phasor representations. At an observation point $\bar{r}$, where permittivity, permeability, and conductivity do not depend on position, we have

$$\bar{\nabla}^2\bar{H}(\bar{r}) + k^2\bar{H}(\bar{r}) - j\omega\mu\sigma\bar{H}(\bar{r}) = 0 \tag{6.99}$$

$$\bar{\nabla}^2\bar{E}(\bar{r}) + k^2\bar{E}(\bar{r}) - j\omega\mu\sigma\bar{E}(\bar{r}) = \frac{1}{\varepsilon}\bar{\nabla}q_v(\bar{r}), \tag{6.100}$$

[48] Using $\bar{\nabla} \cdot \bar{E}(\bar{r},t) = \rho_v(\bar{r})/\varepsilon$.

where $k = \omega\sqrt{\mu\varepsilon}$. In the first equation, we already assume that there is no impressed electric current density at $\bar{r}$. Considering time-harmonic sources, there should not be impressed electric charge density, leading to[49]

$$\bar{\nabla}^2 \bar{H}(\bar{r}) + k^2 \bar{H}(\bar{r}) - j\omega\mu\sigma \bar{H}(\bar{r}) = 0 \tag{6.101}$$

$$\bar{\nabla}^2 \bar{E}(\bar{r}) + k^2 \bar{E}(\bar{r}) - j\omega\mu\sigma \bar{E}(\bar{r}) = 0. \tag{6.102}$$

These equations are similar to the homogeneous forms of Helmholtz equations: i.e. Eqs. (4.279) and (4.280) with zero right-hand sides.

[49]Setting the electric charge density directly to zero may be confusing since there is electric current density (due to conduction). But in fact, given *position-independent* ε and σ, we have

$$\bar{\nabla} \cdot \bar{J}_v(\bar{r},t) = \sigma \bar{\nabla} \cdot \bar{E}(\bar{r},t)$$
$$\longrightarrow -\frac{\partial q_v(\bar{r},t)}{\partial t} = \frac{\sigma}{\varepsilon} q_v(\bar{r},t),$$

leading to $q_v(\bar{r},t) = 0$ in a steady state, even though there is conduction current. On the other hand, if there is an impressed current density, we may have $\bar{\nabla} \cdot \bar{J}_v^{\text{imp}}(\bar{r},t) \neq 0$, indicating the existence of an impressed charge density.

6.6.1 Complex Permittivity

To facilitate solutions of electric and magnetic fields in conducting media in the phasor domain, we define complex permittivity as[50]

$$\varepsilon_c = \varepsilon + \frac{\sigma}{j\omega} = \varepsilon - j\frac{\sigma}{\omega}. \tag{6.103}$$

[50]In this expression, the imaginary part of the permittivity is only due to conduction (finite and nonzero value of σ). In real life, other effects may contribute to the imaginary part of the considered material's permittivity, but they are omitted in this context.

The corresponding complex wavenumber can be written

$$k_c = \omega\sqrt{\mu\varepsilon_c} = \omega\sqrt{\mu\left(\varepsilon + \frac{\sigma}{j\omega}\right)} = \omega\sqrt{\mu\varepsilon}\sqrt{\left(1 + \frac{\sigma}{j\omega\varepsilon}\right)}. \tag{6.104}$$

Also note that

$$k_c^2 = \omega^2\mu\varepsilon\left(1 + \frac{\sigma}{j\omega\varepsilon}\right) = k^2 + \omega^2\mu\varepsilon\frac{\sigma}{j\omega\varepsilon} \tag{6.105}$$

$$= k^2 - j\omega\mu\sigma. \tag{6.106}$$

Then we simply have

$$\bar{\nabla} \times \bar{H}(\bar{r}) = j\omega\varepsilon_c \bar{E}(\bar{r}), \tag{6.107}$$

which is similar to Ampere's law in a source-free and non-conducting position. Obviously, the imaginary part of the complex permittivity is associated with losses due to finite conductance. A relative quantity called the *loss tangent* for a medium at a given frequency is defined as (Figure 6.29)

$$\tan\delta_c = \frac{\sigma}{\omega\varepsilon}. \tag{6.108}$$

The loss tangent is an important quantity since it provides information about the dissipation characteristics of a medium depending on the frequency.

Using the definition of complex wavenumber, we obtain Helmholtz equations in a conducting medium as

$$\nabla^2 \bar{E}(\bar{r}) + k_c^2 \bar{E}(\bar{r}) = 0 \tag{6.109}$$

$$\nabla^2 \bar{H}(\bar{r}) + k_c^2 \bar{H}(\bar{r}) = 0 \tag{6.110}$$

with complex coefficients for the second terms. For various practical reasons, it is useful to define a new variable, the *propagation constant*, as

$$\gamma = jk_c = j\omega\sqrt{\mu\varepsilon_c}. \tag{6.111}$$

Using γ, two other variables are defined: [51]

$$\alpha = \mathrm{Re}\{\gamma\}, \qquad \beta = \mathrm{Im}\{\gamma\}. \tag{6.112}$$

They are called the *attenuation constant* (showing how waves decay) and the *phase constant* (showing how waves propagate). In a non-conducting medium, all these variables become

$$k_c = k = \omega\sqrt{\mu\varepsilon}, \tag{6.113}$$

$$\gamma = jk_c = j\omega\sqrt{\mu\varepsilon}, \qquad \alpha = \mathrm{Re}\{\gamma\} = 0, \qquad \beta = \mathrm{Im}\{\gamma\} = \omega\sqrt{\mu\varepsilon} = k. \tag{6.114}$$

In general, note that $\gamma^2 = -k_c^2$, so the Helmholtz equations become

$$\nabla^2 \bar{E}(\bar{r}) - \gamma^2 \bar{E}(\bar{r}) = 0, \qquad \nabla^2 \bar{H}(\bar{r}) - \gamma^2 \bar{H}(\bar{r}) = 0, \tag{6.115}$$

whose general solutions are hyperbolic functions in terms of γ.

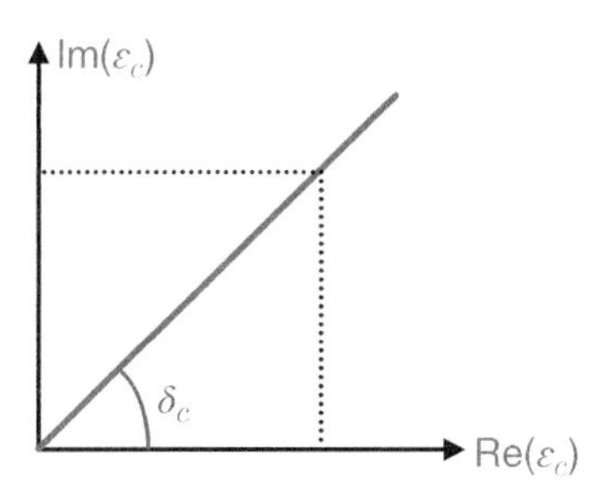

Figure 6.29 A trigonometric demonstration of the loss tangent.

α: attenuation constant
Unit: nepers/meter (Np/m)

β: phase constant
Unit: radians/meter (rad/m)

[51] Note that the unit of the attenuation constant is nepers/meter: i.e. it is not 1/meter. This can be understood by considering that $\exp(-\alpha z)$, where z represents distance, should be unitless. Then αz corresponds to a logarithmic scale with unit neper, but α should be nepers/meter to satisfy this. Note that the neper is based on the natural logarithm, while the more common decibel (dB) is based on the base-10 logarithm. It is possible to convert a value in nepers to dB by multiplying the value by $20\log_{10}(e) \approx 8.69$.

6.6.2 Power and Energy in Conducting Media

As discussed in Section 5.2, electromagnetic waves carry energy, which can be represented by the Poynting vector $\bar{S}(\bar{r},t) = \bar{E}(\bar{r},t) \times \bar{H}(\bar{r},t)$ or the complex Poynting vector $\bar{S}_c(\bar{r}) = \bar{E}(\bar{r}) \times [\bar{H}(\bar{r})]^*/2$ for time-harmonic fields. For a given volume V bounded by a surface S, the total power flux can be written

$$p_{em}(t) = \oint_S \bar{S}(\bar{r},t) \cdot \overline{ds} \tag{6.116}$$

$$= -\frac{\partial w_{e,\mathrm{stored}}(t)}{\partial t} - \frac{\partial w_{m,\mathrm{stored}}(t)}{\partial t} - \int_V \bar{E}(\bar{r},t) \cdot \bar{J}_v(\bar{r},t) dv, \tag{6.117}$$

where $w_{e,\text{stored}}(t)$ is the stored electric potential energy and $w_{m,\text{stored}}(t)$ is the stored magnetic potential energy. In a conducting medium, assuming that the conductivity does not depend on position, the third term can be divided in two as

$$p_{em}(t) = -\sigma \int_V |\bar{E}(\bar{r},t)|^2 dv - \int_V \bar{E}(\bar{r},t) \cdot \bar{J}_v^{\text{imp}}(\bar{r},t) dv, \tag{6.118}$$

where the first term on the right-hand side represents the dissipated power due to losses, while the second term is the generated power due to sources (impressed currents) located inside the medium.

6.6.3 Plane Waves in Conducting Media

In Section 5.3, plane waves are presented as natural time-harmonic solutions of wave and Helmholtz equations. At this stage, plane waves in a conducting medium, which satisfy Eqs. (6.109) and (6.110), can be written[52]

$$\bar{E}(\bar{r}) = \bar{E}_0 \exp(-j\bar{k}_c \cdot \bar{r}), \qquad \bar{H}(\bar{r}) = \hat{a}_k \times \bar{E}_0 \frac{1}{\eta_c} \exp(-j\bar{k}_c \cdot \bar{r}), \tag{6.119}$$

where $\bar{E}_0$ is a constant vector satisfying $\bar{E}_0 \perp \hat{a}_k$ but its value can be a complex number. In the above, the complex wave (intrinsic) impedance can be written

$$\eta_c = \sqrt{\frac{\mu}{\varepsilon_c}} = \sqrt{\frac{\mu}{\left(\varepsilon + \dfrac{\sigma}{j\omega}\right)}} = \sqrt{\frac{\mu}{\varepsilon\,(1 - j\tan\delta_c)}}. \tag{6.120}$$

The complex value of η_c indicates that electric and magnetic fields are generally *not in phase* in conducting media (Figure 6.30).

In terms of the propagation constant, electric and magnetic fields can also be written

$$\bar{E}(\bar{r}) = \bar{E}_0 \exp(-\gamma\hat{a}_k \cdot \bar{r}) = \bar{E}_0 \exp(-\alpha\hat{a}_k \cdot \bar{r}) \exp(-j\beta\hat{a}_k \cdot \bar{r}) \tag{6.121}$$

$$\bar{H}(\bar{r}) = \hat{a}_k \times \bar{E}_0 \frac{1}{\eta_c} \exp(-\gamma\hat{a}_k \cdot \bar{r}) = \hat{a}_k \times \bar{E}_0 \frac{1}{\eta_c} \exp(-\alpha\hat{a}_k \cdot \bar{r}) \exp(-j\beta\hat{a}_k \cdot \bar{r}). \tag{6.122}$$

To proceed further, we now consider a special case involving a plane wave propagating in the z direction with the electric field polarized in the x direction. We have $\bar{E}(\bar{r}) = \hat{a}_x E_0 \exp(-\gamma z)$, leading to

$$\bar{E}(\bar{r}) = \hat{a}_x E_0 \exp(-\alpha z) \exp(-j\beta z), \qquad \bar{H}(\bar{r}) = \hat{a}_y \frac{E_0}{\eta_c} \exp(-\alpha z) \exp(-j\beta z) \tag{6.123}$$

[52] We simply replace real wavenumber k with complex wavenumber k_c so that $\bar{k}_c = \hat{a}_k k_c$.

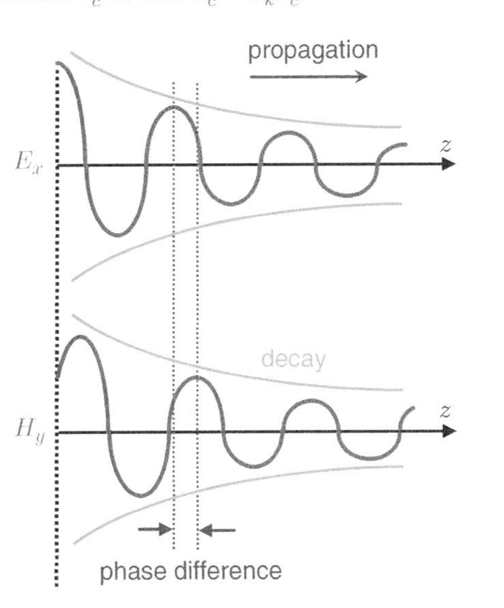

Figure 6.30 Electromagnetic waves decay as they propagate in a conducting medium. This example shows the electric field intensity and magnetic field intensity for a linearly polarized plane wave propagating in the z direction. In addition to the decay, a phase difference occurs between the electric and magnetic fields, depending on the conductivity of the medium.

in the phasor domain. These expressions indicate that the amplitude of the plane wave decays with $\exp(-\alpha z)$ (Figure 6.30), where $\alpha = \text{Re}\{\gamma\} = \text{Re}\{jk_c\}$. An important parameter is the *skin depth*, defined as

$$s_d = 1/\alpha, \tag{6.124}$$

s_d: skin depth
Unit: meter (m)

which indicates the amount of distance for the amplitude decays by $\exp(-1) = 1/e$. In addition to decay, the expressions above indicate propagation with phase constant $\beta = \text{Im}\{\gamma\} = \text{Im}\{jk_c\}$. The velocity of the wave (specifically, the velocity of the wavefront) can be written $u_p = \omega/\beta$, while the wavelength is $\lambda = 2\pi/\beta$.

As mentioned above, the electric field intensity and magnetic field intensity of a plane wave are not in phase in a conducting medium due to the complex value of η_c. We can work on the expression as

$$\eta_c^2 = \mu \exp\left(j\tan^{-1}\left(\frac{\sigma}{\omega\varepsilon}\right)\right) \Big/ \sqrt{\varepsilon^2 + \frac{\sigma^2}{\omega^2}} \tag{6.125}$$

or

$$\eta_c = \sqrt{\mu}\exp\left(\frac{j}{2}\tan^{-1}\left(\frac{\sigma}{\omega\varepsilon}\right)\right) \Big/ \left(\varepsilon^2 + \frac{\sigma^2}{\omega^2}\right)^{1/4}. \tag{6.126}$$

Consequently, the magnetic field intensity can be written in the phasor domain as[53]

$$\bar{H}(\bar{r}) = \hat{a}_y \frac{E_0}{|\eta_c|}\exp(-\alpha z)\exp\left(-j\beta z - j\delta_c/2\right), \tag{6.127}$$

where $|\eta_c| = \eta/[1 + \tan^2\delta_c]^{1/4}$ for $\eta = \sqrt{\mu/\varepsilon}$. At this stage, we now categorize conducting materials in order to derive neat expressions to gain more insight:

- A *low-loss dielectric* can be described as

$$\tan\delta_c = \frac{\sigma}{\omega\varepsilon} \ll 1. \tag{6.128}$$

In this case, the propagation constant can be manipulated as[54]

$$\gamma = j\omega\sqrt{\mu\varepsilon\left(1 + \frac{\sigma}{j\omega\varepsilon}\right)} = j\omega\sqrt{\mu\varepsilon}\left(1 + \frac{\sigma}{j\omega\varepsilon}\right)^{1/2} \tag{6.129}$$

$$\approx j\omega\sqrt{\mu\varepsilon}\left(1 - \frac{j\sigma}{2\omega\varepsilon} + \frac{\sigma^2}{8\omega^2\varepsilon^2}\right), \tag{6.130}$$

[53] We can write the intermediate steps as

$$\bar{H}(\bar{r}) = \hat{a}_y \frac{E_0\left(\varepsilon^2 + \dfrac{\sigma^2}{\omega^2}\right)^{1/4}\exp(-\alpha z)\exp(-j\beta z)}{\sqrt{\mu}\exp\left(\dfrac{j}{2}\tan^{-1}\left(\dfrac{\sigma}{\omega\varepsilon}\right)\right)},$$

leading to

$$\bar{H}(\bar{r}) = \hat{a}_y \frac{E_0(1 + \tan^2\delta_c)^{1/4}}{\eta\exp\left(\dfrac{j}{2}\delta_c\right)}\exp(-\alpha z)\exp(-j\beta z)$$

or

$$\bar{H}(\bar{r}) = \hat{a}_y \frac{E_0(1 + \tan^2\delta_c)^{1/4}}{\eta} \times \exp(-\alpha z)\exp\left(-j\beta z - j\delta_c/2\right).$$

[54] We use an approximation

$$(1 + a)^{1/2} \approx 1 + \frac{a}{2} - \frac{a^2}{8}$$

for small a, based on the Taylor series expansion.

leading to[55]

$$\alpha \approx \omega\sqrt{\mu\varepsilon}\frac{\sigma}{2\omega\varepsilon} = \frac{\sigma}{2}\sqrt{\frac{\mu}{\varepsilon}} \quad (6.131)$$

$$\beta \approx \omega\sqrt{\mu\varepsilon}\left(1 + \frac{\sigma^2}{8\omega^2\varepsilon^2}\right) = \omega\sqrt{\mu\varepsilon}\left(1 + \frac{1}{8}\tan^2\delta_c\right). \quad (6.132)$$

Similarly, the wave (intrinsic) impedance can be approximated as

$$\eta_c = \sqrt{\frac{\mu}{\varepsilon}}\left(1 + \frac{\sigma}{j\omega\varepsilon}\right)^{-1/2} \approx \sqrt{\frac{\mu}{\varepsilon}}\left(1 + \frac{j\sigma}{2\omega\varepsilon}\right) = \eta\left(1 + \frac{j}{2}\tan\delta_c\right), \quad (6.133)$$

where $\eta = \sqrt{\mu/\varepsilon}$. Therefore, for small values of $\tan\delta_c = \sigma/(\omega\varepsilon)$, there is a *slight* phase difference between the electric and magnetic fields. Furthermore, the phase velocity can be found as

$$u_p = \frac{\omega}{\beta} \approx \frac{1}{\sqrt{\mu\varepsilon}\left(1 + \frac{1}{8}\tan^2\delta_c\right)} \approx \frac{1}{\sqrt{\mu\varepsilon}}\left(1 - \frac{1}{8}\tan^2\delta_c\right), \quad (6.134)$$

considering the approximation of the phase constant.[56] Hence, the phase velocity is *slightly* reduced in comparison to a non-conducting medium with permittivity ε and permeability μ.

- A *good conductor* can be described as[57]

$$\tan\delta_c = \frac{\sigma}{\omega\varepsilon} \gg 1. \quad (6.135)$$

Hence, we manipulate and approximate the propagation constant as

$$\gamma = j\omega\sqrt{\mu\varepsilon}\left(1 + \frac{\sigma}{j\omega\varepsilon}\right)^{1/2} \approx j\omega\sqrt{\mu\varepsilon}\left(\frac{\sigma}{j\omega\varepsilon}\right)^{1/2} = j\omega\sqrt{\mu\varepsilon}\frac{\sqrt{\sigma}}{\sqrt{\omega\varepsilon}}\sqrt{-j} \quad (6.136)$$

leading to

$$\gamma \approx \exp(j\pi/2)\sqrt{\omega\mu\sigma}\exp(-j\pi/4) = \sqrt{\frac{\omega\mu\sigma}{2}}(1 + j). \quad (6.137)$$

Then the attenuation and phase constants are approximately the same and can be written

$$\alpha = \beta \approx \sqrt{\frac{\omega\mu\sigma}{2}}. \quad (6.138)$$

[55] It is also possible to further simplify the expressions for γ and β (with less accuracy) as

$$\gamma \approx j\omega\sqrt{\mu\varepsilon}\left(1 - \frac{j\sigma}{2\omega\varepsilon}\right)$$

$$\beta \approx \omega\sqrt{\mu\varepsilon}.$$

[56] We further use

$$\frac{1}{1+a} = \frac{1-a}{(1-a^2)} \approx 1 - a$$

for small a.

[57] Note that a material is categorized as a low-loss dielectric or a good conductor, depending on the value of $\tan\delta_c = \sigma/(\omega\varepsilon)$, not σ alone. For example, typical seawater may have $\sigma \approx 4.8$ S/m and $\varepsilon_r \approx 80$ in a wide range of frequencies. Then $\tan\delta_c > 10$ for $f <$ 100 MHz, while $\tan\delta_c < 0.1$ for $f > 11$ GHz. Using the expressions for good conductors, we can obtain the following data:

$$f = 50 \text{ Hz} \longrightarrow \alpha \approx 0.0308 \text{ Np/m}, \quad s_d \approx 32.5 \text{ m}$$

$$f = 1 \text{ kHz} \longrightarrow \alpha \approx 0.138 \text{ Np/m}, \quad s_d \approx 7.25 \text{ m}$$

$$f = 1 \text{ MHz} \longrightarrow \alpha \approx 4.35 \text{ Np/m}, \quad s_d \approx 0.230 \text{ m}$$

Specifically, at 1 MHz, the amplitude of the electric field intensity of an electromagnetic wave decays from E_0 to E_0/e in approximately 23 cm. Interestingly, for the given values of ε and σ, the attenuation constant and skin depth are fixed as $\alpha \approx 101$ Np/m and $s_d \approx$ 9.89 mm, *independent* of the frequency, if $f > 11$ GHz (when the seawater is a low-loss dielectric). However, one should keep in mind that ε and σ also depend on frequency (any many other parameters, like temperature) in real life. For example, at optical frequencies, the relative permittivity of water drops below 2.0.

Similarly, the complex wave (intrinsic) impedance can approximated as

$$\eta_c = \sqrt{\frac{\mu}{\varepsilon}}\left(1+\frac{\sigma}{j\omega\varepsilon}\right)^{-1/2} \approx \sqrt{\frac{\mu}{\varepsilon}}\left(\frac{\sigma}{j\omega\varepsilon}\right)^{-1/2} = \sqrt{\frac{\mu\omega}{2\sigma}}(1+j). \tag{6.139}$$

This expression indicates that, for good conductors, there is a 45° phase difference (approximately, depending on the conductivity) between the electric and magnetic fields. Obviously, as $\sigma \to \infty$, we have $\eta_c \to 0$. The phase velocity and wavelength in a good conductor can be approximated:

$$u_p = \frac{\omega}{\beta} \approx \frac{\sqrt{2}\omega}{\sqrt{\omega\mu\sigma}} = \sqrt{\frac{2\omega}{\mu\sigma}}, \qquad \lambda = \frac{2\pi}{\beta} \approx \frac{2\sqrt{2}\pi}{\sqrt{\omega\mu\sigma}} = \sqrt{\frac{8\pi^2}{\omega\mu\sigma}}. \tag{6.140}$$

Note that, as $\sigma \to \infty$, we have $u_p \to 0$ and $\lambda \to 0$, but electromagnetic waves cannot penetrate into the medium in this limit case.[58]

6.6.4 Power of Plane Waves in Conducting Media

In Section 5.3.4, electromagnetic energy carried by plane waves in non-conducting media is discussed. Now, we turn our attention to plane waves in conducting media in the context of electromagnetic power and energy. First, considering time-domain representations of the electric field intensity and magnetic field intensity related to an x-polarized plane wave propagating in the z direction as[59]

$$\bar{E}(\bar{r},t) = \hat{a}_x E_0 \exp(-\alpha z)\cos(\omega t - \beta z) \tag{6.141}$$

$$\bar{H}(\bar{r},t) = \hat{a}_y \frac{E_0}{|\eta_c|} \exp(-\alpha z)\cos(\omega t - \beta z - \delta_c/2), \tag{6.142}$$

the Poynting vector can be found as[60]

$$\bar{S}(\bar{r},t) = \bar{E}(\bar{r},t) \times \bar{H}(\bar{r},t) \tag{6.143}$$

$$= \hat{a}_z \frac{E_0^2}{|\eta_c|} \exp(-2\alpha z)\cos(\omega t - \beta z)\cos(\omega t - \beta z - \delta_c/2) \tag{6.144}$$

$$= \hat{a}_z \frac{E_0^2}{2|\eta_c|} \exp(-2\alpha z)\left[\cos(2\omega t - 2\beta z - \delta_c/2) + \cos(\delta_c/2)\right]. \tag{6.145}$$

[58] As a further example, we may consider pure copper, which has a conductivity of $\sigma \approx 6 \times 10^7$ S/m. Assuming that its relative permittivity is unity (indeed, it can be represented by a negative number, which is out of scope), $\tan\delta_c > 1000$ for $f < 1000$ THz: i.e. it can be assumed to be a very good conductor for most frequencies. Then we obtain the following:

$$f = 50 \text{ Hz} \longrightarrow s_d \approx 9.2 \text{ mm}, \quad u_p \approx 2.89 \text{ m/s}$$

$$f = 1 \text{ kHz} \longrightarrow s_d \approx 2.1 \text{ mm}, \quad u_p \approx 12.9 \text{ m/s}$$

$$f = 1 \text{ MHz} \longrightarrow s_d \approx 65\ \mu\text{m}, \quad u_p \approx 408 \text{ m/s}$$

Hence, at 50 Hz, the speed of an electromagnetic wave in a pure copper is even less than 3 m/s. Very small values of the skin depth clearly show how electromagnetic fields are squeezed near the surface of a copper wire.

[59] We assume that E_0 has a real value, while a complex E_0 would only cause an addition phase.

[60] We use

$$\cos(a)\cos(b) = \frac{1}{2}\cos(a+b) + \frac{1}{2}\cos(a-b).$$

It can be seen that the power flux density decreases as the wave propagates due to ohmic losses (when $\sigma > 0$). Next, considering phasor-domain representations of fields as

$$\bar{E}(\bar{r}) = \hat{a}_x E_0 \exp(-\alpha z) \exp(-j\beta z) \tag{6.146}$$

$$\bar{H}(\bar{r}) = \hat{a}_y \frac{E_0}{|\eta_c|} \exp(-\alpha z) \exp(-j\beta z - j\delta_c/2), \tag{6.147}$$

we have the complex Poynting vector as[61]

$$\bar{S}_c(\bar{r}) = \frac{1}{2}\bar{E}(\bar{r}) \times [\bar{H}(\bar{r})]^* = \hat{a}_z \frac{E_0^2}{2|\eta_c|} \exp(-2\alpha z) \exp(j\delta_c/2). \tag{6.148}$$

[61]As intermediate steps, we have

$$\begin{aligned}\bar{S}_c(\bar{r}) &= \frac{1}{2}\hat{a}_x E_0 \exp(-\alpha z) \exp(-j\beta z) \\ &\times \hat{a}_y \frac{E_0}{|\eta_c|} \exp(-\alpha z) \exp(j\beta z + j\delta_c/2) \\ &= \hat{a}_z \frac{E_0^2}{2|\eta_c|} \exp(-\alpha z) \exp(-\alpha z) \\ &\times \exp(-j\beta z) \exp(j\beta z + j\delta_c/2).\end{aligned}$$

The real part of $\bar{S}_c$, which corresponds to the time-average power flux density, can be found as

$$\bar{S}_{\text{avg}}(\bar{r}) = \text{Re}\{\bar{S}_c(\bar{r})\} = \hat{a}_z \frac{E_0^2}{2|\eta_c|} \exp(-2\alpha z) \cos(\delta_c/2). \tag{6.149}$$

For a non-conducting (lossless) medium, we have $|\eta_c| = \eta$, $\delta_c = 0$, and $\alpha = 0$ so that $\bar{S}_{\text{avg}}(\bar{r}) = \hat{a}_z E_0^2/(2\eta)$ is purely real. On the other hand, when the medium is conducting, the imaginary power is nonzero: i.e. we have

$$\text{Im}\{\bar{S}_c(\bar{r})\} = \hat{a}_z \frac{E_0^2}{2|\eta_c|} \exp(-2\alpha z) \sin(\delta_c/2), \tag{6.150}$$

due to phase difference between electric and magnetic fields.

6.6.5 Reflection from PECs

Electromagnetic waves are reflected from conducting objects, which becomes extreme for PECs:[62] i.e. when no fields can enter inside the object. In Section 5.3.5, plane waves are focused in the context of reflection and refraction phenomena for penetrable objects. In this section, we study reflection from perfectly conducting boundaries while considering plane waves as illustrative examples. Given the infinity property of plane waves, boundaries are also assumed to be infinite planes to simplify analysis.

[62]We emphasize again that perfect conductivity is an idealization, but it is a very good model for most metals, depending on the frequency.

Before directly deriving equations for PEC boundaries, we can first consider the equations in Section 5.3.5 to gain insight into what happens as the conductivity of the second medium increases (Figure 6.31). Assuming normal incidence, electric and magnetic field intensities are derived in Eqs. (5.352), (5.353), (5.356), and (5.357) as

$$\bar{E}_1(\bar{r}) = \hat{a}_x E_0 \exp(-jk_1 z) + \hat{a}_x \left(\frac{\eta_{c2} - \eta_1}{\eta_{c2} + \eta_1}\right) E_0 \exp(jk_1 z) \tag{6.151}$$

$$\bar{H}_1(\bar{r}) = \hat{a}_y \frac{E_0}{\eta_1} \exp(-jk_1 z) - \hat{a}_y \left(\frac{\eta_{c2} - \eta_1}{\eta_{c2} + \eta_1}\right) \frac{E_0}{\eta_1} \exp(jk_1 z) \tag{6.152}$$

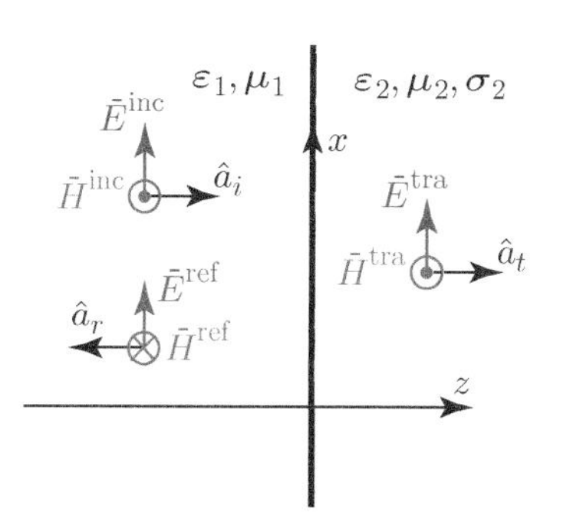

Figure 6.31 Normal incidence of a plane wave on a boundary between two media, where the second medium is conducting.

and

$$\bar{E}_2(\bar{r}) = \hat{a}_x \left(\frac{2\eta_{c2}}{\eta_{c2} + \eta_1}\right) E_0 \exp(-jk_{c2} z) \tag{6.153}$$

$$\bar{H}_2(\bar{r}) = \hat{a}_y \left(\frac{2\eta_{c2}}{\eta_{c2} + \eta_1}\right) \frac{E_0}{\eta_{c2}} \exp(-jk_{c2} z) \tag{6.154}$$

in the phasor domain.[63] In the limit of PEC – i.e. when $\eta_{c2} \to 0$ – we obtain

$$\bar{E}_1(\bar{r}) \to \hat{a}_x E_0 \exp(-jk_1 z) - \hat{a}_x E_0 \exp(jk_1 z) \tag{6.155}$$

$$\bar{H}_1(\bar{r}) \to \hat{a}_y \frac{E_0}{\eta_1} \exp(-jk_1 z) + \hat{a}_y \frac{E_0}{\eta_1} \exp(jk_1 z), \tag{6.156}$$

$\bar{E}_2(\bar{r}) \to 0$, and $\bar{H}_2(\bar{r}) \to 0$, as also derived below. For a general case, however, expressions for the transmitted wave can be written

$$\bar{E}_2(\bar{r}) = \hat{a}_x \left(\frac{2\eta_{c2}}{\eta_{c2} + \eta_1}\right) E_0 \exp(-jz\omega\sqrt{\mu_2 \varepsilon_{c2}}) \tag{6.157}$$

$$\bar{H}_2(\bar{r}) = \hat{a}_y \left(\frac{2}{\eta_{c2} + \eta_1}\right) E_0 \exp(-jz\omega\sqrt{\mu_2 \varepsilon_{c2}}), \tag{6.158}$$

where $\varepsilon_{c2} = \varepsilon_2 - j\sigma_2/\omega$ and $\eta_{c2} = \sqrt{\mu_2/\varepsilon_{c2}}$. When the conductivity increases and the good-conductor assumption can be made as $\sigma_2/(\omega\varepsilon_2) \gg 1$, we further have

$$\bar{E}_2(\bar{r}) \approx \hat{a}_x \left(\frac{2\eta_{c2}}{\eta_{c2} + \eta_1}\right) E_0 \exp(-jz\sqrt{\omega\mu_2\sigma_2/2}) \exp(-z\sqrt{\omega\mu_2\sigma_2/2}) \tag{6.159}$$

$$\bar{H}_2(\bar{r}) \approx \hat{a}_y \left(\frac{2}{\eta_{c2} + \eta_1}\right) E_0 \exp(-jz\sqrt{\omega\mu_2\sigma_2/2}) \exp(-z\sqrt{\omega\mu_2\sigma_2/2}), \tag{6.160}$$

[63] In these expressions, note that η_{c2} and k_{c2} are complex numbers.

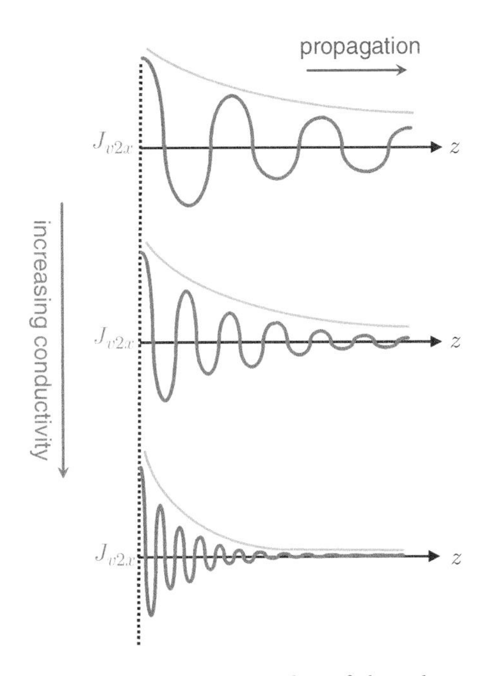

Figure 6.32 Time-domain snapshot of the volume electric current density in the second medium for the scenario in Figure 6.31. As the conductivity of the second medium increases, the volume electric current density decays faster with respect to z. In the limit case of perfect conductivity ($\sigma_2 \to \infty$), $\bar{J}_{v2}$ becomes zero for $z > 0$, while it is perfectly squeezed on the boundary ($z = 0$): i.e. it turns into a surface electric current density. Note that the magnetic field intensity is continuous across the boundary for finite conductivity values, while it becomes discontinuous in the limit case of $\sigma_2 \to \infty$.

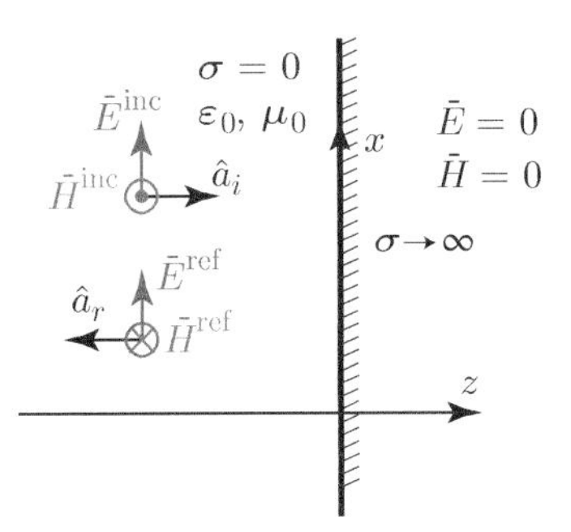

Figure 6.33 Normal incidence of a plane wave onto a planar PEC.

which represent exponentially decaying waves with respect to z. In addition, there is a volume electric current distribution in the second medium as

$$\bar{J}_{v2}(\bar{r}) \approx \hat{a}_x \sigma_2 \left(\frac{2\eta_{c2}}{\eta_{c2} + \eta_1}\right) E_0 \exp(-jz\sqrt{\omega\mu_2\sigma_2/2}) \exp(-z\sqrt{\omega\mu_2\sigma_2/2}). \tag{6.161}$$

As the conductivity increases, the volume electric current density is confined more in the vicinity of the boundary (Figure 6.32). In the limit of $\sigma \to 0$, we have

$$\bar{J}_{v2}(\bar{r}) \to 0 \quad (z \neq 0), \tag{6.162}$$

i.e. it disappears everywhere except at the boundary. On the other hand, exactly at the boundary, the limit should be evaluated carefully. Given the volume electric current density in a region of shrinking skin depth $\Delta z = s_{d2}$, one can show that

$$s_{d2}\bar{J}_{v2}(\bar{r}) \to \bar{J}_s(\bar{r}) = \hat{a}_x 2E_0/\eta_1 \quad (z = 0). \tag{6.163}$$

This corresponds to nothing but the discontinuity in the magnetic field intensity in the limit case, perfectly fitting into the related boundary condition.

Now, we consider a relatively simple case when a plane wave is normally incident onto a planar PEC (Figure 6.33). Consider a plane wave propagating in the z direction with the electric field polarized in the x direction in a vacuum. When half of the space ($z > 0$) is perfectly conducting, there is a reflected plane wave in the $-z$ direction. In the phasor domain, we have

$$\bar{E}^{\rm inc}(\bar{r}) = \hat{a}_x E_0^{\rm inc} \exp(-jk_0 z), \qquad \bar{H}^{\rm inc}(\bar{r}) = \hat{a}_y \frac{E_0^{\rm inc}}{\eta_0} \exp(-jk_0 z) \tag{6.164}$$

and

$$\bar{E}^{\rm ref}(\bar{r}) = \hat{a}_x E_0^{\rm ref} \exp(jk_0 z), \qquad \bar{H}^{\rm ref}(\bar{r}) = -\hat{a}_y \frac{E_0^{\rm ref}}{\eta_0} \exp(jk_0 z), \tag{6.165}$$

respectively. At the interface – i.e. at $z = 0$ – the boundary condition for the electric field intensity requires $\bar{E}^{\rm tot}_{\rm tan}(z = 0) = 0$: the tangential electric field intensity must vanish on the surface, since there is no electric field for $z > 0$. Then we have[64]

$$\bar{E}^{\text{inc}}_{\text{tan}}(z=0) + \bar{E}^{\text{ref}}_{\text{tan}}(z=0) = 0 \longrightarrow E^{\text{inc}}_0 + E^{\text{ref}}_0 = 0 \tag{6.166}$$

or $E^{\text{ref}}_0 = -E^{\text{inc}}_0 = -E_0$. Hence, the total electric field intensity and total magnetic field intensity for $z < 0$ can be written

$$\bar{E}^{\text{tot}}(\bar{r}) = \hat{a}_x E_0 \exp(-jk_0 z) - \hat{a}_x E_0 \exp(jk_0 z) = -\hat{a}_x j2E_0 \sin(k_0 z) \tag{6.167}$$

$$\bar{H}^{\text{tot}}(\bar{r}) = \hat{a}_y \frac{E_0}{\eta_0} \exp(-jk_0 z) + \hat{a}_y \frac{E_0}{\eta_0} \exp(jk_0 z) = \hat{a}_y \frac{2E_0}{\eta_0} \cos(k_0 z). \tag{6.168}$$

Note that

$$\bar{H}^{\text{inc}}(\bar{r}) = \frac{1}{\eta_0} \hat{a}_z \times \bar{E}^{\text{inc}}(\bar{r}), \qquad \bar{H}^{\text{ref}}(\bar{r}) = \frac{1}{\eta_0}(-\hat{a}_z) \times \bar{E}^{\text{inc}}(\bar{r}), \tag{6.169}$$

but there is no such relationship between $\bar{E}^{\text{tot}}(\bar{r})$ and $\bar{H}^{\text{tot}}(\bar{r})$. Obviously, Maxwell's equations must always hold:

$$\bar{\nabla} \times \bar{E}^{\text{tot}}(\bar{r}) = -\hat{a}_y j2E_0 \frac{\partial}{\partial z} \sin(k_0 z) = -\hat{a}_y j2k_0 E_0 \cos(k_0 z) = -j\omega\mu_0 \bar{H}^{\text{tot}}(\bar{r}) \tag{6.170}$$

and

$$\bar{\nabla} \times \bar{H}^{\text{tot}}(\bar{r}) = -\hat{a}_x 2\frac{E_0}{\eta_0} \frac{\partial}{\partial z} \cos(k_0 z) = \hat{a}_x 2k_0 \frac{E_0}{\eta_0} \sin(k_0 z) = j\omega\varepsilon_0 \bar{E}^{\text{tot}}(\bar{r}), \tag{6.171}$$

given that $k_0 = \omega\sqrt{\mu_0\varepsilon_0} = \omega\mu_0/\eta_0 = \omega\varepsilon_0\eta_0$.

In the time domain, the electric field intensity and magnetic field intensity can be written[65]

$$\bar{E}^{\text{inc}}(\bar{r},t) = \hat{a}_x E_0 \cos(\omega t - k_0 z), \qquad \bar{H}^{\text{inc}}(\bar{r},t) = \hat{a}_y (E_0/\eta_0) \cos(\omega t - k_0 z) \tag{6.172}$$

$$\bar{E}^{\text{ref}}(\bar{r},t) = -\hat{a}_x E_0 \cos(\omega t + k_0 z), \qquad \bar{H}^{\text{ref}}(\bar{r},t) = \hat{a}_y (E_0/\eta_0) \cos(\omega t + k_0 z) \tag{6.173}$$

and

$$\bar{E}^{\text{tot}}(\bar{r},t) = -\hat{a}_x 2E_0 \sin(k_0 z) \text{Re}\{j \exp(j\omega t)\} \tag{6.174}$$

$$= \hat{a}_x 2E_0 \sin(k_0 z) \sin(\omega t) \tag{6.175}$$

$$\bar{H}^{\text{tot}}(\bar{r},t) = \hat{a}_y (2E_0/\eta_0) \cos(k_0 z) \text{Re}\{\exp(j\omega t)\} \tag{6.176}$$

$$= \hat{a}_y (2E_0/\eta_0) \cos(k_0 z) \cos(\omega t). \tag{6.177}$$

[64] Note that if the expressions for the reflection/transmission coefficients in Eqs. (5.346) and (5.347) are directly used for $\eta_2 = 0$ (PEC), we have

$$\Gamma = \frac{0 - \eta_1}{0 + \eta_1} = -1, \qquad \tau = \frac{0}{0 + \eta_1} = 0,$$

indicating that there is no transmitted wave ($\tau = 0$), while the reflected wave has the same amplitude as the incident wave with a direction change ($\Gamma = -1$) in the electric field intensity.

[65] Assuming a real value for E_0.

Interestingly, as opposed to plane waves, where $(\omega t + k_0 z)$ is the argument of the trigonometric functions, the ωt and $k_0 z$ terms are separated and located in two different trigonometric functions. Therefore, these waves, which are superpositions of two plane waves propagating in opposite directions, do not possess propagation themselves. Due to their nature, they are called *standing waves*, as previously discussed in Sections 4.4.1 and 5.3.

To further understand standing waves, we can fix the time at certain points and observe the intensity values with respect to position (Figure 6.34). For example, as time progresses with intervals $\omega \Delta t = \pi/4$, we have

$$\bar{E}^{\text{tot}}(z, \omega t = 0) = \bar{E}^{\text{tot}}(z, \omega t = \pi) = \bar{E}^{\text{tot}}(z, \omega t = 2\pi) = 0 \tag{6.178}$$

$$\bar{E}^{\text{tot}}(z, \omega t = \pi/4) = \bar{E}^{\text{tot}}(z, \omega t = 3\pi/4) = \hat{a}_x \sqrt{2} E_0 \sin(k_0 z) \tag{6.179}$$

$$\bar{E}^{\text{tot}}(z, \omega t = \pi/2) = \hat{a}_x 2 E_0 \sin(k_0 z) \tag{6.180}$$

$$\bar{E}^{\text{tot}}(z, \omega t = 5\pi/4) = \bar{E}^{\text{tot}}(z, \omega t = 7\pi/4) = -\hat{a}_x \sqrt{2} E_0 \sin(k_0 z). \tag{6.181}$$

Similarly, for the magnetic field intensity, one can obtain

$$\bar{H}^{\text{tot}}(z, \omega t = 0) = \bar{H}^{\text{tot}}(z, \omega t = 2\pi) = \hat{a}_y 2(E_0/\eta_0) \cos(k_0 z) \tag{6.182}$$

$$\bar{H}^{\text{tot}}(z, \omega t = \pi/4) = \bar{H}^{\text{tot}}(z, \omega t = 7\pi/4) = \hat{a}_y \sqrt{2}(E_0/\eta_0) \cos(k_0 z) \tag{6.183}$$

$$\bar{H}^{\text{tot}}(z, \omega t = \pi/2) = \bar{H}^{\text{tot}}(z, \omega t = 3\pi/2) = 0 \tag{6.184}$$

$$\bar{H}^{\text{tot}}(z, \omega t = 3\pi/4) = \bar{H}^{\text{tot}}(z, \omega t = 5\pi/4) = -\hat{a}_y \sqrt{2}(E_0/\eta_0) \cos(k_0 z) \tag{6.185}$$

$$\bar{H}^{\text{tot}}(z, \omega t = \pi) = -\hat{a}_y 2(E_0/\eta_0) \cos(k_0 z). \tag{6.186}$$

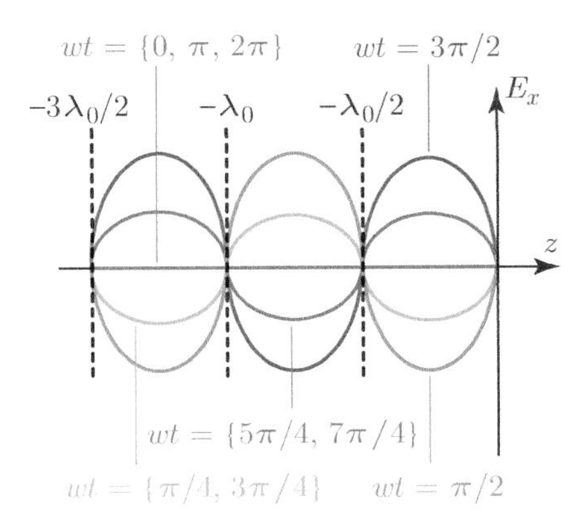

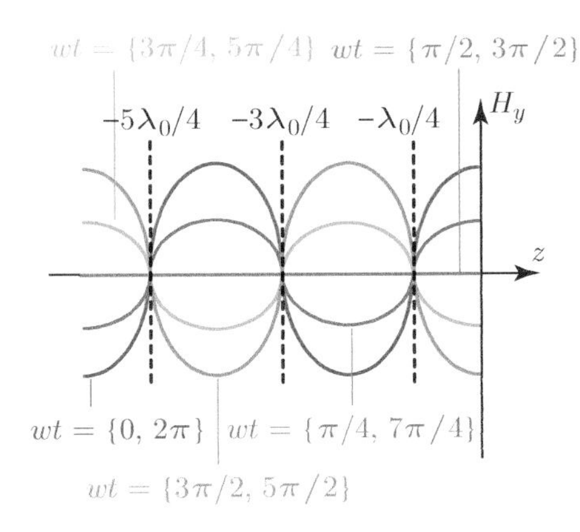

Figure 6.34 Total electric and magnetic field intensities, which represent a standing wave, when a plane wave is normally incident onto a planar PEC.

Therefore, as time progresses, there is no propagation of a wave. Instead, there exists a wave with changing magnitude (including positive and negative values) with respect to time, but it maintains its general shape. The zeros of the electric field intensity occurs when $\sin(k_0 z) = 0$: i.e. at

$$z_E^{\text{zero}} = -n\lambda_0/2 \qquad n = 1, 2, 3, \ldots, \tag{6.187}$$

where λ_0 is the wavelength. On the other hand, the zeros of the magnetic field intensity occurs when $\cos(k_0 z) = 0$: i.e. at

$$z_H^{\text{zero}} = \lambda_0/4 - n\lambda_0/2 \qquad n = 1, 2, 3, \ldots. \tag{6.188}$$

Also note that the tangential component of the magnetic field intensity is discontinuous at the boundary. Given the boundary condition for the magnetic field intensity, we have

$$-\hat{a}_z \times \bar{H}^{\text{tot}}(z=0,t) = \hat{a}_x 2\frac{E_0}{\eta_0}\cos(\omega t) = \bar{J}_s(\bar{r},t). \tag{6.189}$$

Hence, a surface electric current density is induced on the perfectly conducting surface due to the excitation by the plane wave.

But what about the power flux density for the scenario considered above? Given the phasor-domain representations of intensities, the complex Poynting vector can be found:

$$\bar{S}_c^{\text{tot}}(\bar{r}) = \frac{1}{2}\bar{E}^{\text{tot}}(\bar{r}) \times [\bar{H}^{\text{tot}}(\bar{r})]^* = -\frac{1}{2}\hat{a}_x j2E_0\sin(k_0 z) \times \hat{a}_y 2\frac{E_0}{\eta_0}\cos(k_0 z) \tag{6.190}$$

$$= -\hat{a}_z j2\frac{E_0^2}{\eta_0}\sin(k_0 z)\cos(k_0 z) = -\hat{a}_z j\frac{E_0^2}{\eta_0}\sin(2k_0 z). \tag{6.191}$$

Unsurprisingly, the complex Poynting vector (the complex power flux density) is purely imaginary. This indicates that $\bar{S}_{\text{avg}}(\bar{r}) = 0$: i.e. the time-average power flow is zero. In the time domain, we have

$$\bar{S}(\bar{r},t) = \bar{E}(\bar{r},t) \times \bar{H}(\bar{r},t) = \hat{a}_z\frac{4E_0^2}{\eta_0}\sin(k_0 z)\cos(k_0 z)\sin(\omega t)\cos(\omega t) \tag{6.192}$$

$$= \hat{a}_z\frac{E_0^2}{\eta_0}\sin(2k_0 z)\sin(2\omega t). \tag{6.193}$$

Hence, at a given position, the power flux density is periodically in the positive and negative z directions (oscillatory) such that the average value is zero (Figure 6.35). At a given time, the power flux density changes sinusoidally with respect to position; but as time progresses, the magnitude changes to get positive and negative values. Also note that the oscillation rates for the power flux density in both time and position are twice the corresponding rates for the electric field intensity and magnetic field intensity.

In the above, while the power is purely reactive, it is remarkable that there is still a continuous power flow that oscillates with respect to time, and it may be interesting to investigate what this flow corresponds to. Given the electric potential energy and magnetic potential energy, we have

$$dw_{e,\text{stored}}(\bar{r},t) = \frac{\varepsilon_0}{2}|\bar{E}(\bar{r},t)|^2 dv = \frac{\varepsilon_0}{2}4E_0^2\sin^2(k_0 z)\sin^2(\omega t)dv \tag{6.194}$$

$$= 2\varepsilon_0 E_0^2\sin^2(k_0 z)\sin^2(\omega t)dv \tag{6.195}$$

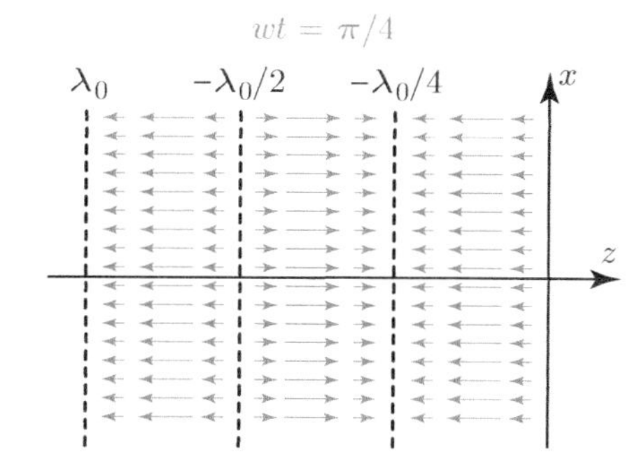

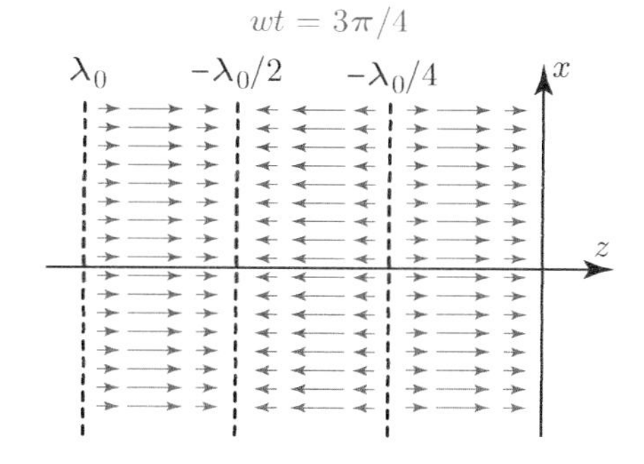

Figure 6.35 Total power flux density at two different instants when a plane wave is normally incident onto a planar PEC at $z = 0$.

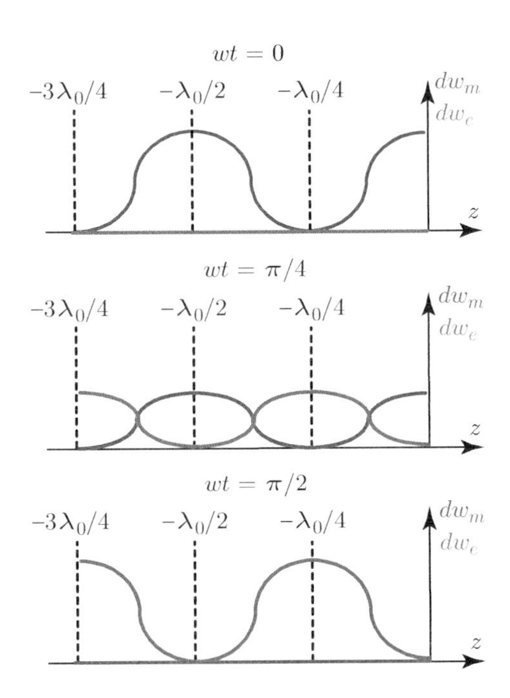

Figure 6.36 Electric potential energy and magnetic potential energy associated with total fields when a plane wave is normally incident onto a planar PEC. At a given position, the stored energy is periodically of electric and magnetic types. This is in agreement with the total power flux density depicted in Figure 6.35, which has an oscillatory nature with respect to time, leading to a zero average power flow.

and

$$dw_{m,\text{stored}}(\bar{r},t) = \frac{\mu_0}{2}|\bar{H}(\bar{r},t)|^2 dv = \frac{\mu_0}{2}4\frac{E_0^2}{\eta_0^2}\cos^2(k_0 z)\cos^2(\omega t)dv \tag{6.196}$$

$$= 2\varepsilon_0 E_0^2 \cos^2(k_0 z)\cos^2(\omega t)dv. \tag{6.197}$$

Hence, at a fixed position, the stored energy is periodically of electric and magnetic types (Figure 6.36). The power flow described by the instantaneous Poynting vector in Eq. (6.193) corresponds to the re-localization of the electric/magnetic energies periodically at different positions.

It becomes slightly more challenging to analyze reflection when the incidence is oblique, as also discussed in Section 5.3.5 for dielectric cases. We now consider perpendicular and parallel polarizations separately, given that an arbitrary plane wave can be a combination of these cases. For perpendicular polarization (Figure 6.37) – i.e. when the electric field intensity is perpendicular to the plane of incidence – we have

$$\bar{E}^{\text{inc}}(\bar{r}) = \hat{a}_y E_0^{\text{inc}} \exp[-jk_0(x\sin\varphi + z\cos\varphi)] \tag{6.198}$$

$$\bar{H}^{\text{inc}}(\bar{r}) = (-\hat{a}_x\cos\varphi + \hat{a}_z\sin\varphi)\frac{E_0^{\text{inc}}}{\eta_0}\exp[-jk_0(x\sin\varphi + z\cos\varphi)] \tag{6.199}$$

for the incident wave and

$$\bar{E}^{\text{ref}}(\bar{r}) = \hat{a}_y E_0^{\text{ref}} \exp[-jk_0(x\sin\varphi - z\cos\varphi)] \tag{6.200}$$

$$\bar{H}^{\text{ref}}(\bar{r}) = (\hat{a}_x\cos\varphi + \hat{a}_z\sin\varphi)\frac{E_0^{\text{ref}}}{\eta_0}\exp[-jk_0(x\sin\varphi - z\cos\varphi)] \tag{6.201}$$

for the reflected wave. According to the boundary condition for the electric field intensity at $z = 0$, the total tangential electric field intensity must vanish on the surface of the PEC,[66] and we have $E_0^{\text{ref}} = -E_0^{\text{inc}} = -E_0$. Then we obtain

$$\bar{E}^{\text{tot}}(\bar{r}) = \hat{a}_y E_0 \exp(-jk_0 x\sin\varphi)\left[\exp(-jk_0 z\cos\varphi) - \exp(jk_0 z\cos\varphi)\right] \tag{6.202}$$

$$= -\hat{a}_y j2E_0\exp(-jk_0 x\sin\varphi)\sin(k_0 z\cos\varphi) \tag{6.203}$$

for the electric field intensity[67] and

$$\bar{H}^{\text{tot}}(\bar{r}) = -\hat{a}_x(2E_0/\eta_0)\cos\varphi\exp(-jk_0 x\sin\varphi)\cos(k_0 z\cos\varphi) \tag{6.204}$$

$$-\hat{a}_z(j2E_0/\eta_0)\sin\varphi\exp(-jk_0 x\sin\varphi)\sin(k_0 z\cos\varphi) \tag{6.205}$$

[66] These expressions are again consistent with using $\eta_2 = 0$ (PEC) in the general expressions in Eqs. (5.395) and (5.396): i.e. we have

$$\Gamma_\perp = \frac{-\eta_1\cos\varphi_t}{+\eta_1\cos\varphi_t} = -1, \qquad \tau_\perp = \frac{0}{\eta_1\cos\varphi_t} = 0,$$

independent of the angle of incidence.

[67] As an intermediate step, we have

$$\bar{E}^{\text{tot}}(\bar{r}) = \hat{a}_y E_0\exp[-jk_0(x\sin\varphi + z\cos\varphi)] - \hat{a}_y E_0\exp[-jk_0(x\sin\varphi - z\cos\varphi)].$$

for the magnetic field intensity[68,69,70]. The time-domain expression for the total electric field intensity can be written[71]

$$\bar{E}^{\text{tot}}(\bar{r},t) = -\hat{a}_y 2E_0 \sin(k_0 z\cos\varphi)\cos(\omega t - k_0 x\sin\varphi + \pi/2) \tag{6.206}$$

$$= \hat{a}_y 2E_0 \sin(k_0 z\cos\varphi)\sin(\omega t - k_0 x\sin\varphi). \tag{6.207}$$

As another important result, the magnetic field intensity at $z = 0$ can be written (in the phasor domain) as

$$\bar{H}^{\text{tot}}(z=0) = -\hat{a}_x(2E_0/\eta_0)\cos\varphi\exp(-jk_0 x\sin\varphi), \tag{6.208}$$

which does not have any z component, as expected due to the boundary condition for the magnetic field intensity (vanishing normal component). Then the induced surface electric current density (at $z = 0$) can be found as

$$\bar{J}_s(\bar{r}) = -\hat{a}_z \times \bar{H}^{\text{tot}}(z=0) = \hat{a}_z \times \hat{a}_x(2E_0/\eta_0)\cos\varphi\exp(-jk_0 x\sin\varphi) \tag{6.209}$$

$$= \hat{a}_y(2E_0/\eta_0)\cos\varphi\exp(-jk_0 x\sin\varphi), \tag{6.210}$$

leading to

$$\bar{J}_s(\bar{r},t) = \hat{a}_y(2E_0/\eta_0)\cos\varphi\cos(\omega t - k_0 x\sin\varphi) \tag{6.211}$$

in the time domain.

Next, we consider parallel polarization: i.e. when the electric field intensity direction is in the plane of incidence (Figure 6.38). In this case, the magnetic field intensity is perpendicular to the plane of incidence, while the electric field intensity (as a vector field) is located in the plane. In the phasor domain, the incident electric field intensity and magnetic field intensity can be written

$$\bar{E}^{\text{inc}}(\bar{r}) = (\hat{a}_x\cos\varphi - \hat{a}_z\sin\varphi)E_0^{\text{inc}}\exp[-jk_0(x\sin\varphi + z\cos\varphi)] \tag{6.212}$$

$$\bar{H}^{\text{inc}}(\bar{r}) = \hat{a}_y\frac{E_0^{\text{inc}}}{\eta_0}\exp[-jk_0(x\sin\varphi + z\cos\varphi)]. \tag{6.213}$$

Similarly, for the reflected plane wave, we have

$$\bar{E}^{\text{ref}}(\bar{r}) = (\hat{a}_x\cos\varphi + \hat{a}_z\sin\varphi)E_0^{\text{ref}}\exp[-jk_0(x\sin\varphi - z\cos\varphi)] \tag{6.214}$$

$$\bar{H}^{\text{ref}}(\bar{r}) = -\hat{a}_y\frac{E_0^{\text{ref}}}{\eta_0}\exp[-jk_0(x\sin\varphi - z\cos\varphi)]. \tag{6.215}$$

[68] As an intermediate step, we have

$$\bar{H}^{\text{tot}}(\bar{r}) = (-\hat{a}_x\cos\varphi + \hat{a}_z\sin\varphi) \times (E_0/\eta_0)\exp[-jk_0(x\sin\varphi + z\cos\varphi)] - (\hat{a}_x\cos\varphi + \hat{a}_z\sin\varphi) \times (E_0/\eta_0)\exp[-jk_0(x\sin\varphi - z\cos\varphi)].$$

[69] Note that $\bar{H}^{\text{tot}}(\bar{r}) = -\bar{\nabla}\times\bar{E}^{\text{tot}}(\bar{r})/(j\omega\mu_0)$ is satisfied.

[70] Note that the complex Poynting vector has both real and imaginary parts.

[71] At time $t = 0$, the position dependency of the total electric field intensity can be found as

$$\bar{E}^{\text{tot}}(\bar{r}, t=0) = -\hat{a}_y 2E_0\sin(k_0 z\cos\varphi)\sin(k_0 x\sin\varphi),$$

i.e. multiplication of two sinusoidal functions in perpendicular directions (Figures 6.39 and 6.40).

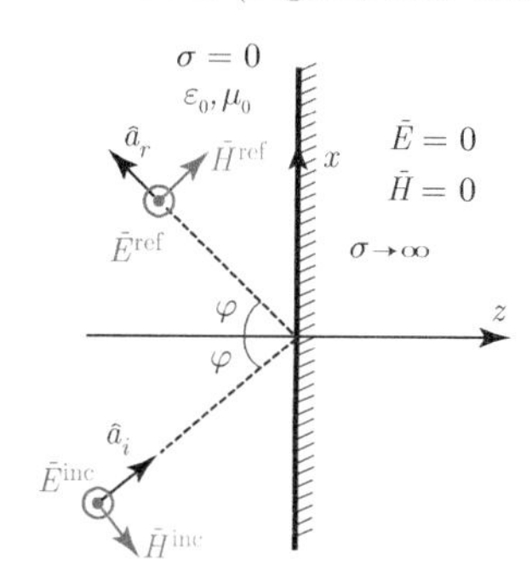

Figure 6.37 Oblique incidence of a plane wave with perpendicular polarization onto a planar PEC.

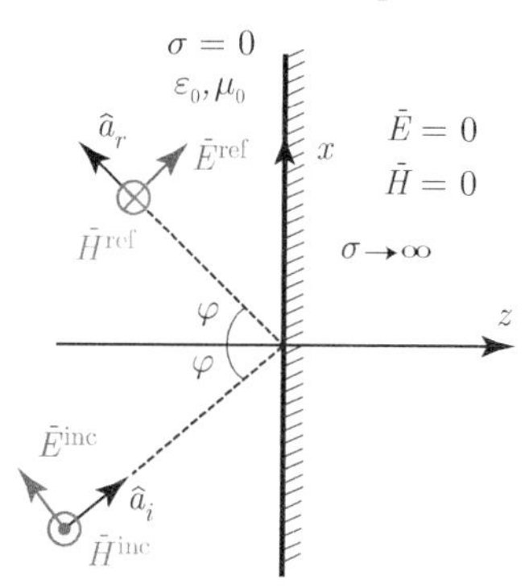

Figure 6.38 Oblique incidence of a plane wave with parallel polarization onto a planar PEC.

Enforcing the boundary condition for the electric field intensity – i.e. the total tangential electric field intensity must vanish at $z = 0$ – we have $E_0^{\rm ref} = -E_0^{\rm inc} = -E_0$, leading to[72,73,74]

$$\bar{E}^{\rm tot}(\bar{r}) = (\hat{a}_x \cos\varphi - \hat{a}_z \sin\varphi)E_0 \exp[-jk_0(x\sin\varphi + z\cos\varphi)]$$

$$- (\hat{a}_x \cos\varphi + \hat{a}_z \sin\varphi)E_0 \exp[-jk_0(x\sin\varphi - z\cos\varphi)] \tag{6.216}$$

$$= -\hat{a}_x j2E_0 \cos\varphi \exp(-jk_0 x\sin\varphi)\sin(k_0 z\cos\varphi)$$

$$- \hat{a}_z 2E_0 \sin\varphi \exp(-jk_0 x\sin\varphi)\cos(k_0 z\cos\varphi) \tag{6.217}$$

and

$$\bar{H}^{\rm tot}(\bar{r}) = \hat{a}_y(E_0/\eta_0)\exp[-jk_0(x\sin\varphi + z\cos\varphi)]$$

$$+ \hat{a}_y(E_0/\eta_0)\exp[-jk_0(x\sin\varphi - z\cos\varphi)] \tag{6.218}$$

$$= \hat{a}_y(2E_0/\eta_0)\exp(-jk_0 x\sin\varphi)\cos(k_0 z\cos\varphi). \tag{6.219}$$

Hence, the time-domain expression for the total magnetic field intensity can be written[75]

$$\bar{H}^{\rm tot}(\bar{r},t) = \hat{a}_y(2E_0/\eta_0)\cos(k_0 z\cos\varphi)\cos(\omega t - k_0 x\sin\varphi). \tag{6.220}$$

To find the induced surface electric current density, we consider the expression for the total magnetic field intensity at $z = 0$ as

$$\bar{H}^{\rm tot}(z=0) = \hat{a}_y(2E_0/\eta_0)\exp(-jk_0 x\sin\varphi) \tag{6.221}$$

in the phasor domain. Then we have

$$\bar{J}_s(\bar{r}) = -\hat{a}_z \times \bar{H}^{\rm tot}(z=0) = \hat{a}_x(2E_0/\eta_0)\exp(-jk_0 x\sin\varphi) \tag{6.222}$$

$$\bar{J}_s(\bar{r},t) = \hat{a}_x(2E_0/\eta_0)\cos(\omega t - k_0 x\sin\varphi) \tag{6.223}$$

in the phasor and time domains, respectively.

6.6.6 Examples

Example 65: At 10 MHz, conductivity values for seawater and lake water are given as 4.8 S/m and 4.8×10^{-3} S/m, respectively, and both have 80 relative permittivity. Find the attenuation and phase constants, as well as the skin depth, in both cases (at 10 MHz).

[72]Note that $\bar{E}^{\rm tot}(\bar{r}) = \bar{\nabla} \times \bar{H}^{\rm tot}(\bar{r})/(j\omega\varepsilon_0)$ is satisfied.

[73]Note that the complex Poynting vector has both real and imaginary parts.

[74]These expressions are again consistent with using $\eta_2 = 0$ (PEC) in the general expressions in Eqs. (5.409) and (5.410): i.e. we have

$$\Gamma_\parallel = \frac{-\eta_1\cos\varphi_i}{+\eta_1\cos\varphi_i} = -1, \qquad \tau_\parallel = \frac{0}{\eta_1\cos\varphi_i} = 0,$$

independent of the angle of incidence.

[75]At time $t = 0$, the position dependency of the magnetic field intensity can be found as

$$\bar{H}^{\rm tot}(\bar{r}, t=0) = \hat{a}_y\frac{2E_0}{\eta_0}\cos(k_0 z\cos\varphi)\cos(k_0 x\sin\varphi),$$

i.e. multiplication of two sinusoidal functions in perpendicular directions.

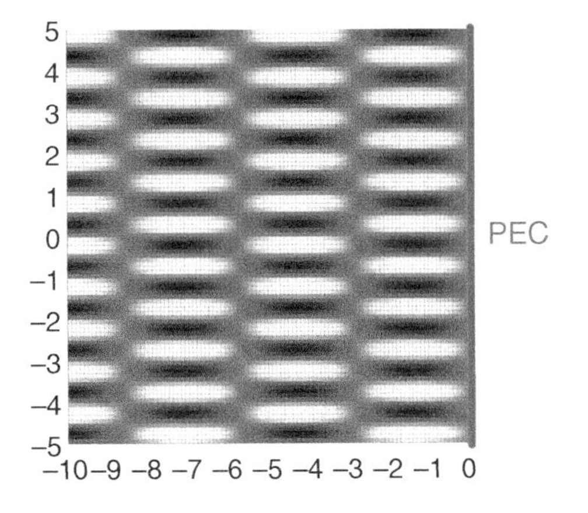

Figure 6.39 A snapshot of the total electric field intensity at $t = 0$ when a plane wave with perpendicular polarization is obliquely incident onto a planar PEC with $\varphi = 80°$ angle. Axes labels show the positions in the plane of incidence in terms the wavelength. Dark and light colors represent dips and peaks, respectively.

Solution: For seawater, we have

$$\tan\delta_c = \frac{\sigma}{\omega\varepsilon} \approx \frac{4.8}{2\pi \times 10^7 \times 80 \times 8.85 \times 10^{-12}} \approx 108 \gg 1, \tag{6.224}$$

which indicates that it is a good conductor. Then we can find the attenuation and phase constants as

$$\alpha = \beta \approx \sqrt{\frac{\omega\mu\sigma}{2}} = \sqrt{\frac{2\pi \times 10^7 \times 4\pi \times 10^{-7} \times 4.8}{2}} \approx 13.8 \text{ Np/m or rad/m}. \tag{6.225}$$

The corresponding skin depth can be found as $s_d = 1/\alpha \approx 7.3$ cm. On the other hand, for lake water, we have

$$\tan\delta_c = \frac{\sigma}{\omega\varepsilon} \approx \frac{4.8 \times 10^{-3}}{2\pi \times 10^7 \times 80 \times 8.85 \times 10^{-12}} \approx 0.108 \ll 1, \tag{6.226}$$

i.e. lake water is a low-loss dielectric. In this case, the attenuation and phase constants can be found separately:

$$\alpha \approx \frac{\sigma}{2}\sqrt{\frac{\mu}{\varepsilon}} \approx \frac{4.8 \times 10^{-3}}{2}\sqrt{\frac{4\pi \times 10^{-7}}{80 \times 8.85 \times 10^{-12}}} \approx 0.101 \text{ Np/m} \tag{6.227}$$

$$\beta \approx \omega\sqrt{\mu\varepsilon} \approx 2\pi \times 10^7 \times \sqrt{4\pi \times 10^{-7} \times 80 \times 8.85 \times 10^{-12}} \approx 1.87 \text{ rad/m}. \tag{6.228}$$

Finally, we obtain the skin depth for lake water as $s_d = 1/\alpha \approx 9.9$ m.

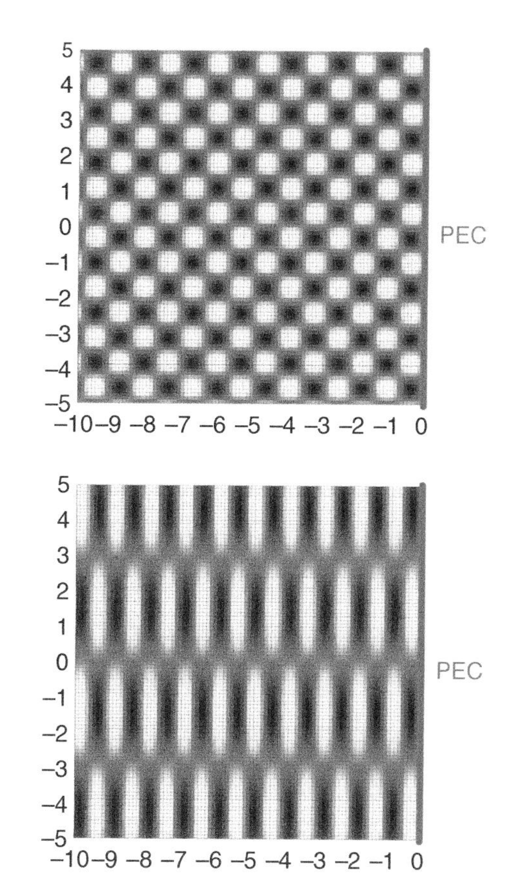

Figure 6.40 Similar to Figure 6.39 when the incidence angle is 45° (top) and 10° (bottom).

Example 66: Consider a plane wave with a phasor-domain expression for the electric field intensity as[76]

$$\bar{E}(\bar{r}) = \hat{a}_x \exp(-\alpha z)\exp(-j\beta z) \quad \text{(V/m)}, \tag{6.229}$$

propagating in seawater ($\sigma = 4.8$, $\varepsilon_r = 80$) at 1 MHz. Find the time-average net power flux into a unit cube centered at the origin (Figure 6.41).

[76] Note that the amplitude of the electric field intensity is given as 1 V/m at $z = 0$.

Solution: First, we can check the characteristics of seawater as

$$\tan\delta_c = \frac{\sigma}{\omega\varepsilon} \approx \frac{4.8}{2\pi \times 10^6 \times 80 \times 8.85 \times 10^{-12}} \approx 1.08 \times 10^3 \gg 1, \tag{6.230}$$

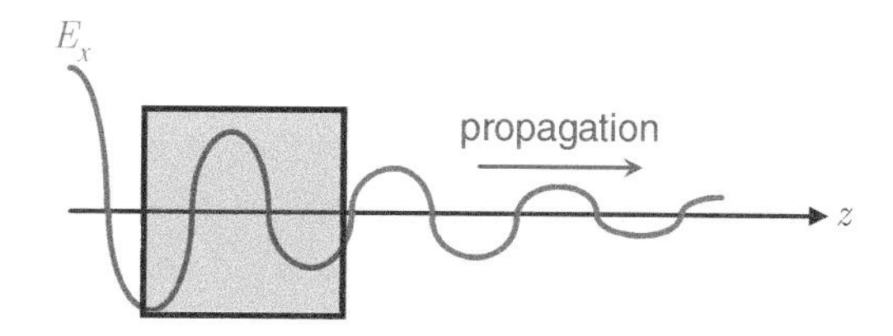

Figure 6.41 A snapshot of the electric field intensity of a plane wave propagating in the z direction in a conducting medium (sea water). A unit cube centered at the origin is considered to investigate the power flux and dissipated power.

which indicates that it is a good conductor at 1 MHz. Then we can obtain the attenuation and propagation constants:

$$\alpha = \beta \approx \sqrt{\frac{\omega\mu\sigma}{2}} = \sqrt{\frac{2\pi \times 10^6 \times 4\pi \times 10^{-7} \times 4.8}{2}} \approx 4.35 \text{ Np/m or rad/m.} \tag{6.231}$$

In addition, we find the complex wave impedance as

$$\eta_c \approx \sqrt{\frac{\mu\omega}{2\sigma}}(1+j) \approx 0.907(1+j) \quad (\Omega). \tag{6.232}$$

Then the expression for the magnetic field intensity can be written[77]

$$\bar{H}(\bar{r}) \approx \hat{a}_y \frac{E_0}{|\eta_c|} \exp(-\alpha z) \exp(-j\beta z - j\delta_c/2) \tag{6.233}$$

$$\approx \hat{a}_y \frac{1}{0.907\sqrt{2}} \exp(-\alpha z) \exp(-j\beta z - j\pi/4) \quad (\text{A/m}). \tag{6.234}$$

At this stage, we can find the complex Poynting vector as

$$\bar{S}_c(\bar{r}) = \frac{1}{2}\bar{E}(\bar{r}) \times [\bar{H}(\bar{r})]^* = \hat{a}_z \frac{E_0^2}{2|\eta_c|} \exp(-2\alpha z) \exp(j\delta_c/2) \tag{6.235}$$

$$\approx \hat{a}_z \frac{1}{2.57} \exp(-8.70z) \exp(j\pi/4) \quad (\text{W/m}^2). \tag{6.236}$$

Then the time-average power flux density can be found as

$$\bar{S}_{\text{avg}}(\bar{r}) = \text{Re}\{\bar{S}_c(\bar{r})\} \approx \hat{a}_z \frac{1}{2.57} \exp(-8.70z) \frac{\sqrt{2}}{2} \approx \hat{a}_z 0.275 \exp(-8.70z). \tag{6.237}$$

Integration of the time-average power flux density on the surface of the unit cube leads to

$$p_{em,\text{avg}} = \int_S \bar{S}_{\text{avg}}(\bar{r}) \cdot \overline{ds} \tag{6.238}$$

$$\approx -[0.275\exp(-8.70z)]_{z=-1/2} + [0.275\exp(-8.70z)]_{z=1/2} \tag{6.239}$$

$$\approx -21.3 \text{ W}. \tag{6.240}$$

Therefore, there is a time-average power flux of 21.3 W into the cube. This corresponds to the dissipated power (specifically, *time-average* power dissipated inside the cube) due to the conductivity of the medium.[78].

[77] Note that $\tan\delta_c \gg 1 \longrightarrow \delta_c \approx \pi/2$, leading to a 45° phase difference between the electric field intensity and magnetic field intensity.

[78] Note that

$$\begin{aligned}
\bar{J}_v(\bar{r},t) \cdot \bar{E}(\bar{r},t) &= \text{Re}\{\bar{J}_v(\bar{r})\exp(j\omega t)\} \cdot \text{Re}\{\bar{E}(\bar{r})\exp(j\omega t)\} \\
&= \frac{1}{4}\bar{J}_v(\bar{r}) \cdot \bar{E}(\bar{r})\exp(j2\omega t) \\
&\quad + \frac{1}{4}[\bar{J}_v(\bar{r})]^* \cdot [\bar{E}(\bar{r})]^* \exp(-j2\omega t) \\
&\quad + \frac{1}{4}[\bar{J}_v(\bar{r})]^* \cdot \bar{E}(\bar{r}) + \frac{1}{4}\bar{J}_v(\bar{r}) \cdot [\bar{E}(\bar{r})]^*.
\end{aligned}$$

Hence, we can define a *complex* dissipated power density $\pi_{J,c}(\bar{r}) = (1/2)[\bar{J}(\bar{r})]^* \cdot \bar{E}(\bar{r}) = (1/2)\sigma(\bar{r})|\bar{E}(\bar{r})|^2$, whose real part corresponds to the time-average dissipated power density.

Example 67: A plane wave propagating in the z direction in a vacuum at 100 MHz is incident normally onto a boundary at $z = 0$ (Figure 6.42). The medium on the right ($z > 0$) is lossy with $\varepsilon_r = 4$ and $\sigma = 10$ S/m. The electric field is polarized in the x direction and has a magnitude of 10 V/m. Find the propagation constants in both media, as well as the electric field intensity of the reflected and transmitted waves (approximately).

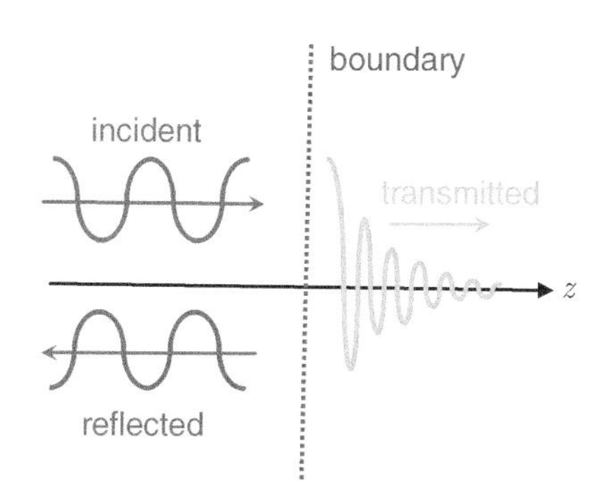

Figure 6.42 A demonstration of the reflection and transmission of a plane wave that is incident normally onto a boundary at $z = 0$ between a vacuum and a lossy medium. Depending on the conductivity, the transmitted wave may quickly decay in the lossy medium.

Solution: First, the propagation constants for the first and second media can be found as

$$\gamma_1 = j\omega\sqrt{\mu_0\varepsilon_0} \approx j\frac{2\pi \times 10^8}{3 \times 10^8} = j\beta_1 = j\frac{2\pi}{3} \text{ rad/m} \tag{6.241}$$

$$\gamma_2 = j\omega\sqrt{\mu_0\varepsilon_2} = j\omega\sqrt{\mu_0\left(\varepsilon + \frac{\sigma}{j\omega}\right)} = j\omega\sqrt{\mu_0\varepsilon_0}\sqrt{4 + \frac{10}{j\omega\varepsilon_0}} \tag{6.242}$$

$$\approx j\frac{2\pi}{3}\sqrt{4 - j1.8 \times 10^2} \approx j\frac{2\pi}{3}\sqrt{-j1.8 \times 10^2} = 2\pi(10 + j10) \text{ rad/m}, \tag{6.243}$$

respectively. Then to find the reflection and transmission coefficients, we calculate the wave impedance in the second medium as[79]

$$\eta_2 = \sqrt{\frac{\mu_0}{\varepsilon_2}} = \sqrt{\frac{\mu_0}{\varepsilon_0\left(4 + \frac{\sigma}{j\omega\varepsilon_0}\right)}} \approx \frac{120\pi}{30 - 30j} = 2\pi(1 + j)\ \Omega. \tag{6.244}$$

We obtain the reflection and transmission coefficients:

$$\Gamma = \frac{\eta_2 - \eta_1}{\eta_2 + \eta_1} \approx \frac{2\pi + j2\pi - 120\pi}{2\pi + j2\pi + 120\pi} = \frac{-59 + j}{61 + j} \tag{6.245}$$

$$\tau = 1 + \Gamma \approx \frac{2 + j2}{61 + j}. \tag{6.246}$$

Consequently, the reflected and transmitted electric field intensity can be written

$$\bar{E}^{\text{ref}}(\bar{r}) \approx \hat{a}_x 10\left(\frac{-59 + j}{61 + j}\right)\exp(j2\pi z/3) \tag{6.247}$$

$$\bar{E}^{\text{tra}}(\bar{r}) \approx \hat{a}_x 10\left(\frac{2 + j2}{61 + j}\right)\exp(-\gamma_2 z). \tag{6.248}$$

[79] Note that the loss tangent of the conducting medium can be found as

$$\tan\delta_c = \frac{\sigma}{\omega\varepsilon} \approx \frac{10}{2\pi \times 10^8 \times 4 \times 8.85 \times 10^{-12}} \approx 450 \gg 1,$$

indicating that it is a good conductor. Therefore, we naturally obtain the phase of η_2 as 45°. In addition, due to the good conductivity, the resulting transmission coefficient is quite small, while the transmitted wave decays quickly.

Note that the transmitted wave is already small at $z = 0$, while it quickly decays for $z > 0$ due to the conductivity.[80]

Example 68: A circularly polarized plane wave propagating in the z direction in a vacuum is incident normally onto a planar PEC located at $z = 0$ (Figure 6.43). The electric field intensity of the incident wave is given in the phasor domain as

$$\bar{E}^{\text{inc}}(\bar{r}) = (\hat{a}_x + \hat{a}_y j)E_0 \exp(-j2z), \tag{6.249}$$

where E_0 (V/m) is a real constant. Find the electric field intensity of the reflected wave, the surface electric current density induced on the PEC, as well as the complex Poynting vector for $z < 0$.

Solution: For normal incidence, the reflected electric field intensity can be written[81]

$$\bar{E}^{\text{ref}}(\bar{r}) = (-\hat{a}_x - \hat{a}_y j)E_0 \exp(j2z). \tag{6.250}$$

The corresponding expressions for the magnetic field intensity of the incident and reflected waves can be written

$$\bar{H}^{\text{inc}}(\bar{r}) = \hat{a}_z \times (\hat{a}_x + \hat{a}_y j)\frac{E_0}{\eta_0}\exp(-j2z) = (-\hat{a}_x j + \hat{a}_y)\frac{E_0}{\eta_0}\exp(-j2z) \tag{6.251}$$

$$\bar{H}^{\text{ref}}(\bar{r}) = -\hat{a}_z \times (-\hat{a}_x - \hat{a}_y j)\frac{E_0}{\eta_0}\exp(j2z) = (-\hat{a}_x j + \hat{a}_y)\frac{E_0}{\eta_0}\exp(j2z). \tag{6.252}$$

Then the total electric field intensity and total magnetic field intensity can be found as[82]

$$\bar{E}^{\text{tot}}(\bar{r}) = (\hat{a}_x + \hat{a}_y j)E_0[\exp(-j2z) - \exp(j2z)] = (-\hat{a}_x j + \hat{a}_y)2E_0 \sin(2z) \tag{6.253}$$

$$\bar{H}^{\text{tot}}(\bar{r}) = (-\hat{a}_x j + \hat{a}_y)\frac{2E_0}{\eta_0}\cos(2z). \tag{6.254}$$

The surface electric current density induced on the PEC can be obtained via the related boundary condition as

$$\bar{J}_s(\bar{r}) = -\hat{a}_z \times \bar{H}^{\text{tot}}(z = 0) = -\hat{a}_z \times (-\hat{a}_x j + \hat{a}_y)\frac{2E_0}{\eta_0} = (\hat{a}_x + \hat{a}_y j)\frac{2E_0}{\eta_0} \tag{6.255}$$

[80] Specifically, the amplitude of the transmitted electric field intensity at $z = 0$ can be found as

$$10\left|\frac{2 + j2}{61 + j}\right| \approx 0.46 \text{ V/m}.$$

Then the exponential part of the transmitted electric field intensity can be written

$$\exp(-\gamma_2 z) = \exp(-2\pi(10 + j10)z) = \exp(-20\pi z)\exp(-j20\pi z).$$

This means the amplitude drops to below 1% of its initial value only 7.5 cm from the boundary.

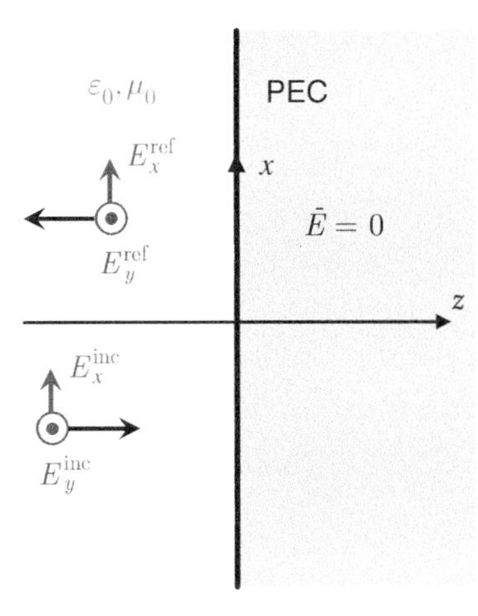

Figure 6.43 Normal incidence of a circularly polarized plane wave onto an interface between a vacuum and a planar PEC.

[81] Note that the incident wave is LHCP, while the reflected wave is RHCP.

[82] The time-domain expression for the electric field intensity can be found as

$$\begin{aligned}\bar{E}^{\text{tot}}(\bar{r}, t) &= \text{Re}\{(-\hat{a}_x j + \hat{a}_y)2E_0 \sin(2z)\exp(j\omega t)\} \\ &= 2E_0 \sin(2z) \\ &\quad \times \text{Re}\{(-\hat{a}_x j + \hat{a}_y)[\cos(\omega t) + j\sin(\omega t)]\} \\ &= \hat{a}_x 2E_0 \sin(2z)\sin(\omega t) \\ &\quad + \hat{a}_y 2E_0 \sin(2z)\cos(\omega t),\end{aligned}$$

i.e. superposition of two standing waves.

in units A/m. Finally, we can find the complex Poynting vector as

$$\bar{S}_c^{\text{tot}}(\bar{r}) = \frac{1}{2}\bar{E}^{\text{tot}}(\bar{r}) \times [\bar{H}^{\text{tot}}(\bar{r})]^* \tag{6.256}$$

$$= \frac{1}{2}(-\hat{a}_x j + \hat{a}_y)2E_0 \sin(2z) \times (\hat{a}_x j + \hat{a}_y)\frac{2E_0}{\eta_0}\cos(2z) \tag{6.257}$$

$$= -\hat{a}_z j2\frac{E_0^2}{\eta_0}\sin(4z), \tag{6.258}$$

which indicates that the time-average power flux density is zero – i.e. $\text{Re}\{\bar{S}_c^{\text{tot}}(\bar{r})\} = 0$ – as expected.

Example 69: A plane wave is obliquely incident onto a planar PEC at $z = 0$ (Figure 6.44). The electric field intensity of the incident wave is given as

$$\bar{E}^{\text{inc}}(\bar{r}) = (\hat{a}_x 3 + \hat{a}_y j5 - \hat{a}_z 4)E_0 \exp[-j(0.8x + 0.6z)], \tag{6.259}$$

where E_0 (V/m) is a real constant. Find the electric field intensity of the reflected wave, as well as the surface electric current density induced on the PEC.

Solution: First, for the exponential part, note that

$$\bar{k}_i \cdot \bar{r} = 0.8x + 0.6y \longrightarrow \bar{k}_i = \hat{a}_i k_0 = \hat{a}_x 4/5 + \hat{a}_z 3/5, \tag{6.260}$$

indicating that $k_0 = 1$ rad/m. The perpendicular and parallel components of the electric field intensity can be written

$$\bar{E}_\perp^{\text{inc}}(\bar{r}) = \hat{a}_y j5E_0 \exp[-j(0.8x + 0.6z)] \tag{6.261}$$

$$\bar{E}_\parallel^{\text{inc}}(\bar{r}) = (\hat{a}_x 3 - \hat{a}_z 4)E_0 \exp[-j(0.8x + 0.6z)]. \tag{6.262}$$

Then for the reflected wave, we have

$$\bar{E}_\perp^{\text{ref}}(\bar{r}) = -\hat{a}_y j5E_0 \exp[-j(0.8x - 0.6z)] \tag{6.263}$$

$$\bar{E}_\parallel^{\text{ref}}(\bar{r}) = (-\hat{a}_x 3 - \hat{a}_z 4)E_0 \exp[-j(0.8x - 0.6z)], \tag{6.264}$$

leading to

$$\bar{E}^{\text{ref}}(\bar{r}) = (-\hat{a}_x 3 - \hat{a}_y j5 - \hat{a}_z 4)E_0 \exp[-j(0.8x - 0.6z)]. \tag{6.265}$$

For the incident and reflected waves, the magnetic field intensity can be written[83]

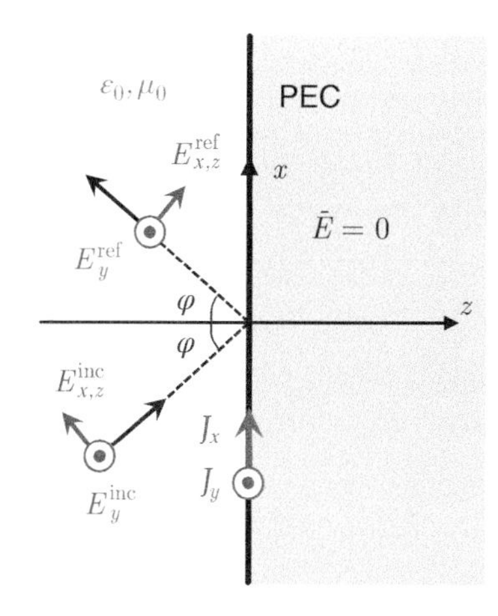

Figure 6.44 Oblique incidence of a plane wave onto an interface between a vacuum and a planar PEC. The incident and reflected waves have both perpendicular and parallel polarizations. The surface electric current density induced on the PEC also has both x and y components.

[83] We use

$$\left(\hat{a}_x\frac{4}{5} + \hat{a}_z\frac{3}{5}\right) \times (\hat{a}_x 3 + \hat{a}_y j5 - \hat{a}_z 4)$$
$$= \hat{a}_z j4 + \hat{a}_y\frac{16}{5} + \hat{a}_y\frac{9}{5} - \hat{a}_x j3$$
$$= -\hat{a}_x j3 + \hat{a}_y 5 + \hat{a}_z j4$$
$$\left(\hat{a}_x\frac{4}{5} - \hat{a}_z\frac{3}{5}\right) \times (-\hat{a}_x 3 - \hat{a}_y j5 - \hat{a}_z 4)$$
$$= -\hat{a}_z j4 + \hat{a}_y\frac{16}{5} + \hat{a}_y\frac{9}{5} - \hat{a}_x j3$$
$$= -\hat{a}_x j3 + \hat{a}_y 5 - \hat{a}_z j4.$$

$$\bar{H}^{\text{inc}}(\bar{r}) = \frac{E_0}{\eta_0}\hat{a}_i \times \bar{E}^{\text{inc}}(\bar{r}) = \frac{E_0}{\eta_0}(-\hat{a}_x j3 + \hat{a}_y 5 + \hat{a}_z j4)\exp[-j(0.8x + 0.6z)] \tag{6.266}$$

$$\bar{H}^{\text{ref}}(\bar{r}) = \frac{E_0}{\eta_0}\hat{a}_r \times \bar{E}^{\text{ref}}(\bar{r}) = \frac{E_0}{\eta_0}(-\hat{a}_x j3 + \hat{a}_y 5 - \hat{a}_z j4)\exp[-j(0.8x - 0.6z)], \tag{6.267}$$

respectively, since

$$\hat{a}_r = \hat{a}_x 4/5 - \hat{a}_z 3/5. \tag{6.268}$$

Note that the tangential component of the magnetic field intensity is nonzero at the boundary.[84] Specifically, we have

$$\bar{H}^{\text{tot}}(z=0) = \frac{E_0}{\eta_0}(-\hat{a}_x j6 + \hat{a}_y 10)\exp(-j0.8x). \tag{6.269}$$

Hence, according to the related boundary condition, there is a surface electric current density on the PEC[85]:

$$\bar{J}_s(\bar{r}) = -\hat{a}_z \times \bar{H}^{\text{tot}}(z=0) = \frac{E_0}{\eta_0}(\hat{a}_x 10 + \hat{a}_y j6)\exp(-j0.8x). \tag{6.270}$$

Example 70: Consider a normal incidence of a plane wave (propagating in the z direction in a vacuum) onto a planar PEC coated with a material of thickness d (Figure 6.45). The wavenumber and intrinsic impedance are denoted by $\{k_m, \eta_m\}$ for the material. Consider the directions shown in the figure: i.e. both incident and reflected (overall from the coated PEC) electric field intensities are defined in the x direction. Then they can be written

$$\bar{E}^{\text{inc}}(\bar{r}) = \hat{a}_x E_0^{\text{inc}}\exp(-jk_0 z), \qquad \bar{E}^{\text{ref}}(\bar{r}) = \hat{a}_x E_0^{\text{ref}}\exp(jk_0 z). \tag{6.271}$$

Derive the effective reflection coefficient $\Gamma_m = E_0^{\text{ref}}/E_0^{\text{inc}}$, as well as the surface electric current density induced on the PEC.

Solution: Given the expressions for the electric field intensities of the incident and reflected waves, we can write the corresponding magnetic field intensities as

$$\bar{H}^{\text{inc}}(\bar{r}) = \hat{a}_y \frac{E_0^{\text{inc}}}{\eta_0}\exp(-jk_0 z), \qquad \bar{H}^{\text{ref}}(\bar{r}) = -\hat{a}_y \frac{E_0^{\text{ref}}}{\eta_0}\exp(jk_0 z). \tag{6.272}$$

[84] Obviously, the normal component of the magnetic field intensity is zero at the boundary.

[85] One can further find the surface electric charge density induced on the PEC in two different ways. First, for the normal component of the electric flux density and the related boundary condition, we have

$$\begin{aligned} q_s(\bar{r}) &= -\hat{a}_z \cdot \bar{D}^{\text{tot}}(z=0) = -\varepsilon_0 \hat{a}_z \cdot \bar{E}^{\text{tot}}(z=0) \\ &= -\varepsilon_0 \hat{a}_z \cdot \bar{E}^{\text{inc}}(z=0) - \varepsilon_0 \hat{a}_z \cdot \bar{E}^{\text{ref}}(z=0) \\ &= 4\varepsilon_0 E_0 \exp(-j0.8x) + 4\varepsilon_0 E_0 \exp(-j0.8x) \\ &= 8\varepsilon_0 E_0 \exp(-j0.8x). \end{aligned}$$

Alternatively, we can employ the continuity equation as ($k = 1$ rad/m means $\omega = 1/\sqrt{\mu_0\varepsilon_0}$)

$$\begin{aligned} q_s(\bar{r}) &= -\frac{1}{j\omega}\nabla \cdot \bar{J}_s(\bar{r}) = \frac{E_0}{j\omega\eta_0}10(j0.8)\exp(-j0.8x) \\ &= 8\varepsilon_0 E_0 \exp(-j0.8x). \end{aligned}$$

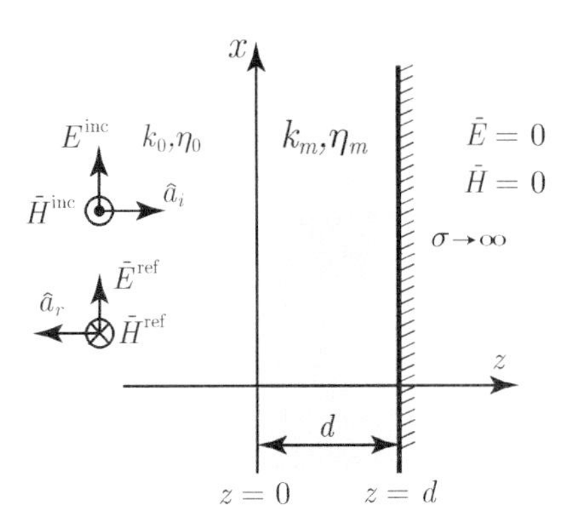

Figure 6.45 Normal incidence of a plane wave on a planar PEC coated with a material of thickness d.

In addition, inside the coating material, we have

$$\bar{E}_m(\bar{r}) = \hat{a}_x E_0^+ \exp(-jk_m z) + \hat{a}_x E_0^- \exp(jk_m z) \quad (6.273)$$

$$\bar{H}_m(\bar{r}) = \hat{a}_y \frac{E_0^+}{\eta_m} \exp(-jk_m z) - \hat{a}_y \frac{E_0^-}{\eta_m} \exp(jk_m z) \quad (6.274)$$

for both right-traveling and left-traveling waves.[86] Applying boundary conditions,[87] we obtain the relationships between the constants as

$$\bar{E}^{\rm inc}(z=0) + \bar{E}^{\rm ref}(z=0) = \bar{E}_m(z=0) \longrightarrow E_0^{\rm inc} + E_0^{\rm ref} = E_0^+ + E_0^- \quad (6.275)$$

$$\bar{H}^{\rm inc}(z=0) + \bar{H}^{\rm ref}(z=0) = \bar{H}_m(z=0) \longrightarrow \frac{E_0^{\rm inc}}{\eta_0} - \frac{E_0^{\rm ref}}{\eta_0} = \frac{E_0^+}{\eta_m} - \frac{E_0^-}{\eta_m} \quad (6.276)$$

$$\bar{E}_m(z=d) = 0 \longrightarrow E_0^- = -E_0^+ \exp(-j2k_m d), \quad (6.277)$$

leading to

$$E_0^{\rm inc} + E_0^{\rm ref} = E_0^+[1 - \exp(-j2k_m d)] \quad (6.278)$$

$$E_0^{\rm inc} - E_0^{\rm ref} = \frac{\eta_0}{\eta_m} E_0^+[1 + \exp(-j2k_m d)]. \quad (6.279)$$

Then, finding suitable expressions[88] for $E_0^{\rm inc}$ and $E_0^{\rm ref}$, we arrive at the effective reflection coefficient as[89,90]

$$\Gamma_m = \frac{E_0^{\rm ref}}{E_0^{\rm inc}} = \frac{\eta_m[1 - \exp(-j2k_m d)] - \eta_0[1 + \exp(-j2k_m d)]}{\eta_m[1 - \exp(-j2k_m d)] + \eta_0[1 + \exp(-j2k_m d)]}. \quad (6.280)$$

To find the induced electric current density on the PEC, we can write the total magnetic field intensity in the material at $z = d$ as

$$\bar{H}_m(z=d) = \hat{a}_y \frac{E_0^+}{\eta_m} \exp(-jk_m d) - \hat{a}_y \frac{E_0^-}{\eta_m} \exp(jk_m d) \quad (6.281)$$

$$= \hat{a}_y \frac{2E_0^+}{\eta_m} \exp(-jk_m d) \quad (6.282)$$

$$= \hat{a}_y \frac{1}{\eta_m} \frac{4E_0^{\rm inc} \exp(-jk_m d)}{[1 - \exp(-j2k_m d)] + \dfrac{\eta_0}{\eta_m}[1 + \exp(-j2k_m d)]} \quad (6.283)$$

$$= \hat{a}_y \frac{2E_0^{\rm inc}}{\eta_0 \cos(k_m d) + j\eta_m \sin(k_m d)}. \quad (6.284)$$

[86] Recall from Section 5.3.11 that these right-traveling and left-traveling waves are actually superpositions of infinitely many plane waves that are created due to infinitely many reflections between the PEC and the a vacuum/material interface.

[87] Both the tangential electric field intensity and tangential magnetic field intensity must be continuous at $z = 0$, while the tangential electric field intensity must vanish at $z = d$.

[88] We have

$$E_0^{\rm inc} = \frac{1}{2} E_0^+[1 - \exp(-j2k_m d)] + \frac{1}{2}\frac{\eta_0}{\eta_m} E_0^+[1 + \exp(-j2k_m d)]$$

$$E_0^{\rm ref} = \frac{1}{2} E_0^+[1 - \exp(-j2k_m d)] - \frac{1}{2}\frac{\eta_0}{\eta_m} E_0^+[1 + \exp(-j2k_m d)].$$

[89] Alternatively, it can be written

$$\Gamma_m = \frac{(\eta_m - \eta_0) - (\eta_m + \eta_0)\exp(-j2k_m d)}{(\eta_m + \eta_0) - (\eta_m - \eta_0)\exp(-j2k_m d)}$$

or

$$\Gamma_m = \frac{-\eta_0 \cos(k_m d) + j\eta_m \sin(k_m d)}{\eta_0 \cos(k_m d) + j\eta_m \sin(k_m d)} = \frac{-\eta_0 + j\eta_m \tan(k_m d)}{\eta_0 + j\eta_m \tan(k_m d)}.$$

[90] Note that in the limit case of $d = 0$, we have $\Gamma_m = -1$, corresponding to a reflection from a planar PEC (without coating). Similarly, for $\eta_m = \eta_0$, we obtain $\Gamma_m = -\exp(-j2k_m d)$, which also corresponds to a reflection from a planar PEC with an additional phase term that accounts for the extra distance $2d$. Furthermore, as two interesting cases to be practiced, we note $d = \lambda_m/2$ and $d = \lambda_m$: i.e. when the thickness of the coating corresponds to the wavelength or half of the wavelength inside the material.

Then we obtain

$$\bar{J}_s(\bar{r}) = -\hat{a}_z \times \bar{H}_m(z = d) = \hat{a}_x \frac{2E_0^{\mathrm{inc}}}{\eta_0 \cos(k_m d) + j\eta_m \sin(k_m d)}. \tag{6.285}$$

Since this is a normal-incidence case, the absence of the normal component of the electric field intensity on the PEC corresponds to zero induced electric charge density. This can be verified by noting the zero divergence (in fact, position independency) of the induced electric current density.

6.7 Capacitance

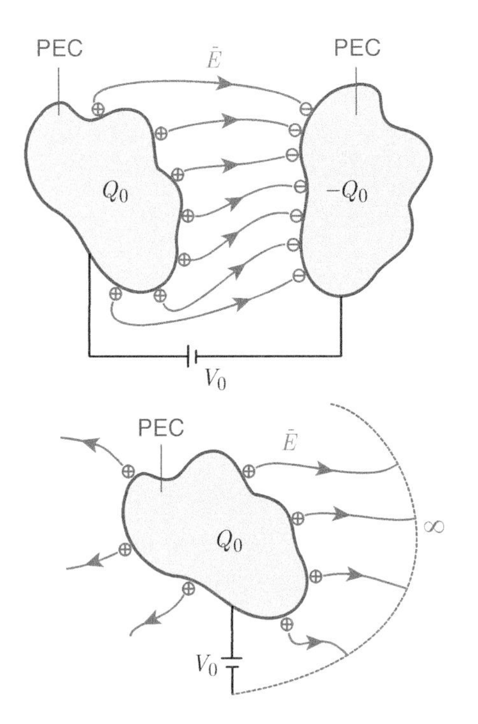

Figure 6.46 Capacitance is typically defined for a pair of two perfectly conducting objects. When voltage is applied between them, electric charges with opposite signs and equal amounts accumulate on PECs. The capacitance is the ratio of electric charges (on one of the PECs) to the applied voltage. It is also possible to define a capacitance for a single PEC if infinity is used as the reference (instead of a second PEC).

C_0: capacitance
Unit: farad (F) (coulombs/volt)

In general terms, capacitance is an important property of an object, corresponding to its ability to store electric charge (Figure 6.46). To make it independent of the electric field and applied potential, it is defined as

$$C_0 = \frac{Q_0(t)}{V_0(t)}, \tag{6.286}$$

where $Q_0(t)$ is the total charge stored by the object and $V_0(t)$ is the related voltage at time t. For a single-body object, $V_0(t)$ can be defined as the electric scalar potential of the object with respect to a reference point: e.g. infinity. In many cases, however, the structure involves two parts (and especially PECs) that are charged negatively as $Q_0(t)$ and $-Q_0(t)$, while $V_0(t)$ is the positive potential difference between them.

In this book, when defining the capacitance of a structure, we do not consider whether charges and voltages change with respect to time. The definition above is valid for both static and dynamic cases. For dynamic cases, however, it can be further manipulated as

$$Q_0(t) = C_0 V_0(t) \tag{6.287}$$

$$\rightarrow I_0(t) = \frac{\partial Q_0(t)}{\partial t} = \frac{\partial [C_0 V_0(t)]}{\partial t}. \tag{6.288}$$

If the *capacitance* itself is constant, we may obtain the standard voltage-current relationship:

$$I_0(t) = C_0 \frac{dV_0(t)}{dt}. \tag{6.289}$$

Note that for a variable C_0, e.g. when it is nonlinear, the capacitance found by using the voltage-current relationship in Eq. (6.289) and the capacitance found by using the charge-voltage ratio in Eq. (6.286) may not be the same. In this book, however, we consider a simple definition (constant capacitance) so that the capacitance of a structure can be found by analyzing an electrostatic case.[91]

A structure's capacitance can be found in various ways, in particular using the following two methods:

- Put Charge ⟶ Find Fields ⟶ Find the Corresponding Voltage
- Put Voltage ⟶ Find Fields ⟶ Find the Corresponding Charge

In relatively simple cases, both methods may work, but there are many cases for which only one method leads to an easy extraction of the capacitance expression.

6.7.1 Capacitance and Electric Potential Energy

When discussing the electric potential energy in Section 3.6.5, the relationship between the stored energy and the capacitance of a simple parallel-plate capacitor (Figure 6.47) is written

$$w_{e,\text{stored}} = \frac{Q_0^2}{2C_0} = \frac{1}{2}C_0V_0^2, \tag{6.290}$$

where V_0 is the electric potential between the plates and Q_0 is the amount of the stored electric charge on one of the plates. In general, a capacitor involves two bodies (usually PECs) with V_0 voltage between them, while equal amounts of positive and negative charges are *distributed* on the bodies with higher and lower potentials, respectively (Figure 6.46). This means external energy must be provided to collect positive/negative charges at the higher/lower potentials. Formally, the required energy to collect charges, which corresponds to the energy stored by the capacitor, can be written[92]

$$w_{e,\text{stored}} = \frac{1}{2}Q_0(V_0 + \Phi_0) + \frac{1}{2}(-Q_0)\Phi_0 = \frac{1}{2}Q_0V_0, \tag{6.291}$$

where Φ_0 is the reference potential: i.e. the potential of the negatively charged body. Hence, we verify the energy-capacitance relationship as

$$w_{e,\text{stored}} = \frac{1}{2}Q_0V_0 = \frac{1}{2}\frac{Q_0^2}{C_0} = \frac{1}{2}C_0V_0^2 \tag{6.292}$$

for any capacitor.[93] The energy-capacitance relationship described in Eq. (6.292) is general: i.e.

[91] In the phasor domain – i.e. assuming steady-state with time-harmonic functions – the voltage/current relationship can be written $I_0 = j\omega C_0 V_0$, where ω is the angular frequency. Hence, capacitance is a negative imaginary impedance: i.e. $Z_0 = V_0/I_0 = -j/(\omega C_0)$ in circuit analysis.

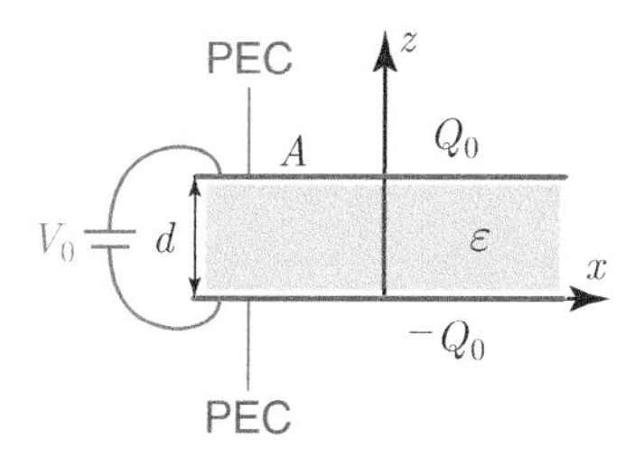

Figure 6.47 A voltage source connected to a parallel-plate capacitor.

[92] Note that this energy is by default a positive value, and it directly corresponds to the energy *stored* by the capacitor. As discussed in Section 3.6.5, a system of two point electric charges involves a negative electric potential energy, if the reference energy (when the charge is located at infinity) is set to zero. This negative energy is a result of the interaction between the charges, without including the energy involved in the formation of each charge. On the other hand, in the case of a capacitor, we consider the energy to build the *entire* system, which always has a positive value.

[93] At this stage, we start to use only static expressions, while these are also valid for dynamic cases. Specifically, we have

$$w_{e,\text{stored}}(t) = \frac{1}{2}Q_0(t)V_0(t) = \frac{1}{2}\frac{[Q_0(t)]^2}{C_0} = \frac{1}{2}C_0[V_0(t)]^2.$$

it can be used even when electric charges do not have simple distributions. Given the stored electric potential energy, we have

$$C_0 = 2w_{e,\text{stored}}/V_0^2 = Q_0^2/(2w_{e,\text{stored}}). \tag{6.293}$$

This can be seen as an alternative method to find the capacitance of a given structure, particularly if the electric potential energy can be found from electric and magnetic fields.

6.7.2 Parallel-Plate Capacitors

As a very common example, we now consider a parallel-plate capacitor[94] consisting of two identical rectangular perfectly conducting plates separated by distance d (Figure 6.47). The area of each plate is A, and the space between the plates is filled with a dielectric of relative permittivity ε_r. Assuming $+Q_0$ and $-Q_0$ on the top and bottom plates, respectively, Gauss' law can be used to derive the electric flux density between the plates as[95]

$$\bar{D}(\bar{r}) = -\hat{a}_z \frac{Q_0}{A}, \tag{6.294}$$

neglecting fringing fields. Then the electric field intensity can be written

$$\bar{E}(\bar{r}) = -\hat{a}_z \frac{Q_0}{\varepsilon_0 \varepsilon_r A}. \tag{6.295}$$

Starting from the total *charge*, we aim to find the *voltage* between the plates. Knowing the electric field intensity in terms of the charge, this can be done via a line integral as

$$V_0 = -\int_0^d \left(-\hat{a}_z \frac{Q_0}{\varepsilon_0 \varepsilon_r A}\right) \cdot \hat{a}_z dz = \frac{Q_0 d}{\varepsilon_0 \varepsilon_r A}. \tag{6.296}$$

The final equality is the one we need to find the capacitance as

$$C_0 = \frac{Q_0}{V_0} = \frac{\varepsilon_0 \varepsilon_r A}{d}. \tag{6.297}$$

Obviously, the capacitance (the ability to store charge) is proportional to the area of the plates, as well as the permittivity of the medium between them, and it is inversely proportional to the distance d.

Alternatively, to find the capacitance, we can start with the voltage between the plates that is directly related to the electric scalar potential distribution. Let a voltage V_0 exist between the

[94] Parallel-plate capacitors have been considered several times in different contexts: i.e. when discussing the superposition principle in Section 1.6.1, equivalent polarization charges in Section 1.7.3, electric potential energy in Section 3.6.5, electric force in Section 3.6.5.4, and electrostatic boundary-value problems in Section 3.6.7, as well as when considering general behaviors of PECs in Section 6.5.4.

[95] As practiced before, we use the expression for the electric flux density due to an infinitely large planar charge distribution for both the bottom and top plates, leading to zero fields outside the capacitor. Hence, we assume the absence of fringing fields and uniformity of charge distributions on the plates. In real life, depending on the geometric parameters, as well as the frequency, more charge tends to accumulate at the edges, leading to different capacitance values.

plates. Assuming that the electric scalar potential values for the bottom ($z = 0$) and top ($z = d$) plates are 0 and V_0, respectively, we must find its distribution between the plates (Figure 6.48). This can be done by solving the related Laplace's equation:

$$\bar{\nabla}^2\Phi(\bar{r}) = \partial^2\Phi(z)/\partial z^2 = 0. \tag{6.298}$$

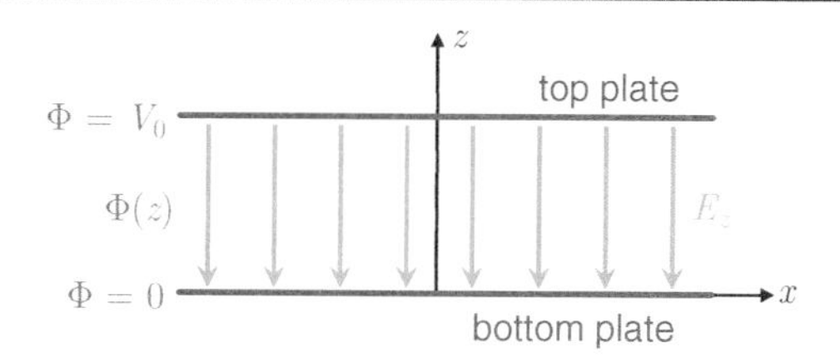

Figure 6.48 The capacitance of a structure can be found by assigning a voltage between the bodies and finding the corresponding charge accumulations. For a parallel-plate capacitor, once 0 and V_0 potential values are assigned to the bottom and top plates, the distribution of the electric scalar potential can be found by solving the related Laplace's equation. In the derivation, assuming that the potential depends only on z (but not on x and y) is the same as assuming uniform distribution of charges on the plates and absence of fringing fields: i.e. assuming infinitely large plates.

This means the potential can be written $\Phi(\bar{r}) = c_1 z + c_2$, where c_1 and c_2 are constants that can be found by enforcing boundary conditions. Specifically, using $\Phi(0) = 0$ and $\Phi(d) = V_0$, we obtain $c_1 = V_0/d$, $c_2 = 0$, and

$$\Phi(\bar{r}) = V_0 z/d. \tag{6.299}$$

Hence, the electric field intensity and electric flux density between the plates can be found:

$$\bar{E}(\bar{r}) = -\bar{\nabla}\Phi(\bar{r}) = -\hat{a}_z\frac{V_0}{d}, \quad \bar{D}(\bar{r}) = -\hat{a}_z\frac{\varepsilon_0\varepsilon_r V_0}{d}. \tag{6.300}$$

Assuming zero fields outside the capacitor (equivalent to the absence of fringing fields: i.e. large dimensions of the plates), the boundary condition for the normal electric flux density gives the surface electric charge density on the top plate as

$$q_{s,\text{top}}(\bar{r}) = -\hat{a}_z \cdot \bar{D}(\bar{r}) = \hat{a}_z \cdot \hat{a}_z\frac{\varepsilon_0\varepsilon_r V_0}{d} = \frac{\varepsilon_0\varepsilon_r V_0}{d}. \tag{6.301}$$

Then we obtain the total charge on the top plate as

$$Q_0 = Aq_{s,\text{top}} = \frac{\varepsilon_0\varepsilon_r A V_0}{d}, \tag{6.302}$$

leading to $C_0 = Q_0/V_0 = \varepsilon_0\varepsilon_r A/d$, as expected.

Since the capacitance of a parallel-plate capacitor is proportional to the permittivity of the material between the plates, one may ask why a dielectric material is helpful to increase the ability to store electric charge. While this can be answered in many different ways, our knowledge of polarization can be helpful to understand what happens when a dielectric is used (Figure 6.49). Assuming that a constant V_0 is applied between the plates, the total charge on the top plate can be written

$$Q_0^{\text{vacuum}} = \frac{\varepsilon_0 A V_0}{d} \quad \text{and} \quad Q_0^{\text{dielectric}} = \frac{\varepsilon_0\varepsilon_r A V_0}{d} \tag{6.303}$$

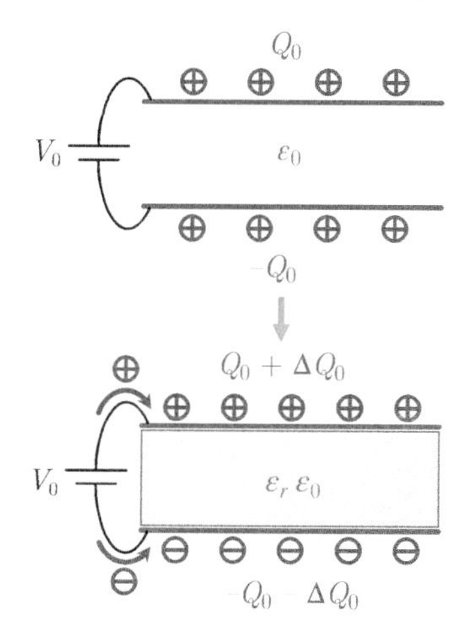

Figure 6.49 The capacitance of a parallel-plate capacitor increases when the space between the plates is filled with a dielectric material. Hence, if the capacitor's voltage is kept constant (via a voltage source), the amount of electric charges accumulated on the plates increases. This can be explained by the attraction of more charge by the dielectric from the voltage source. In an equivalent problem, it can be shown that the amount of attracted charge is the same as the amount of surface polarization charge. If the capacitor is not connected to a voltage source (isolated) such that the amount of charges on the plates is fixed, increased capacitance due to a dielectric leads to decreased voltage between the plates.

when the space between the plates is a vacuum and filled with a dielectric material, respectively. Hence, when a dielectric material is inserted between the plates, the charge on the top plate increases by an amount of

$$\Delta Q_0 = \frac{\varepsilon_0(\varepsilon_r - 1)AV_0}{d}. \tag{6.304}$$

In addition, when there is a dielectric between the plates, the polarization (inside the dielectric) can be written

$$\bar{P}(\bar{r}) = \bar{D}(\bar{r}) - \varepsilon_0 \bar{E}(\bar{r}) = -\hat{a}_z \frac{\varepsilon_0(\varepsilon_r - 1)V_0}{d}. \tag{6.305}$$

Note that the electric field intensity is always $\bar{E}(\bar{r}) = -\hat{a}_z V_0/d$, whether there is a dielectric or not.[96] Hence, on the top surface of the *dielectric*, there is a negative surface polarization charge density

$$q_{ps}(\bar{r}) = \hat{a}_n \cdot \bar{P}(\bar{r}) = \hat{a}_z \cdot (-\hat{a}_z) \frac{\varepsilon_0(\varepsilon_r - 1)V_0}{d}. \tag{6.306}$$

Then the total polarization charge on the top surface of the dielectric can be found:

$$Q_{ps} = -\frac{\varepsilon_0(\varepsilon_r - 1)AV_0}{d}. \tag{6.307}$$

It is not a coincidence that Q_{ps} is just the negative of ΔQ_0: i.e. the amount of *new* electric charge brought (from the voltage source) by inserting a dielectric. Hence, this accumulation can be interpreted as the attraction of more charge from the external source (voltage source) by the negative polarization electric charge in the vicinity of the top plate.

At this stage, we can study relatively complicated cases, especially those involving inhomogeneity and inhomogeneous materials. First, we consider a parallel-plate capacitor filled with two different types of dielectric materials with relative permittivities ε_{r1} and ε_{r2} from $z = 0$ to $z = d_1$ and from $z = d_1$ to $z = d_1 + d_2 = d$, respectively (Figure 6.50). Assuming Q_0 and $-Q_0$ on the top and bottom plates, Gauss' law can be used to derive[97]

$$\bar{D}(\bar{r}) = -\hat{a}_z \frac{Q_0}{A}, \tag{6.308}$$

leading to[98]

$$\bar{E}(\bar{r}) = -\hat{a}_z \frac{Q_0}{\varepsilon_0 A} \begin{cases} 1/\varepsilon_{r1} & (0 \le z < d_1) \\ 1/\varepsilon_{r2} & (d_1 < z \le d_1 + d_2). \end{cases} \tag{6.309}$$

[96] This is not true if the capacitor is isolated such that the amount of electric charge stored by the capacitor is constant, while the voltage between the plates changes when one inserts a dielectric. This scenario is extensively discussed in Section 1.7.3. On the other hand, keeping V_0 constant corresponds to keeping a battery (voltage source) connected to the capacitor so that accumulated charge may increase or decrease.

[97] Note that the expression for the electric flux density is not affected by the existence of the dielectric materials: i.e. its expression is exactly the same as when there is a vacuum between the plates.

[98] Note that $\bar{D}(\bar{r})$ is continuous since there is no surface electric charge density at $z = d_1$, while $\bar{E}(\bar{r})$ is discontinuous due to the jump in the permittivity.

The potential difference between the plates can be found as[99]

$$V_0 = \frac{Q_0 d_1}{\varepsilon_0 \varepsilon_{r1} A} + \frac{Q_0 d_2}{\varepsilon_0 \varepsilon_{r2} A} \tag{6.310}$$

considering a line integral from $z = 0$ to $z = d$. Then the capacitance can be derived as

$$C_0 = \frac{Q_0}{V_0} = \frac{\varepsilon_0 A}{d_1/\varepsilon_{r1} + d_2/\varepsilon_{r2}} \tag{6.311}$$

or

$$\frac{1}{C_0} = \frac{1}{C_1} + \frac{1}{C_2}, \tag{6.312}$$

where

$$C_1 = \frac{\varepsilon_0 \varepsilon_{r1} A}{d_1}, \qquad C_2 = \frac{\varepsilon_0 \varepsilon_{r2} A}{d_2}. \tag{6.313}$$

When they are considered lumped elements, the capacitance of two capacitors that are connected in series is calculated as above: i.e. $1/C_0 = 1/C_1 + 1/C_2$.

But what about the polarization inside this capacitor, which may provide further insight about the scenario? Using $\bar{P}(\bar{r}) = \bar{D}(\bar{r}) - \varepsilon_0 \bar{E}(\bar{r})$, we have

$$\bar{P}(\bar{r}) = -\hat{a}_z \frac{Q_0}{A} \begin{cases} 1 - \varepsilon_0/\varepsilon_{r1} & (0 \le z < d_1) \\ 1 - \varepsilon_0/\varepsilon_{r2} & (d_1 < z \le d_1 + d_2). \end{cases} \tag{6.314}$$

Hence, in the *equivalent* problem, in addition to those near the top and bottom plates, there is a surface polarization charge distribution at the location of the interface between the two dielectric media: i.e. we have[100]

$$q_{ps}^{\text{interface}} = \hat{a}_z \cdot [\bar{P}(z = d_1^-) - \bar{P}(z = d_1^+)] \tag{6.315}$$

$$= -\frac{Q_0}{A}\left(1 - \frac{\varepsilon_0}{\varepsilon_{r1}} - 1 + \frac{\varepsilon_0}{\varepsilon_{r2}}\right) = \frac{\varepsilon_0 Q_0}{A}\left(\frac{1}{\varepsilon_{r1}} - \frac{1}{\varepsilon_{r2}}\right). \tag{6.316}$$

Note that this polarization charge density is responsible for the discontinuity in the normal component of the electric field intensity. However, since it is not a true charge density, the normal component of the electric flux density is continuous across the same interface.

[99] We simply have

$$V_0 = -\int_0^d \bar{E}(\bar{r}) \cdot \hat{a}_z dz$$
$$= -\int_0^{d_1} \left(-\hat{a}_z \frac{Q_0}{\varepsilon_0 \varepsilon_{r1} A}\right) \cdot \hat{a}_z dz$$
$$- \int_{d_1}^{d_1+d_2} \left(-\hat{a}_z \frac{Q_0}{\varepsilon_0 \varepsilon_{r2} A}\right) \cdot \hat{a}_z dz.$$

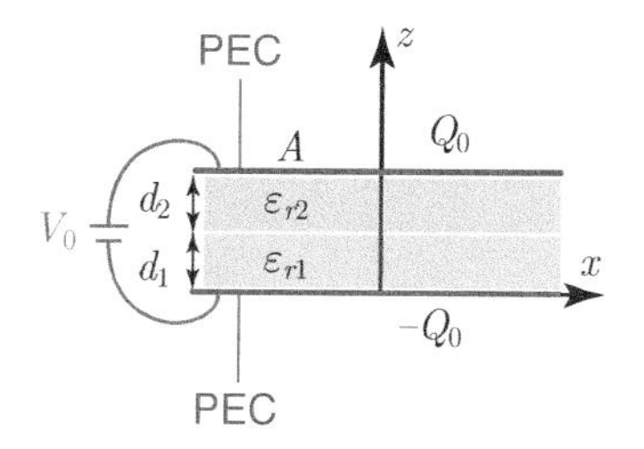

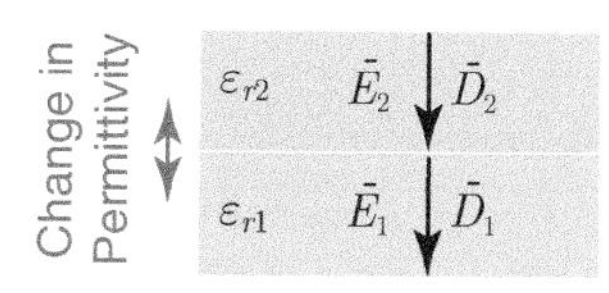

Figure 6.50 A parallel-plate capacitor filled with two different types of dielectric materials that are stacked vertically. The overall structure can be considered two capacitors connected in series. The change in the permittivity (simply a jump in this case) is parallel to the electric field direction (assuming no fringing fields) so that the charge approach is very suitable to find the capacitance. According to the boundary condition, the electric field intensity is continuous across the boundary, which further verifies the suitability of the charge approach. Also note that electric charges are distributed uniformly on the perfectly conducting plates (assuming large plates).

[100] One can show that the total polarization charge, considering the distributions near the top/bottom plates (at $z = 0^+$ and $z = d^-$), as well as the one at $z = d_1$, is zero.

For the capacitance problem above, it may not be easy to start with the electric scalar potential. Specifically, given the derivation of Laplace's equation for the electric scalar potential, we have

$$\bar{\nabla}\varepsilon(\bar{r}) \cdot \bar{E}(\bar{r}) - \varepsilon(\bar{r})\bar{\nabla}^2\Phi(\bar{r}) = 0. \tag{6.317}$$

The jump in the permittivity *in the direction* of the electric field intensity prevents us from eliminating the first term. Hence, it is not trivial to solve this equation, especially given that $\bar{E}(\bar{r}) \neq -\hat{a}_z V_0/d$.

There are also many cases where the potential approach works while the charge approach does not (at least in a trivial way). As a simple example (which is related to one considered above), we consider again a parallel-plate capacitor filled with two different types of dielectric materials with relative permittivities ε_{r1} and ε_{r2}. In this case, however, the dielectrics are placed side by side between the plates, and they cover surfaces of areas A_1 and A_2, respectively, where $A = A_1 + A_2$ is the total surface of each plate (Figure 6.51). The distance between the plates is again d. In this case, we have $\bar{\nabla}^2\Phi(\bar{r}) = 0$, and the solution is obtained simply as

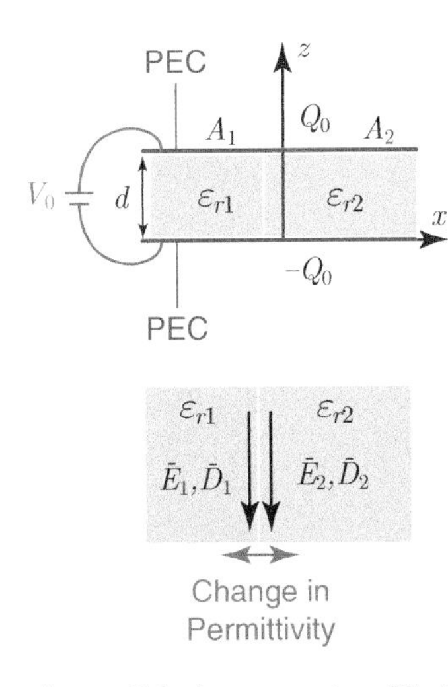

Figure 6.51 A parallel-plate capacitor filled with two different types of dielectric materials that are stacked horizontally. The overall structure can be considered two capacitors connected in parallel. The change in the permittivity (simply a jump in this case) is perpendicular to the electric field direction (assuming no fringing fields) so that the voltage approach is very suitable to find the capacitance. According to the related boundary condition, the electric field intensity must be continuous across the boundary, which further verifies the suitability of the potential approach. Also note that electric charges are not distributed uniformly on the perfectly conducting plates (even when assuming large plates).

$$\Phi(\bar{r}) = \frac{V_0 z}{d}. \tag{6.318}$$

Therefore, the expression for the electric field intensity is the same as the one for the single-dielectric case:

$$\bar{E}(\bar{r}) = -\bar{\nabla}\Phi(\bar{r}) = -\hat{a}_z\frac{V_0}{d}. \tag{6.319}$$

On the other hand, considering the permittivity values of the two different media, we have

$$\bar{D}(\bar{r}) = -\hat{a}_z\frac{\varepsilon_0 V_0}{d}\begin{cases}\varepsilon_{r1} & \text{(Medium 1)}\\ \varepsilon_{r2} & \text{(Medium 2)}.\end{cases} \tag{6.320}$$

Note that across the boundary between the dielectrics, the electric field intensity is now continuous, while the electric flux density is discontinuous; all these are consistent with the boundary conditions for tangential electric fields. Assuming zero fields outside the plates and employing the boundary condition for the normal electric flux density, we have

$$q_{s1}(\bar{r}) = \frac{\varepsilon_0\varepsilon_{r1}V_0}{d}, \qquad q_{s2}(\bar{r}) = \frac{\varepsilon_0\varepsilon_{r2}V_0}{d} \tag{6.321}$$

as the surface electric charge densities in the first and second parts of the top plate. Then the total electric charge on the top plate can be found:

$$Q_0 = A_1 q_{s1} + A_2 q_{s2} = \frac{\varepsilon_0 V_0}{d}(\varepsilon_{r1}A_1 + \varepsilon_{r2}A_2). \tag{6.322}$$

Finally, we obtain the capacitance as

$$C_0 = \frac{Q_0}{V_0} = \frac{\varepsilon_0}{d}(\varepsilon_{r1}A_1 + \varepsilon_{r2}A_2) \tag{6.323}$$

or

$$C_0 = C_1 + C_2, \tag{6.324}$$

where

$$C_1 = \frac{\varepsilon_0\varepsilon_{r1}A_1}{d}, \qquad C_2 = \frac{\varepsilon_0\varepsilon_{r2}A_2}{d}. \tag{6.325}$$

This case can be considered a parallel connection of two capacitors, leading to an overall capacitance as their sum. Also note that this problem's charge approach may not be obvious since the surface electric charge is not distributed uniformly on the plates. The final equations show that they are distributed proportional to the permittivity, while we may not know this before solving the problem.

The examples above involve only two media, while the presented strategies work similarly for completely inhomogeneous media. The selection of the approach can be done by considering the direction of the electric field intensity and the change in the permittivity. In perpendicular cases – i.e. when the permittivity changes perpendicularly to the electric field intensity – the voltage approach works well. In parallel cases, however, the charge approach usually leads to easier solutions. We now consider again a parallel-plate capacitor, but this time, it is filled with a dielectric material with varying permittivity: $\varepsilon = \varepsilon_0 \exp(z/d)$ from $z = 0$ to $z = d$ (Figure 6.52). In this case, the change in permittivity is in the z direction ($\bar{\nabla}\varepsilon(\bar{r})$ is in the z direction), which is parallel to the electric field intensity.[101] Then assuming Q_0 and $-Q_0$ on the top and bottom plates,[102] Gauss' law can be used to derive

$$\bar{D}(\bar{r}) = -\hat{a}_z\frac{Q_0}{A}, \qquad \bar{E}(\bar{r}) = -\hat{a}_z\frac{Q_0}{\varepsilon_0 A\exp(z/d)}. \tag{6.326}$$

Then we also find

$$V_0 = -\int_0^d \left(-\hat{a}_z\frac{Q_0}{\varepsilon_0 A\exp(z/d)}\right)\cdot\hat{a}_z dz = \frac{Q_0 d}{\varepsilon_0 A}[1-\exp(-1)]. \tag{6.327}$$

[101]This means, if we attempt to solve the boundary-value problem, we have

$$\bar{\nabla}\cdot\bar{D}(\bar{r}) = 0 \rightarrow \bar{\nabla}\varepsilon(\bar{r})\cdot\bar{E}(\bar{r}) - \varepsilon(\bar{r})\bar{\nabla}^2\Phi(\bar{r}) = 0,$$

leading to

$$\bar{\nabla}^2\Phi(\bar{r}) = -\bar{\nabla}\varepsilon(\bar{r})\cdot\bar{E}(\bar{r}),$$

as we obtained several times before. Using $\bar{\nabla}\varepsilon(\bar{r}) = \hat{a}_z(\varepsilon_0/d)\exp(z/d)$, we further obtain

$$\bar{\nabla}^2\Phi(\bar{r}) + \hat{a}_z\cdot\bar{E}(\bar{r})\frac{\varepsilon_0}{d}\exp(z/d) = 0,$$

whose solution is not as trivial as the charge approach.

[102]Note that the surface electric charge density is still the total charge divided by the area, since the electric charge is distributed uniformly. The uniform distribution is due to the invariance of the permittivity in the x and y directions: i.e. over the surfaces of the plates.

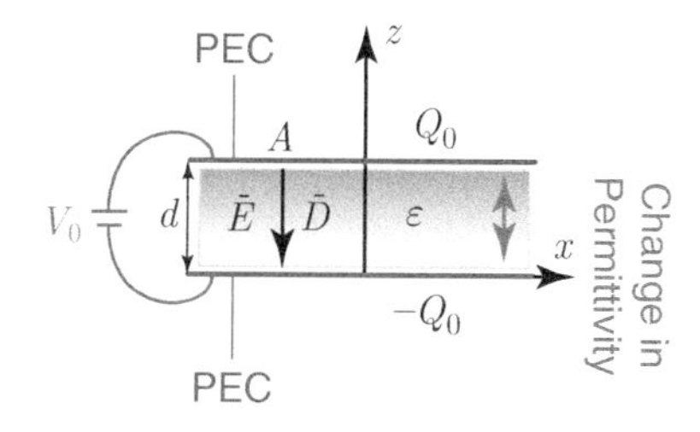

Figure 6.52 A parallel-plate capacitor filled with an inhomogeneous dielectric material having varying permittivity in the vertical direction.

The final equation can be used to derive an expression for the capacitance:

$$C_0 = \frac{Q_0}{V_0} = \frac{\varepsilon_0 A}{d[1-\exp(-1)]}. \tag{6.328}$$

The polarization inside the capacitor can also be found:

$$\bar{P}(\bar{r}) = \bar{D}(\bar{r}) - \varepsilon_0 \bar{E}(\bar{r}) = -[\exp(z/d)-1]\hat{a}_z \frac{Q_0}{A\exp(z/d)} \tag{6.329}$$

$$= \hat{a}_z \frac{Q_0}{A}[\exp(-z/d)-1] = -\hat{a}_z \frac{\varepsilon_0 V_0}{d}\left(\frac{1-\exp(-z/d)}{1-\exp(-1)}\right). \tag{6.330}$$

Hence, in the equivalent problem, there is a volume polarization charge density to represent the dielectric medium[103]

$$q_{pv}(\bar{r}) = -\bar{\nabla}\cdot\bar{P}(\bar{r}) = \hat{a}_z \frac{\varepsilon_0 V_0 \exp(-z/d)}{d^2[1-\exp(-1)]}, \tag{6.331}$$

which is mainly caused by the inhomogeneity of the dielectric material. In addition, a surface polarization charge exists in the vicinity of the top plate[104] (but not in the vicinity of the bottom plate):

$$q_{ps}(\bar{r}) = \hat{a}_z \cdot [\bar{P}(z=d)] = -\frac{\varepsilon_0 V_0}{d}\left(\frac{1-\exp(-1)}{1-\exp(-1)}\right) = -\frac{\varepsilon_0 V_0}{d}. \tag{6.332}$$

Analysis of a parallel-plate capacitor consisting of an dielectric material with an inhomogeneity in a horizontal direction (e.g. x in Figure 3.48) can be considered a further exercise.

6.7.3 Spherical Capacitors

So far, we have discussed parallel-plate capacitors. But now, we consider a spherical capacitor consisting of a perfectly conducting sphere of radius a inside a perfectly conducting spherical surface of radius b, and a homogeneous dielectric of relative permittivity ε_r between them (Figure 6.53). Since the permittivity does not change with respect to position, we should be able to solve this problem using both the charge and voltage approaches. First, using the charge approach, we assume $+Q_0$ charge on the inner PEC and $-Q_0$ charge on the outer PEC. Due to the perfect spherical symmetry, these charges are expected to be distributed uniformly. Specifically, the surface electric charge densities on the inner and outer PECs can be written $Q_0/(4\pi a^2)$ and $-Q_0/(4\pi b^2)$, respectively. Using Gauss' law[105], we obtain

[103] This volume polarization charge density can be considered responsible for the position-dependent electric field intensity in the equivalent problem, despite the constant electric flux density.

[104] Therefore, in the equivalent problem, the total charge at $z=d$ can be found as

$$Q_{0,\text{total}} = Q_{0,\text{true}} + Q_{ps} = C_0 V_0 - \frac{\varepsilon_0 A V_0}{d}$$
$$= \frac{\varepsilon_0 A}{d[1-\exp(-1)]}V_0 - \frac{\varepsilon_0 A V_0}{d}$$
$$= \frac{\varepsilon_0 A\exp(-1)}{d[1-\exp(-1)]}V_0 = \exp(-1)Q_0,$$

using the expression for the capacitance. Hence, the total surface electric charge density in the equivalent problem is $q_{s,\text{total}} = \exp(-1)Q_0/A$. This explains why the electric field intensity jumps from zero (just above the top plate) to $|\bar{E}(\bar{r})| = \exp(-1)Q_0/(\varepsilon_0 A)$ (just below the top plate) considering the expression in Eq.~(6.326). We further note that the discontinuity in the electric field intensity on the bottom plate is $Q_0/(\varepsilon_0 A)$, which is consistent with the fact that there is no surface polarization charge density in the vicinity of the bottom plate (at the location of the bottom surface of the dielectric).

[105] We simply use

$$\oint_S \bar{D}(\bar{r})\cdot\overline{ds} = Q_{\text{enc}} = Q_0$$

for $a < R < b$.

$$\bar{D}(\bar{r}) = \hat{a}_R \frac{Q_0}{4\pi R^2}, \qquad \bar{E}(\bar{r}) = \hat{a}_R \frac{Q_0}{4\pi\varepsilon_0\varepsilon_r R^2} \tag{6.333}$$

for $a < R < b$, given the spherical symmetry (Figure 6.53). Then the potential difference between the PECs can be found as

$$V_0 = -\int_b^a \bar{E}(\bar{r}) \cdot \hat{a}_R dR = -\int_b^a \hat{a}_R \frac{Q_0}{4\pi\varepsilon_0\varepsilon_r R^2} \cdot \hat{a}_R dR \tag{6.334}$$

$$= \frac{Q_0}{4\pi\varepsilon_0\varepsilon_r}\left[\frac{1}{R}\right]_b^a = \frac{Q_0}{4\pi\varepsilon_0\varepsilon_r}\left(\frac{1}{a} - \frac{1}{b}\right) = \frac{Q_0}{4\pi\varepsilon_0\varepsilon_r}\left(\frac{b-a}{ab}\right). \tag{6.335}$$

Hence, the capacitance can be obtained as[106]

$$C_0 = \frac{Q_0}{V_0} = 4\pi\varepsilon_0\varepsilon_r\left(\frac{ab}{b-a}\right). \tag{6.336}$$

This also means the capacitance of a *single* spherical PEC in a dielectric medium can be found:[107]

$$C_0 = 4\pi\varepsilon_0\varepsilon_r a. \tag{6.337}$$

At this stage, we can find the capacitance of a spherical capacitor using the potential approach. As already considered in Section 3.6.7, the solution of Laplace's equation (for a spherically symmetric case) leads to

$$\bar{\nabla}^2\Phi(\bar{r}) = 0 \longrightarrow \Phi(\bar{r}) = -\frac{c_1}{R} + c_2, \tag{6.338}$$

where c_1 and c_2 are constants to be found via boundary values. Setting $\Phi(R = a) = V_0$ and $\Phi(R = b) = 0$, we have $c_1 = bc_2$ and

$$-\frac{c_1}{a} + c_2 = V_0 \longrightarrow c_2\left(1 - \frac{b}{a}\right) = V_0 \tag{6.339}$$

or $c_2 = V_0 a/(a-b)$. This further leads to $c_1 = V_0 ab/(a-b)$. Hence, we obtain the expression for the electric scalar potential as

$$\Phi(\bar{r}) = \frac{V_0 ab}{(b-a)R} - \frac{V_0 a}{(b-a)} = \frac{V_0 a}{(b-a)}\left(\frac{b}{R} - 1\right). \tag{6.340}$$

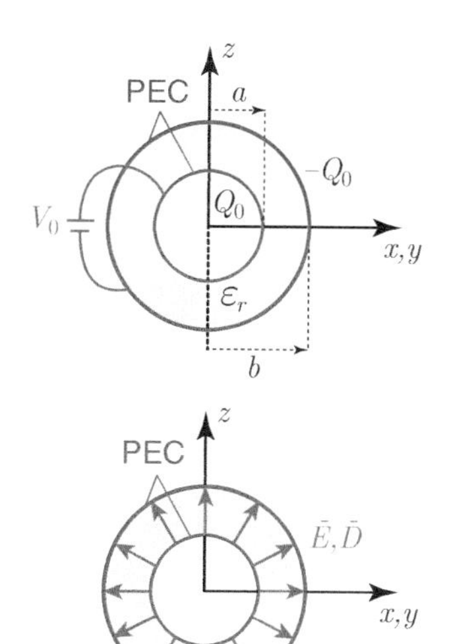

Figure 6.53 A spherical capacitor and the electric field created inside it, if the voltage of the inner PEC is larger than the voltage of the outer PEC.

[106] The capacitance of a structure is usually proportional to the areas of the PECs and inversely proportional to the distance between them, but this relationship may not be always obvious. For a spherical capacitor with a homogeneous dielectric material between PECs, the capacitance is proportional to $a \times b$, which can be seen as a kind of effective area. In addition, it is inversely proportional to $b - a$: i.e. the distance between the PECs.

[107] By evaluating the limit $b \to \infty$.

Next, the electric field intensity can be found via gradient as

$$\bar{E}(\bar{r}) = -\bar{\nabla}\Phi(\bar{r}) = \hat{a}_R\frac{V_0ab}{(b-a)R^2} \quad (a < R < b), \tag{6.341}$$

which is in the radial direction. The electric flux density can be written similarly as[108]

$$\bar{D}(\bar{r}) = \hat{a}_R\frac{\varepsilon_0\varepsilon_r V_0ab}{(b-a)R^2} \quad (a < R < b). \tag{6.342}$$

Knowing the value of the electric flux density $\bar{D}(\bar{r})$, we can find the electric charges accumulated on the perfectly conducting spheres via the related boundary conditions. Given that the electric flux density is zero for $R < a$ and $R > a$, we have

$$q_s(R=a) = \hat{a}_R\cdot\bar{D}(R=a) = \frac{\varepsilon_0\varepsilon_r V_0ab}{(b-a)a^2} = \frac{\varepsilon_0\varepsilon_r V_0 b}{a(b-a)} \tag{6.343}$$

$$q_s(R=b) = -\hat{a}_R\cdot\bar{D}(R=b) = -\frac{\varepsilon_0\varepsilon_r V_0ab}{(b-a)b^2} = -\frac{\varepsilon_0\varepsilon_r V_0 a}{b(b-a)}. \tag{6.344}$$

Hence, as we reach the expression for the electric charge stored by the capacitor, we can find the capacitance as[109]

$$C_0 = \frac{Q_0}{V_0} = \frac{4\pi a^2 q_s(R=a)}{V_0} = 4\pi\varepsilon_0\varepsilon_r\left(\frac{ab}{b-a}\right), \tag{6.345}$$

which is perfectly consistent with Eq. (6.336).

Now, we can consider the energy stored by the spherical capacitor discussed above. Since we already know the expression for the electric field intensity, one can use the expression for the electric potential energy[110] in Eq. (3.179):

$$w_{e,\text{stored}} = \frac{1}{2}\int_\infty \bar{E}(\bar{r})\cdot\bar{D}(\bar{r})dv = \frac{1}{2}\int_{\text{between PECs}} \bar{E}(\bar{r})\cdot\bar{D}(\bar{r})dv \tag{6.346}$$

$$= \frac{1}{2}4\pi\int_a^b \hat{a}_R\frac{V_0ab}{(b-a)R^2}\cdot\hat{a}_R\frac{\varepsilon_0\varepsilon_r V_0ab}{(b-a)R^2}R^2dR \tag{6.347}$$

in terms of voltage V_0. Combining the terms and evaluating the integral, we arrive at[111]

$$w_{e,\text{stored}} = \frac{1}{2}4\pi\varepsilon_0\varepsilon_r\frac{V_0^2(ab)^2}{(b-a)^2}\int_a^b\frac{1}{R^2}dR = \frac{1}{2}4\pi\varepsilon_0\varepsilon_r\frac{V_0^2ab}{(b-a)}. \tag{6.348}$$

[108] Note that this expression describes $\bar{D}(\bar{r})$ in terms of voltage V_0 (as we started by defining the voltage), as opposed to the expression in Eq. (6.333), which is written in terms of Q_0.

[109] Note that, although the surface electric charge densities are different on the inner and outer PECs, the total accumulated charges are the same (with opposite signs):

$$Q_0(R=a) = 4\pi a^2 q_s(R=a) = \frac{4\pi\varepsilon_0\varepsilon_r V_0ab}{(b-a)} = Q_0$$

$$Q_0(R=b) = 4\pi b^2 q_s(R=b) = -\frac{4\pi\varepsilon_0\varepsilon_r V_0ab}{(b-a)} = -Q_0.$$

[110] Note that the energy stored by a capacitor corresponds to the electric potential energy stored by the system: i.e. the energy required to construct the overall system by collecting the electric charges on the plates.

[111] As intermediate steps, we have

$$w_{e,\text{stored}} = \frac{1}{2}4\pi\varepsilon_0\varepsilon_r\frac{V_0^2(ab)^2}{(b-a)^2}\left[-\frac{1}{R}\right]_a^b = \frac{1}{2}4\pi\varepsilon_0\varepsilon_r\frac{V_0^2(ab)^2}{(b-a)^2}\frac{(b-a)}{ab}.$$

The final expression is nothing but $w_{e,\text{stored}} = C_0V_0^2/2$, where $C_0 = 4\pi\varepsilon_0\varepsilon_r ab/(b-a)$ as found above. Hence, the energy stored by a capacitor can be found via the standard expression, if its capacitance is known.

As a second way to find the energy stored, we can use

$$w_{e,\text{stored}} = \frac{1}{2}\int_S q_s(\bar{r})\Phi(\bar{r})ds \tag{6.349}$$

as also shown in Section 3.6.5.3. Given the inner and outer PECs, we obtain

$$w_{e,\text{stored}} = \frac{1}{2}\int_{S,\text{inner}} \frac{Q_0}{4\pi a^2}V_0 ds + \frac{1}{2}\int_{S,\text{outer}} \frac{-Q_0}{4\pi b^2}(0)ds \tag{6.350}$$

given that the voltage of the outer PEC is zero if the voltage of the inner PEC is selected as V_0. Hence, we simply reach the basic expression

$$w_{e,\text{stored}} = \frac{1}{2}4\pi a^2\frac{Q_0}{4\pi a^2}V_0 = \frac{1}{2}Q_0V_0 = \frac{1}{2}C_0V_0^2. \tag{6.351}$$

Similar to the parallel-plate capacitors, we can also study spherical capacitors with inhomogeneous materials between the spherical PECs. As an example, we consider a spherical capacitor, again consisting of two PECs of radii a and $b > a$, while the dielectric material between them has a permittivity given as

$$\varepsilon = \varepsilon_0(1+\sin\theta). \tag{6.352}$$

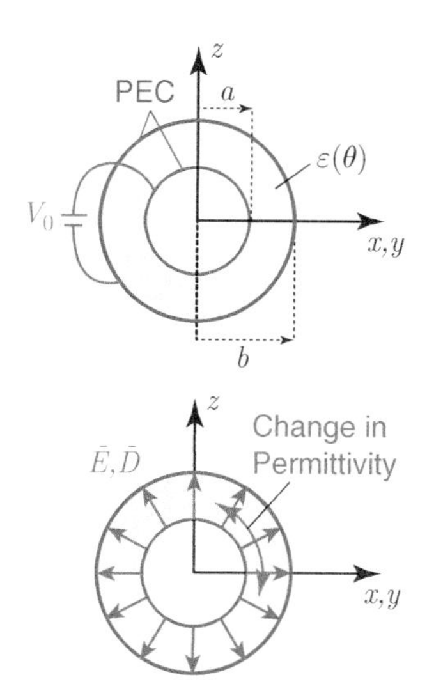

Figure 6.54 A spherical capacitor filled with an inhomogeneous dielectric material with a θ-dependent permittivity. Assuming that the electric field intensity is in the radial direction, Laplace's equation holds.

Assuming that the electric field intensity is only in the radial direction[112] (Figure 6.54), Laplace's equation holds since $\bar{E}(\bar{r}) \perp \bar{\nabla}\varepsilon(\bar{r})$: i.e. the electric field intensity (in the R direction) is perpendicular to the change in the permittivity (in the θ direction). Then we simply have

$$\Phi(\bar{r}) = \frac{V_0 a}{(b-a)}\left(\frac{b}{R}-1\right), \qquad \bar{E}(\bar{r}) = -\bar{\nabla}\Phi(\bar{r}) = \hat{a}_R\frac{V_0ab}{(b-a)R^2} \tag{6.353}$$

exactly as the corresponding expressions for the spherical capacitor with a homogeneous dielectric. On the other hand, the electric flux density can be written

$$\bar{D}(\bar{r}) = \hat{a}_R\varepsilon_0(1+\sin\theta)\frac{V_0ab}{(b-a)R^2}. \tag{6.354}$$

Hence, the surface electric charge density on the inner PEC is obtained as

$$q_s(R=a) = \hat{a}_R\cdot\bar{D}(R=a) = \varepsilon_0(1+\sin\theta)\frac{V_0b}{(b-a)a}, \tag{6.355}$$

[112] This is similar to assuming no fringing fields in a parallel-plate capacitor.

which is obviously nonuniform.[113] To find the total charge Q_0 on the inner PEC, integration is required as[114]

$$Q_0 = \int_0^{2\pi}\int_0^{\pi} \varepsilon_0(1+\sin\theta)\frac{V_0 b}{(b-a)a}R^2\sin\theta d\theta d\phi \tag{6.356}$$

$$= 2\pi\varepsilon_0\frac{V_0 b}{(b-a)a}a^2\int_0^{\pi}(1+\sin\theta)\sin\theta d\theta \tag{6.357}$$

$$= 2\pi\varepsilon_0\frac{V_0 b}{(b-a)a}a^2\left[-\cos\theta+\frac{\theta}{2}+\frac{\sin(2\theta)}{4}\right]_0^{\pi} \tag{6.358}$$

$$= 2\pi\varepsilon_0\frac{V_0 b}{(b-a)a}a^2\left(2+\frac{\pi}{2}\right) = 4\pi\varepsilon_0\frac{V_0 b}{(b-a)a}a^2\left(1+\frac{\pi}{4}\right). \tag{6.359}$$

Hence, the capacitance can be found as

$$C_0 = \frac{Q_0}{V_0} = 4\pi\varepsilon_0\frac{ab}{(b-a)}\left(1+\frac{\pi}{4}\right). \tag{6.360}$$

As a further example, we can find the energy stored in the capacitor for a given voltage V_0. Using the expressions for $\bar{E}(\bar{r})$ and $\bar{D}(\bar{r})$, we obtain

$$w_{e,\text{stored}} = \frac{1}{2}\int_{\infty}\bar{E}(\bar{r})\cdot\bar{D}(\bar{r})dv \tag{6.361}$$

$$= \frac{1}{2}\int_a^b\int_0^{\pi}\int_0^{2\pi}\frac{V_0 ab}{(b-a)R^2}\varepsilon_0(1+\sin\theta)\frac{V_0 ab}{(b-a)R^2}R^2\sin\theta d\theta d\phi dR \tag{6.362}$$

$$= \frac{\varepsilon_0 V_0^2(ab)^2}{2(b-a)^2}\int_a^b\int_0^{\pi}\int_0^{2\pi}\frac{(1+\sin\theta)}{R^2}\sin\theta d\theta d\phi dR \tag{6.363}$$

$$= \frac{\varepsilon_0 V_0^2(ab)^2}{2(b-a)^2}2\pi\left(2+\frac{\pi}{2}\right)\left(\frac{1}{a}-\frac{1}{b}\right) \tag{6.364}$$

$$= 2\pi\varepsilon_0\frac{ab}{(b-a)}\left(1+\frac{\pi}{4}\right)V_0^2 = \frac{1}{2}C_0V_0^2 \tag{6.365}$$

as expected.

[113] This is also why the charge approach is difficult; putting a total charge Q_0 on the inner PEC, it is not obvious how it is distributed.

[114] Note that, although the surface electric charge density has a nonuniform distribution on the inner/outer PECs, the electric field intensity does not depend on θ. For a better understanding, we may consider the polarization as

$$\bar{P}(\bar{r}) = \bar{D}(\bar{r}) - \varepsilon_0\bar{E}(\bar{r})$$
$$= \hat{a}_R\varepsilon_0(1+\sin\theta)\frac{V_0 ab}{(b-a)R^2} - \hat{a}_R\varepsilon_0\frac{V_0 ab}{(b-a)R^2}$$
$$= \hat{a}_R\varepsilon_0\sin\theta\frac{V_0 ab}{(b-a)R^2}.$$

Then in the equivalent problem, the equivalent polarization charges can be found as $q_{pv}(\bar{r}) = -\bar{\nabla}\cdot\bar{P}(\bar{r}) = 0$ and

$$q_{ps}(R=a) = -\varepsilon_0\sin\theta\frac{V_0 b}{(b-a)a}$$
$$q_{ps}(R=b) = \varepsilon_0\sin\theta\frac{V_0 a}{(b-a)a}.$$

Therefore, for the equivalent problem, where the dielectric between PECs is replaced by a vacuum, the *total* surface electric charge density on the inner PEC can be found as

$$q_{s,\text{total}}(R=a) = q_s(R=a) + q_{ps}(R=a)$$
$$= \varepsilon_0 V_0 b/[(b-a)a],$$

which is *uniform* and consistent with the expression for the electric field intensity.

6.7.4 Cylindrical Capacitors

Cylindrical capacitors are frequently used in electrical circuits, making them interesting to analyze. To make the analysis easier, we consider an infinitely long cylindrical capacitor consisting of a perfectly conducting cylinder of radius a inside a perfectly conducting cylindrical surface of radius b, and a dielectric of relative permittivity ε_r between them (Figure 6.55). Note that this infinitely large structure's capacitance is infinite (infinite ability to store charge), while we are concerned about the capacitance per length. Assuming $+Q_{0l}$ (C/m) charge per length on the inner PEC and $-Q_{0l}$ charge per length on the outer PEC, and using Gauss' law, we obtain

$$\bar{D}(\bar{r}) = \hat{a}_\rho \frac{Q_{0l}}{2\pi\rho}, \qquad \bar{E}(\bar{r}) = \hat{a}_\rho \frac{Q_{0l}}{2\pi\varepsilon_0\varepsilon_r\rho}, \tag{6.366}$$

leading to the voltage between the inner and outer PECs as (Figure 6.56)

$$V_0 = -\int_b^a \bar{E}(\bar{r}) \cdot \hat{a}_\rho d\rho = -\int_b^a \hat{a}_\rho \frac{Q_{0l}}{2\pi\varepsilon_0\varepsilon_r\rho} \cdot \hat{a}_\rho d\rho \tag{6.367}$$

$$= -\frac{Q_{0l}}{2\pi\varepsilon_0\varepsilon_r}\left[\ln\rho\right]_b^a = \frac{Q_{0l}}{2\pi\varepsilon_0\varepsilon_r}\ln\left(\frac{b}{a}\right). \tag{6.368}$$

Then the capacitance per length (F/m) can be obtained as

$$C_{0l} = \frac{Q_{0l}}{V_0} = \frac{2\pi\varepsilon_0\varepsilon_r}{\ln(b/a)}. \tag{6.369}$$

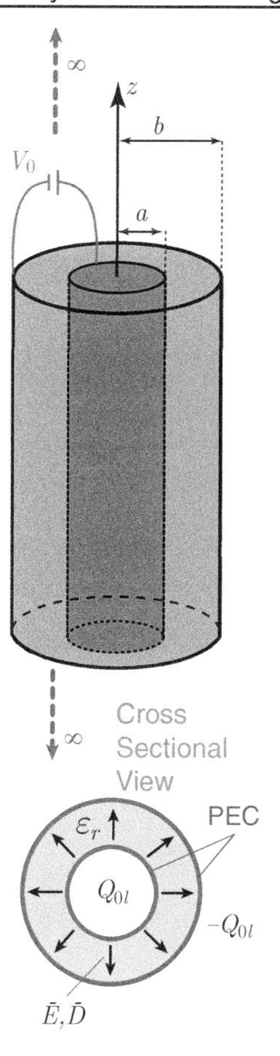

Figure 6.55 A cylindrical capacitor that is assumed to be infinitely long.

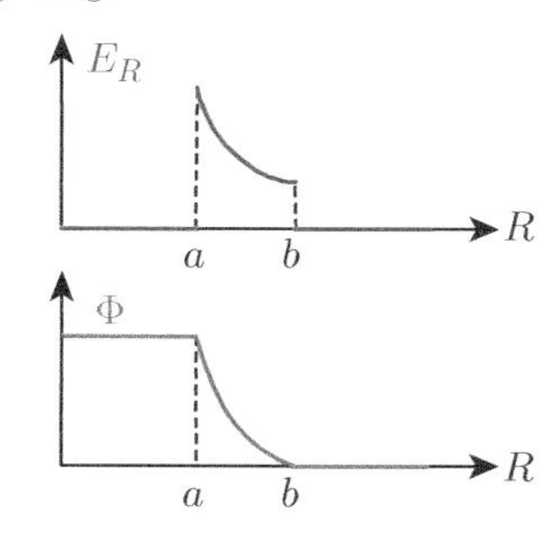

Figure 6.56 The electric field intensity and electric scalar potential with respect to radial variable ρ for an ideal cylindrical capacitor connected to a static voltage.

Alternatively, using the potential approach, we can solve Laplace's equation in a cylindrical coordinate system (with cylindrical symmetry) as

$$\bar{\nabla}^2\Phi(\bar{r}) = 0 \longrightarrow \frac{1}{\rho}\frac{\partial}{\partial\rho}\left(\rho\frac{\partial}{\partial\rho}\Phi(\bar{r})\right) = 0 \longrightarrow \rho\frac{\partial}{\partial\rho}\Phi(\bar{r}) = c_1 \tag{6.370}$$

$$\longrightarrow \frac{\partial}{\partial\rho}\Phi(\bar{r}) = \frac{c_1}{\rho} \longrightarrow \Phi(\bar{r}) = c_1\ln(\rho) + c_2, \tag{6.371}$$

where c_1 and c_2 are constants. Inserting $\Phi(\rho = a) = V_0$ and $\Phi(\rho = b) = 0$, we have $c_2 = -c_1\ln(b)$ and

$$\Phi(\rho = a) = V_0 = c_1\ln(a) - c_1\ln(b) = c_1\ln(a/b) \longrightarrow c_1 = V_0/\ln(a/b). \tag{6.372}$$

Therefore, the expression for the electric scalar potential can be obtained as

$$\Phi(\bar{r}) = V_0 \ln(\rho)/\ln(a/b) - V_0 \ln(b)/\ln(a/b) = V_0 \frac{\ln(\rho/b)}{\ln(a/b)}, \tag{6.373}$$

leading to[115]

$$\bar{E}(\bar{r}) = -\bar{\nabla}\Phi(\bar{r}) = -\hat{a}_\rho \frac{V_0}{\ln(a/b)\rho} = \hat{a}_\rho \frac{V_0}{\ln(b/a)\rho}. \tag{6.374}$$

[115] Hence, we have

$$\bar{D}(\bar{r}) = \varepsilon_0\varepsilon_r \bar{E}(\bar{r}) = \hat{a}_\rho \varepsilon_0\varepsilon_r \frac{V_0}{\ln(b/a)\rho}.$$

Consequently, the surface electric charge density on the inner PEC can be obtained as

$$q_s(\rho = a) = \hat{a}_\rho \cdot \bar{D}(\rho = a) = \varepsilon_0\varepsilon_r \frac{V_0}{\ln(b/a)a}. \tag{6.375}$$

This corresponds to a total electric charge per length:

$$Q_{0l} = 2\pi a q_s(\rho = a) = \varepsilon_0\varepsilon_r 2\pi a \frac{V_0}{\ln(b/a)a} = \frac{2\pi\varepsilon_0\varepsilon_r}{\ln(b/a)} V_0. \tag{6.376}$$

This way, we again obtain the capacitance per length as

$$C_{0l} = \frac{Q_{0l}}{V_0} = \frac{2\pi\varepsilon_0\varepsilon_r}{\ln(b/a)}, \tag{6.377}$$

confirming the expression found via the charge approach. Finally, if the energy stored per length needs to be found, one can use the capacitance per length as

$$w_{e,l} = \frac{1}{2} C_{0l} V_0^2 = \frac{\pi\varepsilon_0\varepsilon_r}{\ln(b/a)} V_0^2. \tag{6.378}$$

6.7.5 Examples

Example 71: Consider an $a \times a \times a$ dielectric material defined in the first octant of the Cartesian coordinate system (Figure 6.57). The permittivity is given as

$$\varepsilon(z) = \varepsilon_0(1 + z/a). \tag{6.379}$$

Find the capacitance between the top and bottom surfaces, assuming that the perfectly conducting plates are placed at these surfaces.

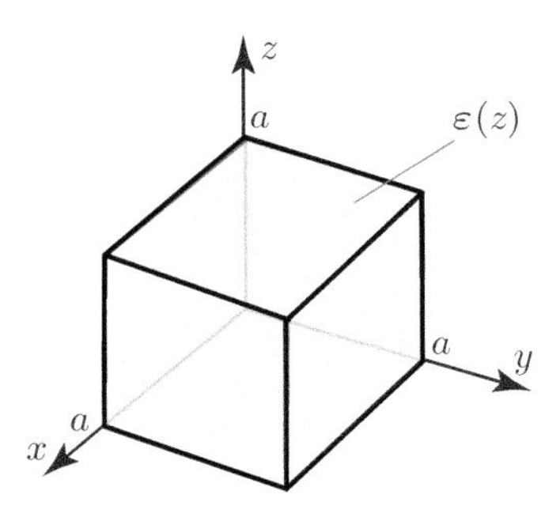

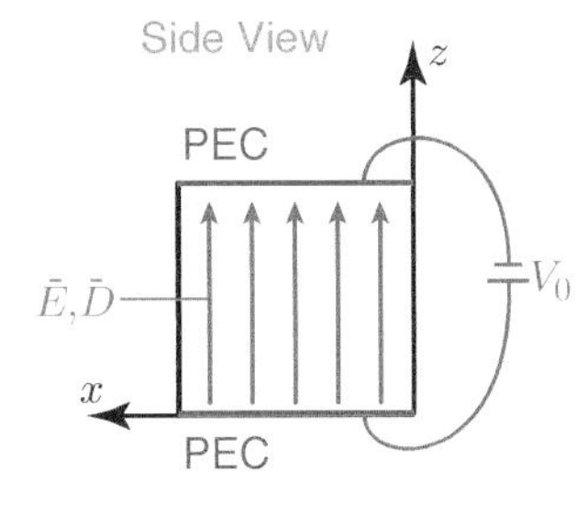

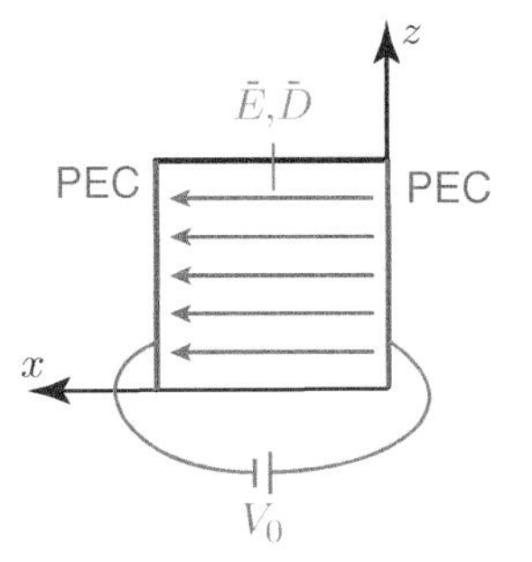

Figure 6.57 A cubic dielectric material with an inhomogeneous permittivity changing in the z direction. The capacitance depends on where the PECs are placed (bottom-top or back-front).

Solution: Neglecting fringing fields, we can assume that the electric field intensity is in the z direction if the bottom/top plate is positively/negatively charged (Figure 6.57). Since the change in permittivity is also in the z direction, we can employ the charge approach. Assuming Q_0 on the bottom plate, we have

$$\bar{D}(\bar{r}) = \hat{a}_z \frac{Q_0}{a^2}, \qquad \bar{E}(\bar{r}) = \frac{\bar{D}(\bar{r})}{\varepsilon(\bar{r})} = \hat{a}_z \frac{Q_0}{\varepsilon_0 a^2(1+z/a)}. \tag{6.380}$$

Hence, the voltage between the plates can be found as[116]

$$V_0 = -\int_a^0 \bar{E}(\bar{r}) \cdot \hat{a}_z dz = \frac{Q_0 \ln(2)}{\varepsilon_0 a}. \tag{6.381}$$

This way, the capacitance can be found as

$$C_0 = Q_0/V_0 = \varepsilon_0 a/\ln(2). \tag{6.382}$$

Example 72: Consider again the $a \times a \times a$ dielectric material defined in the first octant of the Cartesian coordinate system (Figure 6.57), where the permittivity is given as $\varepsilon(z) = \varepsilon_0(1+z/a)$. Find the capacitance between the back and front surfaces, assuming that the perfectly conducting plates are placed at these surfaces.

Solution: In this case, assuming no fringing fields, the electric field intensity is in the x direction so that it is perpendicular to the change in the permittivity. This makes the potential approach suitable since Laplace's equation holds.[117] We have

$$\bar{\nabla}^2 \Phi(\bar{r}) = 0 \longrightarrow \frac{\partial}{\partial x^2}\Phi(x) = 0 \longrightarrow \Phi = c_1 x + c_2, \tag{6.383}$$

where $c_1 = -V_0/a$ and $c_2 = V_0$ if we set $\Phi(x=0) = V_0$ and $\Phi(x=a) = 0$. Then the electric field intensity and electric flux density can be found as

$$\bar{E}(\bar{r}) = -\bar{\nabla}\Phi(\bar{r}) = -\hat{a}_x \frac{\partial}{\partial x}\left(-V_0 \frac{x}{a} + V_0\right) = \hat{a}_x \frac{V_0}{a} \tag{6.384}$$

$$\bar{D}(\bar{r}) = \varepsilon(\bar{r})\bar{E}(\bar{r}) = \hat{a}_x \varepsilon_0 \frac{V_0}{a}\left(1 + \frac{z}{a}\right) = \hat{a}_x \varepsilon_0 V_0 \left(\frac{a+z}{a^2}\right). \tag{6.385}$$

The expression for the electric flux density indicates that the surface electric charge density accumulated at the back plate ($x = 0$) can be written[118]

[116] As intermediate steps, we have

$$V_0 = \int_0^a \bar{E}(\bar{r}) \cdot \hat{a}_z dz = \int_0^a \frac{Q_0}{\varepsilon_0 a^2(1+z/a)} dz = \frac{Q_0}{\varepsilon_0 a}\int_0^a \frac{1}{a+z} dz = \frac{Q_0}{\varepsilon_0 a}\ln\left(\frac{2a}{a}\right).$$

[117] In addition, note that if the charge approach is used, distributions of charges on the plates are not uniform (and may not be obvious).

[118] Using the corresponding boundary condition, we have

$$q_s(x=0) = \hat{a}_x \cdot \bar{D}(x=0) = \varepsilon_0 V_0 \left(\frac{a+z}{a^2}\right).$$

$$q_s(x=0) = \varepsilon_0 V_0 \left(\frac{a+z}{a^2}\right). \tag{6.386}$$

Since we started from the potential, the capacitance can be found if we derive an expression for the total charge accumulated. This can be found via integration as

$$Q_0 = \int_0^a \int_0^a \frac{\varepsilon_0 V_0}{a^2}(a+z)dydz = \frac{\varepsilon_0 V_0}{a}\left[az + \frac{z^2}{2}\right]_0^a = \varepsilon_0 V_0 \frac{3a}{2}, \tag{6.387}$$

leading to

$$C_0 = Q_0/V_0 = 3\varepsilon_0 a/2. \tag{6.388}$$

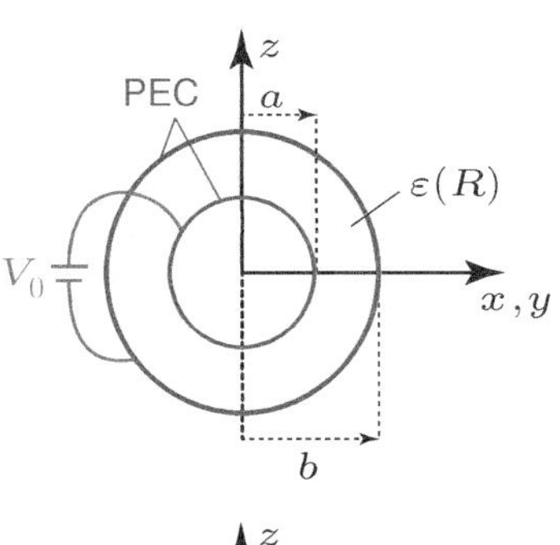

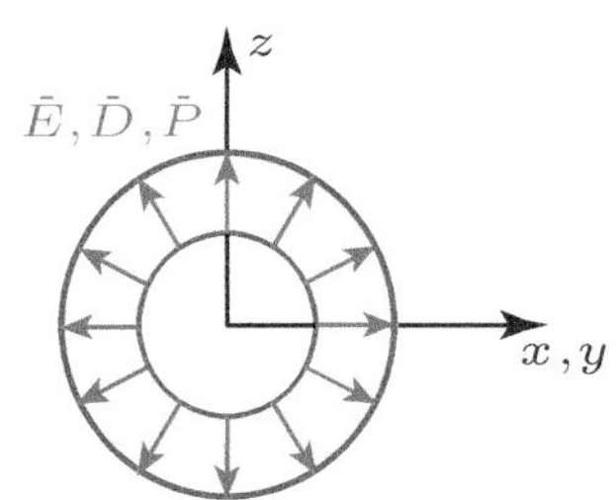

Figure 6.58 A spherical capacitor filled with an inhomogeneous dielectric material with R-dependent permittivity. Assuming that the electric field intensity is in the radial direction, Laplace's equation does not hold.

Example 73: Consider a spherical capacitor consisting of two PECs of radii a and $b > a$, and an inhomogeneous dielectric material between them (Figure 6.58). It is given that the permittivity of the dielectric changes only with respect to the radial variable R (i.e. we have $\varepsilon(\bar{r}) = \varepsilon(R)$). When the inner PEC is charged with Q_0, the polarization inside the dielectric is given as

$$\bar{P}(\bar{r}) = \hat{a}_R \frac{Q_0}{4\pi R^2}\left(1 - \frac{a}{R}\right). \tag{6.389}$$

Find the capacitance of the capacitor.

Solution: First, assuming that the electric field intensity and electric flux density are in the radial direction, we have

$$\bar{D}(\bar{r}) = \hat{a}_R \frac{Q_0}{4\pi R^2}, \qquad \bar{E}(\bar{r}) = \hat{a}_R \frac{Q_0}{4\pi\varepsilon(R)R^2} \tag{6.390}$$

$$\bar{P}(\bar{r}) = \bar{D}(\bar{r}) - \varepsilon_0 \bar{E}(\bar{r}) = \hat{a}_R \frac{Q_0}{4\pi R^2} - \hat{a}_R \frac{\varepsilon_0 Q_0}{4\pi\varepsilon(R)R^2} = \hat{a}_R \frac{Q_0}{4\pi R^2}\left[1 - \frac{\varepsilon_0}{\varepsilon(R)}\right]. \tag{6.391}$$

Then according to the given expression for the polarization, we can write

$$\bar{P}(\bar{r}) = \hat{a}_R \frac{Q_0}{4\pi R^2}\left[1 - \frac{\varepsilon_0}{\varepsilon(R)}\right] = \hat{a}_R \frac{Q_0}{4\pi R^2}\left(1 - \frac{a}{R}\right), \tag{6.392}$$

which can be satisfied if $\varepsilon(R) = \varepsilon_0 R/a$. Note that Laplace's equation does not hold since

$$\bar{\nabla} \cdot \bar{D}(\bar{r}) = 0 \longrightarrow \bar{\nabla} \cdot [\varepsilon(\bar{r})\bar{E}(\bar{r})] = \bar{\nabla}\varepsilon(\bar{r}) \cdot \bar{E}(\bar{r}) + \varepsilon(\bar{r})\bar{\nabla} \cdot \bar{E}(\bar{r}) = 0, \tag{6.393}$$

leading to

$$\bar{\nabla} \cdot \bar{E}(\bar{r}) = -\bar{\nabla}^2 \Phi(\bar{r}) = -\bar{\nabla}\varepsilon(\bar{r}) \cdot \bar{E}(\bar{r})/\varepsilon(\bar{r}) \neq 0. \tag{6.394}$$

In fact, this can be verified by using the expression for the electric field intensity as

$$\bar{E}(\bar{r}) = \hat{a}_R \frac{Q_0 a}{4\pi\varepsilon_0 R^3} \tag{6.395}$$

$$\bar{\nabla} \cdot \bar{E}(\bar{r}) = \frac{1}{R^2}\frac{\partial}{\partial R}\left(R^2 E_R(\bar{r})\right) = \frac{Q_0 a}{4\pi\varepsilon_0}\frac{1}{R^2}\frac{\partial}{\partial R}\left(\frac{1}{R}\right) = -\frac{Q_0 a}{4\pi\varepsilon_0 R^4}, \tag{6.396}$$

while we also have

$$\bar{\nabla}\varepsilon(\bar{r}) = \hat{a}_R \frac{\varepsilon_0}{a} \longrightarrow \bar{\nabla}\varepsilon(\bar{r}) \cdot \bar{E}(\bar{r}) = \frac{Q_0}{4\pi R^3} \longrightarrow -\frac{\bar{\nabla}\varepsilon(\bar{r}) \cdot \bar{E}(\bar{r})}{\varepsilon(\bar{r})} = -\frac{Q_0 a}{4\pi\varepsilon_0 R^4}. \tag{6.397}$$

Obviously, Poisson's equation is valid inside the dielectric:

$$\bar{\nabla}^2 \Phi(\bar{r}) = \frac{Q_0 a}{4\pi\varepsilon_0 R^4}. \tag{6.398}$$

On the other hand, since we already know the electric field intensity, the voltage between the PECs can be found as[119]

$$V_0 = \int_a^b \hat{a}_R \frac{Q_0 a}{4\pi\varepsilon_0 R^3} \cdot \hat{a}_R dR = \frac{Q_0 a}{4\pi\varepsilon_0}\left[-\frac{1}{2R^2}\right]_a^b = \frac{Q_0 a}{8\pi\varepsilon_0}\left(\frac{1}{a^2} - \frac{1}{b^2}\right). \tag{6.399}$$

Then the capacitance can be found as

$$C_0 = \frac{Q_0}{V_0} = \frac{8\pi\varepsilon_0 a b^2}{b^2 - a^2}. \tag{6.400}$$

Example 74: Consider a spherical capacitor consisting of a PEC of radius a inside a PEC of radius $2b$ and a dielectric material from $R = b$ to $R = 2b$ with $b > a$ (Figure 6.59). The permittivity of the dielectric is given as

$$\varepsilon(R) = \varepsilon_0 \frac{b^2}{R^2}, \tag{6.401}$$

while a vacuum exists from $R = a$ to $R = b$. Find the capacitance of the structure.

[119] Note that using the expression for the electric field intensity, we can obtain the electric scalar potential at anywhere inside the dielectric material as

$$\Phi(\bar{r}) = \int_R^b \hat{a}_R \frac{Q_0 a}{4\pi\varepsilon_0 (R')^3} \cdot \hat{a}_R dR' = \frac{Q_0 a}{4\pi\varepsilon_0}\left[-\frac{1}{2(R')^2}\right]_R^b = \frac{Q_0 a}{8\pi\varepsilon_0}\left[\frac{1}{R^2} - \frac{1}{b^2}\right],$$

which becomes

$$\Phi(\bar{r}) = V_0 \frac{a^2}{R^2}\left(\frac{b^2 - R^2}{b^2 - a^2}\right)$$

using

$$Q_0 = C_0 V_0 = \frac{8\pi\varepsilon_0 a b^2}{b^2 - a^2} V_0.$$

Then one can further check the Laplace of the electric scalar potential as

$$\bar{\nabla}^2 \Phi(\bar{r}) = \frac{1}{R^2}\frac{\partial}{\partial R}\left(R^2 \frac{\partial \Phi(R)}{\partial R}\right) = \frac{1}{R^2}\frac{\partial}{\partial R}\left[R^2 \frac{(-2V_0)a^2 b^2}{R^3(b^2 - a^2)}\right] = \frac{2V_0 a^2 b^2}{R^4(b^2 - a^2)} = \frac{Q_0 a}{4\pi\varepsilon_0 R^4},$$

which is the same as Eq. (6.398).

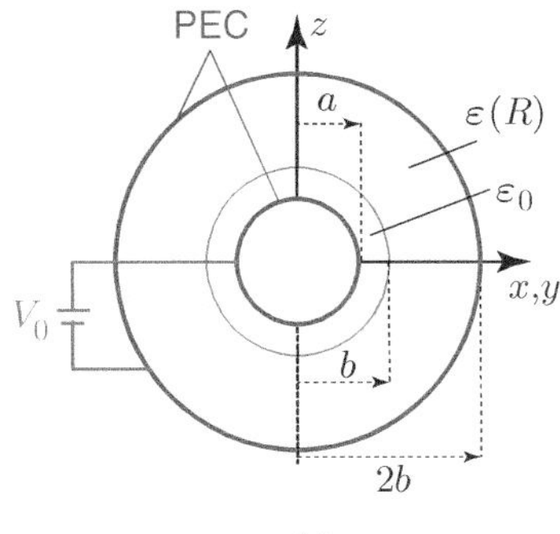

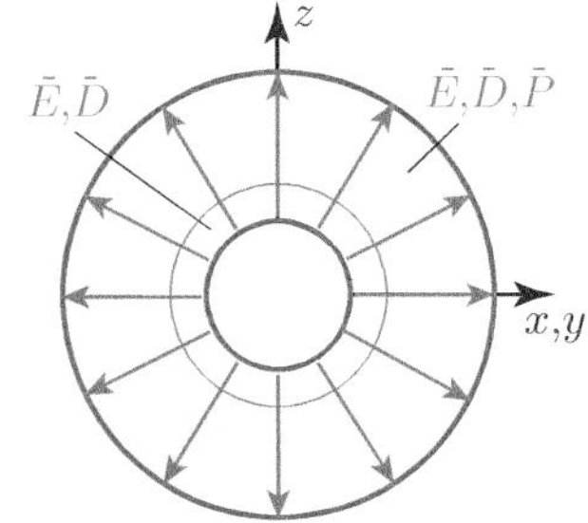

Figure 6.59 A spherical capacitor filled with an inhomogeneous dielectric material from $R = b$ to $R = 2b$ with R-dependent permittivity. A vacuum exists from $R = a$ to $R = b$.

Solution: As in the previous example, the electric field intensity (assumed to be in the radial direction) is parallel to the change in the permittivity such that Laplace's equation does not hold for the electric scalar potential: i.e. $\bar{\nabla}^2\Phi(\bar{r}) \neq 0$. On the other hand, due to the spherical symmetry, we can assume that the surface electric charge density is distributed uniformly on the inner and outer PECs. In addition, Gauss' law leads to

$$\bar{D}(\bar{r}) = \hat{a}_R \frac{Q_0}{4\pi R^2}, \tag{6.402}$$

where Q_0 is the total electric charge on the inner PEC. Depending on the permittivity, the electric field intensity can be written[120]

$$\bar{E}(\bar{r}) = \hat{a}_R \frac{Q_0}{4\pi\varepsilon_0 R^2} \begin{cases} 1 & (a \leq R < b) \\ R^2/b^2 & (b < R \leq 2b). \end{cases} \tag{6.403}$$

Hence, the voltage between the inner and outer PECs can be found as

$$V_0 = -\int_{2b}^{a} \bar{E}(\bar{r}) \cdot \hat{a}_R dR = -\int_{2b}^{b} \frac{Q_0}{4\pi\varepsilon_0 b^2} dR - \int_{b}^{a} \frac{Q_0}{4\pi\varepsilon_0 R^2} dR, \tag{6.404}$$

leading to

$$V_0 = \frac{Q_0}{4\pi\varepsilon_0 b} + \frac{Q_0}{4\pi\varepsilon_0 a} - \frac{Q_0}{4\pi\varepsilon_0 b} = \frac{Q_0}{4\pi\varepsilon_0 a}. \tag{6.405}$$

Hence the capacitance is simply $C_0 = Q_0/V_0 = 4\pi\varepsilon_0 a$.

But what about the polarization inside the dielectric region? It can be written

$$\bar{P}(\bar{r}) = \left(\varepsilon_0 \frac{b^2}{R^2} - \varepsilon_0\right) \bar{E}(\bar{r}) = \left(\varepsilon_0 \frac{b^2}{R^2} - \varepsilon_0\right) \hat{a}_R \frac{Q_0}{4\pi\varepsilon_0 b^2} \qquad (b < R \leq 2b) \tag{6.406}$$

or

$$\bar{P}(\bar{r}) = \hat{a}_R \left(\frac{b^2}{R^2} - 1\right) \frac{Q_0}{4\pi b^2} \qquad (b < R < 2b), \tag{6.407}$$

which also depends on position. Therefore, in the equivalent problem, we have[121]

$$q_{pv}(\bar{r}) = -\bar{\nabla} \cdot \bar{P}(\bar{r}) = \frac{Q_0}{2\pi b^2 R} \qquad (b < R < 2b), \tag{6.408}$$

[120] Note that the electric flux density is continuous across the boundary at $R = b$, while the electric field intensity is not.

[121] This is evaluated as

$$\begin{aligned} -\bar{\nabla} \cdot \bar{P}(\bar{r}) &= -\frac{1}{R^2}\frac{\partial}{\partial R}\left(R^2 P_R(R)\right) \\ &= -\frac{1}{R^2}\frac{\partial}{\partial R}\left[R^2\left(\frac{b^2}{R^2} - 1\right)\frac{Q_0}{4\pi b^2}\right] \\ &= -\frac{1}{R^2}\frac{\partial}{\partial R}\left[(b^2 - R^2)\frac{Q_0}{4\pi b^2}\right] \\ &= \frac{1}{R}\frac{Q_0}{2\pi b^2}. \end{aligned}$$

which is nonzero to represent the inhomogeneity. In addition, an equivalent (surface) polarization charge distribution is required at $R = 2b$ as[122]

$$q_{ps}(R = 2b) = \hat{a}_R \cdot \bar{P}(R = 2b) = \left(\frac{b^2}{4b^2} - 1\right)\frac{Q_0}{4\pi b^2} = -\frac{3Q_0}{16\pi b^2}, \tag{6.409}$$

while there is no need to an equivalent polarization charge distribution at $R = b$: i.e. at the interface between a vacuum and the dielectric.[123]

Example 75: Consider two cylindrical PECs, each with radius a, separated by a distance d from center to center. The PECs are infinitely long and located in a vacuum (Figure 6.60). Find the capacitance per length of the structure.

Solution: Assuming Q_{0l} and $-Q_{0l}$ charges per length are accumulated on the PECs (on the z-x plane), and they are distributed uniformly,[124] the electric field intensity at an arbitrary position between the PECs can be found as

$$\bar{E}(\bar{r}) = \hat{a}_x \frac{Q_{0l}}{2\pi\varepsilon_0 x} + \hat{a}_x \frac{Q_{0l}}{2\pi\varepsilon_0 (d-x)} = \hat{a}_x \frac{Q_{0l}}{2\pi\varepsilon_0}\left(\frac{1}{x} + \frac{1}{d-x}\right), \tag{6.410}$$

where x is the distance from the center of the positively charged PEC. Integration along x from $d-a$ to a, we obtain

$$V_0 = -\int_{d-a}^{a} \bar{E}(\bar{r}) \cdot \hat{a}_x dx = -\frac{Q_{0l}}{2\pi\varepsilon_0}\int_{d-a}^{a}\left(\frac{1}{x} + \frac{1}{d-x}\right)dx \tag{6.411}$$

$$= -\frac{Q_{0l}}{2\pi\varepsilon_0}\left[\ln x - \ln(d-x)\right]_{d-a}^{a} \tag{6.412}$$

$$= -\frac{Q_{0l}}{2\pi\varepsilon_0}\left[\ln a - \ln(d-a) - \ln(d-a) + \ln(a)\right] \tag{6.413}$$

$$= \frac{Q_{0l}}{\pi\varepsilon_0}\ln\left(\frac{d-a}{a}\right). \tag{6.414}$$

Then the capacitance per length can be obtained as

$$C_{0l} = \frac{Q_{0l}}{V_0} = \frac{\pi\varepsilon_0}{\ln(d/a - 1)}. \tag{6.415}$$

Note that if $d \gg a$, then the capacitance per length can be approximated as $C_{0l} \approx \pi\varepsilon_0 / \ln(d/a)$.

[122] One can show that the total equivalent polarization charge is zero as

$$Q_{\text{net}} = \int_V q_{pv}(\bar{r})d\bar{r} + \oint_S q_{ps}(\bar{r})d\bar{r}$$
$$= 4\pi\int_b^{2b} \frac{Q_0}{2\pi b^2 R}R^2 dR - 4\pi(2b)^2\frac{3Q_0}{16\pi b^2}$$
$$= 4\pi\frac{Q_0}{2\pi b^2}\frac{3b^2}{2} - 3Q_0 = 3Q_0 - 3Q_0 = 0.$$

[123] This is related to the fact that the permittivity does not have a jump (discontinuity) at $R = b$: i.e. the permittivity of the dielectric material becomes $\varepsilon(R = b) = \varepsilon_0 b^2/b^2 = \varepsilon_0$ at the interface.

[124] Specifically, we assume that the PECs do not affect each other: i.e. the distribution of charges on a PEC is not affected by the charges on the other PEC.

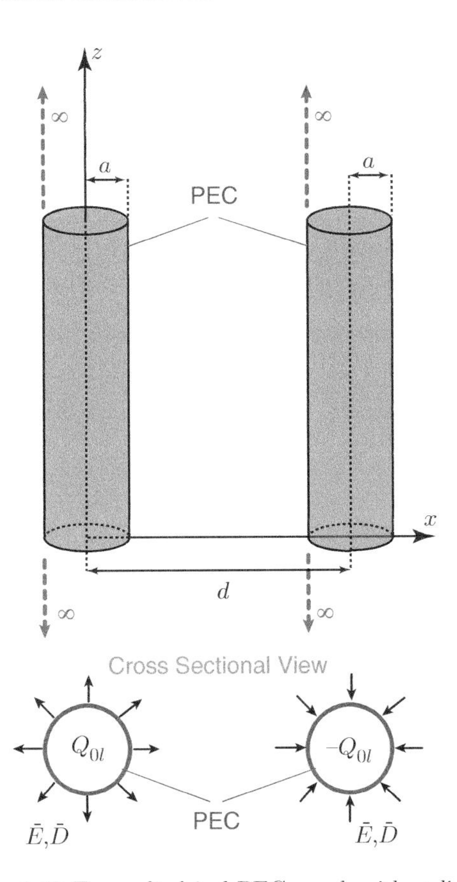

Figure 6.60 Two cylindrical PECs, each with radius a, separated by a distance d from center to center. To find the capacitance per length, we assume that electric charges are distributed uniformly on the PECs. Then the electric field intensity can easily be found via Gauss' law, leading to the expression for the electric scalar potential, as well as the voltage between the PECs.

Example 76: Consider an infinitely long cylindrical capacitor consisting of a perfectly conducting cylinder of radius a inside a perfectly conducting cylindrical surface of radius b aligned in the z direction (Figure 6.61). Between the PECs, a dielectric material with permittivity ε is used from $\phi = 0$ to $\phi = \phi_0$, while the rest is filled with a vacuum. Find the capacitance per length.

Solution: Assuming that the electric field intensity is in the radial (ρ) direction, this problem can also be solved using the potential approach. Setting V_0 potential on the inner PEC and 0 potential on the outer PEC, we have

$$\Phi(\bar{r}) = V_0 \frac{\ln(\rho/b)}{\ln(a/b)} \tag{6.416}$$

$$\bar{E}(\bar{r}) = -\bar{\nabla}\Phi(\bar{r}) = -\hat{a}_\rho \frac{V_0}{\ln(a/b)} \frac{1}{\rho} = \hat{a}_\rho \frac{V_0}{\ln(b/a)} \frac{1}{\rho}, \tag{6.417}$$

exactly as in the case of a cylindrical capacitor with a homogeneous dielectric between the PECs (see Section 6.7.4). Then we can write the electric flux density as

$$\bar{D}(\bar{r}) = \hat{a}_\rho \frac{V_0}{\ln(b/a)} \frac{1}{\rho} \begin{cases} \varepsilon & (0 \leq \phi < \phi_0) \\ \varepsilon_0 & (\phi_0 < z \leq 2\pi). \end{cases} \tag{6.418}$$

For zero fields inside the inner cylinder and using the boundary condition for the normal electric flux density, we have

$$q_s(\rho = a) = \frac{V_0}{\ln(b/a)} \frac{1}{a} \begin{cases} \varepsilon & (0 \leq \phi < \phi_0) \\ \varepsilon_0 & (\phi_0 < z \leq 2\pi), \end{cases} \tag{6.419}$$

as the surface electric charge density on the inner PEC. Then the total charge per length on the inner PEC can be found as

$$Q_{0l} = \phi_0 a \frac{V_0}{\ln(b/a)} \frac{1}{a} \varepsilon + (2\pi - \phi_0) a \frac{\varepsilon_0 V_0}{\ln(b/a)} \frac{1}{a} \varepsilon_0 \tag{6.420}$$

$$= \frac{V_0}{\ln(b/a)} [\phi_0 \varepsilon + 2\pi\varepsilon_0 - \phi_0 \varepsilon_0], \tag{6.421}$$

and we obtain

$$C_{0l} = \frac{Q_{0l}}{V_0} = \frac{1}{\ln(b/a)} [2\pi\varepsilon_0 + \phi_0(\varepsilon - \varepsilon_0)] \tag{6.422}$$

as the capacitance per length. Obviously, this capacitor can be interpreted as a *parallel* connection of two capacitors.

6.8 Resistance

Resistance is an important property of structures, similar to capacitance and inductance. For a given structure, it is generally defined as the voltage between its two sides divided by the current along it. As in the case of capacitance, the resistance of a given structure can be found systematically in various ways. First, one can derive a relationship between the capacitance and resistance. To illustrate this, we consider two PECs with a potential difference V_0 between them (Figure 6.62). The PECs are located in an electrically homogeneous *lossy* medium with permittivity ε and conductivity σ. We assume that charges Q_0 and $-Q_0$ are accumulated on the PECs; consequently, an electric field distribution is formed in the medium. Since the medium is also conducting, there is also an electric current flowing between the PECs. The capacitance can be defined as

$$C_0 = \frac{Q_0}{V_0} = \frac{1}{V_0}\left(\oint_S \bar{D}(\bar{r}) \cdot \overline{ds}\right), \tag{6.423}$$

where the integral is over a Gauss surface enclosing the *positive* PEC. Since the medium is electrically homogeneous, we further have

$$C_0 = \frac{\varepsilon}{V_0}\oint_S \bar{E}(\bar{r}) \cdot \overline{ds} = \frac{\varepsilon}{\sigma V_0}\oint_S \bar{J}_v(\bar{r}) \cdot \overline{ds} = \frac{\varepsilon I_0}{\sigma V_0} = \frac{\varepsilon}{\sigma R_0}, \tag{6.424}$$

where I_0 is the total current flowing outward the Gauss surface and $R_0 = V_0/I_0$ is the resistance between the PECs. Therefore, we obtain a general expression as

$$R_0 C_0 = \frac{\varepsilon}{\sigma} = \tau, \tag{6.425}$$

where τ is the relaxation time of the medium[125].

Including the approach above, there are at least four ways to find a given structure's resistance:

- For a structure involving an electrically homogeneous medium, one can use $R_0C_0 = \varepsilon/\sigma$ if the capacitance is already available.
- If the capacitance is not available, one can start by defining the volume electric current density and then find the electric field intensity, leading to the potential difference between the PECs.
- In some cases, the volume electric current density cannot be defined easily, and it may be easier to start with the potential difference between the PECs and solving Laplace's equation, leading to the full expression for the electric scalar potential, electric field intensity, and volume electric current density.

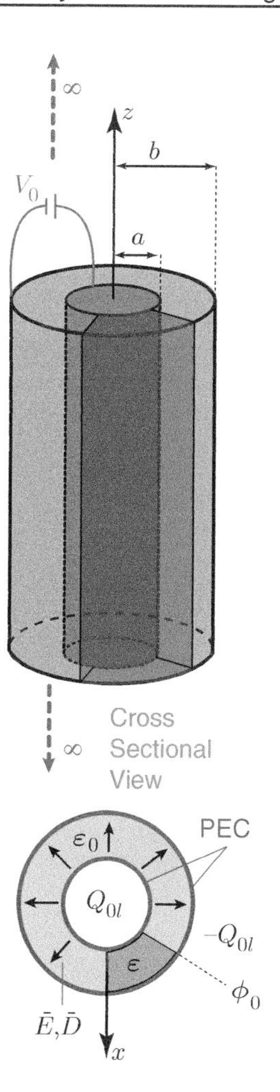

Figure 6.61 A cylindrical capacitor that is assumed to be infinitely long. The space between the PECs is partially filled with a dielectric material.

[125] We emphasize that this equality holds only for an electrically homogeneous medium: i.e. when we have $\varepsilon(\bar{r}) = \varepsilon$ and $\sigma(\bar{r}) = \sigma$.

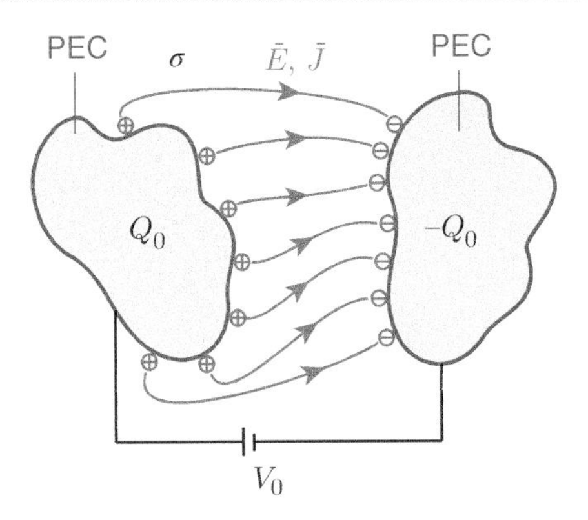

Figure 6.62 Two perfectly conducting bodies located in a conductive medium. When a voltage source is applied, an electric current distribution is created between them, besides the electric field distribution. Therefore, in addition to the capacitance of the system, there is a resistance between the PECs.

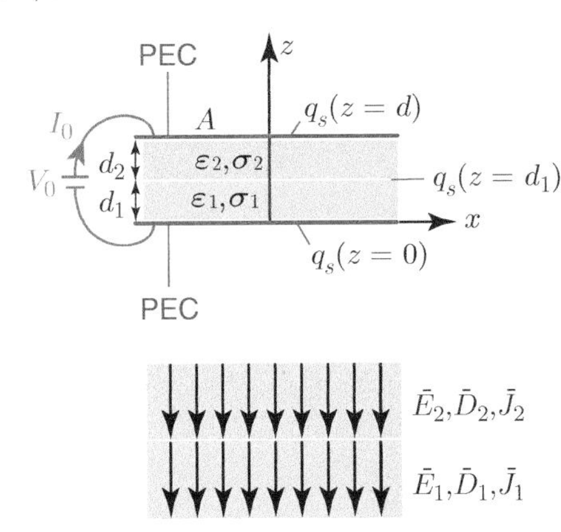

Figure 6.63 A parallel-plate capacitor involving two lossy dielectric slabs with different electrical parameters. An electric current flows though the capacitor due to the lossy materials. At the interface, only the normal component of the volume electric current density is continuous, while both the electric field intensity and electric flux density have jumps.

[126]This is because we have $\bar{\nabla} \cdot \bar{J}_v(\bar{r}) = 0$ for steady-state electric currents: i.e. when Kirchhoff's current law holds. Application of $\bar{\nabla} \cdot \bar{J}_v(\bar{r}) = 0$ at any boundary leads to the continuity of the normal component of the volume electric current density.

[127]And also, we are considering very large plates.

- Series and parallel connections of differential resistors may be helpful to easily derive an overall resistance.

In the following examples, we particularly consider static cases (steady currents). The resistance of a structure can be different in a dynamic case due to possibly changing electrical parameters of the material used (e.g. nonlinearity) and different (particularly nonuniform) distributions of the volume electric current density.

As an example, we consider a static case involving a parallel-plate capacitor with two lossy dielectric slabs (Figure 6.63) – i.e. $\{\varepsilon_1, \sigma_1\}$ for $0 < z < d_1$ and $\{\varepsilon_2, \sigma_2\}$ for $d_1 < z < d_1 + d_2$ – while perfectly conducting plates are located at $z = 0$ and $z = d_1 + d_2$. Due to different relaxation times, there should be charge accumulation at the interface ($z = d_1$). Therefore, for general values of ε and σ, neither the electric field intensity nor the electric flux density is continuous across the interface. The only boundary condition that we can use is related to the volume electric current density, which must be continuous in a steady state[126]

$$\bar{J}_{v0}(\bar{r}) = \bar{J}_{v1}(\bar{r}) = \bar{J}_{v2}(\bar{r}) = -\hat{a}_z I_0 / A, \tag{6.426}$$

where I_0 is the total current flowing through the structure and A is the area of the plates. Note that the volume electric current density is assumed to be uniform since the electrical properties do not change in the x and y directions.[127] Using conductivity values, we arrive at

$$\bar{E}_1(\bar{r}) = -\hat{a}_z I_0/(A\sigma_1), \qquad \bar{E}_2(\bar{r}) = -\hat{a}_z I_0/(A\sigma_2) \tag{6.427}$$

as the electric field intensity values in the first and second dielectric slabs. Assuming V_0 potential difference between the perfectly conducting plates, and assigning V_0 and 0 values at the top plate and the bottom plate, respectively, we have

$$V_0 = -\int_0^{d_1+d_2} \bar{E} \cdot \overline{dl} = |\bar{E}_1| d_1 + |\bar{E}_2| d_2 = \frac{I_0}{A}\left(\frac{d_1}{\sigma_1} + \frac{d_2}{\sigma_2}\right). \tag{6.428}$$

Therefore, the resistance value can be obtained as

$$R_0 = \frac{V_0}{I_0} = \left(\frac{d_1}{A\sigma_1} + \frac{d_2}{A\sigma_2}\right). \tag{6.429}$$

Obviously, this expression consists of two separate terms: i.e. $R_0 = R_1 + R_2$, where $R_1 = d_1/(A\sigma_1)$ and $R_2 = d_2/(A\sigma_2)$, in accordance with a *series* connection of two resistors.

At this stage, we may further investigate the parallel-plate capacitor involving two lossy dielectrics. Updating the electric field intensity values, we obtain[128]

$$\bar{E}_1(\bar{r}) = -\hat{a}_z \frac{\sigma_2 V_0}{(d_1\sigma_2 + d_2\sigma_1)}, \qquad \bar{E}_2(\bar{r}) = -\hat{a}_z \frac{\sigma_1 V_0}{(d_1\sigma_2 + d_2\sigma_1)}. \tag{6.430}$$

In addition, the electric flux density expressions can be written in terms of V_0 as

$$\bar{D}_1(\bar{r}) = -\hat{a}_z \frac{\varepsilon_1\sigma_2 V_0}{(d_1\sigma_2 + d_2\sigma_1)}, \qquad \bar{D}_2(\bar{r}) = -\hat{a}_z \frac{\varepsilon_2\sigma_1 V_0}{(d_1\sigma_2 + d_2\sigma_1)}. \tag{6.431}$$

Hence, using the related boundary conditions[129] (and assuming zero fields outside the structure), the expressions for the surface electric charge densities at the perfectly conducting plates, as well as at the interface, can be found:

$$q_s(z=0) = -\frac{\varepsilon_1\sigma_2 V_0}{(d_1\sigma_2 + d_2\sigma_1)}, \qquad q_s(z=d_1+d_2) = \frac{\varepsilon_2\sigma_1 V_0}{(d_1\sigma_2 + d_2\sigma_1)} \tag{6.432}$$

and

$$q_s(z=d_1) = \frac{\varepsilon_1\sigma_2 V_0}{(d_1\sigma_2 + d_2\sigma_1)} - \frac{\varepsilon_2\sigma_1 V_0}{(d_1\sigma_2 + d_2\sigma_1)} = \frac{(\varepsilon_1\sigma_2 - \varepsilon_2\sigma_1) V_0}{(d_1\sigma_2 + d_2\sigma_1)}. \tag{6.433}$$

Note that $q_s(z=d_1) \neq 0$ only if $\varepsilon_1\sigma_2 - \varepsilon_2\sigma_1 \neq 0$. Specifically, there are electric charges at the interface only if the relaxation times of the two media are different. In other words, such an accumulation may occur due to imbalanced rates of decays of electric charges. Obviously, any accumulation leads to a discontinuous electric flux density.[130]

Finally, the overall structure discussed above can be represented as a series connection of two parallel resistor-capacitor (RC) combinations (Figure 6.64): i.e. $R_1 \parallel C_1 + R_2 \parallel C_2$. It is not a coincidence that

$$R_1C_1 = \frac{d_1}{A\sigma_1}\frac{\varepsilon_1 A}{d_1} = \frac{\varepsilon_1}{\sigma_1} = \tau_1, \qquad R_2C_2 = \frac{d_2}{A\sigma_2}\frac{\varepsilon_2 A}{d_2} = \frac{\varepsilon_2}{\sigma_2} = \tau_2, \tag{6.434}$$

where R_1C_1 and R_2C_2 are well-known to be time constants of the corresponding RC circuits.

As another basic example, we consider a spherical structure consisting of a PEC of radius a inside a PEC of radius b (Figure 6.65). The space between the PECs is filled with an electrically homogeneous lossy material with permittivity ε and conductivity σ. The capacitance of the same structure is found in Section 6.7.3 as $C_0 = 4\pi\varepsilon ab/(b-a)$. Since the medium between the PECs is electrically homogeneous, the resistance can easily be found via $R_0C_0 = \varepsilon/\sigma$ as

$$R_0 = \frac{\varepsilon}{\sigma C_0} = \frac{b-a}{4\pi\sigma ab}. \tag{6.435}$$

[128] Writing I_0 in terms of V_0, we first have

$$\bar{E}_1(\bar{r}) = -\hat{a}_z \frac{V_0}{(d_1/\sigma_1 + d_2/\sigma_2)\,\sigma_1}$$
$$\bar{E}_2(\bar{r}) = -\hat{a}_z \frac{V_0}{(d_1/\sigma_1 + d_2/\sigma_2)\,\sigma_2}.$$

[129] We have

$$q_s(z=0) = \hat{a}_z \cdot \bar{D}_1(z=0)$$
$$q_s(z=d_1+d_2) = -\hat{a}_z \cdot \bar{D}_2(z=d_1+d_2)$$
$$q_s(z=d_1) = -\hat{a}_z \cdot [\bar{D}_1(z=d_1) - \bar{D}_2(z=d_1)].$$

[130] Note these interesting limit cases: (i) if $\sigma_1 = 0$ (lossless dielectric), we have $\bar{E}_2(\bar{r}) = 0$ and $\bar{E}_1(\bar{r}) = -\hat{a}_z V_0/d_1$ – i.e. all voltage drop occurs across the lossless dielectric, while there is no electric field intensity in the lossy part since the overall resistance is infinity and there is no electric current flowing through the structure; (ii) if $\sigma_1 \to 0$ (PEC), we have $\bar{E}_1(\bar{r}) = 0$ and $\bar{E}_2(\bar{r}) = -\hat{a}_z V_0/d_2$ – i.e. all voltage drop occurs across the lossy dielectric, while there is no electric field intensity inside the PEC.

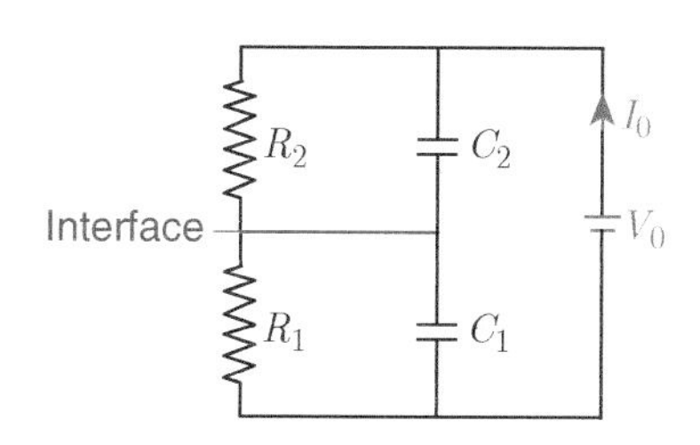

Figure 6.64 The structure in Figure 6.63 can be represented as a series connection of two parallel resistor-capacitor (RC) combinations.

Alternatively, one can start by assuming

$$\bar{J}_v(\bar{r}) = \hat{a}_R \frac{I_0}{4\pi R^2} \tag{6.436}$$

as the volume electric current density.[131] Then the electric field intensity can be obtained as

[131] Due to the symmetry, the electric current is distributed uniformly. Hence, the volume electric current density at R can be found by dividing the total current I_0 by the area of the corresponding spherical surface ($4\pi R^2$).

$$\bar{E}(\bar{r}) = \frac{J_v(\bar{r})}{\sigma} = \hat{a}_R \frac{I_0}{4\pi\sigma R^2}, \tag{6.437}$$

leading to

$$V_0 = -\int_b^a \bar{E}(\bar{r}) \cdot \hat{a}_R dR = \frac{I}{4\pi\sigma}\left[\frac{1}{R}\right]_b^a = \frac{I}{4\pi\sigma}\left(\frac{1}{a} - \frac{1}{b}\right). \tag{6.438}$$

Hence, we obtain the resistance as

$$R_0 = \frac{V_0}{I_0} = \frac{b-a}{4\pi\sigma ab}, \tag{6.439}$$

confirming the previous result. Finally, as the third way, we can follow the potential approach by solving Laplace's equation (see Section 6.7.3), leading to

$$\Phi(\bar{r}) = \frac{V_0 ab}{(b-a)R} - \frac{V_0 a}{(b-a)} = \frac{V_0 a}{(b-a)}\left(\frac{b}{R} - 1\right) \tag{6.440}$$

and

$$\bar{E}(\bar{r}) = -\bar{\nabla}\Phi(\bar{r}) = \hat{a}_R \frac{V_0 ab}{(b-a)R^2}. \tag{6.441}$$

Then the volume electric current density can be written

$$\bar{J}_v(\bar{r}) = \sigma\bar{E}(\bar{r}) = \hat{a}_R \sigma \frac{V_0 ab}{(b-a)R^2} \tag{6.442}$$

in terms of the voltage. The total current flowing from the inner PEC to the outer PEC can be found via integration as[132]

$$I_0 = \int_0^{2\pi}\int_0^{\pi} \sigma \frac{V_0 ab}{(b-a)} \sin\theta d\theta d\phi = 4\pi\sigma \frac{V_0 ab}{(b-a)}. \tag{6.443}$$

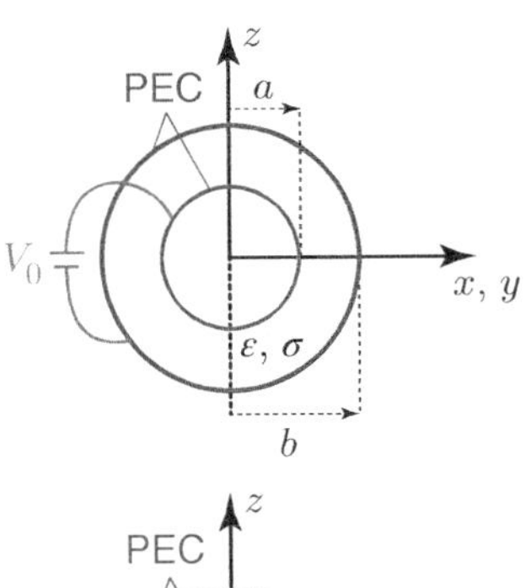

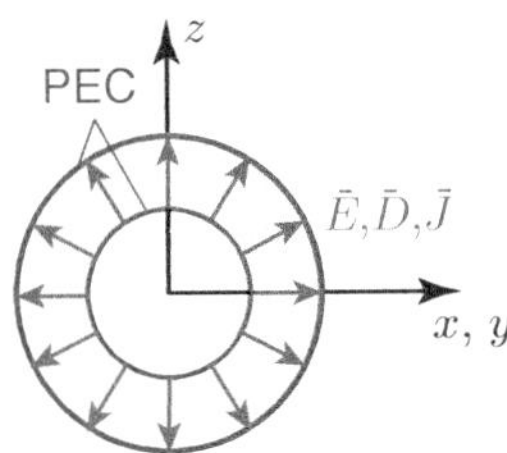

Figure 6.65 A spherical structure with a lossy material between two spherical PECs. When a voltage source is connected, in addition to the electric field intensity and electric flux density, a nonzero volume electric current distribution is created between the PECs.

[132] We have

$$I_0 = \int_0^{2\pi}\int_0^{\pi} \bar{J}_v(\bar{r}) \cdot \hat{a}_R R^2 \sin\theta d\theta d\phi = \int_0^{2\pi}\int_0^{\pi} \sigma \frac{V_0 ab}{(b-a)R^2} R^2 \sin\theta d\theta d\phi.$$

Hence, once again, we obtain the resistance as

$$R_0 = \frac{V_0}{I_0} = \frac{b-a}{4\pi\sigma ab}, \tag{6.444}$$

as expected.

For some resistance problems, the potential approach may be difficult to use because of volume electric charge distributions accumulated due to material inhomogeneities. As an example, we consider again a spherical structure consisting of a spherical PEC of radius a inside a spherical PEC of radius b (Figure 6.66). The space between the PEC is filled with a lossy material with a conductivity depending on the position as

$$\sigma(\bar{r}) = \sigma_0\left(1 + c\frac{a}{R}\right), \tag{6.445}$$

where c is a constant. We assume that the permittivity is ε_0: i.e. it is the same as a vacuum permittivity. As in the examples above, we attempt to find the resistance of the structure in a static case. To see the problem in the potential approach, one can consider the divergence of the electric flux density as[133]

$$\bar{\nabla}\cdot\bar{D}(\bar{r}) = \varepsilon_0\bar{\nabla}\left[1/\sigma(\bar{r})\right]\cdot\bar{J}_v(\bar{r}). \tag{6.446}$$

In the above, we have a nonzero $\bar{\nabla}\cdot\bar{D}(\bar{r})$ that indicates the existence of a volume electric charge distribution:

$$q_v(\bar{r}) = \varepsilon_0\bar{\nabla}\left[1/\sigma(\bar{r})\right]\cdot\bar{J}_v(\bar{r}). \tag{6.447}$$

This occurs particularly when the change in $\varepsilon(\bar{r})/\sigma(\bar{r})$ is not perpendicular to the direction of the electric current.[134] Based on this, the Laplace of the electric scalar potential can be derived as

$$\bar{\nabla}\cdot\bar{D}(\bar{r}) = \varepsilon_0\bar{\nabla}\cdot\bar{E}(\bar{r}) = -\varepsilon_0\bar{\nabla}^2\Phi(\bar{r}) = \varepsilon_0\bar{\nabla}\left[1/\sigma(\bar{r})\right]\cdot\bar{J}_v(\bar{r}), \tag{6.448}$$

leading to[135]

$$\bar{\nabla}^2\Phi(\bar{r}) = -\bar{\nabla}\left[1/\sigma(\bar{r})\right]\cdot\bar{J}_v(\bar{r}). \tag{6.449}$$

This equation may not be trivial to solve in comparison to the following approach.

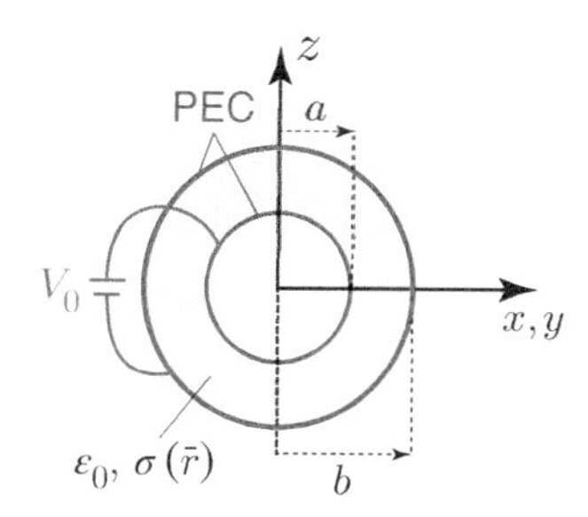

Figure 6.66 A spherical structure with a lossy material between two spherical PECs. The permittivity of the material is ε_0, while its conductivity changes with respect to position.

[133] We have

$$\bar{\nabla}\cdot\bar{D}(\bar{r}) = \bar{\nabla}\cdot[\varepsilon(\bar{r})\bar{E}(\bar{r})] = \bar{\nabla}\cdot\left[\frac{\varepsilon(\bar{r})}{\sigma(\bar{r})}\bar{J}_v(\bar{r})\right]$$
$$= \bar{\nabla}\left[\frac{\varepsilon_0}{\sigma(\bar{r})}\right]\cdot\bar{J}_v(\bar{r}) + \frac{\varepsilon_0}{\sigma(\bar{r})}\bar{\nabla}\cdot\bar{J}_v(\bar{r})$$
$$= \bar{\nabla}\left[\frac{\varepsilon_0}{\sigma(\bar{r})}\right]\cdot\bar{J}_v(\bar{r}),$$

where $\bar{\nabla}\cdot\bar{J}_v(\bar{r})$ is set to zero given steady currents.

[134] In this case, since the permittivity is constant ε_0, the nonzero volume electric charge density is due to the change in $1/\sigma(\bar{r})$ or $\sigma(\bar{r})$ that is not perpendicular to $\bar{J}_v(\bar{r})$.

[135] Hence, it is a Poisson's equation: $\nabla^2\Phi(\bar{r}) \neq 0$.

It is not uncommon that the current method works well to find the resistance of a given structure when the potential approach is relatively difficult to use. In fact, due to the symmetry of the spherical structure considered above, we can directly write the electric current density as

$$\bar{J}_v(\bar{r}) = \hat{a}_R \frac{I_0}{4\pi R^2}, \tag{6.450}$$

assuming I_0 is the total current flowing from the inner PEC to the outer PEC. Then the electric field intensity can be written directly as

$$\bar{E}(\bar{r}) = \hat{a}_R \frac{I_0}{\sigma(\bar{r})4\pi R^2} = \hat{a}_R \frac{I_0}{\sigma_0 (1 + ca/R) 4\pi R^2} = \hat{a}_R \frac{I_0}{4\pi\sigma_0} \frac{1}{R(ca+R)}. \tag{6.451}$$

The potential difference between the PECs can be found via integration as[136]

$$V_0 = -\frac{I_0}{4\pi\sigma_0} \int_b^a \frac{\hat{a}_R}{R(ca+R)} \cdot \hat{a}_R dR = \frac{I_0}{4\pi\sigma_0 ca} \ln\left(\frac{b+cb}{b+ca}\right). \tag{6.452}$$

Finally, we obtain the resistance as[137]

$$R_0 = \frac{V_0}{I_0} = \frac{1}{4\pi\sigma_0 ca} \ln\left(\frac{b+cb}{b+ca}\right). \tag{6.453}$$

As a further investigation, considering ε_0 permittivity of the inhomogeneous lossy material of the spherical structure, we can write the electric flux density as

$$\bar{D}(\bar{r}) = \varepsilon_0 \bar{E}(\bar{r}) = \hat{a}_R \frac{\varepsilon_0 I_0}{4\pi\sigma_0} \frac{1}{R(ca+R)} \tag{6.454}$$

leading to[138]

$$q_v(\bar{r}) = \bar{\nabla} \cdot \bar{D}(\bar{r}) = \frac{1}{R^2}\frac{\partial}{\partial R}\left(R^2 D_R(R)\right) = \frac{1}{R^2}\frac{\partial}{\partial R}\left(\frac{\varepsilon_0 I_0}{4\pi\sigma_0}\frac{R^2}{R(ca+R)}\right) \tag{6.455}$$

$$= \frac{\varepsilon_0 I_0}{4\pi\sigma_0 R^2} \frac{ca}{(ca+R)^2} \quad (a < R < b). \tag{6.456}$$

This already verifies the nonzero charge accumulation due to the varying $\sigma(\bar{r})$: i.e. $q_v(\bar{r}) \neq 0$ for nonzero c. On the other hand, we can further check Eq. (6.448) by considering Eq. (6.456) as

$$\bar{\nabla}\left[1/\sigma(\bar{r})\right] = -\frac{\bar{\nabla}\sigma(\bar{r})}{[\sigma(\bar{r})]^2} = -\hat{a}_R \frac{1}{\sigma_0^2(1+ca/R)^2}\frac{\partial}{\partial R}\left[\sigma_0\left(1+\frac{ca}{R}\right)\right] \tag{6.457}$$

$$= \hat{a}_R \frac{1}{\sigma_0^2(1+ca/R)^2}\frac{\sigma_0 ca}{R^2} = \hat{a}_R \frac{ca}{\sigma_0(ca+R)^2} \tag{6.458}$$

[136] We use

$$\frac{1}{R(ca+R)} = \frac{1/ca}{R} - \frac{1/ca}{(ca+R)}.$$

[137] Note that when $c = 0$, we have

$$R_0(c=0) = \frac{b-a}{4\pi\sigma ab},$$

which is consistent with Eq. (6.444), as the resistance of the same geometry with a homogeneous material between the PECs.

[138] Note that the total electric charge inside the material (volume distribution) can be found:

$$Q_v = 4\pi \int_a^b \frac{\varepsilon_0 I_0}{4\pi\sigma_0 R^2} \frac{ca}{(ca+R)^2} R^2 dR$$
$$= \frac{\varepsilon_0 I_0}{\sigma_0}\left[-\frac{ca}{(ca+R)}\right]_a^b$$
$$= \frac{\varepsilon_0 I_0}{\sigma_0}\left[\frac{c}{(c+1)} - \frac{ca}{(ca+b)}\right].$$

In addition, the electric charges on the inner and outer spheres (surface distributions) can be found:

$$Q_s^{\text{inner}} = \hat{a}_R \cdot \bar{D}(R=a) 4\pi a^2 = \frac{\varepsilon_0 I_0}{\sigma_0} \frac{1}{(c+1)}$$
$$Q_s^{\text{outer}} = -\hat{a}_R \cdot \bar{D}(R=b) 4\pi b^2 = -\frac{\varepsilon_0 I_0}{\sigma_0} \frac{b}{(ca+b)}.$$

One can further show the charge neutrality as $Q_v + Q_s^{\text{inner}} + Q_s^{\text{outer}} = 0$.

and

$$\varepsilon_0\bar{\nabla}\left[1/\sigma(\bar{r})\right]\cdot\bar{J}_v(\bar{r}) = \varepsilon_0\hat{a}_R\frac{ca}{\sigma_0(ca+R)^2}\cdot\hat{a}_R\frac{I_0}{4\pi R^2} = \frac{\varepsilon_0 I_0 ca}{4\pi\sigma_0 R^2(ca+R)^2}$$
$$= q_v(\bar{r}). \tag{6.459}$$

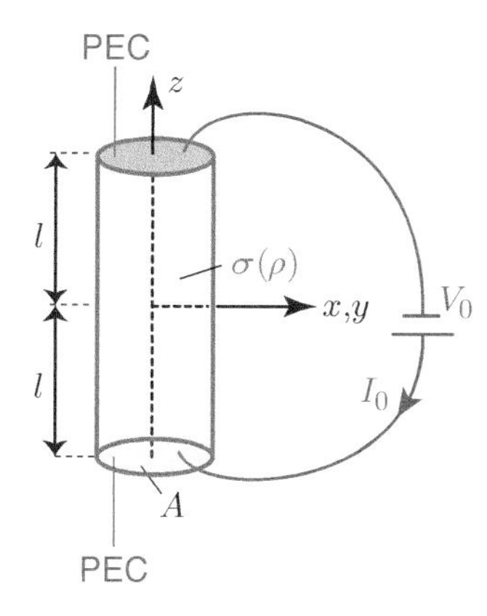

Figure 6.67 A cylindrical structure with an inhomogeneous material between the plates at $z=-l$ and $z=l$.

6.8.1 Examples

Example 77: Consider a cylindrical lossy object aligned along the z axis from $z=-l$ to $z=l$ (Figure 6.67). The radius of the cylinder is a, while the conductivity is given as $\sigma=\sigma_0 l/\rho$. Find the resistance of the structure between its bottom and top surfaces.

Solution: Given that the inhomogeneity of the conductivity is in the ρ direction, this problem can be solved via the potential approach.[139] We have[140]

$$\bar{\nabla}^2\Phi = 0 \longrightarrow \Phi(\bar{r}) = \frac{V_0}{2}\left(1-\frac{z}{l}\right), \tag{6.460}$$

and

$$\bar{E}(\bar{r}) = -\bar{\nabla}\Phi(\bar{r}) = \hat{a}_z V_0/(2l) \tag{6.461}$$
$$\bar{J}_v(\bar{r}) = \sigma(\bar{r})\bar{E}(\bar{r}) = \hat{a}_z\sigma_0(l/\rho)[V_0/(2l)] = \hat{a}_z\sigma_0 V_0/(2\rho). \tag{6.462}$$

Then the total current flowing through the object can be found as

$$I_0 = 2\pi\int_0^a \frac{\sigma_0 V_0}{2\rho}\rho d\rho = \pi\sigma_0 V_0 a. \tag{6.463}$$

Finally, we obtain the resistance as

$$R_0 = \frac{V_0}{I_0} = \frac{1}{\pi\sigma_0 a}. \tag{6.464}$$

[139] If the current approach is used, it is not obvious how the current density is distributed on the cross section. In addition, due to the direction of the inhomogeneity, Laplace's equation holds, which makes the potential approach more suitable.

[140] We assume that $\Phi(z=-l)=V_0$ and $\Phi(z=l)=0$.

Example 78: A resistor is defined as a cylindrical layer with inner radius a and outer radius b (Figure 6.68). The height of the resistor is l, while an inhomogeneous material exists only in the region $a<\rho<b$. The conductivity of the material is defined with respect to z as $\sigma(\bar{r}) = \sigma_0(1+z/l)$, when the bottom of the resistor is at $z=0$. The permittivity of the material is constant ε_0. Find the resistance between the bottom and top surfaces, as well as the dissipated power flux density $\pi_J(\bar{r}) = \bar{J}_v(\bar{r})\cdot\bar{E}(\bar{r})$ and the total dissipated power by the resistor in terms of an applied static voltage V_0.

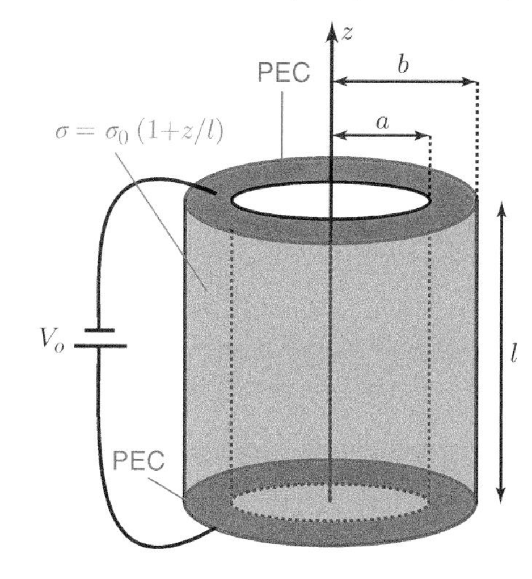

Figure 6.68 A cylindrical layer with an inhomogeneous material.

Solution: Since the conductivity changes in the z direction, the electric current density and electric field intensity can be written[141]

[141] Since the conductivity does not change in the ρ and ϕ directions, we can assume that the current is distributed uniformly on the cross section of the structure: i.e. on the ring of area $\pi(b^2 - a^2)$.

$$\bar{J}_v(\bar{r}) = \hat{a}_z \frac{I_0}{\pi(b^2 - a^2)} \tag{6.465}$$

$$\bar{E}(\bar{r}) = \frac{\bar{J}_v(\bar{r})}{\sigma(\bar{r})} = \hat{a}_z \frac{I_0}{\pi(b^2 - a^2)} \frac{1}{\sigma_0(1 + z/l)} = \hat{a}_z \frac{I_0 l}{\pi(b^2 - a^2)\sigma_0(l + z)}. \tag{6.466}$$

Then the voltage across the structure can be found as

$$V_0 = \int_0^l \hat{a}_z \frac{I_0 l}{\pi(b^2 - a^2)\sigma_0(l + z)} \cdot \hat{a}_z dz = \frac{I_0 l}{\sigma_0 \pi(b^2 - a^2)} \ln(2), \tag{6.467}$$

leading to

$$R_0 = \frac{V_0}{I_0} = \frac{l \ln(2)}{\sigma_0 \pi(b^2 - a^2)}. \tag{6.468}$$

Next, the power flux density can be evaluated as[142]

[142] This expression can be found as

$$\begin{aligned}\pi_J(\bar{r}) &= \hat{a}_z \frac{I_0}{\pi(b^2 - a^2)} \cdot \hat{a}_z \frac{I_0 l}{\sigma_0 \pi(b^2 - a^2)(l + z)} \\ &= \frac{I_0^2 l}{\sigma_0 \pi^2 (b^2 - a^2)^2 (l + z)} \\ &= \left[\frac{I_0^2 l^2 (\ln(2))^2}{\sigma_0^2 \pi^2 (b^2 - a^2)^2}\right] \frac{\sigma}{l(l + z)(\ln(2))^2}.\end{aligned}$$

$$\pi_J(\bar{r}) = \bar{J}_v(\bar{r}) \cdot \bar{E}(\bar{r}) = \frac{V_0^2 \sigma_0}{l(l + z)\ln(4)}. \tag{6.469}$$

Then the dissipated power can be found by integrating this expression in the volume of the structure:

$$p_J = \int_V \bar{J}_v(\bar{r}) \cdot \bar{E}(\bar{r}) dv = \frac{V_0^2 \sigma_0 \pi(b^2 - a^2)}{l \ln(2)}. \tag{6.470}$$

Alternatively, since the resistance is already available, one can simply use $p_J = V_0^2/R_0$.

Example 79: Consider a cylindrical lossy object aligned along the z axis from $z = -l$ to $z = l$ (Figure 6.69). The radius of the cylinder is a. The conductivity of the material depends on the position as

$$\sigma = \sigma_0 \frac{6\rho/a}{1 + 2z/l}. \tag{6.471}$$

Find the resistance of the structure between its bottom and top surfaces.

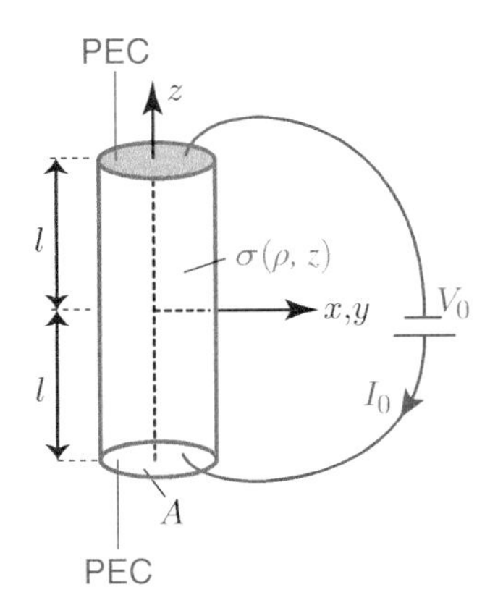

Figure 6.69 A cylindrical structure with an inhomogeneous material between the plates at $z = -l$ and $z = l$.

Solution: For this problem, due to the varying conductivity in the z direction, one cannot use the potential approach since Laplace's equation does not hold for the electric scalar potential. In addition, the distribution of $\bar{J}_v(\bar{r})$ is not obvious as the conductivity also depends on ρ. Using $R_0C_0 = \varepsilon/\sigma$ is also not an option because the lossy material is inhomogeneous. On the other hand, we can still find the resistance of the structure by considering that it consists of smaller pieces (Figure 6.70). First, we can assume that the cylindrical object consists of infinitesimal differential tubes of length $2l$. In addition, each differential tube consists of differential cells. Then we have

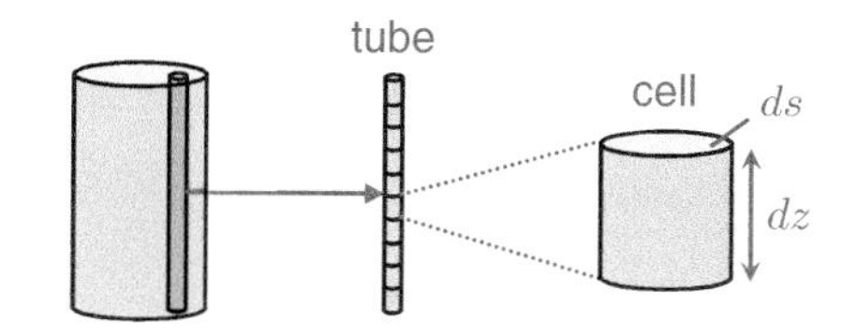

Figure 6.70 A cylindrical structure can be assumed to be formed of infinitely many tubes, each of which can be divided into infinitesimal cells.

$$R_{\text{cell}} = \frac{dl}{\sigma ds} = \frac{a(1+2z/l)dz}{6\rho\sigma_0 ds} \tag{6.472}$$

as the resistance of a cell[143] located at $\bar{r} = (\rho, \phi, z)$ for $\phi \in [0, 2\pi)$. Then the resistance of a tube located at ρ distance from the z axis can be found via integration as

[143]A cell is so small that the derivation in Section 6.1 holds: i.e. its resistance is directly proportional to its length, while it is inversely proportional to the area of its cross section and the conductivity. In addition, for the conductivity, we use Eq. (6.471) directly, since the conductivity of a cell can be assumed to be constant.

$$R_{\text{tube}} = \int_{-l}^{l} \frac{a(1+2z/l)dz}{6\rho\sigma_0 ds} = \frac{a}{6\rho\sigma_0 ds}\left[z + \frac{z^2}{l}\right]_{-l}^{l} = \frac{al}{3\rho\sigma_0 ds}, \tag{6.473}$$

since each tube is a collection of infinitely many cells that are connected in *series*. This also means the *conductance* of a tube can be obtained:

G_0: conductance
Unit: siemens (S) (amperes/volt or 1/ohm)

$$G_{\text{tube}} = \frac{1}{R_{\text{tube}}} = \frac{3\rho\sigma_0 ds}{al}. \tag{6.474}$$

Hence, the conductance of the overall structure is simply an integration:[144]

$$G_0 = \int_0^a \int_0^{2\pi} \frac{3\rho\sigma_0}{al}\rho d\rho d\phi = \frac{2\pi\sigma_0}{al}\left[\rho^3\right]_0^a = \frac{2\pi\sigma_0 a^2}{l}. \tag{6.475}$$

[144]When objects are connected in parallel, the overall conductance is the sum of the conductances of the objects. For two objects, this means

$$G_0 = G_1 + G_2,$$

where $G_0 = 1/R_0$, $G_1 = 1/R_1$, and $G_2 = 1/R_2$.

This leads to the resistance of the overall structure as[145]

$$R_0 = \frac{1}{G_0} = \frac{l}{2\pi\sigma_0 a^2}. \tag{6.476}$$

[145]As an alternative way to analyze this structure, one may initially guess that the volume electric current density can be written $\bar{J}_v(r) = J_0\rho/a$, where J_0 is a constant (A/m^2). This can be verified (but not proven) by checking Faraday's law: i.e. $\bar{\nabla}\times\bar{E}(\bar{r}) = \bar{\nabla}\times[\sigma(\bar{r})\bar{J}_v(\bar{r})] = 0$. Consequently, after writing the electric field intensity as $\bar{E}(\bar{r}) = \bar{J}_v(\bar{r})/\sigma(\bar{r})$, one can obtain V_0 in terms of J_0, leading to the resistance value.

Example 80: A structure of length $4l$ is connected to a static voltage source V_0 (Figure 6.71). The structure involves three different cylindrical parts (with a fixed cross-sectional area of A) connected in series. Conductivity values are given as

$$\sigma(\bar{r}) = \sigma_0 \begin{cases} 1 & (-2l < z < -l) \\ l/|z| & (-l < z < l) \\ 1 & (l < z < 2l), \end{cases} \tag{6.477}$$

where σ_0 is a constant. The permittivity is ε_0 everywhere.

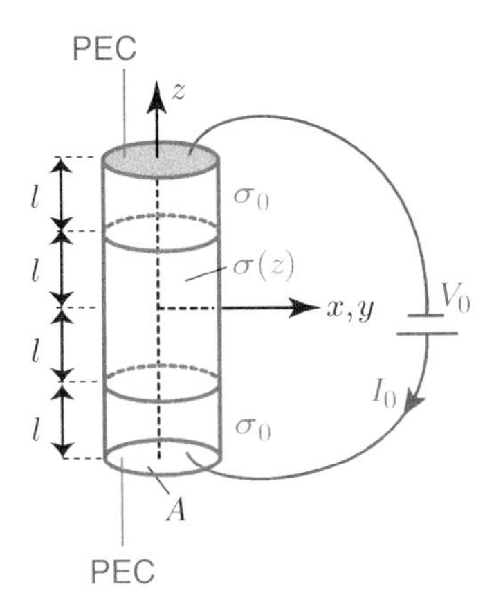

Figure 6.71 A structure of length $4l$, which involves three different cylindrical parts connected to a static voltage source V_0.

- Find the resistance of the structure between its bottom and top surfaces.
- Find the volume electric charge density depending on the position inside the structure.
- Find the electric potential energy stored by the structure.
- Find the total dissipated power.

Solution: Assuming that the total current I_0 is distributed uniformly on the cross section all along the structure, we have

$$\bar{J}_v(\bar{r}) = \hat{a}_z \frac{I_0}{A}, \qquad \bar{E}(\bar{r}) = \hat{a}_z \frac{I_0}{\sigma_0 A} \begin{cases} 1 & (-2l < z < -l) \\ |z|/l & (-l < z < l) \\ 1 & (l < z < 2l). \end{cases} \tag{6.478}$$

Then the voltage between the PECs can be obtained as[146]

$$V_0 = -\int_{2l}^{-2l} \bar{E}(\bar{r}) \cdot \hat{a}_z dz = \int_{-2l}^{2l} \bar{E}(\bar{r}) \cdot \hat{a}_z dz = \frac{3I_0 l}{\sigma_0 A}. \tag{6.479}$$

[146]We have

$$V_0 = \int_{-2l}^{-l} \frac{I_0}{\sigma_0 A} dz + \int_{-l}^{0} -\frac{I_0 z}{\sigma_0 A l} dz + \int_{0}^{l} \frac{I_0 z}{\sigma_0 A l} dz + \int_{l}^{2l} \frac{I_0}{\sigma_0 A} dz = \frac{I_0 l}{\sigma_0 A} + \frac{I_0 l}{2\sigma_0 A} + \frac{I_0 l}{2\sigma_0 A} + \frac{I_0 l}{\sigma_0 A}.$$

Consequently, we obtain the resistance as

$$R_0 = \frac{V_0}{I_0} = \frac{3l}{\sigma_0 A}. \tag{6.480}$$

Given that the permittivity is constant ε_0, one can find the volume electric charge density as[147]

$$q_v(\bar{r}) = \frac{1}{\varepsilon_0} \bar{\nabla} \cdot \bar{E}(\bar{r}) = \frac{I_0}{\sigma_0 \varepsilon_0 A l} \begin{cases} -1 & (-l < z < 0) \\ 1 & (0 < z < l) \\ 0 & \text{otherwise.} \end{cases} \tag{6.481}$$

[147]Note that $Q_{in} = 0$: i.e. the total electric charge inside the structure is zero.

For the electric potential energy, we can use the expression for the electric field intensity as[148,149]

$$w_{e,\text{stored}} = \frac{\varepsilon_0}{2}\int_{\infty}|\bar{E}(\bar{r})|^2 dv = \frac{\varepsilon_0 I_0^2}{2\sigma_0^2 A^2}Al + \frac{\varepsilon_0 I_0^2}{2\sigma_0^2 A^2 l^2}A\int_{-l}^{l} z^2 dz + \frac{\varepsilon_0 I_0^2}{2\sigma_0^2 A^2}Al \tag{6.482}$$

$$= \frac{\varepsilon_0 I_0^2 l}{\sigma_0^2 A} + \frac{\varepsilon_0 I_0^2}{\sigma_0^2 A l^2}\frac{l^3}{3} = \frac{4\varepsilon_0 l}{3\sigma_0^2 A}I_0^2. \tag{6.483}$$

Finally, we can find the total dissipated power, as we already know the electric field intensity and/or the volume electric current density. We have

$$p_J = \int_V \sigma(\bar{r})|\bar{E}(\bar{r})|^2 dv = \frac{I_0^2}{\sigma_0 A^2}2Al + \frac{I_0^2}{\sigma_0 A^2}2A\int_0^l \frac{z^2}{l^2}\frac{l}{z}dz \tag{6.484}$$

$$= \frac{I_0^2}{\sigma_0 A^2}2Al + \frac{I_0^2}{\sigma_0 A^2}Al = \frac{3l}{\sigma_0 A}I_0^2. \tag{6.485}$$

This simply corresponds to $R_0 I_0^2$.

[148] Since there is a steady surface electric current density, we have $\bar{\nabla}\cdot\bar{J}(\bar{r}) = 0$, while an electric charge accumulation occurs due to the inhomogeneity. In addition, the electric charge distribution is static, which can be verified by the continuity equation in Eq. (2.202) as $\bar{\nabla}\cdot\bar{J}(\bar{r},t) = 0 = -\partial q_v(\bar{r},t)/\partial t$. Physically, the accumulation occurs during the transient state: i.e. once the voltage source is connected, electric charge starts to accumulate until the volume electric current density saturates to its final value of $\hat{a}_z I_0/A$ (ideally in infinite time).

[149] Note that if there were no inhomogeneity – i.e. if the conductivity was σ_0 everywhere – the electric potential energy would be

$$w_{e,\text{homogeneous}} = \frac{\varepsilon_0 I_0^2}{2\sigma_0^2 A^2}(4Al) = \frac{2\varepsilon_0 l}{\sigma_0^2 A}I_0^2.$$

This is the same as

$$w_{e,\text{homogeneous}} = \frac{1}{2}C_0 V_0^2 = \frac{1}{2}\frac{\varepsilon_0 A}{4l}V_0^2$$
$$= \frac{1}{2}\frac{\varepsilon_0 A}{4l}R_{0,\text{homogeneous}}^2 I_0^2$$
$$= \frac{1}{2}\frac{\varepsilon_0 A}{4l}\frac{16l^2}{\sigma_0^2 A^2}I_0^2 = \frac{2\varepsilon_0 l}{\sigma_0^2 A}I_0^2,$$

since there is no charge accumulation inside. Note that, for this structure, the inhomogeneity reduces the electric potential energy.

Example 81: The electric scalar potential inside an $a \times a \times a$ square structure (Figure 6.72) is given as

$$\Phi(\bar{r}) = V_0 - V_0\frac{z^2}{a^2}, \tag{6.486}$$

where V_0 (V) is a constant. In addition, the resistance between the top and bottom PECs is defined as R_0 (Ω), while the volume electric charge density inside the structure is q_0 (C/m^3). Find the permittivity (constant ε) and conductivity [z-dependent $\sigma(\bar{r})$: i.e. $\sigma(z)$] of the lossy material (in terms of the given quantities) inside the structure.

Solution: Knowing the distribution of the electric scalar potential, we can find the electric field intensity, as well as the volume electric current density, as

$$\bar{E}(\bar{r}) = -\bar{\nabla}\Phi(\bar{r}) = \hat{a}_z V_0\frac{2z}{a^2}, \qquad \bar{J}_v(\bar{r}) = \sigma(\bar{r})\bar{E}(\bar{r}) = \hat{a}_z\sigma(z)V_0\frac{2z}{a^2}. \tag{6.487}$$

Then the total electric current passing through the structure can be written

$$I_0 = a^2\hat{a}_z\cdot\hat{a}_z\sigma(z)V_0\frac{2z}{a^2} = \sigma(z)V_0 2z \tag{6.488}$$

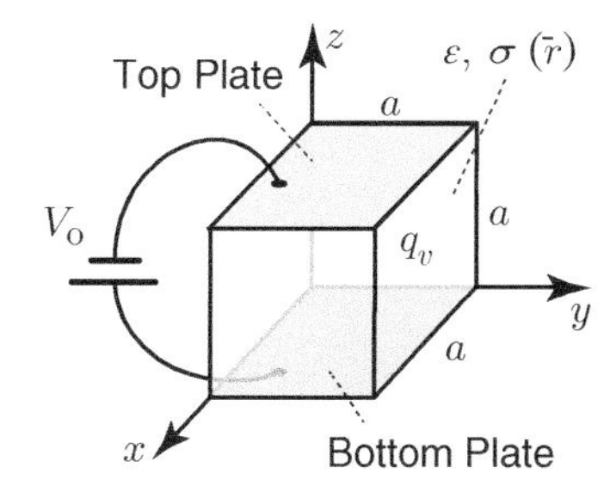

Figure 6.72 A cubic structure consisting of an inhomogeneous material between perfectly conducting plates. A volume electric charge distribution is created inside the structure when a static voltage source V_0 is connected between the PECs.

leading to

$$R_0 = \frac{V_0}{I_0} = \frac{1}{\sigma(z)2z} \longrightarrow \sigma(z) = \frac{1}{2R_0 z}. \tag{6.489}$$

For the permittivity, we can find expressions for the electric flux density and its divergence as

$$\bar{D}(\bar{r}) = \varepsilon \bar{E}(\bar{r}) = \hat{a}_z \varepsilon V_0 \frac{2z}{a^2}, \qquad \bar{\nabla} \cdot \bar{D}(\bar{r}) = \varepsilon V_0 \frac{2}{a^2}. \tag{6.490}$$

The final expression must be equal to the volume electric charge density:

$$\varepsilon V_0 \frac{2}{a^2} = q_0 \longrightarrow \varepsilon = \frac{q_0 a^2}{2V_0}. \tag{6.491}$$

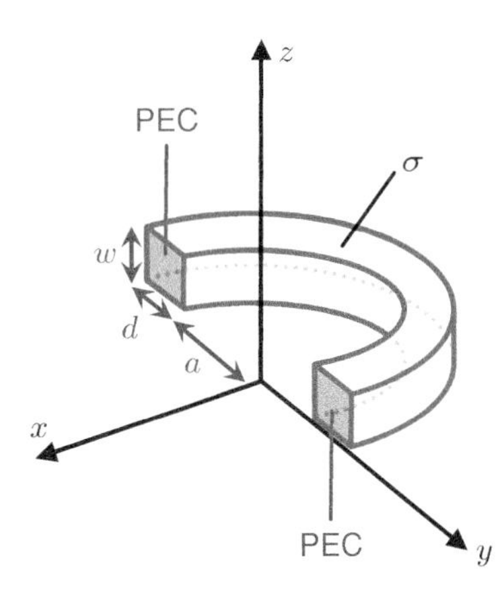

Figure 6.73 An arc-shaped object lying from $\phi = \pi/2$ to $\phi = 3\pi/2$.

Example 82: Consider an arc-shaped object with a $d \times w$ rectangular cross section lying from $\phi = \pi/2$ to $\phi = 3\pi/2$ (Figure 6.73). The inner radius is a, while the conductivity of the object is σ. Find the resistance between the rectangular surfaces.

Solution: For this problem, the distribution of the volume electric current density $\bar{J}_v(\bar{r})$ is not obvious (may not be uniform), and one needs to start with the potential. Assuming V_0 and 0 electric scalar potential values when $\phi = \pi/2$ and $\phi = 3\pi/2$, respectively, we can solve Laplace's equation in a cylindrical coordinate system as[150,151]

$$\bar{\nabla}^2 \Phi(\bar{r}) = 0 \longrightarrow \frac{1}{\rho}\frac{\partial}{\partial \rho}\left(\rho \frac{\partial \Phi(\bar{r})}{\partial \rho}\right) + \frac{1}{\rho^2}\frac{\partial^2 \Phi(\bar{r})}{\partial \phi^2} + \frac{\partial^2 \Phi(\bar{r})}{\partial z^2} = 0 \tag{6.492}$$

$$\longrightarrow \frac{1}{\rho^2}\frac{\partial^2 \Phi(\bar{r})}{\partial \phi^2} = 0 \longrightarrow \Phi(\bar{r}) = c_1\phi + c_2 = -\frac{V_0}{\pi}\phi + \frac{3V_0}{2}. \tag{6.493}$$

Then the electric field intensity can be obtained as

$$\bar{E}(\bar{r}) = -\bar{\nabla}\Phi(\bar{r}) = -\hat{a}_\phi \frac{1}{\rho}\frac{\partial \Phi(\bar{r})}{\partial \phi} = \hat{a}_\phi \frac{V_0}{\pi \rho}, \tag{6.494}$$

leading to

$$\bar{J}_v(\bar{r}) = \sigma(\bar{r})\bar{E}(\bar{r}) = \hat{a}_\phi \frac{\sigma V_0}{\pi \rho}. \tag{6.495}$$

[150] We assume that the electric scalar potential changes only in the angular direction.

[151] After finding the form of the electric scalar potential as $\Phi(\phi) = c_1\phi + c_2$, we have

$$\Phi(3\pi/2) = 0 \longrightarrow c_2 = -c_1(3\pi/2)$$
$$\Phi(\pi/2) = c_1(\pi/2) - c_1(3\pi/2) = -c_1\pi = V_0$$
$$\longrightarrow c_1 = -V_0/\pi$$
$$\longrightarrow c_2 = 3V_0/2.$$

The total current can be found via integration as

$$I_0 = \int_0^w \int_a^{a+d} \hat{a}_\phi \frac{\sigma V_0}{\pi \rho} \cdot \hat{a}_\phi d\rho dz = \frac{\sigma V_0}{\pi} w \ln\left(\frac{a+d}{a}\right). \quad (6.496)$$

Finally, we arrive at the expression for the resistance:[152]

$$R_0 = \frac{V_0}{I_0} = \frac{\pi}{\sigma w \ln(1+d/a)}. \quad (6.497)$$

[152]Note that if $d \ll a$, we have

$$\ln(1+d/a) \approx d/a \longrightarrow R_0 \approx \frac{\pi a}{\sigma w d},$$

i.e. the standard simple expression involving length (πa) divided by conductivity (σ) and cross-sectional area (wd).

Example 83: Consider a cubic parallel-plate structure (Figure 6.74) consisting of two $a \times a$ PECs separated by a distance a. The material between the PECs has permittivity and conductivity

$$\varepsilon(\bar{r}) = \varepsilon_0 \exp(y/a) \quad (6.498)$$

$$\sigma(\bar{r}) = \sigma_0 \exp(z/a), \quad (6.499)$$

where ε_0 (F/m) and σ_0 (S/m) are constants. Find the resistance of the structure. In addition, assuming that V_0 static voltage is applied between the PECs, find the following:

- The surface electric charge density on the PECs
- The volume electric charge density inside the structure
- The density of the electric potential energy stored inside the structure
- The dissipated power flux density inside the structure

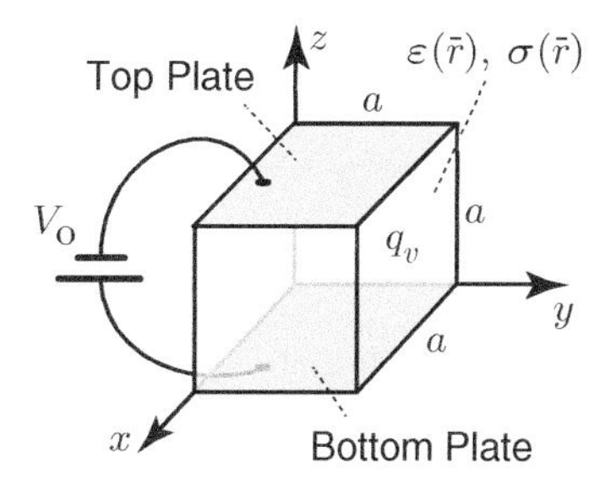

Figure 6.74 A cubic structure consisting of an inhomogeneous material between perfectly conducting plates. A volume electric charge distribution is created inside the structure when a static voltage source V_0 is connected between the PECs.

Solution: Since Laplace's equation does not hold, this problem is difficult to solve via the potential approach. On the other hand, given that the conductivity changes in the z direction, and not in the x and y directions, we have

$$\bar{J}(\bar{r}) = \hat{a}_z \frac{I_0}{a^2}, \qquad \bar{E}(\bar{r}) = \frac{\bar{J}(\bar{r})}{\sigma(\bar{r})} = \hat{a}_z \frac{I_0 \exp(-z/a)}{\sigma_0 a^2}, \quad (6.500)$$

assuming I_0 steady current flowing through the structure. This leads to[153]

[153]We derive

$$V_0 = \frac{I_0}{\sigma_0 a^2} \int_0^a \exp(-z/a) dz$$
$$= \frac{I_0}{\sigma_0 a^2} \left[(-a)\exp(-z/a)\right]_0^a$$
$$= \frac{I_0}{\sigma_0 a^2} [a - a\exp(-1)].$$

$$V_0 = -\int_a^0 \hat{a}_z \frac{I_0 \exp(-z/a)}{\sigma_0 a^2} \cdot \hat{a}_z dz = \frac{I_0}{\sigma_0 a}(1 - 1/e). \quad (6.501)$$

Then the resistance can be written

$$R_0 = \frac{V_0}{I_0} = \frac{1}{\sigma_0 a}(1 - 1/e). \quad (6.502)$$

Next, we can find the electric flux density as[154]

$$\bar{D}(\bar{r}) = \varepsilon(\bar{r})\bar{E}(\bar{r}) = \hat{a}_z \frac{\varepsilon_0 V_0 \exp(y/a - z/a)}{a(1 - 1/e)}. \quad (6.503)$$

At the top/bottom plates (when $z = 0$ and $z = a$), the related boundary conditions give

$$q_{s,\text{top}}(\bar{r}) = -\frac{\varepsilon_0 V_0 \exp(y/a - 1)}{a(1 - 1/e)}, \qquad q_{s,\text{bottom}}(\bar{r}) = \frac{\varepsilon_0 V_0 \exp(y/a)}{a(1 - 1/e)} \quad (6.504)$$

as the surface electric charge density expressions.[155] On the other side, the volume electric charge density can be obtained:

$$q_v(\bar{r}) = \bar{\nabla} \cdot \bar{D}(\bar{r}) = \frac{\varepsilon_0 V_0}{a(1 - 1/e)} \frac{\partial}{\partial z} \exp(y/a - z/a) \quad (6.505)$$

$$= -\frac{\varepsilon_0 V_0}{a^2(1 - 1/e)} \exp(y/a - z/a). \quad (6.506)$$

The density of the electric potential energy can be found as[156]

$$dw_{e,\text{stored}} = \frac{1}{2\varepsilon(\bar{r})}|\bar{D}(\bar{r})|^2 = \frac{\varepsilon_0 V_0^2 \exp(y/a - 2z/a)}{2a^2(1 - 1/e)^2}. \quad (6.507)$$

Similarly, the dissipated power flux density can be obtained as[157]

$$\pi_J(\bar{r}) = \bar{E}(\bar{r}) \cdot \bar{J}(\bar{r}) = \frac{\sigma_0 V_0^2 \exp(-z/a)}{a^2(1 - 1/e)^2}. \quad (6.508)$$

[154] As intermediate steps, we have

$$\bar{D}(\bar{r}) = \hat{a}_z \varepsilon_0 \exp(y/a) \frac{I_0 \exp(-z/a)}{\sigma_0 a^2} = \hat{a}_z \frac{\varepsilon_0 I_0 \exp(y/a - z/a)}{\sigma_0 a^2}.$$

[155] Note that for both the bottom and top plates, the surface electric charge density is not uniformly distributed.

[156] As an intermediate expression, we have

$$\frac{1}{2\varepsilon(\bar{r})}|\bar{D}(\bar{r})|^2 = \frac{1}{2\varepsilon_0 \exp(y/a)} \frac{\varepsilon_0^2 V_0^2 \exp(2y/a - 2z/a)}{a^2(1 - 1/e)^2}.$$

[157] As intermediate expressions, we have

$$\bar{E}(\bar{r}) \cdot \bar{J}(\bar{r}) = \hat{a}_z \frac{I_0 \exp(-z/a)}{\sigma_0 a^2} \cdot \hat{a}_z \frac{I_0}{a^2} = \frac{I_0^2 \exp(-z/a)}{\sigma_0 a^4}.$$

6.9 Inductance

Similar to capacitance and resistance, inductance is an important electromagnetic property of structures, particularly conductors. Theoretically, it measures the ability of an object to store *magnetic charge*. On the other hand, since magnetic charge does not exist, the inductance of an object is defined by considering the magnetic flux that can be generated for a given electric current. To make it independent of the applied current, we have

L_0: inductance
Unit: henry (H) (webers/ampere)

$$L_0 = \frac{f_0(t)}{I_0(t)}, \tag{6.509}$$

where $I_0(t)$ is the total current passing through the structure and

$$f_0(t) = \int_S \bar{B}(\bar{r}, t) \cdot \overline{ds} \tag{6.510}$$

is the generated magnetic flux. This magnetic flux may be confined in the structure (giving *internal* inductance) or exist outside (giving *external* inductance), depending on how the inductance and structure under investigation are defined.

Obviously, as opposed to capacitance, which can be defined in electrostatics, the definition of inductance needs electric current to flow. On the other hand, similar to capacitance, it can be defined for static (in this case, magnetostatic) or dynamic cases. In a dynamic case, we may get a voltage-current relationship by considering the time derivative of the electric current:

$$I_0(t) = \frac{1}{L_0} \int_S \bar{B}(\bar{r}, t) \cdot \overline{ds} \longrightarrow \frac{\partial I_0(t)}{\partial t} = \frac{\partial}{\partial t} \left(\frac{1}{L_0} \int_S \bar{B}(\bar{r}, t) \cdot \overline{ds} \right). \tag{6.511}$$

If the *inductance* itself is constant, we arrive at

$$\frac{\partial I_0(t)}{\partial t} = \frac{1}{L_0} \frac{\partial}{\partial t} \left(\int_S \bar{B}(\bar{r}, t) \cdot \overline{ds} \right) = -\frac{1}{L_0} \Phi_{\mathrm{emf}}(t), \tag{6.512}$$

where Faraday's law is used to rewrite the time derivative of the magnetic flux as the electromotive force. The electromotive force is generally measured as the negative voltage across the structure, leading to

$$V_0(t) = L_0 \frac{dI_0(t)}{dt} \tag{6.513}$$

as the standard voltage-current relationship related to inductance.[158] For a variable L_0 – e.g. when it is nonlinear – the inductance found by using the voltage-current relationship

[158] In the phasor domain – i.e. assuming steady-state with time-harmonic functions – voltage/current relationship can be written $V_0 = j\omega L_0 I_0$, where ω is the angular frequency. Hence, inductance is a positive imaginary impedance: i.e. $Z_0 = V_0/I_0 = j\omega L_0$ in circuit analysis.

in Eq. (6.513) and the inductance found by using the flux-current ratio in Eq. (6.509) may not be the same. On the other hand, similar to the capacitance extractions, we consider a simple definition (constant inductance) so that the inductance of a structure can be found by analyzing a magnetostatic case (steady electric currents).

Starting with a general case, we consider a loop C_1 with N_1 turns carrying a current I_1. The *self* inductance[159] due to the magnetic flux created by the loop through itself can be found as (Figure 6.75)

$$L_{11} = \frac{\Lambda_{11}}{I_1} = \frac{N_1 f_{B1}}{I_1} = \frac{N_1}{I_1} \int_{S_1} \bar{B}_1(\bar{r}) \cdot \overline{ds}, \tag{6.514}$$

where $\Lambda_{11} = N_1 f_{B1}$ is called the *flux linkage* and accounts for the multiplication of the total magnetic flux f_{B1} with the number of turns. If there is another loop C_2 with N_2 turns carrying a current I_2, the mutual inductance between the loops (due to the magnetic interaction between them) can be found as (Figure 6.76)

$$L_{12} = \frac{\Lambda_{12}}{I_1} = \frac{N_2}{I_1} \int_{S_2} \bar{B}_1(\bar{r}) \cdot \overline{ds} \quad \text{or} \quad L_{21} = \frac{\Lambda_{21}}{I_2} = \frac{N_1}{I_2} \int_{S_1} \bar{B}_2(\bar{r}) \cdot \overline{ds}, \tag{6.515}$$

where $L_{12} = L_{21}$. In the former, the magnetic flux density created by C_1 is integrated over S_2 (surface of the second loop), and it is divided by I_1 for normalization while being multiplied by N_2 to account for the overall flux linkage. In the latter, however, the magnetic flux density created by C_2 is integrated over S_1 (surface of the first loop), and it is divided by I_2 for normalization while being multiplied by N_1. If one needs to find the *self* inductance of the *overall* system, we need to add the self inductances of the loops and the mutual inductance between them:

$$L_0 = L_{11} + L_{22} + L_{12} + L_{21}. \tag{6.516}$$

Obviously, self and mutual inductance definitions depend on how we define the overall system and its parts.

In the discussion of the magnetic potential energy in Section 4.3.3, its relation to the inductance for a simple inductor of N turns is written

$$w_{m,\text{stored}} = \frac{1}{2} L_0 I_0^2 = \frac{1}{2} \frac{N^2 f_{B0}^2}{L_0}, \tag{6.517}$$

where f_{B0} is the amount of the magnetic flux and I_0 is the electric current. On the other side, considering two infinitesimal electric currents $I_1\overline{dl}_1$ and $I_2\overline{dl}_2$, the formal definition of the

[159] We avoid using the self inductance and internal inductance interchangeably. Similarly, mutual and external inductances are considered differently in this book:

- Self inductance: Inductance due to a structure's self magnetic interaction.
- Mutual inductance: Inductance due to mutual magnetic interaction between structures.
- Internal inductance: Inductance related to the interior of a structure, which can be part of self or mutual inductance.
- External inductance: Inductance related to the exterior of a structure, which can be part of self or mutual inductance.

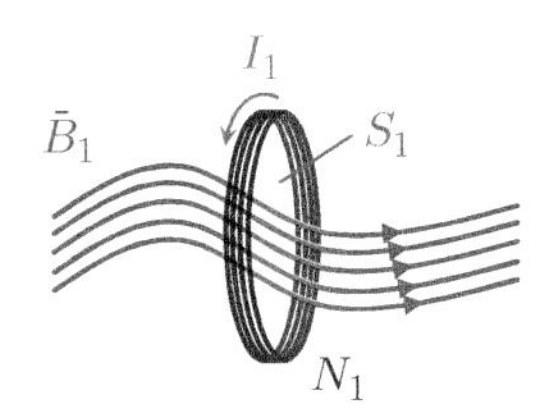

Figure 6.75 The self inductance of a loop C_1 with N_1 turns can be found by considering the magnetic flux created by the loop through itself. The magnetic flux is the integration of the magnetic flux density on the surface of the loop (S_1), and it must be multiplied by N_1 to find the flux linkage and the inductance.

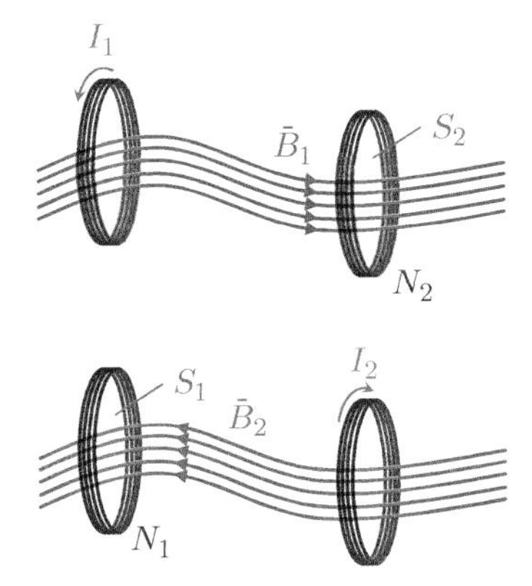

Figure 6.76 The mutual inductance between two loops C_1 and C_2 can be found by considering the magnetic flux created by one of the loops through the other loop. The order does not matter: i.e. we have $L_{12} = L_{21}$.

magnetic potential energy is given in Eq. (4.137) as

$$w_m = \frac{1}{2} I_1 \overline{dl}_1 \cdot \bar{A}_1 + \frac{1}{2} I_2 \overline{dl}_2 \cdot \bar{A}_2, \tag{6.518}$$

where $\bar{A}_1/\bar{A}_2$ represents the magnetic vector potential created at the location of the first/second current by the second/first current. If these infinitesimal electric currents flow in loops of N_1 and N_2 turns (e.g. as in Figure 6.76), we must consider magnetic flux linkages as

$$N_1 \oint_{C_1} \bar{A}_2(\bar{r}) \cdot \overline{dl}_1 = N_1 \oint_{S_1} \bar{\nabla} \times \bar{A}_2(\bar{r}) \cdot \overline{ds}_1 = N_1 \oint_{S_1} \bar{B}_2(\bar{r}) \cdot \overline{ds}_1 = \Lambda_{21} \tag{6.519}$$

$$N_2 \oint_{C_2} \bar{A}_1(\bar{r}) \cdot \overline{dl}_2 = N_2 \oint_{S_2} \bar{\nabla} \times \bar{A}_1(\bar{r}) \cdot \overline{ds}_2 = N_2 \oint_{S_2} \bar{B}_1(\bar{r}) \cdot \overline{ds}_2 = \Lambda_{12}, \tag{6.520}$$

where $A_1(\bar{r})$ and $\bar{A}_2(\bar{r})$ represent the magnetic vector potentials created by the first and second loops, respectively.[160] Then we have

$$w_{m,\text{stored}} = \frac{1}{2} \Lambda_{21} I_1 + \frac{1}{2} \Lambda_{12} I_2 \tag{6.521}$$

as the stored magnetic potential energy due to the *interaction* of two current loops. Inserting $L_{12} = \Lambda_{12}/I_1$ and $L_{21} = \Lambda_{21}/I_2$, we obtain the energy-inductance relationship for a system of two current loops as

$$w_{m,\text{stored}} = \frac{1}{2} I_2 L_{21} I_I + \frac{1}{2} I_1 L_{12} I_2 = \frac{1}{2} (L_{12} + L_{21}) I_1 I_2. \tag{6.522}$$

Note that $L_{12} = L_{21}$ can be *negative*,[161] depending on the selection of $\overline{dl}_1$ and $\overline{dl}_2$. Extending the expression above for a system of multiple current loops $I_1\overline{dl}_1, I_2\overline{dl}_2, \ldots, I_n\overline{dl}_n$ (Figure 6.77), we have

$$w_{m,\text{stored}} = \frac{1}{2} \sum_{k=1}^{n} I_k \sum_{l=1, l \neq k}^{n} L_{kl} I_l, \tag{6.523}$$

where L_{kl} represents the mutual inductance between the kth and lth loops. The expression above is the stored magnetic potential energy due to the interaction of n loops. If the interactions of the loops by themselves (self interactions) are also included, we need to update the expression as

$$w_{m,\text{stored}} = \frac{1}{2} \sum_{k=1}^{n} I_k \sum_{l=1}^{n} L_{kl} I_l, \tag{6.524}$$

where self inductances (L_{kl} for $k = l$) are added.[162]

[160] To be consistent with Section 4.3.3, we define $\bar{A}_1/\bar{A}_2$ as the magnetic vector potential created at the location of the *first/second* current by the *second/first* current in Eq. (6.518). On the other hand, in Eqs. (6.519) and (6.520), $\bar{A}_1(\bar{r})$ and $\bar{A}_2(\bar{r})$ represent the magnetic vector potentials created by the *first* and *second* loops, respectively, at any point $\bar{r}$. This is the reason for the integration of $\bar{A}_2(\bar{r})$ along C_1, while $\bar{A}_1(\bar{r})$ is integrated along C_2. In addition, these line integrals are multiplied by N_1 and N_2, respectively, to account for the corresponding numbers of turns. Also note that Stokes' theorem is employed to convert line integrals to surface integrals, while $\bar{B}(\bar{r}) = \bar{\nabla} \times \bar{A}(\bar{r})$ is used to reveal that these integrals are simply magnetic fluxes.

[161] This does not mean negative total magnetic potential energy stored by the system (which also includes self interactions/inductances), which is always positive.

Figure 6.77 The magnetic potential energy stored by a system of n current loops can be found by considering the mutual inductances between the loops, as well as the self inductances.

[162] Note that the energy in Eq. (6.523) can be negative, depending on the selections of line/surface directions ($\overline{dl}$ and $\overline{ds}$). Eq. (6.524) always gives a positive value as it corresponds to the total energy required to construct the overall system.

All electric current distributions can be considered combinations of smaller electric currents so that the discussion above can be generalized. On the other hand, as opposed to the capacitance and resistance, the inductance of a structure often involves self and mutual parts.[163] For example, given two loops with electric currents I_1 and I_2 (Figure 6.76), we generally have

[163]While they are less common, there are also self/mutual capacitance and resistance definitions.

$$L_{11} = 2w_{m,\text{stored},11}/I_1^2 \tag{6.525}$$

$$L_{22} = 2w_{m,\text{stored},22}/I_2^2 \tag{6.526}$$

$$L_{12} = 2w_{m,\text{stored},12}/[I_1 I_2] = L_{21}, \tag{6.527}$$

where $w_{m,\text{stored},11}$ is the magnetic potential energy stored due to the self interaction of the first loop, $w_{m,\text{stored},22}$ is the magnetic potential energy stored due to the self interaction of the second loop, and $w_{m,\text{stored},12}$ is the magnetic potential energy stored due to the interaction of the two loops. The total inductance of the system can be defined as $L_0 = L_{11} + L_{22} + L_{12} + L_{21}$, which is always positive.[164] For a simple system (e.g. a cylindrical inductor) with electric current I_0, we simply have

[164]Hence, as emphasized above, the total magnetic potential energy stored by a system is always positive.

$$L_0 = \frac{2w_{m,\text{stored}}}{I_0^2} = \frac{N^2 f_{B0}^2}{2w_{m,\text{stored}}}, \tag{6.528}$$

without explicitly considering which part of the inductance is due to self or mutual interactions of individual loops. Considering the system as a whole object, this can be interpreted as the self inductance of the inductor.

As opposed to capacitance and resistance, which can be found in several different ways, it is usually easier to find the inductance of a given structure from the magnetic potential energy.[165] But as the most basic example for which magnetic flux can be used, we now reconsider an ideal inductor with a cylindrical shape of radius a and length l aligned in the z direction (Figure 6.78). As we have found several times, such as in Section 4.3.3.3, the magnetic field intensity inside the structure can be written

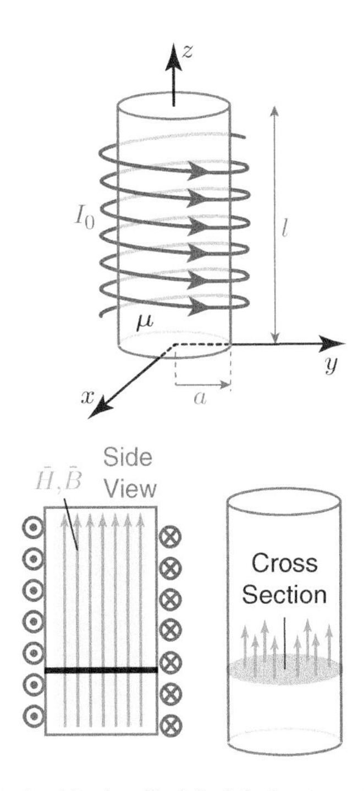

Figure 6.78 An ideal cylindrical inductor with a magnetic material inside. The magnetic flux can be found by integrating the magnetic flux density on the cross section (disk).

[165]In particular because it may not be obvious where to integrate the magnetic flux density.

$$\bar{H}(\bar{r}) = \hat{a}_z I_0 N_l, \tag{6.529}$$

where $N_l = N/l$ is the number of wire turns per length and I_0 is the steady electric current flowing along the wire. If the core material has a permeability of μ, we further have

$$\bar{B}(\bar{r}) = \hat{a}_z \mu I_0 N_l. \tag{6.530}$$

Then the magnetic flux through the inductor can be found as (Figure 6.78)

$$f_{B0} = \int_{\text{cross section}} \bar{B}(\bar{r}) \cdot \overline{ds} = \mu I_0 N_l A, \tag{6.531}$$

where $A = \pi a^2$ is the area of the cross section. The total linkage is the magnetic flux multiplied by the number of turns:

$$\Lambda_0 = N f_{B0} = N \mu I_0 N_l A = \mu I_0 \frac{N^2 A}{l}. \tag{6.532}$$

Finally, the inductance (specifically, the self inductance of an ideal inductor) can be obtained as

$$L_0 = \frac{\Lambda_0}{I_0} = \mu \frac{N^2 A}{l}. \tag{6.533}$$

This gives the magnetic potential energy stored by the inductor as[166]

$$w_{m,\text{stored}} = \frac{1}{2} L_0 I_0^2 = \mu \frac{N^2 A}{2l} I_0^2. \tag{6.534}$$

[166]This is the same as Eq. (4.163) with $A = \pi a^2$.

As another interesting example, we may consider an infinitely long conducting cylinder of radius a in a vacuum. We assume that the cylinder can carry a steady electric current with a constant volume electric current density $\bar{J}_v(\bar{r}) = \hat{a}_z J_0$ (A/m^2) (Figure 6.79). This structure was considered several times previously, such as in Section 2.4 and Chapter 4, but we now consider the *internal* inductance of the structure per length. Note that the *external* inductance is infinite since the magnetic potential energy stored outside the cylinder per length is infinite for a finite value of the electric current.[167] For the internal inductance, recall that the magnetic field intensity can be written

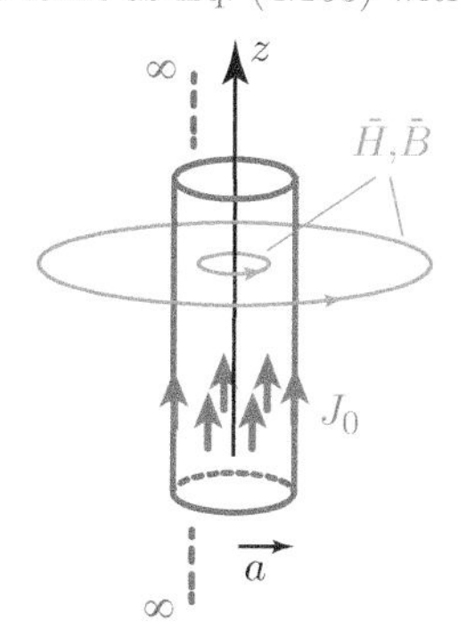

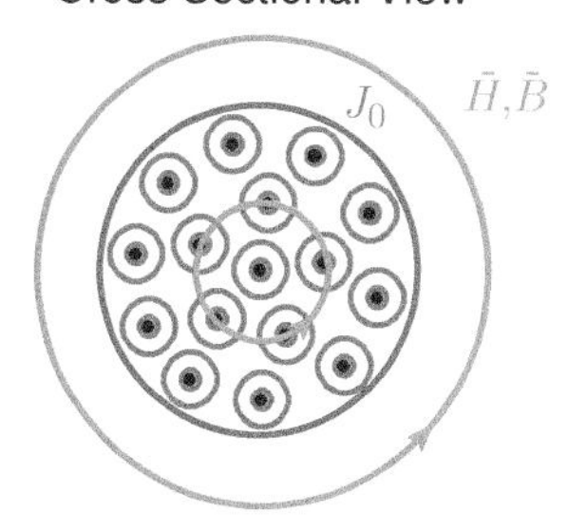

Figure 6.79 A steady volume electric current density $J_v(\bar{r}) = J_0$ (A/m^2) flowing in a cylindrical region of radius a in the z direction. The electric current is assumed to be uniformly distributed over the cross-sectional surface (disk). The internal inductance per length can be found by considering the stored magnetic potential energy.

$$\bar{H}(\bar{r}) = \hat{a}_\phi \frac{I_{\text{enc}}}{2\pi\rho} = \hat{a}_\phi \frac{J_0 \pi \rho^2}{2\pi\rho} = \hat{a}_\phi \frac{J_0 \rho}{2} \quad (\rho < a), \tag{6.535}$$

which can be found via the integral form of Ampere's law. Then the magnetic potential energy stored inside the cylinder per length can be found as

$$w_{ml,\text{internal}} = \frac{1}{2} \int_{\rho<a} \bar{H}(\bar{r}) \cdot \bar{B}(\bar{r}) ds = \frac{J_0^2}{8} \int_0^{2\pi} \int_0^a \rho^2 \rho d\rho d\phi \tag{6.536}$$

$$= \frac{\pi \mu_0 J_0^2 a^4}{16} = \frac{\mu_0 I_0^2}{16\pi}, \tag{6.537}$$

where $I_0 = \pi a^2 J_0$ is the total current flow, assuming that the cylinder is nonmagnetic with permeability μ_0. Then using the relationship between the magnetic potential energy and inductance,

[167]Also note that internal inductance is practically important.

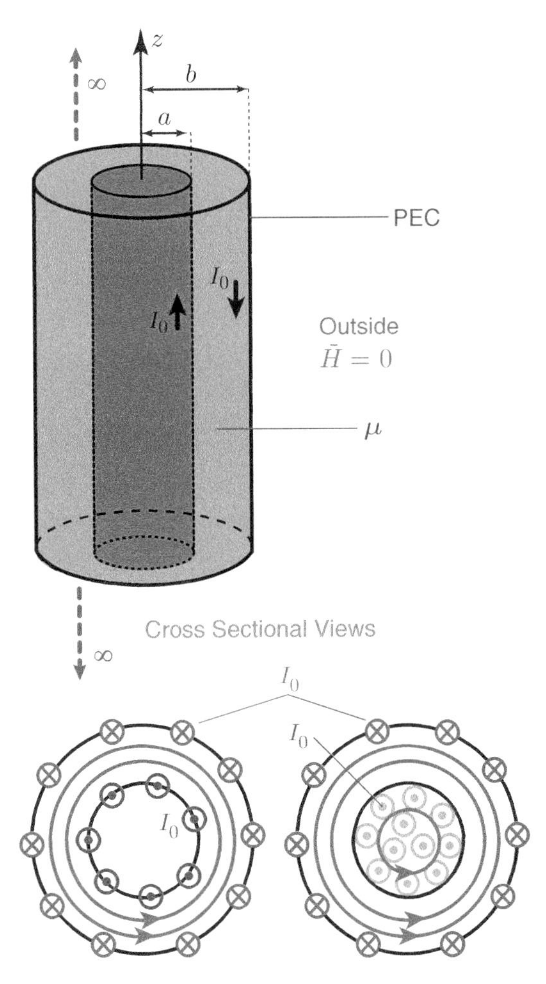

Figure 6.80 An infinitely long coaxial structure consisting of two cylindrical conductors with radii a and $b > a$ aligned in the z direction. Steady electric currents can flow in opposite directions, while two different cases are considered depending on how the electric current is distributed for the inner cylinder.

we have

$$L_{0l,\text{internal}} = \frac{2w_{ml,\text{internal}}}{I_0^2} = \frac{\mu_0}{8\pi} \tag{6.538}$$

as the internal inductance per length.

As further examples, we now consider a coaxial structure consisting of two infinitely long cylindrical conductors with radii a and $b > a$, and a homogeneous magnetic material with permeability μ between them. Steady electric currents can flow in opposite directions on the conductors, while we focus on two different cases (Figure 6.80). In the first case, we assume that all electric currents flow on surfaces. Then the magnetic field intensity is nonzero only between the cylinders: i.e. we have

$$\bar{H}(\bar{r}) = \hat{a}_\phi \frac{I_0}{2\pi\rho} \quad (a < \rho < b), \tag{6.539}$$

which can be found via the integral form of Ampere's law, where I_0 is the total amount of current carried by each cylinder (in opposite directions). This gives the magnetic potential energy stored per length as (see Section 4.3)

$$w_{ml,\text{stored}} = \frac{\mu I_0^2}{4\pi} \ln(b/a). \tag{6.540}$$

Therefore, the inductance (specifically, the *self* inductance of an ideal coaxial structure or the *mutual* inductance between two concentric cylinders) per length can be found:

$$L_{0l} = \frac{\mu}{2\pi} \ln(b/a). \tag{6.541}$$

Next, as the second case for the coaxial structure, we assume that the electric current flow is uniformly distributed on the cross section (disk) of the inner (nonmagnetic) cylinder (Figure 6.80). Then for the same I_0, the magnetic field intensity between the cylinders is the same as one in the previous case, while there is also a nonzero magnetic field intensity inside the inner cylinder:

$$\bar{H}(\bar{r}) = \hat{a}_\phi \frac{I_0\rho}{2\pi a^2} \quad (\rho < a). \tag{6.542}$$

Hence, the magnetic potential energy (per length) is increased with an addition term as (see the exercises in Chapter 4)

$$w_{ml,\text{stored}} = \frac{\mu_0 I_0^2}{16\pi} + \frac{\mu I_0^2}{4\pi} \ln(b/a). \tag{6.543}$$

Consequently, the inductance per length is also increased, and we obtain

$$L_{0l} = \frac{\mu_0}{8\pi} + \frac{\mu}{2\pi}\ln(b/a) \tag{6.544}$$

Obviously, this expression is simply the sum of Eqs. (6.538) and (6.541).

Inductors with toroid shapes are commonly used in electrical circuits (Figure 6.81). As found in the exercises in Chapter 2, the magnetic field intensity and magnetic flux density inside an ideal toroidal inductor can be written

$$\bar{H}(\bar{r}) = \hat{a}_\phi \frac{NI_0}{2\pi\rho}, \qquad \bar{B}(\bar{r}) = \hat{a}_\phi \mu \frac{NI_0}{2\pi\rho} \quad (b < \rho < b+a), \tag{6.545}$$

where I_0 is the steady electric current flowing through wires, N is the number of turns, and μ is the permeability of the core material. The magnetic flux through the cross section can be obtained via integration as

$$f_{B0} = \int_b^{b+a}\int_0^h \mu\frac{NI_0}{2\pi\rho}d\rho dz = \mu\frac{NI_0}{2\pi}h\ln\left(\frac{b+a}{b}\right). \tag{6.546}$$

The flux linkage is N times the magnetic flux

$$\Lambda_0 = Nf_{B0} = \mu\frac{N^2I_0}{2\pi}h\ln\left(\frac{b+a}{b}\right), \tag{6.547}$$

leading to the expression for the inductance as[168]

$$L_0 = \frac{\Lambda_0}{I_0} = \mu\frac{N^2h}{2\pi}\ln\left(\frac{b+a}{b}\right). \tag{6.548}$$

Alternatively, we can find the stored magnetic potential energy in terms of I_0 as

$$w_{m,\text{stored}} = \frac{1}{2}\int_{\text{inside toroid}} \bar{B}(\bar{r})\cdot\bar{H}(\bar{r})dv = \mu\frac{N^2I_0^2}{8\pi^2}2\pi h\int_a^{b+a}\frac{1}{\rho^2}\rho d\rho \tag{6.549}$$

$$= \mu\frac{N^2hI_0^2}{4\pi}\ln[(b+a)/a]. \tag{6.550}$$

Then we have $L_0 = 2w_{m,\text{stored}}/I_0^2$, leading to the same expression for the inductance.

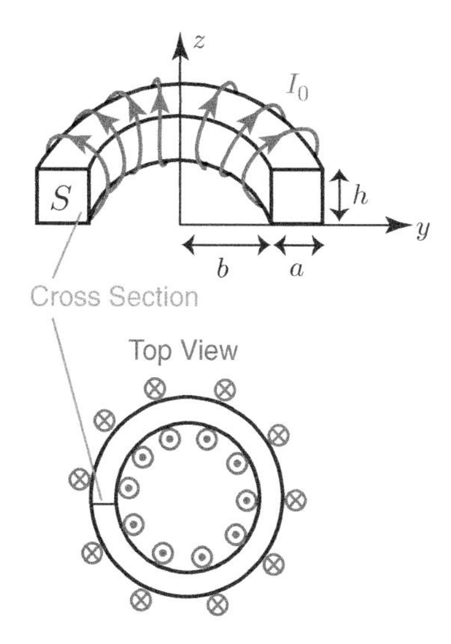

Figure 6.81 An inductor with a toroid shape.

[168] Note that for $b \gg a$, we have

$$L_0 \approx \mu\frac{N^2h}{2\pi}\frac{a}{b} = \mu\frac{N^2(ha)}{2\pi b} = \mu\frac{N^2A}{l},$$

where A is the area of the cross section and l is the length of the inductor. Hence, the expression reduces to the inductance of a cylindrical inductor.

6.9.1 Examples

Example 84: Consider an $a \times b$ rectangular loop carrying a steady electric current I_0 in a vacuum. The loop is located between two infinitely long lines that also carry I_0 in opposite directions (Figure 6.82). The distance between the infinitely long lines is d, while the distance from the left vertical side of the rectangular loop to one of the lines is $x_0 < (d-a)$. Find the magnetic force acting on the loop.

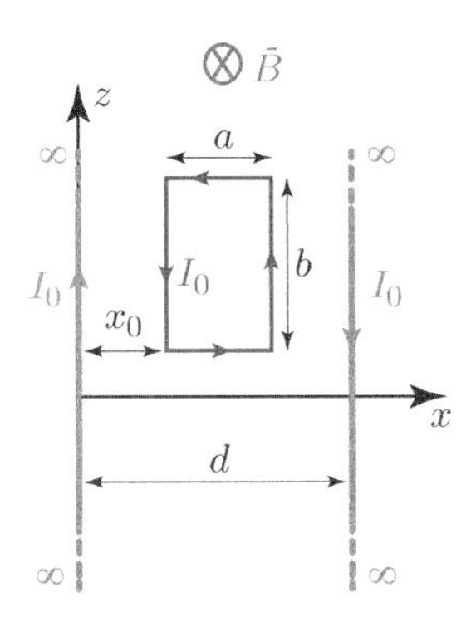

Figure 6.82 A rectangular current-carrying loop between two infinitely long current-carrying wires.

Solution: As considered in the exercises in Chapter 2, the expression for the force acting on the loop can be found via Ampere's law as

$$\bar{F}_m = \hat{a}_x \frac{\mu_0 I_0^2 b}{2\pi}\left[\frac{1}{x_0} + \frac{1}{(d-x_0)} - \frac{1}{(x_0+a)} - \frac{1}{(d-a-x_0)}\right]. \tag{6.551}$$

To solve the problem via the magnetic potential energy, we can first find the mutual inductance between the infinitely long wires and the rectangular loop. Using

$$\bar{B}(\bar{r}) = \hat{a}_y \frac{\mu_0 I_0}{2\pi x} + \hat{a}_y \frac{\mu_0 I_0}{2\pi(d-x)}, \tag{6.552}$$

the magnetic flux through the loop can be found as[169]

$$f_{B0} = \int_S \bar{B}(\bar{r}) \cdot (-\hat{a}_y) dx dz = -\frac{\mu_0 I_0 b}{2\pi}\left[\ln\left(1+\frac{a}{x_0}\right) - \ln\left(1-\frac{a}{d-x_0}\right)\right]. \tag{6.553}$$

Hence, we obtain the mutual inductance between the pair of infinitely long wires and the rectangular loop as

$$L_{0,\text{mutual}} = \frac{f_{B0}}{I_0} = -\frac{\mu_0 b}{2\pi}\left[\ln\left(1+\frac{a}{x_0}\right) - \ln\left(1-\frac{a}{d-x_0}\right)\right]. \tag{6.554}$$

Then the magnetic potential energy stored due to the interaction of the infinitely long wires and the rectangular loop[170] can be written[171]

$$w_{m,\text{mutual}} = L_{0,\text{mutual}} I_0^2 = -\frac{\mu_0 I_0^2 b}{2\pi}\left[\ln\left(1+\frac{a}{x_0}\right) - \ln\left(1-\frac{a}{d-x_0}\right)\right]. \tag{6.555}$$

[169] We have

$$\begin{aligned} f_{B0} &= -\frac{\mu_0 I_0}{2\pi}\int_{x_0}^{x_0+a}\int_0^b \left[\frac{1}{x} + \frac{1}{(d-x)}\right] \\ &= -\frac{\mu_0 I_0 b}{2\pi}\left[\ln(x) - \ln(d-x)\right]_{x_0}^{x_0+a} \\ &= -\frac{\mu_0 I_0 b}{2\pi}\left[\ln\left(\frac{x_0+a}{x_0}\right) - \ln\left(\frac{d-x_0-a}{d-x_0}\right)\right]. \end{aligned}$$

[170] This is not the total magnetic potential energy, which is always positive.

[171] Note that $2L_{0,\text{mutual}}$ is used as the total mutual inductance.

Finally, given x_0 as the virtual displacement variable, we obtain the magnetic force as

$$\bar{F}_m = \bar{\nabla} w_m = -\hat{a}_x \frac{\mu_0 I_0^2 b}{2\pi} \frac{\partial}{\partial x_0} \left[\ln\left(1 + \frac{a}{x_0}\right) - \ln\left(1 - \frac{a}{d - x_0}\right) \right] \tag{6.556}$$

$$= \hat{a}_x \frac{\mu_0 I_0^2 b}{2\pi} \left[\frac{a}{x_0(a + x_0)} - \frac{a}{(d - x_0)(d - x_0 - a)} \right], \tag{6.557}$$

leading to[172]

$$\bar{F}_m = \hat{a}_x \frac{\mu_0 I_0^2 b}{2\pi} \left[\frac{1}{x_0} - \frac{1}{(a + x_0)} + \frac{1}{(d - x_0)} - \frac{1}{(d - x_0 - a)} \right], \tag{6.558}$$

verifying the earlier result.

[172] We use

$$\frac{a}{x_0(a+x_0)} = \frac{a+x_0}{x_0(a+x_0)} - \frac{x_0}{x_0(a+x_0)} = \frac{1}{x_0} - \frac{1}{(a+x_0)}$$

$$\frac{a}{(d-x_0)(d-x_0-a)} = \frac{a-d+x_0}{(d-x_0)(d-x_0-a)} + \frac{d-x_0}{(d-x_0)(d-x_0-a)} = -\frac{1}{(d-x_0)} + \frac{1}{(d-x_0-a)}.$$

Example 85: Consider a coaxial structure consisting of two infinitely long cylindrical conductors with radii a and $b > a$, aligned in the z direction (Figure 6.83). A surface electric current can flow on the outer conductor, while it tends to be uniformly distributed in the form of the volume electric current density in the inner conductor. In addition, a magnetic material with permeability $\mu = \mu_0 \rho / a$ exists between the conductors. Find the inductance of the structure per length.

Solution: Assuming steady electric currents $\pm I_0$ flowing in opposite directions, we can write the magnetic field intensity and magnetic flux density as

$$\bar{H}(\bar{r}) = \hat{a}_\phi \frac{I_0}{2\pi} \begin{cases} \rho/a^2 & (\rho < a) \\ 1/\rho & (a < \rho < b) \\ 0 & (\rho > b) \end{cases} \tag{6.559}$$

and

$$\bar{B}(\bar{r}) = \hat{a}_\phi \frac{\mu_0 I_0}{2\pi} \begin{cases} \rho/a^2 & (\rho < a) \\ 1/a & (a < \rho < b) \\ 0 & (\rho > b). \end{cases} \tag{6.560}$$

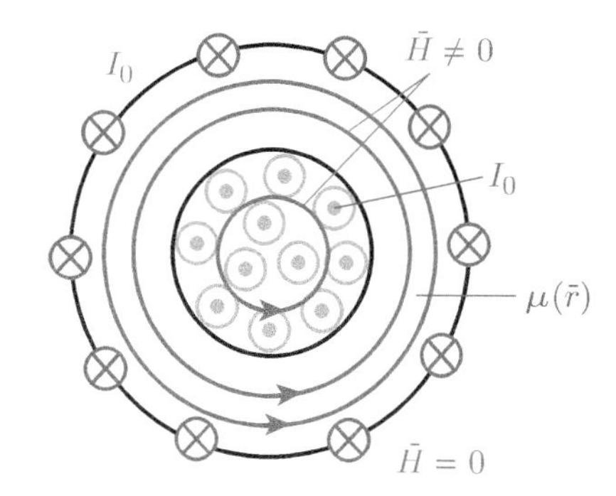

Figure 6.83 An infinitely long coaxial structure consisting of two cylindrical conductors with radii a and $b > a$ aligned in the z direction. A surface electric current distribution can exist on the outer conductor, while it is uniformly distributed on the cross section of the inner conductor. An inhomogeneous magnetic material exists between the conductors.

Then the magnetic potential energy per length can be found as[173,174]

$$w_{ml,\text{stored}} = \frac{1}{2}\int_{\infty} \bar{H}(\bar{r}) \cdot \bar{B}(\bar{r}) ds \tag{6.561}$$

$$= \frac{1}{2}\int_0^a \int_0^{2\pi} \frac{I_0\rho}{2\pi a^2}\frac{\mu_0 I_0 \rho}{2\pi a^2}\rho d\phi d\rho + \frac{1}{2}\int_a^b \int_0^{2\pi} \frac{I_0}{2\pi\rho}\frac{\mu_0 I_0}{2\pi a}\rho d\phi d\rho \tag{6.562}$$

$$= \frac{\mu_0 I_0^2}{16\pi}\left(\frac{4b}{a} - 3\right). \tag{6.563}$$

Consequently, we obtain the inductance per length as

$$L_{0l} = \frac{\mu_0}{8\pi}\left(\frac{4b}{a} - 3\right). \tag{6.564}$$

Example 86: Consider a loop of radius a carrying a steady electric current I_0 located at the origin (Loop 1) and an identical loop carrying an electric current I_0 in the opposite direction located at $z = d$ (Loop 2). The loops are in a vacuum (Figure 6.84). Assuming that $d \gg a$, find the *vertical* magnetic force between the loops.

Solution: As considered in the exercises in Chapter 2, the magnetic force between the loops can be found via Ampere's force law as

$$\bar{F}_{m,z} = \hat{a}_z \frac{3\pi\mu_0 I_0^2 a^4}{2d^4}. \tag{6.565}$$

The same expression can be found by considering the magnetic potential energy. Before proceeding, however, we must emphasize the following points:

- There is a force acting on Loop 1 due to the electric current flowing along it. This force is related to self inductance, and it tries to expand the loop (increase a).
- Similarly, there is a force acting on Loop 2 due to its electric current. This force, which is related to the self inductance of the loop, tries to expand it.
- Loop 1 applies a force to Loop 2, and Loop 2 applies a corresponding force to Loop 1, and these forces try to expand the loops. The forces are related to the mutual inductance between the loops.
- There is also a repulsive force between the loops. This force is also related to the mutual inductance between the loops.

[173]As intermediate steps, we have

$$w_{ml,\text{stored}} = \frac{1}{2}\frac{\mu_0 I_0^2}{4\pi^2 a^4} 2\pi \int_0^a \rho^3 d\rho + \frac{1}{2}\frac{\mu_0 I_0^2}{4\pi^2 a} 2\pi \int_a^b d\rho = \frac{\mu_0 I_0^2}{16\pi} + \frac{\mu_0 I_0^2}{4\pi}\frac{(b-a)}{a}.$$

[174]Note that the material between the conductors is magnetized as

$$\bar{M}(\bar{r}) = \frac{\bar{B}(\bar{r})}{\mu_0} - \bar{H}(\bar{r}) = \hat{a}_\phi \frac{I_0}{2\pi a} - \hat{a}_\phi \frac{I_0}{2\pi\rho} = \hat{a}_\phi \frac{I_0}{2\pi}\left(\frac{1}{a} - \frac{1}{\rho}\right).$$

This magnetization can be represented by equivalent surface and volume current densities as

$$\bar{J}_{ms}(\bar{r}) = -\hat{a}_z \frac{I_0}{2\pi}\left(\frac{1}{a} - \frac{1}{b}\right) \quad (\rho = 2a)$$

$$\bar{J}_{mv}(\bar{r}) = \hat{a}_z \frac{I_0}{2\pi a\rho} \quad (a < \rho < 2a).$$

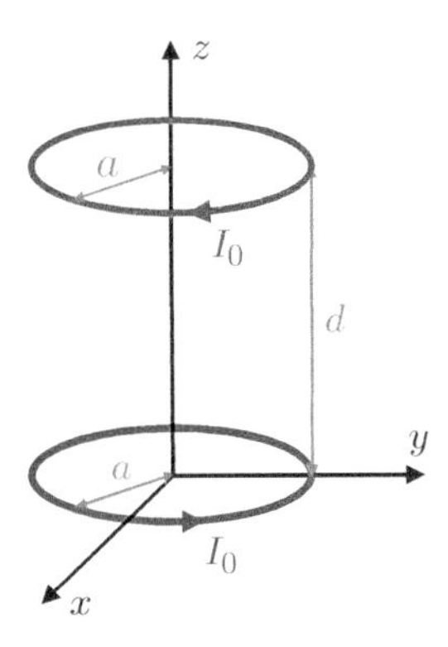

Figure 6.84 A loop over another loop, carrying currents in opposite directions. This structure is commonly known as a Helmholtz coil.

In this question, we are particularly interested in the fourth item: the vertical magnetic force.[175]

First, we can consider the approximate expression for the z-directed magnetic flux density created by Loop 1 at the location of Loop 2 as[176]

$$\bar{B}_{12,z}(\bar{r}) \approx \hat{a}_z \frac{\mu_0 I_0 a^2}{4[a^2+d^2]^{3/2}} \frac{2d^2-a^2}{a^2+d^2} \approx \hat{a}_z \frac{\mu_0 I_0 a^2}{2d^3} \tag{6.566}$$

using a first-order approximation when $d \gg a$. Then the magnetic flux created through Loop 2 can be found as

$$f_{B,12} \approx \int_{S_2} \bar{B}_{12}(\bar{r}) \cdot \overline{ds} = -\frac{\mu_0 I_0 a^2}{2d^3}\pi a^2 = -\frac{\pi \mu_0 I_0 a^4}{2d^3} \tag{6.567}$$

using $\overline{ds} = -\hat{a}_z ds$. Then the mutual inductance between the loops can be found as[177]

$$L_{12} = \frac{f_{B,12}}{I_0} = -\frac{\pi \mu_0 a^4}{2d^3} = L_{21}, \tag{6.568}$$

which is negative. Consequently, for electric currents I_0 flowing along the loops in opposite directions, the stored magnetic potential energy due to the mutual coupling between the loops can be written

$$w_m = \frac{1}{2}(L_{12}+L_{21})I_0^2 = -\frac{\pi \mu_0 I_0^2 a^4}{2d^3}. \tag{6.569}$$

Finally, the vertical magnetic force between the loops can be found by taking d as the virtual displacement variable as

$$F_m = \bar{\nabla} w_m \approx -\hat{a}_d \frac{\pi \mu_0 I_0^2 a^4}{2} \frac{\partial}{\partial d}\left(\frac{1}{d^3}\right) = -\hat{a}_d \frac{\pi \mu_0 I_0^2 a^4}{2} 3d^{-4} = \hat{a}_d \frac{3\pi \mu_0 I_0^2 a^4}{2d^4}, \tag{6.570}$$

as expected. The positive value indicates that the magnetic force tries to separate the loops (a repulsive force).

Example 87: Consider two infinitely long current-carrying cylinders, each of radius a, located in a vacuum (Figure 6.85). Let the cylinders be aligned in the z direction, one at the origin and the other at $x = d$, where $d \gg a$. We also assume that the currents in opposite directions flow through the surfaces of the cylinders. Find the inductance per length.

Solution: Given I_0 steady electric currents flowing through the surfaces of the PECs, and assuming that they are uniformly distributed (the cylinders do not affect the distributions

[175] This corresponds to the assumption that the loops are rigid and they cannot be deformed.

[176] Note that we are considering the z component, as opposed to the radial component, when solving the same problem via Ampere's force law.

[177] The sign of mutual inductance depends on the selection of directions $\overline{dl}$. The corresponding magnetic potential energy is only due to the interaction of the loops: i.e. it is part of the total magnetic potential energy stored by the system.

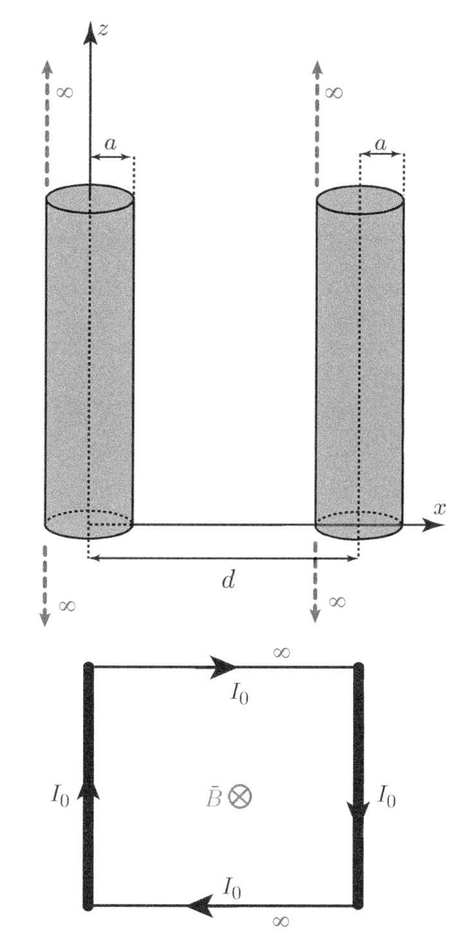

Figure 6.85 Two infinitely long cylindrical PECs. When finding the inductance of the structure, it can be considered a closed loop with connections at infinity.

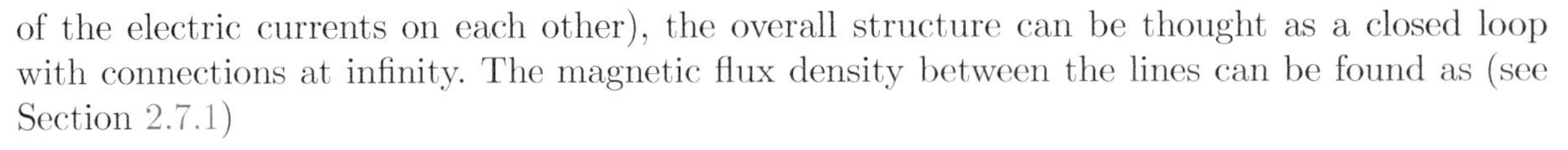
of the electric currents on each other), the overall structure can be thought as a closed loop with connections at infinity. The magnetic flux density between the lines can be found as (see Section 2.7.1)

$$\bar{B}(\bar{r}) = \hat{a}_y \frac{\mu_0 I_0}{2\pi}\left(\frac{1}{x} + \frac{1}{d-x}\right). \tag{6.571}$$

Hence, the magnetic flux per length can be obtained:

$$f_{Bl} = \int_a^{d-a} \hat{a}_y \frac{\mu_0 I_0}{2\pi}\left(\frac{1}{x} + \frac{1}{d-x}\right) \cdot \hat{a}_y dx \tag{6.572}$$

$$= \frac{\mu_0 I_0}{2\pi}\left[\ln\left(\frac{x}{d-x}\right)\right]_a^{d-a} = \frac{\mu_0 I_0}{2\pi}\left[\ln\left(\frac{d-a}{a}\right) - \ln\left(\frac{a}{d-a}\right)\right] \tag{6.573}$$

$$= \frac{\mu_0 I_0}{\pi}\ln\left(\frac{d-a}{a}\right) \approx \frac{\mu_0 I_0}{\pi}\ln(d/a). \tag{6.574}$$

Then the inductance per length can be derived as

$$L_l \approx \frac{\mu_0}{\pi}\ln(d/a). \tag{6.575}$$

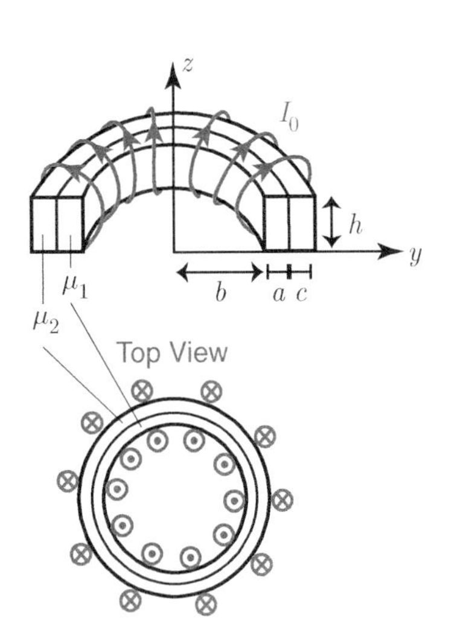

Figure 6.86 A toroidal inductor consisting of a core with two different materials.

Example 88: Consider a toroidal inductor consisting of two different materials with permeability values μ_1 and μ_2 (Figure 6.86). The inner and outer radii of the first material are b and $b+a$, respectively, while its cross section is an $a \times h$ rectangle. Similarly, the inner and outer radii of the second material are $b+a$ and $b+a+c$, respectively, while its cross section is a $c \times h$ rectangle. If there are a total of N turns (with uniform distribution), find the inductance of the structure.

Solution: Assuming a steady electric current I_0 flowing along wires, and using the integral form of Ampere's law, the magnetic field intensity inside the ring can be found as[178]

$$\bar{H}(\bar{r}) = \hat{a}_\phi \frac{NI_0}{2\pi\rho} \quad (b < \rho < b+a+c). \tag{6.576}$$

Then the magnetic flux density can be written

$$\bar{B}_1(\bar{r}) = \hat{a}_\phi \frac{\mu_1 NI_0}{2\pi\rho} \quad (b < \rho < b+a) \tag{6.577}$$

$$\bar{B}_2(\bar{r}) = \hat{a}_\phi \frac{\mu_2 NI_0}{2\pi\rho} \quad (b+a < \rho < b+a+c). \tag{6.578}$$

[178] Note that the magnetic field intensity is continuous (with the same expression everywhere), while there is a jump in the magnetic flux density at $\rho = b + a$.

The magnetic flux through the cross section of the toroid can be evaluated via integration as[179]

$$f_{B0} = \int_{S_1} \bar{B}_1(\bar{r}) \cdot \overline{ds} + \int_{S_2} \bar{B}_2(\bar{r}) \cdot \overline{ds} \tag{6.579}$$

$$= \frac{hNI_0}{2\pi} \left\{ \mu_1 \ln\left(1 + \frac{b}{a}\right) + \mu_2 \ln\left(1 + \frac{c}{b+a}\right) \right\}. \tag{6.580}$$

[179]As intermediate steps, we have

$$f_{B0} = h \int_b^{b+a} \hat{a}_\phi \frac{\mu_1 N I_0}{2\pi\rho} \cdot \hat{a}_\phi d\rho + h \int_{b+a}^{b+a+c} \hat{a}_\phi \frac{\mu_2 N I_0}{2\pi\rho} \cdot \hat{a}_\phi d\rho = h\frac{\mu_1 N I_0}{2\pi} \ln\left(\frac{b+a}{a}\right) + h\frac{\mu_2 N I_0}{2\pi} \ln\left(\frac{b+a+c}{b+a}\right).$$

Finally, the total flux linkage and the inductance can be found as[180]

$$\Lambda_0 = \frac{hN^2 I_0}{2\pi} \left\{ \mu_1 \ln\left(1 + \frac{b}{a}\right) + \mu_2 \ln\left(1 + \frac{c}{b+a}\right) \right\} \tag{6.581}$$

$$L_0 = \frac{\Lambda_0}{I_0} = \frac{hN^2}{2\pi} \left\{ \mu_1 \ln\left(1 + \frac{b}{a}\right) + \mu_2 \ln\left(1 + \frac{c}{b+a}\right) \right\}. \tag{6.582}$$

[180]Note the special case of $\mu_2 = \mu_1$, leading to

$$L_0 = \frac{hN^2}{2\pi}\mu_1 \left\{ \ln\left(\frac{b+a}{a}\right) + \ln\left(\frac{b+a+c}{b+a}\right) \right\} = \frac{hN^2}{2\pi}\mu_1 \ln\left(\frac{b+a+c}{a}\right),$$

which is consistent with Eq. (6.548) given for a toroidal inductor with a single core material.

6.10 Final Remarks

This chapter's focus has been electrical conduction, an important property of all materials, particularly good conductors whose main electromagnetic characteristics are dominated by conduction. By considering this property, in addition to permittivity and permeability, we can fully understand how materials interact with electromagnetic waves. Maxwell's equations, as well as higher-level equations derived from them, introduced in the previous chapters, have been sufficient to analyze conducting objects. We have investigated the general behaviors of perfectly conducting objects in more detail to understand the interesting limit of infinite conductivity, which can be useful to model, represent, and understand diverse physical phenomena in real-life applications. With the knowledge of conduction, we have completed the chapter with three major properties of all structures: capacitance, resistance, and inductance. In the next chapter, we finally focus on the transmission of electromagnetic waves to complete the basics of electromagnetics.

6.11 Exercises

Exercise 69: Consider a static case involving an infinitely large domain (Figure 6.87) defined as $0 < z < d$, filled with a constant volume electric charge density $q_v = q_0$ (C/m^3). There is a PEC at the bottom of the region (at $z = 0$) with a constant electric scalar potential $\Phi(x, y, z = 0) = 0$, while the electric field is zero for $z < 0$. In addition, there is a constant surface electric charge density $q_s = q_0 d$ (C/m^2) at $z = d$. Finally, the electric flux density

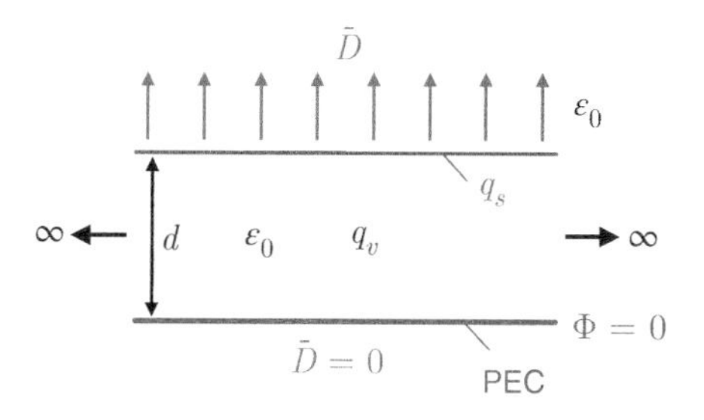

Figure 6.87 An infinitely large structure consisting of a PEC, a volume electric charge distribution ($0 < z < d$), and a surface electric charge distribution at $z = d$.

for $z > d$ is given as $\bar{D}(\bar{r}) = \hat{a}_z 4q_0 d$. Everywhere is a vacuum with ε_0 permittivity. Do the following:

- Find the electric scalar potential $\Phi(z)$ for $0 < z < d$.
- Find the electric flux density for $0 < z < d$.
- Find the surface electric charge density on the PEC.

Exercise 70: Consider a linearly polarized (uniform) plane wave with a phasor-domain expression for the electric field intensity as

$$\bar{E}(\bar{r}) = (\hat{a}_y + 2\hat{a}_z)E_0 \exp(-\alpha \hat{a}_k \cdot \bar{r}) \exp[-j4\pi(y - az)], \tag{6.583}$$

where E_0 (V/m) and a are real constants, propagating in a nonmagnetic lossy medium at 10 MHz. At this frequency, the medium is classified as a good conductor. Find the complete expressions for $\bar{E}(\bar{r})$ and $\bar{H}(\bar{r})$, as well as their time-domain versions.

Exercise 71: Consider a (uniform) plane wave propagating in a nonmagnetic homogeneous medium. The electric field intensity and magnetic field intensity are defined as

$$\bar{E}(\bar{r}) = \hat{a}_x E_0 \exp(-\pi \times 10^3 z) \cos(2\pi \times 10^6 t - \pi \times 10^3 z) \tag{6.584}$$

$$\bar{H}(\bar{r}) \approx \hat{a}_y H_0 \exp(-\pi \times 10^3 z) \cos(2\pi \times 10^6 t - \pi \times 10^3 z - \pi/4) \tag{6.585}$$

where E_0 (V/m) and H_0 (A/m) are constants. Find the conductivity of the medium, H_0 in terms of E_0, as well as the phase velocity and skin depth.

Exercise 72: Wet soil behaves like a good conductor and a low-loss dielectric at 100 kHz and 1 GHz, while the corresponding skin depths are measured as 16 m and 2 m, respectively (interestingly, less when it is a low-loss dielectric). Find (approximately) the relative permittivity and conductivity of wet soil (given that it is nonmagnetic), as well as the phase velocities of uniform plane waves inside wet soil at 100 kHz and 1 GHz.

Exercise 73: A plane wave propagating in a lossy medium is defined with the electric field intensity and magnetic field intensity as

$$\bar{E}(\bar{r}, t) = \hat{a}_x E_0 \exp(-2z) \cos(\omega t - \beta z) \tag{6.586}$$

$$\bar{H}(\bar{r}, t) = \hat{a}_y \frac{E_0}{2} \exp(-2z) \cos(\omega t - \beta z - \pi/3), \tag{6.587}$$

where E_0 (V/m) is a constant. Find the time-average net power flux into a unit cube centered at the origin.

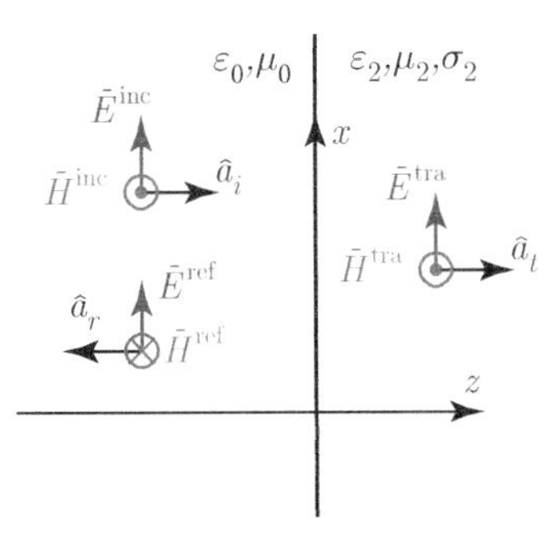

Figure 6.88 Normal incidence of a plane wave onto a boundary between a vacuum and a conducting medium.

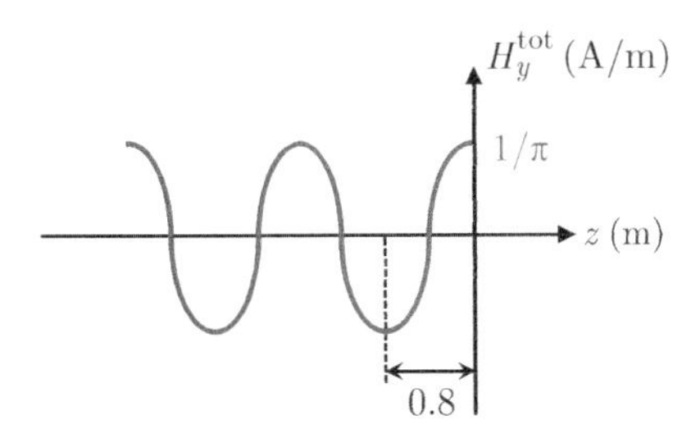

Figure 6.89 A snapshot (plot at a fixed time) of the total magnetic field intensity due to a normal incidence of a plane wave onto a planar PEC at $z = 0$.

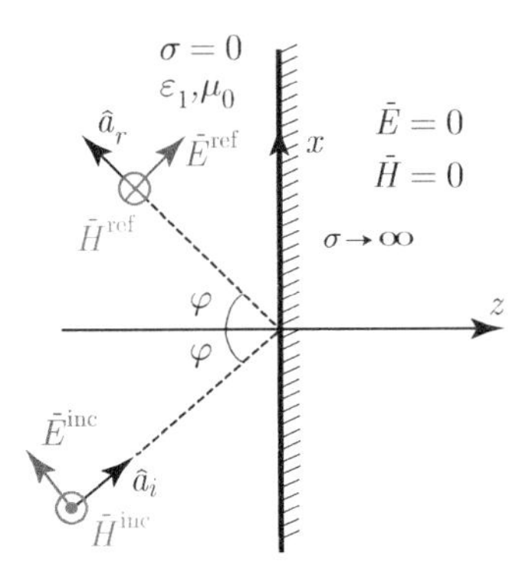

Figure 6.90 Oblique incidence of a plane wave with parallel polarization onto a boundary between a dielectric medium and a PEC.

Exercise 74: A plane wave propagating in the z direction in a vacuum is incident normally onto a boundary at $z = 0$ (Figure 6.88). The medium for $z > 0$ is a lossy dielectric with an intrinsic impedance of $\eta_c \approx 16\pi(1+j)$ Ω. Find a condition for the medium's relative permittivity if it is categorized as a good conductor. Also find the time-average power flux density values for the reflected and transmitted waves if the incident time-average power flux density is 1 W/m^2.

Exercise 75: Consider a normal incidence of a linearly polarized plane wave propagating in the z direction in a vacuum onto a planar PEC at $z = 0$. The total magnetic field intensity for $z < 0$ is plotted for $\omega t = \pi/3$ in Figure 6.89. Find an expression for the total electric field intensity and incident electric field intensity in the time domain.

Exercise 76: A plane wave is incident to a boundary at $z = 0$ between a dielectric medium ($z < 0$) with $\varepsilon_r = 4$ relative permittivity and a PEC (Figure 6.90). The wave has a parallel polarization, and the angle of incidence is 60°, while the surface electric current density induced on the PEC is defined as

$$\bar{J}_s(\bar{r}) = \hat{a}_x 2J_0 \exp(-j4\pi\sqrt{3}x), \tag{6.588}$$

where J_0 (A/m) is a constant. Find expressions for the electric field intensities of the incident and reflected waves.

Exercise 77: Consider a parallel-plate structure consisting of PECs (each with area A) at $z = 0$ and $z = d$, filled with a dielectric material of varying permittivity as $\varepsilon = \varepsilon_0 \exp(x/a)$ (Figure 6.91). Find the capacitance.

Exercise 78: Consider an infinitely long cylindrical PEC of radius a placed at $z = d$ above an infinitely large planar PEC on the x-y plane in a vacuum (Figure 6.92). Find the capacitance of the structure per length.

Exercise 79: Consider a lossy cube of edges $2a$ centered at around the origin. The conductivity of the object is given as

$$\sigma(\bar{r}) = \frac{\sigma_0}{a^2 + z^2}. \tag{6.589}$$

Find the resistance between the top and bottom surfaces: i.e. when planar PECs are placed at these surfaces.

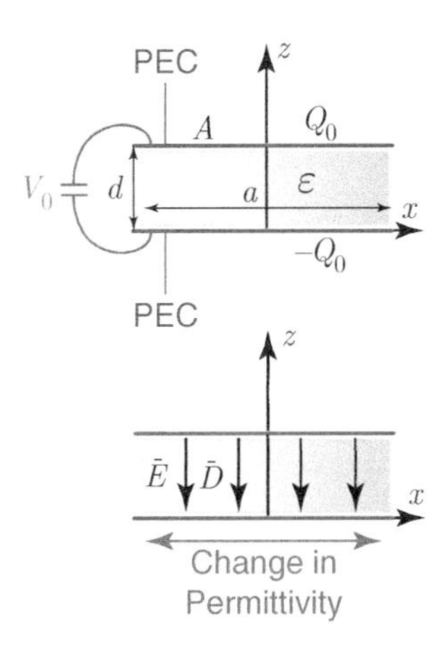

Figure 6.91 A parallel-plate capacitor filled with an inhomogeneous dielectric material having varying permittivity in the horizontal direction.

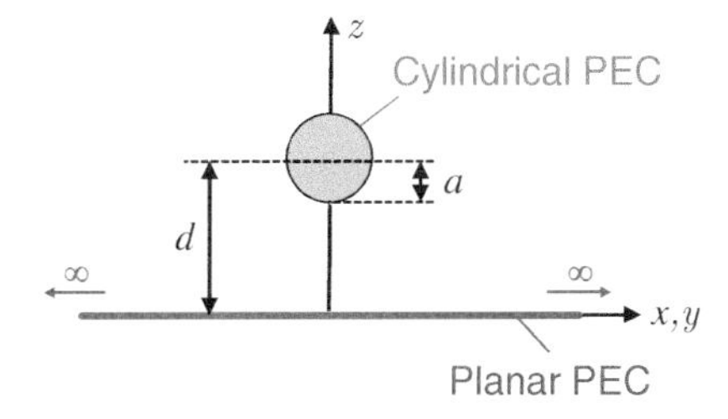

Figure 6.92 An infinitely long cylindrical PEC above an infinitely large planar PEC.

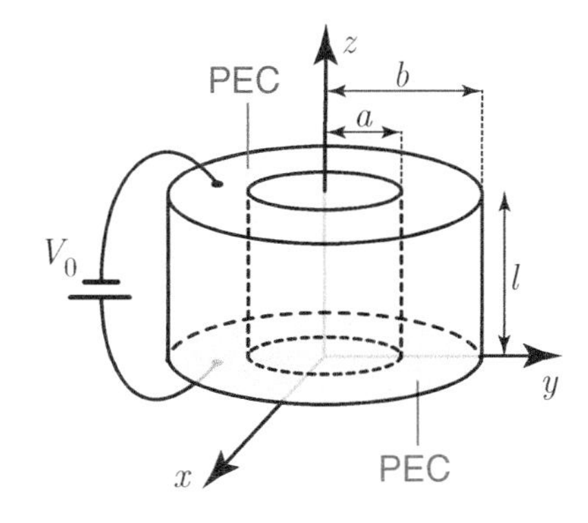

Figure 6.93 A cylindrical structure that is formed by two different materials.

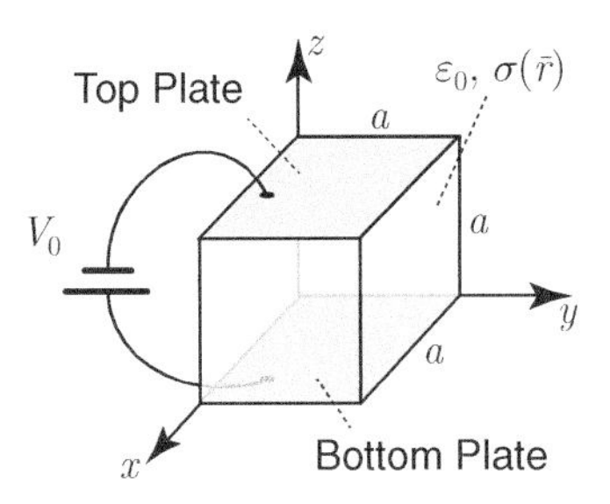

Figure 6.94 A cubic structure consisting of an inhomogeneous material between planar PECs.

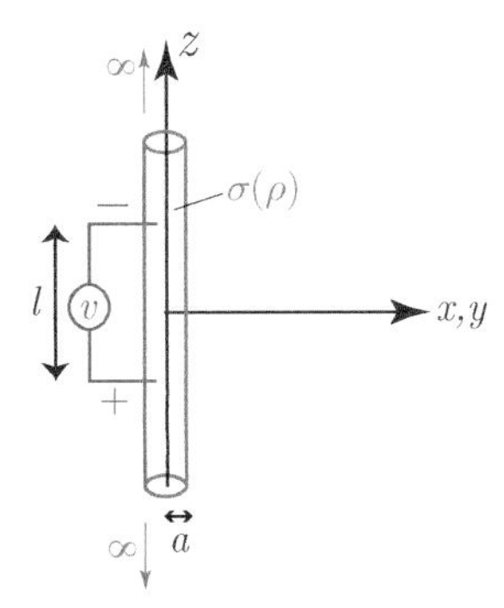

Figure 6.95 An infinitely long cylinder of radius a with conductivity $\sigma = \sigma_0 \exp(\rho/a)$.

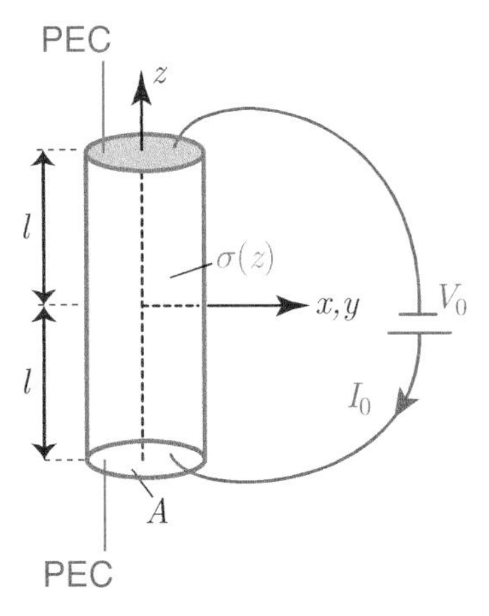

Figure 6.96 A cylindrical structure with an inhomogeneous material between the plates at $z = -l$ and $z = l$.

Exercise 80: Consider a cylindrical structure that is formed by two different materials (Figure 6.93). The conductivity values are given as

$$\sigma(\bar{r}) = \sigma_0 \begin{cases} 1 & (0 < \rho < a) \\ \rho/a & (a < \rho < b), \end{cases} \tag{6.590}$$

where σ_0 (S/m) is a constant. Find the resistance of the structure between the top and bottom plates.

Exercise 81: A cubic object of size $a \times a \times a$ is defined in the first octant between two perfectly conducting plates (Figure 6.94). The permittivity of the object is constant ε_0, while its conductivity depends on the position as

$$\sigma(\bar{r}) = \sigma_0 \left(1 + \frac{z}{a}\right). \tag{6.591}$$

A constant voltage source V_0 is connected between the plates. Do the following, neglecting fringing fields:

- Find the resistance of the object (between the top and bottom plates).
- Find the surface electric charge density on the top and bottom plates.
- Find the volume electric charge density inside the cubic object.
- Find the electric scalar potential inside the cubic object.
- Find the dissipated power.

Exercise 82: An infinitely long cylindrical cable of radius a is aligned in the z direction (Figure 6.95). The conductivity of the cable is given as $\sigma = \sigma_0 \exp(\rho/a)$. In a static case, a voltmeter is connected to the cylinder to measure V_0 (V) across a distance l. Find the magnetic field intensity outside the cable.

Exercise 83: Consider a cylindrical object of length $2l$ and cross-sectional area A (Figure 6.96). The conductivity inside the object is given with respect to position as

$$\sigma(\bar{r}) = \sigma_0 \left(\frac{l}{z}\right)^{2n}, \tag{6.592}$$

where n is an integer. The permittivity is ε_0 everywhere. Find an expression for the resistance between the top and bottom perfectly conducting plates.

Exercise 84: Consider a conic structure aligned in the z direction with θ_0 expansion angle (Figure 6.97). The value of the radial distance R on the surface changes from a to b, while the conductivity is defined as

$$\sigma(\bar{r}) = \sigma_0 \frac{a}{R}. \tag{6.593}$$

Find the resistance between the top and bottom surfaces.

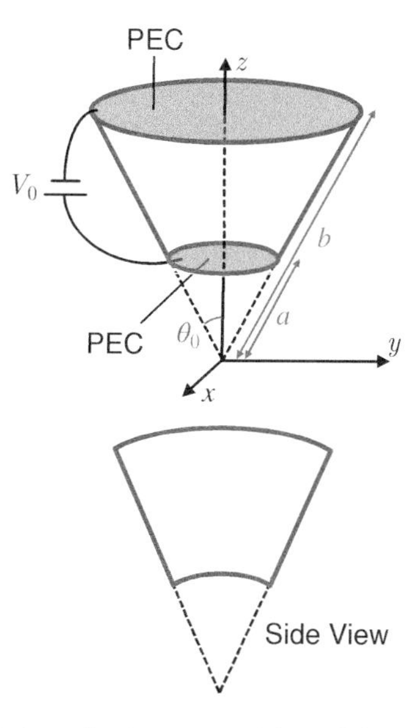

Figure 6.97 A conic structure with an inhomogeneous material between PECs. The top and bottom surfaces are assumed to be curved with fixed radial distances (a for the bottom surface and b for the top surface) to the origin.

Exercise 85: Consider two loops C_1 and C_2 with arbitrary shapes and a center-to-center distance x between them (Figure 6.98). When $I_1 = 10$ A and $I_2 = 0$, it is given that the magnetic flux created by C_1 on C_2 is $f_{12} = 3/x$ (Wb).

- Find the magnetic flux created by C_2 on C_1 when $I_1 = 0$ and $I_2 = 30$ A.
- Find the magnetic force between the loops when $I_1 = 10$ A and $I_2 = 30$ A.

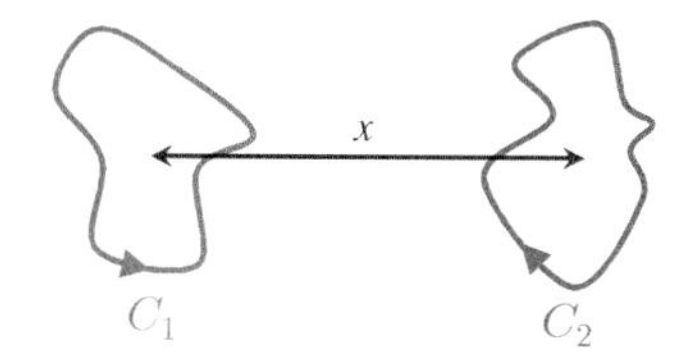

Figure 6.98 Two loops with x distance between them.

Exercise 86: Consider a circular wire of radius a positioned at a distance d from an infinitely long straight wire located in the same plane in a vacuum (Figure 6.99). Find the mutual inductance between the wires.

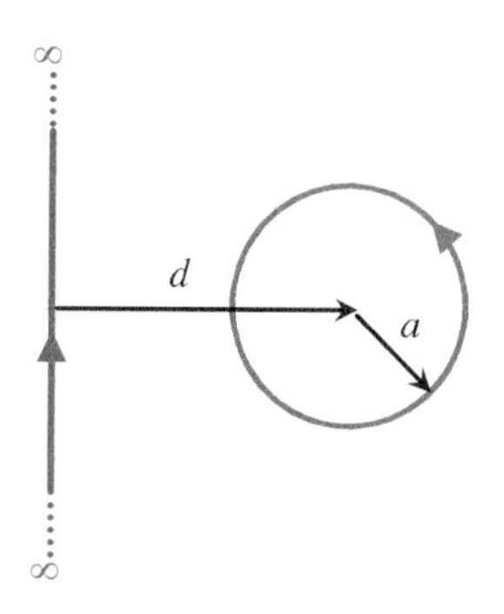

Figure 6.99 A circular wire located at a distance d from an infinitely long wire. The line directions are selected arbitrarily.

Exercise 87: Consider an infinitely long wire carrying a steady electric current flowing in the z direction and a triangular loop that also carries steady electric current I_0 located on the z-x plane in a vacuum (Figure 6.100). Find the magnetic force acting on the loop.

Exercise 88: Consider a cylindrical object of length l and radius a aligned in the z direction from $z = 0$ to $z = l$ (Figure 6.101). Also assume that a steady current I_0 flows in the z direction. The conductivity of the object is given as

$$\sigma(\bar{r}) = \sigma_0 \frac{l}{z}, \tag{6.594}$$

while permittivity and permeability are constants as ε_0 and μ_0.

- Find the dissipated power.
- Find the resistance of the object.
- Find the electric potential energy stored inside the object.
- Find the capacitance of the object.

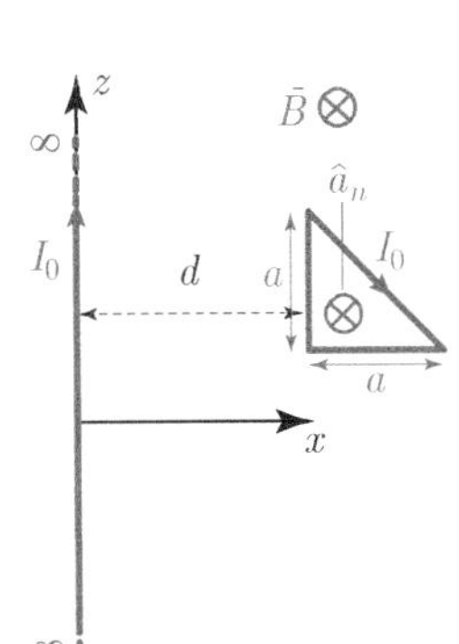

Figure 6.100 An infinitely long wire carrying a steady electric current flowing in the z direction, and a triangular loop (which also carries steady electric current I_0) located on the z-x plane in a vacuum.

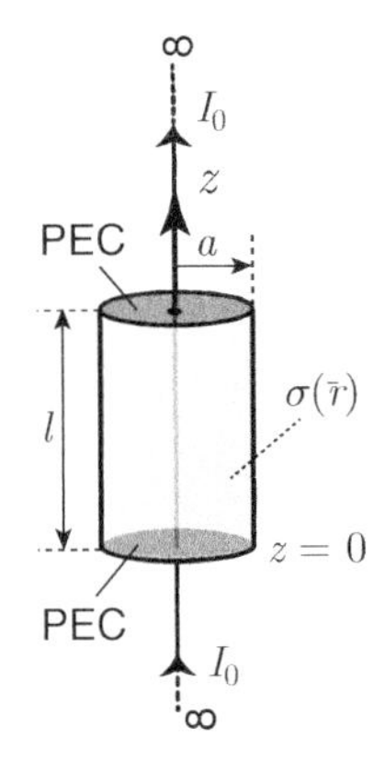

Figure 6.101 A cylindrical object with an inhomogeneous lossy material. A steady electric current I_0 flows through the structure.

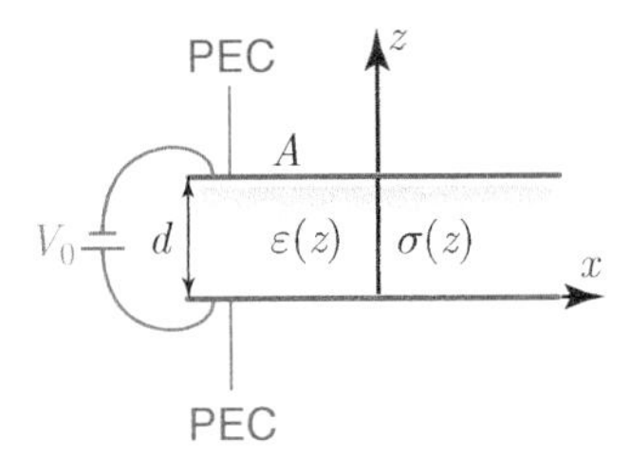

Figure 6.102 A parallel-plate structure filled with an inhomogeneous dielectric material having varying permittivity and conductivity in the vertical direction.

- Find the magnetic potential energy stored inside the object.
- Find the internal inductance of the object.
- Find the Poynting vector inside the object.

6.12 Questions

Question 31: Consider a spherical capacitor consisting of a PEC of radius a inside a perfectly conducting spherical surface of radius c and two kinds of dielectric materials between them: ε_1 from $R = a$ to $R = b$ and ε_2 from $R = b$ to $R = c$ with $a < b < c$. Show that the capacitance can be written

$$C_0 = \frac{4\pi\varepsilon_1\varepsilon_2 abc}{\varepsilon_2(b-a)c + \varepsilon_1(c-b)a}. \tag{6.595}$$

Question 32: Consider a cylindrical lossy object aligned along the z axis from $z = 0$ to $z = l$ with radius a. The resistance between the top and bottom surfaces is given as R_0. If the conductivity is given as $\sigma = \sigma_0 a/z$, show that

$$\sigma_0 = \frac{l^2}{2\pi a^3 R_0}. \tag{6.596}$$

Question 33: Consider an $a \times a \times a$ cubic resistor defined in the first octant of the coordinate system with conductivity

$$\sigma = \sigma_0 \exp(nz/a), \tag{6.597}$$

where σ_0 is constant. Two measurements are performed for the resistance: R_1 is measured between the top and bottom surfaces, while R_2 is measured between the front and back surfaces. Show that the ratio R_2/R_1 can be written solely in terms of n as

$$R_2/R_1 = \frac{n^2}{(1-\exp(-n))(\exp(n)-1)} = \frac{n^2}{\exp(n)+\exp(-n)-2}. \tag{6.598}$$

Question 34: Consider a parallel-plate structure (Figure 6.102) consisting of two PECs of areas A separated by a distance d. The material between the PECs has permittivity and conductivity of

$$\varepsilon(\bar{r}) = \varepsilon_0\left(\frac{d}{z}\right)^n, \qquad \sigma(\bar{r}) = \sigma_0\left(\frac{d}{z}\right)^m, \tag{6.599}$$

where ε_0 and σ_0 are constants, and $(m \geq 0, n \geq 0)$ are real numbers. Note that the bottom plate is at $z = 0$.

- Show that the resistance of the structure is

$$R_0 = d/[A\sigma_0(m+1)]. \quad (6.600)$$

- Find the electric field intensity $\bar{E}(\bar{r})$ and electric flux density $\bar{D}(\bar{r})$ if V_0 voltage is applied between the plates.
- Find the total charge on the bottom plate.
- Find the condition for m and n such that there is no volume electric charge density between the plates.

Question 35: Consider two parallel current-carrying cylinders of infinite length located in a vacuum. The radius of each cylinder is a, while the distance between their centers is d, where $d \gg a$ (e.g. similar to Figure 6.85, while the cylinders are not perfectly conducting). Electric currents flow in opposite directions uniformly through the cross sections of the cylinders.

- Show that the inductance per length of the overall structure can be written[181]

$$L_{0l} = \frac{\mu_0}{\pi} \ln\left(\frac{d}{a}\right) + \frac{\mu_0}{4\pi}. \quad (6.601)$$

- Show that the strength of the magnetic force per length between the cylinders can be written[182,183]

$$F_{0l} = \frac{\mu_0 I_0^2}{2\pi d}, \quad (6.602)$$

 if I_0 corresponds to the steady electric currents flowing through the cylinders (in opposite directions).

Question 36: Consider a spherical structure with two PECs at $R = a$ and $R = b > a$, and an inhomogeneous material between them (Figure 6.66). Find the capacitance and resistance of the structure if the permittivity and conductivity are defined as $\varepsilon(\bar{r}) = \varepsilon_0 R/b$ and $\sigma = \sigma_0 a^2/R^2$.

[181] Also identify the first and second terms.

[182] Is the force repulsive or attractive?

[183] How do the inductance and magnetic force change if the electric currents through the cylinders are assumed to flow in the same direction?

7

Transmission of Electromagnetic Waves

According to Maxwell's equations, electromagnetic radiation occurs naturally: i.e. waves generated by sources (electric charges and currents) propagate away, carrying energy to distant locations even in the absence of any matter (Figure 7.1). These waves naturally interact with dielectric, magnetic, and conducting objects, as described by various phenomena (e.g. reflection, refraction, and diffraction), leading to interruptions in regular wave propagation. But in general, there is no need for any effort to transmit electromagnetic waves. Nevertheless, in real-life applications, it is necessary to establish efficient, reliable, and robust transmission media to *wisely* transmit electromagnetic waves, specifically energy and information (*signals*) carried by them. In this chapter, we consider how these transmission environments operate and how they can be analyzed. We start with the most natural mechanism *wireless transmission* using antennas. Then we study *waveguides*, which can be considered bounded environments to efficiently transmit electromagnetic waves over relatively shorter distances. Finally, *transmission lines* are excellent models of common transmission tools, such as coaxial and twisted pair cables.[1]

Figure 7.1 Sunlight is a collection of electromagnetic waves (some at visible frequencies), which propagate in space before being received directly or indirectly by our eyes or absorbed by plants for photosynthesis. Some events like vision and absorption can be better explained using particle physics: i.e. considering electromagnetic energy carriers (photons) rather than waves.

[1]It should be emphasized that wireless transmission, antennas, waveguides, and transmission lines are broad topics, each of which could easily be the subject of a single book. This chapter provides an overview of these transmission concepts.

7.1 Antennas and Wireless Transmission

Wireless transmission is a fundamental component of modern life (Figure 7.2), particularly since mobile phones have become widespread in the last several decades. By carrying a cellular phone, each person carries at least one antenna, meaning that billions of moving antennas are distributed over the surface of the Earth – and even more are located inside our houses (e.g. for televisions).

Introduction to Electromagnetic Waves with Maxwell's Equations, First Edition. Özgür Ergül.

Companion website: www.wiley.com/go/ergulmax

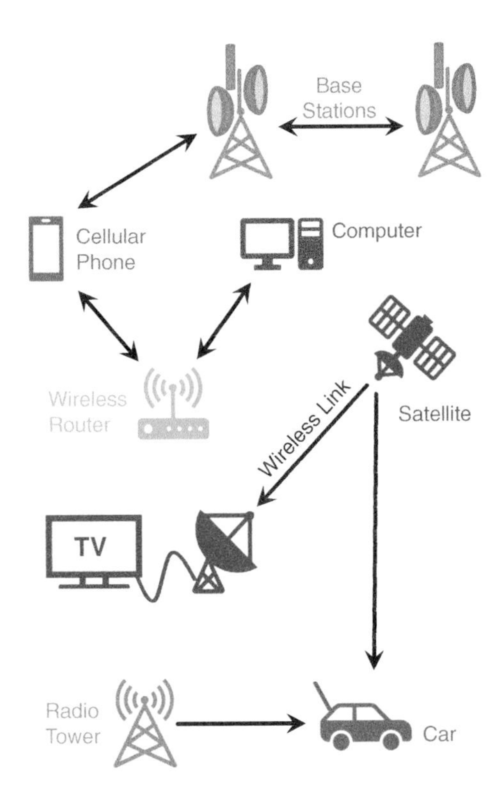

Figure 7.2 Some real-life applications involving wireless communications and, hence, antennas.

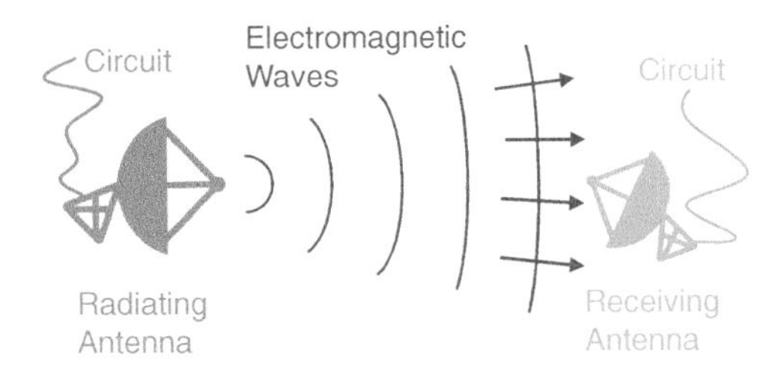

Figure 7.3 Radiating and receiving antennas.

In general, an antenna can be seen as an end point of a wireless transmission environment. Specifically, a *radiating* antenna generates electromagnetic waves by converting bounded electrical energy (provided by a circuit) into electromagnetic energy that propagates away (as an electromagnetic wave) from the antenna (Figure 7.3). Using an antenna, the generated electromagnetic wave is intended to be received by another antenna or antennas, sometimes by itself after being reflected from a target, depending on the application. A *receiving* antenna receives electromagnetic waves and converts them into bounded electrical energy to be transferred to a circuit (Figure 7.3). Radiating or receiving, antennas are generally connected to electronic systems according to their functions. For example, in a cellular phone, electrical signals converted by an antenna and transferred to its circuit (load) are further processed and converted into a voice.

This section provides an overview of antennas: their basic properties, design parameters, and types, as well as the fundamental formulation of wireless transmission. In all discussions, we restrict ourselves to time-harmonic sources/fields that can be studied in the phasor domain. In addition, unless required, concepts are presented for radiating antennas without directly considering receiving antennas. This is thanks to the reciprocity principle, which indicates that radiating and receiving properties of an antenna are identical. Specifically, characteristic properties of an antenna, such as the radiation pattern, directivity, gain, and input impedance remain the same whether it is used as a radiating antenna or a receiving antenna. This interesting principle makes it possible to consider the antenna in one of two modes, depending on the analysis.

7.1.1 Basic Properties of Antennas

Hertzian (ideal) dipoles, studied in Section 5.2.1, can be considered the most basic radiating elements to describe some of the fundamental properties of antennas, particularly their radiation characteristics. As found in Section 5.2.1, the magnetic vector potential and the electric and magnetic fields generated by a Hertzian dipole located at the origin and aligned in the z direction (Figure 7.4) in an empty space[2] (ε, μ) can be written

$$\bar{A}_d(\bar{r}) = (\hat{a}_R \cos\theta - \hat{a}_\theta \sin\theta)\mu I_{DM,h} g(R) \tag{7.1}$$

$$\bar{B}_d(\bar{r}) = \bar{\nabla} \times \bar{A}_d(\bar{r}) = \hat{a}_\phi \mu I_{DM,h} \sin\theta \left(jk + \frac{1}{R}\right) g(R) \tag{7.2}$$

$$\begin{aligned}\bar{E}_d(\bar{r}) = \frac{1}{j\omega\varepsilon\mu}\bar{\nabla} \times \bar{B}_d(\bar{r}) &= -\hat{a}_R j\omega\mu I_{DM,h} g(R)\left[2\cos\theta\left(\frac{j}{kR} + \frac{1}{k^2R^2}\right)\right] \\ &\quad + \hat{a}_\theta j\omega\mu I_{DM,h} g(R)\left[\sin\theta\left(1 - \frac{j}{kR} - \frac{1}{k^2R^2}\right)\right],\end{aligned} \tag{7.3}$$

[2] In this chapter, we assume that antenna(s) are located in empty space with homogeneous material properties, such as a vacuum.

where[3] $I_{DM,h} = I_0 \Delta z$ and

$$g(R) = \frac{\exp(-jkR)}{4\pi R}, \qquad k = \omega\sqrt{\mu\varepsilon} = 2\pi/\lambda. \tag{7.4}$$

In most applications, antennas are used as *far-zone* devices: i.e. they interact with each other and with other objects that are located far away[4] from them, but near-zone interactions (if not within the antenna system itself) are often considered unwanted effects that need to be eliminated. In the far zone ($kR \gg 1$), we have[5]

$$\bar{E}_d^{\text{far}}(\bar{r}) = \hat{a}_\theta j\omega\mu I_{DM,h} \frac{\exp(-jkR)}{4\pi R} \sin\theta \tag{7.5}$$

$$\bar{B}_d^{\text{far}}(\bar{r}) = \hat{a}_\phi jk\mu I_{DM,h} \frac{\exp(-jkR)}{4\pi R} \sin\theta = \frac{\mu}{\eta}\hat{a}_R \times \bar{E}_d^{\text{far}}(\bar{r}), \tag{7.6}$$

where $\eta = \sqrt{\mu/\varepsilon}$. The complex Poynting vector (complex power flux density) corresponding to these fields can be found as

$$\bar{S}_{c,d}^{\text{far}}(\bar{r}) = \frac{1}{2}\bar{E}_d^{\text{far}}(\bar{r}) \times [\bar{H}_d^{\text{far}}(\bar{r})]^* = \hat{a}_R |I_{DM,h}|^2 \frac{\omega k \mu}{32\pi^2 R^2} \sin^2\theta. \tag{7.7}$$

Because it is purely real,[6] the complex Poynting vector directly represents the time-average power flux density of the wave generated by the Hertzian dipole.

At this stage, we will focus on the power flux density to understand the radiation characteristics of a Hertzian dipole as a representative antenna. Considering the expression in Eq. (7.7), we first note that the power flux density decays quadratically with R. This is a natural consequence of the fact that the power radiated by the Hertzian dipole is distributed over larger areas as the observation point gets further away. The *radiation intensity* of an antenna is defined as the power per unit solid angle, which simply corresponds to the power flux density multiplied by the square of the distance, i.e.

$$U^{\text{far}}(\theta,\phi) = R^2 |\bar{S}_c^{\text{far}}(\bar{r})|. \tag{7.8}$$

For the Hertzian dipole, we have

$$U_d^{\text{far}}(\theta,\phi) = R^2 |\bar{S}_{c,d}^{\text{far}}(\bar{r})| = |I_{DM,h}|^2 \frac{\omega k \mu}{32\pi^2} \sin^2\theta, \tag{7.9}$$

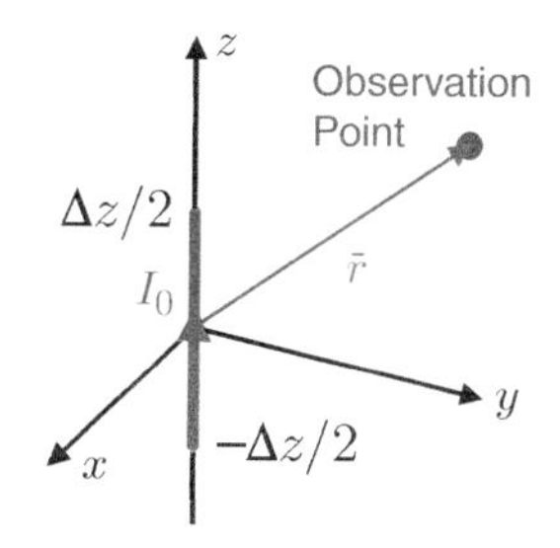

Figure 7.4 A Hertzian dipole located at the origin and oriented in the z direction. Current I_0 is assumed to be constant (position-independent).

[3]Recall that $\bar{I}_{DM,h} = \hat{a}_z I_{DM,h}$ is called the dipole moment.

[4]For a distance R to be a far-zone location, we must have $R \gg \lambda$ (much larger than the wavelength) and $R \gg D$, where D represents the maximum dimension of the antenna.

[5]As a general property, far-zone fields due to a finite source distribution do not contain radial ($\hat{a}_R$) components, but they satisfy the plane-wave property of $\bar{H}^{\text{far}}(\bar{r}) = (1/\eta)\hat{a}_R \times \bar{E}^{\text{far}}(\bar{r})$.

[6]This is also a general property satisfied by all finite sources radiating in a homogeneous space.

U^{far}: radiation intensity
Unit: watts/steradian (W/sr)

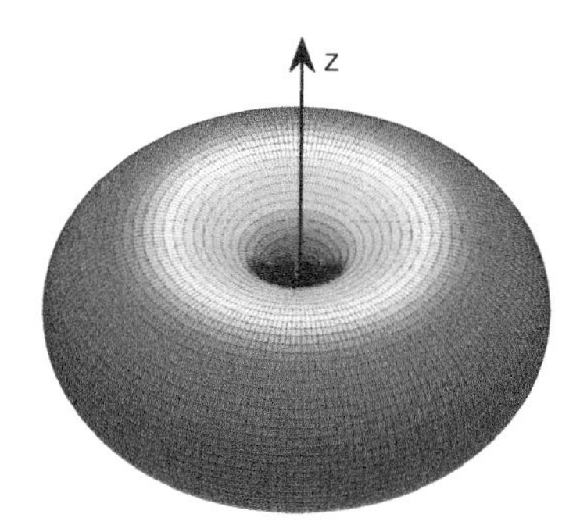

Figure 7.5 A three-dimensional plot of the radiation pattern of a Hertzian dipole oriented in the z direction: i.e. $\tilde{U}_d^{\text{far}}(\theta,\phi) = \sin^2\theta$.

which is free of any distance (R) term.[7] Also note that the integral of the radiation intensity over a unit sphere leads to the total radiated power as

$$p_{em}^{\text{far}} = \int_0^{2\pi} \int_0^{\pi} U^{\text{far}}(\theta, \phi) \sin\theta d\theta d\phi, \tag{7.10}$$

which leads to[8]

$$p_{em,d}^{\text{far}} = \int_0^{2\pi} \int_0^{\pi} U_d^{\text{far}}(\theta, \phi) \sin\theta d\theta d\phi = |I_{DM,h}|^2 \frac{\omega k \mu}{32\pi^2} \frac{8\pi}{3} = |I_{DM,h}|^2 \frac{\omega k \mu}{12\pi} \tag{7.11}$$

for the Hertzian dipole. On the other side, the *radiation pattern* of an antenna corresponds to the normalized angular dependency of the radiation intensity:

$$\tilde{U}^{\text{far}}(\theta, \phi) = \frac{U^{\text{far}}(\theta, \phi)}{\max\{U^{\text{far}}(\theta, \phi)\}}. \tag{7.12}$$

For the Hertzian dipole, we have (Figures 7.5–7.7)

$$\tilde{U}_d^{\text{far}}(\theta, \phi) = U_d^{\text{far}}(\theta, \phi) / \max\{U_d^{\text{far}}(\theta, \phi)\} = \sin^2\theta, \tag{7.13}$$

which shows that the Hertzian dipole (located at the origin and oriented in the z direction) radiates mostly in directions with $\theta = 90°$ (x-y plane), but it has a zero radiation toward the poles ($\theta = 0°$ and $\theta = 180°$).

In general, antennas should radiate in specific directions while suppressing radiation in other directions: e.g. to maximize efficiency and reduce interference with other antennas. The ability of an antenna to direct its radiation is characterized by directivity.[9] Specifically, the *directive gain* of an antenna is defined as the radiation intensity divided by the angle-average of the total radiated power:

$$D(\theta, \phi) = \frac{U^{\text{far}}(\theta, \phi)}{p_{em}^{\text{far}}/(4\pi)} = \frac{4\pi U^{\text{far}}(\theta, \phi)}{p_{em}^{\text{far}}}. \tag{7.14}$$

Note that the angle-average of the total radiated power ($p_{em}^{\text{far}}/(4\pi)$) corresponds to the radiation intensity if the same amount of power was radiated by an isotropic source.[10] For the Hertzian dipole, we obtain the directive gain as

$$D_d(\theta, \phi) = \frac{4\pi |I_{DM,h}|^2 \frac{\omega k \mu}{32\pi^2} \sin^2\theta}{|I_{DM,h}|^2 \frac{\omega k \mu}{12\pi}} = \frac{48\pi^2}{32\pi^2} \sin^2\theta = 1.5 \sin^2\theta. \tag{7.15}$$

[7] This is true for any antenna, since the power flux density of a typical antenna decays quadratically with the distance.

[8] We use

$$\begin{aligned}\int_0^{\pi} \sin^3\theta d\theta &= \int_0^{\pi} (1 - \cos^2\theta) \sin\theta d\theta \\ &= \int_0^{\pi} \sin\theta d\theta - \int_0^{\pi} \cos^2\theta \sin\theta d\theta \\ &= \left[-\cos\theta + (\cos^3\theta)/3\right]_0^{\pi}.\end{aligned}$$

$\tilde{U}^{\text{far}}$: radiation pattern
Unitless

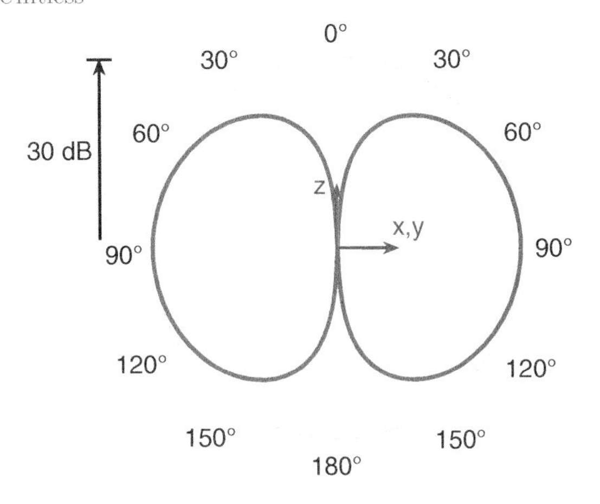

Figure 7.6 Polar plot of the radiation pattern of a Hertzian dipole oriented in the z direction. Logarithmic values are shown in dB with respect to θ. In the plot, 30 dB dynamic range is used, which means the origin corresponds to −30 dB.

[9] Note that the radiation pattern given in Eq. (7.12) already provides information about how the radiation is distributed; but the normalization to unity leads to a loss of information.

D: directive gain
Unitless

[10] An ideal isotropic source radiates equally in all directions.

This means the directive gain of the Hertzian dipole (located at the origin and oriented in the z direction) is 1.5 when[11] $\theta = 0°$, approximately 1.0 when $\theta = 55°$, and 0.0 when $\theta = 90°$. The maximum of the directive gain ($D_{\max}$) is often called *directivity* to characterize the maximum radiation of an antenna.[12] For the Hertzian dipole, the directivity is $D_{d,\max} = \max\{D_d(\theta, \phi)\} = 1.5$.

To gain more insight into the basic properties of antennas, we now consider a short wire antenna (short dipole) with small length $l \ll \lambda$ (Figure 7.8). The wire is an ideal PEC with zero thickness aligned symmetrically along the z axis, and it is fed by a time-harmonic voltage source located at its center. In general, the current distribution on the antenna can be found by fully solving Maxwell's equations (or their higher-order forms), but this can be difficult even for a simple wire without using a numerical solver. Therefore, for analytical calculations, it is common to approximate current distributions on basic antennas (verified by numerical approaches). For a short wire, a simple approximation is a triangular distribution as

$$\bar{I}_{sw}(x = 0, y = 0, z) = \bar{I}_{sw}(z) = \hat{a}_z I_0 \begin{cases} 1 - 2z/l & (0 \leq z \leq l/2) \\ 1 + 2z/l & (-l/2 \leq z \leq 0), \end{cases} \tag{7.16}$$

where I_0 (A) is a constant. To find the electric and magnetic field intensities, we can first obtain the magnetic vector potential via

$$\bar{A}_{sw}(\bar{r}) = \frac{\mu}{4\pi} \int_V \bar{J}_{v,sw}(\bar{r}') \frac{\exp(-jk|\bar{r} - \bar{r}'|)}{|\bar{r} - \bar{r}'|} d\bar{r}'. \tag{7.17}$$

On the other hand, finding first general expressions for $\bar{A}_{sw}(\bar{r})$, $\bar{E}_{sw}(\bar{r})$, and $\bar{H}_{sw}(\bar{r})$, and then evaluating their far-zone expressions may not be trivial.[13] Instead, we can first derive general expressions for the far-zone magnetic vector potential as follows. Considering[14] $|\bar{r}| \gg \lambda$ (or $k|\bar{r}| \gg 1$) and $|\bar{r}| \gg l$, we have

$$\frac{\exp(-jk|\bar{r} - \bar{r}'|)}{4\pi|\bar{r} - \bar{r}'|} \approx \frac{\exp(-jk|\bar{r}|)}{4\pi|\bar{r}|} \exp(jk\hat{a}_r \cdot \bar{r}') = g(R) \exp(jk\hat{a}_r \cdot \bar{r}') \tag{7.18}$$

and[15]

$$\bar{A}^{\text{far}}(\bar{r}) = \mu \int_V \bar{J}_v(\bar{r}') g(R) \exp(jk\hat{a}_r \cdot \bar{r}') d\bar{r}' \tag{7.19}$$

$$= \mu g(R) \int_V \bar{J}_v(\bar{r}') \exp(jk\hat{a}_r \cdot \bar{r}') d\bar{r}' \tag{7.20}$$

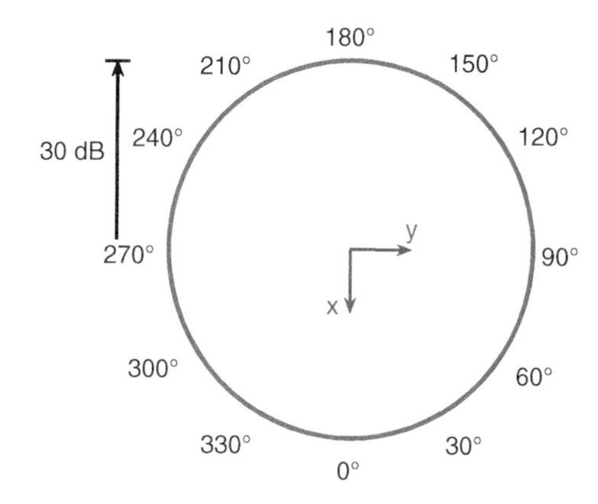

Figure 7.7 Polar plot of the radiation pattern of a Hertzian dipole oriented in the z direction. Logarithmic values are shown in dB with respect to ϕ on the x-y plane. On this plane, the dipole radiates equally in all directions.

[11] In the maximum direction, a Hertzian dipole radiates 1.5 times stronger than an isotropic source.

[12] In another convention used in the literature, $D(\theta, \phi)$ is called *directivity*, and $D_{\max}$ is called *maximum directivity*.

[13] Even if it is not for a short wire, it can be a difficult procedure for more complex systems.

[14] Recall that a far zone should be defined as $|\bar{r}| \gg \lambda$ *and* $|\bar{r}| \gg D$, where D is the maximum dimension of the given structure.

[15] Note that

$$|\bar{r} - \bar{r}'| = \sqrt{(\bar{r} - \bar{r}') \cdot (\bar{r} - \bar{r}')} \approx \sqrt{|\bar{r}|^2 - 2\bar{r} \cdot \bar{r}'}$$
$$\approx |\bar{r}|\sqrt{1 - 2\hat{a}_r \cdot \bar{r}'/|\bar{r}|} \approx |\bar{r}|\,(1 - \hat{a}_r \cdot \bar{r}'/|\bar{r}|)$$
$$= |\bar{r}| - \hat{a}_r \cdot \bar{r}'.$$

This first-order approximation can be used for the phase (exponential part), but the magnitude can be approximated simply as $|\bar{r} - \bar{r}'| \approx |\bar{r}|$.

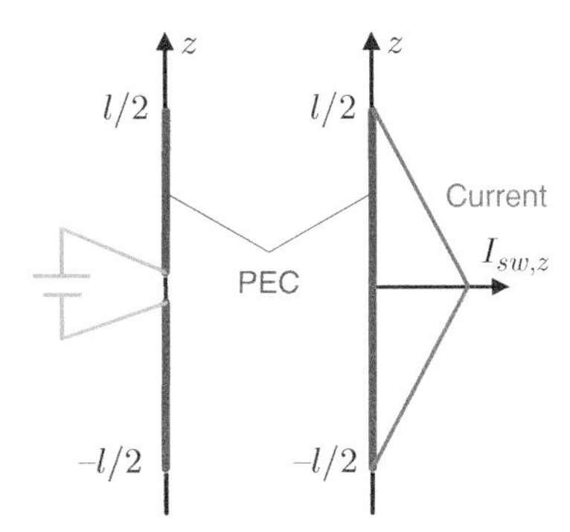

Figure 7.8 A short wire antenna located symmetrically on the z axis and a triangular approximation of the electric current along it.

for the far-zone magnetic vector potential due to an *arbitrary* current distribution[16] $\bar{J}_v(\bar{r})$. One can further show that

$$\bar{B}^{\text{far}}(\bar{r}) = -jk\hat{a}_R \times \bar{A}^{\text{far}}(\bar{r})$$
$$\bar{E}^{\text{far}}(\bar{r}) = -\frac{\eta}{\mu}\hat{a}_R \times \bar{B}^{\text{far}}(\bar{r}) = -j\omega[\bar{A}^{\text{far}}(\bar{r}) - \hat{a}_R\hat{a}_R \cdot \bar{A}^{\text{far}}(\bar{r})]$$

in the far zone. Using the general expression in Eq. (7.20) for the short wire, we have

$$\bar{A}^{\text{far}}_{sw}(\bar{r}) = \mu g(R)\int_{-l/2}^{l/2} \bar{I}_{sw}(z')\exp(jk\hat{a}_r \cdot \bar{r}')d\bar{r}' \tag{7.21}$$

$$= \hat{a}_z\mu I_0 g(R)\int_0^{l/2} (1 - 2z'/l)\exp(jkz'\cos\theta)dz'$$
$$+ \hat{a}_z\mu I_0 g(R)\int_{-l/2}^{0} (1 + 2z'/l)\exp(jkz'\cos\theta)dz' \tag{7.22}$$

$$= \hat{a}_z\mu I_0 g(R)2\int_0^{l/2} (1 - 2z'/l)\cos(kz'\cos\theta)dz' \tag{7.23}$$

leading to

$$\bar{A}^{\text{far}}_{sw}(\bar{r}) = \hat{a}_z\mu I_0 g(R)\frac{4}{k^2 l\cos^2\theta}\big[1 - \cos(k(l/2)\cos\theta)\big]. \tag{7.24}$$

Using a first-order approximation,[17] the magnetic vector potential can further be simplified as

$$\bar{A}^{\text{far}}_{sw}(\bar{r}) = \hat{a}_z\mu I_0 g(R)\frac{4}{k^2 l\cos^2\theta}\frac{(kl)^2\cos^2\theta}{8} = \hat{a}_z\frac{1}{2}\mu I_0 l g(R). \tag{7.25}$$

This expression is remarkably similar to the magnetic vector potential due to a Hertzian dipole: i.e. only multiplication with 1/2 arises due to the triangular distribution of the current. The corresponding electric and magnetic fields can be obtained as

$$\bar{E}^{\text{far}}_{sw}(\bar{r}) = \hat{a}_\theta j\omega\mu I_0 l\frac{\exp(-jkR)}{8\pi R}\sin\theta, \quad \bar{B}^{\text{far}}_{sw}(\bar{r}) = \hat{a}_\phi jk\mu I_0 l\frac{\exp(-jkR)}{8\pi R}\sin\theta. \tag{7.26}$$

[16] This expression is often interpreted as the *Fourier transform* of the electric current distribution. Specifically, the far-zone magnetic vector potential due a current distribution corresponds to its spatial Fourier transform.

[17] Specifically, we use $\cos a \approx 1 - a^2/2$ for small a.

Hence, we have the complex Poynting vector and the radiation intensity as

$$\bar{S}^{\text{far}}_{c,sw}(\bar{r}) = \frac{1}{2}\bar{E}^{\text{far}}_{sw}(\bar{r}) \times [\bar{H}^{\text{far}}_{sw}(\bar{r})]^* = \hat{a}_R|I_0|^2 l^2 \frac{\omega k\mu}{128\pi^2 R^2}\sin^2\theta \quad (7.27)$$

$$U^{\text{far}}_{sw}(\theta,\phi) = R^2|\bar{S}^{\text{far}}_{c,sw}(\bar{r})| = |I_0|^2 l^2 \frac{\omega k\mu}{128\pi^2}\sin^2\theta, \quad (7.28)$$

leading to the total radiated power as[18,19]

$$p^{\text{far}}_{em,sw} = \int_0^{2\pi}\int_0^{\pi} U^{\text{far}}_{sw}(\theta,\phi)\sin\theta d\theta d\phi = |I_0|^2 l^2 \frac{\omega k\mu}{48\pi} = \eta\frac{\pi}{12}\left(\frac{l}{\lambda}\right)^2 |I_0|^2. \quad (7.29)$$

Note that with these expressions, the radiation pattern and directive gain of a short wire antenna are the same as those of a Hertzian dipole.

Next, we consider a class of wire antennas that have lengths comparable to the wavelength. We specifically focus on the commonly used *half-wave* dipoles with $l = \lambda/2$. In such a case, the current on the wire cannot be approximated as a triangular distribution. A more realistic distribution is sinusoidal (Figure 7.9): i.e. we have

$$\bar{I}_{hw}(x=0, y=0, z) = \bar{I}_{hw}(z) = \hat{a}_z I_0 \cos(kz). \quad (7.30)$$

The far-zone magnetic vector potential can be obtained as[20]

$$\bar{A}^{\text{far}}_{hw}(\bar{r}) = \hat{a}_z\mu I_0 g(R)\int_{-\lambda/4}^{\lambda/4}\cos(kz')\exp(-jkz'\cos\theta)dz' \quad (7.31)$$

$$= \hat{a}_z\mu I_0 g(R)\frac{2}{k}\frac{\cos[(\pi/2)\cos\theta]}{\sin^2\theta}. \quad (7.32)$$

Then we can derive the magnetic flux density and electric field intensity in the far zone as

$$\bar{B}^{\text{far}}_{hw}(\bar{r}) = -jk\hat{a}_R \times \bar{A}^{\text{far}}_{hw}(\bar{r}) = \hat{a}_\phi j\mu I_0\frac{\exp(-jkR)}{2\pi R}\frac{\cos[(\pi/2)\cos\theta]}{\sin\theta} \quad (7.33)$$

$$\bar{E}^{\text{far}}_{hw}(\bar{r}) = -\frac{\eta}{\mu}\hat{a}_R \times \bar{B}^{\text{far}}_{hw}(\bar{r}) = \hat{a}_\theta j\eta I_0\frac{\exp(-jkR)}{2\pi R}\frac{\cos[(\pi/2)\cos\theta]}{\sin\theta}. \quad (7.34)$$

[18] Note that

$$\frac{\omega k\mu}{48\pi} = \frac{\omega^2(\sqrt{\mu\varepsilon})\mu}{48\pi} = \frac{4\pi^2 f^2(\sqrt{\mu\varepsilon})\mu}{48\pi} = \frac{4\pi^2(\sqrt{\mu\varepsilon})\mu}{\mu\varepsilon\lambda^2 48\pi} = \frac{\pi(\sqrt{\mu/\varepsilon})}{\lambda^2 12\pi} = \frac{\pi\eta}{\lambda^2 12\pi}.$$

[19] For example, if $I_0 = 1$ A and $l = 0.1\lambda$, we have

$$p^{\text{far}}_{em,sw} \approx 120\pi\frac{\pi}{12}(0.1)^2 \approx 0.99 \text{ W},$$

if the dipole is located in a vacuum.

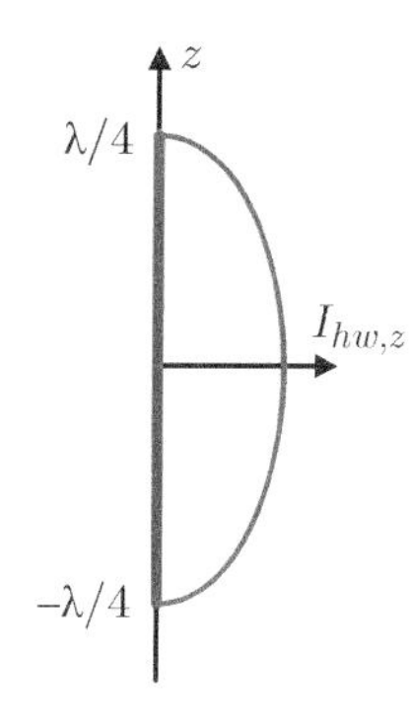

Figure 7.9 Sinusoidal approximation of the electric current along a half-wave dipole antenna.

[20] We have

$$\int_{-\lambda/4}^{\lambda/4}\cos(kz')\exp(-jkz'\cos\theta)dz' = 2\int_0^{\lambda/4}\cos(kz')\cos(kz'\cos\theta)dz'.$$

These expressions lead to the far-zone power flux density as

$$\bar{S}^{\text{far}}_{c,hw}(\bar{r}) = \frac{1}{2}\bar{E}^{\text{far}}_{hw}(\bar{r}) \times [\bar{H}^{\text{far}}_{hw}(\bar{r})]^* = \hat{a}_R|I_0|^2\frac{\eta}{8\pi^2R^2}\frac{\cos^2[(\pi/2)\cos\theta]}{\sin^2\theta}. \tag{7.35}$$

Consequently, the corresponding radiation intensity, radiation pattern, total radiated power, and directive gain of a half-wave dipole antenna can be written[21] (Figures 7.10 and 7.11)

$$U^{\text{far}}_{hw}(\theta,\phi) = R^2|\bar{S}^{\text{far}}_{c,hw}(\bar{r})| = |I_0|^2\frac{\eta}{8\pi^2}\frac{\cos^2[(\pi/2)\cos\theta]}{\sin^2\theta} \tag{7.36}$$

$$\tilde{U}^{\text{far}}_{hw}(\theta,\phi) = U^{\text{far}}_{hw}(\theta,\phi)/\max\{U^{\text{far}}_{hw}(\theta,\phi)\} = \frac{\cos^2[(\pi/2)\cos\theta]}{\sin^2\theta} \tag{7.37}$$

$$p^{\text{far}}_{em,hw} = \int_0^{2\pi}\int_0^{\pi} U^{\text{far}}_{hw}(\theta,\phi)\sin\theta d\theta d\phi \tag{7.38}$$

$$= |I_0|^2\frac{\eta}{8\pi^2}\int_0^{2\pi}\int_0^{\pi}\frac{\cos^2[(\pi/2)\cos\theta]}{\sin^2\theta}\sin\theta d\theta d\phi \approx 1.22|I_0|^2\frac{\eta}{4\pi} \tag{7.39}$$

$$D_{hw}(\theta,\phi) = 4\pi U^{\text{far}}_{hw}(\theta,\phi)/p^{\text{far}}_{em,hw} \approx 1.64\frac{\cos^2[(\pi/2)\cos\theta]}{\sin^2\theta}, \tag{7.40}$$

respectively. Hence, the directivity of a half-wave dipole antenna is 1.64 (larger than a short wire) with maximum radiation toward $\theta = 90°$.

In many cases, antennas cannot be studied as isolated elements without their connections to voltage/current sources and circuits. As discussed below, the compatibility of antennas with other components in antenna systems is extensively considered in the context of important properties, such as gain and input impedance. From a circuit perspective, a related parameter is the *radiation resistance*, which is based on the power consumption of antennas as circuit elements. Specifically, a radiating antenna can be considered a resistor that dissipates power so that the corresponding resistance can be written[22]

$$R_r = 2p^{\text{far}}_{em}/|I_0|^2. \tag{7.41}$$

Then for a short wire antenna and a half-wave dipole, the radiation resistance can be found as[23]

$$R_{r,sw} = \eta\frac{\pi}{6}\left(\frac{l}{\lambda}\right)^2, \qquad R_{r,hw} \approx 2.44\frac{\eta}{4\pi} \approx 0.194\eta. \tag{7.42}$$

For example, at 300 kHz, a 1-meter wire in a vacuum has approximately 0.001λ length, and its radiation resistance corresponds to approximately 2×10^{-4} Ω. Thus it can be extremely difficult to efficiently transfer power from a source to such an antenna.

[21] The total radiated power is found via numerical integration.

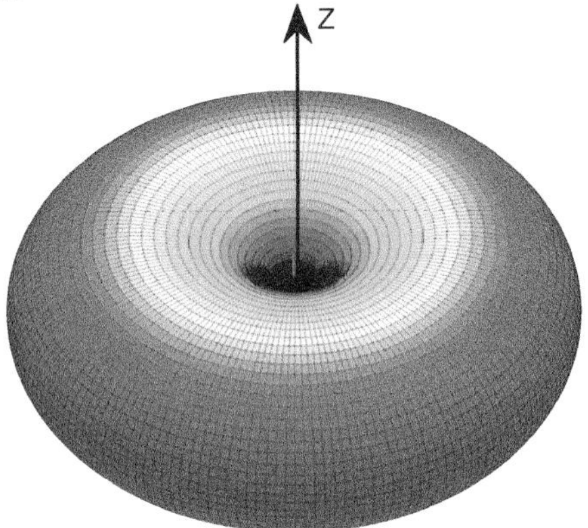

Figure 7.10 A three-dimensional plot of the radiation pattern of a half-wave dipole oriented in the z direction. The plot is slightly different from the radiation pattern of a Hertzian dipole in Figure 7.5.

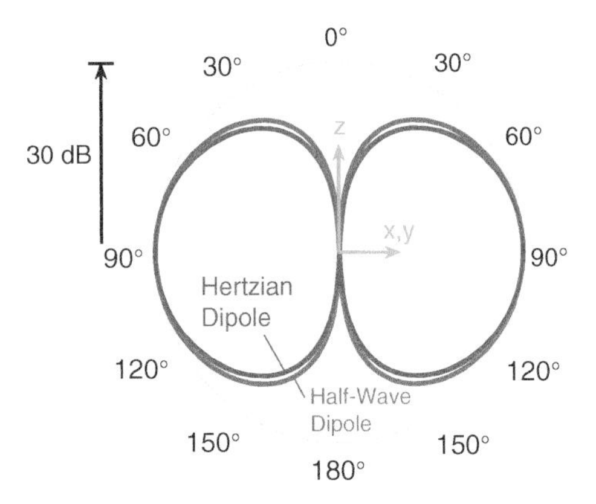

Figure 7.11 Radiation patterns of Hertzian and half-wave dipoles oriented in the z direction with respect to θ.

[22] The factor of two comes from the fact that the overall system (and hence the equivalent circuit) is AC.

[23] Hence, in a vacuum, a short wire antenna has a radiation resistance of $R_{r,sw} \approx 197(l/\lambda)^2$ Ω, and a half-wave dipole has $R_{r,hw} \approx 73.1$ Ω. Using a short-wire approximation for a half-wave dipole would incorrectly lead to $R_{r,hw} \approx 49.3$ Ω.

Gain[24,25] is an important property of antennas defined as

$$G(\theta, \phi) = e_r D(\theta, \phi), \tag{7.43}$$

G: gain
Unitless

where $D(\theta, \phi)$ is the directive gain and $e_r \in [0, 1]$ is the unitless *radiation efficiency*. Therefore, the gain of an antenna is the scaled version of its directive gain. The radiation efficiency of an antenna accounts for conduction losses *inside* the antenna, because the metals and dielectrics used to build the antenna are not perfect in real life. Hence, it can be defined as

$$e_r = p_{em}^{\text{far}} / p_{in} < 1, \tag{7.44}$$

where p_{in} is the real power at the *feedpoint* (connection between the antenna and the feedline to which it is connected): i.e. the power successfully delivered to the antenna from the source (Figure 7.12). Then we can also write the gain in terms of the radiation intensity as

$$G(\theta, \phi) = \frac{p_{em}^{\text{far}}}{p_{in}} D(\theta, \phi) = \frac{p_{em}^{\text{far}}}{p_{in}} \frac{4\pi U^{\text{far}}(\theta, \phi)}{p_{em}^{\text{far}}} = \frac{4\pi U^{\text{far}}(\theta, \phi)}{p_{in}}. \tag{7.45}$$

It should be emphasized that the overall efficiency of an antenna system is not limited to the radiation efficiency. In general, we can define the total efficiency as

$$e_t = e_p e_f e_r, \tag{7.46}$$

where $e_f \in [0, 1]$ accounts for losses due to the feedline (specifically, feedline losses and feedline-to-antenna impedance mismatches) and $e_p \in [0, 1]$ accounts for polarization mismatches.[26,27]

As mentioned above, the value of the radiation resistance is critical in terms of the power that can be transferred from a source to a radiating antenna (or from a receiving antenna to its load). On the other hand, the transfer of power between an antenna and its circuit is fully characterized by its *input impedance* (Figure 7.13). In general, the input impedance of an antenna can be written

$$Z_{in} = R_{in} + jX_{in}, \tag{7.47}$$

where R_{in} and X_{in} are the antenna resistance and reactance, respectively. The reactance is caused by locally stored (non-radiating) energy in the near zone of the antenna, and the resistance can further be divided as

$$R_{in} = R_r + R_c, \tag{7.48}$$

[24] In many cases, the gain of an antenna is given as a single number, which corresponds to the maximum gain over all directions (θ, ϕ).

[25] In antenna engineering, the maximum gain and directivity are often given in dB values. Unlike many other quantities, however, there are two conventions for the dB values of the maximum gain and directivity. For a gain value of G, using the standard $G_{dBi} = 10\log_{10}(G)$ leads to a value in dBi, which indicates that the logarithmic value is with respect to an *isotropic* source. Alternatively, there is a corresponding value in dBd, which can be obtained approximately as $G_{dBd} = 10\log_{10}(G) - 10\log_{10}(1.64) \approx G_{dBi} - 2.15$ dB. This corresponds to the maximum gain with respect to the one achieved by a *half-wave dipole*.

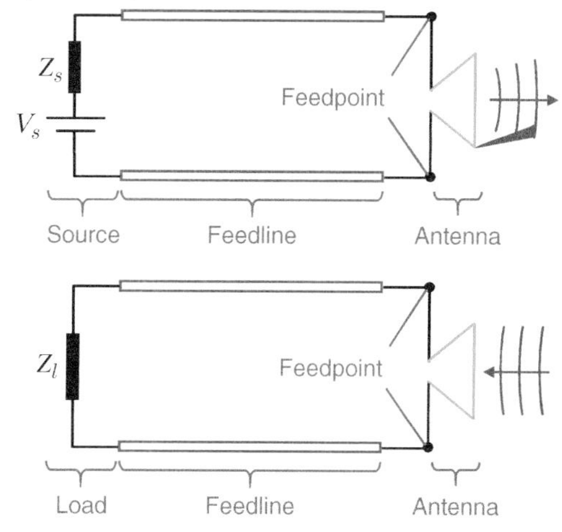

Figure 7.12 Simple representations of radiating (top) and receiving (bottom) antennas when they are in use. The radiating antenna is connected to a voltage source via a feedline (e.g. a coaxial cable). Also using a feedline, the receiving antenna is connected to a load to extract the incoming energy and signal.

[26] In a polarization mismatch, we assume that the radiation from an antenna is received by another antenna, which possibly has a different polarization due to its design or orientation.

[27] According to the current IEEE standards, gain calculated using $e_f e_r$ is called *realized* gain. Similarly, using $e_p e_r$ leads to *partial* gain, and using $e_t = e_p e_f e_r$ leads to *partial realized* gain.

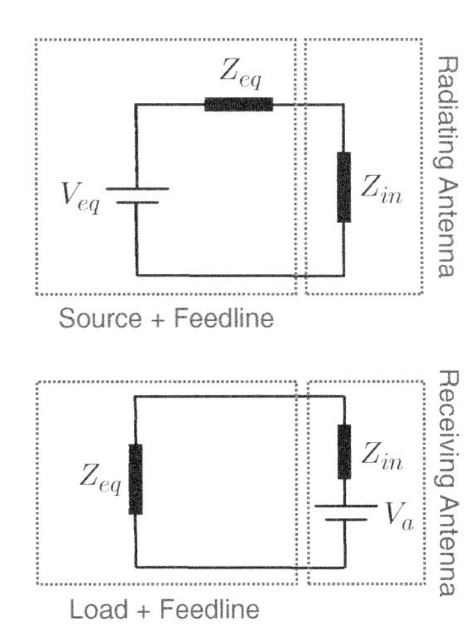

Figure 7.13 Simple circuit representations of radiating (top) and receiving (bottom) antennas when they are in use. Source+feedline or load+feedline can be represented by equivalent lumped elements (V_{eq} and Z_{eq}). The receiving antenna can be represented by a voltage source (V_a) in addition to its input impedance. Note that the input impedance value of an antenna is the same whether it is radiating or receiving.

where R_r is the radiation resistance (as defined above) and R_c is the ohmic resistance due to the finite conductivity of the antenna. The input impedance can be written in terms of power/energy quantities as

$$Z_{in} = 2\frac{p_{em}^{\text{far}} + p_c + j2\omega(w_m - w_e)}{|I_0|^2}, \tag{7.49}$$

where p_c represents the dissipated power due to ohmic losses and w_m/w_e stand for magnetic/electric energies stored in the near zone. To transfer the maximum amount of power from a source to a radiating antenna (or from a receiving antenna to its load), one must match the impedance at the feedpoint (impedance seen by the antenna) to the input impedance as $Z_{eq} = Z_{in}^*$ (Figure 7.13).

Considering again a half-wave dipole, the electric and magnetic energies stored in the near zone are equal to each other (as an advantage) such that we have

$$Z_{in,hw} = 2(p_{em,hw}^{\text{far}} + p_{c,hw})/|I_0|^2 = R_{r,hw} + R_{c,hw}, \tag{7.50}$$

i.e. the input impedance is purely real. This condition of real impedance is called *resonance*, which is a basis for the operation of many antennas. In addition, using a good conductor, conduction losses can be significantly suppressed, leading to an efficient antenna. For example, we can consider a 1.5-meter dipole antenna made of copper with a 0.5 cm radius. At 100 MHz (when $\lambda \approx 3$ m and it is a half-wave dipole), the ohmic resistance of the antenna is even less than 0.1 Ω. Assuming $R_{c,hw} \approx 0.1$ Ω, we have

$$Z_{in,hw} = R_{in,hw} = R_{r,hw} + R_{c,hw} \approx 73.1 + 0.1 = 73.2 \ \Omega. \tag{7.51}$$

Then the radiation efficiency of the antenna can be found as

$$e_{r,hw} = \frac{p_{em,hw}^{\text{far}}}{p_{in,hw}} = \frac{(1/2)R_{r,hw}|I_0|^2}{(1/2)R_{in,hw}|I_0|^2} = \frac{R_{r,hw}}{R_{in,hw}} \approx \frac{73.1}{73.2} \approx 0.999, \tag{7.52}$$

i.e. very high.

For a receiving antenna, an important property is the *effective area* (aperture), defined as

A_e: effective area
Unit: meters^2 (m^2)

$$A_{e,\max} = \frac{p_{rec}}{|\bar{S}_c^{\text{inc}}|}, \tag{7.53}$$

where p_{rec} corresponds to the real power at the feedpoint to be delivered by the antenna to the feedline, and $\bar{S}_c^{\text{inc}}$ is the power flux density of the incident wave.[28]

[28] The incident wave represents the incoming transmission from a radiating antenna and is assumed to be uniform over the entire antenna, assuming that the antennas are far from each other.

In this calculation, it is assumed that there is no polarization mismatch. If possible, the aperture efficiency is defined as the ratio of the effective aperture to the physical area: i.e. $e_{\text{aperture}} = A_{e,\max}/A_{\text{physical}}$, which is generally smaller than one. Due to reciprocity, one can show that

$$A_{e,\max}/G_{\max} = A_e(\theta,\phi)/G(\theta,\phi) = \lambda^2/(4\pi) \tag{7.54}$$

for all antennas.[29] Hence, larger effective area means higher gain.

Considering the equivalent circuit in the case of a radiating antenna (Figure 7.13), it is obvious that the power transferred to the antenna can be maximized by selecting[30] $Z_{eq} = Z_{in}^*$. If this is the case, the real power transferred to the antenna can be written

$$p_{in} = \frac{1}{2}\frac{|V_{eq}|^2}{4R_{in}^2}R_{in} = \frac{|V_{eq}|^2}{8R_{in}}. \tag{7.55}$$

Note that the same amount of power is consumed by feedline losses and the internal resistance of the source ($p_{eq} = p_{in}$). Because $R_{in} = R_r + R_c$ (radiation resistance + ohmic resistance), we have

$$p_{em}^{\text{far}} = \frac{|V_{eq}|^2}{8R_{in}}\frac{R_r}{R_{in}} = \frac{|V_{eq}|^2}{8(R_r+R_c)^2}R_r, \qquad p_c = \frac{|V_{eq}|^2}{8R_{in}}\frac{R_c}{R_{in}} = \frac{|V_{eq}|^2}{8(R_r+R_c)^2}R_c \tag{7.56}$$

as the power radiated and the power lost due to antenna losses, respectively. On the other hand, for a receiving antenna (Figure 7.13) with impedance matching ($Z_{eq} = Z_{in}^*$), we have

$$p_{eq} = p_{rec} = \frac{1}{2}\frac{|V_a|^2}{|Z_{eq}+Z_{in}|^2}R_{eq} = \frac{1}{2}\frac{|V_a|^2}{4R_{in}^2}R_{in} = \frac{|V_a|^2}{8R_{in}} = \frac{|V_a|^2}{8(R_r+R_c)} \tag{7.57}$$

as the power delivered to the feedline, where V_a is the equivalent voltage generated on the antenna by the incident wave. The same amount of power remains on the antenna ($p_{in} = p_{rec}$), and it can be divided into two parts as

$$p_c = \frac{|V_a|^2}{8(R_r+R_c)^2}R_c, \qquad p_{em,\text{rad}}^{\text{far}} = \frac{|V_a|^2}{8(R_r+R_c)^2}R_r, \tag{7.58}$$

which correspond to ohmic losses and the *re-radiation* of the antenna back into space, respectively.

[29] Using the antenna gain, there is no restriction on defining the effective area for ideal (zero-thickness) wire and dipole antennas. For example, for short wire and half-wave dipole antennas, we have

$$A_{e,\max,sw} = G_{\max,sw} \times \lambda^2/(4\pi) \approx 0.12\lambda^2$$
$$A_{e,\max,hw} = G_{\max,hw} \times \lambda^2/(4\pi) \approx 0.13\lambda^2.$$

On the other hand, using these finite values leads to infinite aperture efficiency for ideally zero surface areas. Therefore, for thin antennas, other definitions, like *effective length*, are more commonly used.

[30] For the general case, we have

$$p_{in} = \frac{1}{2}R_{in}|I_{in}|^2 = \frac{1}{2}R_{in}\frac{|V_{eq}|^2}{|Z_{eq}+Z_{in}|^2} = \frac{|V_{eq}|^2}{2}\frac{R_{in}}{(R_{eq}+R_{in})^2+(X_{eq}+X_{in})^2},$$

where I_{in} is the electric current that flows through the circuit. This expression can be maximized by selecting $X_{eq} = -X_{in}$ and finding R_{eq} that satisfy

$$\frac{\partial}{\partial R_{in}}\left[\frac{R_{in}}{(R_{eq}+R_{in})^2}\right] = 0$$
$$\longrightarrow \frac{(R_{eq}+R_{in})^2 - 2(R_{eq}+R_{in})R_{in}}{(R_{eq}+R_{in})^4} = 0$$
$$\longrightarrow R_{eq}^2 = R_{in}^2 \longrightarrow R_{eq} = R_{in}.$$

Hence, p_{in} can be maximized by selecting $Z_{eq} = R_{eq} + jX_{eq} = R_{in} - jX_{in} = Z_{in}^*$.

Using the effective area, one can further find an expression for the equivalent voltage V_a used in the circuit representation of a receiving antenna (Figure 7.13). Assuming a matched load $Z_{eq} = Z_{in}^*$, all power available at the feedpoint can be extracted so that we have

$$p_{eq} = p_{rec} = \frac{|V_a|^2}{8R_{in}}. \tag{7.59}$$

Then using $p_{rec} = A_{e,\max}|\bar{S}_c^{\text{inc}}|$, where $\bar{S}_c^{\text{inc}}$ is the power flux density of the incident wave on the antenna, we obtain

$$|V_a|^2 = 8R_{in}p_{rec} = 8R_{in}A_{e,\max}|\bar{S}_c^{\text{inc}}|. \tag{7.60}$$

This expression can also be written in terms of the gain of the antenna as

$$|V_a|^2 = \frac{2\lambda^2}{\pi}R_{in}G_{\max}|\bar{S}_c^{\text{inc}}|. \tag{7.61}$$

Note that $|V_a|^2$ remains the same for an unmatched load. In such a case, the power transferred to the load can be obtained as[31]

$$p_{eq} = \frac{|V_a|^2}{2|Z_{eq}+Z_{in}|^2}R_{eq} = \frac{\lambda^2}{\pi}G_{\max}|\bar{S}_c^{\text{inc}}|\frac{R_{eq}R_{in}}{|Z_{eq}+Z_{in}|^2} = \frac{4R_{eq}R_{in}}{|Z_{eq}+Z_{in}|^2}p_{rec}. \tag{7.62}$$

[31] Note that the ratio of this power to the maximum that can be delivered (given a matched load) can be written

$$\frac{4R_{eq}R_{in}}{|Z_{eq}+Z_{in}|^2} = 1 - |\Gamma|^2,$$

where

$$\Gamma = \frac{Z_{eq} - Z_{in}^*}{Z_{eq} + Z_{in}}$$

is called the *reflection coefficient* for a power *wave*.

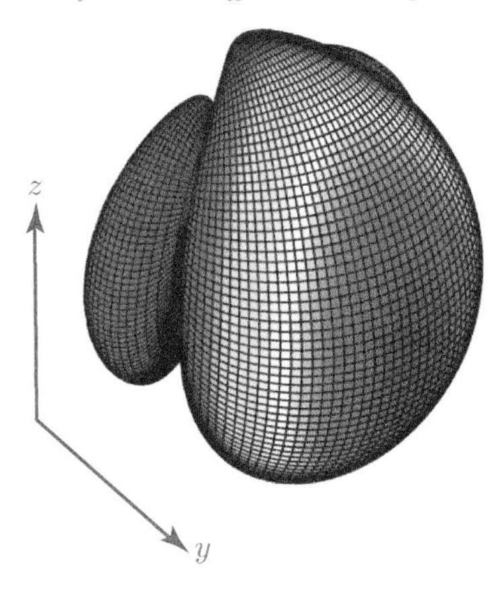

Figure 7.14 A three-dimensional plot of the radiation pattern for the intensity given in Eq. (7.63).

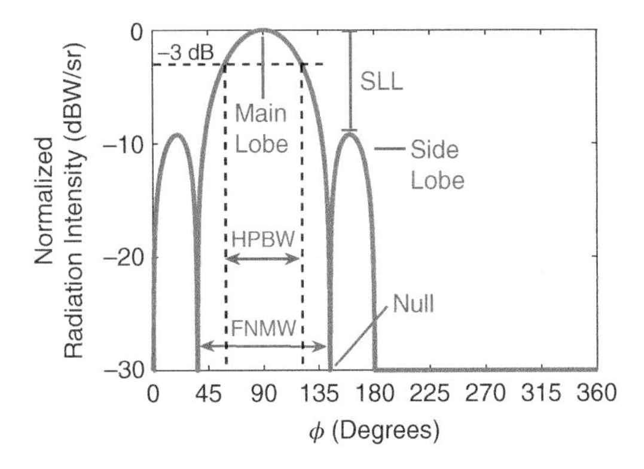

Figure 7.15 A plot of the radiation pattern for the intensity given in Eq. (7.63) with respect to ϕ on the x-y plane. Some important properties of a radiation pattern are also shown.

7.1.2 Antenna Design Parameters

Antenna engineering is a huge scientific and technological area involving the design, simulation, and fabrication of antennas for many applications. Therefore, in addition to the basic properties described above, engineers consider many other parameters while designing antennas. We now consider some of these parameters to gain more insight into the operating principles of these structures.

Unlike simple and relatively small structures, such as half-wave dipoles, many antennas have complicated radiation patterns with multiple lobes (beams) and nulls in different directions. To study important parameters regarding radiation patterns, we consider a hypothetical antenna with a radiation intensity given as (Figures 7.14–7.17)

$$U^{\text{far}}(\theta,\phi) = U_0 \begin{cases} \sin\theta\sin\phi|\sin(\phi-\pi/5)\sin(\phi+\pi/5)| & (0 \le \phi \le \pi) \\ 0 & (\pi < \phi < 2\pi), \end{cases} \tag{7.63}$$

where U_0 (W/sr) is a constant. Looking at the radiation pattern of the antenna, we observe that the maximum radiation occurs in the direction of the y axis: i.e. toward $(\theta,\phi) = (90°, 90°)$. In addition to this *main lobe* (main beam), there are two *side lobes* (local maxima) that occur approximately at $(\theta, \phi) = (90°, 20°)$ and $(\theta, \phi) = (90°, 160°)$. The values of the radiation intensity at the main lobe and at the side lobes can be found to be[32]

$$U^{\text{far}}_{\text{max}} = U^{\text{far}}(90°, 90°) \approx 0.655U_0 \tag{7.64}$$

$$U^{\text{far}}(90°, 20°) = U^{\text{far}}(90°, 160°) \approx 0.0782U_0. \tag{7.65}$$

[32]This means the maximum power flux density at a distance of 100 m can be measured as $|\bar{S}^{\text{far}}_{c,\text{max}}| = U^{\text{far}}_{\text{max}}/R^2 \approx 0.655 \times 10^{-4}U_0$ W/m^2, assuming that 100 m distance is a far zone.

The *side-lobe level* (SLL) of an antenna is defined as the ratio of the radiation intensity at the main lobe (i.e. the maximum radiation intensity) and the radiation intensity at the largest side lobe. For this case, we have (Figure 7.15)

$$\text{SLL} = 0.655U_0/0.0782U_0 \approx 8.38 \approx 9.23 \text{ dB}. \tag{7.66}$$

In practical applications, side lobes are often undesirable as they correspond to wasted power and/or source of unwanted electromagnetic interference. Therefore, we usually aim to maximize the value of SLL.

Considering the hypothetical antenna with the given radiation intensity in Eq. (7.63), we can further find the total radiated power as[33]

[33]A numerical technique is used to evaluate the integral.

$$p^{\text{far}}_{em} = \int_0^{2\pi}\int_0^{\pi} U^{\text{far}}(\theta, \phi)\sin\theta d\theta d\phi \approx 1.21U_0 \quad (\text{W}). \tag{7.67}$$

Hence, we obtain the directive gain as

$$D(\theta, \phi) = 4\pi U^{\text{far}}(\theta, \phi)/p^{\text{far}}_{em}$$
$$\approx \frac{4\pi}{1.21}\begin{cases}\sin\theta\sin\phi|\sin(\phi - \pi/5)\sin(\phi + \pi/5)| & (0 \le \phi \le \pi)\\ 0 & (\pi < \phi < 2\pi),\end{cases} \tag{7.68}$$

leading to $D_{\text{max}} \approx 6.80 \approx 8.33$ dBi as the directivity of the antenna. In many applications, it is desirable to obtain a directive radiation with focused power flow to certain directions.[34] Although directivity provides information about the directional property of an antenna, SLL described above provides further knowledge of side lobes. In addition, *half-power beam width* (HPBW) is defined to measure how sharp the main lobe of an antenna is (Figure 7.15). In general, HPBW can be found on the two principal planes, both containing the maximum radiation direction. Specifically, one can define an E-plane and H-plane that contain electric field and magnetic field components radiated from a given antenna. For the hypothetical antenna with the radiation intensity given in Eq. (7.63), HPBW on the y-z and x-y planes (both containing the maximum radiation direction y) can be found as follows. On the y-z plane (Figure 7.16), the radiation

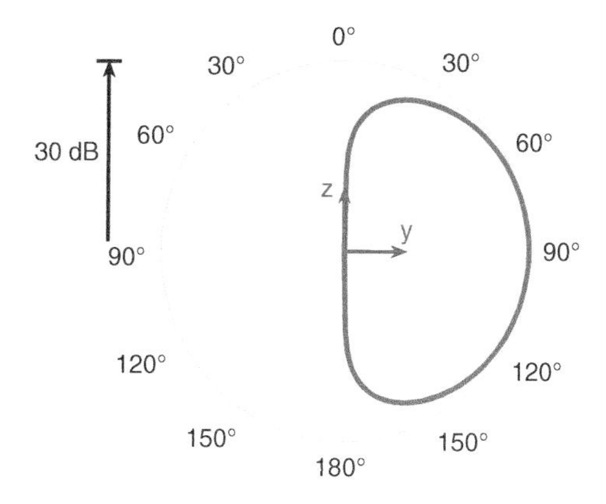

Figure 7.16 Polar plot of the radiation pattern for the intensity given in Eq. (7.63) on the y-z plane. In the plot, a 30 dB dynamic range is used, which means the origin corresponds to −30 dB.

[34]There are also various applications in which isotropic radiation is preferred.

intensity becomes $0.655U_0/2 \approx 0.328U_0$ – i.e. half of the maximum intensity value[35]– when $\theta = 30°$ and $\theta = 150°$. Then on this plane, HPBW can be defined as

$$\text{HPBW}_{zy} = 150° - 30° = 120°. \tag{7.69}$$

On the other hand, for the x-y plane (Figure 7.17), we have $\phi \approx 58.6°$ and $\phi \approx 121.4°$, leading to

$$\text{HPBW}_{xy} \approx 121.4° - 58.6° = 62.8°. \tag{7.70}$$

Hence, the main lobe is narrower on the x-y plane.[36] This does not necessarily indicate that a better (e.g. more directive) radiation is obtain on the x-y plane, since the two side lobes become maximum on this plane, and they are not seen on the y-z plane.

Further investigating the radiation pattern of the hypothetical antenna, we observe that there are directions with zero (null) radiation. Specifically this ideal antenna does not have any radiation: i.e. $U^{\text{far}}(\theta, \phi) = 0$, for $\pi < \phi < 2\pi$. In addition, considering the radiation pattern on the x-y plane, we observe two special directions, *nulls*, that occur at $\phi = 36°$ and $\phi = 144°$. Then the *first-null beam width* of the antenna is defined as[37] FNBW $= 144° - 36° = 108°$.

In the whole discussion until this point, we have examined antennas that operate at single frequencies. On the other hand, many applications in real life require antennas that operate in certain frequency ranges. The parameter that describes the success of an antenna for this purpose is the frequency *bandwidth* (BW), defined (usually) as

$$\text{BW} = f_{\text{max}} : f_{\text{min}} \quad \text{or} \quad \text{BW} = 200 \times \frac{f_{\text{max}} - f_{\text{min}}}{f_{\text{max}} + f_{\text{min}}}\ \%, \tag{7.71}$$

where f_{max} and f_{min} are the maximum and minimum frequencies at which the antenna can be considered to operate (in terms of given requirements). For many antennas, BW values are given as percentages: e.g. less than 10% for a half-wave dipole, but they are represented as ratios particularly for broadband antennas.

7.1.3 Antenna Types

State-of-the-art antenna engineering includes many different types of antennas, each designed and suitable for a particular set of applications. Basically, any piece of a conducting object that can be connected to a generator or a load may be considered an antenna (Figure 7.18), but engineers always work to build an efficient device under the design constraints they are given. Basic geometric designs often outshine others at obtaining successful values for the antenna properties and parameters discussed so far while also considering factors such as cost, weight, and compactness. We now briefly consider some well-known (and time-tested) antenna types and their general properties.

[35] Hence, in a dB scale, we consider radiation intensity values that are approximately 3 dB less than the maximum value.

[36] Note that the z-x plane is not a principal plane: i.e. the maximum radiation (main lobe) is not located on the z-x plane. Consequently, parameters like SLL and HPBW are not (well) defined on this plane. For this particular antenna, radiation is completely zero on the z-x plane.

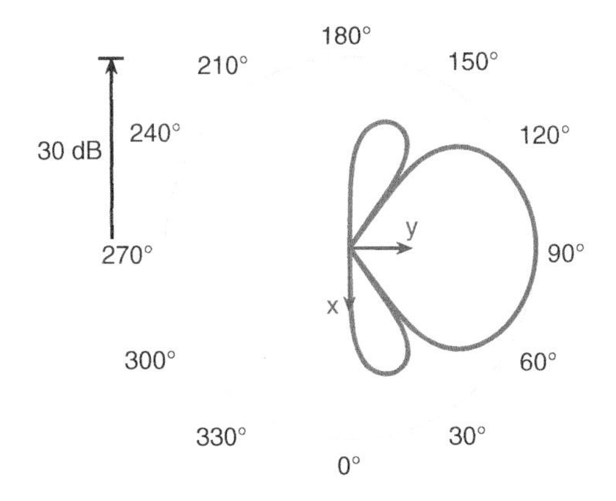

Figure 7.17 Polar plot of the radiation pattern for the intensity given in Eq. (7.63) on the x-y plane using a 30 dB dynamic range.

[37] There is another parameter called *beam efficiency*, which is usually defined as the total power radiated by the main lobe (e.g. directions between the first nulls) relative to the total radiated power.

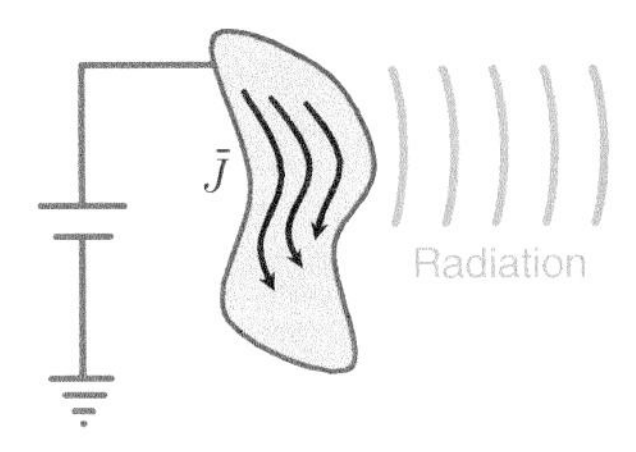

Figure 7.18 When a time-harmonic source is connected to a conducting object, an electric current distribution is generated on/in the object, leading to the radiation of electromagnetic waves as predicted by Maxwell's equations.

There is no single way to categorize antennas; such a classification depends on the context. However, we can group all antennas as resonant, aperture, traveling-wave, and arrays. Note that an antenna type listed in one of these groups may also be used in the other groups, particularly if *type* refers to a particular shape rather than operation principles.

- *Resonant* antennas (Figure 7.19) are based on resonance phenomena: i.e. a resonant antenna operates effectively and efficiently at a resonance frequency that depends on its size and shape. Under a resonance condition, *standing* waves are dominant: e.g. electric currents induced on the antenna possess a standing-wave behavior. This behavior leads to desired antenna properties and parameters. For example, an antenna at a resonance frequency often involves purely real input impedance, balanced electric-magnetic energy stored in the near zone, high antenna gain, etc. Some resonant-type antennas are as follows:

  ✓ Monopole antennas (straight or other variations)

  ✓ Dipole antennas (e.g. half-wave dipoles), folded dipole antennas, turnstile antennas

  ✓ Loop antennas

  ✓ Biconical antennas, bowtie antennas

  ✓ Patch antennas

  Resonant antennas typically operate in narrow bands, as their operations depend on resonance conditions.

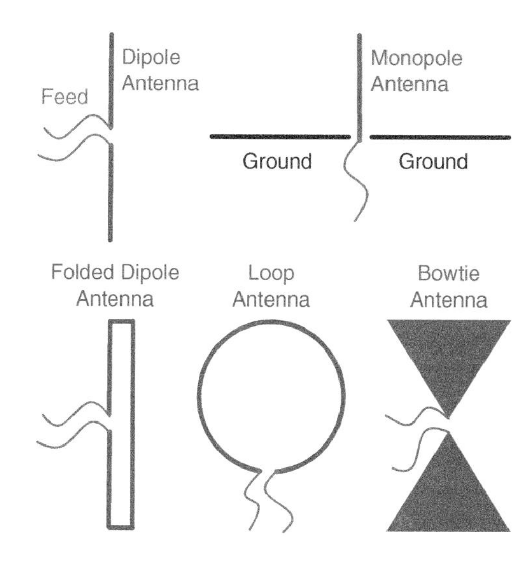

Figure 7.19 Simple illustrations of some common resonant antennas.

- *Aperture* antennas (Figures 7.20–7.22) involve smaller feed parts (that can be resonating by themselves), which excite larger radiating parts (via coupling). They usually operate in wider frequency ranges than resonant antennas, but they may be less compact (depending on the radiating part). Some well-known aperture antennas are as follows:

  ✓ Horn antennas

  ✓ Lens antennas and dielectric-resonator antennas

  ✓ Parabolic antennas

  ✓ Slot antennas (including Vivaldi antennas)

  ✓ Microstrip antennas (e.g. microstrip patch antennas)

  Aperture antennas, especially parabolic antennas, are also quite directional.

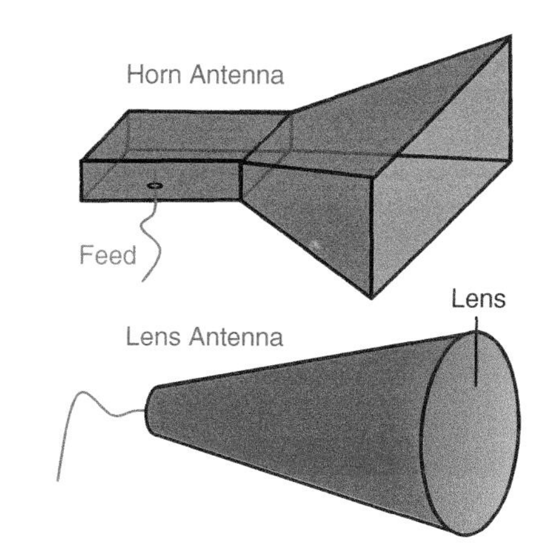

Figure 7.20 Simple illustrations of a horn antenna and a lens antenna.

- *Traveling-wave* antennas (Figure 7.23) do not involve any resonance characteristics and are based on *traveling* waves rather than *standing* (resonance) waves. A very long wire (with

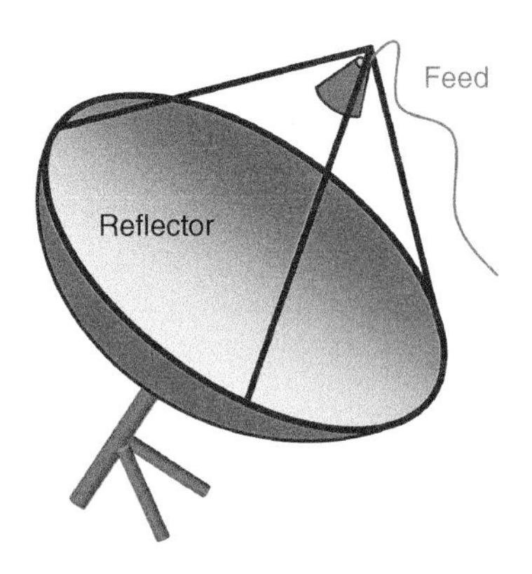

Figure 7.21 A simple illustration of a parabolic reflector antenna. The feed, which illuminates the larger parabolic reflector, is an antenna, such as a horn antenna.

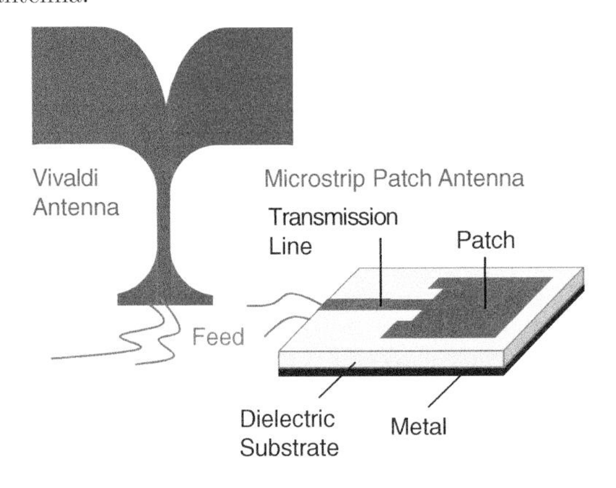

Figure 7.22 Simple illustrations of a Vivaldi antenna and a microstrip patch antenna.

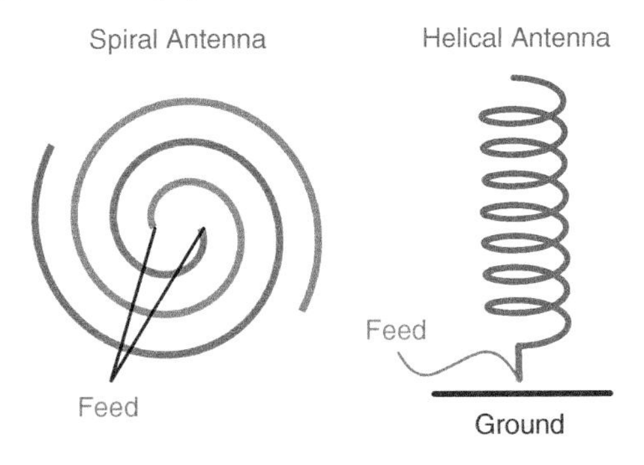

Figure 7.23 Simple illustrations of an Archimedean spiral antenna and a helical antenna.

respect to wavelength) is a traveling-wave antenna (called a Beverage antenna). Rhombic antennas, leaky-wave antennas, helical antennas, and spiral antennas are other well-known examples of traveling-wave antennas. Compared to most resonant antennas, traveling-wave antennas (especially spiral antennas) usually operate in wider frequency ranges, but they are typically less compact and portable.

- *Arrays*, which are considered in more detail below, are arrangements of multiple antennas (elements) for better operations: e.g. for higher directivity values, broadband or multiband operations, beam steering abilities, etc. Array elements may be connected electronically (via a circuit) or integrated into a single body. For example, the well-known *Yagi-Uda* antennas and *log-periodic* antennas (Figure 7.24) can be considered arrays of dipoles/monopoles, while fractal antennas usually consist of fractal-type arrangements of planar patches with varying sizes. Based on the operating principle, an array may be included in one of the three categories above. For example, Yagi-Uda antennas and log-periodic antennas are often considered traveling-wave antennas, because waves travel along them.

7.1.3.1 Antenna Arrays

We now consider some basic concepts of antenna arrays, which are commonly used in wireless systems. We particularly focus on radiation characteristics of arrays: i.e. the well-known principle of the *array factor*. As a starting point, we assume a current distribution $\bar{J}_{v,a}(\bar{r}')$ that represents an antenna located at around the origin (Figure 7.25). We have

$$\bar{A}_a^{\text{far}}(\bar{r}) = \mu\frac{\exp(-jkR)}{4\pi R}\int_V \bar{J}_{v,a}(\bar{r}')\exp(jk\hat{a}_r\cdot\bar{r}')d\bar{r}' \tag{7.72}$$

$$\bar{E}_a^{\text{far}}(\bar{r}) = -j\omega[\bar{A}_a^{\text{far}}(\bar{r}) - \hat{a}_R\hat{a}_R\cdot\bar{A}_a^{\text{far}}(\bar{r})] \tag{7.73}$$

as the far-zone magnetic vector potential and the far-zone electric field intensity, respectively. If the antenna is shifted to a position $\bar{r}_s$, then these expressions should be updated as

$$\bar{A}_{as}^{\text{far}}(\bar{r}) = \mu\frac{\exp(-jkR)}{4\pi R}\exp(jk\hat{a}_r\cdot\bar{r}_s)\int_V \bar{J}_{v,a}(\bar{r}')\exp(jk\hat{a}_r\cdot\bar{r}')d\bar{r}' \tag{7.74}$$

$$= \exp(jk\hat{a}_r\cdot\bar{r}_s)\bar{A}_a^{\text{far}}(\bar{r}) \tag{7.75}$$

$$\bar{E}_{as}^{\text{far}}(\bar{r}) = -j\omega[\bar{A}_{as}^{\text{far}}(\bar{r}) - \hat{a}_R\hat{a}_R\cdot\bar{A}_{as}^{\text{far}}(\bar{r})] = \exp(jk\hat{a}_r\cdot\bar{r}_s)\bar{E}_a^{\text{far}}(\bar{r}). \tag{7.76}$$

Hence, a location shift corresponds to a *phase shift* in the generated far-zone fields.

Consider a total of N antennas located at $\bar{r}_{s,0}, \bar{r}_{s,1}, \ldots, \bar{r}_{s,N-1}$ (Figure 7.26). The total electric field intensity created by them can be obtained as

$$\bar{E}^{\text{far}}_{arr}(\bar{r}) = \sum_{n=0}^{N-1} \zeta_n \bar{E}^{\text{far}}_{as,n}(\bar{r}) = \sum_{n=0}^{N-1} \zeta_n \exp(jk\hat{a}_r \cdot \bar{r}_{s,n}) \bar{E}^{\text{far}}_{a,n}(\bar{r}), \quad (7.77)$$

where ζ_n for $n = 0, 1, \ldots, N-1$ represent excitation coefficients that can be different for different antennas. In the above, $\bar{E}^{\text{far}}_{as,n}(\bar{r})$ is the electric field intensity generated by the nth antenna when it is located at $\bar{r}_{s,n}$, while $\bar{E}^{\text{far}}_{a,n}(\bar{r})$ corresponds to the electric field intensity when the same antenna is located at the origin (without a shift). If all array elements (antennas) are identical with $\bar{E}^{\text{far}}_{a,n}(\bar{r}) = \bar{E}^{\text{far}}_{a}(\bar{r})$ for $n = 0, 1, \ldots, N-1$, the total electric field intensity can be written

$$\bar{E}^{\text{far}}_{arr}(\bar{r}) = \bar{E}^{\text{far}}_{a}(\bar{r}) \sum_{n=0}^{N-1} \zeta_n \exp(jk\hat{a}_r \cdot \bar{r}_{s,n}). \quad (7.78)$$

Obviously, with identical elements (but not necessarily identical excitations), the total electric field intensity is simply a multiplication of the electric field intensity of a single element with a *factor* that depends on element positions and excitations.[38] This is called the *array factor*, which can be written

$$\text{AF}(\theta, \phi) = \sum_{n=0}^{N-1} \zeta_n \exp(jk\hat{a}_r \cdot \bar{r}_{s,n}), \quad (7.79)$$

leading to $\bar{E}^{\text{far}}_{arr}(\bar{r}) = \text{AF}(\theta, \phi) \times \bar{E}^{\text{far}}_{a}(\bar{r})$. Considering that

$$\bar{S}^{\text{far}}_{c}(\bar{r}) = \frac{1}{2}\bar{E}^{\text{far}}(\bar{r}) \times [\bar{H}^{\text{far}}(\bar{r})]^* = \frac{1}{2\eta}\bar{E}^{\text{far}}(\bar{r}) \times [\hat{a}_R \times \bar{E}^{\text{far}}(\bar{r})]^* = \hat{a}_R \frac{|\bar{E}^{\text{far}}(\bar{r})|^2}{2\eta} \quad (7.80)$$

and

$$U^{\text{far}}(\theta, \phi) = R^2 |\bar{S}^{\text{far}}_{c}(\bar{r})| = R^2 \frac{|\bar{E}^{\text{far}}(\bar{r})|^2}{2\eta} \quad (7.81)$$

for a real intrinsic impedance η, the radiation intensity created by an array of identical elements can be written

$$U^{\text{far}}_{arr}(\theta, \phi) = \frac{R^2}{2\eta}|\bar{E}^{\text{far}}_{arr}(\bar{r})|^2 = \frac{R^2}{2\eta}|\text{AF}(\theta, \phi)|^2 |\bar{E}^{\text{far}}_{a}(\bar{r})|^2 = |\text{AF}(\theta, \phi)|^2 U^{\text{far}}_{a}(\theta, \phi), \quad (7.82)$$

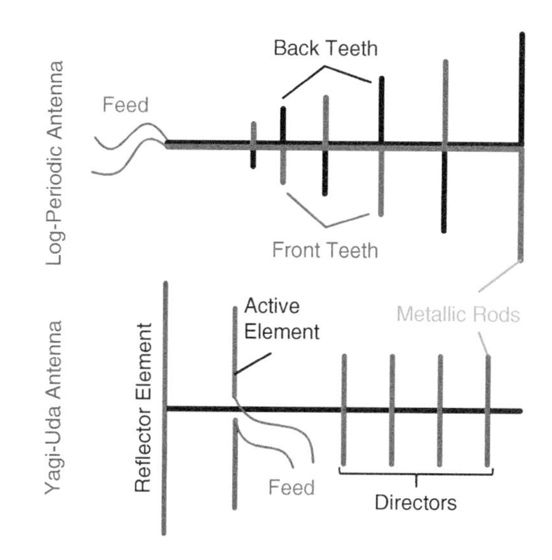

Figure 7.24 Simple illustrations of a wire log-periodic antenna and a Yagi-Uda antenna. These antennas can be considered arrays of dipoles/monopoles, and they are also categorized as traveling-wave antennas.

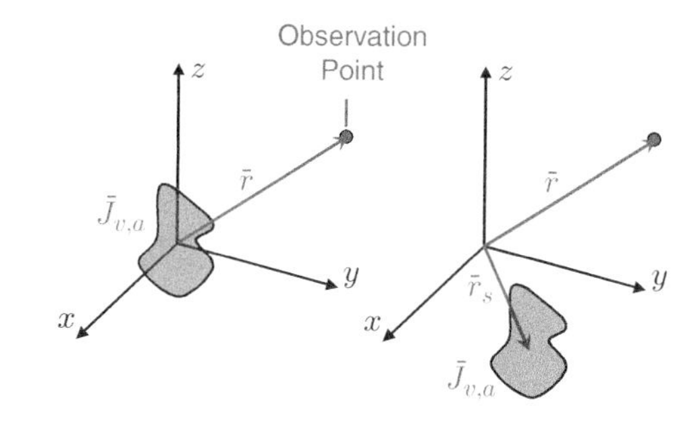

Figure 7.25 A current distribution that represents an antenna located at the origin and its shifted version.

[38] Note the assumption that the antennas do not interact with each other. Specifically, the existence of an antenna does not change the current distributions on the other antennas.

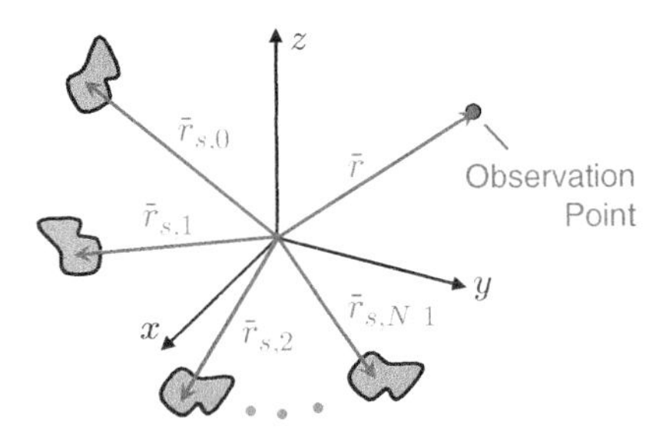

Figure 7.26 A total of N distinct current distributions, each of which represents an antenna located at a specific position.

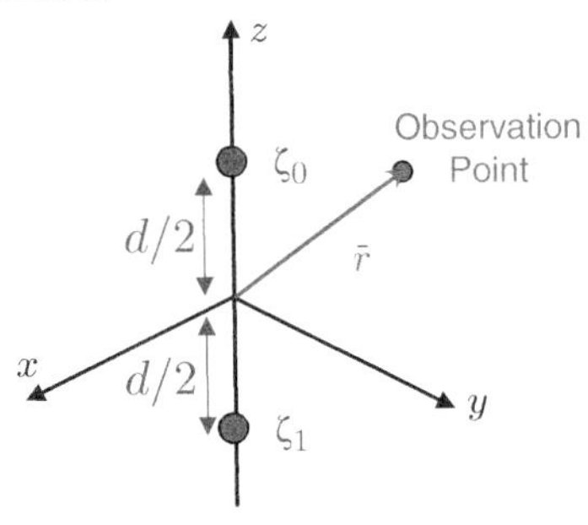

Figure 7.27 Two pointwise elements that represent antennas located at $z = \pm d/2$.

[39] Note that each of these points represents an actual antenna, as if the overall radiation of the antenna emerges from a single point.

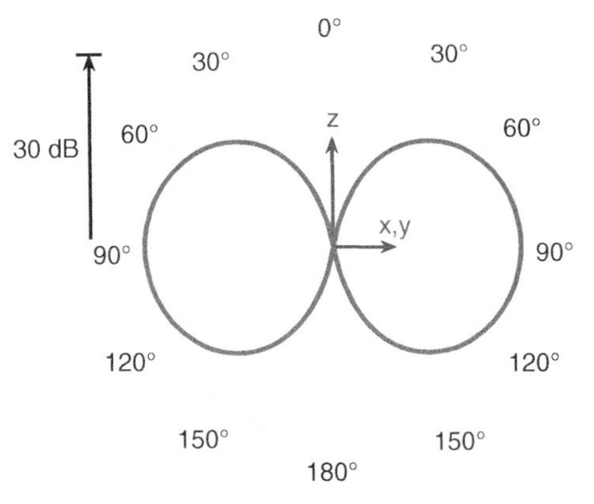

Figure 7.28 Array factor of the two-element configuration in Figure 7.27 when the elements have the same excitation and $d = \lambda/2$.

[40] Cancellation can occur since we assume that the observation point is in the far zone.

where $U_a^{\text{far}}(\theta, \phi)$ is the radiation intensity due to a single element (when it is located at the origin). Note that normalized $|\text{AF}(\theta, \phi)|^2$ directly represents the radiation pattern of an array if the elements are assumed to be isotropic sources (when $\tilde{U}_a^{\text{far}}(\theta, \phi) = 1$).

At this stage, we now consider several fundamental scenarios, starting with a simple case of an array with two elements. Assume that a pair of identical pointwise elements[39] are located at $z = \pm d/2$ (Figure 7.27). Then we have

$$\text{AF}(\theta, \phi) = \zeta_0 \exp(jk\hat{a}_r \cdot \hat{a}_z d/2) + \zeta_1 \exp(-jk\hat{a}_r \cdot \hat{a}_z d/2) \tag{7.83}$$

$$= \zeta_0 \exp\left(\frac{jkd\cos\theta}{2}\right) + \zeta_1 \exp\left(-\frac{jkd\cos\theta}{2}\right), \tag{7.84}$$

where ζ_0 and ζ_1 are excitations. If these excitations are the same ($\zeta_0 = \zeta_1$), we further obtain

$$\text{AF}(\theta, \phi) = 2\zeta_0 \cos\left(\frac{kd\cos\theta}{2}\right). \tag{7.85}$$

Several interesting examples are as follows:

- If the two elements have the same excitation and are ideally isotropic with $U_a^{\text{far}}(\theta, \phi) = 1$, we have

$$U_{arr}^{\text{far}}(\theta, \phi) = \left|\zeta_0 2\cos\left(\frac{kd\cos\theta}{2}\right)\right|^2 = 4|\zeta_0|^2 \cos^2\left(\frac{kd\cos\theta}{2}\right). \tag{7.86}$$

Hence, there is always a maximum radiation toward $\theta = 90°$, while the other characteristics depend on the separation distance between the elements. For example, if $d = \lambda/2$, we further have (Figure 7.28)

$$U_{arr}^{\text{far}}(\theta, \phi) = 4|\zeta_0|^2 \cos^2\left(\frac{\pi\cos\theta}{2}\right), \tag{7.87}$$

indicating zero radiation toward $\theta = \{0°, 180°\}$. This corresponds to a perfect cancellation of the radiation from the two sources in these directions.[40] On the other hand, if $d = \lambda$, the radiation intensity of the array becomes (Figure 7.29)

$$U_{arr}^{\text{far}}(\theta, \phi) = 4|\zeta_0|^2 \cos^2(\pi\cos\theta), \tag{7.88}$$

which means $\theta = \{0°, 180°\}$ also receive maximum radiation (in addition to $\theta = 90°$). Furthermore, using $d = 5\lambda/4$ leads to an interesting radiation intensity with maxima at $\theta \approx \{36.9°, 143.1°\}$ (in addition to $\theta = 90°$) and nulls at $\theta \approx \{66.4°, 113.6°\}$ (Figure 7.30). In general, as distance d becomes larger, the array factor and the radiation intensity/pattern of the array become more oscillatory with multiple lobes and nulls.

- If two isotropic elements (with $U_a^{\text{far}}(\theta, \phi) = 1$) are excited with a 90° phase difference as $\zeta_1 = j\zeta_0$, we have

$$\text{AF}(\theta, \phi) = \zeta_0 \exp\left(\frac{jkd\cos\theta}{2}\right) + \zeta_0 \exp(j\pi/2)\exp\left(-\frac{jkd\cos\theta}{2}\right) \tag{7.89}$$

$$= \zeta_0 \exp\left(j\frac{\pi}{4}\right)\left[\exp\left(\frac{jkd\cos\theta - j\pi/2}{2}\right) + \exp\left(\frac{-jkd\cos\theta + j\pi/2}{2}\right)\right] \tag{7.90}$$

$$= 2\zeta_0 \exp\left(j\frac{\pi}{4}\right)\cos\left(\frac{kd\cos\theta}{2} - \frac{\pi}{4}\right), \tag{7.91}$$

leading to

$$U_{arr}^{\text{far}}(\theta, \phi) = 4|\zeta_0|^2\cos^2\left(\frac{kd\cos\theta}{2} - \frac{\pi}{4}\right). \tag{7.92}$$

Hence, if $d \geq \lambda/4$, we have maxima at directions that satisfy

$$\frac{kd\cos\theta_{\text{max}}}{2} - \frac{\pi}{4} = m\pi \longrightarrow \theta_{\text{max}} = \cos^{-1}\left[\frac{\lambda}{4d}(4m+1)\right], \tag{7.93}$$

while nulls occur when

$$\frac{kd\cos\theta_{\text{zero}}}{2} - \frac{\pi}{4} = \frac{(2m-1)\pi}{2} \longrightarrow \theta_{\text{zero}} = \cos^{-1}\left[\frac{\lambda}{4d}(4m-1)\right] \tag{7.94}$$

for $m = 0, \pm 1, \pm 2, \ldots$ For example, when $d = \lambda/4$, maximum and null radiation occur for $\theta = 0°$ and $\theta = 180°$, respectively (Figure 7.31). If $d = \lambda/2$, however, maximum radiation occurs toward $\theta = 60°$, while zero radiation is obtained for $\theta = 120°$ (Figure 7.32). If the separation distance is further made $d = \lambda$, we obtain (Figure 7.33)

$$\theta_{\text{max}} = \{\cos^{-1}(1/4), \cos^{-1}(-3/4)\} \approx \{75.5°, 138.6°\} \tag{7.95}$$

$$\theta_{\text{zero}} = \{\cos^{-1}(3/4), \cos^{-1}(-1/4)\} \approx \{41.4°, 104.5°\}. \tag{7.96}$$

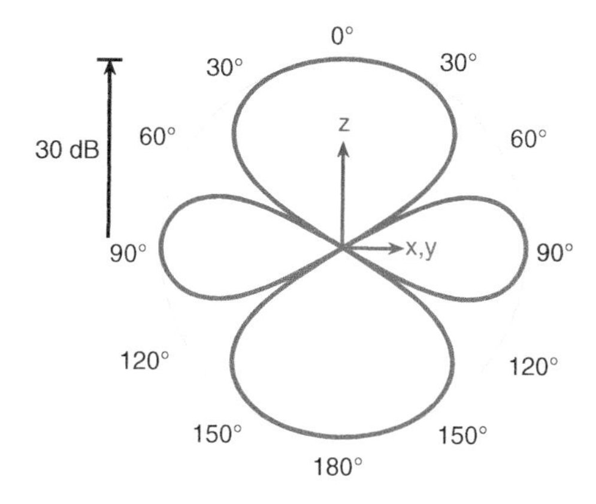

Figure 7.29 Array factor of the two-element configuration in Figure 7.27 when the elements have the same excitation and $d = \lambda$.

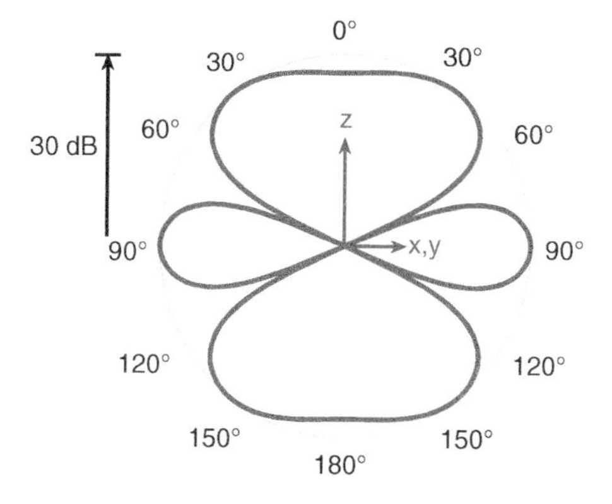

Figure 7.30 Array factor of the two-element configuration in Figure 7.27 when the elements have the same excitation and $d = 5\lambda/4$.

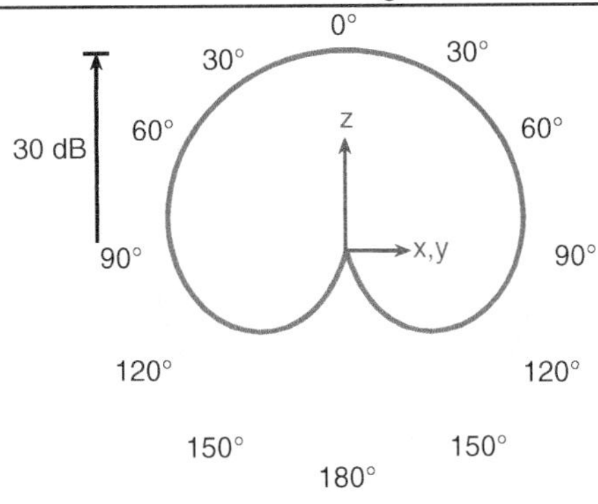

Figure 7.31 Array factor of the two-element configuration in Figure 7.27 when the elements are excited with a 90° phase difference and $d = \lambda/4$.

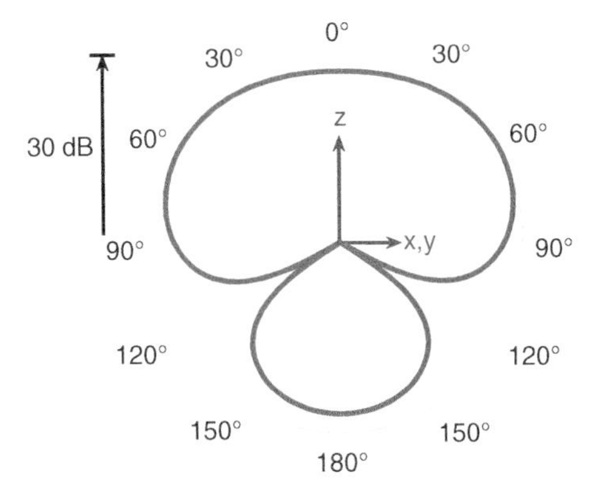

Figure 7.32 Array factor of the two-element configuration in Figure 7.27 when the elements are excited with a 90° phase difference and $d = \lambda/2$.

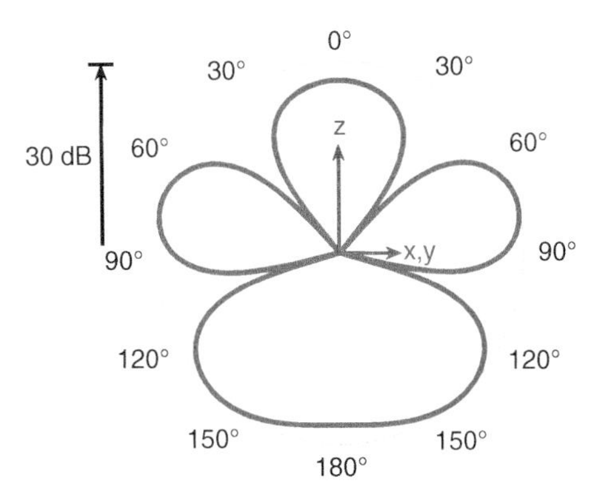

Figure 7.33 Array factor of the two-element configuration in Figure 7.27 when the elements are excited with a 90° phase difference and $d = \lambda$.

- Consider two half-wave dipoles that are excited identically. For a separation distance d, we have $\text{AF} = 2\zeta_0 \cos[(kd/2)\cos\theta]$. If the half-wave dipoles are both oriented in the z direction, we can obtain the radiation intensity of the array as

$$U_{arr}^{\text{far}}(\theta,\phi) = |\text{AF}(\theta,\phi)|^2 U_{hw}^{\text{far}}(\theta,\phi) \tag{7.97}$$

$$= 4|\zeta_0|^2 \cos^2\left(\frac{kd\cos\theta}{2}\right)|I_0|^2 \frac{\eta}{8\pi^2}\frac{\cos^2[(\pi/2)\cos\theta]}{\sin^2\theta} \tag{7.98}$$

$$= |\zeta_0|^2|I_0|^2 \frac{\eta}{2\pi^2}\cos^2\left(\frac{kd\cos\theta}{2}\right)\cos^2\left(\frac{\pi\cos\theta}{2}\right)\frac{1}{\sin^2\theta}. \tag{7.99}$$

As a particular case, we can consider $d = \lambda/2$, leading to

$$U_{arr}^{\text{far}}(\theta,\phi) = |\zeta_0|^2|I_0|^2 \frac{\eta}{2\pi^2}\frac{\cos^4[(\pi/2)\cos\theta]}{\sin^2\theta}, \tag{7.100}$$

which has $\theta_{\text{max}} = 90°$ and $\theta_{\text{zero}} = \{0°, 180°\}$, without any side lobes. On the other hand, if $d = 3\lambda/2$, we have (Figure 7.34)

$$U_{arr}^{\text{far}}(\theta,\phi) = |\zeta_0|^2|I_0|^2 \frac{\eta}{2\pi^2}\frac{\cos^2[(3\pi/2)\cos\theta]\cos^2[(\pi/2)\cos\theta]}{\sin^2\theta} \tag{7.101}$$

with side lobes at $\theta \approx \{52.3°, 127.7°\}$ and nulls at $\theta \approx \{70.5°, 109.5°\}$ (in addition to nulls at $\theta = \{0°, 180°\}$ that occur independent of the value of d).

Next, we focus on *linear* arrays with larger numbers of elements. Consider a total of N identical elements located on the z axis with a separation distance d (Figure 7.35): i.e. at $z = 0, d, 2d, \ldots, (N-1)d$. This type of an array is called an *equally spaced linear array*. For arbitrary excitations $\zeta_0, \zeta_1, \ldots, \zeta_{N-1}$, the array factor can be written

$$\text{AF}(\theta,\phi) = \sum_{n=0}^{N-1} \zeta_n \exp(jknd\cos\theta). \tag{7.102}$$

If the excitations have the same magnitude but a progressive phase distribution (*uniform* array) as $\zeta_n = \zeta_0 \exp(jn\varphi_0)$, the array factor becomes

$$\text{AF}(\theta,\phi) = \zeta_0 \sum_{n=0}^{N-1} \exp(jn\varphi_0)\exp(jknd\cos\theta) \tag{7.103}$$

$$= \zeta_0 \sum_{n=0}^{N-1} \exp[jn(\varphi_0 + kd\cos\theta)]. \tag{7.104}$$

Defining $\varphi = \varphi_0 + kd\cos\theta$ and using

$$\sum_{n=0}^{N-1} a^n = \frac{1-a^N}{1-a} \qquad \text{for} \quad |a| \le 1, \tag{7.105}$$

this expression becomes

$$\mathrm{AF}(\varphi) = \zeta_0 \sum_{n=0}^{N-1} \exp(jn\varphi) = \zeta_0 \frac{1-\exp(jN\varphi)}{1-\exp(j\varphi)} \tag{7.106}$$

$$= \zeta_0 \frac{\exp(jN\varphi/2)}{\exp(j\varphi/2)} \frac{[\exp(-jN\varphi/2) - \exp(jN\varphi/2)]}{[\exp(-j\varphi/2) - \exp(j\varphi/2)]} \tag{7.107}$$

$$= \zeta_0 \exp[j(N-1)\varphi/2] \frac{\sin(N\varphi/2)}{\sin(\varphi/2)}. \tag{7.108}$$

This array factor is often normalized as

$$\widetilde{\mathrm{AF}}(\varphi) = \frac{1}{N}\frac{\sin(N\varphi/2)}{\sin(\varphi/2)}. \tag{7.109}$$

Based on this expression, we first consider general properties in terms of φ (Figure 7.36), without its result in θ (real space). First, note that the maximum value of the array factor is obtained when $\varphi = 2m\pi$ for $m = 0, \pm 1, \pm 2, \ldots$, and there are main lobes centered at these values of φ (with a 2π period). For a main lobe located at $\varphi_{\max}$, the first nulls occur at $\varphi = \varphi_{\max} \pm 2\pi/N$ so that the total width of each main lobe is $4\pi/N$. The nulls tend to repeat with a period of $\Delta\varphi = 2\pi/N$. Therefore, between two consecutive main lobes, there are a total of $2\pi/(2\pi/N) - 1 = N-1$ nulls. Also, if there is no main lobe between two consecutive nulls, there is a side lobe with $2\pi/N$ width. Consequently, between two main lobes are $N-2$ side lobes.[41]

The actual radiation characteristics of a linear array depend on the visibility of the φ dependency of its array factor in real coordinates (Figure 7.37). Specifically, since θ is limited to a range from 0 to π, the *visible region* is defined as

$$\varphi = \varphi_0 + kd\cos\theta \longrightarrow -kd + \varphi_0 \le \varphi \le kd + \varphi_0. \tag{7.110}$$

Therefore, the φ dependency of the array factor is visible only from $\varphi = -kd + \varphi_0$ to $\varphi = kd + \varphi_0$ (from $\theta = \pi$ to $\theta = 0$). Hence, larger values of kd mean larger ranges of the array factor in terms of φ are visible. In terms of θ, main lobes of the array factor occur in general at

$$\theta_{\max} = \cos^{-1}\left(\frac{m\lambda}{d} - \frac{\varphi_0}{kd}\right), \qquad \left(-\frac{d}{\lambda} + \frac{\varphi_0}{2\pi} \le m \le \frac{d}{\lambda} + \frac{\varphi_0}{2\pi}\right). \tag{7.111}$$

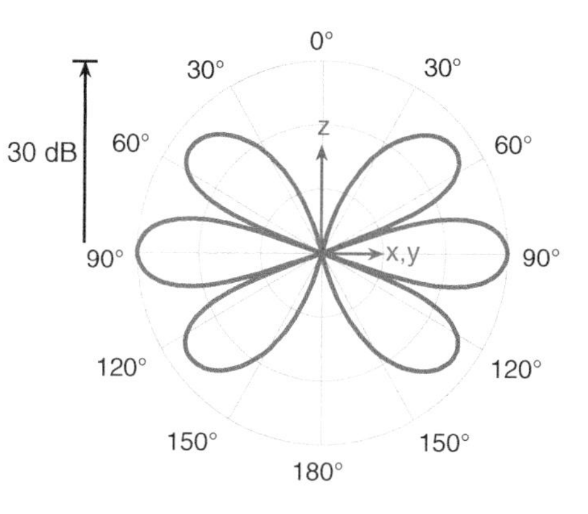

Figure 7.34 Radiation pattern of a pair of identically excited z-oriented half-wave dipoles as arranged in Figure 7.27 with $d = 3\lambda/2$.

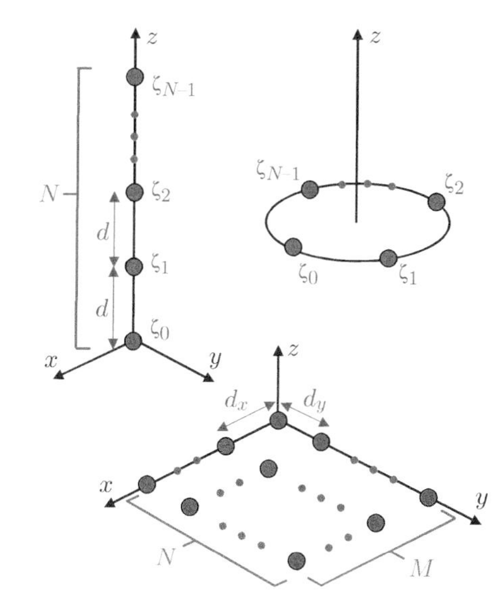

Figure 7.35 Various array configurations of identical elements. A linear array of N equally spaced elements aligned along the z axis; a circular array on the x-y plane; a planar array of $M \times N$ elements located on the x-y plane with equal spacing in the x and y directions.

[41] Hence, as N gets larger, the array factor becomes more oscillatory in terms of φ. In addition, one can show that the SLL (of the array factor) decreases with increasing N.

On the other hand, the lobes that satisfy this expression for $m \neq 0$ are usually called *grating* lobes (lobes that are as large as true main lobes with $m = 0$) as they represent unwanted radiation.[42]

[42]Hence, as d increases, more grating lobes appear in the visible range.

The width of the visible part of AF(φ) depends on kd, and its center is located at φ_0. This allows us to control the radiation (at least, the array factor of the considered array configuration) by selecting φ_0. For example, selecting $\varphi_0 = 0°$ (no phase progress: i.e. all elements are excited identically), we have

$$\theta_{\max} = \cos^{-1}\left(\frac{m\lambda}{d}\right), \qquad \left(-\frac{d}{\lambda} \leq m \leq \frac{d}{\lambda}\right). \tag{7.112}$$

This means there is *always* a main lobe toward $\cos^{-1}(0) = 90°$, which is called *broadside* radiation, while the existence of grating lobes depends on d: e.g. there is no grating lobe if $d < \lambda$. For example, if $d = \lambda$ (in addition to $\varphi_0 = 0°$), leading to $-1 \leq m \leq 1$ or $m = \{-1, 0, 1\}$, we have (Figure 7.37)

$$\theta_{\max} = \cos^{-1}(m) = \{0, 90°, 180°\}. \tag{7.113}$$

Lobes that occur in the alignment direction of the array ($\pm z$ direction) are called *end-fire* radiation. As another value, for $d = 3\lambda/2$ (besides $\varphi_0 = 0°$), one can obtain $-3/2 \leq m \leq 3/2$ or $m = \{-1, 0, 1\}$ (possible m values do not change), and main/grating lobes occur as (Figure 7.38)

$$\theta_{\max} = \cos^{-1}(2m/3) \approx \{48.2°, 90°, 131.8°\}. \tag{7.114}$$

It should be emphasized that there is no linear relationship between φ and θ: i.e. maxima at $\varphi = \{-360°, 0°, 360°\}$ correspond to maxima at $\theta \approx \{131.8°, 90°, 48.2°\}$, respectively. Alternatively, we can consider $\varphi_0 = 180°$, which leads to

$$\theta_{\max} = \cos^{-1}\left(\frac{m\lambda}{d} - \frac{\lambda}{2d}\right), \qquad \left(-\frac{d}{\lambda} + \frac{1}{2} \leq m \leq \frac{d}{\lambda} + \frac{1}{2}\right). \tag{7.115}$$

In this case, for $d = \lambda/2$, we have $0 \leq m \leq 1$ or $m = \{0, 1\}$, and

$$\theta_{\max} = \cos^{-1}(2m - 1) = \{0°, 180°\}. \tag{7.116}$$

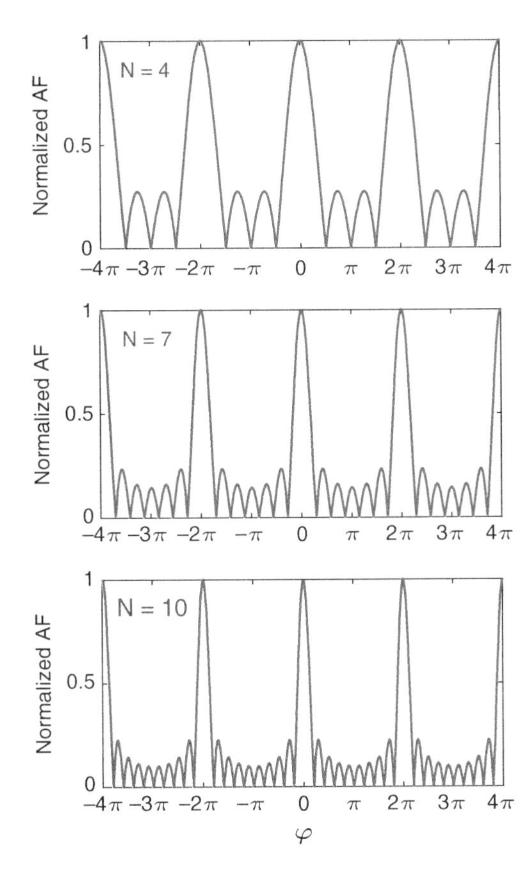

Figure 7.36 Normalized array factor (with respect to φ) of uniform linear arrays aligned along the z axis with $N = 4$, $N = 7$, and $N = 10$ elements.

Hence, we obtain end-fire radiation without any broadside radiation (Figure 7.39). On the other side, using $d = 3\lambda/2$ (in addition to $\varphi = 180°$) leads to $-1 \leq m \leq 2$ or $m = \{-1, 0, 1, 2\}$ (Figure 7.40), and[43]

$$\theta_{\text{max}} = \cos^{-1}(2m/3 - 1/3) \approx \{0°, 70.5°, 109.5°, 180°\}. \tag{7.117}$$

As considered in the examples above, it is possible to avoid grating lobes by selecting a small separation distance. For given array configurations, there can be alternative formulas to describe conditions to eliminate grating lobes. To derive interesting criteria for linear arrays, we consider a common expression for φ_0:[44]

$$\varphi_0 = -kd\cos\theta_0. \tag{7.118}$$

Then the maximum radiation directions can be written

$$\theta_{\text{max}} = \cos^{-1}\left(\frac{m\lambda}{d} + \cos\theta_0\right), \qquad \left(-\frac{d}{\lambda} - \frac{d}{\lambda}\cos\theta_0 \leq m \leq \frac{d}{\lambda} - \frac{d}{\lambda}\cos\theta_0\right). \tag{7.119}$$

This means selecting $m = 0$ gives the main lobe at θ_0, but grating lobes can be avoided if

$$-1 < -\frac{d}{\lambda} - \frac{d}{\lambda}\cos\theta_0 \quad \text{and} \quad \frac{d}{\lambda} - \frac{d}{\lambda}\cos\theta_0 < 1 \tag{7.120}$$

or $d < \lambda/(1 \pm \cos\theta_0)$. Similar to this straightforward derivation for eliminating grating lobes, many criteria can be used to achieve desired radiation characteristics. For example, considering a uniformly excited linear array, a simple form of the Hansen-Woodyard conditions can be written

$$\varphi_0 \approx \pm(kd + \pi/N), \tag{7.121}$$

which is derived (via optimization) to obtain an end-fire radiation with *maximum* directivity in the $\theta = 0°$ or $\theta = 180°$ direction.

Using nonuniform excitations with unequal amplitudes (in addition to different phases) provides additional abilities to control the array factors of arrays. *Binomial* arrays are linear configurations of equally spaced elements with excitation amplitudes given as

$$|\zeta_n| = \binom{N-1}{n} = \frac{(N-1)!}{n!(N-1-n)!}, \qquad (n = 0, 1, \ldots, N-1). \tag{7.122}$$

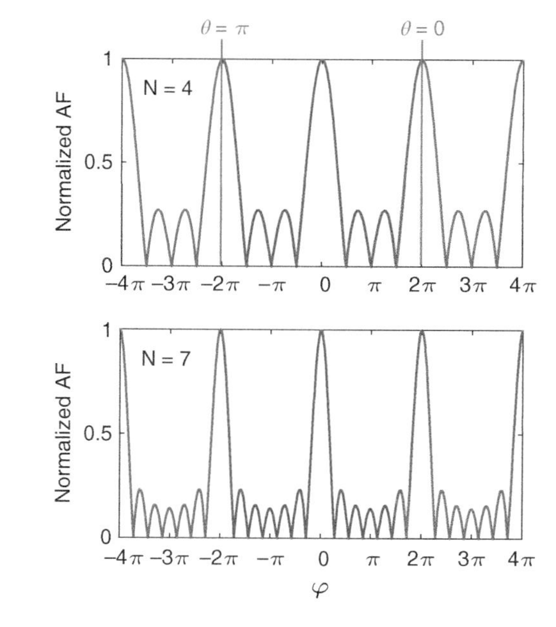

Figure 7.37 Visible regions of array factors of z-directed linear arrays with $N = 4$ and $N = 7$ elements using $d = \lambda$ element spacing and zero phase progress ($\varphi_0 = 0°$). Note that when φ changes from $-360°$ to $360°$, θ changes from 180° to 0°.

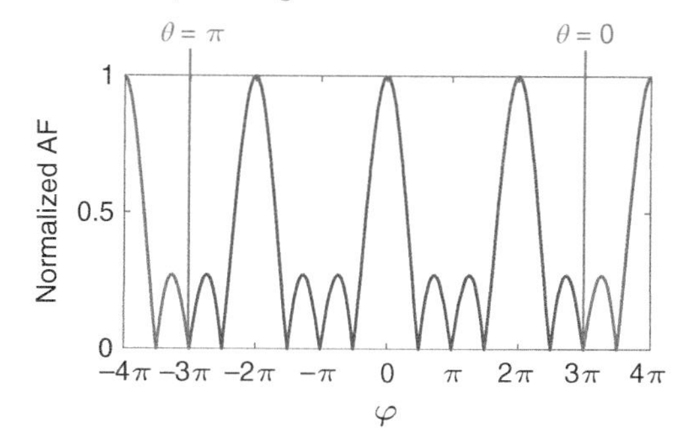

Figure 7.38 Visible region of the array factor of a z-directed linear array with $N = 4$ elements using $d = 3\lambda/2$ element spacing and zero phase progress ($\varphi_0 = 0°$). There are nulls at $\theta = 0°$ and $\theta = 180°$.

[43] Depending on the application, the beams toward 70.5° and 109.5° can be interpreted as grating lobes.

[44] This means we consider linear arrays with excitations $\zeta_n = \zeta_0 \exp(-jknd\cos\theta_0)$: i.e. the progressive phase depends on d itself.

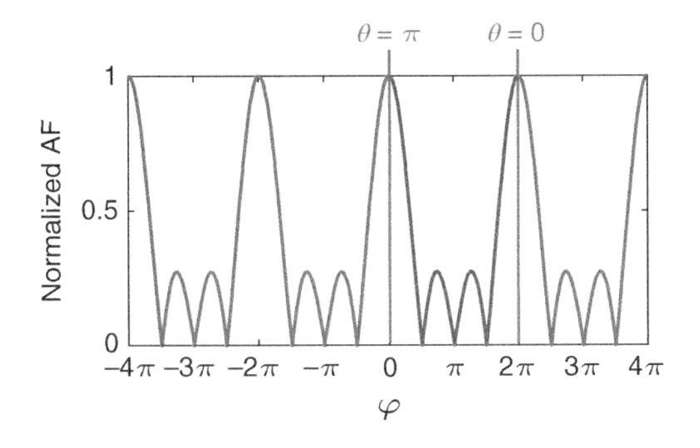

Figure 7.39 Visible region of the array factor of a z-directed linear array with $N = 4$ elements using $d = \lambda/2$ element spacing and $\varphi_0 = 180°$ phase progress. There is no broadside radiation (at $\theta = 90°$).

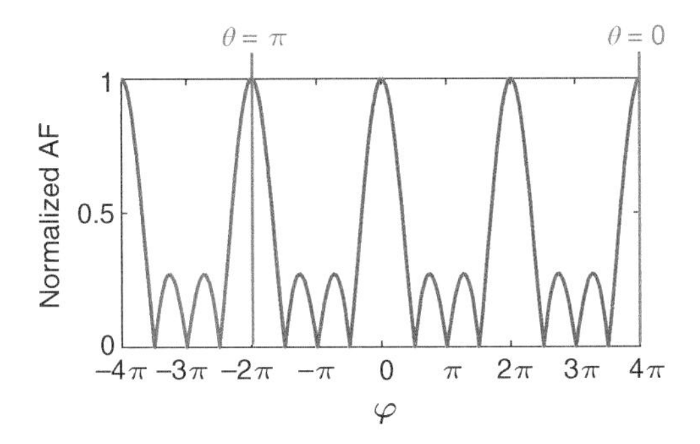

Figure 7.40 Visible region of the array factor of a z-directed linear array with $N = 4$ elements using $d = 3\lambda/2$ element spacing and $\varphi_0 = 180°$ phase progress. There are four main/grating lobes.

Note that a binomial array can be considered a recursive combination of smaller binomial arrays with overlapping elements. For example, for a five-element array, we have $|\zeta_0| = 1$, $|\zeta_1| = 4$, $|\zeta_2| = 6$, $|\zeta_3| = 4$, and $|\zeta_4| = 1$. Without any phase variation, the array factor can be written

$$\mathrm{AF}(\theta, \phi) = \sum_{n=0}^{4} \zeta_n \exp(jknd\cos\theta) \tag{7.123}$$

$$= 1 + 4\exp(jkd\cos\theta) + 6\exp(j2kd\cos\theta)$$

$$+ 4\exp(j3kd\cos\theta) + \exp(j4kd\cos\theta) \tag{7.124}$$

$$= [1 + \exp(jkd\cos\theta)]^4 \tag{7.125}$$

for a separation distance d. Then for the normalized array factor, we have

$$\widetilde{\mathrm{AF}}(\theta, \phi) = \cos^4[(kd/2)\cos\theta]. \tag{7.126}$$

If $d \leq \lambda/2$, this array factor does not have any side lobes (Figure 7.41), as a general property of binomial arrays.

As further examples of linear configurations with nonuniform excitations, we can mention Dolph-Chebyshev arrays. Excitation coefficients in these arrays are selected based on Chebyshev polynomials (specifically, matching array factors to Chebyshev polynomials) such that the side lobes become equal in magnitude, leading to the minimum beam width for a given SLL.

The linear arrangements discussed so far represent a particular class of configurations to build antenna arrays, but there can be many other configurations. For example, planar arrays with two-dimensional arrangements of elements are commonly used in practice. As a particular case, we can consider $M \times N$ identical elements located on the x-y plane with periodicities d_x and d_y in the x and y directions, respectively (Figure 7.35). Assuming that one of the corner elements is at the origin, the positions of the elements can be written

$$\bar{r}_{mn} = (x_m, y_n, 0), \quad x_m = md_x, \quad y_n = nd_y \tag{7.127}$$

for $m = 0, 1, \ldots, M-1$ and $n = 0, 1, \ldots, N-1$. Then using

$$\hat{a}_r \cdot \bar{r}_{mn} = \frac{(\hat{a}_x x + \hat{a}_y y + \hat{a}_z z)}{R} \cdot (\hat{a}_x x_m + \hat{a}_y y_n) \tag{7.128}$$

$$= \frac{x x_m + y y_n}{R} = x_m \sin\theta\cos\phi + y_n \sin\theta\sin\phi, \tag{7.129}$$

the array factor can be written

$$\mathrm{AF}(\theta,\phi)=\sum_{m=0}^{M-1}\sum_{n=0}^{N-1}\zeta_{m,n}\exp[jk(x_m\sin\theta\cos\phi+y_m\sin\theta\sin\phi)] \tag{7.130}$$

$$=\sum_{m=0}^{M-1}\sum_{n=0}^{N-1}\zeta_{m,n}\exp[jk(md_x\sin\theta\cos\phi+nd_n\sin\theta\sin\phi)]. \tag{7.131}$$

If the excitations are *separable* as $\zeta_{m,n}=\zeta_m\zeta_n$, we further have

$$\mathrm{AF}(\theta,\phi)=\sum_{m=0}^{M-1}\zeta_m\exp[jkmd_x\sin\theta\cos\phi]\sum_{n=0}^{N-1}\zeta_n\exp[jknd_n\sin\theta\sin\phi] \tag{7.132}$$

$$=\mathrm{AF}_x(\theta,\phi)\times\mathrm{AF}_y(\theta,\phi), \tag{7.133}$$

which indicates that such a two-dimensional array is nothing but a one-dimensional array of one-dimensional arrays.

As considered for simple two-element configurations, an array factor does not generally represent the overall radiation pattern of an array. Specifically, the array factor should be multiplied with the radiation pattern of an individual element to obtain the overall radiation, which can be quite different from both the array factor and the single-element radiation. As further examples, we now consider five-element uniform (fixed excitation amplitude) linear arrays, whose array factor is given as

$$\mathrm{AF}(\varphi)=\zeta_0\exp(j2\varphi)\frac{\sin(5\varphi/2)}{\sin(\varphi/2)},\qquad \varphi=\varphi_0+kd\cos\theta. \tag{7.134}$$

We focus on two different cases: $\{d_1=\lambda,\varphi_{0,1}=0°\}$ and $\{d_2=\lambda/2,\varphi_{0,2}=180°\}$. In these cases, we have (Figures 7.42–7.44)

$$\varphi_1=2\pi\cos\theta\longrightarrow\mathrm{AF}_1(\theta,\phi)=\zeta_0\exp(j4\pi\cos\theta)\frac{\sin(5\pi\cos\theta)}{\sin(\pi\cos\theta)} \tag{7.135}$$

and

$$\varphi_2=\pi\cos\theta+\pi\longrightarrow\mathrm{AF}_2(\theta,\phi)=\zeta_0\exp(j2\pi\cos\theta)\frac{\cos[\pi(5/2)\cos\theta]}{\cos[\pi(1/2)\cos\theta]}, \tag{7.136}$$

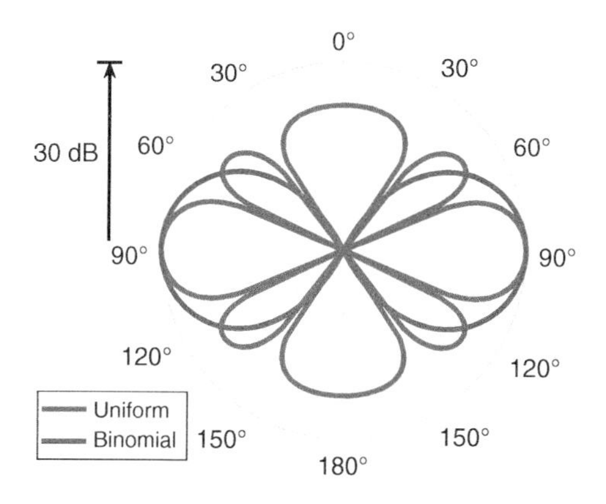

Figure 7.41 Normalized array factors of five-element arrays aligned along the z axis. The elements are separated by $\lambda/2$ distance, and both arrays have zero progressive phase. The uniform array consists of identical excitations, whereas the binomial array uses the amplitudes given in Eq. (7.122).

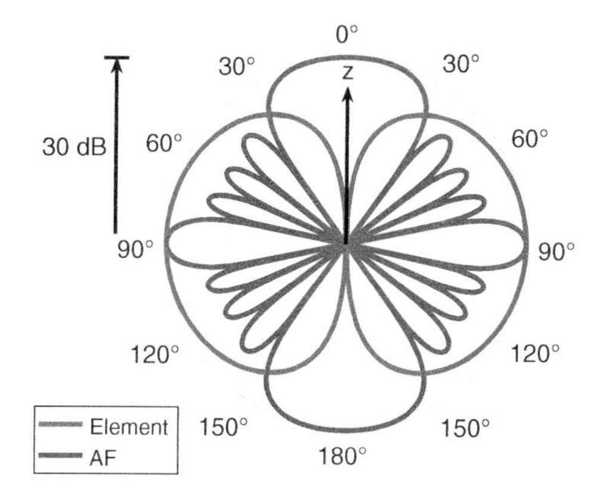

Figure 7.42 Radiation pattern of a Hertzian dipole and the array factor of a five-element linear array with $d=\lambda$ and $\varphi=0°$.

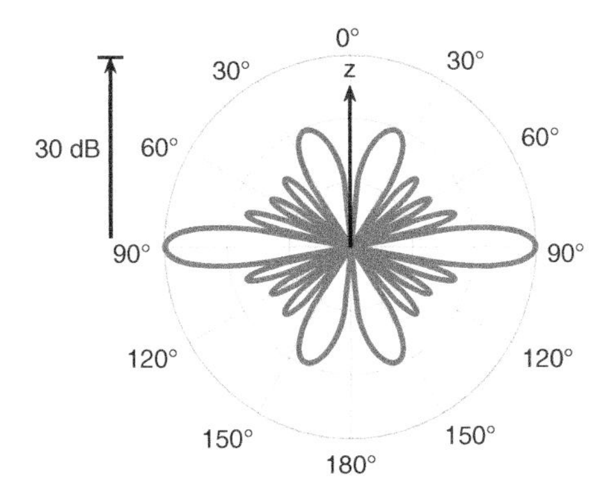

Figure 7.43 Radiation pattern of a five-element array of Hertzian dipoles, when the array factor is as given in Figure 7.42.

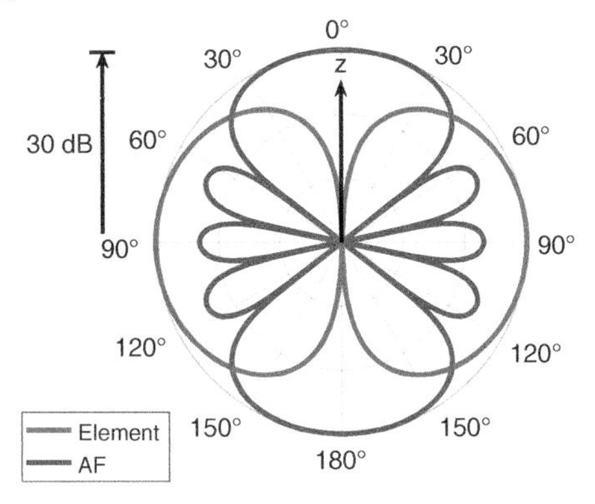

Figure 7.44 Radiation pattern of a Hertzian dipole and the array factor of a five-element linear array with $d = \lambda/2$ and $\varphi_0 = 180°$.

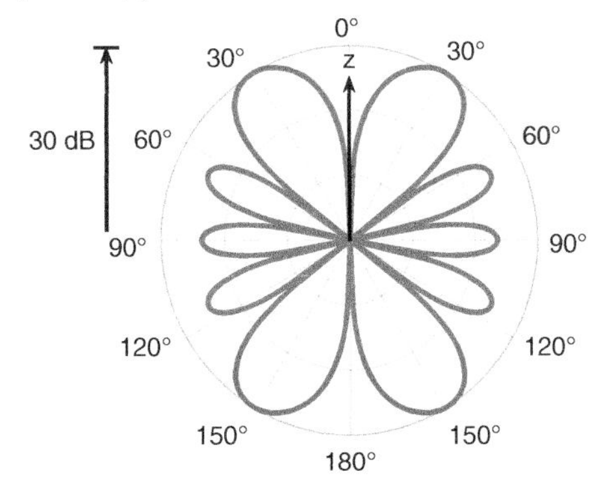

Figure 7.45 Radiation pattern of a five-element array of Hertzian dipoles, when the array factor is as given in Figure 7.44.

respectively. Considering z-directed Hertzian dipoles as array elements, the radiation intensity of a single element at the origin can be written

$$U_d^{\text{far}}(\theta, \phi) = |I_{DM,h}|^2 \frac{\omega k \mu}{32\pi^2} \sin^2\theta. \tag{7.137}$$

Hence, the overall radiation intensities generated by the five-element arrays can be written

$$U_{arr,1}^{\text{far}}(\theta, \phi) = |\text{AF}_1(\theta, \phi)|^2 U_d^{\text{far}}(\theta, \phi) \tag{7.138}$$

$$= |\zeta_0|^2 |I_{DM,h}|^2 \frac{\omega k \mu}{32\pi^2} \frac{\sin^2(\theta)\sin^2(5\pi\cos\theta)}{\sin^2(\pi\cos\theta)} \tag{7.139}$$

and

$$U_{arr,2}^{\text{far}}(\theta, \phi) = |\text{AF}_2(\theta, \phi)|^2 U_d^{\text{far}}(\theta, \phi) \tag{7.140}$$

$$= |\zeta_0|^2 |I_{DM,h}|^2 \frac{\omega k \mu}{32\pi^2} \frac{\sin^2(\theta)\cos^2[\pi(5/2)\cos\theta]}{\cos^2[\pi(1/2)\cos\theta]}. \tag{7.141}$$

Note that the radiation pattern of the first array has a single main lobe at $\theta = 90°$ and many side lobes, despite the fact that the corresponding array factor has maxima at $\theta = \{0°, 90°, 180°\}$ (Figures 7.42 and 7.43). Similarly, for the second array, the array factor has maxima at $\theta = 0°$ and $\theta = 90°$, which become nulls (due to the nulls in the radiation pattern of a Hertzian dipole) in the overall radiation pattern of the array (Figures 7.44 and 7.45).

7.1.4 Friis Transmission Equation

In general, a wireless transmission system needs multiple antennas to transmit and receive electromagnetic waves (Figure 7.46). Given two such antennas (radiating and receiving), the transferred electromagnetic power is fundamentally described by the Friis transmission equation. To understand transmission between two antennas, however, we must first consider the reciprocity theorem. In electromagnetics, reciprocity is written in many ways; but for circuit models (Figure 7.47), we have

$$= V_{a,B}/I_{in,A} = V_{a,A}/I_{in,B}. \tag{7.142}$$

When Antenna A radiates and Antenna B receives, $V_{a,B}$ is the voltage generated on Antenna B and $I_{in,A}$ is the input current on Antenna A. In the reverse case, $V_{a,A}$ is the voltage generated on Antenna A (receiving), while $I_{in,B}$ is the input current on Antenna B (radiating). According to the reciprocity described above, the voltage/current ratios in these cases are equal to each other.

Now, we focus on the case when Antenna A radiates and Antenna B receives. The real power transferred from the generator to Antenna A can be written

$$p_{in,A} = \frac{1}{2} R_{in,A} |I_{in,A}|^2, \tag{7.143}$$

where $R_{in,A}$ is the real part of the input impedance. This power is partially radiated by Antenna A, leading to a power flux density on Antenna B as

$$|\bar{S}^{\text{inc}}_{c,B}| = \frac{1}{4\pi R^2} G_A(\theta_{AB}, \phi_{AB}) p_{in,A} = \frac{1}{4\pi R^2} G_A(\theta_{AB}, \phi_{AB}) \left[\frac{1}{2} R_{in,A} |I_{in,A}|^2\right], \tag{7.144}$$

where $G_A(\theta_{AB}, \phi_{AB})$ is the gain of Antenna A in the direction of Antenna B. The power incident on Antenna B is received by this antenna as

$$p_{rec,B} = \frac{\lambda^2}{4\pi} G_B(\theta_{BA}, \phi_{BA}) |\bar{S}^{\text{inc}}_{c,B}| = \left(\frac{\lambda}{4\pi R}\right)^2 G_B(\theta_{BA}, \phi_{BA}) G_A(\theta_{AB}, \phi_{AB}) p_{in,A} \tag{7.145}$$

$$= \left(\frac{\lambda}{4\pi R}\right)^2 G_A(\theta_{AB}, \phi_{AB}) G_B(\theta_{BA}, \phi_{BA}) \left[\frac{1}{2} R_{in,A} |I_{in,A}|^2\right], \tag{7.146}$$

where $G_B(\theta_{BA}, \phi_{BA})$ is the gain of Antenna B in the direction of Antenna A. Hence, the voltage on Antenna B can be written

$$|V_{a,B}|^2 = 8 R_{in,B} p_{rec,B} \tag{7.147}$$

$$= \left(\frac{\lambda}{2\pi R}\right)^2 G_A(\theta_{AB}, \phi_{AB}) G_B(\theta_{BA}, \phi_{BA}) R_{in,A} R_{in,B} |I_{in,A}|^2, \tag{7.148}$$

where $R_{in,B}$ is the real part of the input impedance of Antenna B, leading to

$$\frac{|V_{a,B}|^2}{|I_{in,A}|^2} = \left(\frac{\lambda}{2\pi R}\right)^2 G_A(\theta_{AB}, \phi_{AB}) G_B(\theta_{BA}, \phi_{BA}) R_{in,A} R_{in,B}. \tag{7.149}$$

This equation is consistent with the reciprocity theorem, as the right-hand side does not depend on which antenna radiates/receives.

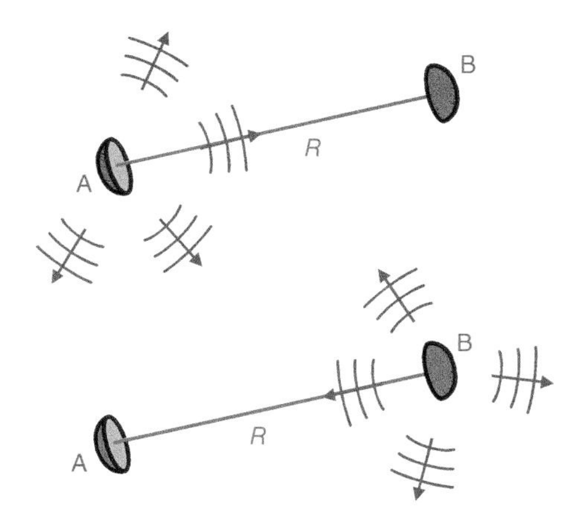

Figure 7.46 Two interacting antennas. In the first case (top), Antenna B receives electromagnetic waves radiated by Antenna A. In the second case (bottom), Antenna A receives while Antenna B radiates.

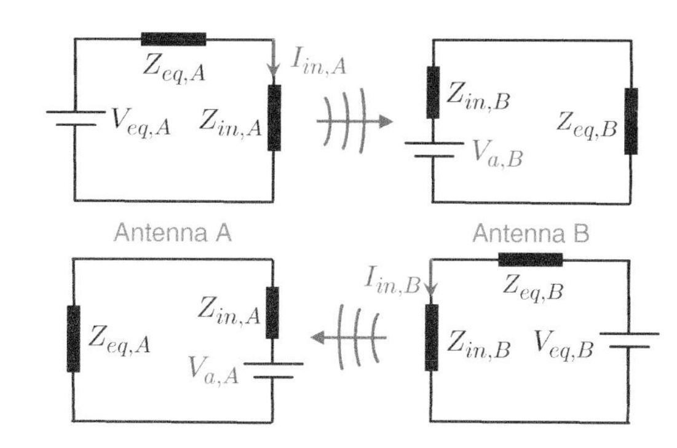

Figure 7.47 Circuit models of two interacting antennas: i.e. Antennas A/B, which operate as radiating/receiving and receiving/radiating one after another.

The Friis transmission equation describes the power transferred from a radiating antenna to a receiving antenna when they are located in each other's *far zone*. Considering Eq. (7.146), we simply have[45]

$$\frac{p_{rec,B}}{p_{in,A}} = \left(\frac{\lambda}{4\pi R}\right)^2 G_A(\theta_{AB}, \phi_{AB}) G_B(\theta_{BA}, \phi_{BA}) \tag{7.150}$$

as the received power (available at the terminals of Antenna B) divided by the input power (delivered to Antenna A by its generator). Note that in the case of an unmatched load connected to Antenna B, this power is partially transferred to its load as

$$\frac{p_{eq,B}}{p_{in,A}} = (1 - |\Gamma_B|^2)\left(\frac{\lambda}{4\pi R}\right)^2 G_A(\theta_{AB}, \phi_{AB}) G_B(\theta_{BA}, \phi_{BA}), \tag{7.151}$$

where Γ_B is the power reflection coefficient.[46] If polarization mismatches exist between the antennas, the received power should be further reduced via multiplication with the polarization efficiency $e_p < 1$. Specifically, considering *both* $G_A(\theta_{AB}, \phi_{AB})$ and $G_B(\theta_{BA}, \phi_{BA})$, we obtain

$$\frac{p_{eq,B}}{p_{in,A}} = |e_p|^2(1 - |\Gamma_B|^2)\left(\frac{\lambda}{4\pi R}\right)^2 G_A(\theta_{AB}, \phi_{AB}) G_B(\theta_{BA}, \phi_{BA}), \tag{7.152}$$

where $e_p = \hat{a}_A \cdot \hat{a}_B$. Here, $\hat{a}_A$ represents the polarization of the radiation by Antenna A, while $\hat{a}_B$ represents the polarization of Antenna B if it was *radiating*. In general, $|e_p|^2$ is called the *polarization loss factor* (PLF).

Finally, in some cases, the Friis equation is written in a logarithmic form (in terms of dB quantities), such as[47]

$$\begin{aligned} \text{dB}(p_{eq,B}) &= \text{dB}(p_{in,A}) + \text{dB}(\text{PLF}) + \text{dB}(1 - |\Gamma_B|^2) + 2 \times \text{dB}[\lambda/(4\pi R)] \\ &\quad + \text{dB}[G_A(\theta_{AB}, \phi_{AB})] + \text{dB}[G_B(\theta_{BA}, \phi_{BA})], \end{aligned} \tag{7.153}$$

$$\text{dB}(\cdot) = 10 \log_{10}(\cdot), \tag{7.154}$$

which can be more suitable when dealing with small/large numbers.

7.1.5 Examples

Example 89: Consider a straight wire antenna with length l aligned from $z = -l/2$ to $z = l/2$. Assume that the current on the antenna can be modeled as

$$\bar{I}_w(x = 0, y = 0, z) = \hat{a}_z I_0 \begin{cases} \sin[k(l/2 - z)] & (0 \leq z \leq l/2) \\ \sin[k(l/2 + z)] & (-l/2 \leq z \leq 0), \end{cases} \tag{7.155}$$

where I_0 (A) is a constant (Figure 7.48).

[45] Once again, the importance of *distance* is clearly observed in the Friis transmission equation: i.e. the electromagnetic power transmitted from an antenna to another antenna decays quadratically with the distance. This is due to the quadratic decay of the power flux density as the observation point moves away from a source (a radiating antenna). As an interesting example, the power of the radiation from the Sun is approximately 3.8×10^{26} W. Despite this huge number, the power flux density on the Earth is only approximately 1.3×10^3 W/m^2, considering 150×10^6 km distance between the Sun and the Earth. Nevertheless, this power flux density on the surface of the Earth leads to an enormous amount of total received power, which demonstrates the importance of solar energy.

[46] It is defined as $\Gamma_B = \dfrac{Z_{eq,B} - Z^*_{in,B}}{Z_{eq,B} + Z_{in,B}}$.

[47] Again, note the importance of the distance. For example, if $p_{in,A} = 1$ W $= 0$ dBW, PFL $= 1 = 0$ dB (no polarization mismatch), $\Gamma_B = 0$ (no impedance mismatch), and $G_A(\theta_{AB}, \phi_{AB}) = G_B(\theta_{AB}, \phi_{AB}) = 1$, then

$$\text{dB}(p_{eq,B}) = 20 \log\left(\frac{\lambda}{4\pi R}\right) \text{ dBW}.$$

If $R = 1000\lambda$, this power is as low as -82 dBW, corresponding to approximately 6.3 nW.

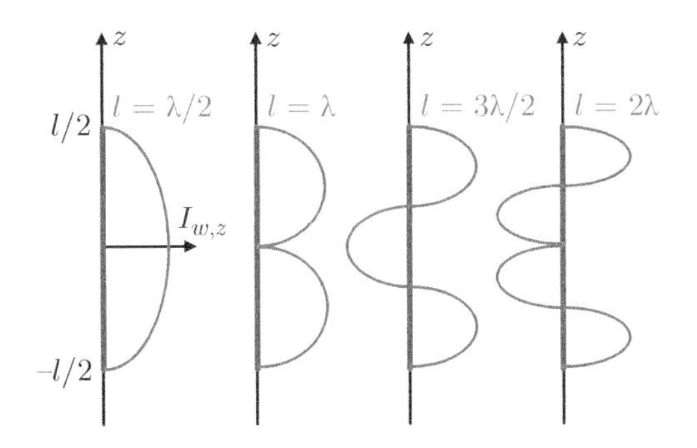

Figure 7.48 A wire antenna aligned along the z axis, and the distribution of the current (z component) for different lengths with respect to wavelength.

- Find the radiation intensity for a general value of l.
- Find the total radiated power and the directive gain of the antenna when $l = (2m+1)\lambda/2$ for $m = 0, 1, \ldots$. Evaluate the expressions for $l = \lambda/2$.

Solution: First, we can find the far-zone magnetic vector potential as[48]

$$\bar{A}_w^{\text{far}}(\bar{r}) = \hat{a}_z \mu I_0 g(R) 2 \int_0^{l/2} \sin[k(l/2 - z')] \cos(kz' \cos\theta) d\bar{r}' \tag{7.156}$$

$$= \hat{a}_z \mu I_0 g(R) \left[\frac{\cos[k(l/2 + z'\cos\theta - z')]}{k(1-\cos\theta)} + \frac{\cos[k(l/2 - z'\cos\theta - z')]}{k(1+\cos\theta)} \right]_0^{l/2} \tag{7.157}$$

$$= \hat{a}_z \mu I_0 g(R) \frac{2}{k \sin^2\theta} [\cos[(kl/2)\cos\theta] - \cos(kl/2)]. \tag{7.158}$$

Hence, the magnetic flux density and electric field intensity in the far zone can be found as[49]

$$\bar{B}_w^{\text{far}}(\bar{r}) = -jk\hat{a}_R \times \bar{A}_w^{\text{far}}(\bar{r}) = \hat{a}_\phi j 2\mu I_0 g(R) \left[\frac{\cos[(kl/2)\cos\theta] - \cos(kl/2)}{\sin\theta} \right] \tag{7.159}$$

$$\bar{E}_w^{\text{far}}(\bar{r}) = -\frac{\eta}{\mu} \hat{a}_R \times \bar{B}_w^{\text{far}}(\bar{r}) = \hat{a}_\theta j 2\eta I_0 g(R) \left[\frac{\cos[(kl/2)\cos\theta] - \cos(kl/2)}{\sin\theta} \right]. \tag{7.160}$$

These expressions lead to the far-zone power flux density and radiation intensity as

$$\bar{S}_{c,w}^{\text{far}}(\bar{r}) = \frac{1}{2} \bar{E}_w^{\text{far}}(\bar{r}) \times [\bar{H}_w^{\text{far}}(\bar{r})]^* \tag{7.161}$$

$$= \hat{a}_R |I_0|^2 \frac{\eta}{8\pi^2 R^2} \left[\frac{\cos[(kl/2)\cos\theta] - \cos(kl/2)}{\sin\theta} \right]^2 \tag{7.162}$$

$$U_w^{\text{far}}(\theta, \phi) = R^2 |\bar{S}_{c,w}^{\text{far}}(\bar{r})| = |I_0|^2 \frac{\eta}{8\pi^2} \left[\frac{\cos[(kl/2)\cos\theta] - \cos(kl/2)}{\sin\theta} \right]^2. \tag{7.163}$$

When $l = (2m+1)\lambda/2$ and $kl = \pi(2m+1)$, the radiation intensity can be written (Figures 7.49 and 7.50)

$$U_w^{\text{far}}(\theta, \phi) = |I_0|^2 \frac{\eta}{8\pi^2} \frac{\cos^2[(\pi/2)(2m+1)\cos\theta]}{\sin^2\theta}. \tag{7.164}$$

This leads to the total radiated power (using a numerical integrator) as

$$p_{em,w}^{\text{far}} = \int_0^{2\pi} \int_0^{\pi} U_w^{\text{far}}(\theta, \phi) \sin\theta d\theta d\phi \tag{7.165}$$

[48] We start with

$$\bar{A}_w^{\text{far}}(\bar{r}) = \hat{a}_z \mu I_0 g(R) \Bigg\{ \int_0^{l/2} \sin[k(l/2 - z')] \exp(-jkz'\cos\theta) d\bar{r}' + \int_{-l/2}^{0} \sin[k(l/2 + z')] \exp(-jkz'\cos\theta) d\bar{r}' \Bigg\},$$

and perform a $z' \to -z'$ transformation for the second integral.

[49] We use

$$\hat{a}_R \times \hat{a}_z = (\hat{a}_x \sin\theta\cos\phi + \hat{a}_y \sin\theta\sin\phi + \hat{a}_z\cos\theta) \times \hat{a}_z = -\hat{a}_y \sin\theta\cos\phi + \hat{a}_x \sin\theta\sin\phi = \sin\theta(-\hat{a}_y\cos\phi + \hat{a}_x\sin\phi) = -\hat{a}_\phi \sin\theta.$$

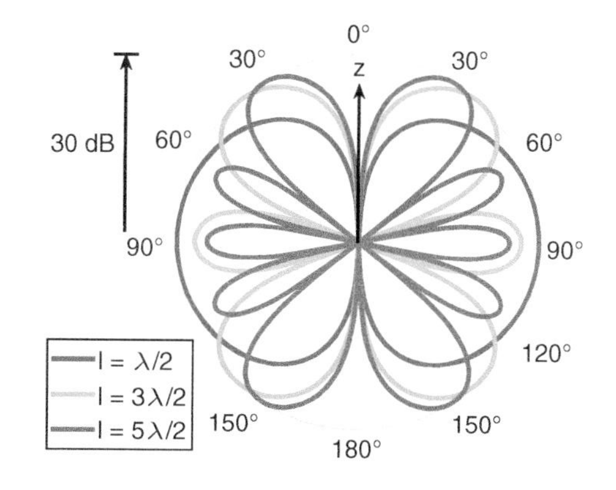

Figure 7.49 Radiation pattern of a wire antenna for different lengths ($l = \lambda/2, 3\lambda/2, 5\lambda/2$), assuming current distributions given in Eq. (7.155). The dynamic range is 30 dB.

$$= |I_0|^2 \frac{\eta}{8\pi} \int_0^{\pi} \frac{1 + \cos[\pi(2m+1)\cos\theta]}{\sin\theta} d\theta \tag{7.166}$$

$$\approx |I_0|^2 \frac{\eta}{8\pi} \left[-\mathrm{Ci}[2(2m+1)\pi] + \ln[2(2m+1)\pi] + 0.577\right], \tag{7.167}$$

where

$$\mathrm{Ci}(a) = -\int_a^{\infty} \frac{\cos b}{b} db \tag{7.168}$$

is called the *cosine integral function*. As a result, we can obtain the directive gain as

$$D_w(\theta,\phi) = \frac{4\pi U_w^{\mathrm{far}}(\theta,\phi)}{p_{em,w}^{\mathrm{far}}} \approx \frac{4\cos^2[(\pi/2)(2m+1)\cos\theta]/\sin^2\theta}{[-\mathrm{Ci}[2(2m+1)\pi] + \ln[2(2m+1)\pi] + 0.577]}. \tag{7.169}$$

When $m = 0$ – i.e. $l = \lambda/2$ – these expressions reduce into

$$p_{em,w}^{\mathrm{far}} \approx |I_0|^2 \frac{\eta}{8\pi} 2.44 \approx 1.22|I_0|^2 \frac{\eta}{4\pi} \tag{7.170}$$

$$D_w(\theta,\phi) \approx \frac{4\cos^2[(\pi/2)\cos\theta]}{2.44\sin^2\theta} \approx 1.64 \frac{\cos^2[(\pi/2)\cos\theta]}{\sin^2\theta}, \tag{7.171}$$

which are perfectly consistent with the earlier expressions found for a half-wave dipole.

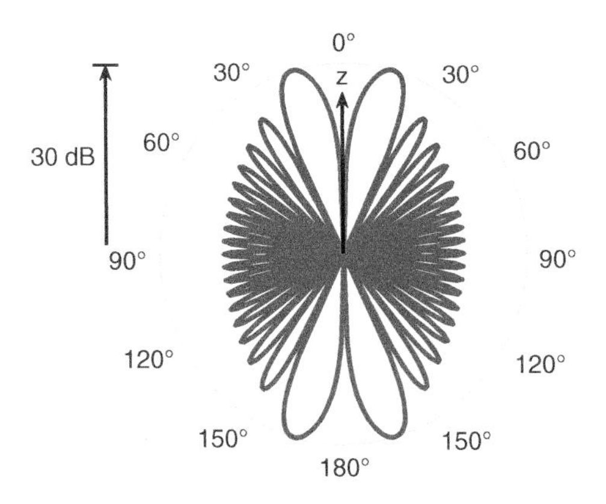

Figure 7.50 Radiation pattern of a wire antenna with $l = 19\lambda$ length, assuming that the current distribution is as given in Eq. (7.155).

Example 90: An antenna with radiation intensity (Figure 7.51)

$$U^{\mathrm{far}}(\theta,\phi) = U_0 \begin{cases} \cos^3\theta & (0 \le \theta \le \pi/2) \\ 0 & (\pi/2 < \theta \le \pi), \end{cases} \tag{7.172}$$

where U_0 (W/sr) is a constant, is connected to a generator (source+feedline) with $V_{eq} = 2$ V voltage and $Z_{eq} = 50 + j75\ \Omega$ impedance. The antenna has $R_r = 50\ \Omega$ radiation resistance, $R_c = 25\ \Omega$ ohmic resistance, and 50 Ω reactance (Figure 7.52).

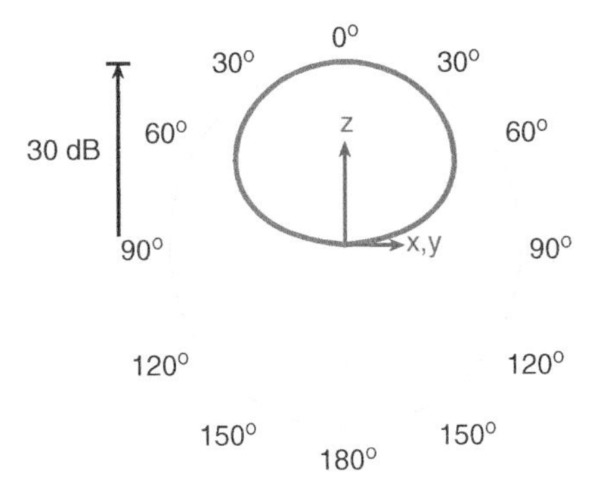

Figure 7.51 The radiation pattern corresponding to the intensity given in Eq. (7.172).

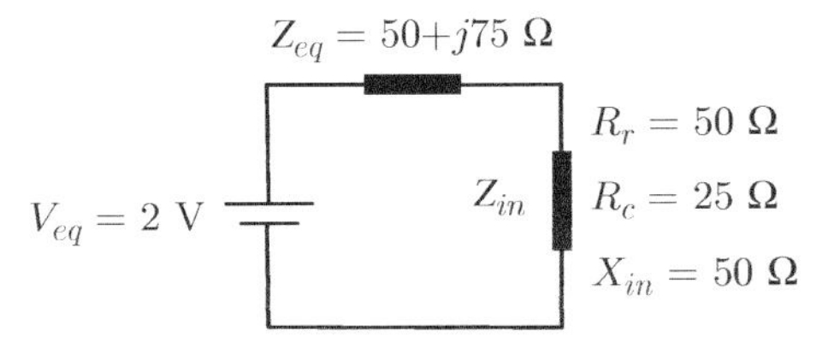

Figure 7.52 Circuit model of an antenna with nonzero reactance and ohmic resistance, connected to a generator.

- Find the directivity of the antenna.
- Find the real power values supplied by the generator, dissipated in the generator, delivered to the antenna, and radiated by the antenna. Also find U_0 and the maximum gain of the antenna.
- Find the power values, U_0, and the maximum gain when the source+feedline is matched to the antenna.

Solution: Considering the radiation intensity, we can obtain the total radiated power as

$$p_{em}^{far} = \int_0^{2\pi}\int_0^{\pi} U^{far}(\theta,\phi)\sin\theta d\theta d\phi = U_0\int_0^{2\pi}\int_0^{\pi/2}\cos^3\theta\sin\theta d\theta d\phi \tag{7.173}$$

$$= 2\pi U_0\left[-\frac{\cos^4\theta}{4}\right]_0^{\pi/2} = \frac{\pi}{2}U_0. \tag{7.174}$$

Then the directive gain can be written

$$D(\theta,\phi) = \frac{4\pi U^{far}(\theta,\phi)}{p_{em}^{far}} = 8\begin{cases}\cos^3\theta & (0\le\theta\le\pi/2)\\ 0 & (\pi/2<\theta\le\pi),\end{cases} \tag{7.175}$$

leading to $D_{max} = 8$ when $\theta = 0°$. For the given values of impedances, we have $Z_{in} = 75 + j50\ \Omega$, and the real power supplied by the generator can be written[50]

$$p_g = \frac{1}{2}\frac{|V_{eq}|^2}{|Z_{eq}+Z_{in}|^2}(R_{eq}+R_{in}) = \frac{1}{2}\frac{4}{|50+j75+75+j50|^2}(50+75) \tag{7.176}$$

$$= \frac{2\times 125}{(125\sqrt{2})^2} = 8\ \text{mW}. \tag{7.177}$$

This power is divided in two as

$$p_{eq} = \frac{R_{eq}}{R_{eq}+R_{in}}p_g = 3.2\ \text{mW}, \qquad p_{in} = \frac{R_{in}}{R_{eq}+R_{in}}p_g = 4.8\ \text{mW}, \tag{7.178}$$

corresponding to the power dissipated in the generator and the power delivered to the antenna, respectively. The power radiated by the antenna can further be found as

$$p_{em}^{far} = \frac{R_r}{R_r+R_c}p_{in} = \frac{50}{50+25}p_{in} = 3.2\ \text{mW}, \tag{7.179}$$

which means $p_c = p_{in} - p_{em}^{far} = 1.6$ mW is the amount of power dissipated by the antenna (ohmic losses). Then we can obtain U_0 as

$$p_{em}^{far} = \frac{\pi}{2}U_0 = 3.2 \longrightarrow U_0 \approx 2.04\ \text{W/sr}. \tag{7.180}$$

Finally, the gain and its maximum value can be found as[51]

$$G(\theta,\phi) = \frac{p_{em}^{far}}{p_{in}}D(\theta,\phi) = \frac{16}{3}\begin{cases}\cos^3\theta & (0\le\theta\le\pi/2)\\ 0 & (\pi/2<\theta\le\pi)\end{cases} \tag{7.181}$$

and $G_{max} = 16/3$.

[50] In circuit analysis, the complex power of a device with impedance $Z_d = R_d + jX_d$ can be found as

$$S_d = \frac{1}{2}V_d(I_d)^* = \frac{1}{2}V_d\frac{(V_d)^*}{(Z_d)^*} = \frac{1}{2}\frac{|V_d|^2}{(Z_d)^*} = \frac{1}{2}Z_d|I_d|^2,$$

where V_d and I_d are voltage and current in the phasor domain. Then the real power can be obtained as

$$p_d = \text{Re}\{S_d\} = \frac{1}{2}\text{Re}\{Z_d\}|I_d|^2 = \frac{1}{2}R_d|I_d|^2$$

or

$$p_d = \text{Re}\{S_d\} = \frac{1}{2}|V_d|^2\text{Re}\left\{\frac{1}{(Z_d)^*}\right\} = \frac{1}{2}\frac{|V_d|^2}{|Z_d|^2}R_d,$$

which corresponds to the time-average power in the time domain.

[51] Hence the radiation efficiency can also be found as $e_r = p_{em}^{far}/p_{in} = 3.2/4.8 \approx 0.667$.

In the case of a matched generator – i.e. when $Z_{eq} = 75 - j50\ \Omega$ – we obtain[52]

$$p_g = \frac{1}{2}\frac{|V_{eq}|^2}{|Z_{eq}+Z_{in}|^2}(R_{eq}+R_{in}) = \frac{1}{2}\frac{4(75+75)}{|75-j50+75+j50|^2} \approx 13.3 \text{ mW} \quad (7.182)$$

$$p_{eq} = \frac{R_{eq}}{R_{eq}+R_{in}}p_g = \frac{75}{150}p_g \approx 6.67 \text{ mW} \quad (7.183)$$

$$p_{in} = \frac{R_{in}}{R_{eq}+R_{in}}p_g = \frac{75}{150}p_g \approx 6.67 \text{ mW} \quad (7.184)$$

$$p_{em}^{\text{far}} = \frac{R_r}{R_r+R_c}p_{in} = \frac{50}{50+25}p_{in} \approx 4.44 \text{ mW} \quad (7.185)$$

as the power values, while $U_0 \approx 2.83$ W/sr. Obviously, the gain of the antenna and its maximum value do not change with the impedance of the generator.

[52]Note that using a matched generator, the real power values supplied by the generator, dissipated in the generator, delivered to the antenna, and radiated by the antenna increase.

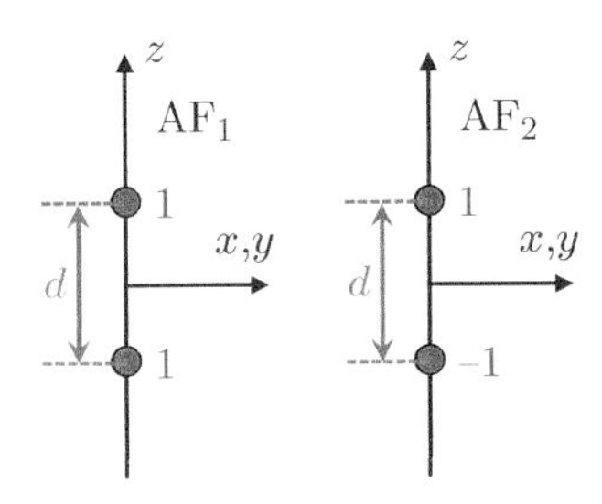

Figure 7.53 Simple arrays of two elements.

Example 91: Consider two simple arrays, each with two elements separated by a distance d, aligned along the z axis (Figure 7.53). The first array with identical excitations has an array factor of AF_1, while the second array with ± 1 excitations has an array factor of AF_2. Find the array factors of the array configurations in Figure 7.54 in terms of AF_1 and AF_2.

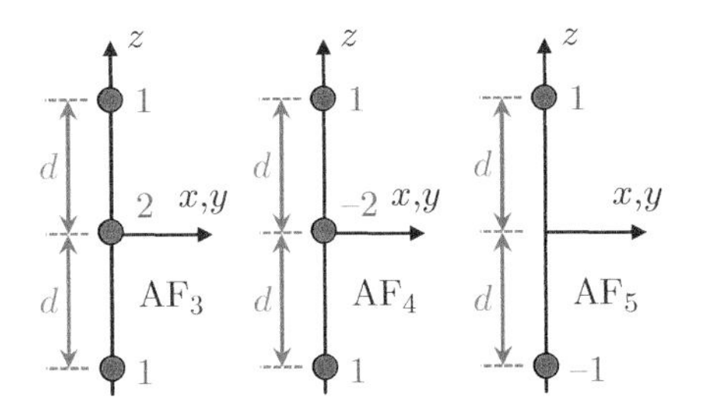

Figure 7.54 Arrays of two or three elements, whose array factors can be derived from the array factors of the configurations in Figure 7.53.

Solution: Array 3 is a two-element array of Array 1 configurations. Specifically, by adding two shifted versions of Array 1 with a separation distance d and identical excitations, one can obtain Array 3 (Figure 7.55). Hence, we have

$$\text{AF}_3 = (\text{AF}_1)^2. \quad (7.186)$$

Similarly, Array 4 can be obtained using two Array 2 configurations with d separation distance and ± 1 excitations for them (Figure 7.56). Therefore, we obtain

$$\text{AF}_4 = (\text{AF}_2)^2. \quad (7.187)$$

Finally, Array 5 is a combination of two Array 1 configurations with d separation distance, if the bottom Array 1 is excited by -1. This means the configuration represented by Array 2 is used to arrange two Array 1 configurations, leading to

$$\text{AF}_5 = \text{AF}_1 \times \text{AF}_2. \quad (7.188)$$

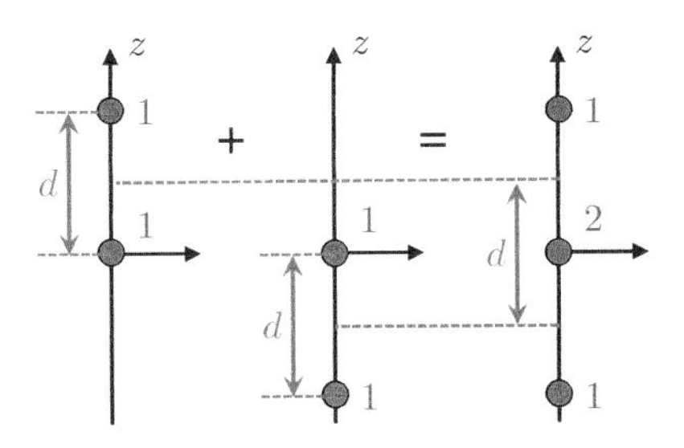

Figure 7.55 Building Array 3 in Figure 7.54 using two Array 1 configurations in Figure 7.53.

Example 92: An linear array of N elements is placed along the z axis with a separation distance $d = \lambda/2$ between consecutive elements (Figure 7.35). The excitations of the elements have different amplitudes but a fixed phase (zero progressive phase).

- Find a value for N and a set of excitation values if we want the array factor to have nulls at $\theta = \{0°, 60°, 120°, 180°\}$.
- Find the array factor if the excitation of the middle element is multiplied by two.

- Find the array factor if the first and last elements are removed from the array.
- Find the excitations if the nulls at 60° and 120° need to be shifted to 45° and 135°.

Solution: For an N-element linear array aligned in the z direction, the array factor can be written[53,54]

$$\mathrm{AF}(\theta,\phi)=\sum_{n=0}^{N-1}\zeta_n\exp(jknd\cos\theta). \tag{7.189}$$

Since $\zeta_n=|\zeta_n|$, we further have

$$\mathrm{AF}(\theta,\phi)=\sum_{n=0}^{N-1}|\zeta_n|\exp(jknd\cos\theta)=\sum_{n=0}^{N-1}|\zeta_n|\chi^n, \tag{7.190}$$

where $\chi=\exp(jkd\cos\theta)$. Hence, in terms of χ, the array factor is simply a polynomial

$$\mathrm{AF}(\chi)=|\zeta_0|+|\zeta_1|\chi+|\zeta_2|\chi^2\ldots+|\zeta_{N-1}|\chi^{N-1}, \tag{7.191}$$

which has a total of $N-1$ roots. Given that $d=\lambda/2$ and the desired nulls are at $\theta=\{0°,60°,120°,180°\}$, the array factor can be in the form of

$$\mathrm{AF}(\chi)\propto(\chi-\chi_1)(\chi-\chi_2)(\chi-\chi_3)(\chi-\chi_4) \tag{7.192}$$

where

$$\chi_1=\exp[j\pi\cos(0°)]=-1,\qquad \chi_2=\exp[j\pi\cos(60°)]=j \tag{7.193}$$

$$\chi_3=\exp[j\pi\cos(120°)]=-j,\qquad \chi_4=\exp[j\pi\cos(180°)]=-1 \tag{7.194}$$

are the roots. Specifically, we have

$$\mathrm{AF}(\chi)\propto(\chi+1)(\chi-j)(\chi+j)(\chi+1)=1+2\chi+2\chi^2+2\chi^3+\chi^4, \tag{7.195}$$

leading to a *possible* selection of excitations as

$$|\zeta_0|=1,\quad|\zeta_1|=2,\quad|\zeta_2|=2,\quad|\zeta_3|=2,\quad|\zeta_4|=1. \tag{7.196}$$

The corresponding array factor in terms of θ can be written (Figure 7.57)

$$\mathrm{AF}(\theta,\phi)=1+2\exp(j\pi\cos\theta)+2\exp(j2\pi\cos\theta)$$
$$+2\exp(j3\pi\cos\theta)+\exp(j4\pi\cos\theta) \tag{7.197}$$
$$=[\exp(j\pi\cos\theta)+1]^2[\exp(j2\pi\cos\theta)+1]. \tag{7.198}$$

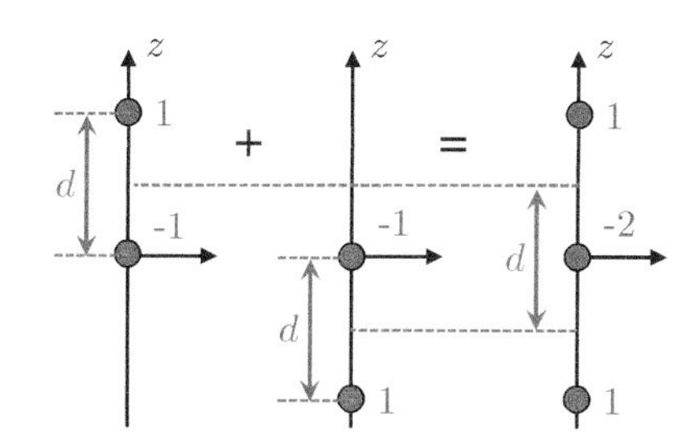

Figure 7.56 Building Array 4 in Figure 7.54 using two Array 2 configurations in Figure 7.53.

[53] We assume that the bottom element is located at the origin. An extra multiplication with a phase factor is required if the array is shifted along the z axis, but this does not change the solution.

[54] The procedure described in this solution is known as the *Schelkunoff polynomial method*.

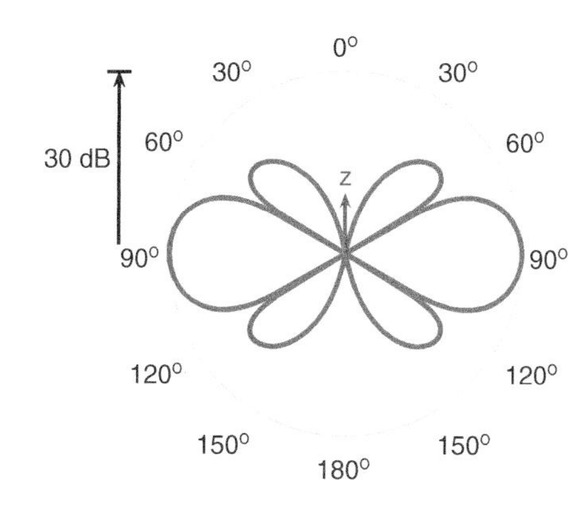

Figure 7.57 Normalized array factor of the designed five-element array with nulls at $\theta=\{0°,60°,120°,180°\}$. The array factor is plotted with respect to θ on a logarithmic scale using a 30 dB dynamic range.

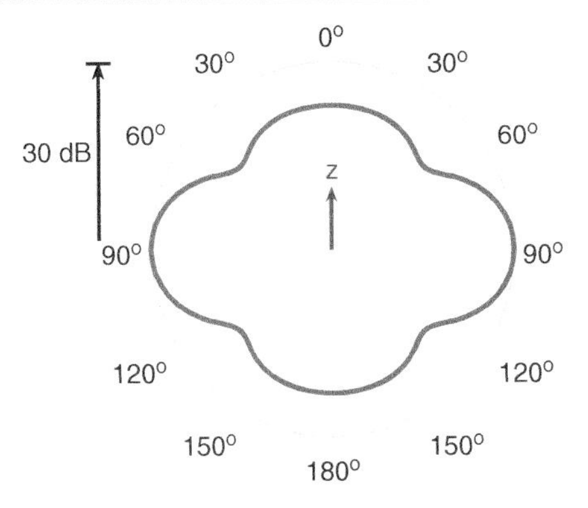

Figure 7.58 Normalized array factor of the designed array when the excitation of the element at the middle is multiplied by two.

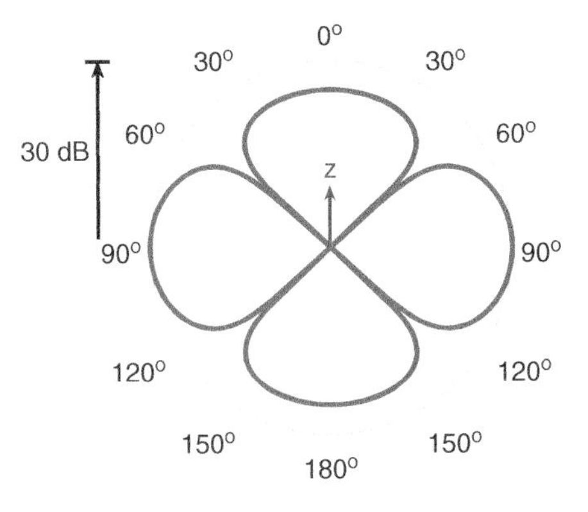

Figure 7.59 Normalized array factor of the designed array when the first and last elements are removed.

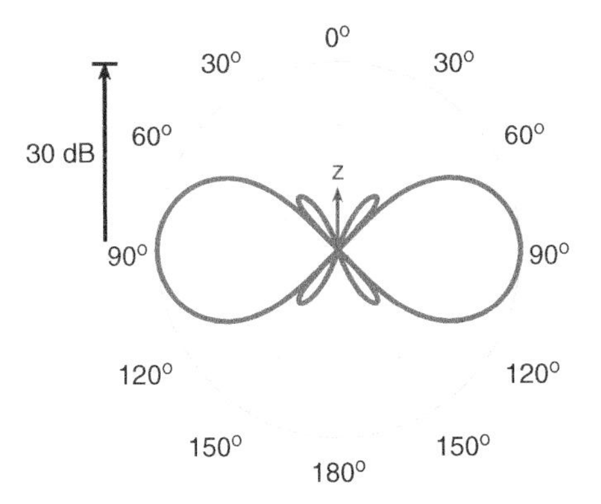

Figure 7.60 Normalized array factor of the designed array with modified excitations for nulls at $\theta = \{0°, 45°, 135°, 180°\}$.

This array factor has nulls at $\theta = \{0°, 60°, 120°, 180°\}$, as desired. If the excitation of the middle element is multiplied by two, we have (Figure 7.58)

$$\begin{aligned} AF(\theta, \phi) &= 1 + 2\exp(j\pi\cos\theta) + 4\exp(j2\pi\cos\theta) \\ &\quad + 2\exp(j3\pi\cos\theta) + \exp(j4\pi\cos\theta) && (7.199) \\ &= [\exp(j\pi\cos\theta) + 1]^2[\exp(j2\pi\cos\theta) + 1], && (7.200) \end{aligned}$$

which does not have any null, as the corresponding polynomial $AF(\chi) \propto 1 + 2\chi + 4\chi^2 + 2\chi^3 + \chi^4$ does not have any root on the *unit circle* with unit magnitude. On the other hand, if we remove the first and last elements of the original array, the array factor becomes (Figure 7.59)

$$\begin{aligned} AF(\theta, \phi) &= 2\exp(j\pi\cos\theta) + 2\exp(j2\pi\cos\theta) + 2\exp(j3\pi\cos\theta) && (7.201) \\ &= 2\exp(j\pi\cos\theta)[1 + \exp(j\pi\cos\theta) + \exp(j2\pi\cos\theta)], && (7.202) \end{aligned}$$

which has nulls at $\theta \approx \{48.2°, 131.8°\}$. Finally, to shift the nulls at $\{60°, 120°\}$ to $\{45°, 135°\}$, the array factor and the excitations should be updated as (Figure 7.60)

$$\begin{aligned} \chi_1 &= \exp[j\pi\cos(0°)] = -1 && (7.203) \\ \chi_2 &= \exp[j\pi\cos(45°)] \approx -0.606 + j0.796 && (7.204) \\ \chi_3 &= \exp[j\pi\cos(135°)] \approx -0.606 - j0.796 && (7.205) \\ \chi_2 &= \exp[j\pi\cos(180°)] = -1 && (7.206) \\ AF(\chi) &\propto (\chi + 1)(\chi + 0.606 - j0.796)(\chi + 0.606 + j0.796)(\chi + 1) && (7.207) \\ &\approx 1 + 3.21\chi + 4.43\chi^2 + 3.21\chi^3 + \chi^4 && (7.208) \end{aligned}$$

and

$$|\zeta_0| = 1, \quad |\zeta_1| \approx 3.21, \quad |\zeta_2| \approx 4.43, \quad |\zeta_3| \approx 3.21, \quad |\zeta_4| = 1. \qquad (7.209)$$

Example 93: Consider Antenna A and Antenna B with radiation intensities given as (Figure 7.61)

$$U_A^{\text{far}}(\theta, \phi) = U_B^{\text{far}}(\theta, \phi) = U_0 \begin{cases} \cos(9\theta) & (0° \le \theta \le 10°) \\ 0 & (10° < \theta \le 180°). \end{cases} \qquad (7.210)$$

The antennas are separated by a distance of 1 km (in each other's far zone), and they are oriented for maximum power transmission (Figure 7.62). Radiation efficiencies are given as unity. Find the received power by Antenna B ($p_{rec,B}$) at 300 MHz, when the input power of Antenna A ($p_{in,A}$) is 50 W, for the following cases:

- One of the antennas is RHCP, and the other is LHCP.
- Both antennas are RHCP.
- One of the antennas is circularly polarized, and the other is linearly polarized.

All polarizations are given for *radiating* modes for both antennas.

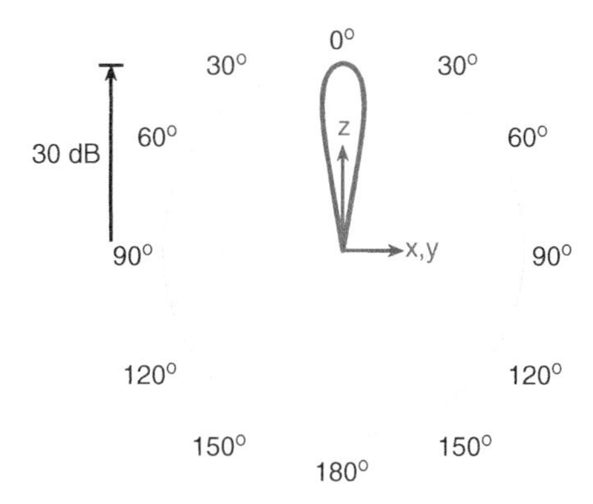

Figure 7.61 Polar plot of the radiation pattern for the intensity given in Eq. (7.210) using a 30 dB dynamic range.

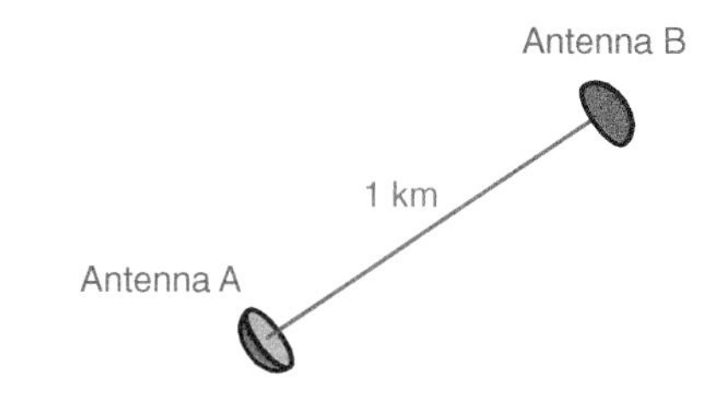

Figure 7.62 Two antennas separated by a distance of 1 km. The antennas are oriented in the maximum gain directions.

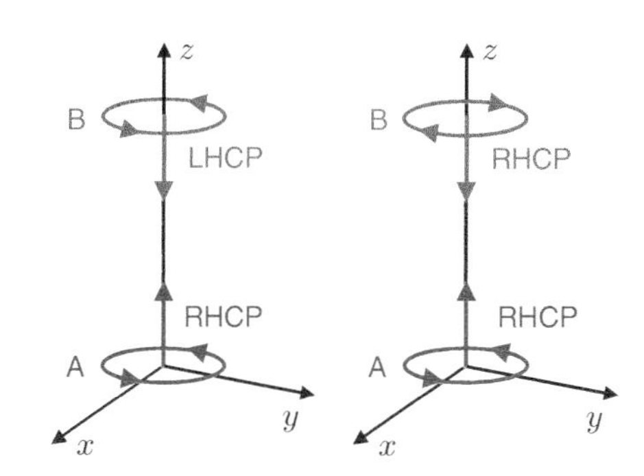

Figure 7.63 Representations of antennas with circular polarizations. Antenna A is oriented upward, and Antenna B is oriented downward.

Solution: Considering the expression for the radiation intensity, we can obtain the total radiated power (when any of the antennas radiates), directive gain, and directivity as

$$p^{\text{far}}_{em,A,B} = \int_0^{2\pi}\int_0^{\pi} U^{\text{far}}_{A,B}(\theta,\phi)\sin\theta d\theta d\phi \tag{7.211}$$

$$= U_0\int_0^{2\pi}\int_0^{\pi/18} \cos(9\theta)\sin\theta d\theta d\phi \tag{7.212}$$

$$= 2\pi U_0\left[\frac{9\sin\theta\sin(9\theta)}{80}+\frac{\cos\theta\cos(9\theta)}{80}\right]_0^{\pi/18} \tag{7.213}$$

$$= U_0\frac{\pi}{40}[9\sin(\pi/18)-1] \quad (\text{W}) \tag{7.214}$$

and

$$D_{A,B}(\theta,\phi) = \frac{160}{[9\sin(\pi/18)-1]}\begin{cases}\cos(9\theta) & (0^\circ \le \theta \le 10^\circ)\\ 0 & (10^\circ < \theta \le 180^\circ)\end{cases} \tag{7.215}$$

$$D_{\max,A} = D_{\max,B} \approx 284.3. \tag{7.216}$$

Since radiation efficiencies are given as unity, we also have $G_{\max,A} = G_{\max,B} \approx 284.3$. Because the antennas are aligned for maximum power transmission, we have

$$p_{rec,B} = p_{in,A}|e_p|^2\left(\frac{\lambda}{4\pi R}\right)^2 G_A(\theta_{AB},\phi_{AB})G_B(\theta_{BA},\phi_{BA}) \tag{7.217}$$

$$\approx 50|e_p|^2\left(\frac{1}{16\pi^2\times 10^6}\right)(284.3)^2 \approx 25.6|e_p|^2 \text{ mW}. \tag{7.218}$$

For different polarization cases, we obtain the following values:

- RHCP/LHCP: Assuming that the antennas are located on the z axis with Antenna A oriented upward and Antenna B oriented downward (Figure 7.63), we can select

$$\hat{a}_A = \frac{1}{\sqrt{2}}(\hat{a}_x - j\hat{a}_y) \qquad \text{(RHCP when radiating toward } +z) \tag{7.219}$$

$$\hat{a}_B = \frac{1}{\sqrt{2}}(\hat{a}_x - j\hat{a}_y) \qquad \text{(LHCP when radiating toward } -z) \tag{7.220}$$

$$e_p = \hat{a}_A \cdot \hat{a}_B = [1 + (j)^2]/\sqrt{2} = 0. \tag{7.221}$$

  Consequently, we obtain $|e_p|^2 = 0$ and $p_{rec,B} = 0$ W.

- RHCP/RHCP: In this case, using the same configuration (Figure 7.63), we have

$$e_p = \hat{a}_A \cdot \hat{a}_B = \frac{1}{2}(\hat{a}_x - j\hat{a}_y) \cdot (\hat{a}_x + j\hat{a}_y) = 1. \tag{7.222}$$

  Therefore, there is no polarization mismatch ($|e_p|^2 = 1$), and we get[55] $p_{rec,B} \approx 25.6$ mW.

- CP/LP: Assuming that Antenna A is circularly polarized as $(\hat{a}_x - j\hat{a}_y)/\sqrt{2}$ and Antenna B is linearly polarized as $\hat{a}_B = (\hat{a}_x + \hat{a}_y)/\sqrt{2}$ (Figure 7.64), we have

$$e_p = \hat{a}_A \cdot \hat{a}_B = \frac{1}{2}(\hat{a}_x - j\hat{a}_y) \cdot (\hat{a}_x + \hat{a}_y) = \frac{1}{2}(1 - j). \tag{7.223}$$

  Hence, we obtain $|e_p|^2 = 1/2$ and $p_{rec,B} \approx 12.8$ mW.

[55] The small value of the received power (25.6 mW) in comparison to the input power (50 W) is mainly due to the 1 km distance between the antennas.

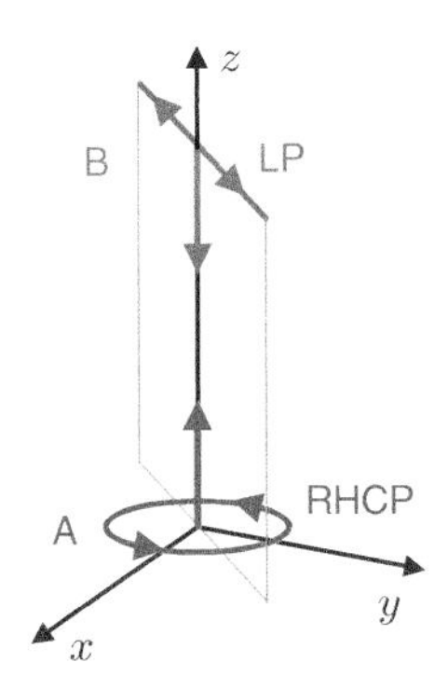

Figure 7.64 Representations of antennas when Antenna A (oriented upward) is circularly polarized and Antenna B (oriented downward) is linearly polarized.

7.2 Waveguides

As discussed in the previous section, wireless transmission requires no particular medium to transfer electromagnetic energy from one point (radiating antenna) to another (receiving antenna). On the other hand, in a boundless medium, the power flux density decays quadratically with the distance, leading to large energy losses as the distance between two points increases. Consequently, various systems and tools have been developed for efficient electromagnetic transmission, at the cost of reduced flexibility compared to wireless systems. Waveguides are structures that guide electromagnetic waves by enclosing them like pipelines (Figure 7.65). Enforcing electromagnetic waves to propagate in a limited region leads to excellent efficiency with minimum loss of energy, but propagation in a waveguide can occur only in certain configurations (called *modes*) that introduce new constraints and challenges.

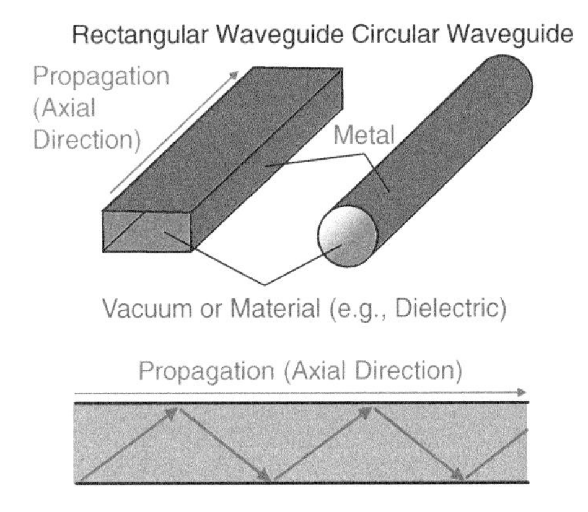

Figure 7.65 Metallic waveguides with rectangular and circular cross sections. Propagation, confined inside a waveguide, occurs in the axial direction. The overall transmission along a waveguide can be interpreted as electromagnetic waves reflecting back and forth from the metallic side surfaces and effectively traveling in the axial direction (for propagating modes).

A natural waveguide is simply a hollow metal pipe (Figure 7.65), which can be in various shapes, most commonly with a rectangular cross section. We examine this kind of rectangular waveguides in this section to illustrate how Maxwell's equations can be employed to analyze electromagnetic problems in closed (internal) regions. While they minimize electromagnetic losses and have excellent shielding properties that reduce electromagnetic interference, metallic waveguides are solid and bulky. Recently, alternative dielectric waveguides based on total-internal-reflection phenomena have become increasingly popular to benefit from the advantages of flexible, lightweight waveguides. Optical fibers are waveguides (Figure 7.66) used at specific optical frequencies.

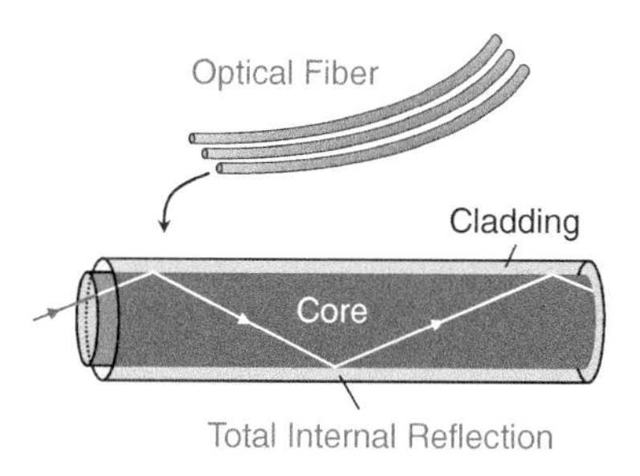

Figure 7.66 Optical fibers are dielectric waveguides.

In real life, a waveguide is excited from one side (e.g. by a probe), which generates electromagnetic waves inside the waveguide. Once transmitted, these waves are received on the other side (e.g. by another probe). In the following analyses, however, our focus is the transmission phenomenon itself: i.e. how modes carry energy. Therefore, when considering a rectangular waveguide, we assume that the structure is infinitely long in the *axial* direction and electromagnetic propagation occurs in this direction (without analyzing how this propagation is initially created). Like previous discussions of wireless transmission, Maxwell's equations are considered in the phasor domain, assuming a steady state.

7.2.1 Transverse and Axial Fields

Consider an infinitely long waveguide made of a PEC aligned in the z direction (Figure 7.67). The permittivity and permeability of the homogeneous material inside the waveguide are[56] ε and μ. The electric field intensity in the phasor domain can be written

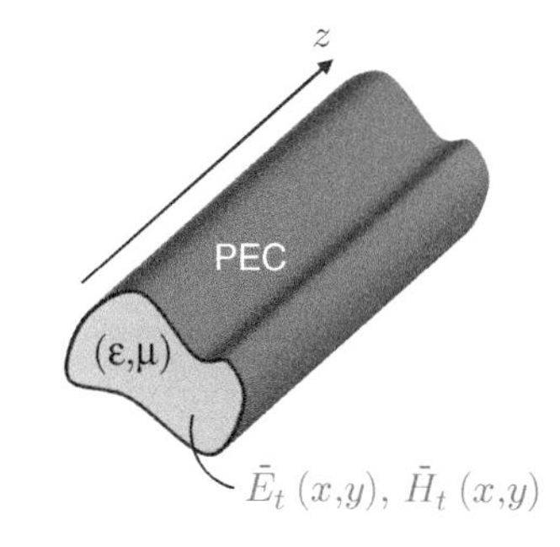

Figure 7.67 A waveguide with an arbitrary cross section. The waveguide is assumed to be made of PEC and infinitely long in the $\pm z$ direction (with a fixed cross section), and the inner material is homogeneous with ε permittivity and μ permeability.

$$\bar{E}(\bar{r}) = \bar{E}_t(x, y)\exp(-\gamma z), \tag{7.224}$$

where $\bar{E}_t(x, y)$ represents the part of the electric field intensity involving x and y dependency, while the z dependency is assumed to be included in the exponential part. Note that $\exp(-\gamma z)$ indicates propagation along the z direction, where $\gamma = \alpha + j\beta$ is the propagation constant to be found.[57] In general, the electric field intensity may have x, y, and z components:

$$\bar{E}(\bar{r}) = \hat{a}_x E_x(\bar{r}) + \hat{a}_y E_y(\bar{r}) + \hat{a}_z E_z(\bar{r}) \tag{7.225}$$

$$\bar{E}_t(x, y) = \hat{a}_x E_{tx}(x, y) + \hat{a}_y E_{ty}(x, y) + \hat{a}_z E_{tz}(x, y) \tag{7.226}$$

and

$$E_x(\bar{r}) = E_{tx}(x, y)\exp(-\gamma z) \tag{7.227}$$

$$E_y(\bar{r}) = E_{ty}(x, y)\exp(-\gamma z) \tag{7.228}$$

$$E_z(\bar{r}) = E_{tz}(x, y)\exp(-\gamma z). \tag{7.229}$$

[56] If a waveguide is filled with a lossy material, complex permittivity ε_c can be used instead of ε in the derivations. Considering lossless filling material and metallic walls (PEC), the overall waveguide becomes an ideal structure with zero loss. In addition, with perfect shielding, fields outside the waveguide are zero.

[57] It should be emphasized that $\gamma \neq jk = j\omega\sqrt{\mu\varepsilon}$ in general, because a waveguide is a bounded region. The relationship between γ and k still needs to be found by applying Maxwell's equations.

Similarly, the magnetic field intensity has a form of

$$\bar{H}(\bar{r}) = \bar{H}_t(x,y)\exp(-\gamma z), \tag{7.230}$$

where

$$\bar{H}(\bar{r}) = \hat{a}_x H_x(\bar{r}) + \hat{a}_y H_y(\bar{r}) + \hat{a}_z H_z(\bar{r}) \tag{7.231}$$
$$\bar{H}_t(x,y) = \hat{a}_x H_{tx}(x,y) + \hat{a}_y H_{ty}(x,y) + \hat{a}_z H_{tz}(x,y) \tag{7.232}$$
$$H_x(\bar{r}) = H_{tx}(x,y)\exp(-\gamma z) \tag{7.233}$$
$$H_y(\bar{r}) = H_{ty}(x,y)\exp(-\gamma z) \tag{7.234}$$
$$H_z(\bar{r}) = H_{tz}(x,y)\exp(-\gamma z). \tag{7.235}$$

Following these general definitions, our purpose is to find relationships between the components of electric and magnetic fields. We can start by employing Faraday's law $\bar{\nabla} \times \bar{E}(\bar{r}) = -j\omega\mu\bar{H}(\bar{r})$ as

$$\bar{\nabla} \times \bar{E}(\bar{r}) = \bar{\nabla} \times \left[\hat{a}_x E_x(\bar{r}) + \hat{a}_y E_y(\bar{r}) + \hat{a}_z E_z(\bar{r})\right] \tag{7.236}$$
$$= -j\omega\mu\left[\hat{a}_x H_x(\bar{r}) + \hat{a}_y H_y(\bar{r}) + \hat{a}_z H_z(\bar{r})\right]. \tag{7.237}$$

Then we arrive at three equations:[58]

$$\left[\frac{\partial E_{tz}(x,y)}{\partial y} + \gamma E_{ty}(x,y)\right] = -j\omega\mu H_{tx}(x,y) \tag{7.238}$$
$$-\left[\frac{\partial E_{tz}(x,y)}{\partial x} + \gamma E_{tx}(x,y)\right] = -j\omega\mu H_{ty}(x,y) \tag{7.239}$$
$$\left[\frac{\partial E_{ty}(x,y)}{\partial x} - \frac{\partial E_{tx}(x,y)}{\partial y}\right] = -j\omega\mu H_{tz}(x,y). \tag{7.240}$$

An application of Ampere's law leads to a similar set of equations:

$$\left[\frac{\partial H_{tz}(x,y)}{\partial y} + \gamma H_{ty}(x,y)\right] = j\omega\varepsilon E_{tx}(x,y) \tag{7.241}$$
$$-\left[\frac{\partial H_{tz}(x,y)}{\partial x} + \gamma H_{tx}(x,y)\right] = j\omega\varepsilon E_{ty}(x,y) \tag{7.242}$$
$$\left[\frac{\partial H_{ty}(x,y)}{\partial x} - \frac{\partial H_{tx}(x,y)}{\partial y}\right] = j\omega\varepsilon E_{tz}(x,y). \tag{7.243}$$

[58] As intermediate steps, we have

$$\begin{aligned}
&\bar{\nabla}\times\left[\hat{a}_x E_x(\bar{r}) + \hat{a}_y E_y(\bar{r}) + \hat{a}_z E_z(\bar{r})\right] \\
&= \hat{a}_z\frac{\partial E_y(\bar{r})}{\partial x} - \hat{a}_y\frac{\partial E_z(\bar{r})}{\partial x} \\
&- \hat{a}_z\frac{\partial E_x(\bar{r})}{\partial y} + \hat{a}_x\frac{\partial E_z(\bar{r})}{\partial y} \\
&+ \hat{a}_y\frac{\partial E_x(\bar{r})}{\partial z} - \hat{a}_x\frac{\partial E_y(\bar{r})}{\partial z} \\
&= \hat{a}_z\frac{\partial E_{ty}(x,y)}{\partial x}\exp(-\gamma z) \\
&- \hat{a}_y\frac{\partial E_{tz}(x,y)}{\partial x}\exp(-\gamma z) \\
&- \hat{a}_z\frac{\partial E_{tx}(x,y)}{\partial y}\exp(-\gamma z) \\
&+ \hat{a}_x\frac{\partial E_{tz}(x,y)}{\partial y}\exp(-\gamma z) \\
&- \hat{a}_y E_{tx}(x,y)\gamma\exp(-\gamma z) \\
&+ \hat{a}_x E_{ty}(x,y)\gamma\exp(-\gamma z) \\
&= -\hat{a}_x j\omega\mu H_{tx}(x,y)\exp(-\gamma z) \\
&- \hat{a}_y j\omega\mu H_{ty}(x,y)\exp(-\gamma z) \\
&- \hat{a}_z j\omega\mu H_{tz}(x,y)\exp(-\gamma z).
\end{aligned}$$

These six equations in Eq. (7.238)–(7.243) can be combined to arrive at expressions that provide *transverse* fields ($E_{tx}(x,y)$, $E_{ty}(x,y)$, $H_{tx}(x,y)$, $H_{ty}(x,y)$) in terms of longitudinal (*axial*) fields[59] ($E_{tz}(x,y)$ and $H_{tz}(x,y)$).

As an example, we can consider the combination of two equations:

$$-\left[\frac{\partial E_{tz}(x,y)}{\partial x}+\gamma E_{tx}(x,y)\right]=-j\omega\mu H_{ty}(x,y) \tag{7.244}$$

$$\left[\frac{\partial H_{tz}(x,y)}{\partial y}+\gamma H_{ty}(x,y)\right]=j\omega\varepsilon E_{tx}(x,y). \tag{7.245}$$

Using the first equation in Eq. (7.244), we have

$$-\gamma E_{tx}(x,y)=-j\omega\mu H_{ty}(x,y)+\frac{\partial E_{tz}(x,y)}{\partial x}. \tag{7.246}$$

Then inserting the expression for $H_{ty}(x,y)$ from the second equation in Eq. (7.245), we have

$$-\gamma E_{tx}(x,y)=-\frac{j\omega\mu}{\gamma}\left[j\omega\varepsilon E_{tx}(x,y)-\frac{\partial H_{tz}(x,y)}{\partial y}\right]+\frac{\partial E_{tz}(x,y)}{\partial x} \tag{7.247}$$

$$=\frac{k^2}{\gamma}E_{tx}(x,y)+\frac{j\omega\mu}{\gamma}\frac{\partial H_{tz}(x,y)}{\partial y}+\frac{\partial E_{tz}(x,y)}{\partial x}, \tag{7.248}$$

where $k=\omega\sqrt{\mu\varepsilon}$ is the wavenumber, as usual. Rearranging the terms, we obtain

$$(\gamma^2+k^2)E_{tx}(x,y)=-\gamma\frac{\partial E_{tz}(x,y)}{\partial x}-j\omega\mu\frac{\partial H_{tz}(x,y)}{\partial y} \tag{7.249}$$

$$\longrightarrow E_{tx}(x,y)=-\frac{1}{h^2}\left[\gamma\frac{\partial E_{tz}(x,y)}{\partial x}+j\omega\mu\frac{\partial H_{tz}(x,y)}{\partial y}\right], \tag{7.250}$$

where[60] $h^2=\gamma^2+k^2$. Similar combinations of equations lead to

$$E_{ty}(x,y)=-\frac{1}{h^2}\left[\gamma\frac{\partial E_{tz}(x,y)}{\partial y}-j\omega\mu\frac{\partial H_{tz}(x,y)}{\partial x}\right] \tag{7.251}$$

$$H_{tx}(x,y)=-\frac{1}{h^2}\left[\gamma\frac{\partial H_{tz}(x,y)}{\partial x}-j\omega\varepsilon\frac{\partial E_{tz}(x,y)}{\partial y}\right] \tag{7.252}$$

$$H_{ty}(x,y)=-\frac{1}{h^2}\left[\gamma\frac{\partial H_{tz}(x,y)}{\partial y}+j\omega\varepsilon\frac{\partial E_{tz}(x,y)}{\partial x}\right]. \tag{7.253}$$

[59] If the axial components are zero, the equations become

$$\gamma E_{ty}(x,y)=-j\omega\mu H_{tx}(x,y)$$
$$\gamma E_{tx}(x,y)=j\omega\mu H_{ty}(x,y)$$
$$\frac{\partial E_{ty}(x,y)}{\partial x}=\frac{\partial E_{tx}(x,y)}{\partial y}$$
$$\gamma H_{ty}(x,y)=j\omega\varepsilon E_{tx}(x,y)$$
$$\gamma H_{tx}(x,y)=-j\omega\varepsilon E_{ty}(x,y)$$
$$\frac{\partial H_{ty}(x,y)}{\partial x}=\frac{\partial H_{tx}(x,y)}{\partial y}.$$

Hence, without axial components $\gamma=j\omega\sqrt{\mu\varepsilon}$ *must* be satisfied, and we further have $\bar{H}(\bar{r})=\hat{a}_z\times\bar{E}(\bar{r})/\eta$. But solutions to these equations (together with z propagation) do not satisfy boundary conditions on waveguide walls, as opposed to solutions to the original set with nonzero axial components.

[60] At this stage, $\gamma\neq jk$ must be satisfied so that $h=\gamma^2+k^2\neq 0$, and we can use it in the denominator.

Consequently, in a given waveguide problem, if we can find axial fields, we can also find transverse fields.[61]

But how do we find axial fields? As we investigate a source-free medium,[62] the Helmholtz equations are satisfied as[63]

$$\bar{\nabla}^2 \bar{E}(\bar{r}) + k^2 \bar{E}(\bar{r}) = 0, \qquad \bar{\nabla}^2 \bar{H}(\bar{r}) + k^2 \bar{H}(\bar{r}) = 0. \tag{7.254}$$

Considering the z dependencies of electric and magnetic fields, these equations can further be written[64,65]

$$\bar{\nabla}^2_{xy} \bar{E}(\bar{r}) + \gamma^2 \bar{E}(\bar{r}) + k^2 \bar{E}(\bar{r}) = \bar{\nabla}^2_{xy} \bar{E}(\bar{r}) + h^2 \bar{E}(\bar{r}) = 0 \tag{7.255}$$

$$\bar{\nabla}^2_{xy} \bar{H}(\bar{r}) + \gamma^2 \bar{H}(\bar{r}) + k^2 \bar{H}(\bar{r}) = \bar{\nabla}^2_{xy} \bar{H}(\bar{r}) + h^2 \bar{H}(\bar{r}) = 0, \tag{7.256}$$

where $\bar{\nabla}^2_{xy} = \partial^2/\partial x^2 + \partial^2/\partial y^2$. While these equations contain all the components of electric and magnetic fields, they are particularly useful to find axial fields $E_z(\bar{r})$ and $H_z(\bar{r})$ that can be used to derive transverse fields.

[61]Note that geometric properties of waveguides have not been considered in the derivations above. Apart from Maxwell's equations, the only information is that there is a wave propagation in the z direction (guided by the waveguide) with a propagation constant of γ. Consequently, the equations above, which are derived directly using Maxwell's equations, are valid not only in waveguides but also in some other homogeneous media.

[62]For a lossy material, the effect of conductivity can be included in the analysis via a complex wavenumber. Otherwise, inside an ideal waveguide is a source-free medium.

[63]Using Helmholtz equations, we include Gauss' laws as further constraints for a complete analysis.

[64]Note that

$$\frac{\partial^2}{\partial z^2}\bar{E}(\bar{r}) = \gamma^2 \bar{E}(\bar{r}), \quad \frac{\partial^2}{\partial z^2}\bar{H}(\bar{r}) = \gamma^2 \bar{H}(\bar{r}).$$

[65]Hence, in the absence of axial fields ($E_z(\bar{r}) = 0$ and $H_z(\bar{r}) = 0$), we have $h = 0$ and

$$\bar{\nabla}^2_{xy} E_x(\bar{r}) = 0, \quad \bar{\nabla}^2_{xy} E_y(\bar{r}) = 0$$

$$\bar{\nabla}^2_{xy} H_x(\bar{r}) = 0, \quad \bar{\nabla}^2_{xy} H_y(\bar{r}) = 0.$$

7.2.2 Rectangular Waveguides

As popular examples, rectangular waveguides with rectangular cross sections are employed in many real-life applications (Figure 7.68). Using the earlier derivations, we can obtain expressions for electric and magnetic fields inside an infinitely long rectangular waveguide with an $a \times b$ cross section. For this purpose, it is useful to assume that fields inside the waveguide can be decomposed into infinitely many configurations, or *modes*. Each mode can be solved separately so that their combinations can provide overall expressions for total fields.

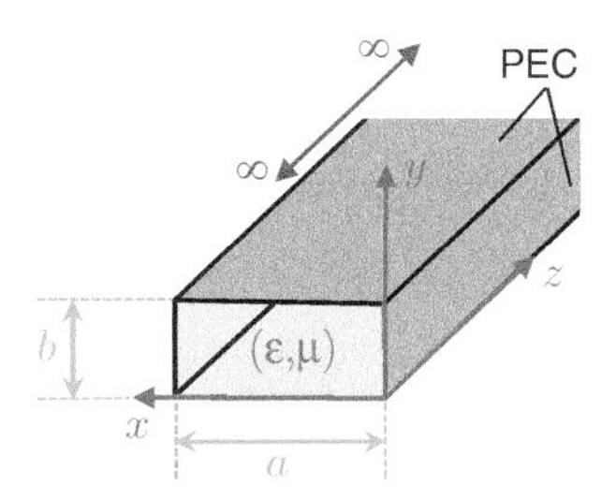

Figure 7.68 A rectangular waveguide with an $a \times b$ cross section. The waveguide is assumed to be made of PEC and infinitely long in the $\pm z$ direction, while the inner material is homogeneous with ε permittivity and μ permeability.

7.2.2.1 Transverse Magnetic Modes

Transverse magnetic (TM) modes are defined as wave configurations with zero axial magnetic field: i.e. $H_z^{\mathrm{TM}}(\bar{r}) = 0$ and $H_{tz}^{\mathrm{TM}}(x, y) = 0$. This leads to

$$E_{tx}^{\mathrm{TM}}(x, y) = -\frac{\gamma}{h^2}\frac{\partial E_{tz}^{\mathrm{TM}}(x, y)}{\partial x}, \qquad E_{ty}^{\mathrm{TM}}(x, y) = -\frac{\gamma}{h^2}\frac{\partial E_{tz}^{\mathrm{TM}}(x, y)}{\partial y} \tag{7.257}$$

$$H_{tx}^{\mathrm{TM}}(x, y) = \frac{j\omega\varepsilon}{h^2}\frac{\partial E_{tz}^{\mathrm{TM}}(x, y)}{\partial y}, \qquad H_{ty}^{\mathrm{TM}}(x, y) = -\frac{j\omega\varepsilon}{h^2}\frac{\partial E_{tz}^{\mathrm{TM}}(x, y)}{\partial x}, \tag{7.258}$$

where $E_{tz}^{\mathrm{TM}}(x,y)$ (the axial electric field intensity) still needs to be found. Using the related Helmholtz equation, we have

$$\bar{\nabla}_{xy}^2 \bar{E}^{\mathrm{TM}}(\bar{r}) + h^2 \bar{E}^{\mathrm{TM}}(\bar{r}) = 0 \tag{7.259}$$

$$\longrightarrow \frac{\partial^2 E_{tz}^{\mathrm{TM}}(x,y)}{\partial x^2} + \frac{\partial^2 E_{tz}^{\mathrm{TM}}(x,y)}{\partial y^2} + h^2 E_{tz}^{\mathrm{TM}}(x,y) = 0. \tag{7.260}$$

Further using a *separation of variables*, a general solution can be obtained as[66]

$$E_{tz}^{\mathrm{TM}}(x,y) = [A\cos(k_x x) + B\sin(k_x x)]\left[C\cos(k_y y) + D\sin(k_y y)\right] \tag{7.261}$$

$$E_z^{\mathrm{TM}}(\bar{r}) = E_{tz}^{\mathrm{TM}}(x,y)\exp(-\gamma z) \tag{7.262}$$

for $k_x^2 + k_y^2 = h^2$. Constants $\{A, B, C, D\}$ in these expressions can be found via boundary conditions (specifically, vanishing *tangential* electric field intensity on the boundaries). First, due to the boundary on the right-hand side ($x = 0$), we have $E_{tz}^{\mathrm{TM}}(0,y) = 0$, which is possible only if $A = 0$. In addition, due to the bottom boundary ($y = 0$), we have $E_{tz}^{\mathrm{TM}}(x,0) = 0$ and $C = 0$. The boundary on the left-hand side ($x = a$) enforces $E_{tz}^{\mathrm{TM}}(a,y) = 0$, which requires

$$\sin(k_x a) = 0 \longrightarrow k_x a = m\pi \qquad (m = 0, 1, 2, \ldots). \tag{7.263}$$

On the other hand, while the selection of $m = 0$ seems to satisfy the required boundary condition, it leads to zero $E_{tz}^{\mathrm{TM}}(x,y)$ everywhere (since we already have $A = 0$). Therefore, we need to exclude $m = 0$ and update the condition as

$$k_x = \frac{m\pi}{a} \qquad (m = 1, 2, \ldots). \tag{7.264}$$

Similarly, for the top boundary ($y = b$) and using $E_{tz}^{\mathrm{TM}}(x,b) = 0$, we derive

$$k_y = \frac{n\pi}{b} \qquad (n = 1, 2, \ldots), \tag{7.265}$$

where $n = 0$ is excluded to avoid null axial field. Consequently, the lowest possible TM mode has $m = 1$ and $n = 1$, which can be written TM_{11} (Figure 7.69). Combining all these results, we arrive at expressions for an axial electric field intensity for a TM mode as

$$E_{tz}^{\mathrm{TM}}(x,y) = A_{mn}\sin\left(\frac{m\pi x}{a}\right)\sin\left(\frac{n\pi y}{b}\right) \tag{7.266}$$

$$E_z^{\mathrm{TM}}(\bar{r}) = A_{mn}\sin\left(\frac{m\pi x}{a}\right)\sin\left(\frac{n\pi y}{b}\right)\exp(-\gamma z), \tag{7.267}$$

[66] Using separation of variables as $\bar{E}_{tz}^{\mathrm{TM}}(x,y) = f(x)g(y)$, we have

$$g(y)\frac{\partial^2 f(x)}{\partial x^2} + f(x)\frac{\partial^2 g(y)}{\partial y^2} + h^2 f(x)g(y) = 0$$

$$\longrightarrow \frac{1}{f(x)}\frac{\partial^2 f(x)}{\partial x^2} + \frac{1}{g(y)}\frac{\partial^2 g(y)}{\partial y^2} + h^2 = 0.$$

This equation can be satisfied using

$$\frac{1}{f(x)}\frac{\partial^2 f(x)}{\partial x^2} = -k_x^2, \qquad \frac{1}{g(y)}\frac{\partial^2 g(y)}{\partial y^2} = -k_y^2,$$

for $h^2 - k_x^2 - k_y^2 = 0$. General solutions for $f(x)$ and $g(y)$ can be obtained as

$$f(x) = A\cos(k_x x) + B\sin(k_x x)$$

$$g(y) = C\cos(k_y y) + D\sin(k_y y).$$

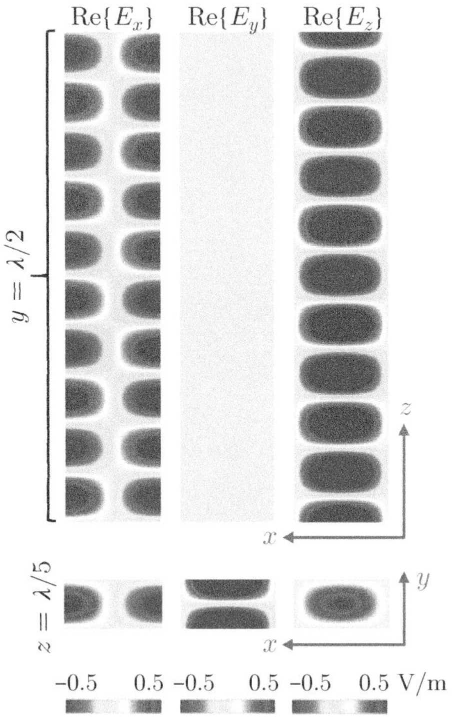

Figure 7.69 The real part of the electric field intensity for the TM_{11} mode in a $2\lambda \times \lambda$ rectangular waveguide (Figure 7.68), where $\lambda = 1$ m. The components $(E_x(\bar{r}), E_y(\bar{r}), E_z(\bar{r}))$ are shown separately with respect to position on the $y = \lambda/2$ plane (with respect to $z \in [0, 6\lambda]$ and $x \in [0, 2\lambda]$) and on the $z = \lambda/5$ plane (with respect to $x \in [0, 2\lambda]$ and $y \in [0, \lambda]$).

where A_{mn} (V/m) is the excitation constant of the TM mode with *mode numbers* (m, n). Finding the axial electric field intensity (and recalling that $H_z^{\text{TM}}(\bar{r}) = 0$ and $H_{tz}^{\text{TM}}(x, y) = 0$), we are able to derive transverse components as

$$E_{tx}^{\text{TM}}(x, y) = -\frac{\gamma}{h^2}\frac{\partial E_{tz}^{\text{TM}}(x, y)}{\partial x} = -\frac{\gamma}{h^2}A_{mn}\left(\frac{m\pi}{a}\right)\cos\left(\frac{m\pi x}{a}\right)\sin\left(\frac{n\pi y}{b}\right) \tag{7.268}$$

$$E_{ty}^{\text{TM}}(x, y) = -\frac{\gamma}{h^2}\frac{\partial E_{tz}^{\text{TM}}(x, y)}{\partial y} = -\frac{\gamma}{h^2}A_{mn}\left(\frac{n\pi}{b}\right)\sin\left(\frac{m\pi x}{a}\right)\cos\left(\frac{n\pi y}{b}\right) \tag{7.269}$$

$$H_{tx}^{\text{TM}}(x, y) = \frac{j\omega\varepsilon}{h^2}\frac{\partial E_{tz}^{\text{TM}}(x, y)}{\partial y} = \frac{j\omega\varepsilon}{h^2}A_{mn}\left(\frac{n\pi}{b}\right)\sin\left(\frac{m\pi x}{a}\right)\cos\left(\frac{n\pi y}{b}\right) \tag{7.270}$$

$$H_{ty}^{\text{TM}}(x, y) = -\frac{j\omega\varepsilon}{h^2}\frac{\partial E_{tz}^{\text{TM}}(x, y)}{\partial x} = -\frac{j\omega\varepsilon}{h^2}A_{mn}\left(\frac{m\pi}{a}\right)\cos\left(\frac{m\pi x}{a}\right)\sin\left(\frac{n\pi y}{b}\right) \tag{7.271}$$

and

$$E_x^{\text{TM}}(\bar{r}) = -\frac{\gamma}{h^2}A_{mn}\left(\frac{m\pi}{a}\right)\cos\left(\frac{m\pi x}{a}\right)\sin\left(\frac{n\pi y}{b}\right)\exp(-\gamma z) \tag{7.272}$$

$$E_y^{\text{TM}}(\bar{r}) = -\frac{\gamma}{h^2}A_{mn}\left(\frac{n\pi}{b}\right)\sin\left(\frac{m\pi x}{a}\right)\cos\left(\frac{n\pi y}{b}\right)\exp(-\gamma z) \tag{7.273}$$

$$H_x^{\text{TM}}(\bar{r}) = \frac{j\omega\varepsilon}{h^2}A_{mn}\left(\frac{n\pi}{b}\right)\sin\left(\frac{m\pi x}{a}\right)\cos\left(\frac{n\pi y}{b}\right)\exp(-\gamma z) \tag{7.274}$$

$$H_y^{\text{TM}}(\bar{r}) = -\frac{j\omega\varepsilon}{h^2}A_{mn}\left(\frac{m\pi}{a}\right)\cos\left(\frac{m\pi x}{a}\right)\sin\left(\frac{n\pi y}{b}\right)\exp(-\gamma z) \tag{7.275}$$

for $\{m, n\} = \{1, 2, \ldots\}$ (e.g. Figures 7.69–7.71). An important expression for the propagation constant in terms of waveguide dimensions, frequency, and material properties can be derived as

$$k_x^2 + k_x^2 = h^2 \longrightarrow h^2 = \gamma^2 + k^2 = k_x^2 + k_y^2 = \left(\frac{m\pi}{a}\right)^2 + \left(\frac{n\pi}{b}\right)^2 \tag{7.276}$$

or

$$\gamma^2 = \left(\frac{m\pi}{a}\right)^2 + \left(\frac{n\pi}{b}\right)^2 - \omega^2\mu\varepsilon \longrightarrow \gamma = j\sqrt{\omega^2\mu\varepsilon - \left(\frac{m\pi}{a}\right)^2 - \left(\frac{n\pi}{b}\right)^2}. \tag{7.277}$$

As discussed below, the propagation constant can be purely real (for propagating modes) or purely imaginary (for evanescent modes), depending on a, b, ω, ε, and μ, even though the waveguide is lossless.

7.2.2.2 Transverse Electric Modes

Transverse electric (TE) modes are defined as wave configurations with zero axial electric field: i.e. $E_z^{\mathrm{TE}}(\bar{r}) = 0$ and $E_{tz}^{\mathrm{TE}}(x,y) = 0$. Hence, we have

$$E_{tx}^{\mathrm{TE}}(x,y) = -\frac{j\omega\mu}{h^2}\frac{\partial H_{tz}^{\mathrm{TE}}(x,y)}{\partial y}, \qquad E_{ty}^{\mathrm{TE}}(x,y) = \frac{j\omega\mu}{h^2}\frac{\partial H_{tz}^{\mathrm{TE}}(x,y)}{\partial x} \tag{7.278}$$

$$H_{tx}^{\mathrm{TE}}(x,y) = -\frac{\gamma}{h^2}\frac{\partial H_{tz}^{\mathrm{TE}}(x,y)}{\partial x}, \qquad H_{ty}^{\mathrm{TE}}(x,y) = -\frac{\gamma}{h^2}\frac{\partial H_{tz}^{\mathrm{TE}}(x,y)}{\partial y}, \tag{7.279}$$

where $H_{tz}^{\mathrm{TE}}(x,y)$ is the nonzero axial magnetic field intensity. Using the associated Helmholtz equation, we also have

$$\bar{\nabla}_{xy}^2 \bar{H}^{\mathrm{TE}}(\bar{r}) + h^2 \bar{H}^{\mathrm{TE}}(\bar{r}) = 0 \tag{7.280}$$

$$\longrightarrow \frac{\partial^2 H_{tz}^{\mathrm{TE}}(x,y)}{\partial x^2} + \frac{\partial^2 H_{tz}^{\mathrm{TE}}(x,y)}{\partial y^2} + h^2 H_{tz}^{\mathrm{TE}}(x,y) = 0. \tag{7.281}$$

Further using a separation of variables, a general solution can be obtained as

$$H_{tz}^{\mathrm{TE}}(x,y) = [A\cos(k_x x) + B\sin(k_x x)]\left[C\cos(k_y y) + D\sin(k_y y)\right] \tag{7.282}$$

$$H_z^{\mathrm{TE}}(\bar{r}) = H_{tz}^{\mathrm{TE}}(x,y)\exp(-\gamma z) \tag{7.283}$$

for $k_x^2 + k_y^2 = h^2$. Once again, constants (A, B, C, and D in A/m) in these expressions can be found via boundary conditions. But as opposed to the derivations for TM modes, *useful* boundary conditions (for the electric field intensity) need derivatives of $H_{tz}^{\mathrm{TE}}(x,y)$. First, when $x = 0$ (given the boundary on the right-hand side), we must have $E_{ty}^{\mathrm{TE}}(x,y) = 0$. Then one can write

$$\left.\frac{\partial H_{tz}^{\mathrm{TE}}(x,y)}{\partial x}\right|_{x=0} = Bk_x\left[C\cos(k_y y) + D\sin(k_y y)\right] = 0, \tag{7.284}$$

leading to $B = 0$. Similarly, for the boundary on the left-hand side (when $x = a$), the same derivative should be identically zero (noting that B is already zero):

$$\left.\frac{\partial H_{tz}^{\mathrm{TE}}(x,y)}{\partial x}\right|_{x=0} = -Ak_x\sin(k_x a)\left[C\cos(k_y y) + D\sin(k_y y)\right] = 0, \tag{7.285}$$

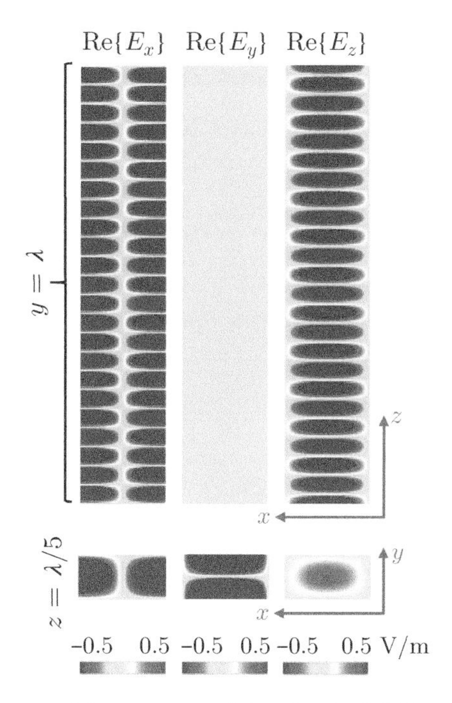

Figure 7.70 The real part of the electric field intensity for the TM$_{11}$ mode in a $4\lambda \times 2\lambda$ rectangular waveguide (Figure 7.68), where $\lambda = 0.5$ m. The components $(E_x(\bar{r}), E_y(\bar{r}), E_z(\bar{r}))$ are shown separately with respect to position on the $y = \lambda$ plane (with respect to $z \in [0, 12\lambda]$ and $x \in [0, 4\lambda]$) and on the $z = \lambda/5$ plane (with respect to $x \in [0, 4\lambda]$ and $y \in [0, 2\lambda]$).

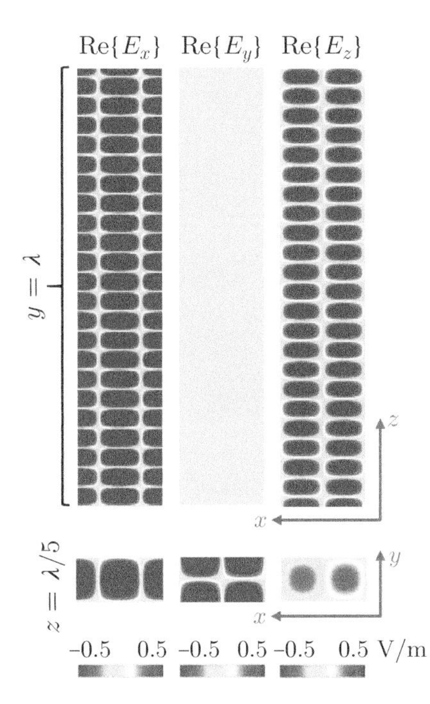

Figure 7.71 The real part of the electric field intensity for the TM_{21} mode in a $4\lambda \times 2\lambda$ rectangular waveguide (Figure 7.68), where $\lambda = 0.5$ m. The components $(E_x(\bar{r}), E_y(\bar{r}), E_z(\bar{r}))$ are shown separately with respect to position on the $y = \lambda$ plane (with respect to $z \in [0, 12\lambda]$ and $x \in [0, 4\lambda]$) and on the $z = \lambda/5$ plane (with respect to $x \in [0, 4\lambda]$ and $y \in [0, 2\lambda]$).

leading to

$$\sin(k_x a) = 0 \longrightarrow k_x = \frac{m\pi}{a} \qquad (m = 0, 1, 2, \ldots). \tag{7.286}$$

Considering the bottom and top boundaries in a similar manner, we obtain $D = 0$ and

$$k_y = \frac{n\pi}{b} \qquad (n = 0, 1, 2, \ldots). \tag{7.287}$$

Inserting these expressions, we derive the axial magnetic field intensity for a TE mode as

$$H_{tz}^{\mathrm{TE}}(x, y) = B_{mn} \cos\left(\frac{m\pi x}{a}\right) \cos\left(\frac{n\pi y}{b}\right) \tag{7.288}$$

$$H_z^{\mathrm{TE}}(\bar{r}) = B_{mn} \cos\left(\frac{m\pi x}{a}\right) \cos\left(\frac{n\pi y}{b}\right) \exp(-\gamma z), \tag{7.289}$$

where B_{mn} (A/m) is the constant of the TE mode with mode numbers (m, n). Then we are able to find transverse components as

$$E_{tx}^{\mathrm{TE}}(x, y) = -\frac{j\omega\mu}{h^2} \frac{\partial H_{tz}^{\mathrm{TE}}(x, y)}{\partial y} = \frac{j\omega\mu}{h^2} B_{mn} \left(\frac{n\pi}{b}\right) \cos\left(\frac{m\pi x}{a}\right) \sin\left(\frac{n\pi y}{b}\right) \tag{7.290}$$

$$E_{ty}^{\mathrm{TE}}(x, y) = \frac{j\omega\mu}{h^2} \frac{\partial H_{tz}^{\mathrm{TE}}(x, y)}{\partial x} = -\frac{j\omega\mu}{h^2} B_{mn} \left(\frac{m\pi}{a}\right) \sin\left(\frac{m\pi x}{a}\right) \cos\left(\frac{n\pi y}{b}\right) \tag{7.291}$$

$$H_{tx}^{\mathrm{TE}}(x, y) = -\frac{\gamma}{h^2} \frac{\partial H_{tz}^{\mathrm{TE}}(x, y)}{\partial x} = \frac{\gamma}{h^2} B_{mn} \left(\frac{m\pi}{a}\right) \sin\left(\frac{m\pi x}{a}\right) \cos\left(\frac{n\pi y}{b}\right) \tag{7.292}$$

$$H_{ty}^{\mathrm{TE}}(x, y) = -\frac{\gamma}{h^2} \frac{\partial H_{tz}^{\mathrm{TE}}(x, y)}{\partial y} = \frac{\gamma}{h^2} B_{mn} \left(\frac{n\pi}{b}\right) \cos\left(\frac{m\pi x}{a}\right) \sin\left(\frac{n\pi y}{b}\right) \tag{7.293}$$

and

$$E_x^{\mathrm{TE}}(\bar{r}) = \frac{j\omega\mu}{h^2} B_{mn} \left(\frac{n\pi}{b}\right) \cos\left(\frac{m\pi x}{a}\right) \sin\left(\frac{n\pi y}{b}\right) \exp(-\gamma z) \tag{7.294}$$

$$E_y^{\mathrm{TE}}(\bar{r}) = -\frac{j\omega\mu}{h^2} B_{mn} \left(\frac{m\pi}{a}\right) \sin\left(\frac{m\pi x}{a}\right) \cos\left(\frac{n\pi y}{b}\right) \exp(-\gamma z) \tag{7.295}$$

$$H_x^{\mathrm{TE}}(\bar{r}) = \frac{\gamma}{h^2} B_{mn} \left(\frac{m\pi}{a}\right) \sin\left(\frac{m\pi x}{a}\right) \cos\left(\frac{n\pi y}{b}\right) \exp(-\gamma z) \tag{7.296}$$

$$H_y^{\mathrm{TE}}(\bar{r}) = \frac{\gamma}{h^2} B_{mn} \left(\frac{n\pi}{b}\right) \cos\left(\frac{m\pi x}{a}\right) \sin\left(\frac{n\pi y}{b}\right) \exp(-\gamma z) \tag{7.297}$$

for $\{m, n\} = \{0, 1, 2, \ldots\}$ (e.g. Figures 7.72 and 7.73), provided that[67] $m + n \neq 0$. Note that for a given pair of m and n, the propagation constant is the same for the corresponding TE and TM modes:

$$\gamma = j\sqrt{\omega^2\mu\varepsilon - \left(\frac{m\pi}{a}\right)^2 - \left(\frac{n\pi}{b}\right)^2} \tag{7.298}$$

is also valid for TE modes.

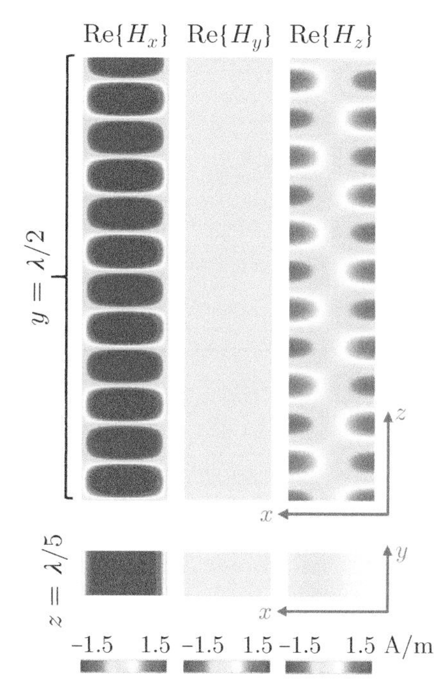

Figure 7.72 The real part of the magnetic field intensity for the TM_{10} mode in a $2\lambda \times \lambda$ rectangular waveguide (Figure 7.68), where $\lambda = 1$ m. The components $(H_x(\bar{r}), H_y(\bar{r}), H_z(\bar{r}))$ are shown separately with respect to position on the $y = \lambda/2$ plane (with respect to $z \in [0, 6\lambda]$ and $x \in [0, 2\lambda]$) and on the $z = \lambda/5$ plane (with respect to $x \in [0, 2\lambda]$ and $y \in [0, \lambda]$).

[67] There are TE_{01} and TE_{10} modes, but there is no TE_{00} mode.

7.2.2.3 Nonexistent Modes

As mentioned above, the TM_{00}, TM_{01}, TM_{10}, and TE_{00} modes do not exist. We now briefly consider these cases. Starting with TM_{01}, we have $h = \pi/b$, $\gamma = j\sqrt{\omega^2\mu\varepsilon - (\pi/b)^2}$, and (using Eqs. (7.267) and (7.272)–(7.275))

$$E_z^{\mathrm{TM}}(\bar{r}) = 0, \quad E_x^{\mathrm{TM}}(\bar{r}) = 0, \quad E_y^{\mathrm{TM}}(\bar{r}) = 0, \quad H_x^{\mathrm{TM}}(\bar{r}) = 0, \quad H_y^{\mathrm{TM}}(\bar{r}) = 0, \tag{7.299}$$

while $H_z^{\mathrm{TM}}(\bar{r}) = 0$ by definition. Hence, simply using the general expressions derived for TM modes, we arrive at zero fields. Similarly, an attempt to derive fields for TM_{10} leads to $h = \pi/a$, $\gamma = j\sqrt{\omega^2\mu\varepsilon - (\pi/a)^2}$, and

$$E_z^{\mathrm{TM}}(\bar{r}) = 0, \quad E_x^{\mathrm{TM}}(\bar{r}) = 0, \quad E_y^{\mathrm{TM}}(\bar{r}) = 0, \quad H_x^{\mathrm{TM}}(\bar{r}) = 0, \quad H_y^{\mathrm{TM}}(\bar{r}) = 0, \tag{7.300}$$

along with $H_z^{\mathrm{TM}}(\bar{r}) = 0$.

Using the general expressions in Eqs. (7.272)–(7.275) for TM_{00} can be problematic since we have $h = 0$, while these final identities for transverse components involve divisions with h^2. Since the Helmholtz equation must still hold for the axial electric field intensity, we have

$$E_z^{\mathrm{TM}}(\bar{r}) = A_{mn}\sin\left(\frac{m\pi x}{a}\right)\sin\left(\frac{n\pi y}{b}\right)\exp(-\gamma z) = 0 \tag{7.301}$$

by inserting $m = n = 0$. Therefore, if it existed, TM_{00} would have zero axial electric and magnetic fields: i.e. it would be *both* TE and TM – the TEM mode. But it is well-known that waveguides, as single-conductor structures, do not support TEM modes (including TM_{00}). Specifically, no electric/magnetic field configuration simultaneously satisfies (i) wave propagation in the z direction, (ii) zero axial component and plane-wave relationship between transverse components, *and* (iii) PEC boundary conditions at planar boundaries. An interesting exception is $a \to \infty$ and $b \to \infty$, for which the waveguide turns into a homogeneous space that can in fact support uniform plane waves (TEM waves).

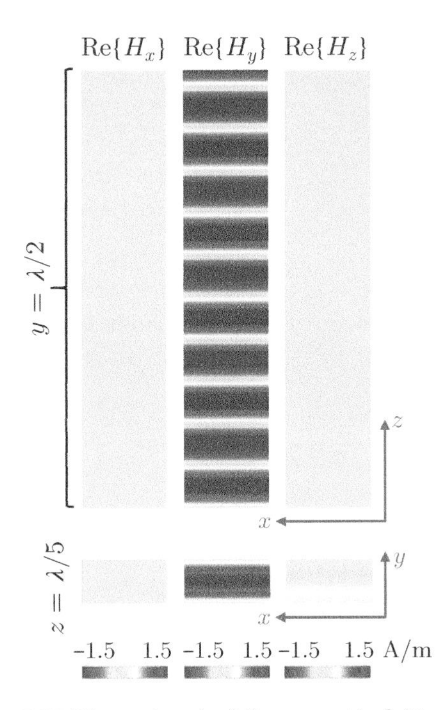

Figure 7.73 The real part of the magnetic field intensity for the TM_{01} mode in a $2\lambda \times \lambda$ rectangular waveguide (Figure 7.68), where $\lambda = 1$ m. The components $(H_x(\bar{r}), H_y(\bar{r}), H_z(\bar{r}))$ are shown separately with respect to position on the $y = \lambda/2$ plane (with respect to $z \in [0, 6\lambda]$ and $x \in [0, 2\lambda]$) and on the $z = \lambda/5$ plane (with respect to $x \in [0, 2\lambda]$ and $y \in [0, \lambda]$).

Finally, considering the possibility of TE_{00}, the magnetic field intensity can be written

$$H_z^{TE}(\bar{r}) = B_{mn} \cos\left(\frac{m\pi x}{a}\right) \cos\left(\frac{n\pi y}{b}\right) \exp(-\gamma z) = B_{00} \exp(-\gamma z). \tag{7.302}$$

It is relatively straightforward to show that such a mode cannot exist since it does not satisfy Gauss' law $\bar{\nabla} \cdot \bar{H}_z^{TE}(\bar{r}) = 0$ if $\gamma \neq 0$. Selecting $B_{00} = 0$ to make Gauss' law valid leads to the TEM mode that cannot exist.

7.2.2.4 Important Properties of Modes

Depending on their mode numbers m and n, waveguide modes may behave quite differently inside a waveguide. For example, depending on the dimensions of a rectangular waveguide, only some of the modes can propagate, but the others attenuate even when there is no material loss. As a starting point, we can go back to the expression for the propagation constant as[68]

$$\gamma_{mn}^2 = \left(\frac{m\pi}{a}\right)^2 + \left(\frac{n\pi}{b}\right)^2 - \omega^2 \mu\varepsilon, \tag{7.303}$$

where we now use subscripts m and n for γ as the propagation constant actually depends on these numbers (and hence, the mode itself). Using $\gamma_{mn} = \alpha_{mn} + j\beta_{mn}$, where α_{mn} and β_{mn} are attenuation and phase constants, respectively, we have several different cases:

- If $(m\pi/a)^2 + (n\pi/b)^2 - \omega^2\mu\varepsilon < 0$, then γ_{mn}^2 is negative, γ_{mn} is purely imaginary, and

$$\beta_{mn} = \sqrt{\omega^2\mu\varepsilon - \left(\frac{m\pi}{a}\right)^2 - \left(\frac{n\pi}{b}\right)^2} \tag{7.304}$$

  is nonzero, while $\alpha_{mn} = 0$. This means mode (m, n) propagates without attenuation (it is a propagating mode: e.g. see Figures 7.69–7.73). Interestingly, the phase constant in Eq. (7.304) is always smaller than the corresponding phase constant in a homogeneous space (Figure 7.74) with the same material properties. This means inside a waveguide, the phase velocity $u_{p,mn} = \omega/\beta_{mn}$ is *faster* (Figure 7.75), and the *guided* wavelength $\lambda_{mn} = 2\pi/\beta_{mn}$ is larger than the corresponding values in a homogeneous space.[69]

- If $(m\pi/a)^2 + (n\pi/b)^2 - \omega^2\mu\varepsilon > 0$, then γ_{mn}^2 is positive, γ_{mn} is purely real, and

$$\alpha_{mn} = \sqrt{\left(\frac{m\pi}{a}\right)^2 + \left(\frac{n\pi}{b}\right)^2 - \omega^2\mu\varepsilon} \tag{7.305}$$

[68] Note that as $a \to \infty$ and $b \to \infty$, then $\gamma_{mn}^2 \to -\omega^2\mu\varepsilon$, $\gamma_{mn} \to j\omega\sqrt{\mu\varepsilon}$, and $\beta_{mn} \to \omega\sqrt{\mu\varepsilon}$.

[69] Hence, a mode can travel faster than the speed of light. However, the information (related to the *group* velocity) cannot be carried faster than the speed of light.

is nonzero, while $\beta_{mn} = 0$. In this case, mode (m, n) attenuates without propagation: i.e. it is an evanescent mode (Figure 7.76). Note that for given values of m and n, the attenuation is higher if the frequency is lower or the dimensions of the waveguide are smaller.

- In a very special case – when $(m\pi/a)^2 + (n\pi/b)^2 - \omega^2\mu\varepsilon = 0$– we have $\gamma = 0$, and there is no propagation or attenuation (Figure 7.76). Such a mode can be interpreted as a standing wave, but it is not caused by reflection. Since the phase velocity is infinity, this special case is also considered an observation of an artificial *zero* refractive index.

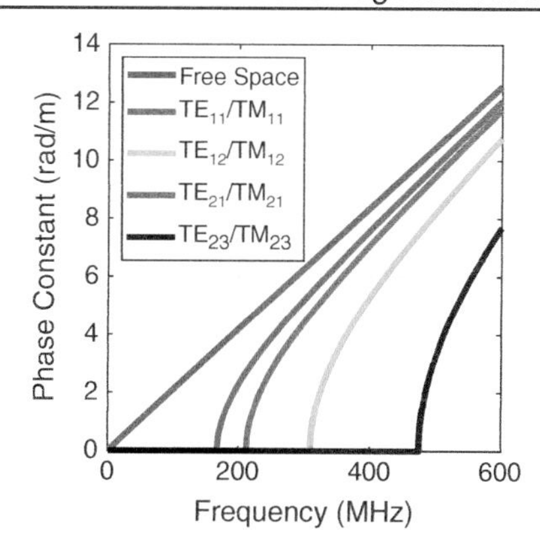

Figure 7.74 Phase constant with respect to frequency for different cases. A rectangular waveguide with a 2×1 m cross section and a vacuum inside is considered from zero frequency to 600 MHz. Given a frequency, the phase constant depends on the mode: i.e. it is smaller for larger mode numbers, but it is always smaller than the corresponding phase constant in a homogeneous space (a vacuum, in this case). Note that for each mode, the phase constant is zero up to the cutoff frequency.

Note that TE and TM modes with the same (m, n) have the same behavior in terms of propagation and attenuation (e.g. propagation/phase/attenuation constants, phase velocity, wavelength, cutoff frequency).

The discussion above shows that, given mode (m, n), there is a cutoff frequency f_c^{mn} at which the mode starts to propagate/attenuate. This frequency can be found in terms of mode numbers and waveguide dimensions as

$$f_c^{mn} = \frac{1}{2\pi\sqrt{\mu\varepsilon}}\sqrt{\left(\frac{m\pi}{a}\right)^2 + \left(\frac{n\pi}{b}\right)^2}. \tag{7.306}$$

If the frequency is larger than the cutoff frequency ($f > f_c^{mn}$), then the related mode (with number (m, n)) propagates, while the same mode attenuates if $f < f_c^{mn}$ (Figure 7.76). One can also deduce that, given a fixed frequency $f < f_c^{mn}$, increasing the dimensions of the waveguide or permittivity/permeability of the inner material can shift f_c^{mn} to lower values such that mode (m, n) may start to propagate.

Obviously, for given dimensions of a waveguide and frequency, some modes may propagate while others (infinitely many, as there is no upper bound for m and n) attenuate (Figure 7.77). Propagating modes (and hence, attenuating modes) can be found by investigating their cutoff frequencies in comparison to the operating frequency (Figure 7.78). It should be emphasized that propagating modes propagate with different velocities: i.e. lower-order modes propagate slower than higher-order modes. On the other hand, since most electromagnetic energy is carried by lower-order modes, the mode with the smallest mode numbers is called the *dominant* (fundamental) mode. In a rectangular waveguide, the cutoff frequencies for TE_{01} and TE_{10} modes can be written

$$f_c^{01} = \frac{1}{2\pi\sqrt{\mu\varepsilon}}\frac{\pi}{b} = \frac{1}{2b\sqrt{\mu\varepsilon}} \tag{7.307}$$

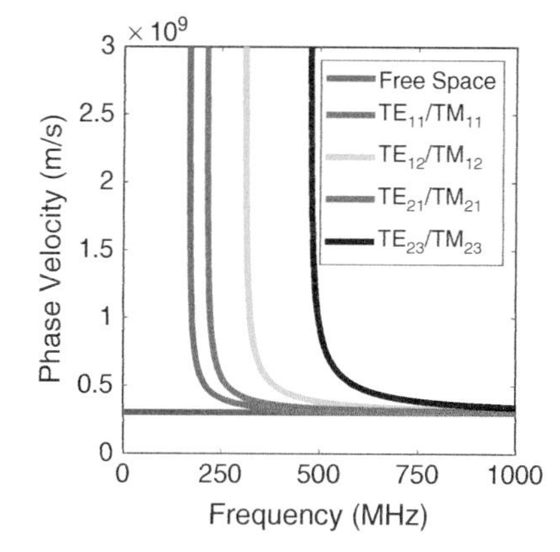

Figure 7.75 Phase velocity with respect to frequency for different cases. A rectangular waveguide with a 2×1 m cross section and a vacuum inside is considered from zero frequency to 1 GHz (see Figure 7.74 for phase constant values up to 600 MHz). Given a frequency, the phase velocity depends on the mode: i.e. it is higher for larger mode numbers, but it is always higher than the corresponding phase velocity in a homogeneous space (a vacuum, in this case). Note that for each mode, the phase velocity goes to infinity at the cutoff frequency.

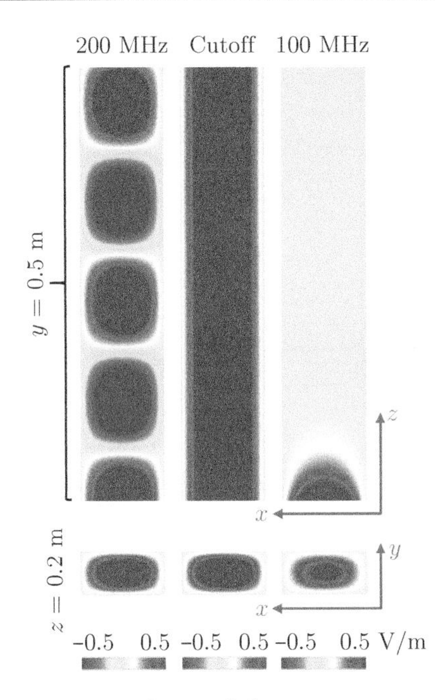

Figure 7.76 The real part of the z component of the electric field intensity for the TM_{11} mode in a 2×1 m rectangular waveguide (a vacuum inside) at three different frequencies: i.e. at 200 MHz (propagating), at approximately 168 MHz (cutoff), and at 100 MHz (evanescent). The values are shown with respect to position on the $y = 0.5$ m plane (with respect to $z \in [0, 6]$ m and $x \in [0, 2]$ m) and on the $z = 0.2$ m plane (with respect to $x \in [0, 2]$ m and $y \in [0, 1]$ m).

and

$$f_c^{10} = \frac{1}{2\pi\sqrt{\mu\varepsilon}}\frac{\pi}{a} = \frac{1}{2a\sqrt{\mu\varepsilon}}, \tag{7.308}$$

respectively. Therefore, depending on the values of a and b, either TE_{01} or TE_{10} is the dominant mode of a rectangular waveguide.[70]

Similar to the wave impedance in a homogeneous space, one can define a wave impedance inside a waveguide. However, such an impedance value, which describes the relationship between amplitudes of electric and magnetic fields, does not directly correspond to intrinsic impedance $\eta = \sqrt{\mu/\varepsilon}$. Instead, it depends on the frequency and mode: i.e. mode numbers m and n. The wave impedance for a TM mode is defined as

$$\eta_{mn}^{\mathrm{TM}} = \frac{E_x^{\mathrm{TM}}(\bar{r})}{H_y^{\mathrm{TM}}(\bar{r})} = -\frac{E_y^{\mathrm{TM}}(\bar{r})}{H_x^{\mathrm{TM}}(\bar{r})} = \frac{\gamma_{mn}}{j\omega\varepsilon} = \frac{1}{j\omega\varepsilon}\sqrt{\left(\frac{m\pi}{a}\right)^2 + \left(\frac{n\pi}{b}\right)^2 - \omega^2\mu\varepsilon}. \tag{7.309}$$

For a TE mode, however, we obtain

$$\eta_{mn}^{\mathrm{TE}} = \frac{E_x^{\mathrm{TE}}(\bar{r})}{H_y^{\mathrm{TE}}(\bar{r})} = -\frac{E_y^{\mathrm{TE}}(\bar{r})}{H_x^{\mathrm{TE}}(\bar{r})} = \frac{j\omega\mu}{\gamma_{mn}} = j\omega\mu\sqrt{\left(\frac{m\pi}{a}\right)^2 + \left(\frac{n\pi}{b}\right)^2 - \omega^2\mu\varepsilon}. \tag{7.310}$$

Using $f_c^{mn} = [1/(2\pi\sqrt{\mu\varepsilon})]\sqrt{(m\pi/a)^2 + (n\pi/b)^2}$, these expressions can be simplified as

$$\eta_{mn}^{\mathrm{TM}} = \eta \begin{cases} [1 - (f_c^{mn}/f)^2]^{1/2} & (f > f_c^{mn}) \\ -j[(f_c^{mn}/f)^2 - 1]^{1/2} & (f < f_c^{mn}) \end{cases} \tag{7.311}$$

$$\eta_{mn}^{\mathrm{TE}} = \eta \begin{cases} [1 - (f_c^{mn}/f)^2]^{-1/2} & (f > f_c^{mn}) \\ j[(f_c^{mn}/f)^2 - 1]^{-1/2} & (f < f_c^{mn}). \end{cases} \tag{7.312}$$

Note that TM and TE modes with the same mode numbers have related wave impedance:

$$\eta_{mn}^{\mathrm{TM}} \times \eta_{mn}^{\mathrm{TE}} = \eta^2 = \mu/\varepsilon. \tag{7.313}$$

Finally, as mentioned above, the phase velocity $u_{p,mn}$ can be faster than the speed of light, but the group velocity must obey the rule that information cannot be transmitted faster than

[70] We can write field expressions for TE_{10} as

$$H_x^{\mathrm{TE}}(\bar{r}) = \gamma B_{10}\left(\frac{a}{\pi}\right)\sin\left(\frac{\pi x}{a}\right)\exp(-\gamma_{10}z)$$

$$E_y^{\mathrm{TE}}(\bar{r}) = -j\omega\mu B_{10}\left(\frac{a}{\pi}\right)\sin\left(\frac{\pi x}{a}\right)\exp(-\gamma_{10}z)$$

$$H_z^{\mathrm{TE}}(\bar{r}) = B_{10}\cos\left(\frac{\pi x}{a}\right)\exp(-\gamma_{10}z)$$

$$E_x^{\mathrm{TE}}(\bar{r}) = 0, \quad H_y^{\mathrm{TE}}(\bar{r}) = 0, \quad \bar{E}_z^{\mathrm{TE}} = 0.$$

Similarly, for TE_{01}, we have

$$E_x^{\mathrm{TE}}(\bar{r}) = j\omega\mu B_{01}\left(\frac{b}{\pi}\right)\sin\left(\frac{n\pi y}{b}\right)\exp(-\gamma_{01}z)$$

$$H_y^{\mathrm{TE}}(\bar{r}) = \gamma_{01}B_{01}\left(\frac{b}{\pi}\right)\sin\left(\frac{n\pi y}{b}\right)\exp(-\gamma_{01}z)$$

$$H_z^{\mathrm{TE}}(\bar{r}) = B_{01}\cos\left(\frac{\pi y}{b}\right)\exp(-\gamma_{01}z)$$

$$H_x^{\mathrm{TE}}(\bar{r}) = 0, \quad E_y^{\mathrm{TE}}(\bar{r}) = 0, \quad \bar{E}_z^{\mathrm{TE}} = 0.$$

the speed of light. To see this, note that

$$u_{g,mn} = \frac{\partial \omega}{\partial \beta_{mn}} = \left(\frac{\partial \beta_{mn}}{\partial \omega}\right)^{-1}, \tag{7.314}$$

where

$$\frac{\partial \beta_{mn}}{\partial \omega} = \frac{\omega\mu\varepsilon}{\sqrt{\omega^2\mu\varepsilon - \left(\frac{m\pi}{a}\right)^2 - \left(\frac{n\pi}{b}\right)^2}}. \tag{7.315}$$

Therefore,

$$u_{g,mn} = \frac{1}{\omega\mu\varepsilon}\sqrt{\omega^2\mu\varepsilon - \left(\frac{m\pi}{a}\right)^2 - \left(\frac{n\pi}{b}\right)^2} < \frac{1}{\sqrt{\mu\varepsilon}}, \tag{7.316}$$

as expected. Note that

$$u_{p,mn} \times u_{g,mn} = 1/(\mu\varepsilon) \tag{7.317}$$

for all m and n, while

$$u_{p,mn} \to 1/\sqrt{\mu\varepsilon}, \qquad u_{g,mn} \to 1/\sqrt{\mu\varepsilon} \tag{7.318}$$

as $a \to \infty$ and $b \to \infty$.

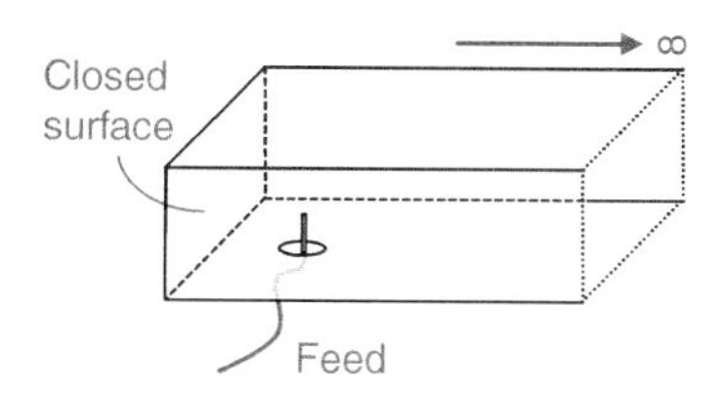

Figure 7.77 Excitation of a waveguide via a probe. The probe may excite all modes while a majority of them are evanescent, and those decay rapidly without propagation. Only propagating modes remain at a distance of a few wavelengths.

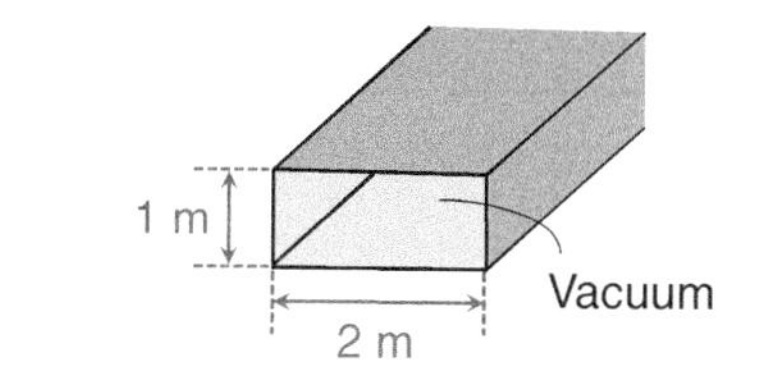

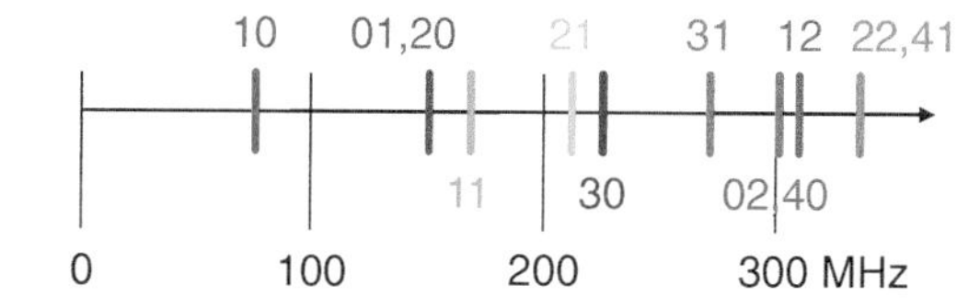

Figure 7.78 Cutoff frequencies of the first 22 modes in a 2 × 1 m rectangular waveguide with a vacuum inside. The fundamental mode is TE_{10}, which is the only mode that can exist below 100 MHz (but above $f_c^{10} \approx 74.95$ MHz). Other than 10 for TE_{10} and 01 for TE_{01}, numbers indicate both TE and TM modes. Note that based on the given dimensions, some modes with different numbers have the same cutoff frequencies: e.g. TE_{01}, TE_{20}, and TM_{20} with a cutoff frequency of approximately 149.8 MHz.

7.2.3 Parallel-Plate Waveguides

As another set of examples, we now briefly focus on parallel-plate waveguides that are formed of parallel conducting plates, as well as the generation of TM modes inside them. For the formulation, we consider a structure consisting of two parallel plates that are infinitely long in the x and z directions (Figure 7.79). A corner of the lower plate is located at the origin such that infinite dimensions are in the $+x$ and $+z$ directions, while we further assume that wave propagation between the plates is in the $+z$ direction. The distance between the plates is constant, b. TM modes satisfy $H_z^{\text{TM}}(\bar{r}) = 0$, leading to

$$E_{tx}^{\text{TM}}(x,y) = -\frac{\gamma}{h^2}\frac{\partial E_{tz}^{\text{TM}}(x,y)}{\partial x}, \qquad E_{ty}^{\text{TM}}(x,y) = -\frac{\gamma}{h^2}\frac{\partial E_{tz}^{\text{TM}}(x,y)}{\partial y} \tag{7.319}$$

$$H_{tx}^{\text{TM}}(x,y) = \frac{j\omega\varepsilon}{h^2}\frac{\partial E_{tz}^{\text{TM}}(x,y)}{\partial y}, \qquad H_{ty}^{\text{TM}}(x,y) = -\frac{j\omega\varepsilon}{h^2}\frac{\partial E_{tz}^{\text{TM}}(x,y)}{\partial x}, \tag{7.320}$$

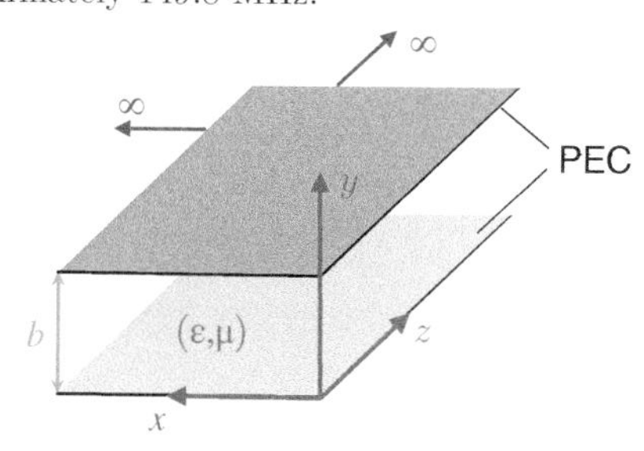

Figure 7.79 A parallel-plate waveguide.

where $h^2 = \gamma^2 + k^2$, similar to those in a rectangular waveguide.[71] In addition, we can use a Helmholtz equation:

$$\bar{\nabla}_{xy}^2 \bar{E}^{\mathrm{TM}}(\bar{r}) + h^2 \bar{E}^{\mathrm{TM}}(\bar{r}) = 0 \tag{7.321}$$

$$\longrightarrow \frac{\partial^2 E_{tz}^{\mathrm{TM}}(x,y)}{\partial x^2} + \frac{\partial^2 E_{tz}^{\mathrm{TM}}(x,y)}{\partial y^2} + h^2 E_{tz}^{\mathrm{TM}}(x,y) = 0. \tag{7.322}$$

Unlike with a rectangular waveguide, however, we can further omit the x dependency, leading to

$$\frac{\partial^2 E_{tz}^{\mathrm{TM}}(y)}{\partial y^2} + h^2 E_{tz}^{\mathrm{TM}}(y) = 0. \tag{7.323}$$

Then a general solution for the axial component of the electric field intensity can be found as

$$E_{tz}^{\mathrm{TM}}(y) = \left[C\cos(k_y y) + D\sin(k_y y)\right] \tag{7.324}$$

$$E_z^{\mathrm{TM}}(\bar{r}) = E_{tz}^{\mathrm{TM}}(y)\exp(-\gamma z) = \left[C\cos(k_y y) + D\sin(k_y y)\right]\exp(-\gamma z), \tag{7.325}$$

where C (V/m) and D (V/m) are constants that need to be determined via boundary conditions. Given that $E_{tz}^{\mathrm{TM}}(y) = 0$ when $y = 0$ and $y = b$, we obtain

$$E_{tz}^{\mathrm{TM}}(y) = D_n \sin\left(\frac{n\pi y}{b}\right) \tag{7.326}$$

$$E_z^{\mathrm{TM}}(\bar{r}) = D_n \sin\left(\frac{n\pi y}{b}\right)\exp(-\gamma_n z), \tag{7.327}$$

and[72]

$$\gamma_n^2 = h_n^2 - \omega^2\mu\varepsilon = \left(\frac{n\pi}{b}\right)^2 - \omega^2\mu\varepsilon \tag{7.328}$$

for each *mode* $n = 1, 2, \ldots$ Using the axial electric field intensity, we can find the transverse components as[73,74]

$$E_{tx}^{\mathrm{TM}}(y) = -\frac{\gamma_n}{h_n^2}\frac{\partial E_{tz}^{\mathrm{TM}}(y)}{\partial x} = 0 \tag{7.329}$$

$$E_{ty}^{\mathrm{TM}}(y) = -\frac{\gamma_n}{h_n^2}\frac{\partial E_{tz}^{\mathrm{TM}}(y)}{\partial y} = -\gamma_n D_n\left(\frac{b}{n\pi}\right)\cos\left(\frac{n\pi y}{b}\right) \tag{7.330}$$

$$H_{tx}^{\mathrm{TM}}(y) = \frac{j\omega\varepsilon}{h_n^2}\frac{\partial E_{tz}^{\mathrm{TM}}(y)}{\partial y} = j\omega\varepsilon D_n\left(\frac{b}{n\pi}\right)\cos\left(\frac{n\pi y}{b}\right) \tag{7.331}$$

[71] Unlike a rectangular waveguide, a parallel-plate waveguide supports the TEM mode. Recall that, without axial electric and magnetic fields, one can write

$$\gamma = j\omega\sqrt{\mu\varepsilon}$$

$$\bar{H}(\bar{r}) = \hat{a}_z \times \bar{E}(\bar{r})/\eta$$

$$\bar{\nabla}_{xy}^2 E_x(\bar{r}) = 0, \quad \bar{\nabla}_{xy}^2 E_y(\bar{r}) = 0$$

$$\bar{\nabla}_{xy}^2 H_x(\bar{r}) = 0, \quad \bar{\nabla}_{xy}^2 H_y(\bar{r}) = 0.$$

But without an x variation (infinity assumption), we further have

$$\bar{\nabla}_{xy}^2 E_x(\bar{r}) = \frac{\partial^2}{\partial y^2}E_x(\bar{r}) = 0$$

$$\longrightarrow E_x(\bar{r}) = (Ay + B)\exp(-\gamma z)$$

$$\bar{\nabla}_{xy}^2 E_y(\bar{r}) = \frac{\partial^2}{\partial y^2}E_y(\bar{r}) = 0$$

$$\longrightarrow E_y(\bar{r}) = (Cy + D)\exp(-\gamma z),$$

where A, B, C, and D are constants. Applying boundary conditions at $y = 0$ and $y = b$ leads to $A = 0$, $B = 0$, and $E_x(\bar{r}) = 0$. In addition, using Gauss' law, the absence of an electric charge between the plates means $C = 0$. Consequently, we obtain

$$E_z^{\mathrm{TEM}}(\bar{r}) = 0, \quad H_z^{\mathrm{TEM}}(\bar{r}) = 0$$

$$E_x^{\mathrm{TEM}}(\bar{r}) = 0, \quad H_y^{\mathrm{TEM}}(\bar{r}) = 0$$

$$E_y^{\mathrm{TEM}}(\bar{r}) = D\exp(-\gamma z)$$

$$H_x^{\mathrm{TEM}}(\bar{r}) = -\frac{D}{\eta}\exp(-\gamma z).$$

Note that the TEM mode does not have a cutoff frequency: i.e. it always exists.

[72] Hence, the cutoff frequency for a TM mode with number n can be written

$$f_c^n = \frac{1}{2\pi\sqrt{\mu\varepsilon}}\frac{n\pi}{b} = \frac{n}{2b\sqrt{\mu\varepsilon}}.$$

[73] In these expressions, the subscript n is omitted for γ and h.

[74] Based on these expressions, the wave impedance of TM_n can be found as

$$\eta_n^{\mathrm{TM}} = -\frac{E_y^{\mathrm{TM}}(\bar{r})}{H_x^{\mathrm{TM}}(\bar{r})} = \frac{\gamma_n}{j\omega\varepsilon} = \frac{1}{j\omega\varepsilon}\sqrt{\left(\frac{n\pi}{b}\right)^2 - \omega^2\mu\varepsilon}$$

$$= \eta\begin{cases}[1-(f_c^n/f)^2]^{1/2} & (f > f_c^n)\\ -j[(f_c^n/f)^2 - 1]^{1/2} & (f < f_c^n).\end{cases}$$

$$H_{ty}^{\mathrm{TM}}(y) = -\frac{j\omega\varepsilon}{h_n^2}\frac{\partial E_{tz}^{\mathrm{TM}}(y)}{\partial x} = 0. \quad (7.332)$$

Also note that

$$E_y^{\mathrm{TM}}(\bar{r}) = -\gamma_n D_n \left(\frac{b}{n\pi}\right) \cos\left(\frac{n\pi y}{b}\right) \exp(-\gamma_n z) \quad (7.333)$$

$$H_x^{\mathrm{TM}}(\bar{r}) = j\omega\varepsilon D_n \left(\frac{b}{n\pi}\right) \cos\left(\frac{n\pi y}{b}\right) \exp(-\gamma_n z). \quad (7.334)$$

Expressions for TE modes can be found in a similar manner.

7.2.4 Examples

Example 94: Consider a rectangular waveguide (Figure 7.80) of dimensions $a = 0.2$ m and $b = 0.1$ m, filled with air ($\varepsilon \approx \varepsilon_0$, $\mu \approx \mu_0$). Find the phase and attenuation constants of the TM_{11} mode at 1 GHz, 2 GHz, and 4 GHz. Find the same constants if the waveguide is filled with a dielectric material having a relative permittivity of $\varepsilon_r = 3$.

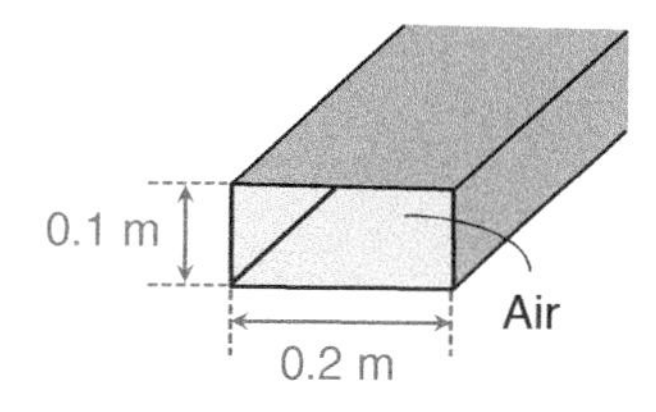

Figure 7.80 A rectangular waveguide filled with air.

Solution: First, we can find the cutoff frequency of the mode as

$$f_c^{11} = \frac{1}{2\pi\sqrt{\mu_0\varepsilon_0}}\sqrt{\left(\frac{\pi}{0.2}\right)^2 + \left(\frac{\pi}{0.1}\right)^2} \approx 1.676 \text{ GHz}. \quad (7.335)$$

When $f = 1$ GHz, the mode is evanescent since $f < f_c^{11}$. The attenuation constant can be found as[75,76]

$$\alpha_{11}(1 \text{ GHz}) = \sqrt{\left(\frac{\pi}{0.2}\right)^2 + \left(\frac{\pi}{0.1}\right)^2 - \omega^2\mu_0\varepsilon_0} \approx 28.2 \text{ Np/m}. \quad (7.336)$$

On the other hand, when $f = 2$ GHz, the mode propagates with a phase constant of[77]

$$\beta_{11}(2 \text{ GHz}) = \sqrt{\omega^2\mu_0\varepsilon_0 - \left(\frac{\pi}{0.2}\right)^2 - \left(\frac{\pi}{0.1}\right)^2} \approx 22.9 \text{ rad/m}. \quad (7.337)$$

If the frequency is further increased to $f = 4$ GHz, the mode is still propagating, but we have[78]

$$\beta_{11}(4 \text{ GHz}) \approx 76.1 \text{ rad/m}. \quad (7.338)$$

[75] The phase constant is automatically zero, in this case: i.e. $\beta_{11}(1 \text{ GHz}) = 0$.

[76] This indicates that the field strength of the mode is multiplied approximately by $\exp(-2.82) \approx 0.0596$ at each 0.1 m distance in the propagation direction.

[77] The corresponding phase velocity and wavelength values can also be obtained as $u_{11}(2 \text{ GHz}) \approx 5.5 \times 10^8$ m/s and $\lambda_{11}(2 \text{ GHz}) \approx 0.275$ m, respectively. Also, note that $\alpha_{11}(2 \text{ GHz}) = 0$.

[78] The corresponding phase velocity and wavelength values can be found as $u_{11}(4 \text{ GHz}) \approx 3.3 \times 10^8$ m/s and $\lambda_{11}(4 \text{ GHz}) \approx 8.26$ cm, respectively. Also, note that $\alpha_{11}(4 \text{ GHz}) = 0$.

If the waveguide is filled with a dielectric material having a relative permittivity of $\varepsilon_r = 3$, the cutoff frequency of TM_{11} is updated as

$$f_{c,d}^{11} = f_c^{11}/\sqrt{3} \approx 967.6 \text{ MHz}. \tag{7.339}$$

Hence, in addition to 2 GHz and 4 GHz, the mode is propagating at 1 GHz. The phase constant can be found as[79]

[79] Hence, we have $u_{11,d}(1 \text{ GHz}) \approx 13.7 \times 10^8$ m/s and $\lambda_{11,d}(1 \text{ GHz}) \approx 0.685$ m.

$$\beta_{11,d}(1 \text{ GHz}) = \sqrt{\omega^2\mu_0\varepsilon - \left(\frac{\pi}{0.2}\right)^2 - \left(\frac{\pi}{0.1}\right)^2} \approx 9.17 \text{ rad/m}. \tag{7.340}$$

Phase constant values at 2 GHz and 4 GHz should also be updated as

$$\beta_{11,d}(2 \text{ GHz}) \approx 63.5 \text{ rad/m}, \quad \beta_{11,d}(4 \text{ GHz}) \approx 141 \text{ rad/m}. \tag{7.341}$$

Example 95: A rectangular waveguide with the given dimensions (Figure 7.81) is filled with a dielectric material having a relative permittivity of $\varepsilon_r = 2.0$. The angular frequency is $\omega = 9 \times 10^8$ rad/s. Assume $1/\sqrt{\mu_0\varepsilon_0} \approx 3 \times 10^8$ m/s.

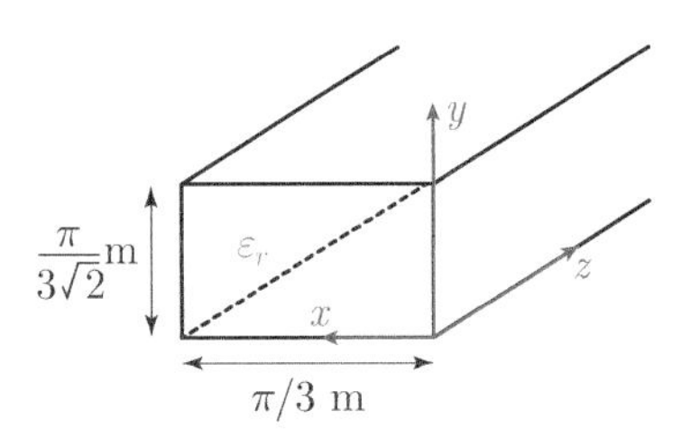

Figure 7.81 A rectangular waveguide filled with a dielectric material.

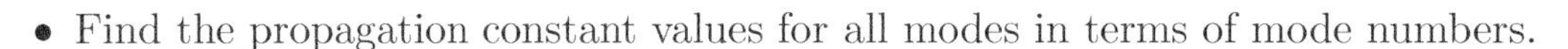

- Find the propagation constant values for all modes in terms of mode numbers.
- Find which mode or modes propagate.
- The z components of the electric field intensity and magnetic field intensity at $z = 0$ are given as

$$E_z(z=0) = \sin(3x)\sin(3\sqrt{2}y) \qquad (\text{V/m}) \tag{7.342}$$

$$H_z(z=0) = \cos(3x) + \cos(3\sqrt{2}y) + \cos(3x)\cos(3\sqrt{2}y) \qquad (\text{A/m}). \tag{7.343}$$

Find the z component of the electric field intensity and magnetic field intensity at $z = \pi$ meter.

Solution: Using the dimensions of the waveguide, we can obtain the expression for the propagation constant as

$$\gamma_{mn} \approx j\sqrt{\frac{2 \times 81 \times 10^{16}}{9 \times 10^{16}} - \left(\frac{m\pi}{\pi/3}\right)^2 - \left(\frac{n\pi}{\pi/(3\sqrt{2})}\right)^2} \tag{7.344}$$

$$= j\sqrt{18 - 9m^2 - 18n^2} \quad (1/\text{m}). \tag{7.345}$$

Propagating modes are those with imaginary propagation constant. Checking the value inside the square root, we have

$$18 - 9m^2 - 18n^2 > 0 \longrightarrow (m, n) = (1, 0), \tag{7.346}$$

i.e. only TE_{10} is a propagating mode. Interestingly, the operating frequency corresponds to the cutoff frequency of TE_{01}: i.e. it is neither propagating nor attenuating, while all other modes are evanescent. Considering the given expressions for the axial components of the electric and magnetic field intensities, we identify the corresponding modes as[80,81]

$$E_z(z = 0): \ TM_{11} \tag{7.347}$$

$$H_z(z = 0): \ TE_{10} + TE_{01} + TE_{11}. \tag{7.348}$$

Among four modes, TE_{11} and TM_{11} are evanescent with an $\alpha_{11} \approx 3.0$ Np/m propagation constant.[82] As discussed above, TE_{01} is neither propagating nor evanescent[83] ($\gamma_{01} \approx 0$), while TE_{10} is a propagating mode with a constant[84] of $\beta_{10} \approx 3$ rad/m. Consequently, at $z = \pi$ meter, the axial fields are expected to be

$$E_z(z = \pi) \approx \sin(3x)\sin(3\sqrt{2}y)\exp(-3\pi) \approx 0 \tag{7.349}$$

$$H_z(z = \pi) \approx \cos(3x)\exp(-j3\pi)$$
$$+ \cos(3\sqrt{2}y) + \cos(3x)\cos(3\sqrt{2}y)\exp(-3\pi) \tag{7.350}$$
$$\approx -\cos(3x) + \cos(3\sqrt{2}y) \qquad \text{(A/m)}. \tag{7.351}$$

Example 96: Consider a parallel-plate waveguide (Figure 7.82) filled with air ($\varepsilon \approx \varepsilon_0$ and $\mu \approx \mu_0$). The distance between the plates is 10 cm. Use approximation $1/\sqrt{\mu_0\varepsilon_0} \approx 3 \times 10^8$ m/s.

- Find the propagation constant for the TE_1 mode at 100 MHz and 10 GHz.
- Find an expression for the electric current density on the top plate due to TE_1 at 100 MHz and 10 GHz.
- Find an expression for the time-average power flux density (Poynting vector) due to TE_1 at 100 MHz and 10 GHz.

Solution: For this parallel-plate waveguide, the propagation constant for mode number n can be written

$$\gamma_n = j\sqrt{\omega^2\mu\varepsilon - \left(\frac{n\pi}{0.1}\right)^2} \approx \sqrt{\frac{\omega^2}{9 \times 10^{16}} - 987n^2}. \tag{7.352}$$

[80] Recall that

$$E_{tz}^{\mathrm{TM}}(x, y) = A_{mn}\sin\left(\frac{m\pi x}{a}\right)\sin\left(\frac{n\pi y}{b}\right)$$
$$E_z^{\mathrm{TM}}(\bar{r}) = E_{tz}^{\mathrm{TM}}(x, y)\exp(-\gamma_{mn}z)$$

and

$$H_{tz}^{\mathrm{TE}}(x, y) = B_{mn}\cos\left(\frac{m\pi x}{a}\right)\cos\left(\frac{n\pi y}{b}\right)$$
$$H_z^{\mathrm{TE}}(\bar{r}) = H_{tz}^{\mathrm{TE}}(x, y)\exp(-\gamma_{mn}z).$$

[81] Hence, we have a mixture of TE_{10}, TE_{01}, TE_{11}, and TM_{11} modes with unit constants (A_{mn} and B_{mn}) at $z = 0$.

[82] Hence, the axial components of the electric and magnetic field intensities *related* to these modes can be written

$$E_z(z) \approx \sin(3x)\sin(3\sqrt{2}y)\exp(-3z) \qquad \text{(V/m)}$$
$$H_z(z) \approx \cos(3x)\cos(3\sqrt{2}y)\exp(-3z) \qquad \text{(A/m)}.$$

[83] The corresponding axial magnetic field intensity *related* to this mode can be written

$$H_z(z) \approx \cos(3\sqrt{2}y) \qquad \text{(A/m)}.$$

[84] The corresponding axial magnetic field intensity *related* to this mode can be written

$$H_z(z) \approx \cos(3x)\exp(-j3z) \qquad \text{(A/m)}.$$

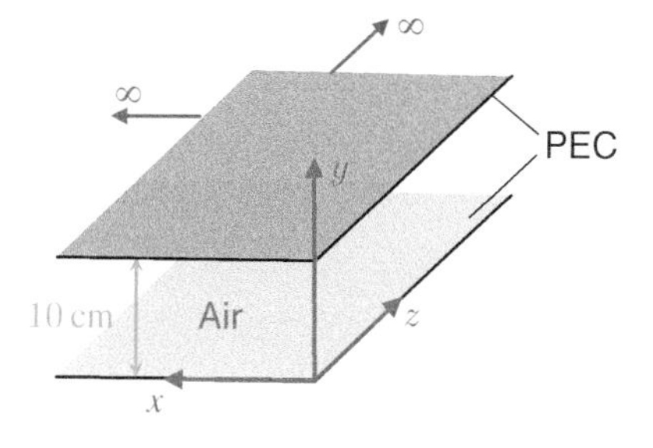

Figure 7.82 A parallel-plate waveguide with a 10 cm distance between the PEC plates and air between them.

Then for the TE_1 mode at 100 MHz, we have

$$\gamma_1(100 \text{ MHz}) \approx j\sqrt{\frac{4\pi^2 \times 10^{16}}{9 \times 10^{16}} - 987} \approx 31.3 \text{ Np/m}, \tag{7.353}$$

which means the mode is evanescent. At 10 GHz, however, we obtain

$$\gamma_1(10 \text{ GHz}) \approx j\sqrt{\frac{4\pi^2 \times 10^{20}}{9 \times 10^{16}} - 987} \approx j207 \text{ rad/m}, \tag{7.354}$$

indicating that the mode propagates with a phase constant of $\beta_1(10 \text{ GHz}) \approx 207$ rad/m. For the electric current density on the top plate, we can first find an expression for the magnetic field intensity as[85]

$$\bar{H}^{\text{TM}}(y = 0.1) = \hat{a}_x j\omega\varepsilon D_1 \left(\frac{0.1}{\pi}\right) \cos\left(\frac{\pi 0.1}{0.1}\right) \exp(-\gamma_1 z) \tag{7.355}$$

$$= -\hat{a}_x \frac{j\omega\varepsilon D_1}{10\pi} \exp(-\gamma_1 z) \qquad (\text{A/m}). \tag{7.356}$$

Then using the boundary condition for the tangential magnetic field intensity, we obtain

$$\bar{J}_s^{\text{TM}}(y = 0.1) = \hat{a}_z \frac{j\omega\varepsilon D_1}{10\pi} \exp(-\gamma_1 z) \qquad (\text{A/m}). \tag{7.357}$$

This means the electric current density on the top plate due to TM_1 can be written

$$\bar{J}_s^{\text{TM}}(y = 0.1, 100 \text{ MHz}) \approx \hat{a}_z j D_1 (1.8 \times 10^{-4}) \exp(-31.3z) \qquad (\text{A/m}) \tag{7.358}$$

$$\bar{J}_s^{\text{TM}}(y = 0.1, 10 \text{ GHz}) \approx \hat{a}_z j D_1 (1.8 \times 10^{-2}) \exp(-j207z) \qquad (\text{A/m}), \tag{7.359}$$

at 100 MHz and 10 GHz, respectively. Finally, we can derive the complex Poynting vector due to TM_1 (considering $b = 0.1$ m) as[86]

$$S_c^{\text{TM}}(\bar{r}) = \frac{1}{2}\bar{E}^{\text{TM}}(\bar{r}) \times [\bar{H}^{\text{TM}}(\bar{r})]^* \tag{7.360}$$

$$= -\hat{a}_y \frac{j\omega\varepsilon|D_1|^2}{10\pi} \sin(20\pi y) \exp(-2\alpha_1 z) - \hat{a}_z \gamma_1 \frac{j\omega\varepsilon|D_1|^2}{100\pi^2} \cos^2(10\pi y) \exp(-2\alpha_1 z) \qquad (\text{W/m}^2), \tag{7.361}$$

[85]Recall that the general expression for the x component of the magnetic field intensity can be written

$$H_x^{\text{TM}}(\bar{r}) = j\omega\varepsilon D_n \left(\frac{b}{n\pi}\right) \cos\left(\frac{n\pi y}{b}\right) \exp(-\gamma_n z),$$

and the y and z components are zero.

[86]For a general TM_n mode, we have

$$E_z^{\text{TM}}(\bar{r}) = D_n \sin\left(\frac{n\pi y}{b}\right) \exp(-\gamma_n z)$$

$$E_y^{\text{TM}}(\bar{r}) = -\gamma_n D_n \left(\frac{b}{n\pi}\right) \cos\left(\frac{n\pi y}{b}\right) \exp(-\gamma_n z)$$

in addition to

$$H_x^{\text{TM}}(\bar{r}) = j\omega\varepsilon D_n \left(\frac{b}{n\pi}\right) \cos\left(\frac{n\pi y}{b}\right) \exp(-\gamma_n z).$$

Hence, the complex Poynting vector can be found as

$$S_c^{\text{TM}}(\bar{r}) = \frac{1}{2}\bar{E}^{\text{TM}}(\bar{r}) \times [\bar{H}^{\text{TM}}(\bar{r})]^*$$

$$= -\hat{a}_y j\omega\varepsilon|D_n|^2 \left(\frac{b}{n\pi}\right) \sin\left(\frac{2n\pi y}{b}\right) \exp(-2\alpha_n z)$$

$$- \hat{a}_z j\omega\varepsilon\gamma_n|D_n|^2 \left(\frac{b}{n\pi}\right)^2 \cos^2\left(\frac{n\pi y}{b}\right) \exp(-2\alpha_n z).$$

Note that the y component is purely imaginary whether $\alpha_n = 0$ or $\alpha_n \neq 0$.

leading to the time-average Poynting vector as

$$S_{\text{avg}}^{\text{TM}}(\bar{r}) = \hat{a}_z \beta_1 \frac{\omega\varepsilon |D_1|^2}{100\pi^2} \cos^2(10\pi y) \exp(-2\alpha_1 z) \qquad (\text{W/m}^2). \tag{7.362}$$

Then at 100 MHz and 10 GHz, we have

$$S_{\text{avg}}^{\text{TM}}(\bar{r}, 100 \text{ MHz}) = 0 \tag{7.363}$$

$$S_{\text{avg}}^{\text{TM}}(\bar{r}, 10 \text{ GHz}) \approx \hat{a}_z 0.12 |D_1|^2 \cos^2(10\pi y) \qquad (\text{W/m}^2). \tag{7.364}$$

7.3 Transmission Line Theory

Waveguides, studied in the previous section, mitigate a significant bottleneck of wireless transmission: quadratic decay of the power density with respect to distance. By minimizing dielectric and metallic losses, waveguides can achieve excellent point-to-point transmission efficiencies (provided that propagating modes exist). In addition, because their structures confine propagating waves and shield them from outside, waveguides avoid the disastrous effects of electromagnetic interference. As another advantage, waveguides can also handle higher power transmission operations. On the other hand, these benefits are balanced by well-known disadvantages: (i) they are not compact and are rigid and bulky; (ii) they are less economical than other transmission tools; and (iii) they have narrow bands of operation. Some of these limitations are alleviated using dielectric waveguides such as optical fibers, but they may not perfectly maintain the advantages mentioned earlier in terms of efficiency and electromagnetic interference. Consequently, various types of waveguides are used in many real-life applications, but they are less preferred in others, where wireless transmission tools or wired media (cables) are used.

As excellent transmission tools, electrical cables of various types can be found everywhere in modern life. These include *twisted-pair* (Figure 7.83) and *coaxial* (Figure 7.84) cables,[87] which are used in household devices, computers, and personal communication tools. Unlike waveguides, cables are usually more compact, flexible, and less expensive, and they offer wider frequency bands of operation. Unfortunately, they can suffer from higher transmission losses (in comparison to waveguides), in addition to having quite limited high-power transmission capabilities, which make them less suitable than waveguides in certain applications.

Given an electromagnetic transmission problem involving a cable, its ideal solution can be achieved by directly solving Maxwell's equations, but this may not be trivial: e.g. when the cable is deformed or bent. Reflections at the end of a cable are well-known phenomena and extremely

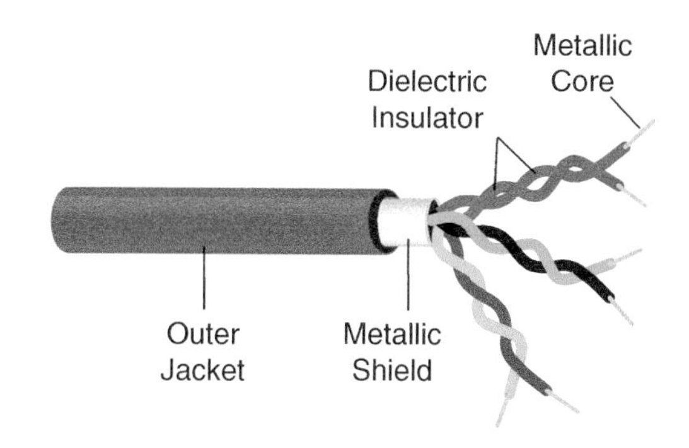

Figure 7.83 A typical twisted-pair cable includes multiple wires that are paired and twisted. Twisting is applied to reduce electromagnetic interference by making the cables equally far from an outer source.

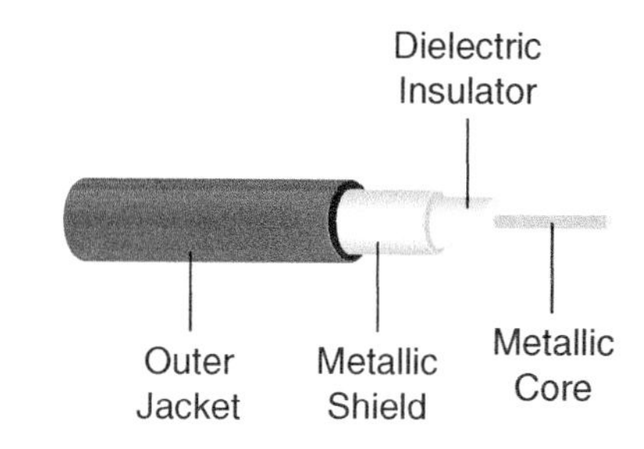

Figure 7.84 A typical coaxial cable includes a metallic core inside a cylindrical shell (shield) and a dielectric insulator between them. Ideally, electromagnetic fields are nonzero only between the conductors (inside the insulator). Coaxial cables support TEM waves.

[87] Coaxial cables are generally more expensive than twisted-pair cables, but they have more effective shielding properties that reduce electromagnetic interference. Also, due to their better efficiency (lower losses), coaxial cables can be longer than twisted-pair cables. Coaxial cables are more efficient than twisted pairs, especially at higher frequencies, making them suitable to transmit data with higher rates.

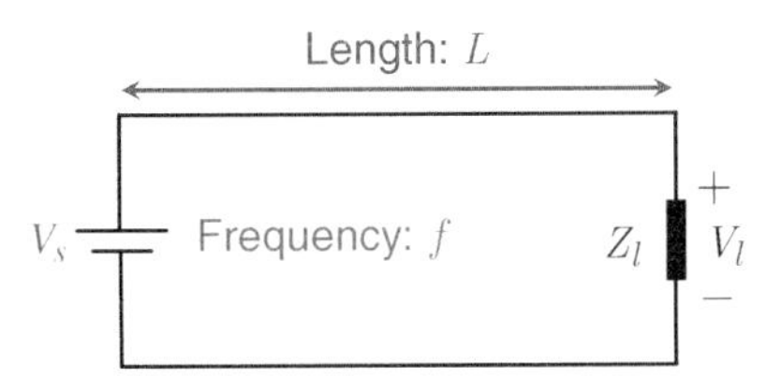

Figure 7.85 In a simple circuit consisting of a source and a load, the connection (cable) between the source and load may need to be modeled as a transmission line, depending on its length (L) with respect to the wavelength (λ). For example, assuming that the velocity of propagation is approximately 3×10^8 m/s (e.g. omitting dielectric effects of the insulator in a coaxial cable), an $L = 1$ m connection corresponds to only $1.67 \times 10^{-7}\lambda$ at 50 Hz, while it becomes λ at 300 MHz. Note that the cable must be 6000 km long to be equal to λ at 50 Hz. When L is comparable to or larger than the wavelength, $V_l \neq V_s$ in the circuit above, even when the connection is lossless.

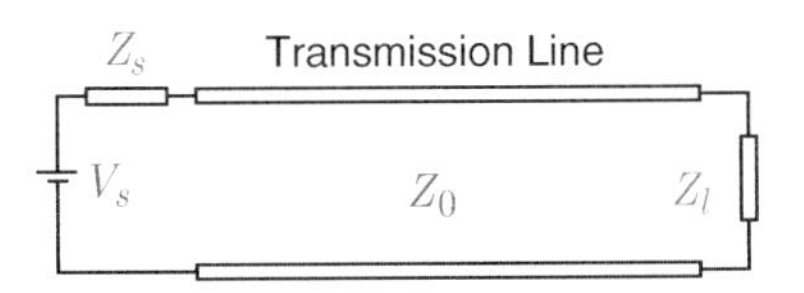

Figure 7.86 A general representation of a transmission line with a characteristic impedance of Z_0 that is connected to a voltage source V_s and a load Z_l. Note that the transmission line is shown with two parallel connections corresponding to positive and negative (ground) conductors in cables.

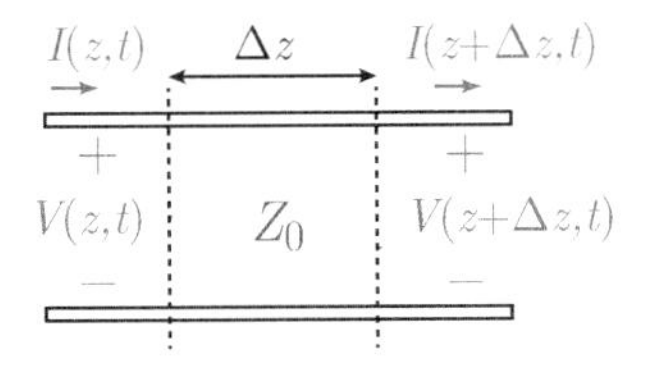

Figure 7.87 A differential portion of a transmission line.

critical in terms of electromagnetic transmission; but creating mathematical models for such reflections together with a *full-wave* model of a cable can be challenging, if not impossible. On the other extreme, a whole cable may be considered a simple connection without any electromagnetic properties, as practiced in circuit theory, but this is possible with acceptable errors only for cables that are sufficiently shorter than the wavelength (Figure 7.85). Transmission line theory comes into play exactly to solve this dilemma by providing a mathematical approach to accurately model electromagnetic transmission along a cable without its full-wave solution. Specifically, using a transmission-line model, the wave behavior of an electromagnetic transmission is still considered in the analysis, facilitated by lumped-element models of tiny segments (to make the analysis feasible). This section briefly considers the fundamentals of transmission lines as excellent mathematical representations of wired transmission environments (cables).

7.3.1 Telegrapher's Equations

In the following derivations, we consider transmission lines connected to sources and loads on different sides (Figure 7.86). A transmission line is mainly characterized by its characteristic impedance Z_0, which depends on the cable that is modeled, as well as the frequency. As a starting point, we examine a differential part of a transmission line (Figure 7.87) that can be modeled using lumped elements (Figure 7.88). The voltage and current along a transmission line can be written in terms of position and time as $V(z,t)$ and $I(z,t)$, respectively. In a Δz section, there is a series combination of an *inductance* $L_0\Delta z$ and a *resistance* $R_0\Delta z$ due to ohmic losses. In addition, between the positive and negative connections, there is a parallel combination of a *capacitance* $C_0\Delta z$ and a *conductance* $G_0\Delta z$ due to leakage (between the positive and negative connections). This lumped-element equivalent circuit can be solved to derive necessary equations – Telegrapher's equations – for the voltage and current.

It should be emphasized that capacitance, inductance, resistance, and conductance values in an equivalent lumped-element model depend on the structure and material properties of the cable, and these values can be found via measurements, simulations, or mathematical modeling. Derivations for static cases are often used to represent elements in lumped-element models. As an example, a coaxial cable's ideal model (Figure 7.89) can be used in an electrostatic scenario to derive the capacitance *per length* as (Section 6.7.4)

$$C_0 = \frac{2\pi\varepsilon}{\ln(b/a)} \quad \text{(F/m)}. \tag{7.365}$$

Similarly, solving the related magnetostatic problem leads to (Section 6.9)

$$L_0 = \frac{\mu}{2\pi}\ln(b/a) \quad \text{(H/m)} \tag{7.366}$$

as the inductance *per length*. If the material between the conductors is lossy, leading to a nonzero volume electric current density for $a < \rho < b$, we can obtain the corresponding resistance per length (R_0) that accounts for such dielectric losses. Similarly, if the conductors are not perfectly conducting and thus cannot be assumed to have zero thicknesses, the corresponding losses can be added to R_0, and the capacitance and inductance calculations may also need to be updated per the changed structure and distribution of charges/currents. Finally, a lossy dielectric material between the conductors may cause a non-negligible leakage current, which can be modeled via conductance per length G_0.

For the lumped-element model of a differential part of a transmission line, an application of Kirchhoff's voltage law leads to[88]

$$V(z+\Delta z, t) - V(z,t) = -R_0 \Delta z I(z,t) - L_0 \Delta z \frac{\partial I(z,t)}{\partial t}, \tag{7.367}$$

which reduces into the first Telegrapher's equation as

$$\frac{\partial V(z,t)}{\partial z} = -R_0 I(z,t) - L_0 \frac{\partial I(z,t)}{\partial t} \tag{7.368}$$

in the limit of $\Delta z \to 0$. For the second equation, we can consider Kirchhoff's current law, leading to[89]

$$I(z+\Delta z, t) - I(z,t) = -G_0 \Delta z V(z,t) - C_0 \Delta z \frac{\partial V(z,t)}{\partial t}. \tag{7.369}$$

Again, in the limit, we obtain the second Telegrapher's equation as

$$\frac{\partial I(z,t)}{\partial z} = -G_0 V(z,t) - C_0 \frac{\partial V(z,t)}{\partial t}. \tag{7.370}$$

Assuming time-harmonic sources, we can obtain expressions in the phasor domain as[90]

$$\frac{\partial V(z)}{\partial z} = -R_0 I(z) - j\omega L_0 I(z) = -(R_0 + j\omega L_0) I(z) \tag{7.371}$$

$$\frac{\partial I(z)}{\partial z} = -G_0 V(z) - j\omega C_0 V(z) = -(G_0 + j\omega C_0) V(z). \tag{7.372}$$

Then these equations can be combined to arrive at transmission-line (second-order Telegrapher's) equations as

$$\frac{\partial^2 V(z)}{\partial z^2} - \gamma^2 V(z) = 0, \qquad \frac{\partial^2 I(z)}{\partial z^2} - \gamma^2 I(z) = 0, \tag{7.373}$$

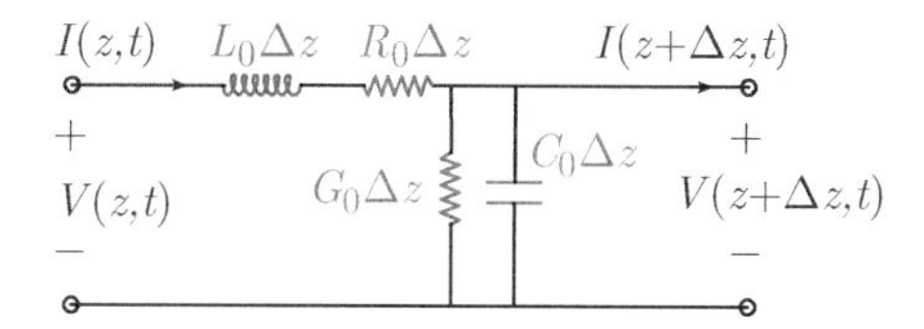

Figure 7.88 Representation of the differential portion of the transmission line in Figure 7.87 using lumped elements. Since we consider only a differential part of the transmission line, it can be modeled via lumped elements, even though the transmission line itself cannot be.

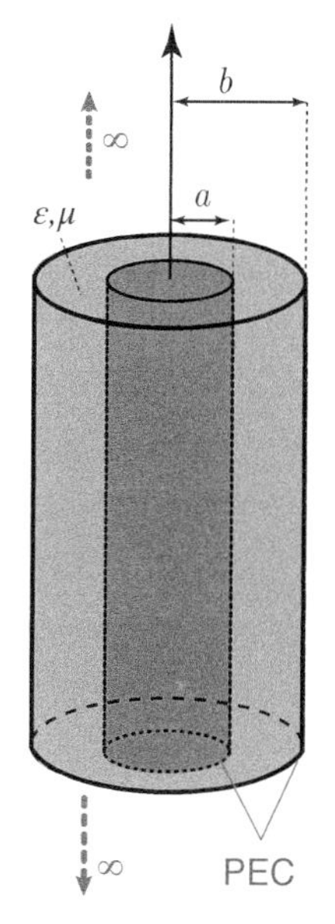

Figure 7.89 An ideal model of a coaxial cable with infinitely long, perfectly conducting cylindrical surfaces with radii $\rho = a$ and $\rho = b$, and a lossless material between them.

[88]Specifically, the voltage drop across the differential part corresponds to the voltage on the inductance and ohmic resistance.

[89]Specifically, the current flowing on the left-hand side of the differential part is divided in three: along the leakage conductance, along the capacitance, and flowing on the right-hand side.

[90]Hence, we assume

$$V(z,t) = \text{Re}\{V(z)\exp(j\omega t)\}$$

$$I(z,t) = \text{Re}\{I(z)\exp(j\omega t)\}.$$

where the propagation constant is defined as[91]

$$\gamma = \sqrt{(R_0 + j\omega L_0)(G_0 + j\omega C_0)}. \tag{7.374}$$

In general, this constant has real and imaginary parts, corresponding to the attenuation constant α and the phase constant β, respectively.[92]

Waves in transmission lines are quite similar to plane waves in unbounded regions. Based on our previous knowledge,[93] a general set of solutions to transmission-line equations can be written

$$V(z) = V_0^+ \exp(-\gamma z) + V_0^- \exp(\gamma z) \tag{7.375}$$

$$I(z) = I_0^+ \exp(-\gamma z) + I_0^- \exp(\gamma z). \tag{7.376}$$

In these expressions, both voltage and current consist of right-traveling (in the positive z direction) and left-traveling (in the negative z direction) waves that are characterized by $\exp(-\gamma z)$ and $\exp(\gamma z)$, respectively. Using the relationship between voltage and current, one can find the ratio of amplitudes. For example, considering the first Telegrapher's equation, we have

$$\frac{\partial V(z)}{\partial z} = -\gamma V_0^+ \exp(-\gamma z) + \gamma V_0^- \exp(\gamma z) = -(R_0 + j\omega L_0) I(z) \tag{7.377}$$

$$= -(R_0 + j\omega L_0)\left[I_0^+ \exp(-\gamma z) + I_0^- \exp(\gamma z)\right], \tag{7.378}$$

leading to

$$-\gamma V_0^+ = -(R_0 + j\omega L_0) I_0^+ \longrightarrow V_0^+ = \frac{(R_0 + j\omega L_0)}{\gamma} I_0^+ \tag{7.379}$$

$$\gamma V_0^- = -(R_0 + j\omega L_0) I_0^- \longrightarrow V_0^- = -\frac{(R_0 + j\omega L_0)}{\gamma} I_0^-. \tag{7.380}$$

Inserting the expression for the propagation constant in these expressions, we obtain the ratio between traveling voltages and currents as

$$V_0^+/I_0^+ = Z_0, \qquad V_0^-/I_0^- = -Z_0, \tag{7.381}$$

where

$$Z_0 = \sqrt{\frac{R_0 + j\omega L_0}{G_0 + j\omega C_0}} \tag{7.382}$$

[91] Then using static expressions for a lossless coaxial cable, we can obtain the propagation constant as

$$\gamma = \sqrt{j\omega L_0)(j\omega C_0)} = j\omega\sqrt{L_0 C_0}$$
$$= j\omega\sqrt{\frac{2\pi\varepsilon}{\ln(b/a)}\frac{\mu}{2\pi}\ln(b/a)}$$
$$= j\omega\sqrt{\mu\varepsilon},$$

which is same as the propagation constant in an unbounded region.

[92] This propagation constant is analogous to the one defined in Section 6.6.1 for electromagnetic waves in conducting media, and it is also similar to the propagation constant defined for waveguides.

[93] Specifically, see Section 4.4.1 for the introduction of wave equations, Section 4.4.3 for Helmholtz equations derived for time-harmonic sources in the phasor domain, Section 5.3 for plane waves as fundamental solutions, and finally Section 6.6.3 for general forms of plane waves in conducting (lossy) media.

Z_0: characteristic impedance
Unit: ohm (Ω) (volts/ampere)

is called the *characteristic impedance* of the transmission line. Note that considering $V_0^+/I_0^+ = Z_0$ and $V_0^-/I_0^- = -Z_0$, we have

$$V(z)/I(z) = Z(z) \neq Z_0 \tag{7.383}$$

except for special cases.[94]

For a lossless transmission line – i.e. when there are no ohmic loss and leakage – both R_0 and G_0 are zero, leading to[95,96]

$$\gamma = \sqrt{(j\omega L_0)(j\omega C_0)} = j\omega\sqrt{L_0 C_0} = j\beta = j\frac{2\pi}{\lambda} = j\frac{\omega}{u_p} \tag{7.384}$$

$$Z_0 = \sqrt{\frac{j\omega L_0}{j\omega C_0}} = \sqrt{\frac{L_0}{C_0}}, \tag{7.385}$$

where λ is the wavelength. Hence, the propagation constant is purely imaginary, indicating that there is no attenuation. In addition, the characteristic impedance is purely real, similar to the intrinsic impedance of a lossless medium (e.g. a vacuum).

7.3.1.1 Transmission Line With a Load

In practice, a transmission line is terminated by a load, which can be considered a *boundary condition* to derive exact expressions for the voltage and current along the line (Figure 7.90). In the context of transmission lines, two variables are defined: the impedance $Z(z) = V(z)/I(z)$ and the reflection coefficient $\Gamma(z)$, both depending on position (z). Now, we examine a transmission line terminated by a load with a known impedance[97] Z_l. For various practical reasons, we use an inverted coordinate variable z' that is zero at the position of the load and increasing in the direction of the transmission line (Figure 7.90). Then the voltage and current with respect to z' can be written

$$V(z') = V_0^+ \exp(\gamma z') + V_0^- \exp(-\gamma z') \tag{7.386}$$

$$I(z') = I_0^+ \exp(\gamma z') + I_0^- \exp(-\gamma z'), \tag{7.387}$$

where the amplitudes of the right-traveling and left-traveling waves are shown with positive and negative superscripts, respectively. The impedance at any position can be written

$$Z(z') = \frac{V(z')}{I(z')} = \frac{V_0^+ \exp(\gamma z') + V_0^- \exp(-\gamma z')}{I_0^+ \exp(\gamma z') + I_0^- \exp(-\gamma z')}. \tag{7.388}$$

[94] Note that Telegrapher's equations mainly provide basic relationships between voltages and currents, but they are based on Maxwell's equations, whose solutions are naturally waves. Using a transmission-line model (and hence, Telegrapher's equations) means a TEM wave propagates along the modeled cable, which is in fact a valid assumption in most cases. For example, a coaxial cable supports the TEM mode at all frequencies (no cutoff frequency). On the other hand, while the TEM mode is always the dominant (fundamental) mode in a coaxial cable, increasing the frequency leads to the appearance of other (undesired) modes (i.e. initially TE_{11}).

[95] An approximation of the attenuation constant for low-loss cases can be found as

$$\gamma = j\omega\sqrt{L_0C_0}\sqrt{\left(1+\frac{R_0}{j\omega L_0}\right)}\sqrt{\left(1+\frac{G_0}{j\omega C_0}\right)}$$
$$\approx j\omega\sqrt{L_0C_0}\left(1+\frac{R_0}{j2\omega L_0}\right)\left(1+\frac{G_0}{j2\omega C_0}\right)$$
$$\longrightarrow \alpha = \mathrm{Re}\{\gamma\} \approx \sqrt{L_0C_0}(R_0/L_0 + G_0/C_0)/2.$$

[96] For a lossless coaxial cable, we have

$$Z_0 = \sqrt{\frac{L_0}{C_0}} = \sqrt{\frac{\frac{\mu}{2\pi}\ln(b/a)}{\frac{2\pi\varepsilon}{\ln(b/a)}}} = \frac{1}{2\pi}\sqrt{\frac{\mu}{\varepsilon}}\ln(b/a).$$

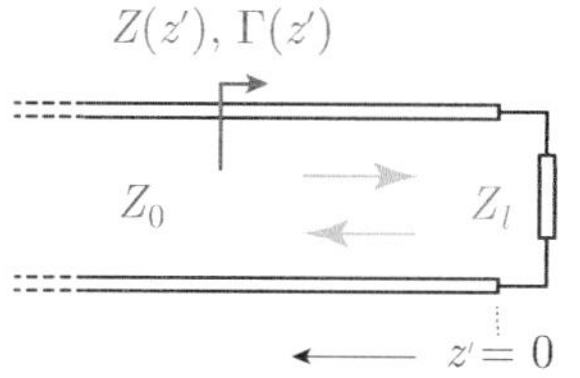

Figure 7.90 Termination of a transmission line by a load.

[97] Load impedance is generally a complex number since the load may be a combination of resistors, capacitors, and inductors. The real part of the impedance is formed by resistors, while the imaginary part is formed by capacitors/inductors. A capacitor with C_l capacitance has an impedance contribution of $Z_C = (j\omega C_l)^{-1}$, whereas an inductor with L_l inductance leads to $Z_L = j\omega L_l$.

Inserting $V_0^+/I_0^+ = Z_0$ and $V_0^-/I_0^- = -Z_0$, we arrive at[98]

$$Z(z') = Z_0 \frac{V_0^+ \exp(\gamma z') + V_0^- \exp(-\gamma z')}{V_0^+ \exp(\gamma z') - V_0^- \exp(-\gamma z')}. \tag{7.389}$$

The reflection coefficient at position z' is the ratio of the left-traveling wave to the right-traveling wave:[99]

$$\Gamma(z') = \frac{V_0^- \exp(-\gamma z')}{V_0^+ \exp(\gamma z')} = \frac{V_0^-}{V_0^+} \exp(-2\gamma z'). \tag{7.390}$$

Note the term $2z'$ in the exponential: it corresponds to the travel distance from z' to the load and then from the load to z' (back).

The equations above can further be manipulated by considering the values at $z' = 0$. First, we have[100]

$$Z(z' = 0) = Z_l = Z_0 \frac{V_0^+ + V_0^-}{V_0^+ - V_0^-}, \tag{7.391}$$

given that the impedance is Z_l at the load location of $z' = 0$. The reflection coefficient at $z' = 0$ can be defined as Γ_l, which can be written

$$\Gamma_l = \Gamma(z' = 0) = V_0^-/V_0^+, \tag{7.392}$$

which allows us to write V_0^- in terms of V_0^+ as $V_0^- = \Gamma_l V_0^+$. Obviously, using Γ_l, the reflection coefficient at an arbitrary position can be written[101]

$$\Gamma(z') = \Gamma_l \exp(-2\gamma z'). \tag{7.393}$$

Furthermore, Z_l and Γ_l can be written in terms of each other. Using $V_0^- = \Gamma_l V_0^+$, we have

$$Z_l = Z_0 \frac{V_0^+ + \Gamma_l V_0^+}{V_0^+ - \Gamma_l V_0^+} = Z_0 \frac{V_0^+(1 + \Gamma_l)}{V_0^+(1 - \Gamma_l)} = Z_0 \frac{1 + \Gamma_l}{1 - \Gamma_l}, \tag{7.394}$$

leading to $Z_l(1 - \Gamma_l) = Z_0(1 + \Gamma_l)$ and

$$\Gamma_l = \frac{Z_l - Z_0}{Z_l + Z_0}. \tag{7.395}$$

Now, again using $V_0^- = \Gamma_l V_0^+$, we can modify the expression for the impedance as

$$Z(z') = Z_0 \frac{V_0^+ \exp(\gamma z') + \Gamma_l V_0^+ \exp(-\gamma z')}{V_0^+ \exp(\gamma z') - \Gamma_l V_0^+ \exp(-\gamma z')} = Z_0 \frac{\exp(\gamma z') + \Gamma_l \exp(-\gamma z')}{\exp(\gamma z') - \Gamma_l \exp(-\gamma z')}. \tag{7.396}$$

[98] Obviously, the impedance at position z' – i.e. $Z(z')$ – is not generally equal to Z_0. To have $Z(z') = Z_0$ all along a transmission line, there should not be any left-traveling wave (reflection) so that $V_0^- = 0$, $I_0^- = 0$, and $Z(z') = V_0^+ \exp(\gamma z')/I_0^+ \exp(\gamma z') = V_0^+/I_0^+ = Z_0$.

[99] Therefore, even when $V_0^+ = V_0^-$, the reflection coefficient is

$$\Gamma(z') = \exp(-2\gamma z') = \exp(-2\alpha z') \exp(-j2\beta z'),$$

which corresponds to attenuation (due to losses) and phase accumulation in $2z'$ distance.

[100] At the load location, the impedance value (the ratio of voltage and current) must be Z_l. Here, we assume that the connection between the transmission line and the load is ideal: e.g. it is too short to be considered a transmission line itself.

[101] Once again

$$\exp(-2\gamma z') = \exp(-2\alpha z') \exp(-j2\beta z')$$

represents attenuation (due to losses) and phase accumulation in $2z'$ distance.

Inserting $\Gamma_l = (Z_l - Z_0)/(Z_l + Z_0)$, and after some mathematical manipulation[102], we arrive at

$$Z(z') = Z_0 \frac{Z_l + Z_0 \tanh(\gamma z')}{Z_0 + Z_l \tanh(\gamma z')}, \qquad \tanh(\gamma z') = \frac{\exp(\gamma z') - \exp(-\gamma z')}{\exp(\gamma z') + \exp(-\gamma z')}. \tag{7.397}$$

For a lossless transmission line with $\gamma = j\beta$, we have[103]

$$\tanh(\gamma z') = \tanh(j\beta z') = j\tan(\beta z') \longrightarrow Z(z') = Z_0 \frac{Z_l + jZ_0 \tan(\beta z')}{Z_0 + jZ_l \tan(\beta z')}. \tag{7.398}$$

7.3.1.2 Special Cases

When a transmission line is connected to a load, a *perfect* match may occur if $Z_l = Z_0$. This leads to

$$Z_l = Z_0 \longrightarrow Z(z') = Z_0 \frac{Z_0 + Z_0 \tanh(\gamma z')}{Z_0 + Z_0 \tanh(\gamma z')} = Z_0 \tag{7.399}$$

$$\Gamma_l = \frac{Z_l - Z_0}{Z_l + Z_0} = 0, \qquad \Gamma(z') = \Gamma_l \exp(-2\gamma z') = 0, \tag{7.400}$$

and

$$Z_l = Z_0 \longrightarrow V_0^- / V_0^+ = \Gamma_l = 0. \tag{7.401}$$

Therefore, there is no reflected wave, and the transmission line behaves like an *infinite* line.[104]

Now, we consider two special cases: when the load of a transmission line is a short circuit or open circuit (Figure 7.91). First, for the short-circuit case, we have

$$Z_l = 0 \longrightarrow \Gamma_l = \frac{Z_l - Z_0}{Z_l + Z_0} = \frac{-Z_0}{Z_0} = -1, \tag{7.402}$$

indicating that a wave reaching the load is completely reflected. The reflection coefficient at an arbitrary position can be written

$$Z_l = 0 \longrightarrow \Gamma(z') = -\exp(-2\gamma z'), \tag{7.403}$$

which becomes

$$Z_l = 0 \longrightarrow \Gamma(z') = -\exp(-j2\beta z') \tag{7.404}$$

[102] We have

$$Z(z') = Z_0 \frac{\exp(\gamma z') + \frac{(Z_l - Z_0)}{(Z_l + Z_0)}\exp(-\gamma z')}{\exp(\gamma z') - \frac{(Z_l - Z_0)}{(Z_l + Z_0)}\exp(-\gamma z')}$$

$$= Z_0 \frac{(Z_l + Z_0)\exp(\gamma z') + (Z_l - Z_0)\exp(-\gamma z')}{(Z_l + Z_0)\exp(\gamma z') - (Z_l - Z_0)\exp(-\gamma z')}$$

$$= Z_0 \frac{Z_l\left(e^{\gamma z'} + e^{-\gamma z'}\right) + Z_0\left(e^{\gamma z'} - e^{-\gamma z'}\right)}{Z_l\left(e^{\gamma z'} - e^{-\gamma z'}\right) + Z_0\left(e^{\gamma z'} + e^{-\gamma z'}\right)}.$$

[103] Hence, the impedance seen at a position is neither the characteristic impedance nor the load impedance in general.

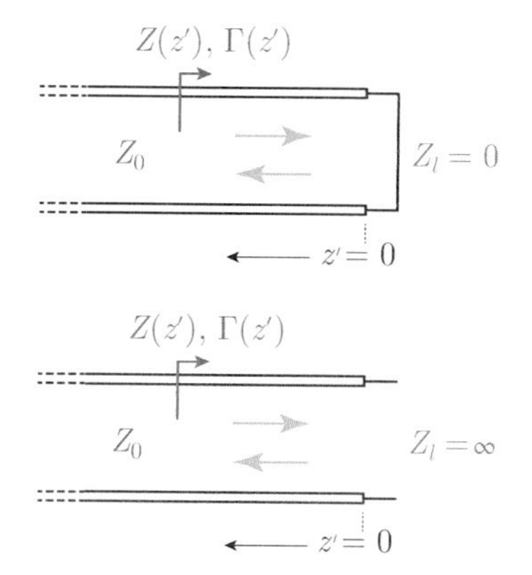

Figure 7.91 Termination of a transmission line by short and open circuits.

[104] This means

$$Z_l = Z_0 \longrightarrow V(z') = V_0^+ \exp(\gamma z'),$$

which becomes $V(z') = V_0^+ \exp(j\beta z')$ if the line is lossless. In the time domain (and for the lossless case), we have

$$Z_l = Z_0 \longrightarrow V(z', t) = V_0^+ \cos(\omega t - \beta z')$$

assuming real V_0^+. Therefore, in the time domain, the voltage and current still have variations with respect to position/time, including zeros at time-dependent locations.

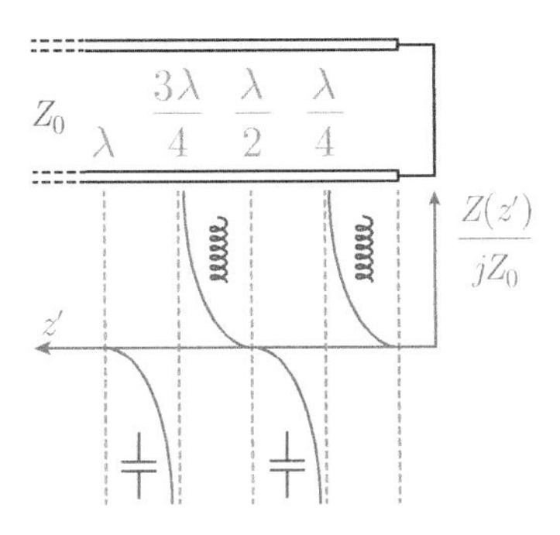

Figure 7.92 Impedance with respect to position along a transmission line terminated by a short circuit. At periodic positions, the impedance is inductive, capacitive, short circuit, or open circuit.

for a lossless case. In addition, we obtain the impedance at any position z' as

$$Z_l = 0 \longrightarrow Z(z') = Z_0 \frac{Z_0 \tanh(\gamma z')}{Z_0} = Z_0 \tanh(\gamma z'). \tag{7.405}$$

Then for a lossless case, we further have

$$Z_l = 0 \longrightarrow Z(z') = jZ_0 \tan(\beta z'), \tag{7.406}$$

which is purely *imaginary*. Considering this expression, we have the following observations (Figure 7.92):

- The impedance is *zero* (short circuit) at $z' = 0, \lambda/2, \lambda, \ldots$, or more generally at $z' = n\lambda/2$ for $n = 0, 1, 2, \ldots$, where $\lambda = 2\pi/\beta$ is the wavelength.
- The impedance is *infinite* (open circuit) at $z' = \lambda/4, 3\lambda/4, \ldots$, or more generally at $z' = \lambda/4 + n\lambda/2$ for $n = 0, 1, 2, \ldots$.
- The impedance is *inductive*[105] (positive imaginary) in the intervals

$$z' \in [0, \lambda/4], [\lambda/2, 3\lambda/4], \ldots. \tag{7.407}$$

- The impedance is *capacitive*[106] (negative imaginary) in the intervals

$$z' \in [\lambda/4, \lambda/2], [3\lambda/4, \lambda], \ldots. \tag{7.408}$$

Interestingly, while the transmission line is terminated by a short circuit, impedance values measured at different positions can be inductive, capacitive, and even infinity (open circuit).

Next, we consider the case when a transmission line is terminated by an open circuit. The reflection coefficient at the load location can be found as[107]

$$Z_l = \infty \longrightarrow \Gamma_l = \frac{Z_l - Z_0}{Z_l + Z_0} \rightarrow \frac{Z_l}{Z_l} = 1, \tag{7.409}$$

indicating again that a wave reaching the load is completely reflected. At an arbitrary position, we have

$$Z_l = \infty \longrightarrow \Gamma(z') = \exp(-2\gamma z'), \tag{7.410}$$

which becomes

$$Z_l = \infty \longrightarrow \Gamma(z') = \exp(-j2\beta z') \tag{7.411}$$

[105] Inductive impedance means $Z(z') = |Z(z')| \exp(j\pi/2)$. Then the relationship between the voltage and current at z' can be written

$$V(z') = Z(z')I(z') = |Z(z')|I(z')\exp(j\pi/2).$$

In the time domain, this corresponds to

$$V(z', t) = |Z(z')|I(z', t + \pi/2),$$

indicating that the current *lags* the voltage by 90°.

[106] Capacitive impedance means $Z(z') = |Z(z')| \exp(-j\pi/2)$, leading to

$$V(z', t) = |Z(z')|I(z', t - \pi/2).$$

Hence, the voltage *lags* the current by 90°.

[107] Note that the reflection coefficient at the load is -1 for a short-circuit termination and $+1$ for an open-circuit termination. For the magnitude $|\Gamma_l| = 1$, both cases correspond to full reflection from the load. This is an expected result since a wave that arrives at the load location has nowhere to go (the transmission line ends), but there is also no resistance to dissipate the incoming electromagnetic power and reduce the amplitude of the wave. On the other hand, the negative sign of the reflection coefficient for a short-circuit termination indicates that the voltage/current wave changes its *sign* when reflecting back. Constructing an analogy with plane waves, short and open circuits correspond to a planar PEC and a planar perfect magnetic conductor (under normal incidence).

for a lossless transmission line. The impedance at any position z' can be found as

$$Z_l = \infty \longrightarrow Z(z') \to Z_0 \frac{Z_l}{Z_l \tanh(\gamma z')} = \frac{Z_0}{\tanh(\gamma z')}. \tag{7.412}$$

Then for a lossless case, we obtain the impedance with respect to position as

$$Z_l = \infty \longrightarrow Z(z') = Z_0/[j \tan(\beta z')] = -jZ_0 \cot(\beta z'), \tag{7.413}$$

which is again purely imaginary (but with a different expression in comparison to a short-circuit termination). Note the following (Figure 7.93):

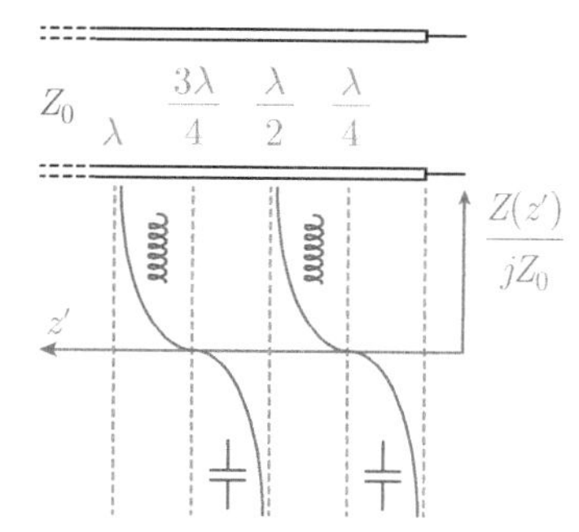

Figure 7.93 Impedance with respect to position along a transmission line terminated by an open circuit. At periodic positions, the impedance is capacitive, inductive, open circuit, or short circuit.

- The impedance is *infinite* (open circuit) at $z' = 0, \lambda/2, \lambda, \ldots$, or more generally at $z' = n\lambda/2$ for $n = 0, 1, 2, \ldots$, where $\lambda = 2\pi/\beta$ is the wavelength.
- The impedance is *zero* (short circuit) at $z' = \lambda/4, 3\lambda/4, \ldots$, or more generally at $z' = \lambda/4 + n\lambda/2$ for $n = 0, 1, 2, \ldots$
- The impedance is *capacitive* (negative imaginary) in the intervals

$$z' \in [0, \lambda/4], [\lambda/2, 3\lambda/4], \ldots. \tag{7.414}$$

- The impedance is *inductive* (positive imaginary) in the intervals

$$z' \in [\lambda/4, \lambda/2], [3\lambda/4, \lambda], \ldots. \tag{7.415}$$

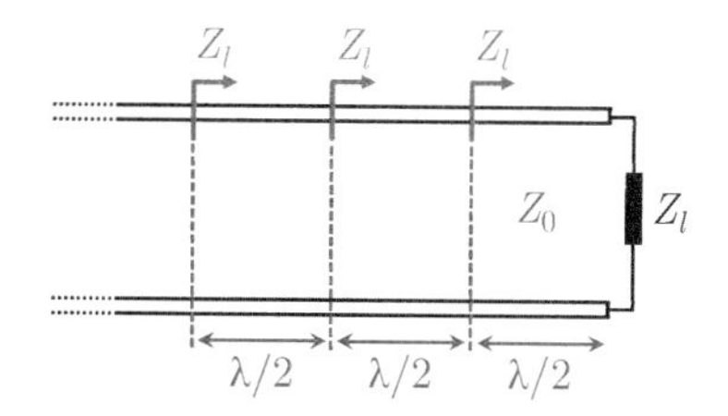

Figure 7.94 The impedance seen along a transmission line is periodic with $\lambda/2$. This means the impedance of the load is exactly seen at $z' = \{\lambda/2, \lambda, 3\lambda/2, \ldots\}$. This is also true when the load is short/open circuit, as observed in Figures 7.92 and 7.93.

Considering the expressions derived above, note the symmetry related to the behavior of impedance values when a transmission line is terminated by short and open circuits. This symmetry can be used to find the unknown characteristic impedance of a transmission line via measurements of short-circuit and open-circuit impedance values (Z_{sc} and Z_{oc}). Specifically, at an arbitrary position (for any arbitrary length of a transmission line), we have

$$Z_{sc}Z_{oc} = jZ_0 \tan(\beta z')[-jZ_0 \cot(\beta z')] = Z_0^2 \tan(\beta z') \cot(\beta z') = Z_0^2, \tag{7.416}$$

leading to

$$Z_0 = \sqrt{Z_{sc}Z_{oc}}. \tag{7.417}$$

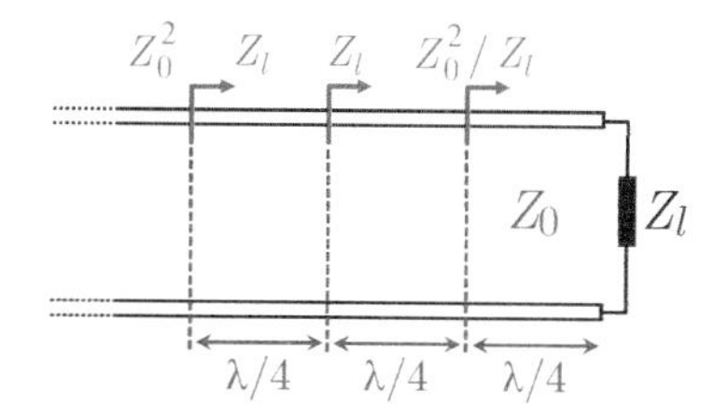

Figure 7.95 The impedance seen along a transmission line is *inverted* at every $\lambda/4$ distance. This means if the impedance of the load is Z_l, the impedance seen at $z' = \lambda/4$ is $Z(\lambda/4) = Z_0^2/Z_l$, while it becomes $Z(\lambda/2) = Z_0^2/Z(\lambda/4) = Z_l$ at $z' = \lambda/2$ due to double inversion. This is consistent with the fact that the impedance is periodic with $\lambda/2$ periodicity, as demonstrated in Figure 7.94.

7.3.1.3 Common Cases

As shown above, when the load is a short or an open circuit, the impedance is periodic with $\lambda/2$ periodicity along the transmission line. This is a common behavior, independent of the load.

Consider a lossless transmission line terminated by a load Z_l. If the length of the line is $L = \lambda/2$, we have

$$Z(L) = Z_0 \frac{Z_l + jZ_0 \tan(2\pi L/\lambda)}{Z_0 + jZ_l \tan(2\pi L/\lambda)} = Z_0 \frac{Z_l + jZ_0 \tan(\pi)}{Z_0 + jZ_l \tan(\pi)} = Z_l. \tag{7.418}$$

Therefore, the impedance is exactly Z_l at the other end of the transmission line: i.e. Z_l is *repeated* at $L = \lambda/2$. This can be generalized to any portion of a longer transmission line; the impedance repeats itself at each $\lambda/2$ distance (Figure 7.94).

Another interesting relationship can be found for $L = \lambda/4$. In this case, we have

$$Z(L) = Z_0 \frac{Z_l + jZ_0 \tan(\pi/2)}{Z_0 + jZ_l \tan(\pi/2)} \rightarrow \frac{Z_0^2}{Z_l}. \tag{7.419}$$

Therefore, the value of the impedance is *inverted* at a distance $\lambda/4$ (Figure 7.95). Obviously, if $Z_l = 0$ and $Z_l = \infty$, we have $Z(\lambda/4) = \infty$ and $Z(\lambda/4) = 0$, respectively.[108]

[108] In other words, a short circuit becomes an open circuit, and vice versa, across $\lambda/4$ distance.

Special transformations of impedances by $\lambda/4$ (inverter) and $\lambda/2$ (repeater) lines can be used to find impedance transformations by longer transmission lines. For example, a transmission line of length $L = 5\lambda/4$ (having a characteristic impedance of Z_0) can be seen as a combination of two repeaters and an inverter ($\lambda/2 + \lambda/2 + \lambda/4$):

$$Z_l \xrightarrow{R} Z_l \xrightarrow{R} Z_l \xrightarrow{I} Z_0^2/Z_l, \tag{7.420}$$

where Z_l is the load connected to the line (Figure 7.96). Alternatively, it can be seen as the concatenation of three inverters and a repeater ($\lambda/4 + \lambda/4 + \lambda/4 + \lambda/2$):

$$Z_l \xrightarrow{I} Z_0^2/Z_l \xrightarrow{I} Z_l \xrightarrow{I} Z_0^2/Z_l \xrightarrow{R} Z_0^2/Z_l. \tag{7.421}$$

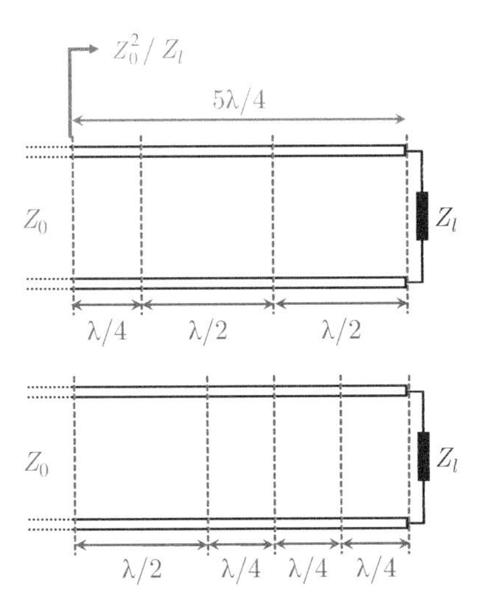

Figure 7.96 A transmission line (or a portion of a transmission line) can be seen as a combination of inverters ($\lambda/4$ segments) and/or repeaters ($\lambda/2$ segments) to find its impedance transformation. In the above, the impedance seen from the left-hand side of the transmission line is Z_0^2/Z_l.

Transmission lines of length $\lambda/4$ can be used to match a load to another transmission line.[109] Consider a load Z_l to be matched *perfectly* to a transmission line with a characteristic impedance of Z_0. A quarter-wavelength transmission line with a characteristic impedance of $Z_0' = \sqrt{Z_0 Z_l}$ can be used so that the impedance of the load is transformed as

[109] This is similar to eliminating reflection between two different dielectric media using a matching layer between them, as discussed in Section 5.3.11.

$$Z_l \xrightarrow{I} \frac{(Z_0')^2}{Z_l} = \frac{Z_0 Z_l}{Z_l} = Z_0, \tag{7.422}$$

and the load becomes perfectly matched.

7.3.2 Voltage and Current Patterns

The voltage and current along transmission lines may have different variations with respect to position, depending on the transmission line properties and connected load. In general, the voltage and current with respect to position can be written

$$V(z') = V_0^+ \exp(\gamma z') + \Gamma_l V_0^+ \exp(-\gamma z') = V_0^+ [\exp(\gamma z') + \Gamma_l \exp(-\gamma z')] \tag{7.423}$$

and

$$I(z') = I_0^+ \exp(\gamma z') - \Gamma_l I_0^+ \exp(-\gamma z') = I_0^+ [\exp(\gamma z') - \Gamma_l \exp(-\gamma z')] \tag{7.424}$$

$$= \frac{V_0^+}{Z_0} [\exp(\gamma z') - \Gamma_l \exp(-\gamma z')]. \tag{7.425}$$

If the transmission line is perfectly matched, we have $\Gamma_l = 0$, leading to

$$V(z') = V_0^+ \exp(\gamma z'), \qquad I(z') = (V_0^+/Z_0) \exp(\gamma z'), \tag{7.426}$$

which becomes

$$V(z') = V_0^+ \exp(j\beta z'), \qquad I(z') = (V_0^+/Z_0) \exp(j\beta z') \tag{7.427}$$

if the transmission line is lossless. Assuming real V_0^+ and Z_0, we can further obtain time-domain expressions as

$$V(z',t) = V_0^+ \cos(\omega t + \beta z'), \qquad I(z',t) = (V_0^+/Z_0) \cos(\omega t + \beta z'), \tag{7.428}$$

indicating voltage and current waves traveling in the $-z'$ direction.

Now, we consider a transmission line terminated via a short circuit (Figure 7.97). We have $\Gamma_l = -1$, and

$$V(z') = V_0^+ [\exp(j\beta z') - \exp(-j\beta z')] = j2V_0^+ \sin(\beta z') \tag{7.429}$$

$$I(z') = \frac{V_0^+}{Z_0} [\exp(j\beta z') + \exp(-j\beta z')] = 2\frac{V_0^+}{Z_0} \cos(\beta z'), \tag{7.430}$$

if the transmission line is lossless.[110] These expressions clearly represent standing waves with $V(z' = 0) = 0$ and $I(z' = 0) = 2V_0^+/Z_0$ at the termination of the transmission line (at $z' = 0$). Note that in the time domain, the voltage and current can be written (Figure 7.97)

$$V(z',t) = -2V_0^+ \sin(\beta z') \sin(\omega t), \qquad I(z',t) = 2\frac{V_0^+}{Z_0} \cos(\beta z') \cos(\omega t), \tag{7.431}$$

assuming real V_0^+ and Z_0.

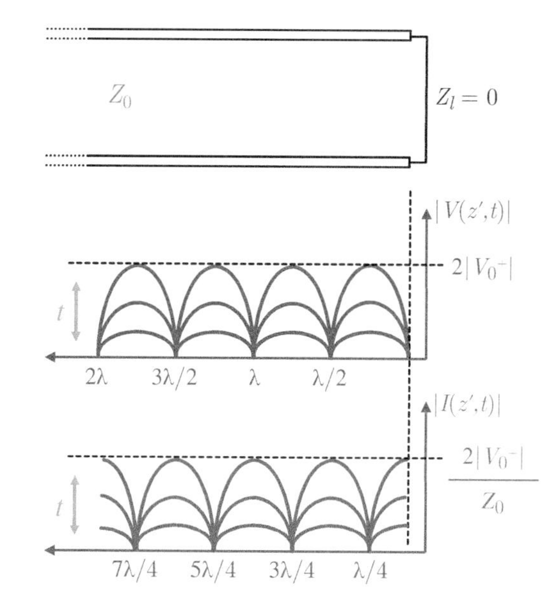

Figure 7.97 Magnitudes of time-domain voltage and current along a transmission line terminated by a short circuit. Both the voltage and current demonstrate standing-wave behaviors without any propagation due to the total reflection of the incoming wave. In addition to $z' = 0$, the voltage is always zero at $z' = \lambda/2, \lambda, 3\lambda/2, \ldots$, which can be interpreted as a short circuit. On the other hand, the current is fixed to zero at open-circuit locations: i.e. at $z' = \lambda/4, 3\lambda/4, 5\lambda/4, \ldots$.

[110] For a general case, we have

$$V(z') = V_0^+ [\exp(\gamma z') - \exp(-\gamma z')]$$

$$I(z') = (V_0^+/Z_0) [\exp(\gamma z') + \exp(-\gamma z')],$$

which include attenuation if $\alpha \neq 0$.

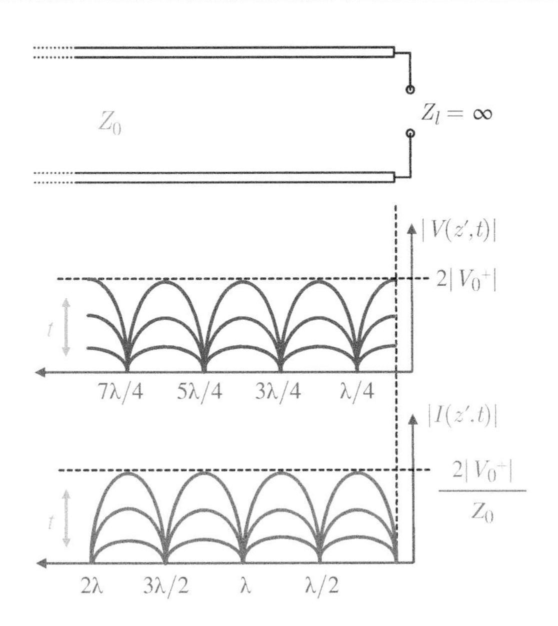

Figure 7.98 Magnitudes of the time-domain voltage and current along a transmission line terminated by an open circuit. Both the voltage and current demonstrate standing-wave behaviors without any propagation due to the total reflection of the incoming wave. In addition to $z' = 0$, the current is always zero at $z' = \lambda/2, \lambda, 3\lambda/2, \ldots$, which can be interpreted as an open circuit. On the other hand, the voltage is fixed to zero at short-circuit locations: i.e. at $z' = \lambda/4, 3\lambda/4, 5\lambda/4, \ldots$.

Next, we consider an open-circuit termination (Figure 7.98). In this case, we have $\Gamma_l = 1$ and

$$V(z') = V_0^+ [\exp(j\beta z') + \exp(-j\beta z')] = 2V_0^+ \cos(\beta z') \tag{7.432}$$

$$I(z') = \frac{V_0^+}{Z_0} [\exp(j\beta z') - \exp(-j\beta z')] = j2\frac{V_0^+}{Z_0} \sin(\beta z') \tag{7.433}$$

for a lossless transmission line.[111] Hence, we again obtain standing waves, leading to (Figure 7.98)

$$V(z', t) = 2V_0^+ \cos(\beta z') \cos(\omega t), \qquad I(z', t) = -\frac{2V_0^+}{Z_0} \sin(\beta z') \sin(\omega t) \tag{7.434}$$

in the time domain, assuming real V_0^+ and Z_0.

In a general case, where the transmission line is terminated by a load Z_l (Figure 7.99), the reflection coefficient is possibly complex as

$$\Gamma_l = |\Gamma_l| \exp(j\varphi_l). \tag{7.435}$$

Then assuming a lossless transmission line, the voltage and current with respect to position can be written

$$V(z') = V_0^+ \exp(j\beta z') [1 + |\Gamma_l| \exp(j\varphi_l - j2\beta z')] \tag{7.436}$$

$$I(z') = \frac{V_0^+}{Z_0} \exp(j\beta z') [1 - |\Gamma_l| \exp(j\varphi_l - j2\beta z')]. \tag{7.437}$$

These expressions describe *mixtures* of standing and traveling waves, but the actual characteristics depend on the value of Γ_l. For $\Gamma_l = 1$ (open circuit), we have purely standing waves as

$$V(z') = V_0^+ \exp(j\beta z') [1 + \exp(-j2\beta z')] \tag{7.438}$$

$$I(z') = \frac{V_0^+}{Z_0} \exp(j\beta z') [1 - \exp(-j2\beta z')], \tag{7.439}$$

verifying the derivation above. Similarly, for $\Gamma_l = -1$ (short circuit), leading to $|\Gamma_l| = 1$ and $\varphi_l = \pi$, we again obtain standing waves as

$$V(z') = V_0^+ \exp(j\beta z') [1 - \exp(-j2\beta z')] \tag{7.440}$$

$$I(z') = \frac{V_0^+}{Z_0} \exp(j\beta z') [1 + \exp(-j2\beta z')]. \tag{7.441}$$

[111] For a general case, we have

$$V(z') = V_0^+ [\exp(\gamma z') + \exp(-\gamma z')]$$

$$I(z') = (V_0^+/Z_0) [\exp(\gamma z') - \exp(-\gamma z')].$$

Obviously, for $\Gamma_l = 0$ (perfect match), the general expressions reduce into those for purely traveling waves.

Considering a lossless transmission line, the magnitudes of the phasor-domain voltage and current for an arbitrary Z_l can be written

$$|V(z')| = |V_0^+|\,|1 + |\Gamma_l| \exp(j\varphi_l - j2\beta z')| \tag{7.442}$$

$$|I(z')| = \frac{|V_0^+|}{Z_0}\,|1 - |\Gamma_l| \exp(j\varphi_l - j2\beta z')|\,. \tag{7.443}$$

For the *voltage pattern* (absolute value of the voltage with respect to position), we can list the following properties:[112]

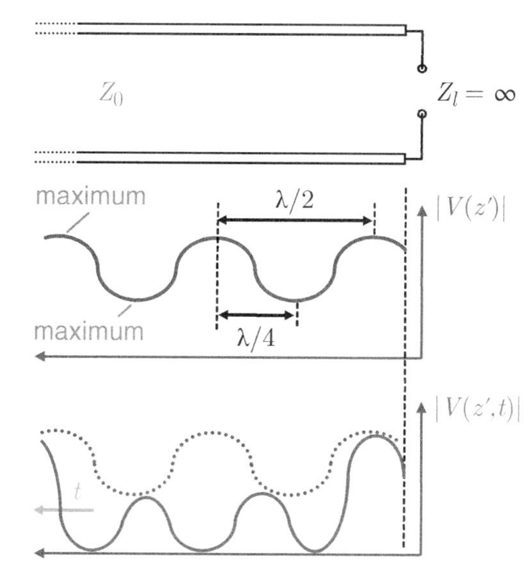

Figure 7.99 Magnitude of the voltage in the phasor domain and time domain along a transmission line terminated by an arbitrary load. In the phasor domain, the voltage has minima and maxima with $\lambda/2$ (maximum-maximum or minimum-minimum) periodicity. The time-domain voltage is a mixture of traveling and standing waves, which may appear as a movement of a wave within an envelope.

- A maximum occurs when $\varphi = 2\beta z'$ or $z' = \varphi/(2\beta)$. Considering the periodicity, locations for maxima can be found as

$$z'_{\max} = \frac{\varphi_l}{2\beta} + \frac{2n\pi}{2\beta} = \frac{\lambda\varphi_l}{4\pi} + \frac{n\lambda}{2} \tag{7.444}$$

  for[113] $n = 0, \pm 1, \pm 2, \ldots$. Note that for $\varphi_l = 0$ and $\varphi_l = \pi$, we have $z' = 0$ and $z' = \lambda/4$ as the locations of the first maximum.

- A minimum occurs when $\varphi = \pi + 2\beta z'$ or $z' = \varphi/(2\beta) - \pi/(2\beta)$. Considering the periodicity, locations for minima can be found as

$$z'_{\min} = \frac{\varphi_l}{2\beta} - \frac{\pi}{2\beta} + \frac{2n\pi}{2\beta} = \frac{\lambda\varphi_l}{4\pi} - \frac{\lambda}{4} + \frac{n\lambda}{2} \tag{7.445}$$

  for $n = 0, \pm 1, \pm 2, \ldots$. One can check this expression for $\varphi_l = 0$ (open circuit) and $\varphi_l = \pi$ (short circuit), leading to $z' = \lambda/4$ and $z' = 0$, respectively, as the locations of the *first* minima.

- The maximum value of the voltage can be written $|V(z')|_{\max} = V_0^+(1 + |\Gamma_l|)$ at the locations specified above. Similarly, the minimum value of the voltage at the listed locations can be written $|V(z')|_{\min} = V_0^+(1 - |\Gamma_l|)$. The ratio of these values is an important quantity called the *voltage standing-wave ratio* (VSWR) as

$$\text{VSWR} = \frac{|V(z')|_{\max}}{|V(z')|_{\min}} = \frac{1 + |\Gamma_l|}{1 - |\Gamma_l|}. \tag{7.446}$$

[112] These properties are described for the phasor-domain voltage.

[113] Obviously, selection of n should make z' positive.

In practice, we want a *small* value of the VSWR, as close as possible to unity. This limit value is possible if $\Gamma_l = 0$: i.e. when the load is perfectly matched to the load. For short-circuit and open-circuit loads, we naturally have VSWR$=\infty$, describing a completely standing wave without any propagation. Interestingly, when the load is imaginary (purely capacitive or inductive), we also have VSWR$=\infty$, assuming that Z_0 is real.[114] In general, the VSWR is between unity and infinity.

[114]This actually means no power can be (permanently) transferred to the load.

We emphasize that all these discussions are valid in the phasor domain.[115] For example, in the time domain, a maximum location for the voltage (derived above) does not mean the voltage at this location is always larger than voltage values at other locations (Figure 7.99).

[115]As an example, the time-domain voltage for a lossless transmission line can be written

$$\begin{aligned} V(z',t) &= \mathrm{Re}\{V_0^+ \exp(j\beta z')\exp(j\omega t)\} \\ &= \mathrm{Re}\{V_0^+|\Gamma_l|\exp(j\varphi_l - j\beta z')\exp(j\omega t)\} \\ &= V_0^+\cos(\omega t + \beta z') \\ &+ V_0^+|\Gamma_l|\cos(\omega t - \beta z' + \varphi_l) \end{aligned}$$

assuming real V_0^+. If $|\Gamma_l| = 1$ – i.e. for a short circuit, an open circuit, or a load with imaginary impedance – this expression turns into

$$\begin{aligned} V(z',t) &= V_0^+[\cos(\omega t + \beta z') + \cos(\omega t - \beta z' + \varphi_l)] \\ &= 2V_0^+\cos\left(\omega t + \frac{\varphi_l}{2}\right)\cos\left(\beta z' - \frac{\varphi_l}{2}\right), \end{aligned}$$

which represents a standing wave.

7.3.3 Examples

Example 97: Consider a lossless transmission line with a characteristic impedance of Z_0 connected to an arbitrary load Z_l (Figure 7.100). There is an incident voltage $V^+(z') = V_0^+\exp(j\beta z')$. Find expressions for the time-average incident and reflected powers along the transmission line, as well the power delivered to the load.

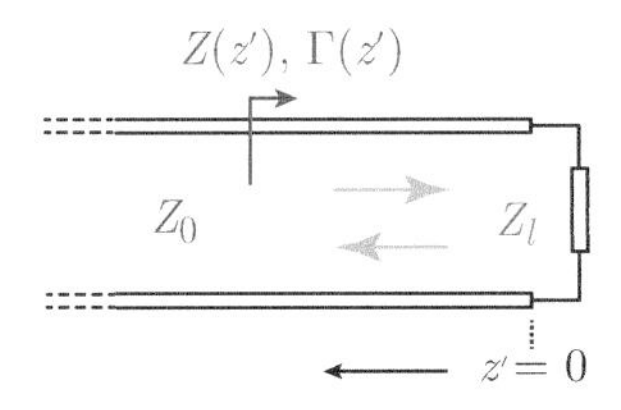

Figure 7.100 Termination of a transmission line by a load.

Solution: The voltage and current for the incident wave can be written

$$V^+(z') = V_0^+\exp(j\beta z'), \qquad I^+(z') = (V_0^+/Z_0)\exp(j\beta z'). \tag{7.447}$$

Hence, the complex power can be obtained as[116]

$$S_c^+(z') = \frac{1}{2}V^+(z')[I^+(z')]^* = \frac{V_0^+(V_0^+)^*}{2(Z_0)^*} = \frac{|V_0^+|^2}{2Z_0}, \tag{7.448}$$

[116]Note that

$$Z_0^* = Z_0 = \sqrt{\frac{L_0}{C_0}}$$

for a lossless transmission line.

which is independent of z'. The real part of the complex power corresponds to the time-average power:

$$S_{\text{avg}}^+(z') = \mathrm{Re}\{S_c^+(z')\} = \frac{|V_0^+|^2}{2Z_0}. \tag{7.449}$$

Using the same procedure for the reflected wave, one can obtain

$$S_{\text{avg}}^-(z') = \frac{|V_0^-|^2}{2Z_0} = |\Gamma_l|^2\frac{|V_0^+|^2}{2Z_0}. \tag{7.450}$$

Then the time-average power delivered to the load can be written[117]

$$S_{\text{avg},l} = S^{+}_{\text{avg}}(z'=0) - S^{-}_{\text{avg}}(z'=0) = \left(1 - |\Gamma_l|^2\right)\frac{|V_0^+|^2}{2Z_0}. \tag{7.451}$$

[117]Note that the time-average power delivered to the load is zero if $\Gamma_l = 1$. This corresponds to a short circuit, an open circuit, or an imaginary load impedance (purely capacitive or inductive). On the other hand, the delivered power is maximized and the reflected power becomes zero for a matched load (when $|\Gamma_l| = 0$). As an example, for a transmission line with 50 Ω characteristic impedance, a resistor with 50 Ω resistance receives all the incoming power without reflection. This power is consumed by the resistor. If the load is inductive as $(50 + j50)$ Ω, the reflection coefficient becomes $|\Gamma_l|^2 = 0.2$ and the time-average power delivered to the load becomes 80% of that for a matched load.

Example 98: An ideal sinusoidal voltage source $V_s(t) = 2\cos(\omega t)$ is connected to an inductor of $L_0 = 1$ nH through a lossless transmission line of length $d = \lambda/4$ at $(25/\pi)$ GHz (Figure 7.101). The characteristic impedance of the line is given as $Z_0 = 50$ Ω. Assume that the phase velocity is 3×10^8 m/s along the line.

- Find the reflection coefficient Γ_l at the load.
- Find the VSWR in the line.
- Find an expression for the voltage across the load (V_l) and the current through the load (I_l) with respect to time.
- Find an expression for the current through the source I_s with respect to time.

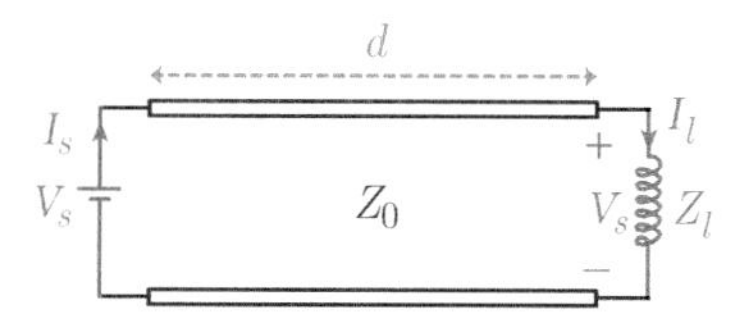

Figure 7.101 A lossless transmission line connected to an inductor.

Solution: First, we can find the impedance of the load as

$$Z_l = j\omega L_0 = j2\pi \times (25/\pi) \times 10^9 \times 1 \times 10^{-9} = j50\ \Omega. \tag{7.452}$$

Then the reflection coefficient at the load can be obtained as[118]

$$\Gamma_l = \frac{Z_l - Z_0}{Z_l + Z_0} = \frac{j50 - 50}{j50 + 50} = \frac{j-1}{j+1} = j. \tag{7.453}$$

[118]In general, $|\Gamma_l| = 1$ if the load is imaginary (lossless) and the characteristic impedance of the transmission line is real (lossless). Note that the *real* part of an impedance corresponds to a loss (resistance) for a lumped element, while it is the *imaginary* part of the characteristic impedance that represents loss in a transmission line.

Using Γ_l, we further obtain the VSWR as

$$\text{VSWR} = \frac{1 + |\Gamma_l|}{1 - |\Gamma_l|} = \frac{1+1}{1-1} \to \infty, \tag{7.454}$$

as expected (since we have a standing wave). Considering the general expressions for the voltage and current along a lossless transmission line, and inserting $\Gamma_l = |\Gamma_l|\exp(j\varphi_l) = 1\exp(j\pi/2)$, we have[119]

[119]Note that

$$\beta = \frac{2\pi}{\lambda} = \frac{\omega}{u_p} = \frac{25/\pi \times 10^9}{3 \times 10^8} = \frac{250}{3\pi}\ \text{rad/m}.$$

$$V(z') = V_0^+ \exp(j\beta z')\left[1 + \exp(j\pi/2 - j2\beta z')\right] \tag{7.455}$$

$$I(z') = \frac{V_0^+}{Z_0}\exp(j\beta z')\left[1 - \exp(j\pi/2 - j2\beta z')\right]. \tag{7.456}$$

Hence, at the load location ($z' = 0$), one can obtain

$$V_l = V_0^+(1+j), \qquad I_l = \frac{V_0^+}{50}(1-j). \tag{7.457}$$

However, V_0^+ is unknown without considering the source side. Using the same expression for the voltage, we have[120]

$$V(z' = \lambda/4) = V_0^+ \exp(j\pi/2)\left[1 + \exp(j\pi/2 - j\pi)\right] = V_0^+(1+j) \tag{7.458}$$

$$= V_s = 2 \text{ V}, \tag{7.459}$$

which indicates that $V_0^+ = 2/(1+j)$ V. Consequently, we can update the expressions for the load voltage/current as[121]

$$V_l = 2 \text{ V}, \qquad I_l = -j/25 \text{ A}, \tag{7.460}$$

leading to[122]

$$V_l(t) = 2\cos(\omega t) \text{ V}, \qquad I_l(t) = \frac{1}{25}\sin(\omega t) \text{ A} \tag{7.461}$$

in the time domain. We finally obtain the source current in the phasor and time domains as[123]

$$I_s = I(z' = \lambda/4) = \frac{V_0^+}{Z_0}\exp(j\pi/2)\left[1 - \exp(j\pi/2 - j\pi)\right] = \frac{j}{25} \text{ A} \tag{7.462}$$

$$I_s(t) = -\frac{1}{25}\sin(\omega t) \text{ A}. \tag{7.463}$$

Example 99: Consider a combination of two loads and two lossless transmission lines (Figure 7.102). The transmission lines are described as follows:

$$\bullet\ Z_{01} = 100\ \Omega, \quad d_1 = \lambda/4, \qquad \bullet\ Z_{02} = 200\ \Omega, \quad d_2 = 5\lambda/8 \tag{7.464}$$

Given the load impedances $Z_{l1} = (100 - j200)\ \Omega$ and $Z_{l2} = (100 + j100)\ \Omega$, do the following:

- In the *second* transmission line, find the positions (measured from the load) of the first maximum and first minimum of the voltage pattern.
- Find the VSWR in both transmission lines.
- Find the input impedance of the overall combination: i.e. Z_{in}.

[120] This can be considered to satisfy the *boundary condition* on the source (left-hand) side. On the load (right-hand) side, the boundary condition was *already* satisfied while deriving the expression for the voltage.

[121] Obviously, $j\omega L_0 = V_l/I_l = 2/(-j/25) = j50\ \Omega$ is satisfied.

[122] Obtaining the same voltage on the source and load is a *special* case.

[123] Considering Figure 7.101, the instantaneous power at $t = t_0$ can be written (in terms of W) as

$$S_l(t_0) = \frac{1}{25}\sin(2\omega t_0), \quad S_s(t_0) = \frac{1}{25}\sin(2\omega t_0),$$

for the load and source, respectively. Interestingly, their sum is not zero, but this is not a conflict. The distance between the source and the load ($\lambda/4$) corresponds to a time delay of $t_d = \pi \times 10^{-7}$ s. Then if the source is considered at $t_0 - t_d$, we have

$$S_s(t_0 - t_d) = \frac{1}{25}\sin(2\omega t_0 - \pi) = -\frac{1}{25}\sin(2\omega t_0),$$

showing the consistency. Also note that the load does not consume the delivered power: i.e. it stores and releases the power periodically.

Solution: Considering the general expressions for maxima and minima, we have

$$z'_{\max} = \frac{\lambda\varphi_l}{4\pi} + \frac{n\lambda}{2} \longrightarrow z'_{\max,\text{first}} = \frac{\lambda\varphi_l}{4\pi} \tag{7.465}$$

$$z'_{\min} = \frac{\lambda\varphi_l}{4\pi} - \frac{\lambda}{4} + \frac{n\lambda}{2} \longrightarrow z'_{\min,\text{first}} = \frac{\lambda\varphi_l}{4\pi} \pm \frac{\lambda}{4} = z'_{\max} \pm \frac{\lambda}{4}, \tag{7.466}$$

where the selection of $\pm$ in $z'_{\min,\text{first}}$ should be based on the value of φ_l. The reflection coefficient at the load can be found as[124]

$$\Gamma_{l2} = \frac{Z_{l2} - Z_{02}}{Z_{l2} + Z_{02}} = \frac{100 + j100 - 200}{100 + j100 + 200} = \frac{-1 + j}{3 + j} = -\frac{1}{5} + j\frac{2}{5}, \tag{7.467}$$

which can be written[125]

$$\Gamma_{l2} = |\Gamma_{l2}| \exp(j\varphi_{l2}) = \frac{1}{\sqrt{5}} \exp[j\pi/2 + j\tan^{-1}(1/2)]. \tag{7.468}$$

Then we obtain[126]

$$z'_{\max,\text{first}} = \frac{\lambda\varphi_l}{4\pi} = \frac{\lambda(\pi/2)}{4\pi} + \frac{\lambda \tan^{-1}(1/2)}{4\pi} = \frac{\lambda}{8} + \frac{\lambda}{4\pi}\tan^{-1}(1/2) \tag{7.469}$$

$$z'_{\min,\text{first}} = z'_{\max,\text{first}} \pm \frac{\lambda}{4} = \frac{3\lambda}{8} + \frac{\lambda}{4\pi}\tan^{-1}(1/2). \tag{7.470}$$

In addition, using Γ_{l2}, we can find the VSWR in the second transmission line as

$$\text{VSWR}_2 = \frac{1 + |\Gamma_{l2}|}{1 - |\Gamma_{l2}|} = \frac{1 + 1/\sqrt{5}}{1 - 1/\sqrt{5}} = \frac{\sqrt{5} + 1}{\sqrt{5} - 1} \approx 2.62. \tag{7.471}$$

To find the VSWR for the first transmission line, we must know the reflection coefficient due to its *effective* load, which is a combination of Z_{l1} and Z_{l2} transformed by the second transmission line. First, the transformation of Z_{l2} by the second transmission line (Figure 7.102) can be found as[127]

$$Z_{in,2} = Z_{02}\frac{Z_{l2} + jZ_{02}\tan(\beta_2 d_2)}{Z_{02} + jZ_{l2}\tan(\beta_2 d_2)} = 200\frac{100 + j100 + j200\tan(5\pi/4)}{200 + j(100 + j100)\tan(5\pi/4)} \tag{7.472}$$

$$= 200\frac{100 + j100 + j200}{200 + j100 - 100} = 200\frac{1 + j3}{1 + j} = (400 + j200)\ \Omega. \tag{7.473}$$

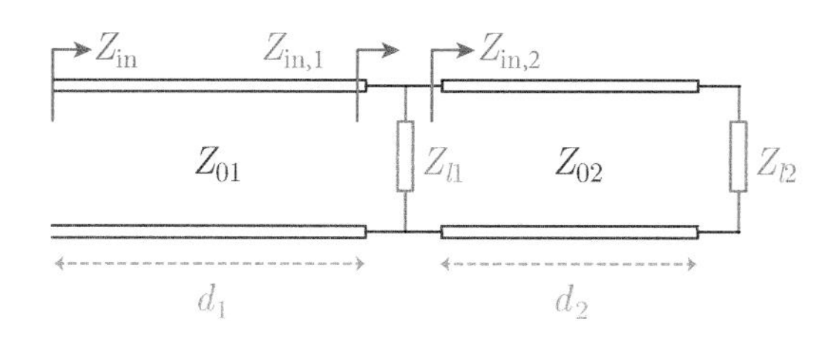

Figure 7.102 A combination of two loads and two lossless transmission lines.

[124]We have

$$\frac{-1+j}{3+j} = \frac{(-1+j)(3-j)}{(3+j)(3-j)} = \frac{-3+j+j3+1}{9-j^2} = \frac{-2+j4}{10} = \frac{-1+j2}{5}.$$

[125]We have

$$|\Gamma_{l2}| = \sqrt{\left(-\frac{1}{5}\right)^2 + \left(\frac{2}{5}\right)^2} = \frac{1}{\sqrt{5}}.$$

$$\varphi_{l2} = \pi - \tan^{-1}[(2/5)/(1/5)] = \pi/2 + \tan^{-1}[(1/5)/(2/5)].$$

[126]Using the minus sign for $z'_{\min,\text{first}}$ would lead to

$$z'_{\min,\text{first}} = -\frac{\lambda}{8} + \frac{\lambda}{4\pi}\tan^{-1}(1/2) < 0,$$

which is invalid. Hence, for this particular case, the first maximum occurs earlier than the first minimum.

[127]Note that we do not need the exact values of β_1 and β_2, since the lengths are given in terms of the wavelength. We have

$$\beta_1 d_1 = \frac{2\pi}{\lambda}\frac{\lambda}{4} = \frac{\pi}{2}, \qquad \beta_2 d_2 = \frac{2\pi}{\lambda}\frac{5\lambda}{8} = \frac{5\pi}{4},$$

while $\tan(\pi/2) \to \infty$ (the first transmission line is an inverter) and $\tan(5\pi/4) = 1$.

using

$$\frac{1+j3}{1+j} = \frac{(1-j)(1+j3)}{(1-j)(1+j)} = \frac{1+j3-j+3}{2} = 2+j. \tag{7.474}$$

Then the effective load of the first transmission line (Figure 7.102) can be obtained as[128]

$$Z_{in,1} = Z_{l1} \parallel Z_{in,2} = (100-j200) \parallel (400+j200) = (160-j120)\ \Omega. \tag{7.475}$$

Hence, we can find the reflection coefficient at the load as

$$\Gamma_{l1} = \frac{Z_{in,1}-Z_{01}}{Z_{in,1}+Z_{01}} = \frac{160-j120-100}{160-j120+100} = \frac{3-j6}{13-j6} \approx 0.37-j0.29. \tag{7.476}$$

Consequently, the VSWR for the first transmission line can be obtained as

$$\text{VSWR}_2 = \frac{1+|\Gamma_{l1}|}{1-|\Gamma_{l1}|} \approx 2.77. \tag{7.477}$$

Finally, we find the input impedance of the overall combination as

$$Z_{in} = \frac{Z_{01}^2}{Z_{in,1}} = \frac{100^2}{(160-j120)} = (40+j30)\ \Omega, \tag{7.478}$$

if the first transmission line is an inverter.

[128] This can be evaluated as

$$\begin{aligned}(100-j200) \parallel (400+j200) &= \frac{(100-j200)(400+j200)}{(100-j200)+(400+j200)} \\ &= 100\frac{(1-j2)(4+j2)}{(1-j2)+(4+j2)} \\ &= 100\frac{(4+j2-j8+4)}{5} \\ &= 40(4-j3)\ \Omega.\end{aligned}$$

Example 100: Consider a lossless transmission line terminated by a load Z_l with a given voltage pattern (Figure 7.103). Assume that the phase velocity is 3×10^8 m/s along the line.

- Find the wavelength and the VSWR in the line.
- Find the magnitudes of the incident and reflected voltage waves.
- Find the reflection coefficient at the load (Γ_l).
- Find the load impedance if the transmission line's characteristic impedance is $Z_0 = 200\ \Omega$.
- If the load consists of a single inductor or capacitor (in addition to a resistor), find the value of the inductance or capacitance.
- Find the impedance value measured at $z' = 8$ cm.

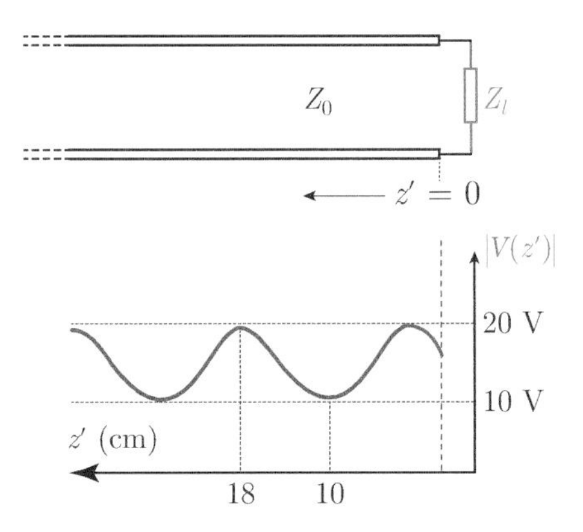

Figure 7.103 A lossless transmission line terminated by a load Z_l. The voltage pattern (phasor-domain voltage with respect to position along the transmission line) is given. By inspecting the pattern, it is obvious that the load is not a short circuit, open circuit, matched load, or load with a purely imaginary impedance.

Solution: Considering the peak-to-peak distance in the voltage pattern (16 cm), we can obtain the wavelength as $\lambda = 0.32$ m. In addition, for the maximum and minimum values of the voltage, we have VSWR = 20/10 = 2. Then the magnitude of the reflection coefficient (at the load) can be obtained as

$$\text{VSWR} = 2 = \frac{1+|\Gamma_l|}{1-|\Gamma_l|} \longrightarrow |\Gamma_l| = 1/3. \tag{7.479}$$

Since the first maximum occurs at $z' = 2$ cm, we can further obtain the phase of the reflection coefficient as

$$\lambda\varphi_l/(4\pi) = 0.02 \longrightarrow \varphi_l = 0.08\pi/\lambda = \pi/4 \text{ rad}, \tag{7.480}$$

leading to $\Gamma_l = |\Gamma_l|\exp(j\varphi_l) = (1/3)\exp(j\pi/4)$ or

$$\Gamma_l = \sqrt{2}(1+j)/6. \tag{7.481}$$

Since the maximum value of the voltage is 20 V, the magnitude of the incident voltage wave can be found as

$$|V_0^+|(1+|\Gamma_l|) = 20 \longrightarrow |V_0^+| = 15 \text{ V}, \tag{7.482}$$

which indicates that $|V_0^-| = |\Gamma_l||V_0^+| = 5$ V is the magnitude of the reflected wave.[129] If the characteristic impedance of the transmission line is 200 Ω, the load impedance can be obtained as[130]

$$Z_l = 200\frac{1+\sqrt{2}/6+j\sqrt{2}/6}{1-\sqrt{2}/6-j\sqrt{2}/6} \approx (277.9 + j147.4)\ \Omega. \tag{7.483}$$

This impedance can be obtained by a series connection of a resistor with 277.9 Ω resistance and an inductor with an inductance of[131]

$$L \approx 147.4/\omega \approx 25 \text{ nH}. \tag{7.484}$$

Finally, the impedance at $z' = 8$ cm, which corresponds to $\lambda/4$, can be obtained as[132] $Z(z' = 8 \text{ cm}) = Z_0^2/Z_l \approx (112.3 - 59.6)\ \Omega$.

[129]This is consistent with the fact that $15 - 5 = 10$ V is the minimum value of the voltage.

[130]Note that if

$$\Gamma_l = \frac{Z_l - Z_0}{Z_l + Z_0},$$

then

$$(Z_l - Z_0) = \Gamma_l(Z_l + Z_0)$$
$$\longrightarrow Z_l - \Gamma_l Z_l = Z_0 + \Gamma_l Z_0$$
$$\longrightarrow Z_l = Z_0\frac{1+\Gamma_l}{1-\Gamma_l}.$$

[131]If the phase velocity is given as 3×10^8 m/s, we can find the frequency and the angular frequency as

$$f = u_p/\lambda = (3/0.32) \times 10^8 \text{ Hz}$$
$$\omega = 2\pi \times (3/0.32) \times 10^8 \approx 5.9 \times 10^9 \text{ rad/s}.$$

[132]So, the impedance at $z' = 8$ cm is capacitive, even though the load is inductive.

7.4 Concluding Remarks

This chapter has summarized various means of electromagnetic transmission. We discussed wireless transmission with an emphasis on antennas, which make a homogeneous space an efficient environment to transmit electromagnetic energy. Despite wireless transmission's significant

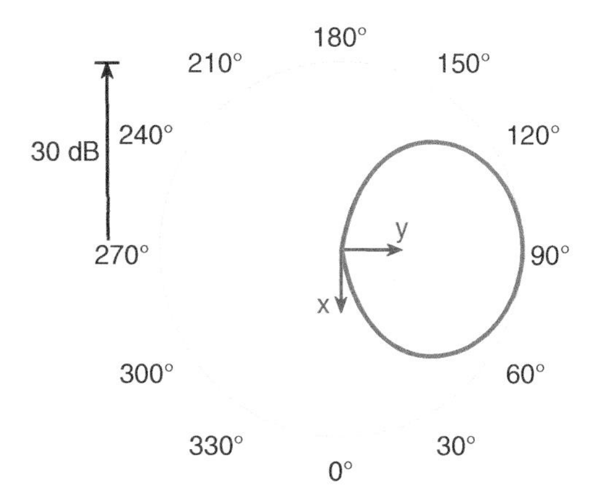

Figure 7.104 Plot of the radiation intensity $U_3^{\text{far}}(\theta, \phi)$ on the x-y plane using a 30 dB dynamic range.

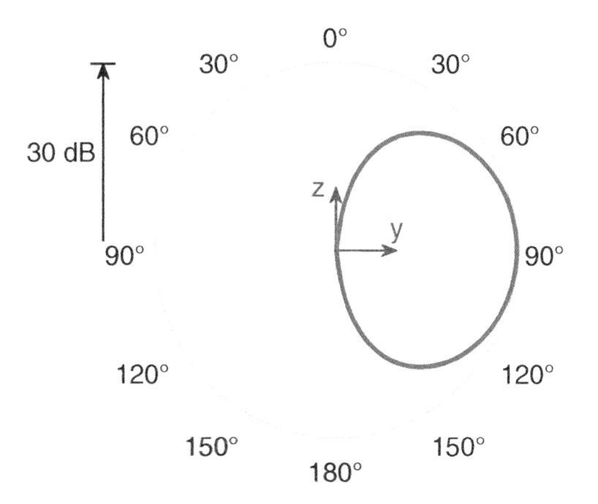

Figure 7.105 Plot of the radiation intensity $U_3^{\text{far}}(\theta, \phi)$ on the y-z plane using a 30 dB dynamic range.

advantages – particularly the fact that it does not require a physical connection – it suffers from inevitable propagation of radiated energy in unwanted directions, leading to interference and quadratic decay of the power density with distance.

Enclosing propagation in closed regions to establish efficient transmission (at least, over relatively short distances) leads to waveguides, which have their own advantages and disadvantages. Specifically, waveguides can provide excellent transmission efficiencies, whereas propagation inside them is quite different from free-space propagation. As shown in detail, electromagnetic energy is transmitted in the form of modes, which have different cutoff frequencies and phase velocities that must be considered carefully when designing a waveguide. Although optical fibers, as a class of waveguides, have become increasingly popular in modern technology (e.g. to build submarine cable systems), standard cables (coaxial and twisted-pair wires) are still the most common tools for wired electromagnetic transmission. Given the wave nature of transmission, such cables are best formulated via transmission line theory without resorting to a full-wave analysis of the entire system.

This book ends with the next chapter, which includes further general information about electromagnetism and common tools to analyze complex problems.

7.5 Exercises

Exercise 89: Consider three antennas with radiation intensities given as (see Figures 7.104 and 7.105 for the third radiation intensity)

$$U_1^{\text{far}}(\theta, \phi) = U_1 \begin{cases} \sin\theta \sin^2\phi & (0 \leq \phi \leq \pi) \\ 0 & (\pi < \phi < 2\pi) \end{cases} \tag{7.485}$$

$$U_2^{\text{far}}(\theta, \phi) = U_2 \begin{cases} \sin^2\theta \sin^3\phi & (0 \leq \phi \leq \pi) \\ 0 & (\pi < \phi < 2\pi) \end{cases} \tag{7.486}$$

$$U_3^{\text{far}}(\theta, \phi) = U_3 \begin{cases} \sin^3\theta \sin^4\phi & (0 \leq \phi \leq \pi) \\ 0 & (\pi < \phi < 2\pi), \end{cases} \tag{7.487}$$

where U_1 (W/sr), U_2 (W/sr), and U_3 (W/sr) are constants.

- Find the directivity values.
- Find constants U_1, U_2, and U_3 if the total radiated power is 5 W for each antenna.
- Find the HPBW values on the x-y and y-z planes.

Exercise 90: Consider an antenna in a vacuum with a radiation intensity given as

$$U^{\text{far}}(\theta, \phi) = U_0 \begin{cases} [\cos(2\theta)]^{2n} \cos\theta & (0 \le \theta \le \pi/2) \\ 0 & (\pi/2 < \theta \le \pi), \end{cases} \quad (7.488)$$

where U_0 (W/sr) is a constant and $n > 0$ is an integer.

- Find the directivity of the antenna.
- Find the directions for zero radiation, as well as the number of lobes.
- Find the maximum effective area of the antenna at 1 GHz if the radiation efficiency is 0.9.

Exercise 91: Consider a two-dimensional array of 4×2 elements on the y-z plane. The separation distance is $\lambda/2$ in the y direction and $\lambda/4$ in the z direction. The excitations of the elements in the first row are given as j, whereas the excitations in the second row are unity (Figure 7.106).

- Find the magnitude of the normalized array factor, and evaluate it on the x-y (Figure 7.108), y-z (Figure 7.109), and z-x cuts (Figure 7.110).
- Using the array factor of the array in Figure 7.106, find the magnitude of the normalized array factor of a modified array with identical excitations and $\lambda/2$ separation distance also in the z direction (Figure 7.107).

Exercise 92: Consider two antennas, Antenna A and Antenna B, with input impedance values $Z_{in,A} = 50\ \Omega$ and $Z_{in,B} = 75\ \Omega$, respectively. The antennas are far from each other. When Antenna A is used as a transmitter, it is connected to a generator (source+feedline) with $Z_{eq,A} = 50\ \Omega$ impedance, whereas Antenna B is used as a receiver and connected to a load+feedline with $Z_{eq,B} = 25\ \Omega$ impedance. Then the output power (on the load of Antenna B) is measured to be 2 W if the generator power (on the Antenna A side) is 20 W. In the following, consider the case when Antenna A and Antenna B become the receiver and the transmitter, respectively.

- Find the output power (on the load of Antenna A) when the generator power (on Antenna B side) is 20 W, if the terminal impedances ($Z_{eq,A}$ and $Z_{eq,B}$) remain the same.
- Find the value of $Z_{eq,A}$ (while keeping other values the same) to maintain 2 W output load power (Antenna A) for 20 W generator power (Antenna B).
- Find the value of $Z_{eq,B}$ (while keeping other values the same) to maintain 2 W output load power (Antenna A) for 20 W generator power (Antenna B).

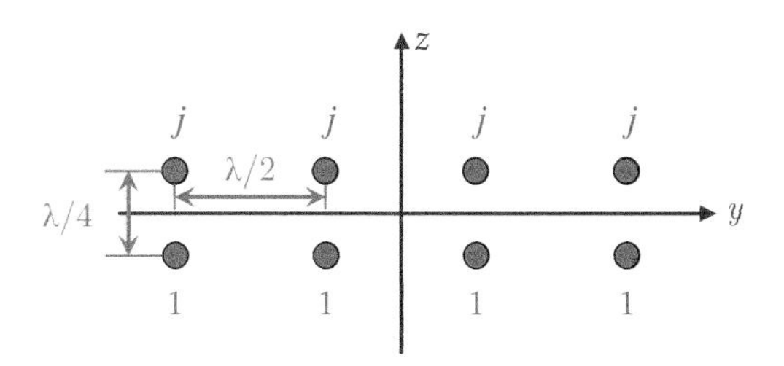

Figure 7.106 A two-dimensional array consisting of 4×2 elements with the given separation distances and excitations.

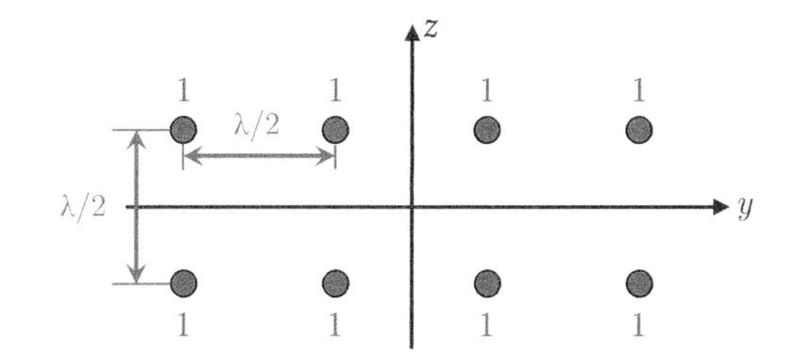

Figure 7.107 A two-dimensional array consisting of 4×2 elements with the given separation distances and identical excitations.

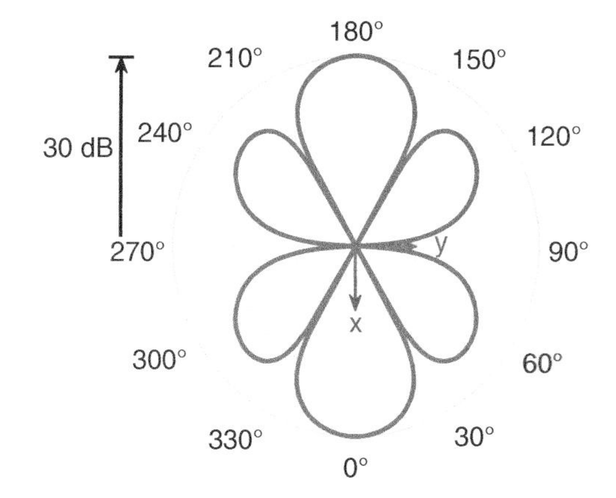

Figure 7.108 Array factor of the array in Figure 7.106 on the x-y plane. The dynamic range is 30 dB.

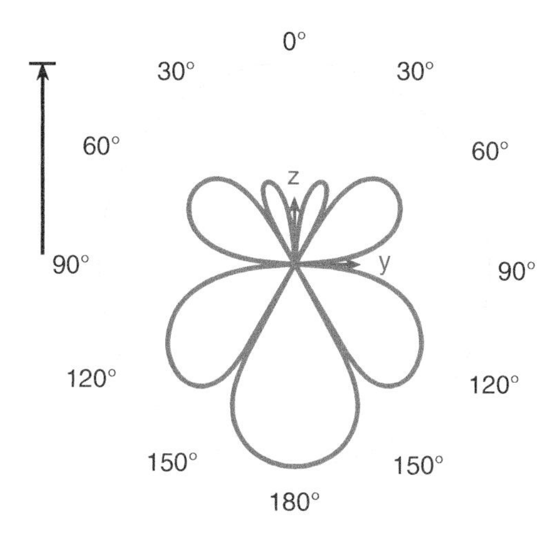

Figure 7.109 Array factor of the array in Figure 7.106 on the y-z plane. The dynamic range is 30 dB.

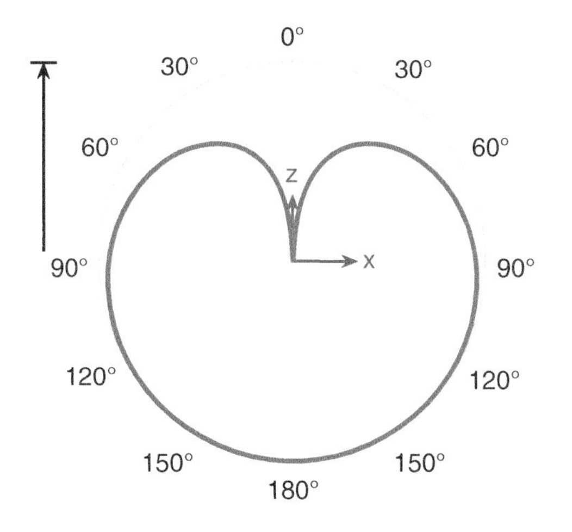

Figure 7.110 Array factor of the array in Figure 7.106 on the z-x plane. The dynamic range is 30 dB.

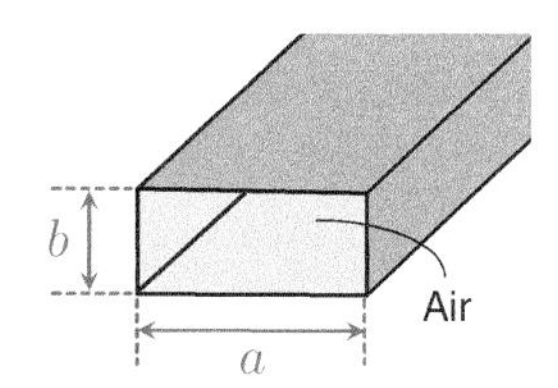

Figure 7.111 An $a \times b$ rectangular waveguide filled with air.

Exercise 93: In an air-filled ($\varepsilon \approx \varepsilon_0$ and $\mu \approx \mu_0$) rectangular waveguide (Figure 7.111), the cutoff frequencies of the TE_{10} and TE_{01} modes are 5.0 GHz and 2.0 GHz, respectively. Find the dimensions of the waveguide, as well as the numbers of the first five modes in terms of frequency (use the approximation $1/\sqrt{\mu_0\varepsilon_0} \approx 3 \times 10^8$ m/s).

Exercise 94: A rectangular waveguide with 2.42×1.12 cm dimensions (Figure 7.112) supports the TE_{10} mode after 6.0 GHz. Find the limits of the relative permittivity of the inner dielectric material if there is no other propagating mode (use the approximation $1/\sqrt{\mu_0\varepsilon_0} \approx 3 \times 10^8$ m/s).

Exercise 95: Consider a rectangular waveguide with 0.15×0.15 m square cross section (Figure 7.113). The waveguide is filled with air ($\varepsilon \approx \varepsilon_0$ and $\mu \approx \mu_0$), while the frequency is set to 1.2 GHz. In the following, use the approximation $1/\sqrt{\mu_0\varepsilon_0} \approx 3 \times 10^8$ m/s.

- Find which modes propagate, as well as their propagation constants.
- Find expressions for the total $H_x(\bar{r})$, $H_y(\bar{r})$, and $H_z(\bar{r})$ inside the waveguide, considering only propagating modes. Assume that all propagating modes are excited equally to simplify expressions.
- Find expressions for the total $E_x(\bar{r})$, $E_y(\bar{r})$, and $E_z(\bar{r})$ inside the waveguide, considering only propagating modes. Assume that all propagating modes are excited equally to simplify expressions.
- Find expressions for the induced electric current density on the waveguide walls, considering only propagating modes. Assume that all propagating modes are excited equally to simplify expressions.

Exercise 96: Consider a transmission line with a characteristic impedance of $Z_0 = 50\ \Omega$ terminated by 10 Ω load (Figure 7.114). Find the impedance seen at $z_1 = 1/8$ m, $z_2 = 1/4$ m, and $z_3 = 1/2$ m, where the position is measured from the load when the wavelength is 1 m.

Exercise 97: A series combination of a 30 Ω resistor and a $1/(80\pi)$ nF capacitor is connected to a lossless transmission line as a load at 1 GHz. The transmission line's characteristic impedance is 50 Ω, while the phase velocity is 3×10^8 m/s. Find the impedance values measured at $z' = 7.5$ cm and $z' = 15$ cm. Also find how the values change if the frequency is increased to 2 GHz.

Exercise 98: Consider a combination of two loads and two lossless transmission lines (Figure 7.115). The characteristic impedances of the transmission lines are given as $Z_{01} = 100\ \Omega$ and $Z_{02} = 200\ \Omega$. Given the load impedances $Z_{l1} = (50 + j50)\ \Omega$ and $Z_{l2} = (25 + j25)\ \Omega$, find the input impedance Z_{in} for the following cases (lengths):

- $d_1 = \lambda/2$, $d_2 = \lambda/4$
- $d_1 = \lambda/4$, $d_2 = \lambda/2$
- $d_1 = \lambda/4$, $d_2 = \lambda/8$

Exercise 99: Consider a 30 cm lossless transmission line with a characteristic impedance of 50 Ω (Figure 7.114). When a short circuit terminates the transmission line, the first voltage maximum is measured at 5 cm from the short circuit. When an unknown load Z_l is connected instead of the short circuit, the VSWR is measured as 9, while the first voltage minimum (again measured from the load) occurs 6 cm away. Find the load impedance Z_l.

Exercise 100: Consider a transmission line with a characteristic impedance of $Z_0 = 300\ \Omega$ terminated by an unknown load Z_l (Figure 7.114). The VSWR is measured as 3.0, while the first voltage minimum from the load occurs at $z' = 0.3$ m. In addition, the distance between two consecutive minima is 0.5 m. Find the load impedance Z_l, as well as the input impedance Z_{in} if the transmission line is 5.25 m long.

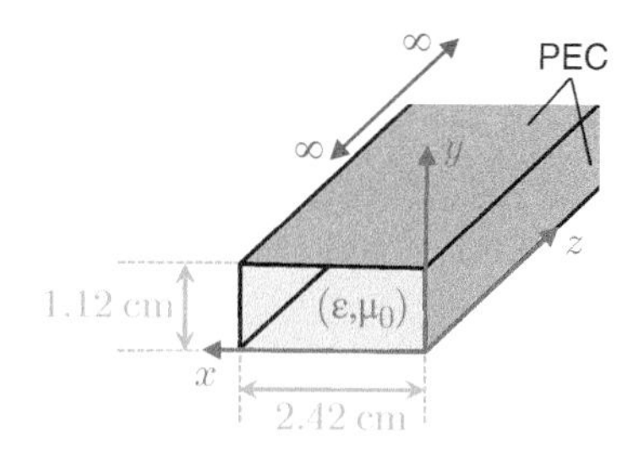

Figure 7.112 A rectangular waveguide with 2.42 × 1.12 cm cross section filled with a dielectric material.

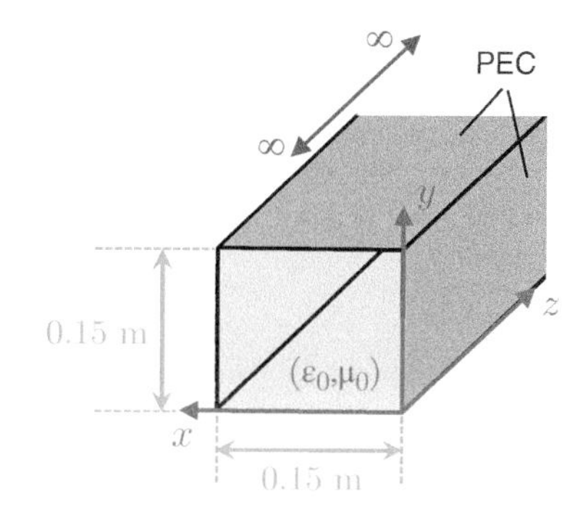

Figure 7.113 A rectangular waveguide with 0.15 × 0.15 m square cross section filled with air.

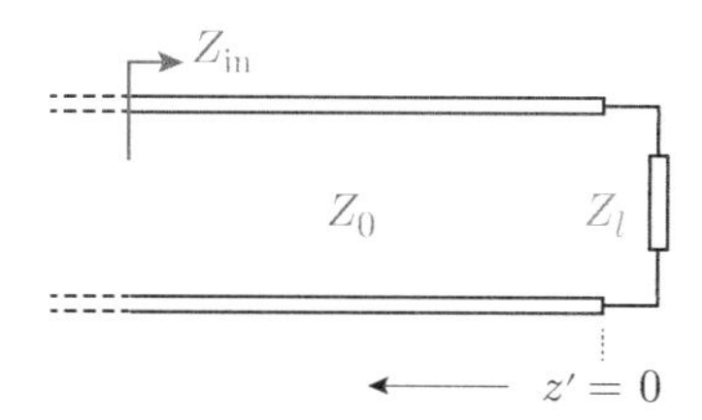

Figure 7.114 Termination of a transmission line by a load.

7.6 Questions

Question 37: Consider an infinitely long cylindrical wire of radius a carrying a uniformly distributed *steady* current I_0 (Figure 7.116). The conductivity is σ_0.

- Find an expression for the static electric field intensity inside the wire.
- Find an expression for the static magnetic field intensity inside the wire.
- Find an expression for the dissipated power per length.
- Show that the Poynting vector can be written

$$\bar{S}(\bar{r}) = -\hat{a}_p \frac{I_0^2 \rho}{2\pi^2 a^4 \sigma_0} \qquad (0 \leq \rho \leq a). \tag{7.489}$$

  Also demonstrate the conservation of energy.

- How can the power flowing from a vacuum into the wire be explained?

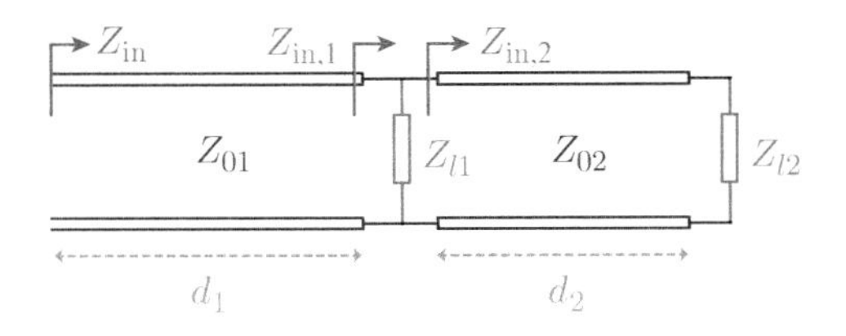

Figure 7.115 A combination of two loads and two lossless transmission lines.

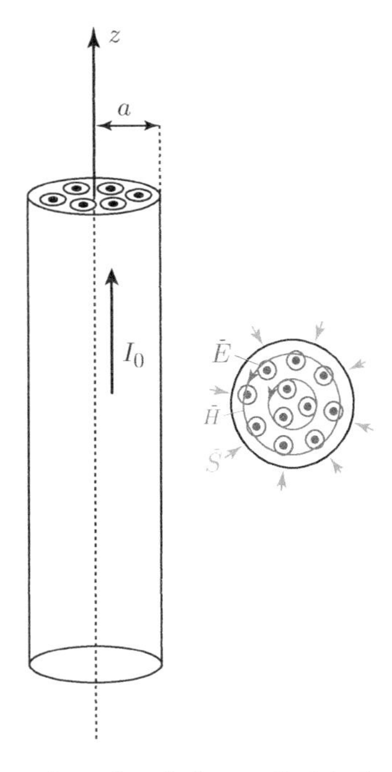

Figure 7.116 An infinitely long cylindrical wire that carries a steady electric current in a vacuum.

Question 38: An unknown current distribution generates an electric field intensity given as

$$\bar{E}(\bar{r}) = jk\eta g(R)\left\{\hat{a}_R\left[\frac{\cos\phi}{jkR} + \frac{\sin^2\theta}{(jkR)^2}\right] + \hat{a}_\theta\left[\sin\theta + \frac{\cos\theta}{jkR}\right] + \hat{a}_\phi\left[\frac{\cos\phi}{jkR} + \frac{\cos\theta\sin\phi}{(jkR)^2}\right]\right\}, \tag{7.490}$$

where $g(R) = \dfrac{\exp(-jkR)}{4\pi R}$ and $k = \omega\sqrt{\mu\varepsilon} = 2\pi/\lambda$.

- Find an expression for the electric field intensity in the far zone.
- Find an expression for the magnetic field intensity in the far zone.
- Find an expression for the complex Poynting vector in the far zone.
- Find the time-average power flux density in the far zone.

Question 39: Consider a lossless antenna ($e_r = 1$) in a vacuum with a radiation intensity given as (Figure 7.117)

$$U^{\text{far}}(\theta,\phi) = U_0\begin{cases}|\cos(4\theta)| & (0 \le \theta \le 5\pi/8)\\ 0 & (5\pi/8 < \theta \le \pi),\end{cases} \tag{7.491}$$

where U_0 (W/sr) is a constant. Find the maximum effective area of the antenna at (a) 60 MHz, (b) 600 MHz, and (c) 6 GHz.

Question 40: Consider a radiating antenna (Antenna A) located at the origin with an electric field intensity given as (Figure 7.118)

$$\bar{E}^{\text{far}}(\bar{r}) = E_0[\hat{a}_\theta\sin\theta\cos\phi - \hat{a}_\phi\sin\phi\cos\theta]g(R)f(\theta,\phi), \tag{7.492}$$

where $f(\theta,\phi)$ is a scalar function and E_0 (V/m) is a constant.

- Find the polarization of a receiving antenna (Antenna B), which minimizes the PLF on the y-z plane.
- Find the polarization of a receiving antenna (Antenna C), which minimizes the PLF on the x-y plane.

- Find the PLF with respect to direction on the z-x plane, if each antenna found above (Antenna B or Antenna C) is used as a receiver.
- Assume that Antenna A is used in a receiving mode to receive circularly polarized waves propagating in the $-z$ direction. Find the PLF when the incident wave is (i) RHCP and (ii) LHCP.

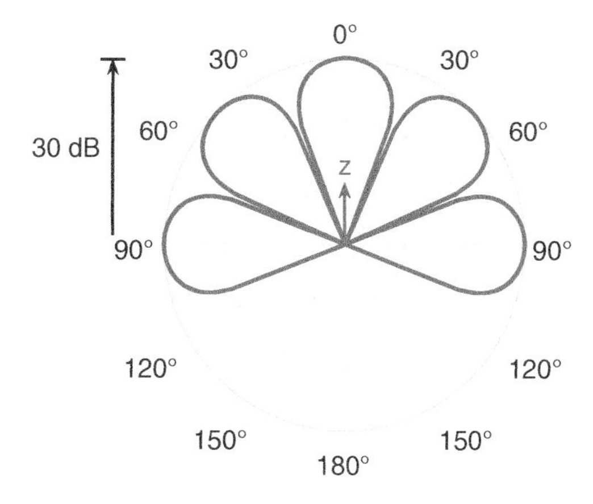

Figure 7.117 Polar plot of the radiation intensity in Eq. (7.491) using a 30 dB dynamic range.

Question 41: Consider an array of five Hertzian dipoles aligned along the z axis. The dipoles are oriented in the z direction, and the separation distance is $d = \lambda/4$. Excitations of the elements have the same amplitude (uniform excitation), while there is phase progress $\varphi_0 = \pi/2$. Find the directions for minimum (zero) and maximum radiation for the overall array.

Question 42: Consider arrays of three elements aligned along the z axis (Figure 7.119). Find the array factors for the following configurations:

- $\zeta_1 = 0,\ \zeta_2 = \zeta_0,\ \zeta_3 = \zeta_0,\ d_1$: NA, $d_2 = \lambda/2,\ d_3 = \lambda/2$
- $\zeta_1 = 0,\ \zeta_2 = \zeta_0,\ \zeta_3 = \zeta_0,\ d_1$: NA, $d_2 = \lambda/4,\ d_3 = \lambda/4$
- $\zeta_1 = 0,\ \zeta_2 = \zeta_0,\ \zeta_3 = -\zeta_0,\ d_1$: NA, $d_2 = \lambda/2,\ d_3 = \lambda/2$
- $\zeta_1 = 0,\ \zeta_2 = \zeta_0,\ \zeta_3 = -\zeta_0,\ d_1$: NA, $d_2 = \lambda/4,\ d_3 = \lambda/4$
- $\zeta_1 = 0,\ \zeta_2 = \zeta_0,\ \zeta_3 = -j\zeta_0,\ d_1$: NA, $d_2 = \lambda/2,\ d_3 = \lambda/2$
- $\zeta_1 = 0,\ \zeta_2 = \zeta_0,\ \zeta_3 = -j\zeta_0,\ d_1$: NA, $d_2 = \lambda/4,\ d_3 = \lambda/4$
- $\zeta_1 = 0,\ \zeta_2 = \zeta_0,\ \zeta_3 = j\zeta_0,\ d_1$: NA, $d_2 = \lambda/8,\ d_3 = \lambda/8$
- $\zeta_1 = 2\zeta_0,\ \zeta_2 = \zeta_0,\ \zeta_3 = \zeta_0,\ d_1 = 0,\ d_2 = \lambda/4,\ d_3 = \lambda/4$

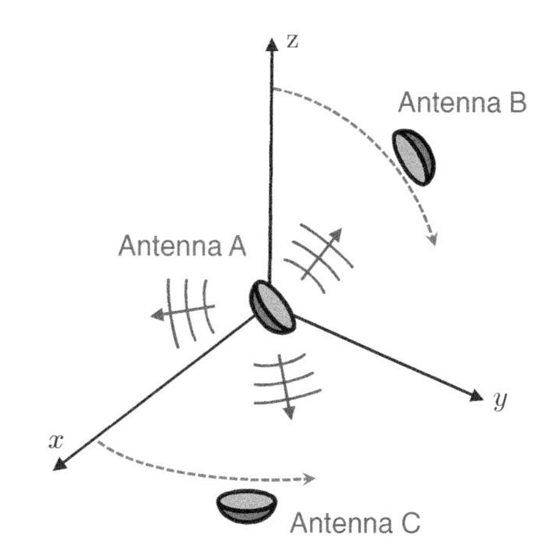

Figure 7.118 A radiating antenna (Antenna A) located at the origin and two receiving antennas (Antennas B and C) located on the y-z and x-y planes, respectively, in the far zone.

Question 43: Four Hertzian dipoles that are oriented in the z direction are arranged as a 2×2 array on the x-y plane with $\lambda/2$ and λ separations in the x and y directions (Figure 7.120). Find the radiation pattern of the array.

Question 44: Consider an $a \times b$ rectangular waveguide filled with a homogeneous material having ε permittivity and μ permeability (Figure 7.121).

- Find the complex Poynting vector for all TM_{mn} and TE_{mn} modes.
- Find the time-average Poynting vector for all TM_{mn} and TE_{mn} modes, and show that it can be nonzero only in the z direction.

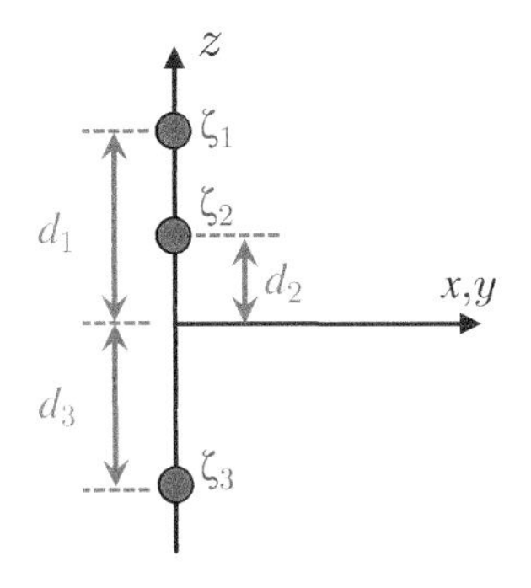

Figure 7.119 Arrangements of three elements on the z axis.

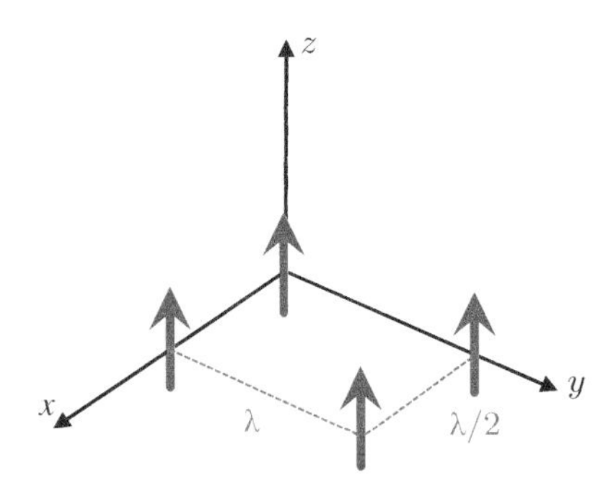

Figure 7.120 An arrangement of four Hertzian dipoles oriented in the z direction.

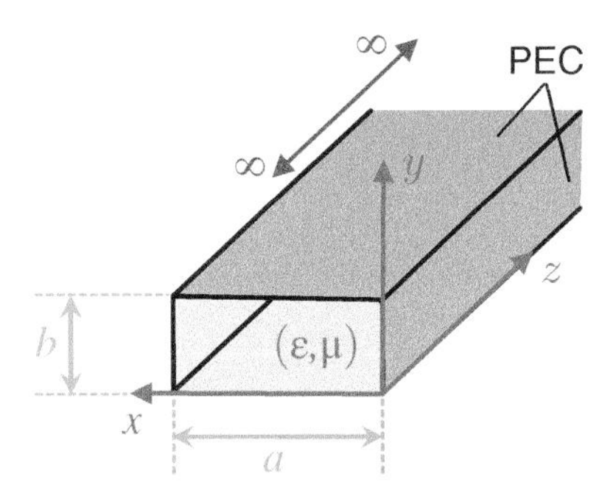

Figure 7.121 A rectangular waveguide with $a \times b$ cross section. The inner material has ε permittivity and μ permeability.

Question 45: Consider an $a \times b$ rectangular waveguide filled with a homogeneous material having ε permittivity and μ permeability (Figure 7.121).

- Find the electric current density on the waveguide walls for all TM_{mn} and TE_{mn} modes.
- Find the electric charge density on the waveguide walls for all TM_{mn} and TE_{mn} modes.

Question 46: Consider a 6.0×4.8 cm rectangular waveguide (Figure 7.122). In the following, use the approximation $1/\sqrt{\mu_0\varepsilon_0} \approx 3 \times 10^8$ m/s.

- Assume that the waveguide is filled with air ($\varepsilon \approx \varepsilon_0$ and $\mu \approx \mu_0$). Find the propagating modes, if any, at 2 GHz.
- Assume that the waveguide is filled with a dielectric material having a relative permittivity of ε_r. Find the range for ε_r to have only one propagating mode at 2 GHz.
- Assume that the waveguide is filled with a dielectric material having a relative permittivity of 3.0. Find the range for the frequency to have only one propagating mode.

Question 47: Consider a lossless transmission line of length L terminated by a load (Figure 7.123). The characteristic impedance of the transmission line is given as 100 Ω, while the phase constant at the operation frequency is $\beta = \pi$ rad/m.

- Find length L if the load impedance is $Z_l = (100 + j100)$ Ω and the second voltage minimum occurs at the input. Also find the corresponding input impedance.
- Find length L if the load impedance is $Z_l = (100 - j100)$ Ω and the second voltage minimum occurs at the input. Also find the corresponding input impedance.
- Find length L to obtain an input impedance with the minimum possible absolute value when the load impedance is $Z_l = (100 + j100)$ Ω.
- Find length L to obtain an input impedance with the maximum possible absolute value when the load impedance is $Z_l = (100 + j100)$ Ω.

Question 48: Consider a combination of two loads and two lossless transmission lines (Figure 7.124). The wavelength is fixed to 1 m in both lines. The characteristic impedances are given as $Z_{01} = 100$ Ω and $Z_{02} = 200$ Ω.

- Find a set of d_2, Z_{l1}, and Z_{l2} to make the VSWR equal to 1.0 and 2.0 in the first and second transmission lines, respectively. Also find the resulting input impedance value if $d_1 = d_2$.

- Find a set of d_2, Z_{l1}, and Z_{l2} to make the VSWR equal to 2.0 and 1.0 in the first and second transmission lines, respectively. Also find the resulting input impedance value if $d_1 = d_2$.
- Let $Z_{l2} = (200 + j200)\ \Omega$. Find d_2 to minimize the VSWR in the first transmission line. Also find the resulting input impedance value if $d_1 = d_2$.

Question 49: For a coaxial cable, the capacitance and inductance per length are given as 75 pF/m and 200 nH/m, respectively. Find (a) the characteristic impedance, (b) the phase velocity inside the cable, (c) the phase constant inside the cable at 500 MHz, (d) the wavelength inside the cable at 500 MHz, and (e) the ratio of the outer/inner conductor radii if the material between them has 1.2 relative permittivity.

Question 50: Consider a transmission line terminated by a load (Figure 7.125). The characteristic impedance of the transmission line is given as $(100 + j100)\ \Omega$. Find the load impedance for the maximum power transfer to the load. Also find the input impedance if the length of the transmission line is $\lambda/3$.

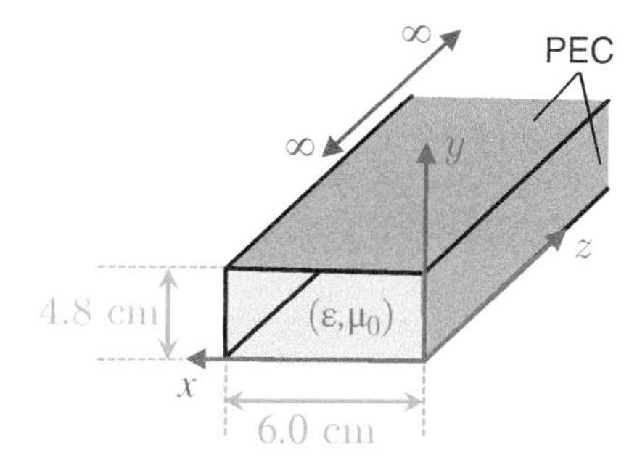

Figure 7.122 A rectangular waveguide with 6.0×4.8 cm cross section. The waveguide is filled with either air or a dielectric material.

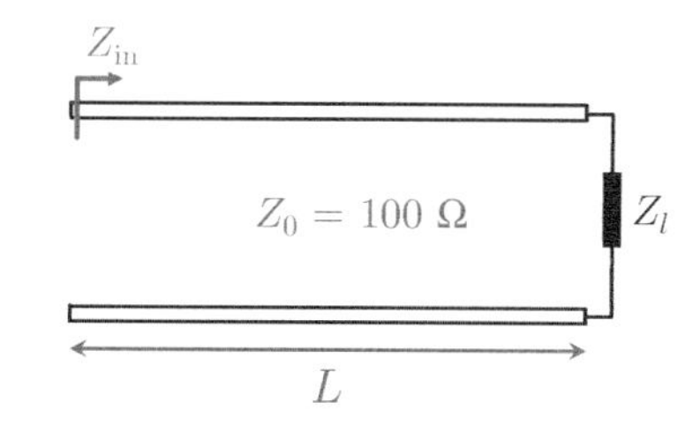

Figure 7.123 A transmission line of length L with a characteristic impedance of 100 Ω terminated by a load.

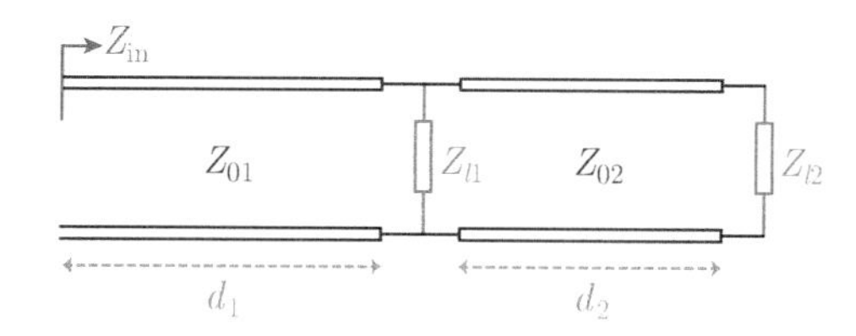

Figure 7.124 A combination of two loads and two lossless transmission lines.

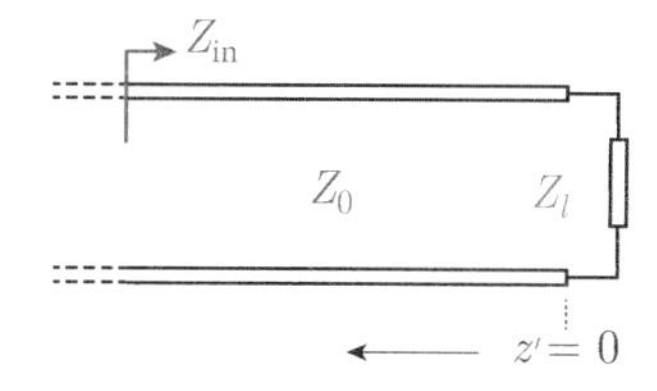

Figure 7.125 Termination of a transmission line by a load.

8

Concluding Chapter

This final chapter concludes the book with essential information about the electromagnetic spectrum, the history of electromagnetism, and examples from nature and technology involving electromagnetic phenomena. In addition, we briefly consider a list of methods and techniques for practical solutions to Maxwell's equations to analyze real-life scenarios.

As discussed in the next section, the electromagnetic spectrum consists of radio waves, microwaves, infrared radiation, visible light, ultraviolet radiation, X-rays, and gamma rays, consecutively, from low to high frequencies (Figure 8.1). All these are electromagnetic waves perfectly satisfy Maxwell's equations, but their interactions with living and nonliving objects are remarkably varied. This is why some electromagnetic waves (e.g. X-rays) were recognized much later than others (e.g. visible light). Electromagnetism became a scientific branch after Faraday's and Maxwell's work, but we should examine its history beginning much earlier. As summarized in Section 8.2, electricity, magnetism, and optics were topics of scientific research for decades, and they were merged into electromagnetism in the nineteenth century. Since then, we have learned a great deal about electromagnetic waves and how to use them to communicate with each other, diagnose diseases, and see the invisible on small and large scales using microscopes, radar, and sensors. Section 8.3 includes sample snapshots of electromagnetism in both nature and state-of-the-art technology.

Maxwell's equations can formulate all electromagnetic scenarios, but their solutions are not trivial for complex structures. Hence, for practical solutions, they are employed on computers; this can be done in various ways, leading to many different methods and techniques in the literature. The final section is devoted to a general overview of these numerical and computational techniques.

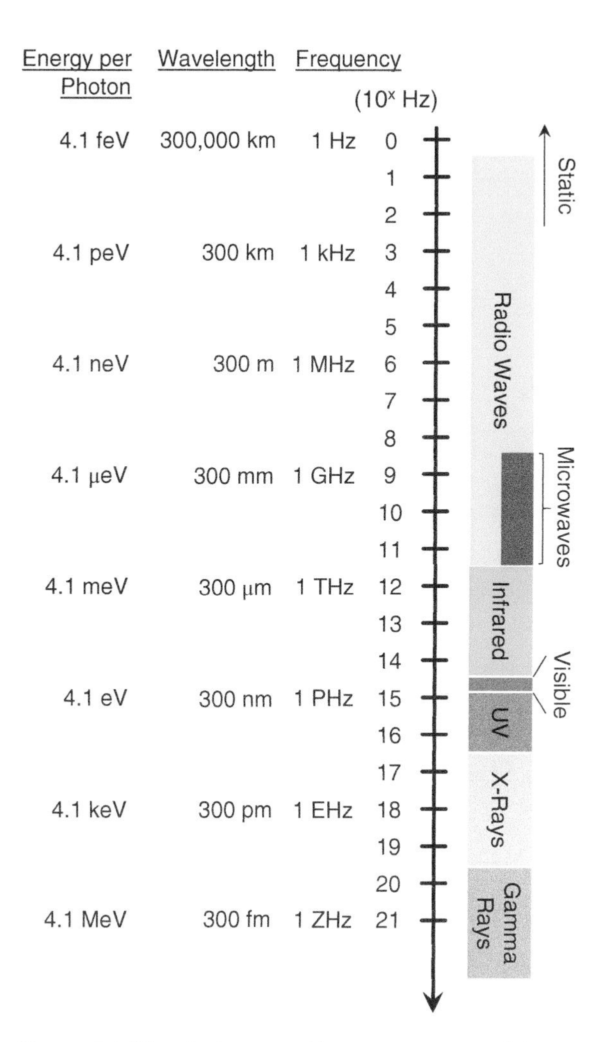

Figure 8.1 The electromagnetic spectrum, containing waves from radio frequencies to gamma rays. The wavelength values are given for a vacuum.

Introduction to Electromagnetic Waves with Maxwell's Equations, First Edition. Özgür Ergül.

Companion website: www.wiley.com/go/ergulmax

8.1 Electromagnetic Spectrum

An electromagnetic wave's behavior is mainly characterized by its frequency f, which is dictated by the source that generates the wave. In a given medium, a frequency value corresponds to a wavelength $\lambda = u_p f$, where u_p is the phase velocity ($\lambda = cf$ is a vacuum). An electromagnetic wave interacts with objects or their parts (matter) in different ways, depending on its wavelength.[1] These interactions lead to reflection, refraction, diffraction, scattering, and absorption that can be observed and/or measured macroscopically as electromagnetic phenomena (Figure 8.2).

[1]For example, microwaves and infrared radiation can interact with molecules, X-rays can interact with atoms, and gamma rays can interact with nuclei.

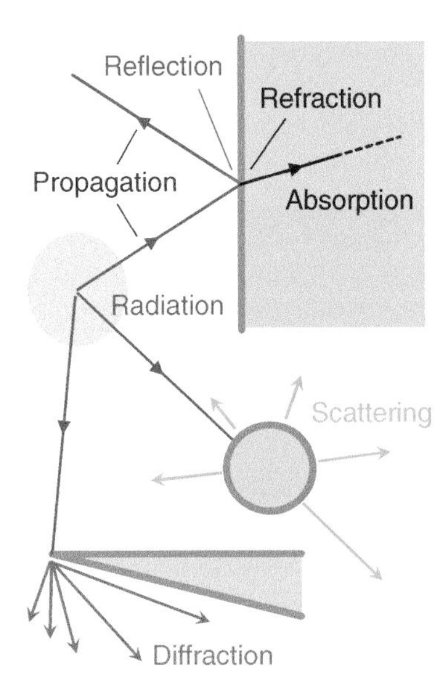

Figure 8.2 A simple demonstration of various electromagnetic phenomena.

The electromagnetic spectrum represents all possible electromagnetic waves in different ranges of frequencies (Figure 8.1). There is no theoretical limit for the lowest possible value of the frequency if we include electrostatics/magnetostatics (with zero frequencies) in the spectrum.[2] For the highest frequency, a theoretical limit can be specified by considering the Planck length the shortest possible wavelength. According to particle theory, which complements wave theory, electromagnetic energy is carried by particles called *photons*. For each frequency, there is a fixed (quantized) amount of energy per photon, which can be found as $E = hf$, where $h = 6.62607015 \times 10^{-34}$ Js (or $h = 4.135667696 \times 10^{-15}$ eVs) is the Planck constant. Hence, as the frequency increases, the energy carried by each photon increases. Increased energy lets photons have individually important and measurable effects on matter. Consequently, to explain electromagnetic phenomena, particle theory is usually preferred at higher frequencies of the spectrum; but photons exist at all frequencies (not only for optical waves) according to this theory.[3] Based on the validity of both theories and experimental observations that identify both characteristics, we emphasize the generally accepted claim that an electromagnetic energy flow (e.g. light) behaves as both a wave and a particle.

[2]Obviously, one may claim that there cannot be an ideal static case; then the lowest possible frequency may correspond to a wavelength as large as the universe.

[3]Photons of low-frequency electromagnetic waves are more difficult to observe than their high-frequency counterparts. Microwave photons were experimentally verified quite recently.

Based on the behaviors of electromagnetic waves, the electromagnetic spectrum is divided into different regimes. We now consider this categorization with generally accepted frequency boundaries.

8.1.1 Radio Waves (3 Hz to 300 GHz)

An extensive range of frequencies from 3 Hz to 300 GHz (3×10^{11} Hz) are categorized as radio waves. This range is further divided into subcategories with the historically given names extremely low frequency (ELF) [3–30 Hz], super-low frequency (SLF) [30–300 Hz], ultra-low frequency (ULF) [0.3–3 kHz], very low frequency (VLF) [3–30 kHz], low frequency (LF) [30–300 kHz], medium frequency (MF) [0.3–3 MHz], high frequency (HF) [3–30 MHz], very high frequency (VHF) [30–300 MHz], ultra-high frequency (UHF) [0.3–3 GHz], super-high frequency (SHF) [3–30 GHz], and extremely high frequency (EHF) [30–300 GHz]. Among these, UHF, SHF, and EHF (from 300 MHz to 300 GHz) are generally called microwaves, as also described below. Note that, in a vacuum, 3 Hz corresponds to approximately 100,000 km wavelength, whereas 300 GHz corresponds to only 1 mm.

In general, radio waves are radiated by accelerating charges: e.g. accelerating electrons on conductors. Antennas are exactly the kind of devices that can generate (and also receive[4]) radio waves for practical purposes: i.e. for wireless communication. Radio waves are suitable for various types of wireless systems. For example, ELF waves, which can also be generated naturally by lightning and disturbances in Earth's magnetic field, are suitable for underwater communication. SLF waves also penetrate seawater, whereas ULF waves can easily penetrate rocks. VLF waves are used for navigation services, LF waves are used for long-distance communications, MF/HF waves are used for radio broadcasting, and VHF waves can be employed for television broadcasting and air-traffic communications. All these applications are possible thanks to their long wavelengths, letting radio waves easily radiate in nature, especially through the atmosphere.

Radio waves (that do not fall into the category of microwaves) can diffract and bend around large obstructions, like mountains (Figure 8.3), while they can partially pass through buildings, vehicles, and living bodies. Therefore, the operation of a radio-wave device does not require a *line of sight*, which is a major advantage for communication systems.[5] In addition, radio waves are generally reflected from the ionosphere, leading to *skywaves* that increase the effective communication distance (Figure 8.4). Energy carried by each photon of a low-frequency radio wave is in the range from 1.24×10^{-14} eV (for 3 Hz) to 1.24×10^{-6} eV (for 300 MHz). Hence, radio waves are non-ionizing radiation: i.e. they do not ionize atoms or molecules, while their absorption (e.g. by human tissue) results in (often negligible) heating. They are substantially reflected by metals and even by metallic cages (such as a Faraday cage) with intervals significantly smaller than the operating wavelength.

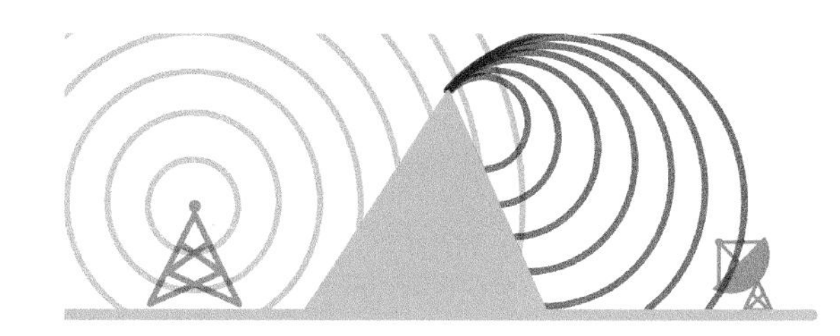

Figure 8.3 An illustration of radio waves diffracted by a hill.

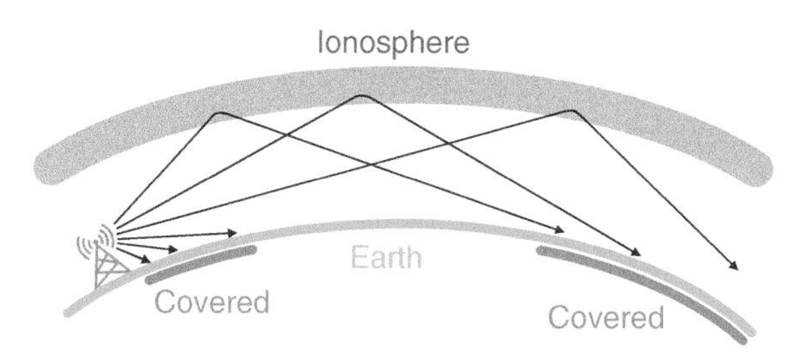

Figure 8.4 Generation of skywaves due to the reflection of radio waves from the ionosphere. The coverage can significantly be increased due to these reflections.

[4]Hence, for a receiving antenna, incoming radio waves affect electrons (on it) and make them oscillate back and forth to generate AC current.

[5]One cannot select and use an arbitrary radio frequency for an application. The radio spectrum is organized and administrated differently in each country, while the International Telecommunications Union institutes global regulations.

8.1.2 Microwaves (300 MHz to 300 GHz)

As mentioned above, microwaves are a subset of radio waves, with frequencies from[6,7] 300 MHz to 300 GHz (UHF, SHF, and EHF) corresponding to wavelengths[8] from approximately 1 m to 1 mm in a vacuum. But due to their considerably different characteristics and the variety of applications specific to these frequencies, microwaves are often studied separately (e.g. microwave engineering). Higher microwave frequencies are further divided into bands:

[6]These limits are not strict. For example, in the literature, microwaves are also defined as a range from 1–100 GHz.

[7]The corresponding energy per photon is in the range from 1.24×10^{-6} eV (for 300 MHz) to 1.24×10^{-3} eV (for 300 GHz). Microwaves are non-ionizing radiation.

[8]Note that the term *microwave* does not indicate a wavelength on the order of *micrometers*. The traditional naming is based on the *shorter* wavelengths of these waves compared to conventional radio waves.

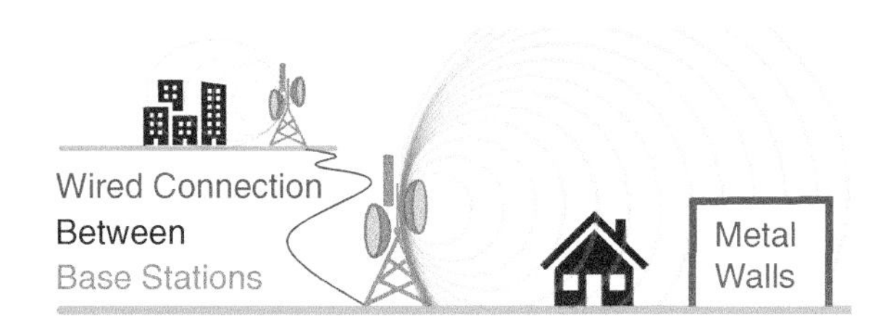

Figure 8.5 Microwaves can usually pass through walls, enabling cellular communication, but they are stopped by metals and conductive structures.

L, S, C, X, Ku, K, Ka, Q, U, V, W, F, and D, named by the IEEE to organize their applications.

Microwaves can be used for communication; their higher frequencies, compared to low-frequency radio waves, make them particularly suitable for such wireless applications. Thanks to their shorter wavelengths, antennas (which should have dimensions comparable to the wavelength) can be smaller and more practical for personal use. In addition, higher frequency ranges provide higher data rates, and more communication bands can be used simultaneously. With these advantages, microwaves are naturally part of modern life in various ways: e.g. they are used in wireless networks (Wi-Fi) and cellular communication systems (mobile phones). On the other hand, unlike radio waves with longer wavelengths, microwaves generally are not reflected by the ionosphere, and their diffraction from large structures is limited. Therefore, microwave transmissions generally require a line of sight[9,10] (Figure 8.5), and they are less preferred for broadcasting.

In addition to various communication systems, microwaves are commonly used in radar, remote sensing, and spectroscopy. While they can penetrate materials, microwaves are absorbed by polar molecules, such as water, leading to their practical use in microwave ovens and for medical treatment.[11] Similar to low-frequency radio waves, microwaves can easily be shielded by metals. In wired systems, waveguides are ideal for transmitting microwaves, as opposed to low-frequency radio waves that can be carried via cables. Microwaves are also used in astronomy (radio astronomy) since some of the electromagnetic energy radiated by stars arrives at the Earth at microwave frequencies. A very interesting phenomenon, cosmic microwave background radiation, was discovered in 1964; it is ancient radiation from the early universe after the Big Bang.[12]

8.1.3 Infrared Radiation (300 GHz to 400 THz)

Infrared radiation includes electromagnetic waves between microwaves and the visible range of the electromagnetic spectrum; it can be defined as the frequency range[13] from 300 GHz (approximately 1 mm wavelength in a vacuum) to 400 THz (approximately 750 nm wavelength in a vacuum). This range is further divided into far-infrared (300 GHz to 30 THz), mid-infrared (30–120 THz), and near-infrared (120–400 THz) frequencies. Far-infrared radiation is also called terahertz radiation, which is sometimes categorized under microwaves, depending on the context.

[9]This is a reason for the *cellular* design of personal communication systems.

[10]Note that line of sight does not mean being visible at *optical* frequencies; its definition depends on the frequency. In the case of microwaves, the lack of ionosphere reflection means altitudes of transmitters/receivers compared to the curvature of the Earth determine the maximum communication distance. On the other hand, especially at lower frequencies, concrete is transparent to microwaves so that building walls and similar structures do not interrupt line of sight. For higher microwave frequencies, these transparencies are generally lost: e.g. even the atmosphere becomes opaque.

[11]In these applications, microwaves are used with high intensity on purpose, but safety precautions are taken (e.g. via shielding) to prevent health issues. Adverse effects of microwaves used in other areas, such as personal communication, are debatable.

[12]This radiation, which may be considered the oldest electromagnetic radiation in the universe, has a peak at around 160 GHz.

[13]The corresponding energy per photon is in the range from 1.24×10^{-3} eV (for 300 GHz) to 1.65 eV (for 400 THz). Similar to radio waves (and microwaves), infrared radiation is non-ionizing.

Infrared waves are naturally generated by electron energy movements and molecular vibrations: i.e. they can be produced by living and nonliving bodies with sufficient temperatures.[14] For example, humans and animals can generate radiation in the mid-infrared range (e.g. at around 30 THz for humans). Similarly, 50% of the Sun's radiation is composed of near-infrared waves, which can easily pass through the atmosphere to heat the Earth. Infrared waves are easily absorbed by molecules and converted (back) into thermal energy. Infrared radiation is a major mechanism for heat transfer,[15] both naturally and artificially. Obviously, the absorption rate depends on the molecule size compared to the wavelength. For example, far-infrared radiation is effectively absorbed by the molecules in the atmosphere, but there are many transparency windows (when the atmosphere is transparent) in the mid-infrared and near-infrared ranges.[16] Thanks to its natural generation, infrared radiation can be used to detect objects in the absence of visible light, leading to the well-known *thermal* camera technology. Infrared radiation is used in astronomy, defense technology, medical imaging, meteorology, security, spectroscopy, and many other areas involving sensing and imaging applications, as well as in fiber optics and for wireless communications.[17]

8.1.4 Visible Range (400 THz to 800 THz)

A very narrow band of the spectrum, between infrared and ultraviolet radiation, includes electromagnetic waves that are visible to the eye: i.e. light.[18,19] The corresponding range of wavelengths is approximately 750–375 nm in a vacuum, while the energy carried by each light photon is from 1.65 eV to 3.31 eV. Light is generally emitted by electrons that move from excited to ground states, as continuously occurs in the Sun.[20] Absorption is the reverse operation: i.e. electrons of atoms and molecules can be excited by incident light, if energy levels match. If the *object* is not specialized, like the eyes of animals or chloroplasts in plants, the absorbed energy may be converted into thermal energy (stored by the object) or reflected in the form an electromagnetic wave (not necessarily similar to the incident wave) when electrons return to their ground states.[21,22]

The visible spectrum is naturally divided into bands represented by colors, as perceived and interpreted by humans and animals (Figure 8.6). The *spectral* colors and corresponding frequency ranges[23] are red (405–480 THz), orange (480–510 THz), yellow (510–530 THz), green (530–600 THz), cyan (600–620 THz), blue (620–680 THz), and violet (680–790 THz). Some colors, like brown, are mixtures of spectral colors: i.e. mixtures of waves with different frequencies. The color of an object that is not itself a light source corresponds to the frequency

[14] In general, thermal electromagnetic radiation is called *black-body radiation*, which occurs at all frequencies depending on the source. However, for living bodies and nonliving objects at room temperature, this radiation mainly occurs in the infrared.

[15] It is the only mechanism in a vacuum: i.e. when there is no conduction or convection.

[16] These windows allow the transmission of infrared radiation by the Sun.

[17] Strong infrared radiation in some industries must be considered with caution as it can damage the human eye.

[18] It is also common to call this region *visible* light, since infrared and ultraviolet radiation may also be considered invisible types of light.

[19] More specifically, the human eye can detect frequencies from 405–790 THz. Animals have varying sensitivity to different frequency ranges, but not all of them correspond to vision. For example, some snakes and carp are sensitive to infrared radiation, but this may not be related to visual activity.

[20] As mentioned above, 50% of the Sun's radiation is infrared, while 40% is in the visible range.

[21] Visible light is non-ionizing since electrons are not completely separated from atoms and molecules.

[22] In the case of an eye, the received energy is converted into electrical signals in optical nerves. In a chloroplast, it is converted into chemical energy bounded in molecules. From an engineering perspective, eyes and chloroplasts can be seen as optical antennas.

[23] These ranges do not mean transitions are sharp; colors in the visible spectrum change smoothly.

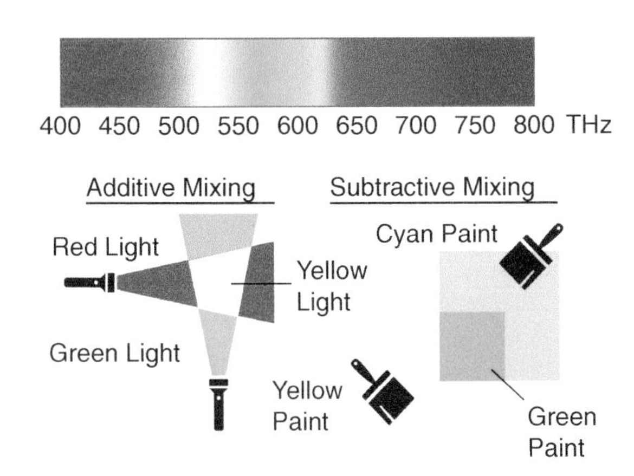

Figure 8.6 While they are well-defined in the electromagnetic spectrum, colors are the result of perception. For example, a mixture of red and green light is seen as yellow light, even though the mixture does not contain yellow frequencies.

or frequencies of the waves reflected from it. White light contains an (ideally equal) mixture of waves at all frequencies in the visible range, and black is an interpretation of the lack of light.

Visible light can easily travel through the atmosphere, but the sky's blue color is the result of scattering of the light from oxygen and nitrogen molecules (combined with our sensitivity to blue). The visible range has been studied extensively in the area of optics, but light was not known to be a form of electromagnetic radiation for a long time. In addition to traditional applications involving visual activities of humans (e.g. signaling) and vision itself, the visible spectrum has the potential to let us build faster communication systems.

8.1.5 Ultraviolet Radiation (800 THz to 30 PHz)

Following the visible range in the spectrum is ultraviolet radiation, which includes electromagnetic waves from 800 THz to 30 PHz (30×10^{15} Hz). These frequencies correspond to wavelengths from approximately 375 nm to only 10 nm in a vacuum, whereas the energy per photon is in the 3.31–124 eV range. Ultraviolet waves can be divided into subcategories:[24] near-ultraviolet (0.8–1.0 PHz), middle ultraviolet (1.0–1.5 PHz), far-ultraviolet (1.5–2.4 PHz), and extreme ultraviolet (2.4–30 PHz).

With their shorter wavelengths and higher energies, compared to radiation at lower frequencies, ultraviolet waves can chemically interact with atoms and molecules, which makes ultraviolet radiation harmful to living bodies. At lower frequencies, ultraviolet radiation can break molecular bonds and make them excited,[25] unstable, or irreversibly damaged.[26] It can be extremely harmful at higher frequencies because it interacts with atoms and ionizes them (separating electrons). As mentioned above, the Sun's radiation is mainly (90%) composed of infrared radiation and visible light, while the remaining 10% is ultraviolet radiation (generated by excited electrons returning to their low-energy states) at different frequencies. Solar electromagnetic waves in the extreme ultraviolet range, which could destroy all life on the Earth's surface, are absorbed by the atmosphere. Specifically, solar ultraviolet radiation is absorbed by nitrogen (higher frequencies), ozone (moderate frequencies), and oxygen (lower frequencies). After this effective filtering, sunlight reaching the Earth's surface includes only 3% ultraviolet radiation, mainly at low frequencies. This remaining ultraviolet radiation is useful for animals and humans: e.g. for the synthesis of vitamin D.[27,28]

As ultraviolet waves interact easily with matter, there are many technological applications for them: e.g. in astronomy, chemistry (mineral analyses), computing, electronics (lithography), forensics, medical imaging, medicine (dermatology, drug discovery), molecular biology (protein analysis, DNA sequencing), optical sensing and tracking, and photography. Ultraviolet lamps are used in many areas for sterilization, decontamination, and purification, because ultraviolet radiation makes microorganisms unable to reproduce (due to DNA/RNA damage).

[24] There are again different types of categorizations: e.g. UV-A, UV-B, and UV-C bands.

[25] Such an excitation of a molecule leads to an interesting phenomenon called *fluorescence*. When returning to the unexcited state, the released energy can be at different frequencies. Hence, depending on its material, an object illuminated by ultraviolet radiation can *glow*: i.e. emit visible light. This emission can be delayed such that the glow continues even after the ultraviolet exposure has stopped.

[26] Sunburn is caused by skin cells with (directly or indirectly) damaged DNA due to excessive exposure to the Sun. Low-frequency ultraviolet radiation that reaches the Earth's surface may not directly damage DNA, but it can still create molecules that damage DNA.

[27] Some plants also use ultraviolet radiation for hormone synthesis.

[28] Ultraviolet light is generally invisible to the human eye, but some animals (insects, birds, and even mammals) can see near-ultraviolet radiation. Some animals have specific receptors in their eyes.

8.1.6 X-Rays (30 PHz to 30 EHz)

Electromagnetic waves from 30 PHz (30×10^{15} Hz) to 30 EHz (30×10^{18} Hz) are called *X-rays*.[29] At these frequencies, wavelengths are very short: i.e. from 10 nm to 10 pm in a vacuum.[30] In addition, the energy per photon is very high, from 124 eV to 124 keV. With these values, X-rays have ionizing characteristics, and their interactions with atoms can vary (depending on the frequency and atoms). In general, X-rays can be categorized[31] as soft X-rays (30 PHz to 3 EHz) and hard X-rays (3–30 EHz). Soft X-rays, as well as hard X-rays with relatively low frequencies, interact with atoms mainly via photoelectric absorption (Figure 8.7). In this phenomenon, an incoming X-ray (or an incoming X-ray photon, in particle theory) is completely absorbed during the ionization of an atom. The separated electron (called a photo-electron) has sufficient energy to ionize more atoms, like a chain reaction. On the other hand, for hard X-rays, energy carried by the wave is partially transferred to an electron (so, ionization still occurs), while the wave continues on its way with lower energy (longer wavelength), possibly changing direction. This is nothing but the scattering of an X-ray from an atomic particle, called *Compton scattering*[32] (Figure 8.7). It is also possible for an incoming X-ray to interact with an electron (Thomson scattering) or an atom (Rayleigh scattering) but scatter without transferring energy or simply pass through an object without interacting with any of its atoms/electrons.

As indicated above, in addition to its frequency, an X-ray's exact behavior depends heavily on matter. For example, in medical imaging, X-rays are absorbed more by bones than soft tissues, leading to visual contrast that forms images. X-rays are generated by accelerated/decelerated electrons with sufficient energy, which occur naturally during cosmic events.[33] In practice, there are various ways to generate X-rays: e.g. with X-ray tubes that consist of accelerated electrons (using high voltages) bombarding a target metal (usually tungsten, copper, or cobalt) atoms. In these devices, the primary challenges are controlling and guiding the X-rays and generating them efficiently.

Due to their high energy, X-rays are very dangerous to living organisms, while they can be used in critical applications under safe conditions. Crystallography, medical imaging (radiography), medicine (cancer treatment), microscopy, security, and spectroscopy are some of the application areas using X-rays. Raymond Gosling and Rosalind Franklin generated their famous DNA images by using X-ray crystallography in 1952.

8.1.7 Gamma Rays (Above 30 EHz)

Electromagnetic waves at frequencies higher than 30 EHz are generally categorized as gamma rays.[34] With their extremely high frequencies, short wavelengths, and high energy, gamma rays are ionizing (as ultraviolet radiation and X-rays), but they can also interact with matter in more

[29] X-rays were discovered by Wilhelm Conrad Röntgen in 1895. As this kind of radiation was *unknown* at that time, he used *X-radiation* to emphasize its *unknown* (X) nature.

[30] To compare, the Bohr radius of a hydrogen atom is larger than 50 pm.

[31] Once again, this categorization and the frequency limits are not strict or fixed in the literature.

[32] Compton scattering is also called inelastic scattering since the wavelength (energy per photon) of the X-ray changes after interacting with an electron, unlike elastic (Rayleigh or Thomson) scattering. Particle theory is more suitable to explain how the frequency of a wave changes after interacting with matter.

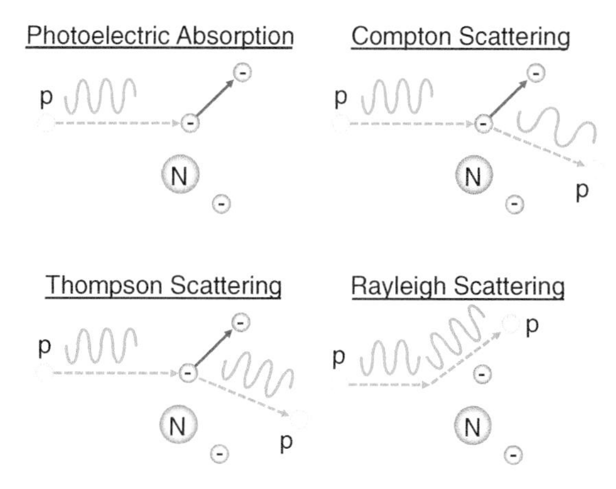

Figure 8.7 Simple illustrations of different kinds of interactions between X-rays and atoms, where an X-ray is represented by a photon.

[33] As the atmosphere completely blocks X-rays, these astronomical observations can be made via satellites.

[34] Gamma rays were discovered by Paul Ulrich Villard in 1900. Villard realized that this type of radiation is quite different from the well-known *radioactive* (alpha and beta) radiation, which are not electromagnetic waves. It took more than 10 years for Ernest Rutherford and other scientists to understand that gamma radiation has electromagnetic characteristics: i.e. it consists of photons (not mass particles) if particle theory is considered.

dramatic ways. First, similar to X-rays, photoelectric interactions may occur: i.e. an incident gamma ray may be absorbed completely, leading to the separation of an electron (called a photo-electron) from its atom. Also similar to X-rays, Compton scattering may occur, particularly for gamma rays with energy per photon less than 10 MeV (greater than 2400 EHz).

For gamma rays with energies greater than 1 MeV, another type of interaction, *pair production*, may occur. In pair production, the incident gamma ray interacts with the nucleus of an atom, and it is converted into *particles* with masses (e.g. an electron/positron pair), which may continue with further interactions (e.g. combination of the positron with an electron to *generate* new gamma rays at different frequencies) that can be understood well with *Feynman diagrams*. A gamma ray with a sufficient energy level (high frequency) may even disintegrate a nucleus (fission). These direct interactions between electromagnetic waves and atoms also explain how gamma rays are generated. Specifically, gamma rays are mainly produced by nuclear reactions during gamma decay, which can occur after fission, fusion, or other types of radioactive decay. Hence, in addition to natural radioactivity, gamma rays can be generated in nuclear reactors.[35]

[35] Interestingly, thunderstorms may produce gamma radiation due to accelerating electrons colliding with atoms in the atmosphere. Solar flares also contain gamma rays, in addition to electromagnetic radiation at lower frequencies.

As they are radiated by astronomical objects and events, such as supernovae, neutron stars, black holes, and quasars,[36] gamma rays are essential for astronomers.[37] In practice, gamma rays are used for chemical sensing, medical imaging (PET), medicine (cancer treatment), and sterilization. Because they are not only ionizing but also interact with the nuclei of atoms, gamma rays are extremely dangerous to living organisms. For safety, gamma rays can be blocked by *dense* materials (to make them interact more with atoms) such as lead or special combinations of concrete, steel, and water.

[36] For example, gamma-ray bursts are important events that provide fundamental information about distant galaxies. Such a burst, which can be generated when a supernova becomes a black hole, produces more than 10^{44} J of energy.

[37] Similar to X-rays, cosmic gamma rays can be observed from orbiting satellites (e.g. the Compton Gamma Ray Observatory), avoiding the atmospheric shield.

8.2 Brief History of Electromagnetism (Electricity, Magnetism, and a Little Optics)

If we consider the history of electromagnetism in chronological order, we may label the middle with James Clerk Maxwell, who completed the *unification* of electricity and magnetism under electromagnetism.[38] He also showed that light is a form of an electromagnetic wave that consists of electric and magnetic fields. Electricity, magnetism, and light had been fundamental topics of much scientific research, as briefly listed below, leading to an accumulation of knowledge that provided a solid basis for Maxwell to employ his own findings to reach unification. Although light was obvious, electricity and magnetism were also known (and sometimes, ironically, confused with each other) even in early civilizations. These mystical forces were naturally observed in certain objects[39] such as amber (electricity) and magnetic iron (magnetism), as well as animals such as catfish and eels. Obviously, natural events, such as lightning, were also examples of electricity, but connections between all these phenomena were not understood. For humanity,

[38] Note that *electromagnetism*, *electromagnetics*, and *electrodynamics* are used synonymously.

[39] Thales of Miletus (seventh and sixth century BCE) and Aristotle (fourth century BCE).

electricity and magnetism remained unknown (and separate) but interesting properties of some objects and a few creatures.

It is well-known that magnetic compasses, which are a truly practical use of magnetism, were used in China in the eleventh century.[40] As these devices became widespread worldwide, electricity and magnetism attracted the interest of more scientists – perhaps these forces could also be used for other purposes. Early scientists particularly focused on explaining magnetism[41] and improving compasses.[42] Gerolamo Cardano, a prolific mathematician of the sixteenth century, was one of the prominent scientists who clearly distinguished electricity and magnetism.[43] Cardano's studies were the basis for William Gilbert,[44] who performed experiments to identify electrical properties of different materials and relationships between electricity and other characteristics: e.g. moisture and temperature. Following Gilbert's studies, Robert Boyle did further experiments on electricity, including its behavior in a vacuum. In 1663, Otto von Guericke build and demonstrated a remarkable invention: the first electric (specifically, electrostatic) generator, based on a friction mechanism.[45]

Scientific experiments on electricity and magnetism became more popular in the eighteenth century; we also have more information about these studies, in comparison to earlier centuries, thanks to the available records. Fascinating properties of electricity attracted the interest of many scientists, including a renowned astronomer, Stephen Gray. His epic experiments starting in 1729 were on the conduction of electricity, which inspired other scientists, including Jean-Antoine Nollet and Charles François de Cisternay du Fay. At that time, although electricity was a clearly observed and somewhat controlled phenomenon, there was no comprehensive explanation of its nature. Du Fay proposed a *two-fluid* theory, which later provided a basis for positive/negative charges, supported by many scientists.[46] Nollet was the inventor of the electroscope, and he performed varied experiments, including measuring the speed of electricity. Similar experiments, which demonstrated this shockingly high speed (and the transmission of information), were carried out by William Watson, who also supported the two-fluid theory of Du Fay. Watson further performed experiments with *iconic* Leyden jars (basic capacitors), invented by Ewald Georg von Kleist and Pieter van Musschenbroek (of Leiden) in 1744–1746. Meanwhile, Von Guericke's electric generator was continuously improved by numerous scientists, including Francis Hauksbee (in 1706), Georg Matthias Bose (in 1730), and Jan Ingenhousz (in 1746). The eighteenth century also saw applications of electricity on living things for good purposes, such as improving the growth of plants, but the results were often questioned.

In the eighteenth century, one remarkable figure was Benjamin Franklin, who contributed significantly to research on electricity and many other fields. Franklin's legendary experiments using a kite really happened in 1752; these were safe versions of experiments that he and other scientists had performed previously using conductive rods.[47] By charging a Leyden jar using a spark from lightning, Franklin proved that lightning is electricity, as had been suspected by many

[40]Described by Shen Kuo in 1088.

[41]Petrus Peregrinus de Maricourt in the thirteenth century.

[42]Flavio Gioja in the fourteenth century.

[43]Gerolamo Cardano in 1550.

[44]William Gilbert also invented the word *electrica*. The word *electricity*, along with many other terms, appears to have been invented by Thomas Browne in the seventeenth century. Athanasius Kircher, also in the seventeenth century, is considered the first user of *electromagnetism*, without an obvious indication that he knew the actual relationship between electricity and magnetism.

[45]In the seventeenth century, studies of optics were carried out independently of electricity and magnetism. At the beginning of the century, Johannes Kepler described various behaviors of light, leading to significant improvements in lens systems for practical use (e.g. telescopes for astronomy). In 1621, Willebrord Snellius put forward Snell's law, which was rediscovered again and again by many scientists, including Ibn Sahl (tenth century), Thomas Harriot (in 1602), and René Descartes (in 1637). In 1662, a famous scientist, Pierre de Fermat, described the *principle of least time* (also known as Fermat's principle), which provides a link between ray and wave optics that was not understood until centuries later. Christiaan Huygens was another prominent scientist who proposed the wave theory of light in 1678. Given the major advances in optics during the seventeenth century, many studies were practiced in the context of astronomy. One interesting figure is Ole Christensen Rømer, who realized, by observing the moons of Jupiter, that the speed of light is not infinite. Rømer's data was used by Huygens to compute this speed as $212,000$ km/s. In 1728, this value was improved by another astronomer, James Bradley, to $283,000$ km/s based on an aberration of light.

[46]Including Christian Gottlieb Kratzenstein, who may be the real Frankenstein of Mary Shelley.

[47]Thomas-François Dalibard did so early in the same year.

scientists. Franklin also extended the two-fluids concept of electricity and used *positive/negative* fluids to describe how electricity behaves, but this was not immediately accepted by others.

The relationship between electricity and magnetism was still unknown in the eighteenth century, while scientists started to realize significant similarities and connections between them.[48] A natural philosopher, Franz Ulrich Theodor Aepinus, put forward an outstanding set of descriptions of electricity (in 1759), including electrical particles and attractions between them, distributions of particles on bodies, and conduction/insulation, as well as magnetism, some of which were verified by experiments. Aepinus's findings were significantly supported by Henry Cavendish, who further put electricity in action by synthesizing water from hydrogen and oxygen. Studies further accelerated toward the end of the eighteenth century. In 1774, Georges-Louis Le Sage showed how electricity could be used to carry information by employing a system of wires, which can be seen as an early telegraph. It was 1784 when Charles-Augustin de Coulomb described forces between charged bodies.[49,50]

Maxwell's studies were in the third quarter of the nineteenth century, and he was *standing on the shoulders of giants*. We now recognize some of these leading scientists of the nineteenth century and their fundamental work:

- Alessandro Giuseppe Antonio Anastasio Volta: In 1800, Volta invented an early electric battery (voltaic pile), which could produce steady electric current. This was a result of lengthy effort by Volta and many other scientists, including Luigi Galvani (the pioneer of bioelectromagnetics), who believed that electricity was mainly an inherent property of living things.[51] Volta's battery clearly demonstrated the generation of electricity via chemical reactions instead of the frictional mechanisms that had long been used.
- Carl Friedrich Gauss: The great mathematician contributed to the development of electromagnetism with his work on the well-known Gauss' law named after him, which was originally formulated by Joseph-Louis Lagrange in 1773. Gauss collaborated with Wilhelm Eduard Weber to construct the first electromagnetic telegraph in 1833.
- Hans Christian Ørsted: In 1820, Ørsted announced the astonishing observation that a nearby electric current could affect the needle of a compass.[52] While Ørsted was already looking for connections between electricity and magnetism, this relatively simple and easy-to-replicate observation attracted the interest of many other scientists, including André-Marie Ampère and Michael Faraday, in performing further research on the unification of two phenomena.
- André-Marie Ampère: In 1821–1823, Ampère observed and defined a set of rules that described attractive and repulsive forces between current-carrying wires.[53] He also predicted the existence of a particle, the electrodynamic molecule, to explain electricity and magnetism.

[48] Henry Elles in 1757.

[49] The inverse square law was realized by Joseph Priestley in 1767.

[50] In the area of optics, Isaac Newton, Jean-Baptiste le Rond d'Alembert, Leonhard Euler, and William Herschel made essential contributions during the eighteenth century. In 1704, Newton published his historically significant book *Opticks*. The corpuscular theory of light used in the book was opposed by Euler, who explained diffraction in 1746 via the wave theory of Huygens. In 1740, d'Alembert theoretically explained refraction, while Herschel's work was mainly on astronomy, leading to the discovery of infrared radiation from the Sun in 1800. Only one year later, ultraviolet radiation from the Sun was shown by Johann Wilhelm Ritter.

[51] In his famous experiments with frog legs, Galvani reached the conclusion that electricity exists in nerves/muscles. Volta opposed this idea by correctly claiming that an electromotive force created by the metal used during the experiments produced the current along a circuit (completed by a leg).

[52] This could be seen as a conceptual demonstration of a galvanometer: i.e. an electromechanical device to measure electric current. A practical galvanometer was invented by Johann Salomo Christoph Schweigger in 1822 and improved by Wilhelm Eduard Weber in 1833. As a related note, Ørsted's observations were also reported by others: e.g. by Gian Domenico Romagnosi in 1802, without a solid explanation.

[53] These studies by Ampère led to Ampere's *force* law. Ampere's (circuital) law, one of the four Maxwell's equations, was derived by Maxwell.

- Georg Simon Ohm: Ohm's work focused on the interaction of electricity and conducting matter, particularly their structural properties.[54] Using a galvanometer, he was able to derive relatively simple equations, published in 1827, to successfully describe the relationship between electromotive force and electric current.
- Michael Faraday: Faraday's initial studies focused on chemistry, as he was supervised by Humphry Davy, who performed important experiments with voltaic piles (electrochemistry). Inspired by Ørsted's work, Faraday built an early electric motor in 1821, which could be seen as an exemplar of *electromagnetic engineering*. His research continued on electricity, magnetism, and optics, with a particular effort to identify relationships between them. A series of experiments he performed in 1831 bore fruit with the discovery of electromagnetic induction,[55] which is the main operating principle for many devices, including early telephones and telegraphs. By studying various substances, Faraday also identified paramagnetism and diamagnetism, which had been observed but remained mysterious for a long time.

With the beginning of the nineteenth century, electromagnetism became a major scientific branch integrating electricity and magnetism, and it was continuously nourished by research in chemistry, physics, and related branches using electricity and magnetism.[56] Scientists found electromagnetism a fruitful area to explore, leading to enthusiastic studies that resulted in various inventions,[57] one after another. Experimental efforts were strongly supported by mathematical studies that provided theoretical bases to understand electromagnetism,[58] while studies of optics were continuing at full speed.[59] Finally, in 1864, Maxwell announced his theory of light, changing the course of history for electromagnetism.

Maxwell performed intensive research on electromagnetism, in addition to his contributions in other topics, starting in the 1850s. In 1855, he published a paper on Faraday's work, providing a simplified approach to describe how electricity and magnetism were related. In a following paper published in 1861, he was able to construct 20 differential equations involving 20 variables to describe electromagnetism. Solutions to these equations described how electromagnetic propagation could occur and predicted its speed, which *appeared* to be the same as the well-known speed of light. Maxwell's papers in 1864 and 1865 concluded that light is an electromagnetic wave consisting of oscillating electric and magnetic fields. The similarity between electricity and light, as well as the mysterious similarity of their speeds, had long been known but unexplained. It was Maxwell who added the missing term – the displacement current in Ampere's law – to complete the picture. Maxwell published his complete work on electromagnetism as a book (*A Treatise on Electricity and Magnetism*) in 1873. Of 20 differential equations derived by Maxwell, 12 were reformulated and reduced into 4 equations by Oliver Heaviside using vector calculus (1884), leading to the *standard* Maxwell's equations that we know today. In addition to constructing a

[54] Similar thermal studies were conducted by Jean Charles Athanase Peltier, who found the so-called *Peltier effect* in 1834. In 1840, James Prescott Joule described his first law, which establishes complete relationships between current, voltage, and power.

[55] Electromagnetic induction was discovered independently by Joseph Henry in 1832. Heinrich Friedrich Emil Lenz was another scientist who studied electromagnetic induction, leading to the well-known Lenz law in electromagnetics (1833). After it was invented by Nicholas Joseph Callan in 1836, many scientists performed research on the *induction coil*.

[56] Gustav Robert Kirchhoff described his well-known circuital rules in 1845. He showed that the speed of electricity is the same as the speed of light in 1857.

[57] For example, William Sturgeon invented electromagnets in 1825.

[58] Siméon Denis Poisson and George Green are among the most famous scientists who used mathematics to analyze electromagnetic problems. Green's milestone essay in 1828 inspired many scientists, including Maxwell. Today, Green's functions are fundamental tools of numerical and computational electromagnetics.

[59] In early 1800s, Thomas Young, Étienne-Louis Malus, David Brewster, Augustin-Jean Fresnel, Joseph Ritter von Fraunhofer, and Alexandre-Edmond Becquerel made numerous experimental and theoretical studies of absorption, reflection, transmission, interference, and polarization of light. Dominique François Jean Arago, Jean Bernard Léon Foucault, and Armand Hippolyte Louis Fizeau studied the speed of light. Foucault and Fizeau reached a quite impressive result of 298,000 km/s in 1849. Arago also discovered a *strange* phenomenon, called Arago's rotation, in 1824, which was explained later by Faraday via electromagnetic induction. Fizeau was involved in the description of the Doppler effect, and Foucault was the inventor of his famous pendulum.

strong basis for modern electromagnetism, Maxwell demonstrated that, when performed systematically, theoretical studies could provide scientific knowledge far beyond experimental studies. He inspired many scientists, including Heinrich Rudolf Hertz, Hendrik Antoon Lorentz, and Albert Einstein.

Supported by the famous scientist Hermann Ludwig Ferdinand von Helmholtz, Hertz proved the existence of electromagnetic waves via a set of experiments during 1886–1889.[60] Hertz's findings triggered many subsequent experiments in the following years, leading to the invention of *radio* by Karl Ferdinand Braun and Guglielmo Giovanni Maria Marconi by the end of the nineteenth century. Meanwhile, scientists became more familiar with electricity and mastered its manipulation, resulting in many practical devices,[61] such as induction motors invented independently by Nikola Tesla in 1887 and Galileo Ferraris in 1888. In 1897, Joseph John Thomson discovered that so-called cathode rays are actually composed of small particles, which had already been conceptualized and named *electrons* by George Johnstone Stoney in 1894. At that time, cathode tubes were subjects or components of many important experiments, including those of Wilhelm Conrad Röntgen, who discovered X-rays in 1895. We should also mention William Thomson (Lord Kelvin), who made significant contributions in the nineteenth century both theoretically and experimentally, leading to the invention of various key components of telegraphy.

[60]Hertz thought that these experiments that demonstrate *invisible* waves were interesting to verify Maxwell's theoretical work. At that time, however, he did not immediately realize the practical importance of electromagnetic propagation.

[61]As a notable example, William Stanley, Jr., invented the first practical AC transformer in 1885. The first practical light bulb is invented in 1878 by Thomas Alva Edison, who also constructed the world's first electrical power distribution system (1882).

At the beginning of the twentieth century, the fundamentals of electromagnetism (including the content of this book) were established for scientists to proceed beyond Maxwell's equations. Considerable research focused on *luminiferous aether*: i.e. the undiscovered medium that enables light propagation, based on the assumption that a wave *must have* a medium to travel.[62] Even though one of the most famous experiments, the Michelson–Morley experiment performed by Albert Abraham Michelson and Edward Williams Morley, failed to prove the existence of aether in 1887, experiments continued for at least 30 years with the same *null* outcome. Lorentz studied the problem mathematically and developed electron theories, still based on aether. He provided explanations for the failure of the aether experiments, leading to the theory of Lorentz–FitzGerald contraction.[63] Lorentz's work, together with extensive contributions by Joseph Larmor, led to the well-known Lorentz transformations (1897–1905), which enable a complete formulation of electromagnetism in different coordinate frames, making *time* a dimensional parameter.[64] In the same period, Jules Henri Poincaré made extensive studies of these transformations and improved them mathematically.[65]

[62]Aether was required to explain the *constant* speed of light, independent of frame, without violating the Galilean invariance of classical mechanics. Note that it was actually classical mechanics that failed to explain the constant speed of light.

[63]Named after Lorentz and George Francis FitzGerald, this theory describes the contraction of a moving object. Einstein showed that this contraction could be explained without assuming the existence of aether: i.e. using special relativity.

[64]This is called Minkowski space, named after Hermann Minkowski.

[65]Poincaré also proposed the existence of gravitational waves in 1905.

On top of these achievements, 1905 was Einstein's *annus mirabilis*; he published four milestone papers on (i) photoelectric effect (supporting the quantum theory by Max Karl Ernst Ludwig Planck), (ii) Brownian motion (atomic theory), (iii) special relativity, and (iv) the equivalence of matter and energy (the famous $E = mc^2$). His theory of special relativity was based on Lorentz transformations, without needing (and, actually, rejecting) the existence of aether. After 1920, various well-known scientists, including Paul Adrien Maurice Dirac, Wolfgang Ernst Pauli, Werner Karl Heisenberg, Enrico Fermi, Arnold Johannes Wilhelm Sommerfeld, Erwin Rudolf Josef Alexander Schrödinger, and Louis Victor Pierre Raymond de Broglie, constructed the basics of quantum electrodynamics.

With few exceptions, scientific and technological advancements in the last 100 years have been the successful results of multidisciplinary research involving contributions from various branches. Electromagnetism has taken an active role in both theoretical and practical studies, leading to a plethora of discoveries and inventions from theories for understanding the universe to technologies that we use every day. Any research and its outcome today would be impossible without the contributions of great scientists, only some of whom could be listed above.

8.3 Electromagnetism in Action

Electromagnetism inherently exists in most of the natural events in the universe, as well as in almost all technological areas developed in human history. We now consider several phenomena and technological areas where electromagnetism can be directly identified as a significant mechanism.

8.3.1 Snapshots from Nature

8.3.1.1 Blue Sky, Bright Sun, Red Sunset

The dominant color of the sky is due to Rayleigh scattering (Figure 8.8), a mechanism named after John William Strutt[66] (Lord Rayleigh). While Rayleigh scattering is mentioned in Section 8.1.6 to describe interactions of X-rays with atoms, it is a general phenomenon, describing elastic scattering of electromagnetic waves from small particles with respect to the wavelength. When a small particle is exposed to an electromagnetic wave, it becomes polarized and behaves like a small dipole that generates secondary waves (scattering) in all directions. If a particle is very small compared to its wavelength, its scattering cross section (ability to scatter electromagnetic power) is inversely proportional to the fourth power of the wavelength. Therefore, at optical frequencies, blue and violet lights (with higher frequencies) are scattered better than yellow and red lights. During the day, air molecules (like nitrogen and oxygen) behave as tiny secondary sources that scatter sunlight (including the entire

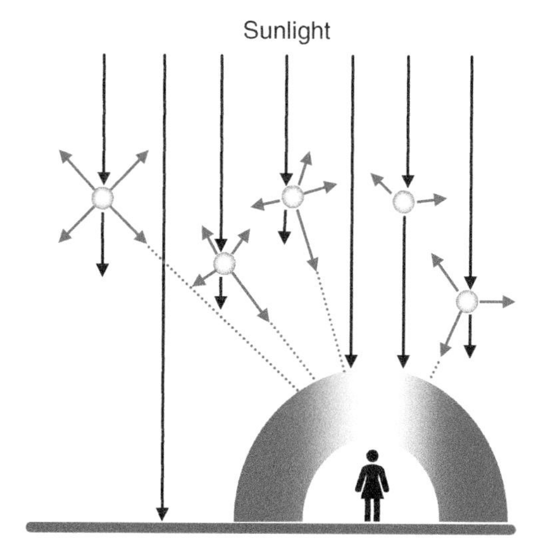

Figure 8.8 Rayleigh scattering is the main mechanism for the blue color of the sky. When sunlight hits an air molecule, all colors scatter; but the intensity of the scattered wave is inversely proportional to the fourth power of the wavelength. Therefore, blue/violet colors dominate, and air molecules behave like tiny blue/violet lamps shining in all directions. This naturally leads to a blue color, except in the close vicinity of the Sun, where direct sunlight (or sunlight with partially eliminated blue/violet) dominates.

[66] Rayleigh explained his scattering theory in 1871. According to particle theory (photons), Rayleigh scattering is elastic: i.e. a photon does not lose energy due to its interaction with an atom. According to wave theory, a light wave makes charges in a particle oscillate, creating radiation at the same frequency.

visible spectrum) in all directions, while blue and violet colors are scattered more than others. Since waves corresponding to the violet range are partially absorbed by oxygen molecules and the human eye is more sensitive to blue than violet, we see the sky as blue with smooth variations.

Rayleigh scattering occurs everywhere in the atmosphere, leading to randomly scattered blue/violet light in all directions. But blue dominates, particularly in directions away from the Sun. In regions close to the Sun, effects of Mie scattering (scattering from larger particles: e.g. water droplets, which are comparable to or larger than wavelength) also become visible, leading to brighter blue colors.[67] In the vicinity of the Sun during the day, we mainly observe light directly from the Sun, after some elimination of the blue/violet range due to Rayleigh scattering (in other directions), leading to white light with yellow hints. During sunset (and sunrise), regions around the Sun have strong yellow and red colors, as a result of the increased effective thickness of the atmosphere from the Sun to our eyes (Figure 8.9). Specifically, due to Rayleigh scattering, blue and violet (and even green) colors are effectively filtered out, leaving yellow and red.[68] Air pollution (various aerosols) and water droplets can also contribute to these spectacular yellow-red scenes by scattering and removing blue-range colors in large areas around the Sun.

[67] In Mie scattering, the maximum scattering occurs in the forward direction; but there is no strong wavelength dependency, so colors are scattered almost equally. The white-gray colors of clouds are also caused by Mie scattering.

[68] Note that a filtered blue light that cannot reach us is scattered and may contribute to blue light reaching an observer at another location.

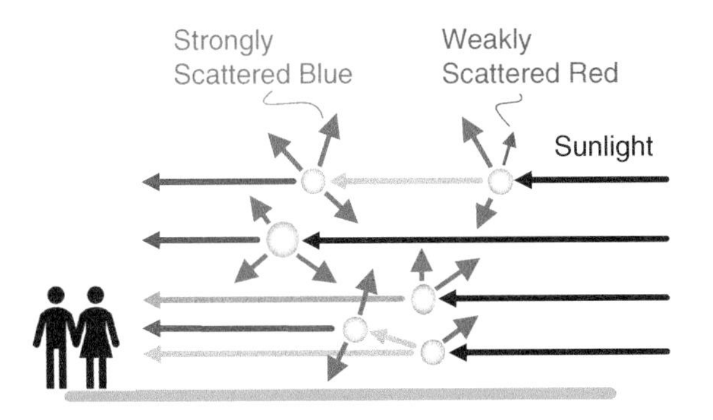

Figure 8.9 During sunrise and sunset, sunlight travels longer distances in the atmosphere, eliminating blue and violet light because they are scattered by air molecules more than other colors of light. This leads to primarily red and yellow colors around the Sun. Blue sky is still visible far from the Sun (due to the scattering of blue light everywhere), as it is during the day.

8.3.1.2 Rainbow in Our Pocket

In an iconic experiment, white light is transmitted through a triangular prism, resulting in optical (rainbow) colors from red to violet being output. As a clear demonstration that white light is composed of electromagnetic waves at different frequencies of the visible spectrum, this experiment also shows that the response of a material generally depends on frequency (*dispersion*). Specifically, if the glass had the same property for all electromagnetic waves, we would obtain the same white light from the prism.

When an object is illuminated by an electromagnetic wave, different phenomena may occur depending on the physical/chemical properties of the object and the frequency of the wave. Specifically, electromagnetic waves can be reflected/diffracted, absorbed (and, possibly, converted into heat), and/or transmitted. For example, in practice, we observe that some metals are good reflectors, wood and soil are good absorbers, and glass and water are good transmitters of optical waves. On the other hand, the transparency of an object does not mean electromagnetic waves simply pass though it without being affected. Depending on the material's properties, which can be represented by its permittivity, permeability, and refractive index, an electromagnetic wave can be significantly slowed, even when it is efficiently transmitted. This is due to the interaction of the electromagnetic wave with the molecules and atoms of the transparent object.

Figure 8.10 White light passing through a glass prism is divided into colors traveling in different directions due to dispersion.

As the result of these interactions, dispersion is caused by the responses of an object to waves at different frequencies.[69] Even though this difference can be small, the results can be dramatic, as in the case of the prism experiment. For glass, the refractive index changes less than 2% in the visible range, increasing from 1.512 (at 405 THz) to 1.534 (at 790 THz), but this is more than sufficient for a palm-sized triangular prism to generate colors from white light. Encountering larger values of the refractive index, the blue-violet side of the spectrum refracts more while passing from air to glass and glass to air. Consequently, with different refractions, we obtain color rays traveling in different directions at the output. Note that white light passing through any type of glass demonstrates dispersion, whereas a prism (or similar) shape leads to a separation of colors via frequency-dependent refraction.[70] Rainbows in the sky are natural results of dispersion of white light through water droplets, combined with internal reflection.

[69]For example, the sizes of atoms and molecules and the distances between them *in terms of wavelength* depend on the frequency. Hence, a material is actually *different* at different frequencies.

[70]For example, when white light is normally incident on a glass slab, waves at different colors travel with different velocities (Figure 8.11), whereas the output is still white light (in steady state), without a separation of colors.

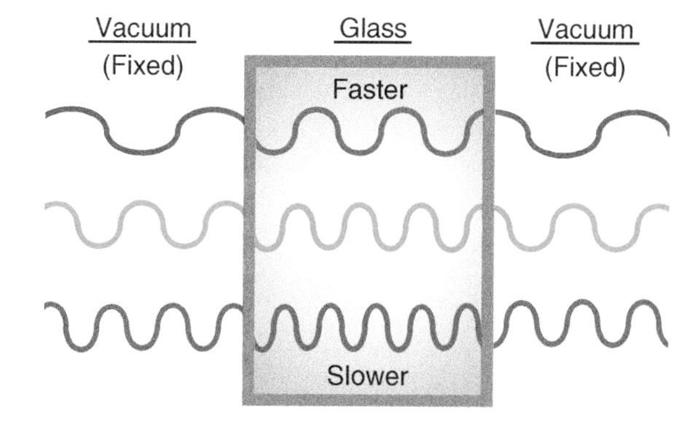

Figure 8.11 Due to *dispersion*, red light is slightly faster than blue light in glass.

8.3.1.3 Green Leaf, Red Apple, Blue Sea

Colors are interesting results of selective reflections from objects (Figure 8.12). Depending on its atoms and molecules, each material in nature has certain responses to electromagnetic waves at different wavelengths. In the visible range, these responses can be categorized as reflection, absorption,[71] and transmission.

For an object that is not a light source itself, what we see corresponds to reflected waves. Under white light (sunlight), a typical leaf is green, since most of the reflected wave is in the range from 530–600 THz. This is because leaves contain green chloroplasts (organelles), which contain green pigments called chlorophyll. These special pigments absorb light to convert it into chemical energy, while this occurs for blue and red sides of the visible spectrum. The unused part, green light, is simply reflected, making leaves look green. In the case of a red apple (or other red objects), all colors are absorbed and converted into heat, except the reflected red color.

The same principle is valid for most of the objects surrounding us, including blue seas. Like glass, water is a transparent material in the visible spectrum, but this does not mean it can transmit light with 100% efficiency. In water with sufficient depth, white light is mostly absorbed by water molecules – except for blue light, which mostly scatters, providing a blue color. Such reflections are increased by impurities (e.g. salt), so alternative colors may arise due to living organisms, dirt, and pollution.

If incident light does not contain a frequency that can be reflected from an object, the result is the absence of color: i.e. black.[72] For example, a red object under blue light (which can be white light passing through a filter) is seen as black. A white object is white since it can reflect all waves in the visible spectrum, while gray is a perception of *reduced* white color. In nature, there is also *structural coloring*, where strong combinations of colors are generated via structural mechanisms (still including reflection/absorption phenomena). Some bugs, animals, plants, and minerals are known to show *iridescence*:[73] angle-dependent colors, which make the world more interesting.

[71]Since electromagnetic waves in the visible spectrum are non-ionizing, their absorption usually results in heating.

[72]Hence, the black color is ideal to absorb sunlight, which is useful in solar *thermal* panels that convert sunlight into heat. Different from these devices, solar photovoltaic cells directly convert sunlight into more useful electrical energy, but usually with less efficiency. Note that the visible spectrum is only half of the total solar energy; hence, extensive research is being conducted to extend the ability of solar cells to collect infrared and ultraviolet frequencies.

[73]Perhaps the most common examples are soap bubbles, which can reflect rainbow colors when viewed from certain angles. Iridescence usually occurs via multiple reflection/diffraction phenomena.

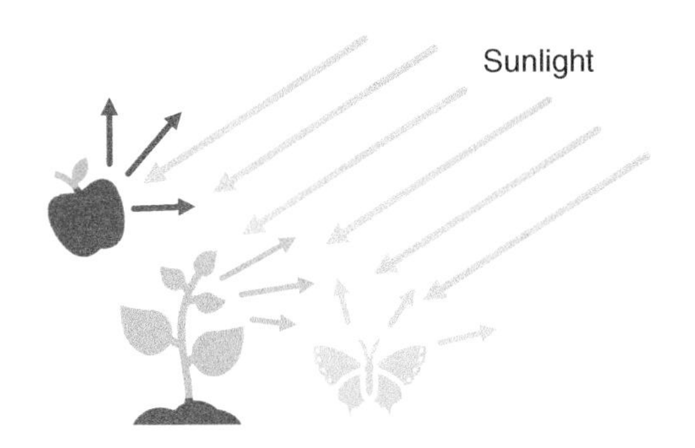

Figure 8.12 Colors of objects are generated by reflections from them.

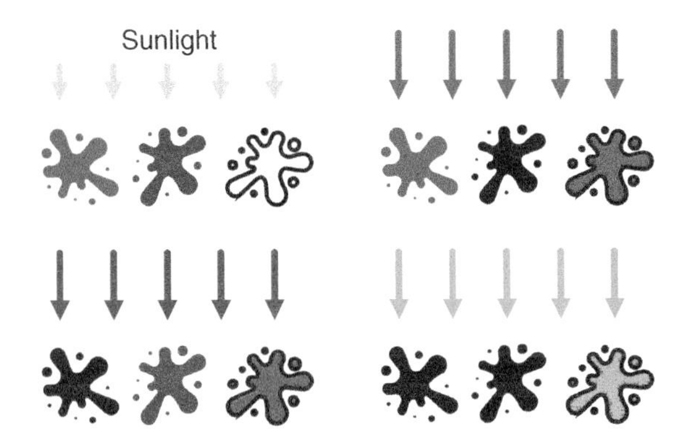

Figure 8.13 Colors depend on illumination, in addition to objects.

8.3.1.4 Electromagnetic Waves from Space

The Earth is literally bombarded by electromagnetic waves from different cosmic events and objects. The most important source is the Sun, which emits electromagnetic waves at many frequencies in the infrared, ultraviolet, and visible ranges, as well as radio waves and X-rays (from solar corona). The Sun can also emit gamma rays via solar flares. As also discussed in Section 8.1, all harmful high-frequency waves (X-rays, gamma rays, and partially ultraviolet radiation) are filtered (absorbed) by the Earth's atmosphere, allowing only useful waves to illuminate the surface of the Earth.

If incoming radiation reaches the inner atmosphere and surface, it mainly consists of visible and infrared frequencies. Some of these waves are reflected by clouds and the ground (particularly icy surfaces: e.g. polar regions), and some of these reflected waves are absorbed by the atmosphere (by *greenhouse gases*, including water vapor and carbon dioxide) before escaping from the Earth. Considering all reflections, we obtain a figure of merit: the albedo of the Earth is 0.3. This means 70% of the solar radiation illuminating the Earth is absorbed, and the rest is reflected into space.[74] Other than specialized processes like vision and photosynthesis, absorbed solar energy (infrared, visible, and ultraviolet radiation) is converted into thermal energy (heat): i.e. molecular vibrations. Heat circulates from object to object, ground to atmosphere, atmosphere to ground; but eventually, all energy is finally released from the Earth, making it balanced in terms of energy input/output.[75]

Similar to the Sun, all stars (and galaxies composed of them) radiate radio, infrared, visible, and ultraviolet waves, as well as X-rays and gamma rays that are filtered by the atmosphere. Since the relationship between heat and electromagnetic radiation is well known, the spectrum of the waves from a star provides fundamental information about its temperature and content. In addition, incoming waves demonstrate frequency shifts (due to the Doppler effect) that let us know the relative velocity of an object with respect to us. As mentioned in Section 8.1.7, high-energy events in the universe result in gamma rays, some of which can be observed via orbiting satellites.[76] All this incoming radiation is on top of background microwave radiation that originates from the early universe. This radiation, which can be continuously observed by radio telescopes, is actually photons from the *surface of last scattering*: i.e. when the universe became transparent due to recombination of protons and electrons, 379,000 years after the Big Bang. The peak frequency is at around 160 GHz, but it was originally a thousand times greater. As these waves have traveled for nearly 13 billion years, expansion of the Universe has *stretched* them, and they arrive at microwave frequencies.

[74] Obviously, these reflections contain the visible range, as astronauts can see the Earth from space.

[75] In recent decades, this balance is considered broken due to the increased greenhouse effect that increases the energy storage capacity of the Earth. From the perspective of physics, the Earth will find a balance, but this balance point may not be suitable for human habitation.

[76] In 2019, gamma rays with a frequency of 2.4×10^{28} Hz (100 TeV energy per photon) were observed, originating from the Crab nebula.

8.3.1.5 Magnetic Earth

Traditional compasses operate based on the Earth's magnetic field: the geomagnetic field[77] caused by the Earth's core (Figure 8.15). The motion of molten metals (iron and nickel) in the Earth's outer core causes electric currents that can generate a magnetic field distribution (dynamo mechanism). In addition to being detectable on the surface of the Earth, this magnetic field is also effective around the Earth, forming the magnetosphere (above the ionosphere) that makes the Earth habitable. The magnetosphere behaves as a shield, protecting the Earth from destructive solar winds (periodic streams of protons and electrons in the form of plasma) and cosmic rays (high-energy protons and nuclei) from outside the solar system. Thanks to this protection, solar winds, which could destroy the atmosphere and erase all living organisms,[78] are visible to us only as the *aurora borealis*.

On the surface of the Earth, the geomagnetic field is in the 25–65 μT range, which is sufficient to direct the needle of a compass. In addition, some animals (particularly birds) can detect the geomagnetic field and use it for navigation. Since it is caused by dynamically changing currents, the geomagnetic field is not constant. The north and south magnetic poles are located close to the geographic poles,[79] but they move slowly over the years[80] (tens of kilometers per year). In addition to these movements, as well as daily fluctuations (tens of meters per day), the geomagnetic field undergoes abrupt reversals. The average periodicity of these irregular events is long compared to a human lifetime, but it is quite short in terms of the Earth's history (183 reversals in the last 83 million years). In addition, the switching process (between two stable states) occurs quickly, estimated to be completed in 100–10,000 years.[81] The last switch occurred 780,000 years ago, and we are currently in a decline era, with a 5–6% decrease every 100 years. Similar to the Earth, the Sun has its own magnetic field distribution, the heliospheric magnetic field, which reverses every 11 years.

8.3.2 Snapshots from Technology

8.3.2.1 Telegraph to Cellular Phones

When scientists started to discover practical, interesting properties of electricity, magnetism, and electromagnetism at the beginning of the nineteenth century, one of their motivations was using them for communication. Before Gauss and Weber constructed the first electromagnetic (electrical) telegraph in 1833,[82] there were already optical and electrostatic telegraphs. Following Gauss and Weber, operational and commercial telegraph systems were constructed by various

[77]The geomagnetic field has been systematically studied since Gauss' early works.

[78]For example, Mars does not have a magnetic field that protects it from solar winds. Astronauts outside spacecraft are in danger from solar winds.

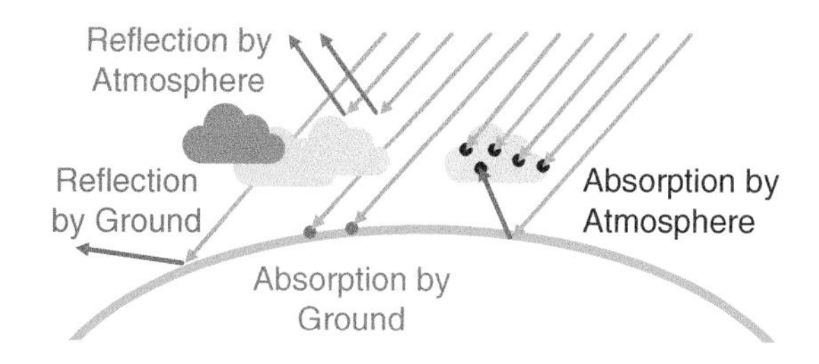

Figure 8.14 Some of the Sun's radiation reaching the Earth is directly (30%) reflected, and the remainder is absorbed by the atmosphere and ground. The absorbed energy is circulated as heat until it is released back to space (balance).

[79]The magnetic south pole is located near the geographic north pole, and the magnetic north pole is located on the opposite side, near the geographic south pole. Hence, the north side of a compass needle is directed north, as it is attracted by the opposite (south) magnetic pole. There are also *geomagnetic* north and south poles, which represent ideal locations if the Earth is modeled by an ideal magnetic dipole. These locations are different than the magnetic north and south poles since the Earth's magnetic field distribution is not perfect.

[80]Since this movement is arbitrary and relatively slow, it has not affected the use of traditional compasses since the eleventh century.

[81]All this information is obtained from rocks by paleomagnetists.

[82]Baron Pavel Lvovitch Schilling demonstrated a similar telegraph system in 1832.

scientists and used for practical purposes, such as railway signaling. In the telegraphy system constructed by Samuel Finley Breese Morse in 1838, only a single wire (instead of multiple wires, as in earlier systems) was sufficient to transmit the signal, using the so-called Morse code. Basically, in this widespread system, a message could be encoded into short/long electric current pulses, which could be transmitted from sender to receiver to be decoded. With the invention and improvement of electromagnetic components, such as induction coils, telegraph systems were quickly refined and became popular all over the world. As early as 1858, transatlantic communication became possible via submarine telegraph cables. To enhance telegraphs, particularly in terms of speed, special electromechanical devices called teleprinters[83] were further invented and used. But in 1876, another invention provided a better means of communication: transmission of the human voice.

After Alexander Graham Bell invented the first telephone, it was quickly accepted, leading to the construction of communication networks. In these early systems, connections were made directly via cables that were switched manually by operators. In addition, early telephones were dependent on their own power generators to operate, but this was improved later by transmitting the required power through communication wires. Regarding their disadvantages,[84] DC signals were replaced by AC signals (retaining DC to power telephones) for information transmission, but the conventional single-wire lines of telegraphs were insufficient to keep high-quality communications. Hence, twisted pair and similar multi-wire links were proposed and used. More complex telephone networks were constructed by using induction coils and amplifiers, allowing for truly long-distance (e.g. transcontinental) communications in the early 1900s.[85] Until 1980, communication via telephones was based on analog systems, involving divisions of transmission bands into small segments (links). With the help of optical fibers, digital and IP telephones came into use; but it took a long time for cellular systems to dominate personal communications.

Today, we are able to communicate with anyone in the world almost instantly. Mobile phones not only enable traditional telephone calls but also provide access to a vast amount of online information through the internet. This is possible using radio waves, which now cover large areas of the Earth's surface. In the last several decades, cellular communication technology has developed quickly, allowing us to access more data (both quantitatively and qualitatively) from our mobile phones, as well as laptops, tablets, and other electronic devices. Basically, a cellular network consists of cells (Figure 8.17), each of which is served by a transceiver (transmitter + receiver) base station using an allocated range of frequencies. Communications between a base station and mobile phones in its cell are wireless, allowing users great freedom of movement. Transceivers are connected to each other, usually via optical fibers, leading to the formation of the global *public switched telephone network*. The establishment of links between users and base stations, allocation of communication channels, maintenance of communications during cell-to-cell movements, and similar practical issues are organized via various protocols. New technologies with rapid advances in cellular communications are grouped as *generations*,[86] such

[83]Teleprinters can be seen as early computers with scanners and printers. Following conventional telegraphy, they are also used in radio communication.

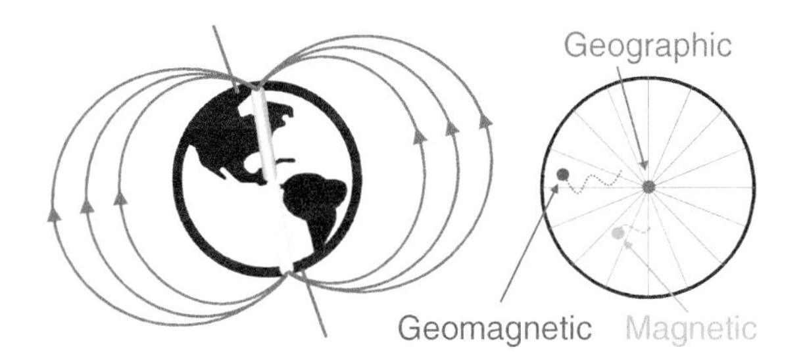

Figure 8.15 The Earth has three kinds of poles: geographic, geomagnetic, and magnetic. The north geographic pole (static), the south geomagnetic pole (tip of the representative magnetic dipole), and the south magnetic pole (where a compass shows the real center of the Earth) are located close to each other. However, the geomagnetic and magnetic poles are not static: i.e. they move over time.

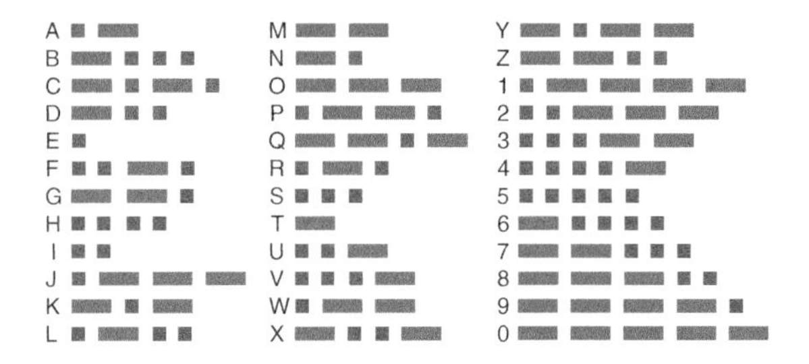

Figure 8.16 International Morse code.

[84]AC signals can be transmitted more efficiently than DC signals.

[85]A transatlantic telephone cable was first used in 1956.

[86]In 1G, cellular phones were wireless but analog. Digital technology was used in 2G, and 3G enabled broadband voice and data transmission. Communication speed has been significantly increased in 4G and 5G.

as 1G, 2G, 3G, 4G, 5G, 6G, and 7G. In general, as the technology develops through generations, the need for higher data rates requires the use of signals with higher frequencies, which puts new constraints on the size of the cells and positioning of base stations.[87]

[87] Radio waves tend to decay faster in the atmosphere as the frequency increases. In addition, their transmission through walls becomes inefficient, and they are diffracted less at higher frequencies. All these are disadvantages in terms of wireless transmission.

8.3.2.2 Home: Where Electromagnetism Happens

Many electrical and electronic household devices use electromagnetism directly or indirectly. As their name suggests, microwave ovens use microwaves directly to heat food[88] (Figure 8.18). A standard microwave oven operates at 2.45 GHz, which is similar to frequencies used for communication.[89] Unlike a mobile phone or Wi-Fi device, however, the total power of the electromagnetic waves used in a microwave oven is around 1200 W, which is sufficient to heat food.[90] Microwaves interact with polar molecules (mainly water[91]) inside foods, which are quickly converted into thermal energy[92]; but microwaves can penetrate effectively only a few inches, which limits the size of the food.

Television is another device that uses electromagnetic waves directly. Several decades ago, most televisions had antennas (particularly dipole or Yagi-Uda), either located on them or outside houses, to capture radio-frequency signals (mainly in the VHF and UHF bands) transmitted by large broadcast towers. Today, more common televisions use dishes that are directed toward *geostationary* satellites to receive 12–18 GHz signals.[93] As an electronic device, a television is like a laboratory of electromagnetic events. For example, early televisions used cathode ray tubes, in which electrons are deflected via magnetic fields. These have been replaced by electrically charged ions in plasma/LCD televisions and by light-emitting diodes in LED/OLED televisions.

Many household devices involve rotational or vibrational movement – electromagnetism in action. Refrigerators, washing machines, vacuum cleaners, electric fans, and mixers have induction motors that convert electrical energy into mechanical energy via electromagnetic induction. Even computers, laptops, and smartphones use small motors to operate properly.

Obviously, all types of light bulbs are electromagnetic devices. In the most standard *incandescent* light bulbs, metal filaments are heated via electricity. Using a suitable material, such as tungsten, electricity is converted into heat and visible light. But note that for traditional light bulbs, the light-conversion efficiency is usually very low: typically, less than 5% of the consumed energy can be released as light, and the rest is converted into heat. The efficiency increases for halogen lamps (using halogen gases in the bulb) and fluorescent lamps that are based on fluorescence (using mercury and phosphor). Today, conventional lamps are continuously replaced by

[88] A major component inside a microwave oven is the *cavity magnetron*, which is a vacuum tube that generates microwaves.

[89] This frequency (exactly 2.45 GHz) is not used for communication, so it can be used for microwave ovens without the possibility of interference.

[90] Hence, the energy per photon may be similar for a mobile phone and a microwave oven, but the total numbers of photons are different.

[91] This is why some paper, plastics, and similar materials that are not fire-resistant *can* be unaffected in a microwave oven.

[92] Hence, there is nothing related to *resonance*; the process is simply the conversion of electromagnetic energy into thermal (vibrational) energy.

[93] There are also cable televisions that use coaxial cables or optical fibers to receive signals, without any wireless transmission.

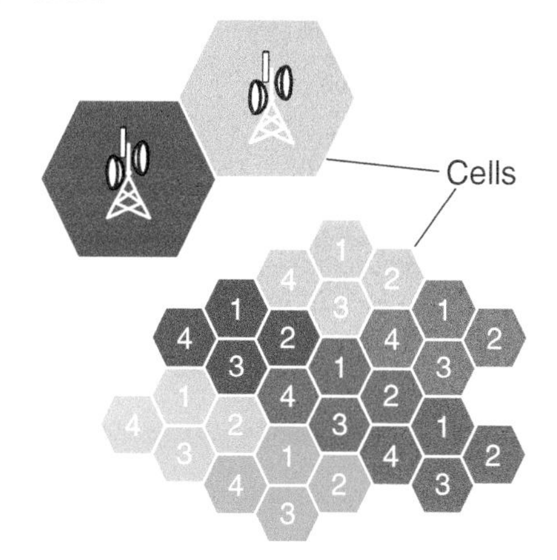

Figure 8.17 Cellular communication systems involve cells, each served by a base station. To avoid interference, nearby cells cannot use the same frequency band. Therefore, various topologies have been developed for reusing frequencies while avoiding interference in cellular networks. In the scheme shown here, with a 1/4 reuse factor, cells with the same number can theoretically use the same frequency band.

more efficient[94] and durable LED lamps, which use light-emitting diodes to produce light. Light that reaches our eyes (either directly or reflected from objects) is converted *back* into electrical signals flowing through our nerves.

[94]The increased efficiency is the reason for the decrease in power consumed by the bulbs.

8.3.2.3 Looking Inside the Body

Medical imaging directly involves electromagnetism. Only a few years after their discovery by Röntgen in 1895, X-rays were used in medical applications, particularly for noninvasive imaging.[95] In the following years, much attention was paid to improving the emission of X-rays for better visualization. X-ray tubes invented by William David Coolidge in 1913 enabled the continuous generation of X-rays and formed the early versions of modern tubes used today (Figure 8.19). When X-rays pass through a body, they are absorbed at different rates, depending on the tissue. For example, with their denser structures, bones absorb more X-rays than fat and muscle tissues. Using a photosensitive plate at the back of the imaged body, it is possible to generate a contrast-type picture based on the transmitted X-rays.

[95]At the same time, clinicians started to realize the hazardous effects of X-rays.

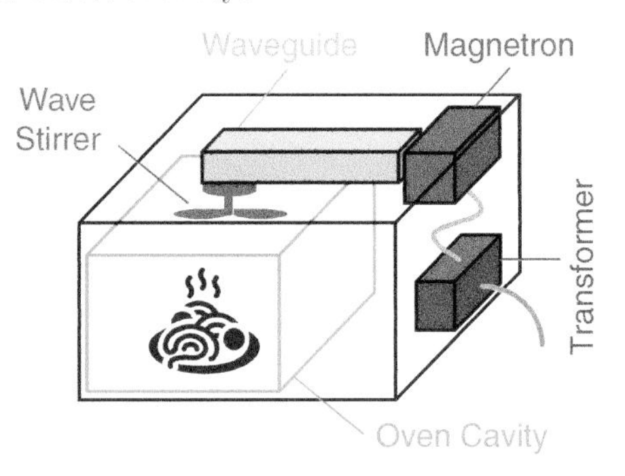

Figure 8.18 A simple illustration of a standard microwave oven. Using high voltage, the magnetron generates microwaves that are transmitted by a waveguide to the oven cavity. A stirrer is used to distribute microwaves uniformly (in addition to a typical turntable that rotates the food). The front side of the oven cavity (lid) is usually covered by a type of Faraday cage that enables perfect isolation while maintaining visual transparency.

In 1946, Felix Bloch and Edward Mills Purcell demonstrated the application of nuclear magnetic resonance on bulk materials, which opened a new door for material imaging. With its application in medical imaging, we now call this technique magnetic resonance imaging (MRI). In 1968, the computed tomography (CT) scan was developed, enabling an extensive visual ability (beyond projectional radiography and fluoroscopy) for medical practitioners. The idea was to illuminate the subject with X-rays from different sides, collecting data and combining it via the tomography technique to generate two-dimensional and three-dimensional images (Figure 8.20). Note that CT is considered an X-ray technique, and the same tomography techniques are also used in other imaging methods, including MRI.

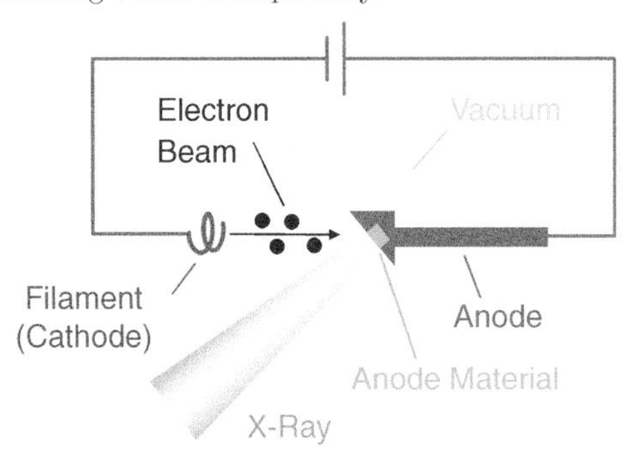

Figure 8.19 A simple schematic of a basic X-ray tube. Due to the high voltage, an electron beam is generated from the cathode filament (e.g. tungsten) to the anode. The beam excites the anode metal (which can also be tungsten), resulting in the emission of X-rays.

In 1971, Paul Christian Lauterbur demonstrated the first MRI images using tomography with nuclear magnetic resonance. MRI body scanners became possible in only 10 years, leading to the first installations in hospitals in the 1980s. Since then, MRI machines have been continuously improved for better visualization: better resolution in less time, four-dimensional (time-dependent) imaging, etc. In addition to MRI and CT, which dominate today's medical imaging, there are various techniques based on interactions of electromagnetic waves and the human body, such as microwave imaging, photoacoustic imaging (using electromagnetic waves at different frequencies, together with acoustic waves), thermography and near-infrared spectroscopy (using infrared radiation), and single-photon emission computed tomography (using gamma rays). Today, we have more than 36,000 MRI machines worldwide as an incredible result of rapid advances in this area. In an MRI machine, a very high static magnetic field (e.g. 7 T) is applied through the subject's body, making it temporarily magnetized. This magnetization is mainly due to hydrogen nuclei in water molecules, which can be found abundantly (but nonuniformly[96]) inside a human body. Due to its spin, each hydrogen atom has a magnetic dipole

[96]Hence, in most MRI applications, water is the visualized material. The distribution of the water provides the required contrast between different parts of the body.

moment: i.e. it can be represented by a magnetic dipole oriented arbitrarily when no external magnetic field is applied. Due to the applied magnetic field inside an MRI machine, magnetic dipoles (spinning hydrogen atoms) tend to orient in the same direction (magnetization). As the critical point, an oriented magnetic dipole can be disturbed by a *special* radio-frequency wave, which leads to the imaging process. Specifically, considering magnetization via a magnetic flux density B, the corresponding Larmor frequency is defined as $f = \gamma B$, where γ is the material-dependent gyromagnetic ratio (42.576 MHz/T for hydrogen). If a radio-frequency wave at the Larmor frequency is incident on a magnetic dipole, the dipole's orientation is temporarily disturbed until it goes back to its enforced orientation (assuming that the static magnetic field still exists). In this *relaxation* from the disturbed state to the aligned state, the magnetic dipole emits a radio-frequency wave at the same (Larmor) frequency, which is a signal to detect for visualization.[97] To generate an image, however, there must be *gradient* magnetic fields, in addition to the main magnetic field. If a whole body were exposed to the same static magnetic field, combinations of all signals from different parts of the body would be mixed. Hence, gradient magnetic fields are applied on top of the main magnetic field to create different magnetization strengths (different Larmor frequencies) for different parts of the body. This allows for the spatial collection of data and the use of tomography techniques to generate slices of the body (without really slicing it).

[97]All these phenomena are mathematically analyzed via Bloch equations.

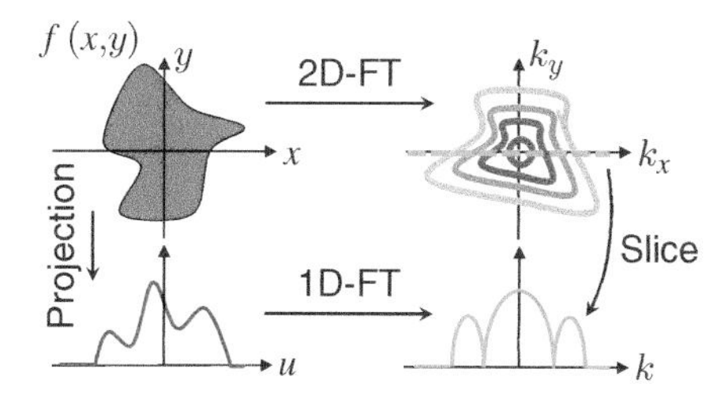

Figure 8.20 According to the central slice theorem in tomography, a one-dimensional Fourier transform of a projection of a two-dimensional function corresponds to a slice in its two-dimensional Fourier transform. Hence, projection data (many projections along different directions) can be used to reconstruct a two-dimensional function without really sampling it. Similarly, two-dimensional projections can be used to reconstruct three-dimensional functions.

8.3.2.4 Seeing the World with Sensors and Radar

Since the visible range forms only a small fraction of the entire electromagnetic spectrum, scientists have developed various techniques to enable *seeing* at other frequencies with the help of technology. Radar is such a detection system that usually operates at radio frequencies[98,99] and can identify objects – e.g. their position, geometry, and velocity – that are beyond human vision.[100] These systems were first developed and used for military purposes (secretly by numerous nations) during World War II. Today, radar is still a major component of defense systems and also used in many other areas,[101] including astronomy, aviation, geology, meteorology, navigation, oceanography, and traffic control (including self-driving cars). Before the war, many scientists were already aware of radio waves' reflection properties and the possibility of capturing them to detect objects under low visibility. But the challenge was to construct a useful and practical device to perform the required detection operations. Robert Watson-Watt is acknowledged as the leading scientist who demonstrated the first practical radar system (operating at low frequencies, e.g. 30 MHz) in 1935. Early prototypes were quickly improved, and a network of radar was constructed all over the UK by 1940. Monopulse radar, which provides effective detection of an object's range and direction, was introduced by Robert Morris Page in 1943. Note that this monopulse technology, with various enhancements, is still used in modern radar systems.

[98]Radar uses the advantages of radio waves, which attenuate less than electromagnetic waves at higher frequencies.

[99]Similar systems using sound waves are called *sonar*.

[100]There is indeed no *distance limitation* for human vision; the *intensity* of the light received by the eye is the limiting factor.

[101]Clouds are mostly transparent to radio waves, so cloudy or rainy weather does not affect the *visual* ability of air traffic systems. Seawater is a good reflector for radio waves, making it possible to measure altitude above seas.

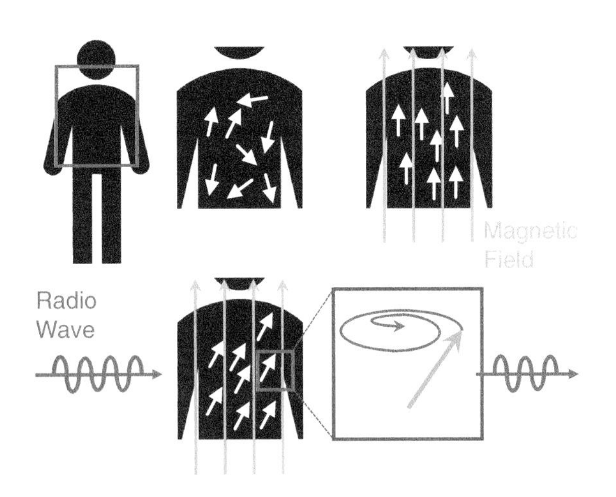

Figure 8.21 A simple illustration of MRI. When the body is placed inside a static magnetic field, magnetic dipoles (mainly hydrogen atoms) are aligned in the same direction and tend to keep their orientation as long as the magnetic field exists. Then a radio-frequency excitation at the Larmor frequency temporarily disturbs these magnetic dipoles, which emit radio-frequency signals during relaxation (while returning to their aligned states).

Since the 1940s, radar technologies have been continuously improved by extensive research in electromagnetics, electronics, computers, and signal processing.

A radar system consists of transmitter and receiver antennas (often the same antenna) connected to an electronic system that controls the antennas, processes received signals, and displays data.[102] The transmitter antenna sends a radio-frequency signal, and the receiver gathers the signals reflected from a target or targets. Depending on the frequency (or the frequency content of a pulse) and the target (size and material), electromagnetic waves generated by a radar system can be transmitted, absorbed, and/or scattered in all directions; only a small fraction of the reflected waves can reach the receiver and be used to detect the target (Figure 8.22).

Many techniques have been developed for different applications of radar, leading to various mechanisms.[103] For example, continuous-wave radar is based on continuous waveforms (rather than pulsed on-off waveforms), which enables better resolution and range at the cost of increased power consumption. Doppler radar is based on the Doppler effect and obtains information about the detected target's velocity, usually in addition to its position and range. This can be done with continuous (and often frequency-modulated) waveforms or pulsed waveforms, leading to the well-known pulse-Doppler radar for the latter. In addition to *monostatic* radar configurations, there are also *bistatic* radar configurations (Figure 8.23) involving separately located transmitters and receivers to facilitate detection. In terms of antenna technologies, phased (electronically scanned) arrays enable electronically controllable beams that can be steered in desired directions without any physical movement. The systematic collection of data via a moving antenna is also a radar technique – synthetic-aperture radar – and can be used to generate two-dimensional and three-dimensional images of targets (e.g. landscapes).

A sensor is a (usually small) device that collects data from the environment: e.g. humidity, light, movement, pressure, temperature, and velocity, as well as biological, chemical, physical, and electrical properties of the surrounding objects. With rapid advances in technology, these devices have become increasingly widespread in modern life. There are many sensing mechanisms, leading to various sensor types, including but not limited to biosensors, chemical sensors, electrochemical sensors, and optical sensors. Optical (or electro-optical) sensors convert light *information* in the infrared and visible ranges into electronic signals. They are often used in automobiles (e.g. parking sensors), lighting and signaling systems, security, smartphones, and traffic controls. Like all sensor types, optical sensors can be passive (passively collecting signals from outside) or active (actively sending signals and collecting reflections). Recently, there has been considerable interest in wireless sensor networks, which can be described as technological systems involving electronically connected sensors for environmental, industrial, and personal applications. In these systems, radio frequencies are not used in sensors, but they are used to connect the sensors effectively.

[102]Usually, antennas and electronic parts of a radar system are enclosed in a *radome* that protects against environmental factors while being transparent to radio-frequency waves.

[103]The frequency is a major parameter in radar systems. The frequencies used in modern radar systems differ significantly, from a few megahertz to several hundred gigahertz. Usually, lower frequencies enable longer ranges, and higher frequencies provide better resolution.

8.3.2.5 Atoms Under the Microscope

For many centuries, microscopes have been major tools of scientists in all areas, particularly biology and physics. These devices resulted from humans' desire to see and explore small-scale worlds, just as we use telescopes to explore distant objects. But as microscopes were continuously improved by manufacturing better lenses and using new illumination techniques, scientists realized there was a limit on the resolution that could be achieved. We now know that this is the fundamental Abbe diffraction limit, named after Ernst Karl Abbe, which defines the minimum *resolvable* distance as

$$d = \lambda/(2\text{NA}), \tag{8.1}$$

where λ is the frequency of the electromagnetic wave used for visualization and NA is the numerical aperture of the microscope (which depends on the medium, where the lens and object are located, and the spot angle). According to this limit, two points separated by a distance smaller than d cannot be distinguished under the microscope (they will be mixed due to blurring), independent of the magnification power. In Eq. (8.1), the microscope's resolution is represented by the numerical aperture, which is usually less than 1.5, even in modern systems. For such a value, we obtain $d = \lambda/3$, corresponding to 125–250 nm in the visible spectrum. Hence, using a *good* light microscope (with NA = 1.5), it is possible (with the violet spectrum) to see details larger than 125 nm,[104] while all smaller details will be blurred (Figure 8.24). Using a remarkable 1000× magnification, a microscope can magnify 125 μm (field of view) to 0.125 m, where 125 nm details are seen as 0.125 mm details in the overall view. Given the numerical aperture's limited value, increasing the frequency is a suitable way to improve the resolution and see smaller objects. For example, ultraviolet microscopes, which can double the resolution, are designed and used for this purpose.[105] Similarly, X-ray microscopes enable nanometer-scale resolution, and the easier penetration of X-rays into materials makes it possible to see internal details (especially of biological structures).

Obviously, outside the visible range, detected signals must be converted into digital pictures, making these microscopes more complicated. Gamma-ray microscopes have been conceptually considered, but they have not been practically constructed, primarily due to difficulty in controlling these rays for microscopy. Instead, non-electromagnetic microscopes are available to obtain very high resolutions. Well-known and common electron microscopes[106] are based on using accelerated electrons to illuminate objects and collecting reflected, scattered, or transmitted electrons or X-rays to create visions. In these microscopes, electrostatic and electromagnetic fields are still used to shape electron beams. Electron microscopes can provide resolutions much higher than X-ray microscopes: e.g. 50 pm resolution is achieved by a transmission electron

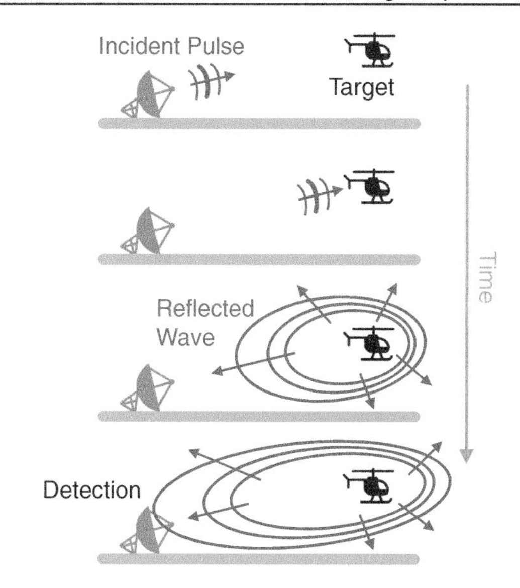

Figure 8.22 A demonstration of a pulse radar. For a target at a distance of 10 km, transmitting a pulse and receiving the reflected waves take less than 33.4 μs. If the target is traveling at 500 km/h, it can move only 5 mm during this period.

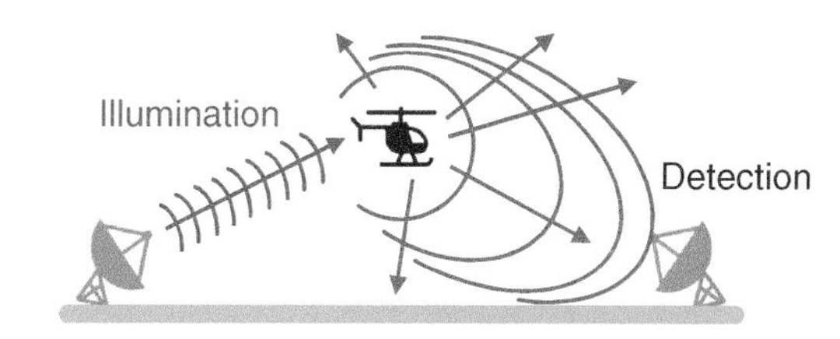

Figure 8.23 A bistatic radar system.

[104] Hence, because most cells are larger than 10 μm, they can be inspected under a light microscope. Some bacteria may also be seen clearly, but not viruses.

[105] Resolution is not the only concern in microscopy; the ability of waves to reflect from an object is also very important. For some structures, infrared microscopes may be preferred, despite their lower resolution.

[106] Two main categories of electron microscopes are transmission electron microscopes (based on transmissions through objects) and scanning electron microscopes (based on reflections and scattering). An important disadvantage is that electron microscopes have special requirements (e.g. very small thicknesses) for specimens.

microscope. With such resolutions, electron microscopes can reach atomic details, where atoms are generally seen as blurry dots.[107,108]

The resolution can also be improved by gathering data in the close vicinity of an object, rather than far-zone observations. Specifically, in the electromagnetic microscopes briefly described above, data is collected in the far zone (with respect to the wavelength) of an object, leading to the Abbe diffraction limit. In near-field scanning optical microscopy, optical waves illuminate the object, which is located very close to the microscope aperture. This way, the resolution is limited by the aperture size itself, which can be on a nanometer scale. In the extreme of measuring from the near zone, scanning force microscopes generate images by literally *touching* samples without a *direct* electromagnetic interaction.[109]

[107]Such an image shows *interactions* of electrons with the atom, rather than the geometric description (if any) of the atom itself.

[108]At atomic levels, it becomes possible to employ quantum effects for visualization. For example, scanning tunneling microscopes are based on quantum tunneling.

[109]Physical contact between two objects is the interaction of atoms (nothing really touches). Therefore, even the atomic force microscope has electromagnetism as its core mechanism.

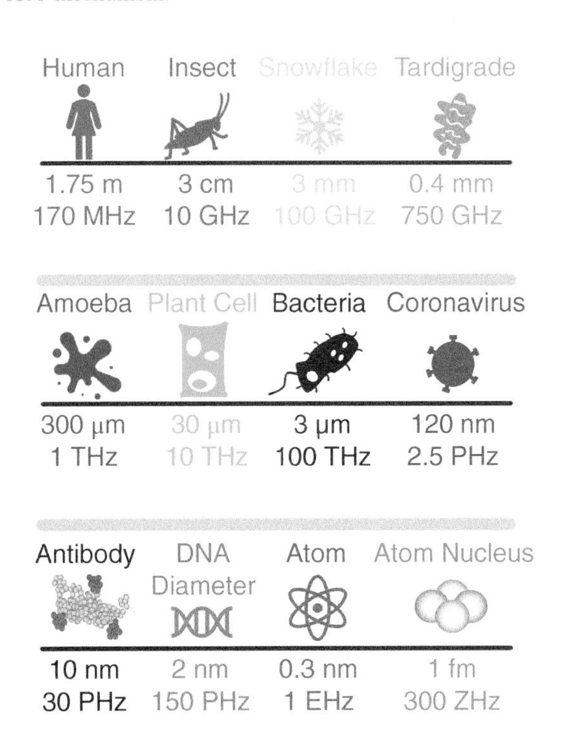

Figure 8.24 Sizes of some objects, and the corresponding frequencies of the electromagnetic waves having the same wavelengths in air.

8.4 How to Solve Maxwell's Equations

In all the examples in this book, solutions to Maxwell's equations are demonstrated on various structures that have ideal geometric properties: e.g. infinitesimality, infinity, and symmetry (perfectly spherical/cylindrical capacitors, resistors, charge distributions, current distributions). While real-life objects do not have such ideal geometries, these assumptions enable solutions to Maxwell's equations to obtain fundamental information about electromagnetic waves' behavior. For example, an infinitely large planar boundary allows us to derive main reflection and transmission properties without considering what happens at edges and corners. While these properties are derived for ideal cases, they are indeed useful for many practical cases: e.g. when an object is very large with respect to the wavelength such that edge/corner effects are negligible. All scientific areas simplify problems to make them analyzable. In electromagnetics, simplifications also include materials (frequency independence, linearity, isotropicity, homogeneity), waves (time-harmonic waves, infinitely large and planar wavefronts), and sources (collection of charges/currents without material support), in addition to geometries (Figures 8.25–8.27). Steady state is also a simplification, considering that transient state effects continue forever.

Under useful assumptions, solutions to Maxwell's equations lead to expressions in terms of given variables. For example, the capacitance of an ideal spherical capacitor was derived (see Eq. (6.336)) as

$$C_0 = 4\pi\varepsilon_0\varepsilon_r \left(\frac{ab}{b-a}\right), \tag{8.2}$$

where a is the radius of the inner PEC, b is the radius of the outer PEC, and ε_r is the relative permittivity of the material between the PECs. This type of expression, which involves variables

(a, b, and ε_r) and constants (π and ε_0) that are ready to be used to find the required quantity (C_0), is generally called an *analytical* expression. These expressions are *exact*: i.e. an analytical expression has a perfect accuracy. But its evaluation can be performed with different accuracy levels, depending on the context. Considering the same example, assume that[110] $a = 1$ cm, $b = 2$ cm, and $\varepsilon_r = 1.0$. Then an engineer may find the capacitance *approximately* as

$$C_0 \approx 4 \times 3 \times 9 \times 10^{-12} \times \frac{0.01 \times 0.02}{0.01} = 2.16 \text{ pF}, \tag{8.3}$$

where $\pi \approx 3$ and $\varepsilon_0 \approx 9 \times 10^{-12}$ F/m are used. While this calculation can be done quickly, it is very approximate. A better one using paper and pen (assuming $\pi \approx 3.1$ and $\varepsilon_0 \approx 8.9 \times 10^{-12}$ F/m) can lead to

$$C_0 \approx 4 \times 3.1 \times 8.9 \times 10^{-12} \times \frac{0.01 \times 0.02}{0.01} = 2.2072 \text{ pF}. \tag{8.4}$$

Using $\pi \approx 3.14159265359$ and $\varepsilon_0 \approx 8.85418782 \times 10^{-12}$ F/m in a standard calculator, we can obtain an even more accurate value as

$$C_0 \approx 4 \times 3.14159265359 \times 8.85418782 \times 10^{-12} \times \frac{0.01 \times 0.02}{0.01} \tag{8.5}$$

$$\approx 2.2253001127 \text{ pF}, \tag{8.6}$$

which is still approximate due to approximations of π and ε_0, as well as the calculation itself[111] (finite precision). Note that in all three of these evaluations, the same analytical formula in Eq. (8.2) is used, but the result depends on how it is evaluated.[112] Depending on the evaluation, different accuracy levels can be achieved, and all results become approximate, as opposed to the symbolic nature of the corresponding analytical formula.

Analytical expressions are derived by analytical (symbolic) derivations, as we have done in most of the examples and exercises in this book. Unfortunately, in electromagnetics, analytical derivations are possible only for a limited set of geometries (e.g. perfect spheres, perfect and infinitely long cylinders, infinitely large planes), material properties, and excitation types. On the other hand, real life cannot be modeled only in terms of these ideal items. Especially in the last several decades, rapid advances in science and technology have required both fast and accurate solutions to increasingly complex electromagnetic problems. Such impressively realistic analyses have been possible by employing *numerical* and *computational* methods and techniques to solve Maxwell's equations on computers.

Computational electromagnetics is an active research branch involving the development and application of numerical and computational methods and techniques for solving electromagnetic problems. While the main aim is to solve Maxwell's equations (or equations

[110] Note that numerical approximations are introduced whenever we insert numbers into the analytical formula. For example, $a = 1$ cm should also be an assumption, since nothing can be exactly 1 cm, considering atomic details.

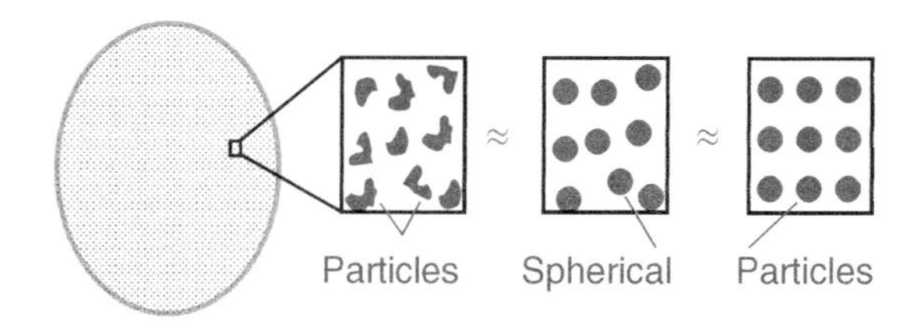

Figure 8.25 Examples of geometric approximations for particles that are very small with respect to the wavelength in a problem involving a large cloud of particles.

[111] Once the result is found, one may prefer to approximate it as $C_0 \approx 2.2253$ pF, again depending on the application.

[112] Approximations in an evaluation depend on the application: e.g. where and why the capacitor will be used. In an application where the capacitor is not critical, a 2.16 pF approximation may work, and more accurate calculations may be unnecessary.

derived from them) and analyze electromagnetic phenomena, computational electromagnetics employs various components from many areas, including but not limited to numerical integration, numerical differentiation, iterative methods, interpolation, algorithms, and high-performance computing. It is an area where fundamental concepts and functions, such as Green's functions, Bessel/Hankel functions, Legendre functions, Lagrange technique, Laguerre polynomials, Riemann sum, multipoles, and many others, are used to build complex algorithms that can solve electromagnetic problems without resorting to structural approximations in analytical approaches.[113]

Depending on the context, different categorizations of methods and techniques can be used in computational electromagnetics. Here, we provide a brief list of properties of different approaches:

- Semi-analytical methods:[114] These methods are conceptually analytical, as they lead to analytical (but not closed-form) formulas for solutions to canonical problems, while evaluations of the derived expressions require approximations (other than approximations *directly* caused by the finite precision). Although this is not strict, semi-analytical methods in computational electromagnetics generally involve infinite series that require approximation. A well-known semi-analytical method is called the *Mie-series* solution (named after Gustav Mie), which provides expressions for scattering from spherical objects. For example, when a PEC sphere of radius a is illuminated by a plane wave, the scattered electric field intensity can be written as

$$E^{\text{sca}}(\bar{r}) = f(R, \theta, \phi, k) \sum_{n=1}^{\infty} g_n(R, \theta, \phi, k), \tag{8.7}$$

  where f and g_n are functions of position and wavenumber $k = \omega\sqrt{\mu\varepsilon}$. Although the expression itself is analytical, its evaluation needs an approximation since the summation cannot be evaluated infinitely, and it must be truncated. On the other hand, as a stable solution technique, the Mie-series method has *convergent* characteristics: i.e. the added terms tend to become smaller and smaller (especially after a threshold depending on the sphere size), providing increasingly accurate results as more terms are added as desired. Note that truncation error can occur at the level of rounding error due to finite precision, so employing a Mie-series solution on a computer may directly correspond to using a full analytical method on a computer (without any truncation effect).

- Full-wave methods: While there is not a consensus on the content of full-wave methods, we can define them as *numerical* methods that solve Maxwell's equations with *controllable* accuracy[115] considering the *wave nature* of electromagnetism (Figures 8.28–8.34). All full-wave

[113]Using a computational method or technique, we accept approximations due to the finite precision of computers, as well as other *numerical* approximations. However, these methods and techniques can be used to analyze complex objects whose geometric (and other types of structural) simplifications (to use an analytical approach) would lead to meaningless results.

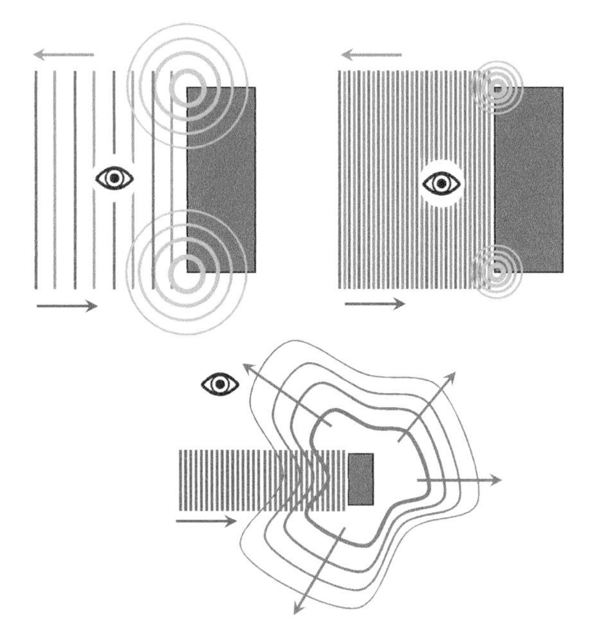

Figure 8.26 The electromagnetic model of an object depends on the scenario. Considering a slab at a low frequency, diffraction at the edges/corners may not be negligible, and computational methods may be needed for the analysis. The same slab at a higher frequency may be modeled as infinite (to be analyzed analytically) if the observation point is located away from the edges. For an observation point located far from the entire slab, the infinite model may fail even at high frequencies.

[114]Depending on how they are used, some semi-analytical methods may be considered analytical methods.

[115]Controllable accuracy is not well-defined, and it is used differently in different contexts.

methods aim to solve general problems involving objects with arbitrary geometries, materials, and excitations, but the underlying methodology puts restrictions on solvable problems. In addition, given a particular problem, one method can be better than other applicable ones in terms of accuracy, efficiency, and/or stability. For this reason, there are many methods in the literature, each providing the best solutions to a particular set of problems. In general, a numerical solution to a problem involving an object needs the *discretization* of domains or boundaries: i.e. dividing the given space into small pieces, on which Maxwell's equations can be reduced into a numerical set of equations. Discretization is important to convert Maxwell's equations into numerical equations that can be solved on computers; but the main form of numerical equations depends on the method used. As further categorized below, some methods generate partial differential equations, and the others lead to integral equations, each having its own advantages and disadvantages. In any case, the virtual elements used for discretization must be small enough to provide sufficient accuracy within the available computational resources. Since Maxwell's equations can be solved in both the time and frequency domains, there are time-domain and frequency-domain methods that provide solutions to a plethora of complex problems.

- Asymptotic techniques: Asymptotic techniques include those that depend on limiting quantities, particularly the wavelength with respect to object size. These techniques are often considered high-frequency techniques, which involve very short wavelengths, but we should also consider the other side: i.e. low-frequency techniques. When an object is very large with respect to the wavelength, its interaction with the incident wave can be approximated with reasonable accuracy. Specifically, the primary mechanisms like reflection, diffraction, and refraction can be modeled by considering incident waves *rays*. This may be necessary for many problems where full-wave models lead to enormous numbers of equations that cannot be solved on computers. Once again, based on the strategy used, various high-frequency techniques have emerged in the literature, and they are used extensively today, particularly when full-wave methods are insufficient. High-frequency techniques do not directly solve Maxwell's equations (or they employ them synthetically), so they do not provide controllable accuracy. Hence, high-frequency techniques are often verified using full-wave methods, which are further verified via semi-analytical methods. High-frequency techniques are also combined with full-wave methods to benefit from both strategies within the same problem. On the other side, low-frequency techniques attack problems involving small objects with respect to the wavelength. In these techniques, quasistatic approximations are employed to avoid full-wave solutions to Maxwell's equations since full-wave methods often suffer from stability issues when used for such problems.

In the following subsections, we consider brief lists of full-wave methods and asymptotic techniques.

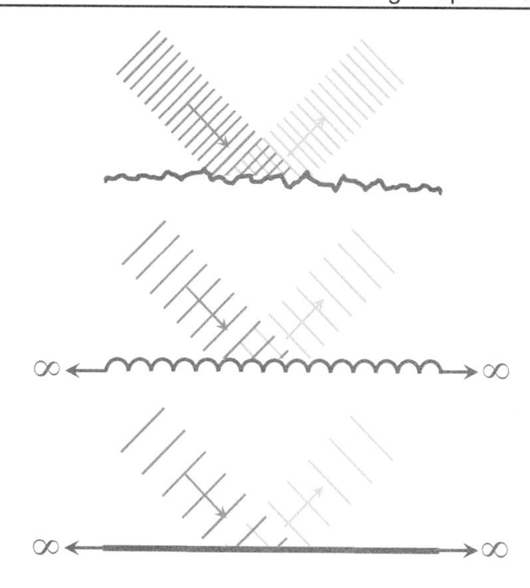

Figure 8.27 At high frequencies, a *rough* surface may need to be modeled directly, using a computational method. At lower frequencies, it can be idealized as an infinitely large regular surface with a periodic texture (enabling the use of more efficient methods) or a planar surface (enabling analytical approaches).

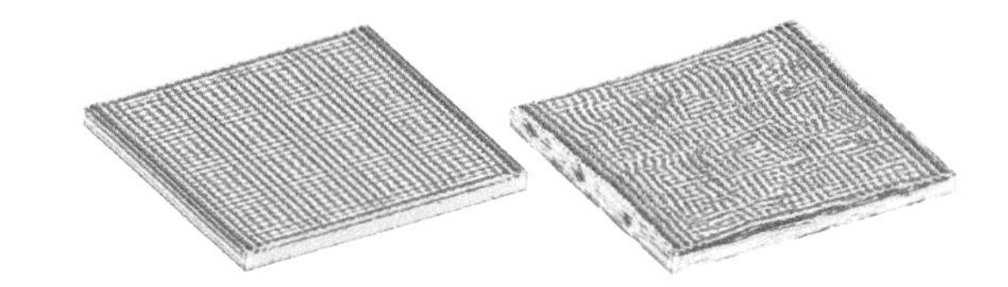

Figure 8.28 Tangential magnetic field intensity on the surfaces of 16×16 μm glass slabs (with perfect and imperfect surfaces) illuminated at an optical frequency. Solutions are obtained via MLFMA (see below).

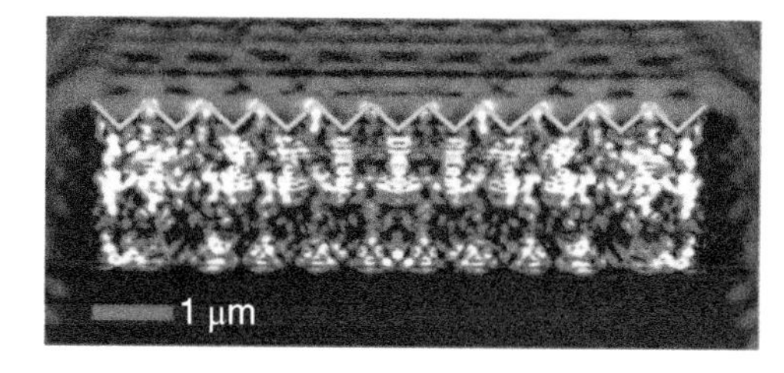

Figure 8.29 Power density inside a silicon solar cell excited at an optical frequency. The solution is obtained via MLFMA (see below).

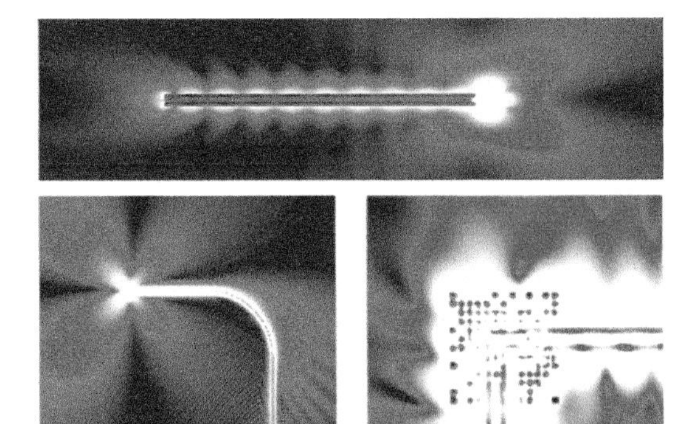

Figure 8.30 Electric field intensity in the vicinity of various nanowire and nanoparticle systems excited via point sources at optical frequencies. Solutions are obtained via MLFMA (see below).

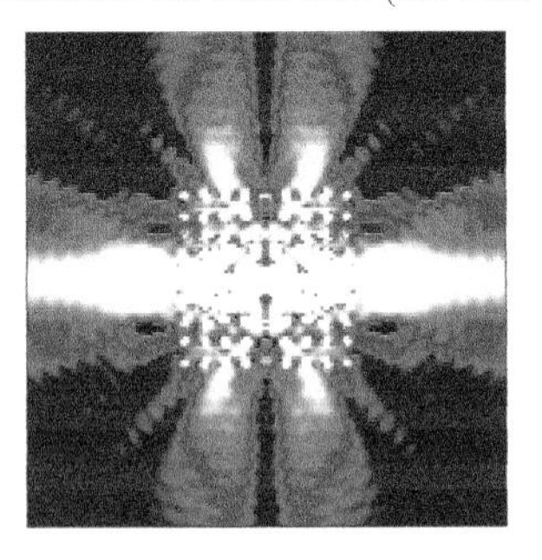

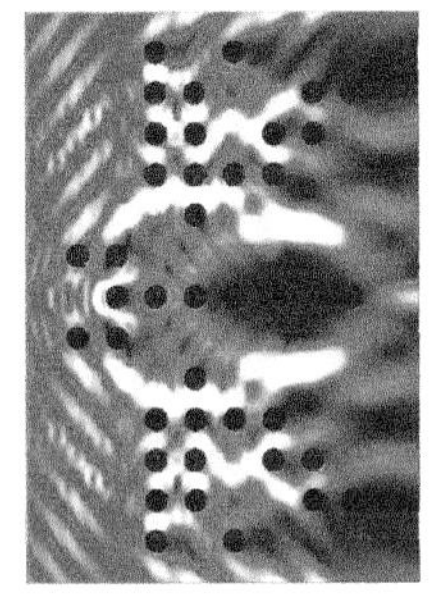

Figure 8.31 Power density in the vicinity of photonic crystals excited by a point source (left) and a beam (right). Solutions are obtained via MLFMA (see below).

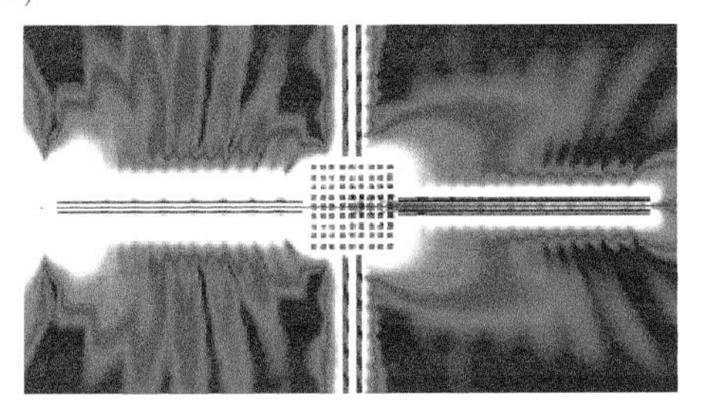

Figure 8.32 Transmission of power through a nano-device. The solution is obtained via MLFMA (see below).

[116]This is, however, a disadvantage when far-zone scattered fields need to be found.

8.4.1 Full-Wave Methods

Full-wave methods, which target numerical solutions to Maxwell's equations with controllable accuracy, can be categorized into differential-equation solvers and integral-equation solvers.

8.4.1.1 Differential-Equation Solvers

Differential-equation solvers are based on direct solutions to Maxwell's equations (their differential forms). Therefore, they are more straightforward, intuitive, and easier to understand and manipulate than integral-equation solvers. However, when using a differential-equation solver for a given problem, its discretization must be performed everywhere: i.e. the whole space must be discretized using small elements (such as a three-dimensional grid). In addition to typically large numbers of discretization elements to handle, this leads to numerical issues for open problems involving distant sources that excite target objects. We now consider two major differential-equation solvers with their advantages and disadvantages.

8.4.1.1.1 Finite-Difference Time-Domain Method (FDTD): FDTD is a very basic and remarkably useful method in computational electromagnetics. While similar finite-difference schemes were well-known in other areas, FDTD became popular in computational electromagnetics with the introduction of Yee cells (grid), named after Kane Shee-Gong Yee. Constructing Yee cells (Figures 8.35 and 8.36), differential forms of Maxwell's equations are written at the edges and surfaces (at finite numbers of positions), where space and time derivatives are handled by using central-difference approximations. With localized equations, solutions are performed in a step-by-step (leapfrog) manner; electric and magnetic fields are solved subsequently in different time steps, as long as desired. Hence, in FDTD, *time* is a solution parameter, and the solution to a problem coincides with the natural flow of time. This makes FDTD powerful to generate visual results, especially for transient-state analyses. Using time as a parameter further enables straightforward analyses of nonlinear materials and complex excitations (e.g. Gaussian pulses) with transient characteristics. In addition, using Maxwell's equations directly, solutions provide immediate access to electric and magnetic fields,[116] without intermediate variables.

As expected, all these advantages are accompanied by disadvantages that make FDTD less suitable for certain problems. First, being a differential-equation solver, FDTD needs the discretization of the whole space (computational domain) of the given problem. If the computational domain is open (unbounded), it must be truncated artificially (Figure 8.35) since it is impossible to discretize an infinitely large space. This truncation is a major issue, as simply stopping a grid causes artificial reflections that contaminate solutions. Various techniques (absorbing boundary conditions) have been developed for smooth truncation of FDTD grids, but this issue is still a significant bottleneck (or at least a weak area) in most existing code. Obviously, the disadvantages of discretization are not limited to open-space problems; constructing a suitable grid to model small geometric features (such as wires) along with smooth and large areas, as well as to accurately model curved structures, especially without exceeding the available computational resources, can be challenging in FDTD (Figure 8.37). Finally, as an inherently time-domain method, FDTD may not always be suitable for steady-state analyses.[117] While sufficient numbers of time steps would theoretically lead to a steady state, this may not work smoothly in numerical computations, often leading to instability issues.[118]

[117]There is also a finite-difference frequency-domain (FDFD) method, but it is used even less frequently than FDTD.

[118]The stability of FDTD solvers has always been an active research area.

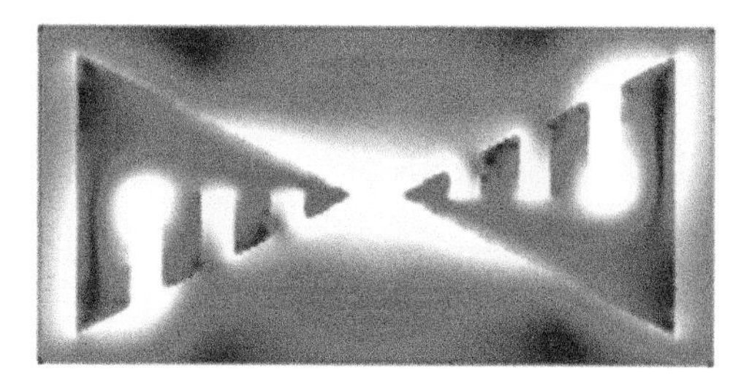

Figure 8.33 Power density around a nanoantenna that harvests solar energy. The solution is obtained via MLFMA (see below).

8.4.1.1.2 Finite Element Method (FEM): FEM has been widely used for decades in many areas, including but not limited to mechanical engineering, fluid dynamics, and thermal modeling, in addition to electromagnetics. In the context of electromagnetics, FEM is mainly used for bounded regions with well-defined boundary conditions. As a differential-equation solver, FEM requires the discretization of the problem space, but it is generally more flexible than FDTD. Specifically, the region of interest is still divided into a finite number of elements, but these elements do not have to be regular like Yee cells. By applying Maxwell's equations at a set of locations (e.g. corners of elements), a large (but sparse) matrix equation is constructed, whose solution provides the required coefficients representing the numerical solution. Minimization of error due to finite numbers of test points (compared to satisfying Maxwell's equations everywhere) is an active research topic in the context of FEM. Using more flexible elements (Figure 8.38), FEM can handle more complex problems, such as geometries with sharp corners, edges, cavities, and curved boundaries, compared to FDTD.

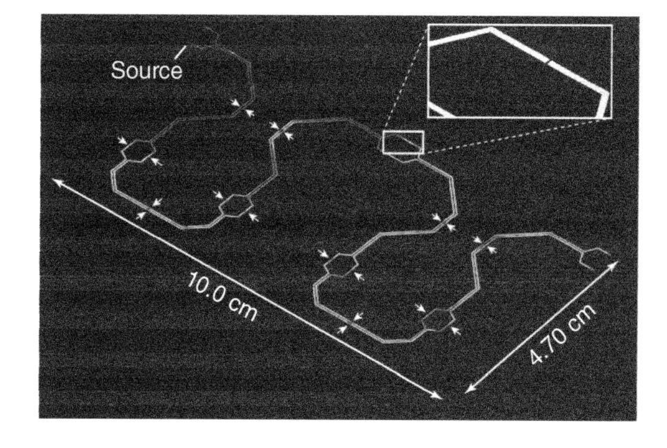

Figure 8.34 Electric current along a microstrip line at a radio frequency. The solution is obtained via MLFMA (see below).

On the other hand, using variously shaped elements within the same problem may make FEM more difficult to implement. More detailed comparisons of FDTD and FEM are topics of numerous papers in the literature, often leading to conclusions that depend on the problem (geometry, material, excitation, required solutions, etc.). In the context of electromagnetics, FEM[119] is usually used in the frequency domain to solve time-harmonic problems, and there are also time-domain versions (usually called TD-FEM or FETD) to analyze transient problems.

[119]Enormous efforts in computational electromagnetics have led to various full-wave solvers that can be found in the literature. In the context of partial differential equations, we can mention the discontinuous Galerkin method (DGM), which has attracted great interest in recent years. Due to its application scheme, DGM in electromagnetics is generally considered a type of FEM, and it is often compared with conventional FEM implementations in terms of accuracy and efficiency.

8.4.1.2 Integral-Equation Solvers

Integral-equation solvers are generally based on representations of original problems in terms of equivalent sources defined at boundaries or in finite volumes. Such a representation enables the use of radiation integrals involving Green's functions to compute electric and magnetic fields from equivalent sources. However, equivalent sources still need to be found by satisfying

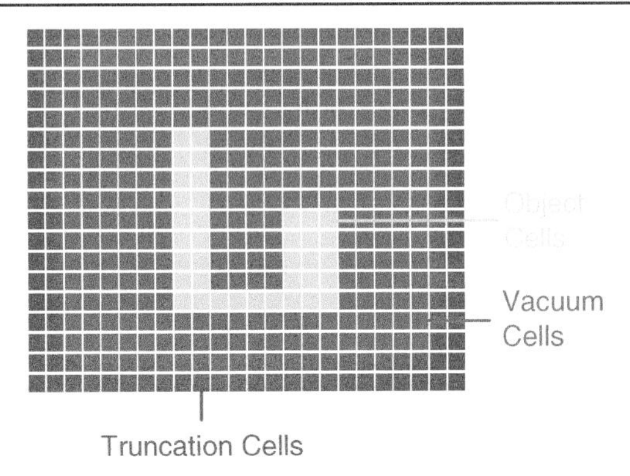

Figure 8.35 A two-dimensional view of an FDTD grid. In addition to cells that represent a vacuum (or host medium) and the target object, there are special cells that truncate the computational domain by artificially absorbing all waves to create the effect of infinite space.

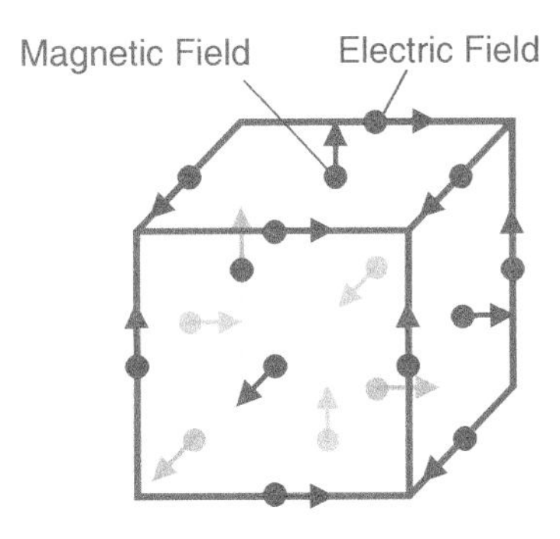

Figure 8.36 A typical Yee cell with locations for testing electric and magnetic fields. Directions show the tested components.

Maxwell's equations in volumes or on surfaces (boundary conditions). For arbitrary geometries, this is possible by discretizing equivalent sources using small elements (*basis* functions), as practiced in FDTD and FEM (discretizing regions).

As a major advantage, integral-equation solvers require the discretization of objects (Figure 8.39), without any discretization of empty spaces. Hence, they are much more suitable for open problems than FDTD and FEM. On the other hand, the constructed matrix equations are *dense* due to the nature of radiation integrals. Hence, acceleration methods are often needed to make them applicable to electrically large problems, for which discretizations lead to large matrix equations.

Integral-equations solvers can be developed in both the frequency and time domains, and they can be based on either surface integral equations or volume integral equations, depending on the problem. Surface integral equations are used for metallic and piecewise homogeneous objects whenever possible, as they require equivalent sources only on boundaries (e.g. surfaces of objects). For inhomogeneous objects, volume integral equations, which require volumetric equivalent sources, become necessary.

8.4.1.2.1 Method of Moments (MoM): In electromagnetics, MoM[120] is used to solve integral equations by minimizing their average errors when satisfying boundary conditions (for surface integral equations) or Maxwell's equations (for volume integral equations). For a given problem, this is achieved by defining a finite number of *testing* functions to test the required conditions. By selecting the same number of testing functions and basis functions (to discretize equivalent sources), the resulting matrix equation becomes square and can be solved directly by Gaussian elimination or iteratively via various techniques. Selecting the testing and basis functions as the same mathematical set, called *Galerkin discretization*, is a popular practice due to its stability. For a PEC object, boundary conditions[121] are satisfied on its surface to find the equivalent surface electric current density.[122] For a homogeneous penetrable object, a typical approach is to define equivalent electric and magnetic currents on the surface. Then the extinction theorem is employed to write surface integral equations, and continuity of tangential electric/magnetic fields is enforced to obtain the equivalent currents (Figure 8.40). When a volume integral equation is used for a homogeneous or an inhomogeneous object, equivalent volumetric currents are defined everywhere inside the object (object is replaced by equivalent sources). For example, the electric flux density can be defined as the unknown, leading to a set of equations to solve and analyze the electromagnetic problem. In all cases, once equivalent currents are found, they can be used to compute electric and magnetic fields *everywhere* in space (Figure 8.41).

[120] Another integral-equation solution method is discrete dipole approximation (DDA), which is frequently used for optical problems.

[121] The tangential electric field intensity must be zero on the surface, and the tangential magnetic field intensity is discontinuous by the amount of the induced electric current density.

[122] This equivalent current density corresponds to the induced current density.

8.4.1.2.2 Acceleration Algorithms: In all full-wave methods, discretization elements should be sufficiently small to reach accurate solutions. For differential-equation solvers based on space discretizations, these elements should also be properly refined at critical locations: e.g. in the vicinity of corners and tips, to capture rapidly changing fields. For integral-equation solvers, this is not directly required since numerical solutions lead to intermediate results (equivalent currents), and fields are computed from them via integration (a smoothing operator) as a post-processing step. Nevertheless, the discretization should be high quality to provide accurate representations of objects and equivalent sources. While there is no strict rule (it depends on the problem), a common practice is selecting discretization elements smaller than $\lambda/10$, where λ is the wavelength. For volume integral equations, this means smaller elements must be used at locations with higher contrast (permittivity and/or permeability). However, for surface integral equations, equivalent sources oscillate with the source frequency such that element size can be determined according to the wavelength of the medium that contains the source. In any case, the number of required elements multiplies with the electrical size of the object, leading to matrix equations involving large numbers of unknowns.[123]

To give an idea of the impressive growth of the computational load in integral-equation solvers, we can consider a sphere of radius λ. Using triangular $\lambda/10$ elements, such a sphere can be discretized using 1000 triangles. If each triangle represents an unknown, a 1000×1000 matrix equation is constructed as a result of the discretization. Using Gaussian elimination, such a matrix equation can be solved in several seconds on a standard laptop, but we may assume it as 1 s. If the frequency is only doubled, the sphere's radius becomes 2λ, making the matrix equation 4000×4000. As the time complexity of Gaussian elimination is cubic, the solution time increases 64-fold: i.e. it becomes 64 s. Similarly, increasing the frequency tenfold makes the solution time 10^6 s, whereas a hundredfold increase leads to 10^{12} s (approximately $31,710$ years) for a single solution. Today, we can solve this kind of large-scale problems involving objects with sizes of hundreds of wavelengths, thanks to acceleration algorithms.

In general, acceleration algorithms do not change the primary characteristics of solutions (still using integral-equation solvers, often together with MoM), but they change how electromagnetic interactions between discretization elements are performed.[124] These algorithms have been topics of considerable research in the last several decades of computational electromagnetics.[125] Among many, we can mention adaptive cross approximation (ACA), the adaptive integral method (AIM), the plane-wave time-domain (PWTD) algorithm, and various algorithms based

[123] The number of unknowns in an integral-equation solver can be much smaller than the corresponding number in a differential-equation solver. But note that the constructed matrix equations are dense for integral-equation solvers.

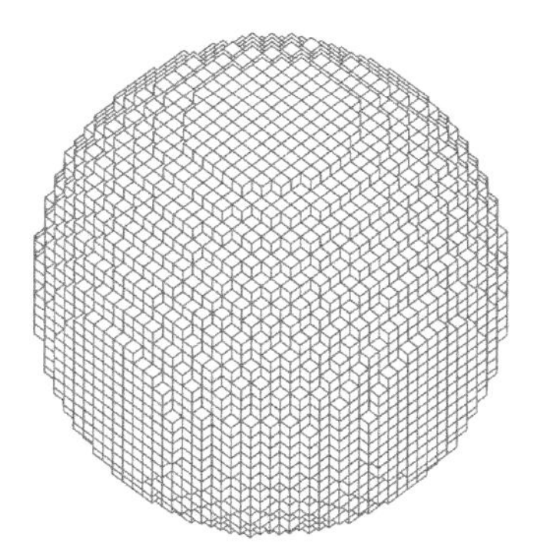

Figure 8.37 An FDTD grid for a sphere using cubic cells. Without using special elements (e.g. curved cells with varying sizes), large numbers of cells may be needed to accurately model the sphere.

[124] Acceleration comes with a relatively controllable relaxation of accuracy. This may raise questions about the full-wave characteristics of the resulting solver.

[125] There are also acceleration algorithms for differential-equation solvers.

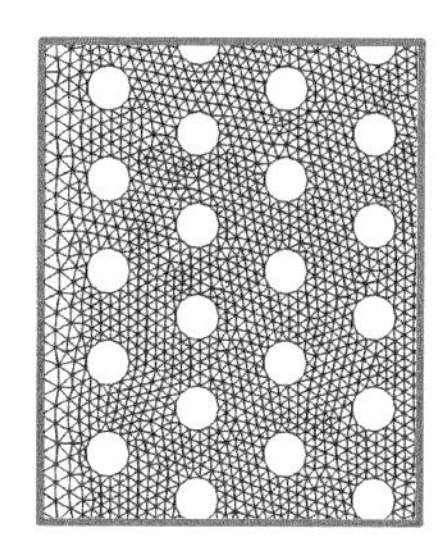

Figure 8.38 Discretization of two-dimensional regions in FEM. Elements can be nonuniform to correctly model curved boundaries or accurately analyze electromagnetic fields at critical locations.

on fast Fourier transform (FFT) as well as the fast multipole method (FMM) and its multilevel version, multilevel fast multipole algorithm (MLFMA). The algorithms are often accompanied by solution techniques, such as the domain decomposition method (DDM) and equivalence principle algorithm (EPA), and they are also supported by algebraic techniques.

8.4.1.2.3 FMM and MLFMA: FMM is considered one of the most important algorithms developed in the twentieth century. It is based on the computation of electromagnetic interactions (between discretization elements: i.e. basis and testing functions) group by group (Figure 8.42). This is achieved by applying two fundamental operations: factorization (via Gegenbauer's addition theorem) and diagonalization (using plane waves). Specifically, using factorization and diagonalization, Green's functions can be written in suitable forms[126] to be used for collective computations of interactions. These interactions correspond to matrix-vector multiplications to be used in iterative solutions to problems. Hence, using FMM, matrix-vector multiplications can be performed *without* directly computing matrix elements.

[126]In these suitable forms, plane waves represent electromagnetic interactions. Specifically, the radiation pattern of a group of discretization elements is expanded in terms of plane waves propagating in different directions. These plane waves are received by other groups and tested by testing functions. Due to the nature of the expansion, FMM interactions can be performed only for groups that are sufficiently far from each other. Nevertheless, there are also *low-frequency* FMM and MLFMA implementations that use other types of expansions suitable for electrically nearby groups.

MLFMA is a multilevel application of FMM, and it can reduce the complexity of a matrix-vector multiplication from quadratic to linear. Recently, by employing MLFMA on high-performance computers, electromagnetic problems involving billions of unknowns have been solved.

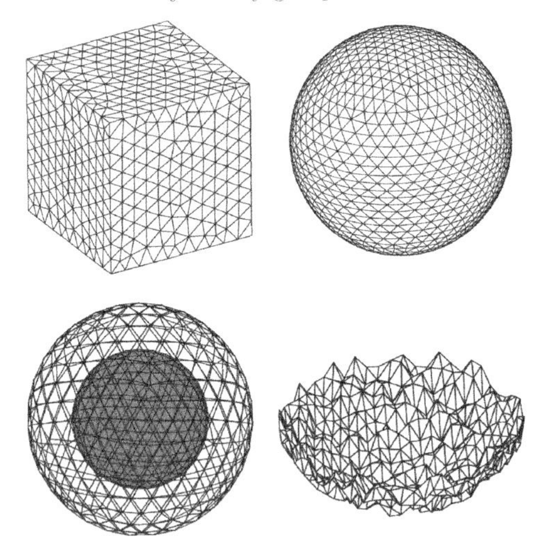

Figure 8.39 Discretization of various objects in surface integral equations. Triangles are very suitable to model three-dimensional surfaces. Discretization is applied only on boundaries: i.e. when electromagnetic parameters change. Bottom-left: a spherical metallic core inside a dielectric layer (both surfaces are discretized).

8.4.2 Asymptotic Techniques

As described above, asymptotic techniques are suitable for objects that are much smaller or larger than the wavelength.[127] In both cases, the application of full-wave methods can be problematic in terms of efficiency, stability, and accuracy. When an object is very large, full-wave methods may require solutions for extremely large matrix equations, which cannot be handled even with acceleration techniques. Similarly, for a very small object in terms of wavelength, full-wave methods often suffer from stability/accuracy issues. Asymptotic techniques can efficiently solve such small-scale and large-scale problems by considering limiting forms of Maxwell's equations to simplify solutions. Consequently, they naturally introduce uncontrollable errors, which tend to diminish as the problem size increases (for high-frequency techniques) or decreases (for low-frequency techniques).

[127]We emphasize that the electrical size of an object (its size with respect to the wavelength), rather than its metric size, is important to decide whether it is small or large. In air, an average human body at 300 kHz is extremely small (smaller than $\lambda/500$), but it is extremely large (larger than 1500λ) at 300 GHz. At the same frequency of 300 GHz, a typical bacterium is an extremely small object.

8.4.2.0.1 Quasistatic Approximations: When an object is very small with respect to the wavelength, quasistatic approximations can be used for its electromagnetic analysis. The procedure is usually based on neglecting one of the time derivatives in Maxwell's equations (based on the fact that these time derivatives have small contributions *across* the given object), from either Ampere's law or Faraday's law, leading to static forms for these equations. Depending on the dropped term, the resulting technique may be called *electro-quasistatic* (modified Faraday's law) or *magneto-quasistatic* (modified Ampere's law). Note that dropping both two-time derivatives leads to a static approximation, where electrostatics and magnetostatics coexist independently.

As an example, we may consider a capacitor in an AC circuit (Figure 8.43). If a full-wave method is used, one can obtain both electric and magnetic fields inside the capacitor. Hence, if the capacitor is modeled by lumped (circuit) elements (if possible), it must be represented by a combination of lumped capacitors and lumped inductors, in addition to lumped resistors to account for losses. Quasistatic approximations lead to simplifications of these combinations, leading to a single lumped capacitor in the static limit. Similarly, with quasistatic approximations, an inductor can be simplified, becoming a single lumped inductor in the static limit. Quasistatic approximations are often used when static limits are not sufficient, whereas static limits can be seen as their most basic forms, as they are frequently used in circuit theory.[128,129]

[128]In circuit theory, whether the circuit itself is DC or AC, static approximations of capacitors, inductors, and resistors are used. For an AC circuit, this does not conflict with the time dependency of current/voltage, since static approximations are used *inside* the components, which can be quite accurate considering their small sizes.

[129]In circuit theory, wires do not have any effect other than providing connections between elements, also considering their short lengths with respect to the wavelength. When they are long, however, transmission-line theory must be used for accurate analyses.

8.4.2.0.2 Geometrical Optics: Geometrical optics (GO) is based on ray-like behaviors of electromagnetic waves when they interact with electrically large objects (Figure 8.44). When an electromagnetic wave is represented by rays, their propagation along straight paths, as well as their reflection and refraction phenomena at interfaces, can easily be shown.[130] In addition, absorption and curved paths due to varying refractive indices can also be demonstrated in GO. On the other hand, some basic mechanisms, like diffraction, cannot be explained directly with this technique. GO is particularly useful for modeling optical systems such as lenses, mirrors, and other components since these structures are typically very large with respect to the wavelength at optical frequencies.

[130]GO has visual advantages for teaching and learning optics, but it can be misleading for understanding the behavior of electromagnetic waves.

Snell's law can be used directly in terms of rays, and it can be considered a special case of Fermat's principles under GO. The use of GO leads to various practical formulas, such as the lensmaker's equation, that can be used to design optical systems. Further manipulations, like paraxial approximation, may further facilitate the applicability of GO. Any application of GO has an approximation error, but it becomes increasingly accurate in the limit of zero wavelength.

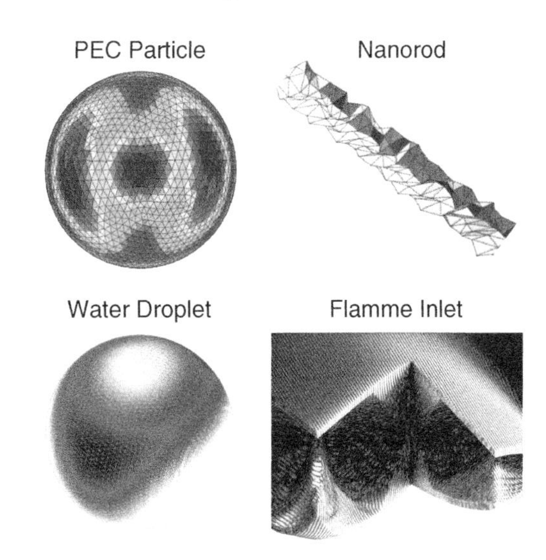

Figure 8.40 Equivalent electric current density on various structures computed using surface integral equations and MoM. Discretizations of the PEC sphere and nanorod are coarse enough to identify triangles.

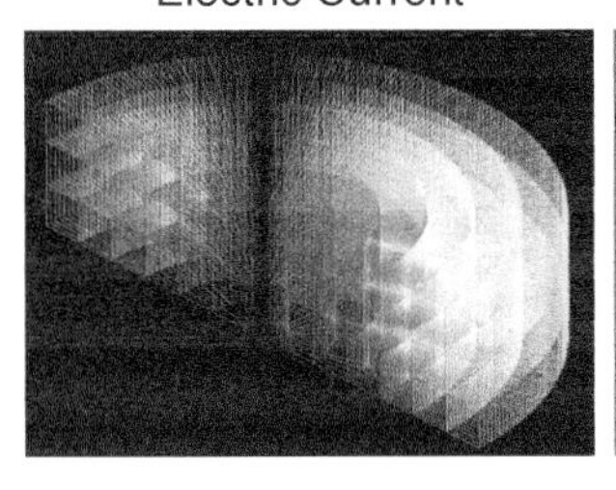

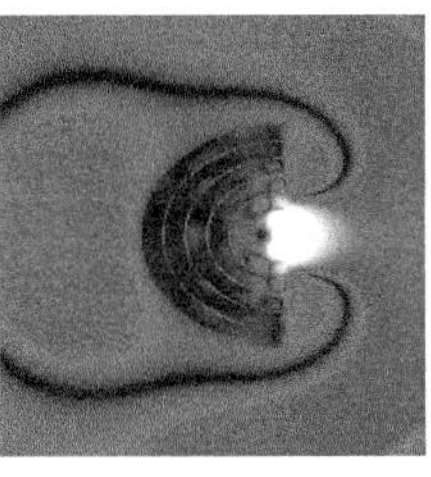

Figure 8.41 Electric current in a complex structure consisting of a network of wires excited by a dipole pair. After the electric current density is found via MoM, the electric field intensity (its imaginary part) is computed in the vicinity of the structure.

[131] GTD can be seen as the generalization of Fermat's principle.

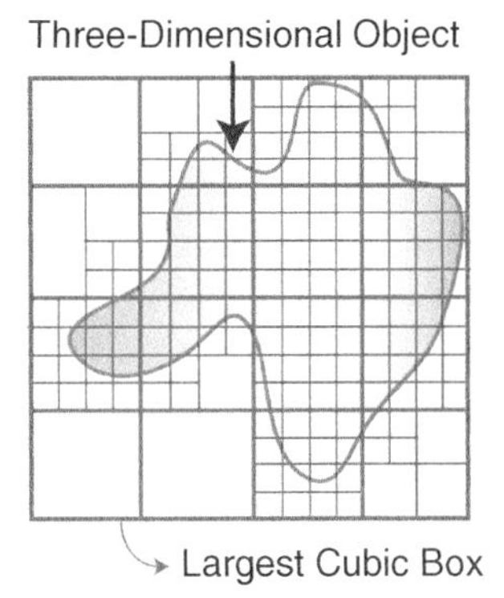

Figure 8.42 A multilevel division of an object into cubic domains.

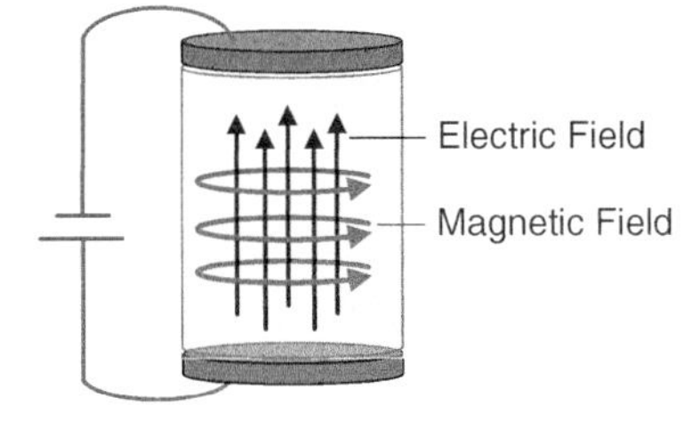

Figure 8.43 When a capacitor is connected to an AC source, the time-dependent electric field leads to a time-dependent magnetic field, which in turn creates an additional electric field. In the static approximation, the magnetic field and, hence, magnetic potential energy and inductance, are ignored, enabling the modeling of the capacitor with a single *lumped capacitor*.

8.4.2.0.3 Uniform Geometrical Theory of Diffraction: It was natural for users of GO to improve it by including diffraction mechanisms (Figure 8.44). One of the early successful attempts was made in 1962 by Joseph Bishop Keller, who proposed the geometrical theory of diffraction (GTD). In GTD, diffracted rays are introduced at edges, corners, and similar locations on objects when illuminated by the incident rays.[131] Definitions of these diffracted rays (e.g. using Huygens' principles, conservation of energy, etc.) are mathematically solid. However, there are numerical problems in GTD; artificially singular values are observed at critical locations at shadow boundaries. Therefore, considerable research has been conducted to improve GTD, leading to various techniques including the uniform geometrical theory of diffraction (UTD). Today, UTD is widely used – often, combined with other full-wave methods or asymptotic techniques – for fast solutions to large-scale problems.

8.4.2.0.4 Physical Optics: Another high-frequency technique is *physical optics* (PO), which considers the wave behavior of electromagnetic fields instead of a full ray approximation (Figure 8.45). Ray optics is still used to identify illuminated (lit) and shadow regions of the given object, whereas radiation integrals are employed to compute scattered fields. Consequently, reflection, transmission, diffraction, and similar phenomena are expected to appear in *scattering* results (similar to full-wave methods) without being manually introduced.

Although it uses waves, PO is an approximate method as it strictly defines lit and shadow regions based on excitation (and a few reflections in improved PO versions) without actually computing them. This corresponds to ignoring electromagnetic interactions between different parts of an object, leading to nonphysical discontinuities in equivalent or induced currents (lit-shadow boundaries). Consequently, diffraction cannot be observed in equivalent currents, which are fixed based on primary sources. Despite issues in the representation of currents, PO may give remarkably accurate results for far-zone scattered fields, thanks to the smoothing nature of the integration, while being uncontrollable. Like all other high-frequency techniques, PO becomes increasingly more accurate as the object becomes large in terms of wavelength. For example, when an object is extremely large, lit-shadow regions can be formed very clearly, exactly as used by PO. However, this is not true for small or moderately large objects. Hence,PO is often combined with full-wave and other high-frequency techniques to build hybrid solvers, such as the shooting and bouncing ray (SBR) technique based on PO and GO, which benefit from the advantages of different methods and techniques.

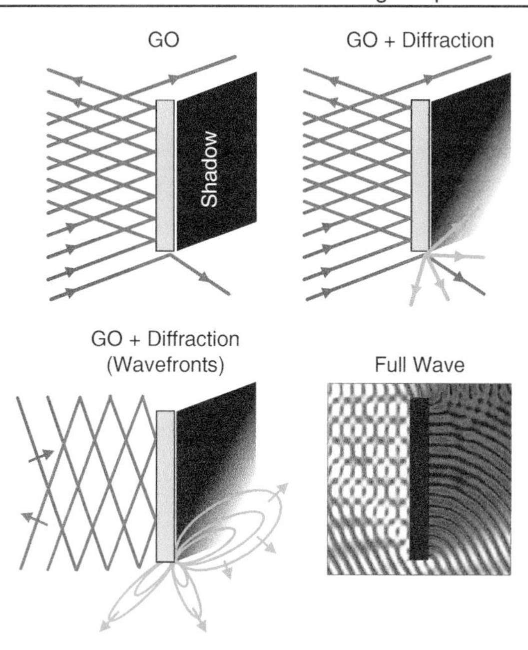

Figure 8.44 Reflection from a $8\lambda \times 8\lambda$ rectangular slab. GO predicts sharp shadowing, but adding diffractions make the solution physically more accurate. A full-wave solution to the problem is shown to illustrate the actual distribution of the electric field intensity.

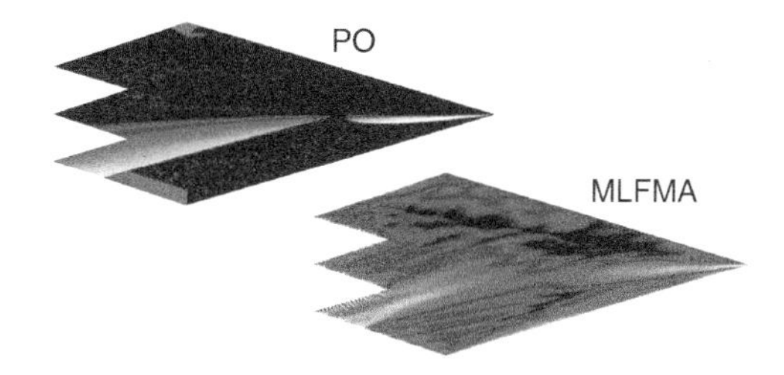

Figure 8.45 Electric current density induced on an airborne target (Flamme) illuminated by a plane wave. Full shadows are observed in the PO solution, but the actual current distribution (obtained via MLFMA) demonstrates the effects of many reflections and diffractions.

Bibliography

Basic Electromagnetics

1. Electromagnetism (Wiley, 2nd Ed., 1991), by I. S. Grant and W. R. Phillips
2. Engineering Electromagnetics (Pearson, 1997), by K. R. Demarest
3. Electromagnetics with Applications (William C Brown Pub., 5th Ed., 1999), by J. D. Kraus and D. A. Fleisch
4. Electromagnetism (Addison-Wesley, 2001), G. L. Pollack and D. R. Stump
5. Electricity, Magnetism, and Light (Academic Press, 2002), by W. M. Saslow
6. Elements of Engineering Electromagnetics (Prentice Hall, 6th Ed., 2004), by N. N. Rao
7. Elements of Electromagnetics (Oxford University Press, 4th Ed., 2006), by M. N. O. Sadiku
8. Electromagnetic Fields (Wiley-IEEE, 2nd Ed., 2007), by R. K. Wangsness
9. A Student's Guide to Maxwell's Equations (Cambridge University Press, 2008), by D. Fleisch
10. Foundations of Electromagnetic Theory (Addison-Wesley, 4th Ed., 2008), by J. R. Reitz, F. J. Milford, and R. W. Christy
11. Electromagnetics (Pearson, 2010), by B. M. Notaros
12. The Feynman Lectures on Physics [Vol. 2] (Basic Books, 2011), by R. P. Feynman, R. B. Leighton, and M. Sands
13. Engineering Electromagnetics (McGraw-Hill, 8th Ed., 2011), by W. H. Hayt and J. A. Buck
14. Fundamentals of Engineering Electromagnetics (Pearson, 2013), by D. K. Cheng
15. MATLAB-Based Electromagnetics (Pearson, 2013), by B. M. Notaros
16. Electricity and Magnetism (Cambridge University Press, 3rd Ed., 2013), by E. M. Purcell and D. J. Morin
17. Engineering Electromagnetics and Waves (Pearson, 2nd Ed., 2014), by U. S. Inan, A. S. Inan, and R. Said

Introduction to Electromagnetic Waves with Maxwell's Equations, First Edition. Özgür Ergül.

Companion website: www.wiley.com/go/ergulmax

18. Fundamentals of Applied Electromagnetics (Pearson, 7th Ed., 2014), by F. T. Ulaby and U. Ravaioli
19. A Student's Guide to Waves (Cambridge University Press, 2015), by D. Fleisch and L. Kinnaman
20. Introduction to Electrodynamics (Cambridge University Press, 4th Ed., 2017), by D. J. Griffiths

Graduate-Level Electromagnetics

1. Electromagnetism (Dover Pub., 1969), by J. C. Slater and N. H. Frank
2. Principles of Electrodynamics (Dover Pub., 1987), by M. Schwartz
3. Electricity and Magnetism: An Introduction to the Theory of Electric and Magnetic Fields (Electret Scientific, 2nd Ed., 1989), by O. D. Jefimenko
4. Field Theory of Guided Waves (IEEE, 2nd Ed., 1996), by R. E. Collin
5. Classical Electrodynamics (Wiley, 3rd Ed., 1998), by J. D. Jackson
6. Classical Electrodynamics (CRC Press, 1998), by J. Schwinger, L. L. Deraad Jr., K. A. Milton, W.-Y. Tsai, and J. Norton
7. Waves and Fields in Inhomogeneous Media (Wiley-IEEE, 1999), by W. C. Chew
8. Scattering of Electromagnetic Waves: Theories and Applications (Wiley-Interscience, 2000), by L. Tsang, J. A. Kong, and K.-H. Ding
9. Time-Harmonic Electromagnetic Fields (Wiley-IEEE, 2nd Ed., 2001), by R. F. Harrington
10. Classical Electricity and Magnetism (Dover Pub., 2nd Ed., 2005), by W. K. H. Panofsky and M. Phillips
11. Electromagnetic Fields (Wiley-IEEE, 2nd Ed., 2007), by J. G. Van Bladel
12. Electromagnetic Theory (Adams Press, 2008), by J. A. Stratton
13. The Classical Electromagnetic Field (Dover Pub., 2010), by L. Eyges
14. Electrodynamics (World Scientific Pub., 2nd Ed., 2011), by H. J. W. Müller-Kirsten
15. The Classical Theory of Fields: Electromagnetism (Springer, 2012), by C. S. Helrich
16. Advanced Engineering Electromagnetics (Wiley, 2nd Ed., 2012), by C. A. Balanis
17. Modern Electrodynamics (Cambridge University Press, 2012), by A. Zangwill
18. Classical Electromagnetism (Dover Pub., 2nd Ed., 2017), by J. Franklin
19. Electromagnetic Wave Propagation, Radiation, and Scattering: From Fundamentals to Applications (Wiley-IEEE, 2nd Ed., 2017), by A. Ishimaru
20. Classical Theory of Electromagnetism (WSPC, 3rd Ed., 2018), by B. Di Bartolo

Specialized Areas

1. Waves and Fields in Optoelectronics (Prentice Hall, 1983), by H. A. Haus
2. Absorption and Scattering of Light by Small Particles (Wiley-VCH, 1998), by C. F. Bohren and D. R. Huffman
3. Electromagnetic Analysis and Design in Magnetic Resonance Imaging (CRC Press, 1998), by J.-M. Jin
4. Principles of Optics: Electromagnetic Theory of Propagation, Interference and Diffraction of Light (Cambridge University Press, 7th Ed., 1999), by M. Born and E. Wolf

5. Wireless Communications: Principles and Practice (Prentice Hall, 2nd Ed., 2002), by T. S. Rappaport
6. Electromagnetic Field Theory and Transmission Lines (Pearson, 2004), by G. S. N. Raju
7. Introduction to RF Propagation (Wiley-Interscience, 2005), by J. S. Seybold
8. Photonic Crystals: Molding the Flow of Light (Princeton University Press, 2nd Ed., 2008), by J. D. Joannopoulos, S. G. Johnson, J. N. Winn, and R. D. Meade
9. Complete Wireless Design (McGraw-Hill, 2nd Ed., 2008), by C. W. Sayre
10. Electromagnetic Compatibility Engineering (Wiley, 2009), by H. W. Ott
11. Optical Physics (Cambridge University Press, 4th Ed., 2010), by A. Lipson, S. G. Lipson, and H. Lipson
12. Principles of Modern Radar: Basic Principles (Scitech Pub., 2010), Edited by M. A. Richards, J. A. Scheer, and W. A. Holm
13. Magnetism and Magnetic Materials (Cambridge University Press, 2010), by J. M. D. Coey
14. Radiowave Propagation: Physics and Applications (Wiley, 2010), by C. Levis, J. T. Johnson, and F. L. Teixeira
15. Microwave Engineering (Wiley, 4th Ed., 2011), by D. M. Pozar
16. Introduction to Laser Technology (Wiley-IEEE, 4th Ed., 2012), by C. B. Hitz, J. Ewing, and J. Hecht
17. Antenna Theory and Design (Wiley, 3rd Ed., 2012), by W. L. Stutzman and G. A. Thiele
18. Transmission Lines: Equivalent Circuits, Electromagnetic Theory, and Photons (Cambridge University Press, 2013), by R. Collier
19. Optics (Pearson, 5th Ed., 2016), by E. Hecht
20. Antenna Theory: Analysis and Design (Wiley, 4th Ed., 2016), by C. A. Balanis

Computational Electromagnetics

1. Field Computation by Moment Methods (Wiley-IEEE, 1993), by R. F. Harrington
2. The Finite Difference Time Domain Method for Electromagnetics (CRC Press, 1993), by K. S. Kunz and R. J. Luebbers
3. High-Frequency Electromagnetic Techniques: Recent Advances and Applications (Wiley-Interscience, 1995), by A. K. Bhattacharyya
4. Computational Methods for Electromagnetics (Wiley-IEEE, 1998), by A. F. Peterson, S. L. Ray, and R. Mittra
5. Fast and Efficient Algorithms in Computational Electromagnetics (Artech House, 2001), by W. C. Chew, J.-M. Jin, E. Michielssen, and J. Song
6. Computational Electrodynamics: The Finite-Difference Time-Domain Method (Artech House, 3rd Ed., 2005), by A. Taflove and S. C. Hagness
7. Fundamentals of Electromagnetics with MATLAB (Scitech Pub., 2nd Ed., 2007), by K. E. Lonngren, S. V. Savov, and R. J. Jost
8. Analytical and Computational Methods in Electromagnetics (Artech House, 2008), by R. Garg
9. Numerical Techniques in Electromagnetics with MATLAB (CRC Press, 3rd Ed., 2009), by M. N. O. Sadiku
10. Advances in Time-Domain Computational Electromagnetics: Beyond Conventional Finite Difference Methods (VDM Verlag, 2010), by S. Wang, F. L. Teixeira, and R. Lee
11. Computational Electromagnetics for RF and Microwave Engineering (Cambridge University Press, 2nd Ed., 2010), by D. B. Davidson
12. Essentials of Computational Electromagnetics (Wiley-IEEE, 2012), by X.-Q. Sheng and W. Song

13. Computational Methods for Electromagnetic Phenomena: Electrostatics in Solvation, Scattering, and Electron Transport (Cambridge University Press, 2013), by W. Cai
14. The Finite Element Method in Electromagnetics (Wiley-IEEE, 3rd Ed., 2014), by J.-M. Jin
15. The Method of Moments in Electromagnetics (CRC, 2nd Ed., 2014), by W. C. Gibson
16. The Multilevel Fast Multipole Algorithm (MLFMA) for Solving Large-Scale Computational Electromagnetics Problems (Wiley-IEEE, 2004), by Ö. Ergül and L. Gürel
17. Higher-Order Techniques in Computational Electromagnetics (Scitech Pub., 2015), by R. D. Graglia and A. F. Peterson
18. Theory and Computation of Electromagnetic Fields (Wiley-IEEE, 2nd Ed., 2015), by J.-M. Jin
19. Boundary Conditions in Electromagnetics (Wiley-IEEE, 2019), by I. V. Lindell and A. Sihvola
20. New Trends in Computational Electromagnetics (Scitech Pub., 2020), Edited by Ö. Ergül

Historically Significant Contributions

1. Opticks: Or, a Treatise of the Reflections, Refractions, Inflections, and Colors of Light (1704), by I. Newton
2. Experiments and Observations on Electricity (1751), by B. Franklin
3. Memories on Electricity and Magnetism (1785–1789), by C.-A. de Coulomb
4. On the Electricity Excited by the Mere Contact of Conducting Substances of Different Kinds (1800), by A. Volta
5. Memoir on the Mathematical Theory of Electrodynamic Phenomena, Uniquely Deduced from Experience (1825), A.-M. Ampère
6. Experimental Researches in Electricity (1839), by M. Faraday
7. The Forces of Matter (1859), by M. Faraday
8. On Physical Lines of Force (1861), by J. C. Maxwell
9. A Dynamical Theory of the Electromagnetic Field (1865), by J. C. Maxwell
10. A Treatise on Electricity and Magnetism (1873), by J. C. Maxwell
11. Electric Waves: Being Researches on the Propagation of Electric Action with Finite Velocity Through Space (1893), by H. R. Hertz
12. Electromagnetic Theory (1893), by O. Heaviside
13. Electricity and Matter (1904), by J. J. Thompson
14. On the Electrodynamics of Moving Bodies (1905), by A. Einstein
15. Two Papers of Henri Poincaré on Mathematical Physics (1921), H. A. Lorentz
16. The Quantum Theory of the Emission and Absorption of Radiation (1927), by P. A. M. Dirac
17. Space-Time Approach to Quantum Electrodynamics (1949), by R. P. Feynman
18. Geometrical Theory of Diffraction (1962), by J. B. Keller
19. Electromagnetic Scattering by Surfaces of Arbitrary Shape (1982), by S. Rao, D. Wilton, and A. Glisson
20. A Fast Algorithm for Particle Simulations (1987), by L. Greengard and V. Rokhlin

On the History of Electromagnetics

1. A History of Electricity and Magnetism (Burndy Library, 1971), by H. W. Meyer
2. The Great Physicists from Galileo to Einstein (Dover Pub., 1988), by G. Gamow
3. A History of the Theories of Aether & Electricity (Dover Pub., 1989), by E. Whittaker
4. The Maxwellians (Cornell University Press, 1994), by B. J. Hunt
5. The Story of Electrical and Magnetic Measurements: From 500 BC to the 1940s (Wiley-IEEE, 1999), by J. F. Keithley
6. Electromagnetics: History, Theory, and Applications (Wiley-IEEE, 1999), by R. S. Elliott
7. Oliver Heaviside: The Life, Work, and Times of an Electrical Genius of the Victorian Age (JHUP, 2002), by P. J. Nahin
8. Empires of Light: Edison, Tesla, Westinghouse, and the Race to Electrify the World (Random House, 2003), by J. Jonnes
9. Electricity and Magnetism: A Historical Perspective (Greenwood, 2006), by B. Baigrie
10. From Falling Bodies to Radio Waves: Classical Physicists and Their Discoveries (Dover Pub., 2007), by E. Segrè
11. Bibliographical History of Electricity and Magnetism (Mottelay Press, 2007), by P. F. Mottelay
12. A History of Optics from Greek Antiquity to the Nineteenth Century (Oxford University Press, 2012), by O. Darrigol
13. The Man Who Changed Everything: The Life of James Clerk Maxwell (Wiley, 2014), by B. Mahon
14. The Invisible Universe: The Story of Radio Astronomy (Springer, 3rd Ed., 2015), by G. Verschuur
15. The Story of Light Science: From Early Theories to Today's Extraordinary Applications (Springer, 2017), by D. F. Vanderwerf
16. Zapped: From Infrared to X-rays, the Curious History of Invisible Light (Little, Brown and Company, 2017), by B. Berman
17. A Brief History of Everything Wireless: How Invisible Waves Have Changed the World (Springer, 2018), by P. Launiainen
18. An Introduction to Classical Electrodynamics (Maricourt Academic Press, 2019), by J. W. Keohane and J. P. Foy
19. Faraday, Maxwell, and the Electromagnetic Field: How Two Men Revolutionized Physics (Prometheus, 2019), by N. Forbes and B. Mahon
20. The Invisible Rainbow: A History of Electricity and Life (Chelsea Green Pub., 2020), by A. Firstenberg

Index

Introduction to Electromagnetic Waves with Maxwell's Equations, First Edition. Özgür Ergül.

Companion website: www.wiley.com/go/ergulmax